Handbook of Filter Synthesis

Handbook of Filter Synthesis

Anatol I. Zverev
Consulting Engineer
Westinghouse Electric Corporation

John Wiley and Sons, Inc.   *New York • London • Sydney*

Library of Congress Catalog Card Number: 67-17352
Printed in the United States of America
ISBN 0 471 98680 1

To my Aunt

Мария Алексеевна Зверева

Preface

This treatment of the electric wave filter is for electronic systems engineers engaged in communication, radar, and any other electronic equipment that depends on selective networks. From the systems engineer's point of view the filter sets the standards of the system. Today he is able to specify almost any type of stable, single-valued analytic function as a subsystem on a block diagram, with reasonable assurance that it can be approximated and built into an operating unit. The exact mathematical technique is so successful that the newer electronic systems are literally packed with synthesized passive and active networks. An exact knowledge of filter performance is therefore essential for the systems engineer. In this book he can find information concerning the performance of all possible types of filters in both the time and frequency domains.

In addition, the filter expert can find here a variety of general and specific information pertinent to his speciality. Almost any type of filter can be designed with the aid of the precalculated data presented.

In the evolution of the electronics industry the first two major developments—radio and the vacuum tube—were followed closely by a third, the electric wave filter. Filter technology was officially born in 1915 when K. Wagner (Germany) and G. Campbell (United States), working independently, proposed the basic concept of the filter. Their results evolved from earlier work on loaded transmission lines and the classical theory of vibrating systems.

Over the years filters have so permeated electronic technology that the modern world is hardly conceivable without them. They direct, channel, integrate, separate, delay, differentiate, and transform all kinds of electric energy and signals. It is appropriate to emphasize the fact that filter technology has not only transformed electronics but has itself been transformed into a theoretical tool of great power. Thus a filter is no longer a mere component neatly packaged in a can. In a much broader and more important sense it is a systems technique, almost a philosophical concept, whose generality has been steadily increasing throughout its fifty-year history.

The generalization of the filter concept began when it was found that filter theory could be used to illuminate problems in mechanical and acoustical systems. By the use of an electromechanical analogy filter theory can be applied to many seemingly unrelated systems in which natural modes of vibration are of interest; for example, loudspeaker design, crystallography, architectural acoustics, airframe behavior, and mechanical systems design. Filter theory shows how to coordinate the action of several resonant elements to obtain uniform transmission over a prescribed frequency range. The concept of an ideal filter with lossless elements, which delivers all of the input energy to its output over the widest possible frequency range, establishes the requirements for broadbanding under prescribed constraints.

Application of filter theory has now gone far beyond these first generalizations. The concepts of exact synthesis techniques for prescribed transfer and immittance functions, of arbitrary functions with realizable rational functions, of time-domain synthesis, matched filters, parametric elements, and various other active devices have added new vitality to an already flourishing technology.

The discovery by Zobel, published in 1923, of a practical method of designing selective filters with an unlimited number of reactances was undoubtedly a work of genius. It was the only known method until about 1940 and the only practical method until the mid-1950s. S. Darlington in the United States and W. Cauer in Germany, both inspired by the work of Norton, published a theory that involved a set of problems relating to modern synthesis procedure. The importance of filter synthesis was not recognized immediately. It could be used to design better low-pass filters but failed to provide such designs in practice because of the extremely heavy burden of computation required. It was not until the advent of relatively cheap computation methods in the 1950s that Cauer-Darlington filters came into widespread use. So many computer-prepared designs have been published that designing an elliptic-function filter involves little more work than copying numbers out of a book, a technique that is actually easier than Zobel's method.

We now synthesize networks and systems by employing a fusion of many theories produced by many authors. In the considerable body of the literature there are many references to Cauer and Darlington but this bibliographical distinction is currently being superseded. It has been assumed that there is little point in listing the names that everyone now takes for granted.

This treatment has several objectives. The first is to present the underlying theory, concepts, and techniques of selective networks. Subject matter of this kind is treated in Chapters 2, 4, and 5. The second objective is the presentation of responses that can be provided by passive, linear, bilateral filtering structures. This subject matter is treated extensively in Chapters 3 and 7. A third objective is to illustrate the first two by the treatment of specialized networks such as crystal filters (Chapter 8) and helical filters (Chapter 9). Chapter 1 is an introduction to the field of selectivity, written with the intent to project the concept and importance of filters in the world of electronics. Chapter 6 provides information pertinent to polynomial filters with monotonic attenuation curves. Information is presented in tables of lowpass element values and normalized coupling coefficients and quality factors. With the aid of data presented in this chapter predistorted filters can readily be designed. All of Chapter 10 is dedicated to techniques of network transformation. These data are not only practical for filter design but can be applied to any other type of electronic circuit design.

The reader will not find an extensive treatment of active networks and microwave structures for themselves they tend to be a specialized field within filter technology. Microwave filters may consist of metal cavities coupled by openings called "irises". They may take the form of printed circuit "stripline" networks that appear to be labyrinthine paths of metal foil and contain no components whatever. Yet designers of these devices still talk about Butterworth, Chebyshev, and Bessel approximations, poles and zeros, and all the other theoretical niceties of network synthesis. In the midst of this diversity there is unity, and the design equations of the microwave engineer are strongly and directly traceable, both historically and ideologically, to the original reflections of Wagner and Campbell fifty years ago.

In addition to being a guide to the solution of filter problems, this book leads into the realm of network synthesis with its specific terminology. It is felt that the wealth of network synthesis theory is still not being fully utilized in design work, and it is hoped that this book will arouse interest in the science of synthesis and help to bridge the gap between the strictly theoretical concepts and the everyday practice in engineering laboratories and scientific establishments.

I wish to express my gratitude to Professor Dr. Fritzsche, Professor Dr. G. Bosse, and C. F. Kurth for permission to use their publications, and to Dr. D. S. Humpherys for bringing network synthesis to the engineering level evidenced by the tables in this book. Special recognition is due to M. Savetman for his valuable help in the preparation of the final manuscript. I also wish to express my appreciation to R. Anderson, R. Ballesteros, H. Blinchikoff, Dr. E. Khu, S. Russell, and C. Vale, the engineers who were instrumental in devising many of the designs and innovations in filter technique, and to Dr. J. Bobis, P. Geffe, R. M. Morrison, and S. I. Rambo who were helpful in creating the proper atmosphere in the Networks Synthesis Department. Many thanks to the Engineering Management of the Surface Division, Westinghouse Electric Corporation, for their continual encouragement.

Anatol I. Zverev

BALTIMORE, MARYLAND
JUNE 1967

Contents

Handbook of Filter Synthesis

1

Filters in Electronics

1.1 TYPES OF FILTERS

Electric wave filters can be classified by several different methods. In terms of the frequency spectrum they may be grouped as audio-frequency, radio-frequency, and microwave filters. In terms of the circuit configuration of the basic elements, filters may take the configuration of a ladder (in the form of T or π) or a lattice (Fig. 1.1), the most general type of network. Classification in terms of the character of the elements is also common: LC filters, filters containing distributed components such as stripline or coaxial filters, and filters comprised of electromechanical, piezoelectric, and magnetostrictive resonators. If a network has an internal source of energy, it may be termed an active filter. An IF amplifier is an example of an active network. Filters with no source of energy within the network are termed passive. Figure 1.2 classifies filter networks according to the character of the elements utilized.

The following five basic types of selective networks are used for frequency discrimination in electronic equipment:

1. The lowpass filter (Fig. 1.3) passes the package of wave energy from zero frequency up to a determined cutoff frequency and rejects all energy beyond that limit. For example, the effective transmission of the human voice requires a frequency band ranging from near 0 to 4000 cps.

2. The highpass filter (Fig. 1.4) prevents the transmission of frequencies below a determined point and appears to be electrically transparent to frequencies beyond this point. The waveguide, used at microwave frequencies, behaves as a typical highpass filter and usually does not pass signals below several hundred megacycles.

3. The bandpass filter (Fig. 1.5) passes the package of waves from certain lower to upper frequency limits and stops all energy outside these two limits. This filter is by far the most important and most commonly used in electronic equipment.

4. The band-reject filter (Fig. 1.6) is used in electronic equipment when a certain unwanted frequency or band of frequencies has to be rejected. Outside of the rejection band or stopband, all frequencies will pass without appreciable attenuation.

5. All-pass filters pass all frequency components of the input signal, but introduce a predictable phase shift for different components of the wave package. A short impulse on the input side of such a filter is modified into a longer frequency-modulated signal at the output. It is evident that the all-pass device can be called a filter only in a limited sense, since in the frequency domain it does not discriminate between the amplitude of the various signals.

From the frequency domain point of view, an ideal filter is one that passes, without attenuation, all frequencies inside certain frequency limits while providing infinite attenuation for all other frequencies.

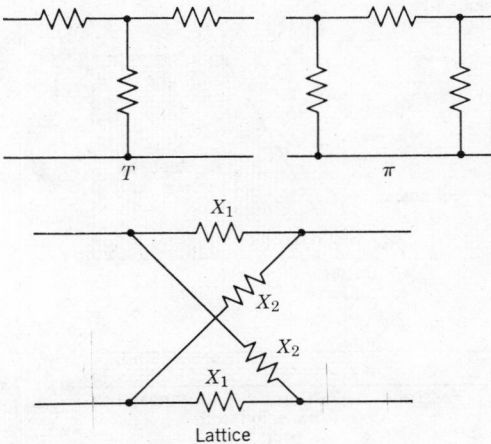

Fig. 1.1. Filter configurations: T, π, and lattice.

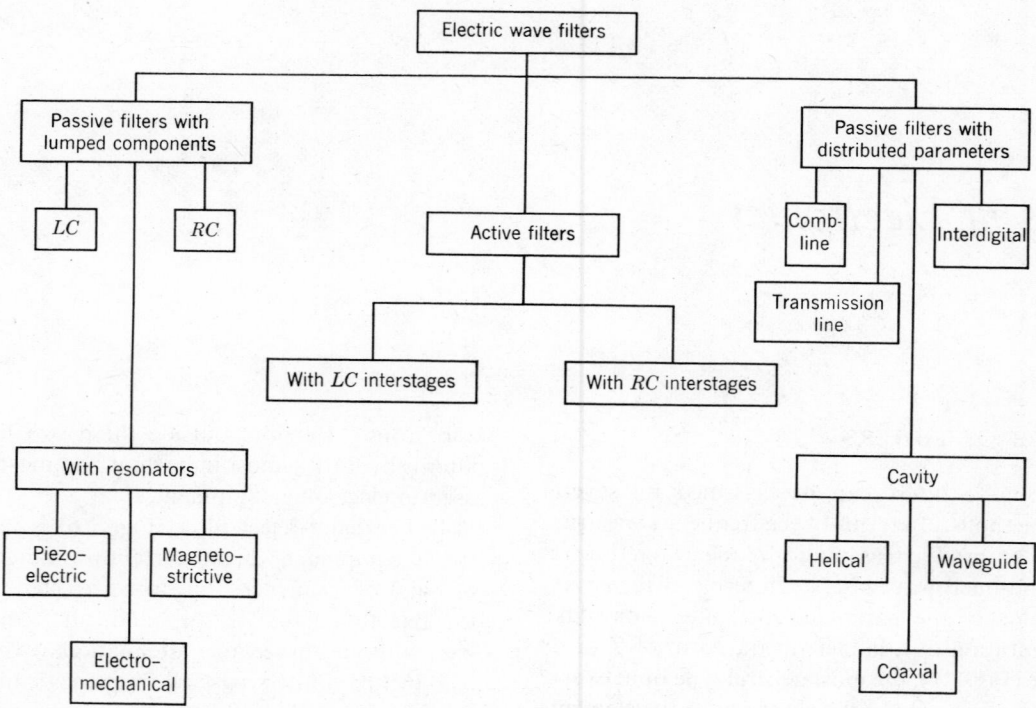

Fig. 1.2. Classification of filters.

The transfer function $|W(j\Omega)|$, the ratio of output to input quantities in the frequency domain is shown for an ideal filter in Fig. 1.7.

From the time domain point of view, an ideal filter is one whose output is identical to its input except for delay τ_o, or

$$e_{\text{out}}(t) = e_{\text{in}}(t - \tau_o) \qquad (1.1.1)$$

Taking the Laplace transform of the above equation and looking at the transfer function in the frequency domain, we obtain the ideal transfer function

$$W(s) = e^{-\tau_o s} \qquad (1.1.2)$$

Letting $s = j\Omega$,

$$W(j\Omega) = e^{-j\Omega\tau_o} \qquad (1.1.3)$$

This function is not frequency selective, since it has unity amplitude; its phase decreases linearly with frequency. These conditions may be realized in practice when the delay approximates a constant for

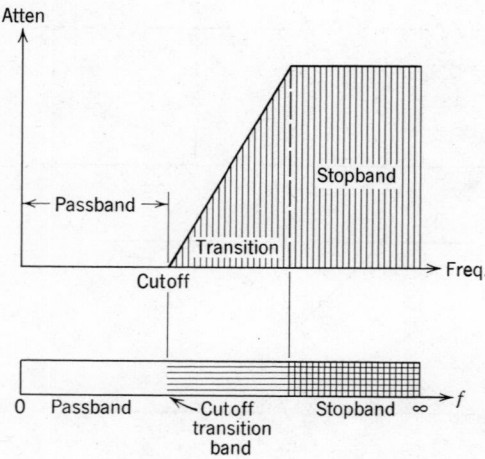

Fig. 1.3. Lowpass filter response.

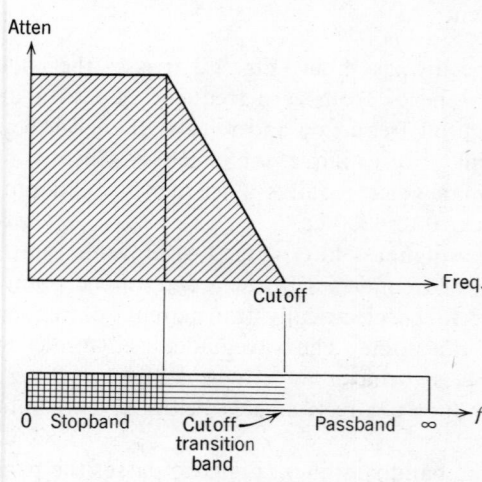

Fig. 1.4. Highpass filter response.

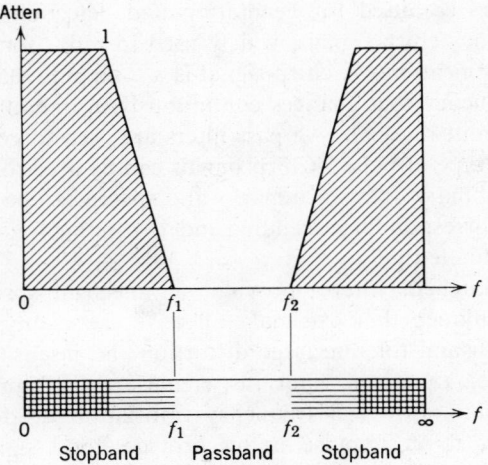

Fig. 1.5. Bandpass filter response.

the range of frequencies over which the attenuation is small.

1.2 FILTER APPLICATIONS

The use of electric wave filters in electronic equipment has increased as equipment has become more complex. Many subsystem operations rely on filters.

Preselector Networks

Preselector networks are required at the input of all sensitive receivers. They separate the desired signal or signals from the unwanted signals. Because the desired signal is usually very low in amplitude, although the undesirable signals including noise may be of appreciably greater magnitude, the preselector

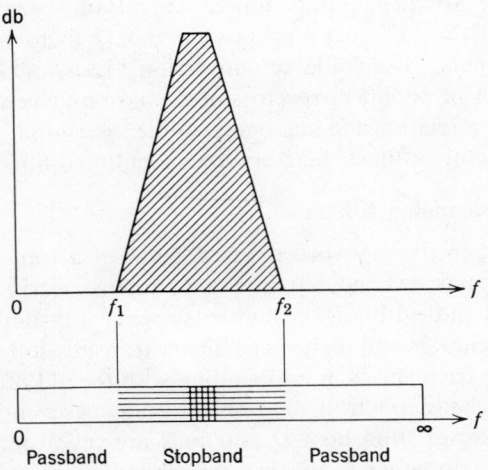

Fig. 1.6. Band-reject filter response.

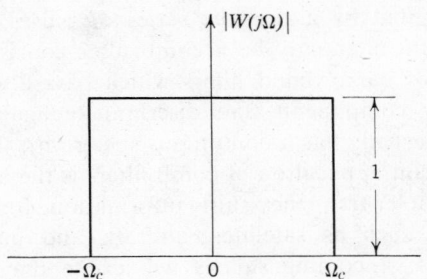

Fig. 1.7. Ideal magnitude response.

network is required to have very low insertion loss for the desired signal and high attenuation for the undesirable signals. The attenuation of desirable signals reduces the signal-to-noise ratio. The preselector filter, covering the entire frequency band of interest, is usually tunable, and provides only a small part of the needed selectivity of the whole receiver.

IF Filters

The next step in signal selection of a typical receiver usually occurs in the IF strip or IF amplifier. This selection and signal discrimination is of very high quality, especially in communication receivers. The bandwidth of the IF filter determines the quality of the system: passband ripple, noise content, and the sharpness of the separation between neighboring transmitted signals. Filters of this kind are usually designed in two or three interstage blocks, separated by tubes or transistors.

SSB Filters

In contrast to the symmetrical preselector filter, the single-sideband (SSB) filter requires a nonsymmetrical attenuation response. Phase-difference networks are sometimes used instead of a filter to eliminate the unwanted sideband. In either case, the main purpose of the network is to suppress the unwanted sideband to such a degree that it does not contribute appreciably to amplitude distortion and instability in the received signal. Insufficient unwanted sideband suppression and insufficient synchronism of the carrier frequency produces undesired beat frequencies.

Comb Filters

When noise is prevalent or jamming is introduced, the extraction of a predetermined signal from a medium can be performed by optimum filters. In general, the input to an optimum filter consists of a mixture of signal and noise. The output of this filter closely approximates the desired signal. For a signal represented by a periodic series of pulses, such an optimum filter may be a comb filter consisting of a chain of narrowband filters which pass discrete frequency components and discriminate against noise (noise usually has a continuous spectrum). The most important application of comb filters is the extraction of doppler-frequency-shift information for passing targets such as satellites, aircraft, and underwater missiles. Incoming signals will excite one of these filters and develop output only in that channel at that time. Evidently, if the input frequency changes, as it usually does when one observes the doppler phenomenon, the output will travel from channel to channel. The number of the channel with maximum output and value of that output is the type of information available in this system.

Multiplexing Filters

Filters can provide multiple use of a broad-spectrum beam between terminal stages of a radio relaying system. It is possible to create up to one thousand telephone channels in one microwave link. In the case of a wire-carrier or power line communication network, the frequency range extends from the audio band up to approximately 200 kc. The use of coaxial lines widens the usable range of frequencies and allows more communication channels to be created. The purpose of multiplexing equipment in power line networks is the channeling of voice communications, telegraph, telemetering, or telecontrol between distant points of road networks, pipeline networks, and power stations. The major requirement of a multiplexing filter is that it obtain the sharpest possible attenuation outside the passband in order to suppress any crosstalk between the channels.

Anti-jam Filtering

Artificially created noise for jamming can completely destroy a radar target signal if no anti-jam features are incorporated into the radar system. To improve target detectability some special equipment features are needed, and the narrowband filter is the key component. The main requirement in this situation is that the filter, operating with pulsed signals, have both selectivity and the ability to minimize overshoot and ringing. To satisfy the requirement, the frequency response curve of the filter usually is of a Bessel or Gaussian shape.

Matched Filters

The new science of correlation techniques and time-domain filtering is based on the matched filter. Such filters are used for generation and detection of the famous chirp signals, widely used in radar for target identification. A chirp signal is a long pulse having a frequency that changes continuously in one direction without reversal. All-pass filters are used to generate this type of pulse. Chirp signals can be used to transmit binary data since marks and spaces can be coded by corresponding ascending and descending frequency-modulated pulses.

Matched filters provide a spectrum-spreading technique; they can make effective use of any bandwidth and tolerate large distortion, be insensitive to noise, tones, or spurious signals, resist jamming, operate with SSB frequency translation or doppler shift, reject impulse noise, provide good signal-to-noise ratio, and require no synchronization. Their areas of application include teletype, signaling field data, and various data-entry systems. Matched filter correlation techniques are very useful in meteor-burst communication systems for minimizing the effect of multipath propagation and external interference.

Frequency Multipliers

Filters utilizing nonlinear reactances find application in all sorts of electronic equipment. In frequency multipliers nonlinear elements and associated idlers are inserted between narrowband filters tuned to the fundamental frequency and one of the harmonic frequencies (which must be extracted). This arrangement can produce any signal harmonic related to the pilot source clock. A chain of parametric frequency multipliers can start with any low frequency which can be maintained constant.

The ideal network with one nonlinear reactive element transforms all the power of the fundamental input frequency into power at certain harmonics. Here the efficiency depends on the Q factor of the elements. Unavoidable conversion losses and the Q factor of nonlinear reactors in the harmonic generator have a relationship analogous to the insertion loss and Q factor of linear reactors in conventional filters.

Broadbanding Filters

A reactive network inserted between a transmitter and a narrowband, low-frequency antenna can improve the bandwidth of the entire system. In solid-state transmitters and high-speed binary transmission at low radio frequencies, broadbanding with the aid of filters is the only practical solution for many cases in which already existing high-Q antennas are involved. The only consequence of this broadbanding is that the filter input impedance, with respect to the transmitter output, varies widely with frequency. This type of system imposes requirements on the available power from the source. However, the efficiency of the transmitter for frequencies in the vicinity of the center frequency can still be as high as 90%.

Impedance Transformation

An impedance-matching network is not always a physical transformer having primary and secondary windings. It may also take the form, for example, of a lowpass ladder, giving a prescribed passband and moderate to high attenuation outside the passband.

Every bandpass filter is potentially an impedance transformer. No matter how the filter is developed, its input and output impedances can be made different from one another. The usual way to obtain an impedance transformation is to introduce Norton's ideal transformer, which consists of three elements in π or T form. This transformation imposes some limitation on the transformation coefficient n (the step-down or step-up ratio). For most filter configurations there exists some maximum value of transformation ratio. At this particular value of n the resulting network may consist of fewer elements than the original filter. If this value is surpassed the filter will become unrealizable because a negative element may be required.

Filters as Coupling Networks

Some signal sources can often be simulated as a current generator with a capacitor across the terminals. A broadband output circuit for such a source can be designed as a filter, driven by an infinite-impedance source with one finite termination. Evidently the output capacitance of the source is utilized as one of the elements required for the filter and consequently the damaging effect of shunt capacitance is controlled. The driving source may also be regarded as a voltage generator plus a reactive element, and therefore a similar situation exists requiring a filter structure for coupling to the following stages.

In some cases the input and output impedances of electronic subassemblies (amplifiers and oscillators, for example) may be represented as a resistor in series with a reactive element. In such cases impedance-matching filters are necessary if optimum broadband performance is to be approached.

Multicouplers

The number of antennas that can be accommodated in a given installation is restricted by mutual disturbance of radiation patterns and intercoupling of signals. In complex electronic systems a large number of individual transmitters and receivers may be operated with a single antenna and still satisfy the system requirements. Special multicoupler networks involving filters are necessary to permit the sharing of antennas by groups of receivers and transmitters.

A desirable multicoupler may consist of minimum-loss type narrowband filters connected together. The number of channels can be large (twenty or more), whereas the passband of each branch is small in comparison with the minimum channel separation. A typical application is found in the 225- to 400-Mc band, with a minimum channel spacing of 2 Mc, an adjacent channel isolation of 60 dB, a channel-center loss of 1 dB or less, and a bandwidth sufficient for typical AM, SSB, or FM signals is achieved.

Harmonic Suppressor

The operation of many systems rely on the purity of a sinusoidal signal and its phase relation to an incoming signal. Harmonic content can easily upset the performance, and therefore the filtering of a pilot frequency source or reference signal is imperative for normal operation.

Coherent Integrators

A set of filters can be used to analyze the return signals from moving targets. If that signal includes the so-called doppler frequency because of a changing relative position to the receiver, the filter can help to determine the location of the target from which the pulse is reflected and also the speed of the target. The coherent integrator is excited by a series of pulses and the energy from this signal is integrated and consequently produces some output. If the frequency content of the pulse is shifting, a narrowband filter with slightly different center frequency responds. Several pulses with the same frequency content are usually necessary to produce a detectable output and must be coherent to produce an additive effect in the resonating system of the filter. After reading the information the energy accumulated in the filter has to be disposed of to make the system free from any electric charge accumulated before the next series of pulses properly excite one of the filters in the bank for a new position of the target.

1.3 ALL-PASS FILTERS

Being an all-pass filter, the lattice network shown in Fig. 1.8*a* has the following properties: along the imaginary (real frequency) axis between $s = -j\Omega$ to

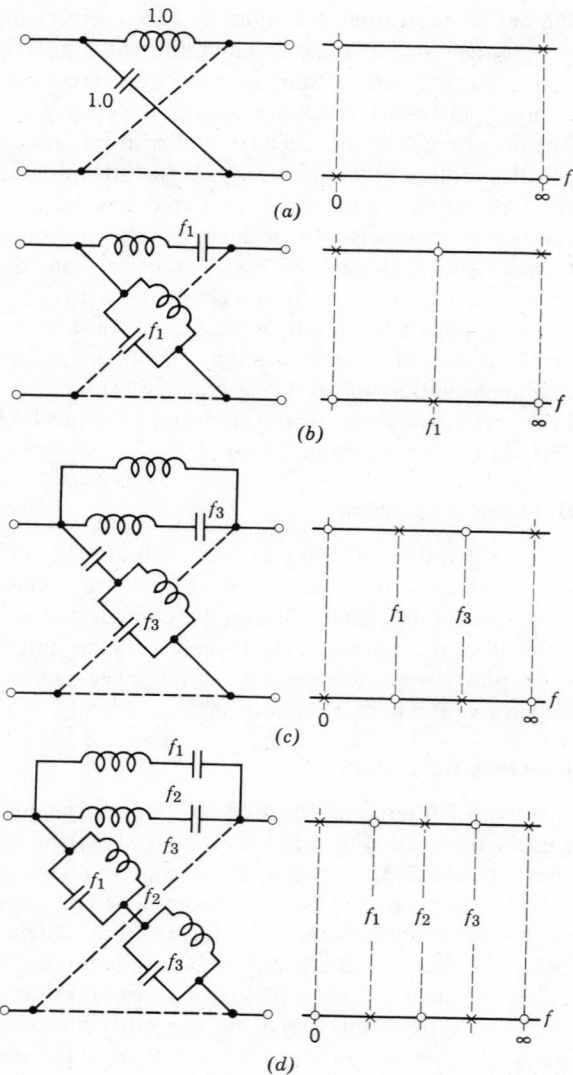

Fig. 1.8. All-pass filters: (*a*) first order, (*b*) second order, (*c*) third order, and (*d*) fourth order. Pole-zero diagrams for the reactances are given.

$s = +j\Omega$ the attenuation is

$$a = \ln \left| \frac{1 + j\Omega}{1 - j\Omega} \right| = 0 \qquad (1.3.1)$$

Thus the filter exhibits zero attenuation, but the phase varies from -180 to $+180°$ as shown by curve *b* in Fig. 1.9.

Along the real axis $s = \sigma$ in the complex frequency plane, which does not correspond to any real frequency, the attenuation is

$$a = \ln \left| \frac{1 + \sigma}{1 - \sigma} \right| \qquad (1.3.2)$$

If $\sigma = \pm 1$, the attenuation $A = \infty$; hence a signal $e^{\pm t}$ will not be transmitted as shown by curve *a* in Fig. 1.9.

It is easy to show that the attenuation of the elementary all-pass lattice has inverse properties along the positive real axis about the point $\sigma = 1$; that is, for any point $\sigma > 1$ there is a corresponding point $0 < 1/\sigma < 1$ where the attenuation is equal.

The all-pass network is one of the most important components in a large number of communication and target-detection systems. It is helpful to mention a few outstanding uses of all-pass filters in order to appreciate its importance:

1. Expansion of signals in the time domain
2. Phase-correction of signals
3. Phase-splitting of signals
4. Intermediate tool in network synthesis
5. Delay of a signal without introducing frequency distortion

In the first case a specified parabolic phase response is obtained. In the second, a network is designed to a specified phase-frequency characteristic in order to obtain over-all linearity. In the third case two all-pass networks with a common input are synthesized to provide, within the given frequency limits, two outputs which are constant in amplitude but whose phases are in quadrature. In the fourth case all-pass networks of any complexity are used only in the initial step of filter design where their properties along the real axis can be translated on the real frequency axis. With the aid of transformations, the resulting lattice networks are made to take the form of well-known lattice or ladder filter types, such as the highpass, lowpass, bandstop, or the bandpass. In the fifth case the all-pass network takes the form of a simple delay line.

1.4 PROPERTIES OF LATTICE FILTERS

A lattice network can be realized with any desired amplitude characteristic, and its characteristic impedance is independent of its transmission properties. If the series arm of the lattice is called X_1 and the parallel arm X_2, as in Fig. 1.1, the characteristic impedance of the lattice is given by

$$Z_0 = \sqrt{X_1 X_2} \qquad (1.4.1)$$

and the transmission constant g by

$$\tanh \frac{g}{2} = \sqrt{\frac{X_1}{X_2}} \qquad (1.4.2)$$

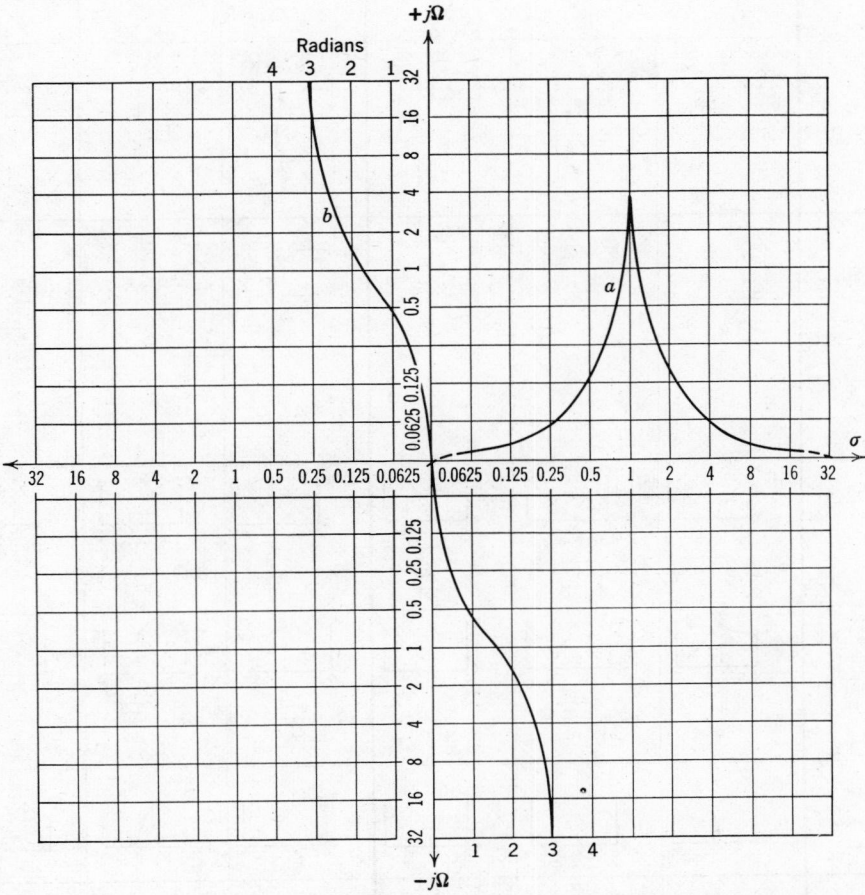

Fig. 1.9. Attenuation and phase relationships of all-pass filter of Fig. 1.8*a*.

The condition of transparency for any lattice filter is very simple. The arm reactances X in the series and parallel arms must be opposite in sign. As soon as the two arm reactances have the same sign (both capacitive or both inductive), the filter stops being transparent.

The amplitude response of the lattice filter depends on only the ratio of the branch reactances and is independent of the input and output impedances. When $X_1 = X_2$, tanh $g/2 = \pm 1$ and a maximum, or peak, of attenuation occurs. The position of the peak can be modified without impedance change by multiplying X_1 by a real positive factor and dividing X_2 by the same factor.

Lattice networks must be designed and built with great care in order to balance their impedances. Once reactance values are chosen to resonate at strategic positions in the passband (such as the cutoffs and the center of the passband) the level of the reactances between these strategic positions must be maintained

in both arms in an exactly prescribed fashion. Furthermore, the peak of attenuation outside the passband is controlled by the level of the impedances of both arms when they are of the same nature (capacitive or inductive), whereas the position of the attenuation peak depends upon the sharpness of the reactance curve at cutoff (which is effectively within the passband).

Therefore the peak's position cannot be controlled by simply changing an inductance or capacitance. Control is very remote and must be applied through the resonances which occur at some frequency in the passband or at cutoff. Although this is an inconvenience, it offers the advantage that the amount of attenuation or the sharpness of the response curve can sometimes be adjusted by a slight variation in the parameters of the resonators. This procedure is often used in crystal filter design but is impractical with conventional coils and capacitors that require precise nominal values and good environmental stability.

A composite lattice has the same basic form as the

	LOWPASS SECTION			HIGHPASS SECTION		
TYPE NUMBER	1	2	3	1	2	3
ARMS OF FULL SECTION Z_1 Z_2	FUNDAMENTAL TYPE 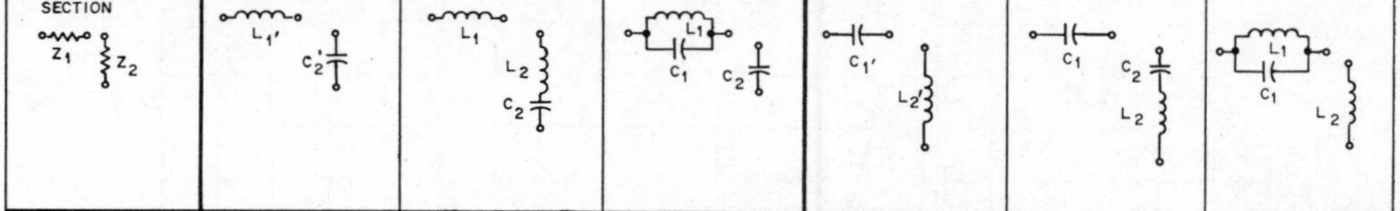 L_1' C_2'	M-DERIVED L_1 L_2 C_2	M-DERIVED L_1 C_1 C_2	FUNDAMENTAL TYPE C_1' L_2'	M-DERIVED C_1 C_2 L_2	M-DERIVED L_1 C_1 L_2
ELEMENT VALUES	$L_1' = \dfrac{R}{\pi f_1}$ $C_2' = \dfrac{1}{\pi f_1 R}$	$L_1 = mL_1'$ $L_2 = \dfrac{1-m^2}{4m}L_1'$ $C_2 = mC_2'$	$L_1 = mL_1'$ $C_1 = \dfrac{1-m^2}{4m}C_2'$ $C_2 = mC_2'$	$C_1' = \dfrac{1}{4\pi f_1 R}$ $L_2' = \dfrac{R}{4\pi f_1}$	$C_1 = \dfrac{C_1'}{m}$ $C_2 = \dfrac{4m}{1-m^2}C_1'$ $L_2 = \dfrac{L_2'}{m}$	$C_1 = \dfrac{C_1'}{m}$ $L_1 = \dfrac{4m}{1-m^2}L_2'$ $L_2 = \dfrac{L_2'}{m}$
$Z_T = \sqrt{Z_1 Z_2}\sqrt{1+\dfrac{Z_1}{4Z_2}}$	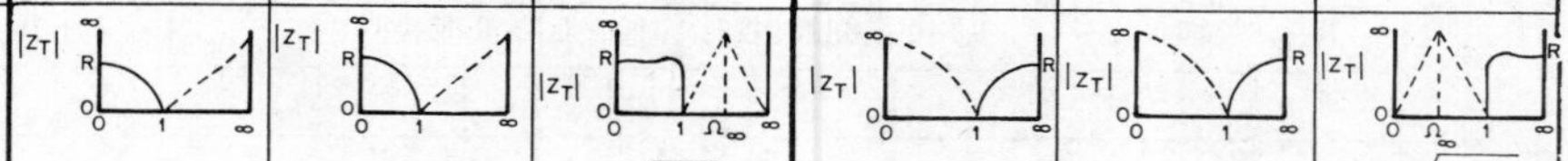 $\lvert Z_T\rvert$ (graph) $Z_T' = R\sqrt{1-\Omega^2}$	$\lvert Z_T\rvert$ (graph) $Z_T = R\sqrt{1-\Omega^2}\;(=Z_T')$	$\lvert Z_T\rvert$ (graph) $Z_T = R\dfrac{\sqrt{1-\Omega^2}}{1-\dfrac{\Omega^2}{a^2}}$	$\lvert Z_T\rvert$ (graph) $Z_T' = R\sqrt{1-\dfrac{1}{\Omega^2}}$	$\lvert Z_T\rvert$ (graph) $Z_T = R\sqrt{1-\dfrac{1}{\Omega^2}}\;(=Z_T')$	$\lvert Z_T\rvert$ (graph) $Z_T = R\dfrac{\sqrt{1-\dfrac{1}{\Omega^2}}}{1-\dfrac{1}{a^2\Omega^2}}$
$Z_\pi = \dfrac{\sqrt{Z_1 Z_2}}{\sqrt{1+\dfrac{Z_1}{4Z_2}}}$	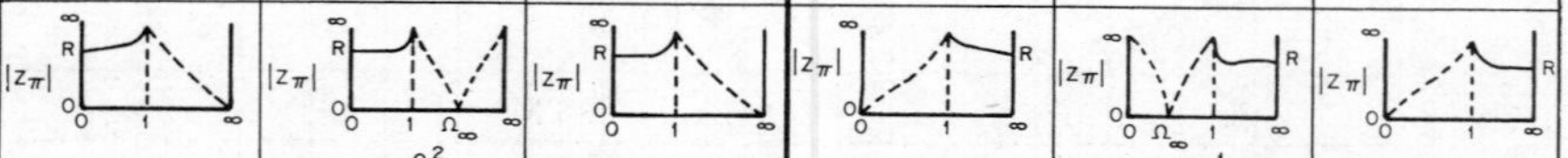 $\lvert Z_\pi\rvert$ (graph) $Z_\pi' = \dfrac{R}{\sqrt{1-\Omega^2}}$	$\lvert Z_\pi\rvert$ (graph) $Z_\pi = R\dfrac{1-\dfrac{\Omega^2}{a^2}}{\sqrt{1-\Omega^2}}$	$\lvert Z_\pi\rvert$ (graph) $Z_\pi = \dfrac{R}{\sqrt{1-\Omega^2}}\;(=Z_\pi')$	$\lvert Z_\pi\rvert$ (graph) $Z_\pi' = \dfrac{R}{\sqrt{1-\dfrac{1}{\Omega^2}}}$	$\lvert Z_\pi\rvert$ (graph) $Z_\pi = R\dfrac{1-\dfrac{1}{a^2\Omega^2}}{\sqrt{1-\dfrac{1}{\Omega^2}}}$	$\lvert Z_\pi\rvert$ (graph) $Z_\pi = \dfrac{R}{\sqrt{1-\dfrac{1}{\Omega^2}}}\;(=Z_\pi')$
ATTENUATION CONSTANT FOR IDEAL ELEMENTS	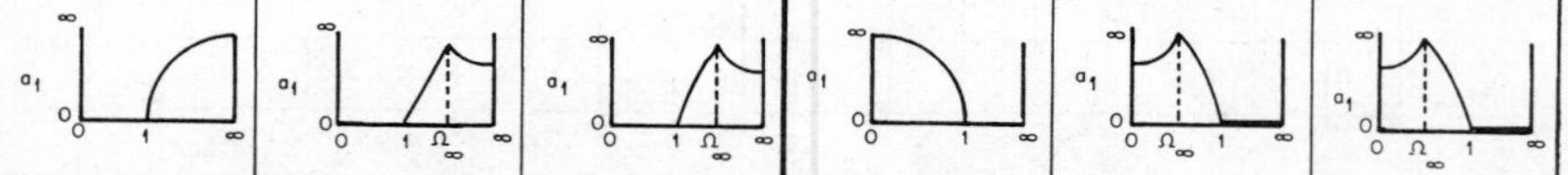 a_1 (graph)	a_1 (graph)	a_1 (graph)	a_1 (graph)	a_1 (graph)	a_1 (graph)
PHASE CONSTANT FOR IDEAL ELEMENTS	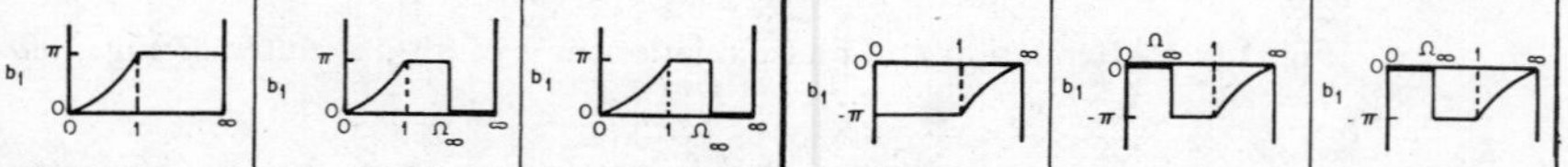 b_1 (graph)	b_1 (graph)	b_1 (graph)	b_1 (graph)	b_1 (graph)	b_1 (graph)
$\dfrac{Z_1}{4Z_2}$ — WITHOUT LOSSES $d=0$	$-\Omega^2$	$\dfrac{a^2-1}{\dfrac{1-a^2}{\Omega^2}}$		$-\dfrac{1}{\Omega^2}$	$\dfrac{a^2-1}{1-a^2\Omega^2}$	
$\dfrac{Z_1}{4Z_2}$ — WITH LOSSES $d\neq0$	$-\Omega^2(1-jd_l)(1-jd_c)$	$\dfrac{\dfrac{a^2-1}{a^2}}{1-\dfrac{1}{\Omega^2(1-jd_l)(1-jd_c)}}$		$-\dfrac{1}{\Omega^2(1-jd_l)(1-jd_c)}$	$\dfrac{a^2-1}{1-a^2\Omega^2(1-jd_l)(1-jd_c)}$	
DEFINITIONS	$\Omega = \dfrac{f}{f_1}$ $\Omega_\infty = \dfrac{f_\infty}{f_1} = a$	$m^2 = 1-\dfrac{1}{a^2}$ $a^2 = \dfrac{1}{1-m^2}$	$d_l = \dfrac{r_l}{\omega L}$ $d_c = \dfrac{1}{R_c\,\omega C}$	$\Omega = \dfrac{f}{f_1}$ $\Omega_\infty = \dfrac{f_\infty}{f_1} = \dfrac{1}{a}$	$m^2 = 1-\dfrac{1}{a^2}$ $a^2 = \dfrac{1}{1-m^2}$	$d_l = \dfrac{r_l}{\omega L}$ $d_c = \dfrac{1}{R_c\,\omega C}$

Fig. 1.10. Lowpass and highpass conventional structures.

elementary lattice. For example, the all-pass filter of the first order (Fig. 1.8*a*) consists of a series and a lattice arm. Higher order all-pass structures as shown in Fig. 1.8*b*, *c*, and *d* still have only two arms, but their schematics are more complicated. The number of reactive components is always the same for both branches, but the nature of the reactances is different at every point on the frequency scale from zero to infinity.

In general, the transmission properties of composite lattice filters outside the passband are controlled by the natural frequencies of the branch reactances inside the passband. Similarly, the flatness of the passband, and consequently the flatness of the input and output impedance of the filter is controlled by the natural frequencies outside the passband. If there is a large number of these frequencies in the passband, the attenuation in the stopband may be high and a more rapid transition from passband to stopband can be obtained.

1.5 FILTER BUILDING BLOCKS

The electric wave filter can be visualized as a combination of simple building blocks called sections, an approach similar to combining the blocks of gain of tubes or transistors. Each of these filter blocks is a certain canonic combination of lumped reactances. At microwave frequencies these reactances are distributed, but for the purpose of analysis they can be reduced to an equivalent schematic with lumped components. A lowpass elementary ladder structure is shown in Fig. 1.10 (type 1). In Fig. 1.15*a* two elementary lowpass lattice structures are shown with their equivalent bridged-T and semilattice circuits. The configuration in the center of Fig. 1.15*a* is a bridged-T schematic, and the form on the extreme right is the so-called semilattice or differential bridge filter. The lowpass filter with a transformer, such as the differential bridge, is a lowpass filter only in a limited sense because it does not pass direct current. Figure 1.15*b* shows the reactance of the lattice arms, and Fig. 1.15*c*, the attenuation curve. Figure 1.10 (highpass type 1) illustrates a highpass elementary structure in ladder form, and Fig. 1.16, the elementary lattice highpass structure and its equivalents. Similarly, an elementary bandpass filter is shown in Fig. 1.11 (bandpass type I_1). It is conventionally known as the *k*-constant type. The lattice bandpass filters of Fig. 1.17 possess a ladder equivalent shown as type IV in Fig. 1.13 and can also be shown in a bridged-T or differential bridge form as is customary in crystal filter

practice. The lattice form of the filter is highly uneconomical. It consists of repetitive elements (shown as dotted lines) and consequently is very seldom used. Being the most general type of building block, however, it theoretically permits the realization of a more universal response than any of its partial equivalents.

The simple lattice-filter building block shown in Fig. 1.17 produces one peak of attenuation at f_∞ that can be realized on either side of the passband, but the ladder equivalent of the given lattice schematic, with given element values, can be realized with a peak of attenuation only on one side, depending on the interrelation of the elements. Under certain conditions one element in the parallel arm of the ladder section can be made to disappear. For example, in the case of filter type IV_2 of Fig. 1.13 the inductance in the arm across the line may be zero, resulting in filter type III in Fig. 1.12. The filter's response in this case will exhibit no attenuation peaks. In filter type IV_2, if the capacitance in the shunt arm is equal to infinity it can be removed from the schematic resulting in type III_2, and the response curve on both sides of the passband will again be monotonic, having no peaks of attenuation.

Concept of Sections for Synthesis

A section (or half section) as a building unit is characteristic of the image parameter theory. In design techniques based on synthesis the concept of sections is fading away and another set of terms (for example, poles and zeros of transfer or transmission functions) is replacing it. Nevertheless, the term sections does appear even when the filter design is based on polynomial development but has no relation at all to the Zobel technique. The concept of sections has proved to be especially helpful in connection with crystal filter design. One section, for instance, could be equivalent to a combination of several zeros of transmission function which describes the more complex filter. One section can also be equivalent to second, third, or higher order polynomial for which the physical components are combined in a schematically separate network.

Subdivision into sections may be dictated by technological reasons, excess of insertion loss (and consequently the necessity to put an amplifier between the filters), and (in the case of crystal filters) the need to eliminate spurious responses. Filters in the form of a single lattice structure, or its equivalent, are unable to satisfy many practical requirements. In fact, being highly dependent on the accuracy of the

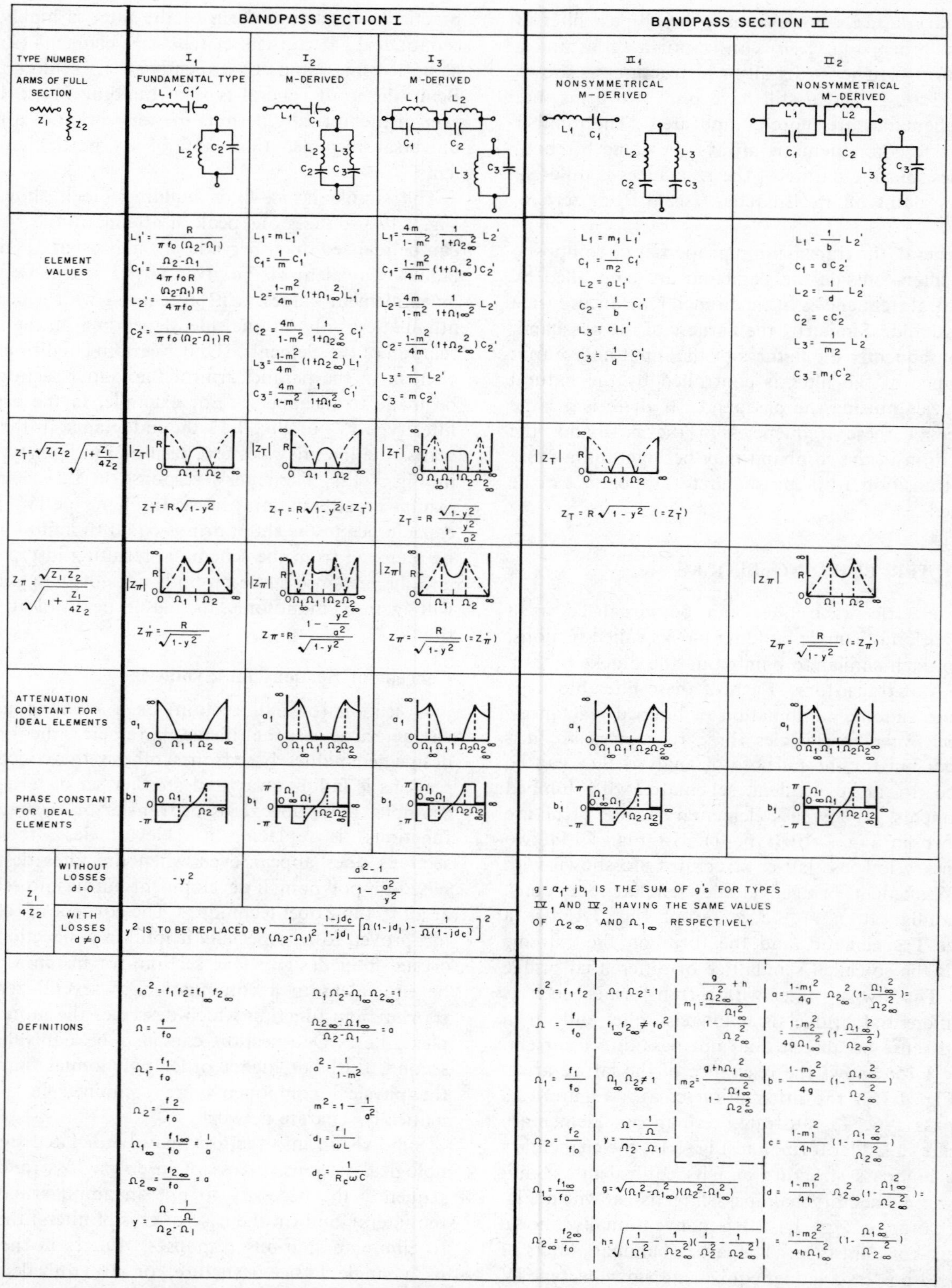

Fig. 1.11. Conventional bandpass structures of the fundamental type (I_1) and *m*-derived type with two attenuation peaks (I_2, I_3, II_1, II_2).

	BANDPASS SECTION I			BANDPASS SECTION II	
TYPE NUMBER	I_1	I_2	I_3	II_1	II_2
ARMS OF FULL SECTION (Z_1, Z_2)	FUNDAMENTAL TYPE L_1' C_1'	M-DERIVED	M-DERIVED	NONSYMMETRICAL M-DERIVED	NONSYMMETRICAL M-DERIVED
ELEMENT VALUES	$L_1' = \dfrac{R}{\pi f_0(\Omega_2-\Omega_1)}$ $C_1' = \dfrac{\Omega_2-\Omega_1}{4\pi f_0 R}$ $L_2' = \dfrac{(\Omega_2-\Omega_1)R}{4\pi f_0}$ $C_2' = \dfrac{1}{\pi f_0(\Omega_2-\Omega_1)R}$	$L_1 = m L_1'$ $C_1 = \dfrac{1}{m}C_1'$ $L_2 = \dfrac{1-m^2}{4m}(1+\Omega_{2\infty}^2)L_1'$ $C_2 = \dfrac{4m}{1-m^2}\dfrac{1}{1+\Omega_{2\infty}^2}C_1'$ $L_3 = \dfrac{1-m^2}{4m}(1+\Omega_{1\infty}^2)L_1'$ $C_3 = \dfrac{4m}{1-m^2}\dfrac{1}{1+\Omega_{1\infty}^2}C_1'$	$L_1 = \dfrac{4m}{1-m^2}\dfrac{1}{1+\Omega_{2\infty}^2}L_2'$ $C_1 = \dfrac{1-m^2}{4m}(1+\Omega_{1\infty}^2)C_2'$ $L_2 = \dfrac{4m}{1-m^2}\dfrac{1}{1+\Omega_{1\infty}^2}L_2'$ $C_2 = \dfrac{1-m^2}{4m}(1+\Omega_{2\infty}^2)C_2'$ $L_3 = \dfrac{1}{m}L_2'$ $C_3 = m C_2'$	$L_1 = m_1 L_1'$ $C_1 = \dfrac{1}{m_2}C_1'$ $L_2 = a L_1'$ $C_2 = \dfrac{1}{b}C_1'$ $L_3 = c L_1'$ $C_3 = \dfrac{1}{d}C_1'$	$L_1 = \dfrac{1}{b}L_2'$ $C_1 = a C_2'$ $L_2 = \dfrac{1}{d}L_2'$ $C_2 = c C_2'$ $L_3 = \dfrac{1}{m_2}L_2'$ $C_3 = m_1 C_2'$
$Z_T = \sqrt{Z_1 Z_2}\sqrt{1+\dfrac{Z_1}{4Z_2}}$	$\|Z_T\|$ (plot) $Z_T = R\sqrt{1-y^2}$	$\|Z_T\|$ (plot) $Z_T = R\sqrt{1-y^2}\,(=Z_T')$	$\|Z_T\|$ (plot) $Z_T = R\dfrac{\sqrt{1-y^2}}{1-\dfrac{y^2}{a^2}}$	$\|Z_T\|$ (plot) $Z_T = R\sqrt{1-y^2}\,(=Z_T')$	
$Z_\pi = \dfrac{\sqrt{Z_1 Z_2}}{\sqrt{1+\dfrac{Z_1}{4Z_2}}}$	$\|Z_\pi\|$ (plot) $Z_\pi = \dfrac{R}{\sqrt{1-y^2}}$	$\|Z_\pi\|$ (plot) $Z_\pi = R\dfrac{1-\dfrac{y^2}{a^2}}{\sqrt{1-y^2}}$	$\|Z_\pi\|$ (plot) $Z_\pi = \dfrac{R}{\sqrt{1-y^2}}\,(=Z_\pi')$		$\|Z_\pi\|$ (plot) $Z_\pi = \dfrac{R}{\sqrt{1-y^2}}\,(=Z_\pi')$
ATTENUATION CONSTANT FOR IDEAL ELEMENTS	a_1 (plot)	a_1 (plot)	a_1 (plot)	a_1 (plot)	a_1 (plot)
PHASE CONSTANT FOR IDEAL ELEMENTS	b_1 (plot)	b_1 (plot)	b_1 (plot)	b_1 (plot)	b_1 (plot)
$\dfrac{Z_1}{4Z_2}$ — WITHOUT LOSSES $d=0$	$-y^2$	$\dfrac{a^2-1}{1-\dfrac{a^2}{y^2}}$		$g = a_1 + jb_1$ IS THE SUM OF g's FOR TYPES II_1 AND II_2 HAVING THE SAME VALUES OF $\Omega_{2\infty}$ AND $\Omega_{1\infty}$ RESPECTIVELY	
$\dfrac{Z_1}{4Z_2}$ — WITH LOSSES $d\neq0$	y^2 IS TO BE REPLACED BY $\dfrac{1}{(\Omega_2-\Omega_1)^2}\dfrac{1-jd_c}{1-jd_l}\left[\Omega(1-jd_l)-\dfrac{1}{\Omega(1-jd_c)}\right]^2$				

DEFINITIONS

Section I:

$$f_0^2 = f_1 f_2 = f_{1\infty} f_{2\infty}$$
$$\Omega = \frac{f}{f_0}$$
$$\Omega_1 = \frac{f_1}{f_0}$$
$$\Omega_2 = \frac{f_2}{f_0}$$
$$\Omega_{1\infty} = \frac{f_{1\infty}}{f_0} = \frac{1}{a}$$
$$\Omega_{2\infty} = \frac{f_{2\infty}}{f_0} = a$$
$$y = \frac{\Omega - \dfrac{1}{\Omega}}{\Omega_2 - \Omega_1}$$

(I_2, I_3):

$$\Omega_1 \Omega_2 = \Omega_{1\infty}\Omega_{2\infty} = 1$$
$$\frac{\Omega_{2\infty} - \Omega_{1\infty}}{\Omega_2 - \Omega_1} = a$$
$$a^2 = \frac{1}{1-m^2}$$
$$m^2 = 1 - \frac{1}{a^2}$$
$$d_l = \frac{r_l}{\omega L}$$
$$d_c = \frac{r_c}{R_c \omega C}$$

Section II:

$$f_0^2 = f_1 f_2 \qquad \Omega_1 \Omega_2 = 1 \qquad m_1 = \frac{\dfrac{g}{\Omega_{2\infty}^2}+h}{1-\dfrac{\Omega_{1\infty}^2}{\Omega_{2\infty}^2}}$$
$$a = \frac{1-m_1^2}{4g}\Omega_{2\infty}^2\left(1-\frac{\Omega_{1\infty}^2}{\Omega_{2\infty}^2}\right) = \frac{1-m_2^2}{4g}\Omega_{2\infty}^2\left(1-\frac{\Omega_{1\infty}^2}{\Omega_{2\infty}^2}\right)$$
$$\Omega = \frac{f}{f_0} \qquad f_{1\infty} f_{2\infty} \neq f_0^2$$
$$\Omega_1 = \frac{f_1}{f_0} \qquad \Omega_1 \Omega_2 \neq 1 \qquad m_2 = \frac{g+h\Omega_{1\infty}^2}{1-\dfrac{\Omega_{1\infty}^2}{\Omega_{2\infty}^2}} \qquad b = \frac{1-m_2^2}{4g}\left(1-\frac{\Omega_{1\infty}^2}{\Omega_{2\infty}^2}\right)$$
$$\Omega_2 = \frac{f_2}{f_0} \qquad y = \frac{\Omega - \dfrac{1}{\Omega}}{\Omega_2 - \Omega_1} \qquad c = \frac{1-m_1^2}{4h}\left(1-\frac{\Omega_{1\infty}^2}{\Omega_{2\infty}^2}\right)$$
$$\Omega_{1\infty} = \frac{f_{1\infty}}{f_0} \qquad g = \sqrt{(\Omega_1^2-\Omega_{1\infty}^2)(\Omega_2^2-\Omega_{1\infty}^2)}$$
$$d = \frac{1-m_1^2}{4h}\Omega_{2\infty}^2\left(1-\frac{\Omega_{1\infty}^2}{\Omega_{2\infty}^2}\right) = \frac{1-m_2^2}{4h\Omega_{1\infty}^2}\left(1-\frac{\Omega_{1\infty}^2}{\Omega_{2\infty}^2}\right)$$
$$\Omega_{2\infty} = \frac{f_{2\infty}}{f_0} \qquad h = \sqrt{\left(\frac{1}{\Omega_1^2}-\frac{1}{\Omega_{2\infty}^2}\right)\left(\frac{1}{\Omega_2^2}-\frac{1}{\Omega_{2\infty}^2}\right)}$$

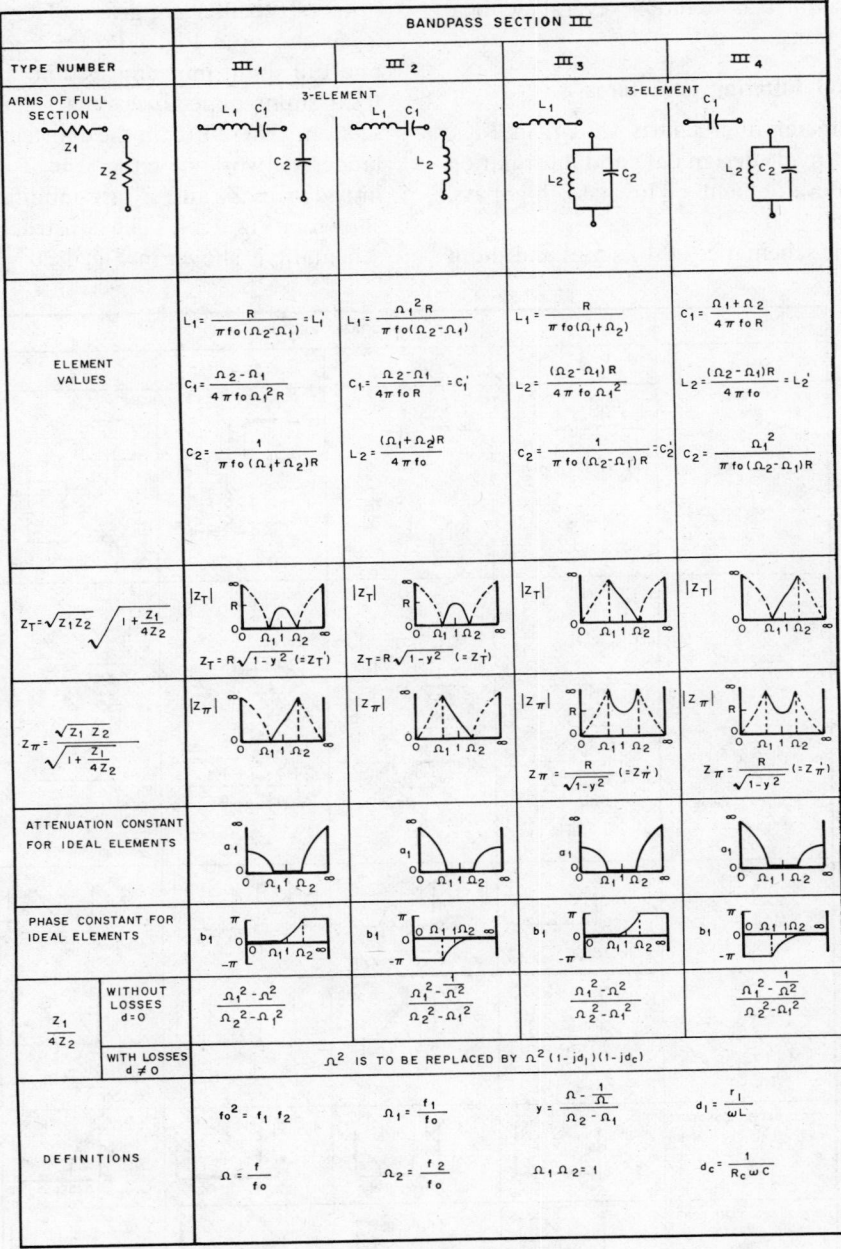

Fig. 1.12. Conventional 3-element bandpass structures.

values of the elements, they cannot provide a large value of ultimate stopband attenuation. Its response is easily degraded by the spurious response characteristic of crystal resonators. Realization in the form of cascaded semilattice sections can solve these problems. Therefore filters may consist of sections, but the meaning of the word is not a conventional one known from image parameter theory, and they are not elementary building blocks as are Zobel sections. A section may include 1, 2, or more crystals in a bridge form and may essentially be a part of ladder *LC* filter (as in the case of very large bandpass filters). Even microwave filters, where the word cavity is usually associated with one resonant circuit or one zero of polynomial, the concept of sections may be employed as a physical division of the structure. In the cases in which coupling is simple and does not produce a transmission zero (or transmission pole) the physical structure may be classified as including several identical or nonidentical sections.

Use of Conventional Filtering Structures

The image parameter filter charts shown in Figs. 1.10 through 1.14 provide design data and information necessary for the development of lowpass, highpass, and bandpass filters.

In the first row the schematics of the series and shunt arms of the filter are given. These impedances Z_1 and Z_2 as shown in Fig. 1.18, are known as the full series and full shunt impedances, and are the total series or total shunt impedance of a full symmetrical T or π section. Therefore, in the construction of a full section ladder network of either the T or π schematic, the impedances Z_1 and Z_2 are modified and connected as shown in Fig. 1.19. The construction of a half-section schematic is shown in Fig. 1.20.

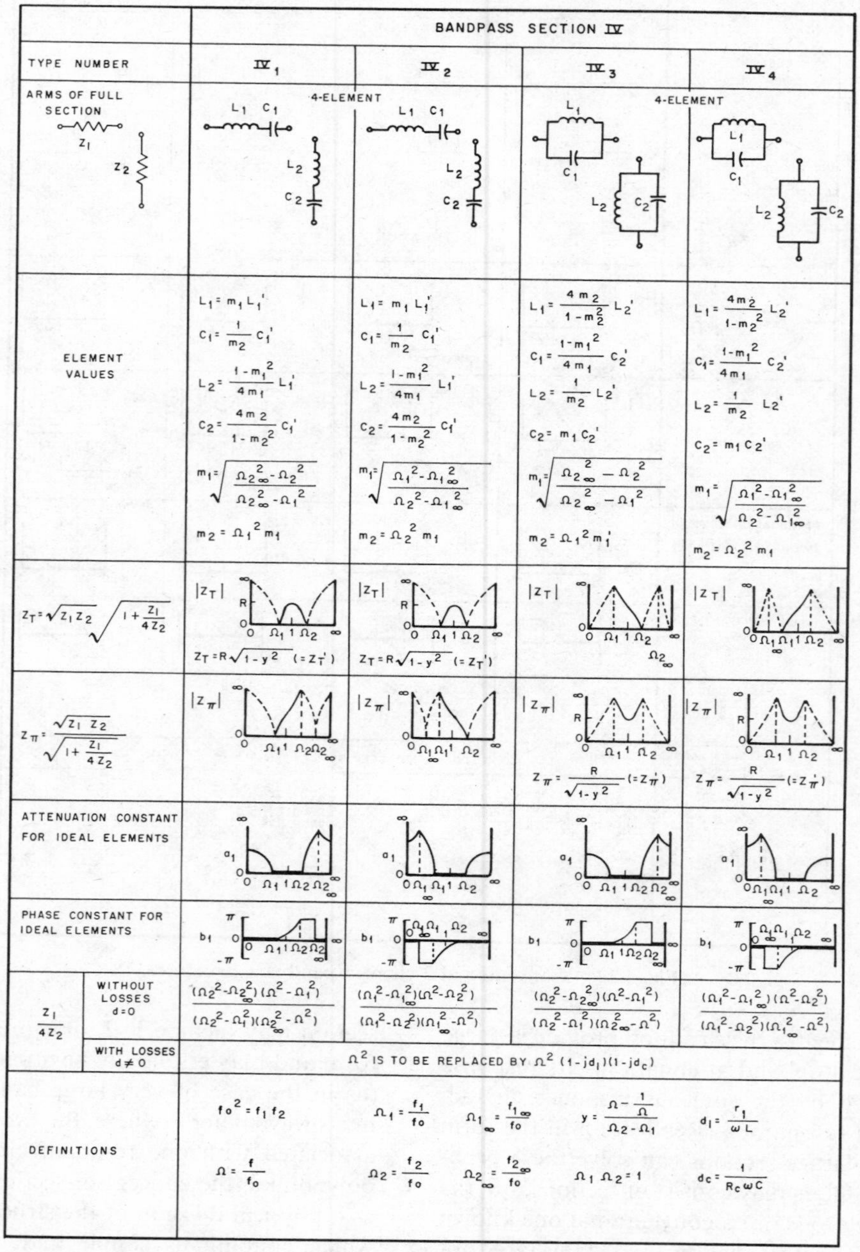

	BANDPASS SECTION IV			
TYPE NUMBER	IV_1	IV_2	IV_3	IV_4
ARMS OF FULL SECTION	4-ELEMENT		4-ELEMENT	
ELEMENT VALUES	$L_1 = m_1 L_1'$ $C_1 = \frac{1}{m_2} C_1'$ $L_2 = \frac{1-m_1^2}{4 m_1} L_1'$ $C_2 = \frac{4 m_2}{1-m_2^2} C_1'$ $m_1 = \sqrt{\frac{\Omega_2^2 - \Omega_\infty^2}{\Omega_\infty^2 - \Omega_1^2}}$ $m_2 = \Omega_1^2 m_1$	$L_1 = m_1 L_1'$ $C_1 = \frac{1}{m_2} C_1'$ $L_2 = \frac{1-m_1^2}{4 m_1} L_1'$ $C_2 = \frac{4 m_2}{1-m_2^2} C_1'$ $m_1 = \sqrt{\frac{\Omega_1^2 - \Omega_{1\infty}^2}{\Omega_2^2 - \Omega_{1\infty}^2}}$ $m_2 = \Omega_2^2 m_1$	$L_1 = \frac{4 m_2}{1-m_2^2} L_2'$ $C_1 = \frac{1-m_1^2}{4 m_1} C_2'$ $L_2 = \frac{1}{m_2} L_2'$ $C_2 = m_1 C_2'$ $m_1 = \sqrt{\frac{\Omega_2^2 - \Omega_\infty^2}{\Omega_2^2 - \Omega_1^2}}$ $m_2 = \Omega_1^2 m_1'$	$L_1 = \frac{4 m_2}{1-m_2^2} L_2'$ $C_1 = \frac{1-m_1^2}{4 m_1} C_2'$ $L_2 = \frac{1}{m_2} L_2'$ $C_2 = m_1 C_2'$ $m_1 = \sqrt{\frac{\Omega_1^2 - \Omega_{1\infty}^2}{\Omega_2^2 - \Omega_{1\infty}^2}}$ $m_2 = \Omega_2^2 m_1$
$Z_T = \sqrt{Z_1 Z_2} \sqrt{1 + \frac{Z_1}{4 Z_2}}$	$|Z_T|$ $Z_T = R\sqrt{1 - y^2} \ (=Z_T')$	$|Z_T|$ $Z_T = R\sqrt{1 - y^2} \ (=Z_T')$	$|Z_T|$	$|Z_T|$
$Z_\pi = \frac{\sqrt{Z_1 Z_2}}{\sqrt{1 + \frac{Z_1}{4 Z_2}}}$	$|Z_\pi|$	$|Z_\pi|$	$|Z_\pi|$ $Z_\pi = \frac{R}{\sqrt{1-y^2}} \ (=Z_\pi')$	$|Z_\pi|$ $Z_\pi = \frac{R}{\sqrt{1-y^2}} \ (=Z_\pi')$
ATTENUATION CONSTANT FOR IDEAL ELEMENTS	a_1	a_1	a_1	a_1
PHASE CONSTANT FOR IDEAL ELEMENTS	b_1	b_1	b_1	b_1
$\frac{Z_1}{4 Z_2}$ — WITHOUT LOSSES $d=0$	$\frac{(\Omega_2^2 - \Omega^2)(\Omega^2 - \Omega_1^2)}{(\Omega_2^2 - \Omega_1^2)(\Omega_2^2 - \Omega^2)}$	$\frac{(\Omega_1^2 - \Omega_\infty^2)(\Omega^2 - \Omega_2^2)}{(\Omega_1^2 - \Omega_2^2)(\Omega_\infty^2 - \Omega^2)}$	$\frac{(\Omega_2^2 - \Omega_\infty^2)(\Omega^2 - \Omega_1^2)}{(\Omega_2^2 - \Omega_1^2)(\Omega_\infty^2 - \Omega^2)}$	$\frac{(\Omega_1^2 - \Omega_\infty^2)(\Omega^2 - \Omega_2^2)}{(\Omega_1^2 - \Omega_2^2)(\Omega_\infty^2 - \Omega^2)}$
WITH LOSSES $d \neq 0$	Ω^2 IS TO BE REPLACED BY $\Omega^2 (1-jd_l)(1-jd_c)$			
DEFINITIONS	$f_0^2 = f_1 f_2$ $\Omega = \frac{f}{f_0}$	$\Omega_1 = \frac{f_1}{f_0}$ $\Omega_2 = \frac{f_2}{f_0}$	$\Omega_{1\infty} = \frac{f_{1\infty}}{f_0}$ $\Omega_{2\infty} = \frac{f_{2\infty}}{f_0}$ $\Omega_1 \Omega_2 = 1$	$y = \frac{\Omega - \frac{1}{\Omega}}{\Omega_2 - \Omega_1}$ $d_l = \frac{r_l}{\omega L}$ $d_c = \frac{1}{R_c \omega C}$

Fig. 1.13.   Conventional 4-element bandpass structures.

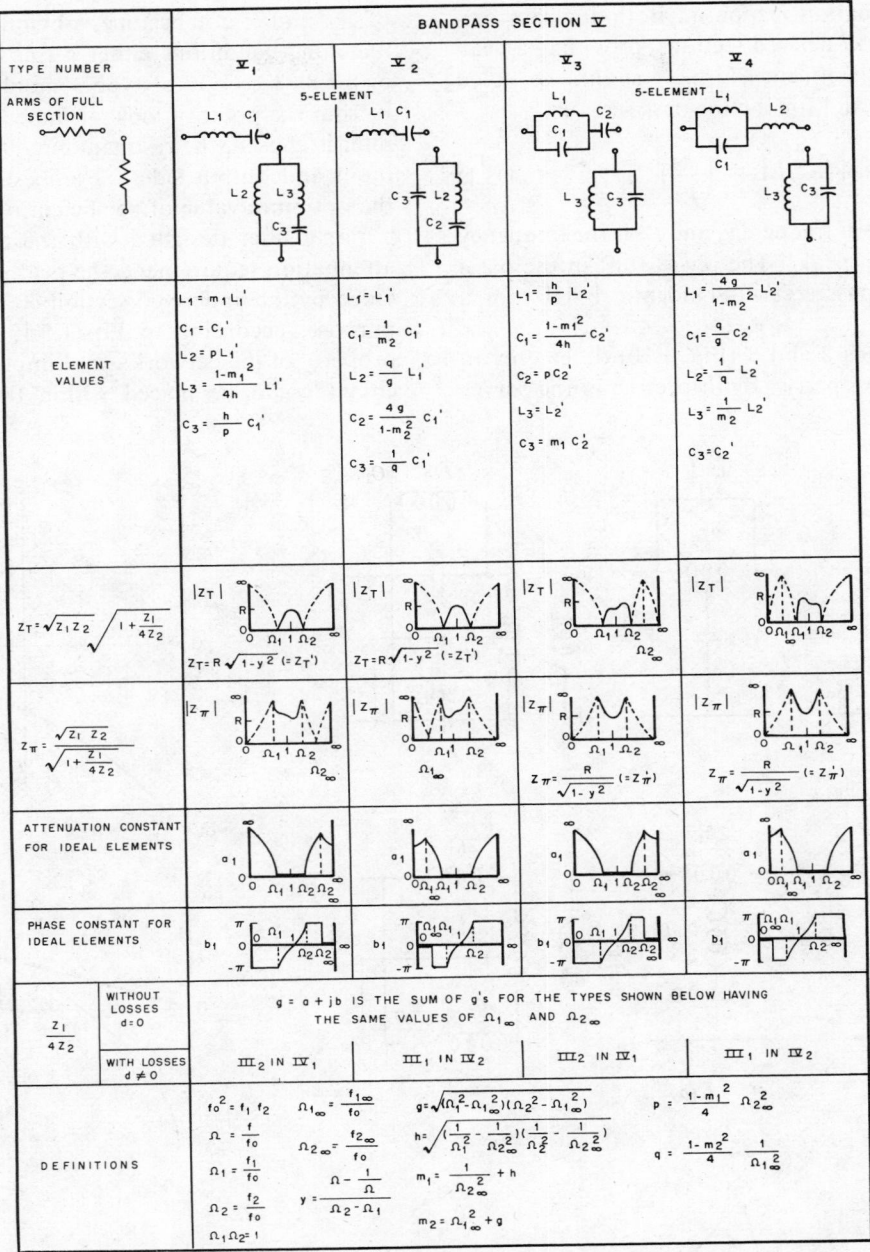

Fig. 1.14. Conventional 5-element bandpass structures.

The element values of the impedance arms are given by the formulas in the second row of the design chart. For the more complicated sections, auxiliary parameters and definitions are given in the last row of the tables. The impedance of the T and π schematics (or from the T side or π side of a half-section schematic) is given in the third and fourth rows of the design charts. The fifth and sixth rows of the charts give the image attenuation and phase response respectively.

These, of course, have been drawn for the case of ideal elements. Finally, the value of the expression $Z_1/4Z_2$ is given in the seventh row of the design charts for both the lossless and lossy cases. This expression is useful in the evaluation of the impedance and transmission of the networks.

Lowpass Sections. Lowpass section 1 of Fig. 1.10 is the fundamental type and is characterized by its monotonic attenuation response. Both lowpass sections 2 and 3 are *m*-derived sections, providing a peak of attenuation at frequency Ω_∞. It should be noted that $0 < m < 1$, and for the lowpass sections

$$m = \sqrt{1 - \left(\frac{f_1}{f_\infty}\right)^2} \qquad (1.5.1)$$

where f_1 is the cutoff frequency, and f_∞ is the frequency of the attenuation peak. The phase shift in the pass-band for all lowpass sections is identical, 180° for a full section.

Lowpass sections 2 and 3 (Fig. 1.10) differ only in their impedance property. By inspection of the curves

of Z_T and Z_π, it becomes obvious that for the best matching conditions either a full π section of type 2 or a full T section of type 3 should be used; that is, if from the point of view of impedance the *m*-derived filter is chosen, the resonant circuits are placed on the input and output sides. For best possible matching the optimum value of coefficient *m* is 0.62.

For a filter designed with the accent on a specific attenuation requirement, the peak of attenuation provided by the *m*-derived section could be used for that purpose, according to Eq. 1.5.1. If the impedance property of the network is not important, the resonant circuit could be placed within the network, and a

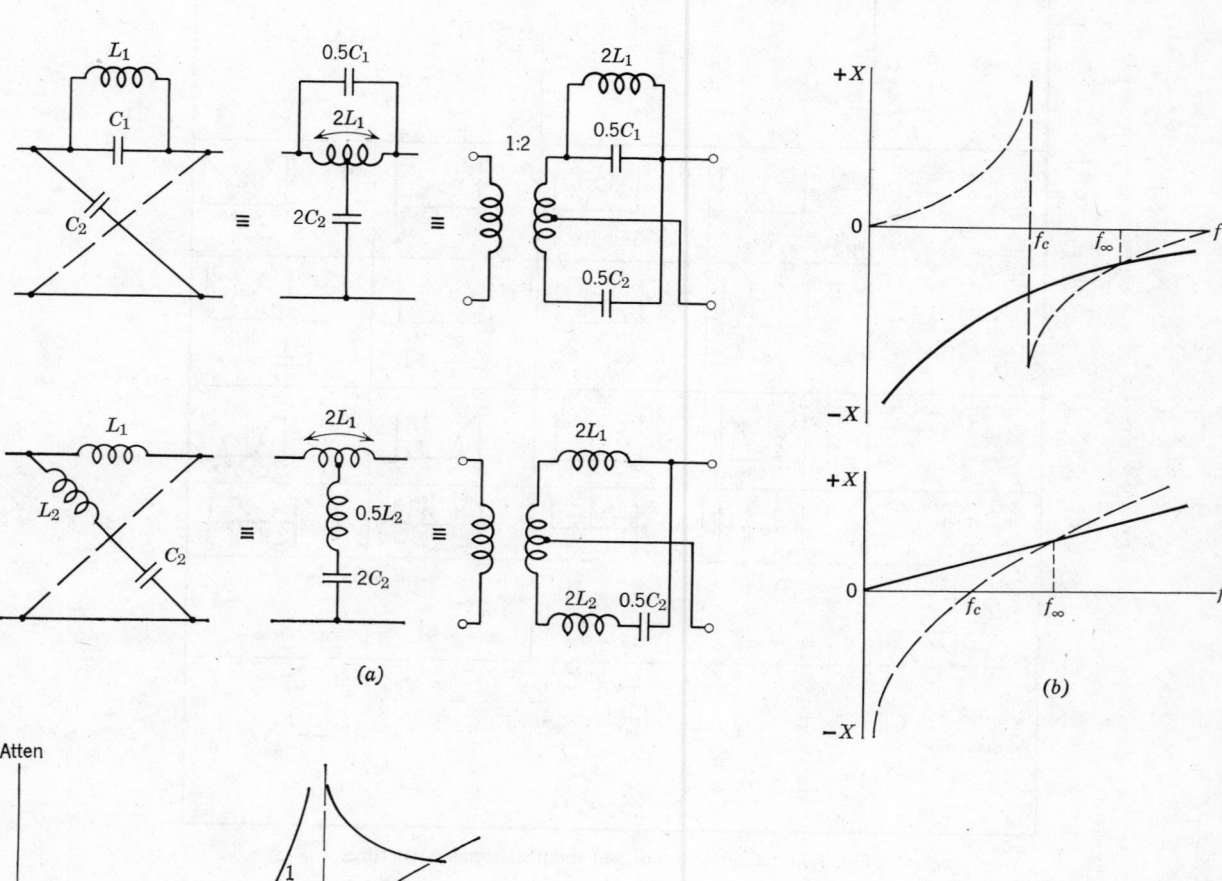

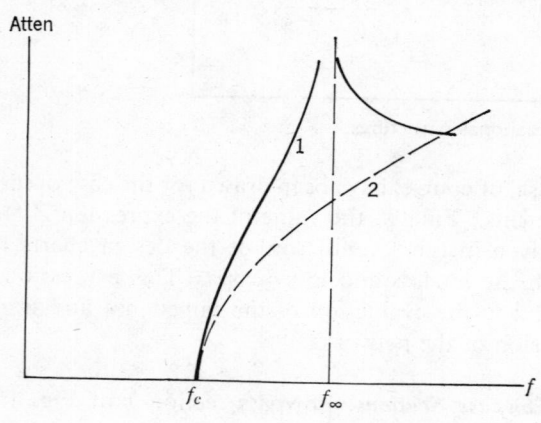

Fig. 1.15. (*a*) Lowpass filters in lattice and equivalent forms. (*b*) reactance of arms X_1 and X_2 (see Fig. 1.1). (*c*) attenuation: (1) when X_1 and X_2 intersect (2) when X_1 and X_2 do not intersect.

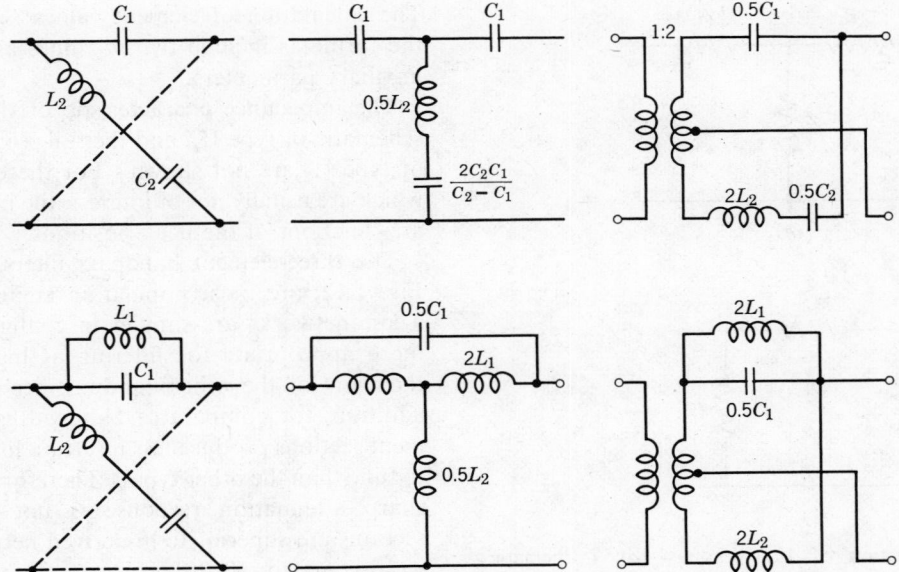

Fig. 1.16. Highpass filters in lattice and equivalent forms.

saving of one element is achieved. For example, with lowpass section 3, the *T* configuration will consist of three capacitors and two inductors. Referring to Fig. 1.19*a*, the schematic of Fig. 1.21*a* is obtained. On the other hand, the *π* schematic of Fig. 1.21*b* consists of three capacitors and one inductor. The elements are obtained by applying Fig. 1.19*b* to the formulas of element values of type 3.

When the required attenuation is complex and the filter must provide more attenuation than one section can offer, several full or half sections can be connected together. The usual way of terminating the filter is to put *m*-derived half sections at the input and output sides. For extremely complex filters *mm*-derived input and output half sections are necessary. The technique of connecting sections in tandem in the construction of a complex ladder is well known and no special treatment will be offered. It should be noted, however, that the amount of attenuation provided by each half-section used can be determined from curve 25 of Chapter 3, Section 14.

Highpass Sections. The design of highpass filters are accomplished in a manner similar to that of the lowpass filter, and therefore no specific treatment is given.

It is necessary to understand that the parameter *m* as applied to highpass types 2 and 3 is

$$m = \sqrt{1 - \left(\frac{f_\infty}{f_1}\right)^2} \qquad (1.5.2)$$

where f_1 is the cutoff frequency and f_∞ is the frequency of the attenuation peak.

Bandpass sections. Bandpass filters constitute the great majority of all filters designed to satisfy certain specific attenuation requirements. The most popular image parameter bandpass sections are those derived from the fundamental type I_1 shown in Fig. 1.11. This fundamental (*k*-constant) type provides a monotonic attenuation response that is symmetrical about the center frequency on a logarithmic scale. The six-element *m*-derived sections I_2 and I_3 (Fig. 1.11) are

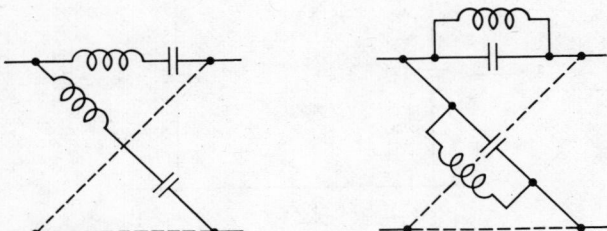

Fig. 1.17. Lattice bandpass filters.

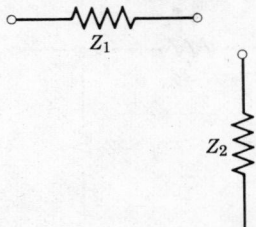

Fig. 1.18. Series and shunt impedance arms, Z_1 and Z_2.

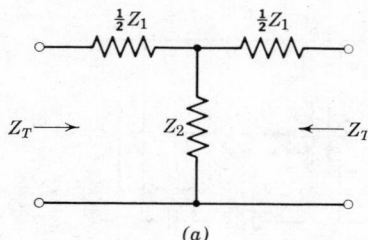

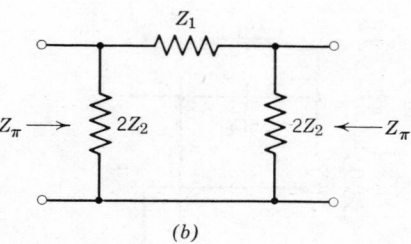

Fig. 1.19. Construction of full sections—(a) *T*-schematic (b) *π*-schematic.

similar to the *m*-derived lowpass sections, except that a bandpass response is obtained with geometric symmetry about the center frequency. One peak of attenuation $f_{1\infty}$ is at a frequency below the passband, and the other peak $f_{2\infty}$, is located above the passband, so that

$$f_{1\infty} f_{2\infty} = f_o^2 \qquad (1.5.3)$$

The position of the peak of "infinite" attenuation is given by

$$m = \sqrt{1 - \left(\frac{f_{1\infty}}{f_0}\right)^2} = \sqrt{1 - \left(\frac{f_0}{f_{2\infty}}\right)^2} \qquad (1.5.4)$$

Bandpass sections I$_2$ and I$_3$ differ only in their impedance characteristics as illustrated in the chart.

Bandpass sections II$_1$ and II$_2$ (shown in Fig. 1.11) yield schematics identical for sections I$_2$ and I$_3$ respectively. However, sections II$_1$ and II$_2$ allow greater flexibility in the positioning of the attenuation peaks; $f_{1\infty}$ is any frequency below the passband, and $f_{2\infty}$ is any frequency above the passband.

$$f_{1\infty} \cdot f_{2\infty} \neq f_o^2 \qquad (1.5.5)$$

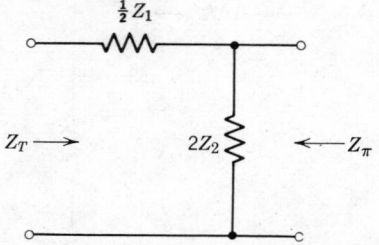

Fig. 1.20. Construction of half-section.

The calculation of element values is more involved; the formulas include two *m*-values as well as several auxiliary parameters.

The impedance characteristic of the full section-*π* schematic of type II$_1$, and the full section-*T* schematic of type II$_2$ are not shown. For these configurations, which are usually not of interest, the impedance curves are functions of the peak positions.

The three-element bandpass filters, sections III of Fig. 1.12, are most popular in engineering practice. These networks are simpler in configuration and are more appropriate for filtering at higher frequencies than any of the other bandpass sections shown. In addition, for comparable bandwidths, the three element sections produce less insertion loss and are easier to tune than the other types. Therefore if an extremely sharp attenuation response is not required, these sections are superior to *m*-derived networks for bandwidths up to 10–20%. A monotonic attenuation curve is provided with less attenuation below the passband for sections III$_1$ and III$_3$ and more attenuation below the passband for sections III$_2$ and III$_4$.

To add the versatility of a peak of attenuation, either above or below the passband, the four-element bandpass sections (type IV of Fig. 1.13) could be used. The peak of attenuation is controlled by the *m*-values defined in the chart, and the position of this peak does

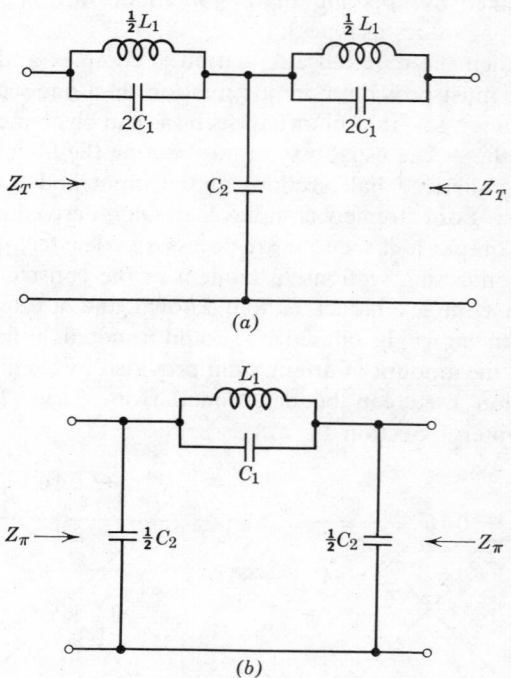

Fig. 1.21. Full sections of lowpass type 3—(a) *T*-schematic (b) *π*-schematic.

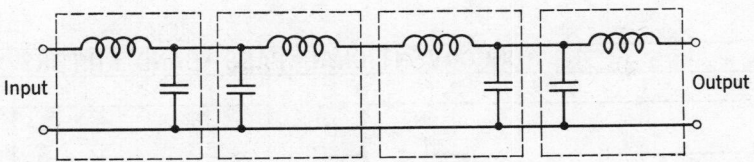

Fig. 1.22. Composite lowpass filter consisting of four half-sections or two full sections.

not upset the impedance characteristic in the passband. These sections are seldom used, since they are not economical in the case of complex networks.

The five-element bandpass sections are of interest for low-frequency applications, when a realization is required with the use of mostly one type of reactances. A full π-section of type V_3 (Fig. 1.14) will consist of three inductors and four capacitors. In the case of networks including many of these sections, the realization will provide a remarkable economy, since fewer inductances than capacitances are required. If, at higher frequencies, the cost of inductances is less than capacitances, types V_1 or V_4 would be desirable.

1.6 HIGHER ORDER FILTERS

A combination of several building blocks like several stages of amplification is an effort to provide the desired filter response. The conventional way to produce a composite filter is to combine many half-sections in tandem or to use a higher order polynomial for synthesis. The sections must be of the same characteristic impedance if they are designed according to image-impedance theory. In the polynomial filter, the composite filter is a chain of components given by a high-order polynomial. The element values of such a filter is the result of continued fraction expansion of the reactance function and appears to be of the same physical structure as a conventional filter. The difference between synthesized and sectional filters is essentially that polynomial filters cannot be subdivided into sections. They are not a combination of sections in the classical sense. Figure 1.22 shows four blocks connected in tandem and is equivalent to four half-sections of a lowpass filter (lowpass type 1 of

Fig. 1.10) which reduces to a filter with three coils and two capacitors. An equivalent polynomial filter would consist also of three coils and two capacitors, the values of the elements being obtained from a fifth-order polynomial. Although the filter designed from the image-parameter approach has a configuration identical to the polynomial network, the values of the elements of these two filters will be different.

The realization of composite lattices, even with such stable components as crystals, is not always easy because of the necessity of achieving an impedance balance. Designers have therefore sought alternate building blocks, such as a combination of the lattice and the tandem form. Instead of complicated branches in one lattice, less complicated blocks may be connected together to overcome inconveniences. The network shown in Fig. 1.23 consists essentially of two semilattice blocks. In combination, they are equivalent to a more complicated semilattice which would require a much more difficult tuning procedure to meet the required performance.

It is interesting to note that the upper arm in each semilattice can be a regular piezoelectric crystal. The differential transformer is replaced by a differential capacitor in parallel with a transformer. This permits, for very narrow bands of frequencies, the input impedance to be kept high while the electrical center can be adjusted for better bridge balance. The lower arm in each bridge is a tunable capacitor that influences the position of the peak of attenuation.

1.7 COIL-SAVING BANDPASS FILTERS

In Fig. 1.24 the simplest and most usable bandpass filter configurations are tabulated. In order to produce

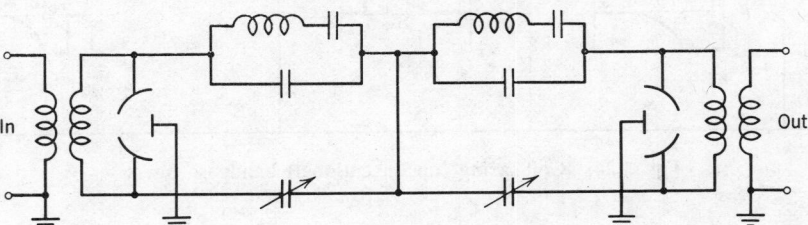

Fig. 1.23. Two-section semi-lattice filter equivalent to two lattices of Fig. 1.25.

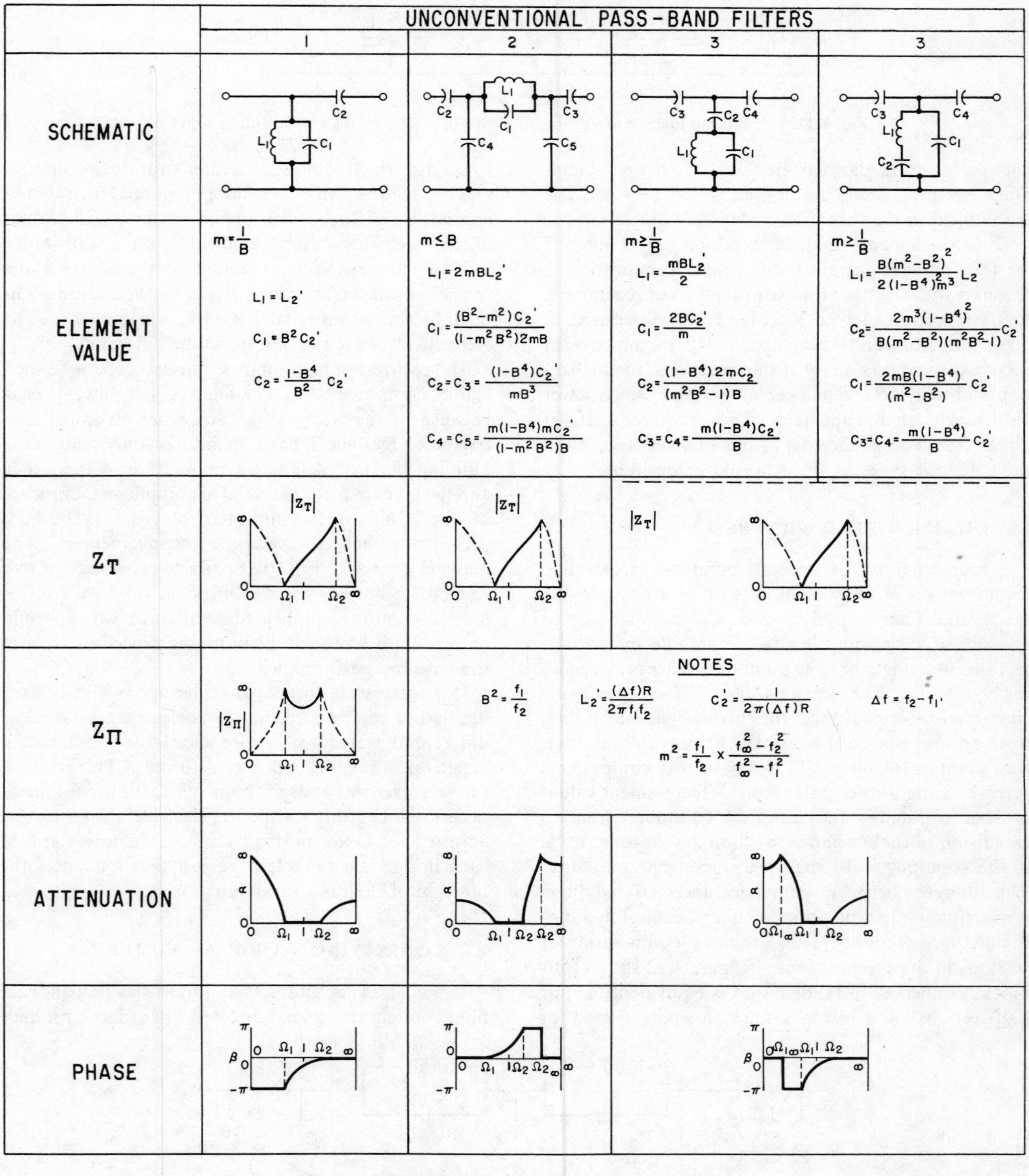

UNCONVENTIONAL PASS-BAND FILTERS

	1	2	3	3
SCHEMATIC				
ELEMENT VALUE	$m = \dfrac{1}{B}$ $L_1 = L_2'$ $C_1 = B^2 C_2'$ $C_2 = \dfrac{1-B^4}{B^2} C_2'$	$m \leq B$ $L_1 = 2mB L_2'$ $C_1 = \dfrac{(B^2 - m^2)C_2'}{(1-m^2 B^2)2mB}$ $C_2 = C_3 = \dfrac{(1-B^4)C_2'}{mB^3}$ $C_4 = C_5 = \dfrac{m(1-B^4)mC_2'}{(1-m^2 B^2)B}$	$m \geq \dfrac{1}{B}$ $L_1 = \dfrac{mB L_2'}{2}$ $C_1 = \dfrac{2BC_2'}{m}$ $C_2 = \dfrac{(1-B^4)2mC_2'}{(m^2 B^2 - 1)B}$ $C_3 = C_4 = \dfrac{m(1-B^4)C_2'}{B}$	$m \geq \dfrac{1}{B}$ $L_1 = \dfrac{B(m^2 - B^2)^2}{2(1-B^4)^2 m^3} L_2'$ $C_2 = \dfrac{2m^3(1-B^4)}{B(m^2 - B^2)(m^2 B^2 - 1)} C_2'$ $C_1 = \dfrac{2mB(1-B^4)}{(m^2 - B^2)} C_2'$ $C_3 = C_4 = \dfrac{m(1-B^4)}{B} C_2'$
Z_T				

NOTES

$$B^2 = \frac{f_1}{f_2} \qquad L_2' = \frac{(\Delta f)R}{2\pi f_1 f_2} \qquad C_2' = \frac{1}{2\pi(\Delta f)R} \qquad \Delta f = f_2 - f_1'$$

$$m^2 = \frac{f_1}{f_2} \times \frac{f_\infty^2 - f_2^2}{f_\infty^2 - f_1^2}$$

	1	2	3	3
Z_Π				
ATTENUATION				
PHASE				

Fig. 1.24.   Coil-saving (unconventional) bandpass filters.

	UNCONVENTIONAL PASS-BAND FILTERS			
	4	5	5	6
SCHEMATIC	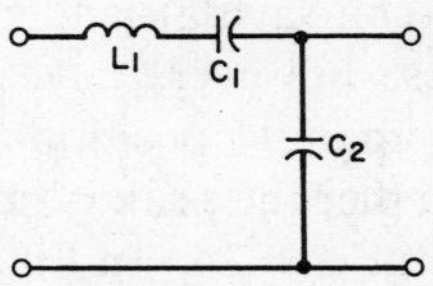	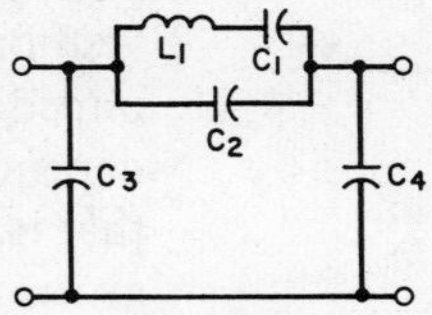	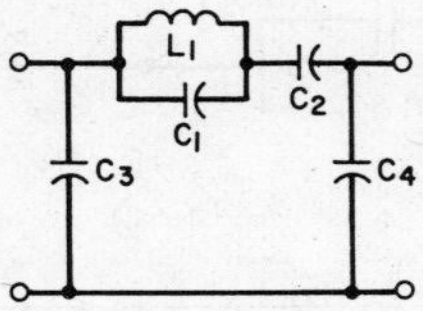	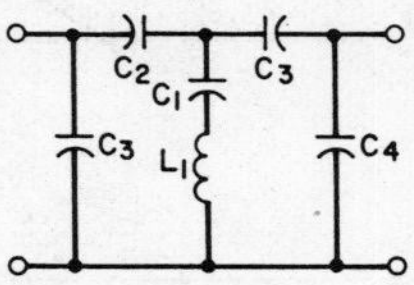
ELEMENT VALUE	$m = B$ $L_1 = L_1' = \frac{R}{2\pi \Delta f}$ $C_1' = \frac{\Delta f}{2\pi f_1 f_2 R}$ $C_1 = \frac{C_1'}{B^2} = \frac{\Delta f}{2\pi f_1 f_2 R B^2}$ $C = \frac{B^2 C_1'}{1-B^4}$	$m \le B$ $L_1 = \frac{2mL_1'}{B}$ $C_1 = \frac{C_1'}{2mB}$ $C_2 = \frac{(B^2-m^2)BC_1'}{(1-B^4)2m}$ $C_3 = C_4 = \frac{mBC_1'}{1-B^4}$	$m \le B$ $L_1 = \frac{2m(1-B^4)^2}{B(1-m^2B^2)^2} L_1'$ $C_1 = \frac{(B^2-m^2)(1-m^2B^2)B}{(1-B^4)^2\, 2m} C_1'$ $C_2 = \frac{(1-m^2B^2)}{2mB(1-B^4)} C_1'$ $C_3 = C_4 = \frac{mB}{(1-B^4)} C_1'$	$m \ge \frac{1}{B}$ $L_1 = \frac{mL_1'}{2B}$ $C_1 = \frac{(m^2-B^2)2BC_1'}{(m^2B^2-1)m}$ $C_2 = C_3 = \frac{(m^2-B^2)BC_1'}{(1-B^4)m}$ $C_4 = C_3 = \frac{B^3C_1}{(1-B^4)m}$
Z_T	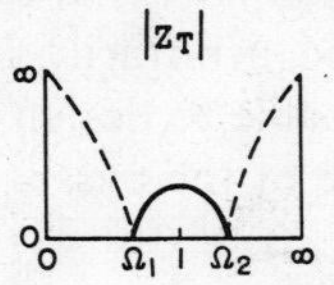			
Z_Π	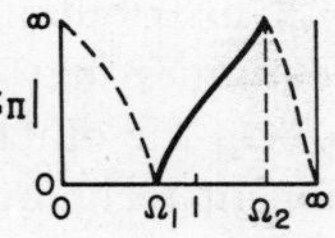		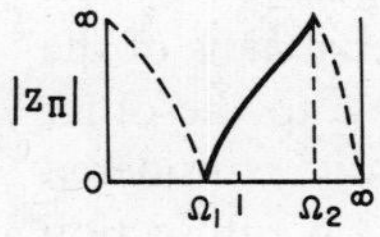	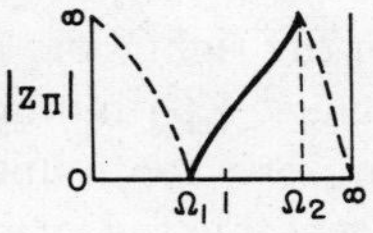
ATTENUATION	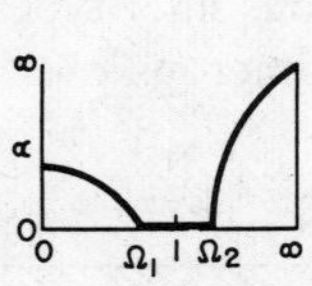		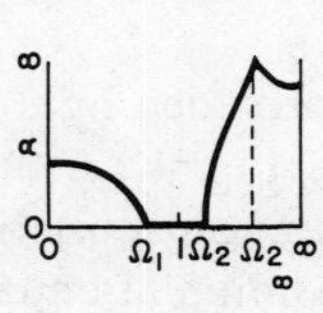	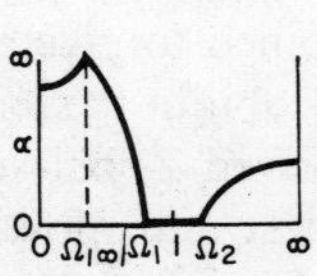
PHASE	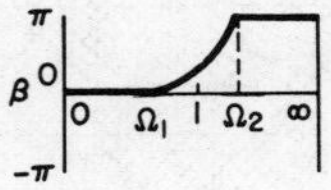		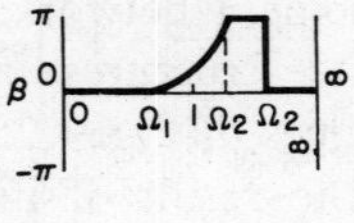	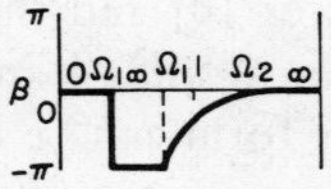

Fig. 1.24. (*Continued*)

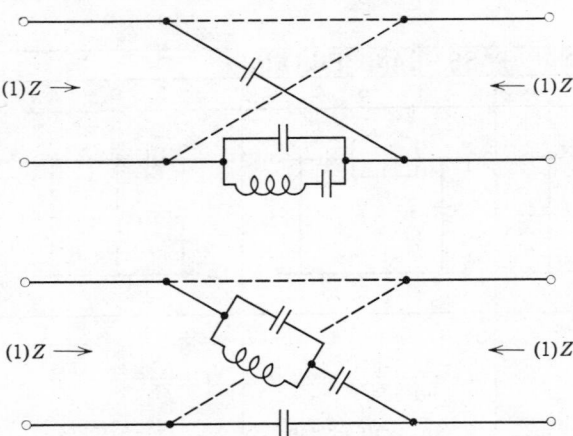

Fig. 1.25. Bandpass lattice networks used to minimize the number of inductors in filter design.

a complete and arbitrary filter using the image-parameter approach, adjacent sections must have exactly equal characteristic impedances for all frequencies.

The simplest three-element half-section with bandpass properties which provide first- and second-class impedances from opposite sides is shown as type 1 in Fig. 1.24. This schematic is identical to the three-element filter type III_4 in Fig. 1.12. From the left side of the half section the impedance is of the second class[1]—the conventional Z_π for bandpass networks. From the right side, however, the impedance is of the first class, Z_t running from one extreme to the other (from zero to infinity) which is usually considered unacceptable for termination purposes. A full section obtained from two of the above half-sections has a second-class impedance when the half sections are combined to give a π-type schematic.

To obtain flexibility similar to that provided by the *m*-derived Zobel sections, tabulated in Fig. 1.13, the simplest and most general bandpass section must be found in which the position of the maximum attenuation can be placed as desired.

With aid of Bartlett's bisectional theorem (see Chapter 10) and the simplest three-element networks (Types 1 and 4 in Fig. 1.24) several structures in lattice form can be obtained. The lattices of Fig. 1.25 used in crystal filter technique are the result of its application. The remaining schematics, types 2, 3, 5, and 6 of Fig. 1.24, are the result of transforming the schematics of Fig. 1.25 back to the ladder form.

[1] Simply stated, the number of intersections with the design value of impedance determines the class of the impedance function.

Conventional sections such as those shown in Fig. 1.13 (with second-class input and output impedances) always have the same form, regardless of where the peaks of attenuation are designed to appear. Unconventional (coil-saving) ladder sections with first-class impedances, however, have different configurations depending upon the position of the attenuation peak—that is, whether the peak is on the right or the left side as shown in Fig. 1.24. In Fig. 1.13 it is seen that if the bandpass ladder is composed of one peaked conventional section, it always requires the same number of inductors and capacitors (three coils and three capacitors). The unconventional section uses only a single coil and still has the desired bandpass characteristic. Unconventional filters of this type have the same attenuation and phase constants as regular or conventional sections, and the attenuation curve can be designed and calculated in exactly the same manner as for a conventional section.

Another property of unconventional filters is that sections having particular desired peaks of attenuation can be placed arbitrarily in the composite network. It is also possible to design a filter having all peaks on the same side of the passband.

1.8 FREQUENCY RANGE OF APPLICATIONS

Microwave filters from 3 Gc and up are designed exclusively with distributed components, but typical microwave techniques may be used at frequencies as low as 20 Mc. On the other hand, the low-frequency, lumped-component technique for filter design can be extended into the microwave region up to 4 Gc. Figure 1.26 indicates the frequency-bandwidth ranges in which various filter types find practical application. Notice the numerous overlap areas where more than

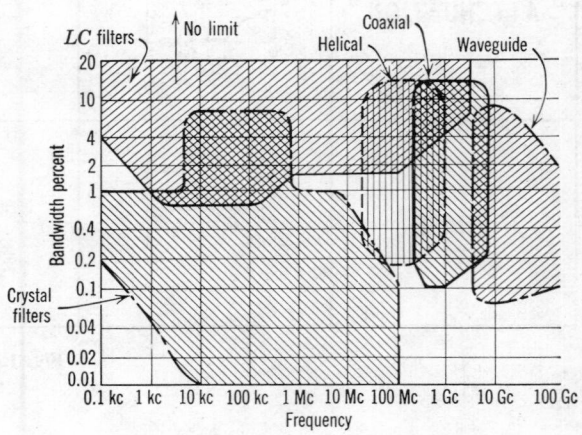

Fig. 1.26. Filter selection guide.

one type may be selected. Since the behavior of distributed circuits operating in a limited frequency band closely approximates that of lumped inductors, capacitors, and resonant circuits, equivalent lumped ladder filter structures can be realized by distributed circuit elements.

Although microwave filter design methods are based on the filters used at lower frequencies, the structures themselves are radically different. Component parameters are distributed and hardware geometry is an integral part of the filter design.

1.9 PHYSICAL ELEMENTS OF THE FILTER

The main elements of a filter are reactances—lumped capacitances and lumped inductances. It is possible to design some filters by using only capacitors and resistors. This combination is especially useful in the case of active networks or active filters. Regular passive filters require both types of reactances.

To a first approximation, lumped inductance and capacitance can be considered as pure reactances, but closer investigation reveals that losses and reactive impurities are also present. The ordinary inductor at relatively low frequencies is wound on a magnetic core (powdered iron or ferramic); other inductors are simply single-layer solenoids wound on a nonmagnetic coil form. In both cases, although the losses are low in comparison with the value of reactance, they cannot be neglected, especially when one designs a very narrow band filter or desires a very sharp response curve.

Losses tend to decrease the rate of attenuation rolloff, increase the attenuation within the passband, and in certain cases, prohibit the realization of narrow bandpass filters. The parasitic effect of distributed capacitance across the coil or series lead inductance in capacitors is damaging in another respect. In the low-pass filter parasitics will create the effect of a parallel-resonant circuit instead of a coil. The filter may thus

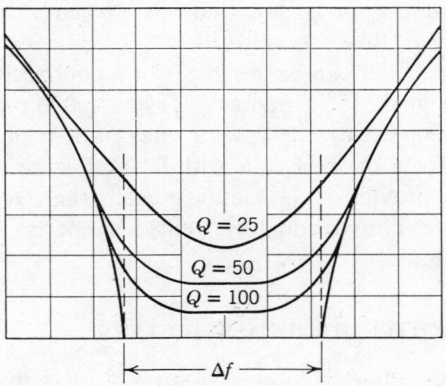

Fig. 1.28. Attenuation characteristics of narrowband bandpass filters having components of different quality factors.

provide unexpected rejection at certain frequencies in the stopband, or even in the passband if the self-resonant frequency of such a coil is sufficiently low. Parasitic reactances in lumped components produce distortion of the amplitude response and may destroy the network response altogether if not neutralized or properly taken into consideration.

Equivalent circuits for an inductor and capacitor are shown, with impurities taken into account in Fig. 1.27. The conventional measure of the quality of any reactance is the quality factor Q, which describes how many times the reactance of a coil or capacitor is greater than the resistance. The most common values of Q of conventional inductors at radio frequencies range from 50 to 300. The Q factor for the capacitor at the same frequencies is usually higher: 500 to 5000. The higher these values are, the better the filter that can be designed. Figure 1.28 illustrates the effect of Q factor on the shape of the response of a bandpass filter.

The demand for quality factor in ordinary lumped components, especially coils, has intensified research to find some substitute for the inductor and capacitor. Historically, the first and most successful substitute was the piezoelectric crystal; next were magnetostrictive components and electromechanical devices. Lumped elements are the oldest filter elements, and they remain the most widely used at low frequencies.

Experience proves that a very good bandpass filter can be made when its components have a Q factor not less than 20 to 25 times $f_0/\Delta f$, where f_o is the center frequency of the filter and Δf is the bandwidth (passband). For the same absolute bandwidth and attenuation characteristic, the filter with the lowest center frequency will need the lowest quality factor. If, for example, $f_o = 10,000$ cycles and $\Delta f = 3000$ cycles, the

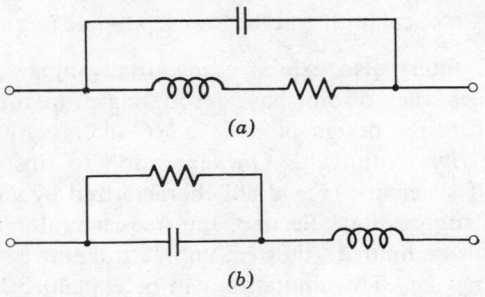

Fig. 1.27. Equivalent circuit of (*a*) inductor and (*b*) capacitor.

Q factor cannot be less than 66. When an element that meets these specifications is used, a very good bandpass filter can be designed for a commercial telephone signal. If the frequency $f_o = 150,000$ cycles and $\Delta f = 3000$ cycles, the Q factor has to be greater than 1000. Even the best coils with ferramic core material cannot provide a Q factor greater than 600 and, therefore, conventional elements cannot be used in such a filter.

1.10 ACTIVE BANDPASS FILTERS

Active filters in most instances are bandpass amplifiers. The main requirement of a bandpass amplifier is to amplify a certain band of frequencies by a certain prescribed amount. Other requirements are the following:

Fidelity Requirements

To ensure distortionless amplification:

1. The amplification within the passband should not fluctuate by more than a certain amount, A_{max}, thus resulting in a tolerable frequency distortion.

2. The phase-shift over the entire passband should not depart from linearity by more than a certain amount, thus ensuring a negligible phase distortion. This type of distortion is important in frequency and phase-modulated systems.

3. For pulse amplifiers the rise time, the overshoot and other similar phenomena should be as small as possible.

Selectivity Requirements

To prevent undesired signals from passing through, frequencies outside the passband should be strongly attenuated relative to frequencies within it. Bandpass amplifiers are in this respect similar to wave filters.

Gain-bandwidth Requirements

The gain-bandwidth product per stage of a multistage amplifier depends on the over-all passband width, the over-all gain, the number of stages, the gain-bandwidth factor, the tube transconductance, and the total shunt capacity across the tuning coil.

In order to justify the added cost of an active device within a filter, some economic or engineering advantages must be evident. Some such advantages may be listed as follows:

1. The active device can furnish gain and hence may already be present in the system.

2. The design of a complex filter is easier if it is made up of several sections separated by buffer amplifiers.

3. The alignment of an active filter with isolated sections is much easier.

4. With the advent of molecular and thin-film circuitry, it is possible to build compact and reliable active devices suitable for this application at a very low production cost.

5. For crystal interstage circuits, either a balanced input or output is usually required which is normally accomplished with a balanced center-tapped coil or transformer. In many circumstances molecular or thin-film devices can be used to perform this same function with an accompanying space and weight savings without a loss in reliability.

6. In multichannel comb filters one crystal resonator can be made to do the work of two if an active operational amplifier is introduced.

1.11 *RC* PASSIVE AND ACTIVE FILTERS

In the low-frequency audio range and even at sonic frequencies, all filters using conventional elements are cumbersome. At frequencies less than 1000 cycles, the Q factor of inductors are so low that a good resonant circuit is difficult to realize. The physical dimensions of coils and capacitors required for resonant circuits became very large and incompatible with modern circuitry using transistors, crystal diodes, and printed circuit techniques. Inductive components, in general, are obstacles in the development of modern low-frequency circuitry, and in filter technology they certainly block the road to progress.

The only way to solve selectivity requirements in this part of the spectrum is to synthesize filters without resonant circuits—*RC* filters as shown in Fig. 1.29. The advantages of *RC* filters are these:

1. Simplicity in manufacturing
2. Small physical dimensions
3. Low cost
4. Negligible sensitivity to external electrical fields
5. Practical for use at the lowest possible frequencies

RC filters also exhibit some disadvantages. For instance, they do not have resonating properties and therefore the design of passive *RC* filters with good selectivity is difficult. One exception to this is the twin-T schematic (Fig. 1.30) characterized by narrow-band suppression. Second, the frequency domain of their use is limited; they are unpractical above several hundred kc. This limitation can be explained by the fact that all capacitors in high-frequency *RC* filters

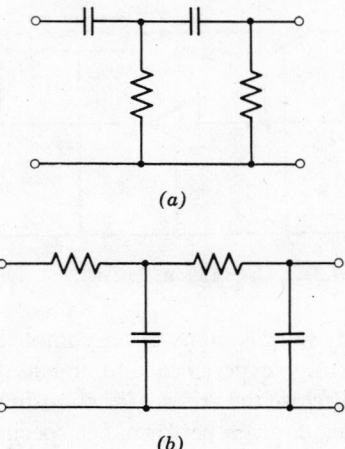

Fig. 1.29. *RC* filters: (*a*) highpass and (*b*) lowpass.

have such values that they cannot be used successfully between the rest of the associated circuitry.

Active *RC* filters overcome the foregoing difficulty of poor discrimination and can be designed with little limitation on their selectivity. In their passband, active *RC* filters can be designed to have no loss, or they may exhibit amplification. At the present time the main disadvantage of highly selective active *RC* filters is the inherent instability associated with the active elements. Also, the active element in the *RC* filter may be said to complicate the schematic, because transistors or tubes, with their required power supplies, must be incorporated. This kind of complication limits, to a certain extent, the domain of application of active filters. On the other hand, this is the only practical solution to the problem of selectivity below 1 kc.

In the most common case of modern application the *RC* filter must be considered as an integral part of an active schematic using transistors whose input impedance is not high, and therefore the influence of loads must be taken into consideration. The operation of *RC* filters when the load impedance is very high is only a limiting case of the more general case of conventional operation.

The exclusion of inductances from *RC* filters is particularly beneficial at low frequencies. It is clear that a large class of *LC* filters can now be displaced by active *RC* networks. Even when they are realized in conventional transistor circuitry, many of the active filters will be smaller and cheaper than their *LC* counterparts, and many are suitable for mass production in thin-film circuitry.

Thus far, two major synthesis techniques have been developed for active filters: (1) those that involve the use of negative impedance converters or (2) operational amplifiers. An ideal negative impedance converter is a network which, if terminated with an impedance *Z* across the output terminals presents an input impedance equal to $-Z$.

Since active filter design is a rather new subject, no general agreement has yet been reached on the best methods to use for practical designs. Various authors have published the configurations shown in Fig. 1.31 *a—e*, with appropriate remarks on the relevant synthesis procedures. In the figures *A* equals amplifier gain and *NIC* is a negative impedance converter.

In Fig. 1.31*e* the block diagram of an *RC* filter with a negative impedance converter is shown. In the operational amplifier shown in Fig. 1.31*b* the internal amplifier component has a large forward voltage gain and 180° phase shift over a limited frequency range. The result is a voltage transfer function that will include the complex conjugate poles and whose network will behave similar to the inductance capacitance combination. *RC* active filters involve ideal voltage amplifiers with shunt feedback or ideal current amplifiers with series feedback. The synthesis technique may incorporate such an attractive feature as the direct control of *DC* gain of the filter. Even imperfections such as the finite input resistance presented by practical amplifiers may be conveniently incorporated into finalized network.

In high-*Q* bandpass filters it is necessary to sort out the various methods according to the sensitivities they exhibit with respect to gain, stability, and mainly *Q* and center frequency. The published sensitivity formulas are sometimes illusory, or missing altogether. Thus the sensitivity of *Q* with respect to gain is 2*Q* for the twin-T method of Fig. 1.32 and 4*Q* in Hakim's circuit (Fig. 1.31*b*). But the latter permits the use of a much more stable amplifier, so that the theoretical advantage of the twin-T cannot be realized in practice.

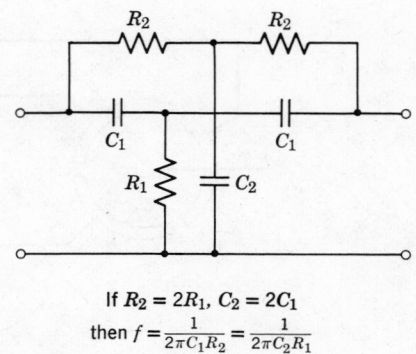

If $R_2 = 2R_1$, $C_2 = 2C_1$
then $f = \dfrac{1}{2\pi C_1 R_2} = \dfrac{1}{2\pi C_2 R_1}$

Fig. 1.30. Twin-T reject network.

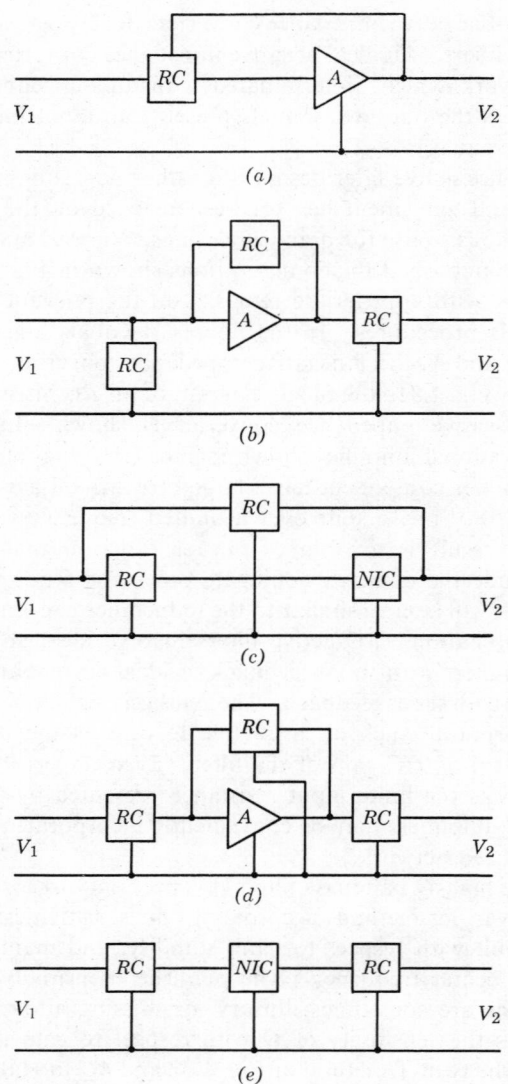

Fig. 1.31. Examples of *RC* filters by (*a*) Sallen and Key, (*b*) Hakim, (*c*) Yanigasawa, (*d*) Dietzold, and (*e*) Linvill.

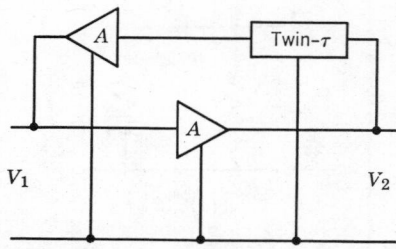

Fig. 1.32. Use of twin-T in active network.

A close study reveals many other complications, indicating that only experience and much thought will eventually disclose the criteria for choosing an optimal design to meet a given performance specification.

Hakim's method gives satisfactory results for obtaining fractional bandwidths as low as 2%. Narrower bands are obtainable but lead to alarming instabilities and alignment procedures that are difficult and expensive.

A method having the dimensions of a genuine breakthrough is shown in Fig. 1.33. Here the mechanical commutators, shown schematically, would be replaced in practice by transistor diode analog-gates driven by conventional logic circuits at an angular velocity, ω_o. The identical *RC* circuits have a lowpass response which is so modified by the sampling action that V_1/V_2 exhibits a related bandpass response centered at ω_o. This method shows promise of yielding figures for Q and stability which have previously been obtainable only with crystal filters. Definite results on the performance of this device will be available soon, but it is too new to be evaluated at present.

Figure 1.34 shows the response of a double-tuned circuit obtained by stagger-tuning two Hakim-type designs, each having a Q of 50. In the figure A represents an operational amplifier with a feedback loop

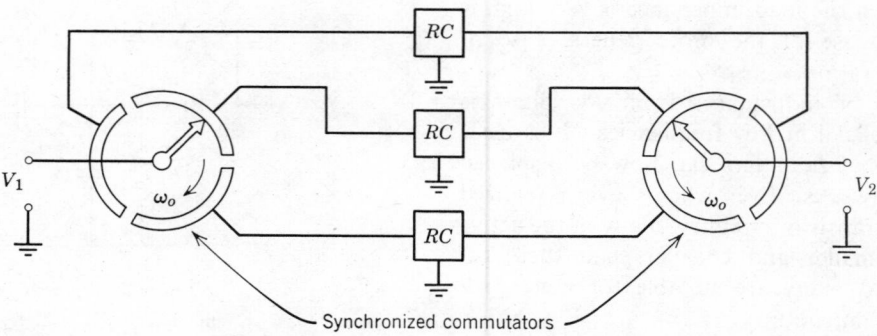

Fig. 1.33. The digital filter.

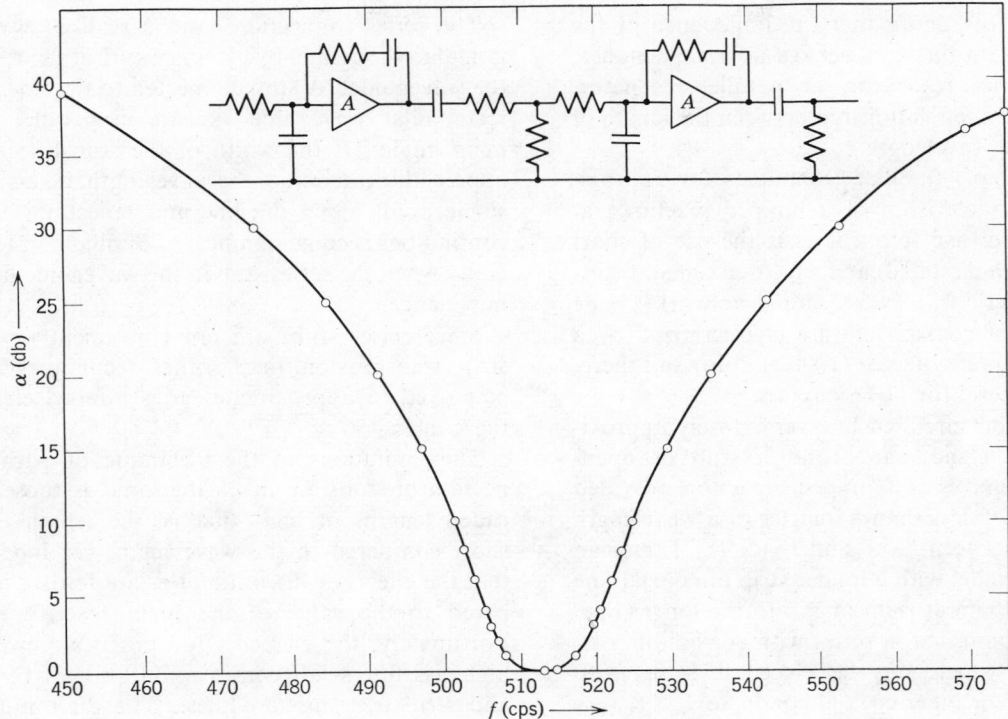

Fig. 1.34 Response of a double-tuned circuit of the type shown in Fig. 1.31(*b*).

adjusted to give a stable gain of two, as recommended by Hakim. More recent calculations suggest that greater stability can probably be obtained by raising this figure somewhat and adjusting the other parameters to suit.

1.12 MICROWAVE FILTERS

In contrast with wave filters, which are designed with lumped components such as coils, capacitors, and crystals, the microwave filter is basically different. They do not include, as a rule, any lumped components. Their components are the distributed reactances of resonators. The simplest example of distributed resonators is a piece of symmetrical line or piece of coaxial line. On the other hand, every distributed component could be approximated in a certain frequency range and in a certain band of frequencies to only one type of reactance, capacitance or inductance, and consequently the synthesis of a filter will be to a certain extent similar to the low-frequency filter design.

It is difficult to describe a microwave filter but in the following text, the term microwave filter will be applied to any filter using coaxial, helical, or waveguide cavities or certain types of structures such as a piece of short line, or strip lines. Filters which include only lumped elements of very small size and are used at microwave frequencies cannot be put in the category of microwave filters, since their design technique is conventional and does not constitute a special distinct category of microwave filter realization.

Filters with Distributed Components

A piece of transmission line may be used in a system as a filter element. It has a resonant frequency determined by the line impedance and the line length. A 3/4 wavelength line is equivalent to 1/4 wave resonator. In the same manner, a 1/2 wavelength line is equivalent to a full wave resonant line.

An electromagnetic wave propagating on a transmission line is continuously attenuated by the lossy elements in the line. When the transmission line is used as a filter component, this effect is apparent in the filter's passband. The characteristic impedance of a transmission line depends upon the geometry of the line.

In Fig. 1.35 two pieces of symmetrical lines are shown. One line is terminated in a short circuit, the other in an open circuit. The input impedance of such lines with negligible losses is expressible as a trigonometric function; the first as the tangent and the second as the cotangent of the same argument. It is apparent from the figure that the impedance of the given transmission line can behave as a capacitance, inductance, series resonator, or parallel resonator, depending upon the relationship between the length of the line and the wavelength.

One of the most familiar techniques for approximating the characteristics of a lumped reactance at UHF and microwave frequencies is the use of short lengths of open-circuited and short-circuited transmission line sections. Two-terminal networks made from sections of coaxial line are characterized by a relatively high quality factor (1000 and up) and therefore are very useful for UHF circuits.

A stub of short-circuited line very closely approximates a lumped inductance and a stub of open-circuited line behaves as a lumped capacitor, provided its length is much less than a quarter of a wavelength (see Fig. 1.35 between $l = 0$ and $l = \lambda/4$). These can be placed in parallel with a longer strip of coaxial line by connecting them at right angles to the longer line. Stubs can be connected in series with coaxial lines by reentrant lines which may be placed in series with either the outer or inner coaxial conductor.

The series connection can be realized with a rectangular waveguide by placing a stub at the top wall of the waveguide. A stub connected to the side wall of a rectangular waveguide appears in parallel with the main guide. If the width of the stub line is not an appreciable fraction of the wavelength, it acts only at a single point along the line and reflections from discontinuities become significant. Similar considerations apply when the series stub in the waveguide has a high impedance.

Since series stubs are not convenient to use with strip transmission lines, other techniques must be employed to approximate series lumped elements in these lines.

The limitations of the technique of parallel connection of stubs are much the same as those for cascaded lengths of line—that is, the lengths must be short compared to the wavelength, yet long enough that the effects of discontinuities are fairly small compared to the value of the main reactive elements. Fortunately, the elements for practical lowpass and highpass filters fall within usable ranges for coaxial and strip transmission lines. The discontinuities at

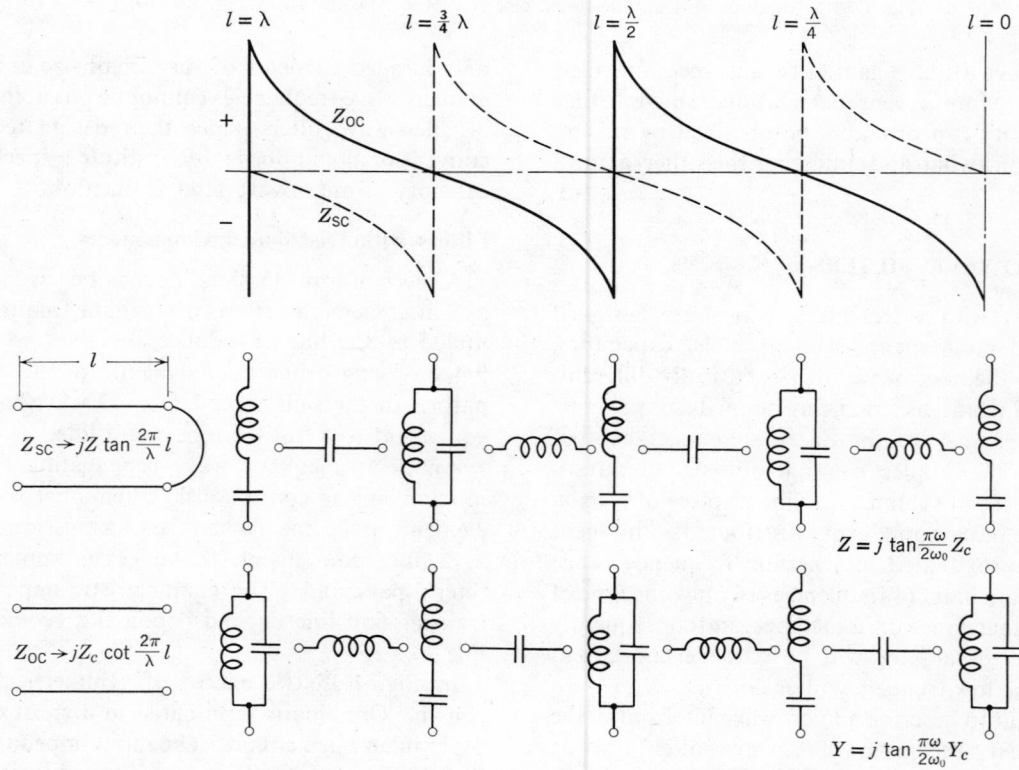

Fig. 1.35. The impedances of short- and open-circuited transmission lines are given by the curves as a function of length. The circuit symbols above the curves indicate the associated lumped equivalent of the line impedance at various points along the two transmission lines.

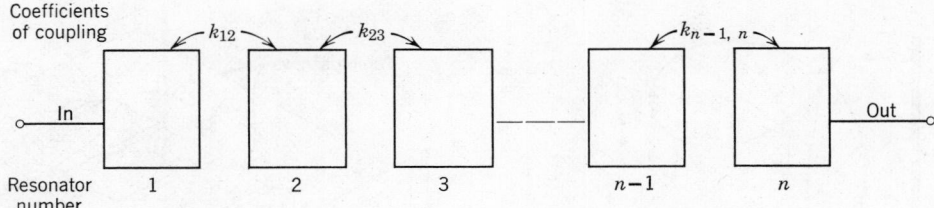

Fig. 1.36. Schematic representation of a microwave filter having *n* cavities.

the junction points in waveguides, however, are frequently so large that lumped-element approximation may not be desirable.

Cavity Filters

Narrow bandpass filters require high-Q resonant circuits and wide variations in characteristic impedance for their constituent elements. Approximation of lumped elements by short lengths of transmission line which have inherently limited quality factor and impedance variation does not lead to practical structures. A similar situation at low frequencies leads to extensive use of coupled resonant circuits and consequently to a narrowband structure. Such structures are also used at VHF, UHF, and microwave frequencies. Here the resonant circuits are usually resonant cavities, since ladder structures can be realized by coupled-cavity resonators. The basic coupled-cavity filter structure is shown in Fig. 1.36.

A newer design, based on quarter-wave coupled cavities, incorporates the susceptances of the coupling elements into the susceptances of the cavity irises to form direct-coupled filters. This design permits filters of greater bandwidths (up to about 20% of resonant frequency f_o), to be realized.

Whether physical quarter-wavelength coupling or direct coupling is used, it is necessary to include quarter-wavelength coupling lines in the analysis. If physical coupling sections are used, the electrical length varies with frequency, producing an error that increases as $(f - f_o)$ increases. The designer generally ignores the frequency sensitivity of the couplings and uses the concept of narrowband approximation. If the coupling is incorporated in the filter elements, the frequency sensitivity is reduced. This reduction is the main reason why direct-coupled filters corresponds to the design equations over a much greater bandwidth than filters coupled by physical quarter-wavelength sections.

A practical coaxial resonator generally consists of a cavity with a quarter-wavelength center post inside. Energy is introduced into the cavity by inductive loops,

capacitive probes, and the like, and the output is taken by similar means. Conventional coaxial resonators are generally used for filters in the range of 200 Mc to 6 Gc. A typical three-cavity coaxial filter is shown in Fig. 1.37.

Coaxial-resonator Q factors are usually very high, making them suitable for most high-quality filters. In the VHF range, however, coaxial-resonator construction becomes bulky compared with active circuit components. The length of a quarter-wave coaxial resonator at 100 Mc, for example, is 75 cm or about 30 in.

The physical explanation of coaxial-cavity filtering may be given with the help of Fig. 1.37. Since maximum current is concentrated at the bottom of the center posts, magnetic flux is at maximum there and can be used for magnetic coupling between cavities. This is accomplished by mounting the coupling loops close to the posts. The induced current into an adjacent resonator cavity produces a magnetic field and excites the resonator. The amount of energy introduced into the second cavity is determined by the size of the loop, its proximity to the center post, and the Q factor of the loop. The coupling loop is a lumped element similar to ordinary coils used at low frequencies.

The helical resonator provides a solution to many difficult filter problems in the VHF range where the conventional coaxial resonator is not applicable. Helical filters fill an important "void" in the

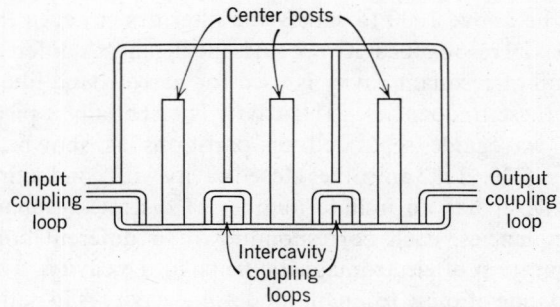

Fig. 1.37. Typical three-cavity coaxial filter.

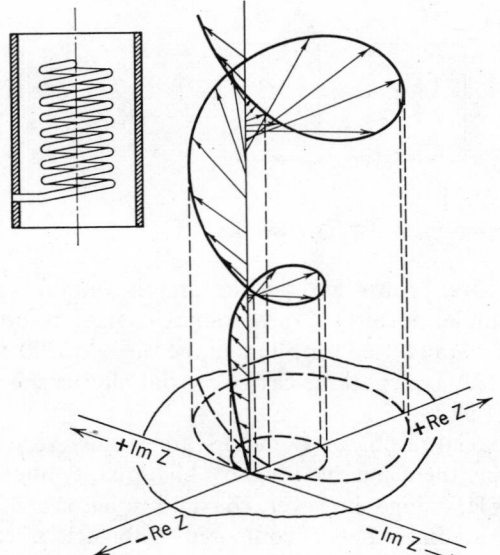

Fig. 1.38. Helical resonator.

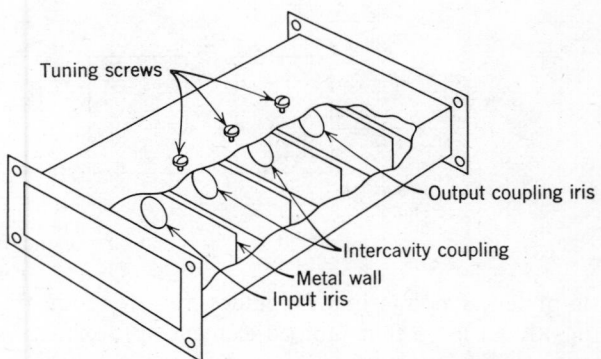

Fig. 1.39. Typical 3-cavity waveguide filter.

frequency spectrum. The resonators are similar to other quarter-wave configurations except that the inner conductor is in the form of a single-layer solenoid (see Fig. 1.38). The Q factor in the VHF range is on the order of 1000. Helical resonators are most useful in the region of 20 to 500 Mc, where lumped elements and crystal resonators cease to be practical but where coaxial cavity resonators are still too large. They can be used equally well for narrowband or wideband filters, up to 20% of center frequency and can be extremely useful even for higher frequencies when a Q factor between several hundred and 1000 is sufficient.

Both the helices and the cavities are made from highly conductive material such as copper, brass, or aluminium. Each dimension of the helix and the cavity is mathematically related to all the other dimensions of the resonator. The design of a helical filter will be treated in Chapter 9.

At higher frequencies, where the quality factor has to be above 1000 to satisfy the filter design, even the coaxial resonant cavity becomes inefficient. A different kind of resonant cavity is used for narrowband filters at these frequencies. This cavity is essentially a piece of waveguide separated by partitions as shown in Fig. 1.39. Like any other closed cavity with conducting walls, it has an infinite number of discrete resonant frequencies, each corresponding to a different configuration of electromagnetic fields in the cavity. For any one of these resonant modes, the cavity is in many ways like an ordinary resonant circuit and its behavior may be described in terms of three ordinary resonant

circuit parameters: center frequency, Q factor, and characteristic impedance. These parameters will, in general, have different values for each of the possible resonant modes.

Stripline Filters

Stripline multiple-coupled resonator bandpass filters are made of half-wavelength strips coupled end-to-end or alternately parallel-coupled as shown in Fig. 1.40.

Parallel coupling offers many important advantages over end coupling since the length of the filter is reduced and symmetrical insertion loss versus frequency response is obtained. The first spurious response occurs at three times the center frequency, and a much larger gap is permitted between parallel adjacent strips. The gap tolerance is also reduced, permitting a broader bandwidth for a given tolerance. The larger gap also permits higher filter power ratings.

Comb- and Interdigital-line Structures

Transmission-line bandpass filters in the form of cascaded cavities or other forms become inconveniently long. A type of filter with resonant lines put side by side, known as a *comb* structure, results in a more

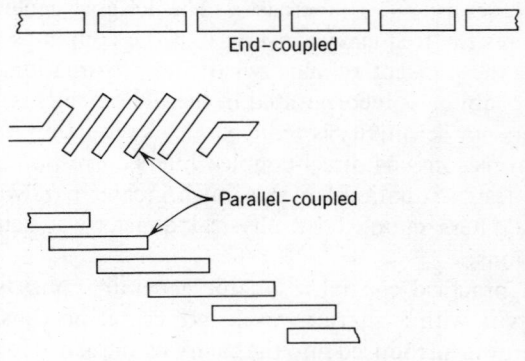

Fig. 1.40. Stripline arrangements.

compact package. Transmission-line elements in a comb-line filters are short-circuited at one end and have lumped capacitance between their other end and the ground. A set of filters consisting of such lines is shown in Fig. 1.41. Two impedance-transforming sections are at the ends of the filter, and the loading capacitances are at the top of the posts. The length of the lines is less than a quarter-wavelength at center frequency. Coupling between resonators is achieved by means of fringing fields between resonator lines. In the arrangement shown the coupling is predominantly magnetic.

If the loading capacitances were not present, the resonating lines would be 1/4 wavelength long. The magnetic and electric coupling effects would then cancel each other, and the comb-line structure would become an all-stop structure. Usually it is desirable to make the loading capacitance in this filter large so that the resonator will be 1/8 wavelength or less. The resulting filter will be small, permitting efficient coupling between the resonators. In this type of filter a second passband response always appears. When these lines are only 1/8 of the wavelength at the fundamental frequency, the second passband will be located at slightly over four times the fundamental center frequency.

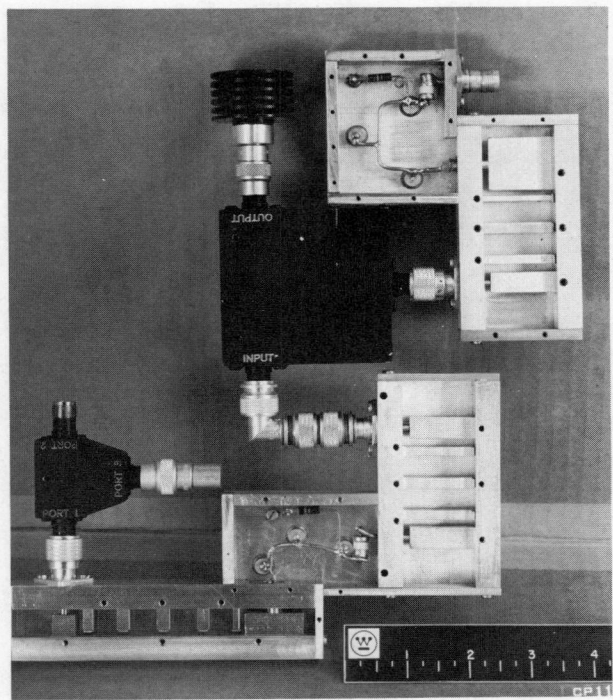

Fig. 1.41. Frequency multiplier utilizing comb-line filtering structures.

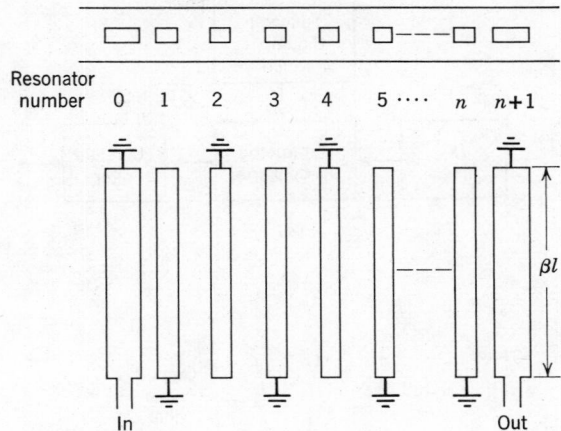

Fig. 1.42. Interdigital bandpass filter with *n* transmission-line elements. Note that alternate ends of the lines are grounded.

Comb-line structures find very broad application in microwave technology. Their attenuation above the primary passband is very high and depends on the electrical length of the resonator lines. Attenuation through the filter is infinite at the frequency for which the resonator lines are 1/4 wavelength long.

The attenuation band of this filter can be very broad because spurious responses can be easily removed. Adequate coupling between resonator elements can be maintained with sizable spacings between physical lines, and this imposes no stringent requirements on tolerances. Filters of this type can be fabricated without the use of dielectric support materials; consequently dielectric losses can be eliminated.

The interdigital filter has the same advantages of the comb-line filter. An example of an interdigital-filter arrangement is shown in Fig. 1.42. When alternate unloaded quarter-wave resonators in a comb filter are turned end-to-end so that the structure has open- and short-circuited ends alternating, the bandstop becomes a bandpass structure, and we have an interdigital filter. Although comb-line filters are somewhat smaller and are capable of broader stopbands above the primary passband, the resonators in the interdigital structure attain higher quality factors. Interdigital filters have very strong stopbands and moderate coupling tolerances.

1.13 PARAMETRIC FILTERS

In Fig. 1.43 a block diagram of a parametric filter is shown. The *LC* filter shown in the left side of the diagram, in a receiver, is a tunable circuit which can cover the frequency range of operation. The output is fed into a parametric up-converter which is basically

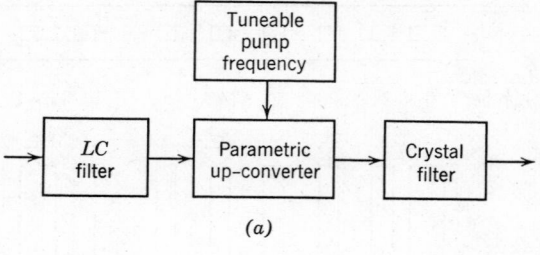

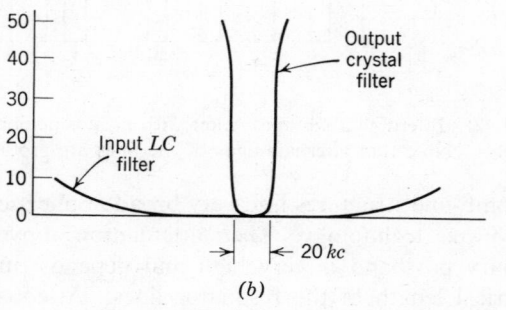

Fig. 1.43. Parametric filter: (a) block diagram and (b) amplitude response.

a mixer and is incidentally a parametric amplifier. The pump signal generator, in essence, is a tunable local oscillator. The pump signal is always much higher in frequency in an up-converter and is higher in power than the incoming signal. Therefore, the mixed

product is of a much higher power level and at a higher frequency.

The signal is now filtered by a relatively narrowband crystal filter. In Fig. 1.43b the response of the filter is shown. From the flat curve at the bottom it is seen that the response of the LC filter is poor compared to the crystal filter. Its prime function is to act as an impedance matching device between antenna and the parametric up-converter. The necessary selectivity of this receiver front end is provided by the crystal filter. The corresponding curve shows that the filter has both a narrow bandwidth and superior selectivity.

The parametric up-converter is not only a mixer that provides a fixed frequency (sum of pump frequency and signal frequency can always be adjusted by changing the frequency of the pump frequency) but also a low noise amplifier. Over the entire tuning range the output frequency is always constant. The higher the output frequency, the greater the gain that can be realized. The only limitations in this direction are the realizability of the output filter and the stability of the pump oscillator.

The parametric filter can handle extreme levels of input signals simultaneously primarily because of the high pump-power level. This ability is in itself no assurance of interference-free reception. However, since both the desired and undesired signals pass through the up-converter without interaction, the undesired signal will eventually be rejected by the output filter.

2

Theory of Effective Parameters

There are two competitive methods of filter design. One was originated by Zobel and is well known as the image-parameter method. The second was originated by Norton and Bennett and is known as the exact method, polynomial method, or insertion-loss method.

The image-parameter theory filter is based on the properties of transmission lines. A simple network with lumped components is described in terms of this continuous structure. Several of such elementary networks with equal characteristic terminal impedances, connected together to produce a chain of ladder networks (or a composite lattice), will possess a transmission constant equal to the sum of all the individual transmission constants of the elementary sections.

In the image-parameter theory, the starting point is the impedance system given by equations and the impedance matrix. The parameters of the network are expressed in terms of the characteristic impedances and the image transmission factor. Image-parameter method of filter design is analytical in nature.

In the effective-parameter theory, the starting point is the effective attenuation and the echo attenuation (reflection). The effective values characterize the behavior of the network. Here, the operational boundry can be more closely and economically approximated. Wave or characteristic impedances play no role in the effective-parameter theory. Filter design technique based on this method is essentially a synthesis.

The image-parameter method seems to be in disrepute among the network theorists partially because of the cut-and-try method that is involved and partly because of the restricted freedom of design. The practical significance of the limitation, however, does not seem to have been adequately explored. Also, very little seems to have been done in applying the results of modern theory in an effort to improve the older

design procedure. Polynomial synthesis deals directly with effective parameters and provides an elegant solution to the approximation problem, but it also involves laborious computation for the determination of the element values. This procedure is now greatly simplified by tables, step-by-step design procedures, and design curves. But the theory behind all this is still one for the specialist. In the more general synthesis procedure, a greater range of designs is possible and greater stopband loss can be obtained at the price of greater ripple factor.

The reflection coefficient is directly related to the insertion loss. To obtain a low reflection coefficient it is usually necessary to design an image-parameter filter for flatter transmission than would otherwise be necessary. The filter synthesized by polynomial methods has no practical limitation, even for an extremely severe requirement, such as 5% reflection or lower.

The more general synthesis procedure is always preferable when maximum stopband loss must be obtained at the expense of higher passband ripple, or where very flat transmission is needed to secure low reflections. The result of image-parameter theory can be regarded as a special case of effective-parameter theory.

The fault of the Zobel design is usually not that it cannot give sufficiently flat transmission, but that the transmission is too flat and discrimination is lost in consequence. The Zobel filter approximated in the Chebyshev sense (with the aid of elliptic functions) still has essential limitations in comparison with the modern filter. In practical terms this means that only one Zobel design is possible, having a particular level of passband ripple for each width of transition region.

In practice, either method of calculation can be used,

depending on the requirements in each specific design case. For a more economical filter and quicker development, especially when the filter is very complex, the use of image-parameter design technique is certainly justified. In certain other cases, when the requirement accents the minimum number of elements to satisfy an exact performance, it is better to use the optimal design of the effective parameter method. This method certainly is worthwhile for mass production.

In this chapter information as to the theory of effective parameters, the elements of the synthesis procedure, will be described. The purpose is to indicate the scientific background of filter design without exhausting the subject matter.

2.1 POWER BALANCE

The typical operating conditions for a filter are shown in Fig. 2.1, where R_1 and R_2 are pure resistances. The output power P_2 in the ideal case of matching (and with purely reactive elements) is equal to the maximum deliverable power P_m of the generator. If V_0 is the source EMF, the maximum deliverable power is $P_m = V_0^2/4R_1$.

Ideal matching is understood as the matching of both pairs of terminals, that is, between the load resistance R_2 and the network output and between the generator internal resistance R_1 the network input. So seldom is ideal matching achieved that in every practical case a complex power distribution is taking place within the network. Several components of the power must be considered.

1. The power P_2, which dissipates in the load.
2. The power P_r, which is reflected back to the generator as a result of mismatching.
3. The power P_L, which is dissipated in the network as a consequence of unavoidable losses in the components.

The sum of these three components must be equal to the maximum deliverable power according to the power balance relation $P_m = P_2 + P_r + P_L$. All

three components are frequency dependent. Although P_m remains constant, the contribution of the variable components is a function of frequency. The equation can be modified so that power balance will be expressed by

$$1 = \frac{P_m}{P_m - P_L}\left(\frac{P_2}{P_m} + \frac{P_r}{P_m}\right) \qquad (2.1.1)$$

It has been assumed that all losses are concentrated in a separate section preceding the purely reactive network.

The effective attenuation is defined as

$$a = \tfrac{1}{2}\ln\frac{P_m}{P_2} \qquad (2.1.2a)$$

where

$$\frac{P_m}{P_2} = e^{2a} = |H|^2$$

and H is the transmission factor. The attenuation a is in nepers. When the decibel unit is used

$$A = 20\log|H|$$

Reflective or echo attenuation is

$$a_e = \tfrac{1}{2}\ln\frac{P_m}{P_r} \qquad (2.1.2b)$$

or

$$\frac{P_m}{P_r} = e^{2a_e} = \left|\frac{1}{\rho}\right|^2 = |T|^2$$

is the squared magnitude of the reflection function. Also, the power loss in the network is

$$a_L = \tfrac{1}{2}\ln\frac{P_m}{P_m - P_L} \qquad (2.1.2c)$$

where

$$\frac{P_m}{P_m - P_L} = e^{2a_L} = k^2$$

is the power loss factor in the loss section. Equation 2.1.1 has, therefore, the following form:

$$1 = e^{2a_L}(e^{-2a} + e^{-2a_e}) = k^2\left(\frac{1}{|H|^2} + |\rho|^2\right) \qquad (2.1.2d)$$

$$1 = k^2(10^{-0.1A} + 10^{-0.1A_e})$$

From this, the transmission function can be found

$$|H| = k\sqrt{1 + \left|\frac{H}{T}\right|^2} = e^a = k\sqrt{1 + |D|^2} \qquad (2.1.3)$$

where $D = H/T$ is known as the discrimination or filtering function or, more often, as the characteristic function of the network.

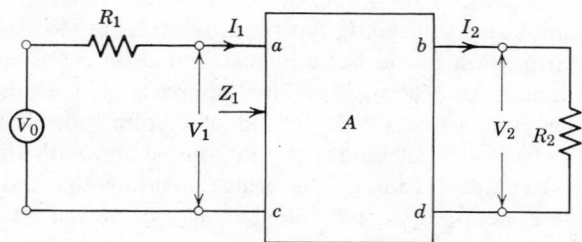

Fig. 2.1. Typical filter operating conditions.

If $P_L = 0$, then $k = 1$, and $a_L = 0$. For a purely reactive network where losses are nonexistant $a = \ln(1 + D^2)$ since

$$|H| = \sqrt{1 + |D|^2} \qquad (2.1.4)$$

and the expression for energy balance will be given in the form

$$e^{-2a} + e^{-2a_e} = 1 \qquad (2.1.5)$$

Equation 2.1.5 can be written

$$e^{2a} = 1 + e^{2(a - a_e)}$$

The filtering factor which characterizes the amplitude discrimination is

$$D = \frac{H}{T}$$

and therefore,

$$a - a_e = \ln |D| \qquad (2.1.6)$$
$$a = -\ln \sqrt{1 - \rho^2} \; \text{Np}$$
$$A = -10 \log (1 - \rho^2) \; \text{dB}$$

The effective transmission factor H and the discrimination D are both functions of frequency f and both are needed to determine quantities such as impedance, admittance, or the chain matrices. When the quantities H and D are expressed as functions of complex frequencies then each of the functions will be a rational function and represented by the ratio of two polynomials in s.

In the filter theory to be developed, we operate with the transmission function $H(s)$ defined as the inverse of the transfer function $W(s)$.

The conditions that $H(s)$ must satisfy for filters having a finite number of lumped components will be given later in the text.

The relationship between echo attenuation, effective attenuation, and the reflection coefficient ρ can be

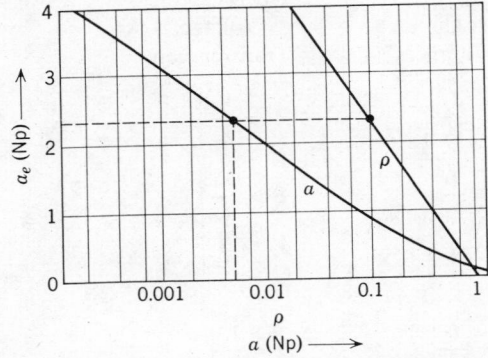

Fig. 2.2. Nomograph relating a_e, a, and ρ.

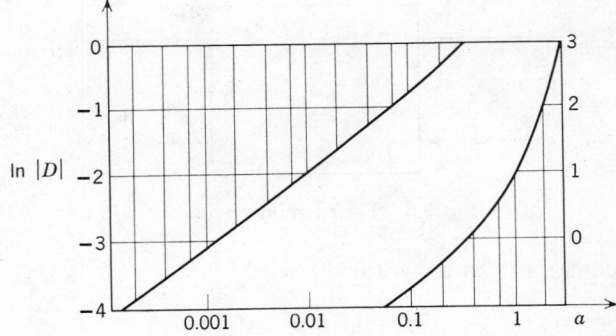

Fig. 2.3. Nomograph relating a and D, $\ln D = \frac{1}{2} \ln (e^{2a \, \max} - 1)$.

easily represented by a nomograph, as shown in Fig. 2.2. The abscissa represents effective attenuation and simultaneously shows the value of the reflection coefficient which must be multiplied by 100 to obtain ρ in percent. The ordinate represents the echo attenuation in nepers. To determine the value of reflection coefficient and the corresponding effective attenuation the procedure is:

1. On the abscissa choose the given value of the reflection coefficient and project that value up to the ρ line.
2. Project that value on the a line parallel to the abscissa.
3. Project the value on the a line down to the abscissa and read the value of effective attenuation in nepers.

For example, the reflection coefficient is 10%, corresponding to a value of the abscissa of 0.1. An intermediate reading on the ordinate indicates 2.3 Np echo attenuation a_e. The intersection of this projection and curve a when projected on the abscissa corresponds to 0.005 Np or an effective attenuation of 0.043 dB.

The relationship between effective attenuation and the filtering function D can also be represented in the form of the nomograph shown in Fig. 2.3. The ordinate is the natural logarithm of D. On the left side negative values of $\ln |D|$ are shown (for small values of D), and on the right side positive values are plotted for larger values of D. The larger values of D correspond to larger values of effective attenuation a in nepers. The use of this chart needs no special explanation.

2.2 TYPES OF GENERAL NETWORK EQUATIONS

In Fig. 2.4 a four-terminal network is shown. According to the Kirchhoff's laws, the following

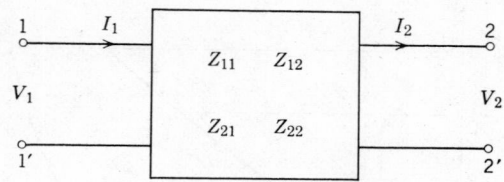

Fig. 2.4.   Four-terminal network.

equations can be written

$$V_1 = I_1 Z_{11} + I_2 Z_{12}$$
$$V_2 = I_1 Z_{21} + I_2 Z_{22}$$ (2.2.1)

or,

$$I_1 = V_1 Y_{11} + V_2 Y_{12}$$
$$I_2 = V_1 Y_{21} + V_2 Y_{22}$$ (2.2.2)

The first pair of equations is known as impedance equations and the second pair as admittance equations.

From the impedance equation it follows that when $I_2 = 0$ (a secondary open circuit),

$$Z_{11} = \frac{V_1}{I_1}$$ primary open-circuit input impedance

$$Z_{21} = \frac{V_2}{I_1}$$ transfer open-circuit impedance   (2.2.3)

When $I_1 = 0$,

$$Z_{22} = \frac{V_2}{I_2}$$ secondary open-circuit input impedance

$$Z_{12} = \frac{V_1}{I_2}$$ transfer open-circuit impedance   (2.2.4)

From the admittance equations it follows that when $V_2 = 0$ (a secondary short circuit):

$$Y_{11} = \frac{I_1}{V_1}$$ primary short-circuit admittance

$$Y_{21} = \frac{I_2}{V_1}$$ transfer short-circuit admittance   (2.2.5)

and when $V_1 = 0$ (a primary short circuit)

$$Y_{22} = \frac{I_2}{V_2}$$ secondary short-circuit admittance

$$Y_{12} = \frac{I_1}{V_2}$$ transfer short-circuit admittance   (2.2.6)

For passive networks generally $Z_{12} = Z_{21}$ and $Y_{12} = Y_{21}$. All of the above values are measurable and can be expressed in terms of network elements.

Table 2.1   Network Equations and Physical Meaning of Network Parameters

	Impedance System	Admittance System	Chain System
Network equations	$V_1 = Z_{11}I_1 + Z_{12}I_2$ $V_2 = Z_{21}I_1 + Z_{22}I_2$	$I_1 = Y_{11}V_1 + Y_{12}V_2$ $I_2 = Y_{21}V_1 + Y_{22}V_2$	$V_1 = AV_2 + BI_2$ $I_1 = CV_2 + DI_2$
Matrix	$\|Z\| = \left\| \begin{matrix} Z_{11} & Z_{12} \\ Z_{21} & Z_{22} \end{matrix} \right\|$	$\|Y\| = \left\| \begin{matrix} Y_{11} & Y_{12} \\ Y_{21} & Y_{22} \end{matrix} \right\|$	$\|M\| = \left\| \begin{matrix} A & B \\ C & D \end{matrix} \right\|$
	Open Circuit Impedances	Short Circuit Admittances	Open Circuit Voltage Transmission
Physical meaning of network parameters	$Z_{11} = \frac{V_1}{I_1}\Big|_{I_2=0}$ $Z_{12} = \frac{V_1}{I_2}\Big|_{I_1=0}$ $Z_{21} = \frac{V_2}{I_1}\Big|_{I_2=0}$ $Z_{22} = \frac{V_2}{I_2}\Big|_{I_1=0}$	$Y_{11} = \frac{I_1}{V_1}\Big|_{V_2=0}$ $Y_{12} = \frac{I_1}{V_2}\Big|_{V_1=0}$ $Y_{21} = \frac{I_2}{V_1}\Big|_{V_2=0}$ $Y_{22} = \frac{I_2}{V_2}\Big|_{V_1=0}$	$A = \frac{V_1}{V_2}\Big|_{I_2=0}$ $B = \frac{V_1}{I_2}\Big|_{V_2=0}$ $C = \frac{I_1}{V_2}\Big|_{I_2=0}$ $D = \frac{I_1}{I_2}\Big|_{V_2=0}$

A different representation of network values can be given by the third pair of equations

$$V_1 = V_2 A + I_2 B$$
$$I_1 = V_2 C + I_2 D \qquad (2.2.7)$$

with a chain matrix given by

$$\|M\| = \begin{Vmatrix} A & B \\ C & D \end{Vmatrix}$$

The determinant of the chain matrices

$$|M| = |AD - BC|$$

has, for all reversible networks, the value one. [Nonreversible networks are those transmitting energy differently in different directions (unilateral) and contain amplifiers or girators.] This form (2.2.7) is especially important and advantageous when several networks are connected in a chain.

The values of A, B, C, and D are given by

$$A = \frac{Z_{11}}{Z_{12}} = \frac{Y_{22}}{Y_{12}}$$

$$B = \frac{1}{Y_{12}} = \frac{Z_{11}Z_{22} - Z_{12}^2}{Z_{12}}$$

$$C = \frac{1}{Z_{12}} = \frac{Y_{11}Y_{22} - Y_{12}^2}{Y_{12}} \qquad (2.2.8)$$

$$D = \frac{Z_{22}}{Z_{12}} = \frac{Y_{11}}{Y_{12}}$$

The values A, B, C, and D can be expressed with measurable parameters or can be described in terms of the constituent network elements. Since the filter's network elements are frequency dependent, the above coefficients are functions of frequency. The exchange of input and output terminals corresponds to an exchange of A and D. With the aid of Eqs. 2.2.8 the open-circuit and short-circuit impedances can be represented as a quotient of two matrix elements so that

$$Z_{11} = \frac{A}{C}, \quad Y_{11} = \frac{D}{B}, \quad Z_{22} = \frac{D}{C}, \quad Y_{22} = \frac{A}{B} \qquad (2.2.9)$$

In Table 2.1, equations describing four-terminal networks are summarized, and the physical meaning of the coefficients are given.

2.3 EFFECTIVE ATTENUATION

Effective attenuation has been defined as half of the natural logarithm of the power ratio (output power P_2,

to maximum available power P_m):

$$e^{2a} = \frac{P_m}{P_2} = \frac{V_0^2}{4R_1} \frac{R_2}{V_2^2} = |H|^2 \qquad (2.3.1)$$

The transmission function H can be represented as

$$H = e^g = e^{a+jb} = \frac{1}{2} \frac{V_0}{V_2} \sqrt{\frac{R_2}{R_1}} \qquad (2.3.2)$$

Often instead of the transmission function H, the natural logarithm of H is used, and called the effective transmission constant g.

$$g = \ln H$$

Its real part is the effective attenuation and its imaginary part is the effective phase angle b.

$$g = a + jb.$$

According to Fig. 2.1, the value of $I_2 = V_2/R_2$. Therefore,

$$e^g = \frac{1}{2} \sqrt{\frac{R_2}{R_1}} \frac{V_0(Z_1 + R_1)}{V_0 Z_1} \left(A + \frac{B}{R_2} \right)$$

From the chain equation (2.2.7) the value of Z_1 can be expressed by

$$Z_1 = \frac{V_1}{I_1} = \frac{A + B/R_2}{C + D/R_2}, \qquad (2.3.3)$$

and by substitution in the previous expression for the transmission function,

$$e^g = \frac{1}{2}\left[A\sqrt{\frac{R_2}{R_1}} + \frac{B}{\sqrt{R_1 R_2}} + C\sqrt{R_1 R_2} + D\sqrt{\frac{R_1}{R_2}} \right] \qquad (2.3.4)$$

For further discussion it is advantageous to introduce normalized impedances and admittances so that all Z_{ik} and Y_{ik} (also R_2 and R_1) are related to R_1.

In normalized form,

$$r_1 = \frac{R_1}{R_1} = 1$$

$$z_{ik} = \frac{Z_{ik}}{R_1} \qquad (2.3.5)$$

$$y_{ik} = \frac{Y_{ik}}{R_1}$$

With new relations

$$a = A\sqrt{\frac{R_2}{R_1}} \qquad c = C\sqrt{R_1 R_2}$$

$$\qquad (2.3.6)$$

$$b = \frac{B}{\sqrt{R_1 R_2}} \qquad d = \sqrt{\frac{R_1}{R_2}}$$

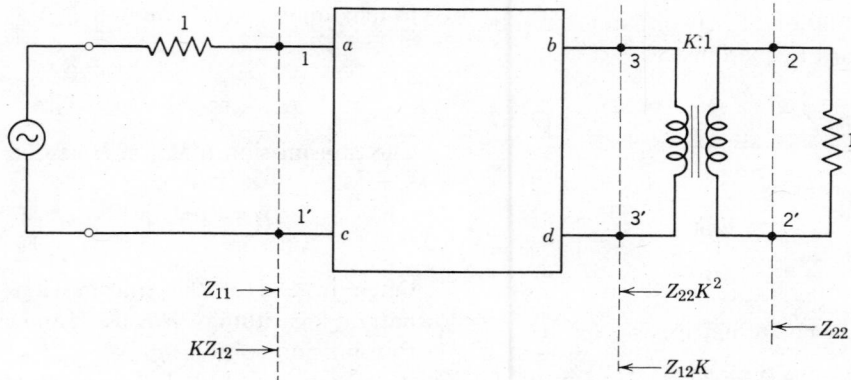

Fig. 2.5.   Network with ideal transformer.

the equation for the transmission function will be

$$e^g = \tfrac{1}{2}(a + b + c + d) = H \qquad (2.3.7)$$

Physically, the normalized functions in the form of (2.3.6) can be interpreted as if a network had an ideal transformer with $K = R_2/R_1$, just before the terminating resistance.   Figure 2.5 shows such a network.

Evidently, from (2.3.6) and (2.3.7), the transmission function H can be expressed by Z_{ik} or Y_{ik} or by the reactive element values of the network.

Let us consider the circuit shown in Fig. 2.1 with R across the load instead of the reactive network A. The transmission coefficient, expressed by the ratio of the maximum available power of the generator and the power consumed by the termination resistor R_2 is

$$\sqrt{\frac{P_m}{P_2}} = e^a = \frac{V_0}{2V_2}\sqrt{\frac{R_2}{R_1}}$$

$$= \frac{1}{2}\left[\sqrt{\frac{R_2}{R_1}} + \sqrt{\frac{R_1}{R_2}} + \frac{\sqrt{R_1 \cdot R_2}}{R}\right]$$

where R_1 is the internal resistance of the generator. Obviously, the amount of the last term of this equation will be less than unity only for negative values of R.   This means amplification exists.   Further pursuing this train of thought, it can be shown that the whole network can be consuming more energy than fed in by the generator.   In this case, we are dealing with an active network.

Before proceeding, it must be pointed out that the scope of this book is restricted to linear time-invariant networks.   By combining a finite number of resistors, capacitors, coils, and ideal transformers having only positive element values we obtain a passive network which will neither produce nor consume more energy

than forced in any port.   The passive network, however, is only one particular type of network.   When extending the element values, especially those of resistors, to the negative scale, we obtain networks able to produce or deliver more energy than is forced in.   An amplifier, for example, can be built out of a single negative resistor.

2.4   REFLECTIVE (ECHO) ATTENUATION

A similar operation can be performed to express the reflection function (2.1.2b)

$$e^{2a_e} = |T|^2 = \frac{P_m}{P_r} \qquad (2.4.1)$$

in terms of the a, b, c, and d parameters.

Since the power

$$P_r = \frac{(V_1 - V_0/2)^2}{R_1}$$

and

$$P_m = \frac{V_0^2}{4R_1}$$

where V_0 is the open-circuit voltage of the generator and R_1 is the resistance of the generator equal to the load resistance and matched with the network impedances),

$$e^{a_e} = |T| = \left|\frac{V_0/2}{V_1 - V_0/2}\right| = \left|\frac{V_0}{2V_1 - V_0}\right| \qquad (2.4.2)$$

The voltage V_1 may be determined from the parameters of the four-terminal network shown in Fig. 2.1:

$$V_1 = V_0 \frac{Z_1}{Z_1 + R_1} = V_0 \frac{Z_1/R_1}{(Z_1/R_1) + 1} \qquad (2.4.3)$$

From Eqs. 2.4.2 and 2.4.3 (and dropping the sign of

the absolute value for T),

$$e^{\alpha_e} = \left| \left(2\frac{Z_1}{Z_1 + R_1} - 1 \right)^{-1} \right| = \left| \frac{Z_1 + R_1}{Z_1 - R_1} \right| = T = \frac{1}{\rho} \quad (2.4.4)$$

It is easy to show that with normalized network coefficients and Eq. 2.3.3, the reflection function is expressible by the values of the network elements. In terms of the a, b, c, and d parameters:

$$T = \frac{a + b + c + d}{a + b - c - d} \quad (2.4.5)$$

2.5 TRANSMISSION FUNCTION AS A FUNCTION OF FREQUENCY PARAMETER, s

For the following discussion, Eq. 2.1.4 will be used:

$$|H| = \sqrt{1 + \left| \frac{H}{T} \right|^2} = \sqrt{1 + |D|^2} \quad (2.5.1)$$

with the network coefficients introduced for both the transmission function and the reflection function, the expression for the filtering function becomes

$$D = \frac{H}{T} = \frac{\frac{1}{2}(a + b + c + d)(a + b - c - d)}{a + b + c + d}$$

$$= \frac{1}{2}(a + b - c - d) \quad (2.5.2)$$

The elements of the chain matrix a, b, c, and d are functions of two-terminal impedances z_{ik} or admittances y_{ik} and consequently are always functions of $j\omega$, or of normalized $j\Omega = s$.

According to Foster's theorem, any combination of pure reactances in complex two-terminal networks can be expressed by a reactance function with the following properties: (1) The reactance function must be a rational function of the frequency parameter with positive coefficients, and (2) the numerator of the function is always an even (odd) function[1] of $s = j\omega$, and

---

[1] When the mathematician speaks of an odd function he means one like x, x^3, x^5, and so on, for which

$$f(s) = -f(-s)$$

For instance, consider the function x^3. If $x = 2$, $x^3 = 8$, and if $x = -2$, $x^3 = -8$. The equation above is therefore satisfied and x^3 is an odd function. The name *odd* function comes from the number in the exponent (see Fig. 2.6).

An even function, on the other hand, is one like x^2, x^4, x^6, and so on (with even exponents) for which

$$f(s) = f(-s)$$

For instance, if $x = 2$, $x^2 = 4$, and if $x = -2$, $x^2 = 4$ also (see Fig. 2.7).

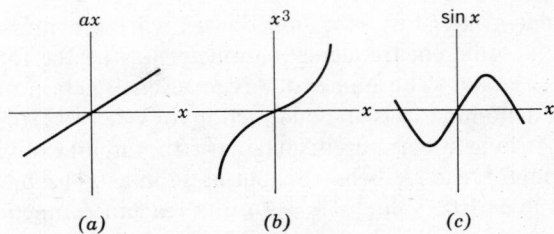

Fig. 2.6. Examples of odd functions.

the respective denominator is always odd (even). Physically, this signifies that the impedance or admittance of the reactive network always increases in value when the frequency is increasing and the quotient of voltage and current is always imaginary.

Figure 2.8 shows an example of a reactance function. The reactance function has the mathematical form

$$X(s) = B\frac{s(s^2 + \omega_3^2)(s^2 + \omega_5^2)}{(s^2 + \omega_2^2)(s^2 + \omega_4^2)} \quad (2.5.3)$$

when $\omega_2^2 < \omega_3^2 < \omega_4^2$

For $s = j\omega$ being odd, the function is purely imaginary and always increases with increasing frequency.

Figure 2.8 shows the plot only for positive ω when $X(s)$ is an odd function. More often, a simpler reactance diagram is used—a pole-zero plot. This pole-zero plot shows only poles and zeros on the positive-frequency axis. Every zero and pole in Fig. 2.8a at finite frequencies represents a pair of zeros or poles in Fig. 2.8b. Such pairs are indicated by circles (zeros) or crosses (poles). Poles and zeros at $\omega = 0$ or $\omega = \infty$ are indicated by half marks, as shown in the diagram. Therefore the figure represents a reactance function with five poles and five zeros.

The poles and zeros of the reactance function alternate. A similar property is exhibited by the reactance function with equal losses in the network elements. This property permits one to extract the lossy elements common to both reactances from a lattice filter and place them outside the lattice in the load circuit without changing the filter's performance. When

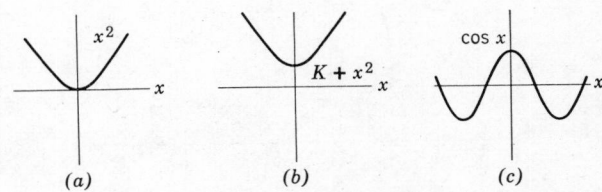

Fig. 2.7. Examples of even functions.

a network is lossy the impedances will be expressed with a different frequency parameter having the form $s' = s + \epsilon$. The numerator is an even function and the denominator is an odd function (or vice versa) in s'.

As long as s' is purely imaginary, the circuit exhibits a purely reactive behavior, but as soon as s' becomes complex (the value of $\epsilon \neq 0$), the reactance function becomes an impedance function.

From the foregoing it can be seen that for a filter with reactive elements, the value of b and c in Eqs. 2.3.7 and 2.4.5 are always odd functions, but a and d are always even functions. The latter are ratios of two impedances or two admittances and are dimensionless quantities.

If the impedances of the components of a four-terminal network are functions of s, the transmission and filtering function of the four-terminal network, according to Eq. 2.3.7, have to be functions of s, and in general are rational functions. $H(s)$ and $D(s)$, being composed of sums and differences of all four functions $a(s)$, $b(s)$, $c(s)$, and $d(s)$ must have identical denominators.

$$H(s) = \frac{E(s)}{P(s)}$$

$$D(s) = \frac{F(s)}{P(s)} \qquad (2.5.4)$$

The following are the conditions that these functions must satisfy if they are to correspond to realizable, reactive networks:

1. $E(s)$, $F(s)$, and $P(s)$ must be real polynomials with positive coefficients.

2. The roots of $E(s)$ must have negative real parts. Such functions are known as Hurwitz polynomials. The negative real part is necessary because it appears

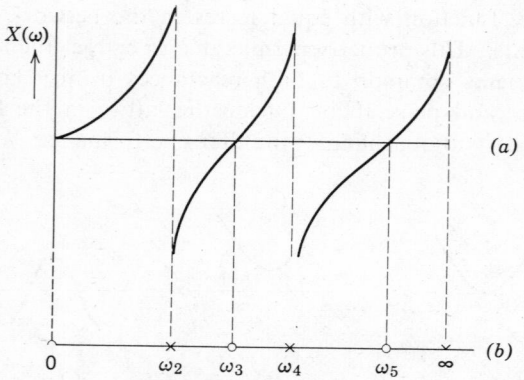

Fig. 2.8. Reactance diagram (*a*), with corresponding pole-zero plot (*b*).

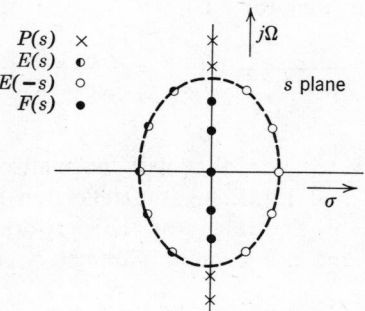

Fig. 2.9. Roots of $P(s)$, $E(s)$, $E(-s)$, and $F(s)$ polynomials of filter with $n = 5$.

as $e^{\alpha t}$ (where α is the real part of the root) in the transient response of the filter and thereby characterizes the amplitude response as a function of time. If these parts are positive, the amplitude of the signal with time would be increasing and this is physically impossible. The output voltage is stabilized after a transition period, and the steady-state condition is reached.

3. $P(s)$ must be an even or odd polynomial whose degree is lower or at most equal to that of $E(s)$.

4. The absolute value of the transmission function must be greater or equal to one for all real ω.

$$|H(s)| \geq 1$$

The physical meaning of this condition is that the power P_2, at the output end, in the case of passive networks, can never be greater than the maximum deliverable power P_m of the source. $H(s)$ is therefore the quotient of two polynomials one of which has zeros in the left half of the s-plane (Hurwitz polynomial) and another polynomial which has an order lower than that of the Hurwitz polynomial. Figure 2.9 shows typical roots of the above polynomials. The order of the transmission function is five.

2.6 POLYNOMIALS OF TRANSMISSION AND FILTERING FUNCTIONS

The functions $E(s)$ and $F(s)$, being rational, can be subdivided into even and odd parts. The previous equations can be written in the following forms:

$$H(s) = \frac{E_e + E_o}{P}$$

$$D(s) = \frac{F_e + F_o}{P} \qquad (2.6.1)$$

With the aid of Eqs. 2.3.7 and 2.5.2 the numerators of Eq. 2.6.1 become:

$$E_e + E_o = \frac{P(a + b + c + d)}{2} = E \quad (2.6.2)$$

$$F_e + F_o = \frac{P(a + b - c - d)}{2} = F \quad (2.6.3)$$

The matrix coefficients a and d are even, and b and c are odd functions. Since $P(s)$, according to condition 3 of Section 2.5 is either an even or odd function, two cases should be considered.

1. $P(s)$ is an even function. From (2.6.2),

$$\frac{P(a + d)}{2} = E_e \quad (2.6.4)$$

$$\frac{P(c + b)}{2} = E_o \quad (2.6.5)$$

From (2.6.3),

$$\frac{P(a - d)}{2} = F_e \quad (2.6.6)$$

$$\frac{P(b - c)}{2} = F_o \quad (2.6.7)$$

Following from the foregoing equations:

$$a = \frac{E_e + F_e}{P}, \quad b = \frac{E_o + F_o}{P}$$
$$c = \frac{E_o - F_o}{P}, \quad d = \frac{E_e - F_e}{P} \quad (2.6.8)$$

2. $P(s)$ is an odd function. From (2.6.2),

$$\frac{P(a + d)}{2} = E_0$$
$$\frac{P(b + c)}{2} = E_e \quad (2.6.9a)$$

From (2.6.3),

$$\frac{P(a - d)}{2} = F_0$$
$$\frac{P(b - c)}{2} = F_e \quad (2.6.9b)$$

With (2.6.9a) and (2.6.9b), the network coefficients can be expressed as:

$$a = \frac{E_0 + F_o}{P}, \quad b = \frac{E_e + F_e}{P}$$
$$c = \frac{E_e - F_e}{P}, \quad d = \frac{E_0 - F_o}{P} \quad (2.6.10)$$

Table 2.2 Network Impedance Parameters as Functions of Polynomials E, F, and P

z_{11}	$\dfrac{E_e + F_e}{E_0 - F_0}$	$\dfrac{E_0 + F_0}{E_e - F_e}$
z_{22}	$\dfrac{E_e - F_e}{E_0 - F_0}$	$\dfrac{E_0 - F_0}{E_e - F_e}$
z_{12}	$\dfrac{P}{E_0 - F_0}$	$\dfrac{P}{E_e - F_e}$
y_{11}	$\dfrac{E_e - F_e}{E_0 + F_0}$	$\dfrac{E_0 - F_0}{E_e + F_e}$
y_{22}	$\dfrac{E_e + F_e}{E_0 + F_0}$	$\dfrac{E_0 + F_0}{E_e + F_e}$
y_{12}	$\dfrac{P}{E_0 + F_0}$	$\dfrac{P}{E_e + F_e}$

With the aid of (2.2.8), the following can be obtained

$$z_{11} = \frac{a}{c} \quad z_{22} = \frac{d}{c} \quad z_{12} = \frac{1}{c} \quad (2.6.11)$$

$$y_{11} = \frac{d}{b} \quad y_{22} = \frac{a}{b} \quad y_{12} = \frac{1}{b} \quad (2.6.12)$$

Therefore the impedances and admittances can be calculated as functions of the frequency parameters from the transmission function $H(s)$ as given in Table 2.2. The network is fully determined by one of these reactances together with the complete set of zeros of $P(s)$. Therefore the circuit elements values can be calculated by partial or full removal of residues at these poles.

2.7 FILTER NETWORKS

From general filter theory it is known that a given attenuation response may be realized by a number of different network configurations of the four-terminal type. In ladder form two different element arrangements can be mentioned, for example, the T and π shown in Figs. 2.10 and 2.11. They exhibit equal transmission characteristics, but the physical differences are apparent—they are reciprocals in impedance.

For the frequencies where the π arrangement presents a low-impedance input or output, the T arrangement displays high-impedance behavior. The following is a criterion for impedance reciprocity of two circuits: When in two comparable circuits with equal attenuation characteristics the open-circuit

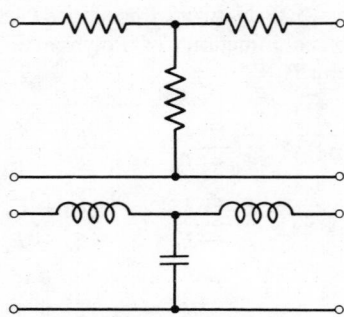

Fig. 2.10. Ladder *T* lowpass network.

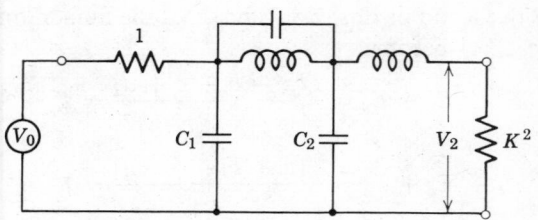

Fig. 2.12. Antimetric (*n* = 4) lowpass filter.

impedance of one equals the short-circuit impedance of the other (within a constant factor), these two circuits are reciprocal. Mathematically, this criterion is expressed by

$$\frac{Z_{ik}}{Y_k} = 1$$

Likewise, with normalized impedance and admittance

$$\frac{z_{ik}}{y_{ik}} = 1 \qquad (2.7.1)$$

From Table 2.2 we conclude that Z_{ik} will become Y_{ik} when F_o and F_e, in one network, are the negatives of those in the other. This means that $F(s)$ is calculated for one network with $-F(s)$ from the other. The exchange has no influence on the attenuation response because the attenuation is expressed in terms of the square of $F(s)$ according to the equation

$$|H|^2 = 1 + |D|^2$$

It is therefore possible to use the expressions in Table 2.2 for impedance-reciprocal schematics.

In the most general case the load resistance R_2 does not match the internal resistance of the generator. Since all resistances are normalized with respect to R_1, R_1 will be set equal to 1, and R_2/R_1 is then equal to K^2, as shown in Fig. 2.12 for a general lowpass

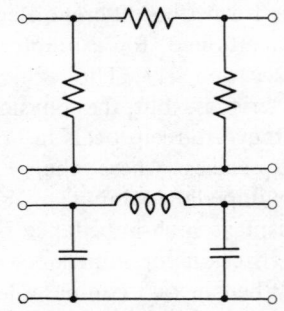

Fig. 2.11. Ladder *π* lowpass network.

filter. For frequency $\Omega = 0$ the shunt arms C_1 and C_2 present infinite impedances and the series branch reactance equals zero. Then the schematic in Fig. 2.12 is equivalent to that in Fig. 2.13. The transmission function at the point $s = j\Omega = 0$ is

$$H(0) = \frac{V_0}{2V_2}\sqrt{\frac{R_2}{R_1}} = \tfrac{1}{2}K\frac{V_0}{V_2}$$

$$= \frac{1}{2}\frac{1+K^2}{K^2}K = \tfrac{1}{2}(K + K^{-1}) \quad (2.7.2)$$

For $K = 1$ then $H(0) = 1$. Equation 2.7.2 shows, however, that for $\Omega = 0$, attenuation can appear as a result of mismatching. The following theorem can be stated for four-terminal networks: If the effective attenuation at $\Omega = 0$ is zero the network is symmetric and its effective load resistance is equal to the source resistance ($K = 1$).

Let us analyze Eq. 2.5.1. When $H(0) = 1$ then $D(0) = 0$. But $D(0)$ is equal to zero when $F(s)$ is an odd function. Therefore, when $F(s)$ is an odd function the filter is always symmetric ($R_2 = R_1$). When $F(s)$ is an even function or some general rational function the filter is antimetric ($R_1 \neq R_2$).

From Eq. 2.7.2 it follows that

$$K_{1,2} = H(0) \pm \sqrt{H^2(0) - 1}$$

Each of the solutions are valid.

If $\Omega = 0$ and $I_2 = 0$ the ratio $V_0/V_2 = 1$. For the same position on the frequency scale ($\Omega = 0$) when $V_2 = 0$

$$\frac{I_1}{I_2} = 1$$

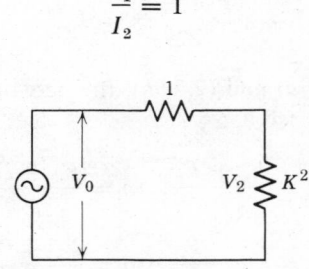

Fig. 2.13. Equivalent circuit of Fig. 2.12 for $\Omega = 0$.

Regarding Fig. 2.12 the situation is easily recognizable. According to Eq. 2.3.6,

$$a = \sqrt{\frac{R_2}{R_1}} A = KA$$

$$d = \sqrt{\frac{R_1}{R_2}} D = K^{-1}D$$

Therefore for $\Omega = 0$,

$$K = a(0) \quad \text{and} \quad K^{-1} = d(0) \qquad (2.7.3)$$

The functions $a(s)$ and $d(s)$ have been expressed previously by Eqs. 2.6.8 and 2.6.10.

2.8 VOLTAGE AND CURRENT SOURCES

Sometimes the network is either connected to a source whose internal impedance is approximately zero or one so high that it can be assumed as infinite.

In Fig. 2.14 the very low-impedance source is shown. The condition is described as:

$V_1 = $ constant (constant voltage source)

Figure 2.15 shows a very high-impedance source. The condition is described as:

$I_1 = $ constant (constant current source)

In practice, naturally, both cases are only approximations to the actual conditions. For a voltage source, the internal impedance of the source is zero. Therefore from Eq. 2.2.6

$$I_2 = \frac{V_2}{R_2}$$

The voltage V_1 will be expressed as:

$$V_1 = AV_2 + B \frac{V_2}{R_2} \qquad (2.8.1)$$

For a network with a current source

$$I_1 = CI_2R_2 + I_2D \qquad (2.8.2)$$

with normalized values (with respect to R_2) the matrix

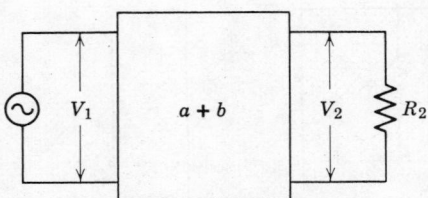

Fig. 2.14. Filter with constant voltage source.

parameters are

$$a = A = \frac{y_{22}}{y_{12}} \qquad b = \frac{B}{R_2} = \frac{1}{y_{12}}$$

$$c = CR_1 = \frac{1}{z_{12}} \qquad d = D = \frac{z_{22}}{z_{12}} \qquad (2.8.3)$$

From (2.8.1) and (2.8.2)

$$\frac{V_1}{V_2} = a + b = \frac{E_v(s)}{P_v(s)} = H_v(s) \qquad (2.8.4)$$

where $H_v(s)$ is the voltage-transmission function. For the current-transmission function, the expression will be

$$\frac{I_1}{I_2} = c + d = \frac{E_i(s)}{P_i(s)} = H_v(s) \qquad (2.8.5)$$

The effective current and voltage attenuation is

$$a = \ln \tfrac{1}{2} |H_v + H_i|$$

Here also, a and d must be even function of s, and b and c must be odd functions of s. Separating the odd and even parts of E_v and E_i we obtain

$$V(s) = \frac{E_{ev} + E_{ov}}{P_v} \qquad (2.8.6a)$$

$$I(s) = \frac{E_{ei} + E_{oi}}{P_i} \qquad (2.8.6b)$$

For an even P function

$$a = \frac{E_{ev}}{P_v} \qquad b = \frac{E_{ov}}{P_v}$$

$$c = \frac{E_{oi}}{P_i} \qquad d = \frac{E_{ei}}{P_i} \qquad (2.8.7)$$

For the constant-voltage generator, by the use of (2.8.3) and (2.8.7),

$$y_{22} = \frac{E_{ev}}{E_{ov}}$$

$$y_{12} = \frac{P_v}{E_{ov}} \qquad (2.8.8)$$

and for a constant-current generator (2.8.3) and (2.8.7) show that

$$z_{22} = \frac{E_{ei}}{E_{oi}}$$

$$z_{12} = \frac{P_i}{E_{oi}} \qquad (2.8.9)$$

For an odd P function the values of z_{ik} and y_{ik} can be calculated in exactly the same way. It has to be remembered that in the case of a constant-current generator the power can be said to be flowing in the

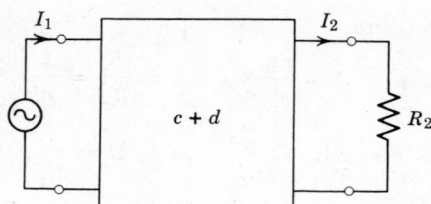

Fig. 2.15.   Filter with constant current source.

opposite direction.   An imaginary generator on the right side of Fig. 2.15 has internal resistance R_2. Therefore the network can be viewed as being open-circuit terminated.   A constant-voltage source of Fig. 2.14 can be also viewed as operating in the opposite direction with a generator having R_2 as its internal resistance and with shorted opposite terminals.   This is especially important to remember in the case of catalog usage.

There are two more possible modes of operation: (1) the output side is open circuited; the corresponding transmission constant is:

$$g_o = \ln \frac{V_0}{V_2} = \ln \frac{Z_1 + R_1}{Z_{12}}$$

(2) the output side is a short circuit; the transmission constant in this condition is

$$g_s = \ln \frac{I_0}{I_2} = \ln \frac{Y_1 + 1/R_1}{Y_{12}}$$

Figure 2.16 shows these conditions.

2.9   THE FUNCTION $D(s)$ AS AN APPROXIMATION FUNCTION

The filtering function (or discrimination function) $D(s)$ is the ratio $D(s) = F(s)/P(s)$, where $F(s)$ and $P(s)$ are real even or odd polynomials in s, with real coefficients.   Function $D(s)$ is less restricted than the function $H(s)$.   This determines the attenuation response of the filter.

If H is an effective-transmission factor and a is the

effective attenuation, the value of $|H|^2$ can be represented by the value of the discrimination D as

$$|H|^2 = e^{2a} = 1 + |D|^2$$

and

$$a = \tfrac{1}{2} \ln (1 + |D|^2)$$

For large values of D, $a = \ln |D|$, for small values of D, $a = \frac{1}{2}|D|^2$ (see Fig. 2.3).

Because $s = j\omega$, the relation $|E(s)|^2 = E(s) E(-s)$ holds and similar ones hold for F and P, and the following important relation can be written

$$E(s) E(-s) = F(s) F(-s) + P(s) P(-s) \quad (2.9.1)$$

The polynomial $E(s)$ can be obtained by determining the roots of the polynomial $F(s) F(-s) + P(s) P(-s)$ and multiplying all those roots having negative real parts to form a new polynomial—the Hurwitz polynomial $E(s)$.

When the filtering function is odd

$$D(s) = -D(-s)$$

the filter which corresponds to that function is symmetric.   When the filtering function is even

$$D(s) = D(-s)$$

the filter is antimetric.

In the passband of the filter the reflected power P_r must approach zero; and in the stopband the output-transmitted power P_2 must approach zero.   That is, in the passband the transmission function $|H(s)|$ must approach one; and in the stopband it must approach infinity.   The last will be satisfied since in Eq. 2.1.4 $D(s) \to \infty$.   The discrimination D as a function of frequency behaves more logically than $|H(s)|$.   In the passband the discrimination D approaches zero and in the stopband it approaches infinity (as does the attenuation of the filter).   The filtering function $D(s)$ determines the effective-transmission function $H(s)$ within a constant factor.   However, it is generally easier to satisfy less stringently the conditions for realizability, and thus simply approximate the values of

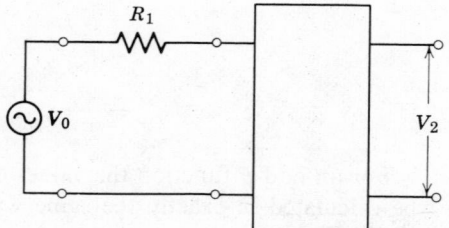
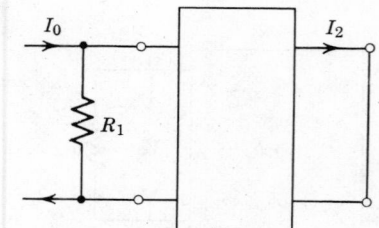

Fig. 2.16.   Open and short circuit operating conditions.

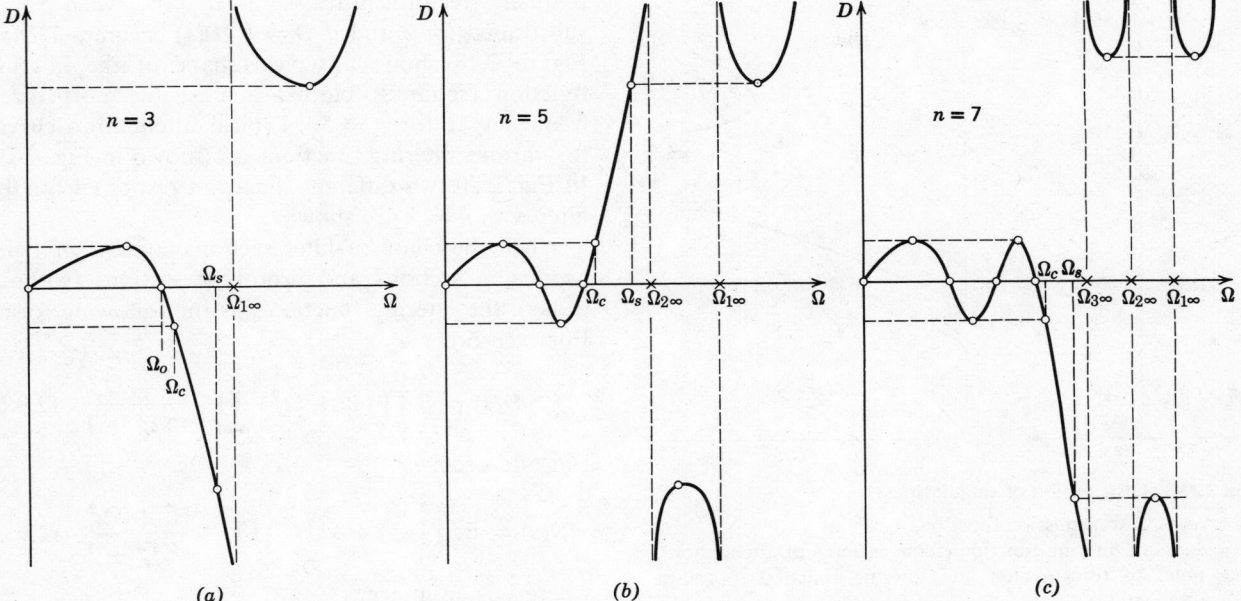

Fig. 2.17. Filtering function for filters with $n = 3$, 5, and 7.

zero and infinity. Using $H(s)\,H(-s)$ only $D(s)\,D(-s)$ can be determined; not $D(s)$ alone. The components of $D(s)$ and $D(-s)$ may be freely chosen.

The filtering function $D(s)$ forms the basic design quantity for filters whose components are tabulated numerically in Chapter 5. The zeros and poles of the filtering function are distributed on the real-frequency axis, the zeros within the passband to provide the frequencies where the effective attenuation vanishes, the poles within the stopband to provide the frequencies where the effective attenuation are infinite. The distribution of zeros and poles of $D(s)$ is arranged to ensure equal maximum in the passband and equal minimum in the stopband. The following are several filtering functions which can be used to realize filters with poles of attenuation.

For $n = 2$ $D(s) = B\,\dfrac{s^2 + \Omega_1^{\,2}}{s^2\Omega_1^{\,2} + 1}$

For $n = 3$ $D(s) = Bs\,\dfrac{s^2 + \Omega_1^{\,2}}{s^2\Omega_1^{\,2} + 1}$

For $n = 4$ $D(s) = B\,\dfrac{(s^2 + \Omega_1^{\,2})(s^2 + \Omega_2^{\,2})}{(s^2\Omega_1^{\,2} + 1)(s^2\Omega_2^{\,2} + 1)}$

For $n = 5$ $D(s) = Bs\,\dfrac{(s^2 + \Omega_1^{\,2})(s^2 + \Omega_2^{\,2})}{(s^2\Omega_1^{\,2} + 1)(s^2\Omega_2^{\,2} + 1)}$

where n is the order of the function and simultaneously the number of series and shunt arms of the ladder

structure, and B is a constant. In Fig. 2.17 the plot of filtering function for filters with poles and zeros of transmission at finite frequencies is shown.

The expression for the filtering function can be generalized: for even-order filters

$$D(s) = B \cdot \prod_{i=1}^{n/2} \frac{s^2 + \Omega_i^{\,2}}{s^2\Omega_i^{\,2} + 1} \qquad (2.9.2)$$

and for odd-order filters

$$D(s) = Bs \cdot \prod_{i=1}^{(n-1)/2} \frac{s^2 + \Omega_i^{\,2}}{s^2\Omega_i^{\,2} + 1} \qquad (2.9.3)$$

Equations 2.9.2 and 2.9.3 describe only networks which either have all attenuation poles at finite frequencies or those which have only one pole at infinity. A network with all its poles at infinity is not covered by this type of function. For networks with severe requirements for the stopband, realization without peaks of attenuation leads to more elements in the filter, compared with the realization including peaks at finite frequencies. But from certain points of view, peakfree filter responses may be very important, especially for use at higher frequencies.

For even-order filters without peaks of attenuation (no poles of $D(s)$ on the $j\omega$ axis) for $s = j\Omega$,

$$D(s) = B \prod_{i=1}^{n/2} (s^2 + \Omega_i^{\,2}) \qquad (2.9.4)$$

For odd-order filters without peaks of attenuation (no

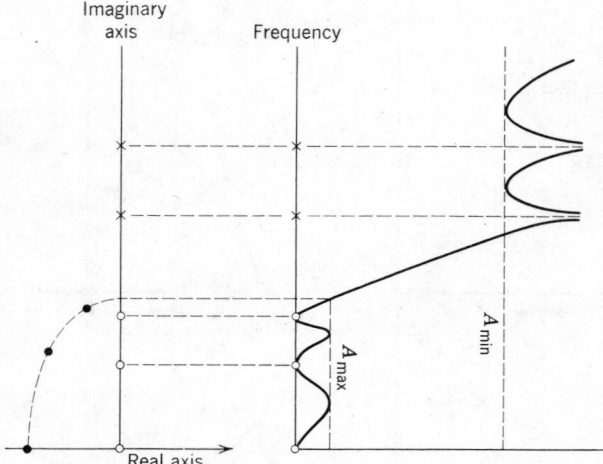

Fig. 2.18. Distribution of singularities:

● = zeros of transmission function,
○ = zeros of filtering function (same as zeros of attenuation),
× = poles of transmission and filtering function (poles of attenuation).

poles on the $j\omega$ axis)

$$D(s) = Bs \prod_{i=1}^{(n-1)/2} (s^2 + \Omega_i^2) \qquad (2.9.5)$$

If the filtering functions have to produce a Chebyshev response, both the odd and even functions have to be Chebyshev polynomials. The odd function, Eqs. 2.9.3 and 2.9.5, is equal to zero at $s = j\Omega = 0$. This signifies that the attenuation has to be zero.

Functions shown by (2.9.2) and (2.9.3) exhibit

frequency-reciprocal response—that is, when s^{-1} is substituted for s and if $B = 1$, $D(s)$ becomes $1/D(s)$. Figure 2.18 shows a typical shape of the filtering function (squared) plotted against the normalized frequency Ω for $n = 5$. Typical attenuation curves for various filtering functions are shown in Fig. 2.19. In Fig. 2.20 two different shapes of responses for the filter with $n = 7$ are shown.

When the chain of filter sections consists of pole-producing sections and prototype sections (without peaks), the filtering function has the following form: For even-order n

$$D(s) = B \prod_{i=1}^{k/2} (s^2 + \Omega_i^2) \prod_{i=1}^{n/2 - k/2} \frac{s^2 + \Omega_i^2}{s^2 \Omega_i^2 + 1} \qquad (2.9.6)$$

For odd-order n

$$D(s) = Bs \prod_{i=1}^{k/2} (s^2 + \Omega_i^2) \prod_{i=1}^{(n-1)/2 - k/2} \frac{s^3 + \Omega_i^2}{s^2 \Omega_i^2 + 1} \qquad (2.9.7)$$

where k is an integer.

In Table 2.3, filter networks and their filtering functions are given. Only the lowpass circuits are shown since the others can easily be derived from the lowpass model with the aid of frequency transformations. Figure 2.21 shows further canonic realization possibilities for one order, $n = 5$.

It is therefore possible by prescribing peak frequencies Ω_i and a constant multiplier B in the D-function to describe the effective attenuation of the filter. Then, by considering D as given, $F(s)$ and $P(s)$ can be determined.

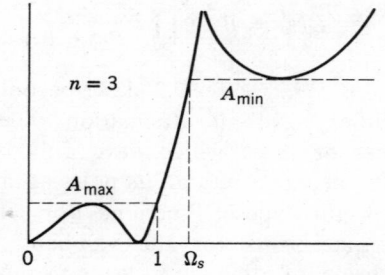

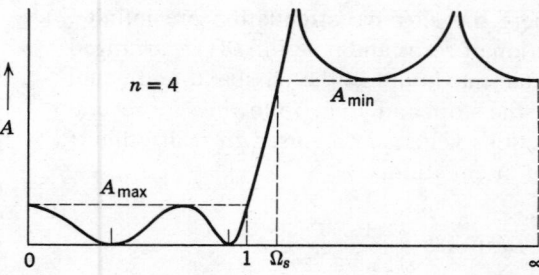

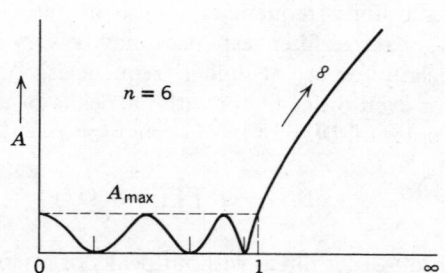

Fig. 2.19. Typical Chebyshev response curves.

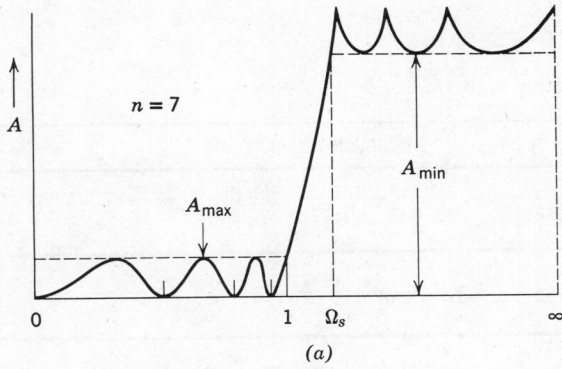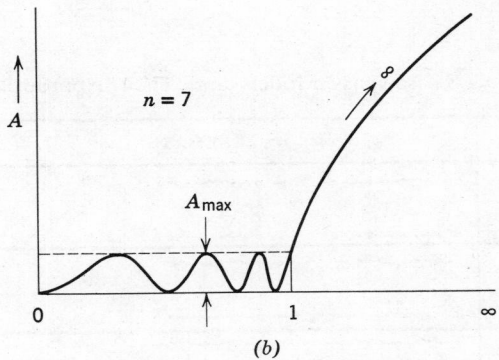

Fig. 2.20. Filters with equal ripple in passbands: (*a*) has equal minima in the stopband, and (*b*) has monotonic response in the stopband.

In conclusion, it has been generally shown that the effective parameters of a four-terminal network having a finite number of lumped linear elements, can be expressed with the help of three real polynomials P, E, F of the complex frequency $s = \sigma + j\Omega$:

$$D = \frac{F}{P}, \quad H = \frac{E}{P}, \quad Z_{in} = \frac{E + F}{E - F} \quad (2.9.8)$$

For reactive networks the polynomials are connected in the manner shown by Eq. 2.9.1.

Polynomials P, E, and F are the simplest and most universal characteristic of the network, since knowing these polynomials makes it possible to find all its parameters, both effective and image (see Section 2.12) and also to determine the elements of the schematic.

The problems of synthesis of an electric wave filter, according to effective attenuation, consists in finding a filtering system with the minimum number of elements necessary for the satisfaction of requirements, not to exceed the given A_{max} in the passband and not to be less than the given A_{min} in the stopband.

This problem can be separated into three parts:

1. The problem of the best approximation of the given requirements with the help of a rational function $D = F/P$ to ensure the minimum number of elements in the system.
2. Determining polynomial $E(s)$ by Eq. 2.9.1.
3. Determining the filter system according to the constructed polynomials P, E, and F.

The problem of determining polynomial $E(s)$ presents the greatest difficulties in designing filters according to effective parameters.

The roots of polynomial $E(s)$ equal the roots of the system's characteristic equation and correspond to the frequencies of the free oscillations which may occur in a charged network during the transition processes (see Section 2.5). The Hurwitz polynomial $E(s)$ is the characteristic polynomial of the network. Since in a passive network the free oscillations should be damped, the real part of the roots of $E(s)$ should be negative (see second condition of realizability).

This condition makes it possible to determine the polynomial $E(s)$ by the known roots of the right-hand part of Eq. 2.9.1: all the roots with a negative real part correspond to polynomial $E(s)$, and with a positive real part to polynomial $E(-s)$.

Thus the whole problem is reduced to finding the roots of the right-hand part of (2.9.1).

2.10 EXAMPLES OF TRANSMISSION FUNCTION APPROXIMATION

Before any transmission function $H(s)$ is written, an approximation process has to take place. During this process the elements of the rational function are found from the prescribed requirements. For example, one may require a transmission function for a filter which is supposed to have monotonic frequency response and at zero frequency be maximally flat. The attenuation function which satisfies this type of requirement is

$$a = \tfrac{1}{2}\ln\left[1 + \left(\frac{\omega}{\omega_c}\right)^{2n}\right] = \tfrac{1}{2}\ln(1 + \Omega^{2n}) \quad (2.10.1)$$

ω_c is a cutoff frequency where the attenuation is 0.35 Np (or 3 dB).

This function, which has a logarithmic argument, is usually called a power function or Butterworth function. Its order, $2n$, determines the sharpness of the attenuation characteristic. The following approximations are true:

$a \approx \tfrac{1}{2}(\Omega)^{2n}$ for ω larger than ω_c (in the stopband)

$a \approx n \ln |\Omega|$ for ω smaller than ω_c (in the passband)

Table 2.3 Lowpass Models and Their Appropriate Filtering Functions, D

	SCHEMATIC	N	FILTERING FUNCTION
1	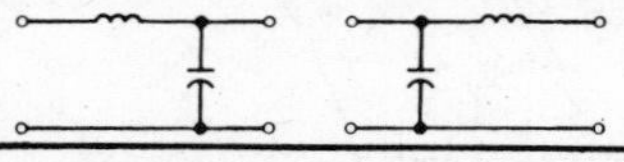	2	$D = B(s^2 + \Omega_1^2)$
2	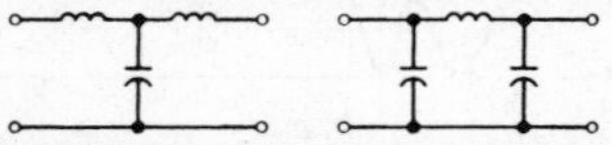	3	$D = BS(s^2 + \Omega_1^2)$
3	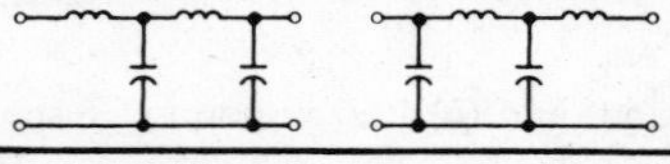	4	$D = B(s^2 + \Omega_1^2)(s^2 + \Omega_2^2)$
4	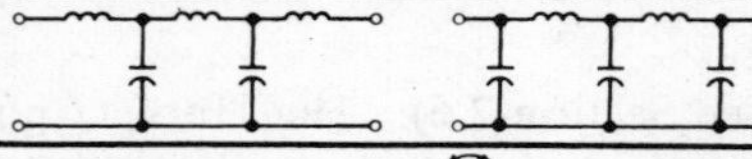	5	$D = BS(s^2 + \Omega_1^2)(s^2 + \Omega_2^2)$
5	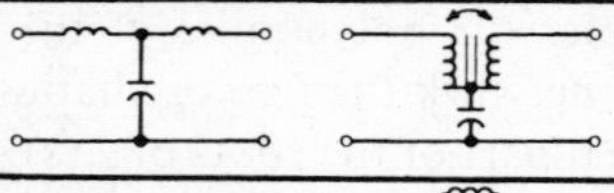	2	$D = B\dfrac{(s^2 + \Omega_1^2)}{(s^2\Omega_1 + 1)}$
6	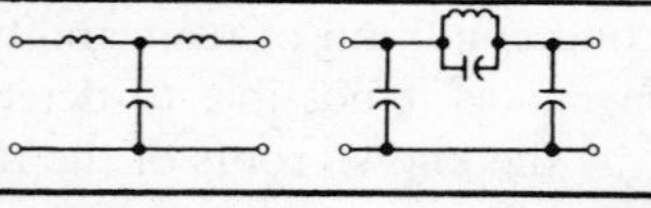	3	$D = BS\dfrac{s^2 + \Omega_1^2}{s^2\Omega_1^2 + 1}$
7	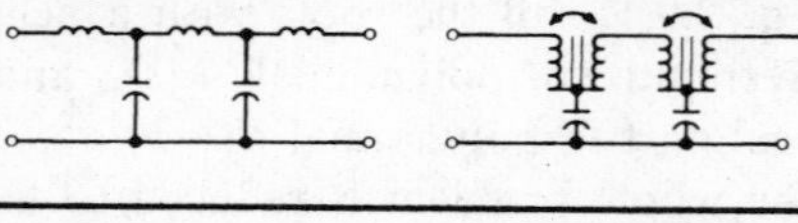	4	$D = B\dfrac{(s^2 + \Omega_1^2)(s^2 + \Omega_2^2)}{(s^2\Omega_1^2 + 1)(s^2\Omega_2^2 + 1)}$
8	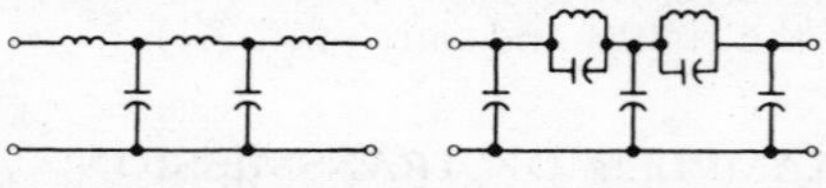	5	$D = BS\dfrac{(s^2 + \Omega_1^2)(s^2 + \Omega_2^2)}{(s^2\Omega_1^2 + 1)(s^2\Omega_2^2 + 1)}$
9	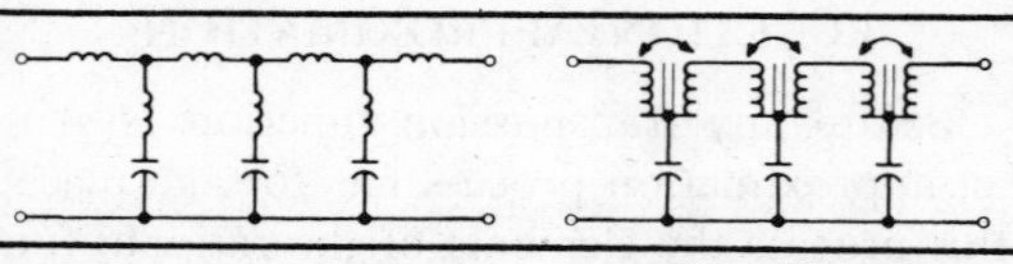	6	$D = B\dfrac{(s^2 + \Omega_1^2)(s^2 + \Omega_2^2)(s^2 + \Omega_3^2)}{(s^2\Omega_1^2 + 1)(s^2\Omega_2^2 + 1)(s^2\Omega_3^2 + 1)}$
10	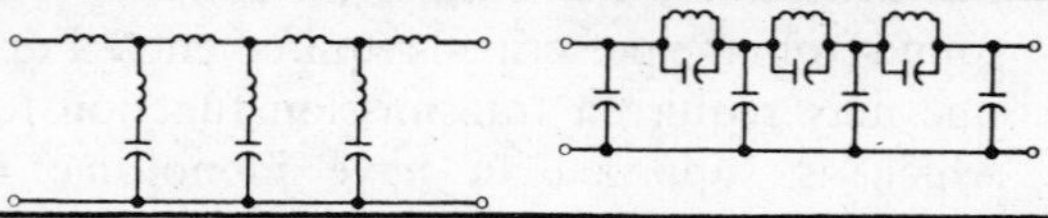	7	$D = BS\dfrac{(s^2 + \Omega_1^2)(s^2 + \Omega_2^2)(s^2 + \Omega_3^2)}{(s^2\Omega_2^2 + 1)(s^2\Omega_2^2 + 1)(s^2\Omega_3^2 + 1)}$
11	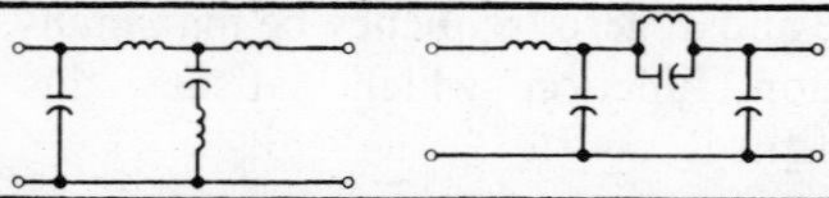	4	$D = B\dfrac{(s^2 + \Omega_1^2)(s^2 + \Omega_2^2)}{(s^2\Omega_3^2 + 1)}$
12	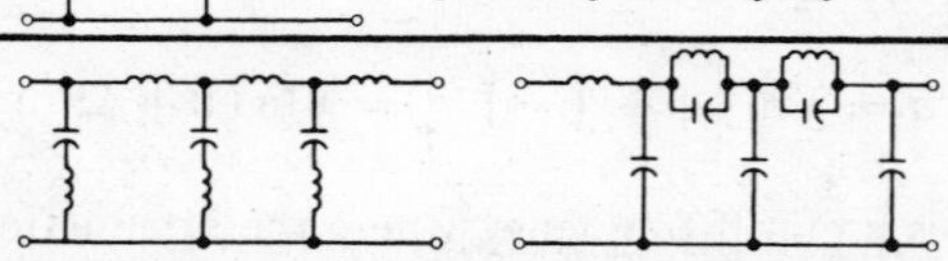	6	$D = B\dfrac{(s^2 + \Omega_1^2)(s^2 + \Omega_2^2)(s^2 + \Omega_3^2)}{(s^2\Omega_4^2 + 1)(s^2\Omega_5^2 + 1)}$
13	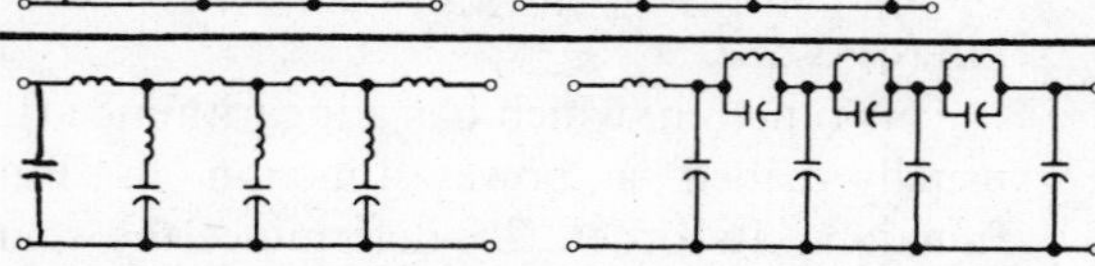	8	$D = B\dfrac{(s^2 + \Omega_1^2)(s^2 + \Omega_2^2)(s^2 + \Omega_3^2)(s^2 + \Omega_4^2)}{(s^2\Omega_5^2 + 1)(s^2\Omega_6^2 + 1)(s^2\Omega_7^2 + 1)}$

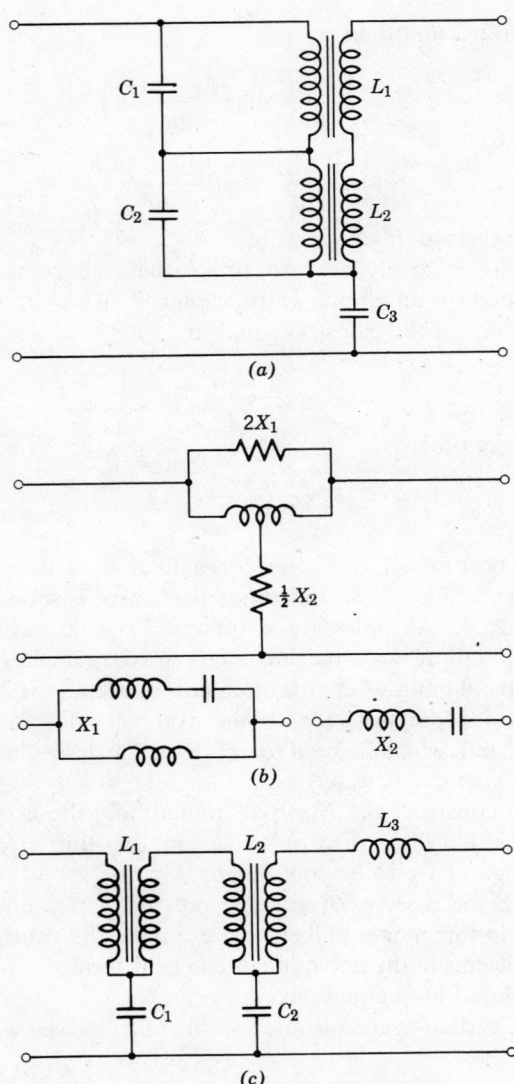

Fig. 2.21.   Canonic realization of filter with $n = 5$.

To find the transmission function, the equation which defines the transmission function will be used

$$a(s) = \tfrac{1}{2} \ln [H(s) \, H(-s)]$$

Therefore,

$$H(s) \, H(-s) = 1 + \left(\frac{-s^2}{\omega_c{}^2} \right)^n = 1 + D^2$$

$$D = \pm (\Omega)^n$$

where $s = j\omega$ and $s^2 = -\omega^2$.

All $2n$ zeros are on a circle of radius ω_c all $2n$ poles are at infinity (see Fig. 2.22). For $n = 3$ the function $H(s)$ with three zeros and three poles (at infinity) will be constructed from the following three multipliers

(when $H(0) = 1$)

$$\begin{aligned}
& \frac{s}{\omega_c} + 1 \\
& \frac{s}{\omega_c} + j\frac{\sqrt{3}}{2} + \frac{1}{2} \\
& \frac{s}{\omega_c} - j\frac{\sqrt{3}}{2} + \frac{1}{2}
\end{aligned} \qquad (2.10.2)$$

$$H(s) = \left(\frac{s}{\omega_c} \right)^3 + 2 \left(\frac{s}{\omega_c} \right)^2 + 2\frac{s}{\omega_c} + 1$$

In this example the zeros of $H(s)$ can be found in elementary fashion, but in most cases this is not so. Finding the zeros for transmission function of higher orders is a very time-consuming and complicated calculating process, which computers can perform most effectively.

Let us consider another requirement; the transmission function has to provide an effective attenuation in the passband $(0 - \omega_c)$ with equal ripples. The amplitude of the ripple is $A_{\max}$. Outside the passband the function exhibits a monotonic response. The condition is satisfied with the following function:

$$H(s) \, H(-s) = 1 + D(s) \, D(-s) = 1 + D^2$$

or

$$|H(s)|^2 = 1 + \left[\frac{\cos (n \cos^{-1} s/j\omega_c)}{(d^n - d^{-n})/2} \right]^2 \qquad (2.10.3)$$

where $H(s) \, H(-s)$ is a polynomial in s. The expression

$$\cos (n \cos^{-1} x) = C_n(x)$$

is a Chebyshev polynomial in x of nth order.

From the theory of Chebyshev functions it is known that in the range $-1 < x < +1$ the values of this polynomial oscillates between $+1$ and -1. The

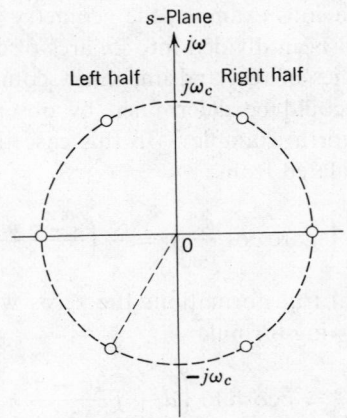

Fig. 2.22.   Position of zeros of Butterworth filter with $n = 3$.

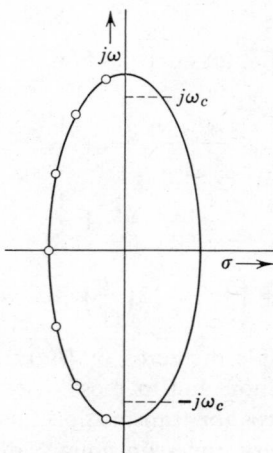

Fig. 2.23. Position of zeros of Chebyshev filter with $n = 7$.

parameter d determines the maximum loss A_{max} inside of the passband

$$|\omega| \leq \omega_c$$

at frequencies for which the Chebyshev polynomial in (2.10.3) take the values ± 1. Therefore,

$$e^{2a_{max}} = 1 + \left(\frac{2}{d^n - d^{-n}}\right)^2 = \left(\frac{d^n + d^{-n}}{d^n - d^{-n}}\right)^2$$

and the value of d will be given as

$$d^2 = \sqrt[n]{\tanh\left(\frac{a_{max}}{2}\right)}$$

For small attenuation this value is approximately

$$d^2 = \sqrt[n]{\frac{a_{max}}{2}}$$

For the transmission function $H(s)$ the position of zeros in the right side of Eq. 2.10.3 have to be found.

In the previous example, the geometry was simple: the unit had been divided into $2n$ arcs of equal length and the values of real and imaginary components for every zero could be determined by observation (see the Butterworth example). In this case the zeros s_{ov} can be calculated from:

$$\cos^2\left(n \arccos \frac{s_{ov}}{j\omega_c}\right) = -\left(\frac{d^n - d^{-n}}{2}\right)^2$$

After several transformations the zeros will be given by the following formula

$$\frac{s_{ov}}{\omega_c} = j \cosh \ln\left(d + j\frac{2v - 1}{2n}\pi\right)$$

The real and imaginary quantities can be further

separated and then

$$\frac{s_{ov}}{\omega_c} = \frac{1 - d^2}{2d} \sin\left(\frac{2v - 1}{2n}\pi\right)$$
$$+ j\frac{1 + d^2}{2d} \cos\left(\frac{2v - 1}{2n}\pi\right)$$

where v is an integer.

It is relatively easy to prove that the zeros are situated on an ellipse whose center is at $s = 0$. One half-axis of the ellipse is equal to

$$\omega_c \frac{1 - d^2}{2d}$$

and the other is

$$\omega_c \frac{1 + d^2}{2d}$$

The position of the zeros for a filter with $n = 7$ is shown in Fig. 2.23. Its attenuation curve is shown in Fig. 2.24. All poles are at infinity. The intersection of the ellipse with the imaginary s-axis indicates the passband limit, where the attenuation reaches the 3-dB point in comparison with the average value in the passband, which is equal to $a_{max}/2$. (The design cutoff is ω_c where $a = a_{max}$.)

To construct the $H(s)$ polynomial only the n-zeros in the left half of s-plane (as in the Butterworth example) have to be considered.

For the case of $H(s)$ in the previous case and for $H(s)$ in this more cumbersome example, the values of coefficients in the polynomials can be determined from the closed-loop equations.

A certain generalization will take place when

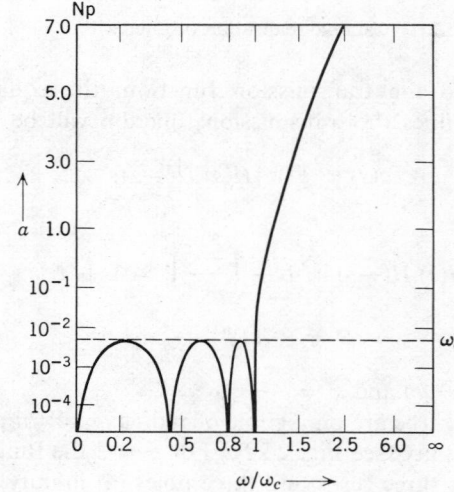

Fig. 2.24. Attenuation curve of filter in Fig. 2.23.

instead of a monotonic response curve, the attenuation above cutoff (in lowpass filters) has to have equal minima. $H(s)$, in that case, is a rational function (Zolotarev fraction). Even this case can be handled with explicit formulas. But in contrast with the Chebyshev polynomial filter, the expressions for poles and zeros will include elliptic functions instead of trigonometric functions.

2.11 SIMPLEST POLYNOMIAL FILTERS IN ALGEBRAIC FORM

The first order filter network shown in Fig. 2.25 consisting simply of a shunt capacitance connected between a source that has an impedance R_1 and a load R_2 is a convenient starting point. The quantity H which defines the behavior of the network is

$$H = 1 + j\omega C \frac{R_1 R_2}{R_1 + R_2}$$

or simply,

$$H = 1 + j\omega CR$$

The presence of j in this expression shows that there is a phase shift introduced by the filter.

The effective attenuation in dB is

$$A = 20 \log |H| = 10 \log (1 + \omega^2 C^2 R^2)$$

In Fig. 2.25 a filter of the second order is also given. The quantity H is

$$H = 1 + j\omega \left(C \frac{R_1 R_2}{R_1 + R_2} + \frac{L}{R_1 + R_2} \right) - \omega^2 LC \frac{R_2}{R_1 + R_2}$$

This expression can be simplified by substituting

$$\frac{R_1 R_2}{R_1 + R_2} = R_p, \quad R_1 + R_2 = R_s, \quad \text{and} \quad \frac{R_2}{R_1 + R_2} = K$$

thus yielding

$$H = 1 + j\omega \left(CR_p + \frac{L}{R_s} \right) - \omega^3 LCK$$

To obtain the expression for effective attenuation we must eliminate j by multiplication by the complex conjugate and considering

$$|H|^2 = 1 + \omega^2 \left[\left(CR_p + \frac{L}{R_s} \right)^2 - 2LCK \right] + \omega^4 L^2 C^2 K^2$$

which is the expanded form of the Eq. 2.1.4.

The effective attenuation is (see 2.1.2a)

$$A = 20 \log |H|$$

To save writing the transmission factor, H can be expressed by

$$|H|^2 = 1 + |D|^2 = 1 + a_1 \omega^2 + a_2 \omega^4$$

where a_1 and a_2 are free to choose, and we can do it in such a way that we get a Butterworth or Chebyshev response.

Mathematically, a Butterworth response is $1 + \omega^{2n}$ which here means that the value of a_1 must be made equal to zero.

A first-order filter cannot have a maximally flat response. In the second-order filter the coefficient of ω^2 must be equal to zero so that

$$A = 10 \log (1 + \omega^4 L^2 C^2 K^2) = 10 \log |H|^2$$

when

$$CR_p + \frac{L}{R_s} = 2LCK$$

Adding one more capacitor to the previous circuit we obtain a third-order filter for which the transmission factor H is

$$H = 1 + j\omega \left[(C_1 + C_1)R_p + \frac{L}{R_s} \right] - \omega^2 \frac{L}{R_s} (C_1 R_1 + C_2 R_2) - j\omega^3 LC_1 C_2 R_p$$

The absolute value can be found by the same procedure we used before.

$$|H|^2 = 1 + \omega^2 \left\{ \left[(C_1 + C_2)R_p + \frac{L}{R_s} \right]^2 - \frac{2L(C_1 R_1 + C_2 R_2)}{R_s} \right\}$$
$$+ \omega^4 \left\{ \frac{L^2(C_1 R_1 + C_2 R_2)^2}{R_s^2} - 2LC_1 C_2 \left[(C_1 + C_2)R_p + \frac{L}{R_s} \right] R_1 \right\}$$
$$+ \omega^6 L^2 C_1^2 C_2^2 R_p^2$$

$$|H|^2 = 1 + a_1 \omega^2 + a_2 \omega^4 + a_6 \omega^6 = 1 + |D|^2$$

and the Butterworth response can be found by equating a_1 and a_2 to zero.

$$A = 10 \log (1 + \omega^6 L^2 C_1^2 C_2^2 R_p^2)$$

The ratio of input to output for the all-zero transmission function filter always takes the general form

$$|H|^2 = a_0 + a_1 \omega^2 + a_2 \omega^4 + \cdots a_n \omega^{2n} \quad (2.11.1)$$

where n is the order of the network and $a_1, a_2, \ldots, a_n$ depend on the values of resistances, capacitances and

Order	Network	Transmission Factor H
1ST order	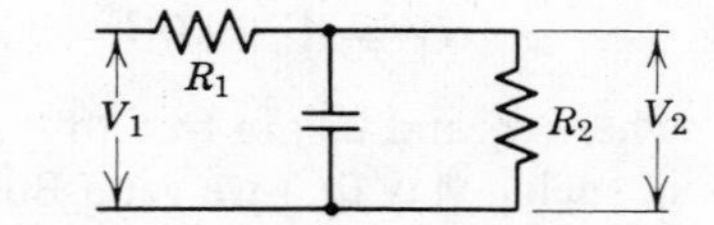	$1 + j\omega C \dfrac{R_1 R_2}{R_1 + R_2} = 1 + j\omega CR$
2ND order	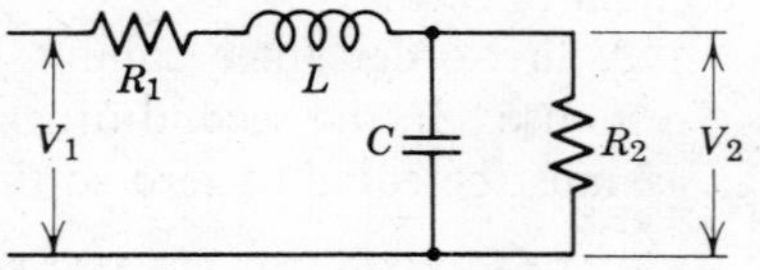	$1 + j\omega(CR_p + \dfrac{L}{R_s}) - \omega^2 LCK$
2ND order	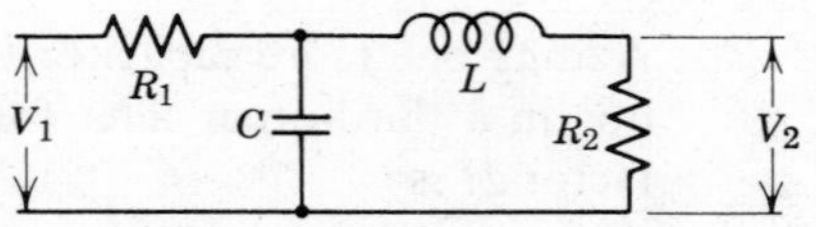	$1 + j\omega(CR_p + \dfrac{L}{R_s}) - \omega^2 LCK$
3RD order	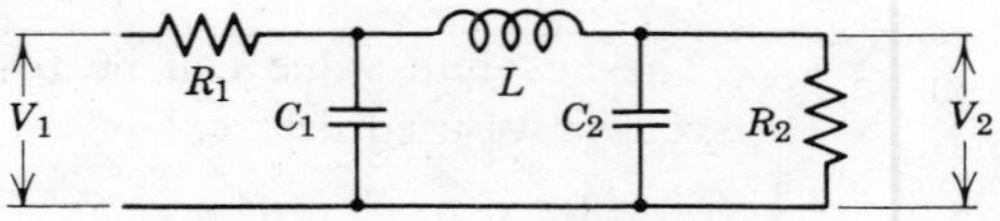	$1 + j\omega[(C_1 + C_2)R_p + \dfrac{L}{R_s}]$ $- \omega^2 \dfrac{L}{R_s}(C_1 R_1 + C_2 R_2)$ $- j\omega^3 LC_1 C_2 R_p$

Fig. 2.25. Simplest polynomial filters.

inductances. Since a lowpass filter has no attenuation at zero frequency, the term a_0 is always unity.

2.12 INTRODUCTION TO IMAGE-PARAMETER THEORY

Image-parameter theory is not of extreme interest from the point of view of filter design, since we have available precalculated parameters which facilitate the evaluation of circuit elements in terms of effective parameters. However, this theory permits the simplification of the physical analysis and the study of a multitude of filtering structures available for filter design. Moreover, the entire image-parameter technique can be incorporated in a modern synthesis based on the effective-parameter approach in which the most versatile attenuation responses (filters with general

parameters) can be readily designed by combining both methods and using both types of terminology.

A network of high order may be designed by adding simple building blocks (Fig. 1.18) generally known as sections or half-sections. Zobel based his theory on structures known as k-constant sections which are merely elementary LC filters of the lowest possible order with reciprocal impedances (Fig. 1.10 and 1.11).

With the aid of an ingenious operation named derivation in m (see Figs. 1.10 and 1.11), Zobel derived the complex structures which are sometimes more advantageous. This theory contributed tremendously to the progress in filter technology. Its historical importance is well established, but the derivation in m does not seem to be either simple or general. It is not simple because it does not provide structures in rising order of complexity and it is not general because it does not provide all possible structures. The available catalogs, textbooks, and handbooks always include a very limited number of half-sections. The structural analysis which is the essence of the image-parameter theory does not make conspicuous certain T and π networks. Usually they escape treatment.

The complete and logical presentation of Zobel (image parameters) filters has to be based on several appropriate uses of attenuation and impedance functions. This method has been exposed by Cauer and Piloty in a series of remarkable papers.

For every type of filter (lowpass, highpass, bandpass, bandstop) there are two image impedances, z_1 and z_2 (z_T and z_π) that can be used for synthesis. There are always a multitude of image-attenuation functions which, being associated with z_1 and z_2, define the filter. The simplest (lowest degree) attenuation function q compatible with z_1 and z_2 will generate the elementary filter. In a great majority of cases the elementary filter is a half-section and consists of two branches (two arms). In other cases the elementary filter consists of three branches.

Relation Between the Impedance Matrix and the Image Parameters

The image-parameter filter can be represented by an impedance function and an attenuation function. The impedance Z is generally related to a resistance R, which is the nominal design resistance. Therefore $z = Z/R$. Similarly the frequency ω is related to the fixed reference frequency ω_r. In the lowpass and highpass case, ω_r is the cutoff frequency ω_c. In the case of bandpass filters ω_r is situated between ω_1 and ω_2 in the geometric center of the passband ω_m, or the arithmetic center ω_a (as in narrowband crystal filters)

$$\omega_m = \sqrt{\omega_1 \omega_2}$$

For all normalized cases,

$$\Omega = \frac{\omega}{\omega_r} \quad \text{and} \quad s = \frac{j\omega}{\omega_r} \tag{2.12.1}$$

The relations which exist between the characteristic quantities of a reactive network and the transmission parameters (such as phase and attenuation) can take different equivalent forms (2.2.1), (2.2.2), and (2.2.7). It is convenient to use the impedance equations.

The relations between the voltages and currents at the input and the output of an ordinary four-terminal network, as we know, can be shown in the form (2.2.1) with the impedance matrix,

$$\frac{Z}{R} = \begin{Vmatrix} z_{11} & z_{12} \\ z_{12} & z_{22} \end{Vmatrix} \tag{2.12.2}$$

In the case of a reactive four-terminal network, z_{11} and z_{12} are open-circuit reactances of four-terminal networks. z_{12} is an odd rational function of the variable $s = j\Omega$, but it is not necessarily a reactance. Conventionally, we call it the transfer impedance. The values z_{11}, z_{12}, and z_{22} must satisfy certain conditions to represent a realizable reactive four-terminal network.

The characteristic impedance is a main design parameter in image-parameter theory. It is determined by

$$z_0 = \sqrt{z_{11} z_{22} - z_{12}^2} \tag{2.12.3}$$

The image impedances z_1 and z_2 can be obtained from the equation which describes the input impedance of the four-terminal network terminated in z_2 for z_1 and in z_1 for z_2 (see Fig. 2.1). The expression for these image-input impedances will be:

$$z_1 = z_0 \sqrt{\frac{z_{11}}{z_{22}}} \; ; \qquad z_2 = z_0 \sqrt{\frac{z_{22}}{z_{11}}} \tag{2.12.4}$$

The image-transmission constant is defined by

$$g_1 = a_1 + jb_1 = \tfrac{1}{2} \ln \frac{V_1 I_1}{V_2 I_2}$$

when the four-terminal network in Fig. 2.1 is terminated by its image impedances. Value a_1 represents the image attenuation, and b_1 is the image phase. One One can verify simply that

$$\frac{V_1 I_1}{V_2 I_2} = e^{2g_1} = \left(\frac{z_0 + z_{11} z_{22}}{z_{12}} \right)^2$$

The formulas giving coth g_1 is the most interesting.

$$q = \coth g_1 = \frac{\sqrt{z_{11} z_{22}}}{z_0} \tag{2.12.5}$$

which is the attenuation q function (quantum function) expressed in terms of networks parameters. The relations shown in (2.12.3), (2.12.4), and (2.12.5) express the image parameters as a function of the elements of the impedance matrix and are valuable for any four-terminal network.

Effective Parameters in Terms of Image Parameters

The effective parameters define the conditions under which a four-terminal network will work if it is inserted between two actual load resistances (see Eq. 2.3.2). The effective-input impedances (when the second pair of terminals are loaded by R_2 as shown in Fig. 2.1) will be

$$z_{e1} = \frac{z_{11} + z_0^2}{1 + z_{22}} \; ; \qquad z_{e2} = \frac{z_{22} + z_0^2}{1 + z_{11}} \tag{2.12.6}$$

If a source of emf V_0 with internal impedance R_1, is connected to the network terminated on the output side by R_2, the output voltage when R_1 equals R_2 will be $V_0/2V_2 = e^g$. The effective-transmission constant is $g = a + jb$; a is the effective-attenuation constant and b is the corresponding effective-phase constant. The effective-transmission factor H is given by the equation

$$H = e^g = \frac{(1 + z_{11})(1 + z_{22}) - z_{12}{}^2}{2z_{12}}$$

$$= \frac{z_{11} + z_{22} + z_0{}^2 + 1}{2z_{12}} \qquad (2.12.7)$$

The matching of a four-terminal network can be characterized by the reflection (echo) constant $g_r = a_e + jb_e$.

$$e^{g_r} = T = \frac{z_{in} + 1}{z_{in} - 1} = \frac{z_{11} + z_{22} + z_0{}^2 + 1}{z_{11} - z_{22} + z_0{}^2 - 1} \qquad (2.12.8)$$

a_e is the attenuation of the reflected wave; b_e characterizes the phase of the reflected signal. Equation 2.12.8 shows that a_e for a reactive network does not depend on the terminals considered. In fact, the change of z_{11} to z_{22} does not modify the modulus of the rational function.

With the aid of image parameters q and z_0, the impedance matrix z and the effective parameters can be expressed. The calculation presents no difficulty. We find:

$$\frac{Z}{R} = \left\| \begin{matrix} z_{11} & z_{12} \\ z_{12} & z_{22} \end{matrix} \right\| = \left\| \begin{matrix} z_1 q & z_0\sqrt{q^2 - 1} \\ z_0\sqrt{q^2 - 1} & z_2 q \end{matrix} \right\| \qquad (2.12.9)$$

$$e^g = H = \frac{z_0{}^2 + 1 + q(z_1 + z_2)}{2z_0\sqrt{q^2 - 1}} \qquad (2.12.10)$$

$$e^{g_r} = T = \frac{z_0{}^2 + 1 + q(z_1 + z_2)}{z_0{}^2 - 1 + q(z_1 - z_2)} \qquad (2.12.11)$$

$$D = \frac{H}{T} = \frac{z_0{}^2 - 1 + q(z_1 - z_2)}{2z_0 q^2 - 1} \qquad (2.12.12)$$

$$z_{in} = \frac{z_0{}^2 + qz_1}{1 + qz_2} \qquad (2.12.13)$$

The values H, T, z_{in} are rational functions of the variable s. These formulas show also that there are no fundamental difficulties in the determination of the effective and echo attenuation when the image impedance and the image attenuation are known.

It is sufficient to evaluate the rational function D with the aid of image parameters z_0 and q and then evaluate the effective parameters of the network.

$$a = \tfrac{1}{2}\ln\left(1 + |D|^2\right) \qquad a_e = \tfrac{1}{2}\ln\left(1 + \frac{1}{|D|^2}\right) \qquad (2.12.14)$$

2.13 BRIDGE NETWORKS

Filter schematics basically take the form of one of two different types: ladder or bridge (see Fig. 1.1). The stopband is obtained in the ladder schematic from the relatively high impedances in the series arms and the low impedances in the shunt arms. Current from the source divides from the series line and returns through the shunt arms as it progresses through the filter. As a consequence of the dividing action, only part of the source signal appears at the load.

Operation of the bridge schematic is quite different from the ladder. In the bridge the source and load are inserted in the diagonal arms. The stopband is achieved when the four impedances are equal; then the bridge is said to be balanced—the better the balance the larger the rejection.

When the matrix of the reactance network is found, the first problem of synthesis is solved. The second part is to find a schematic of the network which corresponds to that matrix. In general there are many such equivalents but usually for a specific problem one of them is more suitable than any other.

From many possible schematics the bridge network and the ladder networks are the most useful. Although, in the majority of filter problems, especially with conventional elements, ladder networks are best, let us analyze the realization with the bridge schematic.

Every symmetrical four-terminal network can be realized by a bridge, which is sometimes called a lattice, consisting of two series and two diagonal arms. The lattice schematic (Fig. 2.26) is equivalent to the bridged-T schematic as shown in Fig. 2.27. The bridged-π schematic, better known as a *differential bridge*, is shown in Fig. 2.28. Both of these require only two reactive arms and one ideal transformer, and consequently are more economical than the lattice. The two terminal networks Z_1 and Z_2 of the bridge are

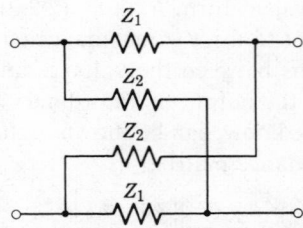

Fig. 2.26. Lattice schematic.

reactances and are uniquely determined by the matrix. The realization of two-terminal networks can be accomplished by partial fraction or continuous fraction development of the appropriate reactance functions. The effective parameters of the bridge can be expressed by

$$e^{-g} = \frac{P}{E} = \frac{Z_2/R - Z_1/R}{(Z_1/R + 1)(Z_2/R + 1)} \quad (2.13.1)$$

and

$$\frac{1}{\rho} = e^{-g_e} = \frac{F}{E} = \frac{Z_1 Z_2/R^2 - 1}{(Z_1/R + 1)(Z_2/R + 1)} \quad (2.13.2)$$

The sum and difference of the foregoing equations include, for the sum Z_1 only, and for the difference Z_2 only. Therefore,

$$\frac{F + P}{E} = \frac{Z_2/R - 1}{Z_2/R + 1} \quad (2.13.3)$$

$$\frac{F - P}{E} = \frac{Z_1/R - 1}{Z_1/R + 1} \quad (2.13.4)$$

$$\frac{Z_2}{R} = \frac{E + (F + P)}{E - (F + P)} \quad (2.13.5)$$

$$\frac{Z_1}{R} = \frac{E + (F - P)}{E - (F - P)} \quad (2.13.6)$$

When the network is symmetrical the filtering function $D = F/P$ is an odd function. Therefore either F is odd and P even or vice versa. For odd F and even P the determining condition is

$$E(s) E(-s) = P^2(s) - F^2(s) = (P + F)(P - F) \quad (2.13.7)$$

For even F and odd P the analogy holds. In both instances the zeros of $(P + F)$ and $(P - F)$ are mirror images about the imaginary axis. The $2n$ zeros of $E(s) E(-s)$ have to be equally divided into $(P + F)$ and $(P - F)$. Arm reactances Z_1 and Z_2, in sum, have the same order n as the matrix. The bridge equivalent schematic in the form of T or π are canonical schematics and require for their realization exactly n elements and one ideal transformer.

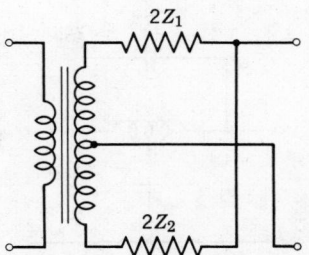

Fig. 2.28. Differential bridge (semilattice).

2.14 EXAMPLES OF REALIZATION IN THE BRIDGE FORM

Let us consider the same example of Section 2.10 for the Butterworth type of response

$$H(s) H(-s) = 1 + D(s) D(-s) = 1 + \left(\frac{s}{j\omega_0}\right)^6 \quad (2.14.1)$$

Consequently $D = \pm(s/\omega_0)^3$. We limit ourselves to the negative part and obtain

$$F = -s^3, \quad P = 1, \quad E = (s + 1)(s^2 + s + 1)$$

Substituting in the equation for effective parameters the following will be obtained:

$$\frac{F + P}{E} = \frac{(1 - s^3)}{(s + 1)(s^2 + s + 1)} = \frac{1 - s}{1 + s} \quad (2.14.2)$$

We recognize that by the confluent placement of the zeros of the numerator and zeros of the denominator the order of the quotation $(F + P)/E$ is diminished.

Using Eqs. 2.13.3 and 2.13.4 both impedances could be written as functions of s. Impedance Z_1 consists of one parallel circuit with resonant frequency ω_0. Impedance Z_2 consists of one capacitor.

$$\frac{Z_1}{R} = \frac{s}{s^2 + 1} \qquad \frac{Z_2}{R} = \frac{1}{s}$$

The schematic is shown in Fig. 2.29, the element values being

$$L_0 = \frac{R}{\omega_0}$$

$$C_0 = \frac{1}{\omega_0 R}$$

when both load resistors are equal to R. The ideal transformer evidently can be merged with the coil of the parallel-resonant circuit. Bridge schematics are difficult for the practical realization of filters, especially

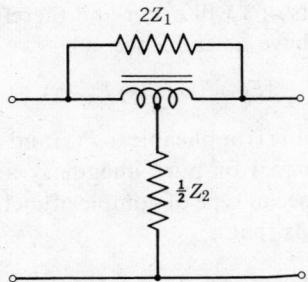

Fig. 2.27. Bridged-T schematic.

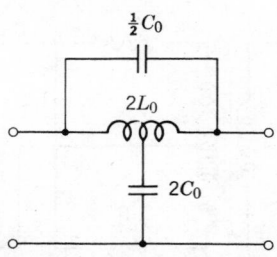

Fig. 2.29. Butterworth filter with $n = 3$ in bridged-T form.

when high-attenuation requirements are imposed in the stopband. The explanation for these difficulties is very simple. The attenuation effect is produced by the balancing of the bridge and for high attenuation the balance must be very accurate. The inverse of the filtering function is

$$\frac{1}{D} = \frac{P}{F} = \frac{Z_2/R - Z_1/R}{Z_1 Z_2/R^2 - 1} \qquad (2.14.3)$$

Differentiation with respect to one of the reactances, for example Z_1, will give

$$\frac{\partial(1/D)}{\partial(Z_1)} = \frac{1 - Z_2^2/R^2}{R(Z_1 Z_2/R^2 - 1)^2} \qquad (2.14.4)$$

The pole of attenuation occurs when

$$\frac{1}{D} = 0$$

For the condition Z_1 must be equal to Z_2, and therefore,

$$\frac{\partial(1/D)}{\partial Z_1} = \frac{1}{R(1 - Z_1^2/R^2)}$$

For a small change in $1/D$ we have

$$\Delta(1/D) \approx \frac{\Delta Z_1}{Z_1} = \frac{1}{(R/Z_1 - Z_1/R)} \qquad (2.14.5)$$

When the attenuation peak is positioned on the imaginary axis (real frequency) then Z_1 is a reactance X_1. Therefore,

$$\Delta(1/D) \approx \frac{\Delta X_1}{X_1} = \frac{j}{(R/X_1 + X_1/R_1)} \qquad (2.14.6)$$

When

$$\frac{X_1}{R} = 1$$

the value in the parenthesis of Eq. 2.14.6 equals two. If, for example, the attenuation has to be a minimum of 60 dB

$$\left|\frac{1}{D}\right| \leq 10^{-3}$$

so the ratio $\Delta Z_1/Z_1 \leq 2 \times 10^{-3}$ is required. The sig-

nificance of these numbers is that both reactances in bridge arms can maximally deviate one from the other by about two parts per thousand. Such requirement is often unobtainable. Therefore for filters where high attenuation in the stopband is required only ladder filters are recommended.

2.15 HURWITZ POLYNOMIAL

For calculation of normalized z_{ik} or y_{ik}, as we also know, the function $E(s)$ is needed. According to (2.1.4) the square magnitude of the transmission function is equal to

$$|H(s)|^2 = 1 + |D(s)|^2$$

and

$$\frac{E(s)}{P(s)} = 1 + \left|\frac{F(s)}{P(s)}\right|^2$$

From the previous discussion it can be seen that: (1) $E(s)$ is the Hurwitz polynomial, and (2) the polynomials $F(s)$ and $P(s)$ are odd or even. This signifies that $E(s)$ can contain odd and even powers of s, but $F(s)$ and $P(s)$ can contain only even or only odd powers of s.

For $s = j\Omega$, terms which are even powers of s are always real and those which are odd powers of s are always imaginary. Consequently the polynomials $F(s)$ and $P(s)$ are either pure real or pure imaginary, but $E(s)$ can be complex on the $j\Omega$-axis (see Fig. 2.9).

The zeros of the Hurwitz polynomial are the frequencies at which the network oscillates when excited by pulses. The zeros of the denominator polynomial of $H(s)$ signify the frequencies of infinite attenuation poles; when the order of the polynomial is lower than that of the Hurwitz polynomial, then one or more poles at attenuation are at infinity. In the limiting case the denominator reduces to a constant and $H(s)$ becomes a polynomial with all poles of attenuation at infinity.

If we consider $E(s) E(-s)$, then $E(-s)$ is the complex conjugate value of $E(+s)$, since in the real part only even exponents of s will occur and therefore the negative sign will have no influence.

$$|E(s)|^2 = E(s) E(-s)$$

Similar reasoning is applicable to $P(s)$ and $F(s)$ although they are pure real or pure imaginary and therefore, represents a special type of complex functions. Consequently it holds that

$$\frac{E(s) E(-s)}{P(s) P(-s)} = 1 + \frac{F(s) F(-s)}{P(s) P(-s)} \qquad (2.15.1)$$

or

$$\frac{E(s)\,E(-s)}{P(s)\,P(-s)} = \frac{P(s)\,P(-s) + F(s)\,F(-s)}{P(s)\,P(-s)} \quad (2.15.2)$$

The squared magnitude will be (see Eq. 2.9.1)

$$|E(s)|^2 = E(s)\,E(-s) = P(s)\,P(-s) + F(s)\,F(-s)$$

This permits us to make some simplifications on the $j\omega$-axis

For odd F	$F(s)\,F(-s) = -F^2(s)$	(2.15.3)
For even F	$F(s)\,F(-s) = F^2(s)$	(2.15.4)
For odd P	$P(s)\,P(-s) = -P^2(s)$	(2.15.5)
For even P	$P(s)\,P(-s) = P^2(s)$	(2.15.6)

Substituting with the equation for the squared magnitude of $E(s)$

1. For odd F and even P one can obtain

$$|E(s)|^2 = P^2(s) - F^2(s) = [P(s) + F(s)][P(s) - F(s)]$$

2. For even F and odd P:

$$|E(s)|^2 = -P^2(s) + F^2(s)$$
$$= [-P(s) + F(s)][P(s) + F(s)]$$

3. If P and F are both even:

$$|E(s)|^2 = P^2(s) + F^2(s)$$

For 1 and 2, the roots of the equation

$$P(s) + F(s) = 0 \quad (2.15.7)$$

have to be found, and for 3, the roots of the equation

$$|E(s)|^2 = P^2(s) + F^2(s) = 0 \quad (2.15.8)$$

must be found.

According to the conditions of realization, the roots s_i of $E(s)$ can only have negative real parts. Then,

$$E(s) = B \prod_{i=1}^{n} (s - s_i) \quad (2.15.9)$$

For imaginary $s = j\omega$

$$a = \tfrac{1}{2} \ln\,[H(s)H(-s)]$$

$$jb = \tfrac{1}{2} \ln \frac{H(s)}{H(-s)}$$

or also

$$j \tan b = \frac{H(s) - H(-s)}{H(s) + H(-s)}$$

When $H(s)$ is a polynomial it has to be a Hurwitz polynomial. In such a case

$$j \tan b = \frac{E(s) - E(-s)}{E(s) + E(-s)}$$

Since $E = E_e + E_o$

$$\tan b = \frac{E_o}{jE_e}$$

This $H(s)$ is a typical reactance function: The quotient of the odd and even parts of a Hurwitz polynomial is a reactance function. The sum of the numerator and denominator of a reactance function is a Hurwitz polynomial.

2.16 THE SMALLEST REALIZABLE NETWORKS

In the previous pages we have shown that every reactance ladder network can be represented as a chain of partial networks with prescribed attenuation poles (see Fig. 1.18). Here, the smallest possible partial networks will be described and the formula for element values will be given.

The smallest partial network can be of the following orders:

1. Order one with an attenuation pole at $s = 0$ or $s = \infty$.

2. Order two with a pair of attenuation poles at opposite but equal frequencies on the real or imaginary s-axis.

3. Order four with four attenuation peaks at conjugate-complex frequencies.

The four-terminal network of order one with an attenuation pole at $s = \infty$ and $H(0) = 1$ has the polynomials:

$$E = 1 + a_1 s, \quad F = \pm a_1 s, \quad \text{and}$$
$$P = 1 \text{ with } a_1 > 0 \quad (2.16.1)$$

The appropriate schematic can be devised immediately for $F = +a_1 s$. The network consists of one series inductance with $L = 2a_1 R$. For $F = -a_1 s$, these polynomials are obtained with one shunt capacitor (in the parallel arm), whose value $C = 2(a_1/R)$. Both schematics are shown in Fig. 2.30.

On the other hand, if the first-order network has a pole at $s = 0$ and $H(\infty) = 1$, the same three polynomials are

$$E = 1 + a_1 s, \quad F = \pm 1, \quad \text{and}$$
$$P = a_1 s \text{ with } a_1 > 0 \quad (2.16.2)$$

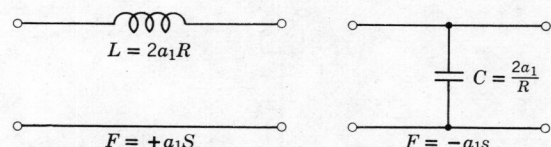

$$L = 2a_1 R$$
$$C = \frac{2a_1}{R}$$
$$F = +a_1 S \qquad F = -a_1 s$$

Fig. 2.30. First order networks with attenuation pole at $s = \infty$.

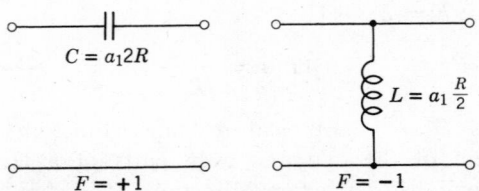

Fig. 2.31. First-order networks with attenuation pole at $s = 0$.

In this case, for $F = +1$ the four-terminal network will have one series capacitor with value $C = a_1/2R$. For $F = -1$ the network will consist of one shunt coil with inductance $L = a_1R/2$. Both schematics are shown in Fig. 2.31.

The network of second order with attenuation poles at frequencies $\pm s_\infty$, and with unobstructed passage for direct current, $H(0) = 1$, has the polynomials

$$E = 1 + a_1s + a_2s^2, \quad F = b_1s + b_2s^2, \quad \text{and}$$
$$P = 1 - s^2/s_\infty^2 \qquad (2.16.3)$$

The polynomials have to satisfy the realizability conditions. Therefore a_1 and a_2 have to be positive, and because the determinant condition must hold,

$$(a_2{}^2 - b_2{}^2)s_\infty{}^4 = 1 \qquad (2.16.4)$$

The realizable schematic cannot be determined directly from the polynomials, but they can be found relatively easily when the impedance or admittance matrix are established. It can be proved that the simplest second-order network in Fig. 2.32 will have one capacitor of value

$$C = \frac{a_1 - b_1}{R} \qquad (2.16.5)$$

and one shunt inductance

$$L_1 = \frac{a_2 + b_2}{a_1 - b_1}R \qquad (2.16.6)$$

with an ideal transformer having a transformation ratio K following the partial network, where

$$K = -s_\infty{}^2(a_2 + b_2) = -1/s_\infty{}^2(a_2 - b_2) \quad (2.16.7)$$

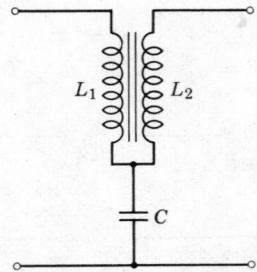

Fig. 2.32. Second-order polynomial filter in canonic form.

Consequently, the output inductance value L_2 has the value

$$L_2 = \frac{a_2 - b_2}{a_1 - b_1}R \qquad (2.16.8)$$

The network is shown in Fig. 2.32 corresponding to the above E, F, and P polynomials. The K of the transformer between the strongly coupled windings is positive when $s_\infty{}^2$ is negative, or when the attenuation poles are on imaginary s-axis. The K is negative when the attenuation poles are on real s-axis.

From Eqs. 2.16.4 and 2.16.7 the coefficient K can be expressed as

$$K = \pm \frac{a_2 + b_2}{a_2 - b_2} \qquad (2.16.9)$$

The positive sign corresponds to imaginary and the minus sign to real attenuation poles. The schematic in Fig. 2.32 is canonic, since it realizes the filter of second order with two reactances (when, naturally, the coil with a tap is regarded as one element).

When one replaces the strongly coupled coils by their equivalent T-network, the noncanonic network shown in Fig. 2.33 is obtained. The values of the inductances will be as follows:

$$L_1 = \frac{a_2 + b_2}{a_1 - b_1}R\left(1 \mp \sqrt{\frac{a_2 - b_2}{a_2 + b_2}}\right)$$

$$L_2 = \frac{a_2 - b_2}{a_1 - b_1}R\left(1 \mp \sqrt{\frac{a_2 + b_2}{a_2 - b_2}}\right)$$

$$L_3 = \pm R\frac{\sqrt{a_2{}^2 - b_2{}^2}}{a_1 - b_1}$$

The plus sign is for attenuation poles on the imaginary s-axis. The value of the capacitor remains unchanged.

One of the inductances in the above circuit is always negative. The network in this form can be realized only when the attenuation pole is on the imaginary axis. In this case the inductance L_3, which is part of a resonant circuit responsible for a pole, is positive. The

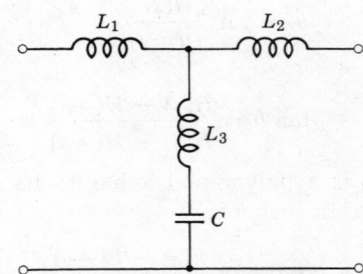

Fig. 2.33. Noncanonic network equivalent to Fig. 2.32.

negative inductance in the series arm must be combined with a positive one in the next filter section.

In Fig. 2.34 another equivalent network is shown. The admittance matrix may be represented as sum of several components. The values of the elements included in the network will be given by the following formulas

$$L = (a_1 + b_1)R$$

$$C_1 = \frac{a_2 - b_2}{(a_1 + b_1)R}\left(1 \mp \sqrt{\frac{a_2 + b_2}{a_2 - b_2}}\right)$$

$$C_2 = \frac{a_2 + b_2}{(a_1 + b_1)R}\left(1 \mp \sqrt{\frac{a_2 - b_2}{a_2 + b_2}}\right)$$

$$C_3 = \pm \frac{\sqrt{a_2{}^2 - b_2{}^2}}{(a_1 + b_1)R}$$

$$C_1C_2 + C_1C_3 + C_3C_2 = 0$$

Even this network is realizable only for an imaginary attenuation pole (upper sign). One of the shunt capacitors is negative and must be combined with a positive capacitor in the neighboring circuitry.

In Fig. 2.35 three symmetric networks of order two are shown. In this case when the transformation coefficient $K = -1$ the element values are as follows:

$$L = R(a_1 + b_1)$$

$$C = \frac{a_1 - b_1}{R}$$

For the network in the center, with $K = 1$

$$L = \frac{a_2 R}{2a_1}$$

$$C = \frac{2a_1}{R}$$

$$a_1 - b_1 = 0$$

For a network with $K = 1$, having a parallel-resonant

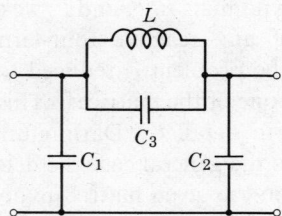

Fig. 2.34. Second-order polynomial filter equivalent to realizations in Figs. 2.32 and 2.33.

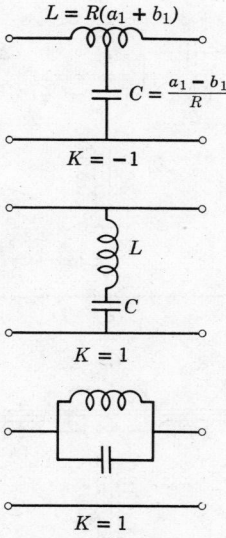

Fig. 2.35. Symmetric second-order networks.

circuit in the line as shown on the right of Fig. 2.35

$$L = 2a_1 R$$

$$C = \frac{a_2}{2a_1 R}$$

$$a_1 + b_1 = 0$$

2.17 FOURTH-ORDER NETWORKS

A network of fourth order with $H(0) = 1$ has the following polynomials

$$E = 1 + a_1 s + a_2 s^2 + a_3 s^3 + a_4 s^4$$

$$F = b_1 s + b_2 s^2 + b_3 s^3 + b_4 s^4$$

$$P = 1 + d_2 s^2 + d_4 s^4$$

which must satisfy the realizability conditions.

The formulas for the element values of a realizable network as in Fig. 2.36 can be obtained with the aid of an impedance matrix which can be decomposed in three parts. The first two correspond to a network of second order. Solving for element values the following equations can be obtained:

$$C_a = \frac{a_1 - b_1}{R}$$

$$L_1 = \frac{a_4 + b_4}{a_3 - b_3} R$$

$$L_2 = \frac{a_4 - b_4}{a_3 - b_3} R$$

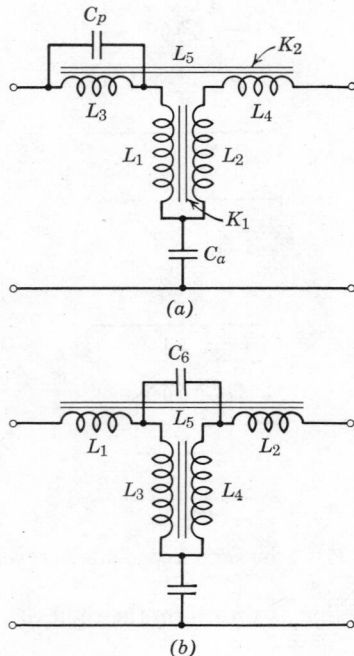

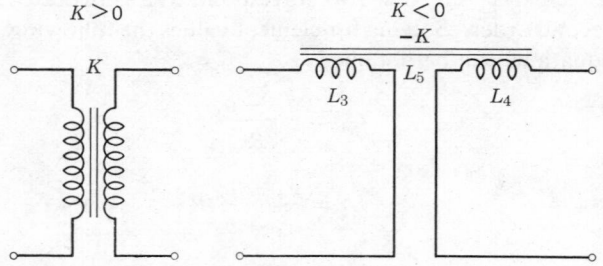

Fig. 2.36. Fourth-order networks.

The transformation coefficient K_1 here is

$$K_1 = \frac{a_4 + b_4}{d_4} = \frac{d_4}{a_4 - b_4} = \pm\sqrt{\frac{a_4 + b_4}{a_4 - b_4}}$$

Because of the determinant condition

$$a_4{}^2 - b_4{}^2 = d_4{}^2$$

K_1 is positive when all four attenuation poles are complex, real, or imaginary, and it is negative when one pair of poles is real and one pair imaginary.

The third matrix corresponds to a network that consists of parallel-resonant circuits in the parallel arm (across the line) followed by an ideal transformer.

In the most interesting cases of complex attenuation poles, the coefficient of transformation K_2 of this transformer is always negative and therefore the equivalent

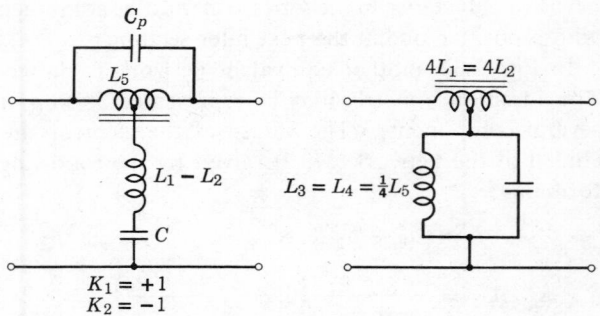

Fig. 2.37. Equivalent schematic of transformer with $K > 0$ and $K < 0$.

Fig. 2.38. Fourth-order symmetrical network.

schematic of the transformer shown in Fig. 2.37 is useful.

The total inductance of this transformer (in series)

$$L_5 = 2\left(\frac{a_2 - d_2}{a_1 - b_1} - \frac{a_4 - d_4}{a_3 - b_3}\right)$$

The resonant frequency of the parallel circuit is

$$\omega_p = \frac{a_1 - b_1}{a_3 - b_3}$$

As a measure of the transformation coefficient, the difference between inductances in the transformer can be used.

$$L_3 - L_4 = \frac{K_2 + 1}{K_2 - 1} L_5 = 2\left(\frac{b_2}{a_1 - b_1} - \frac{b_4}{a_3 - b_3}\right)$$

The total network which corresponds to all three parts of the total impedance matrix (in Fig. 2.36a) is shown in Fig. 2.36b. Both schematics are canonic since the mutually coupled coils are considered as one inductance.

When the network is symmetric, then the result of both transformations is equal to one absolute value. If b_4 is positive, and all poles are complex, real, or imaginary, then $K_1 = +1$ and $K_2 = -1$. The networks shown in Fig. 2.38 are realizations that have this property.

2.18 FIFTH-ORDER NETWORKS

Using the polynomials E, F, and P, we can determine the elements of any reactive four-terminal matrix. Consequently, the problem is reduced to determining the circuit from one of the matrices. This problem has been examined in detail by Darlington, Cauer, and many others. In the general case the determination of the network from the given matrix involves enormous arithmetic calculations which have to be performed with great accuracy. However, in the case of symmetrical filters (for odd $n \leq 9$) it is possible to derive comparatively simple formulas which express the values of the elements in the network directly through the roots of the polynomials E, F, and P. For example, for $n = 5$, Cauer gives the following formula for the normalized inductances L' and capacitances C' of a lowpass ladder filter shown in Fig. 2.39.

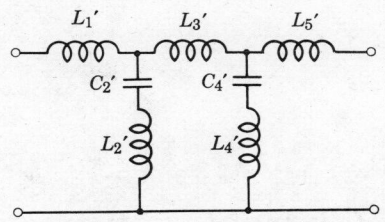

Fig. 2.39. Fifth-order realization.

$$L_1' = a_s - a_2\epsilon_1, \qquad L_2' = \frac{\epsilon_1 a_2}{2} \frac{2a_1 - \epsilon_1 - \epsilon_2}{a_1 - \epsilon_1}$$

$$L_3' = a_2(2a_1 - \epsilon_1 - \epsilon_2), \quad L_4' = \frac{\epsilon_2 a_2}{2} \frac{2a_1 - \epsilon_1 - \epsilon_2}{a_1 - \epsilon_2}$$

$$L_5' = a_3 - a_2\epsilon_2, \qquad C_2 = \frac{\epsilon_1}{L_1'}, \quad C_4' = \frac{\epsilon_2}{L_4'}$$

In these formulas the auxiliary values a_1, a_2, and a_3 are expressed by the roots of the polynomial $E(s)$ and the values of ϵ_1 and ϵ_2 by the roots of the polynomial $P(s)$.

The actual values of L and C are determined from the formulas:

$$L = \frac{R}{2\pi f_r} L' \qquad C = \frac{1}{2\pi f_r R} C'$$

where R is the internal impedance of the generator.

3

Filter Characteristics in the Frequency Domain

There are several parameters which can characterize a filter's performance. Among the most important are the attenuation, phase, and group delay. All of these parameters may be expressed as a function of frequency, plotted in the form of curves, and may be calculated or measured in the steady-state condition. Besides responses for steady-state signals, such as a sinusoidal signal, there are several important transient responses for nonstationary signals. They characterize the dynamic properties of the filter, revealing performance in the time domain for different types of existing functions.

3.1 AMPLITUDE RESPONSES

There are many different shapes of amplitude-versus-frequency responses of filters which may be described by analytic functions. There are also established design procedures to approximate these types of responses. The following are the main distinctive varieties of responses:

1. Butterworth response (B) with "maximally flat" passband. Filters having this response are sometimes called power-term filters.
2. Chebyshev response (C) with "equal ripple" attenuation in the passband.
3. Inverse Chebyshev response (IC) with equal minima of attenuation in the stopband.
4. Chebyshev complete response (CC) with equal ripple attenuation in the passband and equal minima of attenuation in the stopband. Filters having this response are sometimes called elliptic-integral filters or filters with Cauer parameters.
5. Gaussian response.
6. Bessel response with "maximally flat delay."

Filters having this response are also known as Thomson filters.
7. Equal-ripple-delay filter response.
8. Legendre filter response.
9. Synchronously tuned-filter response.
10. Minimum insertion-loss filters.

The first four response types belong to the Chebyshev family.

Usually, gain is cheap whereas selectivity is very expensive. For this reason a tremendous effort is made to find the most rational way to design a network with the minimum number of expensive components while satisfying the fundamental requirements of attenuation, phase, and so forth. The mathematical problem posed by these considerations consists primarily in finding a network whose transmission or transfer function fits the appropriate polynomials. In most cases the polynomials that provide the best and most economical solution are of the Chebyshev type.

The next three responses (5, 6, and 7) belong to the Gaussian family in the sense that they may all be conveniently used to satisfy phase and group-delay requirements. The Gaussian response is represented by the well-known exponential formula. The importance of this response shape cannot be fully appreciated from the point of view of frequency discrimination since the rate of increase in attenuation for Gaussian, Bessel, and equal-ripple-delay filters is very low. Gaussian-response filters should be regarded as compromise designs for pulsed systems where, except for the frequency discrimination, it is the truthful reproduction of the pulse shape that is important.

The amplitude responses of an equal-ripple group-delay approximation depends upon a special design parameter, characterized by the height of the group-delay ripples. This filter, for a given number of zeros in transmission function will produce different curves of attenuation which depend upon the magnitude of the group-delay ripples.

3.2 PHASE- AND GROUP-DELAY RESPONSES

A specific amplitude response given by attenuation versus frequency does not describe the complete transmission property of the filter. Indeed, the filter discrimination factor is somewhat superficial because it describes only the gain or loss of the network. The phase characteristic which is included in the effective-transmission factor H becomes very important when considering radar and communication systems especially those using pulse signaling. More and more of the questions a filter designer must answer pertain to the time delay, rise time, amount of overshoot, rate of decay, quasi (transient) oscillations. A flat group delay is desirable because this signifies that all frequencies will be delayed the same amount while going through the filter. If the various frequencies are not delayed equally, dispersion results and the output for a pulsed input does not retain its identity.

In a reactive filtering network the attenuation behavior is described by both the denominator and the numerator polynomial of the transmission function $H(s) = E(s)/P(s)$. The group delay, on the contrary, is determined exclusively by the numerator (characteristic) polynomial $E(s)$. Consequently the computation procedure begins with an approximation of polynomial $E(s)$ alone. Resulting attenuation curves in the stopband can be modified by the polynomial $P(s)$.

3.3 GROUP DELAY OF AN IDEALIZED FILTER

The logarithm of transmission factor H of the network consists of two components: attenuation a and the phase b. The sum of both is a transmission constant which has been defined as:

$$g = a + jb = \ln H \qquad (3.3.1)$$

If the first real parameter describes the absolute value of the signal in the steady-state condition, the second imaginary parameter introduces a completely different realm of time. Both parameters are equally important when one describes the nonstationary performance and dynamic property of the network in modern electronic systems.

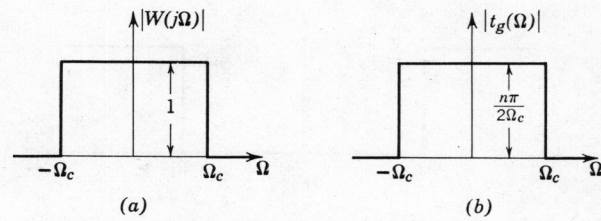

Fig. 3.1. Idealized magnitude and group-delay responses.

According to the definition, group delay is the derivative of phase. If the phase of the transmission function is b, then the group delay $t_g = db/d\omega$, where b is the phase constant of a filter and ω is angular frequency. For the ideal case, the phase is a linear function of frequency, group delay is constant, and consequently the network does not produce phase distortion at any frequency.

Because the ideal pulse-transmission system is not frequency selective, the function which describes it is not a particularly desirable network function. In the desirable situation the group delay approximates a constant in the passband of the filter when the attenuation is small. But, nevertheless, with the knowledge of the ideal time response, the limitations of the physical network can appear more understandable.

Let us assume that we have an ideal filter which satisfies the ideal frequency and time-domain requirements. Such a filter has a rectangular response for its magnitude as in Fig. 3.1a, and its phase angle increases linearly and is given by:

$$b = \frac{n\pi\Omega}{2} \qquad \text{for} \quad |j\Omega| = 1 \qquad (3.3.2)$$

$$b = 0 \qquad \text{for} \quad |j\Omega| = 0 \qquad (3.3.3)$$

The group-delay response is therefore rectangular as illustrated in Fig. 3.1b. The value n in (3.3.2) is the order of the approximating polynomial, and every term is s gives the maximum phase shift of 90°.

3.4 GROUP-DELAY—ATTENUATION RELATIONSHIP

The majority of filters at high frequencies are polynomial filters with no peaks of attenuation in the stopband. In the theory of synthesis, the transfer functions representing these polynomial filters are called all-pole functions. It is customary in filter design practice, however, to express the transmission property by a loss and phase parameter. The real part a and the imaginary part b can be related by the Hilbert

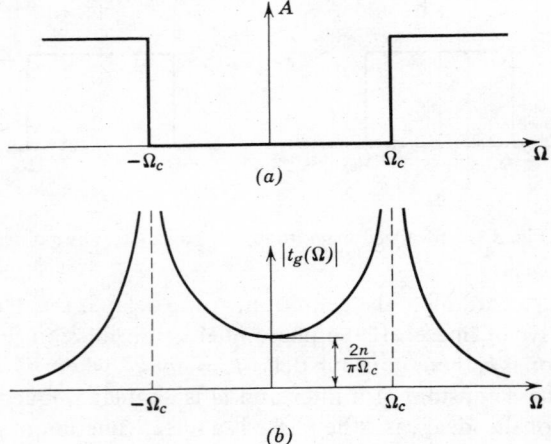

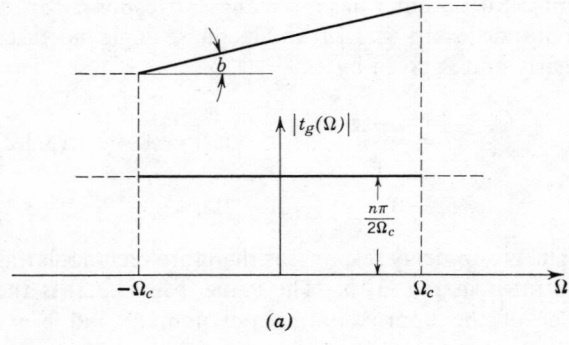

Fig. 3.2. Hilbert transform of attenuation response into group delay.

transform if $H(j\omega)$ is conveniently restricted. $H(j\omega)$ can have neither poles nor zeros in the right half-plane. Fortunately all polynomial filters represented by the transmission function satisfy this necessary requirement.

Let us assume that the response curve has a sharp cutoff as a result of the corresponding group-delay responses given by the Hilbert transform as shown in

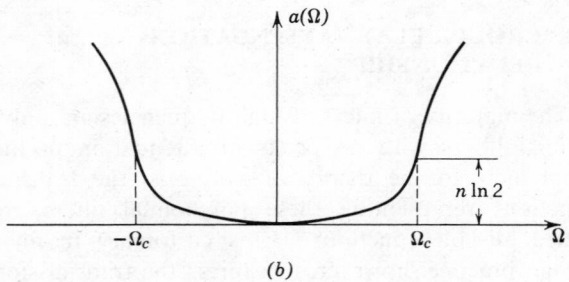

Fig. 3.3. Transformation of phase response into amplitude response.

Fig. 3.2*a* and *b*. The figure shows once more that a sharp cutoff transforms itself into an extremely non-constant delay. Moreover, the time delay becomes infinite at the ideal cutoff discontinuity. If the cutoff is smooth, then the delay would decrease to some finite value, hence the conclusion: a polynomial filter cannot be made to approximate a function with sharp attenuation cutoffs and at the same time have a constant group delay. The two requirements are not compatible.

If, on the other hand, we know a phase function *b* which increases linearly from zero to certain negative and positive values depending upon the complexity of the function, by the Hilbert transform the corresponding attenuation can be found. The group delay and attenuation are shown in Fig. 3.3*a* and 3.3*b* respectively. As can be seen from these figures, the linear phase shift is accompanied by a rounded attenuation response. The amplitude characteristic belonging to the Gaussian family of responses (synchronously tuned, Bessel, and equal-ripple group-delay) all exhibit rounded amplitude response characteristics. However, the group-delay response characteristics are essentially linear.

3.5 THE CHEBYSHEV FAMILY OF RESPONSE CHARACTERISTICS

As indicated, the Butterworth, Chebyshev, inverse Chebyshev, and Cauer-parameter filters are closely related to each other and can be developed along similar lines.

Butterworth Filter

The Butterworth lowpass insertion-loss function is expressed by

$$A = 10 \log \left| 1 + \left(\frac{\omega}{\omega_c}\right)^{2n} \right| = 10 \log (1 + \Omega^{2n})$$
$$A = 10 \log (1 + D^2) \, \text{dB} \qquad (3.5.1)$$

or,

$$a = \tfrac{1}{2} \ln (1 + \Omega^{2n}) = \tfrac{1}{2} \ln (1 + D^2) \, \text{Np} \quad (3.5.2)$$

where the filter discrimination factor $D = \Omega^n$, and the normalized frequency $\Omega = \omega/\omega_c$ is the ratio of the given frequency ω to the cutoff frequency ω_c. The first ten Butterworth polynomials are given in Table 3.1.

The attenuation for these filters is 3 dB at Ω_c. The insertion-loss function has the flattest possible shape at the center of the passband (maximally flat) and is a

Table 3.1 Butterworth Polynomials

1. $1 + s$

2. $1 + 1.4142s + s^2$

3. $1 + 2.0000s + 2.000s^2 + s^3$

4. $1 + 2.6131s + 3.4142s^2 + 2.6131s^3 + s^4$

5. $1 + 3.2361s + 5.2361s^2 + 5.2361s^3 + 3.2361s^4 + s^5$

6. $1 + 3.8637s + 7.4641s^2 + 9.1416s^3 + 7.4641s^4$
   $+ 3.8637s^5 + s^6$

7. $1 + 4.4940s + 10.0978s^2 + 14.5920s^3 + 14.5920s^4$
   $+ 10.0978s^5 + 4.4940s^6 + s^7$

8. $1 + 5.1528s + 13.1371s^2 + 21.8462s^3 + 25.6884s^4$
   $+ 21.8462s^5 + 13.1371s^6 + 5.1528s^7 + s^8$

9. $1 + 5.7588s + 16.5817s^2 + 31.1634s^3 + 41.9864s^4$
   $+ 41.9864s^5 + 31.1634s^6 + 16.5817s^7 + 5.7588s^8$
   $+ s^9$

10. $1 + 6.3925s + 20.4317s^2 + 42.8021s^3 + 64.8824s^4$
    $+ 74.2334s^5 + 64.8824s^6 + 42.8021s^7 + 20.4317s^8$
    $+ 6.3925s^9 + s^{10}$

monotonically increasing function. Above the cutoff frequency the loss approaches a line drawn from Ω_c which is linear and increasing on a logarithmic frequency scale at a rate of $6n$ dB per octave.

Butterworth filters require n reactive elements (n appears as an exponent in the foregoing equations). Maximally flat response curves for various values of n are given in Section 3.14, curve 1. For $n = 2$ the shape corresponds to that obtained for the critically coupled condition of the familiar synchronously tuned circuit.

The Butterworth approximation is useful for many applications, however, its main advantage is its mathematical simplicity. The Butterworth response was derived on the assumption that behavior at zero frequency was far more important than behavior at any other frequency. This leads to a class of filters with good phase response and tolerably good amplitude response but with very poor characteristics around the cutoff frequency. The Butterworth function is unsuitable for applications that require uniform transmission of frequencies in the passband and sharp rise at cutoff. The response to a unit impulse input has overshoot which increases with increasing n, exceeding 11% for $n > 4$.

Chebyshev Function

By changing the approximation conditions, it is possible to obtain much better characteristics near cutoff. One may decide, for instance, that all frequencies in the passband are equally important, and that it is desirable to minimize the maximum deviation from the ideal response. This type of approximation is called a Chebyshev approximation.

Figure 3.4 shows an example of a third-order low-pass filter response with an attenuation typical of the so-called Chebyshev filters where the abscissa is the normalized frequency Ω. The insertion loss function for this response is

$$A = 10 \log |1 + \epsilon^2 C_n{}^2(\Omega)| = 10 \log (1 + D^2)$$

where $D = \epsilon C_n(\Omega)$, and the parameter ϵ, the ripple factor is given by

$$\epsilon = \sqrt{10^{0.1 A_{max}} - 1}$$

and $C_n(\Omega)$ is the Chebyshev polynomial of the first kind and of order n. The first 12 orders of C_n are given in Table 3.2. In the passband ($-1 \leq \Omega \leq 1$), the attenuation response varies between the values of zero and A_{max}. The maximum passband insertion loss is $A_{max} = 10 \log (1 + \epsilon^2)$. At frequencies slightly above Ω_c (the passband limit), the attenuation will surpass A_{max} for the first time. A transition range follows and the stopband begins with frequency Ω_s. Here the attenuation A_{min} is reached for the first time. If the filter is designed according to a Butterworth or Chebyshev polynomial, the attenuation curve will rise monotonically.

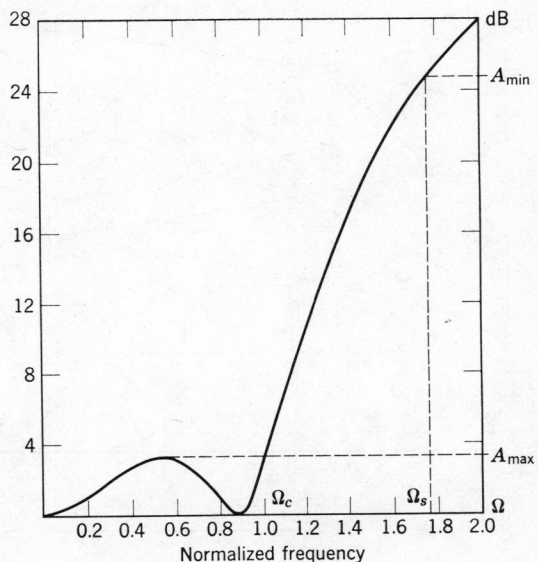

Fig. 3.4. Lowpass Chebyshev amplitude response, $n = 3$.

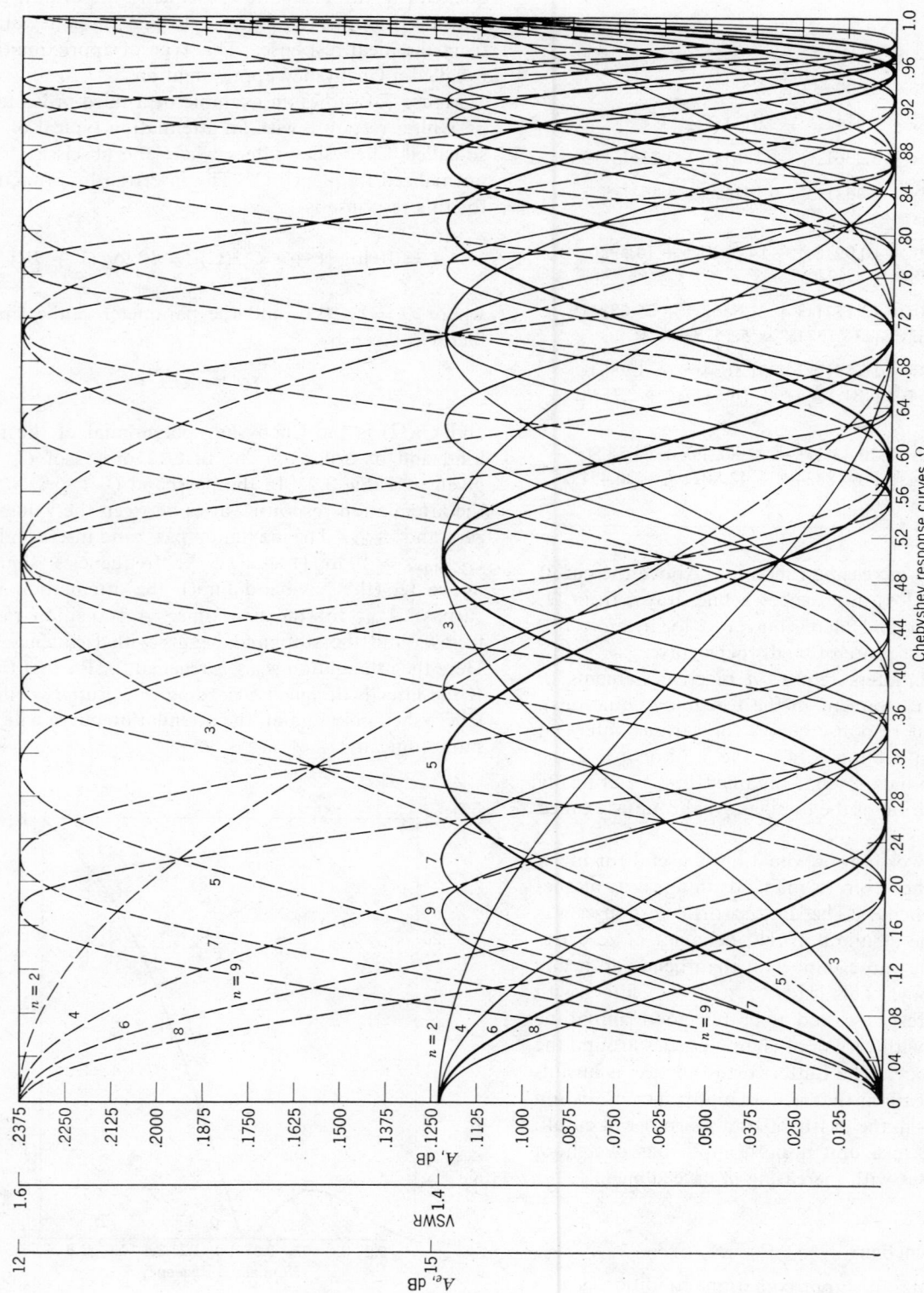

Fig. 3.5. Chebyshev response curves in passband.

Chebyshev response curves, Ω

A, dB

VSWR

A_e, dB

Table 3.2 Chebyshev Polynomials

0. 1

1. Ω

2. $2\Omega^2 - 1$

3. $4\Omega^3 - 3\Omega$

4. $8\Omega^4 - 8\Omega^2 + 1$

5. $16\Omega^5 - 20\Omega^3 + 5\Omega$

6. $32\Omega^6 - 48\Omega^4 + 18\Omega^2 - 1$

7. $64\Omega^7 - 112\Omega^5 + 56\Omega^3 - 7\Omega$

8. $128\Omega^8 - 256\Omega^6 + 160\Omega^4 - 32\Omega^2 + 1$

9. $256\Omega^9 - 576\Omega^7 + 432\Omega^5 - 120\Omega^3 + 9$

10. $512\Omega^{10} - 1280\Omega^8 + 1120\Omega^6 - 400\Omega^4 + 50\Omega^2 - 1$

11. $1024\Omega^{11} - 2816\Omega^9 + 2816\Omega^7 - 1232\Omega^5 + 220\Omega^3 - 11\Omega$

12. $2048\Omega^{12} - 6144\Omega^{10} + 6912\Omega^8 - 3584\Omega^6 + 840\Omega^4 - 72\Omega^2 - 1$

The Chebyshev response ($n = 2$) corresponds to that obtained with the overcoupled conditions of the familiar double-tuned circuit. In general, this shape has a number of ripples of equal height—a number equal to the number of resonant circuits used. The equal-ripple filter for a given bandwidth has the greatest attenuation outside of the passband of any monotonic stopband or all-pole filter. The rate of increase depends not only upon the number of poles or resonators but also upon a special design parameter—that is, the height of the ripples; the attenuation rate is higher for larger passband ripples. In Fig. 3.5 the family of Chebyshev amplitude responses in the passband is shown. The lower set of curves belongs to the filters designed with low ripples (higher echo attenuation), and the upper set belongs to the similar filters designed for a slightly higher ripple level (0.23 dB). A set of curves for attenuation outside the ripple band is shown in Section 3.14, curves 3, 5, and 7.

The Chebyshev function is exceedingly useful in applications where the magnitude of the transfer function is of primary concern. This approximation gives more constant magnitude response throughout the passband but no improvement in decreasing the overshoot of the impulse response. The class of Chebyshev functions is optimum in the sense that of all possible transmission functions with poles at infinity (all-zero functions) it has the lowest complexity for yielding a prescribed maximum deviation in the passband and the fastest possible rate of cutoff outside the passband. As a consequence, the transition range for reaching a prescribed attenuation A_{min} is a minimum and the attenuation in the stopband is never less than that prescribed attenuation. No other polynomial possessing these optimum properties exists.

Zolotarev Function

The Chebyshev polynomial is a particular example of the Chebyshev rational function. The Zolotarev function is another, when the poles are placed in such a way that all the minima of attenuation of the function in the stopband are identical in their absolute value.

The Chebyshev polynomial includes the restriction that all the zeros of the transfer function lie at infinity. In other words, the reciprocal of the transfer function is required to be a polynomial. On the other hand, the Zolotarev rational functions are not so restricted. Rather, their transfer function takes the form

$$|W(\Omega)|^2 = \frac{1}{1 + \epsilon^2 C_n{}^2(\Omega)}$$

and the attenuation becomes,

$$A = 10 \log [1 + \epsilon^2 C_n{}^2(\Omega)]$$
$$A = 10 \log (1 + D^2) \qquad (3.5.3)$$

where $C_n(\Omega)$ is chosen so that it has an equiripple attenuation in the passband and the stopband. Here the filter discrimination factor $D = \epsilon C_n(\Omega)$. Depending upon whether it is even or odd, $C_n(\Omega)$ has one of two forms:

$$C_{2n}(\Omega) = \frac{A(\Omega^2 - \Omega_1{}^2)(\Omega^2 - \Omega_3{}^2) \cdots (\Omega^2 - \Omega_{2n-1}^2)}{(\Omega^2 - \Omega_2{}^2)(\Omega^2 - \Omega_4{}^2) \cdots (\Omega^2 - \Omega_{2n}{}^2)}$$

$$(3.5.4)$$

or,

$$C_{2n+1}(\Omega) = \frac{B(\Omega^2 - \Omega_1{}^2)(\Omega^2 - \Omega_3{}^2) \cdots (\Omega^2 - \Omega_{2n-1}^2)}{(\Omega^2 - \Omega_2{}^2)(\Omega_2{}^2 - \Omega_4{}^2) \cdots (\Omega - \Omega_{2n}{}^2)}$$

$$(3.5.5)$$

In the passband, $-1 \leq \Omega \leq 1$, $C_n(\Omega)$ must lie between the limits -1 and $+1$. In the stopband, $C_n(\Omega)$ should take the maximum possible absolute values for the given degree of n.

Parameters $\Omega_1 \cdots \Omega_{2n-1}$ are in the passband, whereas $\Omega_2 \cdots \Omega_{2n}$ are in the stopband. Moreover, the following relation also holds:

$$\Omega_1 \Omega_2 = \Omega_{2n-1} \Omega_{2n} = \Omega_s \Omega_c \qquad (3.5.6)$$

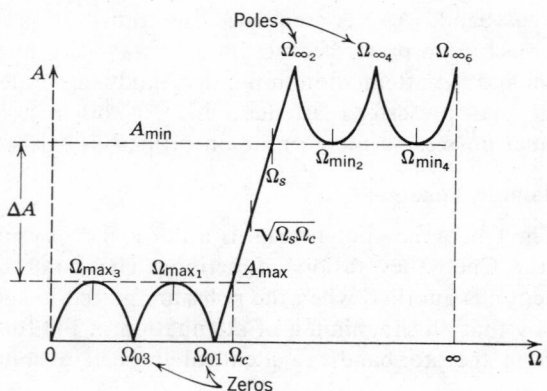

Fig. 3.6. Zolotarev type of amplitude response.

where Ω_c is the passband ripple bandwidth and Ω_s corresponds to the first frequency attenuated by $A_{\min}$. In other words, the poles of C_n are the reciprocals of its zeros. The integer n determines the complexity of the function, and specifically, it is equal to the number of Ω^2 zeros (or poles or a suitable combination of the two types of critical frequencies) that must be specified. Because of the reciprocal relationship between the zeros and the poles of the function its value at any Ω_1

in the range $0 < \Omega < 1$ is the reciprocal of its value at $1/\Omega_1$ in the range $1 < \Omega < \infty$. If the critical frequencies can be found so that rational function has equiripples in the passband, it will automatically have equiripples in the stopband. In Fig. 3.6 the amplitude response (with $n = 5$) according to Zolotarev type of rational function from Chebyshev family is shown. $\Omega_c = 1$ is used as the normalized cutoff frequency which corresponds to $A_{\max}$ in the amplitude response before entrance into the transition region.

Filters with General Parameters

Under certain conditions all of the filters of the Chebyshev family can degenerate one into another. There is, however, one type of filter, the so-called general-parameter filter which is a part of the Chebyshev family but cannot be directly related by applying the limiting conditions.

The general-parameter filter may have explicit solutions though it might, in certain cases, be explicit only up to the determination of the filtering function. For those cases in which various requirements regarding the attenuation in the stopband also exists, the solution is not explicit to the extent that it can be

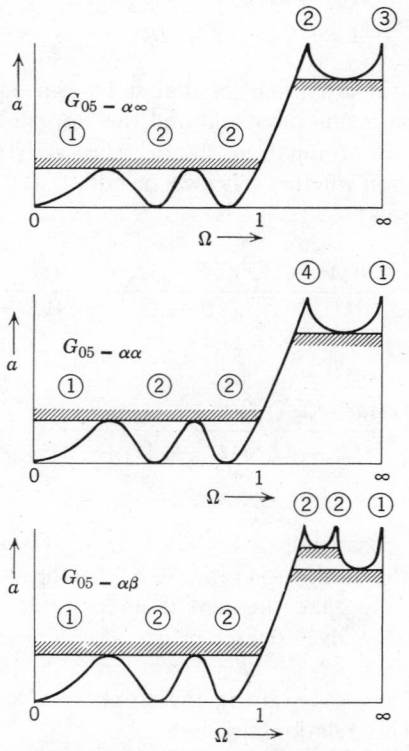

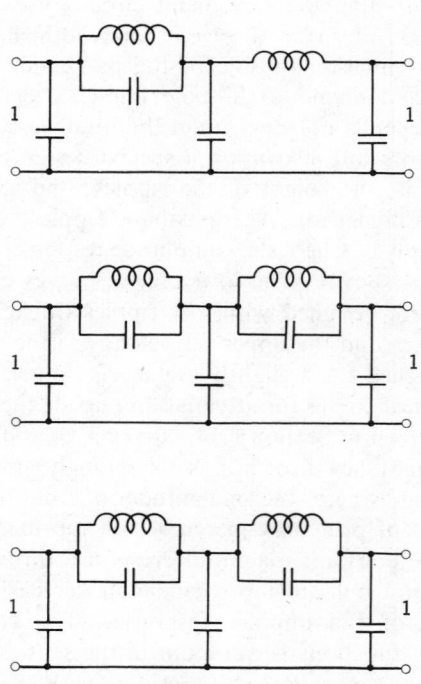

Fig. 3.7. Response shapes and realization of a filter with $n = 5$. The numbers in the circles indicate the position and the number of poles or zeros of the transmission function.

tabulated. When the parameter of the filtering function for those cases have been determined the attenuation curve in the passband can be improved with conventional predistortion methods. It is evident that Zolotarev's rational function cannot be used. Instead the Chebyshev rational function has to be applied. The difficulties of determining the roots of this very important general case is a recognized obstacle in engineering practice, but surmountable by the use of iteration techniques. In the filter with general parameters the typical Chebyshev response will remain only in the passband, but outside the passband the attenuation curve has a restricted character which could be advantageous from the point of view of economy in the filter construction.

In Fig. 3.7 simple (fifth order) filters with restricted behavior are shown in order to demonstrate the problems involved in their design. These filters are certainly more difficult to tabulate. Nevertheless, when the position of the attenuation peaks are known, the filtering function which provides a Chebyshev response in the passband can be found. This possibility was demonstrated by Fetzer. In the first filter shown, the peaks are at α and infinity; in the second filter, both peaks are at the same frequency; and in the third one, at α and β.

Properties of the Chebyshev Family of Filters

The universal normalized response of the Chebyshev family is shown in Fig. 3.6. The transition interval between passband and stopband is represented by ΔA which can be simply related to the transmission function. All design parameters, such as value and position of A_{max} and A_{min} as well as frequencies Ω_c and Ω_s are under control of the designer.

If the stopband attenuation A_{min} is increased to infinity, the value of Ω_s will then have to go to infinity, and the defined restricted stopband will be compressed and moved to infinite frequency, in which case the Chebyshev rational function reduces to a Chebyshev polynomial. To design a lowpass prototype filter which exhibits any response of the Chebyshev family we need not go through the complete synthesis procedure. Tables of the element values for normalized lowpass filters are available.

Properties of Chebyshev filters are tabulated in Fig. 3.8. The responses for inverse Chebyshev filters are shown in row 1, column *a* of the figure. The corresponding locus of the transmission function zeros, shown in column *b* of row 1 changes from the deformed half circle to a semicircle which is characteristic of Butterworth filters.

The responses in row 3 of Fig. 3.8 belong to the filters with Zolotarev function approximation (Chebyshev in the passband and stopband). The amplitudes of ripple and amount of attenuation guaranteed in the stopband are changing in definite steps from one extreme value to another. The corresponding transmission function zeros remain on a circle, as shown in column *b*.

In intermediate cases when the ripples in the passband and ripples in the stopband are arbitrary as in row 4 column *a*, (more ripple may be accepted in the passband in order to get more attenuation in the stopband), the locus of zeros is deformed into an ellipse as shown in row 4 column *b*, of the figure.

In the case of an arbitrary design with small ripples in the passband and guaranteed minimum in the stopband, the half-circle locus is deformed as shown in row 2 column *b*. In this case, the half-axis of the ellipse along the real negative direction is longer than the axis along the imaginary direction.

Row 5 illustrates the degeneration of typical Chebyshev responses when limiting conditions are gradually applied. The filter with no ripples in the passband and no poles of attenuation outside it degenerates into a power-term filter with maximally flat amplitude response and with the zeros of the transmission function located on a semicircle.

The plot of zeros and poles for the Chebyshev family of filters indicates that when a curve enters into the transition frequency region, it is sharper when the transfer-function poles are concentrated closer to the imaginary *s*-plane axis. In the proximity of the poles the effective-phase angle changes more rapidly and the group delay has high peaks.

The curves in column *c* of Fig. 3.8 are curves of normalized group delay; it may be observed that relatively constant group delay in the passband can be reached only when we avoid using sharp-attenuation curves. A transmission function of the Chebyshev family is thus optimum only from the point of view of the attenuation requirement but not from the point of view of the group delay.

3.6 GAUSSIAN FAMILY OF RESPONSE CHARACTERISTICS

There is a demand in pulse communication systems and other related areas for filters whose impulse responses have the two properties: (1) freedom from ringing or overshoots, and (2) symmetry about the time for which the response is a maximum. A filter that satisfies the above is called a Gaussian filter. The

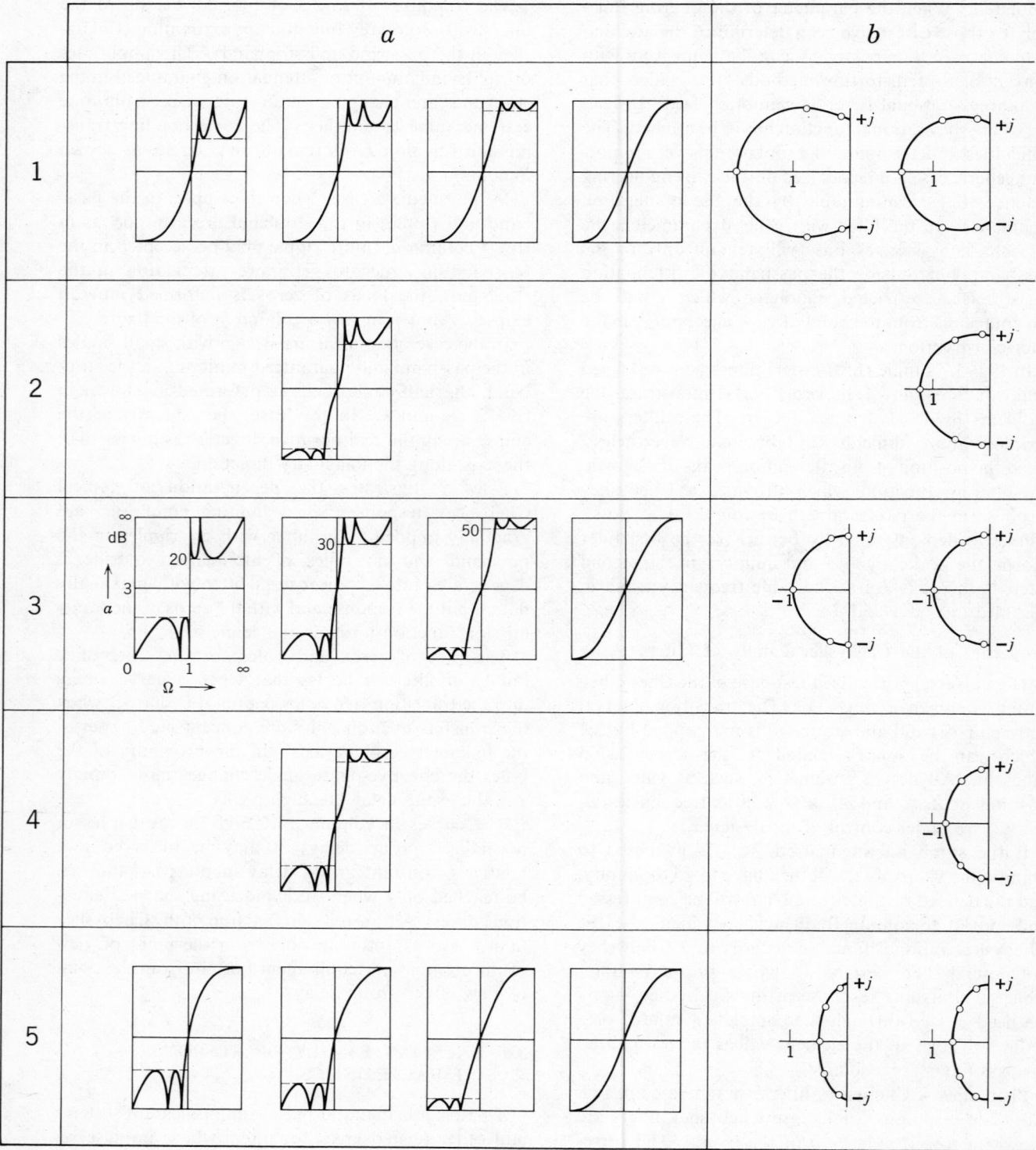

Fig. 3.8. Tabulated properties of the

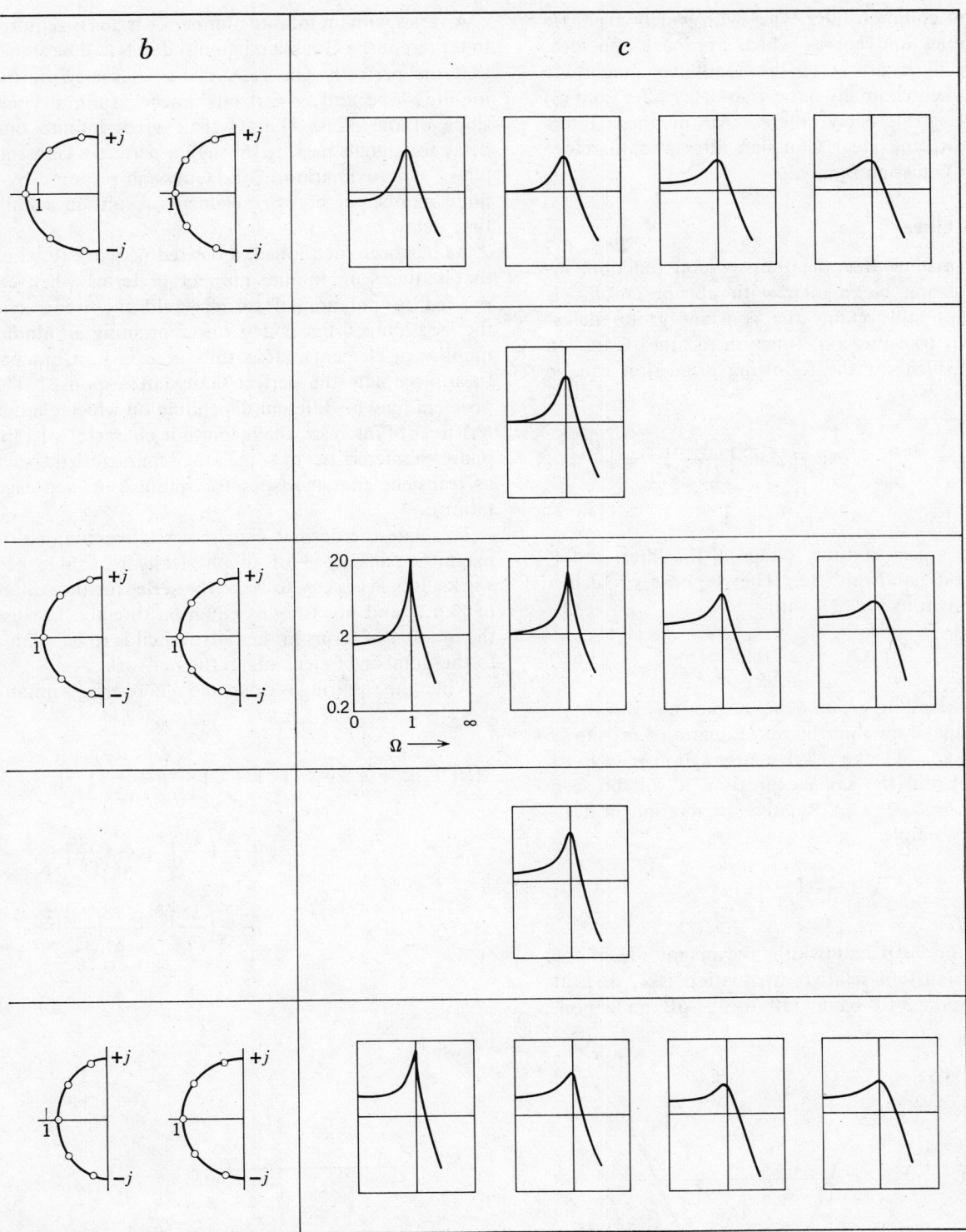

Chebyshev family of filters for $n = 5$.

three most common filter types with widely available design tables and curves which approach the ideal Gaussian filter are: (1) the Gaussian magnitude filter, (2) the maximally flat group-delay filter, and (3) equiripple group-delay filters. All of these filters approximate the ideal Gaussian filter and therefore belong to Gaussian family.

Gaussian Filter

Let us assume that the transmission function, instead of being rectangular with abrupt cutoffs, is smooth but still retains the constant group delay. One such transmission function is the Gaussian function which for the following discussion can be defined as

$$|H(j\Omega)| = e^{\ln \Omega^2} = \exp\left[0.3466\left(\frac{\Omega}{\Omega_{3\,dB}}\right)^2\right] = \left|\frac{V_p}{V}\right|$$
(3.6.1)

where V_p is peak output voltage of the filter, and V is output at bandwidth Ω. The response of such a filter is shown in Fig. 3.9 with

$$|H(j\Omega_c)| = 2.71828 = e$$
(3.6.2)

The corresponding group-delay response is shown on the right side of the same figure. Equation 3.6.1 shows that when $\Omega = \Omega_c$ the relative attenuation is 1 Np or 8.68 dB. From the same expression it will be seen that $\Omega_{3\,dB} = 0.588\,\Omega_c$. Relative attenuation in decibels is very simple.

$$dB = 3\left(\frac{\Omega}{\Omega_{3\,dB}}\right)^2$$
(3.6.3)

At twice the 3-dB bandwidth, the magnitude of the perfectly Gaussian relative attenuation is 12 dB; at three times the 3-dB bandwidth it is 27 dB, and so on.

A series with an infinite number of terms is required to represent the Gaussian magnitude. It can be shown that the perfectly Gaussian phase characteristic has infinite slope and is perfectly linear. Infinite linear slope of the phase characteristic gives infinite time delay for signals passing through a perfectly Gaussian filter. Approximation of the Gaussian response by a finite number of network elements results in a finite time delay.

As has been mentioned, a desired network function that requires an infinite number of terms when expressed in polynomial form would require a synthesized lumped linear network containing an infinite number of elements. It is thus necessary in practice to approximate the perfect Gaussian response. The problem may be different depending on which characteristic is of interest: the magnitude characteristic, the phase characteristic, or some other characteristic (such as transient characteristics for impulse or step excitation).

To relate the desired Gaussian relative-attenuation magnitude of Eq. 3.6.1 to physically realizable networks, it is necessary to write the series for the square of (3.6.1) and not for this equation directly, because the number of terms in the polynomial is to be related to the number of elements in the network.

When the squaring is done and a convergent infinite series is used

$$|H(j\Omega)|^2 = e^{2(\Omega/\Omega_c)^2} = 1 + 2\left(\frac{\Omega}{\Omega_c}\right)^2 + \frac{2^2}{2!}\left(\frac{\Omega}{\Omega_c}\right)^4$$
$$+ \frac{2^3}{3!}\left(\frac{\Omega}{\Omega_c}\right)^6 + \frac{2^4}{4!}\left(\frac{\Omega}{\Omega_c}\right)^8$$
$$+ \frac{2^5}{5!}\left(\frac{\Omega}{\Omega_c}\right)^{10} + \frac{2^6}{6!}\left(\frac{\Omega}{\Omega_c}\right)^{12}$$
$$+ \cdots$$
(3.6.4)

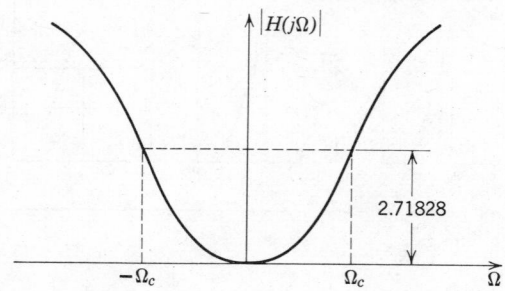
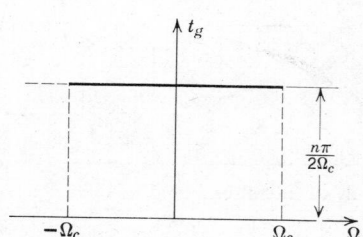

Fig. 3.9. Gaussian amplitude and group-delay response.

It can be seen that a two-element filter can satisfy the first three terms of (3.6.4), a three-element filter the first four terms, and so forth. In general, an n-element filter can satisfy the first $(n + 1)$ terms of (3.6.4).

There are a number of different ways of approximation to a desired curve. The Gaussian magnitude shapes obtainable with n-elements within the 3-dB bandwidth are shown in curve 9, Section 3.14. The part outside the 3-dB band is also given for filters containing up to 10 elements. From these curves it is a simple matter to determine the number of network elements required to satisfy a specified condition. The number of elements used in the filter determines how far up on the relative attenuation curve the approximation satisfies the perfect Gaussian response. For example, a four-element filter can approximate the Gaussian response within 1 dB down to about the 11-dB point, six elements down to about 18 dB; eight elements to about the 26-dB point; 10 elements to about the 34-dB point; and so forth.

How far down must the perfect Gaussian response be satisfied? This is an important question that depends on the requirements of a specific system.

The phase response of the ideal Gaussian filter is linear, and no overshoot will be produced as a result of rapid signal changes. Realizable Gaussian magnitude filters with a finite number of elements yield, in many instances, not a sufficiently linear phase response in the frequency domain which results in an inconvenience in the time domain.

Bessel Filter

The best approximation in the sense that it produces maximally flat group delay in the frequency domain is accomplished with the Bessel polynomial, known also as the Thomson filter and shown in Fig. 3.10. Accordingly, this approximation results in better responses in the time domain and approaches the ideal Gaussian curve as the degree of approximation is increased. The transfer function and the realized lowpass filter can be expressed by the operator

$$W(s) = e^{-st_0} = (\cosh s + \sinh s)^{-1} \quad (3.6.5)$$

where t_0 is a fixed delay time.

Using identity $e^s = \cosh s + \sinh s$, a new function,

$$f(s) = \frac{1/\sinh s}{1 + (\cosh s/\sinh s)} \quad (3.6.6)$$

can be formed. Now, if a continued fraction expansion

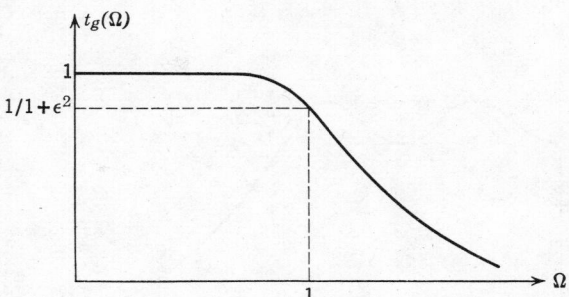

Fig. 3.10. Group-delay response of Bessel amplitude-response approximation.

of coth s is used for coth s, the resultant approximation of e^s will give a maximally flat delay. The phase response in the finite-transfer-pole Bessel filter is more linear than that of the Gaussian magnitude type with an equal number of poles. The skirt selectivity of the passband is sharper for the Bessel filter, but the attenuation slope near cutoff for both filters is not very great. Both types have very poor attenuation characteristics especially with wide passbands. But from the point of view of group-delay distortion or phase characteristics, their responses in comparison with Chebyshev or Butterworth filters are remarkably good.

Linear Phase Filters with an Equiripple Error

As the Chebyshev filter (equiripple) is a better approximation to the ideal magnitude filter, it follows that an equiripple approximation of the phase will give better results than the maximally flat phase approximation.

As in the case of the maximally flat magnitude approximation, the maximally flat delay approximation does not make as efficient use of the polynomials as does the equiripple approximation. For a given degree of polynomial n, the equiripple approximation approximates a constant over a longer interval than does the maximally flat delay approximation. The steady-state amplitude response of the equiripple delay filter is somewhat better than the filter of comparable complexity designed by the Bessel approximation.

Typical group-delay characteristics of equiripple-error linear-phase filters are illustrated in Fig. 3.11. In the realizable filter, the number of maximum phase deviations of ϵ degrees shown in Fig. 3.12 is $(n + 1)$ where n is the degree of the polynomial and also the minimum number of reactive components required to construct the lowpass filter. Because a closed form

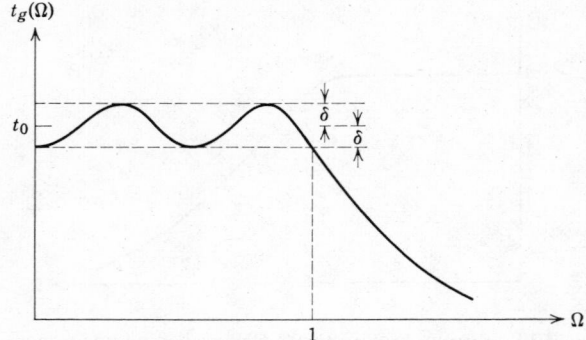

Fig. 3.11. Typical equiripple group-delay response.

solution to this problem has not been found, an iterative procedure was devised. The roots of the polynomial are found by the iterative scheme, and curves are plotted for two values of ϵ ($\epsilon = 0.05°$, $0.5°$) and for $n = 2, 3, \ldots, 10$ in Section 3.14. The polynomials are the ones commonly used for the Bessel filter that is, the phase characteristic approximates a linear function of unit slope. The degree of the polynomial used in this approximation is finite, and consequently the linear phase is approximated over only a finite frequency interval (see Fig. 3.12).

From the root locations all of the frequency- and time-domain properties can be determined. The two most useful frequency-domain characteristics are the phase and the magnitude. The phase however, is the defining feature of this particular filter and hence, is known. On the other hand, the magnitude characteristics, must be calculated from the transmission

zero locations. In Section 3.14 the reader can find the amplitude- and group-delay responses of these filters calculated for different levels of phase error in the passband. As is customary for such representations, a unity 3-dB radian-frequency bandwidth normalization has been used.

From a comparison of the linear phase attenuation curves to the ideal Gaussian attenuation curve several conclusions can be drawn, some of which are:

1. The linear-phase filter, for small ϵ, is approximately Gaussian within the 3-dB passband.
2. The attenuation of the linear-phase filter is higher than that of the Gaussian filter for frequencies just above the passband. This region is sometimes called the transition region.
3. The attenuation of the linear-phase filter approaches $20\,n$ dB per decade for frequencies far removed from the cutoff. This is less than the ideal Gaussian.

As the ripple factor ϵ is increased, it is noted that:

1. The attenuation characteristics become rather "lumpy" and the response of each individual transmission zero can be detected.
2. The attenuation increases rapidly for frequencies above the interval for which a linear phase is being approximated. This is especially true for large ϵ and large n.

In general, it can be concluded that a linear-phase filter does not also have a marked cutoff frequency. Its transition region is broad.

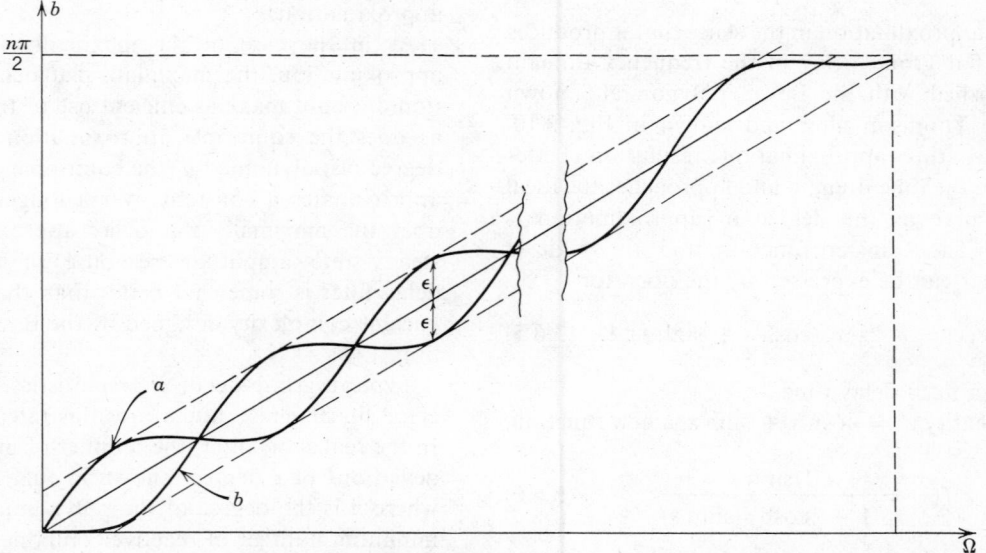

Fig. 3.12. A typical phase characteristic when a linear function is approximated in an equiripple manner for (*a*) *n* odd and (*b*) *n* even.

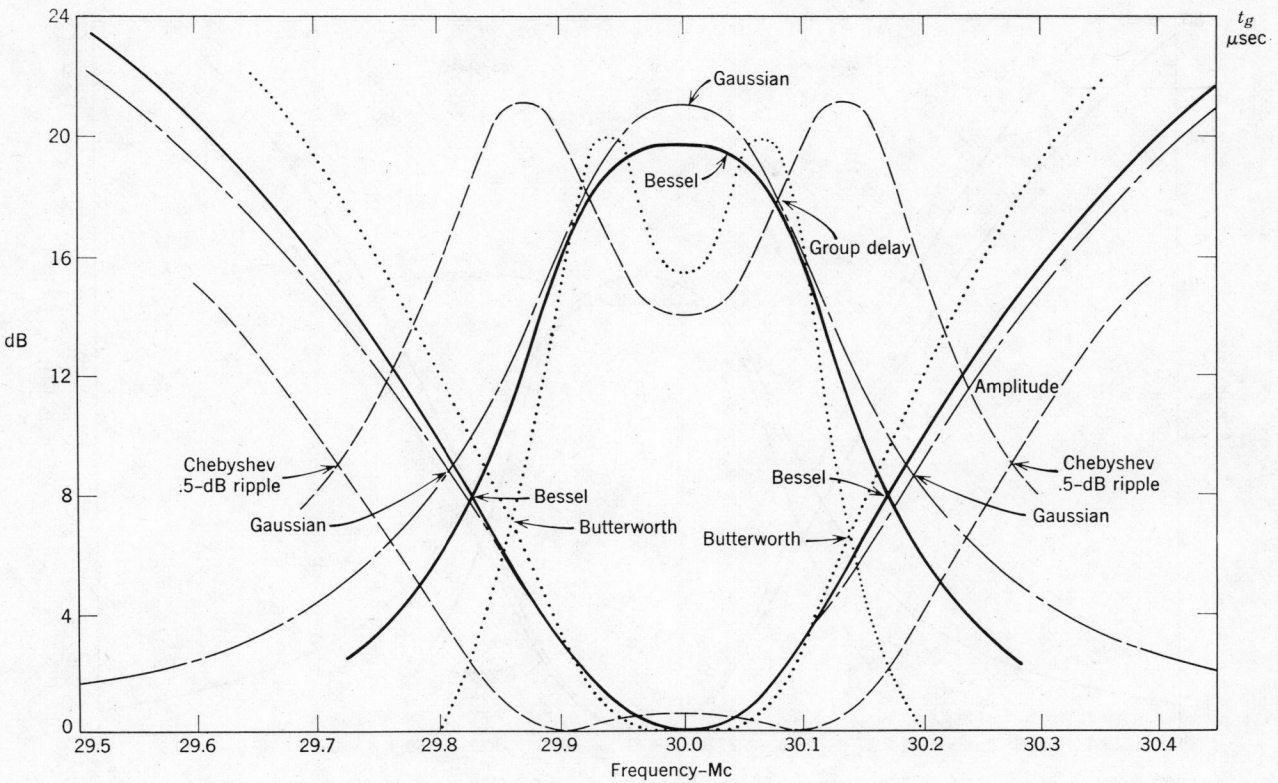

Fig. 3.13. Amplitude and group-delay response of ———— Bessel, — — — Gaussian, – – – – Chebyshev-.5-dB ripple, ... Butterworth.

From the curves of Section 3.14 a filter designer can rapidly determine which value of n and ϵ will most closely satisfy his particular frequency-domain needs.

Comparison of Group-Delay Characteristics in the Frequency Domain

For purposes of comparison, the group delay of Bessel, Gaussian, Chebyshev, and Butterworth filters of second degree are plotted in Fig. 3.13. It should be noted that the group delay of the Bessel filter is flat at the center of the passband, whereas the Gaussian curve drops off more rapidly, and the Butterworth has a pronounced peak at the cutoff frequency. In Fig. 3.14, the curves of the Gaussian and Bessel responses for $n = 4$ are plotted.

Although the Bessel filter phase characteristic is greatly superior to that of the Chebyshev filter, its disadvantages preclude its use in most cases. The Bessel amplitude response approximates a parabolic curve, so that its voltage standing-wave ratio (VSWR) increases rapidly as the frequency deviates from center frequency. Over most of the 3-dB bandwidth the VSWR and reflection coefficient are high. As a result, reflection interactions with a slightly mismatched load or generator may seriously affect the otherwise good phase-shift response of this type of filter. For a given selectivity, Bessel filters need more elements than the Chebyshev type. In fact, no matter how many elements are used, this type of filter cannot provide appreciably greater selectivity than a parabolic attenuation-response curve. The multiple-resonator structure for Bessel filters is highly unsymmetrical, with a very large variation in the coupling elements from one end to the other. This large variation makes the design difficult, and values of the couplings are more likely to be in error relative to each other.

In most filter design problems the difficulty is mainly that of obtaining a group delay as nearly constant as possible. In view of the relations already mentioned for a lowpass filter, for example, in order to obtain nearly constant group delay, the attenuation must increase gradually in the passband in a Gaussian fashion. This is illustrated in Fig. 3.15 where the frequency response of attenuation and of group delay are shown for some of the filter types mentioned.

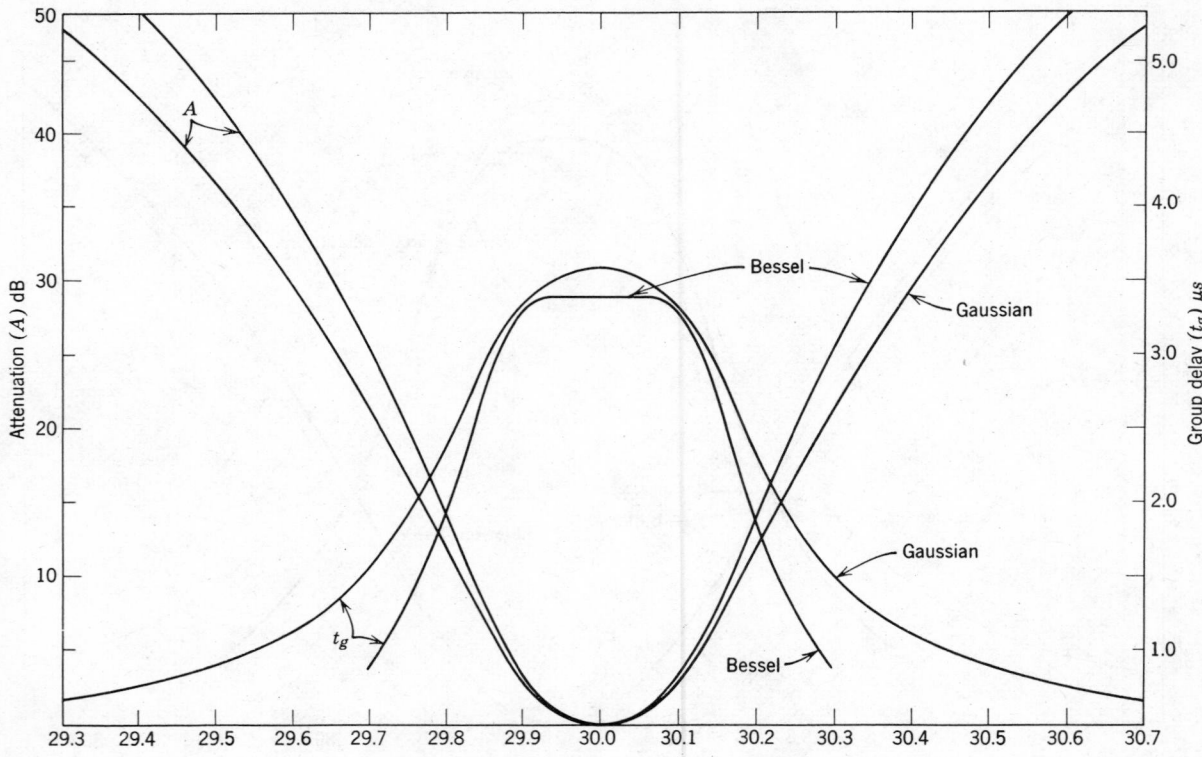

Fig. 3.14. Gaussian and Bessel responses for $n = 4$.

3.7 A FILTER WITH TRANSITIONAL MAGNITUDE CHARACTERISTICS

There are several applications for compromise filters which have frequency-domain attenuation characteristics midway between those of the Butterworth filter and those of the linear-phase family of filters. Some of the approximations give a smooth attenuation and group-delay characteristics which resemble the nonefficient Butterworth magnitude and Bessel group-delay characteristics. Some approximations propose a lowpass prototype filter which makes more efficient use of each pole of the transfer function as do the Chebyshev and equiripple group-delay filters. It is known that a filter with Gaussian magnitude characteristics possesses an almost linear-phase characteristic. Hence, the transitional filter should exhibit a linear-phase characteristic near the origin. At the end of the approximation interval the magnitude characteristic, like the Chebyshev filter, will exhibit a sharp break to give a high discrimination for those frequencies outside the passband. In the tables in Section 3.14, the normalized transitional filter is characterized as a filter whose attenuation charac-

teristics in decibels approximate the reference function

$$R(\Omega, \epsilon, k) = (10k \log 2 \pm \epsilon)\Omega^2 - (\pm\epsilon) \qquad \text{dB}$$

$$(3.7.1)$$

in an equiripple manner to give an error function which has n peak error deviations of ϵ dB within the interval $(0, 1)$ and also has a zero at $\Omega = 1$. The minus sign in (3.7.1) is for n odd. k is a constant related to the number of dB for which the approximation is Gaussian. For example, if $k = 4$, the approximation will be Gaussian within the range of 0–12 dB.

3.8 LEGENDRE FILTERS

The Butterworth response is monotonic and the amplitude decreases with increasing frequency. The Chebyshev response is monotonic only outside of the passband but equiripple in the passband. A Chebyshev response has better cutoff characteristics than the Butterworth, but by being in the same family with the Butterworth it generates into a Butterworth response if no ripples are allowed. Allowance of ripples, even a fraction of 1 dB, makes a big difference in the stopband.

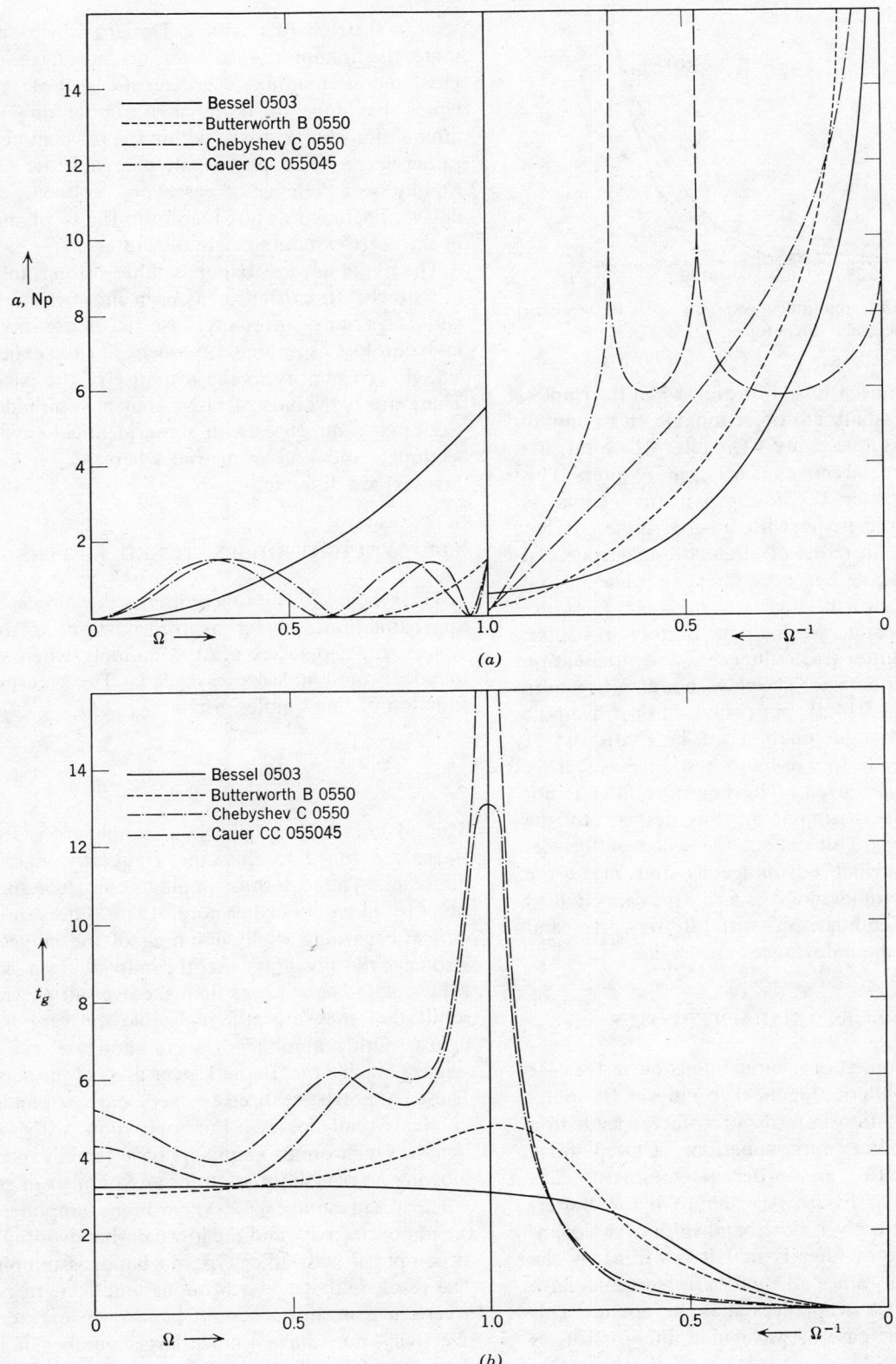

Fig. 3.15. (a) Effective attenuation of lowpass filters of order $n = 5$; (b) Group delay of lowpass filters of order $n = 5$.

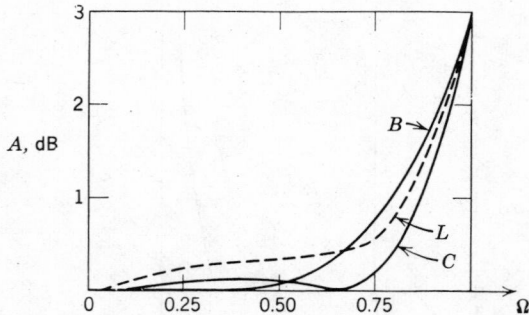

Fig. 3.16. In-band amplitude response of Butterworth, Chebyshev, and Legendre filters for $n = 3$.

There may be occasions, however, when the ripples, no matter how small, are objectionable and a monotonic response is a necessity. The filter which satisfies these kinds of requirements is a Legendre filter. This amplitude characteristic is approximately equal to Chebyshev characteristic with 0.1-dB ripple. It has been found that in terms of attenuation sharpness at cutoff the three-pole Legendre filter is equivalent to a four-pole Butterworth filter. A five-pole Legendre filter is equivalent to a nine-pole Butterworth filter. A three-pole Butterworth filter has the attenuation rate of 10 dB per octave. The corresponding Legendre filter has a rate of 14 dB per octave. For a five-pole filter the respective attenuation will be 18 dB and 31 dB. In Fig. 3.16 the in-band responses of three different filters are given. The Legendre filter is not symmetrical. It is similar in this respect to the Gaussian family of filters, but the value of the elements are somewhat advantageous and make the filter easier to manufacture. As a rule the capacitances are larger in comparison with Butterworth and Chebyshev and the inductances are smaller.

3.9 MINIMUM-LOSS CHARACTERISTICS

A filter may consist of identical units or, in the case of microwave filters, identical cavities. Its polynomial representation in terms of Q-factor leads to a minimum-loss filter approximation. Closed form solution exists for even-order polynomials. The resultant 3-dB bandwidth is wider than the Butterworth design for a given skirt bandwidth. An advantage of this type of filter is that it is ordinarily the simplest to build. Since all cavities (or sections) have the same Q, all susceptances may be equal. This means that in the case of waveguide filters, all irises may be identical.

The disadvantage of this identical-unit filter can be

seen in their characteristics. Designs which approximate the minimum-loss filter do not have regular passband-attenuation characteristics. For a filter which has four or more cavities, the dips in the attenuation characteristic within the passband become rather large. The phase-shift characteristic changes rapidly near the edge of passband, producing a large delay. There is less attenuation in the stopband than in the corresponding Chebyshev filter.

The minimum-loss bandpass filter belongs in a class of its own. Its existence has been shown by S. Cohn, and it permits, in many cases, a more favorable insertion loss at midband frequency at the expense of a slight irregularity in the response of the passband. There are two kinds of filters that have nonidentical sections: symmetric, with some identical cavities or sections; and nonsymmetric, where all sections (or cavities) are different.

3.10 SYNCHRONOUSLY TUNED FILTERS

The synchronously tuned filter is the simplest of all approximations, as far as transmission function is concerned, and arises most commonly when several identical amplifiers are cascaded. The transmission function of this simple form is

$$H(s) = \frac{(s + a)^n}{a^n} \qquad (3.10.1)$$

The advantage of synchronous amplifiers is that all stages are tuned to the same frequency, namely to midband. They are thus simple to construct and very easy to align. Furthermore, these filters are not critical regarding slight detuning of the stages, and also have the advantage that the individual stages have values of Q factor lower than the over-all Q with the result that they are still realizable for very narrow bands. Other amplifiers, in common use, require a stage Q much larger than the over-all Q of the network. These cannot be realized for very narrow bands due to the circuit losses. For operation with pulsed signals, synchronous amplifiers have the advantage of showing no overshoot at all for any number of stages.

The disadvantages of synchronous amplifiers are the poor selectivity and the low gain-bandwidth factor as compared with other types of bandpass amplifiers. The result is that a synchronous amplifier, to realize a certain gain and a certain bandwidth or a certain selectivity, must have a much larger number of stages than would be necessary for other types. The synchronous amplifier can often not satisfy the given gain-bandwidth or selectivity requirements at all, however large the number of stages may be, whereas other types can (in theory) always be designed for many such requirements.

For the foregoing consideration, synchronous amplifiers are sometimes used for modest selectivity requirements which when the gain-bandwidth product is well below the theoretical maximum, although even in this case the use of other types may be advantageous.

Having examined the characteristics of the various filters, several conclusions can be made. For example:

1. Butterworth, Legendre, Chebyshev, and minimum insertion-loss filter attenuation characteristics have a sharp increase in attenuation as the frequency increases slightly above cutoff. This is in sharp contrast to the remaining filters where the attenuation cutoff is smooth.

2. The general attenuation-delay relationship predicted by the Hilbert transforms are being approached by several of the filters.

3.11 ARITHMETICALLY SYMMETRICAL BANDPASS FILTERS

In certain transmission systems, specifically in data-transmission systems, bandpass filters are required to exhibit an essentially symmetrical characteristic on an arithmetic scale, in both the attenuation and group delay. The conventional lowpass-to-bandpass transformation $s \to p + 1/p$ (where p is the bandpass frequency variable) leads to bandpass filters exhibiting attenuation symmetry about $f_m = \sqrt{f_1 f_2}$ on a logarithmic scale, whereas the original lowpass filter had arithmetic symmetry at about $f = 0$. The two kinds of symmetry differ negligibly in the narrow band where $f_m \approx f_a = (f_1 + f_2)/2$, but the difference becomes noticable when the band of interest exceeds approximately 5% of center frequency. If the lowpass had linear phase, then this property is lost in the bandpass filter (with exception of the narrow band case). The difficulty here is a fundamental one; exact arithmetic symmetry in filters with lumped components is not physically possible.

Arithmetic symmetry is exhibited by transmission-line filters having a periodic infinite number of pass-bands. A filter with lumped parameters cannot exhibit arithmetic symmetry, but this characteristic may be approximated over a limited bandwidth. The designer is faced with two simultaneous approximations instead of the usual one: (1) the approximation to the ideal filter characteristic and (2) the approximation of

symmetry. Usually, symmetry outside of the passband is of no consequence. Hence its approximation can be restricted to the passband.

Approximating the symmetry of attenuation characteristics leads to a symmetrical-delay characteristic. Restricting the discussion to ladder structures, the delay can be obtained uniquely from the attenuation characteristic and symmetry in one will lead to symmetry in the other. The relationship for deviation from symmetry in one characteristic is simply related to deviation in the other.

Let us now consider a normalized lowpass filter with passband

$$0 < \Omega < 1$$

and stopband

$$1 < k < \Omega < \infty$$

The attenuation characteristic is shown in Fig. 3.17a. Now we apply the periodic transformation

$$s = j\omega = \frac{d}{\tanh(\pi p/2\omega_0)}$$

or,

$$\Omega = -\frac{d}{\tan(\pi\omega/2\omega_0)}$$

Here

$$d = \tan\frac{\pi\omega_1}{2\omega_0} = -\tan\frac{\pi\omega_2}{2\omega_0}$$

where ω_1 and ω_2 are the lower and upper ends respectively of the first passband. These periodic transformations lead to a filter of an infinite number of passbands, each with exact arithmetic symmetry and

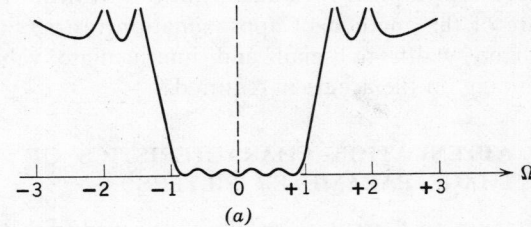

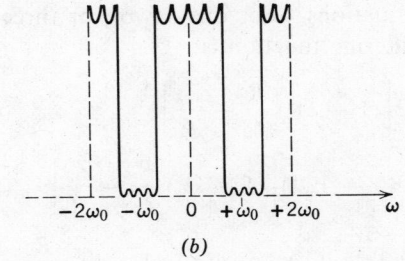

Fig. 3.17. (a) Normalized filter characteristic, (b) periodic bandpass characteristic.

centered at $\pm\omega_0$, $\pm3\omega_0$, $\pm5\omega_0$, and so on as shown in Fig. 3.17b.

A lumped-element approximation can be obtained for this transformation by approximating the hyperbolic tangent function by a rational function. The disadvantages of this method is the excessive number of elements required and the one or more spurious passbands generated. In order to prevent this, a slightly different philosophy by G. Szentirmai can be applied. The poles and zeros for all passbands except those at $\pm\omega_0$ are discarded and a filtering function $D^*(p)$ is formed out of those that remain.

From the exact arithmetic symmetry obtained by the periodic transformation only an approximation exists, since the infinite number of superfluous passbands have been discarded.

Nevertheless $D^*(p)$ is a good approximation and can be improved by choosing a new filtering function as follows

$$D(p) = D^*(p)n(p)$$

where

$$n(p) = 1 + ap + bp^2$$

The quadratic factor is inserted to correct the dissymmetry of $D^*(p)$ in both the passband and the transmission bands of the filter. The correction is achieved by plotting the deviation from symmetry of $D^*(p)$ as a function of ω and choosing the coefficients a and b, so that $n(p)$ matches the deviation curve. From here on, all that remains is to realize $D(p)$ in a network, and this is a straightforward calculation.

A second simpler method may be applied to polynomial filters. The numerical approximation can be avoided and explicit formulas can be provided. The nature of this method of approximation restricts the passband width to small and intermediate values depending on the accuracy required.

3.12 ATTENUATION CHARACTERISTICS OF IMAGE PARAMETER FILTERS

Attenuation formulas for filters designed with the aid of image-parameter theory are obtained with the following equations. For one-, two-, or three-π sections, the filtering function is:

$$\pm D = -\Omega\left[(1 - \Omega^2)\frac{R}{Z_\pi} - \frac{Z_\pi}{R}\right] \qquad (3.12.1)$$

$$\pm D = -\Omega(2 - 4\Omega^2)\left[(1 - \Omega^2)\frac{R}{Z_\pi} - \frac{Z_\pi}{R}\right] \qquad (3.12.2)$$

$$\pm D = -\Omega(3 - 4\Omega^2)(1 - 4\Omega^2)\left[(1 - \Omega^2)\frac{R}{Z_\pi} - \frac{Z_\pi}{R}\right] \qquad (3.12.3)$$

Equation 3.12.1 describes the filtering function for a single-π section and (3.12.2) and (3.12.3) refer to two- and three-π sections respectively. In each case, $2 \geq (Z_\pi/R) \geq 1$. The curve of effective attenuation for the three-π filter is shown in Fig. 3.18.

The corresponding expressions for T-sections are

$$\pm D = \Omega\left[(1 - \Omega^2)\frac{Z_T}{R} - \frac{R}{Z_T}\right] \qquad (3.12.4)$$

$$\pm D = \Omega(2 - 4\Omega^2)\left[(1 - \Omega^2)\frac{Z_T}{R} - \frac{R}{Z_T}\right] \qquad (3.12.5)$$

$$\pm D = \Omega(3 - 4\Omega^2)(1 - 4\Omega^2)\left[(1 - \Omega^2)\frac{Z_T}{R} - \frac{R}{Z_T}\right]$$
$$(3.12.6)$$

Equation 3.12.4 gives the filtering function for a single-T section; (3.12.5) and (3.12.6) give the functions for two- and three-T section filters respectively.

These expressions describe the peaks and valleys of attenuation in essentially the same way as the familiar curves of the Chebyshev filter with one important exception. The values of the ripples are not equal.

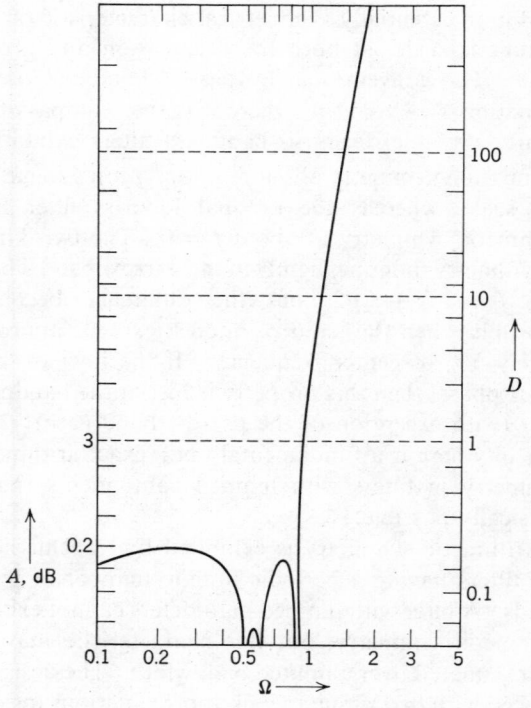

Fig. 3.18. Curve of effective attenuation for three-π section Zobel lowpass filter having optimum termination.

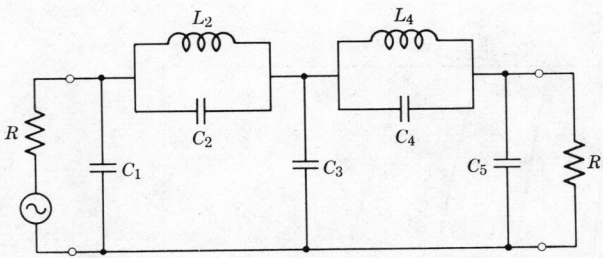

Fig. 3.19. Two *m*-derived Zobel lowpass sections.

For the case of the two *m*-derived sections shown in Fig. 3.19, the insertion loss in terms of the equivalent lattice reactances is given by

$$A = 10 \log\left[1 + \left(\frac{UV}{U-V}\right)^2\right] = 10 \log\left(1 + D^2\right)$$

(3.12.7)

The quantity D can be computed from

$$D = \frac{-(m_1 + m_2)(1 + m_1 m_2)}{r(1 - m_1^2)(1 - m_2^2)}$$

$$\times \frac{\Omega[\Omega^2 - (1 - r^2)][\Omega^2 - (1 + m_1 m_2)^{-1}]}{[\Omega^2 - (1 - m_1^2)^{-1}][\Omega^2 - (1 - m_2^2)^{-1}]}$$

(3.12.8)

Equations 3.12.7 and 3.12.8 give the location of the poles and zeros. The two peaks of attenuation appear in familiar form as the two factors in the denominator or when $U = V$ in (3.12.7). It should be noted that for $r < 1$ there are two finite zeros in the passband; one determined by the product of the *m* values and the other determined solely by r (the ratio of the design resistance to the terminating resistance). For unity ratio, when load resistance and design resistance are equal, the second zero moves back to zero frequency. The expression for D shows that if either m_1 or m_2 is given the value of unity, then the corresponding peak of attenuation moves to infinite frequency. For two constant k sections, both *m*'s are equal to unity and the passband zero determined by the *m*'s occurs at 0.707 times the cutoff frequency. The most interesting possibility indicated by Eq. 3.12.8 is that of controlling the passband ripples by proper choice of the design resistance (relative to the terminations). The zero determined by r can be located freely without regard to the other design constants at the point in the passband that gives the minimum ripple. This is the point that makes the two ripple peaks of equal amplitude. In

Fig. 3.20, the solid curves show a single passband zero and a single ripple occurring when the load and source have equal resistances. The dashed curve shows what happens when the terminating resistances are increased 20%, corresponding to $r = 0.833$, for this particular design. A second passband zero is brought in at $1 - r^2$ or at 0.553 times the cutoff frequency. This reduces the ripple amplitude by more than 3 to 1. As the terminating resistance is made even greater than the design resistance, the two ripples can be made equal and the amplitude reduction is then more than 4 to 1. The passband ripple can be controlled by selecting the terminating resistance in much the same way that the stopband peaks and valleys can be controlled by choice of the *m* values.

In the case of the Zobel filter with three constant-*k* sections.

$$D = \Omega(3 - 4\Omega^2)(1 - 4\Omega^2)\left[(1 - \Omega^2)\frac{1}{r} - r\right]$$

(3.12.9)

When one of the factors in Eq. 3.12.9 disappears, the value of D vanishes, and with it vanishes the effective attenuation.

In the case of a π-section filter, one zero will appear when the expression in brackets vanishes. For the T-section filter, one zero will be at

$$(1 - \Omega^2)r - \frac{1}{r} = 0 \qquad \text{or} \qquad R_1 = R\sqrt{1 - \Omega^2}$$

(3.12.10)

and three zeros will be independent of the load resistance, as shown in Fig. 3.21. One zero will be at

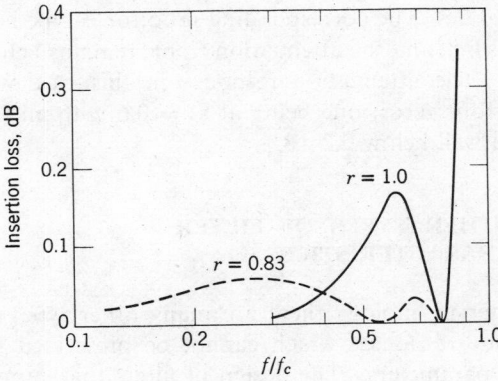

Fig. 3.20. Passband insertion loss of lowpass Zobel filter with two sections.

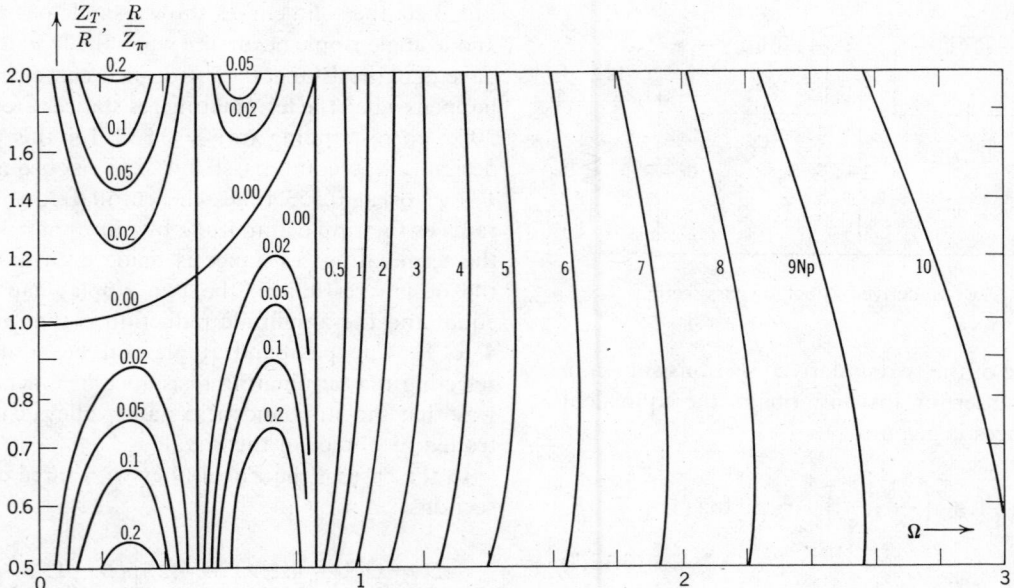

Fig. 3.21. Effective attenuation of Zobel filters as a function of frequency and design resistance.

$\Omega = 0$, the second zero will occur when

$$\Omega = \sqrt{\tfrac{1}{4}} = \pm 0.5$$

and the third at

$$\Omega = \sqrt{\tfrac{3}{4}} = \pm 0.87$$

It is possible with the aid of Eq. 3.12.9 to calculate attenuation as a function of normalized frequency for different values of load and design impedances.

From Fig. 3.21 it is evident that from the point of view of attenuation ripples for the lower part of the passband, the best ratio for Z_T/R is 1.1. In the upper part of the passband the best ratio is 1.5. For practical reasons, only a single value of this ratio can be used, such as 1.25. The corresponding ratio for π-type networks is 0.8, and the attenuation ripple remains below 0.2 dB. The attenuation response in this case will exhibit four zeros, one being at $\Omega = 0.6$ with all the maxima well below 0.2 dB.

3.13   OTHER TYPES OF FILTER CHARACTERISTICS

In filter technology there are many other types of amplitude responses which cannot be prescribed by specific parameters. The design of filters that supply each response requires consideration of physical rather than mathematical factors. Among these filters are:

1. Single-sideband filters, which are basically unsymmetrical (sharper on one side of the passband than on the other).
2. Extremely narrow bandpass filters designed to transmit virtually one frequency only.
3. Very wideband-response filters and impedance-matching networks.
4. Filters with restricted or unrestricted attenuation in a restricted band of frequencies.

Each of these types has different design techniques, different requirements for element values, and different component technology. An example is that of filters designed to reject unwanted frequencies on one side of the passband. Here the sharpness of the response curve on this particular side has to be much higher than on the opposite side of the passband where there are only limited attenuation requirements. The use of a symmetrical Chebyshev type for such an application would be wasteful of components and would complicate auxiliary problems such as insertion loss in the passband.

The Chebyshev approximation is impractical for very narrow-bandpass filters because of the high-Q requirements for the components. Very wideband-response filters, on the other hand, need high-Q resonators only when high rates of cutoff attenuation are required. Insertion loss is very low and does not present any problems for components in such wideband applications. In order to realize bandpass filters (with a 10% bandwidth, for example), a tandem combination of low- and highpass filters, is often used instead of direct bandpass synthesis. This technique can simplify the circuit and separate the problem of selectivity from the problem of insertion loss in the proximity of cutoff. It also avoids excessive requirements on inductive and other components used in the circuit.

3.14 PLOTS OF THE ATTENUATION AND GROUP-DELAY CHARACTERISTICS

The following curves are plots of the attenuation and group-delay characteristics for the filter types discussed in this chapter, and are valuable reference material for typical filtering problems. Note that for all curves $\Omega_{3\,\mathrm{dB}} = 1$.

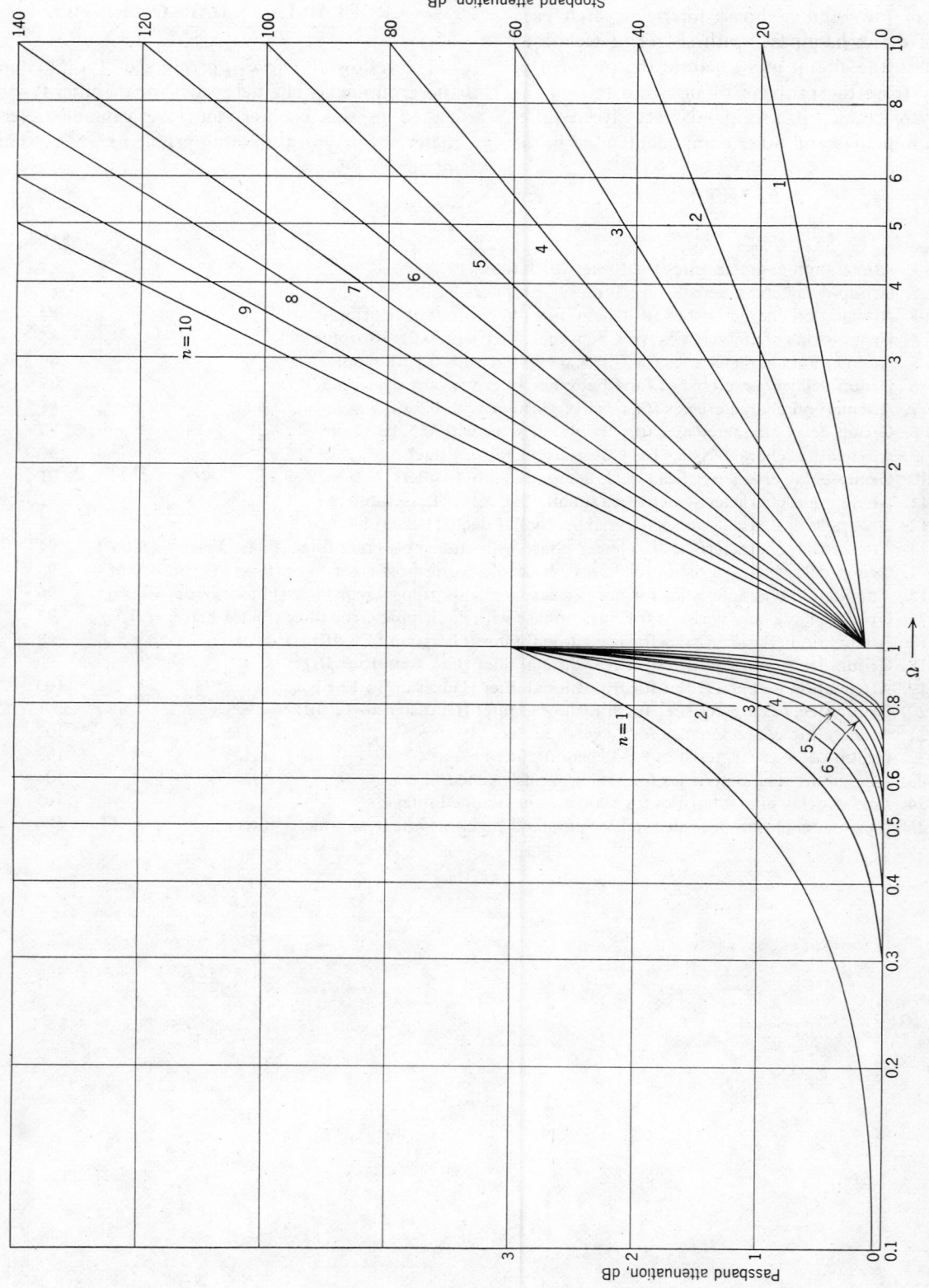

Curve 1. Attenuation characteristics for Butterworth filters.

82

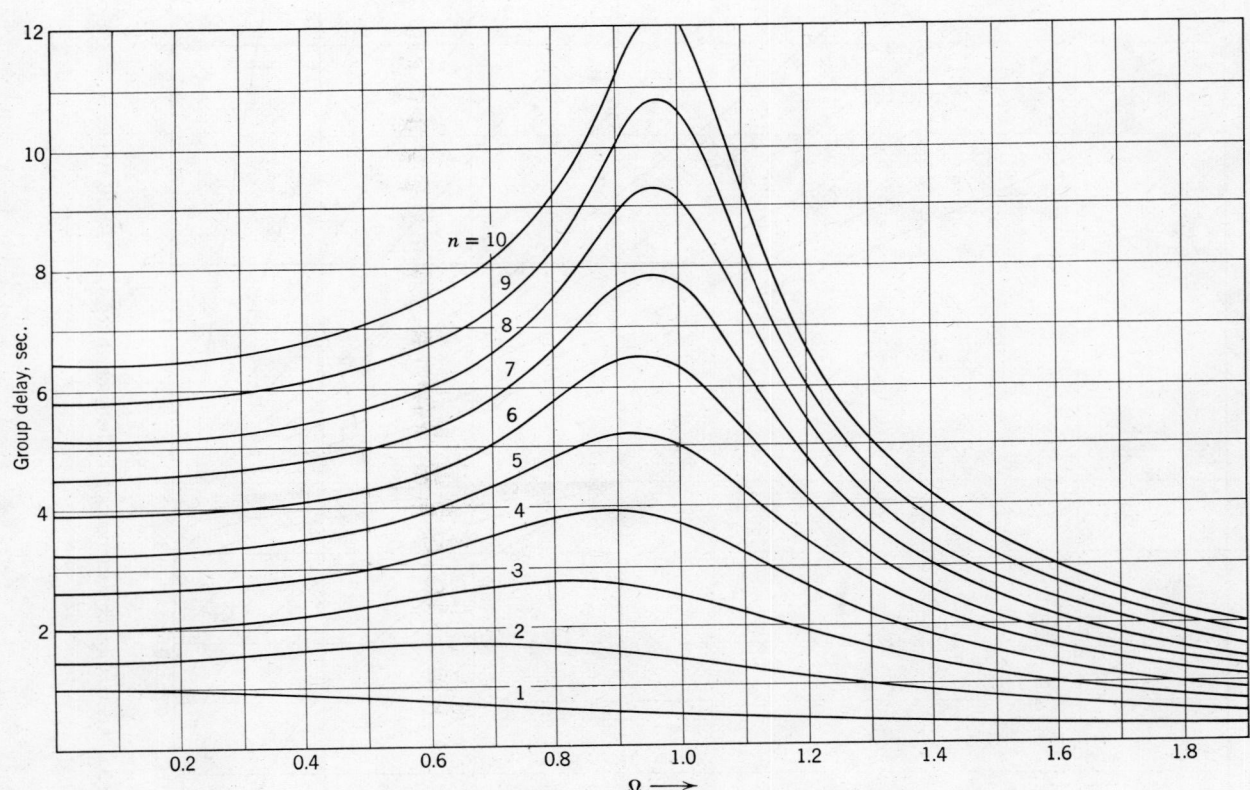

Curve 2. Group-delay characteristics for Butterworth filters.

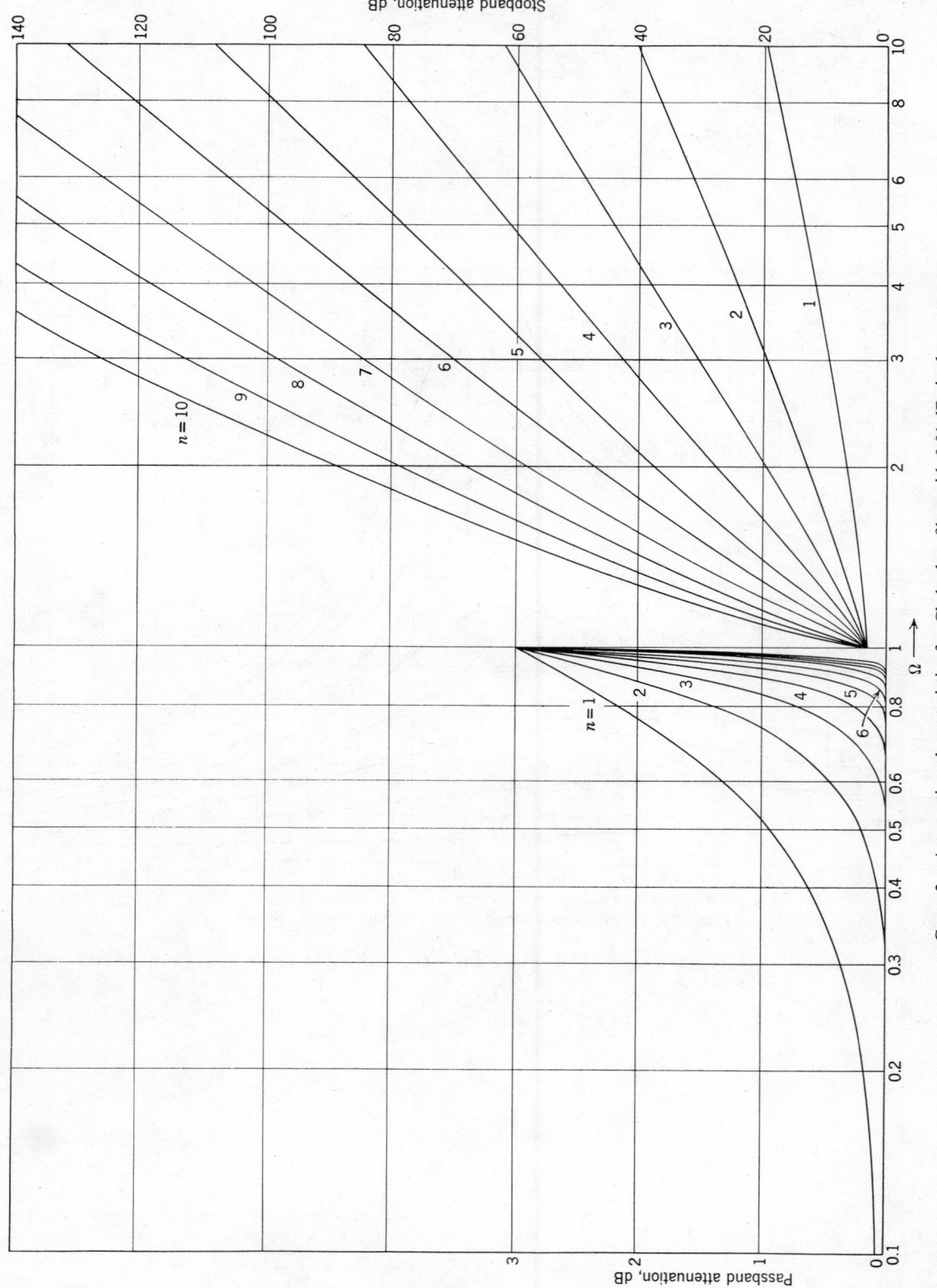

Curve 3. Attenuation characteristics for Chebyshev filter with 0.01 dB ripple.

84

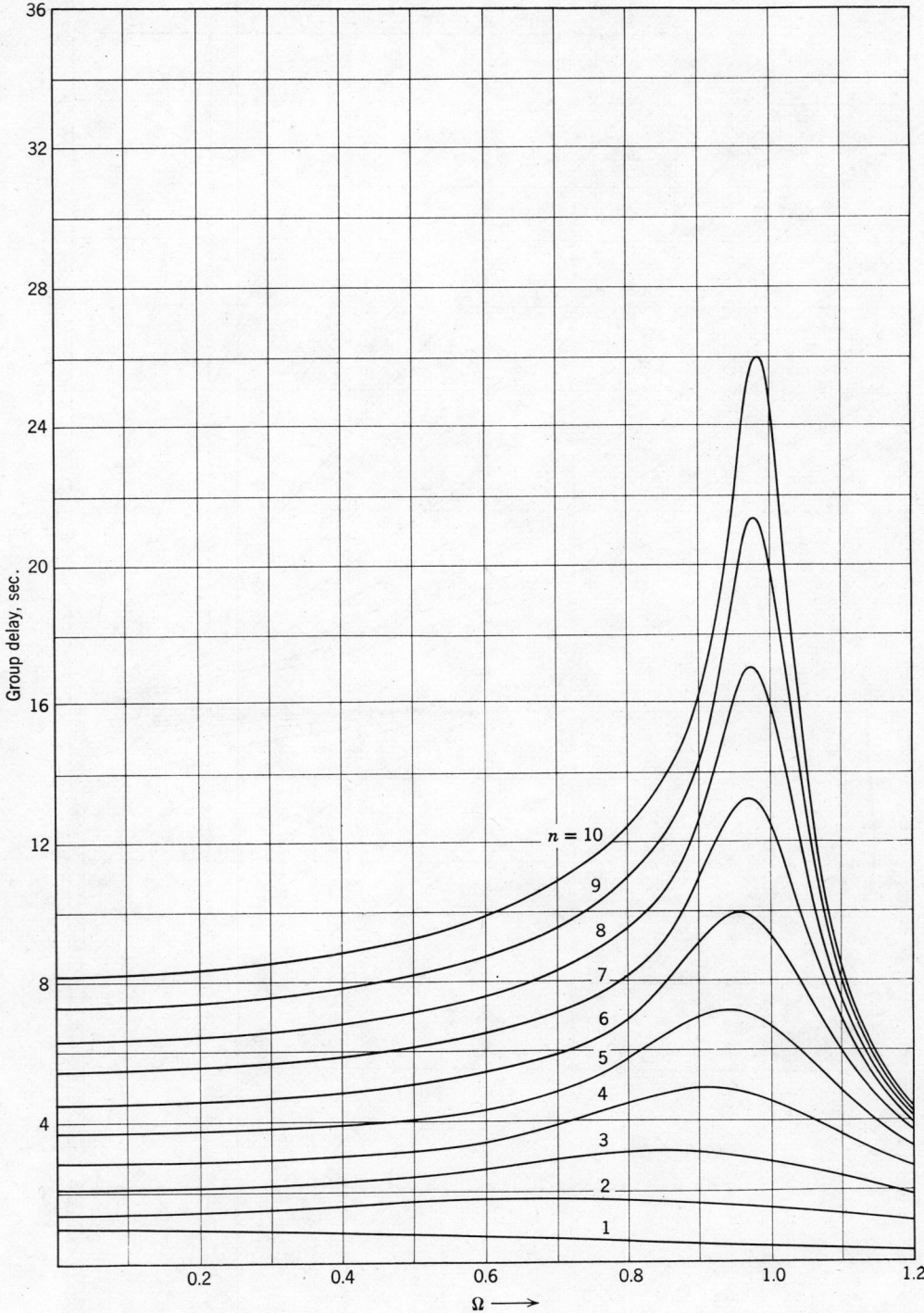

Curve 4. Group-delay characteristics for Chebyshev filter with 0.01 dB ripple.

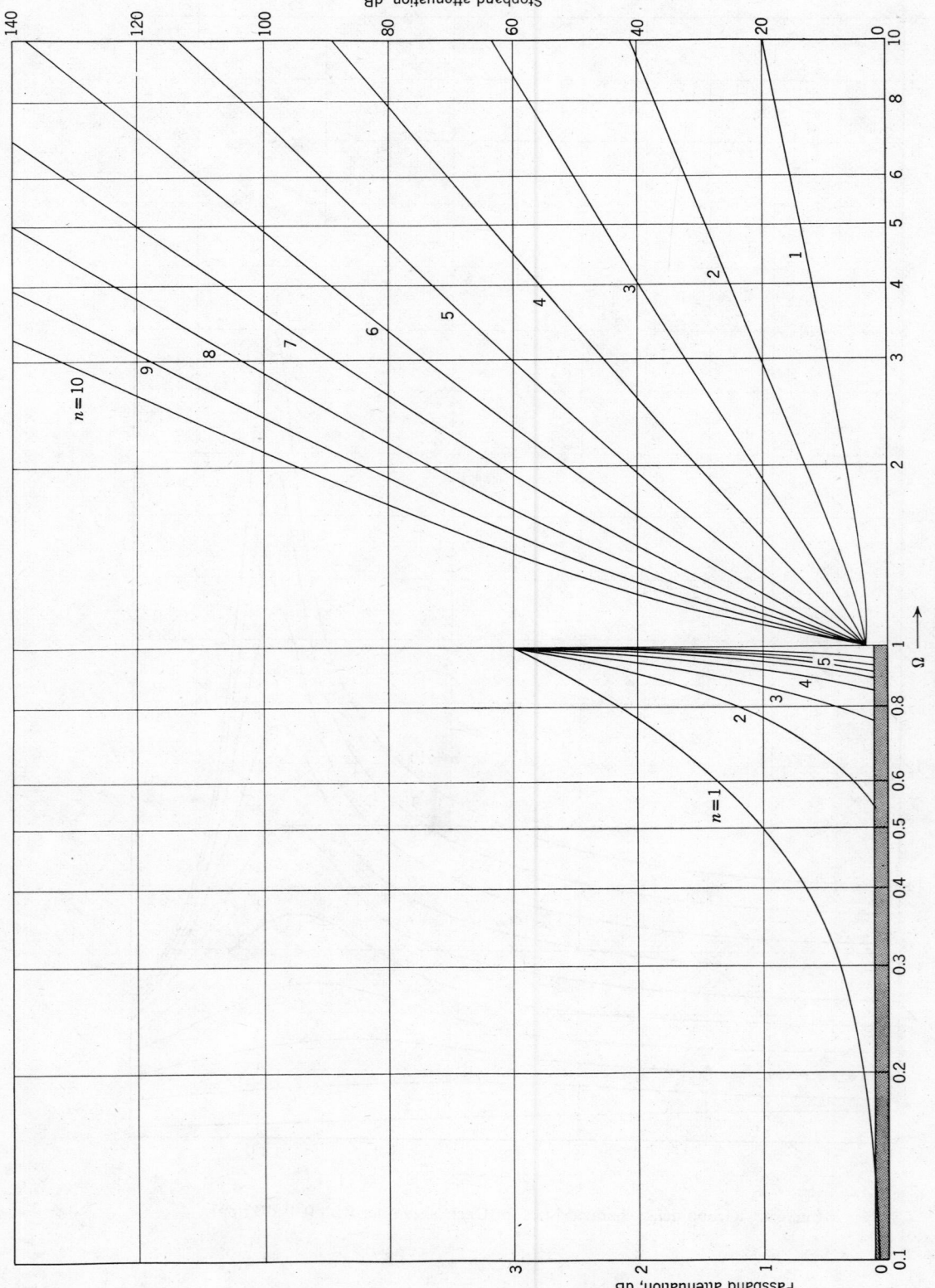

Curve 5. Attenuation characteristics for Chebyshev filter with 0.1 dB ripple.

86

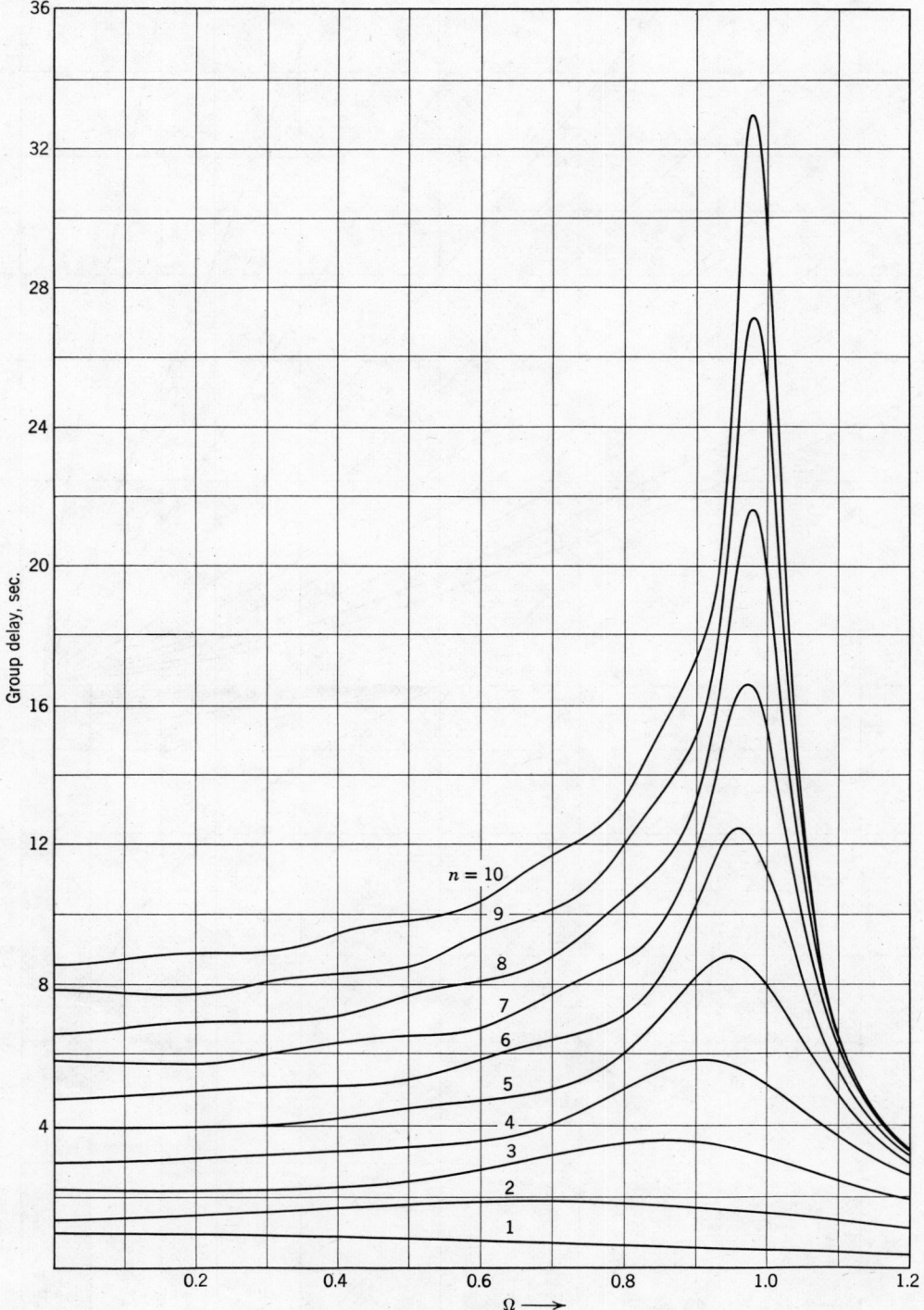

Curve 6. Group-delay characteristics for Chebyshev filter with 0.1 dB ripple.

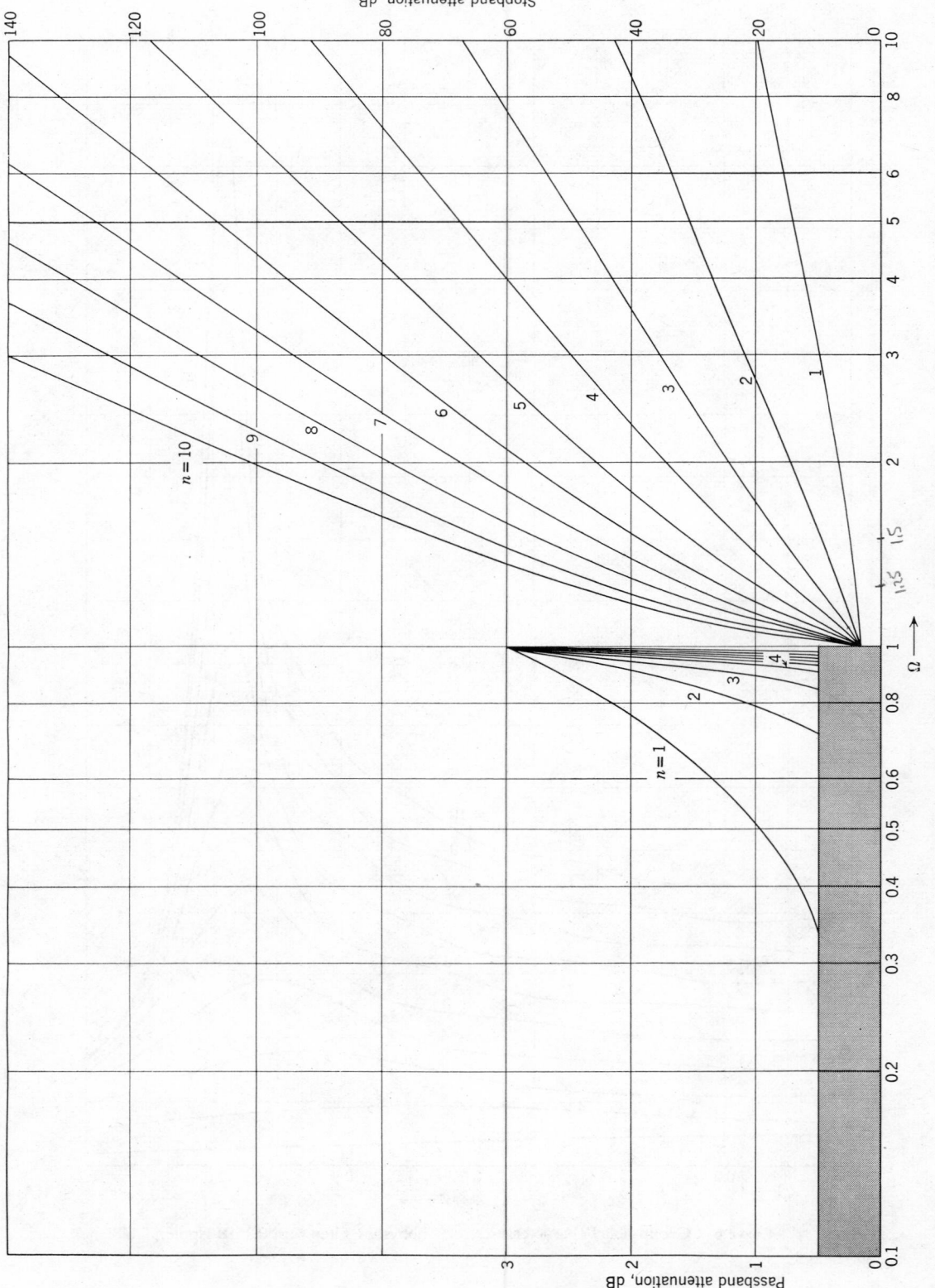

Curve 7. Attenuation characteristics for Chebyshev filter with 0.5 dB ripple.

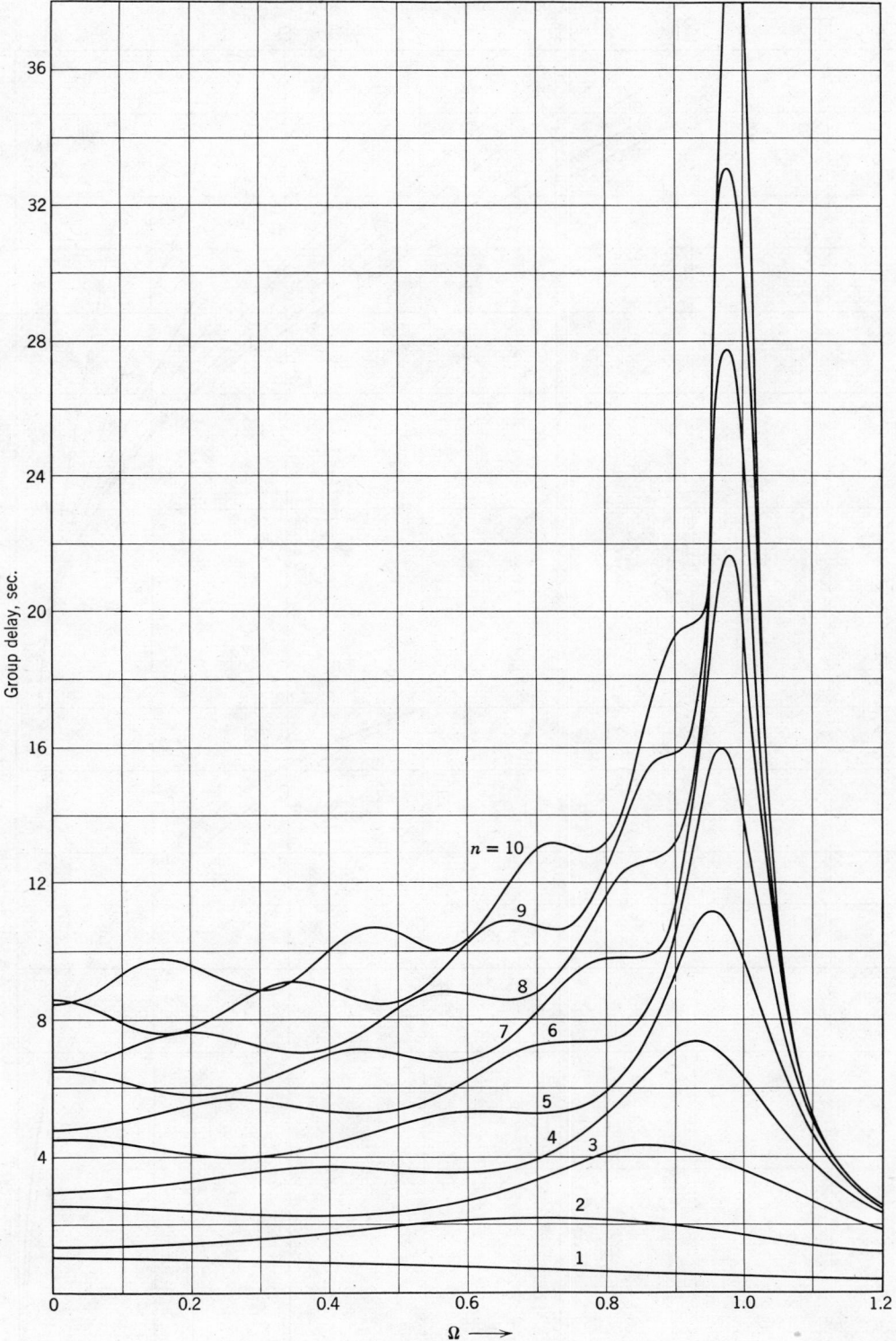

Curve 8. Group-delay characteristics for Chebyshev filter with 0.5 dB ripple.

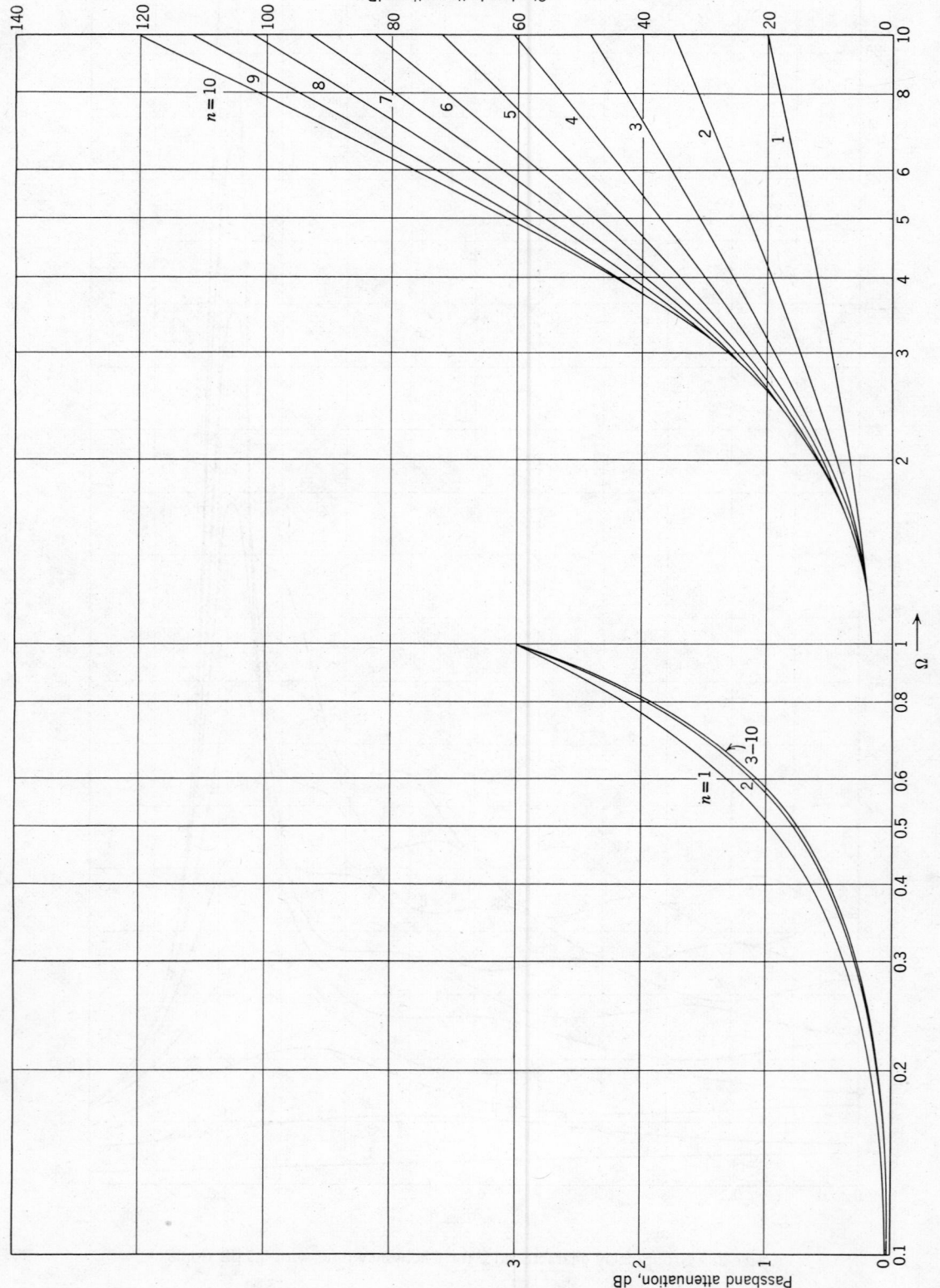

Curve 9. Attenuation characteristics for Gaussian magnitude filters.

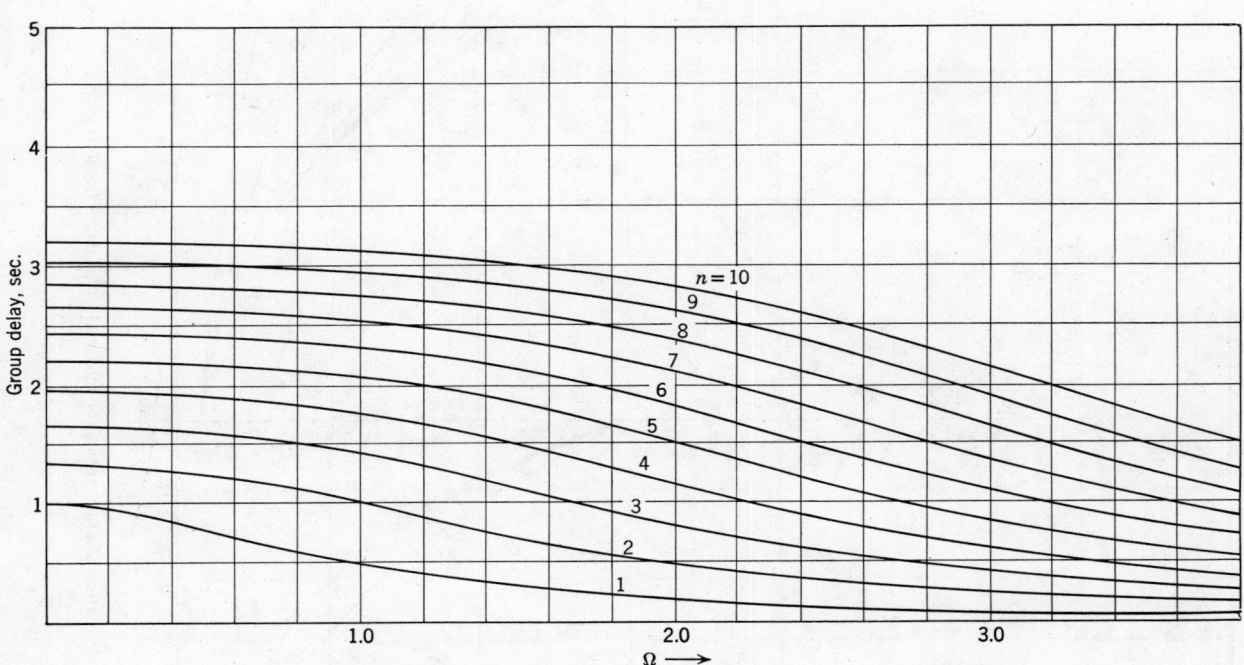

Curve 10. Group-delay characteristics for Gaussian magnitude filters.

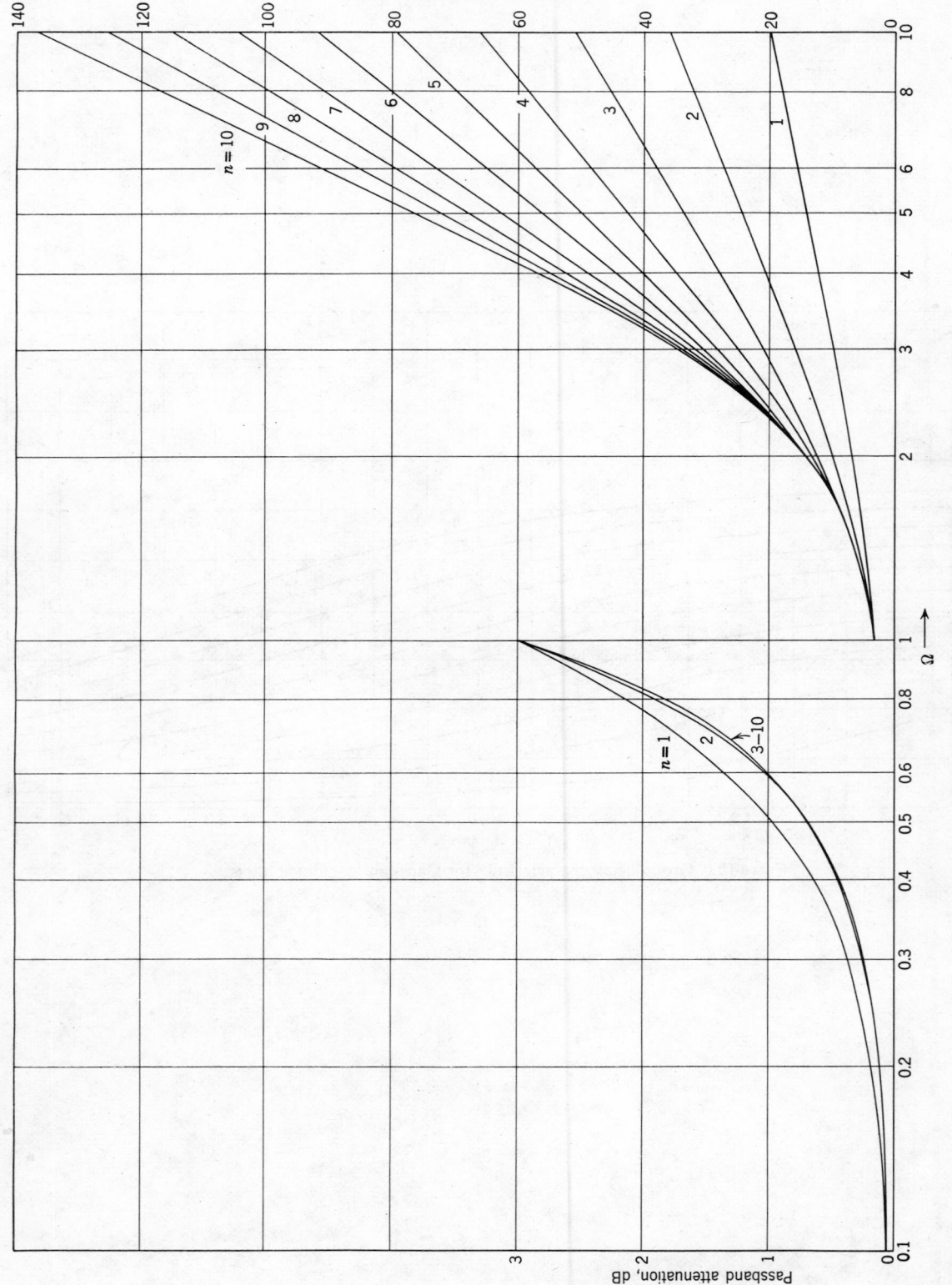

Curve 11. Attenuation characteristics for maximally flat delay (Bessel) filters.

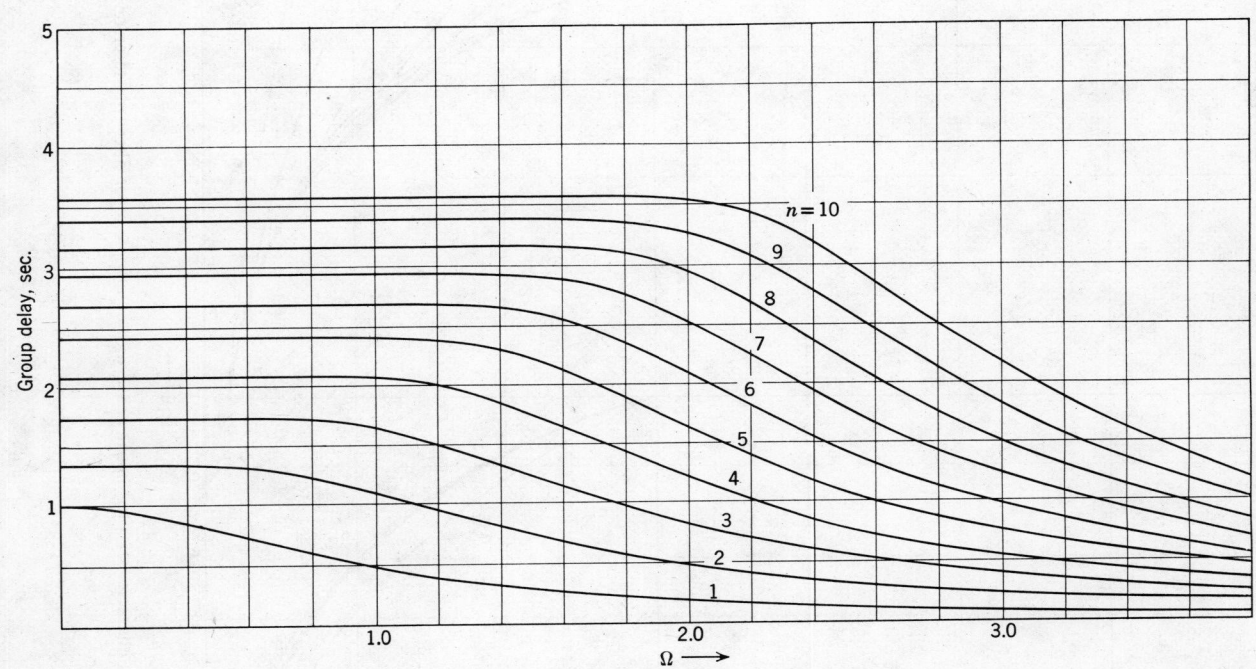

Curve 12. Group-delay characteristics for maximally flat delay (Bessel) filters.

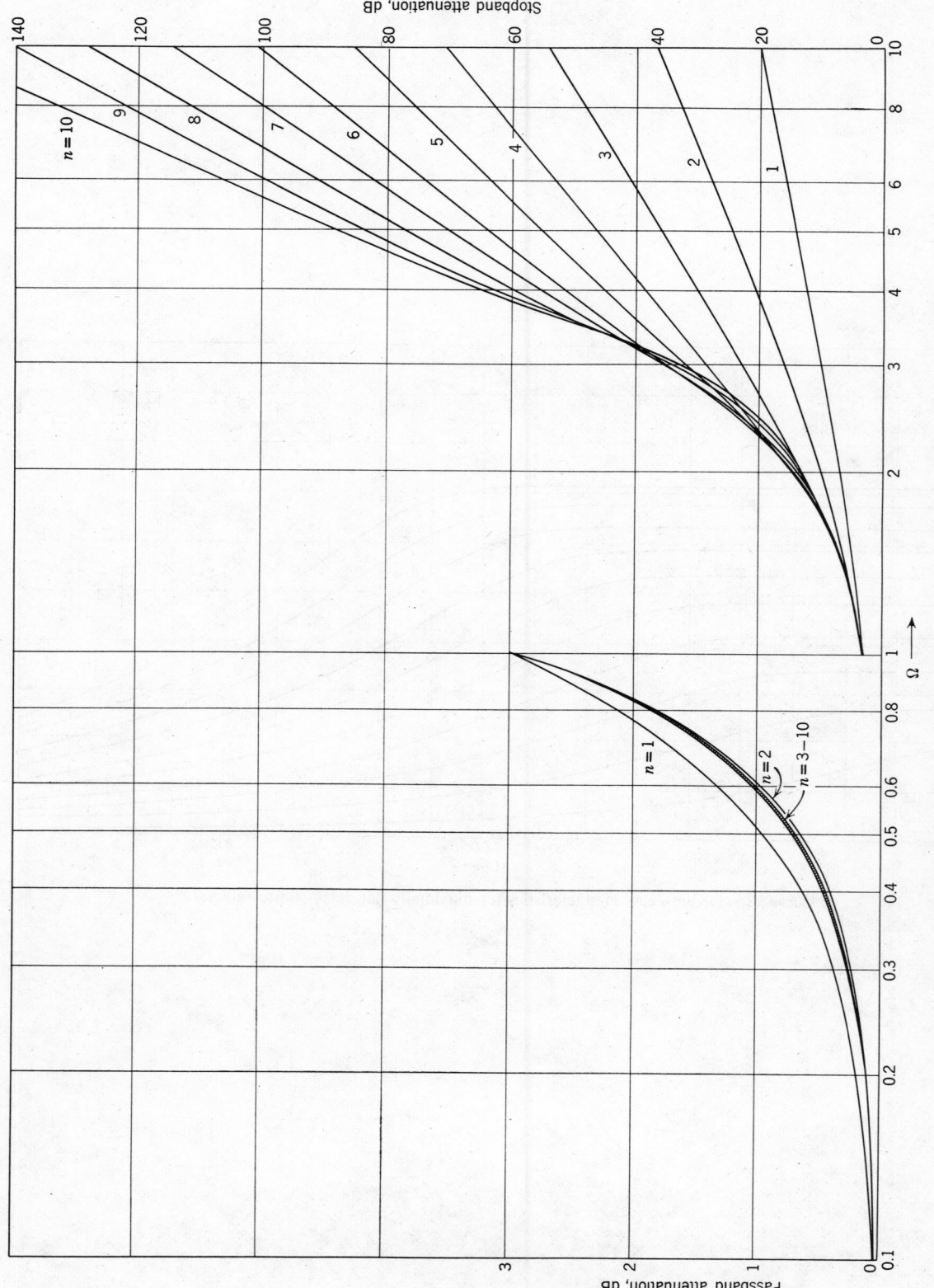

Curve 13. Attenuation characteristics for linear phase with equiripple error filter (phase error = 0.05°).

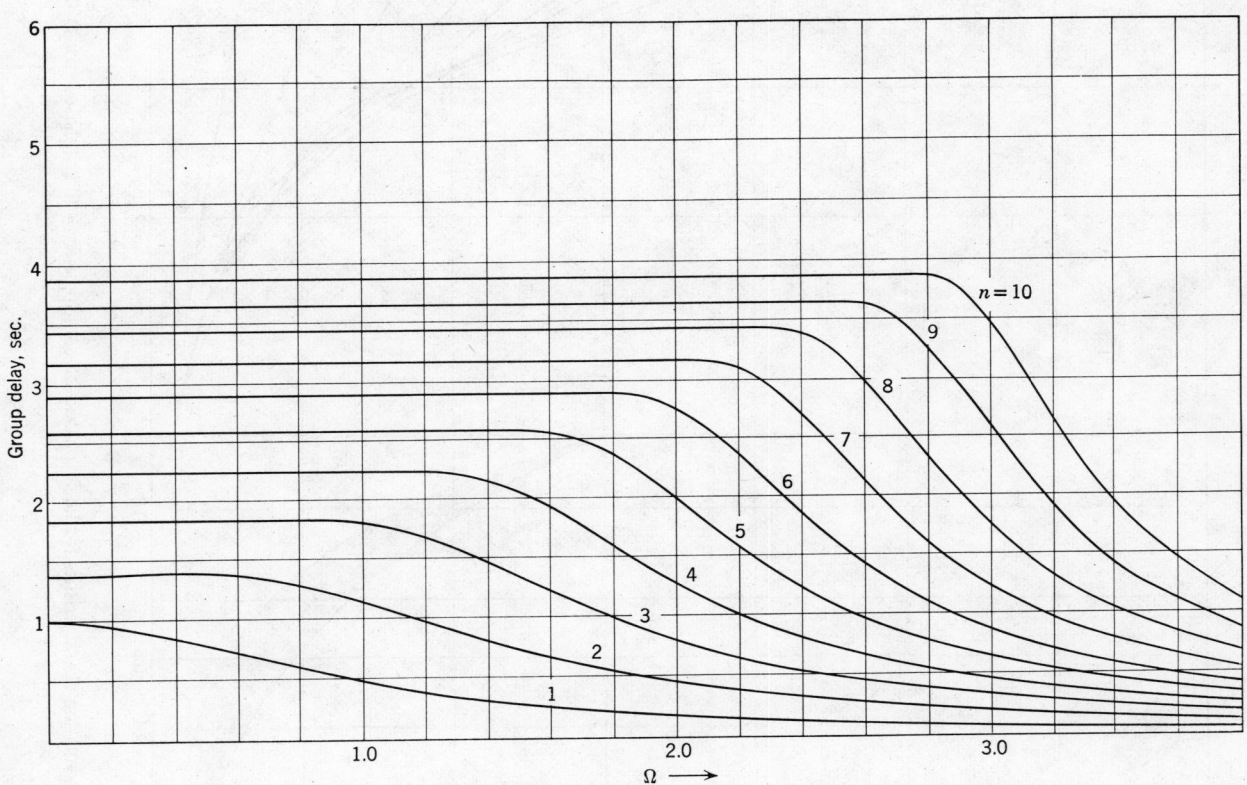

Curve 14. Group-delay characteristics for linear phase with equiripple error filter (phase error = 0.05°).

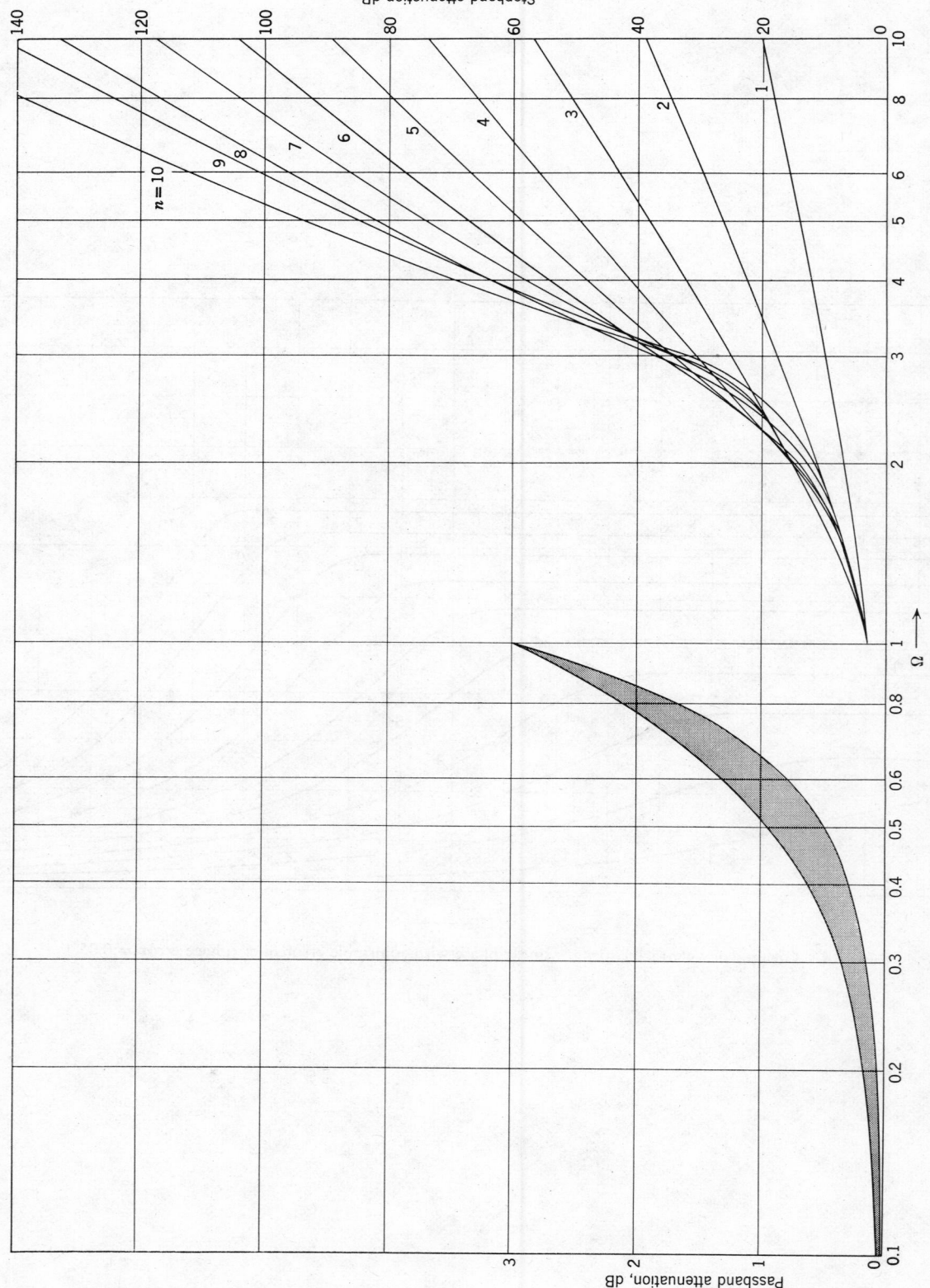

Curve 15. Attenuation characteristics for linear phase with equiripple error filter (phase error = 0.5°).

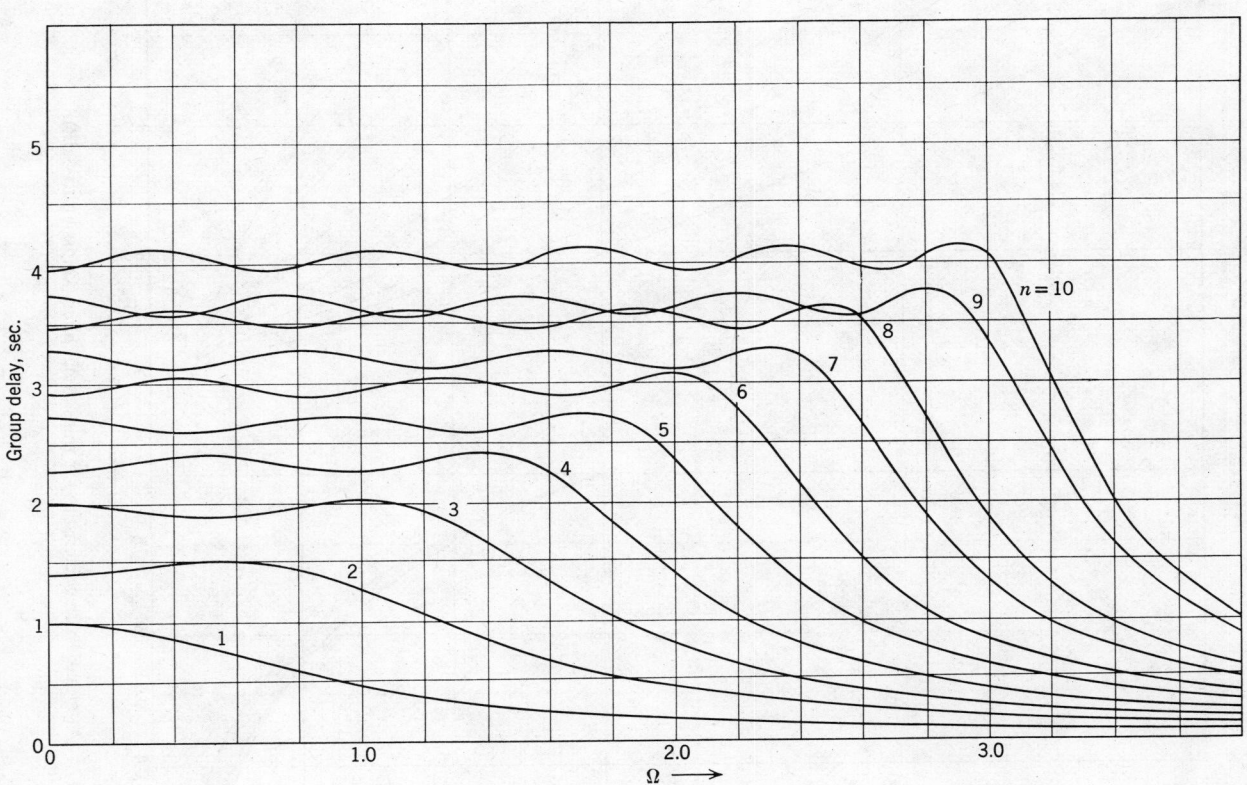

Curve 16. Group-delay characteristics for linear phase with equiripple error filter (phase error = 0.5°).

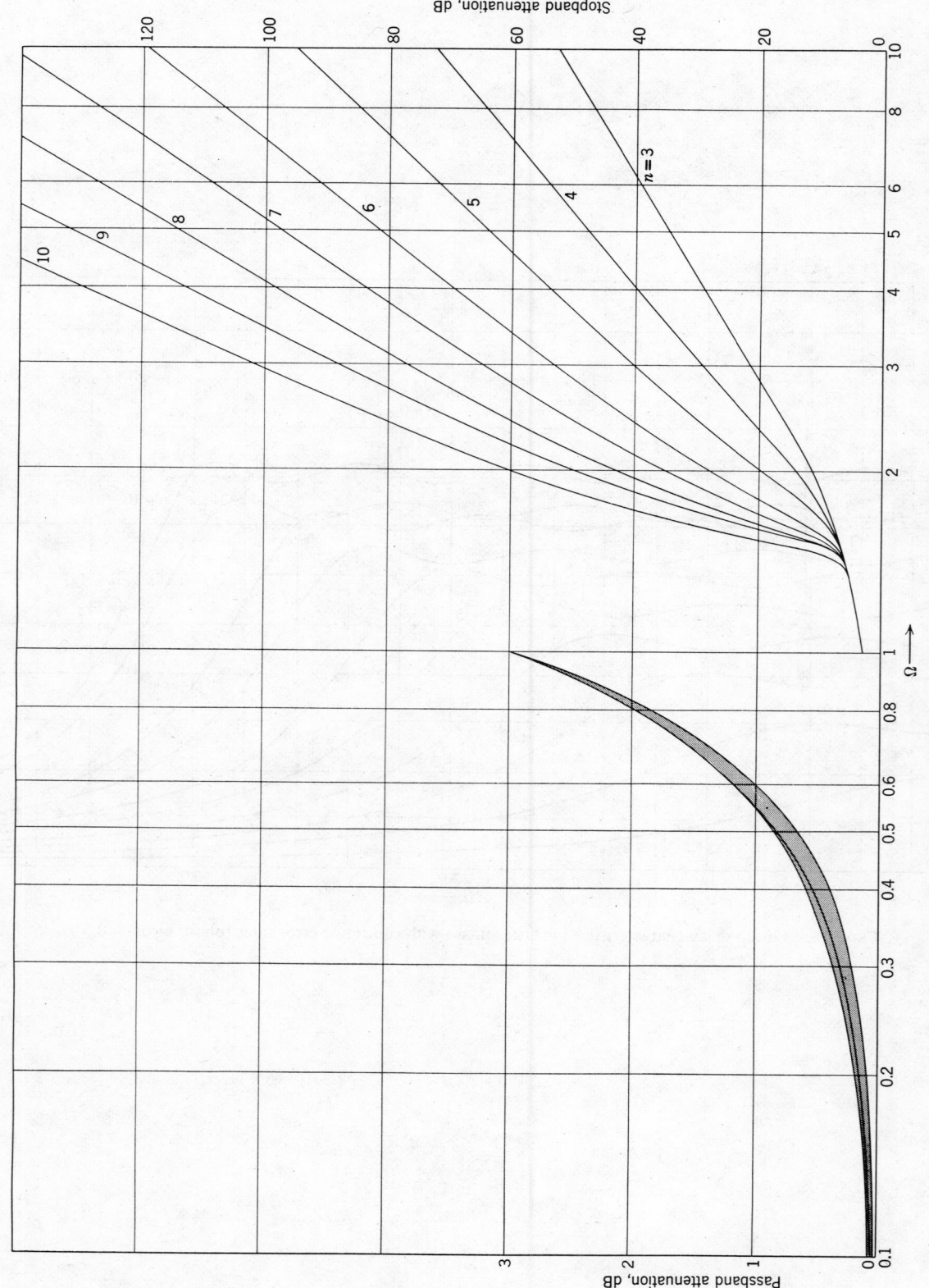

Curve 17. Attenuation characteristics for transitional filter (Gaussian to 6 dB).

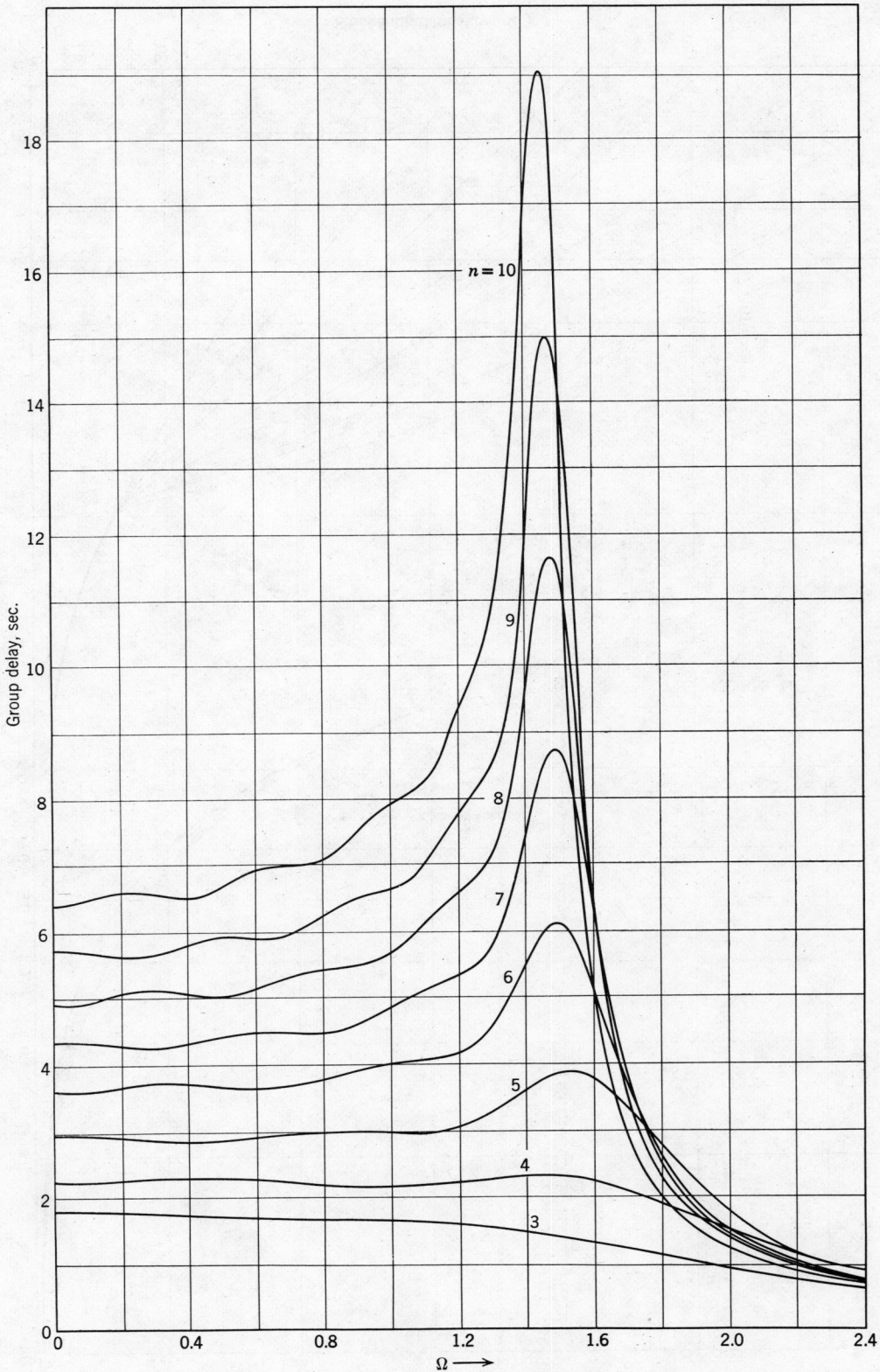

Curve 18. Group-delay characteristics for transitional filter (Gaussian to 6 dB).

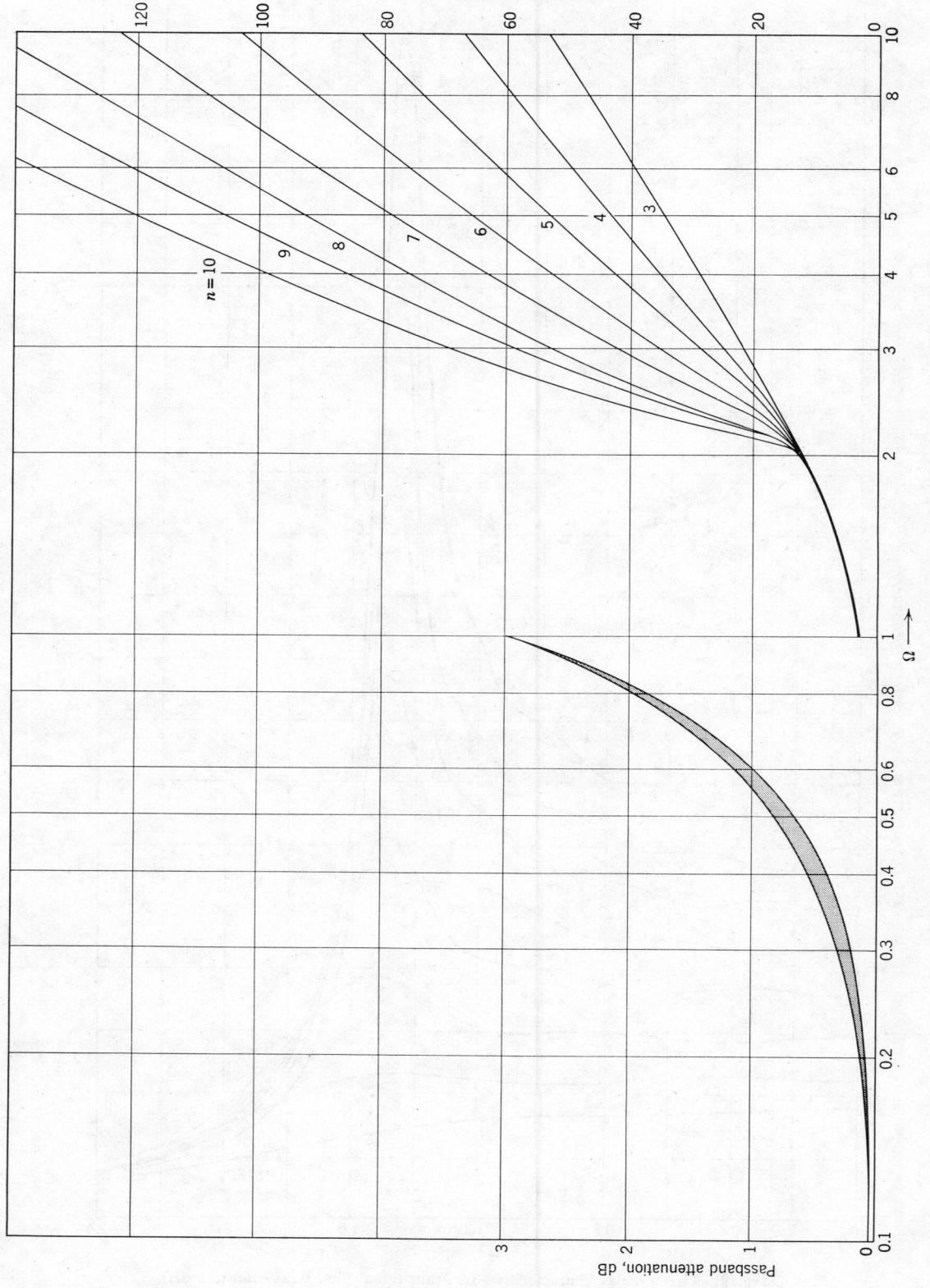

Curve 19. Attenuation characteristics for transitional filter (Gaussian to 12 dB).

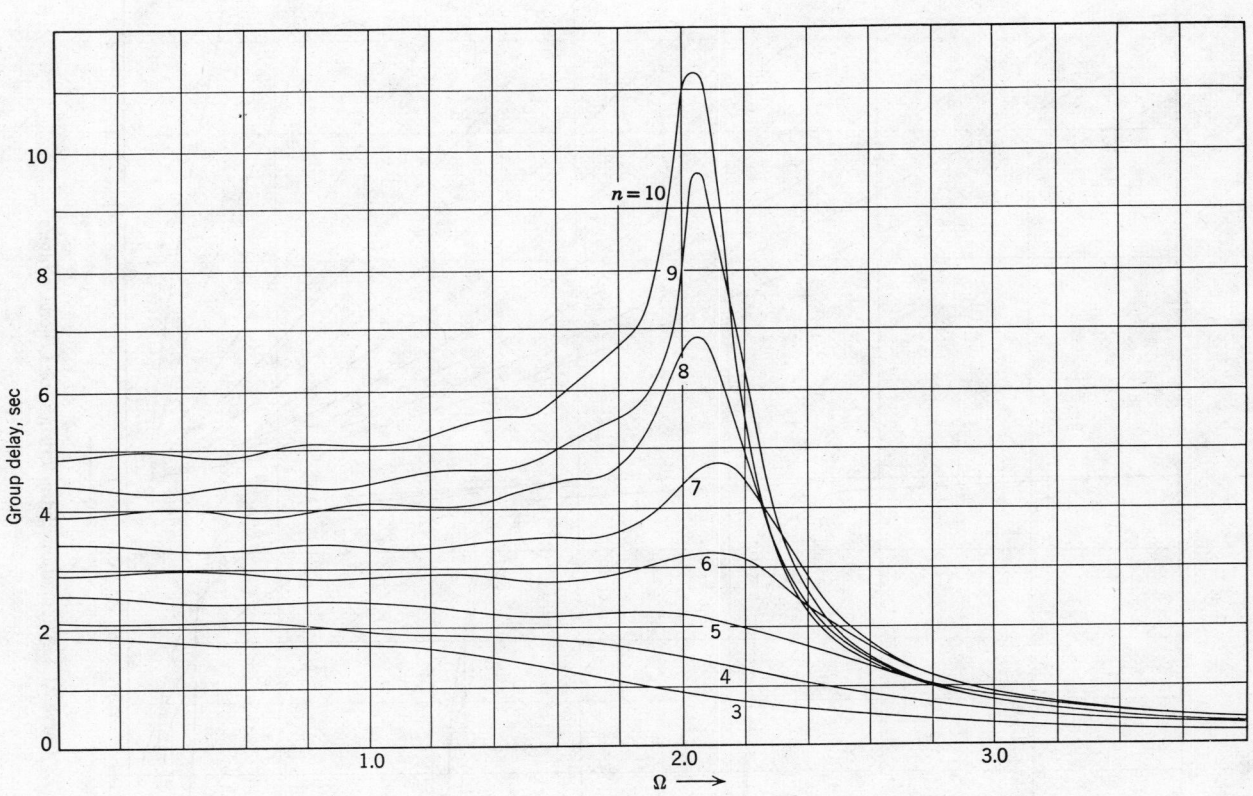

Curve 20. Group-delay characteristics for transitional filter (Gaussian to 12 dB).

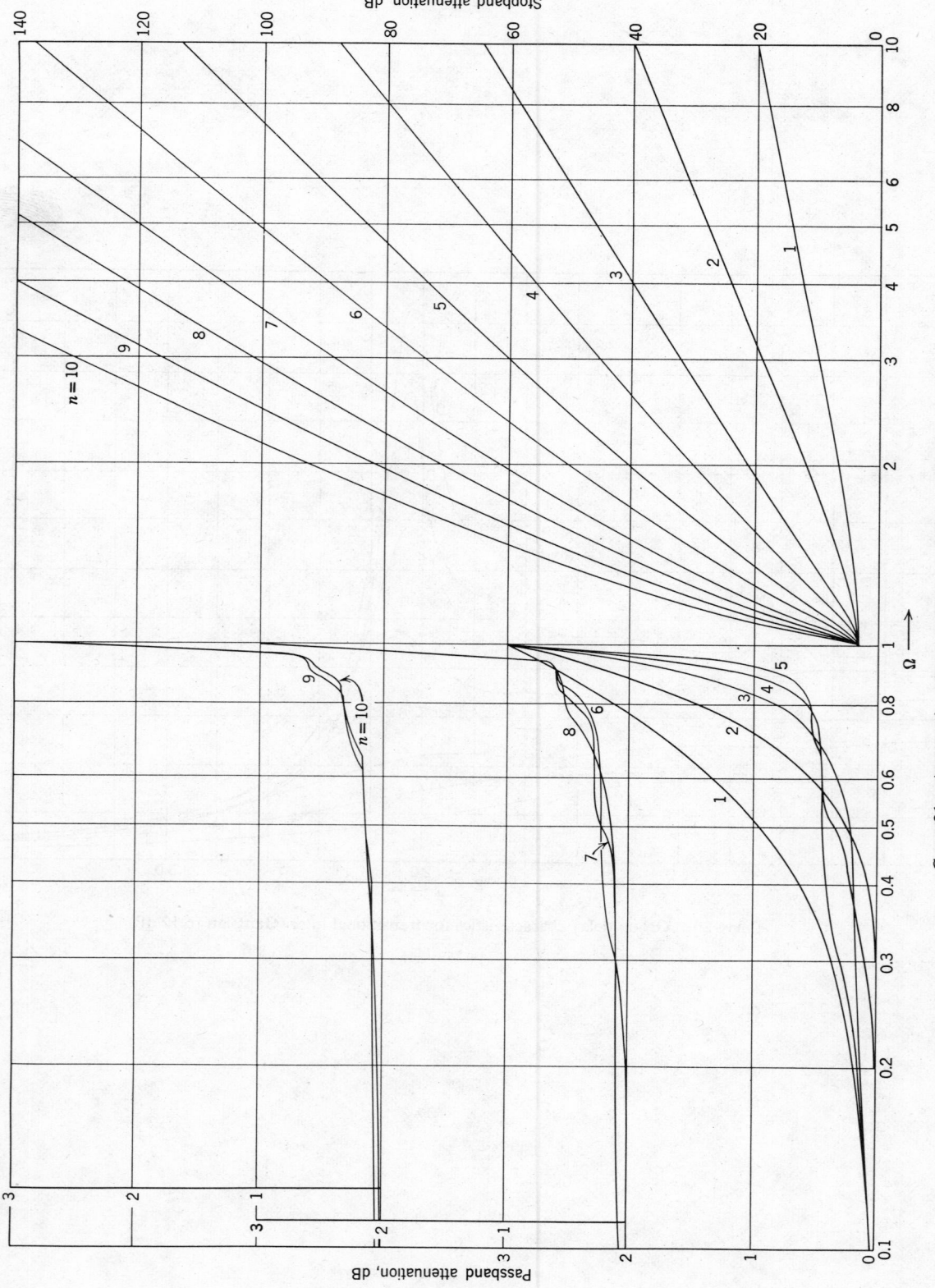

Curve 21. Attenuation characteristics for Legendre filters.

Stopband attenuation, dB

Passband attenuation, dB

$\Omega \longrightarrow$

$n = 10$

102

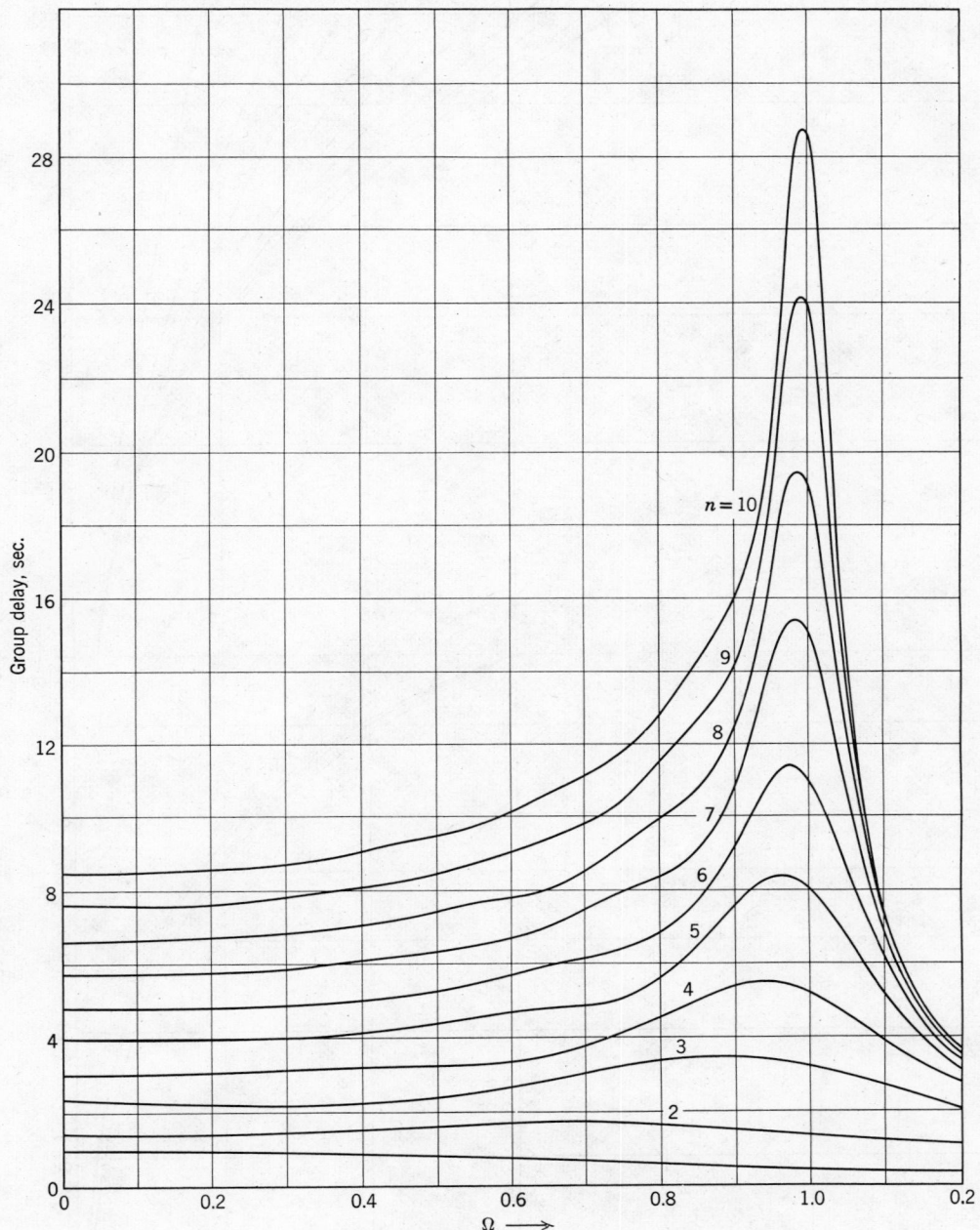

Curve 22. Group-delay characteristics for Legendre filters.

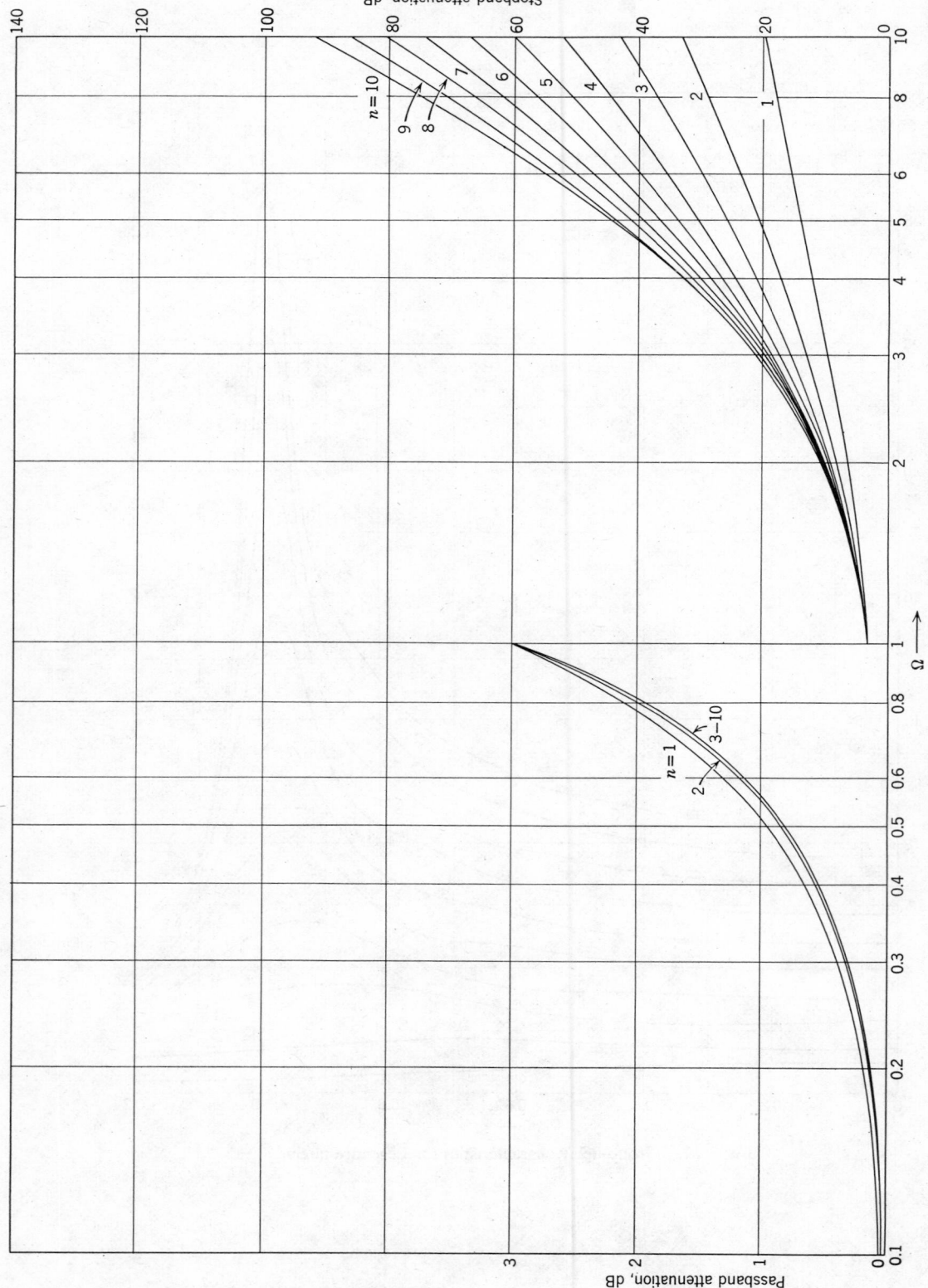

Curve 23. Attenuation characteristics for synchronously tuned filters.

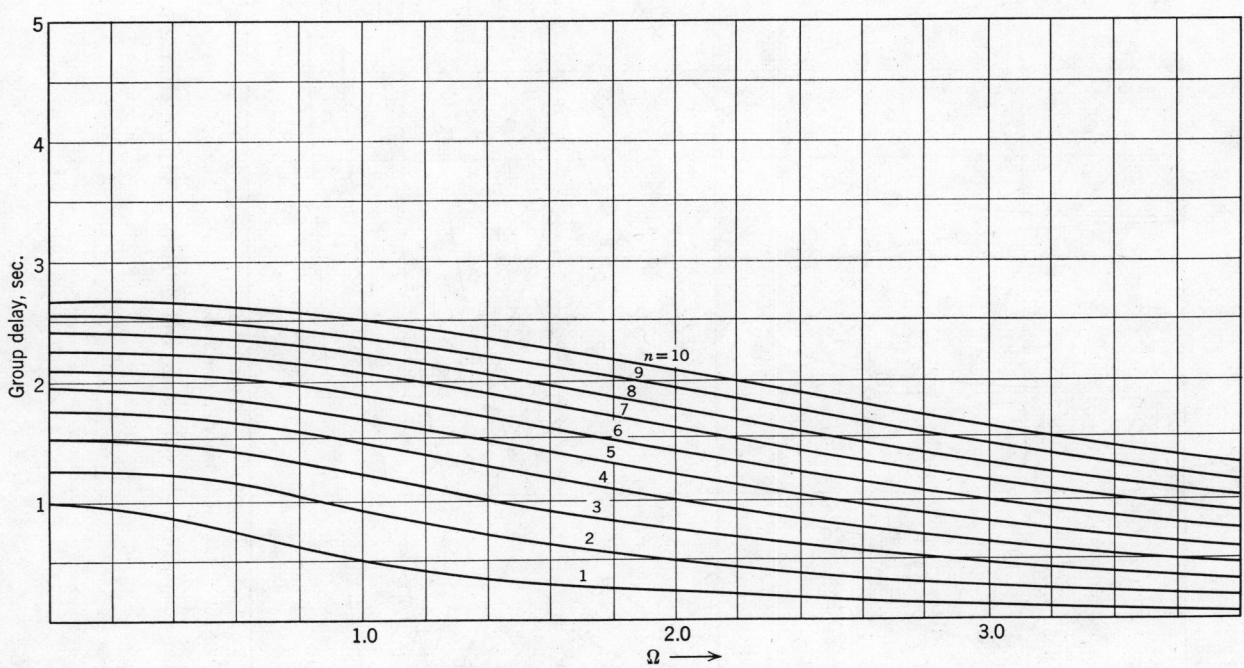

Curve 24. Group-delay characteristics for synchronously tuned filters.

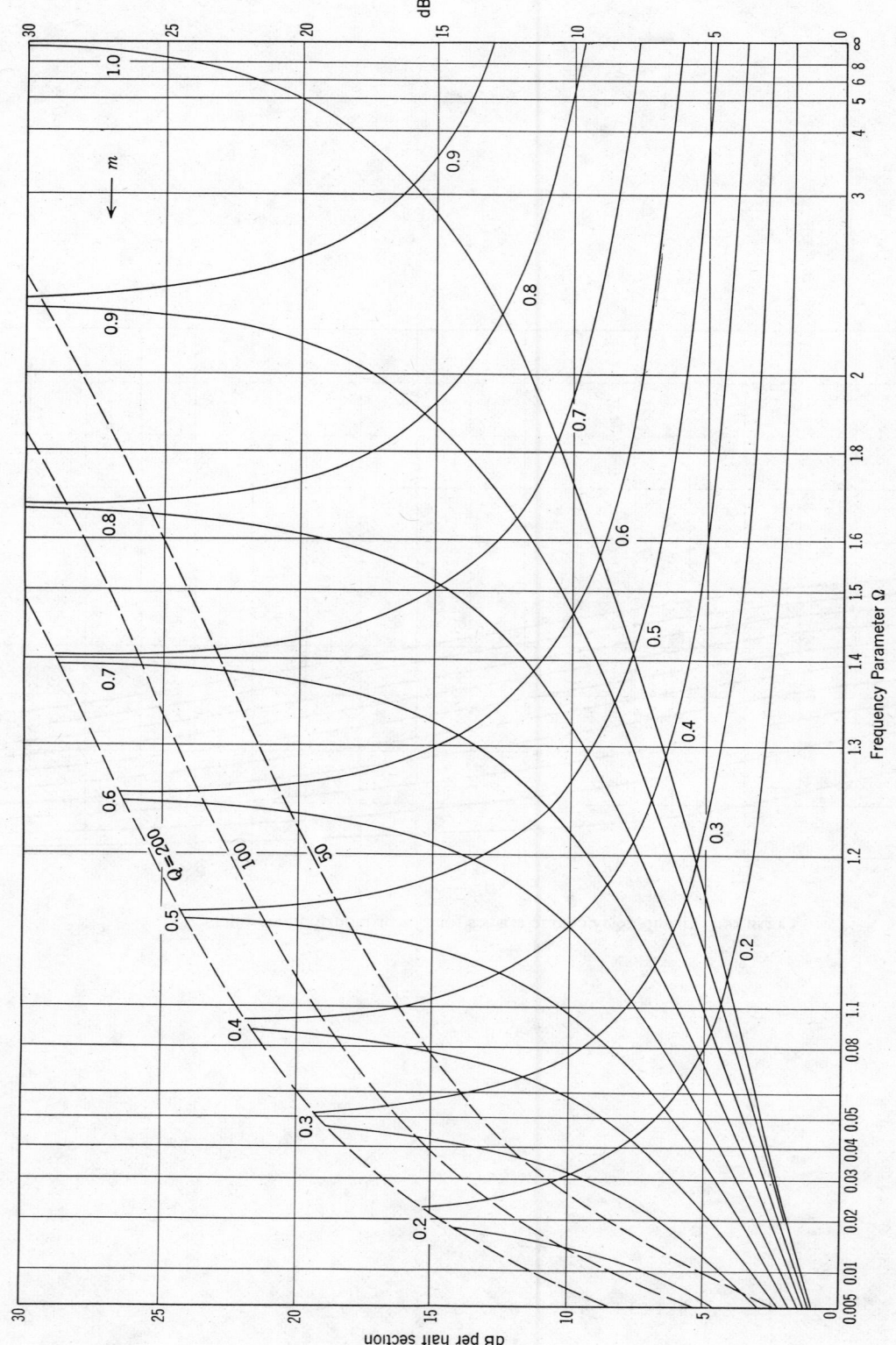

Curve 25. Image attenuation of *m*-derived lowpass and highpass image parameter filters.

4

Elliptic Functions and
Elements of Realization

In polynomial filters with equiripples in the passband and in which all the poles of attenuation lie at infinity the discrimination function D is a Chebyshev polynomial. The roots of the characteristic polynomial $E(s)$ are situated in the complex frequency plane on an ellipse (see Section 2.10) and expressed by hyperbolic functions. In the similar equiripple case in which the attenuation in the stopband does not go below certain specified limits the discrimination function D is a Zolotarev function. The roots of the characteristic polynomial are also on an ellipse and can be evaluated from zeros of the Zolotarev rational function and expressed by elliptic functions (see Section 3.5). Unlike general Chebyshev functions, the Zolotarev function is completely defined by its zeros and poles. The zeros of the function can be determined by means of tables that give the parameters of the Zolotarev function.

A great majority of filters are based on the transmission function of the lowpass prototype and the translation of lowpass properties approximated in a Chebyshev manner into any other type of response is the subject of the following discussion. Image-parameter filters can be synthesized in a nonelementary fashion when we use the Cauer method, based on the symmetrical lattice structure. The fact that the parametric relations in the lattice are independent permits the stopband property and passband property to be designed independently. Location of the control frequencies to obtain a Chebyshev approximation is the main problem of the nonelementary synthesis of image-parameter filters. The problem of synthesis of filters based on the direct use of effective parameters is begun by letting the required transmission function be, for example, a Zolotarev rational function.

Then, the desired filter is specified by any three of the four parameters (a_{max}, a_{min}, Ω_s/Ω_c, and n). Both image-parameter and effective-parameter filters use a similar theory and technique in the choice of design parameters. The optimum solution, in both cases, involves the Chebyshev approximation below and above cutoff, and the use of an elliptic function for conformal transformation of the s-plane into the u-plane where the Chebyshev approximation is taking place.

4.1 DOUBLE PERIODIC ELLIPTIC FUNCTIONS

Representation of the filtering function D and the complex frequency s, involves the Jacobi elliptic functions, which are doubly periodic functions of a complex variable u. The fundamental property of both the circular and the hyperbolic functions is that they are periodic; the former have a single real period π or 2π and the latter have a single imaginary period $j\pi$ or $j2\pi$. The elliptic function combines the property of circular and hyperbolic functions and has real and imaginary periods. The ratio of these two periods is an imaginary or complex number. In contrast to trigonometric functions, which have the property of being repeated in strips of the complex plane, the properties of doubly periodic functions are repeated in parallelograms, or, in especially interesting cases, in rectangles.

The most useful functions are the simplest ones; those that assume only one value in each periodic rectangle. Function theory shows that such a function does not exist. The simplest possible doubly periodic functions are those with two poles and two zeros in the periodic rectangle. The elliptic functions of Jacobi, $sn(u)$, $cn(u)$, and $dn(u)$ are functions of this sort in that they have, in each periodic rectangle, two simple poles with the necessary opposite and equal residues

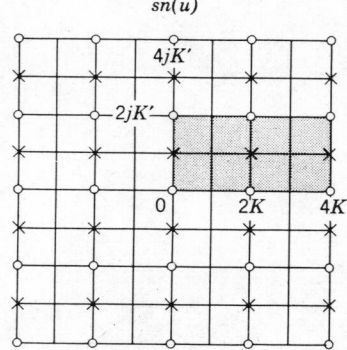 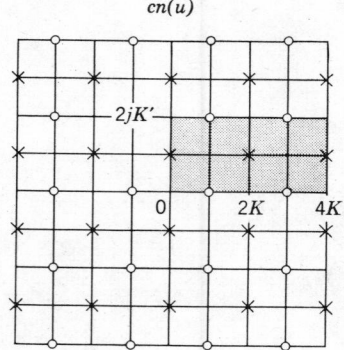 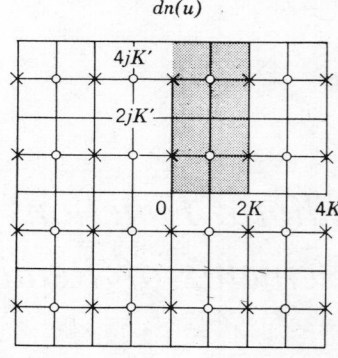

Fig. 4.1. Poles and zeros of different elliptic functions.

and two simple zeros. They may be represented as the ratio of two entire functions.

The application of Jacobi functions to the approximation problem originated with Cauer in 1931. He achieved an equiripple approximation to constant attenuation in the stopband of image-parameter filters and a constant passband image impedance. Darlington used them (1939) to obtain equiripple attenuation characteristics in both the passband and stopbands of an effective-parameter filter.

Figure 4.1 shows how the poles and zeros of functions $sn(u)$, $cn(u)$, and $dn(u)$ are distributed in the complex plane. (The explicit and picturesque three-dimensional representation of this function may be found in Jahnke and Emde.) The function $sn(u)$ has, in the direction of the real axis, a period equal to $4K$, and in the direction of the imaginary axis a period of $j2K'$. Table 4.1 summarizes the periods of the elliptic functions.

Both periods may be represented as functions of one parameter (k or k') by complete elliptic integrals of the first kind.

$$K = \int_0^{\pi/2} \frac{d\phi}{\sqrt{1 - k^2 \sin^2 \phi}} \qquad (4.1.1)$$

$$K' = \int_0^{\pi/2} \frac{d\phi}{\sqrt{1 - k'^2 \sin^2 \phi}} \qquad (4.1.2)$$

where ϕ is the amplitude of the elliptic integral and k' is the complimentary modulus.

$$k' = \sqrt{1 - k^2} \quad \text{or} \quad k^2 + k'^2 = 1 \quad (4.1.3)$$

$$1 > k > 0$$

$$k' = \sin \theta \qquad (4.1.4)$$

$$k = \cos \theta \qquad (4.1.5)$$

where θ is a modular angle as shown in Fig. 4.2. For

Table 4.1

Function	Real Period	Imaginary Period
$sn(u, k)$	$4K$	$j2K'$
$cn(u, k)$	$4K$	$2(K + jK')$
$dn(u, k)$	$2K$	$j4K'$

k'^2 much less than one, or when θ (modular angle) is larger than $89.95°$,

$$K \approx \ln \frac{4}{k'} + \frac{1}{4}\left(\ln \frac{4}{k'} - 1\right)k'^2 \qquad (4.1.6)$$

and

$$K' \approx \frac{\pi}{2}\left(1 + \frac{k'^2}{4} + \cdots\right) \qquad (4.1.7)$$

For the elliptic functions $sn(u; k)$ and $cn(u; k)$, one of the periods is equal to the four-fold elliptic integral K of modulus k. The period of $dn(u; k)$ is only one half of that and equal to $2K$—that is, $dn(u + 2vK; k) = dn(u; k)$.

The values of the elliptic function of particular interest and difficulty are those where the modulus k approaches unity (for example, $k = 0.9999995$). The same approximations are valid for the case when k^2 is much less than one ($k^2 \ll 1$), with primed values

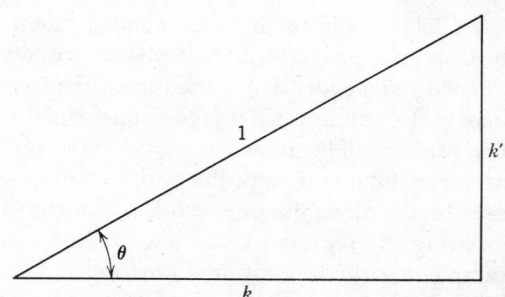

Fig. 4.2. Relationship between modulus k and k'.

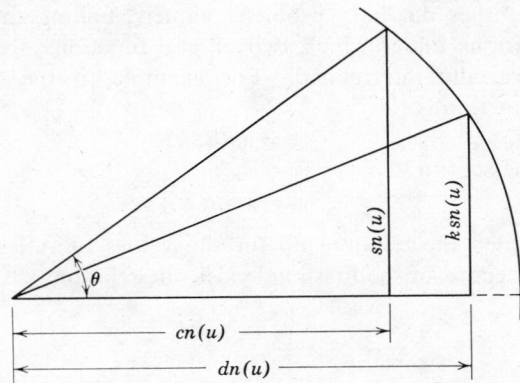

Fig. 4.3. Relationship between elliptic functions.

exchanged with nonprimed values. The relations between elliptic functions are analogous to those between trigonometric functions, as shown in Fig. 4.3.

$$sn^2(u) + cn^2(u) = 1 \qquad (4.1.8)$$

$$dn^2(u) + k^2 sn^2(u) = 1 \qquad (4.1.9)$$

Since elliptic functions are functions of two variables, the notation $sn(u; k)$ is sometimes used instead of $sn(u)$ to distinguish functions with different k. Figure 4.4 shows the behavior of all the above functions. If the argument u is real and the argument k remains between $0 < k < 1$, the function $sn(u; k)$ and $cn(u; k)$ will oscillate between -1 and $+1$. The value of the elliptic functions for any complex variable can be expressed by the value of the same functions for real variables.

If one of the periods becomes infinitely large, the elliptic function degenerates into trigonometric or hyperbolic functions and the periodic rectangle becomes a strip. For example, if k' approaches 0 and

k approaches 1, then $K' = 2\pi$; $K = \infty$ and

$$\lim_{k \to 1} sn(u) = \tanh(u)$$

$$\lim_{k \to 1} cn(u) = \frac{1}{\cosh(u)} \qquad (4.1.10)$$

$$\lim_{k \to 1} dn(u) = \frac{1}{\cosh(u)}$$

If $k \to 0$ and $k' \to 1$, then $K \to \pi/2$ and $K' \to \infty$. All zeros which are on the real axis move to ∞, and the following relations are true:

$$\lim_{k \to 0} sn(u) = \sin(u)$$

$$\lim_{k \to 0} cn(u) = \cos(u) \qquad (4.1.11)$$

$$\lim_{k \to 0} dn(u) = 1$$

4.2 MAPPING OF *s*-PLANE INTO *u*-PLANE

In passing in the rectangular contour from $u = jK'$ counterclockwise, the function $s = sn(u; k)$ increases from $s = -\infty$ to $s = +\infty$. The above function, indeed, maps the rectangle $ABCD$ of Fig. 4.5 in the u-plane in the upper half of the s-plane.

The characteristic points of the rectangle corresponds to the values of function s shown in Table 4.2.

In the case where $u = jw$ is variable, the function

$$s = sn(jw; K) = j\frac{sn(w; K')}{cn(w; K')}$$

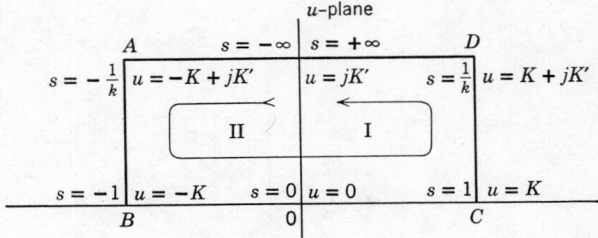

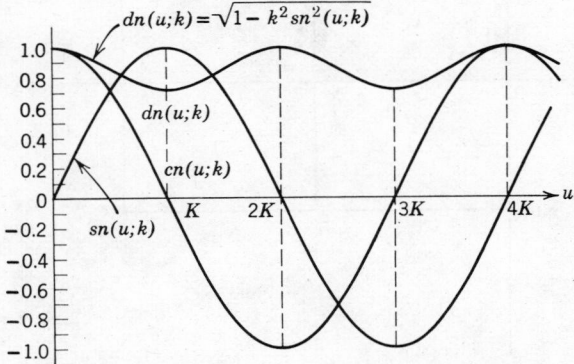

Fig. 4.4. $sn(u; k)$, $cn(u; k)$, and $dn(u; k)$ for the case when $k = 0.5$ ($K = K' = 1.8541$).

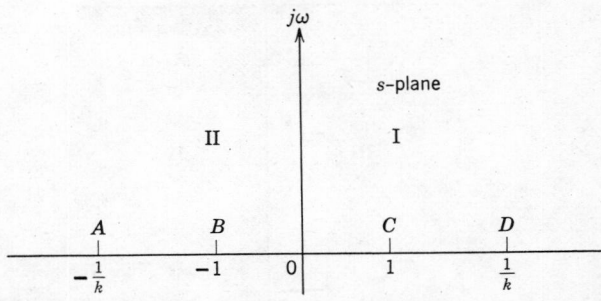

Fig. 4.5. Conformal transformation of s-plane into rectangle in u-plane.

Table 4.2

s	u	s	u	s	u	s	u
$-\infty$	jK'	-1	$-K$	$\frac{1}{k}$	$K + jK'$	0	0
$-\frac{1}{k}$	$-K + jK'$	$+1$	K	∞	jK'		

is imaginary, and the line between 0 and jK' of the periodic rectangle corresponds to the positive imaginary half-axis of the s-plane. We can write also:

$$dn(k) = k' \qquad dn(K + jK') = 0$$

$$cn(K + jK') = \frac{-jk'}{k} \qquad cn(jK') = \infty \qquad (4.2.1)$$

$$dn(jK') = \infty$$

This leads one to observe that the cn and dn functions map the fourth quadrant of the s-plane into a rectangle in the u-plane with vertices at 0, K, $K + jK'$, and jK (see Figs. 4.6 and 4.7).

4.3 FIRST BASIC TRANSFORMATION OF ELLIPTIC FUNCTIONS

The solution of approximation problems with the aid of elliptic functions can be based on the application of the first basic transformation of the nth power.

This transformation is, in reality, a solution for another auxiliary problem, namely, finding the conditions under which two elliptic functions are algebraically interrelated. For example, if the elliptic function is

$$s = sn(u; k)$$

the second is

$$x = sn(u; k_1)$$

When the relationship for the values of full elliptic integrals of the first kind exist, the relation

$$\frac{K'}{K} = n \frac{K_1'}{K_1}$$

holds true where n is a number.

To obtain the desired Chebyshev behavior of the discrimination D, D and s will be represented as elliptic functions of the variable z (normalized to $1/4$ of the period $-K$) instead of the variable u. In other words,

$$u = zK = (x + jy)K \qquad (4.3.1)$$

Figure 4.8a shows the s-plane where the thick line indicates the transition frequencies, Figure 4.8b shows the folding of the s-plane, transforming the s-plane into a z-plane, and Figure 4.8c shows the transformed s-plane with all values related to the s-plane inside of the rectangle.

The following function mathematically describes the foregoing procedure:

$$s = 0 + j\Omega = j \frac{dn(zK, k)}{dn(x_1K, k)} \qquad (4.3.2)$$

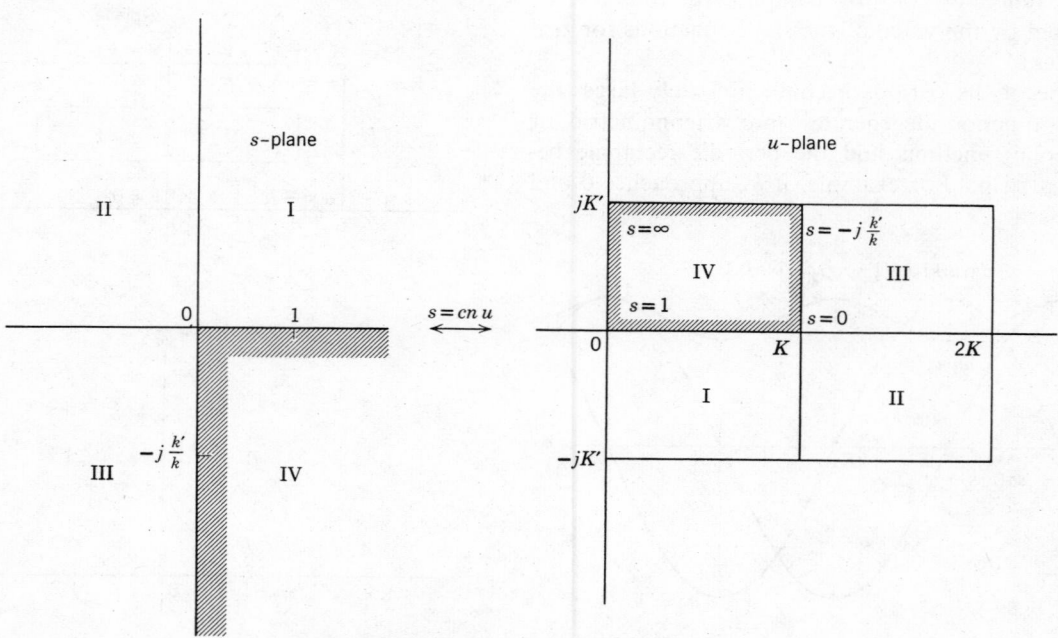

Fig. 4.6. Mapping of s-plane into u-plane by $s = cn\, u$.

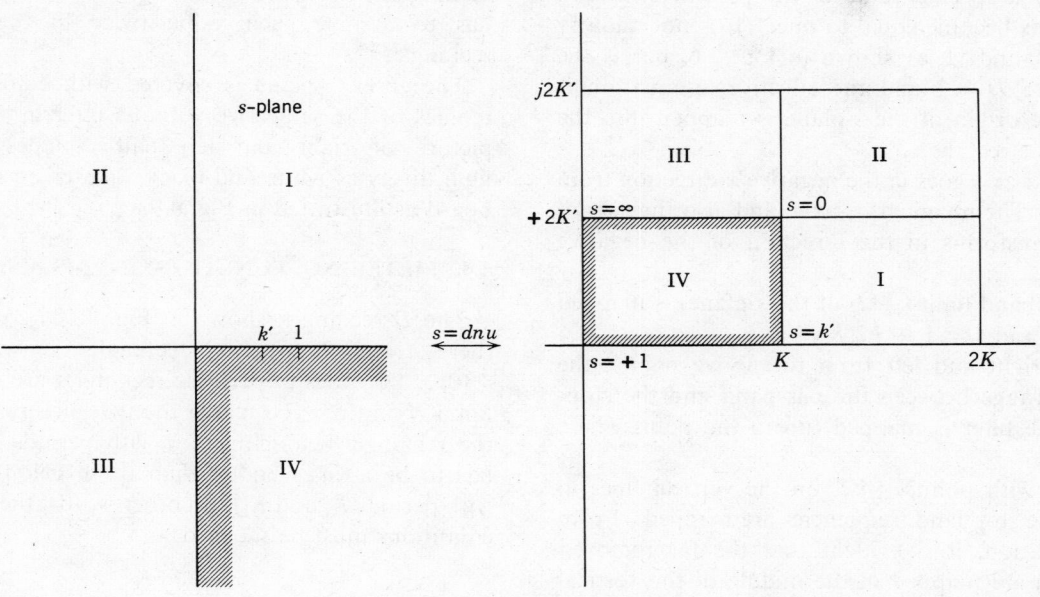

Fig. 4.7. Mapping of s-plane into u-plane by $s = dn\ u$.

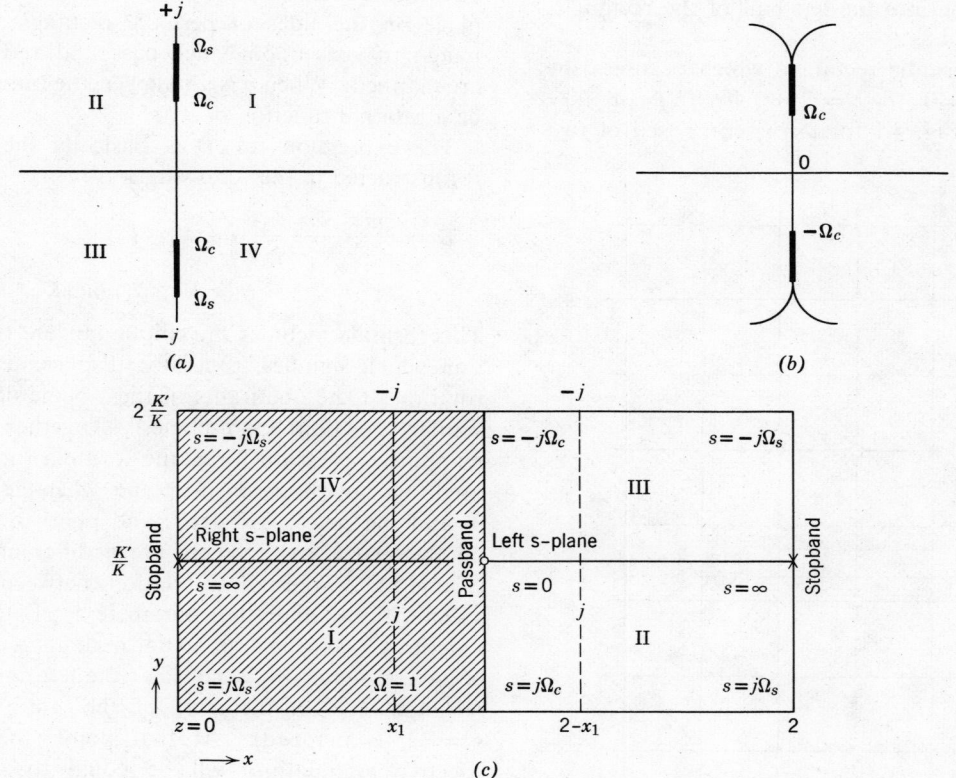

Fig. 4.8. Consecutive steps in transformation of s-plane into u-plane.

where x_1 is a specific value of the parameter z with which Ω has become equal to one. (It is not midway between Ω_s and Ω_c as shown in Fig. 3.6, but is the point where $D = 1$ and the effective attenuation is 3 dB). The origin of the s-plane is mapped into the center of the rectangle.

The real s-axis goes in the negative x-direction from the origin. The imaginary s-axis (and also the axis of real Ω) transforms in the direction of the negative y-axis.

The passband limit ($\pm\Omega_c$ of the s-plane) is mapped into $z = 1$ and $z = 1 + j\,2(K'/K)$.

On the right and left from the above points, the transition range between the passband and the stopband of the filter is mapped (up to the point where $s = \pm j\Omega_s$).

Starting with points $\pm j\Omega_s$ on the vertical lines in Fig. 4.8, the stopband frequencies are mapped. From this illustration, it is evident that the far removed frequencies are mapped in the middle of the vertical lines and that this is another place where the real and imaginary axis meet together.

The mapping function of (4.3.2) translates the right half of the s-plane into the left half of the rectangle, as shown in Fig. 4.8.

A complete periodic rectangle which occupies the region from $z = 0$ to $z = 2 + j\,4(K'/K)$, or $u = 2K + j\,4K'$ (see Fig. 4.1 for dn) is composed of two

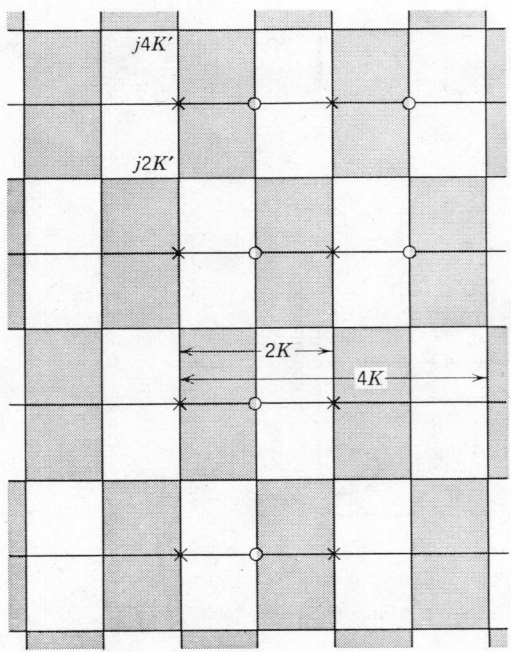

Fig. 4.9. Location of periodic rectangle in u-plane.

rectangles shown in the figure. Every elliptic function has to assume each value twice in the periodic rectangle.

The entire z-plane is covered with conjugate rectangles of the same size without interruptions. The picture of right and left half s-planes alternate figuratively as white and black squares on a checkerboard as illustrated in Fig. 4.9.

4.4   FILTERING FUNCTION IN z-PLANE

The D rectangle, shown in Fig. 4.10a consists of n (here $n = 5$) half-period rectangles shown in Fig. 4.10b. The sides of periodic rectangles are dependent upon K and K'. To match the two pictures together, the rectangle belonging to s with periods K and K' has to be n-times higher than those belonging to D with periods K_1 and K_1'. In other words, the following conditions must be satisfied:

$$n\,\frac{K_1'}{K_1} = \frac{K'}{K} \qquad (4.4.1)$$

When $j\Omega$ runs along its axis one time, the property of the function D will be repeated n times. This takes place on the sides of periodic rectangle parallel to imaginary axis upon which passband and stopbands are mapped. When n is an integer, the function D will be a rational function of s.

The expression for D is basically the same but reconstructed in the following way:

$$D = \frac{sn(x_1K_1;\ k_1)}{cn(x_1K_1;\ k_1)}\,dn(zK_1;\ k_1) =$$
$$tn(x_1K_1;\ k_1)dn(xK_1;\ k_1) \quad (4.4.2)$$

This formula includes the elliptic tangent transformation which signifies, generally, displacement without rotation. (The quadrants of the s-plane occupies the similar position in u-plane.) Together with the mapping expression for s, the formula for D can be used to locate the zeros in z-plane (when $dn(zK_1;\ k_1) = 0$) and the poles. All zeros and poles in z-plane lie on the real Ω-axis. There are n different values of Ω_0 and Ω_∞. To establish relations between the filter's specifications and the parameters of the elliptic functions, two special points of periodic rectangle can be investigated. First, consider the left corner of the rectangle ($z = 0$, the start of the stopband where $s = j\Omega_s$ is mapped). At that point, the value of effective attenuation will be equal to a_{min}. The corresponding value of the discrimination D_{min} is reached for the first time. Because $dn(0) = 1$,

$$s = [dn(x_1K;\ k)]^{-1} \qquad (4.4.3)$$

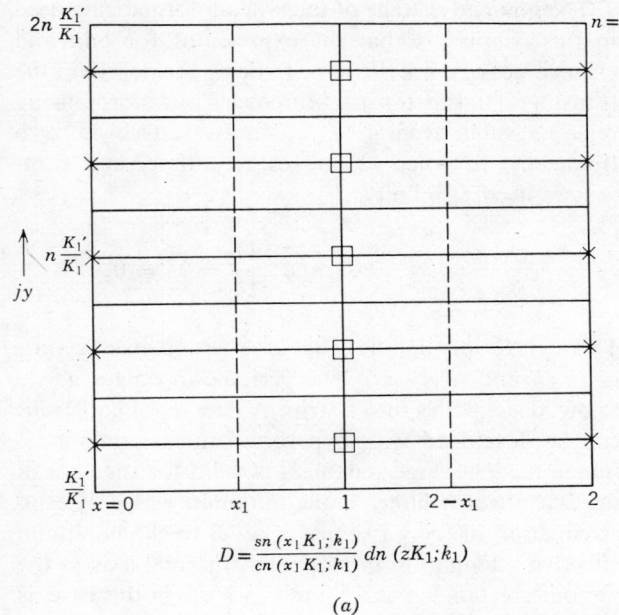

$$D = \frac{sn\,(x_1 K_1; k_1)}{cn\,(x_1 K_1; k_1)}\,dn\,(zK_1; k_1)$$

(a)

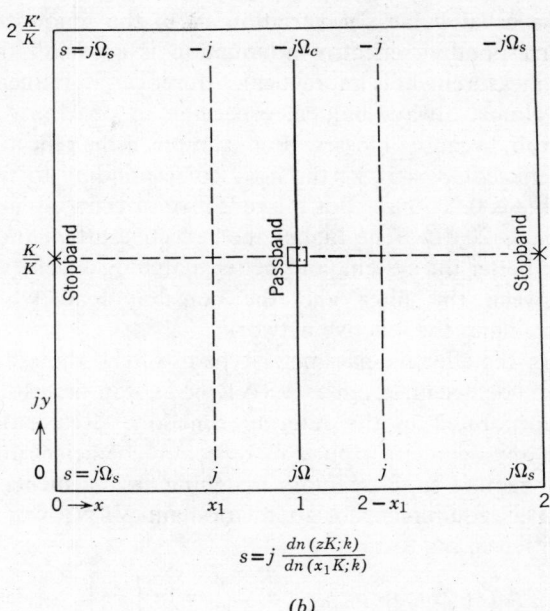

$$s = j\,\frac{dn\,(zK; k)}{dn\,(x_1 K; k)}$$

(b)

Fig. 4.10. Mapping of fifth-order filter in *u*-plane.

and according to (4.4.2)

$$D_{\min} = \frac{sn(x_1 K_1;\ k_1)}{cn(x_1 K_1;\ k_1)} = tn(x_1 K_1;\ k_1) \quad (4.4.4)$$

or

$$e^{a_{\min}} = \sqrt{1 + D_{\min}^2} = [cn(x_1 K_1;\ k_1)]^{-1} \quad (4.4.5)$$

Second, consider the middle of the lower side, when $z = 1$, $s = j\Omega_c$, and the passband starts. At that point, the effective attenuation is equal to $a_{\max}$. The corresponding value $D_{\max}$ is reached for the first time. Here $dn(K; k) = k_1{'}$; therefore,

$$\Omega_c = \frac{k'}{dn(x_1 K;\ k)} \quad (4.4.6)$$

$$D_{\max} = \frac{sn(x_1 K_1;\ k_1)k_1{'}}{cn(x_1 K_1;\ k_1)} \quad (4.4.7)$$

The ratios of two limiting frequencies describe the sharpness of the filter and are equal to

$$\frac{\Omega_s}{\Omega_c} = \frac{1}{k'} \quad (4.4.8)$$

where k' is a basic selectivity parameter for the filter. The corresponding insertion-power ratio or effective attenuation for the filter is

$$a_{\max} = \ln\left(\sqrt{1 + D_{\max}^2}\right)$$

$$= \ln\left[\frac{dn(x_1 K_1;\ k_1)}{cn(x_1 K_1;\ k_1)}\right] \quad \text{Np} \quad (4.4.9)$$

Therefore the difference between the specified attenuation in the passband and the required $a_{\min}$ in the stopband can be represented by

$$e^{-\Delta a} = dn(x_1 K_1;\ k_1) \quad (4.4.10)$$

and

$$e^{-\Delta a_e} = dn[(1 - x_1)K_1;\ k_1] \quad (4.4.11)$$

It is relatively easy to show that the sum of the two differences in the transition band

$$a_{\min} - a_{\max} = \Delta_a$$

and

$$a_{e\ \min} - a_{e\ \max} = \Delta a_e$$

can be represented by

$$\Delta a + \Delta a_e = \ln\left(\frac{1}{k_1{'}}\right) \quad (4.4.12)$$

since

$$dn(x_1 K_1;\ k_1)\,dn[(1 - x_1)K_1;\ k_1] = k_1{'} \quad (4.4.13)$$

This value is a discrimination parameter given by the filter's specification in one form or another, and for all practical cases, is much less than 1 ($k_1{'} \ll 1$).

Instead of specifying the maximum effective attenuation or value of the ripples in the passband, the value of $a_{e\ \min}$ echo (reflective) attenuation, the reflection coefficient ρ, or the VSWR can be specified.

The relationship between effective attenuation and reflection coefficient in reactive networks is

$$a_{\max} = -\ln\sqrt{1 - \rho^2} \quad (4.4.14)$$

For usual values of variation in ρ, the amount of corresponding effective attenuation is so small that its measurement is impractical. Moreover, its influence is almost always hidden, especially in proximity of cutoff, because of losses. For example, if the reflection coefficient $\rho = 10\%$, the a_{max} corresponding to that value is 0.005 Np. But the reflective or echo attenuation is 20 dB. The higher the reflective attenuation, the better the system, and better matching is achieved between the filter and the constant load which terminates the reactive network.

In the effective-parameter type of filters, the reflection coefficient, given as VSWR or as ρ in percent, is incorporated in the filtering function. Realization for any degree of ripples of reflection coefficient does not require supplementary matching arrangements.

The requirement for attenuation and VSWR can be written in the form

$$\Delta a + \Delta a_e \approx a_{min} + a_{e\,min} = \ln\left(\frac{1}{k_1'}\right) \quad (4.4.15)$$

The difference of the above quantities is

$$\Delta a - \Delta a_e = \ln\frac{dn[(1-x_g)K_1;\ k]}{dn(x_gK_1;\ k_1)} \quad (4.4.16)$$

As an intermediate value for the evaluation of limiting frequencies for optimum design, the product of Ω_s and Ω_c can be found:

$$\Omega_s \cdot \Omega_c = \frac{dn[(1-x_1)K;\ k]}{dn(x_1K;\ k)} \quad (4.4.17)$$

The passband echo attenuation $a_{e\,min}$ in the case of lossy filters may formally be evaluated from the equation

$$e^{2a} + a^{2a_e} = 1$$

The filtering function D, at the 3-dB point where $\Omega = 1$ is

$$D_1 = \frac{sn(x_1K_1;\ k_1)}{cn(x_1K_1;\ k_1)}\,dn(x_1K_1;\ k_1),$$

or

$$D_1 = D_{3\,dB} = \left(\frac{1-e^{-2\Delta a}}{1-e^{-2\Delta a_e}}\right)^{1/2} \quad (4.4.18)$$

and finally

$$D_1 = \left(\frac{1-e^{-2\Delta a}}{1-e^{-2\Delta a_e}}\right)^{1/2} = \left(\frac{1-e^{-2a_{min}}}{1-e^{-2a_{e\,min}}}\right)^{1/2} \quad (4.4.19)$$

This expression shows that D_1 has the exact value 1 when $a_{min} = a_{e\,min}$; however the deviation of D_1 from this exact value in all other practical cases is negligible.

The main advantage of the type of normalizing used in this chapter is that the expressions for odd and even values of n are the same (n being the degree of the transmission function). Moreover, this normalizing remains valid even if Ω_c compresses itself at zero frequency, or when Ω_s moves to infinity and compresses itself at infinity:

$$\frac{D_{max}}{D_{min}} = k_1' = 0 \quad \text{and} \quad \frac{\Omega_c}{\Omega_s} = k' = 0.$$

This expression signifies that one period has become large (K and $K_1 \to \infty$). The periodic rectangle, as we know, degenerates into a strip. A new set of equations can be developed with hyperbolic and trigonometric functions. The same statement is valid for the case of the Butterworth filter. If the minimum value of echo attenuation in the passband is equal to the minimum effective attenuation in stopband, the real axis in the periodic rectangle has a point x_1 which, in this case, is equal to one-half. According to definition, it is a reference frequency for normalization where the filtering function $D = 1$ (3-dB point).

The product of the two limiting frequencies for this specific case is equal to one; also

$$\Omega_c \cdot \Omega_s = 1$$

and corresponds to the normalizing scheme by Cauer and others. In the tables in Chapter 5, the cutoff frequency is equal to one ($\Omega_c = 1$).

4.5 GRAPHICAL REPRESENTATION OF PARAMETERS

Both the sharpness parameter k' and the discrimination parameter k_1' are given in the specifications for the filter. Such a pair of values can be used which relate two ratios K'/K and K_1'/K_1 by an integer n.

One can find this n value from the formulas of first basic transformation of nth power. Special nomographs constructed by Kawakami (see Chapter 5) correlates the design parameters for different-order networks. The numerical calculation of poles and zeros (frequencies of maximum and minimum of the effective attenuation) using the Jacobi elliptic function is, in many cases, relatively difficult.

A curve of attenuation $\Delta A + \Delta A_e \approx A_{min} + A_{e\,min}$ is shown in Fig. 4.11 as a function of sharpness. The guaranteed attenuation in the stopband is A_{min}, and $A_{e\,min}$ is the value corresponding to the maximum admissible reflection factor ρ in the passband. The

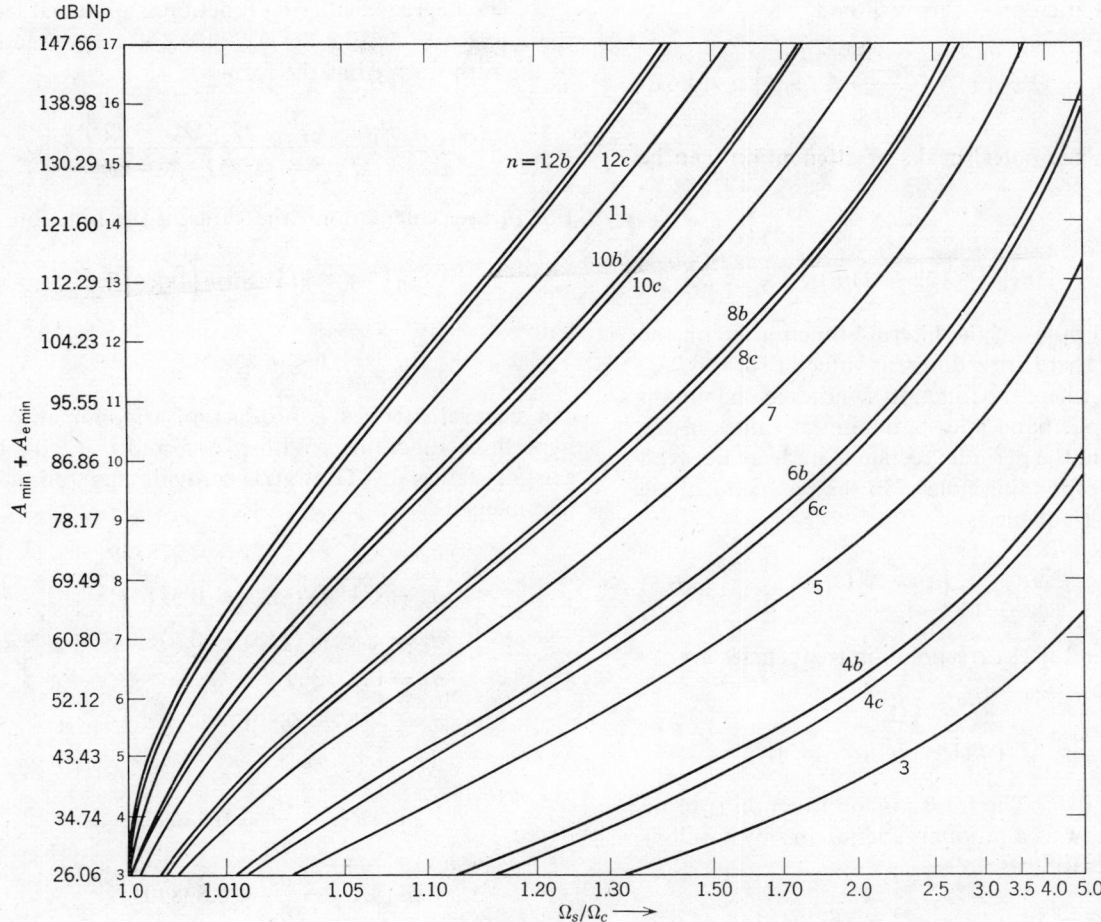

Fig. 4.11. Curves of attenuation versus sharpness.

parameter n for the curves indirectly determines the requirements for the elements of the filter. The relationship between k', $A_{\min}$, and ρ is uniquely indicated.

As an example of the use of Fig. 4.11 for insertion-loss estimating, the following requirements are formulated. At normalized frequency $\Omega_s/\Omega_c = 1.25$, the minimum attenuation $A_{\min}$ in the stopband has to be greater than 43 dB. At the same time, the filter in the passband has to have a reflection factor $\rho = 10\%$ or less which corresponds to $A = 0.043$ dB ($A_e = 20$ dB). The abscissa $\Omega_s/\Omega_c = 1.25$ intersects the ordinate $A = 63$ dB between the curves of order $6b$ and 7. This means that all three requirements (for Ω_s/Ω_c, $A_{\min}$ and ρ), with a filter of order $6b$, cannot be satisfied, and the choice of the seventh order will provide a reserve. Every point on the curve for order 7 between $1.16 \leq 1/k' \leq 1.25$ (same as $63 < A_{\min} + A_{e\,\min} < 72$) will satisfy the requirements and will be sufficient for the given problem.

4.6 CHARACTERISTIC VALUES OF $D(s)$

To determine the rational function $D(s)$, it is necessary to find the specific values of s where function D has poles and zeros. According to Eq. 4.4.2 the value of D is zero where $dn(zK_1; k_1) = 0$. In the periodic rectangle the zero values are at

$$z_{0v} = 1 + j\,\frac{2v-1}{n}\,\frac{K'}{K} \qquad (4.6.1)$$

where $1 \leq v \leq n$. This position in the periodic rectangle corresponds to the frequency

$$\frac{s_{0v}}{j} = \Omega_{0v} = \frac{dn\{K + j[(2v-1)/n]K'; k\}}{dn(x_1 K; k)} \qquad (4.6.2)$$

which is a function of a complex argument. If the real and imaginary parts of the function of the complex variable are separated, the value of real frequency where the attenuation is zero will be

$$\Omega_{0v} = \Omega_c sn\left(1 - \frac{2v - 1}{n} K_1'; k'\right) \quad (4.6.3)$$

Analogously, the poles (peaks of attenuation) can be found

$$\frac{s_{\infty v}}{j} = \Omega_{\infty v} = \frac{\Omega_s}{sn\{[1 - (2v - 1)/n]K'; k'\}} \quad (4.6.4)$$

All zeros and poles of the filtering function lie on the real Ω axis. There are n different value of Ω_0 and Ω_∞.

The places where the function D takes its maximum value in the passband and its minimum values in the stopband lie in the periodic rectangle midway between zeros or corresponding poles. In the passband, these points are at frequencies

$$\Omega_{max} = \Omega_c sn\left[\left(1 - \frac{2v}{n}\right)K'; k'\right] \quad (4.6.5)$$

In the stopband the corresponding frequencies are

$$\Omega_{min} = \frac{\Omega_s}{[sn(1 - 2v/n)K'; k']} \quad (4.6.6)$$

where $1 < v < n$. The transmission function is to be represented now as a rational function in s by developing $1 + D^2$ as follows:

$$|H|^2 = 1 + D^2 = e^{2a} = e^{g(-s)}e^{g(s)} \quad (4.6.7)$$

with zeros which lie in the left half of the s-plane.

4.7   AN EXAMPLE OF FILTER DESIGN

Consider a lowpass filter that satisfies the following conditions:

1. Amplitude ripples in the passband, from 0 to 10 kc should not exceed 0.05 Np, or 0.43 dB.
2. Effective attenuation in the stopband, from 15.6 kc to infinity is to be less than 6 Np, or 52 dB.
3. The filter is to be connected between source and load impedances of 600 ohms.

SOLUTION

1. The slope of selectivity is $k' = 0.641$ $(1/k' = 1.56)$, which corresponds to the modular angle $\theta = 39°52'$. Assuming $\theta = 40°$, then $k' = \sin \theta = 0.642788$ and $K'/K = 0.93$; $f_s = 15.557$ kc.
2. Total attenuation,

$$a_{min} + a_{e\,min} = 52 + 10 = 62 \text{ dB.}$$

3. The degree of filtering function is obtained from the curve $n = 5$ (Fig. 4.11). The Zolotarev fraction of the fifth degree has the form

$$D = \frac{B_1\Omega(\Omega_{01}^2 - \Omega^2)(\Omega_{02} - \Omega^2)}{(\Omega_{\infty1}^2 - \Omega^2)(\Omega_{\infty2}^2 - \Omega^2)}$$

For further calculations, the value of the function

$$sn\left(\frac{v}{5} K'; k'\right) = sn'\left(\frac{v}{5} K'\right)$$

with

$$1 \le v \le 5$$

can be evaluated as a product of trigonometric or hyperbolic functions. With $n = 5$ and $\theta = 40°$, the existing tables by Glowatzki provide the following parameters:

$$a_1 = \sqrt{k'}\, sn'(\tfrac{1}{5}K') = 0.278149$$

$$a_2 = \sqrt{k'}\, sn'(\tfrac{2}{5}K') = 0.511671$$

$$a_3 = \sqrt{k'}\, sn'(\tfrac{3}{5}K') = 0.676806$$

$$a_4 = 0.771369$$

$$a_5 = \sqrt{k'} = 0.801740$$

$$sn'(0) = 0$$

$$sn'(\tfrac{1}{5}K') = \frac{a_1}{a_5} = 0.346931$$

$$sn'(\tfrac{2}{5}K') = \frac{a_2}{a_5} = 0.638201$$

$$sn'(\tfrac{3}{5}K') = \frac{a_3}{a_5} = 0.844171$$

$$sn'(\tfrac{4}{5}K') = \frac{a_4}{a_5} = 0.962119$$

$$sn'(K') = 1$$

Parameters of Filtering Function

The values of limiting frequencies Ω_c and Ω_s must now be found. With $\Delta a - \Delta a_e = 42$, the value of normalizing cutoff frequency x_1 in rectangle is equal to 0.8. This gives the product $\Omega_c \cdot \Omega_s = 1.38$ with known

$$k' = \frac{\Omega_c}{\Omega_s} = 0.642788$$

$$\Omega_c = \sqrt{1.39 \times 0.642788} = \sqrt{0.893475} = 0.945238$$

and

$$\Omega_s = \sqrt{\frac{1.39}{0.642788}} = \sqrt{2.16245} = 1.47053$$

Also

$$\Omega_{01} = \frac{a_4}{a_5} \cdot \Omega_c = 0.909431$$

$$\Omega_{02} = \frac{a_2}{a_5} \cdot \Omega_c = 0.591443$$

$$\Omega_{\infty 1} = \frac{\Omega_s}{a_4/a_5} = 1.52843$$

$$\Omega_{\infty 2} = \frac{\Omega_s}{a_2/a_5} = 2.35019$$

$$\Omega_{\max 1} = \Omega_c \frac{a_1}{a_5} = 0.327932$$

$$\Omega_{\max 2} = \Omega_c \frac{a_3}{a_5} = 0.797942$$

$$\Omega_{\max 3} = \Omega_c \cdot 1 = 0.945238$$

The minimum attenuation in the stopband appears at

$$\Omega_{\min 1} = \frac{\Omega_s}{1} = 1.47053$$

$$\Omega_{\min 2} = \frac{\Omega_s}{a_3/a_5} = \frac{1.47053}{0.844171} = 1.74198$$

$$\Omega_{\min 3} = \frac{\Omega_s}{a_1/a_5} = \frac{1.47053}{0.346931} = 4.23868$$

In the Glowatzki tables, normalization is relative to the geometric mean frequency $\Omega_c\Omega_s = 1$. Therefore in order to determine the poles and zeros in the case of normalization relative to actual cutoff where the attenuation is equal to $a_{\max}$ it is necessary to use the formula in the form

$$\Omega_{0v} = \frac{a_m}{a_n} \quad \text{and} \quad \Omega_{\infty v} = \frac{1}{a_m a_n}$$

The corresponding values of zeros in the passband and poles in the stopband are

$$\Omega_{01} = \frac{a_4}{a_5} = 0.962119$$

$$\Omega_{02} = \frac{a_2}{a_5} = 0.638201$$

$$\Omega_{\infty 1} = \frac{1}{a_4 a_5} = 1.61697$$

$$\Omega_{\infty 2} = \frac{1}{a_4 a_5} = 2.43767$$

The expression for D can be substituted in the formula

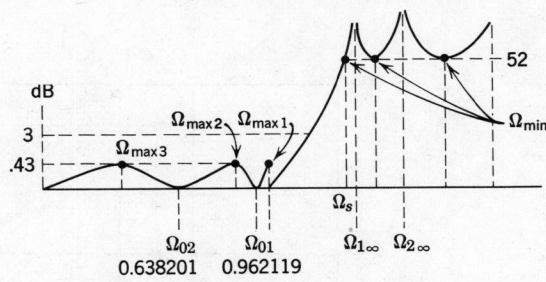

Fig. 4.12. Example of practical realization ($n = 5$).

for attenuation,

$$e^{2a} = 1 + (e^{2a_{\max}} - 1)\,|D|^2$$

(when $\Omega = 1$, the coefficient B_1 can be evaluated) and the response curve plotted. Figure 4.12 shows the response with imposed requirements met in both passband and stopband.

Transmission Function $H(s)$

Another rational function to be determined (after the problem of approximation is solved) is the transmission function H with its complex zeros. The regularity found in the transformed frequency plane (z-plane) enables us to deduce the following formulas for the zeros of H.

$$1 \pm jD = 0$$

or

$$dn(z_0 K_1; k_1) = \pm \frac{j}{tn(x_1 K_1; k_1)}$$

where z_0 values are complex. After separating the real and imaginary parts the values of the zeros will be

$$z_{0v} = \pm x_1 + \frac{j(2v - 1)K_1'}{K_1} = \pm x_1 + j\frac{2v - 1}{n}\frac{K'}{K}$$

These zeros are all in the periodic rectangle, but are not on the line passing through the center where the zeros of filtering are located. They are on two parallel lines passing distance x_1 from the center on both sides of the center. Figure 4.13 shows 10 zeros of the transmission factor and 5 zeros of the filtering factor for the filter with $n = 5$.

Taking the negative values of x_1, the zeros can be expressed by mapping the formula

$$s_{0v} = \frac{j\,dn\{-x_1 K + j[(2v - 1)/n]K'; k\}}{dn(x_1 K; k)}$$

The real and imaginary parts must be separated and the final expressions for s_{0v} can be written in the following form:

$$s_{0v} = \frac{-\sqrt{1 - 1/\Omega_s{}^2}\sqrt{1 - \Omega_c{}^2} \cdot sn\{[(2v - 1)/n]K'\} + jcn\{[(2v - 1)/n]K'\}\,dn\{[(2v - 1)/n]K'\}}{1 - \left(\dfrac{sn\{[(2v - 1)/n]K'\}}{\Omega_s}\right)^2}$$

where $sn(zK)$ is an abbreviation for $sn(zK; k)$.

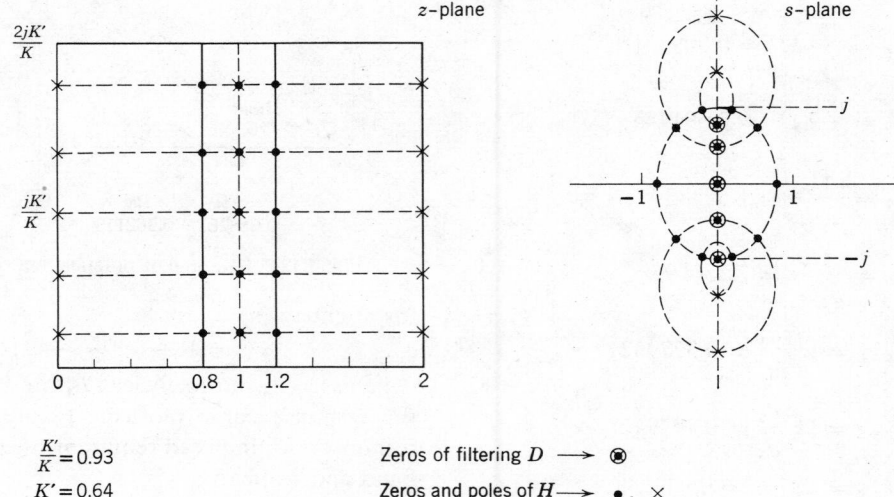

$\dfrac{K'}{K} = 0.93$ Zeros of filtering $D \longrightarrow \circledast$

$K' = 0.64$ Zeros and poles of $H \longrightarrow \bullet , \times$

Fig. 4.13. Poles and zeros in s- and u-planes.

The values of s_{0v}, for positive x_1, will be the same. With the foregoing expression the zeros can be evaluated in the form

$$s_{0v} = -U \pm jV$$

Each pair of zeros is a conjugate complex pair, always with a negative real part. In the case of odd-order filters one of the zeros is negative real.

$$s_r = -\sqrt{\frac{1 - \Omega_c^2}{1 - 1/\Omega_s^2}}$$

The construction of the $E(s)$ polynomial for the transmission function H is an involved procedure. For the given example the zeros are

$$s_{01} = -0.087394 \pm j1.016566$$

$$s_{02} = -0.297797 \pm j0.701139$$

$$s_r = -0.43783$$

The polynomial will include three multipliers

$$s^2 + 2sU_1 + C_1 \qquad (V = 1)$$

$$s^2 + 2sU_2 + C_2 \qquad (V = 2)$$

$$s + 0.43783$$

where

$$C_1 = U_1^2 V_1^2$$

$$C_2 = U_2^2 + V_2^2$$

so that

$$E(s) = (s^2 + 2sU_1 + C_1)(s^2 + 2sU_2 + C_2)(s + C_3)$$

Finally the transmission function for the given

example will be written in the following form:

$$e^q = H = B\,\frac{s^5 + E_4 s^4 + E_3 s^3 + E_2 s^2 + E_0}{s^4 + P_2 s^2 + P_0} = \frac{E}{P}$$

The denominator in both rational functions H and D for the given example with $n = 5$ consists of two factors, $s^2 + (1.52848)^2$ and $s^2 + (2.35019)^2$ which are directly related to the peaks of attenuation. The denominators are the same in both expressions which signifies that the poles of both functions coincide in the s-plane and in the periodic rectangle.

$$\pm \frac{D}{j} = B\,\frac{s^5 + F_3 s^3 + F_1 s}{s^4 + P_2 s^2 + P_0} = \frac{F}{P}$$

where B is a constant which determines the component of transmission function for $s = 0$ and

$$P = s^4 + 7.85964 s^2 + 12.90403$$

The next problem is the realization of the network with aid of E and F. Figure 4.14 shows one of the realizations suitable for the frequency range of interest.

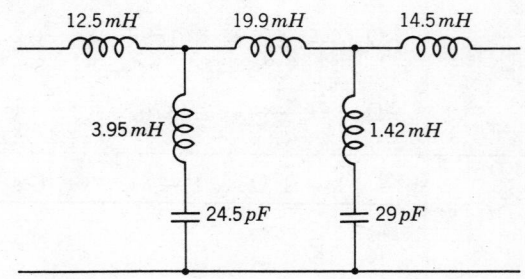

Fig. 4.14. Schematic of filter corresponding to Fig. 4.12.

4.8 CONSIDERATION OF LOSSES

In the previous discussion on circuit theory, all network elements were pure reactances. In reality physical inductances and capacitances have resistive as well as reactive components as has been shown in Section 1.9.

The equivalent circuit of an inductance shown in Fig. 1.23 can be once more approximated by a series circuit which includes the ideal inductance and a series resistor which simplifies the synthesis technique. By similar reasoning the equivalent circuit of a capacitor in Fig. 1.23 may be approximated by the combination of a pure capacitor with a parallel resistor. These equivalent circuits are shown in Fig. 4.15. Therefore, for a lossy coil, the impedance will be expressed by

$$Z_L = j\omega L + R_L \qquad (4.8.1)$$

and for capacitor by

$$\frac{1}{Z_C} = j\omega C + \frac{1}{R_C} \qquad (4.8.2)$$

The impedance reciprocity which is recognizable from this expression is valid even for the nonideal case.

For further discussion, the following normalization will be introduced:

$$r_L = \frac{Z_L}{R_r} = \frac{j\omega L \omega_r}{R_r \omega_r} + \frac{R_L}{R_r} = j\frac{\omega}{\omega_r}\frac{\omega_r L}{R_r} + \frac{R_L}{R_r} \quad (4.8.3)$$

$$r_L = \frac{\omega_r L}{R_r}\left(j\Omega + \frac{\omega_L}{\omega_r L}\right) \qquad (4.8.4)$$

where R_r is the reference resistance (usually the internal source resistance) and ω_r is the reference frequency (usually the center frequency for bandpass filters or the cutoff frequency for lowpass filters). The capacitor's impedance will be

$$r_C = \frac{Z_C}{R_r} = \left[\omega_r C R_r\left(j\Omega + \frac{1}{\omega_r C R_C}\right)\right]^{-1} \quad (4.8.5)$$

Ordinarily the value $\omega_r L/R_L$ is called the quality factor of the coil Q_L and the value $\omega_r C R_c$ is the corresponding quality factor of the capacitor Q_C. The reciprocal of a quality factor is a loss factor

$$\epsilon_L = \frac{1}{Q_L}$$

$$\epsilon_C = \frac{1}{Q_C}$$

With normalized element value such as

$$L' = \frac{\omega_r L}{R_r}$$

$$C' = \omega_r C R_r$$

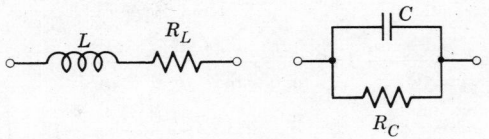

Fig. 4.15. Equivalent circuit of lossy inductor and capacitor.

The corresponding impedances in normalized form will be as follows

$$r_L = \frac{1}{j\Omega + \epsilon_L} \qquad (4.8.6)$$

$$r_C = \frac{1}{C(j\Omega + \epsilon_C)} \qquad (4.8.7)$$

ϵ_C and ϵ_L here are frequency independent constants. It will be assumed that the losses in every network elements are equal ($\epsilon_L = \epsilon_C$). This permits us to substitute for the frequency variable $j\Omega$ a new complex variable

$$s + \epsilon = s' \qquad (4.8.8)$$

Every coil now will possess $r_L = s'L$ and every capacitor $r_C = 1/s'C$.

4.9 INTRODUCTION OF LOSSES BY FREQUENCY TRANSFORMATION

In the case of a bandpass circuit, the lossy coil of the lowpass model will be transformed into a series-resonant circuit and the lossy capacitor will be transformed into a parallel-resonant circuit. In such networks, the circuits will be approximately as shown in Fig. 4.16. The normalized impedance with respect to R_r of the series circuit is

$$z_s = \frac{j\omega L}{R_r}\left(1 - \frac{\omega_m^2}{\omega^2}\right) + \frac{R_s}{R_r} \qquad (4.9.1)$$

The same impedance normalized with respect to the bandwidth $(\omega_2 - \omega_1)$ will take the form

$$z_s = j\frac{\omega_2 - \omega_1}{R_r}\frac{\omega_m}{\omega_2 - \omega_1}\left(\frac{\omega}{\omega_m} - \frac{\omega_m}{\omega}\right) + \frac{R_s}{R_r}$$

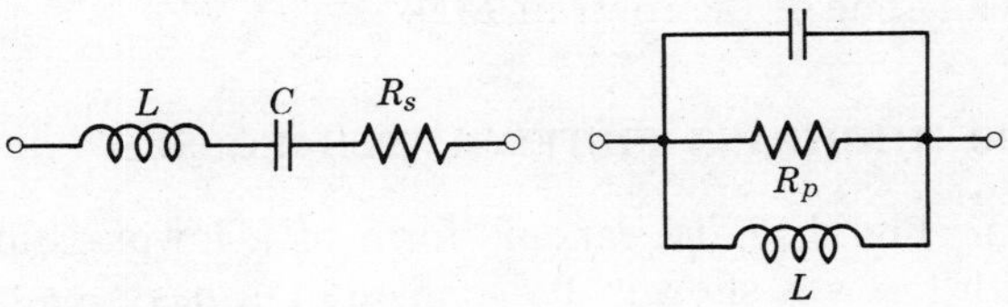

Fig. 4.16. Lossy resonant circuits.

or

$$z_s = \frac{\omega_2 - \omega_1}{R_r} L\left[j\,\frac{\omega_m}{\omega_2 - \omega_1}\left(\frac{\omega}{\omega_m} - \frac{\omega_m}{\omega}\right) + \frac{R_s R_r}{(\omega_2 - \omega_1)R_r L}\right] \quad (4.9.2)$$

Let us call the expression outside the brackets the normalized inductance

$$l = \frac{\omega_2 - \omega_1}{R_r} L \quad (4.9.3)$$

The bandpass transformation for a series circuit with losses will then be

$$z_s = l(j\Omega + \epsilon) = l(s + \epsilon) = ls' \quad (4.9.4)$$

In these cases, the value of ϵ corresponds to

$$\epsilon = \frac{R_s}{(\omega_2 - \omega_1)L} \quad (4.9.5)$$

and the Q for the circuit is

$$Q = \frac{\omega_m L}{R_s} \quad (4.9.6)$$

Equation 4.9.5 can be modified, using this definition of Q,

$$\epsilon = \frac{\omega_m}{\omega_2 - \omega_1}\frac{R_s}{\omega_m L} = \frac{\omega_m}{\omega_2 - \omega_1}\frac{1}{Q} \quad (4.9.7)$$

In a similar way, a parallel circuit will be described by the relation

$$z_p = \frac{1}{(j\Omega + \epsilon)C} = \frac{1}{(s + \epsilon)C} = \frac{1}{Cs'}$$

In both cases, ϵ is not simply the reciprocal quality factor of the circuit but the reciprocal product of the relative bandwidth and the quality of the circuit. Losses can be taken into account by the transformation $s' = s + \epsilon$ to the extent that the quality factors for the series and parallel arms are equal.

When the bandwidth is very narrow and the quality factor is by necessity very high, the value of ϵ has to be, as has been mentioned, sufficiently small to be able to satisfy the necessary condition that it be smaller in magnitude than the smallest magnitude of the real part of any one of the roots of $E(s)$.

4.10 HIGHPASS FILTERS WITH LOSSES

In Fig. 4.17 the canonic form of a lowpass filter with losses is shown. Its configuration has been obtained with the aid of the transformation $s' = s + \epsilon$.

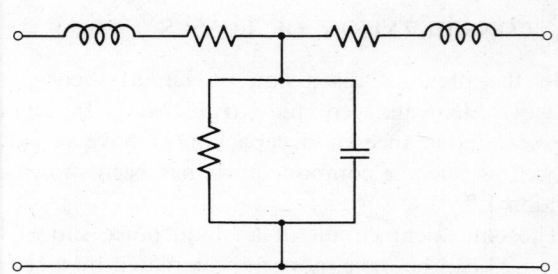

Fig. 4.17. Canonic lowpass filter with losses.

For a highpass filter the frequency parameter is the reciprocal of that for a lowpass filter.

$$s_H = \frac{1}{s_L} = \frac{\omega_r}{j\omega} \quad (4.10.1)$$

where s_H is the highpass normalizing frequency parameter and s_L is the lowpass normalizing frequency parameter.

When the above transformation is introduced, every element of the circuit will have the complex-frequency parameter

$$s_H' = \frac{\omega_r}{j\omega} + \epsilon \quad (4.10.2)$$

Therefore the normalized impedance of the capacitor will be

$$z_c = \frac{1}{C}\left(\frac{\omega_r}{j\omega} + \epsilon\right) \quad (4.10.3)$$

and that of the inductance will be

$$z_l = \frac{L}{(\omega_r/j\omega + \epsilon)} \quad (4.10.4)$$

The corresponding equivalent circuits are in Fig. 4.18.

This picture of lossy elements is different from the usual representation of those elements. We are accustomed to thinking of losses in the capacitor as resistance in parallel with a pure capacitance and of lossy coils as pure inductances in series with a resistance.

To preserve the usual equivalent circuit and develop the network as shown in Fig. 4.19, the transformation to be introduced is to have the form

$$s_H' = \frac{1}{s + \epsilon} \quad (4.10.5)$$

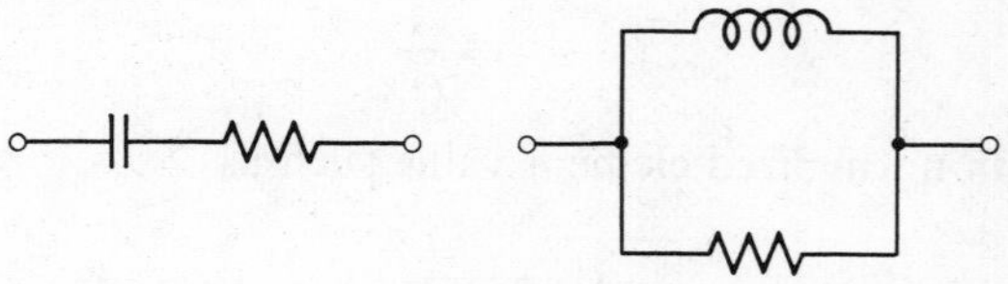

Fig. 4.18. Equivalent lossy inductor and capacitor pertaining to highpass filter.

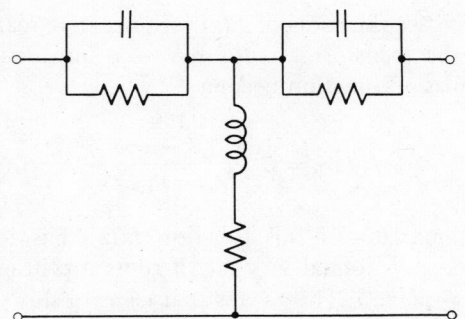

Fig. 4.19. Canonic highpass filter with losses.

instead of

$$s_H' = \frac{1}{s} + \epsilon.$$

The transmission function H, when this transformation is used, must be the same as that obtained by using the transformation in Eq. 4.10.2. To prove the validity of the transformation, the expression which it represents can be replaced by a power series, so that

$$s_H' = \frac{1}{s} - \epsilon\frac{1}{s^2} + \epsilon^2\frac{1}{s^3} - \epsilon^3\frac{1}{s^4} + \cdots + (-1)^{n-1}\epsilon^{n-1}\frac{1}{s^n}$$

(4.10.6)

or with $s = j\Omega$

$$s_H' = \frac{1}{s} + \epsilon\frac{1}{\Omega^2} + \epsilon^2\frac{1}{s^3} + \cdots + (-1)^{n-1}\epsilon^{n-1}\frac{1}{s^n}$$

(4.10.7)

Both parts of this expansion (real and imaginary) converge for $\epsilon^2/\Omega^2 < 1$.

This signifies that in the passband of a highpass filter, between $1 < \Omega < \infty$, the series is convergent. However, in the stopband, where the normalized frequency is between 0 and 1, the series may diverge. This region is of little interest since we cannot predistort to alter the filter's characteristics in this region.

If the ϵ parameter is small enough, the high-order terms in (4.10.7) can be neglected. We may approximate the series by

$$s_H' = \frac{1}{s} + \epsilon\frac{1}{\Omega^2} \qquad \text{for } 1 < \Omega < \infty \quad (4.10.8)$$

At $\Omega = 1$ (the cutoff frequency), the series is approximately equal to

$$s_H' = \frac{1}{s} + \epsilon \tag{4.10.9}$$

In lowpass filters,

$$s_L' = s + \epsilon$$

Thus even at the cutoff frequency of the filter, the losses are taken into account by use of the transformation in Eq. 4.10.5.

When Ω increases, the value of ϵ/Ω^2 tends to diminish. Physically this signifies that the influence of losses tend to disappear and the response curve will follow the ideal transfer response of a lossless filter. When ϵ is small, but of the same order of magnitude as the smallest real part of one of the roots of $E(s)$, the response will produce supplementary ripples since the transformation $s' = s + \epsilon$ is calculated with a constant ϵ value. The effect of assuming constant ϵ can cause a distortion of the passband when the usable passband of a highpass filter is very broad.

It is interesting to note the behavior of the attenuation at frequencies below cutoff. The power series is not convergent in the lowest part of the bandstop range (certainly not at $\Omega = 0$). Therefore let us set $s = 0$ in the expression for s_H'. When $\Omega = 0$

$$s_H' = \frac{1}{\epsilon} \tag{4.10.10}$$

In contrast with the transformation for lossless elements, this value is not infinite.

In lowpass filters the attenuation in the stopband reaches the calculated values, except at the peaks, which have finite values of attenuation and are spread over a finite range of frequencies. Here, in the highpass filter, the actual attenuation in the stopband is always below the theoretical value. The magnitude of the attenuation at $\Omega = 0$ is easy to find by setting $s' = 1/\epsilon$, and introducing this into the transmission function.

Similar reasoning is valid for bandstop filters. Evidently Ω has a different meaning, namely, the reciprocal of the frequency transformation for a bandpass filter. The expression used for a bandstop filter transformation is

$$s_S' = \frac{1}{s + \epsilon} \tag{4.10.11}$$

For the details of these transformations see Chapter 5.

Equation 4.10.11 can be represented by a power series and the value of loss coefficient will be

$$\epsilon = \frac{\omega_r}{\omega_2 - \omega_1}\frac{1}{Q}$$

where Q is the quality factor of the resonant circuit.

4.11 TRANSMISSION FUNCTIONS WITH LOSSES

It has been shown that the filtering function and the transmission function are simply related in lossless filters, and how it is possible to obtain the transmission function from the filtering function. For lossy filters, similar steps can be followed to obtain $H(s)$. In the given transmission function

$$H(s) = \frac{E(s)}{P(s)} \qquad (4.11.1)$$

if the polynomial $P(s)$ has zeros at finite frequency, then the function $H(s)$ has poles at real frequencies. When the components of the ladder network are composed of lossy elements, the attenuation does not reach infinite value at the poles, and the ripples of the attenuation response in the passband (in the case of Chebyshev filter) are "smeared." These have to be taken into consideration, and therefore a filter with lossy elements has to be designed with the aid of a special approximation technique. It it is desired to achieve a required transmission function $H(s)$ despite the presence of losses, a transformation from lossy to lossless "frequency" $s = s' - \epsilon$ should be made and a predistorted transmission function obtained before the realization process is begun. When the function $H(s)$ is given, the effective transmission function with losses can be obtained from it by introducing the transformation $H'(s) = kH(s)$. The filter must have a theoretical response which agrees with $|H(s)|$ (in spite of losses), but this response will be superimposed on a constant (frequency-independent) insertion loss k. Introducing the expression for s as a new variable in the expression for $H(s)$ yields the following relation

$$H''(s') = kH(s) = \frac{E(s' - \epsilon)}{P(s' - \epsilon)} \qquad (4.11.2)$$

$P(s)$ can no longer be an odd or even polynomial, that violates the condition for realizability for an odd or even $P(s)$.

It has been mentioned that the compensation of losses imposes considerable requirements on the filter design, and can be accomplished only approximately by introducing Eq. 4.11.2 into $E(s)$ when ϵ is not too large. The passband response will reflect the compensation of losses and its shape will be very close to the exact analytical curve. The response in the stopband will show the error in approximation, but the places where it is evident are only at the attenuation maxima (instead of attenuation poles). In most of the practical cases these phenomena are of no consequence, since the distortion in the passband is the only one of concern to the equipment designer.

Let us assume that the transmission function is given

$$H(s) = B \prod_{i=1}^{n} \frac{(s - s_i)}{P(s)} \qquad (4.11.3)$$

where s_i are the roots of $E(s)$ with negative real parts. With the transformation $s = s' - \epsilon$ introduced the transmission function becomes

$$H'(s') = B \prod_{i=1}^{n} \frac{(s' - \epsilon - s_i)}{P(s')} \qquad (4.11.4)$$

The numerator of this function $E'(s')$ has to be a Hurwitz polynomial in s'. (All roots are in left half of the s'-plane.) This means that ϵ has to stay smaller than the magnitude of the smallest real part of any root s_i.

The function $P(s)$ stays the same except that s' is substituted for s. With the function

$$H'(s') = \frac{E'(s')}{P(s')} \qquad (4.11.5)$$

realization can be accomplished while first regarding s' as a purely imaginary frequency parameter. $P(s')$ remains an even or odd function of s'. We must also prove that $H'(s')$ is realizable, and that $H'(s') \geq 1$ for all the imaginary values of s'.

To determine whether H' remains greater than or equal to one for all imaginary s', the absolute minimum of $|H'(s')|$ for imaginary s' has to be found. Usually it is less than one. To make $H'(s')$ realizable, its value is multiplied by a factor k, which is larger than one. If, for example $M = |H'(s')|_{\min}$ at $s_m' = j\Omega_m$, the expression for the realizable function with $k = M^{-1}$ is

$$H''(s') = kH'(s') = kH(s) \qquad (4.11.6)$$

The factor k is introduced to satisfy the realizability condition.

In Fig. 4.20 the response curves of a fourth-order lowpass filter with and without losses are shown. The figure also illustrates the effect of predistortion. The echo attenuation is drastically reduced, and the flat loss becomes higher proportionally to the amount of predistortion, but the shape of the ideal response is achieved.

In the equation for power balance, the insertion loss was represented by the term $2a_L$ where a_L is the frequency independent attenuation in the passband. Therefore the value k in Eq. 4.11.6 has to be equal to $k = e^{2a_L}$. The rest of the function $H(s)$ is a realizable function with the condition $|H(s)| \geq 1$ satisfied for all imaginary s or real Ω. Equal losses result in shift of the imaginary axis to the left in the s-plane. The properties of lossless filters at frequencies s are the same as those of lossy filters at frequency $s - \epsilon$.

It has been shown that for the calculation of

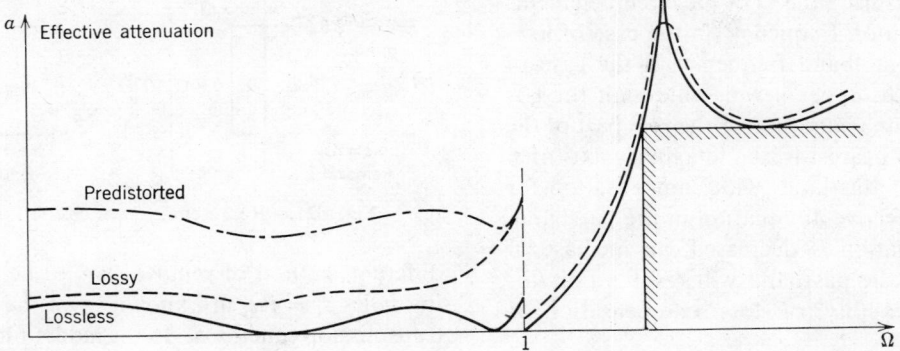

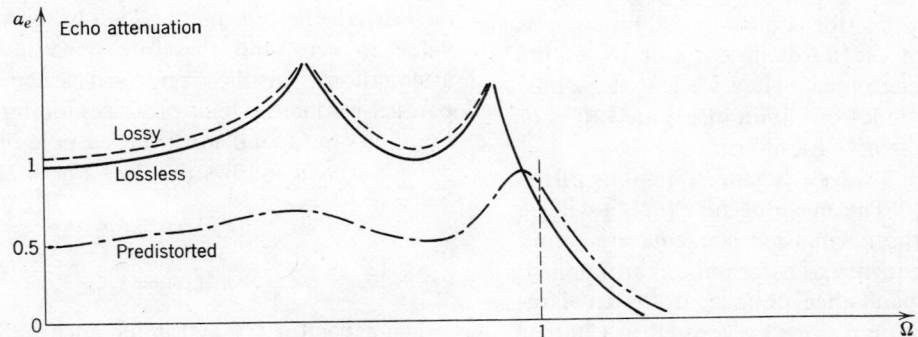

Fig. 4.20. Effective parameters of low pass Chebyshev filter ($n = 4$).

the input or output impedances (consequently admittances), the function $F(s)$ is necessary. Impedance or admittance will now be determined as a function of the new parameter $s' = s + \epsilon$. Consequently, the polynomial $F'(s')$ must be determined. On the basis of the relation

$$\left|\frac{E(s)}{P(s)}\right|^2 = 1 + \left|\frac{F(s)}{P(s)}\right|^2$$

it must be true that

$$|H''(s')|^2 = \left|\frac{E''(s')}{P(s')}\right|^2 = 1 + \left|\frac{F'(s)}{P(s')}\right|^2 \quad (4.11.7)$$

Since $H''(s')$ has a minimum at $s' = j\Omega_m$ and the absolute value of the minimum was changed to one by multiplication of $H'(s')$ by k, the polynomial $F'(s')$ is

$$F'(s')F'(-s') = |E''(s')|^2 - |P(s')|^2 \quad (4.11.8)$$

The right side of this relation is known, and the equation has twice as many roots, s_i', as the order of the polynomial $E''(s')$. These roots appear as pairs, such that every root of one pair differs from the other root of the same pair only by sign.

4.12 CONCLUSIONS ON CONSIDERATION OF LOSSES

The addition of losses in the network has the following consequences:

1. The peaks of attenuation are finite. In the complex-frequency plane, the poles are moved into the left half-plane, from the imaginary axis. On the imaginary axis only finite attenuation can be obtained.

2. The transition from passband to stopband is rounded as a result of the influence of the zeros closest to the imaginary axis. (This can be seen in Fig. 4.20.)

3. The minimum value $H = 1$, which corresponds to zero attenuation, is not obtainable; some insertion loss independent of frequency is always present. In order to compensate for the influence of losses, especially in the passband of the filter, all poles and zeros of $H(s)$ have to be moved toward the right (less ϵ). But that action will violate the realization conditions. Therefore only zeros of $H(s)$ are shifted to the right, and finite attenuation in the stopband will be the consequence. The action of losses brings the zeros to their original position. The quantity ϵ can be

determined by measuring the Q of the circuit elements to be used at the cutoff frequency (in the case of low-pass filters) or at midband frequency in the case of bandpass filters. An upper permissable limit for ϵ is given by the absolute magnitude of the real part of the root of $E(s)$ that is nearest to the imaginary axis of s. Reactance filters of this kind, with compensation for losses, increases effective attenuation in the passband. If the echo attenuation is decreased, an increase of reflected power in the passband will result. For this reason the applicability of loss compensation is limited.

4.13 REALIZATION PROCESS

As we know, when the transmission function is given, the value of the normalized z_{ik} or y_{ik} of the network can be determined. Here we will show how to find (from impedance or admittances) the values of the reactive components of a filter.

The z_{ik} or y_{ik} are always reactance functions of the parameters s or s'. The meaning of s (or s') will be determined after the normalized elements are found and appropriate frequency transformations are applied. Bandpass and stopband filters designed by the usual frequency-transformation process are treated in Chapter 5 and are frequency-symmetric filters. When z_{ik} or y_{ik} are given as reactance functions, the values of network elements can be found from a set of n-algebraic equations.

The pole-removal process is much more demonstrative and for the type of reactance functions used in filters is particularly useful. If the reactance function $z(s)$ or $y(s)$ is given, and the circuit has to be of the ladder type, and inductor or capacitor can be removed in the following manner:

$$z(s) = sL' + z'(s) \tag{4.13.1}$$

or

$$y(s) = sC' + y'(s) \tag{4.13.2}$$

$z'(s)$ or $y'(s)$ are then one-order lower-remainder functions, from which a capacitor or an inductance can be removed.

$$\frac{1}{z'(s)} = sC_1' + \frac{1}{z''(s)} \tag{4.13.3}$$

$$\frac{1}{y'(s)} = sL_1' + \frac{1}{y''(s)} \tag{4.13.4}$$

This process leads to a lowpass filter with either a π- or T-type input, having no poles of attenuation. The elements are fully determined by z_{11} and z_{22} or y_{11} and y_{22}. For a filter with poles of attenuation, a

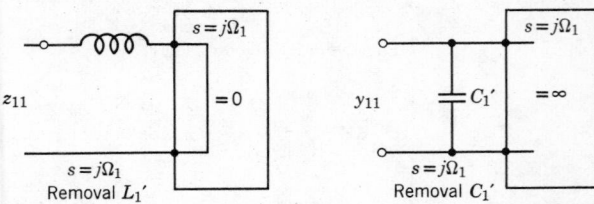

Fig. 4.21.   Representation of circuit for removal.

different method of removal has to be used. In fact, the poles $s_i = j\Omega_i$ are known from the approximated transmission functions. In the ladder filters the attenuation poles will require either a parallel circuit in the series arms of the ladder or a series circuit in the shunt arms of the ladder. In schematic 10 of Table 2.3, when $j\Omega = s_1$, the first resonant circuit brings the impedance value to zero and therefore constitutes a pole of attenuation. In the reciprocal schematic the first parallel-resonant circuit produces an impedance equal to infinity at Ω_1 and constitutes a pole of attenuation.

From this it follows that (see Fig. 4.21):

$$z_{11}(s_1) = s_1 L_1'$$

or

$$y_{11}(s_1) = s_1 C_1'$$

The capacitor C_1' and inductance L_1' can be calculated from the above relations. The remainder functions will be z'_{11} and y_1' (see Fig. 4.22).

$$z_{11}(s) = sL_1' + z_{11}'(s)$$
$$y_{11}(s) = sC_1' + y_{11}'(s)$$

Both of these remainder functions have to have a zero at $s = s_1$, since

$$z_{11}'(s) = z_{11}(s) - sL_1'$$
$$y_{11}'(s) = y_{11}(s) - sC_1'$$

The first pole circuit can be removed from this function. It is represented by the expression

$$x_p = \frac{K_1 s}{s^2 + \Omega^2}$$

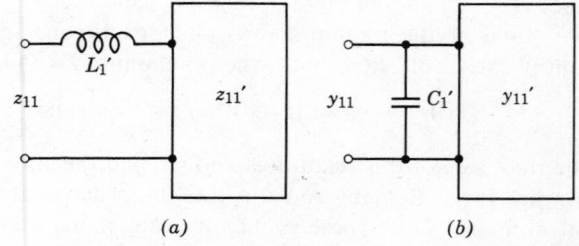

Fig. 4.22.   Definition of remainder functions (*a*) to determine z_{11}' from known z_{11} (*b*) to determine y_{11}' from known y_{11}.

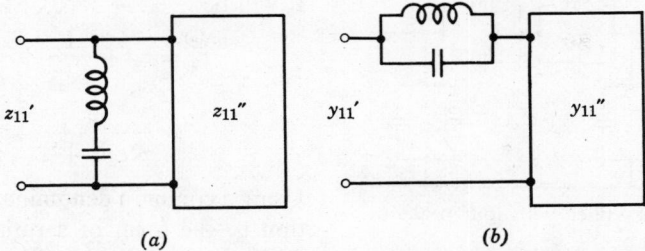

Fig. 4.23. (*a*) Removal of resonant circuit from $1/z_{11}'$. (*b*) Removal of parallel resonant circuit from $1/y_{11}'$.

where K_1 can be a capacitor or an inductor and x_p is the reactance, impedance, or admittance. When the first pole is expressed by the foregoing formula the remainder function will be (see Fig. 4.23)

$$\frac{1}{z_{11}'(s)} = \frac{K_1 s}{s^2 + \Omega_1{}^2} + \frac{1}{z_{11}''(s)}$$

or

$$\frac{1}{y_{11}'(s)} = \frac{K_1 s}{s^2 + \Omega_1{}^2} + \frac{1}{y_{11}''(s)}$$

The remainder function z'' or y'' must be two orders lower than z' or y'. The coefficient K_1 and the new coefficients for z'' and y'' are obtainable in this way by comparison with z' and y'. The elements of the rest of the ladder are in the remainder function, and their values can be found in a similar fashion.

Finally, we note that for the *T*-network shown in Table 2.3, schematic 10, the last inductance in the series arm does not exist in the primary open circuit impedance, and must be calculated from either the primary short-circuit or secondary open-circuit impedance. The same problem exists with the reciprocal π-schematic and its last shunt capacitor. When the filter is complicated it is reasonable to start the removal from both sides. The elements in the center of the network have to be at the same value, and that fact alone is a good checking point.

4.14 BANDPASS FILTER WITH A MINIMUM NUMBER OF INDUCTORS

Bandpass filters which produce one pole of attenuation in every arm of a ladder are conventionally called Zig-Zag filters. They were originated by Laurent and developed originally on the basis of image-parameter theory by quantizing the attenuation in half-sections. Being technologically superior to conventional filters they deserve some special attention.

Figures 4.24 and 4.25 show two equivalent bandpass filters. The first is the coil-saving Zig-Zag and the second is the conventional bandpass filter developed from the normalized lowpass schematic shown in Table 2.3, schematic No. 12.

Both have similar attenuation characteristics. The network shown in Fig. 4.25, being a result of a lowpass-bandpass frequency transformation, must have a symmetric amplitude-frequency response and consequently must have a symmetric pole distribution. (A nonsymmetric response having an image-parameter structure similar to this is shown in Fig. 1.11, bandpass II_1 and II_2.) A necessary condition for realization is that the filter have two poles of attenuation: one in its upper and one in its lower stopband.

The same condition must apply to the Zig-Zag configuration in Fig. 4.24. The poles produced by the series arms are always in the upper part of the stopband. The poles produced by the shunt arms of the

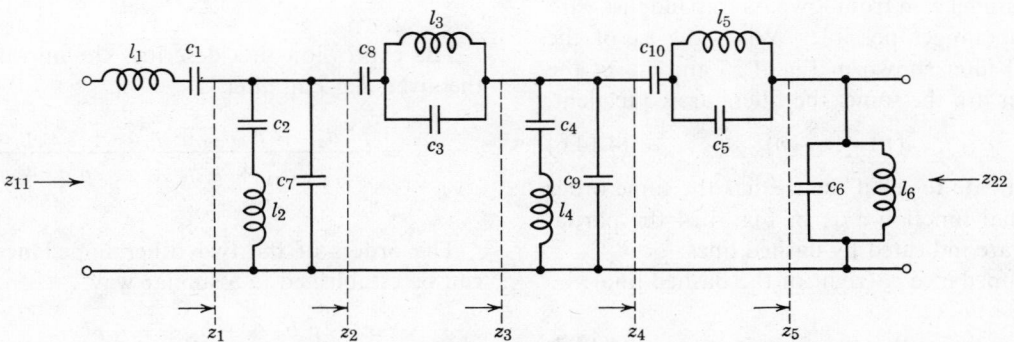

Fig. 4.24. Coil-saving bandpass filter with partial impedances shown.

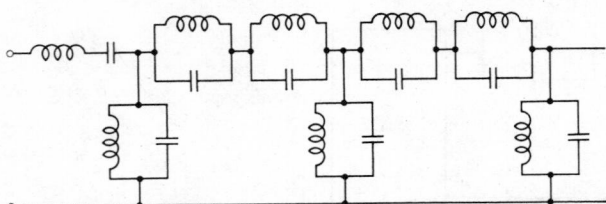

Fig. 4.25. Conventional bandpass filter with four peaks of attenuation.

ladder produce attenuation peaks only below the passband.

In Table 2.3 of filtering functions, for filter No. 12 with $n = 6$ there is

$$D = B \cdot \frac{(s^2 + \Omega_1{}^2)(s^2 + \Omega_2{}^2)(s^2 + \Omega_3{}^2)}{(s^2\Omega_4{}^2 + 1)(s^2\Omega_5{}^2 + 1)} \quad (4.14.1)$$

Let us use the frequency transformation for a bandpass filter

$$s = j\Omega = a\left(\Omega_B - \frac{1}{\Omega_B}\right) \quad (4.14.2)$$

From Table 2.2 one of the tabulated functions can be used for element evaluation. Let us limit ourselves to z_{ik}, the sixth-order filter which, according to Fig. 4.25, will have impedance functions

$$z_{11} = \frac{s^6K_6 + s^4K_4 + s^2K_2 + K_0}{s^5K_5 + s^3K_3 + sK_1} \quad (4.14.3)$$

$$z_{22} = \frac{s^4M_4 + s^2M_2 + M_0}{s^5M_5 + s^3M_3 + sM_1} \quad (4.14.4)$$

$$z_{12} = \frac{s^4N_4 + s^2N_2 + N_0}{s^5N_5 + s^3N_3 + sN_1} \quad (4.14.5)$$

With Eq. 4.14.2 the values of the impedance may be found as a function of $p = j\Omega_B$; that is, as a function of the normalized parameter p (with a reference frequency equal to geometric midband frequency). This is very important since, in the case of Zig-Zag filters, the separate frequency transformation in the series and parallel arm from lowpass to bandpass with (4.14.2) is no longer possible. When the z_{ik} of the conventional filter shown in Fig. 4.25 and z_{ik}' of the Zig-Zag filter are the same, the filters are equivalent.

$$z_{ik}(p) = z_{ik}'(p) \quad (4.14.6)$$

Let us investigate to see if the z_{ik}' has the same order as the rational function z_{ik}. In Fig. 4.24 the partial impedances are indicated by dashed lines.

For the impedance z_5 (right of the dashed line)

$$z_5 = \frac{pl_6}{p^2c_6l_6 + 1} \quad (4.14.7)$$

z_4 will be

$$z_4 = z_5 + \frac{1}{pc_{10}} + \frac{pl_5}{p^2c_5l_5 + 1}$$

$$= \frac{pl_6}{p^2c_6l_6 + 1} + \frac{1}{pc_{10}} + \frac{pl_5}{p^2c_5l_5 + 1} \quad (4.14.8)$$

Using a common denominator, and bringing this equation to the form of a rational function we obtain a numerator of fourth degree and a denominator of fifth degree. Including the next series-resonant circuit (c_4 and l_4) and capacitor to the left (c_9) for the construction of the reactance function we will find that $1/z_3$ will be a rational function of 7/6 order.

The series arm on the left of z_3 also contributes 2/3 to the reactance function and consequently z_2 will be of degree 8/9, since both functions contain a common factor $1/p$

$$\frac{\text{Order}}{\text{Order}} = \frac{0}{1}\left(\frac{2}{2} + \frac{6}{6}\right) = \frac{0}{1}\left(\frac{8}{8}\right) = \frac{8}{9}$$

The next circuit on the left of z_2 is of degree 2/3 and consequently $1/z_1$ is 11/10 order. The series-resonant circuit l_1 and c_1 has an impedance degree 2/1 and therefore the whole network will have a degree of 12/11. z_{11} is a reactance function of twelfth degree in the numerator and of eleventh degree in the denominator.

For a filter developed in conventional form, the impedances are

$$z_{11} = \frac{\sum_{i=0}^{6} A_{2i}p^{2i}}{\sum_{i=0}^{6} A_{2i-1}p^{2i-1}} = \frac{12}{11} \quad (4.14.9)$$

$$z_{22} = \frac{\sum_{i=0}^{4} B_{2i+1}p^{2i+1}}{\sum_{i=1}^{6} B_{2i-1}p^{2i-2}} \quad (4.14.10)$$

$$z_{12} = \frac{\sum_{i=0}^{4} D_{2i+1}p^{2i+1}}{\sum_{i=1}^{6} D_{2i-2}p^{2i-2}} \quad (4.14.11)$$

The expression that describes the impedance z_{11} of the given Zig-Zag filter is

$$z_{11} = \frac{p^{12}a_{12} + p^{10}a_{10} + p^8a_8 + \dots p^2a_2 + a_0}{p^{11}a_{11} + p^9a_9 + \dots p^3a_3 + pa_1} \quad (4.14.12)$$

The orders of the two other impedance functions can be established in a similar way

$$z_{22} = \frac{p^9b_9 + p^7b_7 + \dots pb_1}{p^{10}b_{10} + p^8b_8 + \dots p^2b_2 + b_0} \quad (4.14.13)$$

and

$$z_{12} = \frac{p^9 d_9 + p^7 d_7 + p d_1}{p^{10} d_{10} + p^8 d_8 + p^2 d_2 + d_0} \quad (4.14.14)$$

All three functions have the same order as the corresponding impedance functions for standard development. The condition of equality is fulfilled when the coefficients a, b, and d of the corresponding functions are equal, is $a_v = A_v$, $b_v = B_v$, $d_v = D_v$.

4.15 THE ELEMENTS OF A COIL-SAVING NETWORK

Circuits similar to that shown in Fig. 4.24 are shown in Table 4.3. They all have a series-resonant circuit in the series arms on the input side. On the output side the filter begins with a shunt arm which is a parallel-resonant circuit. The networks are basically antimetric and are represented by even-order transmission functions (4, 6, 8, . . .). The number of inductors saved by

this development is equal to the number of normalized attenuation poles in the lowpass model. For the filter considered here, we will use the example shown in Fig. 4.24 to demonstrate the removal process. First of all, it has to be shown that the number of unknowns correspond to the number of known conditions. In this network there are 16 unknown elements. The order of the impedance polynomial is 12. It is necessary to find at least four complementary independent conditions. These conditions are given in the transmission function by the position of the poles (resonant frequencies).

The known resonant frequencies of the pole-producing circuits are

$$\Omega_2 = (\sqrt{l_2 c_2})^{-1} \quad (4.15.1) \qquad \Omega_4 = (\sqrt{l_4 c_4})^{-1} \quad (4.15.3)$$

$$\Omega_3 = (\sqrt{l_3 c_3})^{-1} \quad (4.15.2) \qquad \Omega_5 = (\sqrt{l_5 c_5})^{-1} \quad (4.15.4)$$

Ω_2 and Ω_4 are in the upper stopband, Ω_3 and Ω_5 in the lower. There are two more resonant circuits that

Table 4.3 Lowpass models and corresponding coil-saving bandpass filters

LP MODEL	COIL SAVING B. P.	NORMALIZED FILTERING FUNCTION
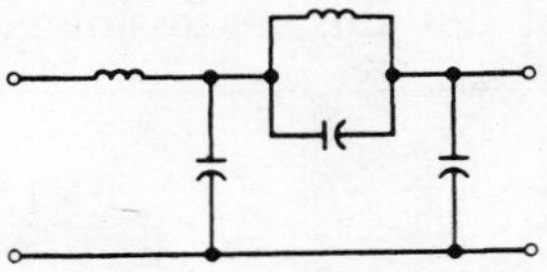	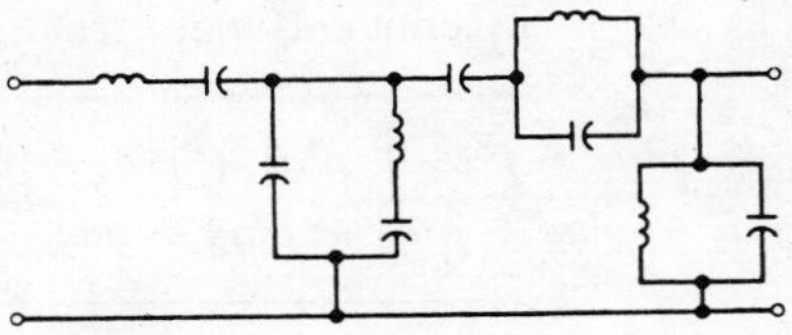	$D = B \dfrac{(s^2 + \Omega_1{}^2)(s^2 + \Omega_2{}^2)}{s^2 \Omega_3{}^2 + 1}$
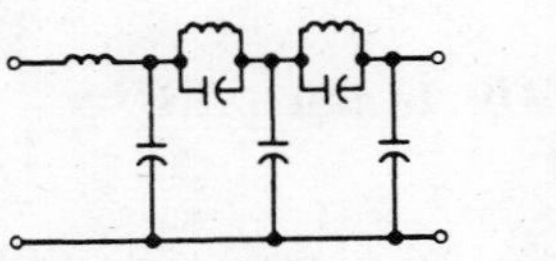	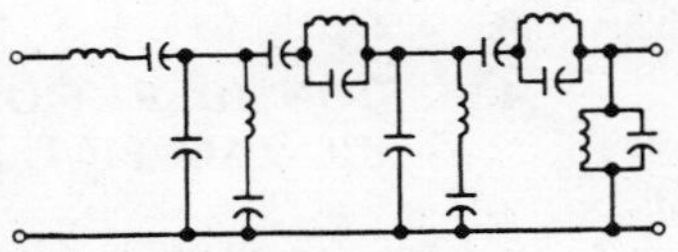	$D = B \dfrac{(s^2 + \Omega_1{}^2)(s^2 + \Omega_2{}^2)(s^2 + \Omega_3{}^2)}{(s^2 \Omega_4{}^2 + 1)(s^2 \Omega_5{}^2 + 1)}$
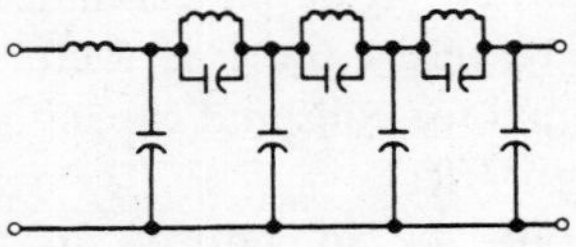	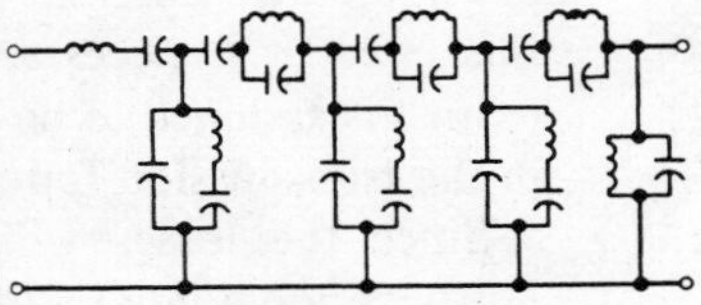	$D = B \dfrac{(s^2 + \Omega_1{}^2)(s^2 + \Omega_2{}^2)(s^2 + \Omega_3{}^2)(s^2 + \Omega_4{}^2)}{(s^2 \Omega_5{}^2 + 1)(s^2 \Omega_6{}^2 + 1)(s^2 \Omega_7{}^2 + 1)}$
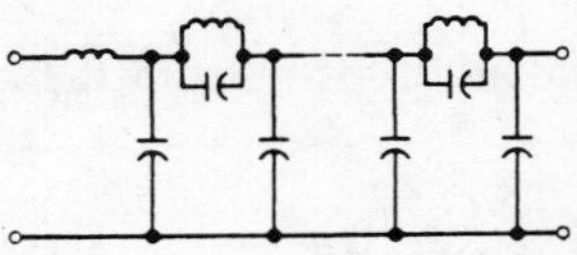	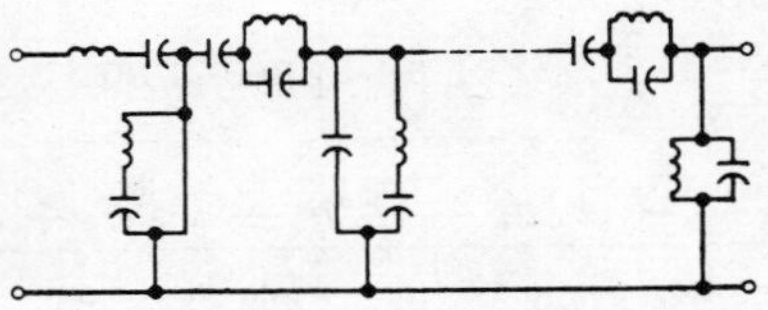	$D = B \dfrac{\prod\limits_{i=1}^{n} (s^2 + \Omega_i{}^2)}{\prod\limits_{v=1}^{n-1} (s^2 \Omega_v{}^2 + 1)}$

cannot produce attenuation peaks and whose frequencies are unknown

$$\gamma_1 = (\sqrt{l_1 c_1})^{-1} \qquad (4.15.5)$$

$$\gamma_6 = (\sqrt{l_6 c_6})^{-1} \qquad (4.15.6)$$

The direction of removal (from left to right or backwards) is not important and therefore let us proceed from the z_{11} side of Fig. 4.24. The function $z_{11}(p)$ is as given in Eq. 4.14.12.

$$z_{11} = pl_1 + \frac{1}{pc_1} + z_1 \qquad (4.15.7)$$

where z_1 is as shown in Fig. 4.24.

Since a series circuit in the shunt arm l_2 and c_2 produces a zero in z_1 at $p^2 = -\Omega_2^2$

$$z_{11} = \frac{p^2 l_1 + c_1^{-1}}{p}$$
$$+ \frac{(p^2\Omega_2^{-2} + 1)(p^8 b_8 + p^6 b_6 + p^4 b_4 + p^2 b_2 + b_0)}{p^{11}a_{11} + p^9 a_9 + p^7 a_7 + p^5 a_5 + p^3 a_3 + pa_1}$$
$$(4.15.8)$$

Combining these two terms over a common denominator, and comparing with the numerator of the original equation for z_{11} (4.14.12)

$$z_{11} = \frac{p^{12}a_{12} + p^{10}a_{10} + p^8 a_8 + p^6 a_6 + p^4 a_4 + p^2 a_2 + a_0}{p^{11}a_{11} + p^9 a_9 + p^7 a_7 + p^5 a_5 + p^3 a_3 + pa_1} \qquad (4.15.9)$$

the value of l_1 must be

$$l_1 = \frac{a_{12}}{a_{11}} \qquad (4.15.10)$$

The rest of unknowns may be expressed as follows:

$$b_0 = a_0 - a_1 c_1^{-1} \qquad (4.15.11)$$

$$b_2 = a_2 - a_3 c_1^{-1} - a_1 l_1 - b_0 \Omega_2^{-2} \qquad (4.15.12)$$

$$b_4 = a_4 - a_5 c_1^{-1} - a_3 l_1 - b_2 \Omega_2^{-2} \qquad (4.15.13)$$

$$b_6 = a_6 - a_7 c_1^{-1} - a_5 l_1 - b_4 \Omega_2^{-2} \qquad (4.15.14)$$

$$b_8 = a_8 - a_9 c_1^{-1} - a_7 l_1 - b_6 \Omega_2^{-2}$$
$$= a^{10}\Omega_2^2 - a_{11}c_1^{-1}\Omega_2^2 - a_9 l_1 \Omega_2^2 \qquad (4.15.15)$$

At $p^2 = -\Omega_2^2$ when the first shunt arm is at resonance, and hence presents a short circuit to the input

$$z_{11} = \frac{c_1^{-1} - \Omega_2^2 l_1}{j\Omega_2} \qquad (4.15.16)$$

From this it follows that

Thus both first circuit elements and the remainder function z_1 are known and further removal can proceed.

$$\frac{1}{z_1} = pc_7 + \frac{pc_2}{p^2\Omega_2^{-2} + 1} + \frac{1}{z_2} \qquad (4.15.18)$$

Thus $1/z_2$ must have a zero at $p^2 = -\Omega_3^2$ so that:

$$\frac{1}{z_1} = pc_7 + \frac{pc_2}{p^2\Omega_2^{-2} + 1}$$
$$+ \frac{(p^2\Omega_3^{-2} + 1)(p^7 d_7 + p^5 d_5 + p^3 d_3 + pd_1)}{p^8 b_8 + p^6 b_6 + p^4 b_4 + p^2 b_2 + b_0}$$
$$(4.15.19)$$

The remaining removal process presents no new difficulties.

Having determined all circuit elements of the filter it should be remembered that Ω_2, Ω_4, Ω_3, and Ω_1 can be interchanged if so desired. If the calculations lead to negative values for the circuit elements the transmission function cannot be realized in a Zig-Zag configuration.

The disadvantage of this design procedure is that it requires, as a rule, more accuracy in arithmetic computations in comparison with frequency transformation in conventional structures, but this disadvantage is compensated for by the economy in inductive components.

4.16   CONSIDERATION OF LOSSES IN ZIG-ZAG FILTERS

Coil-saving filters cannot be calculated by the method of regular frequency transformation. As a consequence, the effect of losses in circuit elements has to be considered in different fashion. For an estimate of this effect it is necessary to insert the loss factor ϵ' (which is assumed to be equal for coils and capacitors) in the transmission function $H'(s)$.

Since the losses will only be introduced in the Hurwitz polynomial a transformation of the following kind is necessary:

$$p = p' - \epsilon' \qquad (4.16.1)$$

For an approximative evaluation of ϵ' from the ϵ of

$$c_1^{-1} = \frac{\Omega_2^{12}a_{12} - \Omega_2^{10}a_{10} + \Omega_2^8 a_8 - \Omega_2^6 a_6 + \Omega_2^4 a_4 - \Omega_2^2 a_2 + a_0}{-\Omega_2^{10}a_{11} + \Omega_2^8 a_9 - \Omega_2^6 a_7 + \Omega_2^4 a_5 - \Omega_2^2 a_3 - a_1} + \frac{\Omega_2^2 a_{12}}{a_{11}} \qquad (4.15.17)$$

the normalized function, the following can be formulated:

$$\frac{f_m}{\Delta f}\frac{(p'^2 + 1)}{p'} + \epsilon = \frac{f_m}{\Delta f}\frac{(p + \epsilon')^2 + 1}{p + \epsilon'} \quad (4.16.2)$$

A power series development then provides the following approximation:

$$\frac{f_m}{\Delta f}\left[\frac{p^2 + 1}{p} + \epsilon\frac{\Delta f}{f_m}\right] \approx \frac{f_m}{\Delta f}\left[\frac{p^2 + 1}{p} + \left(1 - \frac{1}{p^2}\right)\epsilon'\right] \quad (4.16.3)$$

Therefore when $p^2 = -\omega^2/\omega_m^2$;

$$\epsilon\frac{\Delta f}{f_m} = \left(1 + \frac{\omega_m^2}{\omega^2}\right)\epsilon' \quad (4.16.4)$$

In relatively narrow-band filters the value of ω is always approximately equal to ω_m. Consequently, for such filters the coil quality (or capacitors quality) is approximately

$$Q = \frac{1}{\epsilon'} = \frac{1}{\epsilon}\frac{2f_m}{\Delta f} \quad (4.16.5)$$

4.17 REALIZATION PROCEDURE

The first step of realization is the selection of suitable reactances from the open circuit and/or short circuit X's shown in Table 2.2, which have the degree of the polynomial E.

The reactances X_o and X_s can be calculated from the polynomials E and F, which are already determined. (See also Table 4.4.)

Table 4.4 Looking at the Network from Its Input Terminals

	P even	P odd
X_{1o}	$\dfrac{E_e - F_e}{E_o + F_o}$	$\dfrac{E_o - F_o}{E_e + F_e}$
X_{1s}	$\dfrac{E_o - F_o}{E_e + F_e}$	$\dfrac{E_e - F_e}{E_o + F_o}$

Looking Back from the Output Terminals

	P even	P odd
X_{2o}	$\dfrac{E_e + F_e}{E_o + F_o}$	$\dfrac{E_o + F_o}{E_e + F_e}$
X_{2s}	$\dfrac{E_o - F_o}{E_e - F_e}$	$\dfrac{E_e - F_e}{E_o - F_o}$

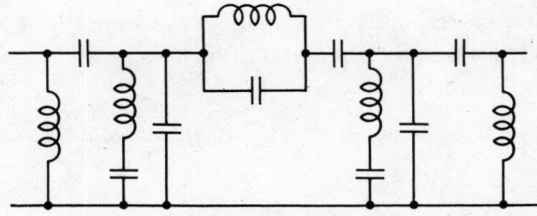

Fig. 4.26. Example of bandpass filter.

The network is completely determined by one of these reactances together with the complete set of poles of the transmission $H(s) = [E(s)/P(s)]$, or zeros of $P(s)$. The circuit elements can be calculated by partial or full removal of the residues at these poles.

Certain limiting rules must be observed in drawing the circuit configuration for the filter. These limitations may be taken into consideration as soon as the set of poles has been determined, before actual calculation.

The poles at finite, nonzero frequencies of Fig. 4.26 are easily seen, since they correspond to parallel-resonant circuits in the series arms and to series-resonant circuits in the shunt arms.

To check the poles at $\Omega = 0$, all the series inductances are replaced by short circuits and all the shunting capacitors by open circuits. For example, the filter of Fig. 4.26 has three singular poles at zero frequency, because, after performing this operation, the remaining circuit will be as shown in Fig. 4.27.

Each circuit element means a singular pole at $\Omega = 0$. To check those at $\Omega = \infty$, all the series capacitors are replaced by short circuits and all the shunting inductances by open circuits. The result for the above network is shown in Fig. 4.28.

This means there is one pole at $\Omega = \infty$. Let us assume that the transmission function $H(s)$ is given

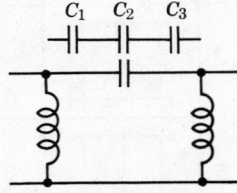

Fig. 4.27. Circuit corresponding to poles at $\Omega = 0$.

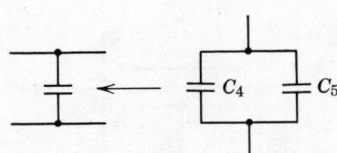

Fig. 4.28. Poles at $\Omega = \infty$.

and the corresponding schematic is shown in Fig. 4.26. This leads to a discrimination function

$\Omega_{2\infty}$, and $\Omega_{3\infty}$ ($\pm j\Omega_{1\infty}$, etc.). These are the nonzero roots of the denominator.

$$D(s) = B\frac{(s^2 + \Omega_1{}^2)(s^2 + \Omega_2{}^2)(s^2 + \Omega_3{}^2)(s^2 + \Omega_4{}^2)(s^2 + \Omega_5{}^2)}{s^3(s^2 + \Omega_{1\infty}{}^2)(s^2 + \Omega_{2\infty}{}^2)(s^2 + \Omega_{3\infty}{}^2)}$$

where B is a constant. The poles in this rational expression are in the denominator. That is,

1. Three poles at $\Omega = 0$ (s^3 in denominator).
2. Two poles at each of the real frequencies $\Omega_{1\infty}$,

3. One pole at $\Omega = \infty$.

The removal has been done in the following way:

1. Full removal of a pole at $\Omega = 0$, which results in a shunting inductance (see Fig. 4 26).

Table 4.5 Steps of removal in example of Fig. 4.26, where B is susceptance

	POLE	REMOVAL	REMAINDER FUNCTION
1	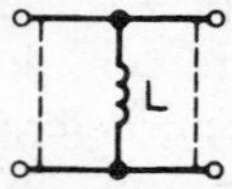 L	$L = \dfrac{1}{SB}\Big\|_{S=0}$	$B_1 = B - \dfrac{1}{SL}$
2	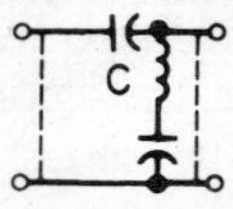 C	$C = \dfrac{1}{SX_1}\Big\|_{S^2 = -\Omega_{1\infty}^2}$	$X_2 = X_1 - \dfrac{1}{SC}$
3	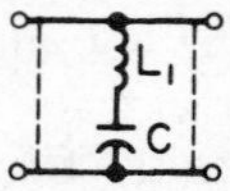 L_1 C	$L = \dfrac{SX_2}{S^2 + \Omega_{1\infty}^2}\Big\|_{S^2 = -\Omega_{1\infty}^2}$ $C = \dfrac{1}{L_1\Omega_{1\infty}^2}$	$\dfrac{1}{X_3} = \dfrac{1}{X_2} - \dfrac{S/L_1}{S^2 + \Omega_{1\infty}^2}$
4	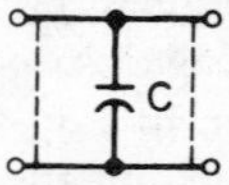 C	$C = \dfrac{B_3}{S}\Big\|_{S^2 = -\Omega_{3\infty}^2}$	$B_4 = B_3 - SC$
5	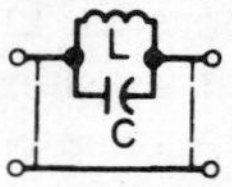 L C	$C = \dfrac{SB_4}{S^2 + \Omega_{3\infty}^2}\Big\|_{S^2 = -\Omega_{3\infty}^2}$ $L = \dfrac{1}{C\Omega_{3\infty}^2}$	$\dfrac{1}{B_5} = \dfrac{1}{B_4} - \dfrac{S/C}{S^2 + \Omega_{3\infty}^2}$
6	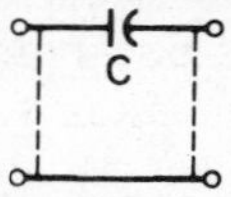 C	$C = \dfrac{1}{SX}\Big\|_{S = -\Omega_{2\infty}^2}$	$X_7 = X_6 - \dfrac{1}{SC}$
7	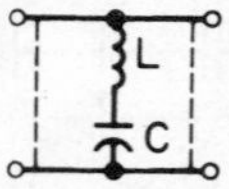 L C	$L = \dfrac{SX_7}{S^2 + \Omega_{2\infty}^2}\Big\|_{S^2 = -\Omega_{2\infty}^2}$ $C = \dfrac{1}{L\Omega_{2\infty}^2}$	$\dfrac{1}{X_8} = \dfrac{1}{X_7} - \dfrac{S/L}{S^2 + \Omega_{2\infty}^2}$
8	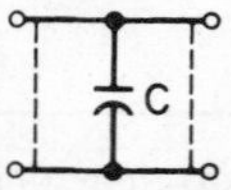 C	$C = \dfrac{B}{S}\Big\|_{S=\infty}$	$B_9 = B_8 - SC$
9	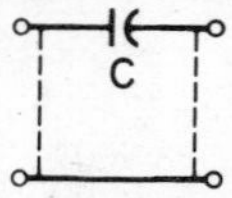 C	$C = \dfrac{1}{SX_9}\Big\|_{S=0}$	$X_{10} = X_9 - \dfrac{1}{SC}$
10	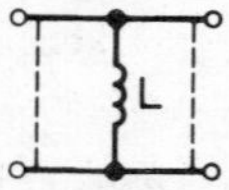 L	$L = \dfrac{1}{SB_{10}}\Big\|_{S=0}$	——

Table 4.6 Simplest pole removals

Pole frequency configuration		Remainder function
	$C=\left.\dfrac{1}{sX}\right|_{s=0}$	$X_1 = X - \dfrac{1}{sC}$
	$L=\left.\dfrac{1}{sB}\right|_{s=0}$	$B_1 = B - \dfrac{1}{sL}$
	$L=\left.\dfrac{X}{s}\right|_{s=\infty}$	$X_1 = X - sL$
	$C=\left.\dfrac{B}{s}\right|_{s=\infty}$	$B_1 = B - sL$

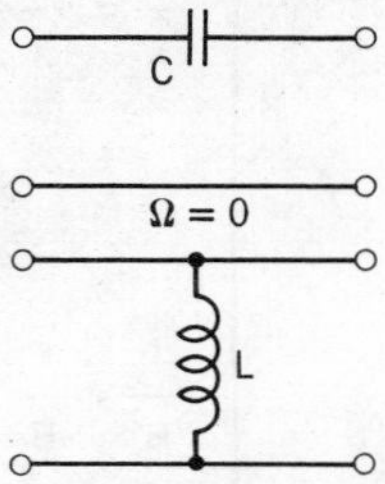

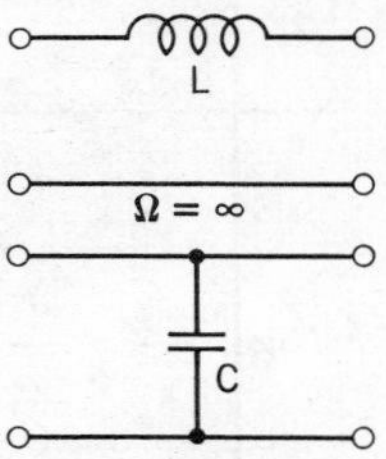

2. Partial removal of pole C_1 at $\Omega = 0$, in preparation for full removal of the poles at $\Omega = \pm\Omega_{1\infty}$. This partial pole removal is shown in Fig. 4.27 as a fraction of the total capacitance.

3. Full removal of pole at $\Omega = \Omega_{1\infty}$, resulting in the resonant circuit across the line.

4. Partial removal of pole at $\Omega = \infty$ in preparation for full removal of the poles at $\Omega = \pm\Omega_{3\infty}$. This partial pole C_4 is a fraction of the total capacitance, as shown in Fig. 4.28.

5. Full removal of the poles at $\Omega = \pm\Omega_{3\infty}$ with a parallel-resonant circuit in the series arm.

6. Partial removal of a pole $\Omega = 0$, which is a series capacitor C_2 in preparation for full removal of the poles at $\Omega = \pm\Omega_{2\infty}$.

7. Full removal of the poles on $\Omega = \pm\Omega_{2\infty}$ with a series-resonant circuit across the line.

8. Full, final removal of the remainder of the pole at $\Omega = \infty$ with a capacitor C_5 across the line.

9. Full removal of the remainder of a pole at $\Omega = 0$ with a series capacitance C_3.

10. Full removal of a pole at $\Omega = 0$ with a shunting inductance. Table 4.5 corresponds to the steps outlined above. Some other rules have to be observed and remembered.

For example, the full removal of an attenuation pole by use of an inductance or capacitance can be done only at $\Omega = 0$ or $\Omega = \infty$. Partial pole removals (inductance, capacitance) take place at finite frequencies

in preparation for full pole removal in the form of a resonant circuit. In Table 4.6 the simplest pole removals are tabulated.

In Table 4.7 partial pole removals from left to right are tabulated. The same reactances are being removed, but at finite frequencies. The remainder function is the same as in Fig. 4.26 but partial removal is used as a means to obtain full removal of the following pole.

4.18 NUMERICAL EXAMPLE OF REALIZATION

The Cauer-Chebyshev filter of degree $n = 3$ with $\rho = 20\%$ and $\theta = 21°$ ($\Omega_s = 2.790428$) provides a polynomial D

$$D(s) = \frac{s^3 + 0.273333s}{0.0755665s^2 + 0.276463} = \frac{F(s)}{P(s)}$$

which enables us to make the polynomial $F(s) + P(s)$, of the following form:

$$F(s) + P(s) = s^3 + 0.0755665s^2 + 0.273333s + 0.276463$$

From this expression the roots of Hurwitz polynomial are determined so that

$$E(s) = s^3 + 0.992009s^2 + 0.7625195s + 0.276463$$

For the realization from left to right we choose

$$X_{10} = \frac{E_e - F_e}{E_0 + F_o} = \frac{0.992009s^2 + 0.276463}{2s^3 + 1.035852s} = \frac{1}{B_1(s)}$$

Table 4.7   Partial pole removals from left to right

(circuit: C, L_1, C_1)	$C = \dfrac{1}{SX}\Big\|_{s^2=-\Omega_\infty^2}$	$L_1 = \dfrac{SX_1}{s^2+\Omega_\infty^2}\Big\|_{s^2=-\Omega_\infty^2}$ $C_1 = \dfrac{1}{L_1\,\Omega_\infty^2}$	$\dfrac{1}{X_2} = \dfrac{1}{X_1} - \dfrac{S/L_1}{s^2-\Omega_\infty^2}$
(circuit: L, C_1, L_1)	$L = \dfrac{1}{SB}\Big\|_{s^2=-\Omega_\infty^2}$	$L_1 = \dfrac{1}{C_1\,\Omega_\infty^2}$ $C_1 = \dfrac{SB_1}{s^2+\Omega_\infty^2}\Big\|_{s^2=-\Omega_\infty^2}$	$\dfrac{1}{B_2} = \dfrac{1}{B} - \dfrac{S/C_1}{s^2+\Omega_\infty^2}$
(circuit: L, L_1, C_1)	$L = \dfrac{X}{s}\Big\|_{s^2=-\Omega_\infty^2}$	$L_1 = \dfrac{SX_1}{s^2+\Omega_\infty^2}\Big\|_{s^2=-\Omega_\infty^2}$ $C_1 = \dfrac{1}{L_1\,\Omega_\infty^2}$	$\dfrac{1}{X_2} = \dfrac{1}{X_1} - \dfrac{S/L_1}{s^2+\Omega_\infty^2}$
(circuit: C, C_1, L_1)	$C = \dfrac{B}{s}\Big\|_{s^2=-\Omega_\infty^2}$	$C_1 = \dfrac{SB_1}{s^2+\Omega_\infty^2}\Big\|_{s^2=-\Omega_\infty^2}$	$\dfrac{1}{B_2} = \dfrac{1}{B_1} - \dfrac{S/C_1}{s^2+\Omega_\infty^2}$

There are two poles of attenuation at $\Omega_2 = 1.91273$ and a simple attenuation pole at $\Omega = \infty$. The circuit configuration is given in Fig. 4.29.

According to the rules for removal for $s^2 = -\Omega_2^2 = -3.658536$, which are given in Table 4.5 (case 4)

$$C_1 = \frac{B}{s}\Big|_{s^2=-\Omega_2^2} = \frac{-6.281220}{-3.352838} = 1.873404$$

The remainder function is

$$B_2 = B_1 - sC_1 = \frac{0.141566s^3 + 0.517925s}{0.992009s^2 + 0.276463}$$

Full removal of the poles at $s^2 = -\Omega_2^2 = -3.658536$

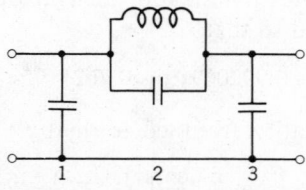

Fig. 4.29.   Third-order lowpass filter with one attenuation peak.

yields a parallel-resonant circuit in the series arm

$$C_2 = \frac{sB_2}{s^2 + \Omega_2^2}\Big|_{s^2=-\Omega_2^2} = \frac{-0.517925}{-3.352838} = 0.154473$$

$$L_2 = \frac{1}{C_2\Omega_2^2} = 1.769457$$

The remainder function is

$$\frac{1}{B_3} = \frac{1}{B_2} - \frac{s/C_2}{s^2 + \Omega_2^2} = \frac{0.075564s^2 + 0.276463}{0.141566s(s^2 + 3.658536)}$$

$$= \frac{0.075564(s^2 + 3.658536)}{0.141566s(s^2 + 3.658536)}$$

The unnormalizing process does not present any new difficulty.

4.19   FULL AND PARTIAL REMOVAL FOR A FIFTH-ORDER FILTER

In the following text a further explanation of the removal process will be given. The order of the filter

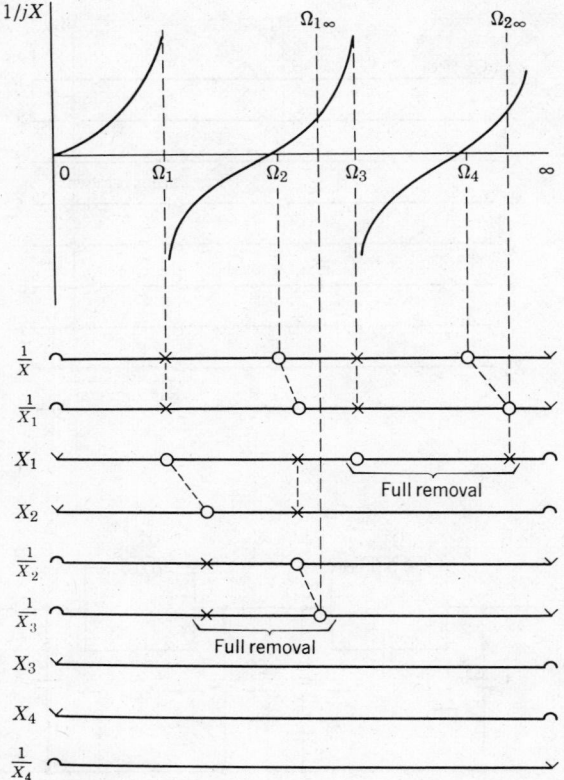

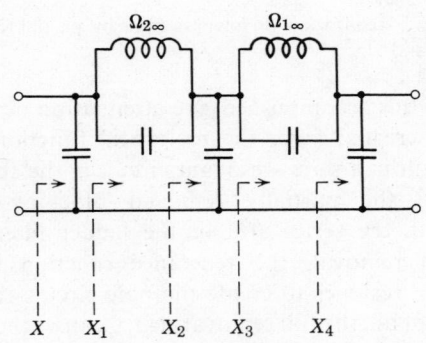

Fig. 4.30. Realization of lowpass filter by partial removal only at $\Omega = \infty$.

is five, with poles of attenuation at $\pm\Omega_{1\infty}$ and $\pm\Omega_{2\infty}$ and ∞. The reactance function is shown in Fig. 4.30. In the lower part of the same figure, a realization of the reactance function is shown. As we know, the reactance function is characterized by position of its zeros and poles. We will first try to remove the pole $\pm j\Omega_{2\infty}$.

As a preparatory step we subtract from $1/X$ a part of the function at $\Omega = \infty$, which is part of one pole common to the reactance function and the attenuation.

In the remainder reactance function $1/X_1$, a zero Ω_4 has to move to $\Omega_{2\infty}$ as shown in Fig. 4.30. X_1 now has a pair of poles at $\pm j\Omega_{2\infty}$ that can be removed and hence reduce the order of the function by two. The remainder function X_2 is now only a third order.

This process now has to be repeated. From $1/X_2$ a part of the residual function is removed again at $\Omega = \infty$, so that the residual function $1/X_3$ has a zero at $\Omega_{1\infty}$. After removal of a pair of poles at $\pm j\Omega_{1\infty}$ from X_3 the remainder function X_4 will be first order. Full removal from $1/X_4$ finally provides the last pole at $\Omega = \infty$.

The removal process is completed. All of the steps of the removal and the network are shown in Fig. 4.30. The positions of the appropriate reactance mentioned above are shown also. From this simple example it is evident that the position of the poles cannot be exchanged in the foregoing circuit. If, for example, the pole at $\pm j\Omega_{1\infty}$ is removed first then in $1/X$ the zero at Ω_4 would be forced to travel further to the right than $\Omega_{2\infty}$. Its position would allow no further partial removal at $\Omega = \infty$, thus necessitating that Ω_4 equal $\Omega_{2\infty}$.

The possibility of realization without mutual coupling is dependent on the two following conditions.

1. At least one pole of attenuation has to be confluent with a pole of the reactance function. The residue of this pole has to be partially removed at $\Omega = 0$ or $\Omega = \infty$. In the case shown in Fig. 4.30, that pole is at $\Omega = \infty$.

2. Looking from the position of the reactance pole at which the partial removal is taking place, every pole of attenuation is followed by a zero of reactance on the low frequency side.

In Fig. 4.31 realizable and unrealizable distributions of attenuation poles are shown. When both conditions are fulfilled, there exists at least one succession of pole removals which leads to a realizable network without mutual inductances. At every partial removal only as

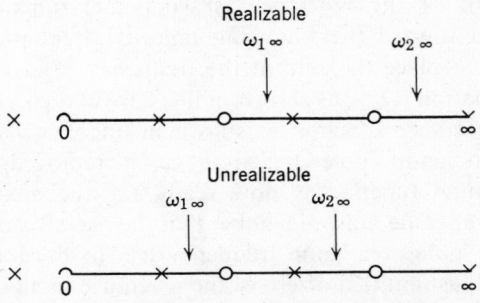

Fig. 4.31. Realizable and unrealizable distribution of attenuation poles.

much can be removed so that the other poles of attenuation still lie above the corresponding zeros of the reactance function.

If these conditions are not fulfilled, the realization is still possible, but with mutual inductances since the negative elements can be removed simply by using coupled coils.

The method just described allows even more generalizations. The full removals described earlier have been accomplished with the aid of the preliminary partial removals at $s = 0$ or $s = \infty$. But the partial removals can be accomplished even at finite frequencies; for example, at the frequency corresponding to a finite attenuation pole. Instead of a coil or capacitor, the partial removal will consist of a resonant circuit that constitutes a pole of attenuation. Since this circuit only partially removes the attenuation pole, full removal of this pole will be performed later in the circuit. Naturally, this signifies more (redundant) elements and can be justified only by some other reasoning; such as the possibility of realizing the network without mutual inductance, the desire to redistribute the reactance power circulating in the components, or the creation of conditions which allow including magnetostrictive and piezoelectric resonators.

Figure 4.32 shows an example of a reactance function similar to the one shown in Fig. 4.30. The partial removal is accomplished at $\Omega_{2\infty}$. Both blocking-resonant circuits in the series arm are tuned to the same frequency $\Omega_{2\infty}$ which is an attenuation pole of the filter. The partial removal can be accomplished simultaneously at two frequencies to create two attenuation poles. The following full removal will furnish the circuit elements for the poles. The circuit configuration is the familiar bandpass schematic with four parallel-resonant circuits as shown in Fig. 4.25.

The first operation shown in Fig. 4.32 is the partial removal of a reactance zero at $\Omega = \infty$, as a preparatory step for a partial removal of the resonant circuit. A zero of the reactance (susceptance) function is moved toward the following pole of attenuation in order to place the zero at the frequency of a pole of attenuation ($\Omega_{2\infty}$, as shown in line 2 from top). From the reactance function as shown in line 3, a part of the attenuation pole $\Omega_{2\infty}$ circuit can be removed. The remaining function is now ready for the next full removal. One must remember that the partial removal of the pole at a finite frequency has to be done in such a fashion that a zero of the reactance function on line 3 is moved to the same frequency as the attenuation pole $\Omega_{1\infty}$.

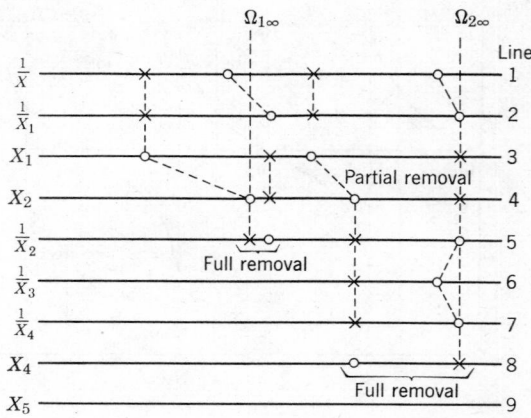

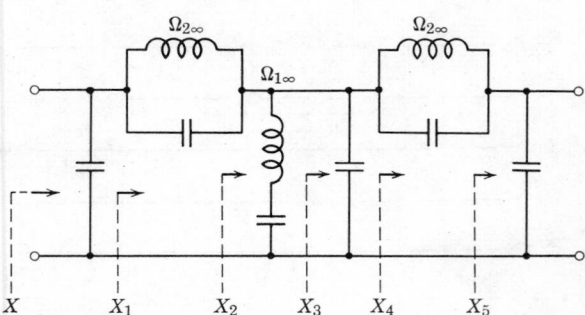

Fig. 4.32. Realization of lowpass filter by partial removals at $\Omega = \infty$ and $\Omega_{2\infty}$.

With this accomplished, the attenuation pole circuit can be created from the reciprocal functions. This will result in a series-resonant circuit in the shunt arm next to the partially removed blocking-resonant circuit in the series arm of the ladder filter. After the full removal, the reactance function still has sufficient residues to create the pole circuit at $\Omega_{2\infty}$.

First of all, the full removal at $\Omega_{1\infty}$ causes the shifting of the remaining finite frequency zero of the reactance function in line 6 toward the left, and hence satisfies the realizability condition (covering the attenuation pole by reactance zero from below).

The same reactance function therefore permits the partial removal from $1/X_3$, (on line 6) which produces one capacitor. In Fig. 4.32, on line 7, the remainder function will have a reactance zero which coincides with the attenuation pole position $\Omega_{2\infty}$. The reactance function can now have the pole at $\Omega_{2\infty}$ removed, as shown in line 8. The reactance on line 9 has a pole at zero frequency and zero at infinity. These are the properties of a capacitor across the line as shown in Fig. 4.32.

Table 4.8 Steps in reactance removal for the bandpass filter

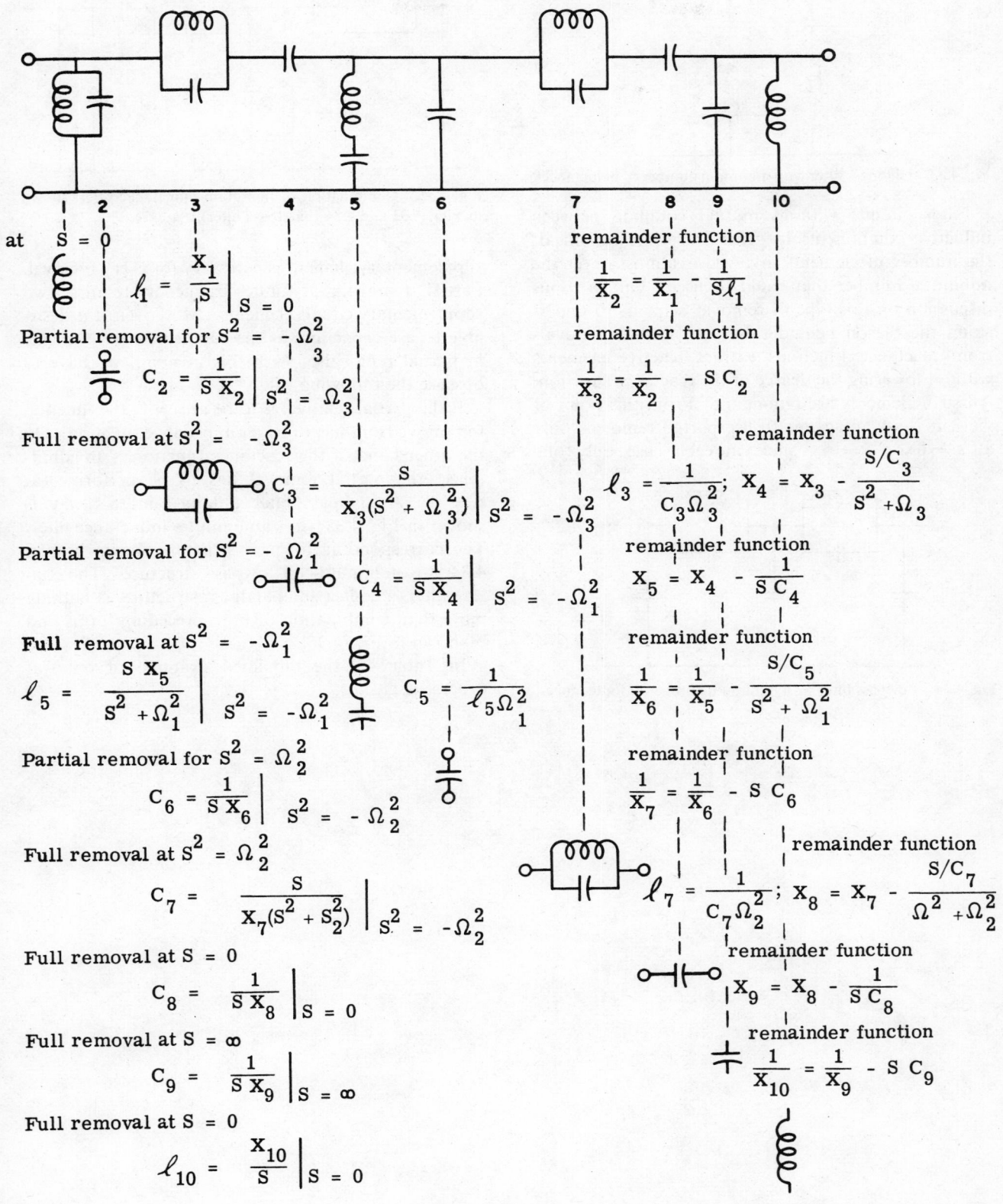

at $S = 0$

$$\ell_1 = \frac{X_1}{S}\Bigg|_{S=0}$$

remainder function

$$\frac{1}{X_2} = \frac{1}{X_1} - \frac{1}{S\ell_1}$$

Partial removal for $S^2 = -\Omega_3^2$

$$C_2 = \frac{1}{S X_2}\Bigg|_{S^2 = \Omega_3^2}$$

remainder function

$$\frac{1}{X_3} = \frac{1}{X_2} - S C_2$$

Full removal at $S^2 = -\Omega_3^2$

$$C_3 = \frac{S}{X_3(S^2 + \Omega_3^2)}\Bigg|_{S^2 = -\Omega_3^2}$$

remainder function

$$\ell_3 = \frac{1}{C_3 \Omega_3^2}; \quad X_4 = X_3 - \frac{S/C_3}{S^2 + \Omega_3}$$

Partial removal for $S^2 = -\Omega_1^2$

$$C_4 = \frac{1}{S X_4}\Bigg|_{S^2 = -\Omega_1^2}$$

remainder function

$$X_5 = X_4 - \frac{1}{S C_4}$$

Full removal at $S^2 = -\Omega_1^2$

$$\ell_5 = \frac{S X_5}{S^2 + \Omega_1^2}\Bigg|_{S^2 = -\Omega_1^2}$$

$$C_5 = \frac{1}{\ell_5 \Omega_1^2}$$

remainder function

$$\frac{1}{X_6} = \frac{1}{X_5} - \frac{S/C_5}{S^2 + \Omega_1^2}$$

Partial removal for $S^2 = \Omega_2^2$

$$C_6 = \frac{1}{S X_6}\Bigg|_{S^2 = -\Omega_2^2}$$

remainder function

$$\frac{1}{X_7} = \frac{1}{X_6} - S C_6$$

Full removal at $S^2 = \Omega_2^2$

$$C_7 = \frac{S}{X_7(S^2 + S_2^2)}\Bigg|_{S^2 = -\Omega_2^2}$$

remainder function

$$\ell_7 = \frac{1}{C_7 \Omega_2^2}; \quad X_8 = X_7 - \frac{S/C_7}{\Omega^2 + \Omega_2^2}$$

Full removal at $S = 0$

$$C_8 = \frac{1}{S X_8}\Bigg|_{S=0}$$

remainder function

$$X_9 = X_8 - \frac{1}{S C_8}$$

Full removal at $S = \infty$

$$C_9 = \frac{1}{S X_9}\Bigg|_{S=\infty}$$

remainder function

$$\frac{1}{X_{10}} = \frac{1}{X_9} - S C_9$$

Full removal at $S = 0$

$$\ell_{10} = \frac{X_{10}}{S}\Bigg|_{S=0}$$

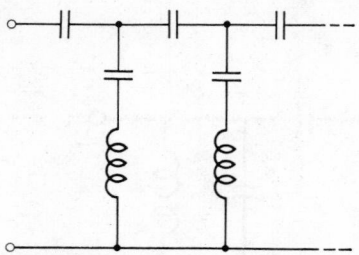

Fig. 4.33. Highpass filter with minimum number of inductances.

Filter circuits without mutual coupling between inductive components in general are not canonical; the number of elements involved is almost never the minimum number that could be used. Our previous discussion on this type of removal helps us to understand the reason behind this fact. Partial removals from reactance functions extract reactive elements without lowering the degree of the reactance function. All such elements are redundant. From this point of view, it is advantageous to use partial removals only at $s = 0$ or $s = \infty$, since in each case only one

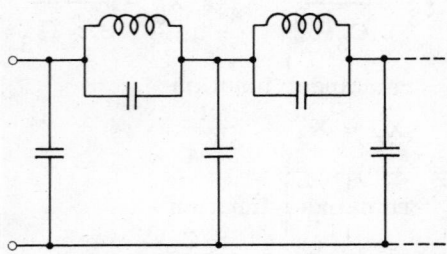

Fig. 4.34. Lowpass filter with minimum number of inductances.

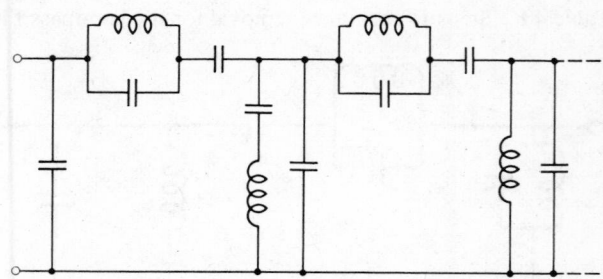

Fig. 4.35. Combination of minimum inductance filters shown in Figs. 4.33 and 4.34 (bandpass filter).

supplementary element is necessary for every removal. Partial removals at finite frequency require two supplementary circuit elements. Moreover, it is desirable to use capacitors as the supplementary elements in partial removals. With this in mind, we have to observe the following rule:

If the partial removal is to be at $s = 0$, the function to remove from has to be an impedance function. On the other hand, if the reactance function is an admittance, the partial removal has to be performed at $s = \infty$. A highpass filter obtained in this way is shown in Fig. 4.33; it is a minimum inductance filter. The corresponding lowpass structure shown in Fig. 4.34 is a minimum-coil lowpass structure. The Zig-Zag filter, a coil-saving bandpass structure, is nothing more than combination of both preceding forms and is shown in Fig. 4.35.

In Table 4.8 the tabulated example of removal process is given.

5

The Catalog of Normalized Lowpass Filters

5.1 INTRODUCTION TO THE CATALOG

This chapter will show that modern filter design methods can be used by engineers, and that it is not a restricted field for specialists. Frequently, requirements for extreme numerical exactness make the modern design techniques prohibitive. In the simplification of the modern design approach, it is desirable that the element values be precalculated and normalized in a convenient manner and related to the desired filter characteristics. It is also desirable that these relationships be in catalog form, at least for the types of filters most commonly used. For a number of simpler cases, the effective filter parameters have already been published by many authors. The information presented in this chapter is developed along the lines established by Glowatzki, Saal, and Ulbrich. Extensive data will be presented for the Cauer-Chebyshev filter developed with the aid of Zolatarev fractions having equal minima of attenuation in the stopband. In Chapter 6, information in the form of tables will be presented for the lowpass model of the Chebyshev polynomial filter, as well as other all-pole types.

Principles of Tabulation

The properties of the lowpass model can be determined by the following effective parameters:

1. The order of the transmission function which sets the number of necessary reactive elements necessary for the realization.
2. The minimum attenuation in the stopband above a certain limiting frequency, $A_{\min}$.
3. The maximum level of ripples in the passband, $A_{\max}$.

In Cauer-Chebyshev filters, the most important parameter is the sharpness of the required response curve. There is a table of useful parameters for the design of filters published by Glowatzki as a function of filter sharpness (Ω_s/Ω_c). In modern tabulation, originated also by Glowatzki, instead of normalized frequency, the modular angle θ was found to be more practical as an input function for the tables. The filter parameters can be classified and tabulated in accordance with the following:

1. Type of response—Butterworth (B), Chebyshev (C), and Cauer-Chebyshev (CC)
2. Order of complexity—n
3. Reflection coefficient ρ, or ripple factor ($A_{\max}$)
4. Modular angle θ.

Responses types (B) and (C) are relatively simple. They do not have modular angles and, therefore, the tables are short. To have (CC) responses without modular angles is senseless since the controllable sharpness of the response is a function of the modular angle. For example, the filter under catalog number CC 09 05 67 means a Cauer-Chebyshev filter of order $n = 9$ with $\rho = 5\%$ (reflection) and having a modular angle of 67°. The sharpness of the response is then (see Eq. 4.4.8)

$$\Omega_s = \frac{1}{k'} = \frac{1}{\sin \theta} = 1.086360$$

Simplest Tabulation

The simplest tabulated symmetric and antimetric polynomial filters are shown in Table 5.1, where only the type and order of response are included in the code number. For the odd order (symmetric) filters, the realizations having no transformers are easy to obtain. In the case indicated CC 04a, and for all the cases of antimetric filters with magnetically coupled inductances, the tabulated Cauer parameters in the

Table 5.1 Table of Simplest Polynomial Filters

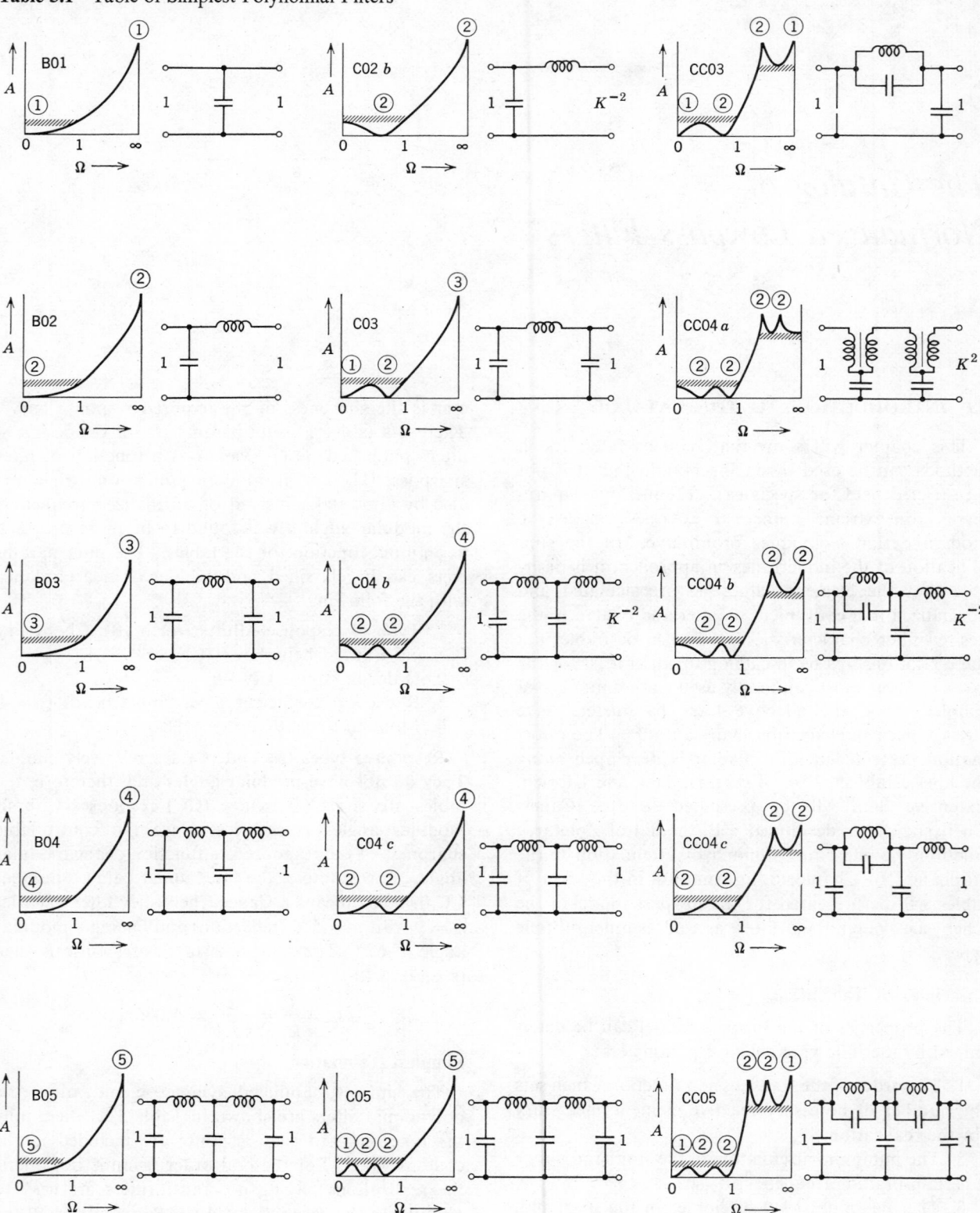

Numbers in the circles show the position of zeros in the passband and poles in the stopband.

Glowatzki tables are still valid, and the filter exhibits neither zero attenuation at $\Omega = 0$ nor a pole of attenuation $\Omega = \infty$. To realize such filters without magnetic coupling, a transformation must be performed.

Unlike the schematics for odd orders of filters, where the normalized loading resistances are equal ($r_1 = r_2 = 1$), schematics for even-order filters require nonequal load resistances, such as

$$r_2 = K^2 = \frac{1 - \rho}{1 + \rho}$$

In case b, the last pole is moved toward infinity. In this way, the need for magnetic coupling between inductances may be avoided. Filters of this kind are indicated by the letter b, for example CC 08 10b. To realize filter characteristics with equal input and output load impedances similar to that of odd-order filter, more transformations must be made. Sharpness in the transition region will be somewhat greater. In Table 5.1, the filters of even order and yet having equal terminations are shown (CC 04c). Instead of having finite attenuation at zero frequency, these filters exhibit $A = 0$ at $\Omega = 0$. The numbers shown in the circles in Table 5.1 show the position and the number of poles and zeros of the function.

Estimate of Filter Complexity with Nomograph

Figures 5.1, 5.2, and 5.3 show the nomographs published by Kawakami, which are useful in determining the required degree of transmission function in order to satisfy a given design condition. They avoid direct reference to any elliptic function parameter. The maximum value of ripples in the passband (A_{max}) is given at the right side of the nomograph. A straight line is drawn from the value of A_{max} permitted through the desired value of attenuation in the stopband (A_{min}).

The line runs up to the third vertical line and is then rotated to run parallel to the Ω-scale. The desired amount of attenuation at a given frequency will be guaranteed if the filtering function is of the order found at the intersection of the vertical line erected from the Ω-scale value and the line which runs parallel to the Ω-scale. If the crossing appears between two curves as shown in Fig. 5.4, the order that must be chosen is the largest integer, written above the curve.

Normalizing

The advantages of using normalized parameters in developing filter networks are well known. The most important advantage is the convenience of the numerical values obtained (from 0.01 up to 100). The resulting normalized network can be used as a basis for a multiplicity of real schematics having specific reference values. Two reference values are generally sufficient: (1) impedance level R_r and (2) reference frequency ω_r.

Normalized values included in the tables are dimensionless and defined as follows:

1. Normalized frequency

$$\Omega = \frac{f}{f_r}$$

2. Normalized resistance

$$r = \frac{R}{R_r}$$

3. Normalized inductance

$$L' = \frac{\omega_r L}{R_r}$$

4. Normalized capacitance

$$C' = \omega_r C R_r$$

The normalized cutoff frequency is $\Omega_c = 1$ for all lowpass filters. The stopband limit is given as Ω_s or f_s. The numerical value of corresponding frequencies is related by the following expression:

$$\frac{\Omega_s}{\Omega_r} = \frac{f_s}{f_r}$$

As a rule, the normalized input impedance is equal to unity, signifying that the reference resistance is equal to the source resistance.

Table Values

For the reader's convenience, tables for normalized lowpass models, tabulated according to ρ values, have been placed in Section 5.8.

The table values consist of three groups of information: (a) Operating parameters, (b) Pole-zero location in the complex plane, and (c) Element values.

The attenuation of a lowpass filter is completely described in terms of the following operating parameters.

Maximum peak ripple attenuation in the passband, A_{max}

Minimum attenuation in the stopband, A_{min}

Normalized stopband limit,

$$\Omega_s = \frac{\omega_s}{\omega_r} = \frac{f_s}{f_r}$$

For the transfer function of the third order ($n = 3$),

which is to be used throughout for an example, the definitive relation between A_{max}, A_{min}, and Ω_s is

$$\left(a_{min} + \tfrac{1}{2}\ln\frac{8}{a_{max}}\right)\ln\frac{8}{\Omega_s - 1} \approx 5n = 15 \quad (5.1.1)$$

where a is in Np and A is in dB. This formula is approximate, and valid only for small A_{max}, large A_{min}, and small Ω_s. It is evident that with two given parameters, it is possible to evaluate the third.

The filtering network is usually part of a more complex system and is preceded and followed by some other network or subsystem. The filter designer is often interested not only in attenuation or phase characteristics but also in the behavior of the input impedance in the passband, or consequently, the value of the reflection coefficient ρ.

The effective peak ripple attenuation of the reactive networks and the reflection coefficient are interconnected by the following expression:

$$a_{max} = -\ln\sqrt{1 - \rho^2} \quad (5.1.2)$$

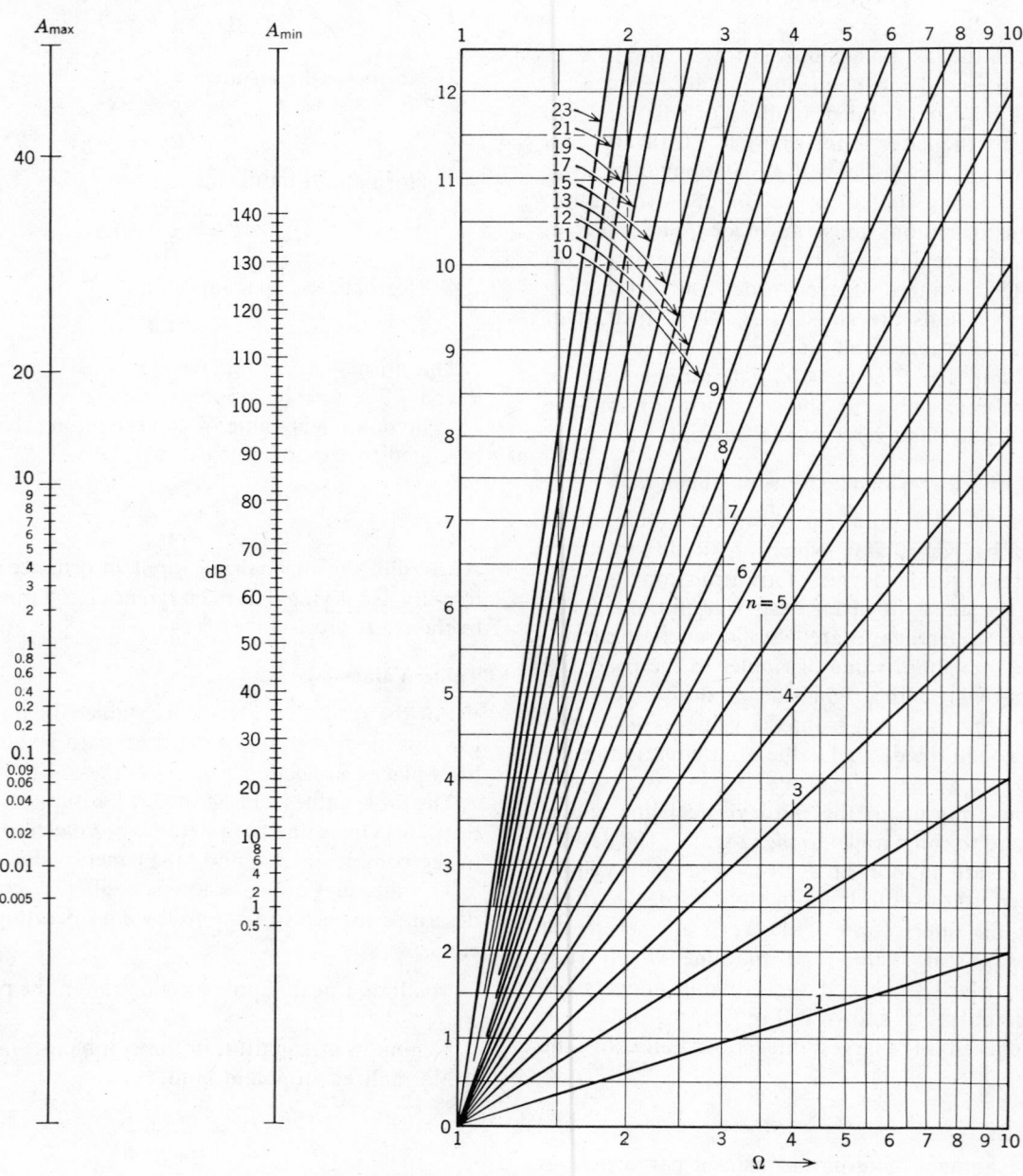

Fig. 5.1. Nomograph for Butterworth filters.

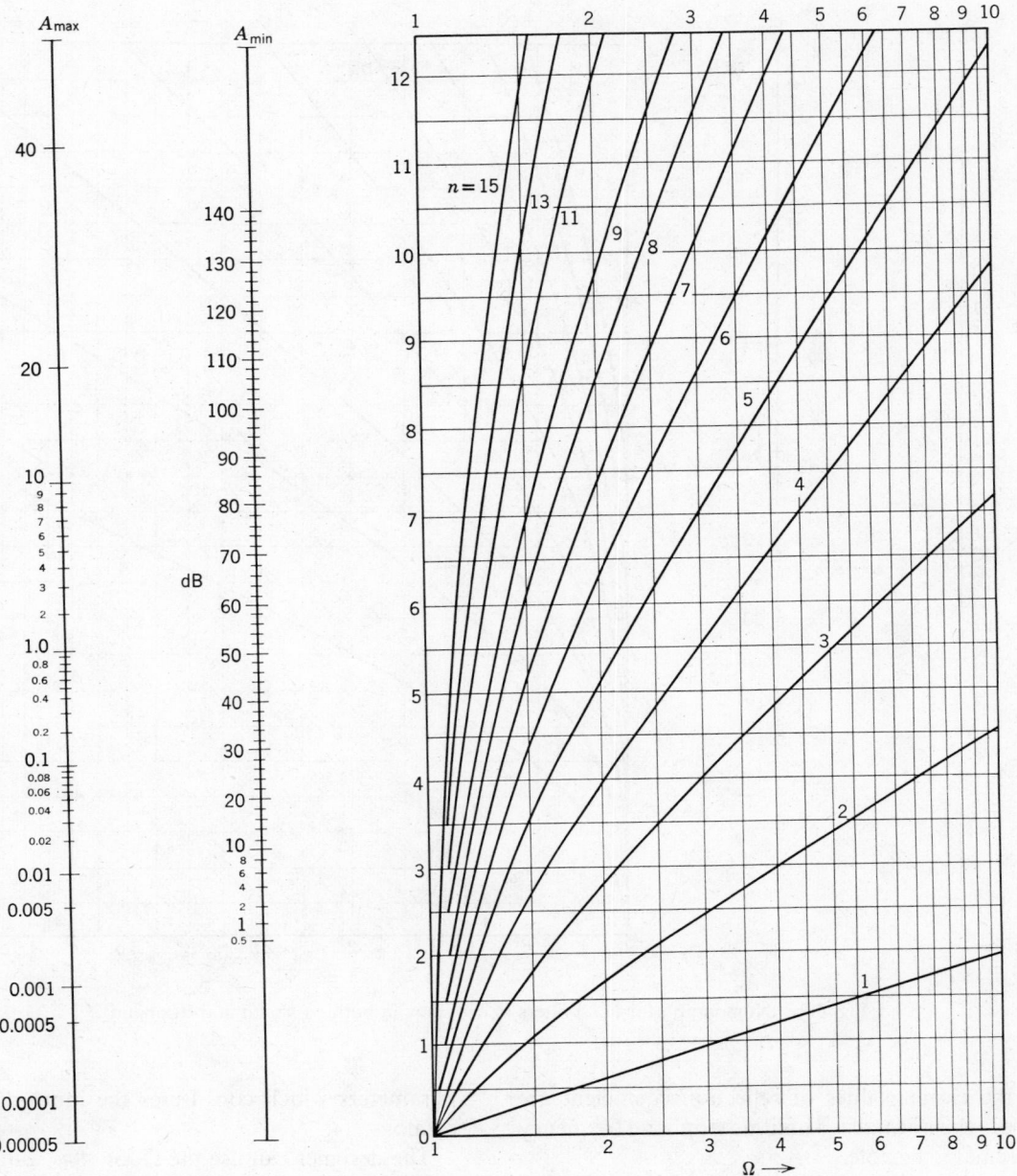

Fig. 5.2. Nomograph for Chebyshev Filters.

The reflection coefficients, usually specified for electronic equipment, correspond to a very small value of peak ripple attenuation *a*. Checking the correct realization of the network by means of the attenuation properties in the passband is, therefore, not easily achieved. A much more sensitive criterion for correct design and construction is the input impedance in the passband. In view of this consideration, it has been found useful to use the maximum permissible reflection coefficient in the passband as a characteristic and as a check quantity for a filter. Furthermore, it is sensible to use only integers for ρ values. Therefore the reflection coefficient is the basic parameter in the following tabulation.

In Table 5.2, the numerical value of attenuation for some practical values of reflection coefficients are shown along with the corresponding values of VSWR. When the order of filter *n* has been determined and

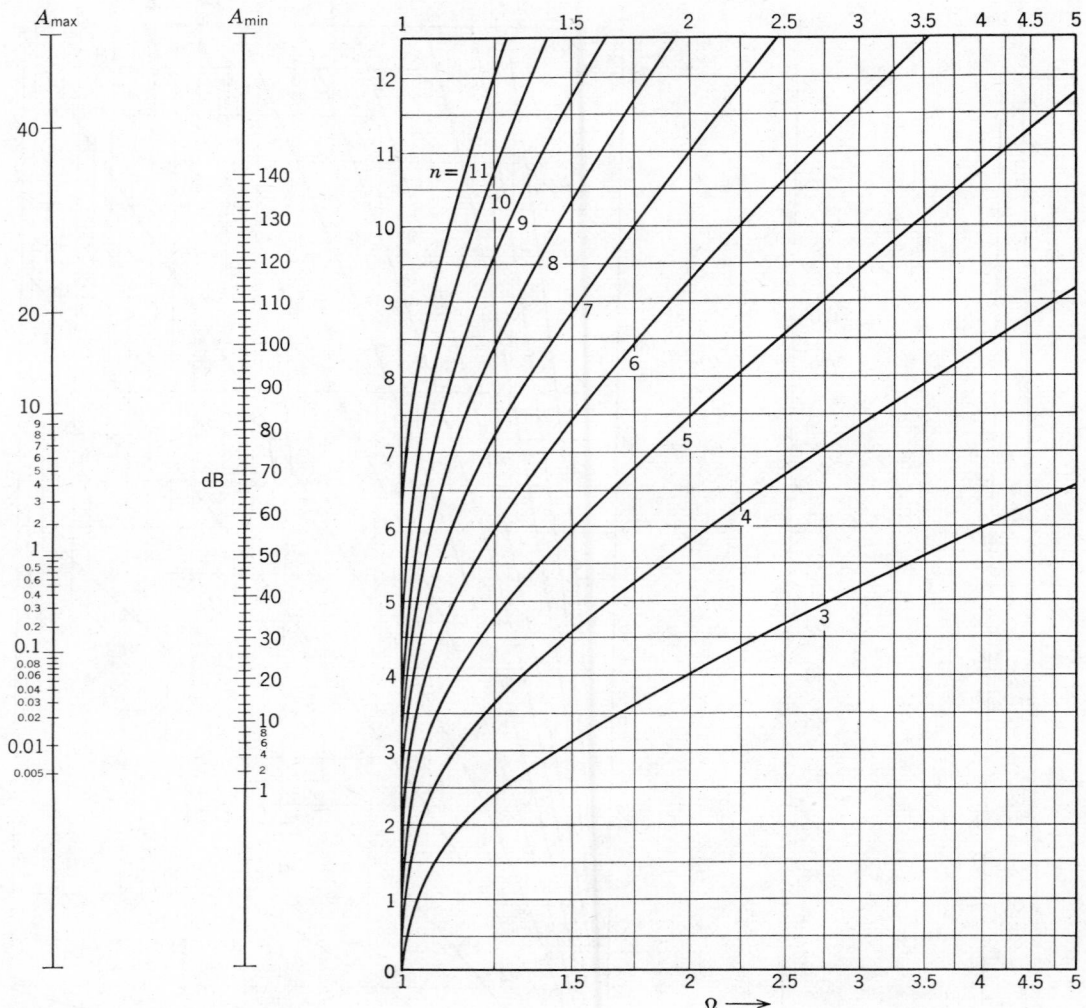

Fig. 5.3. Nomograph for (CC) filters (Chebyshev in both passband and stopband).

one of the eleven values of reflection coefficient are prescribed, the normalized limiting stopband frequency Ω_s still remains flexible.

For practical design work, it is desirable to have the sharpness tabulation progress in very fine steps especially in the proximity of unity. That fineness is reached when one uses the expression

$$\Omega_s = \frac{1}{\sin \theta} \quad \theta < 90° \qquad (5.1.3)$$

where θ is permitted to increase gradually. The density of values when angle θ becomes large is evident.

The relationship shown in (5.1.1) states the existence of a rigid numerical interrelation between the operating

parameters which constitutes the input data of filter catalogs.

The designer can use the Ω_s or A_{min} values in order to find the row of data for a specific filter design.

Pole-zero Information

The complex transmission function for the three-pole filter under consideration can be expected by

$$H(s) = B \frac{s^3 + a_2 s^2 + a_1 s + a_0}{s^2 + b_0} = \frac{1}{W(s)} \qquad (5.1.4)$$

with normalized complex frequency $s = \sigma + j\omega$, and real constants a_v and b_0. B is a relatively unimportant constant easily determined from boundary conditions at $\Omega = 0$.

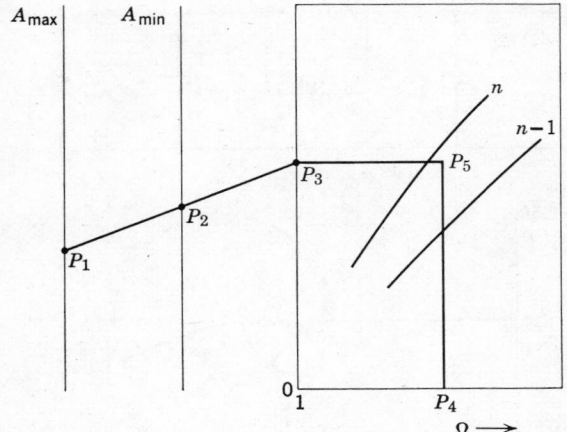

Fig. 5.4. Use of nomographs.

The same expression in factorized form is

$$H(s) = B \frac{(s + \sigma_0)(s + \sigma_1 - j\Omega_1)(s + \sigma_1 + j\Omega_1)}{(s - j\Omega_2)(s + j\Omega_2)}$$

$$= \frac{1}{W(s)} \quad (5.1.5)$$

where σ_0 and σ_1 are the real parts of the zeros, Ω_1 is the imaginary part of the conjugate zero, and Ω_2 is the imaginary part of the imaginary pole (same as the real pole of attenuation). The geometric representation of this frequency in the complex plane (s-plane) involves the pole-zero coordinates (P-Z data). With the aid of known pole and zero data, the transmission function is fully determined. With this information the steady-state filter properties, such as attenuation, phase, and group delay, as well as the dynamic properties, such as step response, can be determined

Table 5.2 Relation Between Reflection Coefficient, Attenuation, VSWR, and Echo Attenuation

ρ, %	a, Np	A, dB	VSWR	A_e, dB
1	0.00005	0.00043	1.020	39.9556
2	0.0002	0.0017	1.041	33.8754
3	0.0005	0.004	1.062	30.4010
4	0.0008	0.007	1.083	27.7952
5	0.0013	0.011	1.105	26.0580
8	0.0032	0.028	1.174	21.7150
10	0.005	0.044	1.222	19.9778
15	0.0113	0.098	1.353	16.5034
20	0.0207	0.18	1.500	13.8976
25	0.032	0.28	1.667	11.7261
50	0.14	1.25	3.000	4.7773

in an elementary fashion. For this reason, the transmission or transfer function parameters σ_0, σ_1, Ω_1, and Ω_2 are given in the tables.

In the pole-zero diagram there are only finite effective zeros of the transmission function. By adding poles on the opposite side of the $j\omega$-axis corresponding to the given zeros (image of zeros) an all-pass pole-zero diagram will appear. Because of symmetry with respect to the $j\omega$-axis no attenuation is possible, but the amount of phase shift or group delay that the all pass produces is doubled in comparison with the lowpass filter.

The Element Values

Filters having equal terminations constitute the great majority of all designs. On the other hand, open-circuit operation is very popular in certain systems. Most of the known tables include information for equally terminated cases. In this collection, the tabulation has been extended to open-circuit conditions.

In the three-zero configuration a relatively simple case exists, and it is possible to use some simple basic relations which are derived from the analysis of pole-zero locations in connection with coefficients:
For $K^2 = 1$,

$$C_1' = \frac{1}{\sigma_1} = C_3' \qquad C_2' = \frac{\sigma_0 - 2\sigma_1}{4\sigma_0\sigma_1} \quad (5.1.6)$$

For $K^2 = \infty$,

$$C_{1,3}' = \frac{\Omega_2^2 \mp \sigma_0^2}{2\sigma_0\Omega_2^2} \qquad C_2' = \frac{1}{\Omega_2^2 L_2'}$$

$$L_2' = \frac{8\sigma_1}{\sigma_1 + \Omega_1^2 + 2\sigma_0\sigma_1 + \sigma_0^2} \quad (5.1.7)$$

It can be noticed that, in the filter network, the pole-zero parameters are interconnected. For the specific case of $n = 3$

$$\Omega_2^2 = \frac{\sigma(\sigma_1^2 + \Omega_1^2)}{\sigma_0 - 2\sigma_1} \quad (5.1.8)$$

This relation can be used for checking purposes.

In the case of $K^2 = \infty$ (open-termination operation) when the attenuation in the passband is small, some element values are negative, as can be seen from Eqs. (5.1.7). The limiting case is the filter having one reactance equal to zero. In the catalog, the element values are in the following order:

$$C_1' = C_3', C_2', L_2' \text{ for } K^2 = 1$$
$$C_1', C_2', L_2', C_3' \text{ for } K^2 = \infty$$

and can be directly read from the table. The normalized element values for the dual schematic are given from the bottom of each table.

In the tables for $n = 3$, negative element values begin to appear when θ approaches 19° for a 1% reflection coefficient. For 2% reflection, the corresponding value of θ is 26°. In the schematic at the top of the filter catalog, the capacity C_1' is the limiting element. In the dual schematic, at the bottom of the tables, the inductance L_1' is the limiting element.

This leads to the natural conclusion that filters with negative element values cannot be realized without mutual coupling. To avoid transformers in the filter, some modification of requirements can be met, or some realization procedure can be modified. First of all, if the order of pole removal is changed, the effect could be sufficient to realize the filter without mutual inductance. If the goal is not achieved, the amount of permissible reflection in the passband must be increased. It is interesting to note that some elements could completely disappear from the schematic at a certain value of the reflection coefficient and a certain value of θ. Figure 5.5 shows the influence of pole removal order. If the first pole is at Ω_4 and the second is at Ω_2, the negative values for the input capacitor will begin to appear at larger values of θ. The case of the reverse order of poles given is illustrated by the dashed line and suggests a very big difference between two cases, especially for a low reflection coefficient.

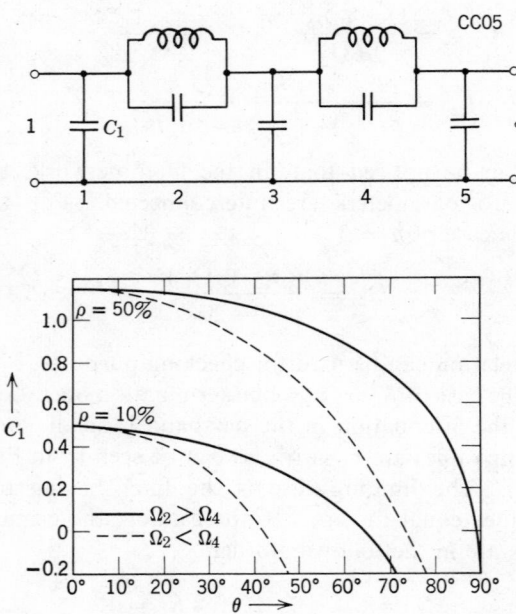

Fig. 5.5. Influence of pole removal order.

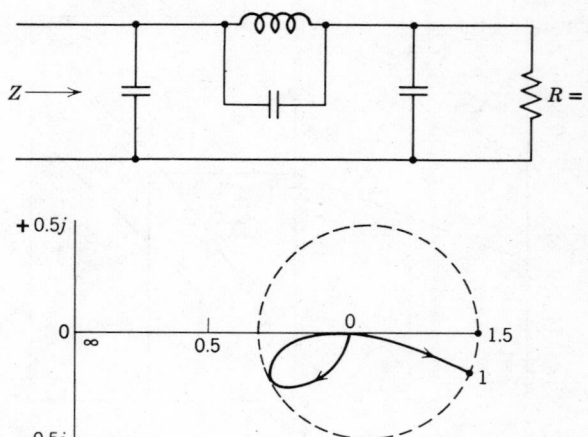

Fig. 5.6. Typical frequency behavior of input impedance.

Impedance in the Passband

Figure 5.6 shows the typical frequency behavior of the input impedance. When the frequency is zero, the filter exhibits a nominal impedance (normalized value $R = 1$). When the frequency increases, the reactive components will appear negative (reactance), then the curve of impedance will reach the circle of maximum allowable reflection coefficient. Further increase of frequency brings the input impedance back to the nominal value and finally to the second maximum at $\Omega = 1$ (catalog number of the filter is CC 03 20 30). All maximum values lie on the circle which outlines the given filter.

Figure 5.7 shows a more complicated filter, namely

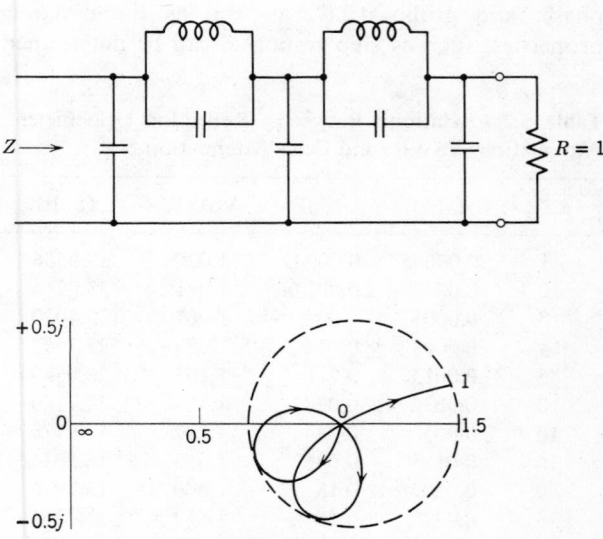

Fig. 5.7. Impedance characteristic in passband.

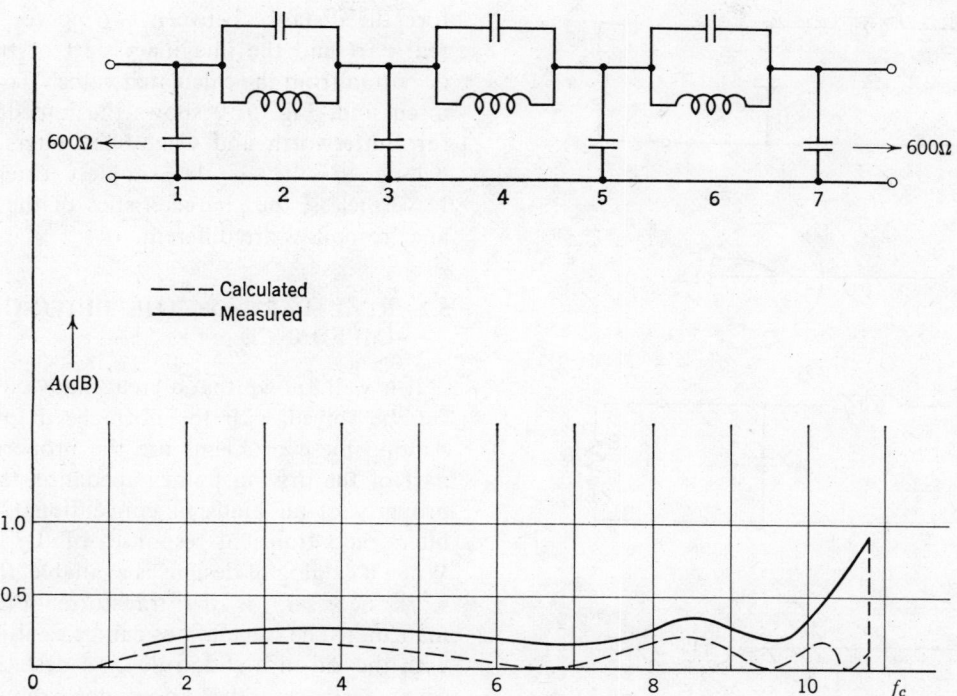

Fig. 5.8. Schematic and attenuation in the passband of seventh order lowpass.

CC 05 20 30. Its impedance behaves in the same fashion except that it exhibits more ripples in the passband. The attenuation reaches maximum value three times in the passband. The filter of higher complexity, shown in Fig. 5.8, is of the seventh order, and an equally-terminated design of a similar structure.

Its attenuation characteristic in the passband is shown below the schematic. Figure 5.9 shows the corresponding behavior of the input admittance. The calculated and measured values are different. On the base of purely reactive network theory, if the calculated values exhibit equal ripple attenuation up to cutoff

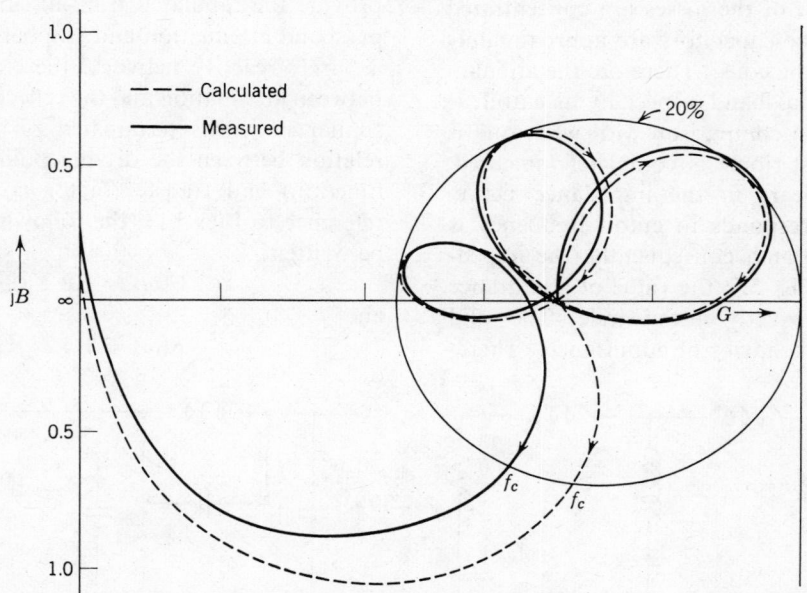

Fig. 5.9. Admittance characteristic inside and outside the passband.

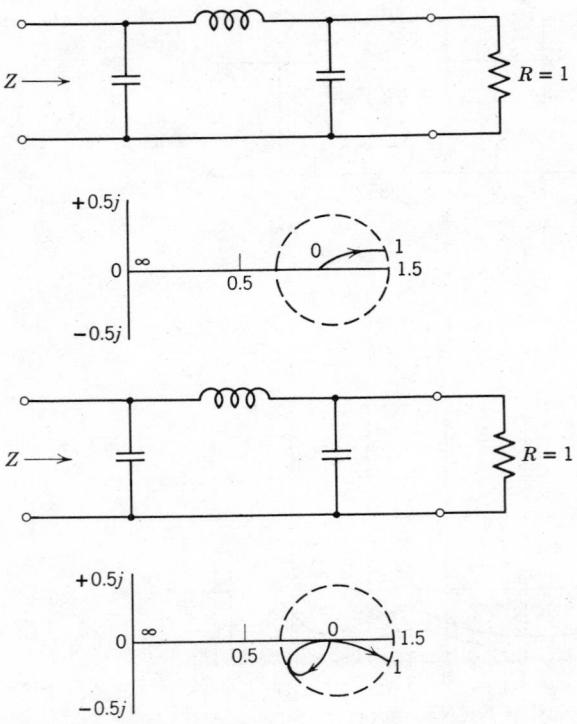

Fig. 5.10. Impedance characteristics of Butterworth and Chebyshev filters.

frequency and the impedance varies up to the circumference of equal reflection values, the measured values will deviate since the components are not purely reactive but include the impurity of losses. In the low frequency domain (where these kind of filters find application) most of the losses are concentrated in the coils because the capacitors are approximately ten times better than the coils. Therefore the attenuation response in the passband, especially at cutoff, is very much distorted in comparison with what one is led to expect; the last ripple is completely smeared. A similar effect appears in the impedance curve. The point which corresponds to cutoff frequency is now out of the circle and, consequently, the impedance is different. In Fig. 5.9, the value of admittance at cutoff frequency is exactly on the circle. The solid line shows the actual behavior of admittance. There-

fore the distance between two curves illustrates the real part and the imaginary part of the admittance deviation from the calculated value. To complete this discussion, Fig. 5.10 shows the impedance behavior for Butterworth and Chebyshev filters. Both filters belong to one of the simplest categories $n = 3$. Nevertheless, the characteristics of the input impedance responses are different.

5.2 REAL PART OF THE DRIVING POINT IMPEDANCE

It is well known that a great many circuit problems can be solved with the normalized lowpass model. Among these problems are the property of the real part of the driving point impedance, the attenuation property of all kinds of conventional filters, crystal filters, and transient responses of the lowpass filter. When a catalog of designs is available, the only design work necessary is the transformation and denormalizing. The calculations can be easily accomplished with the aid of a slide rule or a desk calculator. In some instances, the knowledge about the filter's behavior in the time domain is even more important than its steady-state performance, especially in the application of impulse techniques, radar, and telemetry.

Design of Two-terminal Networks Having a Prescribed Driving Point Impedance

The filter catalog has been computed primarily to provide the tabulated transmission property, namely passband attenuation and stopband attenuation. For a purely reactive network, there is a simple relation between attenuation and the reflection coefficient. For similar networks terminated on only one side, the relation between the driving point impedance (input function) and transfer factor is also simple. With reference to Fig. 5.11, the following relationships can be written:

$$G(\omega) = \text{Re } Y_1(j\Omega)$$

and $\qquad\qquad\qquad\qquad\qquad$ (5.2.1)

$$R(\omega) = \text{Re } Z(j\Omega)$$

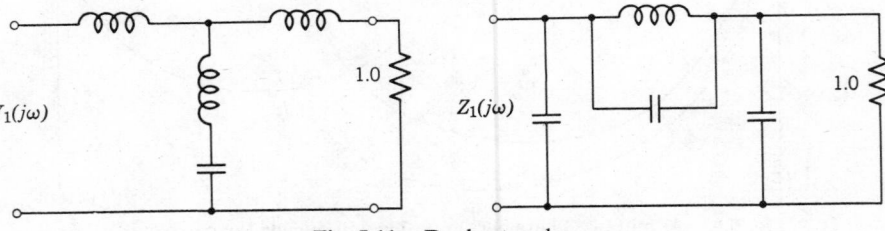

Fig. 5.11. Dual networks.

The power passing through the network in terms of effective voltage and current is expressed by

$$P_1 = |V_1|^2 \, G(\Omega)$$

and

$$P_1 = |I_1|^2 \, R(\Omega) \qquad (5.2.2)$$

No losses are anticipated in the network and, therefore, the input and output power are equal

$$P_1 = P_2$$

Therefore the output power in both cases will be

$$P_2 = |V_2|^2 \cdot 1$$
$$P_2 = |I_2|^2 \cdot 1$$

The corresponding transfer factors are as follows:

$$|H_v(j\Omega)|^2 = \left| \frac{V_1}{V_2} \right|^2 = R(\Omega)$$
$$|H_i(j\Omega)|^2 = \left| \frac{I_1}{I_2} \right|^2 = G(\Omega) \qquad (5.2.3)$$

The transmission function for the network having a voltage or current source of input power can be expressed by the square of the voltages in one case and currents in the other, or by the real part of the driving point impedance function. As a consequence of this property, it is possible to insert a finite reactance in the shunt arm of the network (with $R_s = 0$) without effecting the real part of the input function (driving point impedance). Attention must be paid to the type of source resistance and the type of the schematic. When the source is of zero internal impedance, or a so-called voltage source, the series branch (*T*-schematic) must be connected with the source. When the source resistance is equal to infinity (so-called current source) the first reactance facing the source in the network must be a parallel branch (*π*-schematic).

Parameters of the Real Part of the Driving Point Impedance

It is advantageous to have some established formal relationship between two-terminal networks to use the tabulated filter values to obtain the necessary design characteristics. Let us introduce the attenuation tolerance scheme for lowpass filters (see Fig. 5.12). The appearance of the impedance tolerance scheme in Fig. 5.13 is similar to that of the transmission function. The details of the responses are shown in Fig. 5.12 and 5.13. The squared magnitudes equation will take the

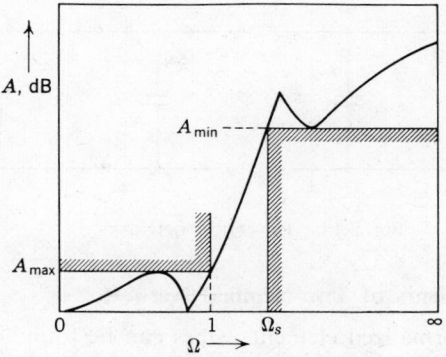

Fig. 5.12. Tolerance scheme for lowpass filters.

familiar form

$$|H(j\Omega)|^2 = e^{2a} = 10^{2A/20} \qquad (5.2.4)$$

The ripple parameter in the passband of the filter $A_{\max}$ is similar to the ripple parameter of the real part of the impedance given by the following relation:

$$\delta = \sqrt{e^{2a} - 1} \qquad (5.2.5)$$

or for small values of $a_{\max}$ (in Np),

$$\delta \approx 2a \qquad (5.2.6)$$

In the stopband, the deviation parameter will be $\Delta = e^{a_{\min}}$. So the filter parameters, such as $A_{\max}$ (consequently ρ) and $A_{\min}$ can be translated into ripples of the real part of the driving point impedance δ and Δ. With aid of the filter catalog, the two-terminal network having a prescribed behavior of the real part of the driving point impedance can be easily designed. The entire input function can be theoretically represented by algebraic equations in a similar way, but the direct development for impedance (or admittance) from continuous fraction expansion when the element values are known is much simpler.

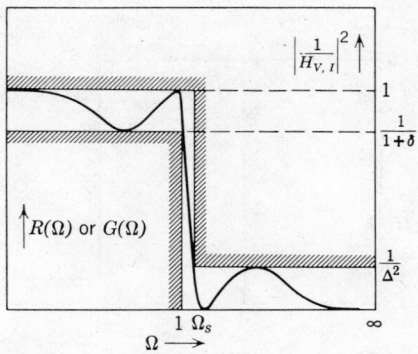

Fig. 5.13. Impedance tolerance scheme.

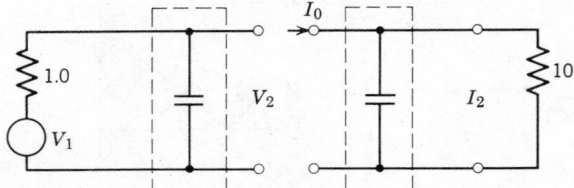

Fig. 5.14.   Reversible networks.

The Elements of Two-terminal Networks

The normalized element values can be found from filter catalog when we consider that, for reversible networks, the open-circuit transmission factor equals the current transmission factor so that

$$H_0 = \frac{V_0}{V_2} = H_i = \frac{I_0}{I_2} \tag{5.2.7}$$

Figure 5.14 shows the schematic for the foregoing conditions. Table values from Section 5.8 can be used when the input side and output side are reversed. (See the following numerical example.)

With the aid of frequency transformation techniques, the behavior of the real part of the driving point impedance of the lowpass model can be transformed into the highpass, passband, and stopband type characteristic. The physical sense behind this transformation is that reactive elements of the lowpass schematic are to be substituted by other simple reactances or combination of reactances.

Numerical Example

The problem here is to find a network with the real part of the driving point impedance practically constant (with negligible deviation from the nominal prescribed value) in a band of frequencies from 0 to 8 kc, as shown in Fig. 5.15.

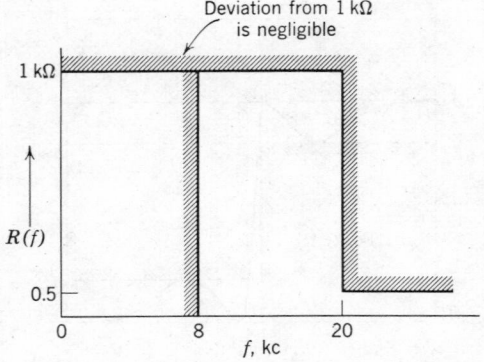

Fig. 5.15.   Prescribed real part limitations.

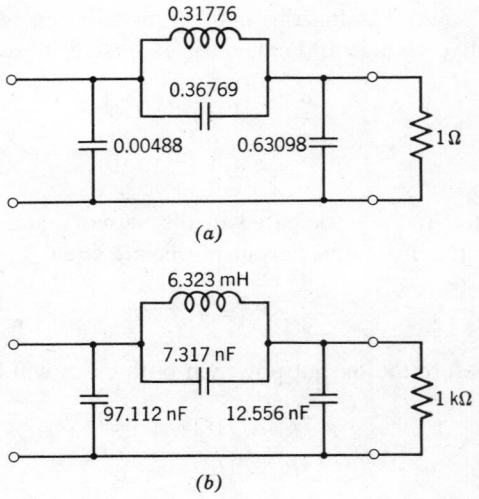

Fig. 5.16.   Normalized and unnormalized two-terminal network.

SOLUTION:

1. The parameters of a two-terminal network in terms of a four-terminal filter are:

$$\delta\sqrt{e^{2a} - 1} = \frac{\rho}{\sqrt{1 - \rho^2}} \approx \rho$$

for small values of ρ. If the value of $\rho = 2\%$, then the amount of deviation from the nominal value one is 0.1%.

2. For $\Omega_s = 20/8 = 2.5$, from the catalog, for $\rho = 2\%$, the closest tabulated values belong to the filter of type CC 03 02 23 with $\Omega_s = 2.5593$, $A_{\min} = 13.72$. Figure 5.16a shows the normalized structure. For example, $\omega_r = 50.240$, and the reference impedance $R_r = 1$ kΩ. Therefore $L_r = R_r/\omega_r = 19.9$ mH, $C_r = (\omega_r R_r)^{-1} = 19.9$ nF, $L_v = L_v{'}L_r$, and $C_v = C_v{'}C_r$. The schematic of Fig. 5.16b gives the element values of the designed network.

5.3   LOWPASS FILTER DESIGN

Sometimes it is desirable to have more exact knowledge about the attenuation, phase, and group delay characteristics than those given in the tables. Because the catalog includes the eigenvalues, it is possible to determine these more exact characteristics in an elementary fashion. The transmission function of a network with lumped elements can always be represented by a rational function. The zeros are in the numerator, and the poles are given in the denominator. In the third-order case, taken as an example, this expression is given by Eqs. 5.1.4 and

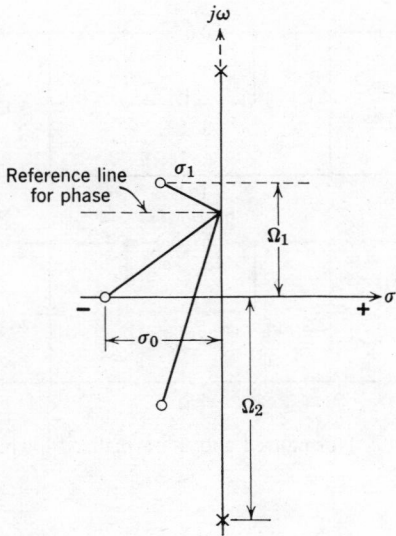

Fig. 5.17. Pole-zero diagram for lowpass filter ($n = 3$).

5.1.5 where σ_0, σ_1, Ω_1, and Ω_2 are the values determined in the *s*-plane shown in Fig. 5.17. Poles are symbolized by ($\times$) and zeros by (0). For $s_0 = -\sigma_0$ and $s_1 = -\sigma_1 \pm j\Omega$, the transmission function is equal to zero. For $s_2 = \pm j\Omega_2$ the transfer function is infinitely large.

These fixed complex numbers are characteristic to the transfer function. Knowledge of pole-zero data permits us to find all steady-state and dynamic properties of the network. Attenuation, phase, and group–delay characteristics can be constructed directly from Eq. 5.1.5 for $H(s)$. In the steady-state condition, the argument of complex frequency is purely imaginary signifying prolonged sinusoidal oscillation. The value of attenuation can be described by the distances between the reference frequency $j\Omega$ and the poles and zeros. Figure 5.17 shows these distances for illustrative purposes. The entire magnitude is evaluated, when the pole and zero distances corresponding to Eqs. 5.1.4 and 5.1.5 are multiplied.

In communication techniques, it is customary to express the transmission function in terms of attenuation so that

$$H(j\Omega) = e^{a(\Omega)}e^{j\beta(\Omega)} \quad \text{for } a \text{ in Np}$$
$$H(j\Omega) = 10^{A(\Omega)/20}e^{j\beta(\Omega)} \quad \text{for } A \text{ in dB} \quad (5.3.1)$$

With these, the attenuation will be

$$A(\Omega) = \log B + \log |j\Omega - j\Omega_2| + \log |j\Omega + j\Omega_2|$$
$$- \log |j\Omega + \sigma_0| - \log |j\Omega + \sigma_1 - j\Omega_1|$$
$$- \log |j\Omega + \sigma_1 + j\Omega_1| \quad (5.3.2)$$

The meaning of these expressions is that the attenuation is fully determined except for the constant $\log B$, by the sum of the logarithms of the phasors (complex numbers).

The phase can be determined even simpler than the attenuation. The expression for the phase is given by corresponding lines with corresponding signs (with positive signs for zeros and negative signs for poles). See Fig. 5.17—the reference line.

$$b(\Omega) = \text{arc}\,[-H(j\Omega)]$$

or

$$b(\Omega) = K - \tan^{-1}\frac{\Omega_0}{\sigma_0} + \tan^{-1}\frac{\Omega - \Omega_1}{\sigma_1}$$

Here $K = 0$ for $|\Omega| < \Omega_2$ and $K = \pm\pi$ for $|\Omega| > \Omega_2$. The phase is determined within a constant by the sum of the arc tangents.

The group delay $t_g(\Omega)$ is given by the familiar expression represented as a derivative of the phase relative to the frequency:

$$t_g(\Omega) = \frac{db}{d\Omega} = \frac{\sigma_0}{\sigma_0^2 + \Omega^2} + \frac{\sigma_1}{\sigma_1^2 + (\Omega - \Omega_1^2)}$$
$$+ \frac{\sigma_1}{\sigma_1^2 + (\Omega + \Omega_1)^2} \quad (5.3.3)$$

This expression shows that the total group delay is a sum of partial curves similar to resonance response curves. The purely imaginary places (such as Ω_2) correspond to infinitely large group-delay spikes and therefore are not included.

In conclusion it can be stated that the knowledge of pole-zero geometry and elementary rules of design procedure permits the calculation of the steady-state property of lowpass models of any complexity, if it is linear and has a lumped element equivalent schematic.

Numerical Example Number 1

A lowpass filter is to satisfy the requirements imposed by the tolerance scheme shown in Fig. 5.18. Figure 5.19 shows the normal operating condition.

According to Table 5.2, a reflection factor $\rho = 3\%$ corresponds to $A_{\max} = 0.004$ dB. The requirements dictate $A_{\min} = 18$ dB at $\Omega_s = 2.7$. The table with $\Omega_s = 2.67$ and $A_{\min} = 18.31$, (CC 03 03 22), provides a set of necessary design values for the required filter.

Figure 5.20*a* shows the normalized network with the reference values $\omega_r = 2\pi \cdot 15{,}000$ and $R_r = 5 \times 10^3$. The reference inductor is $L_r = R_r/\omega_r = 53.1$ mH, the circuit inductance is $L_v = L_v'L_r$. The

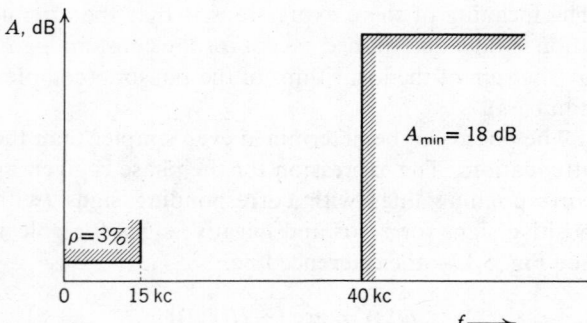

Fig. 5.18.   The attenuation requirements for the filter.

reference capacitor is $C_r = 1/\omega_r R_r = 2.12$ nF, and the circuit capacitor is $C_v = C_v' C_r$. Figure 5.20b shows the unnormalized network. The resonant frequency for the LC resonator must equal 45.9 kc.

From the same catalog, the following information will be used for group delay response calculation:

$$\sigma_0 = 2.290$$
$$\sigma_1 = 0.673$$
$$\Omega_1 = 1.842$$

$$tg(\Omega) = \frac{2.290}{(2.290)^2 + \Omega^2} + \frac{0.673}{(0.673)^2 + (\Omega - 1.842)^2}$$
$$+ \frac{0.673}{(0.673)^2 + (\Omega + 1.842)^2}$$

The curve of total magnitude can be represented as a sum of three curves.

The partial curves are marked by several characteristic points, for example when

$$\Omega - \Omega_1 = 0$$
$$t_1 = \frac{1}{\sigma_1}$$

for

$$\Omega - \Omega_1 = \sigma_1$$
$$t_1 = \frac{1}{2\sigma_1}$$

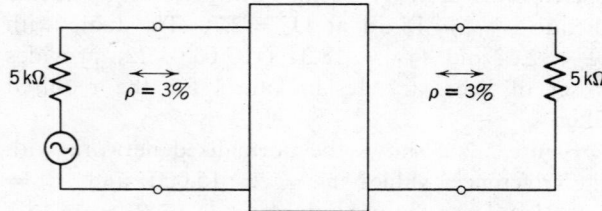

Fig. 5.19.   Normal operating conditions.

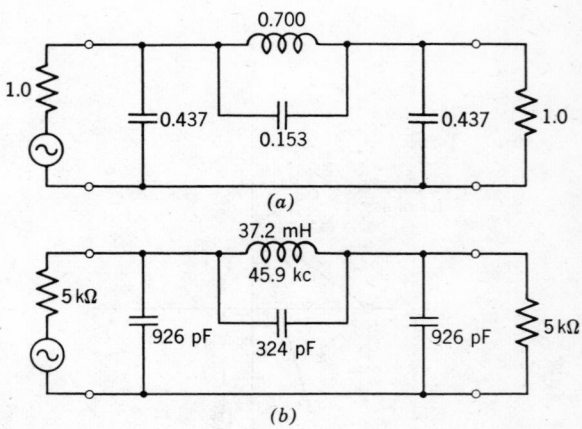

Fig. 5.20.   Normalized and unnormalized lowpass filters.

for

$$\Omega - \Omega_1 = 2\sigma_1$$
$$t_1 = \frac{1}{5\sigma_1}$$

and, in general, for

$$\Omega - \Omega_v = k\sigma_v$$
$$t_v = \frac{1}{1 + k^2}\frac{1}{\sigma_v}$$

Practically, it is sufficient to have such points to distinguish the partial curves.

Figure 5.21 shows the partial and total curves. It is evident that the group delay inside of the passband is gradually increasing to about 30% of its initial value at the beginning of the scale.

The actual group delay can be found with the aid of the reference time

$$t_r = \frac{1}{\omega_r} \tag{5.3.4}$$

and all normalized $t_g(\Omega)$ values must be multiplied by this reference value. The physical time units are shown on the left-side scale of Fig. 5.21.

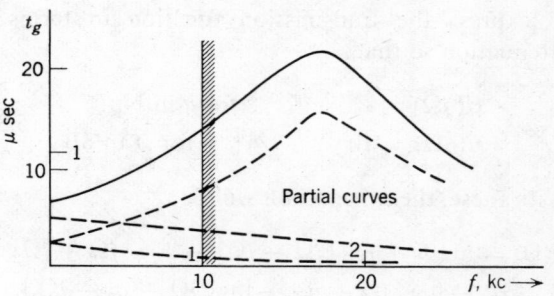

Fig. 5.21.   Partial and total curves of group delay.

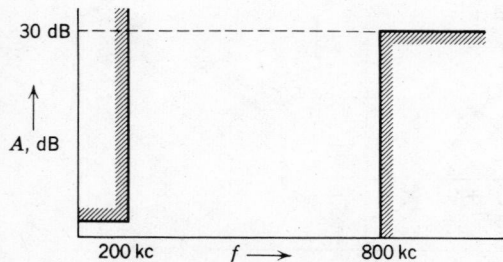

Fig. 5.22. Tolerance scheme for broadband amplifier.

Numerical Example Number 2

For a second example, consider a broadband amplifier operating between 0 and 200 kc with a guaranteed attenuation above 800 kc of no less than 40 dB (see Fig. 5.22). According to the catalog, considering $\rho = 2\%$, the filter CC 03 02 14 will satisfy the requirements with $\Omega_s = 4.13$ and $A_{min} = 26.7$ dB. The load impedances are not equal, and represent an extreme case of impedance transformation (see Fig. 5.23). The problem can be solved with the aid of the circuit transfer concept.

The network to insert is shown in Fig. 5.24 being calculated with the reference values:

$$\omega_r = 2\pi \, 2 \times 10^5 \quad R_r = 1500 \, \Omega$$

Figure 5.24 shows the schematic with unnormalized element values. Note the reversed position for C_1 and C_3.

5.4 DESIGN OF HIGHPASS FILTERS

Highpass filters are characterized by their property to pass all frequencies above a certain point and stop frequencies below a certain point. Figure 5.25 shows the attenuation response of highpass filters. The highpass filter is exactly reciprocal in its performance to the lowpass filter. Nevertheless, the steady-state properties of highpass filters and their pole-zero locations can be obtained from the lowpass data. The lowpass-highpass transformation is especially simple when the normalized cutoff frequencies in both cases are equal to one. Here,

$$\Omega_{HP} = \frac{1}{\Omega_{LP}} < 1 \tag{5.4.1}$$

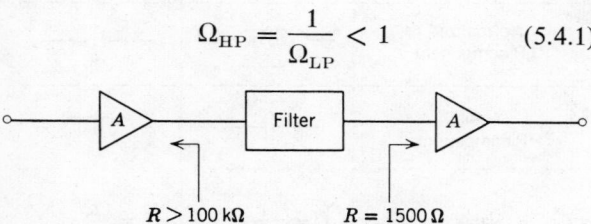

Fig. 5.23. Block diagram of broadband amplifier.

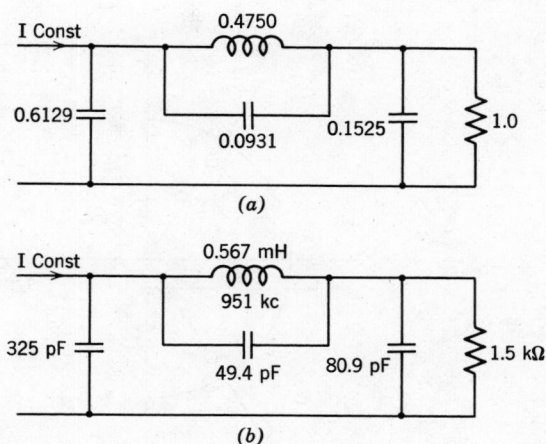

Fig. 5.24. Normalized and unnormalized value of coupling filter.

Therefore all characteristics of highpass networks are known. The transformation for complex frequency values is made with the following expression:

$$s_{HP} = \frac{1}{s_{LP}} \tag{5.4.2}$$

Highpass pole-zero information is obtained from the normalized lowpass pole-zero information by a simple reflection through the unit circle.

Because in network theory only real or conjugate complex pole-zero data are involved it is sufficient to obtain the angle and find the reciprocal magnitude as a distance of pole or zero location from the origin of the coordinate system as shown in Fig. 5.26. It is important when one uses the transformation of the lowpass pole-zero data, to consider the zeros at infinity.

It should be noted, for completeness, that with the aid of potential analogy the poles and zeros can be developed and thought of as sources and sinks. In special cases of electrical networks the continuity must be fulfilled. The number of electrical sources

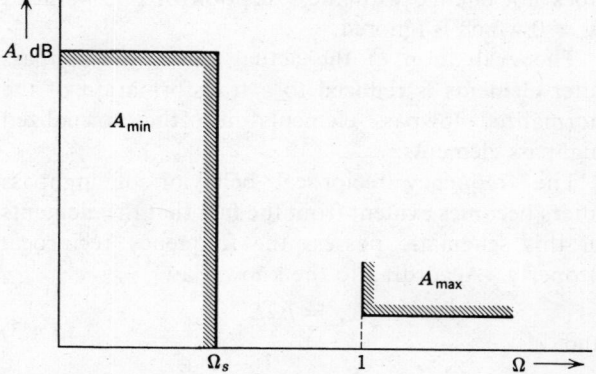

Fig. 5.25. Highpass filter attenuation response specification.

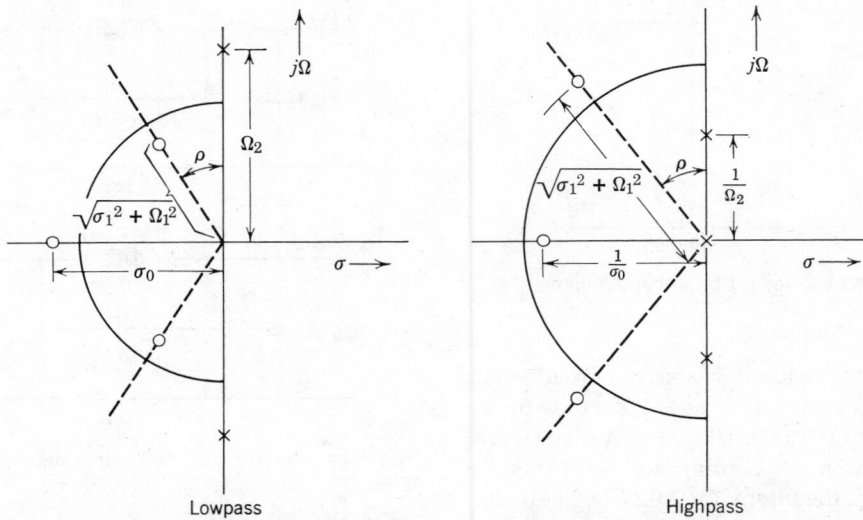

Fig. 5.26. Pole-zero diagram of lowpass and highpass filters.

(poles) have to be equal to the number of discrete sinks (zeros). In the catalog for the third-order filter, only two zeros are evident. The third one must lie, by necessity, at infinity. Using the equations of reciprocity (5.4.1) and (5.4.2), it is possible to transform the lowpass information into highpass information. The steady-state property can be evaluated in exactly the same fashion as in the lowpass filter. Here, we can illustrate the use of the tables for a design example. In some special cases, the lowpass and highpass schematics exhibit the same group-delay responses inspite of the fundamental differences in their attenuation responses.

In the Butterworth case, all zeros are on the circle and poles are at infinity (polynomial filter). The imaging about the unit circle in the pole-zero plane is according to Eq. 5.4.2. The pole positions are unchanged, and all the zeros migrate to the origin of the coordinate system. Therefore the group delay does not change with the exception of one point at $\omega = 0$ which is ignored.

The evaluation of the actual values of highpass filter elements is reduced to a transformation of the normalized lowpass elements into the normalized highpass elements.

The frequency reciprocal behavior of highpass filters becomes evident from the fact that the elements of this schematic possess the frequency reciprocal property. According to the known law

$$X_L = j\Omega L$$

and

$$X_C = \frac{1}{j\Omega C} \qquad (5.4.3)$$

the coils must be substituted by capacitors and vice versa, with the cutoff frequency unchanged. Considering Eqs. 5.4.1 and 5.4.3

$$C_{HP} = \frac{1}{L_{LP}}$$

and

$$L_{HP} = \frac{1}{C_{LP}} \qquad (5.4.4)$$

In Table 5.3 the relationship between the normalized elements and actual physical components of highpass filters are shown.

Numerical Example Number 1

A highpass filter is to satisfy the requirements shown in the tolerance scheme of Fig. 5.27. The sharpness of the response curve given as

$$\Omega_s = \frac{1}{\Omega_{s_{HP}}} = 10.$$

Table 5.3 Lowpass to Highpass Transformation

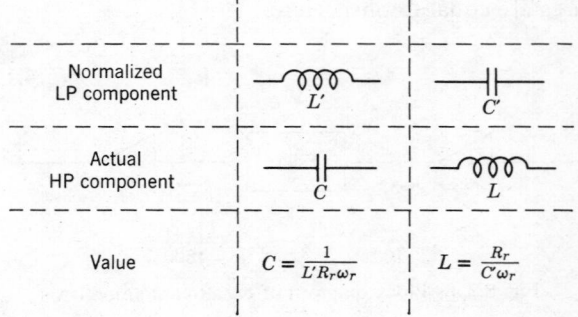

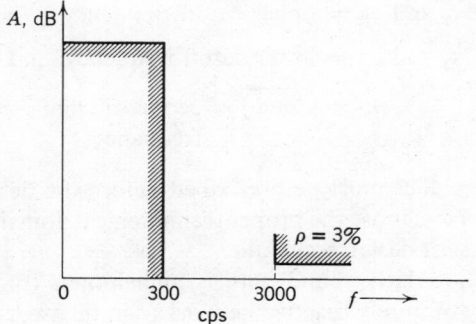

Fig. 5.27. Tolerance scheme for numerical example number 1.

Consequently, the filter CC 03 03 06 will provide the following:

$$\Omega_s = 9.5668 \quad \text{and} \quad A_{\min} = 52.40 \text{ dB}.$$

Figure 5.28 shows the normalized and unnormalized schematics. The necessary lowpass to highpass transformation is accomplished with the aid of the design expressions of Table 5.3.

Numerical Example Number 2

A highpass filter is required to pass frequencies from 150 kc and up. The peak ripple attenuation in the passband must be no greater than 0.05 dB, and in the stopband the attenuation must be no less than 50 dB. The problem is to find maximum value of Ω_s for a filter with four reactive elements.

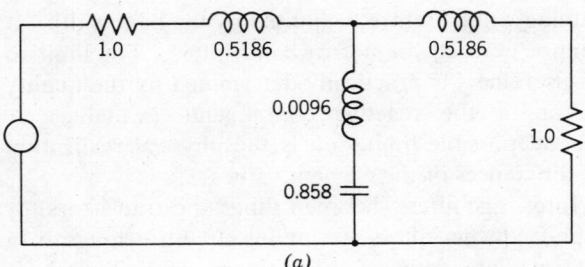

(a)

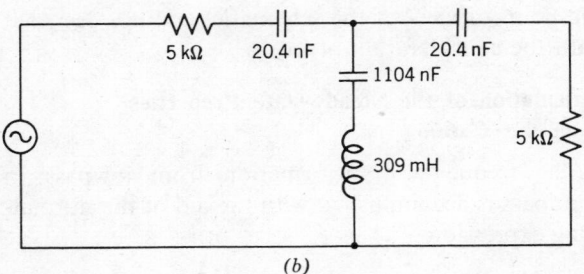

(b)

Fig. 5.28. Normalized and unnormalized highpass filter.

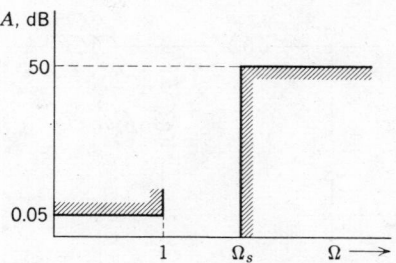

Fig. 5.29. Tolerance scheme for numerical example number 2.

In the case of an extreme impedance transformation, and without losing accuracy in calculations, the load impedance can be accepted as nonexistent (short circuit). In this condition, the data from the catalog can still be used. With the known recalculation procedure, the tolerance response can be drawn as shown in Fig. 5.29 which is the familiar lowpass limits. Figure 5.30 shows the normal operating conditions. From the catalog with $\rho = 3\%$ the corresponding model filter CC 03 03 07 will provide $\Omega_s = 8.2055$ and $A_{\min} = 48.38$ dB. The maximum unnormalized limiting frequency f_s is

$$f_{\text{HP}} = \frac{f_r}{\Omega_{\text{LP}}} = 18.28 \ kc$$

Because the current transfer function is prescribed, the model schematic must be of the form shown in Fig. 5.31. The physical values of the actual network are evaluated with the following reference values:

$$R_r = 20 \text{ k}\Omega$$

$$\omega_r = 2\pi \cdot 150 \times 10^3 = 9.42 \times 10^5$$

$$L_r = \frac{R_r}{\omega_r} = 21.2 \text{ mH}$$

$$C_r = \frac{1}{R_r \omega_r} = 53.1 \text{ pF}$$

The necessary lowpass to highpass transformation is

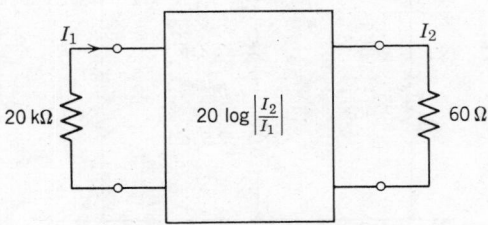

Fig. 5.30. Normal operating conditions.

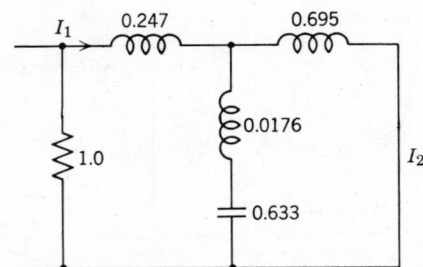

Fig. 5.31.   Model schematic (lowpass prototype).

accomplished with the aid of the design expressions

$$L_{HP} = \frac{L_r}{C'} \quad \text{and} \quad C_{HP} = \frac{C_r}{L'}$$

Figure 5.32 shows the realizable network.

As a checking point for the calculated elements, the following formula can be used.

$$f_\infty = \frac{1}{2\pi\sqrt{L_2 C_2}} = \frac{f_r}{\Omega_\infty} = 15.8 \text{ kc} \quad (5.4.5)$$

The value of Ω_∞, in normalized frequency, is given in the table (zero transmission).

5.5   DESIGN OF *LC* BANDPASS FILTERS

A conventional bandpass filter can be in the narrow-band or broadband category. There is a remarkable difference between narrow and wide bandwidth filters, and the design procedure is also different. The bandpass filter is certainly the most important type of filter. The catalog of lowpass prototype filters provides the necessary information for design, and the problem now is to investigate the method in which the given information can be used to design a bandpass filter.

Bandpass filter design terminology includes the concept of relative bandwidth

$$\frac{\omega_2 - \omega_1}{\sqrt{\omega_1 \omega_2}} = \frac{f_2 - f_1}{\sqrt{f_1 f_2}} = \frac{\Delta f}{f_m} = \frac{1}{a} \quad (5.5.1)$$

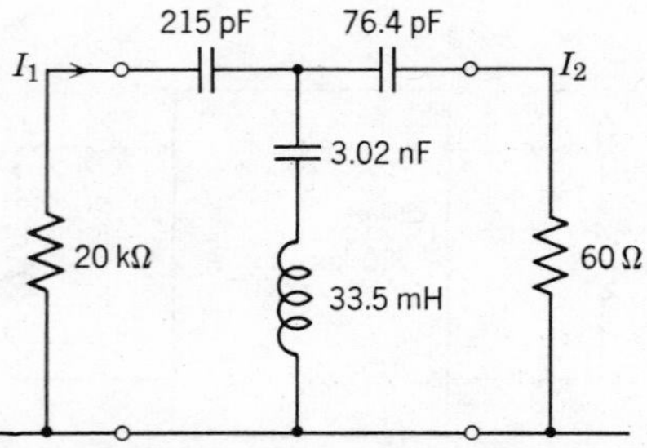

Fig. 5.32.   The realized network (highpass filter).

where f_2 and ω_2 = upper cutoff frequency,

f_1 and ω_1 = lower cutoff frequency, and

$\omega_m = \sqrt{\omega_2 \omega_1}$ = geometric mid-frequency.

These values must be prescribed before the design is started to choose the proper technological domain and the easiest design technique.

If the relative bandwidth is higher than 10%, the theory of purely reactive networks can be used. Both design methods (namely, the image-parameter method and the effective-parameter method) can be applied, and the choice is dependent upon the designer's decision and the specific problem involved. If the relative bandwidth is narrower than 10%, the best method is to use predistortion techniques or the theory of effective parameters with lossy elements. If the relative passband is less than 1%, the special theory of small bandwidth crystal filters or intermediate crystal filters must be used. The above speculation does not establish the limits but shows the direction in which the design can go.

When the quality factor of available components is high, it is quite reasonable to use the coil-saving bandpass filter, and design filters with bandwidth narrower than 10% according to the theory of purely reactive networks. In both design methods it is possible to transform the prototype bandpass filter into the Zig-Zag or a similar configuration.

It is known that the filter catalog is based on the theory of reactive networks, and the application of the catalog will therefore be limited by the bandwidth. It cannot be used for narrowband filters. The limit to its usefulness is practically determined by the quality factor of the reactive components available. A second possible limitation is the physical realization of differences of the element values.

Interstage filters (between tubes and transistors) or high-frequency filters as coupling circuits belongs to an intermediate category of a more general class of bandpass filters. They are usually very simple from the point of view of the schematic, and are designed with the consideration of losses.

Calculation of the Steady-state Properties from the Catalog

The frequency transformation from lowpass to bandpass is accomplished with the aid of the normalizing expression

$$\Omega = a\left(\Omega_B - \frac{1}{\Omega_B}\right)$$

where Ω is the frequency scale of the lowpass model, and Ω_B is the bandpass frequency which corresponds to this lowpass frequency.

It is evident that for every value of normalized frequency Ω there are two corresponding values of normalized frequency Ω_B and $1/\Omega_B$ which are geometrically symmetric since

$$\Omega_B \frac{1}{\Omega_B} = 1$$

At the cutoff frequency $\Omega = 1$,

$$\Omega_c' = 1 = a\left(\Omega_{BC} - \frac{1}{\Omega_{BC}}\right) = a(\Omega_{B2} - \Omega_{B1}) \quad (5.5.2)$$

The bandpass characteristic has two cutoff frequencies; namely, Ω_{B1} the upper and Ω_{B2} the lower, as illustrated in Fig. 5.33.

The constant a can be determined from the expression

$$a = (\Omega_{B2} - \Omega_{B1})^{-1} \quad (5.5.3)$$

If the upper and lower cutoff frequencies are normalized with respect to the reference f_r, which is

$$f_r = \sqrt{f_1 f_2} = f_m \text{ (midband frequency)}$$

then,

$$\Omega_{B2} = \frac{f_2}{f_m} = \sqrt{\frac{f_2}{f_1}} \qquad \Omega_{B1} = \sqrt{\frac{f_1}{f_2}} \quad (5.5.4)$$

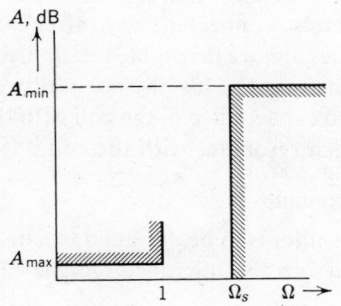

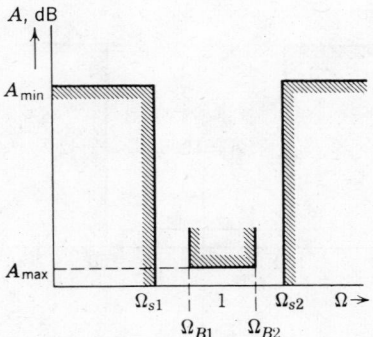

Fig. 5.33. Typical attenuation specifications for lowpass and bandpass filters.

The constant of transformation

$$a = \left(\frac{f_2}{f_m} - \frac{f_1}{f_m}\right)^{-1} = \frac{f_m}{f_2 - f_1} = \frac{f_m}{\Delta f} \quad (5.5.5)$$

is a reciprocal to the relative bandwidth.

Normalized frequencies in the stopband such as Ω_s can be obtained in the similar way

$$\Omega_s = a(\Omega_{s2} - \Omega_{s1}), \quad (5.5.6)$$

and the relationship between the stopband and passband limit is

$$\frac{\Omega_s}{\Omega_c} = \frac{\Omega_{s2} - \Omega_{s1}}{\Omega_{B2} - \Omega_{B1}} \quad (5.5.7)$$

which is a measure of the sharpness or selectivity of the response curve.

In general, any normalized frequency in the passband of the filter Ω_B can be found as a function of Ω, which belongs to the lowpass prototype and which runs from 0 to ∞.

$$\Omega_B = f(\Omega)$$

The solution is a quadratic expression

$$\Omega_B = \pm \frac{\Omega}{2a} + \sqrt{1 + \left(\frac{\Omega}{2a}\right)^2} \quad (5.5.8)$$

For

$$\frac{\Omega}{2a} \ll 1$$

as in the case of narrow bandpass filters, the expression for passband frequencies will be reduced to the appropriate expression

$$\Omega_B \approx 1 \pm \frac{\Omega}{2a}$$

The last formula suggests that in the case of very narrowband filters the response curve is approaching arithmetic symmetry and the passband frequencies are expressed as a simple deviation from the center ($\Omega_B = 1$) by certain number of half bandwidths $1/2a$. Evidently this approximation holds for narrowband realizations. Nevertheless in normalized diagrams, the attenuation is shown as equal on both sides of the passband. The response curve will look very distorted if, instead of a logarithmic frequency scale, a linear scale is used. The attenuation on opposite sides of the passband are related as follows

$$A(\Omega) = A\left(\frac{1}{\Omega}\right)$$

When the requirement is formulated (as usually) in arithmetic terms such as so many dB for $f_m \pm X$

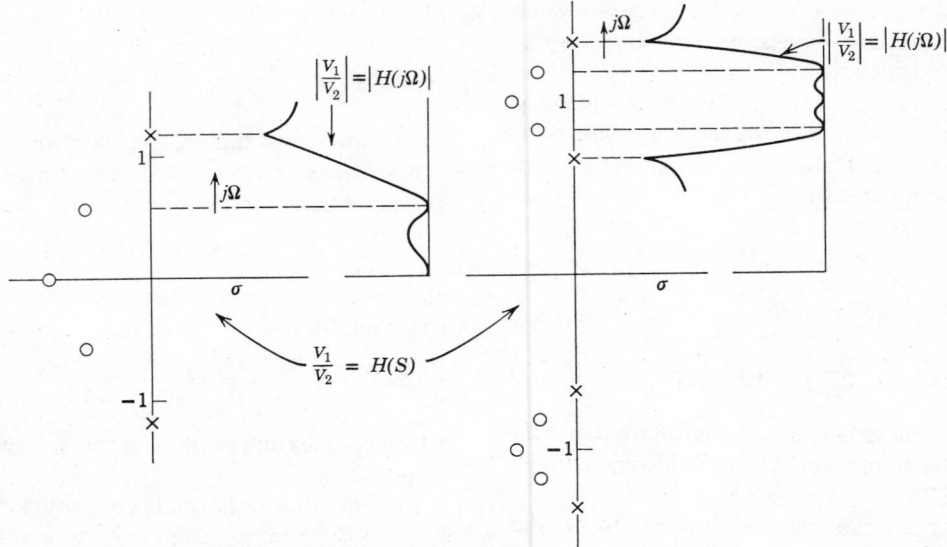

Fig. 5.34. Lowpass and bandpass pole-zero plane.

(deviation) the attenuation at $f_m + X$ will be lower than at $f_m - X$.

It is easy to understand this, from the fact that the geometric center is always lower than the arithmetic center. For example, when $\Omega_1 = 1$ and $\Omega_2 = 25$, the geometric frequency is $\Omega_m = \sqrt{1 \times 25} = 5$ and the arithmetic center is $\Omega_{am} = (1 + 25)/2 = 13$.

This property of transformed bandpass characteristic must be considered when one designs the filter attenuation characteristic with geometry symmetry. Pole-zero information of the lowpass filter can be transformed into corresponding information for the bandpass filter by an expression similar to that for frequency:

$$s_B = \frac{s}{2a} \pm j\sqrt{1 - \left(\frac{s}{2a}\right)^2} \qquad (5.5.9)$$

To every point in the complex frequency plane for lowpass filters, correponds two bandpass eigenvalues. All points in the lowpass pole-zero diagram must be considered. An infinite lowpass value produces a transformed zero at infinity in the bandpass pole-zero plane. Figure 5.34 shows a plot of poles and zeros.

For the narrowband case, a similar significant simplification is applicable:

$$s_B = \frac{s}{2a} \pm j \qquad \text{when} \quad \frac{s}{2a} \ll 1$$

The expression shows that the original pole-zero geometry of the lowpass prototype is displaced two times (about $\pm j$). The evaluation of the physical bandpass elements consists of translation of the normalized lowpass into a normalized bandpass filter. Table 5.4 relates the lowpass and bandpass data for element values.

The resonant lowpass circuit values are of special practical importance to the corresponding bandpass circuits and are translated into a combination of a series resonant circuit with a parallel resonant circuit. The equivalent circuit, consisting of two parallel resonant circuits connected one after another (in series), provides a more favorable relationship between coils and capacitors and has the possibility of lumping the distributed capacitance of the coil with the physical capacitor which resonates with the coil.

Numerical Example

A bandpass filter is to be designed to satisfy the conditions outlined in the tolerance scheme of Fig. 5.35.

$$R_1 = R_2 = 20 \text{ k}\Omega$$

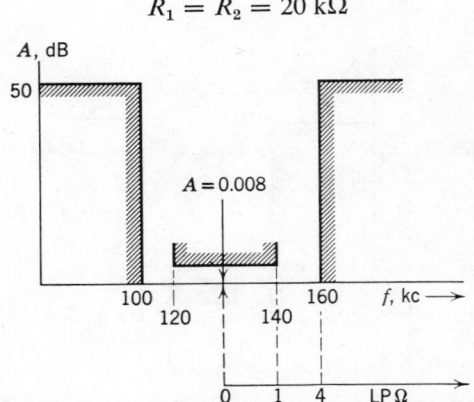

Fig. 5.35. Tolerance scheme for numerical example.

Table 5.4 Lowpass to Bandpass Transformation

LP	BP Schematic	BP Values
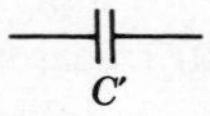	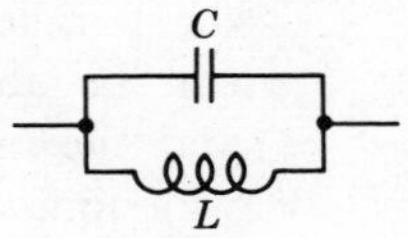	$C = aC' \dfrac{1}{R_r \omega_r}$ $L = \dfrac{1}{aC'} \dfrac{R_r}{\omega_r}$
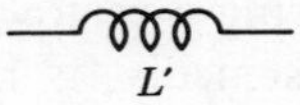		$L = aL' \dfrac{R_r}{\omega_r}$ $C = \dfrac{1}{aL'} \dfrac{1}{R_r \omega_r}$ $L, C \rightarrow$ are unnormalized BP values
		$c_+ = \dfrac{1}{l_-} = aC'(1 + \Omega_-{}^2)$ $c_- = \dfrac{1}{l_+} = aC'(1 + \Omega_+{}^2)$ $l_+ = \dfrac{1}{c_-} = aL'(1 + \Omega_-{}^2)$ $l_- = \dfrac{1}{c_+} = aL'(1 + \Omega_+{}^2)$ where $\Omega_\pm = \sqrt{1 + \left(\dfrac{\Omega_\infty}{2a}\right)^2} \pm \dfrac{\Omega_\infty}{2a}$ $l_- + l_+ = \dfrac{1}{aC'}$ $c_+, l_+, c_-,$ and l_- are normalized BP values

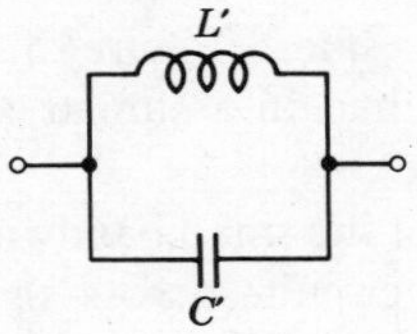

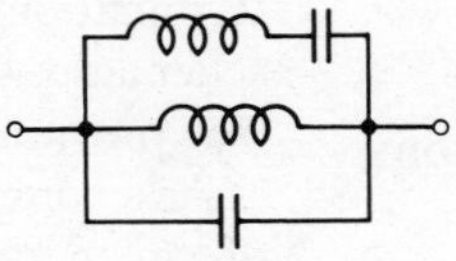

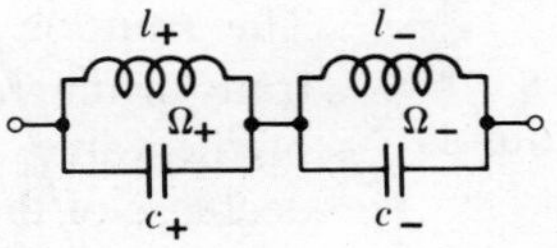

C' and L' are normalized LP values

with

$$\frac{f_2 - f_1}{\sqrt{f_2 f_1}} \approx 0.15 = 15\%$$

The filter can be designed as a purely reactive network since the available coils at the specified frequency have a good quality factor and the realized network will closely approximate the theoretical characteristic. Most critical is the upper stopband limit. With the approximate formula

$$\Omega_B \approx 1 \mp \frac{\Omega}{2a}$$

the information for the equivalent lowpass filter can be calculated. This is done below the bandpass tolerance scheme in Fig. 5.35.

The requirement will be satisfied with the lowpass filter catalog number CC 03 04 07. The corresponding transmission properties are

$$A_{max} = 0.008 \text{ dB}$$
$$A_{min} = 50.88 \text{ dB}$$
$$\Omega_s = 8.21$$

Figure 5.36 shows the normalized schematic.

Using the known reference value, the network components will be calculated.

$$L_r = \frac{R_r}{\omega_r} = 24.5 \text{ mH}$$

$$C_r = \frac{1}{\omega_r R_r} = 61.2 \text{ pF}$$

The inductance in the parallel branch is

$$L_1 = L_3 = \frac{1}{aC'} L_r = 6.52 \text{ mH}$$

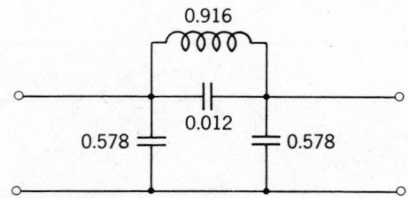

Fig. 5.36. Normalized schematic (according to catalog).

where

$$a = \frac{f_m}{\Delta f} = 6.5$$

and the capacitance across the line is

$$C_1 = C_3 = aC' \times C_r = 230 \text{ pF}.$$

The upper resonant frequency (for the attenuation peak) is

$$\Omega_+ = \sqrt{1 + \left(\frac{\Omega_\infty}{2a}\right)^2} + \frac{\Omega_\infty}{2a} = 1.976$$

The lower resonant frequency (for the attenuation peak) is

$$\Omega_- = \sqrt{1 + \left(\frac{\Omega_\infty}{2a}\right)^2} = \frac{\Omega_\infty}{2a} = 0.508$$

The unnormalized element values and frequencies for corresponding antiresonant circuits will be found with the following relations:

$$C_- = \frac{1}{l_+} C_r = aC'(1 + \Omega_+^2)C_r$$
$$= 0.383 \times 61.2 = 23.4 \text{ pF}$$

$$C_+ = \frac{1}{l_-} C_r = aC'(1 + \Omega_-^2)C_r$$
$$= 0.098 \times 61.2 = 6.0 \text{ pF}$$

$$f_{2\infty} = f_r\Omega_+ = 256.9 \text{ kc}$$
$$f_{1\infty} = f_r\Omega_- = 66.0 \text{ kc}$$

$$L_- = \frac{1}{C_+} L_r = 2.614 \times 15.9 = 41.6 \text{ mH}$$

$$L_+ = \frac{1}{C_-} L_r = 10.191 \times 15.9 = 162 \text{ mH}$$

All these elements are a part of actual network shown in Fig. 5.37. The schematic is very practical and theoretically flexible, resulting from the direct translation of elements. Tuning is simplified by using only the parallel resonant circuits. Impedance matching can be accomplished by tapping the output coils, and the impedance transformation can be made in wide limits. A dc path is available through the input and output coils.

Extreme Impedance Transformation

A bandpass filter having a relative bandwidth less than 10% is to be designed. Its output load is a finite, but the input is considered to be a current source (very high impedance) as shown in Fig. 5.38. The starting point is still the normalized lowpass with termination on one side. Figure 5.39 shows the schematic which resulted in a similar transformation (see previous example).

As a consequence of the small bandwidth, the values of capacitors in the circuits across the line are too large and the corresponding inductances are too small. The solution, in this case, is the use of an auto-transforming arrangement which permits an increase of the value of inductances and at the same time a decrease of the values of capacitors down to practical values (the resonant frequency of the combinations are the same).

Let us impose the unloaded quality factors of circuits I, II, and III equal Q_0 and the unloaded quality of circuit IV $= Q_4$ as in Fig. 5.39. Then (approximately)

$$\frac{1}{Q_t} = \frac{3}{Q_0} + \frac{1}{Q_4} \qquad (5.5.10)$$

When Q_t is known, the value of Q_4 is then

$$Q_4 = \frac{1}{1/Q_t - 3/Q_0} \qquad (5.5.11)$$

Appropriate tapping of the last coil can bring the agreement between this value of Q_4 (which is reduced

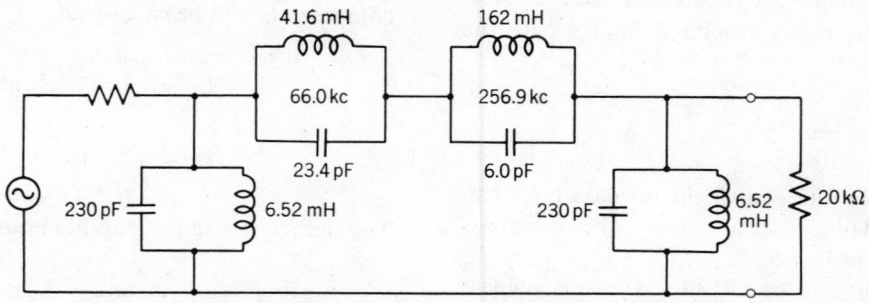

Fig. 5.37. Example of bandpass ($n = 3$) realization.

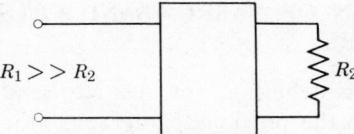

Fig. 5.38. Block diagram of extreme impedance transformation.

to the resistance across the coil) to the load resistance R_2. Figure 5.40 shows the final schematic without numerical values for this example, since the mechanics of the transformation have already been shown.

Bandpass Filter Transformation

The bandpass circuit computed as an example of reactance transformation, even in its modified form as shown in Fig. 5.37 produces inconveniences. It is found to be impractical with certain conditions and especially at frequencies higher even than 10 kc. The damaging phenomena consists of the stray capacitance between the junction of the parallel circuits in the series arm and ground. But the schematic can be once more modified in such a way that this capacitance is taken into consideration and thereby made harmless.

Let us consider the following example: The catalog filter CC 03 20 21 has the center frequency $f_0 = 5$ Mc and the pass band between 4.8 and 5.2 Mc. Being equally terminated with 150 ohms require normalized circuit elements for lowpass filter Fig. 5.41 as follows:

$$C_1' = C_3' = 1.121$$
$$C_2' = 0.09247$$
$$\Omega_2 = 3.195134$$

The transformation constant $a = 10.5529$. Attenuation poles of the bandpass filter calculated with tabulated formula will be

$$\Omega_+ = 1.162781$$
$$\Omega_- = 0.860007$$

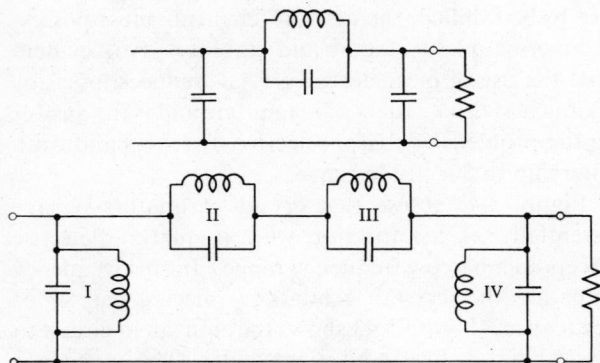

Fig. 5.39. Actual schematic for the example of Fig. 5.38.

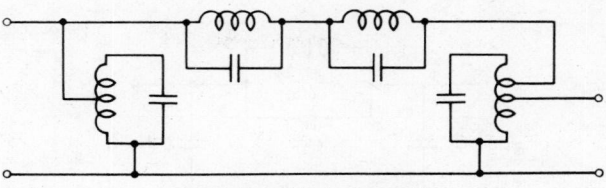

Fig. 5.40. Final schematic.

and the reference values for design are

$$L_r = 4.778 \ \mu\text{H}$$
$$C_r = 212.4 \ \text{pF}$$

Figure 5.42 shows the schematic of a modified bandpass section. The important design parameter in this case is the value

$$t_0 = 1 + \frac{C_2'}{C_3'}$$

The maximum value of normalized shunt capacitor to insert between the resonant circuits is:

$$c_{\max} = aC_2' \frac{(t_0 - 1)\Omega_-^2}{[(t_0 - 1)\Omega_-^2 + 1]} (t_0\Omega_-^2 + 1)(\Omega_+^2 - 1)$$

The actual capacitor c_i has to be chosen smaller than that value. The transformation ratio K is thereby fixed and can be obtained from the equation

$$K^2 \left[(\Omega_+^2 - 1) - \frac{c_i}{aC_2'} \right] - K(\Omega_+^2 - 1)(1 - \Omega_-^2) - (1 - \Omega_-^2) = 0$$

where K is greater than zero.

The normalized element values of the filter are given by the set of following expressions:

$$l_1 = \left[a\left(C_1' + C_2' \frac{K - 1}{\Omega_+} \right) \right]^{-1}$$

$$c_1 = a\left(C_1' + C_2' \frac{K - 1}{K} \Omega_-^2 \right)$$

$$l_2 = (ac_2')^{-1} \frac{K}{1 + K\Omega_+^2}$$

$$c_2 = \frac{\Omega_-^2}{L_2}$$

$$c_3 = (KL_2')^{-1}$$

$$l_3 = \frac{\Omega_+^2}{C_3'}$$

$$c_4 = \frac{a}{K_2} [C_3' - C_2'(K - 1)\Omega_+^2]$$

$$l_4 = \frac{K_2}{a[C_3' - C_2'(K - 1)]}$$

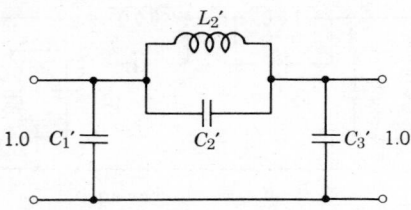

Fig. 5.41.   Normalized lowpass.

In the given example the value t_0 will be

$$t_0 = 13.122851$$

and the capacitance c_{max}

$$c_{max} = 0.332.$$

With reference capacitor $C_r = 212.4$ pF one would obtain

$$C_{max} = C_r \cdot 0.332 = 70.5 \text{ pF}$$

If it is decided to choose $C_i = 30$ pF, the normalized intermediate capacitor will be $c_i = 0.141243$ and therefore the equation for the transformation coefficient can be used.

$$K^2 - 0.442K - 1.156 = 0$$
$$K = 1.3634$$

The normalized element values are therefore

Element	Normalized	Unnormalized	
L_1	0.0827	0.3952 μH	
C_1	12.022	2554 pF	$f_1 = 5.01$ Mc
L_2	0.491	2.348 μH	
C_2	1.505	320 pF	$f_2 = 5.809$ Mc
C_3	1.492	317 pF	
L_3	0.9058	4.328 μH	$f_3 = 4.300$ Mc
C_4	6.106	1297 pF	
L_4	0.162	0.774 μH	$f_4 = 5.023$ Mc
C_i		30 pF	
R_2	K^2R_1	279 Ω	

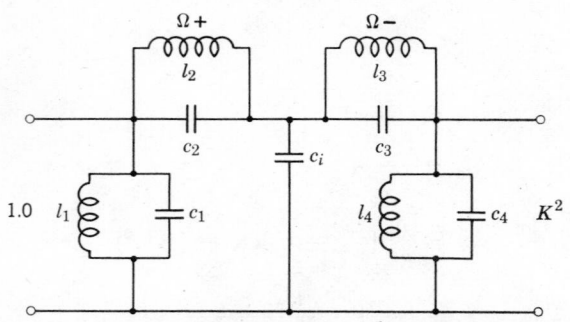

Fig. 5.42.   Schematic of modified bandpass (from Fig. 5.41).

5.6   DESIGN OF NARROWBAND CRYSTAL FILTERS

The basic definition for relative bandwidth was discussed in the previous paragraphs along with the lowpass-bandpass transformation scheme. In dealing with extremely sharp filters another general parameter, that of selectivity, must be added to bandpass theory. The numerical definition of selectivity is

$$S = \frac{f_c}{f_\infty - f_c} \qquad (5.6.1)$$

where f_c is the cutoff frequency and f_∞ is the frequency of the closest pole outside of the passband. Design experience indicates that when S is less than 60, the filter can be realized with coils and capacitors (LC filters), but if S is greater than 60, the filter by necessity is a crystal filter. Because the bandpass filter is the type most used, the catalog information relating to lowpass networks usually must be transformed into bandpass information. All transformation formulas used in previous paragraphs are applicable for quartz crystal filter design.

Crystal resonators as filter elements are somewhat different from LC resonators. Crystals are characterized by their excellent frequency stability and low electrical losses. The equivalent element values are dependent on the types of crystal, cut, frequency range, and packaging.

Crystals in the Filter Network

Many filters in the high-frequency domain consist of lumped inductances and capacitors. All lumped inductances, however, inherit two disadvantages: 1. the loss resistance, which cannot be neglected and 2. an unavoidable distributed capacitance. Since the resistance of the coil can never be eliminated completely, the bandwidth of the filter cannot be made extremely narrow or sharp. When these requirements are to be fulfilled, the network elements must possess a superior quality factor and stability. It is evident that the use of piezoelectric crystals reduces losses by as much as three orders of magnitude and is the answer to the problem of sharp, selective, narrow bandwidth filters up to 30–40 Mc range.

Figure 5.43 shows that crystal resonators behave essentially as a capacitor with a quartz dielectric except in a narrow frequency range. In this frequency interval, the crystal exhibits a mechanical series resonance. Figure 5.43 shows the equivalent circuit to possess certain fixed equivalent values. The impedance level is between 10 and 100 MΩ. The ratio of parallel

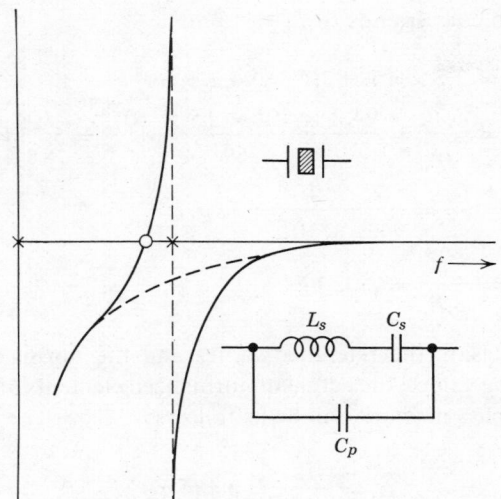

Fig. 5.43. Impedance diagram of crystal resonator.

capacitance to series equivalent capacitance usually is higher than 150. The information to be presented here will be limited to ladder filters, and the passband and stopband characteristics will be related directly to the corresponding resonant frequencies in the series (or parallel arms) of the filter. It is evident that most pronounced rejection, or a pole of effective attenuation, can be produced only by a crystal in the parallel arm.

The attention of the designer must be concentrated on the fact that the losses of inductance play a most critical role in narrowband structures. It is desirable that the series arm of the bandpass filter be a parallel combination of one series resonant circuit with one parallel resonant circuit. Both circuits are to be tuned to the center frequency. This type of schematic is obtained as the original form of the bandpass schematic after it has been correctly transformed from the low-pass prototype. The configuration requires a high quality series circuit with very high impedance level which is determined from $\sqrt{L/C}$. Evidently, a crystal element is best suited for this circuit. The parallel resonant circuit can be regarded as a *neutralizing capacitance*.

The following examples will show that only small θ values obtained from the catalog are reasonable to use. The relatively distant attenuation peak is not recommended, on the ground that the parallel resonant circuit will not exhibit a sufficient quality factor.

Numerical Example of Design

A crystal filter is to be designed according to the requirements set forth in the tolerance scheme shown

in Fig. 5.44. The input impedance must be equal or greater than 18 kΩ, and the output impedance must be 1 kΩ. The problem is to find:

1. Bandwidth parameter and selectivity.
2. Attenuation at ± 50 kc.
3. The values of the filter elements. A crystal is to be used in the series arm ($C_p/C_s \approx 200$; $L_s = 15$ mH).

SOLUTION

1. $a = \dfrac{f_m}{f_2 - f_1} = \dfrac{15{,}000}{50} = 300$

$S = \dfrac{f_c}{f_\infty - f_c} = \dfrac{15{,}025}{150} = 100$

The distance $f_\infty - f_c$ is shown as equal to 150 kc. Both numbers indicate that a crystal filter is the only possible solution to realize the given bandwidth and selectivity parameters.

2. The equation relating A_{max}, A_{min} and Ω_s for third-order filter is

$$\left(A_{min} + \tfrac{1}{2}\ln\frac{8}{A_{max}}\right)\ln\frac{8}{\Omega_s - 1} \approx 5n = 15$$

with $n = 3$, $\Omega_s = \tfrac{1}{2}(100)/\tfrac{1}{2}(40) = 2.5$, and $A_{max} = 0.0032$ Np (which corresponds to 8% reflection).

$$A_{min} = \frac{15}{\ln\dfrac{8}{\Omega_s - 1}} - \tfrac{1}{2}\ln\frac{8}{A_{max}} = 5.5 \text{ Np} = 48 \text{ dB}$$

3. Using the value of a, the bandwidth parameter, the following relations can be solved (see Fig. 5.45).
For the parallel circuit:

$$C_{2p} = aC_2{'}\frac{1}{R_r\omega_r}, \qquad L_{2p} = \frac{1}{aC_2{'}}\frac{R_r}{\omega_r}$$

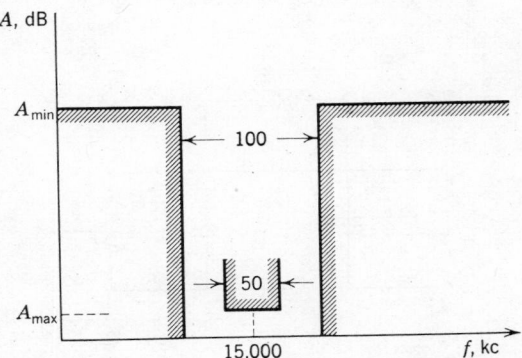

Fig. 5.44. Tolerance scheme for numerical example.

For the series circuit:

$$L_{2s} = aL_2{}' \frac{R_r}{\omega_r}$$

$$C_{2s} = \frac{1}{aL_2{}'} \frac{1}{R_r\omega_r}$$

For the combination of the two circuits, the ratio of capacitors will be

$$\frac{C_{2p}}{C_{2s}} = C_2{}'L_2{}'a^2$$

The available crystal has the ratio of $C_p/C_s = 200$, therefore with

$$\frac{C_{2p}}{C_{2s}} = 200$$

the resonant frequency

$$\omega_2{}^2 = \frac{1}{L_2{}'C_2{}'} = \frac{a^2C_{2s}}{C_{2p}} = 450$$

or

$$\omega_2 = 21.2$$

4. From the catalog, the set of values which corresponds to filter CC 03 08 03, with $K^2 = \infty$, will satisfy the above conditions.

The normalized element values are as follows:

$$C_1{}' = 0.3855$$
$$C_2{}' = 0.0023$$
$$C_3{}' = 0.9213$$
$$L_2{}' = 0.8954$$

Reversing the schematic as shown in Fig. 5.45 and using the reference values f_r and R_r, and the value of the equivalent motional inductance of the crystal, $L_s = 15$ mH, the following can be calculated

$$l_s = L_2{}'a = 0.8954 \times 300 = 269$$

which corresponds to $L_s = 15$ mH.

$$\omega_r = 2\pi \times 15 \times 10^6 = 94.2 \times 10^6$$

$$R_r = \frac{\omega_r L_s}{l_s} = \frac{94.2 \times 10^6 \times 15 \times 10^{-3}}{269} = 5.25 \, k\Omega$$

$$L_r = \frac{R_r}{\omega_r} = \frac{5.25 \times 10^3}{94.2 \times 10^6} = 56 \, \mu H$$

$$C_r = \frac{1}{\omega_r R_r} = 2.02 \, pF$$

5. Using the reference values and the normalized catalog values, the actual unnormalized elements of the bandpass network will be as follows:

$$L_1 = \frac{L_r}{C_1{}'a} = 0.484 \, \mu H$$

$$C_1 = C_1{}'a \, C_r = 234 \, pF$$

$$L_3 = \frac{L_r}{C_3{}'a} = 0.203 \, \mu H$$

$$C_3 = C_3{}'aC_r = 558 \, pF$$

$$L_{2p} = \frac{L_r}{C_2{}'a} = 81 \, pH$$

$$C_{2p} = C_2{}'aC_r = 1.4 \, pF$$

$$L_{2s} = L_2{}'aL_r = 15 \, mH \qquad \text{crystal}$$

$$C_{2s} = \frac{1}{C_2{}'a} C_r = 2.9 \, pF$$

This example shows the design procedure up to the point of actual realization. In the construction of the engineering model of the filter there may be certain difficulties in the realization coils on the input and output sides of the filter. It will be more practical to use auto-transformers instead of straightforward coils and employ an impedance transformation. The coils can be made of the same value, and be optimized from point of view of quality factor. Figure 5.46

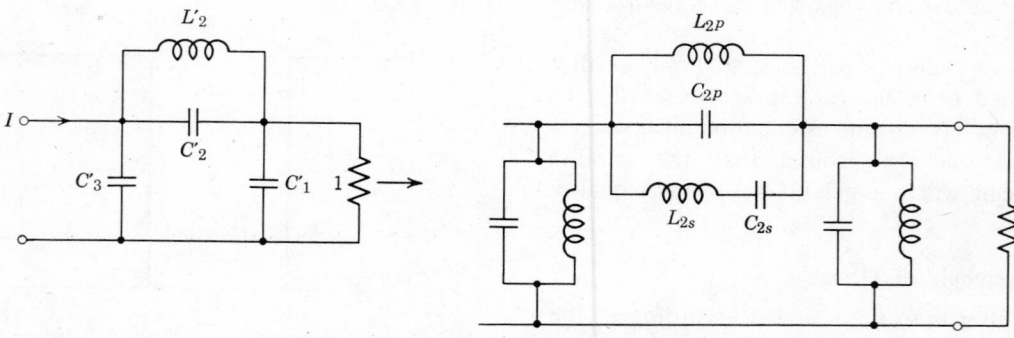

Fig. 5.45.   Lowpass prototype and bandpass frequency transformation.

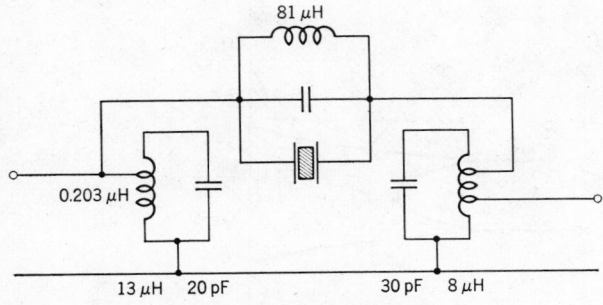

Fig. 5.46. Bandpass crystal filter.

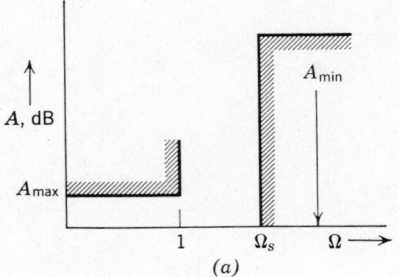

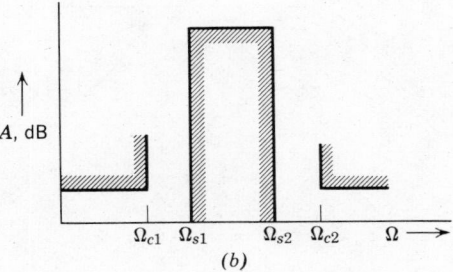

Fig. 5.47. Attenuation requirements for bandstop filter.

shows the configuration of the filter after the appropriate modification. Instead of 0.2 μH, the input coil is now 13 μH which is easier to realize at the given frequency. The output coil is 8 μH, also realizable, and the same type of core material can be used. All circuits are tuned to resonate at the same frequency. The coil in parallel with the crystal serves in neutralizing the damaging parallel capacitance of the crystal.

Theoretically the signal source impedance R_1 is assumed to be infinity, and under this condition the open-circuit transmission factor is identical with the current transmission factor. But practically, it is sufficient that the source impedance is large in comparison with the impedance at resonance of the lossy input circuit.

The partial compensation of losses similar to that mentioned for the case of regular bandpass filter can be used. The inductive components with high Q (quality factors of 250 can be realized) permits an almost complete agreement between theory and practice for this specific filter.

5.7 DESIGN OF BANDSTOP FILTERS

Bandstop filters are characterized by having the reverse properties of bandpass filters: Signals are suppressed within a certain finite band of frequencies and passed at frequencies both above and below that band. The bandstop filter presents little attenuation except in its transition region and, of course, in its stopband. Figure 5.47b shows the attenuation requirements for bandstop filters. Once again, the starting point in the development procedure is the lowpass prototype as shown in Fig. 5.47a. A straight forward transformation will provide a symmetric bandstop response curve.

The bandstop parameter is defined in similar fashion as the bandpass parameter,

$$a = \frac{f_0}{f_2 - f_1} \qquad (5.7.1)$$

where f_1 is the lower cutoff frequency, f_2 is the upper cutoff frequency, and f_0 is the geometric mid-frequency. When the relative bandwidth $1/a$ is only a small percent, the losses because of resistances in the inductors and other realization difficulties must be taken into consideration. When bandwidth is made even smaller and is only a fraction of one percent of the center frequency, electromechanical resonators, such as crystals must be used to eliminate the influence of losses. As a rule, bandstop filters offer much less realization freedom than the bandpass realizations.

The design problem consists of simply obtaining the necessary information for bandstop filters by transforming the information tabulated in the catalog of normalized lowpass filters. The transformation can be specified as follows:

$$\Omega = \frac{\Omega_{BS}}{a(\Omega_{BS}^2 - 1)} \qquad (5.7.2)$$

Figure 5.48 shows frequency scales that indicate the

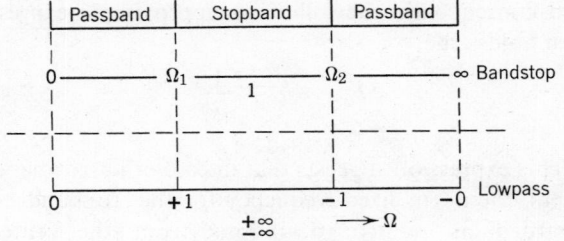

Fig. 5.48. Frequency scales.

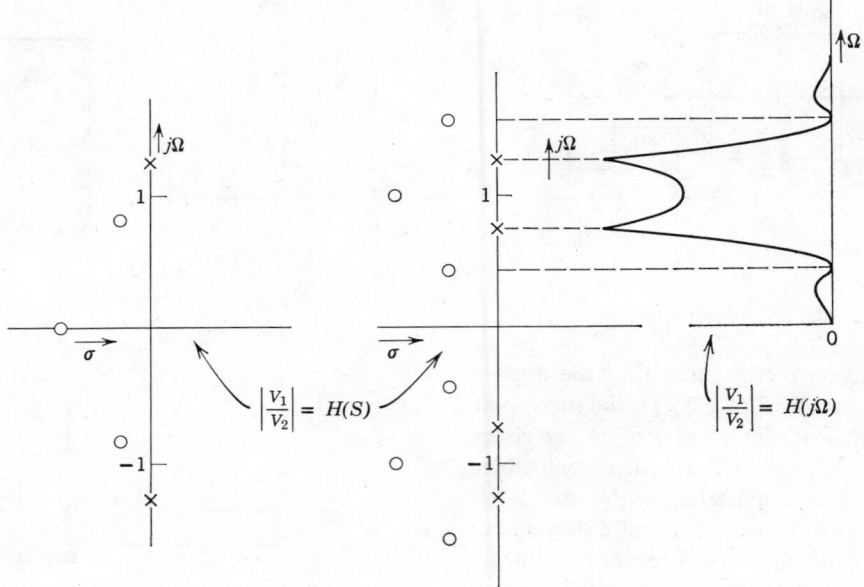

Fig. 5.49. Two typical pole-zero plots (lowpass and bandstop filters).

data of bandstop frequencies from $\Omega_{BS} = 0$ to 1 which can be obtained immediately from the appropriate lowpass data. For higher frequencies the mirror image data must be used. It is evident that the normalized frequency Ω_{BS} of bandstop filters appear in a form analogous to that of the normalized frequency of bandpass filters, Ω_{BP}.

When Eq. 5.7.2 is solved, the following expression for Ω_{BS} is obtained

$$\Omega_{BS} = \sqrt{1 + \left(\frac{1}{2a\Omega}\right)^2} \pm \frac{1}{2a\Omega} \qquad (5.7.3)$$

with $0 < \Omega < \infty$. Using this equation, the lowpass information from the catalog bandstop behavior can be specified. For large Ω values (practically for all frequencies above the passband of the lowpass prototype)

$$\frac{1}{2a\Omega} \ll 1$$

and consequently the following approximate expression holds true

$$\Omega_{BS} \approx 1 \mp \frac{1}{a\Omega} \qquad (5.7.4)$$

This expression suggests that in case of narrowband filters, the normalized frequency for the stopband be regarded as relative deviations from the center frequency.

The exact Eq. 5.7.3 for normalized frequencies indicates a reciprocal quadratic equation

$$A(\Omega_{BS}) = A\left(\frac{1}{\Omega_{BS}}\right) \qquad (5.7.5)$$

which has been shown in Fig. 5.47. When the attenuation specifications are given for equal arithmetic distances from the center frequency, such as

$$\Omega_{s2} - 1 = 1 - \Omega_{s1}$$

then the upper limiting frequency, Ω_{s2} is the critical frequency which must be considered in the remaining calculations, following in complete analogy with bandpass filters.

Pole-zero information for bandstop filters can be developed from the lowpass pole-zero information given in the catalog. s, which belongs to the complex frequency plane, is to be substituted for $j\Omega$. From Eq. 5.7.3

$$s_{BS} = \frac{1}{2as} \pm j\sqrt{1 - \left(\frac{1}{2as}\right)^2} \qquad (5.7.6)$$

From this expression, it is evident that for every s_{BS}, corresponds two bandstop eigenvalues. All lowpass information, even that pertaining to infinity, must be taken into account. Figure 5.49 shows two typical pole-zero plots. They can serve to illustrate the transformation procedure.

For narrow bandstop filters

$$s_{BS} \approx \frac{1}{2as} \pm j$$

since the expression under the radical of Eq. 5.7.6 is

$$\frac{1}{2as} \ll 1$$

The original lowpass pole-zero geometric appears reflected, but twice displaced on $+j$. The knowledge of pole-zero distribution permits the evaluation of the steady-state properties of bandstop filter: its phase, attenuation, and group delay.

The actual element values remain unknown; denormalization proceeds with given reference values which can be represented as a set of design equations. From the physical point of view, the problem of lowpass to bandstop transformation repre-

sents the case when the lowpass inductive element must be substituted by one reactive element with places of "transparance" at $\Omega = 0$ and $\Omega = \infty$. The reactance with this property is a parallel resonant circuit. Consequently, for the dual lowpass filter, the capacitor must be substituted by a reactance having to block frequencies at $\Omega = 0$ and $\Omega = \infty$. The reactance with that property is a series resonant circuit.

In Table 5.5, the relationship between lowpass and bandstop elements is shown along with the relations which provide the element values. Every lowpass structure not dependent on mutual coupling can be transformed into a bandstop structure. But in transforming the parallel resonant circuit from the lowpass model, it is more sensible to use the alternative equivalent BS schematic which includes two parallel resonant circuits in series than the schematic consisting

Table 5.5 Lowpass to Bandstop Transformation

LP	BS	BS Values

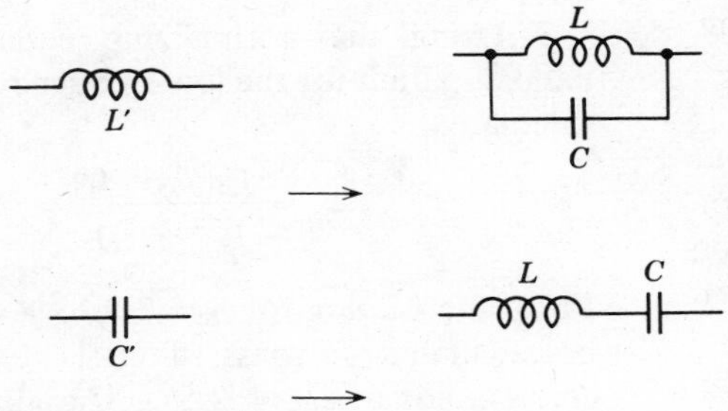

$$L = \frac{L'}{a}\frac{R_r}{\omega_r}$$

$$C = \frac{a}{L'}\frac{1}{R_r\omega_r}$$

$$L = \frac{a}{C'}\frac{R_r}{\omega_r}$$

$$C = \frac{C'}{a}\frac{1}{R_r\omega_r}$$

Where L' and C' are normalized values

Where L, C are unnormalized values

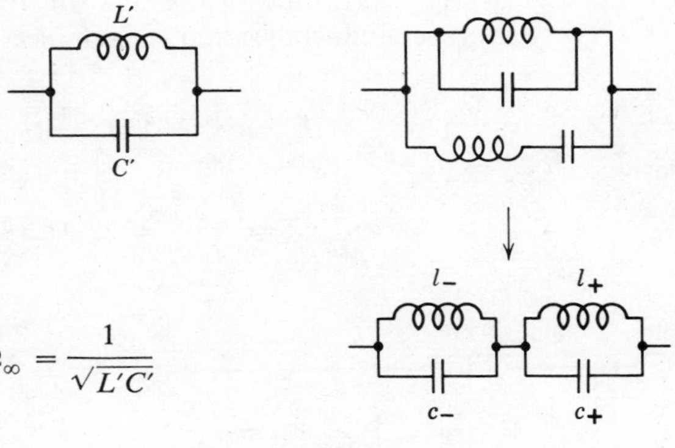

$$\Omega_\infty = \frac{1}{\sqrt{L'C'}}$$

$$c_-' = \frac{a}{L'}(1 + \Omega_+{}^2)$$

$$c_+' = \frac{a}{L'}(1 + \Omega_-{}^2)$$

$$l_-' = \frac{1}{c_-'}$$

$$l_+' = \frac{1}{c_+'}$$

Where c_-', c_+', l_-', l_+' are normalized values

$$l_-' + l_+' = \frac{L'}{a}$$

$$\Omega_\pm = \sqrt{1 + \left(\frac{1}{2\Omega_\infty a}\right)^2} \pm \frac{1}{2\Omega_\infty a}$$

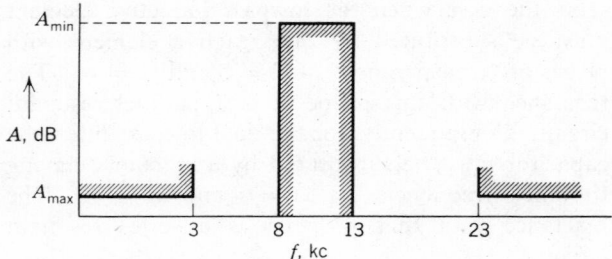

Fig. 5.50.   Prescribed attenuation.

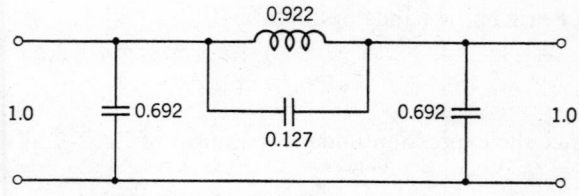

Fig. 5.51.   Lowpass prototype for design example.

of a parallel combination of a series and parallel resonant circuit.

The unnormalized element values for this schematic are

$$C_+ = \frac{1}{\omega_r^2 L_-} = \frac{a}{L'}(1 + \Omega_-^2)\frac{1}{\omega_r R_r}$$

$$C_- = \frac{1}{\omega_r^2 L_+} = \frac{a}{L'}(1 + \Omega_+^2)\frac{1}{\omega_r R}$$

where

$$\Omega_\pm = \sqrt{1 + \left(\frac{1}{a\Omega_\infty}\right)^2} \pm \frac{1}{a\Omega_\infty}$$

and L' is the normalized inductance from the catalog.

Numerical Example Number 1

A bandstop filter is to be designed consisting of four resonant circuits. Figure 5.50 shows the attenuation requirements. The level of guaranteed attenuation in the stopband $A_{\min}$ must be found if the reflection coefficient in the passband is allowed to be 8%.

Figures 5.51 and 5.52 show the lowpass prototype. The transformed canonic schematic of such a filter is shown having an input and output resistance equal to 8 kΩ. The steps to follow for the solution are:

1. Calculate the stopband parameter a.

$$a = \frac{\sqrt{f_2 f_1}}{f_2 - f_1} = \frac{\sqrt{23 \times 3}}{23 - 3} = \frac{8.31}{20} = 0.416.$$

With the coefficient of such magnitude, the design based on the theory of lossless reactive networks is possible.

2. The required response is geometrically symmetric and, therefore,

$$\Omega_{S2}\Omega_{S1} = \Omega_2\Omega_1$$

Knowing that the critical side is the upper side, the modified lower limit is

$$f_{S1} = \frac{f_2 f_1}{f_{S2}} = \frac{23 \cdot 3}{13} = 5.31$$

This figure indicates the importance of distinguishing between geometric and arithmetic symmetry. The lower limit of the stopband must be moved even lower in order to ensure the necessary amount of attenuation at 30 kc.

3. Using the normalizing equation (5.7.2), the bandstop limit for the low-frequency prototype can be calculated.

$$\Omega_S = \frac{f_{s1}(f_2 - f_1)}{f_{s1}^2 - f_2 f_1} = 2.60$$

From the catalog for $n = 3$ and $\rho = 8\%$, the closest precalculated lowpass filter is CC 03 08 23 which corresponds to $\Omega_S = 2.56$. The guaranteed attenuation in the stopband between 8 and 13 kc is $A_{\min} = 25.6$ dB.

4. From the lowpass normalized schematic shown in Fig. 5.51, and using known relations for the reference inductance and capacitance,

$$L_r = \frac{R_r}{\omega_r} = 153.2 \ mH$$

$$C_r = \frac{1}{R_r\omega_r} = 2.394 \ nF,$$

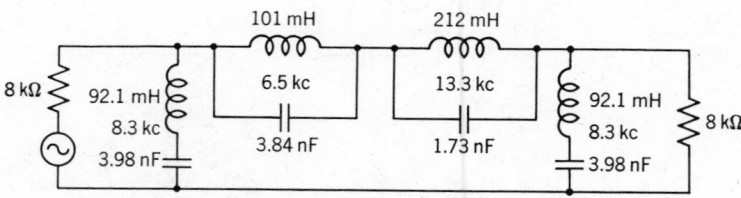

Fig. 5.52.   Bandstop realization.

the element values of the bandstop schematic can be evaluated with the formulas for the shunt circuits.

$$L_1 = L_3 = \frac{a}{C_1'} L_r = 92.1 \text{ mH}$$

$$C_1 = C_3 = \frac{C_1'}{a} C_r = 3.98 \text{ nF}$$

The normalized frequencies where the peaks of attenuation will take place can be calculated with the equations from Table 5.5.

$$\Omega_+ = \sqrt{1 + \left(\frac{1}{2\Omega_\infty a}\right)^2} + \frac{1}{2\Omega_\infty a} = 1.599$$

$$\Omega_- = \sqrt{1 + \left(\frac{1}{2\Omega_\infty a}\right)^2} - \frac{1}{2\Omega_\infty a} = 0.777$$

The values of elements which constitute the parallel resonant circuits are:

$$C_- = \frac{a}{L'}(1 + \Omega_+^2)C_r = 3.84 \text{ nF}$$

$$L_- = \frac{1}{C_-'} L_r = 101 \text{ mH}$$

$$C_+ = \frac{a}{L'}(1 + \Omega_-^2)C_r = 1.73 \text{ nF}$$

$$L_+ = \frac{1}{C_+'} L_r = 212 \text{ mH}$$

The resonant frequencies of the circuits in Fig. 5.52 are

$$_+ = \Omega + f_r = 13.3 \text{ kc} \quad \text{and} \quad f_- = \Omega - f_r = 6.5 \text{ kc}.$$

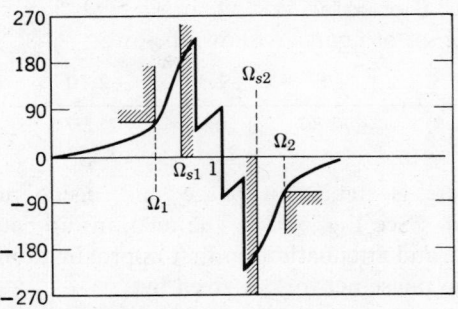

Fig. 5.53. Bandstop phase response curve.

Numerical Example Number 2

The phase response for a bandstop filter with bandwidth parameter $a = 1$ will be calculated.

1. By using the frequency transformation process, the bandstop filter from lowpass prototype CC 03 08 25 is developed. The load impedances are unequal, and the design is accomplished with $K^2 = \infty$. With catalog data, $\Omega_s = 2.37$, and $\Omega_2(=\Omega_\infty) = 2.70$, the following set of data is generated by using the equation for Ω_{BS}.

Ω	0	1	2.37	2.70	∞
Ω_{BS}	0	0.618	0.811	0.832	1
	∞	1.618	1.233	1.202	

2. Pole-zero data (σ_0, σ_1, Ω_1, Ω_2) for the lowpass prototype are used to find the phase of the lowpass model

$$b(\Omega) = \tan^{-1}\frac{\Omega}{\sigma_0} + \tan^{-1}\frac{\Omega - \Omega_1}{\sigma_1}$$
$$+ \tan^{-1}\frac{\Omega + \Omega_1}{\sigma_1} - K$$

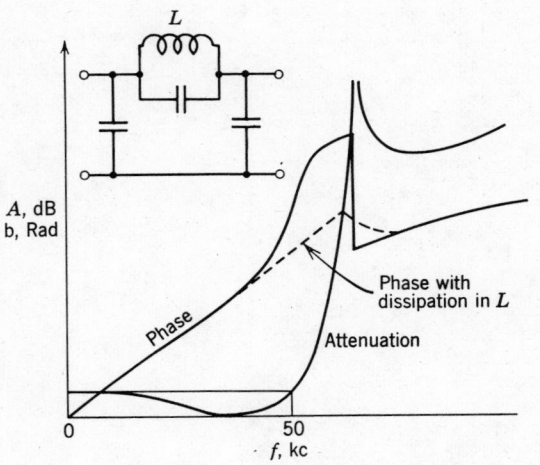

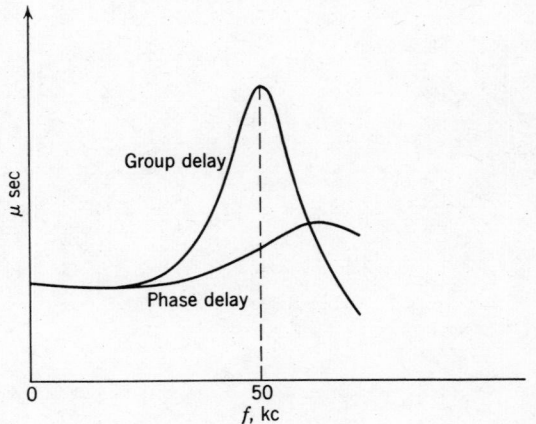

Fig. 5.54. Attenuation, phase, group, and phase delay responses of a lowpass filter ($n = 3$).

with $K = 0$ or π for $\Omega < \Omega_2$ or $\Omega > \Omega_2$. A corresponding set of figures is shown below.

Ω	0	± 1	± 2.37	± 2.70	$\pm \infty$
$b(\Omega)$	0	± 72.80	$\pm 202.4°$	$\pm 212.7°$	$\pm 90°$
	∞		$\pm 22.4°$	$\pm 32.7°$	

3. There is significant phase shift inside of the passband. (See Fig. 5.53). The relationship between the phase and attenuation (in first approximation) for minimum phase network is given by

$$b \approx \frac{da}{d\omega}$$

The phase response can also be obtained graphically from bandstop pole-zero diagram. Figure 5.54 illustrates the total performance of a simple prototype filter with $n = 3$.

5.8 CATALOG OF NORMALIZED LOWPASS MODELS

The information in this section is presented in tabular form, describing a series of lowpass models by

1. Order of polynomial n
2. Reflection coefficient, ρ (proportional to A_{max} as given in Table 2.2).
3. Modular angle θ.
4. Minimum attenuation A_{min}, above Ω_s.

The filters are of the CC type, Chebyshev in both bands, and pay tribute to Cauer and Chebyshev. Every filter could be classified by

$$CC \quad n \quad \rho \quad \theta \quad K^2.$$

The even-order filters ($n = 4, 6$) are of the type b characteristic (see Table 5.1).

Given are filters of complexity $n = 3$ through $n = 7$, with $\rho = 1, 2, 3, 4, 5, 8, 10, 15, 20, 25,$ and 50%.

Besides the normal element value information, other data are presented pertinent to the transmission function. With this, the phase and group delay, and transient response can be calculated.

The tables also include component information for open-circuit (or short-circuit) loads.

$\rho = 1\%$

$n = 3$

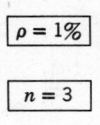
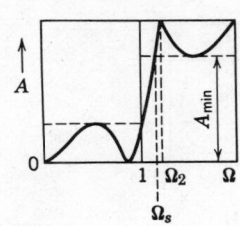
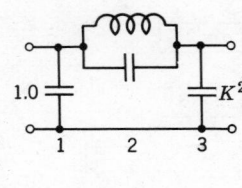

θ	Ω_s	A_{min} (db)	σ_0	σ_1	Ω_1	Ω_2	$K^2 = 1.0$			$K^2 = \infty$			
							$C'_1 = C'_3$	C'_2	L'_2	C'_1	C'_2	L'_2	C'_3
C	∞	∞	2.83850	1.41925	2.60631	∞	0.3523	0.0000	0.6446	0.17615	0.00000	0.45559	0.49845
1	57.2987	89.57	2.84045	1.41737	2.60634	66.1616	0.3521	0.0004	0.6441	0.17570	0.00050	0.45498	0.49841
2	28.6537	71.51	2.84634	1.41174	2.60647	33.0839	0.3513	0.0014	0.6427	0.17436	0.00202	0.45312	0.49830
3	19.1073	60.94	2.85622	1.40239	2.60665	22.0595	0.3501	0.0032	0.6403	0.17212	0.00457	0.45002	0.49812
4	14.3356	53.44	2.87016	1.38935	2.60681	16.5483	0.3484	0.0057	0.6369	0.16896	0.00819	0.44568	0.49789
5	11.4737	47.62	2.88828	1.37266	2.60687	13.2424	0.3462	0.0090	0.6326	0.16488	0.01296	0.44010	0.49763
6	9.5668	42.86	2.91074	1.35237	2.60671	11.0392	0.3436	0.0131	0.6273	0.15983	0.01894	0.43329	0.49735
7	8.2055	38.83	2.93771	1.32858	2.60619	9.4661	0.3404	0.0180	0.6210	0.15381	0.02624	0.42426	0.49710
8	7.1853	35.34	2.96944	1.30136	2.60517	8.2868	0.3368	0.0237	0.6138	0.14676	0.03501	0.41599	0.49691
9	6.3925	32.26	3.00620	1.27084	2.60348	7.3700	0.3326	0.0304	0.6057	0.13865	0.04540	0.40552	0.49683
10	5.7588	29.51	3.04832	1.23713	2.60091	6.6370	0.3280	0.0381	0.5966	0.12942	0.05764	0.39384	0.49690
11	5.2408	27.01	3.09616	1.20041	2.59726	6.0377	0.3230	0.0468	0.5865	0.11902	0.07200	0.38099	0.49721
12	4.8097	24.74	3.15015	1.16083	2.59233	5.5386	0.3174	0.0566	0.5755	0.10738	0.08883	0.36690	0.49784
13	4.4454	22.64	3.21080	1.11861	2.58589	5.1166	0.3114	0.0678	0.5637	0.09440	0.10855	0.35190	0.49888
14	4.1336	20.71	3.27865	1.07398	2.57772	4.7552	0.3050	0.0803	0.5509	0.08000	0.13172	0.33575	0.50044
15	3.8637	18.91	3.35433	1.02721	2.56759	4.4423	0.2981	0.0943	0.5373	0.06407	0.15903	0.31863	0.50268
16	3.6280	17.23	3.43854	0.97858	2.55529	4.1688	0.2908	0.1101	0.5228	0.04648	0.19140	0.30063	0.50574
17	3.4203	15.67	3.53204	0.92842	2.54063	3.9277	0.2831	0.1277	0.5076	0.02709	0.22997	0.28187	0.50982
18	3.2361	14.20	3.63569	0.87710	2.52344	3.7137	0.2751	0.1475	0.4916	0.00571	0.27625	0.26247	0.51512
19	3.0716	12.83	3.75038	0.82501	2.50359	3.5224	0.2666	0.1697	0.4749	-0.01781	0.33219	0.24263	0.52191
20	2.9238	11.55	3.87710	0.77255	2.48096	3.3505	0.2579	0.1946	0.4577	-0.04371	0.40032	0.22253	0.53049
21	2.7904	10.36	4.01686	0.72016	2.45553	3.1951	0.2490	0.2227	0.4399	-0.07225	0.48403	0.20237	0.54116
22	2.6695	9.25	4.17071	0.66828	2.42731	3.0541	0.2398	0.2542	0.4217	-0.10368	0.58771	0.18242	0.55432
23	2.5593	8.22	4.33972	0.61737	2.39637	2.9256	0.2304	0.2897	0.4033	-0.13830	0.71715	0.16292	0.57037
24	2.4586	7.27	4.52490	0.56784	2.36285	2.8079	0.2210	0.3298	0.3846	-0.17644	0.88000	0.14413	0.58976
25	2.3662	6.40	4.72726	0.52009	2.32698	2.6999	0.2115	0.3749	0.3659	-0.21848	1.08641	0.12627	0.61299
θ	Ω_s	A_{min} (db)	σ_0	σ_1	Ω_1	Ω_2	$L'_1 = L'_3$	L'_2	C'_2	L'_1	L'_2	C'_2	L'_3

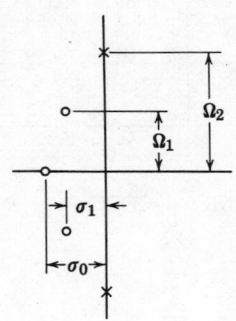
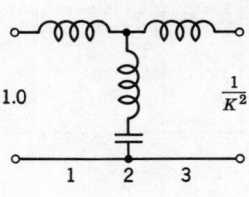

$\rho = 2\%$

$n = 3$

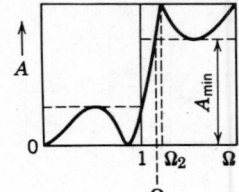

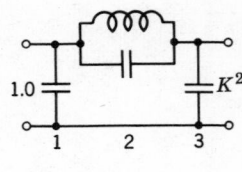

Θ	Ω$_s$	A$_{Min}$ (dB)	σ$_0$	σ$_1$	Ω$_1$	Ω$_2$	K² = 1.0			K² = ∞			
							C'$_1$=C'$_3$	C'$_2$	L'$_2$	C'$_1$	C'$_2$	L'$_2$	C'$_3$
C	∞	∞	2.21299	1.10649	2.10310	∞	0.4519	0.0000	0.7837	0.22594	0.00000	0.57324	0.61780
1	57.2987	95.59	2.21398	1.10556	2.10313	66.1616	0.4517	0.0003	0.7833	0.22558	0.00040	0.57274	0.61776
2	28.6537	77.53	2.21697	1.10277	2.10324	33.0839	0.4511	0.0012	0.7821	0.22452	0.00160	0.57123	0.61762
3	19.1073	66.96	2.22197	1.09813	2.10342	22.0595	0.4501	0.0026	0.7802	0.22274	0.00361	0.56870	0.61739
4	14.3356	59.46	2.22900	1.09164	2.10364	16.5483	0.4486	0.0047	0.7774	0.22024	0.00646	0.56517	0.61707
5	11.4737	53.64	2.23811	1.08332	2.10388	13.2424	0.4468	0.0074	0.7738	0.21702	0.01017	0.56064	0.61669
6	9.5668	48.88	2.24934	1.07318	2.10410	11.0392	0.4446	0.0107	0.7695	0.21306	0.01478	0.55510	0.61624
7	8.2055	44.85	2.26276	1.06125	2.10425	9.4661	0.4419	0.0146	0.7643	0.20834	0.02034	0.54856	0.61575
8	7.1853	41.36	2.27844	1.04755	2.10429	8.2868	0.4389	0.0192	0.7584	0.20286	0.02692	0.54101	0.61521
9	6.3925	38.28	2.29646	1.03211	2.10415	7.3700	0.4355	0.0245	0.7516	0.19659	0.03458	0.53247	0.61468
10	5.7588	35.53	2.31692	1.01497	2.10379	6.6370	0.4316	0.0305	0.7441	0.18951	0.04341	0.52293	0.61416
11	5.2408	33.03	2.33995	0.99616	2.10311	6.0377	0.4274	0.0373	0.7358	0.18159	0.05354	0.51241	0.61367
12	4.8097	30.75	2.36567	0.97574	2.10206	5.5386	0.4227	0.0449	0.7267	0.17280	0.06508	0.50091	0.61327
13	4.4454	28.65	2.39424	0.95376	2.10054	5.1166	0.4177	0.0533	0.7169	0.16311	0.07820	0.48844	0.61299
14	4.1336	26.70	2.42582	0.93028	2.09847	4.7552	0.4122	0.0626	0.7062	0.15248	0.09310	0.47502	0.61286
15	3.8637	24.89	2.46059	0.90537	2.09577	4.4423	0.4064	0.0729	0.6948	0.14086	0.10999	0.46069	0.61296
16	3.6280	23.19	2.49878	0.87910	2.09234	4.1688	0.4002	0.1602	0.6827	0.12821	0.12917	0.44546	0.61334
17	3.4203	21.60	2.54060	0.85156	2.08809	3.9277	0.3936	0.1802	0.6698	0.11446	0.15097	0.42936	0.61406
18	3.2361	20.10	2.58632	0.82286	2.08293	3.7137	0.3866	0.2021	0.6562	0.09956	0.17581	0.41243	0.61520
19	3.0716	18.68	2.63621	0.79309	2.07676	3.5224	0.3793	0.2261	0.6419	0.08343	0.20418	0.39474	0.61686
20	2.9238	17.34	2.69059	0.76238	2.06949	3.3505	0.3717	0.2526	0.6270	0.06599	0.23670	0.37636	0.61916
21	2.7904	16.07	2.74979	0.73085	2.06106	3.1951	0.3637	0.2818	0.6113	0.04716	0.27413	0.35733	0.62217
22	2.6695	14.86	2.81416	0.69865	2.05136	3.0541	0.3553	0.3140	0.5951	0.02682	0.31741	0.33776	0.62606
23	2.5593	13.72	2.88411	0.66592	2.04035	2.9256	0.3467	0.3496	0.5783	0.00488	0.36769	0.31776	0.63098
24	2.4586	12.63	2.96003	0.63284	2.02796	2.8079	0.3378	0.3889	0.5609	-0.01879	0.42642	0.29743	0.63707
25	2.3662	11.60	3.04238	0.59956	2.01415	2.6999	0.3287	0.4323	0.5430	-0.04433	0.49539	0.27692	0.64455
26	2.2812	10.63	3.13161	0.56627	1.99891	2.6003	0.3193	0.4804	0.5248	-0.07190	0.57685	0.25638	0.65362
27	2.2027	9.70	3.22820	0.53315	1.98221	2.5083	0.3098	0.5337	0.5062	-0.10165	0.67361	0.23593	0.66451
28	2.1301	8.83	3.33262	0.50039	1.96409	2.4231	0.3001	0.5928	0.4872	-0.13377	0.78924	0.21580	0.67746
29	2.0627	8.02	3.44537	0.46818	1.94457	2.3438	0.2902	0.6583	0.4681	-0.16845	0.92814	0.19613	0.69276
30	2.0000	7.25	3.56694	0.43668	1.92373	2.2701	0.2804	0.7309	0.4489	-0.20591	1.09595	0.17706	0.71070
Θ	Ω$_s$	A$_{Min}$ (dB)	σ$_0$	σ$_1$	Ω$_1$	Ω$_2$	L'$_1$=L'$_3$	L'$_2$	C'$_2$	L'$_1$	L'$_2$	C'$_2$	L'$_3$

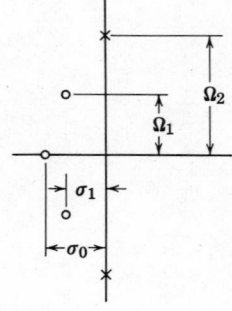

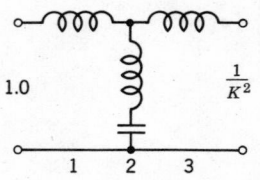

$\rho = 3\%$

$n = 3$

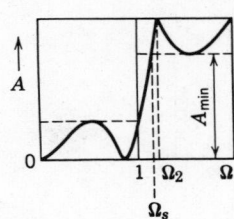

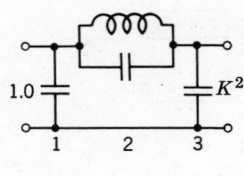

θ	Ω_s	A_{min} (db)	σ_0	σ_1	Ω_1	Ω_2	$K^2 = 1.0$			$K^2 = \infty$			
							$C'_1=C'_3$	C'_2	L'_2	C'_1	C'_2	L'_2	C'_3
C	∞	∞	1.90393	0.95196	1.86246	∞	0.5252	0.0000	0.8704	0.26261	0.00000	0.65512	0.69780
1	57.2987	99.12	1.90460	0.95134	1.86248	66.1616	0.5250	0.0003	0.8700	0.26231	0.00035	0.65468	0.69775
2	28.6537	81.05	1.90661	0.94950	1.86258	33.0839	0.5245	0.0011	0.8690	0.26138	0.00140	0.65334	0.69760
3	19.1073	70.48	1.90998	0.94642	1.86275	22.0595	0.5236	0.0024	0.8672	0.25982	0.00316	0.65108	0.69733
4	14.3356	62.98	1.91471	0.94212	1.86297	16.5483	0.5223	0.0042	0.8647	0.25764	0.00564	0.64794	0.69697
5	11.4737	57.16	1.92083	0.93660	1.86322	13.2424	0.5206	0.0066	0.8615	0.25483	0.00886	0.64390	0.69651
6	9.5668	52.40	1.92837	0.92987	1.86350	11.0392	0.5186	0.0096	0.8576	0.25137	0.01284	0.63896	0.69597
7	8.2055	48.38	1.93735	0.92193	1.86377	9.4661	0.5162	0.0131	0.8529	0.24728	0.01763	0.63314	0.69536
8	7.1853	44.89	1.94781	0.91281	1.86402	8.2868	0.5134	0.0172	0.8476	0.24258	0.02325	0.62641	0.69467
9	6.3925	41.81	1.95980	0.90251	1.86420	7.3700	0.5103	0.0219	0.8415	0.23709	0.02975	0.61879	0.69395
10	5.7588	39.05	1.97336	0.89104	1.86430	6.6370	0.5067	0.0272	0.8348	0.23097	0.03720	0.61028	0.69317
11	5.2408	36.55	1.98857	0.87844	1.86427	6.0377	0.5029	0.0332	0.8273	0.22416	0.04565	0.60089	0.69238
12	4.8097	34.27	2.00549	0.86471	1.86407	5.5386	0.4986	0.0398	0.8192	0.21663	0.05520	0.59061	0.69158
13	4.4454	32.17	2.02418	0.84989	1.86365	5.1166	0.4940	0.0471	0.8103	0.20836	0.06592	0.57947	0.69082
14	4.1336	30.22	2.04475	0.83400	1.86298	4.7552	0.4891	0.0552	0.8007	0.19931	0.07793	0.56744	0.69010
15	3.8637	28.41	2.06727	0.81707	1.86201	4.4423	0.4837	0.0641	0.7905	0.18949	0.09137	0.55458	0.68947
16	3.6280	26.71	2.09187	0.79913	1.86067	4.1688	0.4780	0.0738	0.7795	0.17884	0.10638	0.54887	0.68896
17	3.4203	25.11	2.11864	0.78023	1.85893	3.9277	0.4720	0.0844	0.7679	0.16733	0.12316	0.52634	0.68861
18	3.2361	23.60	2.14773	0.76041	1.85672	3.7137	0.4656	0.0960	0.7556	0.15494	0.14190	0.51099	0.68845
19	3.0716	22.17	2.17926	0.73971	1.85398	3.5224	0.4589	0.1085	0.7426	0.14161	0.16286	0.49488	0.68855
20	2.9238	20.82	2.21340	0.71819	1.85068	3.3505	0.4518	0.1222	0.7290	0.12731	0.18636	0.47803	0.68897
21	2.7904	19.53	2.25031	0.69590	1.84673	3.1951	0.4444	0.1371	0.7147	0.11198	0.21273	0.46046	0.68977
22	2.6695	18.31	2.29017	0.67292	1.84210	3.0541	0.4366	0.1532	0.6998	0.09558	0.24243	0.44222	0.69100
23	2.5593	17.14	2.33318	0.64930	1.83673	2.9256	0.4286	0.1707	0.6843	0.07800	0.27596	0.42339	0.69278
24	2.4586	16.02	2.37956	0.62513	1.83056	2.8079	0.4202	0.1898	0.6683	0.05622	0.31394	0.40399	0.69516
25	2.3662	14.95	2.42953	0.60050	1.82355	2.6999	0.4116	0.2105	0.6517	0.03915	0.35715	0.38412	0.69828
26	2.2812	13.94	2.48333	0.57548	1.81566	2.6003	0.4027	0.2331	0.6345	0.01771	0.40646	0.36385	0.70223
27	2.2027	12.96	2.54124	0.55017	1.80685	2.5083	0.3935	0.2576	0.6169	-0.00519	0.46300	0.34328	0.70716
28	2.1301	12.03	2.60351	0.52467	1.79708	2.4231	0.3841	0.2844	0.5988	-0.02967	0.52813	0.32250	0.71317
29	2.0627	11.14	2.67046	0.49910	1.78634	2.3438	0.3745	0.3137	0.5803	-0.05581	0.60347	0.30164	0.72045
30	2.0000	10.29	2.74236	0.47354	1.77461	2.2701	0.3646	0.3456	0.5615	-0.08375	0.69107	0.28080	0.72916
31	1.9416	9.49	2.81955	0.44813	1.76190	2.2012	0.3547	0.3805	0.5424	-0.11361	0.79338	0.26013	0.73946
32	1.8871	8.72	2.90234	0.42297	1.74821	2.1368	0.3445	0.4188	0.5230	-0.14554	0.91350	0.23974	0.75158
33	1.8361	8.00	2.99104	0.39817	1.73356	2.0765	0.3343	0.4607	0.5034	-0.17967	1.05512	0.21980	0.76572
34	1.7883	7.31	3.08600	0.37385	1.71800	2.0199	0.3240	0.5067	0.4838	-0.21617	1.22293	0.20843	0.78211
35	1.7434	6.67	3.18752	0.35011	1.70156	1.9666	0.3137	0.5572	0.4640	-0.25522	1.42261	0.18175	0.80097
θ	Ω_s	A_{min} (db)	σ_0	σ_1	Ω_1	Ω_2	$L'_1=L'_3$	L'_2	C'_2	L'_1	L'_2	C'_2	L'_3

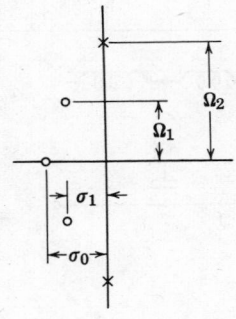

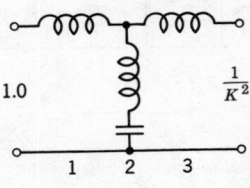

$\rho = 4\%$

$n = 3$

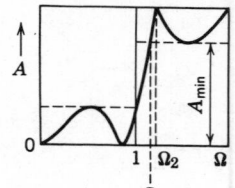

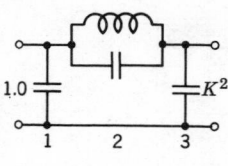

θ	Ω_s	A_{MIN} (db)	σ_0	σ_1	Ω_1	Ω_2	$K^2 = 1.0$			$K^2 = \infty$			
							$C'_1=C'_3$	C'_2	L'_2	C'_1	C'_2	L'_2	C'_3
C	∞	∞	1.70606	0.85303	1.71261	∞	0.5862	0.0000	0.9321	0.29307	0.00000	0.71971	0.75912
1	57.2987	101.62	1.70653	0.85255	1.71260	66.1615	0.5860	0.0002	0.9318	0.29280	0.00032	0.71931	0.75908
2	28.6537	83.55	1.70806	0.85117	1.71270	33.0839	0.5855	0.0010	0.9308	0.29195	0.00127	0.71807	0.75891
3	19.1073	72.98	1.71061	0.84888	1.71285	22.0595	0.5846	0.0022	0.9291	0.29053	0.00287	0.71598	0.75862
4	14.3356	65.48	1.71419	0.84567	1.71306	16.5483	0.5834	0.0039	0.9268	0.28855	0.00512	0.71308	0.75822
5	11.4737	59.66	1.71882	0.84155	1.71332	13.2424	0.5818	0.0062	0.9239	0.28599	0.00804	0.70934	0.75772
6	9.5668	54.90	1.72451	0.83652	1.71361	11.0392	0.5799	0.0089	0.9202	0.28286	0.01164	0.70478	0.75712
7	8.2055	50.88	1.73128	0.83060	1.71391	9.4661	0.5776	0.0122	0.9159	0.27914	0.01596	0.69940	0.75642
8	7.1853	47.39	1.73916	0.82377	1.71423	8.2868	0.5750	0.0160	0.9110	0.27483	0.02101	0.69317	0.75563
9	6.3925	44.31	1.74817	0.81606	1.71452	7.3700	0.5720	0.0203	0.9053	0.26993	0.02683	0.68615	0.75478
10	5.7588	41.55	1.75835	0.80747	1.71478	6.6370	0.5687	0.0253	0.8991	0.26440	0.03347	0.67827	0.75385
11	5.2408	39.05	1.76973	0.79801	1.71499	6.0377	0.5651	0.0307	0.8921	0.25825	0.04147	0.66958	0.75287
12	4.8097	36.77	1.78236	0.78770	1.71510	5.5386	0.5611	0.0369	0.8845	0.25148	0.04939	0.66007	0.75185
13	4.4454	34.67	1.79629	0.77654	1.71511	5.1166	0.5567	0.0436	0.8763	0.24405	0.05879	0.64976	0.75081
14	4.1336	32.72	1.81156	0.76455	1.71497	4.7552	0.5520	0.0510	0.8674	0.23595	0.06925	0.63862	0.74976
15	3.8637	30.91	1.82824	0.75175	1.71467	4.4423	0.5470	0.0591	0.8579	0.22717	0.08086	0.62669	0.74874
16	3.6280	29.20	1.84639	0.73816	1.71416	4.1688	0.5416	0.0679	0.8477	0.21768	0.09372	0.61397	0.74776
17	3.4203	27.60	1.86608	0.72380	1.71340	3.9277	0.5359	0.0775	0.8368	0.20746	0.10796	0.60046	0.74685
18	3.2361	26.09	1.88739	0.70869	1.71237	3.7137	0.5298	0.0878	0.8254	0.19649	0.12369	0.58618	0.74603
19	3.0716	24.66	1.91041	0.69285	1.71103	3.5224	0.5234	0.0991	0.8133	0.18473	0.14111	0.57115	0.74535
20	2.9238	23.31	1.93523	0.67633	1.70933	3.3505	0.5167	0.1113	0.8006	0.17217	0.16039	0.55540	0.74486
21	2.7904	22.01	1.96196	0.65915	1.70725	3.1951	0.5097	0.1244	0.7872	0.15876	0.18176	0.53892	0.74456
22	2.6695	20.78	1.99070	0.64134	1.70473	3.0541	0.5023	0.1386	0.7733	0.14446	0.20548	0.52175	0.74453
23	2.5593	19.60	2.02157	0.62295	1.70174	2.9256	0.4947	0.1540	0.7588	0.12924	0.23184	0.50394	0.74482
24	2.4586	18.48	2.05471	0.60402	1.69824	2.8079	0.4867	0.1705	0.7437	0.11304	0.26123	0.48551	0.74548
25	2.3662	17.40	2.09027	0.58459	1.69419	2.6999	0.4784	0.1884	0.7280	0.09582	0.29408	0.46649	0.74658
26	2.2812	16.36	2.12838	0.56472	1.68955	2.6003	0.4698	0.2078	0.7118	0.09316	0.33087	0.44698	0.74820
27	2.2027	15.36	2.16922	0.54446	1.68429	2.5083	0.4610	0.2287	0.6951	0.05811	0.37224	0.42700	0.75043
28	2.1301	14.41	2.21296	0.52387	1.67838	2.4231	0.4519	0.2513	0.6778	0.03748	0.41889	0.40660	0.75332
29	2.0627	13.49	2.25979	0.50301	1.67178	2.3438	0.4425	0.2757	0.6602	0.01559	0.47170	0.38590	0.75701
30	2.0000	12.61	2.30990	0.48196	1.66447	2.2701	0.4329	0.3023	0.6420	-0.00765	0.53173	0.36495	0.76159
31	1.9416	11.77	2.36350	0.46077	1.65644	2.2012	0.4231	0.3310	0.6235	-0.03233	0.60022	0.34385	0.76719
32	1.8871	10.96	2.42081	0.43953	1.64765	2.1368	0.4131	0.3622	0.6046	-0.05854	0.67868	0.32270	0.77393
33	1.8361	10.19	2.48207	0.41831	1.63812	2.0765	0.4029	0.3962	0.5854	-0.08637	0.76895	0.30161	0.78196
34	1.7883	9.44	2.54749	0.39719	1.62782	2.0199	0.3925	0.4331	0.5659	-0.11593	0.87326	0.28069	0.79143
35	1.7434	8.74	2.61734	0.37626	1.61677	1.9666	0.3821	0.4734	0.5462	-0.14734	0.99435	0.26003	0.80250
36	1.7013	8.06	2.69185	0.35558	1.60499	1.9165	0.3715	0.5173	0.5263	-0.18070	1.13541	0.23980	0.81536
37	1.6616	7.42	2.77127	0.33525	1.59249	1.8692	0.3608	0.5653	0.5063	-0.21617	1.30056	0.22008	0.83019
38	1.6243	6.82	2.85585	0.31533	1.57931	1.8245	0.3502	0.6177	0.4863	-0.25388	1.49458	0.20100	0.84720
39	1.5890	6.25	2.94583	0.29591	1.56548	1.7823	0.3395	0.6751	0.4663	-0.29396	1.72348	0.18266	0.86659
40	1.5557	5.71	3.04147	0.27704	1.55106	1.7423	0.3288	0.7380	0.4464	-0.33657	1.99444	0.16517	0.88856
θ	Ω_s	A_{MIN} (db)	σ_0	σ_1	Ω_1	Ω_2	$L'_1=L'_3$	L'_2	C'_2	L'_1	L'_2	C'_2	L'_3

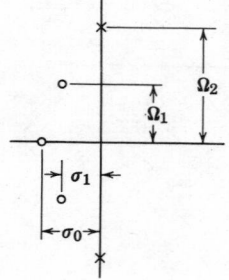

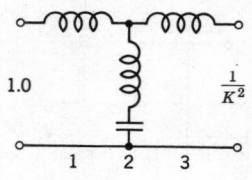

ρ = 5%

n = 3

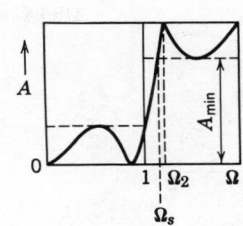

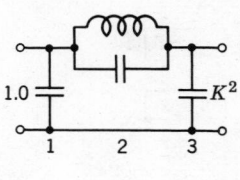

Θ	Ωs	A_Min	σ₀	σ₁	Ω₁	Ω₂	K²=1.0			K²=∞			
							$C'_1=C'_3$	C'_2	L'_2	C'_1	C'_2	L'_2	C'_3
1	57.2987	103.56	1.56379	0.78132	1.60724	66.1616	0.6395	0.0002	0.9786	0.3196	0.0003	0.7733	0.8092
2	28.6537	85.50	1.56503	0.78023	1.60733	33.0839	0.6390	0.0009	0.9776	0.3188	0.0012	0.7721	0.8090
3	19.1073	74.93	1.56708	0.77840	1.60748	22.0595	0.6381	0.0021	0.9761	0.3175	0.0027	0.7702	0.8087
4	14.3356	67.43	1.56997	0.77585	1.60768	16.5483	0.6370	0.0037	0.9739	0.3156	0.0048	0.7675	0.8083
5	11.4737	61.61	1.57370	0.77257	1.60793	13.2424	0.6354	0.0059	0.9711	0.3132	0.0075	0.7639	0.8077
6	9.5668	56.85	1.57828	0.76857	1.60822	11.0392	0.6336	0.0085	0.9676	0.3103	0.0180	0.7596	0.8071
7	8.2055	52.82	1.58373	0.76384	1.60854	9.4661	0.6314	0.0116	0.9636	0.3069	0.0148	0.7546	0.8063
8	7.1853	49.33	1.59006	0.75840	1.60887	8.2868	0.6289	0.0152	0.9589	0.3029	0.0195	0.7487	0.8055
9	6.3925	46.25	1.59730	0.75225	1.60922	7.3700	0.6261	0.0193	0.9536	0.2983	0.0248	0.7421	0.8045
10	5.7588	43.49	1.60546	0.74539	1.60955	6.6370	0.6229	0.0240	0.9477	0.2932	0.0309	0.7346	0.8035
11	5.2408	41.00	1.61458	0.73784	1.60986	6.0377	0.6194	0.0291	0.9411	0.2875	0.0378	0.7264	0.8024
12	4.8097	38.71	1.62467	0.72959	1.61013	5.5386	0.6155	0.0349	0.9339	0.2813	0.0454	0.7175	0.8012
13	4.4454	36.61	1.63579	0.72065	1.61033	5.1166	0.6113	0.0412	0.9261	0.2744	0.0540	0.7077	0.8000
14	4.1336	34.66	1.64795	0.71104	1.61046	4.7552	0.6068	0.0482	0.9177	0.2670	0.0634	0.6972	0.7987
15	3.8637	32.85	1.66121	0.70077	1.61048	4.4423	0.6020	0.0558	0.9087	0.2589	0.0739	0.6860	0.7974
16	3.6280	31.14	1.67561	0.68984	1.61037	4.1688	0.5968	0.0640	0.8991	0.2502	0.0854	0.6739	0.7961
17	3.4203	29.54	1.69120	0.67828	1.61011	3.9277	0.5913	0.0729	0.8888	0.2408	0.0980	0.6612	0.7949
18	3.2361	28.03	1.70803	0.66609	1.60967	3.7137	0.5855	0.0826	0.8780	0.2308	0.1120	0.6477	0.7936
19	3.0716	26.60	1.72616	0.65329	1.60903	3.5224	0.5793	0.0930	0.8665	0.2201	0.1272	0.6334	0.7925
20	2.9238	25.24	1.74565	0.63990	1.60815	3.3505	0.5728	0.1043	0.8545	0.2087	0.1440	0.6185	0.7914
21	2.7904	23.95	1.76659	0.62595	1.60701	3.1951	0.5661	0.1164	0.8418	0.1965	0.1625	0.6028	0.7905
22	2.6695	22.71	1.78903	0.61144	1.60558	3.0541	0.5590	0.1294	0.8286	0.1836	0.1828	0.5865	0.7897
23	2.5593	21.53	1.81308	0.59641	1.60383	2.9256	0.5515	0.1434	0.8148	0.1699	0.2052	0.5695	0.7891
24	2.4586	20.40	1.83881	0.58089	1.60172	2.8079	0.5438	0.1585	0.8004	0.1553	0.2298	0.5519	0.7887
25	2.3662	19.31	1.86632	0.56489	1.59923	2.6999	0.5358	0.1747	0.7855	0.1399	0.2571	0.5337	0.7887
26	2.2812	18.27	1.89572	0.54847	1.59633	2.6003	0.5275	0.1921	0.7700	0.1236	0.2873	0.5149	0.7889
27	2.2027	17.26	1.92713	0.53164	1.59298	2.5083	0.5189	0.2108	0.7540	0.1063	0.3208	0.4955	0.7896
28	2.1301	16.30	1.96066	0.51445	1.58916	2.4231	0.5100	0.2309	0.7375	0.0880	0.3580	0.4757	0.7908
29	2.0627	15.37	1.99644	0.49694	1.58485	2.3438	0.5009	0.2526	0.7205	0.0687	0.3997	0.4555	0.7924
30	2.0000	14.47	2.03461	0.47915	1.58000	2.2701	0.4915	0.2760	0.7031	0.0483	0.4463	0.4348	0.7947
31	1.9416	13.61	2.07532	0.46114	1.57461	2.2012	0.4819	0.3012	0.6852	0.0268	0.4986	0.4139	0.7977
32	1.8871	12.77	2.11873	0.44294	1.56865	2.1368	0.4720	0.3284	0.6669	0.0040	0.5577	0.3927	0.8014
33	1.8361	11.97	2.16500	0.42462	1.56210	2.0765	0.4619	0.3578	0.6482	-0.0201	0.6244	0.3714	0.8061
34	1.7883	11.20	2.21430	0.40622	1.55494	2.0199	0.4516	0.3896	0.6291	-0.0456	0.7003	0.3500	0.8117
35	1.7434	10.46	2.26682	0.38781	1.54717	1.9666	0.4411	0.4241	0.6097	-0.0725	0.7868	0.3286	0.8185
Θ	Ωs	A_Min	σ₀	σ₁	Ω₁	Ω₂	$L'_1=L'_3$	L'_2	C'_2	L'_1	L'_2	C'_2	L'_3

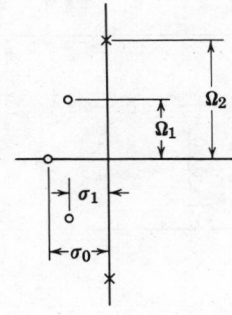

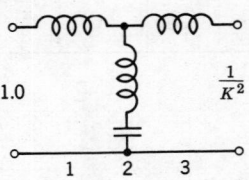

173

ρ = 8%

n = 3

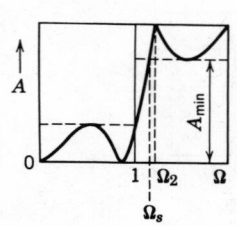

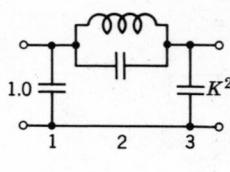

Θ	Ωs	A_Min (dB)	σ_o	σ_1	Ω1	Ω2	K² = 1.0			K² = ∞			
							$C'_1=C'_3$	C'_2	L'_2	C'_1	C'_2	L'_2	C'_3
1	57.2987	107.66	1.29040	0.64484	1.41365	66.1616	0.7750	0.0002	1.0684	0.3873	0.0003	0.8982	0.9218
2	28.6537	89.59	1.29118	0.64417	1.41373	33.0839	0.7745	0.0009	1.0676	0.3867	0.0010	0.8971	0.9216
3	19.1073	79.03	1.29249	0.64304	1.41386	22.0595	0.7737	0.0019	1.0662	0.3855	0.0023	0.8954	0.9213
4	14.3356	71.53	1.29434	0.64147	1.41405	16.5483	0.7726	0.0034	1.0642	0.3839	0.0041	0.8930	0.9280
5	11.4737	65.70	1.29671	0.63945	1.41428	13.2424	0.7712	0.0054	1.0617	0.3819	0.0064	0.8898	0.9202
6	9.5668	60.95	1.29963	0.63698	1.41455	11.0392	0.7694	0.0078	1.0587	0.3794	0.0093	0.8860	0.9194
7	8.2055	56.92	1.30309	0.63406	1.41486	9.4661	0.7674	0.0106	1.0551	0.3764	0.0127	0.8815	0.9185
8	7.1853	53.43	1.30710	0.63070	1.41521	8.2868	0.7650	0.0139	1.0509	0.3730	0.0166	0.8763	0.9175
9	6.3925	50.35	1.31168	0.62690	1.41559	7.3700	0.7624	0.0176	1.0462	0.3691	0.0212	0.8704	0.9164
10	5.7588	47.59	1.31684	0.62265	1.41599	6.6370	0.7594	0.0218	1.0409	0.3647	0.0263	0.8638	0.9151
11	5.2408	45.10	1.32258	0.61797	1.41640	6.0377	0.7561	0.0265	1.0351	0.3599	0.0320	0.8565	0.9137
12	4.8097	42.81	1.32892	0.61284	1.41682	5.5386	0.7525	0.0317	1.0287	0.3546	0.0384	0.8485	0.9123
13	4.4454	40.71	1.33589	0.60729	1.41724	5.1166	0.7486	0.0374	1.0218	0.3488	0.0455	0.8399	0.9107
14	4.1336	38.76	1.34349	0.60130	1.41765	4.7552	0.7443	0.0436	1.0143	0.3425	0.0532	0.8305	0.9090
15	3.8637	36.94	1.35174	0.59488	1.41803	4.4423	0.7398	0.0504	1.0063	0.3356	0.0618	0.8205	0.9073
16	3.6280	35.24	1.36067	0.58804	1.41839	4.1688	0.7349	0.0577	0.9977	0.3283	0.0711	0.8098	0.9055
17	3.4203	33.64	1.37030	0.58078	1.41869	3.9277	0.7298	0.0656	0.9886	0.3205	0.0812	0.7984	0.9036
18	3.2361	32.13	1.38066	0.57310	1.41894	3.7137	0.7243	0.0741	0.9789	0.3121	0.0922	0.7863	0.9016
19	3.0716	30.70	1.39176	0.56502	1.41912	3.5224	0.7185	0.0832	0.9687	0.3032	0.1042	0.7736	0.8997
20	2.9238	29.33	1.40365	0.55653	1.41921	3.3505	0.7124	0.0930	0.9579	0.2937	0.1172	0.7602	0.8977
21	2.7904	28.04	1.41636	0.54765	1.41920	3.1951	0.7060	0.1035	0.9467	0.2836	0.1313	0.7462	0.8957
22	2.6695	26.80	1.42992	0.53838	1.41908	3.0541	0.6993	0.1147	0.9348	0.2730	0.1466	0.7315	0.8937
23	2.5593	25.61	1.44436	0.52873	1.41883	2.9256	0.6923	0.1267	0.9225	0.2618	0.1631	0.7162	0.8918
24	2.4586	24.47	1.45974	0.51871	1.41844	2.8079	0.6851	0.1394	0.9096	0.2500	0.1811	0.7002	0.8899
25	2.3662	23.38	1.47609	0.50833	1.41788	2.6999	0.6775	0.1531	0.8962	0.2375	0.2007	0.6837	0.8881
26	2.2812	22.32	1.49346	0.49760	1.41714	2.6003	0.6696	0.1676	0.8823	0.2244	0.2219	0.6665	0.8864
27	2.2027	21.31	1.51191	0.48653	1.41620	2.5083	0.6614	0.1831	0.8679	0.2106	0.2450	0.6488	0.8848
28	2.1301	20.33	1.53149	0.47514	1.41505	2.4231	0.6530	0.1997	0.8530	0.1961	0.2701	0.6305	0.8834
29	2.0627	19.39	1.55225	0.46344	1.41366	2.3438	0.6442	0.2173	0.8376	0.1808	0.2976	0.6117	0.8822
30	2.0000	18.47	1.57428	0.45145	1.41203	2.2701	0.6352	0.2362	0.8217	0.1649	0.3276	0.5923	0.8812
31	1.9416	17.59	1.59762	0.43919	1.41012	2.2012	0.6259	0.2563	0.8054	0.1481	0.3605	0.5725	0.8805
32	1.8871	16.73	1.62236	0.42667	1.40794	2.1368	0.6164	0.2777	0.7885	0.1305	0.3966	0.5523	0.8801
33	1.8361	15.90	1.64858	0.41392	1.40544	2.0765	0.6066	0.3007	0.7713	0.1121	0.4363	0.5316	0.8801
34	1.7883	15.10	1.67635	0.40095	1.40263	2.0199	0.5965	0.3252	0.7536	0.0928	0.4801	0.5106	0.8805
35	1.7434	14.32	1.70578	0.38780	1.39948	1.9666	0.5862	0.3515	0.7355	0.0726	0.5285	0.4892	0.8814
36	1.7013	13.56	1.73694	0.37449	1.39599	1.9165	0.5757	0.3797	0.7171	0.0514	0.5823	0.4676	0.8829
37	1.6616	12.83	1.76995	0.36105	1.39212	1.8692	0.5650	0.4099	0.6982	0.0292	0.6421	0.4458	0.8849
38	1.6243	12.12	1.80491	0.34750	1.38788	1.8245	0.5540	0.4424	0.6791	0.0059	0.7088	0.4238	0.8877
39	1.5890	11.43	1.84194	0.33388	1.38326	1.7823	0.5429	0.4773	0.6596	-0.0185	0.7835	0.4018	0.8912
40	1.5557	10.77	1.88114	0.32022	1.37823	1.7423	0.5316	0.5149	0.6398	-0.0441	0.8674	0.3798	0.8955
Θ	Ωs	A_Min (dB)	σ_o	σ_1	Ω1	Ω2	$L'_1=L'_3$	L'_2	C'_2	L'_1	L'_2	C'_2	L'_3

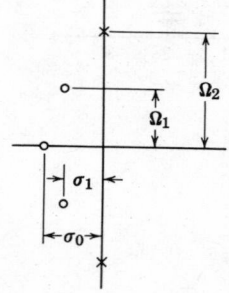

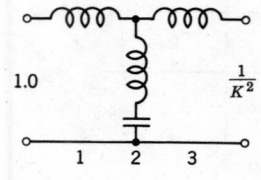

174

$\rho = 10\%$

$n = 3$

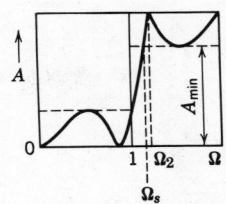

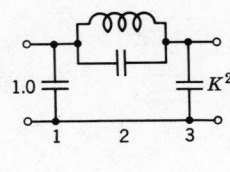

θ	Ω_s	A_{min} (db)	σ_0	σ_1	Ω_1	Ω_2	$K^2 = 1.0$			$K^2 = \infty$			
							$C'_1 = C'_3$	C'_2	L'_2	C'_1	C'_2	L'_2	C'_3
1.	57.2987	109.61	1.17193	0.58568	1.33407	66.1616	0.8533	0.0002	1.1036	0.4265	0.0002	0.9623	0.9786
2.	28.6537	91.55	1.17256	0.58514	1.33415	33.0839	0.8528	0.0008	1.1028	0.4259	0.0010	0.9613	0.9784
3.	19.1073	80.98	1.17362	0.58425	1.33427	22.0595	0.8521	0.0019	1.1015	0.4248	0.0021	0.9597	0.9780
4.	14.3356	73.48	1.17511	0.58300	1.33445	16.5483	0.8510	0.0033	1.0997	0.4233	0.0038	0.9574	0.9775
5.	11.4737	67.66	1.17703	0.58140	1.33467	13.2424	0.8496	0.0052	1.0973	0.4214	0.0060	0.9544	0.9768
6.	9.5668	62.90	1.17938	0.57944	1.33493	11.0392	0.8479	0.0075	1.0944	0.4191	0.0086	0.9508	0.9760
7.	8.2055	58.88	1.18218	0.57713	1.33524	9.4661	0.8459	0.0102	1.0910	0.4163	0.0118	0.9465	0.9750
8.	7.1853	55.39	1.18542	0.57446	1.33558	8.2868	0.8436	0.0134	1.0871	0.4132	0.0155	0.9416	0.9740
9.	6.3925	52.31	1.18911	0.57144	1.33596	7.3700	0.8410	0.0170	1.0826	0.4095	0.0197	0.9360	0.9727
10.	5.7588	49.55	1.19326	0.56807	1.33636	6.6370	0.8380	0.0211	1.0776	0.4055	0.0244	0.9297	0.9714
11.	5.2408	47.05	1.19788	0.56435	1.33679	6.0377	0.8348	0.0256	1.0721	0.4010	0.0297	0.9228	0.9699
12.	4.8097	44.77	1.20299	0.56027	1.33724	5.5386	0.8313	0.0306	1.0661	0.3960	0.0356	0.9152	0.9683
13.	4.4454	42.66	1.20858	0.55585	1.33770	5.1166	0.8274	0.0361	1.0596	0.3906	0.0421	0.9070	0.9666
14.	4.1336	40.72	1.21468	0.55108	1.33817	4.7552	0.8233	0.0420	1.0525	0.3848	0.0492	0.8981	0.9647
15.	3.8637	38.90	1.22129	0.54597	1.33864	4.4423	0.8188	0.0485	1.0449	0.3785	0.0570	0.8886	0.9628
16.	3.6280	37.20	1.22844	0.54051	1.33909	4.1688	0.8140	0.0555	1.0368	0.3717	0.0655	0.8785	0.9608
17.	3.4203	35.59	1.23613	0.53472	1.33953	3.9277	0.8090	0.0630	1.0282	0.3644	0.0747	0.8677	0.9586
18.	3.2361	34.08	1.24439	0.52859	1.33994	3.7137	0.8036	0.0712	1.0190	0.3567	0.0847	0.8562	0.9564
19.	3.0716	32.65	1.25324	0.52212	1.34032	3.5224	0.7979	0.0798	1.0094	0.3485	0.0955	0.8441	0.9542
20.	2.9238	31.29	1.26269	0.51532	1.34065	3.3505	0.7920	0.0892	0.9992	0.3397	0.1071	0.8314	0.9518
21.	2.7904	29.99	1.27277	0.50820	1.34093	3.1951	0.7857	0.0991	0.9885	0.3305	0.1197	0.8180	0.9494
22.	2.6695	28.75	1.28351	0.50075	1.34114	3.0541	0.7791	0.1097	0.9774	0.3208	0.1333	0.8041	0.9470
23.	2.5593	27.56	1.29493	0.49299	1.34127	2.9256	0.7722	0.1210	0.9657	0.3105	0.1480	0.7895	0.9446
24.	2.4586	26.42	1.30706	0.48491	1.34132	2.8079	0.7651	0.1330	0.9535	0.2996	0.1638	0.7743	0.9422
25.	2.3662	25.32	1.31993	0.47653	1.34126	2.6999	0.7576	0.1458	0.9408	0.2883	0.1809	0.7585	0.9397
26.	2.2812	24.27	1.33358	0.46784	1.34109	2.6003	0.7499	0.1594	0.9276	0.2763	0.1993	0.7421	0.9374
27.	2.2027	23.25	1.34803	0.45887	1.34079	2.5083	0.7418	0.1739	0.9139	0.2638	0.2192	0.7251	0.9350
28.	2.1301	22.27	1.36334	0.44961	1.34036	2.4231	0.7335	0.1893	0.8998	0.2506	0.2407	0.7076	0.9327
29.	2.0627	21.32	1.37953	0.44007	1.33977	2.3438	0.7249	0.2056	0.8852	0.2369	0.2640	0.6895	0.9306
30.	2.0000	20.40	1.39666	0.43027	1.33901	2.2701	0.7160	0.2230	0.8701	0.2225	0.2892	0.6709	0.9285
31.	1.9416	19.51	1.41477	0.42021	1.33808	2.2012	0.7068	0.2415	0.8545	0.2074	0.3166	0.6518	0.9267
32.	1.8871	18.65	1.43392	0.40991	1.33695	2.1368	0.6974	0.2612	0.8385	0.1917	0.3464	0.6322	0.9250
33.	1.8361	17.82	1.45415	0.39937	1.33561	2.0765	0.6877	0.2821	0.8220	0.1752	0.3789	0.6121	0.9235
34.	1.7883	17.00	1.47553	0.38862	1.33405	2.0199	0.6777	0.3044	0.8051	0.1580	0.4143	0.5917	0.9223
35.	1.7434	16.21	1.49812	0.37767	1.33225	1.9666	0.6675	0.3282	0.7878	0.1401	0.4530	0.5708	0.9213
36.	1.7013	15.45	1.52199	0.36653	1.33020	1.9165	0.6570	0.3536	0.7701	0.1213	0.4955	0.5495	0.9208
37.	1.6616	14.70	1.54721	0.35522	1.32789	1.8692	0.6463	0.3806	0.7520	0.1017	0.5421	0.5280	0.9206
38.	1.6243	13.98	1.57384	0.34376	1.32531	1.8245	0.6354	0.4095	0.7335	0.0813	0.5935	0.5061	0.9208
39.	1.5890	13.27	1.60198	0.33217	1.32243	1.7823	0.6242	0.4405	0.7147	0.0599	0.6504	0.4841	0.9216
40.	1.5557	12.59	1.63171	0.32047	1.31926	1.7423	0.6129	0.4737	0.6955	0.0377	0.7133	0.4618	0.9229
41.	1.5243	11.93	1.66312	0.30869	1.31578	1.7044	0.6013	0.5092	0.6760	0.0144	0.7833	0.4395	0.9249
42.	1.4945	11.28	1.69630	0.29684	1.31198	1.6684	0.5895	0.5474	0.6562	-0.0099	0.8613	0.4171	0.9276
43.	1.4663	10.65	1.73135	0.28496	1.30786	1.6343	0.5776	0.5885	0.6362	-0.0353	0.9486	0.3947	0.9310
44.	1.4396	10.05	1.76838	0.27307	1.30340	1.6018	0.5655	0.6328	0.6159	-0.0619	1.0466	0.3724	0.9353
45.	1.4142	9.46	1.80750	0.26119	1.29862	1.5710	0.5532	0.6805	0.5954	-0.0896	1.1569	0.3502	0.9406
θ	Ω_s	A_{min} (db)	σ_0	σ_1	Ω_1	Ω_2	$L'_1 = L'_3$	L'_2	C'_2	L'_1	L'_2	C'_2	L'_3

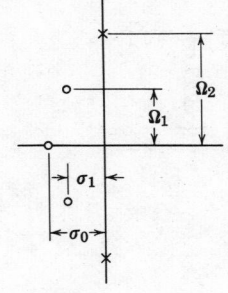

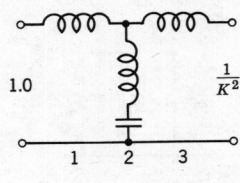

$\rho = 15\%$

$n = 3$

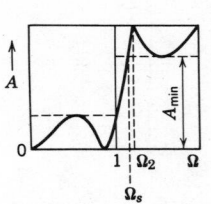

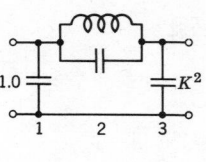

θ	Ω_s	A_min (db)	σ_0	σ_1	Ω_1	Ω_2	K²=1.0			K²=∞			
							$C'_1=C'_3$	C'_2	L'_2	C'_1	C'_2	L'_2	C'_3
1.	57.2987	113.19	0.97227	0.48595	1.20782	66.1616	1.0285	0.0002	1.1468	0.5141	0.0002	1.0843	1.0878
2.	28.6537	95.13	0.97271	0.48560	1.20789	33.0839	1.0281	0.0008	1.1461	0.5136	0.0008	1.0834	1.0875
3.	19.1073	84.56	0.97343	0.48502	1.20800	22.0595	1.0273	0.0018	1.1449	0.5126	0.0019	1.0819	1.0871
4.	14.3356	77.06	0.97444	0.48420	1.20816	16.5483	1.0262	0.0032	1.1433	0.5113	0.0034	1.0798	1.0865
5.	11.4737	71.24	0.97574	0.48316	1.20836	13.2424	1.0249	0.0050	1.1411	0.5096	0.0053	1.0771	1.0858
6.	9.5668	66.48	0.97734	0.48188	1.20861	11.0392	1.0232	0.0072	1.1386	0.5076	0.0076	1.0738	1.0849
7.	8.2055	62.45	0.97923	0.48037	1.20890	9.4661	1.0212	0.0098	1.1355	0.5051	0.0104	1.0699	1.0838
8.	7.1853	58.96	0.98143	0.47863	1.20922	8.2868	1.0189	0.0129	1.1320	0.5023	0.0137	1.0654	1.0826
9.	6.3925	55.88	0.98393	0.47665	1.20959	7.3700	1.0163	0.0163	1.1280	0.4991	0.0174	1.0603	1.0812
10.	5.7588	53.12	0.98674	0.47445	1.20999	6.6370	1.0134	0.0202	1.1235	0.4955	0.0215	1.0546	1.0797
11.	5.2408	50.63	0.98986	0.47201	1.21042	6.0377	1.0102	0.0245	1.1186	0.4915	0.0262	1.0483	1.0780
12.	4.8097	48.34	0.99330	0.46934	1.21088	5.5386	1.0067	0.0293	1.1132	0.4872	0.0313	1.0414	1.0761
13.	4.4454	46.24	0.99706	0.46644	1.21137	5.1166	1.0029	0.0345	1.1073	0.4824	0.0369	1.0339	1.0742
14.	4.1336	44.29	1.00116	0.46331	1.21188	4.7552	0.9988	0.0402	1.1010	0.4773	0.0431	1.0258	1.0720
15.	3.8637	42.47	1.00559	0.45995	1.21241	4.4423	0.9944	0.0463	1.0941	0.4717	0.0498	1.0171	1.0698
16.	3.6280	40.77	1.01038	0.45636	1.21296	4.1688	0.9897	0.0529	1.0869	0.4658	0.0571	1.0078	1.0674
17.	3.4203	39.17	1.01551	0.45254	1.21352	3.9277	0.9847	0.0601	1.0791	0.4594	0.0650	0.9979	1.0648
18.	3.2361	37.66	1.02102	0.44850	1.21409	3.7137	0.9794	0.0677	1.0709	0.4527	0.0734	0.9875	1.0622
19.	3.0716	36.22	1.02690	0.44422	1.21466	3.5224	0.9738	0.0759	1.0623	0.4455	0.0825	0.9764	1.0594
20.	2.9238	34.86	1.03317	0.43972	1.21523	3.3505	0.9679	0.0846	1.0531	0.4379	0.0923	0.9648	1.0565
21.	2.7904	33.56	1.03984	0.43500	1.21579	3.1951	0.9617	0.0939	1.0436	0.4299	0.1028	0.9526	1.0535
22.	2.6695	32.32	1.04692	0.43005	1.21634	3.0541	0.9552	0.1037	1.0335	0.4215	0.1141	0.9398	1.0505
23.	2.5593	31.13	1.05443	0.42488	1.21687	2.9256	0.9484	0.1142	1.0230	0.4126	0.1261	0.9264	1.0473
24.	2.4586	29.99	1.06238	0.41949	1.21737	2.8079	0.9413	0.1253	1.0121	0.4033	0.1390	0.9124	1.0440
25.	2.3662	28.89	1.07079	0.41387	1.21784	2.6999	0.9339	0.1371	1.0006	0.3935	0.1528	0.8979	1.0407
26.	2.2812	27.84	1.07967	0.40804	1.21828	2.6003	0.9262	0.1496	0.9888	0.3833	0.1675	0.8829	1.0373
27.	2.2027	26.82	1.08905	0.40200	1.21867	2.5083	0.9182	0.1628	0.9765	0.3726	0.1833	0.8672	1.0339
28.	2.1301	25.83	1.09894	0.39574	1.21900	2.4231	0.9100	0.1767	0.9637	0.3614	0.2001	0.8510	1.0304
29.	2.0627	24.88	1.10937	0.38928	1.21928	2.3438	0.9014	0.1915	0.9505	0.3497	0.2182	0.8343	1.0269
30.	2.0000	23.96	1.12036	0.38260	1.21948	2.2701	0.8926	0.2071	0.9369	0.3376	0.2375	0.8171	1.0234
31.	1.9416	23.06	1.13194	0.37573	1.21962	2.2012	0.8834	0.2236	0.9228	0.3249	0.2582	0.7993	1.0199
32.	1.8871	22.20	1.14412	0.36866	1.21966	2.1368	0.8740	0.2411	0.9083	0.3117	0.2804	0.7810	1.0165
33.	1.8361	21.35	1.15695	0.36139	1.21962	2.0765	0.8643	0.2596	0.8934	0.2980	0.3043	0.7623	1.0130
34.	1.7883	20.53	1.17044	0.35393	1.21947	2.0199	0.8544	0.2792	0.8780	0.2837	0.3299	0.7430	1.0097
35.	1.7434	19.73	1.18464	0.34629	1.21921	1.9666	0.8441	0.2999	0.8623	0.2689	0.3575	0.7233	1.0064
36.	1.7013	18.95	1.19957	0.33847	1.21883	1.9165	0.8336	0.3218	0.8461	0.2535	0.3872	0.7031	1.0032
37.	1.6616	18.20	1.21528	0.33048	1.21833	1.8692	0.8229	0.3450	0.8296	0.2375	0.4194	0.6825	1.0001
38.	1.6243	17.46	1.23180	0.32233	1.21768	1.8245	0.8118	0.3697	0.8126	0.2209	0.4541	0.6615	0.9972
39.	1.5890	16.73	1.24918	0.31402	1.21689	1.7823	0.8005	0.3959	0.7953	0.2036	0.4918	0.6401	0.9945
40.	1.5557	16.03	1.26745	0.30556	1.21594	1.7423	0.7890	0.4237	0.7776	0.1857	0.5327	0.6184	0.9920
41.	1.5243	15.34	1.28668	0.29697	1.21483	1.7044	0.7772	0.4532	0.7595	0.1671	0.5772	0.5964	0.9898
42.	1.4945	14.67	1.30691	0.28824	1.21354	1.6684	0.7652	0.4847	0.7411	0.1478	0.6258	0.5740	0.9879
43.	1.4663	14.02	1.32819	0.27940	1.21206	1.6343	0.7529	0.5183	0.7224	0.1278	0.6790	0.5514	0.9863
44.	1.4396	13.38	1.35059	0.27046	1.21040	1.6018	0.7404	0.5541	0.7033	0.1070	0.7372	0.5286	0.9850
45.	1.4142	12.75	1.37416	0.26142	1.20853	1.5710	0.7277	0.5924	0.6840	0.0854	0.8013	0.5057	0.9842
46.	1.3902	12.14	1.39898	0.25231	1.20645	1.5415	0.7148	0.6334	0.6643	0.0630	0.8719	0.4826	0.9839
47.	1.3673	11.54	1.42511	0.24313	1.20416	1.5135	0.7017	0.6774	0.6444	0.0398	0.9501	0.4595	0.9841
48.	1.3456	10.96	1.45262	0.23390	1.20164	1.4868	0.6884	0.7246	0.6243	0.0156	1.0367	0.4364	0.9849
49.	1.3250	10.40	1.48160	0.22464	1.19890	1.4613	0.6749	0.7754	0.6039	-0.0094	1.1332	0.4133	0.9864
50.	1.3054	9.84	1.51212	0.21536	1.19593	1.4369	0.6613	0.8301	0.5834	-0.0355	1.2409	0.3903	0.9885
θ	Ω_s	A_min (db)	σ_0	σ_1	Ω_1	Ω_2	$L'_1=L'_3$	L'_2	C'_2	L'_1	L'_2	C'_2	L'_3

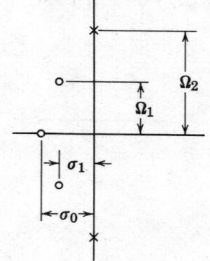

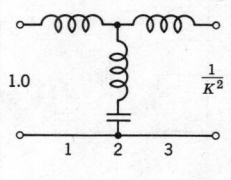

176

$\rho = 20\%$

$n = 3$

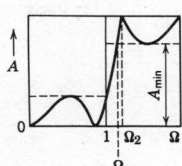

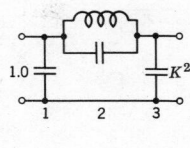

θ	Ω_s	A_{min} (db)	σ_0	σ_1	Ω_1	Ω_2	$K^2 = 1.0$			$K^2 = \infty$			
							$C'_1=C'_3$	C'_2	L'_2	C'_1	C'_2	L'_2	C'_3
1.	57.2987	115.77	0.84082	0.42027	1.13143	66.1616	1.1893	0.0002	1.1540	0.5946	0.0002	1.1713	1.1717
2.	28.6537	97.70	0.84114	0.42001	1.13150	33.0839	1.1889	0.0008	1.1533	0.5940	0.0008	1.1704	1.1715
3.	19.1073	87.13	0.84169	0.41958	1.13160	22.0595	1.1881	0.0018	1.1522	0.5932	0.0018	1.1690	1.1710
4.	14.3356	79.63	0.84246	0.41899	1.13175	16.5483	1.1870	0.0032	1.1507	0.5920	0.0031	1.1670	1.1704
5.	11.4737	73.81	0.84345	0.41822	1.13194	13.2424	1.1856	0.0050	1.1488	0.5904	0.0049	1.1645	1.1696
6.	9.5668	69.05	0.84466	0.41728	1.13218	11.0392	1.1839	0.0072	1.1464	0.5885	0.0071	1.1614	1.1686
7.	8.2055	65.03	0.84609	0.41617	1.13245	9.4661	1.1819	0.0098	1.1436	0.5862	0.0096	1.1577	1.1675
8.	7.1853	61.54	0.84776	0.41489	1.13276	8.2868	1.1796	0.0128	1.1404	0.5836	0.0126	1.1535	1.1662
9.	6.3925	58.46	0.84965	0.41344	1.13311	7.3700	1.1770	0.0162	1.1367	0.5807	0.0160	1.1487	1.1647
10.	5.7588	55.70	0.85177	0.41182	1.13350	6.6370	1.1740	0.0200	1.1326	0.5773	0.0199	1.1434	1.1630
11.	5.2408	53.20	0.85413	0.41003	1.13392	6.0377	1.1708	0.0243	1.1281	0.5737	0.0241	1.1374	1.1611
12.	4.8097	50.92	0.85673	0.40807	1.13438	5.5386	1.1672	0.0290	1.1231	0.5696	0.0288	1.1310	1.1591
13.	4.4454	48.82	0.85957	0.40593	1.13486	5.1166	1.1634	0.0342	1.1177	0.5653	0.0340	1.1239	1.1570
14.	4.1336	46.87	0.86266	0.40363	1.13538	4.7552	1.1592	0.0398	1.1119	0.5605	0.0396	1.1163	1.1546
15.	3.8637	45.05	0.86600	0.40115	1.13592	4.4423	1.1547	0.0458	1.1057	0.5554	0.0457	1.1082	1.1521
16.	3.6280	43.35	0.86959	0.39851	1.13649	4.1688	1.1500	0.0524	1.0990	0.5500	0.0523	1.0994	1.1495
17.	3.4203	41.75	0.87345	0.39569	1.13709	3.9277	1.1449	0.0594	1.0919	0.5441	0.0595	1.0902	1.1467
18.	3.2361	40.23	0.87759	0.39270	1.13770	3.7137	1.1395	0.0669	1.0844	0.5379	0.0671	1.0803	1.1437
19.	3.0716	38.80	0.88199	0.38954	1.13833	3.5224	1.1338	0.0749	1.0764	0.5314	0.0753	1.0700	1.1407
20.	2.9238	37.44	0.88668	0.38621	1.13897	3.3505	1.1278	0.0834	1.0681	0.5244	0.0841	1.0590	1.1374
21.	2.7904	36.14	0.89167	0.38272	1.13963	3.1951	1.1215	0.0925	1.0593	0.5171	0.0935	1.0475	1.1340
22.	2.6695	34.90	0.89695	0.37905	1.14029	3.0541	1.1149	0.1021	1.0500	0.5094	0.1035	1.0355	1.1305
23.	2.5593	33.71	0.90254	0.37521	1.14096	2.9256	1.1080	0.1123	1.0404	0.5013	0.1142	1.0229	1.1269
24.	2.4586	32.57	0.90845	0.37120	1.14162	2.8079	1.1008	0.1231	1.0303	0.4928	0.1256	1.0098	1.1232
25.	2.3662	31.47	0.91469	0.36702	1.14228	2.6999	1.0933	0.1345	1.0199	0.4839	0.1377	0.9961	1.1193
26.	2.2812	30.41	0.92127	0.36268	1.14294	2.6003	1.0855	0.1466	1.0090	0.4746	0.1506	0.9819	1.1153
27.	2.2027	29.39	0.92820	0.35817	1.14358	2.5083	1.0773	0.1593	0.9976	0.4649	0.1643	0.9672	1.1113
28.	2.1301	28.41	0.93550	0.35349	1.14420	2.4231	1.0689	0.1728	0.9859	0.4548	0.1789	0.9519	1.1071
29.	2.0627	27.45	0.94318	0.34864	1.14480	2.3438	1.0602	0.1869	0.9738	0.4443	0.1944	0.9362	1.1029
30.	2.0000	26.53	0.95125	0.34364	1.14538	2.2701	1.0512	0.2019	0.9612	0.4333	0.2110	0.9199	1.0985
31.	1.9416	25.63	0.95973	0.33847	1.14592	2.2012	1.0420	0.2176	0.9483	0.4219	0.2285	0.9030	1.0942
32.	1.8871	24.76	0.96863	0.33313	1.14643	2.1368	1.0324	0.2343	0.9349	0.4101	0.2473	0.8857	1.0897
33.	1.8361	23.92	0.97799	0.32764	1.14689	2.0765	1.0225	0.2518	0.9212	0.3978	0.2672	0.8679	1.0853
34.	1.7883	23.09	0.98780	0.32199	1.14730	2.0199	1.0123	0.2702	0.9070	0.3851	0.2885	0.8496	1.0808
35.	1.7434	22.29	0.99810	0.31619	1.14766	1.9666	1.0019	0.2897	0.8925	0.3719	0.3112	0.8308	1.0762
36.	1.7013	21.51	1.00890	0.31023	1.14796	1.9165	0.9912	0.3103	0.8776	0.3582	0.3355	0.8116	1.0717
37.	1.6616	20.74	1.02024	0.30412	1.14819	1.8692	0.9802	0.3320	0.8623	0.3441	0.3614	0.7919	1.0672
38.	1.6243	20.00	1.03213	0.29786	1.14835	1.8245	0.9689	0.3549	0.8466	0.3294	0.3892	0.7718	1.0627
39.	1.5890	19.27	1.04460	0.29147	1.14842	1.7823	0.9573	0.3791	0.8305	0.3142	0.4191	0.7512	1.0583
40.	1.5557	18.56	1.05768	0.28493	1.14841	1.7423	0.9455	0.4047	0.8141	0.2985	0.4511	0.7303	1.0540
41.	1.5243	17.86	1.07140	0.27825	1.14830	1.7044	0.9334	0.4318	0.7973	0.2823	0.4856	0.7089	1.0497
42.	1.4945	17.18	1.08579	0.27145	1.14810	1.6684	0.9210	0.4605	0.7801	0.2655	0.5228	0.6872	1.0456
43.	1.4663	16.52	1.10089	0.26452	1.14778	1.6343	0.9084	0.4909	0.7627	0.2481	0.5629	0.6651	1.0416
44.	1.4396	15.86	1.11673	0.25747	1.14735	1.6018	0.8955	0.5232	0.7448	0.2301	0.6064	0.6427	1.0378
45.	1.4142	15.22	1.13336	0.25031	1.14679	1.5710	0.8823	0.5576	0.7267	0.2115	0.6535	0.6200	1.0341
46.	1.3902	14.60	1.15082	0.24304	1.14611	1.5415	0.8689	0.5942	0.7082	0.1923	0.7048	0.5971	1.0307
47.	1.3673	13.98	1.16915	0.23567	1.14528	1.5135	0.8553	0.6331	0.6895	0.1725	0.7607	0.5739	1.0276
48.	1.3456	13.38	1.18840	0.22821	1.14432	1.4868	0.8415	0.6747	0.6705	0.1519	0.8217	0.5505	1.0248
49.	1.3250	12.79	1.20862	0.22067	1.14320	1.4613	0.8274	0.7192	0.6511	0.1307	0.8886	0.5270	1.0223
50.	1.3054	12.22	1.22988	0.21306	1.14192	1.4369	0.8131	0.7668	0.6316	0.1087	0.9621	0.5034	1.0202
51.	1.2868	11.65	1.25221	0.20539	1.14048	1.4137	0.7986	0.8179	0.6118	0.0860	1.0431	0.4797	1.0185
52.	1.2690	11.10	1.27570	0.19766	1.13887	1.3914	0.7839	0.8728	0.5918	0.0625	1.1327	0.4560	1.0173
53.	1.2521	10.56	1.30040	0.18990	1.13709	1.3702	0.7690	0.9319	0.5716	0.0382	1.2320	0.4323	1.0166
54.	1.2361	10.03	1.32639	0.18211	1.13512	1.3498	0.7539	0.9958	0.5512	0.0130	1.3426	0.4088	1.0165
55.	1.2208	9.51	1.35374	0.17431	1.13297	1.3303	0.7387	1.0648	0.5306	-0.0131	1.4662	0.3854	1.0171
56.	1.2062	9.01	1.38253	0.16652	1.13064	1.3117	0.7233	1.1397	0.5100	-0.0401	1.6046	0.3622	1.0184
57.	1.1924	8.51	1.41284	0.15873	1.12811	1.2938	0.7078	1.2210	0.4892	-0.0681	1.7605	0.3393	1.0205
58.	1.1792	8.03	1.44478	0.15098	1.12540	1.2767	0.6921	1.3097	0.4684	-0.0971	1.9366	0.3168	1.0235
59.	1.1666	7.57	1.47842	0.14328	1.12249	1.2603	0.6764	1.4065	0.4476	-0.1272	2.1364	0.2947	1.0274
60.	1.1547	7.11	1.51387	0.13565	1.11939	1.2446	0.6606	1.5127	0.4268	-0.1584	2.3640	0.2731	1.0323
θ	Ω_s	A_{min} (db)	σ_0	σ_1	Ω_1	Ω_2	$L'_1=L'_3$	L'_2	C'_2	L'_1	L'_2	C'_2	L'_3

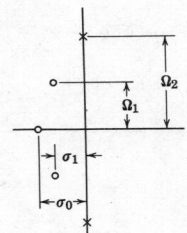

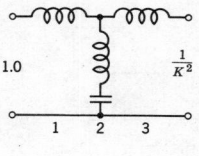

$\rho = 25\%$

$n = 3$

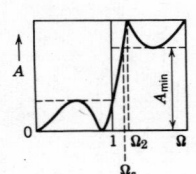

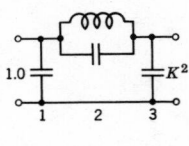

θ	Ω_s	A_{MIN} (db)	σ_0	σ_1	Ω_1	Ω_2	$K^2=1.0$ $C'_1=C'_3$	C'_2	L'_2	$K^2=\infty$ C'_1	C'_2	L'_2	C'_3
1	57.2987	117.81	0.74343	0.37160	1.07910	66.1616	1.3451	0.0002	1.1412	0.6725	0.0002	1.2347	1.2432
2	28.6537	99.74	0.74369	0.37140	1.07916	33.0839	1.3446	0.0008	1.1406	0.6720	0.0007	1.2339	1.2429
3	19.1073	89.18	0.74413	0.37107	1.07926	22.0595	1.3438	0.0018	1.1396	0.6712	0.0017	1.2325	1.2425
4	14.3356	81.67	0.74475	0.37060	1.07941	16.5483	1.3427	0.0032	1.1382	0.6700	0.0030	1.2307	1.2418
5	11.4737	75.85	0.74554	0.37000	1.07959	13.2424	1.3413	0.0050	1.1364	0.6685	0.0046	1.2282	1.2410
6	9.5668	71.10	0.74651	0.36927	1.07981	11.0392	1.3396	0.0072	1.1342	0.6667	0.0067	1.2253	1.2399
7	8.2055	67.07	0.74767	0.36840	1.08007	9.4661	1.3375	0.0099	1.1316	0.6646	0.0091	1.2218	1.2387
8	7.1853	63.58	0.74900	0.36740	1.08037	8.2868	1.3351	0.0129	1.1286	0.6621	0.0120	1.2178	1.2373
9	6.3925	60.50	0.75052	0.36626	1.08071	7.3700	1.3324	0.0164	1.1252	0.6593	0.0152	1.2132	1.2357
10	5.7588	57.74	0.75223	0.36499	1.08109	6.6370	1.3294	0.0202	1.1214	0.6562	0.0188	1.2081	1.2339
11	5.2408	55.24	0.75412	0.36359	1.08150	6.0377	1.3260	0.0246	1.1172	0.6527	0.0228	1.2025	1.2319
12	4.8097	52.96	0.75620	0.36205	1.08195	5.5386	1.3224	0.0293	1.1126	0.6489	0.0272	1.1963	1.2298
13	4.4454	50.86	0.75848	0.36038	1.08243	5.1166	1.3184	0.0345	1.1076	0.6447	0.0321	1.1896	1.2275
14	4.1336	48.91	0.76095	0.35858	1.08294	4.7552	1.3141	0.0401	1.1022	0.6402	0.0374	1.1824	1.2250
15	3.8637	47.09	0.76362	0.35664	1.08348	4.4423	1.3095	0.0462	1.0964	0.6354	0.0431	1.1746	1.2223
16	3.6280	45.39	0.76650	0.35456	1.08406	4.1688	1.3046	0.0528	1.0902	0.6303	0.0493	1.1663	1.2195
17	3.4203	43.79	0.76958	0.35235	1.08466	3.9277	1.2994	0.0598	1.0836	0.6248	0.0560	1.1575	1.2165
18	3.2361	42.28	0.77288	0.35000	1.08528	3.7137	1.2938	0.0673	1.0767	0.6189	0.0632	1.1482	1.2133
19	3.0716	40.84	0.77640	0.34752	1.08593	3.5224	1.2880	0.0754	1.0693	0.6127	0.0708	1.1383	1.2099
20	2.9238	39.48	0.78014	0.34490	1.08660	3.3505	1.2818	0.0839	1.0615	0.6062	0.0790	1.1279	1.2064
21	2.7904	38.18	0.78410	0.34215	1.08730	3.1951	1.2753	0.0930	1.0534	0.5993	0.0877	1.1169	1.2028
22	2.6695	36.94	0.78830	0.33926	1.08800	3.0541	1.2685	0.1026	1.0448	0.5920	0.0970	1.1055	1.1989
23	2.5593	35.75	0.79275	0.33624	1.08873	2.9256	1.2614	0.1128	1.0359	0.5844	0.1068	1.0935	1.1950
24	2.4586	34.61	0.79744	0.33308	1.08947	2.8079	1.2540	0.1236	1.0265	0.5764	0.1173	1.0810	1.1908
25	2.3662	33.51	0.80238	0.32979	1.09021	2.6999	1.2463	0.1349	1.0168	0.5681	0.1285	1.0680	1.1866
26	2.2812	32.45	0.80759	0.32635	1.09097	2.6003	1.2382	0.1469	1.0067	0.5594	0.1403	1.0544	1.1822
27	2.2027	31.43	0.81307	0.32279	1.09172	2.5083	1.2299	0.1595	0.9962	0.5503	0.1528	1.0404	1.1777
28	2.1301	30.44	0.81884	0.31909	1.09248	2.4231	1.2212	0.1729	0.9853	0.5409	0.1660	1.0258	1.1730
29	2.0627	29.49	0.82489	0.31525	1.09324	2.3438	1.2123	0.1869	0.9741	0.5311	0.1801	1.0108	1.1683
30	2.0000	28.57	0.83124	0.31128	1.09399	2.2701	1.2030	0.2016	0.9624	0.5209	0.1950	0.9952	1.1634
31	1.9416	27.67	0.83791	0.30717	1.09473	2.2012	1.1934	0.2172	0.9504	0.5103	0.2108	0.9792	1.1584
32	1.8871	26.80	0.84490	0.30293	1.09546	2.1368	1.1836	0.2335	0.9380	0.4993	0.2275	0.9626	1.1533
33	1.8361	25.95	0.85223	0.29855	1.09617	2.0765	1.1734	0.2507	0.9252	0.4879	0.2453	0.9456	1.1481
34	1.7883	25.12	0.85991	0.29405	1.09686	2.0199	1.1629	0.2687	0.9121	0.4761	0.2641	0.9281	1.1429
35	1.7434	24.32	0.86795	0.28941	1.09752	1.9666	1.1521	0.2878	0.8986	0.4639	0.2841	0.9101	1.1376
36	1.7013	23.54	0.87637	0.28464	1.09816	1.9165	1.1411	0.3078	0.8847	0.4512	0.3054	0.8916	1.1322
37	1.6616	22.77	0.88519	0.27974	1.09876	1.8692	1.1297	0.3288	0.8704	0.4382	0.3280	0.8727	1.1267
38	1.6243	22.02	0.89442	0.27471	1.09932	1.8245	1.1180	0.3510	0.8558	0.4247	0.3520	0.8533	1.1213
39	1.5890	21.29	0.90409	0.26956	1.09984	1.7823	1.1061	0.3744	0.8409	0.4107	0.3777	0.8335	1.1158
40	1.5557	20.58	0.91421	0.26428	1.10031	1.7423	1.0938	0.3990	0.8255	0.3963	0.4050	0.8133	1.1103
41	1.5243	19.88	0.92480	0.25887	1.10072	1.7044	1.0813	0.4251	0.8099	0.3815	0.4343	0.7927	1.1048
42	1.4945	19.19	0.93589	0.25335	1.10107	1.6684	1.0685	0.4525	0.7939	0.3661	0.4656	0.7716	1.0993
43	1.4663	18.52	0.94750	0.24771	1.10136	1.6343	1.0554	0.4815	0.7775	0.3503	0.4991	0.7502	1.0938
44	1.4396	17.86	0.95966	0.24195	1.10158	1.6018	1.0420	0.5122	0.7608	0.3340	0.5351	0.7284	1.0884
45	1.4142	17.22	0.97240	0.23608	1.10173	1.5710	1.0284	0.5447	0.7438	0.3172	0.5737	0.7062	1.0831
46	1.3902	16.58	0.98574	0.23010	1.10179	1.5415	1.0145	0.5792	0.7265	0.2998	0.6154	0.6838	1.0779
47	1.3673	15.96	0.99972	0.22402	1.10176	1.5135	1.0003	0.6158	0.7089	0.2819	0.6604	0.6610	1.0728
48	1.3456	15.35	1.01436	0.21784	1.10164	1.4868	0.9858	0.6547	0.6910	0.2635	0.7091	0.6379	1.0678
49	1.3250	14.75	1.02972	0.21157	1.10142	1.4613	0.9711	0.6961	0.6728	0.2445	0.7620	0.6146	1.0631
50	1.3054	14.16	1.04583	0.20520	1.10109	1.4369	0.9562	0.7402	0.6543	0.2248	0.8194	0.5910	1.0585
51	1.2868	13.58	1.06272	0.19875	1.10065	1.4137	0.9410	0.7873	0.6355	0.2046	0.8820	0.5673	1.0541
52	1.2690	13.01	1.08045	0.19223	1.10009	1.3914	0.9255	0.8378	0.6165	0.1837	0.9505	0.5434	1.0501
53	1.2521	12.45	1.09906	0.18563	1.09941	1.3702	0.9099	0.8918	0.5973	0.1622	1.0255	0.5194	1.0463
54	1.2361	11.90	1.11860	0.17897	1.09860	1.3498	0.8940	0.9498	0.5778	0.1400	1.1080	0.4953	1.0429
55	1.2208	11.37	1.13914	0.17226	1.09766	1.3303	0.8779	1.0123	0.5582	0.1171	1.1991	0.4712	1.0398
56	1.2062	10.84	1.16072	0.16551	1.09657	1.3117	0.8615	1.0797	0.5383	0.0935	1.2999	0.4471	1.0372
57	1.1924	10.32	1.18341	0.15872	1.09534	1.2938	0.8450	1.1526	0.5183	0.0690	1.4119	0.4231	1.0351
58	1.1792	9.81	1.20728	0.15190	1.09395	1.2767	0.8283	1.2316	0.4981	0.0438	1.5367	0.3992	1.0335
59	1.1666	9.31	1.23239	0.14508	1.09242	1.2603	0.8114	1.3175	0.4779	0.0178	1.6765	0.3755	1.0326
60	1.1547	8.83	1.25884	0.13825	1.09073	1.2446	0.7944	1.4111	0.4575	0.0091	1.8337	0.3521	1.0323
θ	Ω_s	A_{MIN} (db)	σ_0	σ_1	Ω_1	Ω_2	$L'_1=L'_3$	L'_2	C'_2	L'_1	L'_2	C'_2	L'_3

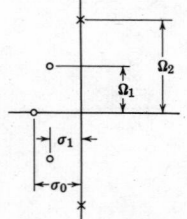

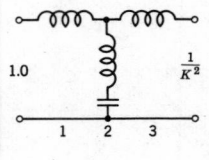

178

ρ = 50%

n = 3

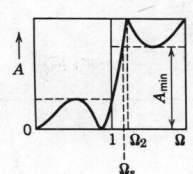 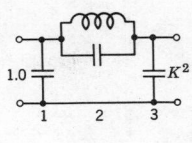

θ	Ω_s	A_{min} (db)	σ_o	σ_1	Ω_1	Ω_2	$K^2 = 1.0$			$K^2 = \infty$			
							$C'_1=C'_3$	C'_2	L'_2	C'_1	C'_2	L'_2	C'_3
1.	57.2987	124.80	0.45326	0.22658	0.95083	66.1616	2.2062	0.0002	0.9486	1.1031	0.0002	1.3267	1.5775
2.	28.6537	106.73	0.45338	0.22649	0.95089	33.0839	2.2056	0.0010	0.9482	1.1026	0.0007	1.3261	1.5771
3.	19.1073	96.17	0.45359	0.22635	0.95098	22.0595	2.2046	0.0022	0.9475	1.1018	0.0016	1.3250	1.5765
4.	14.3356	88.66	0.45388	0.22615	0.95110	16.5483	2.2032	0.0039	0.9465	1.1008	0.0028	1.3234	1.5757
5.	11.4737	82.84	0.45426	0.22589	0.95127	13.2424	2.2014	0.0060	0.9452	1.0994	0.0043	1.3215	1.5746
6.	9.5668	78.09	0.45472	0.22557	0.95147	11.0392	2.1991	0.0087	0.9437	1.0977	0.0062	1.3191	1.5733
7.	8.2055	74.06	0.45526	0.22520	0.95170	9.4661	2.1965	0.0118	0.9418	1.0957	0.0085	1.3162	1.5717
8.	7.1853	70.57	0.45589	0.22477	0.95197	8.2868	2.1935	0.0155	0.9397	1.0934	0.0111	1.3130	1.5699
9.	6.3925	67.49	0.45661	0.22428	0.95228	7.3700	2.1900	0.0196	0.9373	1.0908	0.0141	1.3092	1.5679
10.	5.7588	64.73	0.45741	0.22373	0.95262	6.6370	2.1862	0.0243	0.9346	1.0879	0.0174	1.3051	1.5656
11.	5.2408	62.23	0.45830	0.22313	0.95299	6.0377	2.1819	0.0294	0.9317	1.0847	0.0211	1.3005	1.5631
12.	4.8097	59.95	0.45928	0.22246	0.95340	5.5386	2.1773	0.0351	0.9284	1.0812	0.0252	1.2955	1.5603
13.	4.4454	57.85	0.46035	0.22174	0.95385	5.1166	2.1722	0.0413	0.9249	1.0773	0.0296	1.2901	1.5574
14.	4.1336	55.90	0.46151	0.22096	0.95433	4.7552	2.1668	0.0480	0.9211	1.0732	0.0344	1.2842	1.5541
15.	3.8637	54.08	0.46276	0.22012	0.95484	4.4423	2.1609	0.0553	0.9170	1.0687	0.0397	1.2779	1.5507
16.	3.6280	52.38	0.46411	0.21922	0.95539	4.1688	2.1547	0.0630	0.9126	1.0640	0.0453	1.2711	1.5470
17.	3.4203	50.78	0.46555	0.21826	0.95597	3.9277	2.1480	0.0714	0.9080	1.0589	0.0513	1.2639	1.5431
18.	3.2361	49.26	0.46708	0.21724	0.95659	3.7137	2.1409	0.0803	0.9031	1.0535	0.0577	1.2563	1.5389
19.	3.0716	47.83	0.46871	0.21616	0.95724	3.5224	2.1335	0.0898	0.8979	1.0478	0.0646	1.2483	1.5346
20.	2.9238	46.47	0.47045	0.21503	0.95791	3.3505	2.1256	0.0998	0.8924	1.0419	0.0719	1.2398	1.5300
21.	2.7904	45.17	0.47228	0.21383	0.95863	3.1951	2.1173	0.1105	0.8866	1.0355	0.0796	1.2309	1.5251
22.	2.6695	43.93	0.47422	0.21256	0.95937	3.0541	2.1087	0.1217	0.8806	1.0289	0.0878	1.2216	1.5201
23.	2.5593	42.74	0.47627	0.21124	0.96014	2.9256	2.0996	0.1336	0.8743	1.0220	0.0964	1.2118	1.5148
24.	2.4586	41.59	0.47842	0.20986	0.96095	2.8079	2.0902	0.1462	0.8677	1.0147	0.1055	1.2016	1.5093
25.	2.3662	40.50	0.48069	0.20841	0.96178	2.6999	2.0803	0.1594	0.8608	1.0072	0.1152	1.1910	1.5035
26.	2.2812	39.44	0.48307	0.20690	0.96264	2.6003	2.0701	0.1732	0.8537	0.9993	0.1253	1.1800	1.4976
27.	2.2027	38.42	0.48556	0.20533	0.96353	2.5083	2.0594	0.1878	0.8462	0.9911	0.1360	1.1685	1.4914
28.	2.1301	37.43	0.48818	0.20369	0.96445	2.4231	2.0484	0.2031	0.8385	0.9826	0.1473	1.1567	1.4850
29.	2.0627	36.48	0.49092	0.20199	0.96539	2.3438	2.0370	0.2192	0.8306	0.9738	0.1591	1.1443	1.4785
30.	2.0000	35.55	0.49379	0.20023	0.96636	2.2701	2.0251	0.2360	0.8223	0.9647	0.1715	1.1316	1.4717
31.	1.9416	34.65	0.49678	0.19840	0.96736	2.2012	2.0129	0.2536	0.8138	0.9552	0.1845	1.1185	1.4646
32.	1.8871	33.78	0.49991	0.19650	0.96838	2.1368	2.0003	0.2720	0.8050	0.9454	0.1982	1.1049	1.4574
33.	1.8361	32.93	0.50318	0.19454	0.96942	2.0765	1.9873	0.2914	0.7960	0.9353	0.2126	1.0910	1.4500
34.	1.7883	32.10	0.50660	0.19252	0.97049	2.0199	1.9739	0.3116	0.7867	0.9249	0.2277	1.0766	1.4424
35.	1.7434	31.30	0.51016	0.19043	0.97158	1.9666	1.9602	0.3327	0.7771	0.9141	0.2435	1.0618	1.4346
36.	1.7013	30.51	0.51387	0.18827	0.97268	1.9165	1.9460	0.3549	0.7672	0.9030	0.2602	1.0466	1.4266
37.	1.6616	29.74	0.51774	0.18604	0.97381	1.8692	1.9314	0.3781	0.7571	0.8916	0.2776	1.0310	1.4184
38.	1.6243	28.99	0.52178	0.18374	0.97495	1.8245	1.9165	0.4023	0.7467	0.8799	0.2960	1.0150	1.4100
39.	1.5890	28.26	0.52598	0.18138	0.97611	1.7823	1.9012	0.4277	0.7360	0.8678	0.3153	0.9985	1.4014
40.	1.5557	27.54	0.53036	0.17894	0.97729	1.7423	1.8855	0.4543	0.7251	0.8554	0.3356	0.9817	1.3927
41.	1.5243	26.83	0.53493	0.17644	0.97848	1.7044	1.8694	0.4822	0.7140	0.8426	0.3569	0.9645	1.3837
42.	1.4945	26.14	0.53968	0.17387	0.97968	1.6684	1.8529	0.5114	0.7025	0.8295	0.3794	0.9469	1.3747
43.	1.4663	25.46	0.54464	0.17123	0.98089	1.6343	1.8361	0.5420	0.6908	0.8161	0.4030	0.9289	1.3654
44.	1.4396	24.79	0.54980	0.16851	0.98211	1.6018	1.8188	0.5741	0.6789	0.8023	0.4280	0.9106	1.3560
45.	1.4142	24.14	0.55518	0.16573	0.98334	1.5710	1.8012	0.6078	0.6667	0.7881	0.4543	0.8918	1.3464
46.	1.3902	23.49	0.56078	0.16288	0.98457	1.5415	1.7832	0.6433	0.6542	0.7736	0.4822	0.8727	1.3367
47.	1.3673	22.86	0.56662	0.15995	0.98581	1.5135	1.7649	0.6805	0.6415	0.7588	0.5116	0.8532	1.3268
48.	1.3456	22.23	0.57270	0.15695	0.98705	1.4868	1.7461	0.7198	0.6285	0.7435	0.5428	0.8334	1.3168
49.	1.3250	21.62	0.57904	0.15388	0.98828	1.4613	1.7270	0.7611	0.6153	0.7279	0.5759	0.8132	1.3067
50.	1.3054	21.01	0.58565	0.15074	0.98951	1.4369	1.7075	0.8947	0.6018	0.7119	0.6110	0.7926	1.2965
51.	1.2868	20.41	0.59255	0.14753	0.99074	1.4137	1.6876	0.8508	0.5882	0.6956	0.6484	0.7718	1.2861
52.	1.2690	19.82	0.59974	0.14424	0.99196	1.3914	1.6674	0.8995	0.5742	0.6788	0.6882	0.7505	1.2757
53.	1.2521	19.24	0.60725	0.14088	0.99317	1.3702	1.6468	0.9511	0.5601	0.6616	0.7307	0.7290	1.2651
54.	1.2361	18.66	0.61509	0.13745	0.99436	1.3498	1.6258	1.0058	0.5457	0.6441	0.7761	0.7071	1.2545
55.	1.2208	18.09	0.62327	0.13396	0.99554	1.3303	1.6044	1.0640	0.5310	0.6261	0.8249	0.6850	1.2438
56.	1.2062	17.53	0.63183	0.13039	0.99669	1.3117	1.5827	1.1260	0.5162	0.6077	0.8772	0.6626	1.2331
57.	1.1924	16.97	0.64077	0.12675	0.99783	1.2938	1.5606	1.1921	0.5011	0.5889	0.9336	0.6399	1.2223
58.	1.1792	16.42	0.65013	0.12304	0.99894	1.2767	1.5382	1.2627	0.4859	0.5697	0.9945	0.6169	1.2114
59.	1.1666	15.88	0.65992	0.11926	1.00001	1.2603	1.5153	1.3384	0.4704	0.5499	1.0604	0.5937	1.2006
60.	1.1547	15.33	0.67017	0.11542	1.00106	1.2446	1.4922	1.4198	0.4547	0.5298	1.1320	0.5703	1.1897
θ	Ω_s	A_{min} (db)	σ_o	σ_1	Ω_1	Ω_2	$L'_1=L'_3$	L'_2	C'_2	L'_1	L'_2	C'_2	L'_3

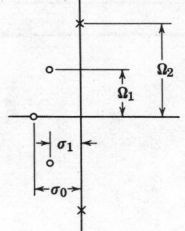 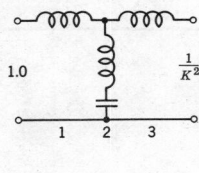

179

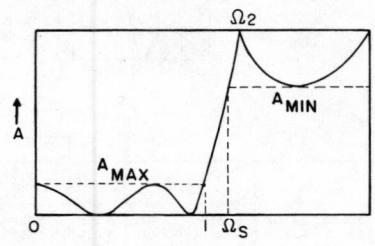

$n = 4$

$\rho = 1\%$

θ	Ω_S	A_{MIN}	σ_1	σ_3	Ω_1	Ω_2	Ω_3
C	∞	∞	-0.6686748	-1.6143238	1.8599986	∞	0.7704367
6.0	10.350843	77.25	-0.6571195	-1.6200284	1.8559428	11.367741	0.7827807
7.0	8.876727	71.89	-0.6529580	-1.6220847	1.8544662	9.747389	0.7872678
8.0	7.771760	67.25	-0.6481638	-1.6244551	1.8527544	8.532615	0.7924650
9.0	6.912894	63.14	-0.6427402	-1.6271388	1.8508036	7.588226	0.7983804
10.0	6.226301	59.47	-0.6366912	-1.6301349	1.8486098	6.833109	0.8050238
11.0	5.664999	56.15	-0.6300211	-1.6334426	1.8461680	6.215646	0.8124058
12.0	5.197666	53.12	-0.6227349	-1.6370611	1.8434724	5.701423	0.8205383
13.0	4.802620	50.32	-0.6148384	-1.6409900	1.8405170	5.266618	0.8294343
14.0	4.464371	47.74	-0.6063379	-1.6452288	1.8372945	4.894214	0.8391082
15.0	4.171563	45.32	-0.5972405	-1.6497773	1.8337969	4.571732	0.8495753
16.0	3.915678	43.07	-0.5875541	-1.6546358	1.8300154	4.289813	0.8608524
17.0	3.690200	40.94	-0.5772876	-1.6598051	1.8259400	4.041300	0.8729572
18.0	3.490065	38.94	-0.5664508	-1.6652865	1.8215599	3.820626	0.8859090
19.0	3.311272	37.04	-0.5550545	-1.6710820	1.8168631	3.623399	0.8997281
20.0	3.150622	35.24	-0.5431111	-1.6771945	1.8118364	3.446101	0.9144362
21.0	3.005526	33.53	-0.5306338	-1.6836281	1.8064656	3.285888	0.9300562
22.0	2.873864	31.89	-0.5176378	-1.6903880	1.8007350	3.140431	0.9466124
23.0	2.753885	30.32	-0.5041398	-1.6974809	1.7946279	3.007807	0.9641301
24.0	2.644133	28.82	-0.4901584	-1.7049154	1.7881263	2.886413	0.9826361
25.0	2.543380	27.38	-0.4757143	-1.7127018	1.7812110	2.774903	1.0021581
26.0	2.450592	25.99	-0.4608307	-1.7208530	1.7738615	2.672139	1.0227251
27.0	2.364885	24.66	-0.4455335	-1.7293846	1.7660564	2.577149	1.0443670
28.0	2.285502	23.37	-0.4298515	-1.7383149	1.7577733	2.489103	1.0671144
29.0	2.211792	22.12	-0.4138170	-1.7476661	1.7489892	2.407283	1.0909986
30.0	2.143189	20.92	-0.3974658	-1.7574639	1.7396807	2.331070	1.1160510
31.0	2.079202	19.76	-0.3808380	-1.7677383	1.7298243	2.259921	1.1423029
32.0	2.019399	18.64	-0.3639779	-1.7785243	1.7193970	2.193363	1.1697851
33.0	1.963403	17.55	-0.3469347	-1.7898615	1.7083773	2.130982	1.1985269
34.0	1.910879	16.50	-0.3297626	-1.8017953	1.6967452	2.072410	1.2285556
35.0	1.861534	15.48	-0.3125208	-1.8143764	1.6844838	2.017322	1.2598955
36.0	1.815103	14.49	-0.2952740	-1.8276612	1.6715803	1.965429	1.2925670
37.0	1.771354	13.53	-0.2780913	-1.8417118	1.6580267	1.916475	1.3265853
38.0	1.730076	12.61	-0.2610468	-1.8565948	1.6438216	1.870229	1.3619590
39.0	1.691083	11.72	-0.2442178	-1.8723817	1.6289715	1.826485	1.3986889
40.0	1.654204	10.86	-0.2276842	-1.8891467	1.6134918	1.785057	1.4367671

θ	Ω_S	A_{MIN}	σ_1	σ_3	Ω_1	Ω_2	Ω_3

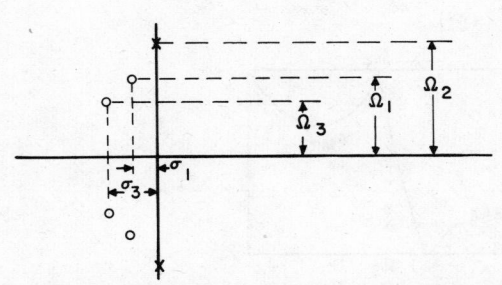

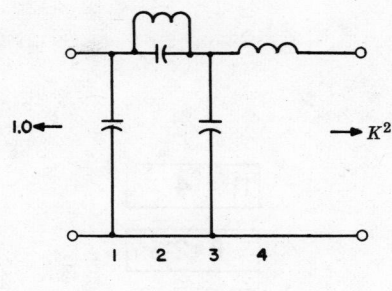

	K² = 0.9802					K² = ∞			
C_1	C_2	L_2	C_3	L_4	C_1	C_2	L_2	C_3	L_4
0.4380	0.000000	0.9687	0.9270	0.4293	0.21901	0.000000	0.59110	0.81280	0.76029
0.4306	0.008635	0.8962	0.9229	0.4304	0.20631	0.01348	0.57392	0.81208	0.76599
0.4279	0.01180	0.8917	0.9214	0.4308	0.20165	0.01854	0.56771	0.81188	0.76805
0.4248	0.01549	0.8865	0.9196	0.4313	0.19622	0.02450	0.56053	0.81170	0.77044
0.4212	0.01972	0.8806	0.9177	0.4318	0.19001	0.03144	0.55238	0.81155	0.77315
0.4173	0.02451	0.8740	0.9155	0.4324	0.18298	0.03942	0.54327	0.81146	0.77618
0.4129	0.02987	0.8667	0.9131	0.4331	0.17510	0.04855	0.53319	0.81146	0.77954
0.4080	0.03583	0.8587	0.9105	0.4338	0.16633	0.05892	0.52213	0.81158	0.78322
0.4028	0.04242	0.8500	0.9077	0.4345	0.15663	0.07068	0.51010	0.81186	0.78722
0.3971	0.04967	0.8406	0.9046	0.4353	0.14595	0.08398	0.49711	0.81235	0.79155
0.3909	0.05761	0.8304	0.9014	0.4362	0.13424	0.09903	0.48314	0.81310	0.79619
0.3843	0.06630	0.8196	0.8970	0.4372	0.12142	0.11606	0.46821	0.81419	0.80113
0.3772	0.07577	0.8081	0.8943	0.4382	0.10741	0.13536	0.45233	0.81568	0.80638
0.3696	0.08600	0.7958	0.8904	0.4392	0.09212	0.15730	0.43550	0.81768	0.81190
0.3615	0.09729	0.7829	0.8864	0.4403	0.07543	0.18233	0.41775	0.82028	0.81769
0.3530	0.1095	0.7693	0.8822	0.4415	0.05723	0.21099	0.39910	0.82362	0.82372
0.3440	0.1227	0.7549	0.8778	0.4427	0.03736	0.24399	0.37959	0.82785	0.82995
0.3344	0.1371	0.7398	0.8732	0.4439	0.01562	0.28224	0.35925	0.83316	0.83635
0.3244	0.1527	0.7241	0.8686	0.4452	-0.00819	0.32689	0.33814	0.83977	0.84286
0.3138	0.1696	0.7077	0.8637	0.4465	-0.03436	0.37944	0.31633	0.84797	0.84941
0.3027	0.1881	0.6905	0.8588	0.4479	-0.06320	0.44188	0.29390	0.85810	0.85592
0.2910	0.2082	0.6727	0.8537	0.4493	-0.09514	0.51688	0.27095	0.87057	0.86229
0.2788	0.2301	0.6543	0.8485	0.4507	-0.13067	0.60804	0.24762	0.88592	0.86841
0.2660	0.2541	0.6352	0.8433	0.4521	-0.17046	0.72038	0.22405	0.90482	0.87412
0.2526	0.2804	0.6155	0.8380	0.4535	-0.21533	0.86098	0.20043	0.92814	0.87927
0.2386	0.3092	0.5952	0.8327	0.4549	-0.26637	1.04002	0.17695	0.95702	0.88364
0.2240	0.3409	0.5744	0.8274	0.4562	-0.32505	1.27260	0.15386	0.99296	0.88704
0.2088	0.3759	0.5530	0.8221	0.4575	-0.39335	1.58170	0.13142	1.03802	0.88921
0.2014	0.4327	0.5344	0.8169	0.4588	-0.47406	2.00348	0.10991	1.09506	0.88989
0.1765	0.4575	0.5089	0.8118	0.4599	-0.57118	2.59714	0.08965	1.16816	0.88882
0.1595	0.5053	0.4863	0.8068	0.4609	-0.69072	3.46428	0.07093	1.26346	0.88572
0.1418	0.5586	0.4634	0.8021	0.4617	-0.84211	4.78961	0.05405	1.39048	0.88035
0.1235	0.6183	0.4403	0.7975	0.4624	-1.04096	6.93444	0.03926	1.56499	0.87251
0.1045	0.6854	0.4171	0.7933	0.4629	-1.31522	10.67602	0.02678	1.81508	0.86204
0.0850	0.7609	0.3939	0.7893	0.4631	-1.72028	17.91955	0.01673	2.19631	0.84888
0.0649	0.8463	0.3708	0.7858	0.4630	-2.38352	34.31970	0.00914	2.83622	0.83308
L_1	L_2	C_2	L_3	C_4	L_1	L_2	C_2	L_3	C_4

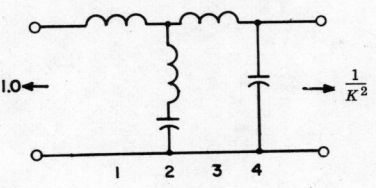

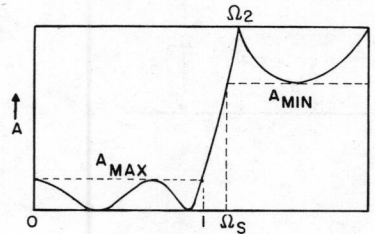

θ	Ω_S	A_{MIN}	σ_1	σ_3	Ω_1	Ω_2	Ω_3
C	∞	∞	-0.5445514	-1.3146634	1.6068271	∞	0.6655696
6.0	10.350843	83.28	-0.5377497	-1.3182283	1.6046598	11.367741	0.6730071
7.0	8.876727	77.91	-0.5352967	-1.3195150	1.6038718	9.747389	0.6757071
8.0	7.771760	73.27	-0.5324685	-1.3209991	1.6029589	8.532615	0.6788319
9.0	6.912894	69.17	-0.5292659	-1.3226807	1.6019196	7.588226	0.6823855
10.0	6.226301	65.50	-0.5256902	-1.3245594	1.6007519	6.833109	0.6863726
11.0	5.664999	62.17	-0.5217425	-1.3266354	1.5994538	6.215646	0.6907983
12.0	5.197666	59.14	-0.5174243	-1.3289083	1.5980227	5.701423	0.6956683
13.0	4.802620	56.35	-0.5127372	-1.3313784	1.5964560	5.266618	0.7009891
14.0	4.464371	53.76	-0.5076831	-1.3340455	1.5947507	4.894214	0.7067674
15.0	4.171563	51.35	-0.5022640	-1.3369099	1.5929033	4.571732	0.7130108
16.0	3.915678	49.09	-0.4964822	-1.3399717	1.5909101	4.289813	0.7197274
17.0	3.690200	46.97	-0.4903402	-1.3432314	1.5887669	4.041300	0.7269259
18.0	3.490065	44.96	-0.4838408	-1.3466896	1.5864691	3.820626	0.7346158
19.0	3.311272	43.06	-0.4769873	-1.3503469	1.5840118	3.623399	0.7428069
20.0	3.150622	41.26	-0.4697831	-1.3542046	1.5813894	3.446101	0.7515101
21.0	3.005526	39.55	-0.4622322	-1.3582638	1.5785961	3.285888	0.7607366
22.0	2.873864	37.91	-0.4543388	-1.3625262	1.5756251	3.140431	0.7704984
23.0	2.753885	36.34	-0.4461077	-1.3669941	1.5724696	3.007807	0.7808084
24.0	2.644133	34.84	-0.4375444	-1.3716698	1.5691219	2.886413	0.7916801
25.0	2.543380	33.39	-0.4286547	-1.3765566	1.5655737	2.774903	0.8031275
26.0	2.450592	32.00	-0.4194452	-1.3816583	1.5618162	2.672139	0.8151658
27.0	2.364885	30.67	-0.4099234	-1.3869793	1.5578399	2.577149	0.8278105
28.0	2.285502	29.37	-0.4000974	-1.3925250	1.5536346	2.489103	0.8410782
29.0	2.211792	28.13	-0.3899764	-1.3983017	1.5491895	2.407283	0.8549861
30.0	2.143189	26.92	-0.3795705	-1.4043168	1.5444931	2.331070	0.8695521
31.0	2.079202	25.75	-0.3688912	-1.4105790	1.5395334	2.259921	0.8847950
32.0	2.019399	24.61	-0.3579512	-1.4170983	1.5342977	2.193363	0.9007342
33.0	1.963403	23.51	-0.3467645	-1.4238864	1.5287727	2.130982	0.9173896
34.0	1.910879	22.44	-0.3353470	-1.4309566	1.5229447	2.072410	0.9347818
35.0	1.861534	21.40	-0.3237164	-1.4383243	1.5167997	2.017322	0.9529319
36.0	1.815103	20.39	-0.3118921	-1.4460071	1.5103234	1.965429	0.9718611
37.0	1.771354	19.41	-0.2998960	-1.4540247	1.5035015	1.916475	0.9915910
38.0	1.730076	18.45	-0.2877522	-1.4623998	1.4963199	1.870229	1.0121429
39.0	1.691083	17.52	-0.2754873	-1.4711577	1.4887650	1.826485	1.0335377
40.0	1.654204	16.61	-0.2631307	-1.4803266	1.4808241	1.785057	1.0557959
41.0	1.619289	15.73	-0.2507146	-1.4899380	1.4724856	1.745777	1.0789365
42.0	1.586200	14.87	-0.2382739	-1.5000265	1.4637399	1.708493	1.1029772
43.0	1.554811	14.03	-0.2258466	-1.5106300	1.4545795	1.673069	1.1279335
44.0	1.525009	13.21	-0.2134731	-1.5217894	1.4450000	1.639380	1.1538182
45.0	1.496692	12.42	-0.2011966	-1.5335488	1.4350002	1.607311	1.1806408
46.0	1.469765	11.65	-0.1890621	-1.5459547	1.4245836	1.576760	1.2084069
47.0	1.444142	10.90	-0.1771163	-1.5590558	1.4137580	1.547632	1.2371177
48.0	1.419745	10.18	-0.1654066	-1.5729020	1.4025371	1.519839	1.2667692
49.0	1.396501	9.48	-0.1539802	-1.5875437	1.3909400	1.493303	1.2973522
50.0	1.374345	8.81	-0.1428834	-1.6030309	1.3789919	1.467949	1.3288516
θ	Ω_S	A_{MIN}	σ_1	σ_3	Ω_1	Ω_2	Ω_3

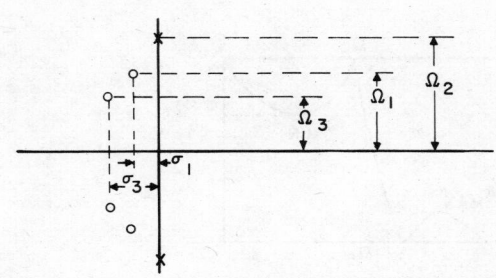

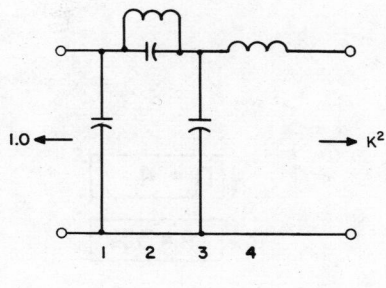

K²=0.9608					K²=∞				
C_1	C_2	L_2	C_3	L_4	C_1	C_2	L_2	C_3	L_4
0.5379	0.000000	1.041	1.084	0.5168	0.26893	0.000000	0.70935	0.95315	0.87995
0.5313	0.007510	1.030	1.080	0.5177	0.25839	0.01114	0.69465	0.95190	0.88436
0.5290	0.01025	1.026	1.078	0.5180	0.25454	0.01527	0.68933	0.95148	0.88595
0.5263	0.01344	1.022	1.077	0.5184	0.25008	0.02010	0.68320	0.95103	0.88779
0.5232	0.01708	1.017	1.075	0.5188	0.24498	0.02568	0.67624	0.95054	0.88989
0.5197	0.02119	1.011	1.073	0.5193	0.23924	0.03204	0.66845	0.95004	0.89223
0.5158	0.02577	1.004	1.071	0.5198	0.23284	0.03923	0.65984	0.94953	0.89482
0.5116	0.03085	0.9973	1.068	0.5204	0.22575	0.04730	0.65040	0.94904	0.89766
0.5070	0.03643	0.9897	1.065	0.5210	0.21797	0.05632	0.64013	0.94859	0.90076
0.5021	0.04254	0.9814	1.062	0.5217	0.20946	0.06637	0.62903	0.94819	0.90410
0.4967	0.04920	0.9725	1.059	0.5224	0.20019	0.07753	0.61710	0.94787	0.90769
0.4910	0.05643	0.9630	1.056	0.5232	0.19014	0.08992	0.60435	0.94767	0.91154
0.4848	0.06426	0.9529	1.052	0.5240	0.17927	0.10364	0.59076	0.94761	0.91562
0.4783	0.07272	0.9421	1.049	0.5249	0.16753	0.11886	0.57635	0.94774	0.91995
0.4714	0.08184	0.9307	1.045	0.5258	0.15488	0.13574	0.56112	0.94810	0.92452
0.4640	0.09166	0.9187	1.040	0.5267	0.14127	0.15449	0.54507	0.94874	0.92932
0.4563	0.1022	0.9061	1.036	0.5278	0.12664	0.17534	0.52822	0.94973	0.93433
0.4481	0.1136	0.8929	1.032	0.5288	0.11091	0.19859	0.51057	0.95112	0.93955
0.4395	0.1257	0.8790	1.027	0.5299	0.09401	0.22460	0.49214	0.95301	0.94497
0.4305	0.1388	0.8645	1.022	0.5311	0.07584	0.25378	0.47296	0.95548	0.95055
0.4210	0.1529	0.8495	1.017	0.5322	0.05630	0.28666	0.45304	0.95864	0.95628
0.4112	0.1680	0.8338	1.012	0.5334	0.03527	0.32388	0.43241	0.96262	0.96213
0.4008	0.1842	0.8174	1.007	0.5347	0.01259	0.36622	0.41112	0.96757	0.96806
0.3900	0.2016	0.8005	1.001	0.5360	-0.01190	0.41469	0.38922	0.97366	0.97401
0.3788	0.2204	0.7830	0.9954	0.5373	-0.03841	0.47053	0.36674	0.98111	0.97995
0.3670	0.2406	0.7649	0.9897	0.5386	-0.06718	0.53532	0.34378	0.99017	0.98581
0.3548	0.2624	0.7462	0.9838	0.5399	-0.09850	0.61113	0.32039	1.00114	0.99150
0.3421	0.2859	0.7270	0.9779	0.5413	-0.13272	0.70062	0.29669	1.01438	0.99694
0.3289	0.3114	0.7072	0.9719	0.5426	-0.17027	0.80731	0.27277	1.03036	1.00203
0.3152	0.3390	0.6869	0.9658	0.5440	-0.21169	0.93593	0.24877	1.04963	1.00665
0.3010	0.3690	0.6660	0.9597	0.5453	-0.25764	1.09289	0.22484	1.07289	1.01065
0.2863	0.4016	0.6446	0.9535	0.5466	-0.30897	1.28707	0.20113	1.10102	1.01389
0.2710	0.4371	0.6228	0.9474	0.5478	-0.36676	1.53097	0.17784	1.13515	1.01620
0.2552	0.4760	0.6006	0.9412	0.5490	-0.43244	1.84262	0.15516	1.17676	1.01739
0.2389	0.5187	0.5780	0.9352	0.5501	-0.50790	2.24865	0.13330	1.22779	1.01728
0.2220	0.5655	0.5550	0.9292	0.5511	-0.59572	2.78950	0.11250	1.29091	1.01566
0.2046	0.6171	0.5317	0.9233	0.5520	-0.69949	3.52866	0.09299	1.36978	1.01234
0.1866	0.6742	0.5082	0.9175	0.5527	-0.82441	4.56971	0.07497	1.46970	1.00712
0.1681	0.7374	0.4845	0.9120	0.5533	-0.97822	6.08976	0.05866	1.59852	0.99984
0.1491	0.8077	0.4606	0.9066	0.5537	-1.17304	8.40941	0.04425	1.76847	0.99037
0.1296	0.8861	0.4368	0.9015	0.5538	-1.42892	12.15229	0.03185	1.99971	0.97863
0.1095	0.9738	0.4130	0.8967	0.5538	-1.78157	18.65124	0.02157	2.32810	0.96458
0.08899	1.072	0.3894	0.8923	0.5534	-2.30140	31.14538	0.01341	2.82419	0.94827
0.06800	1.183	0.3660	0.8883	0.5527	-3.14911	59.13418	0.00732	3.64881	0.92982
0.04660	1.307	0.3430	0.8846	0.5517	-4.78850	140.46249	0.00319	5.26589	0.90942
0.02481	1.448	0.3204	0.8815	0.5503	-9.33868	555.83990	0.00083	9.79467	0.88733
L_1	L_2	C_2	L_3	C_4	L_1	L_2	C_2	L_3	C_4

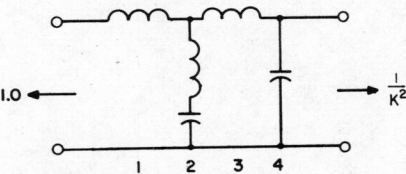

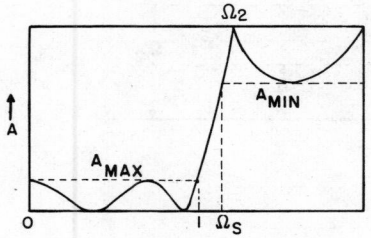

$$n = 4$$

$$\rho = 3\%$$

θ	Ω_S	A_{MIN}	σ_1	σ_3	Ω_1	Ω_2	Ω_3
c	∞	∞	-0.4797504	-1.1582199	1.4815623	∞	0.6136832
6.0	10.350843	86.80	-0.4747726	-1.1609403	1.4801063	11.367741	0.6192316
7.0	8.876727	81.44	-0.4729762	-1.1619227	1.4795772	9.747389	0.6212446
8.0	7.771760	76.79	-0.4709043	-1.1630563	1.4789644	8.532615	0.6235735
9.0	6.912894	72.69	-0.4685573	-1.1643410	1.4782669	7.588226	0.6262209
10.0	6.226301	69.02	-0.4659355	-1.1657769	1.4774836	6.833109	0.6291901
11.0	5.664999	65.70	-0.4630394	-1.1673641	1.4766131	6.215646	0.6324842
12.0	5.197666	62.66	-0.4598697	-1.1691027	1.4756539	5.701423	0.6361072
13.0	4.802620	59.87	-0.4564269	-1.1709928	1.4746045	5.266618	0.6400633
14.0	4.464371	57.28	-0.4527118	-1.1730347	1.4734628	4.894214	0.6443570
15.0	4.171563	54.87	-0.4487253	-1.1752284	1.4722269	4.571732	0.6489933
16.0	3.915678	52.61	-0.4444682	-1.1775744	1.4708944	4.289813	0.6539776
17.0	3.690200	50.49	-0.4399417	-1.1800731	1.4694629	4.041300	0.6593157
18.0	3.490065	48.49	-0.4351468	-1.1827248	1.4679295	3.820626	0.6650139
19.0	3.311272	46.59	-0.4300849	-1.1855303	1.4662913	3.623399	0.6710787
20.0	3.150622	44.79	-0.4247576	-1.1884902	1.4645449	3.446101	0.6775175
21.0	3.005526	43.07	-0.4191663	-1.1916055	1.4626868	3.285888	0.6843376
22.0	2.873864	41.43	-0.4133131	-1.1948771	1.4607130	3.140431	0.6915472
23.0	2.753885	39.86	-0.4071999	-1.1983064	1.4586192	3.007807	0.6991548
24.0	2.644133	38.36	-0.4008290	-1.2018948	1.4564010	2.886413	0.7071695
25.0	2.543380	36.92	-0.3942031	-1.2056442	1.4540534	2.774903	0.7156008
26.0	2.450592	35.53	-0.3873251	-1.2095566	1.4515710	2.672139	0.7244589
27.0	2.364885	34.19	-0.3801980	-1.2136345	1.4489481	2.577149	0.7337543
28.0	2.285502	32.90	-0.3728257	-1.2178808	1.4461784	2.489103	0.7434983
29.0	2.211792	31.65	-0.3652120	-1.2222988	1.4432556	2.407283	0.7537026
30.0	2.143189	30.44	-0.3573616	-1.2268924	1.4401723	2.331070	0.7643796
31.0	2.079202	29.27	-0.3492795	-1.2316659	1.4369213	2.259921	0.7755423
32.0	2.019399	28.13	-0.3409714	-1.2366246	1.4334943	2.193363	0.7872042
33.0	1.963403	27.03	-0.3324435	-1.2417742	1.4298832	2.130982	0.7993795
34.0	1.910879	25.95	-0.3237031	-1.2471217	1.4260788	2.072410	0.8120830
35.0	1.861534	24.91	-0.3147581	-1.2526746	1.4220718	2.017322	0.8253301
36.0	1.815103	23.90	-0.3056174	-1.2584418	1.4178524	1.965429	0.8391366
37.0	1.771354	22.91	-0.2962909	-1.2644332	1.4134104	1.916475	0.8535191
38.0	1.730076	21.94	-0.2867899	-1.2706602	1.4087352	1.870229	0.8684946
39.0	1.691083	21.00	-0.2771267	-1.2771355	1.4038160	1.826485	0.8840805
40.0	1.654204	20.08	-0.2673153	-1.2838737	1.3986419	1.785057	0.9002945
41.0	1.619289	19.19	-0.2573711	-1.2908908	1.3932018	1.745777	0.9171547
42.0	1.586200	18.31	-0.2473112	-1.2982052	1.3874846	1.708493	0.9346792
43.0	1.554811	17.46	-0.2371546	-1.3058369	1.3814798	1.673069	0.9528861
44.0	1.525009	16.62	-0.2269222	-1.3138086	1.3751773	1.639380	0.9717930
45.0	1.496692	15.80	-0.2166369	-1.3221450	1.3685676	1.607311	0.9914172
46.0	1.469765	15.01	-0.2063239	-1.3308735	1.3616425	1.576760	1.0117750
47.0	1.444142	14.23	-0.1960103	-1.3400240	1.3543953	1.547632	1.0328817
48.0	1.419745	13.47	-0.1857255	-1.3496288	1.3468211	1.519839	1.0547508
49.0	1.396501	12.73	-0.1755006	-1.3597226	1.3389174	1.493303	1.0773941
50.0	1.374345	12.00	-0.1653688	-1.3703425	1.3306843	1.467949	1.1008207
51.0	1.353215	11.30	-0.1553645	-1.3815275	1.3221254	1.443712	1.1250372
52.0	1.333055	10.61	-0.1455235	-1.3933185	1.3132479	1.420528	1.1500469
53.0	1.313814	9.95	-0.1358816	-1.4057570	1.3040629	1.398341	1.1758494
54.0	1.295444	9.30	-0.1264748	-1.4188854	1.2945862	1.377098	1.2024406
55.0	1.277901	8.68	-0.1173380	-1.4327460	1.2848378	1.356750	1.2298128
θ	Ω_S	A_{MIN}	σ_1	σ_3	Ω_1	Ω_2	Ω_3

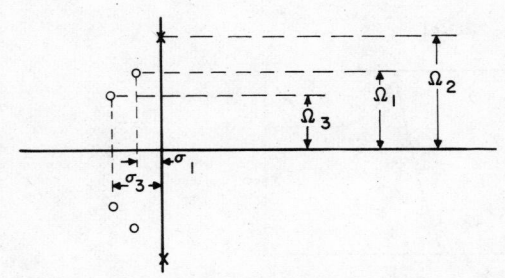

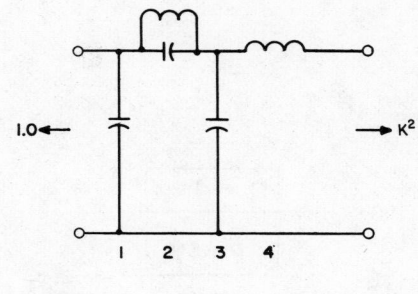

K²=0.9417					K²= ∞				
C₁	C₂	L₂	C₃	L₄	C₁	C₂	L₂	C₃	L₄
0.6105	0.000000	1.118	1.187	0.5749	0.30526	0.000000	0.78980	1.04335	0.95411
0.6044	0.006985	1.108	1.183	0.5757	0.29580	0.00997	0.77633	1.04180	0.95784
0.6022	0.009532	1.104	1.182	0.5760	0.29236	0.01364	0.77147	1.04126	0.95919
0.5997	0.01249	1.100	1.180	0.5764	0.28837	0.01793	0.76585	1.04066	0.96075
0.5968	0.01586	1.095	1.178	0.5767	0.28383	0.02287	0.75948	1.04000	0.96252
0.5935	0.01965	1.090	1.176	0.5772	0.27872	0.02847	0.75235	1.03929	0.96451
0.5899	0.02399	1.084	1.174	0.5776	0.27303	0.03477	0.74447	1.03854	0.96670
0.5860	0.02856	1.077	1.172	0.5781	0.26675	0.04181	0.73583	1.03776	0.96910
0.5817	0.03369	1.070	1.169	0.5787	0.25987	0.04963	0.72644	1.03696	0.97172
0.5771	0.03930	1.062	1.166	0.5793	0.25237	0.05828	0.71628	1.03616	0.97455
0.5721	0.04539	1.054	1.163	0.5799	0.24423	0.06783	0.70537	1.03538	0.97760
0.5667	0.05199	1.045	1.160	0.5806	0.23543	0.07833	0.69370	1.03463	0.98085
0.5610	0.05912	1.036	1.156	0.5813	0.22595	0.08988	0.68127	1.03394	0.98432
0.5550	0.06679	1.026	1.152	0.5821	0.21576	0.10254	0.66808	1.03333	0.98799
0.5485	0.07503	1.015	1.148	0.5829	0.20484	0.11644	0.65414	1.03283	0.99188
0.5417	0.08387	1.004	1.144	0.5838	0.19314	0.13169	0.63944	1.03245	0.99596
0.5345	0.09334	0.9923	1.140	0.5847	0.18064	0.14843	0.62399	1.03228	1.00025
0.5270	0.1035	0.9800	1.135	0.5856	0.16729	0.16683	0.60779	1.03230	1.00473
0.5191	0.1143	0.9671	1.131	0.5866	0.15305	0.18707	0.59086	1.03259	1.00940
0.5107	0.1259	0.9536	1.126	0.5876	0.13786	0.20940	0.57320	1.03319	1.01424
0.5020	0.1382	0.9396	1.121	0.5886	0.12166	0.23407	0.55483	1.03417	1.01925
0.4929	0.1514	0.9250	1.115	0.5897	0.10439	0.26141	0.53575	1.03559	1.02440
0.4834	0.1655	0.9098	1.110	0.5909	0.08596	0.29180	0.51598	1.03752	1.02969
0.4735	0.1805	0.8940	1.105	0.5920	0.06630	0.32570	0.49556	1.04007	1.03508
0.4632	0.1966	0.8777	1.099	0.5932	0.04528	0.36367	0.47450	1.04334	1.04056
0.4524	0.2138	0.8608	1.093	0.5944	0.02281	0.40639	0.45284	1.04744	1.04609
0.4412	0.2322	0.8434	1.087	0.5957	-0.00126	0.45470	0.43061	1.05252	1.05163
0.4297	0.2518	0.8254	1.081	0.5970	-0.02709	0.50962	0.40788	1.05876	1.05713
0.4176	0.2729	0.8068	1.074	0.5982	-0.05487	0.57245	0.38468	1.06633	1.06255
0.4051	0.2956	0.7878	1.068	0.5995	-0.08482	0.64481	0.36109	1.07548	1.06783
0.3922	0.3199	0.7682	1.061	0.6008	-0.11720	0.72876	0.33718	1.08649	1.07288
0.3788	0.3461	0.7480	1.055	0.6021	-0.15232	0.82694	0.31305	1.09969	1.07763
0.3649	0.3743	0.7274	1.048	0.6034	-0.19058	0.94281	0.28878	1.11547	1.08197
0.3506	0.4048	0.7063	1.041	0.6047	-0.23244	1.08087	0.26451	1.13434	1.08582
0.3358	0.4378	0.6848	1.035	0.6059	-0.27847	1.24717	0.24035	1.15690	1.08904
0.3205	0.4735	0.6627	1.028	0.6071	-0.32939	1.44987	0.21645	1.18389	1.09150
0.3047	0.5124	0.6403	1.021	0.6083	-0.38610	1.70022	0.19298	1.21627	1.09305
0.2884	0.5548	0.6175	1.014	0.6093	-0.44976	2.01396	0.17011	1.25523	1.09355
0.2716	0.6011	0.5943	1.008	0.6103	-0.52186	2.41366	0.14801	1.30233	1.09282
0.2543	0.6518	0.5709	1.001	0.6112	-0.60440	2.93237	0.12689	1.35964	1.09069
0.2365	0.7074	0.5471	0.9944	0.6120	-0.70007	3.61984	0.10693	1.42990	1.08700
0.2182	0.7688	0.5232	0.9879	0.6126	-0.81258	4.55336	0.08834	1.51692	1.08159
0.1994	0.8365	0.4991	0.9816	0.6131	-0.94725	5.85743	0.07128	1.62611	1.07430
0.1801	0.9115	0.4749	0.9755	0.6134	-1.11191	7.74162	0.05592	1.76539	1.06503
0.1603	0.9949	0.4507	0.9697	0.6134	-1.31861	10.57807	0.04239	1.94694	1.05368
0.1400	1.088	0.4266	0.9641	0.6132	-1.58692	15.07306	0.03079	2.19043	1.04021
0.1193	1.192	0.4026	0.9588	0.6128	-1.95080	22.68776	0.02115	2.52995	1.02463
0.0982	1.308	0.3788	0.9539	0.6120	-2.47490	36.81599	0.01346	3.03027	1.00702
0.0766	1.439	0.3553	0.9493	0.6109	-3.29914	66.75732	0.00766	3.83143	0.98749
0.0547	1.587	0.3322	0.9452	0.6094	-4.79337	145.51403	0.00362	5.30341	0.96626
0.0324	1.754	0.3097	0.9415	0.6075	-8.35859	462.24884	0.00118	8.84731	0.94355
L₁	L₂	C₂	L₃	C₄	L₁	L₂	C₂	L₃	C₄

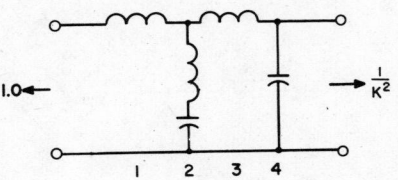

185

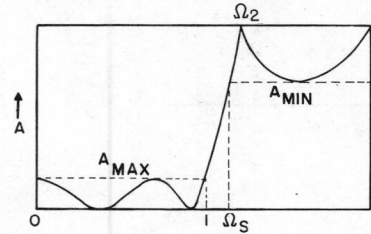

$$\boxed{\textsf{n = 4}}$$

$$\boxed{\textsf{ρ = 4\%}}$$

θ	Ω_S	A_{MIN}	σ_1	σ_3	Ω_1	Ω_2	Ω_3
c	∞	∞	−0.4367918	−1.0545087	1.4019778	∞	0.5807182
6.0	10.350843	89.30	−0.4328080	−1.0567584	1.4009044	11.367741	0.5852342
7.0	8.876727	83.94	−0.4313698	−1.0575711	1.4005143	9.747389	0.5868720
8.0	7.771760	79.29	−0.4297107	−1.0585089	1.4000626	8.532615	0.5887664
9.0	6.912894	75.19	−0.4278307	−1.0595721	1.3995485	7.588226	0.5909195
10.0	6.226301	71.52	−0.4257302	−1.0607606	1.3989713	6.833109	0.5933335
11.0	5.664999	68.20	−0.4234091	−1.0620747	1.3983300	6.215646	0.5960111
12.0	5.197666	65.17	−0.4208678	−1.0635145	1.3976235	5.701423	C.5989550
13.0	4.802620	62.37	−0.4181066	−1.0650801	1.3968506	5.266618	0.6021684
14.0	4.464371	59.78	−0.4151257	−1.0667719	1.3960101	4.894214	0.6056548
15.0	4.171563	57.37	−0.4119254	−1.0685901	1.3951004	4.571732	0.6094179
16.0	3.915678	55.11	−0.4085063	−1.0705349	1.3941200	4.289813	0.6134618
17.0	3.690200	52.99	−0.4048687	−1.0726069	1.3930671	4.041300	0.6177909
18.0	3.490065	50.99	−0.4010132	−1.0748065	1.3919398	3.820626	0.6224098
19.0	3.311272	49.09	−0.3969403	−1.0771341	1.3907359	3.623399	0.6273236
20.0	3.150622	47.29	−0.3926508	−1.0795904	1.3894531	3.446101	0.6325376
21.0	3.005526	45.57	−0.3881455	−1.0821762	1.3880890	3.285888	0.6380577
22.0	2.873864	43.93	−0.3834251	−1.0848922	1.3866408	3.140431	0.6438898
23.0	2.753885	42.37	−0.3784907	−1.0877395	1.3851056	3.007807	0.6500404
24.0	2.644133	40.86	−0.3733434	−1.0907191	1.3834802	2.886413	0.6565164
25.0	2.543380	39.42	−0.3679846	−1.0938323	1.3817611	2.774903	0.6633250
26.0	2.450592	38.03	−0.3624156	−1.0970808	1.3799446	2.672139	0.6704739
27.0	2.364885	36.69	−0.3566380	−1.1004661	1.3780268	2.577149	0.6779712
28.0	2.285502	35.40	−0.3506538	−1.1039902	1.3760033	2.489103	0.6858253
29.0	2.211792	34.15	−0.3444650	−1.1076554	1.3738696	2.407283	0.6940453
30.0	2.143189	32.94	−0.3380739	−1.1114642	1.3716206	2.331070	0.7026406
31.0	2.079202	31.77	−0.3314832	−1.1154196	1.3692513	2.259921	0.7116211
32.0	2.019399	30.63	−0.3246960	−1.1195250	1.3667558	2.193363	0.7209972
33.0	1.963403	29.52	−0.3177155	−1.1237840	1.3641282	2.130982	0.7307800
34.0	1.910879	28.45	−0.3105455	−1.1282011	1.3613622	2.072410	0.7409808
35.0	1.861534	27.41	−0.3031905	−1.1327811	1.3584510	2.017322	0.7516116
36.0	1.815103	26.39	−0.2956551	−1.1375296	1.3553874	1.965429	0.7626850
37.0	1.771354	25.40	−0.2879449	−1.1424529	1.3521640	1.916475	0.7742141
38.0	1.730076	24.43	−0.2800659	−1.1475579	1.3487728	1.870229	0.7862124
39.0	1.691083	23.49	−0.2720250	−1.1528528	1.3452056	1.826485	0.7986941
40.0	1.654204	22.57	−0.2638299	−1.1583465	1.3414539	1.785057	0.8116739
41.0	1.619289	21.66	−0.2554893	−1.1640491	1.3375087	1.745777	0.8251669
42.0	1.586200	20.78	−0.2470129	−1.1699720	1.3333610	1.708493	0.8391887
43.0	1.554811	19.92	−0.2384115	−1.1761280	1.3290015	1.673069	0.8537553
44.0	1.525009	19.08	−0.2296972	−1.1825311	1.3244208	1.639380	0.8688829
45.0	1.496692	18.26	−0.2208835	−1.1891972	1.3196097	1.607311	0.8845881
46.0	1.469765	17.45	−0.2119853	−1.1961439	1.3145588	1.576760	0.9008874
47.0	1.444142	16.66	−0.2030191	−1.2033907	1.3092596	1.547632	0.9177974
48.0	1.419745	15.88	−0.1940032	−1.2109590	1.3037037	1.519839	0.9353344
49.0	1.396501	15.13	−0.1849573	−1.2188724	1.2978836	1.493303	0.9535141
50.0	1.374345	14.38	−0.1759034	−1.2271567	1.2917930	1.467949	0.9723518
51.0	1.353215	13.66	−0.1668647	−1.2358397	1.2854269	1.443712	0.9918615
52.0	1.333055	12.95	−0.1578665	−1.2449517	1.2787821	1.420528	1.0120560
53.0	1.313814	12.25	−0.1489357	−1.2545249	1.2718573	1.398341	1.0329467
54.0	1.295444	11.58	−0.1401004	−1.2645936	1.2646540	1.377098	1.0545428
55.0	1.277901	10.91	−0.1313901	−1.2751937	1.2571763	1.356750	1.0768512
θ	Ω_S	A_{MIN}	σ_1	σ_3	Ω_1	Ω_2	Ω_3

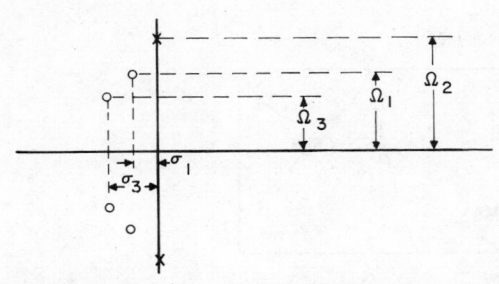

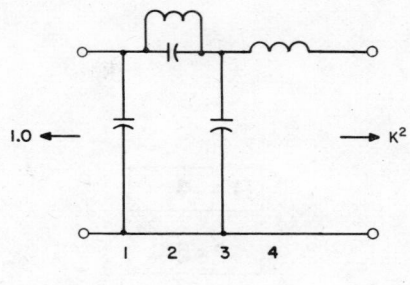

	K² = 0.9231					K² = ∞			
C_1	C_2	L_2	C_3	L_4	C_1	C_2	L_2	C_3	L_4
0.6706	0.000000	1.170	1.267	0.6190	0.33528	0.000000	0.85248	1.11080	1.00792
0.6647	0.006670	1.160	1.263	0.6197	0.32653	0.00921	0.83980	1.10903	1.01121
0.6626	0.009100	1.157	1.262	0.6200	0.32335	0.01260	0.83522	1.10841	1.01240
0.6601	0.01192	1.153	1.260	0.6203	0.31967	0.01655	0.82993	1.10771	1.01377
0.6574	0.01513	1.148	1.259	0.6206	0.31547	0.02108	0.82394	1.10693	1.01533
0.6543	0.01874	1.143	1.256	0.6210	0.31076	0.02621	0.81723	1.10608	1.01707
0.6508	0.02276	1.137	1.254	0.6214	0.30553	0.03196	0.80981	1.10517	1.01900
0.6471	0.02720	1.131	1.252	0.6219	0.29976	0.03837	0.80169	1.10420	1.02112
0.6430	0.03207	1.124	1.249	0.6224	0.29344	0.04547	0.79285	1.10319	1.02343
0.6385	0.03738	1.117	1.246	0.6229	0.28657	0.05330	0.78330	1.10214	1.02592
0.6337	0.04315	1.109	1.243	0.6235	0.27912	0.06189	0.77304	1.10107	1.02860
0.6286	0.04938	1.100	1.240	0.6241	0.27109	0.07131	0.76206	1.10000	1.03146
0.6232	0.05610	1.091	1.236	0.6248	0.26246	0.08160	0.75037	1.09893	1.03452
0.6174	0.06332	1.082	1.232	0.6255	0.25320	0.09283	0.73796	1.09788	1.03776
0.6112	0.07106	1.072	1.228	0.6262	0.24330	0.10508	0.72485	1.09689	1.04118
0.6047	0.07934	1.061	1.224	0.6270	0.23274	0.11843	0.71102	1.09596	1.04479
0.5979	0.08820	1.050	1.220	0.6278	0.22148	0.13298	0.69648	1.09512	1.04858
0.5907	0.09764	1.038	1.215	0.6287	0.20950	0.14884	0.68123	1.09440	1.05255
0.5831	0.1077	1.026	1.211	0.6296	0.19677	0.16615	0.66529	1.09383	1.05669
0.5752	0.1184	1.013	1.206	0.6305	0.18325	0.18505	0.64864	1.09345	1.06099
0.5669	0.1299	1.000	1.201	0.6315	0.16890	0.20572	0.63130	1.09330	1.06546
0.5582	0.1420	0.9862	1.195	0.6325	0.15368	0.22836	0.61328	1.09342	1.07008
0.5492	0.1549	0.9717	1.190	0.6335	0.13752	0.25322	0.59459	1.09387	1.07484
0.5398	0.1687	0.9567	1.184	0.6346	0.12039	0.28058	0.57524	1.09469	1.07972
0.5300	0.1833	0.9412	1.178	0.6357	0.10220	0.31078	0.55525	1.09597	1.08472
0.5198	0.1989	0.9252	1.172	0.6368	0.08289	0.34421	0.53464	1.09777	1.08980
0.5092	0.2155	0.9086	1.166	0.6380	0.06237	0.38136	0.51343	1.10018	1.09496
0.4983	0.2332	0.8914	1.160	0.6392	0.04055	0.42279	0.49164	1.10330	1.10016
0.4869	0.2520	0.8738	1.153	0.6404	0.01730	0.46921	0.46932	1.10725	1.10536
0.4751	0.2721	0.8556	1.147	0.6416	-0.00748	0.52146	0.44650	1.11217	1.11054
0.4629	0.2936	0.8369	1.140	0.6428	-0.03397	0.58059	0.42323	1.11820	1.11565
0.4503	0.3166	0.8176	1.133	0.6441	-0.06232	0.64788	0.39957	1.12554	1.12063
0.4372	0.3412	0.7979	1.126	0.6453	-0.09274	0.72495	0.37557	1.13440	1.12542
0.4237	0.3676	0.7777	1.119	0.6466	-0.12548	0.81382	0.35130	1.14503	1.12997
0.4098	0.3960	0.7570	1.112	0.6478	-0.16081	0.91707	0.32686	1.15773	1.13418
0.3954	0.4265	0.7359	1.105	0.6491	-0.19909	1.03802	0.30234	1.17288	1.13798
0.3806	0.4594	0.7142	1.098	0.6503	-0.24073	1.18096	0.27783	1.19090	1.14126
0.3653	0.4949	0.6922	1.090	0.6514	-0.28623	1.35158	0.25347	1.21234	1.14392
0.3496	0.5334	0.6698	1.083	0.6526	-0.33621	1.55741	0.22939	1.23784	1.14583
0.3333	0.5752	0.6469	1.076	0.6536	-0.39145	1.80870	0.20572	1.26822	1.14686
0.3166	0.6206	0.6237	1.068	0.6546	-0.45292	2.11951	0.18263	1.30451	1.14687
0.2995	0.6701	0.6002	1.061	0.6555	-0.52187	2.50959	0.16027	1.34800	1.14571
0.2818	0.7243	0.5764	1.054	0.6563	-0.59993	3.00718	0.13884	1.40038	1.14324
0.2636	0.7837	0.5524	1.047	0.6570	-0.68924	3.65366	0.11849	1.46385	1.13930
0.2450	0.8490	0.5282	1.040	0.6575	-0.79269	4.51127	0.09940	1.54139	1.13374
0.2259	0.9211	0.5038	1.033	0.6579	-0.91428	5.67662	0.08175	1.63709	1.12643
0.2063	1.001	0.4794	1.026	0.6581	-1.05973	7.30519	0.06568	1.75674	1.11726
0.1862	1.089	0.4549	1.019	0.6580	-1.23741	9.65829	0.05131	1.90884	1.10613
0.1657	1.188	0.4305	1.013	0.6577	-1.46024	13.19899	0.03875	2.10640	1.09301
0.1448	1.298	0.4062	1.007	0.6571	-1.74907	18.80380	0.02804	2.37038	1.07788
0.1234	1.421	0.3822	1.002	0.6563	-2.14002	28.27906	0.01921	2.73702	1.06078
L_1	L_2	C_2	L_3	C_4	L_1	L_2	C_2	L_3	C_4

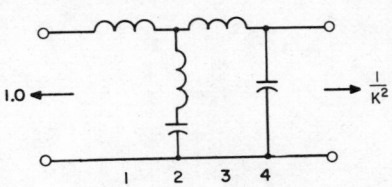

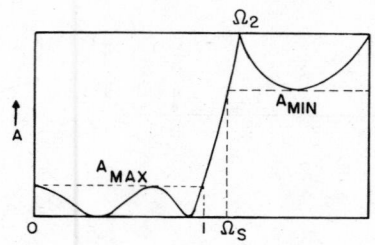

$$n = 4$$

$$\rho = 5\%$$

θ	Ω_S	A_{MIN}	σ_1	σ_3	Ω_1	Ω_2	Ω_3
C	∞	∞	-0.4050275	-0.9778230	1.3452476	∞	0.5572198
6.0	10.350843	91.24	-0.4016789	-0.9797660	1.3444158	11.367741	0.5610746
7.0	8.876727	85.88	-0.4004697	-0.9804681	1.3441135	9.747389	0.5624723
8.0	7.771760	81.24	-0.3990745	-0.9812783	1.3437635	8.532615	0.5640888
9.0	6.912894	77.13	-0.3974934	-0.9821970	1.3433651	7.588226	0.5659256
10.0	6.226301	73.46	-0.3957264	-0.9832241	1.3429179	6.833109	0.5679847
11.0	5.664999	70.14	-0.3937735	-0.9843599	1.3424210	6.215646	0.5702682
12.0	5.197666	67.11	-0.3916348	-0.9856046	1.3418737	5.701423	0.5727782
13.0	4.802620	64.31	-0.3893103	-0.9869583	1.3412750	5.266618	0.5755174
14.0	4.464371	61.73	-0.3868003	-0.9884214	1.3406240	4.894214	0.5784885
15.0	4.171563	59.31	-0.3841047	-0.9899941	1.3399195	4.571732	0.5816946
16.0	3.915678	57.06	-0.3812237	-0.9916767	1.3391603	4.289813	0.5851389
17.0	3.690200	54.93	-0.3781575	-0.9934696	1.3383450	4.041300	0.5888249
18.0	3.490065	52.93	-0.3749063	-0.9953733	1.3374723	3.820626	0.5927565
19.0	3.311272	51.03	-0.3714704	-0.9973882	1.3365405	3.623399	0.5969377
20.0	3.150622	49.23	-0.3678500	-0.9995149	1.3355479	3.446101	0.6013728
21.0	3.005526	47.51	-0.3640455	-1.0017540	1.3344925	3.285888	0.6060665
22.0	2.873864	45.88	-0.3600572	-1.0041063	1.3333724	3.140431	0.6110236
23.0	2.753885	44.31	-0.3558857	-1.0065726	1.3321853	3.007807	0.6162494
24.0	2.644133	42.80	-0.3515315	-1.0091538	1.3309287	2.886413	0.6217494
25.0	2.543380	41.36	-0.3469952	-1.0118510	1.3296002	2.774903	0.6275294
26.0	2.450592	39.97	-0.3422776	-1.0146654	1.3281970	2.672139	0.6335958
27.0	2.364885	38.63	-0.3373795	-1.0175982	1.3267160	2.577149	0.6399549
28.0	2.285502	37.34	-0.3323020	-1.0206511	1.3251539	2.489103	0.6466138
29.0	2.211792	36.09	-0.3270460	-1.0238258	1.3235074	2.407283	0.6535797
30.0	2.143189	34.88	-0.3216130	-1.0271242	1.3217728	2.331070	0.6608602
31.0	2.079202	33.71	-0.3160044	-1.0305485	1.3199459	2.259921	0.6684634
32.0	2.019399	32.57	-0.3102219	-1.0341012	1.3180227	2.193363	0.6763979
33.0	1.963403	31.46	-0.3042674	-1.0377851	1.3159984	2.130982	0.6846724
34.0	1.910879	30.39	-0.2981431	-1.0416034	1.3138683	2.072410	0.6932966
35.0	1.861534	29.35	-0.2918515	-1.0455595	1.3116273	2.017322	0.7022800
36.0	1.815103	28.33	-0.2853955	-1.0496574	1.3092698	1.965429	0.7116332
37.0	1.771354	27.34	-0.2787783	-1.0539018	1.3067901	1.916475	0.7213667
38.0	1.730076	26.37	-0.2720036	-1.0582975	1.3041820	1.870229	0.7314921
39.0	1.691083	25.42	-0.2650754	-1.0628501	1.3014390	1.826485	0.7420210
40.0	1.654204	24.50	-0.2579985	-1.0675662	1.2985544	1.785057	0.7529658
41.0	1.619289	23.60	-0.2507782	-1.0724527	1.2955210	1.745777	0.7643394
42.0	1.586200	22.71	-0.2434205	-1.0775176	1.2923314	1.708493	0.7761550
43.0	1.554811	21.85	-0.2359321	-1.0827699	1.2889780	1.673069	0.7884265
44.0	1.525009	21.00	-0.2283206	-1.0882194	1.2854527	1.639380	0.8011682
45.0	1.496692	20.17	-0.2205946	-1.0938772	1.2817475	1.607311	0.8143949
46.0	1.469765	19.36	-0.2127634	-1.0997558	1.2778540	1.576760	0.8281216
47.0	1.444142	18.57	-0.2048380	-1.1058689	1.2737640	1.547632	0.8423639
48.0	1.419745	17.78	-0.1968300	-1.1122318	1.2694691	1.519839	0.8571374
49.0	1.396501	17.02	-0.1887528	-1.1188613	1.2649613	1.493303	0.8724580
50.0	1.374345	16.27	-0.1806209	-1.1257762	1.2602328	1.467949	0.8883416
51.0	1.353215	15.53	-0.1724505	-1.1329968	1.2552764	1.443712	0.9048038
52.0	1.333055	14.81	-0.1642592	-1.1405457	1.2500856	1.420528	0.9218602
53.0	1.313814	14.10	-0.1560663	-1.1484474	1.2446548	1.398341	0.9395254
54.0	1.295444	13.41	-0.1478925	-1.1567284	1.2389798	1.377098	0.9578137
55.0	1.277901	12.73	-0.1397604	-1.1654174	1.2330578	1.356750	0.9767381
θ	Ω_S	A_{MIN}	σ_1	σ_3	Ω_1	Ω_2	Ω_3

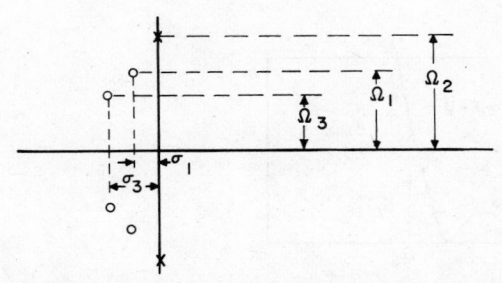

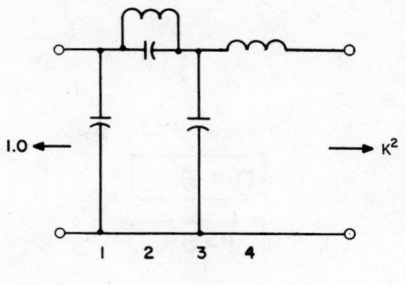

K²=0.9048					K²=∞				
C_1	C_2	L_2	C_3	L_4	C_1	C_2	L_2	C_3	L_4
0.7231	0.000000	1.207	1.334	0.6543	0.36157	0.000000	0.90444	1.16498	1.04995
0.7174	0.006461	1.198	1.330	0.6549	0.35333	0.00867	0.89233	1.16304	1.05291
0.7154	0.008813	1.194	1.329	0.6552	0.35034	0.01185	0.88795	1.16236	1.05398
0.7130	0.01154	1.190	1.327	0.6555	0.34687	0.01556	0.88290	1.16158	1.05522
0.7103	0.01464	1.186	1.325	0.6558	0.34293	0.01980	0.87717	1.16072	1.05662
0.7073	0.01814	1.181	1.323	0.6561	0.33851	0.02460	0.87076	1.15976	1.05820
0.7040	0.02202	1.176	1.321	0.6565	0.33360	0.02997	0.86368	1.15873	1.05993
0.7003	0.02630	1.170	1.318	0.6569	0.32819	0.03594	0.85592	1.15763	1.06184
0.6963	0.03100	1.163	1.316	0.6574	0.32227	0.04254	0.84748	1.15646	1.06391
0.6920	0.03612	1.156	1.313	0.6579	0.31584	0.04980	0.83836	1.15523	1.06616
0.6874	0.04166	1.148	1.310	0.6584	0.30888	0.05774	0.82856	1.15396	1.06857
0.6824	0.04766	1.140	1.306	0.6590	0.30139	0.06642	0.81808	1.15265	1.07115
0.6771	0.05411	1.132	1.303	0.6596	0.29334	0.07588	0.80692	1.15132	1.07390
0.6715	0.06103	1.122	1.299	0.6603	0.28473	0.08616	0.79508	1.14998	1.07682
0.6655	0.06845	1.113	1.295	0.6610	0.27554	0.09733	0.78256	1.14865	1.07991
0.6592	0.07637	1.103	1.291	0.6617	0.26575	0.10945	0.76935	1.14734	1.08316
0.6526	0.08482	1.092	1.286	0.6624	0.25534	0.12260	0.75547	1.14607	1.08658
0.6456	0.09383	1.081	1.282	0.6632	0.24428	0.13685	0.74091	1.14486	1.09016
0.6383	0.1034	1.069	1.277	0.6641	0.23257	0.15232	0.72568	1.14374	1.09391
0.6306	0.1136	1.057	1.272	0.6649	0.22015	0.16911	0.70977	1.14274	1.09780
0.6226	0.1244	1.044	1.267	0.6658	0.20702	0.18735	0.69320	1.14187	1.10185
0.6143	0.1359	1.030	1.262	0.6668	0.19313	0.20719	0.67596	1.14118	1.10605
0.6055	0.1481	1.017	1.256	0.6677	0.17844	0.22880	0.65807	1.14071	1.11038
0.5964	0.1611	1.002	1.250	0.6687	0.16291	0.25238	0.63954	1.14049	1.11485
0.5870	0.1748	0.9872	1.244	0.6698	0.14651	0.27816	0.62037	1.14057	1.11943
0.5772	0.1894	0.9717	1.238	0.6708	0.12916	0.30642	0.60057	1.14101	1.12412
0.5670	0.2049	0.9558	1.232	0.6719	0.11082	0.33749	0.58017	1.14186	1.12890
0.5564	0.2213	0.9393	1.226	0.6730	0.09142	0.37172	0.55919	1.14320	1.13375
0.5455	0.2388	0.9223	1.219	0.6742	0.07088	0.40959	0.53763	1.14511	1.13865
0.5341	0.2573	0.9048	1.212	0.6753	0.04911	0.45163	0.51554	1.14767	1.14358
0.5224	0.2771	0.8868	1.205	0.6765	0.02602	0.49848	0.49294	1.15100	1.14850
0.5103	0.2982	0.8683	1.198	0.6777	0.00149	0.55094	0.46988	1.15520	1.15338
0.4978	0.3206	0.8492	1.191	0.6789	-0.02462	0.60995	0.44638	1.16042	1.15818
0.4848	0.3446	0.8297	1.184	0.6801	-0.05244	0.67668	0.42250	1.16681	1.16286
0.4715	0.3702	0.8098	1.177	0.6813	-0.08216	0.75258	0.39830	1.17457	1.16735
0.4577	0.3976	0.7893	1.169	0.6825	-0.11399	0.83946	0.37385	1.18392	1.17160
0.4435	0.4270	0.7684	1.161	0.6838	-0.14816	0.93958	0.34921	1.19511	1.17554
0.4289	0.4586	0.7470	1.154	0.6849	-0.18498	1.05582	0.32448	1.20845	1.17908
0.4138	0.4926	0.7252	1.146	0.6861	-0.22477	1.19185	0.29974	1.22432	1.18215
0.3983	0.5293	0.7030	1.138	0.6872	-0.26797	1.35247	0.27511	1.24314	1.18464
0.3823	0.5689	0.6804	1.130	0.6883	-0.31507	1.54395	0.25071	1.26546	1.18645
0.3659	0.6118	0.6574	1.123	0.6893	-0.36670	1.77461	0.22665	1.29194	1.18745
0.3490	0.6584	0.6341	1.115	0.6903	-0.42363	2.05572	0.20310	1.32339	1.18752
0.3317	0.7091	0.6105	1.107	0.6912	-0.48682	2.40270	0.18018	1.36082	1.18653
0.3138	0.7645	0.5866	1.099	0.6919	-0.55751	2.83707	0.15806	1.40552	1.18434
0.2956	0.8251	0.5624	1.092	0.6926	-0.63730	3.38948	0.13691	1.45915	1.18080
0.2768	0.8917	0.5380	1.084	0.6931	-0.72827	4.10454	0.11689	1.52386	1.17578
0.2576	0.9650	0.5135	1.076	0.6934	-0.83325	5.04892	0.09815	1.60256	1.16914
0.2379	1.046	0.4889	1.069	0.6936	-0.95609	6.32519	0.08085	1.69917	1.16078
0.2177	1.136	0.4643	1.062	0.6935	-1.10223	8.09682	0.06513	1.81926	1.15058
0.1971	1.235	0.4397	1.055	0.6932	-1.27963	10.63514	0.05108	1.97085	1.13848
L_1	L_2	C_2	L_3	C_4	L_1	L_2	C_2	L_3	C_4

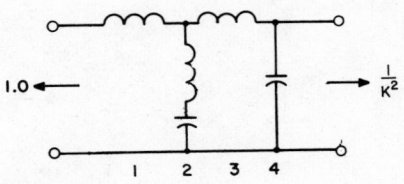

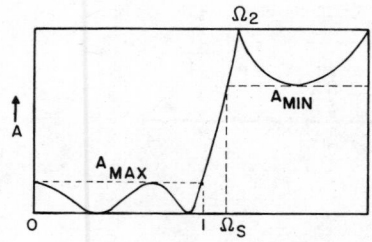

$$n = 4$$

$$\rho = 8\%$$

θ	Ω_S	A_{MIN}	σ_1	σ_3	Ω_1	Ω_2	Ω_3
C	∞	∞	-0.3420766	-0.8258460	1.2391832	∞	0.5132865
11.0	5.664999	74.24	-0.3342909	-0.8306545	1.2376691	6.215646	0.5226765
12.0	5.197666	71.21	-0.3328093	-0.8315710	1.2373756	5.701423	0.5244804
13.0	4.802620	68.41	-0.3311984	-0.8325682	1.2370546	5.266618	0.5264483
14.0	4.464371	65.83	-0.3294578	-0.8336462	1.2367053	4.894214	0.5285817
15.0	4.171563	63.41	-0.3275876	-0.8348053	1.2363273	4.571732	0.5308827
16.0	3.915678	61.16	-0.3255876	-0.8360460	1.2359199	4.289813	0.5333535
17.0	3.690200	59.03	-0.3234577	-0.8373685	1.2354822	4.041300	0.5359961
18.0	3.490065	57.03	-0.3211976	-0.8387732	1.2350136	3.820626	0.5388132
19.0	3.311272	55.13	-0.3188073	-0.8402605	1.2345130	3.623399	0.5418072
20.0	3.150622	53.33	-0.3162866	-0.8418309	1.2339797	3.446101	0.5449810
21.0	3.005526	51.61	-0.3136355	-0.8434849	1.2334125	3.285888	0.5483376
22.0	2.873864	49.97	-0.3108537	-0.8452231	1.2328103	3.140431	0.5518802
23.0	2.753885	48.41	-0.3079411	-0.8470461	1.2321720	3.007807	0.5556120
24.0	2.644133	46.90	-0.3048978	-0.8489545	1.2314962	2.886413	0.5595366
25.0	2.543380	45.46	-0.3017235	-0.8509492	1.2307815	2.774903	0.5636580
26.0	2.450592	44.07	-0.2984183	-0.8530309	1.2300264	2.672139	0.5679799
27.0	2.364885	42.73	-0.2949822	-0.8552005	1.2292293	2.577149	0.5725068
28.0	2.285502	41.44	-0.2914151	-0.8574590	1.2283885	2.489103	0.5772430
29.0	2.211792	40.19	-0.2877171	-0.8598076	1.2275020	2.407283	0.5821933
30.0	2.143189	38.98	-0.2838883	-0.8622475	1.2265679	2.331070	0.5873626
31.0	2.079202	37.80	-0.2799290	-0.8647799	1.2255840	2.259921	0.5927562
32.0	2.019399	36.67	-0.2758393	-0.8674063	1.2245479	2.193363	0.5983796
33.0	1.963403	35.56	-0.2716195	-0.8701283	1.2234572	2.130982	0.6042387
34.0	1.910879	34.49	-0.2672700	-0.8729478	1.2223093	2.072410	0.6103396
35.0	1.861534	33.44	-0.2627914	-0.8758667	1.2211013	2.017322	0.6166887
36.0	1.815103	32.42	-0.2581843	-0.8788871	1.2198301	1.965429	0.6232928
37.0	1.771354	31.43	-0.2534494	-0.8820115	1.2184926	1.916475	0.6301591
38.0	1.730076	30.46	-0.2485875	-0.8852424	1.2170854	1.870229	0.6372951
39.0	1.691083	29.51	-0.2435999	-0.8885830	1.2156047	1.826485	0.6447086
40.0	1.654204	28.59	-0.2384877	-0.8920363	1.2140466	1.785057	0.6524080
41.0	1.619289	27.68	-0.2332524	-0.8956060	1.2124071	1.745777	0.6604021
42.0	1.586200	26.80	-0.2278959	-0.8992962	1.2106816	1.708493	0.6686998
43.0	1.554811	25.93	-0.2224201	-0.9031112	1.2088658	1.673069	0.6773110
44.0	1.525009	25.08	-0.2168274	-0.9070558	1.2069544	1.639380	0.6862455
45.0	1.496692	24.25	-0.2111204	-0.9111357	1.2049426	1.607311	0.6955141
46.0	1.469765	23.43	-0.2053025	-0.9153568	1.2028247	1.576760	0.7051277
47.0	1.444142	22.63	-0.1993771	-0.9197259	1.2005952	1.547632	0.7150979
48.0	1.419745	21.84	-0.1933483	-0.9242503	1.1982480	1.519839	0.7254368
49.0	1.396501	21.07	-0.1872210	-0.9289383	1.1957771	1.493303	0.7361569
50.0	1.374345	20.30	-0.1810004	-0.9337992	1.1931760	1.467949	0.7472714
51.0	1.353215	19.56	-0.1746925	-0.9388430	1.1904383	1.443712	0.7587940
52.0	1.333055	18.82	-0.1683044	-0.9440811	1.1875572	1.420528	0.7707386
53.0	1.313814	18.10	-0.1618436	-0.9495260	1.1845259	1.398341	0.7831199
54.0	1.295444	17.38	-0.1553190	-0.9551915	1.1813378	1.377098	0.7959529
55.0	1.277901	16.68	-0.1487403	-0.9610927	1.1779861	1.356750	0.8092528
56.0	1.261143	15.99	-0.1421184	-0.9672467	1.1744644	1.337251	0.8230354
57.0	1.245134	15.31	-0.1354654	-0.9736718	1.1707666	1.318559	0.8373164
58.0	1.229837	14.65	-0.1287947	-0.9803884	1.1668870	1.300636	0.8521119
59.0	1.215222	13.99	-0.1221209	-0.9874188	1.1628208	1.283446	0.8674376
60.0	1.201256	13.34	-0.1154601	-0.9947871	1.1585638	1.266954	0.8833094
θ	Ω_S	A_{MIN}	σ_1	σ_3	Ω_1	Ω_2	Ω_3

190

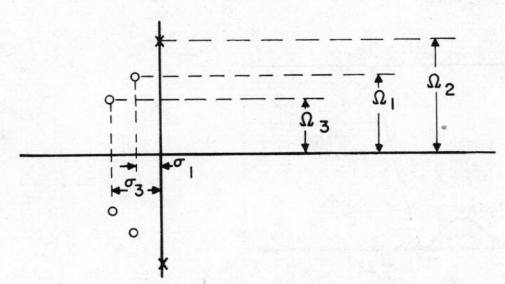

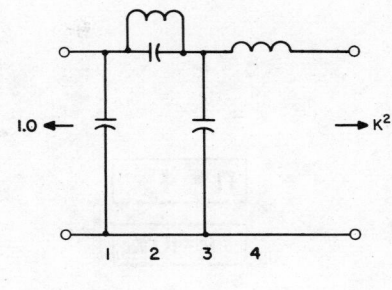

K²=0.8519					K²=∞				
C_1	C_2	L_2	C_3	L_4	C_1	C_2	L_2	C_3	L_4
0.8562	0.000000	1.271	1.493	0.7294	0.42811	0.000000	1.02333	1.28418	1.13759
0.8379	0.02083	1.242	1.479	0.7312	0.40349	0.02625	0.98618	1.27667	1.14543
0.8344	0.02487	1.237	1.477	0.7316	0.39875	0.03142	0.97911	1.27529	1.14692
0.8306	0.02929	1.231	1.474	0.7320	0.39358	0.03711	0.97142	1.27382	1.14855
0.8264	0.03409	1.224	1.471	0.7324	0.38797	0.04335	0.96312	1.27225	1.15031
0.8220	0.03930	1.218	1.468	0.7328	0.38191	0.05014	0.95420	1.27059	1.15221
0.8173	0.04491	1.210	1.464	0.7333	0.37540	0.05752	0.94466	1.26885	1.15423
0.8122	0.05093	1.202	1.460	0.7338	0.36843	0.06552	0.93450	1.26703	1.15639
0.8069	0.05739	1.194	1.457	0.7344	0.36099	0.07416	0.92372	1.26515	1.15868
0.8012	0.06428	1.185	1.453	0.7349	0.35307	0.08349	0.91232	1.26320	1.16111
0.7952	0.07163	1.176	1.448	0.7355	0.34466	0.09353	0.90030	1.26121	1.16366
0.7889	0.07945	1.166	1.444	0.7362	0.33575	0.10434	0.88766	1.25918	1.16635
0.7823	0.08776	1.155	1.439	0.7368	0.32633	0.11596	0.87441	1.25713	1.16917
0.7753	0.09658	1.145	1.434	0.7375	0.31638	0.12845	0.86053	1.25506	1.17212
0.7681	0.1059	1.133	1.429	0.7383	0.30589	0.14187	0.84605	1.25299	1.17520
0.7605	0.1158	1.121	1.424	0.7390	0.29484	0.15629	0.83094	1.25094	1.17841
0.7525	0.1263	1.109	1.418	0.7398	0.28322	0.17179	0.81522	1.24893	1.18174
0.7443	0.1373	1.097	1.412	0.7406	0.27099	0.18846	0.79890	1.24697	1.18520
0.7357	0.1490	1.083	1.406	0.7415	0.25815	0.20641	0.78196	1.24509	1.18877
0.7268	0.1613	1.070	1.400	0.7423	0.24467	0.22574	0.76443	1.24331	1.19245
0.7175	0.1744	1.055	1.394	0.7432	0.23052	0.24659	0.74630	1.24165	1.19624
0.7079	0.1881	1.041	1.387	0.7442	0.21566	0.26911	0.72758	1.24015	1.20013
0.6980	0.2027	1.026	1.381	0.7451	0.20008	0.29348	0.70828	1.23884	1.20412
0.6877	0.2180	1.010	1.374	0.7461	0.18373	0.31989	0.68840	1.23776	1.20819
0.6770	0.2343	0.9939	1.367	0.7471	0.16657	0.34858	0.66796	1.23695	1.21233
0.6660	0.2514	0.9774	1.360	0.7481	0.14856	0.37981	0.64698	1.23645	1.21654
0.6547	0.2696	0.9603	1.352	0.7492	0.12964	0.41389	0.62546	1.23632	1.22079
0.6430	0.2888	0.9428	1.345	0.7502	0.10977	0.45120	0.60343	1.23661	1.22506
0.6309	0.3091	0.9249	1.337	0.7513	0.08887	0.49216	0.58090	1.23740	1.22935
0.6184	0.3307	0.9064	1.329	0.7524	0.06688	0.53729	0.55790	1.23875	1.23362
0.6056	0.3536	0.8875	1.321	0.7535	0.04371	0.58718	0.53447	1.24076	1.23784
0.5924	0.3779	0.8682	1.313	0.7546	0.01927	0.64257	0.51063	1.24352	1.24199
0.5788	0.4038	0.8484	1.304	0.7557	-0.00653	0.70431	0.48642	1.24714	1.24603
0.5648	0.4314	0.8281	1.296	0.7568	-0.03383	0.77347	0.46188	1.25175	1.24991
0.5504	0.4608	0.8075	1.287	0.7580	-0.06276	0.85132	0.43707	1.25750	1.25360
0.5357	0.4922	0.7864	1.279	0.7591	-0.09348	0.93944	0.41203	1.26457	1.25704
0.5205	0.5259	0.7649	1.270	0.7601	-0.12617	1.03976	0.38684	1.27314	1.26016
0.5049	0.5620	0.7430	1.261	0.7612	-0.16106	1.15471	0.36157	1.28346	1.26291
0.4888	0.6007	0.7207	1.252	0.7622	-0.19839	1.28733	0.33629	1.29580	1.26521
0.4724	0.6425	0.6980	1.243	0.7632	-0.23846	1.44147	0.31110	1.31048	1.26697
0.4555	0.6876	0.6749	1.234	0.7641	-0.28165	1.62208	0.28609	1.32789	1.26811
0.4382	0.7363	0.6516	1.225	0.7650	-0.32837	1.83558	0.26138	1.34849	1.26853
0.4204	0.7892	0.6270	1.216	0.7658	-0.37916	2.09037	0.23707	1.37285	1.26812
0.4022	0.8468	0.6039	1.207	0.7665	-0.43465	2.39762	0.21330	1.40164	1.26677
0.3836	0.9096	0.5797	1.197	0.7671	-0.49563	2.77238	0.19020	1.43571	1.26435
0.3644	0.9783	0.5553	1.188	0.7675	-0.56311	3.23526	0.16792	1.47609	1.26076
0.3449	1.054	0.5307	1.180	0.7679	-0.63833	3.81496	0.14658	1.52412	1.25585
0.3248	1.137	0.5059	1.171	0.7680	-0.72291	4.55227	0.12635	1.58146	1.24953
0.3043	1.229	0.4811	1.162	0.7680	-0.81897	5.50637	0.10736	1.65031	1.24166
0.2834	1.331	0.4562	1.153	0.7678	-0.92933	6.76536	0.08973	1.73358	1.23214
0.2620	1.444	0.4313	1.145	0.7673	-1.05784	8.46416	0.07360	1.83520	1.22090
L_1	L_2	C_2	L_3	C_4	L_1	L_2	C_2	L_3	C_4

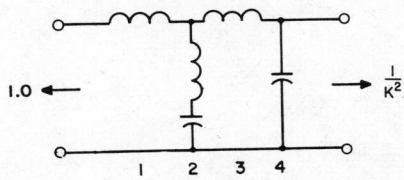

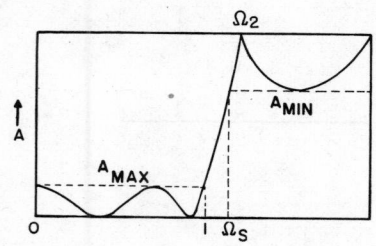

$$\boxed{n = 4}$$

$$\boxed{\rho = 10\%}$$

θ	Ω_S	A_{MIN}	σ_1	σ_3	Ω_1	Ω_2	Ω_3
c	∞	∞	-0.3138480	-0.7576961	1.1948459	∞	0.4949214
11.0	5.664999	76.20	-0.3073267	-0.7618501	1.1937989	6.215646	0.5029786
12.0	5.197666	73.16	-0.3060851	-0.7626421	1.1935957	5.701423	0.5045257
13.0	4.802620	70.37	-0.3047348	-0.7635040	1.1933732	5.266618	0.5062131
14.0	4.464371	67.78	-0.3032756	-0.7644359	1.1931311	4.894214	0.5080421
15.0	4.171563	65.37	-0.3017073	-0.7654381	1.1928689	4.571732	0.5100144
16.0	3.915678	63.11	-0.3000298	-0.7665109	1.1925861	4.289813	0.5121317
17.0	3.690200	60.99	-0.2982429	-0.7676546	1.1922821	4.041300	0.5143958
18.0	3.490065	58.98	-0.2963462	-0.7688696	1.1919564	3.820626	0.5168087
19.0	3.311272	57.09	-0.2943397	-0.7701563	1.1916083	3.623399	0.5193726
20.0	3.150622	55.28	-0.2922231	-0.7715150	1.1912371	3.446101	0.5220898
21.0	3.005526	53.57	-0.2899962	-0.7729463	1.1908420	3.285888	0.5249626
22.0	2.873864	51.93	-0.2876586	-0.7744507	1.1904222	3.140431	0.5279936
23.0	2.753885	50.36	-0.2852103	-0.7760287	1.1899769	3.007807	0.5311857
24.0	2.644133	48.86	-0.2826509	-0.7776809	1.1895050	2.886413	0.5345417
25.0	2.543380	47.41	-0.2799803	-0.7794080	1.1890055	2.774903	0.5380647
26.0	2.450592	46.02	-0.2771983	-0.7812106	1.1884773	2.672139	0.5417580
27.0	2.364885	44.68	-0.2743045	-0.7830896	1.1879192	2.577149	0.5456250
28.0	2.285502	43.39	-0.2712989	-0.7850457	1.1873299	2.489103	0.5496694
29.0	2.211792	42.14	-0.2681812	-0.7870799	1.1867081	2.407283	0.5538951
30.0	2.143189	40.93	-0.2649513	-0.7891933	1.1860522	2.331070	0.5583061
31.0	2.079202	39.76	-0.2616091	-0.7913868	1.1853607	2.259921	0.5629067
32.0	2.019399	38.62	-0.2581544	-0.7936617	1.1846318	2.193363	0.5677015
33.0	1.963403	37.52	-0.2545873	-0.7960193	1.1838637	2.130982	0.5726952
34.0	1.910879	36.44	-0.2509077	-0.7984609	1.1830544	2.072410	0.5778928
35.0	1.861534	35.40	-0.2471156	-0.8009882	1.1822018	2.017322	0.5832997
36.0	1.815103	34.38	-0.2432112	-0.8036029	1.1813038	1.965429	0.5889214
37.0	1.771354	33.38	-0.2391947	-0.8063067	1.1803578	1.916475	0.5947639
38.0	1.730076	32.41	-0.2350662	-0.8091018	1.1793614	1.870229	0.6008332
39.0	1.691083	31.47	-0.2308262	-0.8119904	1.1783117	1.826485	0.6071358
40.0	1.654204	30.54	-0.2264751	-0.8149750	1.1772059	1.785057	0.6136788
41.0	1.619289	29.64	-0.2220136	-0.8180582	1.1760409	1.745777	0.6204691
42.0	1.586200	28.75	-0.2174424	-0.8212431	1.1748133	1.708493	0.6275145
43.0	1.554811	27.88	-0.2127623	-0.8245329	1.1735197	1.673069	0.6348229
44.0	1.525009	27.03	-0.2079747	-0.8279313	1.1721562	1.639380	0.6424027
45.0	1.496692	26.20	-0.2030807	-0.8314423	1.1707189	1.607311	0.6502627
46.0	1.469765	25.38	-0.1980819	-0.8350702	1.1692036	1.576760	0.6584122
47.0	1.444142	24.57	-0.1929803	-0.8388201	1.1676060	1.547632	0.6668610
48.0	1.419745	23.78	-0.1877780	-0.8426971	1.1659212	1.519839	0.6756193
49.0	1.396501	23.01	-0.1824776	-0.8467075	1.1641444	1.493303	0.6846979
50.0	1.374345	22.24	-0.1770820	-0.8508576	1.1622703	1.467949	0.6941081
51.0	1.353215	21.49	-0.1715947	-0.8551549	1.1602937	1.443712	0.7038617
52.0	1.333055	20.75	-0.1660197	-0.8596076	1.1582088	1.420528	0.7139710
53.0	1.313814	20.03	-0.1603615	-0.8642245	1.1560098	1.398341	0.7244491
54.0	1.295444	19.31	-0.1546252	-0.8690156	1.1536907	1.377098	0.7353095
55.0	1.277901	18.60	-0.1488168	-0.8739921	1.1512453	1.356750	0.7465662
56.0	1.261143	17.91	-0.1429430	-0.8791660	1.1486675	1.337251	0.7582340
57.0	1.245134	17.22	-0.1370115	-0.8845511	1.1459508	1.318559	0.7703278
58.0	1.229837	16.55	-0.1310308	-0.8901622	1.1430892	1.300636	0.7828635
59.0	1.215222	15.88	-0.1250104	-0.8960159	1.1400764	1.283446	0.7958571
60.0	1.201256	15.22	-0.1189613	-0.9021304	1.1369067	1.266954	0.8093252
θ	Ω_S	A_{MIN}	σ_1	σ_3	Ω_1	Ω_2	Ω_3

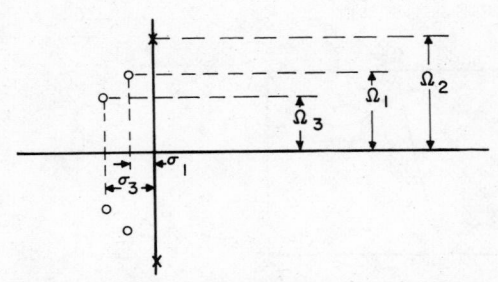

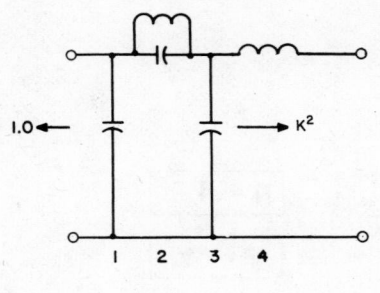

K²=0.8182					K²=∞				
C_1	C_2	L_2	C_3	L_4	C_1	C_2	L_2	C_3	L_4
0.9332	0.000000	1.292	1.580	0.7636	0.46662	0.000000	1.08381	1.34325	1.17766
0.9151	0.02047	1.265	1.566	0.7652	0.44340	0.02469	1.04820	1.33515	1.18457
0.9116	0.02443	1.259	1.563	0.7656	0.43895	0.02954	1.04143	1.33365	1.18588
0.9079	0.02876	1.254	1.560	0.7659	0.43408	0.03486	1.03406	1.33204	1.18732
0.9038	0.03347	1.247	1.557	0.7663	0.42881	0.04069	1.02611	1.33031	1.18887
0.8994	0.03856	1.241	1.554	0.7667	0.42313	0.04702	1.01756	1.32848	1.19053
0.8948	0.04405	1.234	1.550	0.7671	0.41702	0.05389	1.00842	1.32655	1.19231
0.8898	0.04994	1.226	1.547	0.7676	0.41049	0.06131	0.99869	1.32453	1.19422
0.8845	0.05625	1.218	1.543	0.7681	0.40353	0.06931	0.98836	1.32241	1.19623
0.8789	0.06298	1.209	1.539	0.7686	0.39612	0.07793	0.97744	1.32022	1.19837
0.8730	0.07015	1.200	1.534	0.7692	0.38827	0.08718	0.96593	1.31794	1.20062
0.8668	0.07776	1.191	1.530	0.7697	0.37996	0.09710	0.95382	1.31560	1.20299
0.8603	0.08585	1.181	1.525	0.7703	0.37119	0.10774	0.94113	1.31321	1.20547
0.8534	0.09441	1.171	1.520	0.7710	0.36194	0.11913	0.92784	1.31076	1.20807
0.8462	0.1035	1.160	1.515	0.7716	0.35220	0.13133	0.91396	1.30828	1.21079
0.8388	0.1131	1.149	1.509	0.7723	0.34196	0.14438	0.89949	1.30578	1.21362
0.8310	0.1232	1.137	1.503	0.7730	0.33121	0.15835	0.88443	1.30326	1.21656
0.8228	0.1339	1.125	1.498	0.7738	0.31993	0.17331	0.86878	1.30075	1.21961
0.8144	0.1451	1.112	1.492	0.7745	0.30811	0.18932	0.85255	1.29825	1.22277
0.8056	0.1570	1.099	1.485	0.7753	0.29572	0.20648	0.83573	1.29579	1.22604
0.7965	0.1695	1.085	1.479	0.7762	0.28275	0.22488	0.81834	1.29339	1.22941
0.7871	0.1828	1.071	1.472	0.7770	0.26917	0.24464	0.80038	1.29106	1.23287
0.7773	0.1967	1.057	1.465	0.7779	0.25497	0.26586	0.78184	1.28884	1.23642
0.7672	0.2113	1.042	1.458	0.7788	0.24011	0.28871	0.76274	1.28674	1.24007
0.7568	0.2268	1.027	1.451	0.7797	0.22458	0.31333	0.74309	1.28479	1.24379
0.7460	0.2431	1.011	1.443	0.7806	0.20833	0.33992	0.72289	1.28304	1.24757
0.7349	0.2603	0.9944	1.436	0.7816	0.19133	0.36868	0.70215	1.28151	1.25142
0.7234	0.2785	0.9776	1.428	0.7826	0.17354	0.39987	0.68089	1.28025	1.25532
0.7116	0.2977	0.9604	1.420	0.7836	0.15492	0.43376	0.65912	1.27929	1.25925
0.6994	0.3180	0.9427	1.412	0.7846	0.13543	0.47069	0.63685	1.27870	1.26320
0.6869	0.3394	0.9246	1.403	0.7856	0.11500	0.51104	0.61411	1.27852	1.26715
0.6740	0.3621	0.9061	1.395	0.7867	0.09357	0.55527	0.59091	1.27883	1.27108
0.6607	0.3862	0.8871	1.386	0.7877	0.07109	0.60391	0.56728	1.27969	1.27497
0.6471	0.4117	0.8676	1.377	0.7888	0.04747	0.65760	0.54326	1.28119	1.27878
0.6330	0.4389	0.8478	1.368	0.7898	0.02262	0.71710	0.51887	1.28341	1.28249
0.6186	0.4678	0.8275	1.359	0.7909	-0.00356	0.78332	0.49415	1.28647	1.28606
0.6038	0.4985	0.8068	1.350	0.7919	-0.03117	0.85735	0.46915	1.29050	1.28945
0.5886	0.5314	0.7857	1.341	0.7930	-0.06035	0.94052	0.44391	1.29562	1.29262
0.5730	0.5665	0.7642	1.331	0.7940	-0.09125	1.03447	0.41849	1.30200	1.29550
0.5570	0.6041	0.7423	1.322	0.7950	-0.12404	1.14119	0.39296	1.30984	1.29805
0.5406	0.6445	0.7200	1.312	0.7959	-0.15892	1.26319	0.36737	1.31934	1.30019
0.5238	0.6880	0.6974	1.303	0.7969	-0.19613	1.40359	0.34182	1.33077	1.30186
0.5065	0.7348	0.6744	1.293	0.7977	-0.23595	1.56632	0.31639	1.34443	1.30298
0.4888	0.7855	0.6511	1.283	0.7985	-0.27872	1.75644	0.29117	1.36068	1.30345
0.4707	0.8404	0.6275	1.273	0.7993	-0.32482	1.98044	0.26626	1.37994	1.30319
0.4521	0.9000	0.6036	1.263	0.7999	-0.37473	2.24679	0.24179	1.40273	1.30209
0.4331	0.9651	0.5794	1.253	0.8005	-0.42904	2.56672	0.21787	1.42968	1.30004
0.4136	1.036	0.5551	1.244	0.8009	-0.48846	2.95523	0.19463	1.46154	1.29694
0.3937	1.114	0.5305	1.234	0.8012	-0.55388	3.43273	0.17221	1.49925	1.29265
0.3733	1.200	0.5058	1.224	0.8013	-0.62641	4.02744	0.15074	1.54399	1.28707
0.3525	1.295	0.4809	1.215	0.8013	-0.70748	4.77911	0.13036	1.59725	1.28007
L_1	L_2	C_2	L_3	C_4	L_1	L_2	C_2	L_3	C_4

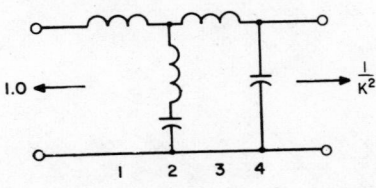

193

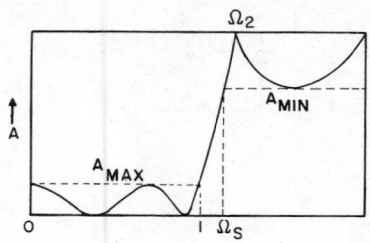

$$n = 4$$

$$\rho = 15\%$$

θ	Ω_S	A_{MIN}	σ_1	σ_3	Ω_1	Ω_2	Ω_3
C	∞	∞	-0.2648393	-0.6393787	1.1235473	∞	0.4653885
11.0	5.664999	79.77	-0.2601411	-0.6425506	1.1231569	6.215646	0.4715308
12.0	5.197666	76.74	-0.2592458	-0.6431558	1.1230803	5.701423	0.4727092
13.0	4.802620	73.95	-0.2582719	-0.6438144	1.1229963	5.266618	0.4739941
14.0	4.464371	71.36	-0.2572191	-0.6445267	1.1229044	4.894214	0.4753865
15.0	4.171563	68.94	-0.2560873	-0.6452929	1.1228046	4.571732	0.4768874
16.0	3.915678	66.69	-0.2548762	-0.6461132	1.1226964	4.289813	0.4784982
17.0	3.690200	64.56	-0.2535855	-0.6469880	1.1225797	4.041300	0.4802200
18.0	3.490065	62.56	-0.2522151	-0.6479175	1.1224540	3.820626	0.4820543
19.0	3.311272	60.66	-0.2507646	-0.6489021	1.1223190	3.623399	0.4840025
20.0	3.150622	58.86	-0.2492338	-0.6499421	1.1221744	3.446101	0.4860664
21.0	3.005526	57.14	-0.2476224	-0.6510379	1.1220197	3.285888	0.4882475
22.0	2.873864	55.51	-0.2459300	-0.6521899	1.1218543	3.140431	0.4905476
23.0	2.753885	53.94	-0.2441563	-0.6533986	1.1216780	3.007807	0.4929688
24.0	2.644133	52.43	-0.2423011	-0.6546644	1.1214900	2.886413	0.4955130
25.0	2.543380	50.99	-0.2403639	-0.6559879	1.1212899	2.774903	0.4981824
26.0	2.450592	49.60	-0.2383444	-0.6573696	1.1210770	2.672139	0.5009794
27.0	2.364885	48.26	-0.2362423	-0.6588102	1.1208507	2.577149	0.5039063
28.0	2.285502	46.97	-0.2340572	-0.6603101	1.1206101	2.489103	0.5069656
29.0	2.211792	45.72	-0.2317887	-0.6618702	1.1203547	2.407283	0.5101602
30.0	2.143189	44.51	-0.2294364	-0.6634911	1.1200835	2.331070	0.5134928
31.0	2.079202	43.33	-0.2270000	-0.6651738	1.1197956	2.259921	0.5169665
32.0	2.019399	42.20	-0.2244791	-0.6669189	1.1194902	2.193363	0.5205845
33.0	1.963403	41.09	-0.2218733	-0.6687275	1.1191661	2.130982	0.5243501
34.0	1.910879	40.02	-0.2191822	-0.6706005	1.1188223	2.072410	0.5282668
35.0	1.861534	38.97	-0.2164055	-0.6725390	1.1184576	2.017322	0.5323384
36.0	1.815103	37.95	-0.2135429	-0.6745441	1.1180707	1.965429	0.5365688
37.0	1.771354	36.96	-0.2105939	-0.6766172	1.1176604	1.916475	0.5409621
38.0	1.730076	35.99	-0.2075583	-0.6787595	1.1172251	1.870229	0.5455227
39.0	1.691083	35.04	-0.2044357	-0.6809726	1.1167632	1.826485	0.5502552
40.0	1.654204	34.12	-0.2012260	-0.6832580	1.1162733	1.785057	0.5551645
41.0	1.619289	33.21	-0.1979288	-0.6856174	1.1157534	1.745777	0.5602555
42.0	1.586200	32.32	-0.1945440	-0.6880527	1.1152017	1.708493	0.5655338
43.0	1.554811	31.45	-0.1910715	-0.6905660	1.1146161	1.673069	0.5710050
44.0	1.525009	30.60	-0.1875112	-0.6931594	1.1139945	1.639380	0.5766751
45.0	1.496692	29.77	-0.1838630	-0.6958355	1.1133346	1.607311	0.5825504
46.0	1.469765	28.95	-0.1801272	-0.6985967	1.1126338	1.576760	0.5886375
47.0	1.444142	28.14	-0.1763038	-0.7014461	1.1118896	1.547632	0.5949437
48.0	1.419745	27.35	-0.1723867	-0.7043867	1.1110991	1.519839	0.6014762
49.0	1.396501	26.57	-0.1683957	-0.7074221	1.1102592	1.493303	0.6082429
50.0	1.374345	25.81	-0.1643121	-0.7105561	1.1093669	1.467949	0.6152522
51.0	1.353215	25.05	-0.1601430	-0.7137928	1.1084186	1.443712	0.6225128
52.0	1.333055	24.31	-0.1558896	-0.7171369	1.1074108	1.420528	0.6300340
53.0	1.313814	23.58	-0.1515528	-0.7205935	1.1063396	1.398341	0.6378256
54.0	1.295444	22.86	-0.1471343	-0.7241682	1.1052009	1.377098	0.6458981
55.0	1.277901	22.15	-0.1426359	-0.7278671	1.1039904	1.356750	0.6542623
56.0	1.261143	21.44	-0.1380597	-0.7316972	1.1027037	1.337251	0.6629299
57.0	1.245134	20.75	-0.1334082	-0.7356661	1.1013358	1.318559	0.6719132
58.0	1.229837	20.07	-0.1286845	-0.7397822	1.0998819	1.300636	0.6812251
59.0	1.215222	19.39	-0.1238921	-0.7440550	1.0983366	1.283446	0.6908795
60.0	1.201256	18.72	-0.1190352	-0.7484947	1.0966947	1.266954	0.7008906
θ	Ω_S	A_{MIN}	σ_1	σ_3	Ω_1	Ω_2	Ω_3

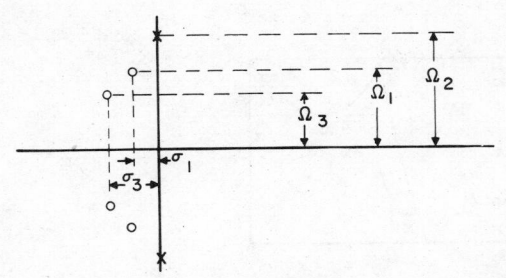
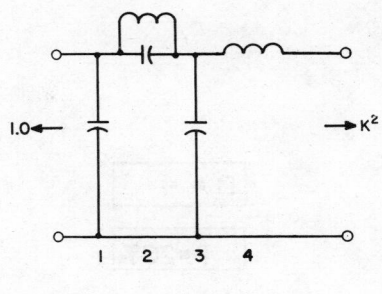

K²=0.7391					K²=∞				
C₁	C₂	L₂	C₃	L₄	C₁	C₂	L₂	C₃	L₄
1.106	0.000000	1.306	1.767	0.8174	0.55296	0.000000	1.19778	1.45588	1.24446
1.088	0.02021	1.281	1.753	0.8188	0.53201	0.02222	1.16475	1.44672	1.24980
1.084	0.02411	1.276	1.750	0.8191	0.52800	0.02656	1.15846	1.44500	1.25081
1.081	0.02837	1.271	1.747	0.8194	0.52363	0.03131	1.15163	1.44315	1.25192
1.077	0.03300	1.265	1.744	0.8197	0.51890	0.03648	1.14426	1.44116	1.25312
1.072	0.03801	1.259	1.740	0.8200	0.51380	0.04210	1.13633	1.43903	1.25440
1.068	0.04339	1.252	1.737	0.8204	0.50833	0.04818	1.12786	1.43678	1.25578
1.063	0.04916	1.245	1.733	0.8207	0.50249	0.05473	1.11884	1.43439	1.25725
1.057	0.05533	1.238	1.729	0.8211	0.49627	0.06176	1.10927	1.43189	1.25881
1.052	0.06191	1.230	1.724	0.8216	0.48967	0.06930	1.09915	1.42926	1.26046
1.046	0.06890	1.222	1.720	0.8220	0.48268	0.07736	1.08849	1.42653	1.26220
1.040	0.07633	1.213	1.715	0.8225	0.47531	0.08597	1.07727	1.42368	1.26403
1.033	0.08419	1.204	1.710	0.8230	0.46753	0.09516	1.06551	1.42073	1.26595
1.026	0.09251	1.195	1.704	0.7235	0.45935	0.10495	1.05320	1.41768	1.26796
1.019	0.1013	1.185	1.699	0.8240	0.45076	0.11537	1.04034	1.41454	1.27006
1.012	0.1106	1.175	1.693	0.8246	0.44175	0.12646	1.02693	1.41132	1.27225
1.004	0.1203	1.164	1.687	0.8252	0.43232	0.13826	1.01297	1.40802	1.27453
0.9961	0.1306	1.153	1.681	0.8258	0.42245	0.15079	0.99847	1.40465	1.27690
0.9877	0.1414	1.141	1.674	0.8264	0.41214	0.16412	0.98343	1.40123	1.27935
0.9790	0.1528	1.129	1.668	0.8271	0.40137	0.17830	0.96784	1.39775	1.28190
0.9700	0.1648	1.117	1.661	0.8278	0.39014	0.19337	0.95170	1.39424	1.28452
0.9607	0.1774	1.104	1.654	0.8285	0.37843	0.20941	0.93503	1.39070	1.28723
0.9510	0.1906	1.091	1.646	0.8292	0.36623	0.22648	0.91782	1.38715	1.29002
0.9410	0.2045	1.077	1.639	0.8299	0.35352	0.24466	0.90007	1.38360	1.29288
0.9307	0.2191	1.063	1.631	0.8307	0.34029	0.26405	0.88179	1.38007	1.29582
0.9201	0.2344	1.048	1.623	0.8315	0.32653	0.28474	0.86299	1.37657	1.29883
0.9092	0.2505	1.033	1.615	0.8323	0.31221	0.30685	0.84366	1.37313	1.30190
0.8979	0.2674	1.018	1.607	0.8331	0.29731	0.33050	0.82381	1.36976	1.30504
0.8862	0.2852	1.002	1.598	0.8339	0.28181	0.35584	0.80345	1.36649	1.30822
0.8742	0.3040	0.9861	1.589	0.8348	0.26569	0.38303	0.78258	1.36334	1.31145
0.8619	0.3237	0.9695	1.580	0.8357	0.24892	0.41227	0.76123	1.36034	1.31472
0.8493	0.3445	0.9525	1.571	0.8365	0.23146	0.44377	0.73938	1.35753	1.31802
0.8363	0.3664	0.9351	1.562	0.8374	0.21329	0.47777	0.71706	1.35493	1.32133
0.8229	0.3895	0.9172	1.552	0.8384	0.19437	0.51456	0.69429	1.35258	1.32464
0.8092	0.4139	0.8990	1.543	0.8393	0.17466	0.55447	0.67106	1.35054	1.32795
0.7951	0.4397	0.8803	1.533	0.8402	0.15412	0.59789	0.64741	1.34884	1.33122
0.7806	0.4670	0.8613	1.523	0.8411	0.13268	0.64525	0.62336	1.34754	1.33444
0.7658	0.4959	0.8418	1.512	0.8421	0.11030	0.69710	0.59892	1.34670	1.33759
0.7506	0.5267	0.8220	1.502	0.8430	0.08692	0.75405	0.57412	1.34639	1.34065
0.7350	0.5593	0.8018	1.492	0.8439	0.06246	0.81683	0.54900	1.34669	1.34357
0.7191	0.5941	0.7812	1.481	0.8448	0.03684	0.88631	0.52359	1.34768	1.34634
0.7027	0.6311	0.7602	1.470	0.8457	0.00996	0.96356	0.49792	1.34947	1.34891
0.6859	0.6707	0.7388	1.460	0.8466	-0.01826	1.04982	0.47205	1.35217	1.35125
0.6688	0.7131	0.7171	1.448	0.8475	-0.04797	1.14664	0.44601	1.35591	1.35329
0.6512	0.7587	0.6951	1.437	0.8483	-0.07928	1.25589	0.41967	1.36085	1.35501
0.6332	0.8076	0.6727	1.426	0.8491	-0.11236	1.37989	0.39369	1.36714	1.35632
0.6148	0.8604	0.6499	1.414	0.8498	-0.14740	1.52149	0.36754	1.37501	1.35717
0.5959	0.9175	0.6269	1.403	0.8505	-0.18460	1.68428	0.34150	1.38469	1.35749
0.5767	0.9794	0.6036	1.391	0.8511	-0.22423	1.87280	0.31564	1.39644	1.35720
0.5569	1.047	0.5800	1.380	0.8516	-0.26657	2.09283	0.29008	1.41061	1.35620
0.5367	1.120	0.5561	1.368	0.8520	-0.31199	2.35181	0.26490	1.42757	1.35442
L₁	L₂	C₂	L₃	C₄	L₁	L₂	C₂	L₃	C₄

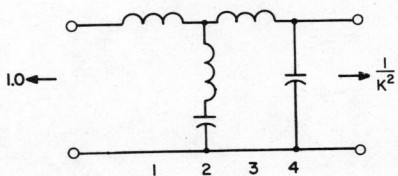

n = 4

ρ = 20%

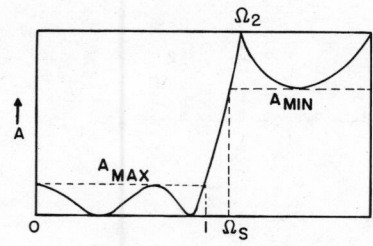

θ	Ω_S	A_MIN	σ_1	σ_3	Ω_1	Ω_2	Ω_3
c	∞	∞	−0.2315236	−0.5589474	1.0798035	∞	0.4472693
11.0	5.664999	82.35	−0.2278270	−0.5615502	1.0797609	6.215646	0.4523695
12.0	5.197666	79.32	−0.2271222	−0.5620470	1.0797514	5.701423	0.4533475
13.0	4.802620	76.52	−0.2263554	−0.5625877	1.0797405	5.266618	0.4544137
14.0	4.464371	73.93	−0.2255264	−0.5631725	1.0797281	4.894214	0.4555688
15.0	4.171563	71.52	−0.2246350	−0.5638016	1.0797140	4.571732	0.4568138
16.0	3.915678	69.26	−0.2236809	−0.5644753	1.0796981	4.289813	0.4581496
17.0	3.690200	67.14	−0.2226639	−0.5651938	1.0796802	4.041300	0.4595772
18.0	3.490065	65.14	−0.2215837	−0.5659574	1.0796600	3.820626	0.4610977
19.0	3.311272	63.24	−0.2204402	−0.5667663	1.0796374	3.623399	0.4627122
20.0	3.150622	61.44	−0.2192330	−0.5676209	1.0796121	3.446101	0.4644221
21.0	3.005526	59.72	−0.2179618	−0.5685214	1.0795839	3.285888	0.4662286
22.0	2.873864	58.08	−0.2166264	−0.5694683	1.0795524	3.140431	0.4681332
23.0	2.753885	56.51	−0.2152263	−0.5704620	1.0795174	3.007807	0.4701374
24.0	2.644133	55.01	−0.2137613	−0.5715028	1.0794785	2.886413	0.4722427
25.0	2.543380	53.57	−0.2122310	−0.5725911	1.0794353	2.774903	0.4744510
26.0	2.450592	52.18	−0.2106351	−0.5737275	1.0793876	2.672139	0.4767640
27.0	2.364885	50.84	−0.2089731	−0.5749124	1.0793348	2.577149	0.4791835
28.0	2.285502	49.54	−0.2072447	−0.5761464	1.0792766	2.489103	0.4817117
29.0	2.211792	48.29	−0.2054495	−0.5774300	1.0792124	2.407283	0.4843506
30.0	2.143189	47.08	−0.2035871	−0.5787638	1.0791417	2.331070	0.4871025
31.0	2.079202	45.91	−0.2016570	−0.5801485	1.0790640	2.259921	0.4899697
32.0	2.019399	44.77	−0.1996587	−0.5815848	1.0789786	2.193363	0.4929548
33.0	1.963403	43.67	−0.1975920	−0.5830734	1.0788850	2.130982	0.4960603
34.0	1.910879	42.59	−0.1954562	−0.5846151	1.0787824	2.072410	0.4992892
35.0	1.861534	41.55	−0.1932510	−0.5862107	1.0786702	2.017322	0.5026441
36.0	1.815103	40.53	−0.1909759	−0.5878612	1.0785474	1.965429	0.5061284
37.0	1.771354	39.54	−0.1886303	−0.5895675	1.0784133	1.916475	0.5097452
38.0	1.730076	38.57	−0.1862139	−0.5913308	1.0782669	1.870229	0.5134979
39.0	1.691083	37.62	−0.1837262	−0.5931520	1.0781073	1.826485	0.5173902
40.0	1.654204	36.69	−0.1811666	−0.5950324	1.0779333	1.785057	0.5214259
41.0	1.619289	35.79	−0.1785348	−0.5969733	1.0777439	1.745777	0.5256089
42.0	1.586200	34.90	−0.1758302	−0.5989761	1.0775378	1.708493	0.5299436
43.0	1.554811	34.03	−0.1730524	−0.6010424	1.0773138	1.673069	0.5344344
44.0	1.525009	33.18	−0.1702010	−0.6031737	1.0770704	1.639380	0.5390860
45.0	1.496692	32.34	−0.1672757	−0.6053718	1.0768061	1.607311	0.5439035
46.0	1.469765	31.52	−0.1642759	−0.6076387	1.0765194	1.576760	0.5488921
47.0	1.444142	30.72	−0.1612014	−0.6099764	1.0762083	1.547632	0.5540573
48.0	1.419745	29.92	−0.1580519	−0.6123872	1.0758712	1.519839	0.5594052
49.0	1.396501	29.14	−0.1548212	−0.6148735	1.0755061	1.493303	0.5649419
50.0	1.374345	28.38	−0.1515269	−0.6174380	1.0751106	1.467949	0.5706741
51.0	1.353215	27.62	−0.1481511	−0.6200837	1.0746827	1.443712	0.5766087
52.0	1.333055	26.88	−0.1446998	−0.6228137	1.0742197	1.420528	0.5827533
53.0	1.313814	26.15	−0.1411730	−0.6256316	1.0737192	1.398341	0.5891156
54.0	1.295444	25.42	−0.1375708	−0.6285412	1.0731782	1.377098	0.5957042
55.0	1.277901	24.71	−0.1338938	−0.6315468	1.0725938	1.356750	0.6025279
56.0	1.261143	24.01	−0.1301423	−0.6346530	1.0719627	1.337251	0.6095961
57.0	1.245134	23.31	−0.1263172	−0.6378650	1.0712816	1.318559	0.6169190
58.0	1.229837	22.63	−0.1224193	−0.6411884	1.0705466	1.300636	0.6245073
59.0	1.215222	21.95	−0.1184499	−0.6446297	1.0697540	1.283446	0.6323724
60.0	1.201256	21.28	−0.1144105	−0.6481957	1.0688996	1.266954	0.6405266
θ	Ω_S	A_MIN	σ_1	σ_3	Ω_1	Ω_2	Ω_3

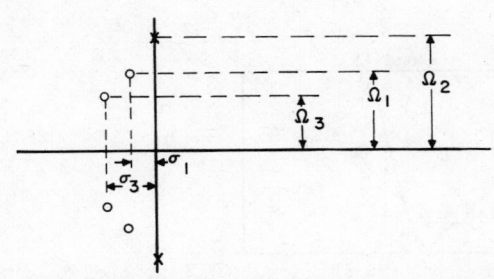

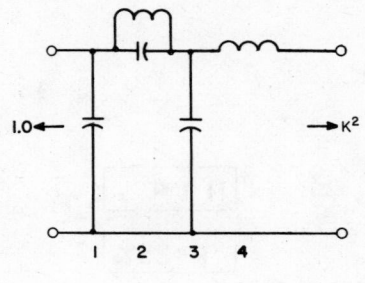

K²=0.6667					K²=∞				
C_1	C_2	L_2	C_3	L_4	C_1	C_2	L_2	C_3	L_4
1.265	0.000000	1.291	1.936	0.8434	0.63253	0.000000	1.27782	1.54262	1.28323
1.260	0.006028	1.284	1.932	0.8437	0.61292	0.02076	1.24651	1.53268	1.28756
1.258	0.008216	1.281	1.930	0.8439	0.60918	0.02480	1.24056	1.53081	1.28838
1.255	0.01074	1.278	1.928	0.8440	0.60509	0.02921	1.23409	1.52878	1.28928
1.253	0.01362	1.275	1.926	0.8442	0.60068	0.03402	1.22710	1.52660	1.29025
1.250	0.01685	1.271	1.924	0.8443	0.59592	0.03923	1.21959	1.52427	1.29129
1.247	0.02043	1.267	1.921	0.8445	0.59082	0.04485	1.21156	1.52179	1.29240
1.243	0.02436	1.263	1.918	0.8448	0.58539	0.05090	1.20302	1.51916	1.29359
1.239	0.02866	1.258	1.915	0.8450	0.57960	0.05738	1.19395	1.51639	1.29485
1.235	0.03333	1.253	1.912	0.8453	0.57347	0.06431	1.18437	1.51348	1.29619
1.231	0.03837	1.247	1.908	0.8456	0.56698	0.07171	1.17427	1.51043	1.29760
1.226	0.04380	1.241	1.904	0.8459	0.56014	0.07959	1.16365	1.50725	1.29908
1.221	0.04961	1.234	1.900	0.8462	0.55294	0.08798	1.15251	1.50394	1.30064
1.216	0.05581	1.227	1.895	0.8465	0.54537	0.09689	1.14085	1.50050	1.30227
1.210	0.06242	1.220	1.891	0.8469	0.53744	0.10634	1.12867	1.49694	1.30397
1.204	0.06944	1.213	1.886	0.8473	0.52913	0.11637	1.11597	1.49326	1.30575
1.198	0.07689	1.205	1.881	0.8477	0.52044	0.12700	1.10276	1.48947	1.30760
1.191	0.08476	1.196	1.875	0.8481	0.51136	0.13826	1.08902	1.48558	1.30952
1.184	0.09309	1.187	1.870	0.8485	0.50190	0.15018	1.07477	1.48159	1.31151
1.177	0.1019	1.178	1.864	0.8490	0.49203	0.16279	1.06001	1.47750	1.31358
1.169	0.1111	1.169	1.858	0.8494	0.48176	0.17615	1.04472	1.47333	1.31571
1.161	0.1209	1.159	1.851	0.8499	0.47107	0.19030	1.02893	1.46908	1.31792
1.153	0.1311	1.148	1.845	0.8505	0.45996	0.20527	1.01262	1.46476	1.32019
1.145	0.1419	1.138	1.838	0.8510	0.44842	0.22114	0.99580	1.46039	1.32253
1.136	0.1532	1.126	1.831	0.8516	0.43644	0.23796	0.97847	1.45596	1.32494
1.127	0.1651	1.115	1.824	0.8521	0.42400	0.25580	0.96063	1.45150	1.32740
1.117	0.1775	1.103	1.816	0.8527	0.41109	0.27473	0.94228	1.44701	1.32993
1.108	0.1906	1.091	1.808	0.8533	0.39771	0.29484	0.92344	1.44251	1.33251
1.097	0.2043	1.078	1.800	0.8540	0.38383	0.31623	0.90409	1.43801	1.33515
1.087	0.2186	1.065	1.792	0.8546	0.36944	0.33899	0.88426	1.43352	1.33783
1.076	0.2337	1.051	1.784	0.8553	0.35453	0.36326	0.86393	1.42908	1.34056
1.065	0.2495	1.038	1.775	0.8560	0.33907	0.38916	0.84312	1.42468	1.34332
1.054	0.2661	1.023	1.766	0.8567	0.32305	0.41686	0.82183	1.42036	1.34612
1.042	0.2835	1.009	1.757	0.8574	0.30644	0.44652	0.80008	1.41614	1.34894
1.030	0.3017	0.9936	1.748	0.8581	0.28921	0.47834	0.77786	1.41205	1.35177
1.017	0.3208	0.9782	1.738	0.8589	0.27135	0.51256	0.75519	1.40810	1.35461
1.005	0.3409	0.9624	1.728	0.8597	0.25282	0.54942	0.73208	1.40434	1.35744
0.9916	0.3621	0.9462	1.718	0.8604	0.23359	0.58924	0.70855	1.40080	1.36026
0.9781	0.3843	0.9296	1.708	0.8612	0.21362	0.63236	0.68460	1.39751	1.36304
0.9643	0.4077	0.9127	1.698	0.8620	0.19287	0.67918	0.66027	1.39453	1.36576
0.9501	0.4323	0.8953	1.687	0.8628	0.17129	0.73017	0.63555	1.39189	1.36842
0.9356	0.4583	0.8776	1.676	0.8637	0.14885	0.78589	0.61049	1.38966	1.37098
0.9207	0.4867	0.8595	1.665	0.8645	0.12547	0.84698	0.58510	1.38790	1.37343
0.9054	0.5147	0.8411	1.654	0.8653	0.10109	0.91420	0.55941	1.38667	1.37573
0.8897	0.5454	0.8222	1.643	0.8661	0.07565	0.98849	0.53346	1.38605	1.37785
0.8737	0.5779	0.8030	1.631	0.8670	0.04906	1.07092	0.50728	1.38614	1.37976
0.8573	0.6124	0.7834	1.620	0.8678	0.02124	1.16282	0.48091	1.38703	1.38141
0.8405	0.6491	0.7635	1.608	0.8686	-0.00794	1.26578	0.45440	1.38885	1.38277
0.8233	0.6881	0.7432	1.596	0.8694	-0.03857	1.38176	0.42781	1.39172	1.38377
0.8057	0.7298	0.7226	1.583	0.8702	-0.07082	1.51316	0.40120	1.39580	1.38436
0.7878	0.7743	0.7016	1.571	0.8710	-0.10483	1.66297	0.37462	1.40127	1.38449
L_1	L_2	C_2	L_3	C_4	L_1	L_2	C_2	L_3	C_4

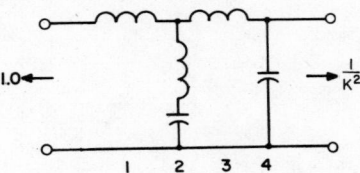

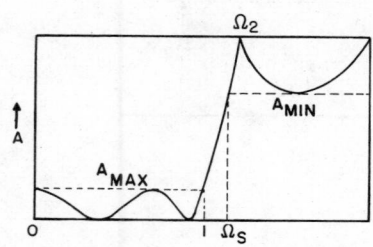

θ	Ω_S	A_{MIN}	σ_1	σ_3	Ω_1	Ω_2	Ω_3
c	∞	∞	-0.2062835	-0.4980125	1.0495570	∞	0.4347407
11.0	5.664999	84.39	-0.2032345	-0.5002303	1.0497315	6.215646	0.4391769
12.0	5.197666	81.36	-0.2026531	-0.5006536	1.0497638	5.701423	0.4400273
13.0	4.802620	78.56	-0.2020203	-0.5011144	1.0497985	5.266618	0.4409542
14.0	4.464371	75.98	-0.2013362	-0.5016129	1.0498356	4.894214	0.4419584
15.0	4.171563	73.56	-0.2006004	-0.5021492	1.0498750	4.571732	0.4430405
16.0	3.915678	71.31	-0.1998127	-0.5027235	1.0499166	4.289813	0.4442013
17.0	3.690200	69.18	-0.1989731	-0.5033361	1.0499602	4.041300	0.4454417
18.0	3.490065	67.18	-0.1980811	-0.5039872	1.0500058	3.820626	0.4467627
19.0	3.311272	65.28	-0.1971367	-0.5046770	1.0500531	3.623399	0.4481651
20.0	3.150622	63.48	-0.1961394	-0.5054058	1.0501020	3.446101	0.4496500
21.0	3.005526	61.76	-0.1950891	-0.5061739	1.0501525	3.285888	0.4512185
22.0	2.873864	60.12	-0.1939855	-0.5069816	1.0502041	3.140431	0.4528719
23.0	2.753885	58.56	-0.1928282	-0.5078293	1.0502569	3.007807	0.4546113
24.0	2.644133	57.05	-0.1916169	-0.5087172	1.0503105	2.886413	0.4564382
25.0	2.543380	55.61	-0.1903513	-0.5096459	1.0503648	2.774903	0.4583539
26.0	2.450592	54.22	-0.1890311	-0.5106156	1.0504194	2.672139	0.4603599
27.0	2.364885	52.88	-0.1876558	-0.5116269	1.0504742	2.577149	0.4624579
28.0	2.285502	51.59	-0.1862251	-0.5126801	1.0505287	2.489103	0.4646494
29.0	2.211792	50.34	-0.1847387	-0.5137758	1.0505828	2.407283	0.4669364
30.0	2.143189	49.13	-0.1831960	-0.5149145	1.0506361	2.331070	0.4693206
31.0	2.079202	47.95	-0.1815967	-0.5160967	1.0506882	2.259921	0.4718041
32.0	2.019399	46.82	-0.1799404	-0.5173230	1.0507386	2.193363	0.4743888
33.0	1.963403	45.71	-0.1782266	-0.5185941	1.0507871	2.130982	0.4770771
34.0	1.910879	44.64	-0.1764548	-0.5199106	1.0508331	2.072410	0.4798712
35.0	1.861534	43.59	-0.1746246	-0.5212732	1.0508761	2.017322	0.4827735
36.0	1.815103	42.57	-0.1727354	-0.5226827	1.0509157	1.965429	0.4857867
37.0	1.771354	41.58	-0.1707868	-0.5241399	1.0509512	1.916475	0.4889135
38.0	1.730076	40.61	-0.1687783	-0.5256457	1.0509820	1.870229	0.4921566
39.0	1.691083	39.66	-0.1667093	-0.5272009	1.0510075	1.826485	0.4955192
40.0	1.654204	38.73	-0.1645794	-0.5288066	1.0510269	1.785057	0.4990045
41.0	1.619289	37.83	-0.1623879	-0.5304639	1.0510396	1.745777	0.5026156
42.0	1.586200	36.94	-0.1601344	-0.5321738	1.0510446	1.708493	0.5063563
43.0	1.554811	36.07	-0.1578183	-0.5339377	1.0510410	1.673069	0.5102303
44.0	1.525009	35.22	-0.1554390	-0.5357567	1.0510280	1.639380	0.5142414
45.0	1.496692	34.38	-0.1529960	-0.5376323	1.0510044	1.607311	0.5183939
46.0	1.469765	33.56	-0.1504888	-0.5395661	1.0509693	1.576760	0.5226922
47.0	1.444142	32.76	-0.1479168	-0.5415597	1.0509213	1.547632	0.5271410
48.0	1.419745	31.96	-0.1452795	-0.5436148	1.0508592	1.519839	0.5317451
49.0	1.396501	31.18	-0.1425764	-0.5457334	1.0507816	1.493303	0.5365099
50.0	1.374345	30.42	-0.1398070	-0.5479175	1.0506869	1.467949	0.5414409
51.0	1.353215	29.66	-0.1369709	-0.5501695	1.0505737	1.443712	0.5465439
52.0	1.333055	28.92	-0.1340676	-0.5524917	1.0504401	1.420528	0.5518253
53.0	1.313814	28.18	-0.1310967	-0.5548869	1.0502844	1.398341	0.5572917
54.0	1.295444	27.46	-0.1280579	-0.5573580	1.0501044	1.377098	0.5629501
55.0	1.277901	26.75	-0.1249510	-0.5599082	1.0498980	1.356750	0.5688081
56.0	1.261143	26.04	-0.1217757	-0.5625410	1.0496628	1.337251	0.5748737
57.0	1.245134	25.35	-0.1185321	-0.5652603	1.0493964	1.318559	0.5811554
58.0	1.229837	24.66	-0.1152201	-0.5680703	1.0490961	1.300636	0.5876625
59.0	1.215222	23.98	-0.1118399	-0.5709758	1.0487589	1.283446	0.5944047
60.0	1.201256	23.30	-0.1083918	-0.5739819	1.0483817	1.266954	0.6013925
θ	Ω_S	A_{MIN}	σ_1	σ_3	Ω_1	Ω_2	Ω_3

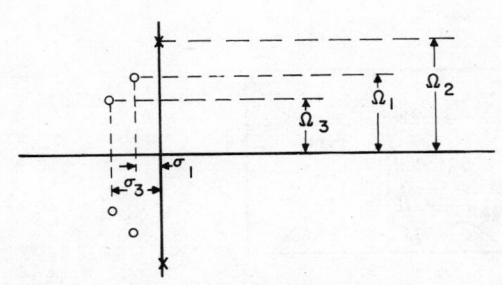
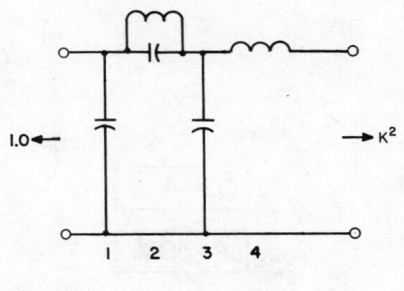

K²=0.600					K²=∞				
C_1	C_2	L_2	C_3	L_4	C_1	C_2	L_2	C_3	L_4
1.420	0.000000	1.260	2.100	0.8519	0.70993	0.000000	1.33510	1.61737	1.30465
1.401	0.02091	1.238	2.084	0.8529	0.69118	0.01983	1.30513	1.60678	1.30825
1.397	0.02494	1.234	2.081	0.8531	0.68760	0.02367	1.29943	1.60478	1.30893
1.393	0.02933	1.229	2.077	0.8533	0.68370	0.02788	1.29324	1.60262	1.30968
1.389	0.03411	1.224	2.074	0.8535	0.67948	0.03245	1.28655	1.60029	1.31048
1.384	0.03926	1.219	2.070	0.8538	0.67495	0.03740	1.27936	1.59779	1.31135
1.380	0.04479	1.213	2.066	0.8540	0.67009	0.04273	1.27168	1.59513	1.31228
1.374	0.05072	1.207	2.061	0.8543	0.66491	0.04846	1.26350	1.59231	1.31326
1.369	0.05706	1.201	2.057	0.8546	0.65940	0.05459	1.25483	1.58933	1.31431
1.363	0.06379	1.194	2.052	0.8549	0.65356	0.06115	1.24566	1.58619	1.31542
1.357	0.07095	1.187	2.047	0.8553	0.64740	0.06813	1.23600	1.58290	1.31660
1.350	0.07853	1.179	2.041	0.8556	0.64089	0.07555	1.22583	1.57946	1.31783
1.344	0.08655	1.171	2.035	0.8560	0.63405	0.08344	1.21518	1.57587	1.31912
1.336	0.09502	1.163	2.030	0.8563	0.62688	0.09180	1.20402	1.57213	1.32048
1.329	0.1039	1.155	2.023	0.8567	0.61935	0.10066	1.19237	1.56825	1.32189
1.321	0.1133	1.146	2.017	0.8571	0.61148	0.11004	1.18023	1.56424	1.32337
1.313	0.1232	1.136	2.010	0.8576	0.60326	0.11995	1.16759	1.56008	1.32491
1.305	0.1336	1.127	2.003	0.8580	0.59469	0.13042	1.15445	1.55580	1.32651
1.296	0.1445	1.117	1.996	0.8585	0.58575	0.14148	1.14082	1.55139	1.32817
1.287	0.1560	1.106	1.988	0.8590	0.57645	0.15316	1.12670	1.54686	1.32989
1.278	0.1680	1.096	1.981	0.8595	0.56677	0.16548	1.11208	1.54222	1.33167
1.268	0.1806	1.084	1.973	0.8600	0.55672	0.17849	1.09696	1.53746	1.33351
1.258	0.1937	1.073	1.965	0.8605	0.54629	0.19222	1.08136	1.53261	1.33540
1.247	0.2075	1.061	1.956	0.8611	0.53546	0.20672	1.06526	1.52765	1.33736
1.237	0.2220	1.049	1.947	0.8616	0.52424	0.22203	1.04868	1.52261	1.33937
1.226	0.2371	1.036	1.938	0.8622	0.51262	0.23820	1.03160	1.51748	1.34143
1.214	0.2530	1.023	1.929	0.8628	0.50058	0.25529	1.01404	1.51228	1.34355
1.203	0.2696	1.010	1.920	0.8634	0.48811	0.27336	0.99599	1.50702	1.34572
1.191	0.2870	0.9963	1.910	0.8640	0.47521	0.29249	0.97746	1.50171	1.34793
1.178	0.3052	0.9823	1.900	0.8647	0.46187	0.31275	0.95845	1.49635	1.35019
1.165	0.3243	0.9679	1.890	0.8653	0.44807	0.33423	0.93897	1.49096	1.35250
1.152	0.3443	0.9531	1.879	0.8660	0.43381	0.35703	0.91901	1.48555	1.35484
1.139	0.3652	0.9380	1.869	0.8667	0.41905	0.38126	0.89857	1.48015	1.35722
1.125	0.3873	0.9225	1.858	0.8674	0.40704	0.40704	0.87768	1.47475	1.35963
1.111	0.4104	0.9066	1.847	0.8681	0.38804	0.43451	0.85632	1.46938	1.36206
1.096	0.4347	0.8904	1.836	0.8688	0.37174	0.46384	0.83451	1.46407	1.36451
1.082	0.4603	0.8739	1.824	0.8695	0.35488	0.49519	0.81226	1.45882	1.36696
1.066	0.4872	0.8570	1.812	0.8702	0.33745	0.52878	0.78957	1.45367	1.36943
1.051	0.5156	0.8397	1.800	0.8710	0.31942	0.56484	0.76644	1.44863	1.37188
1.035	0.5455	0.8221	1.788	0.8717	0.30076	0.60363	0.74290	1.44374	1.37431
1.019	0.5771	0.8041	1.776	0.8724	0.28144	0.64547	0.71896	1.43903	1.37671
1.002	0.6106	0.7858	1.763	0.8732	0.26143	0.69070	0.69462	1.43453	1.37907
0.9847	0.6460	0.7671	1.750	0.8739	0.24070	0.73975	0.66991	1.43028	1.38137
0.9672	0.6836	0.7481	1.737	0.8747	0.21919	0.79309	0.64484	1.42633	1.38358
0.9493	0.7236	0.7287	1.724	0.8754	0.19688	0.85129	0.61943	1.42272	1.38569
0.9309	0.7662	0.7090	1.711	0.8761	0.17370	0.91501	0.59371	1.41952	1.38768
0.9122	0.8116	0.6890	1.697	0.8768	0.14961	0.98504	0.56770	1.41677	1.38951
0.8931	0.8602	0.6686	1.684	0.8775	0.12453	1.06231	0.54144	1.41455	1.39117
0.8735	0.9123	0.6479	1.670	0.8781	0.09839	1.14793	0.51496	1.41295	1.39260
0.8535	0.9683	0.6269	1.656	0.8787	0.07112	1.24327	0.48829	1.41204	1.39378
0.8331	1.029	0.6056	1.642	0.8793	0.04261	1.34995	0.46149	1.41195	1.39466
L_1	L_2	C_2	L_3	C_4	L_1	L_2	C_2	L_3	C_4

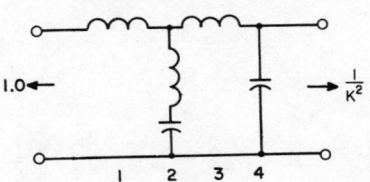

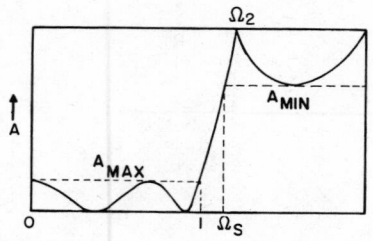

n = 4

ρ = 50%

θ	Ω_S	A_MIN	σ_1	σ_3	Ω_1	Ω_2	Ω_3
c	∞	∞	-0.1282831	-0.3097029	0.9744071	∞	0.4036127
11.0	5.664999	91.38	-0.1267466	-0.3109331	0.9750421	6.215646	0.4065895
12.0	5.197666	88.35	-0.1264533	-0.3111680	0.9751629	5.701423	0.4071596
13.0	4.802620	85.55	-0.1261341	-0.3114238	0.9752944	5.266618	0.4077808
14.0	4.464371	82.96	-0.1257888	-0.3117006	0.9754363	4.894214	0.4084536
15.0	4.171563	80.55	-0.1254174	-0.3119984	0.9755889	4.571732	0.4091783
16.0	3.915678	78.30	-0.1250196	-0.3123173	0.9757521	4.289813	0.4099554
17.0	3.690200	76.17	-0.1245955	-0.3126576	0.9759259	4.041300	0.4107855
18.0	3.490065	74.17	-0.1241447	-0.3130194	0.9761103	3.820626	0.4116691
19.0	3.311272	72.27	-0.1236672	-0.3134027	0.9763054	3.623399	0.4126068
20.0	3.150622	70.47	-0.1231627	-0.3138079	0.9765111	3.446101	0.4135991
21.0	3.005526	68.75	-0.1226312	-0.3142349	0.9767275	3.285888	0.4146468
22.0	2.873864	67.11	-0.1220724	-0.3146842	0.9769546	3.140431	0.4157505
23.0	2.753885	65.55	-0.1214861	-0.3151557	0.9771923	3.007807	0.4169111
24.0	2.644133	64.04	-0.1208721	-0.3156498	0.9774408	2.886413	0.4181292
25.0	2.543380	62.60	-0.1202301	-0.3161667	0.9777000	2.774903	0.4194058
26.0	2.450592	61.21	-0.1195600	-0.3167065	0.9779699	2.672139	0.4207418
27.0	2.364885	59.87	-0.1188615	-0.3172697	0.9782505	2.577149	0.4221381
28.0	2.285502	58.58	-0.1181344	-0.3178563	0.9785419	2.489103	0.4235956
29.0	2.211792	57.33	-0.1173783	-0.3184668	0.9788440	2.407283	0.4251155
30.0	2.143189	56.12	-0.1165930	-0.3191014	0.9791568	2.331070	0.4266988
31.0	2.079202	54.94	-0.1157782	-0.3197604	0.9794804	2.259921	0.4283468
32.0	2.019399	53.81	-0.1149335	-0.3204441	0.9798146	2.193363	0.4300607
33.0	1.963403	52.70	-0.1140587	-0.3211529	0.9801596	2.130982	0.4318418
34.0	1.910879	51.63	-0.1131535	-0.3218872	0.9805153	2.072410	0.4336914
35.0	1.861534	50.58	-0.1122174	-0.3226474	0.9808816	2.017322	0.4356110
36.0	1.815103	49.56	-0.1112501	-0.3234339	0.9812585	1.965429	0.4376021
37.0	1.771354	48.57	-0.1102513	-0.3242470	0.9816461	1.916475	0.4396664
38.0	1.730076	47.60	-0.1092205	-0.3250874	0.9820441	1.870229	0.4418056
39.0	1.691083	46.65	-0.1081573	-0.3259554	0.9824527	1.826485	0.4440213
40.0	1.654204	45.72	-0.1070613	-0.3268517	0.9828717	1.785057	0.4463156
41.0	1.619289	44.82	-0.1059321	-0.3277767	0.9833010	1.745777	0.4486904
42.0	1.586200	43.93	-0.1047691	-0.3287310	0.9837406	1.708493	0.4511479
43.0	1.554811	43.06	-0.1035720	-0.3297153	0.9841904	1.673069	0.4536901
44.0	1.525009	42.21	-0.1023401	-0.3307302	0.9846502	1.639380	0.4563195
45.0	1.496692	41.37	-0.1010731	-0.3317765	0.9851199	1.607311	0.4590386
46.0	1.469765	40.55	-0.0997703	-0.3328548	0.9855993	1.576760	0.4618500
47.0	1.444142	39.74	-0.0984313	-0.3339660	0.9860884	1.547632	0.4647564
48.0	1.419745	38.95	-0.0970553	-0.3351109	0.9865868	1.519839	0.4677608
49.0	1.396501	38.17	-0.0956419	-0.3362905	0.9870944	1.493303	0.4708662
50.0	1.374345	37.40	-0.0941904	-0.3375056	0.9876110	1.467949	0.4740761
51.0	1.353215	36.65	-0.0927002	-0.3387575	0.9881361	1.443712	0.4773939
52.0	1.333055	35.90	-0.0911706	-0.3400471	0.9886696	1.420528	0.4808232
53.0	1.313814	35.17	-0.0896009	-0.3413756	0.9892111	1.398341	0.4843682
54.0	1.295444	34.44	-0.0879905	-0.3427445	0.9897601	1.377098	0.4880330
55.0	1.277901	33.73	-0.0863385	-0.3441551	0.9903164	1.356750	0.4918221
56.0	1.261143	33.02	-0.0846443	-0.3456089	0.9908793	1.337251	0.4957404
57.0	1.245134	32.33	-0.0829070	-0.3471075	0.9914483	1.318559	0.4997930
58.0	1.229837	31.64	-0.0811259	-0.3486528	0.9920228	1.300636	0.5039854
59.0	1.215222	30.95	-0.0793002	-0.3502466	0.9926021	1.283446	0.5083237
60.0	1.201256	30.28	-0.0774290	-0.3518912	0.9931856	1.266954	0.5128141
θ	Ω_S	A_MIN	σ_1	σ_3	Ω_1	Ω_2	Ω_3

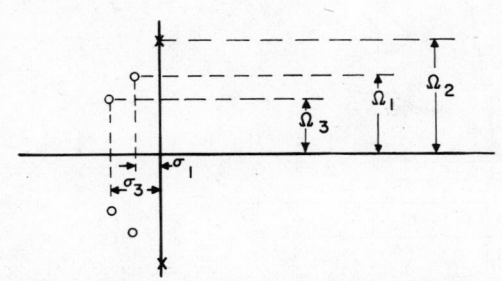

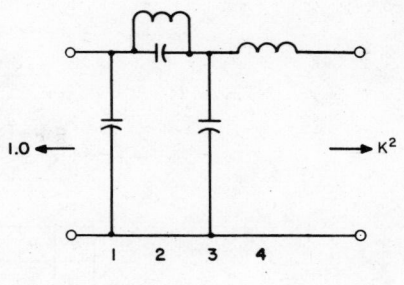

K²=0.3333					K²=∞				
C_1	C_2	L_2	C_3	L_4	C_1	C_2	L_2	C_3	L_4
2.283	0.000000	1.011	3.034	0.7611	1.14159	0.000000	1.40190	1.98851	1.25691
2.259	0.02600	0.9956	3.013	0.7616	1.12377	0.01879	1.37723	1.97504	1.25856
2.254	0.03099	0.9926	3.008	0.7617	1.12038	0.02241	1.37255	1.97249	1.25887
2.249	0.03644	0.9893	3.004	0.7618	1.11669	0.02636	1.36745	1.96972	1.25921
2.244	0.04235	0.9858	2.999	0.7619	1.11270	0.03065	1.36195	1.96673	1.25958
2.238	0.04872	0.9820	2.994	0.7621	1.10842	0.03528	1.35604	1.96352	1.25998
2.232	0.05557	0.9779	2.988	0.7622	1.10383	0.04026	1.34972	1.96010	1.26040
2.225	0.06289	0.9736	2.982	0.7623	1.09895	0.04559	1.34299	1.95645	1.26085
2.218	0.07070	0.9690	2.976	0.7625	1.09377	0.05128	1.33586	1.95260	1.26133
2.211	0.07900	0.9641	2.969	0.7627	1.08828	0.05734	1.32832	1.94852	1.26184
2.203	0.08781	0.9590	2.962	0.7628	1.08249	0.06377	1.32037	1.94424	1.26237
2.195	0.09712	0.9536	2.955	0.7630	1.07640	0.07059	1.31202	1.93974	1.26294
2.186	0.1070	0.9480	2.947	0.7632	1.07001	0.07780	1.30326	1.93503	1.26353
2.177	0.1173	0.9421	2.939	0.7634	1.06331	0.08542	1.29409	1.93011	1.26415
2.168	0.1282	0.9359	2.931	0.7636	1.05630	0.09344	1.28452	1.92499	1.26479
2.158	0.1397	0.9295	2.922	0.7638	1.04899	0.10189	1.27454	1.91965	1.26547
2.148	0.1518	0.9228	2.913	0.7641	1.04136	0.11079	1.26415	1.91411	1.26617
2.137	0.1644	0.9159	2.903	0.7643	1.03343	0.12013	1.25336	1.90837	1.26691
2.126	0.1776	0.9087	2.893	0.7645	1.02518	0.12994	1.24216	1.90242	1.26767
2.115	0.1915	0.9012	2.883	0.7648	1.01662	0.14023	1.23056	1.89627	1.26846
2.103	0.2060	0.8935	2.873	0.7651	1.00774	0.15102	1.21855	1.88993	1.26927
2.091	0.2211	0.8855	2.862	0.7653	0.99855	0.16234	1.20613	1.88339	1.27012
2.078	0.2369	0.8772	2.851	0.7656	0.98903	0.17419	1.19331	1.87665	1.27099
2.065	0.2535	0.8687	2.839	0.7659	0.97919	0.18661	1.18009	1.86973	1.27189
2.052	0.2707	0.8600	2.827	0.7662	0.96903	0.19961	1.16646	1.86261	1.27282
2.038	0.2888	0.8509	2.815	0.7665	0.95854	0.21322	1.15242	1.85531	1.27377
2.023	0.3076	0.8417	2.802	0.7668	0.94771	0.22748	1.13798	1.84782	1.27475
2.009	0.3272	0.8321	2.789	0.7672	0.93655	0.24242	1.12314	1.84016	1.27576
1.994	0.3477	0.8223	2.776	0.7675	0.92506	0.25806	1.10789	1.83231	1.27680
1.978	0.3691	0.8122	2.762	0.7678	0.91322	0.27444	1.09224	1.82430	1.27785
1.962	0.3914	0.8019	2.748	0.7682	0.90104	0.29161	1.07619	1.81611	1.27894
1.946	0.4147	0.7913	2.734	0.7686	0.88850	0.30962	1.05974	1.80776	1.28005
1.929	0.4390	0.7804	2.719	0.7689	0.87562	0.32850	1.04288	1.79925	1.28118
1.912	0.4644	0.7693	2.704	0.7693	0.86237	0.34832	1.02562	1.79058	1.28233
1.894	0.4909	0.7580	2.689	0.7697	0.84876	0.36914	1.00797	1.78175	1.28351
1.876	0.5187	0.7463	2.673	0.7701	0.83477	0.39102	0.98991	1.77279	1.28470
1.857	0.5477	0.7344	2.657	0.7705	0.82041	0.41404	0.97146	1.76368	1.28591
1.838	0.5781	0.7223	2.641	0.7709	0.80567	0.43828	0.95261	1.75443	1.28714
1.819	0.6099	0.7099	2.624	0.7713	0.79053	0.46382	0.93337	1.74506	1.28838
1.799	0.6432	0.6972	2.607	0.7717	0.77500	0.49078	0.91373	1.73557	1.28963
1.779	0.6782	0.6842	2.590	0.7721	0.75905	0.51926	0.89370	1.72597	1.29089
1.758	0.7150	0.6710	2.573	0.7726	0.74269	0.54940	0.87328	1.71627	1.29216
1.737	0.7536	0.6576	2.555	0.7730	0.72589	0.58133	0.85247	1.70647	1.29343
1.715	0.7943	0.6439	2.536	0.7734	0.70866	0.61522	0.83128	1.69659	1.29470
1.693	0.8371	0.6299	2.518	0.7739	0.69097	0.65125	0.80970	1.68664	1.29596
1.671	0.8824	0.6157	2.499	0.7743	0.67282	0.68962	0.78775	1.67663	1.29720
1.648	0.9302	0.6012	2.480	0.7747	0.65418	0.73059	0.76543	1.66657	1.29843
1.624	0.9808	0.5864	2.461	0.7752	0.63505	0.77441	0.74273	1.65649	1.29964
1.600	1.035	0.5714	2.441	0.7756	0.61539	0.82140	0.71967	1.64639	1.30081
1.576	1.092	0.5561	2.421	0.7760	0.59520	0.87191	0.69626	1.63630	1.30194
1.551	1.152	0.5406	2.400	0.7764	0.57445	0.92638	0.67250	1.62624	1.30302
L_1	L_2	C_2	L_3	C_4	L_1	L_2	C_2	L_3	C_4

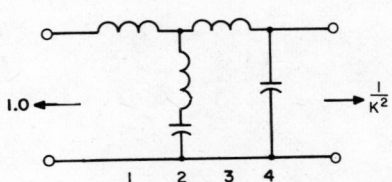

201

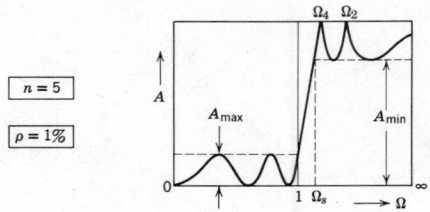

$n = 5$

$\rho = 1\%$

θ	Ω_K	A_{MIN}	σ_0	σ_1	σ_3	Ω_1	Ω_2	Ω_3	Ω_4
c	∞	∞	1.26940	-.39226	-1.02697	1.5368	∞	0.9498	∞
2.0	28.6537	153.87	1.27041	-0.39155	-1.02676	1.5364	48.7389	0.9507	30.1274
3.0	19.1073	136.26	1.27168	-0.39067	-1.02650	1.5359	32.4927	0.9519	20.0893
4.0	14.3356	123.75	1.27345	-0.38943	-1.02614	1.5352	24.3697	0.9536	15.0716
5.0	11.4737	114.05	1.27574	-0.38785	-1.02567	1.5342	19.4959	0.9557	12.0620
6.0	9.5668	106.12	1.27855	-0.38592	-1.02508	1.5330	16.2468	0.9583	10.0565
7.0	8.2055	99.41	1.28189	-0.38364	-1.02437	1.5317	13.9260	0.9614	8.6247
8.0	7.1853	93.60	1.28576	-0.38103	-1.02354	1.5301	12.1854	0.9649	7.5516
9.0	6.3925	88.46	1.29017	-0.37809	-1.02258	1.5283	10.8316	0.9690	6.7175
10.0	5.7588	83.87	1.29514	-0.37481	-1.02147	1.5262	9.7486	0.9735	6.0507
11.0	5.2408	79.70	1.30068	-0.37121	-1.02021	1.5240	8.8625	0.9785	5.5057
12.0	4.8097	75.90	1.30679	-0 36730	-1.01879	1.5215	8.1241	0.9840	5.0520
13.0	4.4454	72.40	1.31350	-0.36308	-1.01720	1.5189	7.4993	0.9900	4.6684
14.0	4.1336	69.15	1.32083	-0.35856	-1.01542	1.5160	6.9638	0.9965	4.3401
15.0	3.8637	66.12	1.32878	-0.35375	-1.01345	1.5129	6.4997	1.0034	4.0559
16.0	3.6280	63.28	1.33739	-0.34865	-1.01125	1.5095	6.0936	1.0109	3.8076
17.0	3.4203	60.61	1.34668	-0.34328	-1.00882	1.5060	5.7353	1.0188	3.5888
18.0	3.2361	58.09	1.35667	-0.33764	-1.00614	1.5022	5.4168	1.0273	3.3946
19.0	3.0716	55.70	1.36738	-0.33175	-1.00319	1.4982	5.1318	1.0362	3.2212
20.0	2.9238	53.43	1.37886	-0.32562	-0.99994	1.4939	4.8753	1.0457	3.0654
21.0	2.7904	51.26	1.39114	-0.31925	-0.99638	1.4895	4.6433	1.0556	2.9246
22.0	2.6695	49.19	1.40425	-0.31267	-0.99248	1.4848	4.4323	1.0661	2.7970
23.0	2.5593	47.21	1.41823	-0.30587	-0.98820	1.4799	4.2397	1.0770	2.6807
24.0	2.4586	45.30	1.43314	-0.29888	-0.98354	1.4747	4.0631	1.0885	2.5743
25.0	2.3662	43.47	1.44901	-0.29171	-0.97845	1.4693	3.9007	1.1004	2.4767
26.0	2.2812	41.71	1.46591	-0.28437	-0.97289	1.4638	3.7507	1.1128	2.3868
27.0	2.2027	40.01	1.48390	-0.27687	-0.96685	1.4579	3.6119	1.1257	2.3038
28.0	2.1301	38.36	1.50304	-0.26923	-0.96028	1.4519	3.4829	1.1391	2.2270
29.0	2.0627	36.77	1.52340	-0.26146	-0.95315	1.4456	3.3629	1.1529	2.1556
30.0	2.0000	35.23	1.54507	-0.25357	-0.94542	1.4392	3.2508	1.1671	2.0892
31.0	1.9416	33.73	1.56814	-0.24559	-0.93704	1.4325	3.1460	1.1818	2.0274
32.0	1.8871	32.28	1.59271	-0.23751	-0.92797	1.4256	3.0476	1.1968	1.9695
33.0	1.8361	30.86	1.61889	-0.22936	-0.91818	1.4184	2.9553	1.2123	1.9154
34.0	1.7883	29.49	1.64680	-0.22116	-0.90761	1.4111	2.8683	1.2281	1.8646
35.0	1.7434	28.14	1.67659	-0.21291	-0.89622	1.4036	2.7864	1.2441	1.8170
36.0	1.7013	26.83	1.70839	-0.20463	-0.88396	1.3959	2.7089	1.2605	1.7722
37.0	1.6616	25.55	1.74240	-0.19634	-0.87079	1.3879	2.6356	1.2770	1.7299
38.0	1.6243	24.30	1.77880	-0.18805	-0.85666	1.3798	2.5662	1.2938	1.6901
39.0	1.5890	23.08	1.81780	-0.17978	-0.84152	1.3715	2.5003	1.3106	1.6525
40.0	1.5557	21.89	1.85965	-0.17153	-0.82534	1.3631	2.4377	1.3275	1.6170
41.0	1.5243	20.72	1.90463	-0.16333	-0.80807	1.3544	2.3781	1.3443	1.5833
42.0	1.4945	19.57	1.95305	-0.15518	-0.78967	1.3456	2.3213	1.3610	1.5515
43.0	1.4663	18.45	2.00525	-0.14711	-0.77012	1.3366	2.2672	1.3775	1.5213
44.0	1.4396	17.36	2.06164	-0.13913	-0.74940	1.3275	2.2154	1.3937	1.4926
45.0	1.4142	16.28	2.12266	-0.13126	-0.72749	1.3182	2.1660	1.4094	1.4654
46.0	1.3902	15.23	2.18882	-0.12350	-0.70438	1.3088	2.1187	1.4246	1.4396
47.0	1.3673	14.20	2.26069	-0.11588	-0.68010	1.2992	2.0733	1.4390	1.4150
48.0	1.3456	13.20	2.33894	-0.10840	-0.65467	1.2895	2.0299	1.4527	1.3916
49.0	1.3250	12.22	2.42429	-0.10110	-0.62814	1.2797	1.9881	1.4653	1.3693
50.0	1.3054	11.27	2.51758	-0.09398	-0.60057	1.2698	1.9480	1.4768	1.3481
51.0	1.2868	10.35	2.61976	-0.08706	-0.57207	1.2598	1.9095	1.4870	1.3279
52.0	1.2690	9.45	2.73188	-0.08035	-0.54275	1.2498	1.8724	1.4958	1.3087
53.0	1.2521	8.59	2.85511	-0.07388	-0.51276	1.2396	1.8366	1.5029	1.2903
54.0	1.2361	7.76	2.99077	-0.06767	-0.48228	1.2294	1.8021	1.5084	1.2728
55.0	1.2208	6.98	3.14032	-0.06172	-0.45151	1.2192	1.7689	1.5120	1.2561
56.0	1.2062	6.23	3.30533	-0.05605	-0.42066	1.2090	1.7368	1.5136	1.2402
57.0	1.1924	5.52	3.48754	-0.05068	-0.39000	1.1988	1.7057	1.5133	1.2250
58.0	1.1792	4.87	3.68883	-0.04562	-0.35975	1.1886	1.6757	1.5109	1.2104
59.0	1.1666	4.26	3.91120	-0.04088	-0.33018	1.1785	1.6467	1.5065	1.1966
60.0	1.1547	3.70	4.15678	-0.03647	-0.30152	1.1685	1.6185	1.5002	1.1834
θ	Ω_K	A_{MIN}	σ_0	σ_1	σ_3	Ω_1	Ω_2	Ω_3	Ω_4

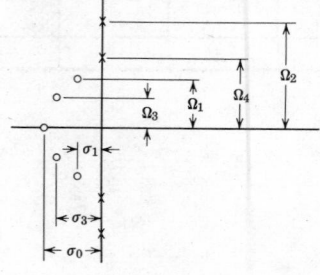

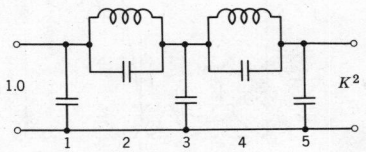

K²=1.0							K²=∞						
c_1	c_2	L_2	c_3	c_4	L_4	c_5	c_1	c_2	L_2	c_3	c_4	L_4	c_5
0.4869	0.0000	1.050	1.225	0.0000	1.050	0.4869	0.2434	0.0000	0.6651	0.9495	0.0000	1.0882	0.9561
0.4865	0.0004	1.0492	1.2248	0.0010	1.0481	0.4859	0.2428	0.0006	0.6644	0.9487	0.0010	1.0869	0.9557
0.4861	0.0009	1.0487	1.2239	0.0023	1.0462	0.4847	0.2421	0.0014	0.6635	0.9477	0.0023	1.0852	0.9551
0.4856	0.0016	1.0480	1.2228	0.0042	1.0435	0.4830	0.2411	0.0025	0.6622	0.9462	0.0041	1.0828	0.9543
0.4849	0.0025	1.0471	1.2213	0.0066	1.0402	0.4809	0.2398	0.0040	0.6606	0.9442	0.0064	1.0707	0.9533
0.4841	0.0036	1.0460	1.2195	0.0095	1.0360	0.4782	0.2382	0.0058	0.6586	0.9419	0.0092	1.0760	0.9520
0.4831	0.0049	1.0448	1.2173	0.0130	1.0312	0.4751	0.2362	0.0079	0.6562	0.9391	0.0125	1.0716	0.9505
0.4819	0.0064	1.0433	1.2149	0.0171	1.0255	0.4715	0.2340	0.0103	0.6534	0.9359	0.0164	1.0664	0.9488
0.4806	0.0081	1.0417	1.2121	0.0217	1.0192	0.4674	0.2315	0.0131	0.6503	0.9323	0.0209	1.0606	0.9468
0.4792	0.0101	1.0398	1.2091	0.0269	1.0120	0.4628	0.2286	0.0163	0.6467	0.9283	0.0259	1.0542	0.9446
0.4776	0.0122	1.0378	1.2057	0.0328	1.0042	0.4577	0.2255	0.0198	0.6429	0.9239	0.0315	1.0470	0.9422
0.4758	0.0146	1.0355	1.2020	0.0393	0.9956	0.4521	0.2220	0.0237	0.6386	0.9191	0.0377	1.0392	0.9395
0.4739	0.0172	1.0331	1.1980	0.0465	0.9862	0.4461	0.2182	0.0280	0.6340	0.9139	0.0445	1.0307	0.9367
0.4718	0.0200	1.0305	1.1938	0.0543	0.9761	0.4394	0.2141	0.0328	0.6290	0.9083	0.0520	1.0215	0.9336
0.4696	0.0230	1.0277	1.1893	0.0629	0.9652	0.4323	0.2096	0.0380	0.6237	0.9023	0.0601	1.0116	0.9303
0.4672	0.0262	1.0246	1.1845	0.0723	0.9537	0.4247	0.2048	0.0436	0.6179	0.8959	0.0689	1.0011	0.9267
0.4646	0.0297	1.0214	1.1794	0.0824	0.9413	0.4165	0.1996	0.0497	0.6118	0.8892	0.0784	0.9899	0.9230
0.4619	0.0334	1.0180	1.1742	0.0934	0.9283	0.4077	0.1941	0.0563	0.6053	0.8821	0.0887	0.9781	0.9190
0.4590	0.0374	1.0144	1.1686	0.1053	0.9145	0.3984	0.1882	0.0634	0.5985	0.8746	0.0998	0.9656	0.9148
0.4559	0.0416	1.0106	1.1629	0.1182	0.8999	0.3886	0.1819	0.0712	0.5912	0.8668	0.1117	0.9524	0.9104
0.4527	0.0460	1.0066	1.1570	0.1321	0.8847	0.3781	0.1752	0.0795	0.5836	0.8587	0.1246	0.9385	0.9058
0.4493	0.0507	1.0023	1.1509	0.1471	0.8687	0.3671	0.1681	0.0884	0.5756	0.8502	0.1383	0.9240	0.9010
0.4458	0.0557	0.9979	1.1446	0.1633	0.8520	0.3555	0.1605	0.0981	0.5673	0.8415	0.1531	0.9089	0.8960
0.4420	0.0609	0.9933	1.1381	0.1808	0.8346	0.3432	0.1525	0.1084	0.5585	0.8324	0.1690	0.8931	0.8908
0.4381	0.0664	0.9884	1.1316	0.1996	0.8164	0.3303	0.1441	0.1196	0.5494	0.8231	0.1860	0.8766	0.8854
0.4341	0.0722	0.9833	1.1249	0.2200	0.7976	0.3168	0.1352	0.1316	0.5399	0.8135	0.2042	0.8595	0.8798
0.4298	0.0783	0.9780	1.1182	0.2421	0.7781	0.3025	0.1258	0.1446	0.5301	0.8036	0.2238	0.8418	0.8740
0.4254	0.0847	0.9724	1.1114	0.2660	0.7579	0.2876	0.1158	0.1586	0.5198	0.7935	0.2449	0.8234	0.8681
0.4207	0.0914	0.9666	1.1046	0.2919	0.7371	0.2719	0.1053	0.1737	0.5092	0.7833	0.2675	0.8044	0.8620
0.4159	0.0985	0.9605	1.0979	0.3201	0.7156	0.2555	0.0942	0.1899	0.4982	0.7728	0.2919	0.7848	0.8558
0.4109	0.1058	0.9542	1.0912	0.3508	0.6934	0.2382	0.0825	0.2076	0.4868	0.7622	0.3182	0.7645	0.8494
0.4057	0.1136	0.9475	1.0846	0.3843	0.6707	0.2202	0.0702	0.2266	0.4751	0.7514	0.3467	0.7436	0.8428
0.4003	0.1217	0.9406	1.0781	0.4210	0.6473	0.2012	0.0572	0.2473	0.4629	0.7405	0.3775	0.7221	0.8362
0.3947	0.1302	0.9333	1.0719	0.4613	0.6234	0.1814	0.0434	0.2699	0.4504	0.7295	0.4109	0.7000	0.8295
0.3889	0.1391	0.9256	1.0659	0.5056	0.5990	0.1605	0.0289	0.2944	0.4375	0.7185	0.4472	0.6773	0.8227
0.3829	0.1485	0.9176	1.0602	0.5546	0.5740	0.1387	0.0135	0.3212	0.4243	0.7075	0.4868	0.6541	0.8158
0.3767	0.1583	0.9092	1.0550	0.6090	0.5486	0.1157	-0.0027	0.3506	0.4107	0.6965	0.5302	0.6302	0.8089
0.3702	0.1686	0.9003	1.0502	0.6696	0.5228	0.0916	-0.0199	0.3828	0.3967	0.6856	0.5778	0.6059	0.8021
0.3635	0.1795	0.8909	1.0461	0.7373	0.4966	0.0663	-0.0382	0.4184	0.3823	0.6747	0.6303	0.5810	0.7952
0.3566	0.1910	0.8809	1.0426	0.8135	0.4701	0.0396	-0.0576	0.4577	0.3676	0.6641	0.6884	0.5556	0.7885
0.3495	0.2031	0.8704	1.0399	0.8996	0.4433	0.0115							
0.3422	0.2159	0.8592	1.0382	0.9975	0.4164	-0.0181							
0.3346	0.2296	0.8472	1.0376	1.1094	0.3894	-0.0495							
0.3268	0.2441	0.8345	1.0383	1.2381	0.3625	-0.0826							
0.3187	0.2596	0.8210	1.0405	1.3872	0.3356	-0.1178							
L_1	L_2	C_2	L_3	L_4	C_4	L_5	L_1	L_2	C_2	L_3	L_4	C_4	L_5

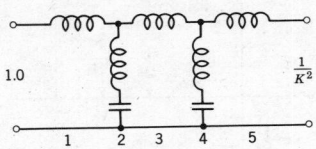

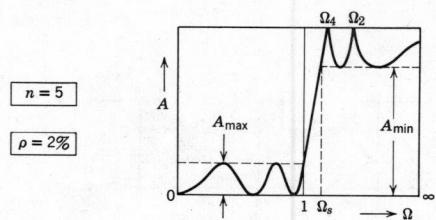

θ	Ω_K	A_{MIN}	σ_0	σ_1	σ_3	Ω_1	Ω_2	Ω_3	Ω_4
C	∞	∞	1.05686	-0.32659	-0.85502	1.3838	∞	0.8552	∞
2.0	28.6537	159.89	1.05754	-0.32612	-0.85489	1.3835	48.7389	0.8558	30.1274
3.0	19.1073	142.28	1.05839	-0.32555	-0.85474	1.3832	32.4927	0.8566	20.0893
4.0	14.3356	129.78	1.05959	-0.32474	-0.85453	1.3827	24.3697	0.8577	15.0716
5.0	11.4737	120.07	1.06113	-0.32371	-0.85425	1.3821	19.4959	0.8592	12.0620
6.0	9.5668	112.14	1.06302	-0.32245	-0.85390	1.3815	16.2468	0.8609	10.0565
7.0	8.2055	105.44	1.06527	-0.32096	-0.85349	1.3806	13.9260	0.8630	8.6247
8.0	7.1853	99.62	1.06787	-0.31925	-0.85300	1.3797	12.1854	0.8655	7.5516
9.0	6.3925	94.49	1.07083	-0.31732	-0.85244	1.3786	10.8316	0.8682	6.7175
10.0	5.7588	89.89	1.07416	-0.31517	-0.85180	1.3774	9.7486	0.8713	6.0507
11.0	5.2408	85.73	1.07787	-0.31280	-0.85107	1.3761	8.8625	0.8747	5.5057
12.0	4.8097	81.92	1.08195	-0.31022	-0.85025	1.3746	8.1241	0.8784	5.0520
13.0	4.4454	78.42	1.08643	-0.30744	-0.84933	1.3730	7.4993	0.8825	4.6684
14.0	4.1336	75.17	1.09130	-0.30444	-0.84832	1.3713	6.9638	0.8869	4.3401
15.0	3.8637	72.14	1.09658	-0.30124	-0.84719	1.3694	6.4997	0.8916	4.0559
16.0	3.6280	69.30	1.10228	-0.29785	-0.84594	1.3674	6.0936	0.8967	3.8076
17.0	3.4203	66.63	1.10841	-0.29426	-0.84457	1.3653	5.7353	0.9021	3.5888
18.0	3.2361	64.11	1.11499	-0.29048	-0.84306	1.3631	5.4168	0.9079	3.3946
19.0	3.0716	61.72	1.12203	-0.28651	-0.84140	1.3607	5.1318	0.9140	3.2212
20.0	2.9238	59.45	1.12954	-0.28237	-0.83959	1.3581	4.8753	0.9204	3.0654
21.0	2.7904	57.28	1.13754	-0.27805	-0.83761	1.3554	4.6433	0.9272	2.9246
22.0	2.6695	55.21	1.14606	-0.27357	-0.83544	1.3526	4.4323	0.9344	2.7970
23.0	2.5593	53.23	1.15510	-0.26892	-0.83308	1.3497	4.2397	0.9419	2.6807
24.0	2.4586	51.33	1.16470	-0.26412	-0.83052	1.3466	4.0631	0.9497	2.5743
25.0	2.3662	49.50	1.17488	-0.25916	-0.82772	1.3433	3.9007	0.9580	2.4767
26.0	2.2812	47.73	1.18566	-0.25407	-0.82469	1.3400	3.7507	0.9665	2.3868
27.0	2.2027	46.03	1.19708	-0.24884	-0.82140	1.3364	3.6119	0.9754	2.3038
28.0	2.1301	44.39	1.20916	-0.24347	-0.81783	1.3328	3.4829	0.9847	2.2270
29.0	2.0627	42.79	1.22194	-0.23799	-0.81396	1.3290	3.3629	0.9943	2.1556
30.0	2.0000	41.25	1.23545	-0.23239	-0.80977	1.3250	3.2508	1.0043	2.0892
31.0	1.9416	39.75	1.24974	-0.22668	-0.80525	1.3209	3.1460	1.0146	2.0274
32.0	1.8871	38.30	1.26486	-0.22087	-0.80036	1.3167	3.0476	1.0253	1.9695
33.0	1.8361	36.88	1.28084	-0.21497	-0.79509	1.3123	2.9553	1.0362	1.9154
34.0	1.7883	35.50	1.29775	-0.20899	-0.78940	1.3078	2.8683	1.0476	1.8646
35.0	1.7434	34.16	1.31565	-0.20293	-0.78327	1.3032	2.7864	1.0592	1.8170
36.0	1.7013	32.85	1.33459	-0.19681	-0.77668	1.2984	2.7089	1.0711	1.7722
37.0	1.6616	31.57	1.35466	-0.19062	-0.76959	1.2934	2.6356	1.0834	1.7299
38.0	1.6243	30.31	1.37592	-0.18438	-0.76197	1.2884	2.5662	1.0959	1.6901
39.0	1.5890	29.09	1.39848	-0.17811	-0.75379	1.2832	2.5003	1.1086	1.6525
40.0	1.5557	27.89	1.42242	-0.17179	-0.74502	1.2778	2.4377	1.1216	1.6170
41.0	1.5243	26.71	1.44786	-0.16546	-0.73562	1.2723	2.3781	1.1348	1.5833
42.0	1.4945	25.56	1.47491	-0.15911	-0.72557	1.2667	2.3213	1.1482	1.5515
43.0	1.4663	24.43	1.50371	-0.15275	-0.71483	1.2610	2.2672	1.1618	1.5213
44.0	1.4396	23.32	1.53441	-0.14640	-0.70336	1.2552	2.2154	1.1754	1.4926
45.0	1.4142	22.23	1.56718	-0.14005	-0.69113	1.2492	2.1660	1.1892	1.4654
46.0	1.3902	21.15	1.60221	-0.13373	-0.67812	1.2431	2.1187	1.2029	1.4396
47.0	1.3673	20.10	1.63970	-0.12744	-0.66428	1.2369	2.0733	1.2167	1.4150
48.0	1.3456	19.06	1.67990	-0.12119	-0.64959	1.2305	2.0299	1.2303	1.3916
49.0	1.3250	18.04	1.72307	-0.11500	-0.63403	1.2241	1.9881	1.2438	1.3693
50.0	1.3054	17.04	1.76952	-0.10886	-0.61758	1.2175	1.9480	1.2570	1.3481
51.0	1.2868	16.06	1.81959	-0.10280	-0.60022	1.2109	1.9095	1.2700	1.3279
52.0	1.2690	15.09	1.87367	-0.09681	-0.58195	1.2041	1.8724	1.2825	1.3087
53.0	1.2521	14.14	1.93219	-0.09093	-0.56276	1.1973	1.8366	1.2945	1.2903
54.0	1.2361	13.20	1.99566	-0.08514	-0.54267	1.1903	1.8021	1.3060	1.2728
55.0	1.2208	12.29	2.06464	-0.07947	-0.52170	1.1833	1.7689	1.3167	1.2561
56.0	1.2062	11.40	2.13977	-0.07393	-0.49989	1.1762	1.7368	1.3266	1.2402
57.0	1.1924	10.52	2.22177	-0.06854	-0.47730	1.1691	1.7057	1.3356	1.2250
58.0	1.1792	9.67	2.31146	-0.06329	-0.45399	1.1619	1.6757	1.3435	1.2104
59.0	1.1666	8.85	2.40977	-0.05821	-0.43007	1.1546	1.6467	1.3502	1.1966
60.0	1.1547	8.05	2.51773	-0.05331	-0.40565	1.1473	1.6185	1.3556	1.1834
θ	Ω_K	A_{MIN}	σ_0	σ_1	σ_3	Ω_1	Ω_2	Ω_3	Ω_4

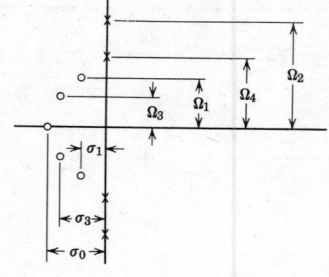

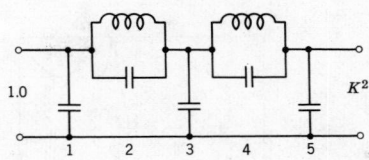

K²=1.0							K²=∞						
c_1	c_2	L_2	c_3	c_4	L_4	c_5	c_1	c_2	L_2	c_3	c_4	L_4	c_5
0.5848	0.0000	1.1690	1.3690	0.0000	1.1691	0.5848	0.2924	0.0000	0.7796	1.0841	0.0000	1.2193	1.0614
0.5845	0.0003	1.1689	1.3683	0.0009	1.1679	0.5839	0.2919	0.0005	0.7790	1.0833	0.0009	1.2181	1.0617
0.5841	0.0008	1.1684	1.3675	0.0021	1.1662	0.5828	0.2913	0.0012	0.7782	1.0823	0.0020	1.2164	1.0611
0.5836	0.0014	1.1677	1.3663	0.0037	1.1638	0.5813	0.2904	0.0022	0.7770	1.0809	0.0036	1.2142	1.0602
0.5830	0.0022	1.1669	1.3648	0.0059	1.1607	0.5794	0.2893	0.0034	0.7756	1.0790	0.0057	1.2113	1.0592
0.5822	0.0032	1.1658	1.3629	0.0085	1.1569	0.5770	0.2879	0.0049	0.7738	1.0768	0.0082	1.2078	1.0578
0.5813	0.0044	1.1646	1.3608	0.0116	1.1524	0.5742	0.2863	0.0067	0.7717	1.0742	0.0112	1.2036	1.0563
0.5803	0.0057	1.1632	1.3582	0.0152	1.1473	0.5709	0.2844	0.0088	0.7692	1.0712	0.0146	1.1988	1.0545
0.5791	0.0073	1.1616	1.3554	0.0194	1.1414	0.5673	0.2822	0.0111	0.7665	1.0678	0.0186	1.1933	1.0525
0.5778	0.0090	1.1598	1.3522	0.0240	1.1349	0.5631	0.2798	0.0138	0.7634	1.0640	0.0230	1.1872	1.0502
0.5763	0.0110	1.1578	1.3487	0.0292	1.1277	0.5586	0.2771	0.0168	0.7599	1.0598	0.0279	1.1805	1.0477
0.5747	0.0131	1.1556	1.3449	0.0349	1.1198	0.5536	0.2742	0.0200	0.7562	1.0552	0.0334	1.1731	1.0450
0.5729	0.0154	1.1532	1.3408	0.0412	1.1113	0.5481	0.2710	0.0236	0.7521	1.0503	0.0394	1.1651	1.0420
0.5710	0.0179	1.1506	1.3364	0.0481	1.1020	0.5422	0.2675	0.0276	0.7477	1.0449	0.0459	1.1565	1.0388
0.5690	0.0206	1.1479	1.3317	0.0556	1.0921	0.5359	0.2637	0.0319	0.7429	1.0392	0.0530	1.1473	1.0354
0.5668	0.0235	1.1449	1.3267	0.0637	1.0815	0.5290	0.2597	0.0365	0.7378	1.0331	0.0606	1.1374	1.0318
0.5645	0.0266	1.1418	1.3213	0.0725	1.0702	0.5218	0.2553	0.0415	0.7324	1.0267	0.0689	1.1269	1.0279
0.5620	0.0299	1.1384	1.3157	0.0820	1.0583	0.5140	0.2507	0.0469	0.7266	1.0199	0.0778	1.1157	1.0238
0.5594	0.0334	1.1349	1.3099	0.0921	1.0457	0.5058	0.2458	0.0527	0.7206	1.0127	0.0873	1.1040	1.0194
0.5566	0.0371	1.1311	1.3037	0.1030	1.0324	0.4971	0.2405	0.0589	0.7141	1.0052	0.0975	1.0916	1.0149
0.5537	0.0411	1.1272	1.2974	0.1147	1.0185	0.4879	0.2350	0.0656	0.7074	0.9973	0.1084	1.0786	1.0101
0.5506	0.0453	1.1230	1.2907	0.1273	1.0038	0.4782	0.2291	0.0727	0.7003	0.9891	0.1200	1.0650	1.0051
0.5473	0.0497	1.1187	1.2838	0.1407	0.9986	0.4681	0.2229	0.0803	0.6928	0.9806	0.1324	1.0508	0.9999
0.5439	0.0543	1.1141	1.2767	0.1551	0.9727	0.4574	0.2163	0.0884	0.6850	0.9718	0.1457	1.0360	0.9945
0.5404	0.0592	1.1094	1.2694	0.1705	0.9561	0.4462	0.2094	0.0971	0.6769	0.9626	0.1597	1.0205	0.9888
0.5367	0.0643	1.1044	1.2619	0.1869	0.9389	0.4344	0.2021	0.1063	0.6684	0.9532	0.1747	1.0045	0.9830
0.5328	0.0697	1.0992	1.2542	0.2045	0.9210	0.4222	0.1945	0.1162	0.6596	0.9434	0.1907	0.9879	0.9769
0.5287	0.0753	1.0938	1.2464	0.2234	0.9025	0.4093	0.1864	0.1267	0.6504	0.9334	0.2077	0.9706	0.9706
0.5245	0.0812	1.0882	1.2383	0.2436	0.8834	0.3959	0.1780	0.1380	0.6409	0.9231	0.2258	0.9529	0.9642
0.5202	0.0874	1.0823	1.2302	0.2652	0.8637	0.3820	0.1691	0.1500	0.6310	0.9126	0.2452	0.9345	0.9575
0.5156	0.0938	1.0762	1.2219	0.2884	0.8433	0.3674	0.1599	0.1628	0.6208	0.9018	0.2658	0.9155	0.9507
0.5109	0.1006	1.0698	1.2136	0.3134	0.8224	0.3522	0.1501	0.1764	0.6102	0.8908	0.2877	0.8960	0.9437
0.5060	0.1076	1.0632	1.2052	0.3403	0.8009	0.3364	0.1399	0.1911	0.5993	0.8795	0.3112	0.8759	0.9365
0.5009	0.1150	1.0563	1.1968	0.3693	0.7788	0.3199	0.1291	0.2067	0.5880	0.8681	0.3363	0.8552	0.9292
0.4956	0.1227	1.0491	1.1883	0.4005	0.7561	0.3027	0.1179	0.2235	0.5763	0.8565	0.3632	0.8340	0.9216
0.4901	0.1308	1.0416	1.1799	0.4344	0.7329	0.2849	0.1061	0.2415	0.5643	0.8448	0.3920	0.8122	0.9140
0.4845	0.1392	1.0338	1.1716	0.4711	0.7092	0.2663	0.0937	0.2609	0.5518	0.8329	0.4230	0.7899	0.9062
0.4786	0.1480	1.0257	1.1633	0.5110	0.6849	0.2469	0.0807	0.2817	0.5391	0.8209	0.4564	0.7671	0.8983
0.4726	0.1572	1.0172	1.1552	0.5546	0.6602	0.2267	0.0671	0.3042	0.5259	0.8088	0.4924	0.7437	0.8902
0.4663	0.1668	1.0084	1.1473	0.6022	0.6351	0.2057	0.0527	0.3284	0.5124	0.7966	0.5313	0.7198	0.8821
0.4598	0.1769	0.9991	1.1397	0.6544	0.6095	0.1837	0.0377	0.3547	0.4985	0.7844	0.5735	0.6955	0.8739
0.4531	0.1875	0.9895	1.1323	0.7119	0.5835	0.1609	0.0218	0.3833	0.4842	0.7722	0.6195	0.6706	0.8657
0.4462	0.1986	0.9793	1.1254	0.7754	0.5572	0.1370	0.0051	0.4144	0.4695	0.7601	0.6696	0.6453	0.8574
0.4390	0.2103	0.9687	1.1188	0.8459	0.5306	0.1121	-0.0125	0.4483	0.4545	0.7480	0.7244	0.6196	0.8492
0.4317	0.2226	0.9575	1.1129	0.9244	0.5037	0.0861	-0.0310	0.4855	0.4391	0.7361	0.7847	0.5934	0.8409
0.4240	0.2355	0.9457	1.1075	1.0123	0.4766	0.0588							
0.4162	0.2492	0.9332	1.1929	1.1112	0.4494	0.0302							
0.4081	0.2637	0.9201	1.0991	1.2231	0.4221	0.0003							
0.3997	0.2791	0.9062	1.0964	1.3503	0.3949	-0.0311							
0.3911	0.2955	0.8915	1.0947	1.4960	0.3677	-0.0642							
L_1	L_2	c_2	L_3	L_4	c_4	L_5	L_1	L_2	c_2	L_3	L_4	c_4	L_5

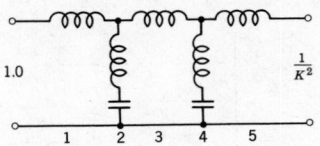

205

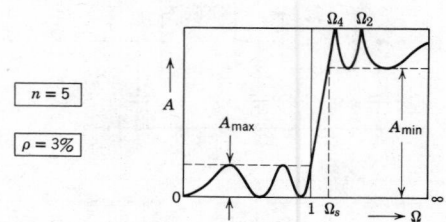

$n = 5$

$\rho = 3\%$

θ	Ω_K	A_{MIN}	σ_0	σ_1	σ_3	Ω_1	Ω_2	Ω_3	Ω_4
C	∞	∞	0.94218	-0.29115	-0.76224	1.3067	∞	0.8076	∞
2.0	28.6537	163.42	0.94272	-0.29079	-0.76215	1.3065	48.7389	0.8080	30.1274
3.0	19.1073	145.80	0.94340	-0.29034	-0.76204	1.3062	32.4927	0.8087	20.0893
4.0	14.3356	133.30	0.94435	-0.28972	-0.76189	1.3059	24.3697	0.8096	15.0716
5.0	11.4737	123.60	0.94558	-0.28891	-0.76169	1.3055	19.4959	0.8108	12.0620
6.0	9.5668	115.67	0.94708	-0.28793	-0.76144	1.3050	16.2468	0.8122	10.0565
7.0	8.2055	108.96	0.94886	-0.28678	-0.76114	1.3044	13.9260	0.8138	8.6247
8.0	7.1853	103.14	0.95093	-0.28545	-0.76079	1.3037	12.1854	0.8158	7.5516
9.0	6.3925	98.01	0.95328	-0.28394	-0.76038	1.3030	10.8316	0.8180	6.7175
10.0	5.7588	93.41	0.95591	-0.28227	-0.75992	1.3021	9.7485	0.8205	6.0507
11.0	5.2408	89.25	0.95885	-0.28042	-0.75940	1.3012	8.8625	0.8232	5.5057
12.0	4.8097	85.45	0.96208	-0.27841	-0.75882	1.3001	8.1241	0.8262	5.0520
13.0	4.4454	81.94	0.96562	-0.27623	-0.75816	1.2989	7.4993	0.8294	4.6684
14.0	4.1336	78.69	0.96947	-0.27389	-0.75744	1.2977	6.9638	0.8330	4.3401
15.0	3.8637	75.66	0.97363	-0.27138	-0.75663	1.2964	6.4997	0.8368	4.0559
16.0	3.6280	72.83	0.97812	-0.26871	-0.75575	1.2949	6.0936	0.8408	3.8076
17.0	3.4203	70.16	0.98295	-0.26589	-0.75477	1.2934	5.7353	0.8452	3.5888
18.0	3.2361	67.63	0.98812	-0.26292	-0.75370	1.2918	5.4168	0.8498	3.3946
19.0	3.0716	65.24	0.99364	-0.25979	-0.75253	1.2900	5.1318	0.8547	3.2212
20.0	2.9238	62.97	0.99953	-0.25652	-0.75126	1.2882	4.8573	0.8599	3.0654
21.0	2.7904	60.81	1.00579	-0.25311	-0.74986	1.2863	4.6433	0.8654	2.9246
22.0	2.6695	58.74	1.01244	-0.24955	-0.74834	1.2842	4.4323	0.8711	2.7970
23.0	2.5593	56.75	1.01950	-0.24586	-0.74668	1.2821	4.2397	0.8772	2.6807
24.0	2.4586	54.85	1.02697	-0.24204	-0.74489	1.2799	4.0631	0.8835	2.5743
25.0	2.3662	53.02	1.03487	-0.23808	-0.74293	1.2775	3.9007	0.8901	2.4767
26.0	2.2812	51.26	1.04323	-0.23401	-0.74082	1.2751	3.7507	0.8970	2.3868
27.0	2.2027	49.55	1.05205	-0.22981	-0.73852	1.2725	3.6119	0.9042	2.3038
28.0	2.1301	47.91	1.06137	-0.22550	-0.73604	1.2699	3.4829	0.9117	2.2270
29.0	2.0627	46.32	1.07119	-0.22108	-0.73335	1.2671	3.3629	0.9195	2.1556
30.0	2.0000	44.77	1.08156	-0.21655	-0.73044	1.2642	3.2508	0.9275	2.0892
31.0	1.9416	43.28	1.09249	-0.21192	-0.72731	1.2612	3.1460	0.9359	2.0274
32.0	1.8871	41.82	1.10402	-0.20720	-0.72392	1.2581	3.0476	0.9446	1.9695
33.0	1.8361	40.41	1.11616	-0.20238	-0.72027	1.2549	2.9553	0.9535	1.9154
34.0	1.7883	39.03	1.12897	-0.19748	-0.71634	1.2516	2.8683	0.9628	1.8646
35.0	1.7434	37.68	1.14247	-0.19251	-0.71210	1.2482	2.7864	0.9723	1.8170
36.0	1.7013	36.37	1.15672	-0.18745	-0.70755	1.2447	2.7089	0.9821	1.7722
37.0	1.6616	35.09	1.17174	-0.18233	-0.70265	1.2411	2.6356	0.9922	1.7299
38.0	1.6243	33.84	1.18759	-0.17715	-0.69738	1.2373	2.5662	1.0026	1.6901
39.0	1.5890	32.61	1.20433	-0.17192	-0.69173	1.2335	2.5003	1.0132	1.6525
40.0	1.5557	31.41	1.22202	-0.16663	-0.68567	1.2295	2.4377	1.0240	1.6170
41.0	1.5243	30.23	1.24071	-0.16130	-0.67917	1.2255	2.3781	1.0352	1.5833
42.0	1.4945	29.08	1.26049	-0.15594	-0.67221	1.2213	2.3213	1.0465	1.5515
43.0	1.4663	27.94	1.28143	-0.15054	-0.66477	1.2170	2.2672	1.0581	1.5213
44.0	1.4396	26.83	1.30362	-0.14512	-0.65680	1.2127	2.2154	1.0698	1.4926
45.0	1.4142	25.74	1.32716	-0.13968	-0.64829	1.2082	2.1660	1.0817	1.4654
46.0	1.3902	24.66	1.35215	-0.13424	-0.63921	1.2036	2.1187	1.0938	1.4396
47.0	1.3673	23.60	1.37873	-0.12879	-0.62952	1.1989	2.0733	1.1060	1.4150
48.0	1.3456	22.56	1.40703	-0.12334	-0.61921	1.1941	2.0299	1.1183	1.3916
49.0	1.3250	21.53	1.43719	-0.11791	-0.60823	1.1893	1.9881	1.1306	1.3693
50.0	1.3054	20.52	1.46940	-0.11250	-0.59656	1.1843	1.9480	1.1430	1.3481
51.0	1.2868	19.52	1.50384	-0.10711	-0.58418	1.1792	1.9095	1.1553	1.3279
52.0	1.2690	18.54	1.54072	-0.10176	-0.57106	1.1741	1.8724	1.1675	1.3087
53.0	1.2521	17.57	1.58031	-0.09645	-0.55718	1.1688	1.8366	1.1796	1.2903
54.0	1.2361	16.61	1.62286	-0.09119	-0.54251	1.1635	1.8021	1.1915	1.2728
55.0	1.2208	15.67	1.66870	-0.08600	-0.52705	1.1581	1.7689	1.2031	1.2561
56.0	1.2062	14.74	1.71818	-0.08087	-0.51078	1.1526	1.7368	1.2144	1.2402
57.0	1.1924	13.83	1.77170	-0.07583	-0.49371	1.1471	1.7057	1.2252	1.2250
58.0	1.1792	12.93	1.82973	-0.07087	-0.47584	1.1415	1.6757	1.2355	1.2104
59.0	1.1666	12.04	1.89278	-0.06601	-0.45719	1.1358	1.6467	1.2452	1.1966
60.0	1.1547	11.18	1.96146	-0.06125	-0.43778	1.1300	1.6185	1.2541	1.1834
θ	Ω_K	A_{MIN}	σ_0	σ_1	σ_3	Ω_1	Ω_2	Ω_3	Ω_4

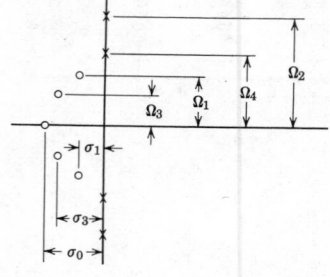

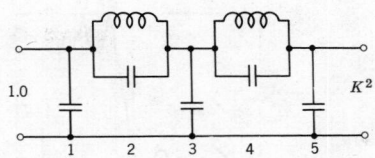

K²=1.0							K²=∞						
c_1	c_2	L_2	c_3	c_4	L_4	c_5	c_1	c_2	L_2	c_3	c_4	L_4	c_5
0.6560	0.0000	1.2362	1.4610	0.0000	1.2360	0.6560	0.3280	0.0000	0.8571	1.1694	0.0000	1.2984	1.1251
0.6556	0.0003	1.2358	1.4599	0.0008	1.2348	0.6551	0.3275	0.0005	0.8565	1.1686	0.0008	1.2971	1.1246
0.6553	0.0007	1.2353	1.4590	0.0020	1.2332	0.6541	0.3270	0.0011	0.8557	1.1676	0.0019	1.2956	1.1240
0.6549	0.0013	1.2347	1.4578	0.0035	1.2309	0.6526	0.3262	0.0020	0.8547	1.1663	0.0034	1.2934	1.1231
0.6543	0.0021	1.2338	1.4563	0.0056	1.2280	0.6508	0.3251	0.0031	0.8533	1.1645	0.0053	1.2906	1.1220
0.6535	0.0030	1.2328	1.4544	0.0080	1.2244	0.6486	0.3239	0.0044	0.8516	1.1624	0.0077	1.2872	1.1207
0.6527	0.0041	1.2316	1.4521	0.0110	1.2202	0.6459	0.3224	0.0061	0.8497	1.1598	0.0105	1.2831	1.1191
0.6517	0.0054	1.2302	1.4496	0.0144	1.2153	0.6428	0.3207	0.0079	0.8474	1.1569	0.0137	1.2784	1.1173
0.6505	0.0069	1.2286	1.4467	0.0183	1.2097	0.6393	0.3187	0.0101	0.8448	1.1536	0.0174	1.2731	1.1152
0.6493	0.0085	1.2268	1.4434	0.0226	1.2035	0.6354	0.3165	0.0125	0.8419	1.1499	0.0216	1.2672	1.1129
0.6478	0.0103	1.2248	1.4398	0.0275	1.1967	0.6311	0.3141	0.0152	0.8387	1.1458	0.0262	1.2607	1.1104
0.6463	0.0123	1.2226	1.4359	0.0329	1.1892	0.6264	0.3115	0.0181	0.8352	1.1413	0.0313	1.2535	1.1076
0.6446	0.0145	1.2203	1.4317	0.0388	1.1810	0.6212	0.3085	0.0214	0.8314	1.1365	0.0368	1.2458	1.1045
0.6428	0.0169	1.2177	1.4271	0.0452	1.1722	0.6156	0.3054	0.0249	0.8272	1.1312	0.0429	1.2374	1.1013
0.6409	0.0194	1.2150	1.4223	0.0522	1.1628	0.6096	0.3020	0.0288	0.8228	1.1257	0.0495	1.2284	1.0978
0.6388	0.0222	1.2121	1.4171	0.0598	1.1528	0.6032	0.2983	0.0329	0.8181	1.1197	0.0566	1.2188	1.0941
0.6365	0.0251	1.2090	1.4116	0.0679	1.1420	0.5964	0.2944	0.0374	0.8130	1.1134	0.0642	1.2086	1.0901
0.6342	0.0282	1.2057	1.4058	0.0767	1.1307	0.5891	0.2902	0.0422	0.8076	1.1067	0.0724	1.1978	1.0859
0.6317	0.0315	1.2022	1.3997	0.0861	1.1187	0.5813	0.2858	0.0474	0.8019	1.0997	0.0812	1.1864	1.0814
0.6290	0.0351	1.1985	1.3934	0.0962	1.1061	0.5731	0.2811	0.0529	0.7959	1.0923	0.0906	1.1744	1.0768
0.6262	0.0388	1.1946	1.3867	0.1069	1.0928	0.5645	0.2761	0.0587	0.7896	1.0845	0.1006	1.1618	1.0719
0.6232	0.0427	1.1905	1.3798	0.1184	1.0790	0.5555	0.2708	0.0650	0.7829	1.0765	0.1113	1.1486	1.0668
0.6201	0.0469	1.1861	1.3726	0.1307	1.0644	0.5459	0.2652	0.0717	0.7760	1.0681	0.1226	1.1349	1.0614
0.6169	0.0512	1.1816	1.3651	0.1438	1.0493	0.5359	0.2594	0.0788	0.7687	1.0593	0.1347	1.1205	1.0558
0.6135	0.0558	1.1769	1.3574	0.1577	1.0336	0.5255	0.2532	0.0864	0.7611	1.0503	0.1475	1.1055	1.0501
0.6099	0.0606	1.1720	1.3495	0.1725	1.0172	0.5145	0.2467	0.0944	0.7531	1.0409	0.1610	1.0900	1.0440
0.6062	0.0656	1.1669	1.3413	0.1883	1.0002	0.5031	0.2399	0.1029	0.7449	1.0312	0.1754	1.0739	1.0378
0.6024	0.0709	1.1615	1.3329	0.2051	0.9827	0.4912	0.2328	0.1120	0.7363	1.0213	0.1907	1.0572	1.0314
0.5984	0.0764	1.1560	1.3243	0.2231	0.9645	0.4788	0.2253	0.1216	0.7273	1.0110	0.2069	1.0399	1.0247
0.5942	0.0822	1.1502	1.3155	0.2422	0.9457	0.4659	0.2175	0.1318	0.7181	1.0005	0.2241	1.0221	1.0178
0.5898	0.0883	1.1441	1.3065	0.2626	0.9264	0.4524	0.2093	0.1426	0.7084	0.9896	0.2424	1.0037	1.0108
0.5853	0.0946	1.1379	1.2974	0.2843	0.9065	0.4384	0.2007	0.1541	0.6985	0.9786	0.2618	0.9848	1.0036
0.5806	0.1012	1.1314	1.2881	0.3076	0.8860	0.4239	0.1917	0.1664	0.6882	0.9673	0.2824	0.9653	0.9961
0.5758	0.1080	1.1246	1.2787	0.3325	0.8649	0.4088	0.1824	0.1794	0.6776	0.9557	0.3043	0.9453	0.9885
0.5707	0.1152	1.1176	1.2693	0.3591	0.8433	0.3932	0.1726	0.1932	0.6666	0.9439	0.3275	0.9247	0.9807
0.5655	0.1227	1.1103	1.2597	0.3877	0.8211	0.3769	0.1623	0.2080	0.6552	0.9319	0.3524	0.9037	0.9727
0.5601	0.1305	1.1027	1.2501	0.4184	0.7984	0.3600	0.1516	0.2237	0.6435	0.9197	0.3788	0.8820	0.9645
0.5545	0.1386	1.0949	1.2404	0.4515	0.7753	0.3425	0.1404	0.2405	0.6314	0.9074	0.4071	0.8599	0.9562
0.5487	0.1471	1.0867	1.2307	0.4872	0.7516	0.3244	0.1287	0.2584	0.6190	0.8949	0.4374	0.8373	0.9477
0.5428	0.1560	1.0782	1.2211	0.5257	0.7274	0.3055	0.1164	0.2776	0.6062	0.8822	0.4698	0.8141	0.9390
0.5366	0.1653	1.0693	1.2115	0.5675	0.7028	0.2860	0.1036	0.2982	0.5930	0.8694	0.5046	0.7905	0.9303
0.5302	0.1750	1.0601	1.2021	0.6129	0.6777	0.2657	0.0901	0.3203	0.5794	0.8565	0.5420	0.7664	0.9214
0.5236	0.1851	1.0505	1.1928	0.6623	0.6523	0.2446	0.0761	0.3441	0.5655	0.8436	0.5824	0.7419	0.9124
0.5168	0.1958	1.0405	1.1837	0.7164	0.6264	0.2227	0.0613	0.3697	0.5511	0.8306	0.6261	0.7169	0.9033
0.5097	0.2069	1.0301	1.1748	0.7757	0.6002	0.2000	0.0458	0.3974	0.5364	0.8175	0.6735	0.6914	0.8942
0.5025	0.2185	1.0192	1.1663	0.8409	0.5737	0.1763	0.0296	0.4274	0.5213	0.8045	0.7250	0.6655	0.8850
0.4950	0.2308	1.0077	1.1581	0.9130	0.5469	0.1517	0.0125	0.4600	0.5057	0.7915	0.7813	0.6393	0.8758
0.4872	0.2437	0.9957	1.1503	0.9930	0.5199	0.1261	-0.0054	0.4955	0.4898	0.7786	0.8429	0.6126	0.8665
0.4792	0.2573	0.9832	1.1431	1.0822	0.4927	0.0994	-0.0243	0.5343	0.4735	0.7659	0.9106	0.5857	0.8574
0.4709	0.2716	0.9699	1.1365	1.1820	0.4654	0.0715	-0.0443	0.5770	0.4567	0.7533	0.9854	0.5584	0.8483
0.4624	0.2868	0.9560	1.1306	1.2943	0.4381	0.0424							
0.4536	0.3030	0.9414	1.1255	1.4214	0.4107	0.0119							
0.4445	0.3201	0.9259	1.1214	1.5661	0.3835	-0.0200							
0.4352	0.3385	0.9095	1.1185	1.7318	0.3564	-0.0535							
0.4255	0.3581	0.8922	1.1168	1.9229	0.3296	-0.0887							
L_1	L_2	c_2	L_3	L_4	c_4	L_5	L_1	L_2	c_2	L_3	L_4	c_4	L_5

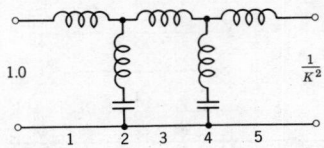

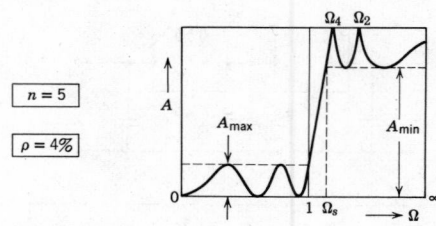

$$n = 5$$

$$\rho = 4\%$$

θ	Ω_K	A_{MIN}	σ_0	σ_1	σ_3	Ω_1	Ω_2	Ω_3	Ω_4
C	∞	∞	0.86460	-0.26718	-0.69948	1.2572	∞	0.7770	∞
2.0	28.6537	165.92	0.86506	-0.26687	-0.69941	1.2571	48.7389	0.7774	30.1274
3.0	19.1073	148.30	0.86564	-0.26650	-0.69932	1.2569	32.4927	0.7780	20.0893
4.0	14.3356	135.80	0.86644	-0.26598	-0.69920	1.2566	24.3697	0.7787	15.0716
5.0	11.4737	126.10	0.86748	-0.26531	-0.69904	1.2563	19.4959	0.7797	12.0620
6.0	9.5668	118.17	0.86876	-0.26449	-0.69884	1.2559	16.2468	0.7809	10.0565
7.0	8.2055	111.46	0.87027	-0.26353	-0.69861	1.2555	13.9260	0.7824	8.6247
8.0	7.1853	105.65	0.87202	-0.26241	-0.69833	1.2549	12.1854	0.7841	7.5516
9.0	6.3925	100.51	0.87402	-0.26116	-0.69802	1.2543	10.8316	0.7859	6.7175
10.0	5.7588	95.91	0.87625	-0.25976	-0.69766	1.2537	9.7486	0.7881	6.0507
11.0	5.2408	91.75	0.87874	-0.25821	-0.69725	1.2529	8.8625	0.7904	5.5057
12.0	4.8097	87.95	0.88148	-0.25653	-0.69679	1.2521	8.1241	0.7930	5.0520
13.0	4.4454	84.44	0.88447	-0.25470	-0.69628	1.2512	7.4993	0.7958	4.6684
14.0	4.1336	81.19	0.88773	-0.25273	-0.69571	1.2502	6.9638	0.7988	4.3401
15.0	3.8637	78.17	0.89125	-0.25063	-0.69509	1.2492	6.4997	0.8021	4.0559
16.0	3.6280	75.33	0.89505	-0.24839	-0.69440	1.2481	6.0936	0.8055	3.8076
17.0	3.4203	72.66	0.89912	-0.24602	-0.69364	1.2469	5.7353	0.8093	3.5888
18.0	3.2361	70.14	0.90348	-0.24352	-0.69281	1.2456	5.4168	0.8132	3.3946
19.0	3.0716	67.75	0.90814	-0.24089	-0.69190	1.2443	5.1318	0.8175	3.2212
20.0	2.9238	65.47	0.91310	-0.23813	-0.69090	1.2428	4.8753	0.8219	3.0654
21.0	2.7904	63.31	0.91837	-0.23524	-0.68982	1.2413	4.6433	0.8266	2.9246
22.0	2.6695	61.24	0.92396	-0.23224	-0.68864	1.2397	4.4323	0.8315	2.7970
23.0	2.5593	59.25	0.92988	-0.22911	-0.68736	1.2381	4.2397	0.8367	2.6807
24.0	2.4586	57.35	0.93614	-0.22587	-0.68597	1.2363	4.0631	0.8422	2.5743
25.0	2.3662	55.52	0.94276	-0.22252	-0.68446	1.2345	3.9007	0.8478	2.4767
26.0	2.2812	53.76	0.94975	-0.21905	-0.68283	1.2326	3.7507	0.8538	2.3868
27.0	2.2027	52.06	0.95712	-0.21548	-0.68106	1.2306	3.6119	0.8600	2.3038
28.0	2.1301	50.41	0.96489	-0.21180	-0.67915	1.2285	3.4829	0.8664	2.2270
29.0	2.0627	48.82	0.97308	-0.20803	-0.67708	1.2263	3.3629	0.8731	2.1556
30.0	2.0000	47.28	0.98170	-0.20415	-0.67484	1.2240	3.2508	0.8801	2.0892
31.0	1.9416	45.78	0.99077	-0.20019	-0.67243	1.2217	3.1460	0.8873	2.0274
32.0	1.8871	44.32	1.00031	-0.19613	-0.66983	1.2193	3.0476	0.8948	1.9695
33.0	1.8361	42.91	1.01036	-0.19199	-0.66703	1.2168	2.9553	0.9025	1.9154
34.0	1.7883	41.53	1.02093	-0.18776	-0.66401	1.2142	2.8683	0.9105	1.8646
35.0	1.7434	40.18	1.03204	-0.18346	-0.66077	1.2115	2.7864	0.9188	1.8170
36.0	1.7013	38.87	1.04374	-0.17909	-0.65727	1.2087	2.7089	0.9273	1.7722
37.0	1.6616	37.59	1.05606	-0.17464	-0.65352	1.2058	2.6356	0.9361	1.7299
38.0	1.6243	36.34	1.06902	-0.17013	-0.64949	1.2029	2.5662	0.9451	1.6901
39.0	1.5890	35.11	1.08266	-0.16557	-0.64516	1.1998	2.5003	0.9544	1.6525
40.0	1.5557	33.91	1.09704	-0.16094	-0.64051	1.1967	2.4377	0.9639	1.6170
41.0	1.5243	32.73	1.11220	-0.15627	-0.63553	1.1934	2.3781	0.9737	1.5833
42.0	1.4945	31.58	1.12818	-0.15155	-0.63020	1.1901	2.3213	0.9837	1.5515
43.0	1.4663	30.44	1.14504	-0.14680	-0.62449	1.1867	2.2672	0.9939	1.5213
44.0	1.4396	29.33	1.16285	-0.14200	-0.61837	1.1832	2.2154	1.0043	1.4926
45.0	1.4142	28.23	1.18168	-0.13718	-0.61183	1.1797	2.1660	1.0150	1.4654
46.0	1.3902	27.15	1.20160	-0.13233	-0.60485	1.1760	2.1187	1.0258	1.4396
47.0	1.3673	26.09	1.22269	-0.12747	-0.59739	1.1722	2.0733	1.0368	1.4150
48.0	1.3456	25.05	1.24505	-0.12259	-0.58943	1.1684	2.0299	1.0479	1.3916
49.0	1.3250	24.02	1.26878	-0.11770	-0.58095	1.1645	1.9881	1.0592	1.3693
50.0	1.3054	23.00	1.29400	-0.11282	-0.57191	1.1605	1.9480	1.0706	1.3481
51.0	1.2868	22.00	1.32084	-0.10794	-0.50462	1.1564	1.9095	1.0821	1.3279
52.0	1.2690	21.01	1.34944	-0.10307	-0.49100	1.1522	1.8724	1.0936	1.3087
53.0	1.2521	20.04	1.37997	-0.09822	-0.47665	1.1480	1.8366	1.1051	1.2903
54.0	1.2361	19.07	1.41261	-0.09339	-0.46155	1.1436	1.8021	1.1166	1.2728
55.0	1.2208	18.12	1.44757	-0.08860	-0.44570	1.1392	1.7689	1.1280	1.2561
56.0	1.2062	17.18	1.48509	-0.08384	-0.56230	1.1347	1.7368	1.1392	1.2402
57.0	1.1924	16.25	1.52532	-0.07914	-0.55207	1.1302	1.7057	1.1503	1.2250
58.0	1.1792	15.33	1.56889	-0.07448	-0.54122	1.1256	1.6757	1.1611	1.2104
59.0	1.1666	14.42	1.61583	-0.06989	-0.52971	1.1209	1.6467	1.1716	1.1966
60.0	1.1547	13.53	1.66664	-0.06537	-0.51751	1.1161	1.6185	1.1817	1.1834
θ	Ω_K	A_{MIN}	σ_0	σ_1	σ_3	Ω_1	Ω_2	Ω_3	Ω_4

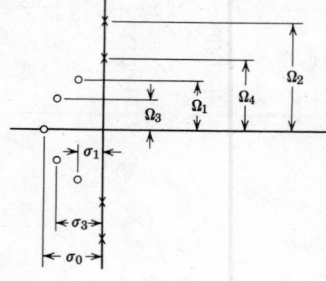

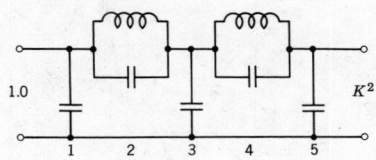

K²=1.0							K²=∞						
c_1	c_2	L_2	c_3	c_4	L_4	c_5	c_1	c_2	L_2	c_3	c_4	L_4	c_5
0.7148	0.0000	1.2800	1.5300	0.0000	1.2800	0.7148	0.3574	0.0000	0.9173	1.2328	0.0000	1.3548	1.1697
0.7145	0.0003	1.2795	1.5297	0.0008	1.2786	0.7140	0.3570	0.0005	0.9167	1.2320	0.0008	1.3535	1.1692
0.7142	0.0007	1.2790	1.5288	0.0019	1.2770	0.7130	0.3565	0.0010	0.9160	1.2311	0.0018	1.3520	1.1686
0.7137	0.0013	1.2784	1.5276	0.0034	1.2748	0.7116	0.3557	0.0018	0.9150	1.2297	0.0033	1.3499	1.1678
0.7132	0.0020	1.2775	1.5260	0.0054	1.2720	0.7098	0.3548	0.0029	0.9137	1.2280	0.0051	1.3471	1.1666
0.7124	0.0029	1.2765	1.5241	0.0077	1.2685	0.7076	0.3536	0.0042	0.9121	1.2258	0.0074	1.3437	1.1653
0.7116	0.0040	1.2753	1.5218	0.0106	1.2644	0.7051	0.3522	0.0057	0.9102	1.2233	0.0100	1.3398	1.1637
0.7106	0.0052	1.2739	1.5192	0.0139	1.2597	0.7021	0.3506	0.0074	0.9080	1.2204	0.0131	1.3352	1.1618
0.7095	0.0067	1.2723	1.5162	0.0176	1.2543	0.6987	0.3488	0.0094	0.9056	1.2172	0.0167	1.3300	1.1597
0.7083	0.0082	1.2706	1.5129	0.0218	1.2484	0.6949	0.3467	0.0117	0.9028	1.2135	0.0206	1.3242	1.1574
0.7069	0.0100	1.2686	1.5093	0.0265	1.2418	0.6908	0.3445	0.0142	0.8998	1.2095	0.0250	1.3178	1.1548
0.7054	0.0119	1.2665	1.5053	0.0317	1.2345	0.6862	0.3420	0.0169	0.8964	1.2051	0.0299	1.3108	1.1520
0.7038	0.0140	1.2641	1.5009	0.0374	1.2267	0.6812	0.3393	0.0199	0.8928	1.2003	0.0352	1.3032	1.1489
0.7020	0.0163	1.2616	1.4963	0.0435	1.2182	0.6758	0.3363	0.0232	0.8888	1.1952	0.0410	1.2949	1.1456
0.7001	0.0188	1.2589	1.4913	0.0502	1.2091	0.6700	0.3332	0.0268	0.8846	1.1897	0.0473	1.2861	1.1420
0.6981	0.0214	1.2560	1.4860	0.0575	1.1994	0.6638	0.3298	0.0306	0.8801	1.1838	0.0540	1.2767	1.1382
0.6959	0.0242	1.2529	1.4804	0.0652	1.1891	0.6572	0.3261	0.0347	0.8752	1.1775	0.0613	1.2667	1.1342
0.6936	0.0272	1.2496	1.4744	0.0736	1.1781	0.6502	0.3222	0.0392	0.8701	1.1709	0.0691	1.2561	1.1299
0.6911	0.0304	1.2461	1.4682	0.0826	1.1666	0.6427	0.3181	0.0439	0.8647	1.1640	0.0774	1.2449	1.1254
0.6885	0.0338	1.2425	1.4616	0.0921	1.1544	0.6348	0.3137	0.0490	0.8589	1.1567	0.0863	1.2331	1.1206
0.6858	0.0374	1.2386	1.4548	0.1024	1.1416	0.6265	0.3091	0.0544	0.8529	1.1490	0.0958	1.2208	1.1157
0.6829	0.0412	1.2345	1.4476	0.1132	1.1283	0.6178	0.3042	0.0601	0.8465	1.1410	0.1058	1.2078	1.1105
0.6799	0.0452	1.2303	1.4402	0.1248	1.1143	0.6087	0.2991	0.0662	0.8399	1.1327	0.1165	1.1943	1.1050
0.6768	0.0494	1.2258	1.4324	0.1372	1.0997	0.5991	0.2936	0.0727	0.8329	1.1240	0.1279	1.1802	1.0993
0.6734	0.0538	1.2211	1.4244	0.1503	1.0845	0.5890	0.2879	0.0796	0.8256	1.1150	0.1399	1.1655	1.0934
0.6700	0.0584	1.2162	1.4162	0.1642	1.0687	0.5786	0.2820	0.0869	0.8181	1.1057	0.1526	1.1503	1.0873
0.6664	0.0632	1.2112	1.4077	0.1790	1.0524	0.5676	0.2757	0.0946	0.8101	1.0960	0.1661	1.1345	1.0810
0.6626	0.0683	1.2059	1.3989	0.1947	1.0354	0.5562	0.2691	0.1028	0.8019	1.0861	0.1803	1.1181	1.0744
0.6587	0.0736	1.2003	1.3899	0.2114	1.0179	0.5444	0.2623	0.1115	0.7934	1.0759	0.1954	1.1012	1.0676
0.6546	0.0792	1.1946	1.3807	0.2291	0.9998	0.5321	0.2551	0.1206	0.7845	1.0653	0.2114	1.0837	1.0606
0.6504	0.0850	1.1886	1.3713	0.2479	0.9812	0.5193	0.2475	0.1303	0.7753	1.0545	0.2283	1.0657	1.0534
0.6460	0.0910	1.1824	1.3616	0.2679	0.9619	0.5060	0.2397	0.1406	0.7658	1.0434	0.2462	1.0471	1.0460
0.6414	0.0973	1.1760	1.3518	0.2893	0.9421	0.4922	0.2315	0.1515	0.7559	1.0320	0.2651	1.0280	1.0384
0.6367	0.1039	1.1693	1.3418	0.3119	0.9218	0.4779	0.2229	0.1630	0.7457	1.0204	0.2852	1.0084	1.0305
0.6318	0.1108	1.1624	1.3317	0.3361	0.9009	0.4631	0.2140	0.1752	0.7352	1.0085	0.3065	0.9882	1.0225
0.6267	0.1179	1.1552	1.3214	0.3620	0.8796	0.4477	0.2047	0.1881	0.7243	0.9964	0.3291	0.9675	1.0143
0.6215	0.1254	1.1478	1.3110	0.3896	0.8576	0.4318	0.1950	0.2019	0.7131	0.9841	0.3531	0.9463	1.0059
0.6160	0.1331	1.1401	1.3005	0.4191	0.8352	0.4153	0.1848	0.2165	0.7015	0.9715	0.3786	0.9246	0.9973
0.6104	0.1413	1.1320	1.2899	0.4507	0.8123	0.3983	0.1742	0.2320	0.6895	0.9588	0.4058	0.9024	0.9886
0.6046	0.1497	1.1237	1.2792	0.4847	0.7889	0.3806	0.1632	0.2485	0.6772	0.9459	0.4348	0.8796	0.9796
0.5986	0.1585	1.1151	1.2686	0.5213	0.7650	0.3624	0.1516	0.2661	0.6645	0.9328	0.4657	0.8566	0.9705
0.5924	0.1677	1.1061	1.2579	0.5608	0.7407	0.3435	0.1396	0.2849	0.6515	0.9195	0.4988	0.8329	0.9613
0.5860	0.1773	1.0968	1.2473	0.6034	0.7159	0.3239	0.1270	0.3049	0.6380	0.9061	0.5342	0.8088	0.9519
0.5793	0.1874	1.0871	1.2367	0.6497	0.6908	0.3036	0.1139	0.3264	0.6242	0.8926	0.5723	0.7842	0.9424
0.5725	0.1978	1.0771	1.2262	0.6999	0.6652	0.2827	0.1001	0.3494	0.6100	0.8791	0.6133	0.7592	0.9328
0.5654	0.2088	1.0666	1.2159	0.7547	0.6393	0.2609	0.0858	0.3742	0.5954	0.8654	0.6576	0.7338	0.9230
0.5581	0.2203	1.0556	1.2058	0.8145	0.6131	0.2384	0.0707	0.4008	0.5804	0.8517	0.7055	0.7080	0.9132
0.5506	0.2324	1.0442	1.1959	0.8802	0.5866	0.2150	0.0549	0.4296	0.5649	0.8380	0.7574	0.6818	0.9033
0.5428	0.2450	1.0323	1.1864	0.9526	0.5598	0.1907	0.0384	0.4608	0.5491	0.8243	0.8140	0.6552	0.8934
0.5348	0.2583	1.0198	1.1772	1.0327	0.5327	0.1655	0.0210	0.4946	0.5328	0.8106	0.8758	0.6282	0.8834
0.5265	0.2724	1.0068	1.1684	1.1215	0.5056	0.1393	0.0028	0.5314	0.5161	0.7971	0.9436	0.6010	0.8735
0.5179	0.2872	0.9931	1.1601	1.2207	0.4783	0.1120	-0.0164	0.5717	0.4990	0.7836	1.0183	0.5734	0.8635
0.5091	0.3029	0.9787	1.1525	1.3319	0.4509	0.0836	-0.0366	0.6158	0.4814	0.7704	1.1009	0.5456	0.8537
0.4999	0.3195	0.9636	1.1456	1.4573	0.4235	0.0539	-0.0579	0.6645	0.4634	0.7573	1.1927	0.5175	0.8440
0.4905	0.3372	0.9477	1.1395	1.5994	0.3962	0.0229	-0.0805	0.7183	0.4449	0.7446	1.2953	0.4893	0.8344
L_1	L_2	c_2	L_3	L_4	c_4	L_5	L_1	L_2	c_2	L_3	L_4	c_4	L_5

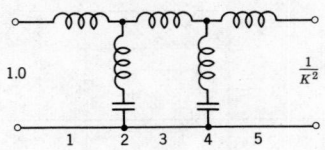

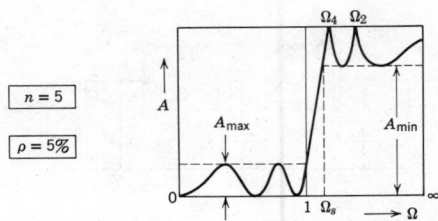

$n = 5$

$\rho = 5\%$

θ	Ω_K	A_{MIN}	σ_0	σ_1	σ_3	Ω_1	Ω_2	Ω_3	Ω_4
C	∞	∞	0.80639	-0.24919	-0.65238	1.2218	∞	0.7551	∞
2.0	28.6537	167.86	0.80679	-0.24892	-0.65232	1.2216	48.7389	0.7554	30.1274
3.0	19.1073	150.25	0.80730	-0.24860	-0.65225	1.2214	32.4927	0.7559	20.0893
4.0	14.3356	137.74	0.80801	-0.24815	-0.65215	1.2212	24.3697	0.7566	15.0716
5.0	11.4737	128.04	0.80893	-0.24757	-0.65202	1.2210	19.4959	0.7575	12.0620
6.0	9.5668	120.11	0.81005	-0.24686	-0.65186	1.2207	16.2468	0.7586	10.0565
7.0	8.2055	113.40	0.81138	-0.24602	-0.65167	1.2203	13.9260	0.7599	8.6247
8.0	7.1853	107.59	0.81292	-0.24505	-0.65144	1.2199	12.1854	0.7613	7.5516
9.0	6.3925	102.45	0.81468	-0.24396	-0.65118	1.2194	10.8316	0.7630	6.7175
10.0	5.7588	97.86	0.81664	-0.24274	-0.65088	1.2188	9.7486	0.7649	6.0507
11.0	5.2408	93.69	0.81883	-0.24140	-0.65055	1.2182	8.8625	0.7670	5.5057
12.0	4.8097	89.89	0.82124	-0.23993	-0.65017	1.2176	8.1241	0.7693	5.0520
13.0	4.4454	86.39	0.82387	-0.23834	-0.64975	1.2168	7.4993	0.7718	4.6684
14.0	4.1336	83.14	0.82673	-0.23663	-0.64928	1.2160	6.9638	0.7744	4.3401
15.0	3.8637	80.11	0.82982	-0.23479	-0.64877	1.2152	6.4997	0.7774	4.0559
16.0	3.6280	77.27	0.83315	-0.23284	-0.64820	1.2143	6.0936	0.7805	3.8076
17.0	3.4203	74.60	0.83672	-0.23077	-0.64758	1.2133	5.7353	0.7838	3.5888
18.0	3.2361	72.08	0.84055	-0.22859	-0.64690	1.2123	5.4168	0.7873	3.3946
19.0	3.0716	69.69	0.84463	-0.22629	-0.64616	1.2112	5.1318	0.7911	3.2212
20.0	2.9238	67.41	0.84897	-0.22387	-0.64534	1.2100	4.8753	0.7950	3.0654
21.0	2.7904	65.25	0.85358	-0.22135	-0.64446	1.2088	4.6433	0.7992	2.9246
22.0	2.6695	63.18	0.85847	-0.21872	-0.64350	1.2075	4.4323	0.8036	2.7970
23.0	2.5593	61.20	0.86364	-0.21598	-0.64245	1.2061	4.2397	0.8082	2.6807
24.0	2.4586	59.29	0.86911	-0.21313	-0.64131	1.2047	4.0631	0.8131	2.5743
25.0	2.3662	57.46	0.87488	-0.21019	-0.64008	1.2032	3.9007	0.8181	2.4767
26.0	2.2812	55.70	0.88097	-0.20714	-0.63875	1.2016	3.7507	0.8234	2.3868
27.0	2.2027	54.00	0.88739	-0.20400	-0.63731	1.2000	3.6119	0.8289	2.3038
28.0	2.1301	52.35	0.89414	-0.20076	-0.63575	1.1983	3.4829	0.8347	2.2270
29.0	2.0627	50.76	0.90126	-0.19742	-0.63407	1.1965	3.3629	0.8407	2.1556
30.0	2.0000	49.22	0.90873	-0.19400	-0.63225	1.1947	3.2508	0.8469	2.0892
31.0	1.9416	47.72	0.91660	-0.19049	-0.63029	1.1928	3.1460	0.8533	2.0274
32.0	1.8871	46.27	0.92486	-0.18690	-0.62818	1.1908	3.0476	0.8600	1.9695
33.0	1.8361	44.85	0.93355	-0.18322	-0.62590	1.1887	2.9553	0.8669	1.9154
34.0	1.7883	43.47	0.94267	-0.17947	-0.62345	1.1866	2.8683	0.8741	1.8646
35.0	1.7434	42.13	0.95225	-0.17565	-0.62081	1.1844	2.7864	0.8815	1.8170
36.0	1.7013	40.81	0.96232	-0.17175	-0.61798	1.1821	2.7089	0.8891	1.7722
37.0	1.6616	39.53	0.97290	-0.16778	-0.61493	1.1797	2.6356	0.8970	1.7299
38.0	1.6243	38.28	0.98402	-0.16375	-0.61166	1.1773	2.5662	0.9051	1.6901
39.0	1.5890	37.05	0.99571	-0.15966	-0.60815	1.1748	2.5003	0.9135	1.6525
40.0	1.5557	35.85	1.00800	-0.15552	-0.60438	1.1722	2.4377	0.9220	1.6170
41.0	1.5243	34.67	1.02092	-0.15132	-0.60034	1.1695	2.3781	0.9308	1.5833
42.0	1.4945	33.52	1.03453	-0.14707	-0.59601	1.1668	2.3213	0.9399	1.5515
43.0	1.4663	32.38	1.04885	-0.14277	-0.59137	1.1640	2.2672	0.9491	1.5213
44.0	1.4396	31.27	1.06395	-0.13844	-0.58641	1.1611	2.2154	0.9586	1.4926
45.0	1.4142	30.17	1.07986	-0.13407	-0.58110	1.1581	2.1660	0.9683	1.4654
46.0	1.3902	29.09	1.09665	-0.12967	-0.57542	1.1551	2.1187	0.9782	1.4396
47.0	1.3673	28.03	1.11439	-0.12524	-0.56935	1.1519	2.0733	0.9883	1.4150
48.0	1.3456	26.99	1.13313	-0.12079	-0.56287	1.1487	2.0299	0.9985	1.3916
49.0	1.3250	25.95	1.15297	-0.11632	-0.55595	1.1455	1.9881	1.0090	1.3693
50.0	1.3054	24.94	1.17399	-0.11184	-0.54858	1.1421	1.9480	1.0195	1.3481
51.0	1.2868	23.93	1.19628	-0.10735	-0.54072	1.1387	1.9095	1.0302	1.3279
52.0	1.2690	22.94	1.21995	-0.10286	-0.53235	1.1352	1.8724	1.0410	1.3087
53.0	1.2521	21.96	1.24513	-0.09838	-0.52344	1.1316	1.8366	1.0519	1.2903
54.0	1.2361	20.99	1.27194	-0.09390	-0.51398	1.1280	1.8021	1.0628	1.2728
55.0	1.2208	20.04	1.30055	-0.08944	-0.50393	1.1243	1.7689	1.0738	1.2561
56.0	1.2062	19.09	1.33113	-0.08500	-0.49326	1.1205	1.7368	1.0847	1.2402
57.0	1.1924	18.15	1.36386	-0.08058	-0.48196	1.1167	1.7057	1.0956	1.2250
58.0	1.1792	17.23	1.39898	-0.07620	-0.47000	1.1128	1.6757	1.1064	1.2104
59.0	1.1666	16.31	1.43673	-0.07186	-0.45737	1.1088	1.6467	1.1170	1.1966
60.0	1.1547	15.40	1.47740	-0.06757	-0.44404	1.1048	1.6185	1.1274	1.1834
θ	Ω_K	A_{MIN}	σ_0	σ_1	σ_3	Ω_1	Ω_2	Ω_3	Ω_4

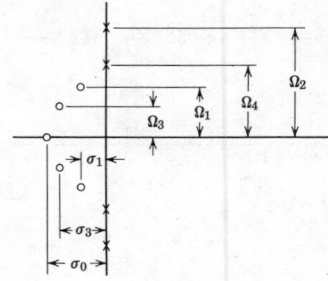

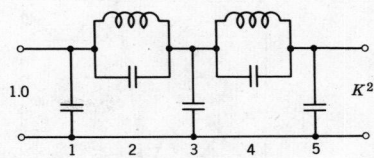

			$K^2=1.0$							$K^2=\infty$			
C_1	C_2	L_2	C_3	C_4	L_4	C_5	C_1	C_2	L_2	C_3	C_4	L_4	C_5
0.7664	0.0000	1.3100	1.5880	0.0000	1.3100	0.7664	0.3832	0.0000	0.9671	1.2835	0.0000	1.3983	1.2042
0.7661	0.0003	1.3099	1.5877	0.0008	1.3091	0.7656	0.3828	0.0004	0.9666	1.2827	0.0008	1.3971	1.2038
0.7658	0.0007	1.3095	1.5868	0.0018	1.3075	0.7646	0.3823	0.0010	0.9659	1.2818	0.0018	1.3955	1.2031
0.7654	0.0012	1.3088	1.5855	0.0033	1.3054	0.7633	0.3816	0.0017	0.9649	1.2804	0.0032	1.3934	1.2023
0.7648	0.0020	1.3080	1.5839	0.0052	1.3026	0.7615	0.3807	0.0027	0.9637	1.2787	0.0049	1.3907	1.2011
0.7641	0.0029	1.3070	1.5820	0.0076	1.2993	0.7594	0.3796	0.0039	0.9621	1.2766	0.0071	1.3874	1.1997
0.7632	0.0039	1.3058	1.5796	0.0103	1.2953	0.7569	0.3783	0.0054	0.9603	1.2741	0.0097	1.3835	1.1981
0.7623	0.0051	1.3044	1.5770	0.0135	1.2907	0.7540	0.3768	0.0070	0.9582	1.2713	0.0127	1.3789	1.1962
0.7612	0.0065	1.3028	1.5739	0.0172	1.2855	0.7507	0.3750	0.0089	0.9558	1.2680	0.0161	1.3738	1.1941
0.7600	0.0080	1.3011	1.5706	0.0213	1.2797	0.7470	0.3731	0.0110	0.9531	1.2644	0.0200	1.3681	1.1917
0.7586	0.0098	1.2991	1.5669	0.0259	1.2733	0.7429	0.3710	0.0134	0.9502	1.2604	0.0242	1.3618	1.1891
0.7572	0.0116	1.2970	1.5628	0.0309	1.2663	0.7384	0.3686	0.0160	0.9470	1.2561	0.0289	1.3548	1.1863
0.7556	0.0137	1.2947	1.5584	0.0364	1.2586	0.7335	0.3660	0.0188	0.9435	1.2513	0.0341	1.3473	1.1831
0.7538	0.0159	1.2922	1.5536	0.0424	1.2504	0.7283	0.3633	0.0219	0.9397	1.2462	0.0396	1.3392	1.1798
0.7519	0.0183	1.2895	1.5485	0.0489	1.2416	0.7226	0.3603	0.0253	0.9356	1.2408	0.0457	1.3305	1.1762
0.7499	0.0209	1.2866	1.5431	0.0559	1.2321	0.7165	0.3570	0.0289	0.9312	1.2349	0.0522	1.3212	1.1723
0.7478	0.0236	1.2836	1.5374	0.0635	1.2221	0.7101	0.3536	0.0328	0.9265	1.2287	0.0592	1.3113	1.1682
0.7455	0.0266	1.2803	1.5313	0.0716	1.2115	0.7032	0.3499	0.0370	0.9216	1.2222	0.0667	1.3009	1.1639
0.7431	0.0297	1.2768	1.5249	0.0802	1.2002	0.6959	0.3460	0.0414	0.9163	1.2153	0.0747	1.2898	1.1593
0.7406	0.0330	1.2732	1.5182	0.0895	1.1884	0.6883	0.3419	0.0462	0.9108	1.2080	0.0833	1.2782	1.1545
0.7379	0.0365	1.2694	1.5112	0.0994	1.1760	0.6802	0.3375	0.0513	0.9050	1.2004	0.0923	1.2660	1.1495
0.7350	0.0402	1.2653	1.5038	0.1099	1.1630	0.6717	0.3329	0.0566	0.8988	1.1924	0.1020	1.2532	1.1442
0.7321	0.0441	1.2611	1.4962	0.1210	1.1494	0.6628	0.3281	0.0623	0.8924	1.1841	0.1122	1.2399	1.1387
0.7290	0.0482	1.2567	1.4882	0.1329	1.1353	0.6534	0.3230	0.0684	0.8857	1.1755	0.1231	1.2260	1.1329
0.7257	0.0524	1.2520	1.4800	0.1454	1.1205	0.6437	0.3176	0.0748	0.8786	1.1666	0.1346	1.2115	1.1269
0.7223	0.0569	1.2472	1.4715	0.1588	1.1052	0.6335	0.3120	0.0816	0.8713	1.1573	0.1467	1.1964	1.1207
0.7187	0.0617	1.2421	1.4627	0.1729	1.0893	0.6229	0.3061	0.0888	0.8637	1.1477	0.1595	1.1809	1.1143
0.7150	0.0666	1.2369	1.4537	0.1879	1.0729	0.6118	0.2999	0.0963	0.8557	1.1377	0.1731	1.1647	1.1076
0.7112	0.0718	1.2314	1.4444	0.2038	1.0559	0.6003	0.2935	0.1043	0.8475	1.1275	0.1875	1.1480	1.1007
0.7072	0.0772	1.2257	1.4348	0.2206	1.0383	0.5884	0.2867	0.1128	0.8389	1.1170	0.2026	1.1308	1.0936
0.7030	0.0828	1.2198	1.4250	0.2384	1.0201	0.5760	0.2797	0.1217	0.8300	1.1061	0.2186	1.1130	1.0863
0.6987	0.0887	1.2136	1.4150	0.2574	1.0015	0.5631	0.2723	0.1312	0.8208	1.0950	0.2355	1.0947	1.0788
0.6942	0.0948	1.2073	1.4048	0.2774	0.9822	0.5498	0.2647	0.1411	0.8112	1.0836	0.2534	1.0758	1.0710
0.6896	0.1012	1.2007	1.3943	0.2988	0.9625	0.5360	0.2567	0.1517	0.8013	1.0719	0.2722	1.0565	1.0630
0.6847	0.1078	1.1938	1.3837	0.3214	0.9422	0.5217	0.2484	0.1628	0.7911	1.0600	0.2922	1.0366	1.0549
0.6798	0.1148	1.1867	1.3729	0.3455	0.9214	0.5070	0.2397	0.1746	0.7806	1.0478	0.3134	1.0162	1.0465
0.6746	0.1220	1.1794	1.3619	0.3712	0.9001	0.4917	0.2306	0.1870	0.7697	1.0354	0.3357	0.9953	1.0379
0.6693	0.1295	1.1717	1.3508	0.3985	0.8782	0.4759	0.2212	0.2002	0.7585	1.0227	0.3595	0.9738	1.0291
0.6637	0.1374	1.1638	1.3396	0.4278	0.8559	0.4595	0.2114	0.2142	0.7469	1.0098	0.3847	0.9519	1.0202
0.6580	0.1456	1.1556	1.3282	0.4590	0.8331	0.4426	0.2012	0.2290	0.7350	0.9967	0.4115	0.9295	1.0017
0.6521	0.1541	1.1472	1.3168	0.4924	0.8099	0.4252	0.1905	0.2447	0.7227	0.9834	0.4399	0.9067	0.9923
0.6460	0.1630	1.1384	1.3053	0.5283	0.7862	0.4072	0.1794	0.2614	0.7100	0.9699	0.4703	0.8833	0.9826
0.6397	0.1722	1.1292	1.2937	0.5669	0.7620	0.3885	0.1678	0.2791	0.6970	0.9563	0.5027	0.8595	0.9728
0.6332	0.1819	1.1198	1.2822	0.6085	0.7375	0.3693	0.1558	0.2981	0.6836	0.9425	0.5374	0.8353	0.9728
0.6265	0.1920	1.1099	1.2706	0.6535	0.7125	0.3494	0.1432	0.3183	0.6697	0.9285	0.5745	0.8106	0.9629
0.6195	0.2025	1.0997	1.2591	0.7022	0.6871	0.3288	0.1300	0.3398	0.6555	0.9145	0.6143	0.7855	0.9528
0.6124	0.2135	1.0891	1.2478	0.7550	0.6614	0.3075	0.1163	0.3630	0.6409	0.9003	0.6572	0.7599	0.9426
0.6050	0.2251	1.0780	1.2365	0.8126	0.6354	0.2855	0.1019	0.3878	0.6259	0.8861	0.7035	0.7430	0.9323
0.5973	0.2372	1.0665	1.2254	0.8756	0.6090	0.2628	0.0869	0.4144	0.6104	0.8718	0.7536	0.7077	0.9219
0.5894	0.2498	1.0545	1.2145	0.9446	0.5824	0.2392	0.0712	0.4432	0.5946	0.8575	0.8079	0.6810	0.9114
0.5813	0.2632	1.0420	1.2039	1.0206	0.5556	0.2147	0.0548	0.4743	0.5783	0.8432	0.8671	0.6540	0.9008
0.5729	0.2772	1.0289	1.1937	1.1047	0.5285	0.1894	0.0375	0.5080	0.5615	0.8290	0.9317	0.6267	0.8903
0.5642	0.2920	1.0152	1.1838	1.1980	0.5013	0.1631	0.0194	0.5447	0.5443	0.8148	1.0027	0.5990	0.8797
0.5552	0.3076	1.0008	1.1744	1.3020	0.4740	0.1357	0.0004	0.5847	0.5266	0.8007	1.0808	0.5711	0.8691
0.5460	0.3242	0.9858	1.1656	1.4187	0.4467	0.1073	-0.0197	0.6285	0.5085	0.7868	1.1672	0.5430	0.8586
0.5364	0.3417	0.9700	1.1575	1.5502	0.4194	0.0777	-0.0408	0.6768	0.4899	0.7731	1.2633	0.5147	0.8482
0.5265	0.3605	0.9533	1.1501	1.6993	0.3921	0.0468	-0.0631	0.7301	0.4708	0.7597	1.3707	0.4862	0.8380
0.5163	0.3805	0.9358	1.1437	1.8693	0.3651	0.0145	-0.0867	0.7893	0.4512	0.7465	1.4914	0.4576	0.8280
0.5057	0.4020	0.9174	1.1382	2.0645	0.3382	-0.0191	-0.1117	0.8554	0.4312	0.7338	1.6279	0.4290	0.8182
0.4948	0.4251	0.8979	1.1340	2.2902	0.3117	-0.0545	-0.1382	0.9295	0.4107	0.7215	1.7833	0.4004	0.8087
L_1	L_2	C_2	L_3	L_4	C_4	L_5	L_1	L_2	C_2	L_3	L_4	C_4	L_5

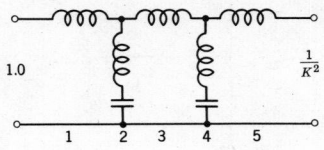

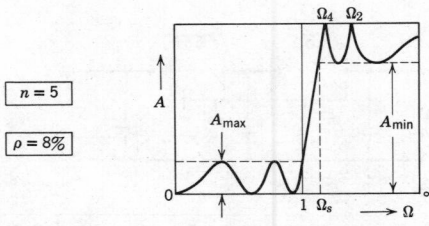

$n = 5$

$\rho = 8\%$

θ	Ω_K	A_{MIN}	σ_0	σ_1	σ_3	Ω_1	Ω_2	Ω_3	Ω_4
c	∞	∞	0.68878	−0.21285	−0.55724	1.1548	∞	0.7137	∞
2.0	28.6537	171.96	0.68909	−0.21265	−0.55720	1.1547	48.7389	0.7140	30.1274
3.0	19.1073	154.34	0.68948	−0.21241	−0.55715	1.1546	32.4927	0.7144	20.0893
4.0	14.3356	141.84	0.69002	−0.21208	−0.55708	1.1545	24.3697	0.7149	15.0716
5.0	11.4737	132.14	0.69072	−0.21165	−0.55700	1.1543	19.4959	0.7156	12.0620
6.0	9.5668	124.21	0.69158	−0.21112	−0.55689	1.1542	16.2468	0.7165	10.0565
7.0	8.2055	117.50	0.69259	−0.21050	−0.55677	1.1539	13.9260	0.7175	8.6247
8.0	7.1853	111.69	0.69376	−0.20978	−0.55662	1.1537	12.1854	0.7186	7.5516
9.0	6.3925	106.55	0.69510	−0.20897	−0.55645	1.1534	10.8316	0.7200	6.7175
10.0	5.7588	101.96	0.69660	−0.20807	−0.55626	1.1530	9.7486	0.7214	6.0507
11.0	5.2408	97.79	0.69826	−0.20707	−0.55604	1.1527	8.8625	0.7231	5.5057
12.0	4.8097	93.99	0.70009	−0.20598	−0.55579	1.1522	8.1241	0.7249	5.0520
13.0	4.4454	90.48	0.70209	−0.20480	−0.55552	1.1518	7.4993	0.7268	4.6684
14.0	4.1336	87.24	0.70426	−0.20353	−0.55522	1.1513	6.9638	0.7290	4.3401
15.0	3.8637	84.21	0.70661	−0.20216	−0.55489	1.1508	6.4997	0.7313	4.0559
16.0	3.6280	81.37	0.70913	−0.20071	−0.55452	1.1502	6.0936	0.7337	3.8076
17.0	3.4203	78.70	0.71184	−0.19917	−0.55412	1.1496	5.7353	0.7363	3.5888
18.0	3.2361	76.18	0.71473	−0.19753	−0.55368	1.1490	5.4168	0.7391	3.3946
19.0	3.0716	73.79	0.71782	−0.19581	−0.55319	1.1483	5.1318	0.7421	3.2212
20.0	2.9238	71.51	0.72109	−0.19400	−0.55267	1.1476	4.8753	0.7452	3.0654
21.0	2.7904	69.35	0.72457	−0.19211	−0.55210	1.1469	4.6433	0.7485	2.9246
22.0	2.6695	67.28	0.72825	−0.19013	−0.55148	1.1461	4.4323	0.7519	2.7970
23.0	2.5593	65.30	0.73214	−0.18807	−0.55080	1.1452	4.2397	0.7556	2.6807
24.0	2.4586	63.39	0.73625	−0.18592	−0.55007	1.1444	4.0631	0.7594	2.5743
25.0	2.3662	61.56	0.74058	−0.18370	−0.54928	1.1434	3.9007	0.7634	2.4767
26.0	2.2812	59.80	0.74514	−0.18139	−0.54842	1.1425	3.7507	0.7676	2.3868
27.0	2.2027	58.10	0.74993	−0.17901	−0.54750	1.1414	3.6119	0.7719	2.3038
28.0	2.1301	56.45	0.75497	−0.17654	−0.54650	1.1404	3.4829	0.7765	2.2270
29.0	2.0627	54.86	0.76027	−0.17401	−0.54542	1.1393	3.3729	0.7812	2.1556
30.0	2.0000	53.32	0.76583	−0.17139	−0.54425	1.1381	3.2508	0.7861	2.0892
31.0	1.9416	51.82	0.77166	−0.16871	−0.54300	1.1370	3.1460	0.7912	2.0274
32.0	1.8871	50.37	0.77777	−0.16595	−0.54165	1.1357	3.0476	0.7965	1.9695
33.0	1.8361	48.95	0.78414	−0.16313	−0.54019	1.1344	2.9553	0.8020	1.9154
34.0	1.7883	47.57	0.79090	−0.16023	−0.53862	1.1331	2.8683	0.8077	1.8646
35.0	1.7434	46.23	0.79794	−0.15728	−0.53694	1.1317	2.7864	0.8136	1.8170
36.0	1.7013	44.91	0.80532	−0.15425	−0.53512	1.1303	2.7089	0.8197	1.7722
37.0	1.6616	43.63	0.81305	−0.15117	−0.53318	1.1288	2.6356	0.8259	1.7299
38.0	1.6243	42.38	0.82114	−0.14803	−0.53109	1.1273	2.5662	0.8324	1.6901
39.0	1.5890	41.15	0.82963	−0.14483	−0.52884	1.1257	2.5003	0.8391	1.6525
40.0	1.5557	39.95	0.83852	−0.14157	−0.52644	1.1240	2.4377	0.8460	1.6170
41.0	1.5243	38.77	0.84784	−0.13826	−0.52386	1.1223	2.3781	0.8531	1.5833
42.0	1.4945	37.62	0.85761	−0.13490	−0.52109	1.1206	2.3213	0.8604	1.5515
43.0	1.4663	36.48	0.86786	−0.13150	−0.51813	1.1188	2.2672	0.8679	1.5213
44.0	1.4396	35.37	0.87862	−0.12805	−0.51495	1.1170	2.2154	0.8756	1.4926
45.0	1.4142	34.27	0.88991	−0.12455	−0.51156	1.1151	2.1660	0.8835	1.4654
46.0	1.3902	33.19	0.90177	−0.12102	−0.50792	1.1131	2.1187	0.8916	1.4396
47.0	1.3673	32.13	0.91424	−0.11745	−0.50403	1.1111	2.0733	0.8999	1.4150
48.0	1.3456	31.08	0.92736	−0.11384	−0.49987	1.1090	2.0299	0.9084	1.3916
49.0	1.3250	30.05	0.94117	−0.11021	−0.49543	1.1069	1.9881	0.9170	1.3693
50.0	1.3054	29.03	0.95571	−0.10655	−0.49068	1.1047	1.9480	0.9259	1.3481
51.0	1.2868	28.02	0.97105	−0.10286	−0.48560	1.1025	1.9095	0.9349	1.3279
52.0	1.2690	27.03	0.98724	−0.09915	−0.48019	1.1002	1.8724	0.9441	1.3087
53.0	1.2521	26.04	1.00435	−0.09543	−0.47441	1.0979	1.8366	0.9535	1.2903
54.0	1.2361	25.07	1.02246	−0.09169	−0.46825	1.0955	1.8021	0.9630	1.2728
55.0	1.2208	24.11	1.04164	−0.08795	−0.46169	1.0930	1.7689	0.9726	1.2561
56.0	1.2062	23.16	1.06198	−0.08410	−0.45469	1.0905	1.7368	0.9824	1.2402
57.0	1.1924	22.21	1.08360	−0.08044	−0.44724	1.0880	1.7057	0.9923	1.2250
58.0	1.1792	21.28	1.10660	−0.07669	−0.43932	1.0854	1.6757	1.0022	1.2104
59.0	1.1666	20.35	1.13113	−0.07295	−0.43089	1.0827	1.6467	1.0122	1.1966
60.0	1.1547	19.43	1.15731	−0.06923	−0.42193	1.0800	1.6185	1.0222	1.1834
θ	Ω_K	A_{MIN}	σ_0	σ_1	σ_3	Ω_1	Ω_2	Ω_3	Ω_4

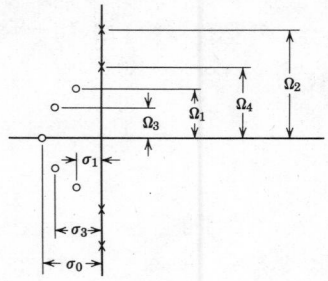

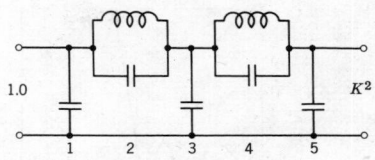

$K^2=1.0$							$K^2=\infty$						
c_1	c_2	L_2	c_3	c_4	L_4	c_5	c_1	c_2	L_2	c_3	c_4	L_4	c_5
0.8973	0.0000	1.3590	1.7270	0.0000	1.3590	0.8973	0.4486	0.0000	1.0809	1.3946	0.0000	1.48729	1.2765
0.8970	0.0003	1.3589	1.7257	0.0008	1.3581	0.8965	0.4483	0.0004	1.0805	1.3939	0.0007	1.48613	1.2760
0.8967	0.0007	1.3584	1.7248	0.0018	1.3566	0.8955	0.4478	0.0009	1.0798	1.3929	0.0017	1.48462	1.2754
0.8962	0.0012	1.3578	1.7235	0.0032	1.3546	0.8942	0.4472	0.0016	1.0789	1.3916	0.0030	1.48261	1.2745
0.8957	0.0019	1.3570	1.7218	0.0050	1.3520	0.8925	0.4464	0.0024	1.0777	1.3899	0.0046	1.47991	1.2733
0.8950	0.0027	1.3560	1.7197	0.0073	1.3489	0.8905	0.4454	0.0035	1.0763	1.3878	0.0067	1.4767	1.2719
0.8942	0.0038	1.3548	1.7173	0.0099	1.3452	0.8880	0.4442	0.0048	1.0746	1.3854	0.0091	1.4729	1.2702
0.8932	0.0049	1.3535	1.7145	0.0130	1.3409	0.8852	0.4429	0.0063	1.0727	1.3826	0.0119	1.4684	1.2683
0.8922	0.0063	1.3519	1.7113	0.0165	1.3360	0.8820	0.4413	0.0080	1.0704	1.3794	0.0151	1.4634	1.2661
0.8910	0.0077	1.3502	1.7078	0.0205	1.3306	0.8784	0.4396	0.0099	1.0680	1.3758	0.0187	1.4578	1.2637
0.8896	0.0094	1.3483	1.7039	0.0249	1.3245	0.8745	0.4377	0.0120	1.0652	1.3719	0.0227	1.4517	1.2610
0.8882	0.0112	1.3462	1.6996	0.0297	1.3180	0.8701	0.4356	0.0143	1.0622	1.3676	0.0271	1.4449	1.2580
0.8866	0.0132	1.3440	1.6950	0.0350	1.3108	0.8654	0.4333	0.0168	1.0589	1.3629	0.0319	1.4376	1.2548
0.8849	0.0153	1.3415	1.6900	0.0407	1.3031	0.8603	0.4308	0.0195	1.0554	1.3579	0.0371	1.4297	1.2514
0.8831	0.0176	1.3389	1.6846	0.0469	1.2949	0.8549	0.4282	0.0225	1.0516	1.3525	0.0428	1.4212	1.2477
0.8811	0.0201	1.3361	1.6789	0.0536	1.2860	0.8490	0.4253	0.0257	1.0475	1.3467	0.0488	1.4121	1.2437
0.8790	0.0228	1.3331	1.6729	0.0608	1.2766	0.8428	0.4222	0.0291	1.0432	1.3406	0.0554	1.4025	1.2395
0.8768	0.0256	1.3299	1.6664	0.0685	1.2667	0.8362	0.4190	0.0328	1.0386	1.3341	0.0623	1.3923	1.2350
0.8744	0.0286	1.3265	1.6597	0.0767	1.2562	0.8292	0.4155	0.0367	1.0337	1.3272	0.0698	1.3815	1.2303
0.8719	0.0318	1.3230	1.6526	0.0854	1.2451	0.8218	0.4118	0.0409	1.0285	1.3200	0.0777	1.3702	1.2254
0.8692	0.0351	1.3192	1.6451	0.0947	1.2335	0.8140	0.4080	0.0453	1.0231	1.3125	0.0861	1.3583	1.2202
0.8665	0.0387	1.3153	1.6373	0.1046	1.2213	0.8058	0.4039	0.0500	1.0174	1.3046	0.0950	1.3458	1.2147
0.8636	0.0424	1.3111	1.6292	0.1151	1.2086	0.7973	0.3996	0.0550	1.0114	1.2964	0.1044	1.3328	1.2091
0.8605	0.0463	1.3068	1.6208	0.1262	1.1954	0.7883	0.3951	0.0603	1.0051	1.2878	0.1144	1.3192	1.2031
0.8573	0.0504	1.3023	1.6120	0.1379	1.1816	0.7790	0.3903	0.0658	0.9986	1.2780	0.1249	1.3051	1.1970
0.8540	0.0547	1.2975	1.6029	0.1503	1.1672	0.7692	0.3854	0.0717	0.9917	1.2696	0.1360	1.2904	1.1906
0.8505	0.0593	1.2926	1.5935	0.1634	1.1523	0.7591	0.3802	0.0779	0.9846	1.2600	0.1477	1.2752	1.1839
0.8469	0.0640	1.2875	1.5838	0.1773	1.1369	0.7485	0.3748	0.0844	0.9772	1.2501	0.1601	1.2595	1.1770
0.8431	0.0689	1.2821	1.5738	0.1919	1.1210	0.7376	0.3691	0.0912	0.9695	1.2399	0.1731	1.2432	1.1699
0.8392	0.0741	1.2765	1.5635	0.2074	1.1045	0.7262	0.3632	0.0984	0.9614	1.2293	0.1868	1.2264	1.1626
0.8351	0.0795	1.2708	1.5529	0.2237	1.0876	0.7144	0.3571	0.1060	0.9531	1.2185	0.2012	1.2090	1.1550
0.8308	0.0851	1.2648	1.5421	0.2409	1.0701	0.7022	0.3507	0.1140	0.9445	1.2073	0.2164	1.1911	1.1472
0.8264	0.0909	1.2586	1.5309	0.2590	1.0521	0.6896	0.3440	0.1224	0.9356	1.1958	0.2324	1.1728	1.1391
0.8219	0.0970	1.2521	1.5195	0.2782	1.0336	0.6766	0.3370	0.1312	0.9264	1.1840	0.2493	1.1539	1.1309
0.8172	0.1034	1.2455	1.5079	0.2985	1.0146	0.6631	0.3298	0.1405	0.9168	1.1720	0.2670	1.1344	1.1224
0.8123	0.1100	1.2386	1.4960	0.3199	0.9950	0.6492	0.3223	0.1502	0.9070	1.1596	0.2857	1.1145	1.1137
0.8072	0.1169	1.2314	1.4838	0.3426	0.9751	0.6348	0.3145	0.1605	0.8968	1.1470	0.3054	1.0941	1.1048
0.8020	0.1240	1.2240	1.4715	0.3667	0.9546	0.6200	0.3064	0.1713	0.8863	1.1341	0.3262	1.0732	1.0957
0.7966	0.1315	1.2164	1.4589	0.3922	0.9336	0.6048	0.2979	0.1827	0.8754	1.1210	0.3481	1.0519	1.0864
0.7910	0.1392	1.2084	1.4461	0.4192	0.9122	0.5890	0.2892	0.1947	0.8643	1.1076	0.3713	1.0300	1.0768
0.7853	0.1473	1.2002	1.4332	0.4479	0.8904	0.5728	0.2801	0.2074	0.8527	1.0939	0.3958	1.0077	1.0671
0.7793	0.1557	1.1918	1.4200	0.4785	0.8681	0.5561	0.2706	0.2207	0.8408	1.0800	0.4218	0.9849	1.0571
0.7731	0.1644	1.1830	1.4068	0.5111	0.8453	0.5389	0.2608	0.2348	0.8286	1.0659	0.4493	0.9616	1.0470
0.7668	0.1735	1.1739	1.3934	0.5459	0.8221	0.5212	0.2506	0.2497	0.8160	1.0516	0.4785	0.9379	1.0367
0.7602	0.1830	1.1645	1.3798	0.5830	0.7986	0.5030	0.2399	0.2654	0.8030	1.0370	0.5096	0.9138	1.0262
0.7534	0.1929	1.1548	1.3662	0.6229	0.7746	0.4842	0.2289	0.2821	0.7896	1.0223	0.5426	0.8893	1.0155
0.7464	0.2032	1.1447	1.3525	0.6657	0.7502	0.4648	0.2174	0.2998	0.7759	1.0074	0.5779	0.8643	1.0046
0.7392	0.2139	1.1343	1.3387	0.7117	0.7255	0.4449	0.2055	0.3186	0.7617	0.9923	0.6155	0.8389	0.9936
0.7318	0.2251	1.1235	1.3249	0.7614	0.7004	0.4244	0.1931	0.3386	0.7471	0.9771	0.6559	0.8131	0.9824
0.7241	0.2369	1.1123	1.3112	0.8151	0.6750	0.4033	0.1801	0.3599	0.7321	0.9618	0.6992	0.7870	0.9711
0.7161	0.2491	1.1006	1.2974	0.8733	0.6492	0.3815	0.1666	0.3827	0.7167	0.9463	0.7457	0.7604	0.9597
0.7069	0.2620	1.0885	1.2838	0.9368	0.6232	0.3590	0.1525	0.4070	0.7008	0.9307	0.7960	0.7336	0.9481
0.6995	0.2755	1.0759	1.2702	1.0060	0.5970	0.3358	0.1378	0.4331	0.6844	0.9151	0.8503	0.7063	0.9364
0.6907	0.2897	1.0628	1.2568	1.0819	0.5705	0.3119	0.1225	0.4612	0.6676	0.8994	0.9093	0.6788	0.9246
0.6817	0.3046	1.0491	1.2436	1.1654	0.5438	0.2872	0.1065	0.4915	0.6503	0.8836	0.9736	0.6510	0.9127
0.6723	0.3203	1.0348	1.2306	1.2576	0.5170	0.2617	0.0897	0.5242	0.6325	0.8679	1.0438	0.6229	0.9007
0.6627	0.3369	1.0199	1.2180	1.3599	0.4900	0.2352	0.0721	0.5596	0.6142	0.8522	1.1209	0.5946	0.8888
0.6527	0.3545	1.0044	1.2057	1.4740	0.4630	0.2079	0.0536	0.5982	0.5953	0.8366	1.2058	0.5660	0.8768
0.6424	0.3732	0.9880	1.1939	1.6018	0.4359	0.1795	0.0342	0.6404	0.5759	0.8211	1.2999	0.5373	0.8648
0.6317	0.3931	0.9709	1.1827	1.7459	0.4089	0.1501	0.0138	0.6867	0.5559	0.8057	1.4046	0.5084	0.8528
L_1	L_2	c_2	L_3	L_4	c_4	L_5	L_1	L_2	c_2	L_3	L_4	c_4	L_5

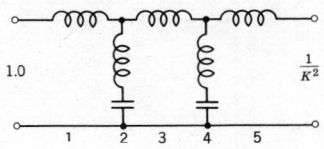

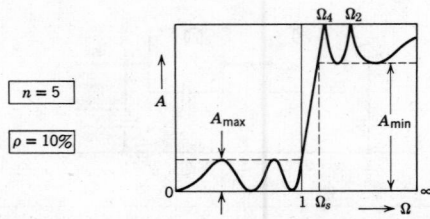

$n = 5$

$\rho = 10\%$

θ	Ω_K	A_{MIN}	σ_0	σ_1	σ_3	Ω_1	Ω_2	Ω_3	Ω_4
C	∞	∞	0.63505	-0.19624	-0.51376	1.1266	∞	0.6963	∞
1.0	57.2987	204.02	0.63511	-0.19619	-0.51375	1.1266	97.4775	0.6963	60.2470
2.0	28.6537	173.91	0.63531	-0.19607	-0.51373	1.1265	48.7389	0.6965	30.1274
3.0	19.1073	156.30	0.63565	-0.19586	-0.51369	1.1265	32.4927	0.6969	20.0893
4.0	14.3356	143.80	0.63613	-0.19557	-0.51364	1.1264	24.3697	0.6974	15.0716
5.0	11.4737	134.10	0.63674	-0.19520	-0.51357	1.1262	19.4959	0.6980	12.0620
6.0	9 5668	126.17	0.63749	-0.19475	-0.51349	1.1261	16.2468	0.6987	10.0565
7.0	8.2055	119.46	0.63838	-0.19421	-0.51338	1.1259	13.9260	0.6996	8.6247
8.0	7.1853	113.64	0.63941	-0.19359	-0.51327	1.1257	12.1854	0.7007	7.5516
9.0	6.3925	108.51	0.64059	-0.19289	-0.51313	1.1255	10.8316	0.7019	6.7175
10.0	5.7588	103.91	0.64190	-0.19211	-0.51297	1.1252	9.7486	0.7032	6.0507
11.0	5.2408	99.75	0.64336	-0.19125	-0.51280	1.1250	8.8625	0.7047	5.5057
12.0	4.8097	95.94	0.64496	-0.19031	-0.51260	1.1247	8.1241	0.7063	5.0520
13.0	4.4454	92.44	0.64671	-0.18928	-0.51238	1.1243	7.4993	0.7080	4.6684
14.0	4.1336	89.19	0.64861	-0.18818	-0.51213	1.1239	6.9638	0.7100	4.3401
15.0	3.8637	86.16	0.65067	-0.18700	-0.51187	1.1236	6.4997	0.7120	4.0559
16.0	3.6280	83.32	0.65288	-0.18574	-0.51157	1.1231	6.0936	0.7142	3.8076
17.0	3.4203	80.65	0.65524	-0.18440	-0.51125	1.1227	5.7353	0.7165	3.5888
18.0	3.2361	78.13	0.65777	-0.18298	-0.51089	1.1222	5.4168	0.7190	3.3946
19.0	3.0716	75.74	0.66047	-0.18148	-0.51050	1.1217	5.1318	0.7217	3.2212
20.0	2.9238	73.47	0.66333	-0.17991	-0.51008	1.1211	4.8753	0.7245	3.0654
21.0	2.7904	71.30	0.66637	-0.17827	-0.50962	1.1206	4.6433	0.7274	2.9246
22.0	2.6695	69.23	0.66958	-0.17655	-0.50912	1.1199	4.4323	0.7305	2.7970
23.0	2.5593	67.25	0.67297	-0.17475	-0.50858	1.1193	4.2397	0.7338	2.6807
24.0	2.4586	65.35	0.67656	-0.17288	-0.50799	1.1186	4.0631	0.7372	2.5743
25.0	2.3662	63.52	0.68033	-0.17094	-0.50735	1.1179	3.9007	0.7408	2.4767
26.0	2.2812	61.75	0.68430	-0.16893	-0.50666	1.1172	3.7507	0.7446	2.3868
27.0	2.2027	60.05	0.68847	-0.16685	-0.50592	1.1164	3.6119	0.7485	2.3038
28.0	2.1301	58.41	0.69285	-0.16470	-0.50511	1.1156	3.4829	0.7526	2.2270
29.0	2.0627	56.82	0.69745	-0.16248	-0.50424	1.1148	3.3629	0.7568	2.1556
30.0	2.0000	55.27	0.70228	-0.16019	-0.50330	1.1139	3.2508	0.7612	2.0892
31.0	1.9416	53.77	0.70734	-0.15784	-0.50229	1.1130	3.1460	0.7658	2.0274
32.0	1.8871	52.32	0.71264	-0.15542	-0.50120	1.1120	3.0476	0.7706	1.9695
33.0	1.8361	50.90	0.71819	-0.15294	-0.50003	1.1110	2.9553	0.7755	1.9154
34.0	1.7883	49.52	0.72400	-0.15040	-0.49877	1.1100	2.8683	0.7806	1.8646
35.0	1.7434	48.18	0.73009	-0.14779	-0.49742	1.1089	2.7864	0.7859	1.8170
36.0	1.7013	46.87	0.73645	-0.14513	-0.49596	1.1078	2.7089	0.7914	1.7722
37.0	1.6616	45.59	0.74312	-0.14241	-0.49439	1.1067	2.6356	0.7970	1.7299
38.0	1.6243	44.33	0.75009	-0.13963	-0.49271	1.1055	2.5662	0.8029	1.6901
39.0	1.5890	43.10	0.75739	-0.13680	-0.49091	1.1042	2.5003	0.8089	1.6525
40.0	1.5557	41.90	0.76503	-0.13392	-0.48897	1.1030	2.4377	0.8151	1.6170
41.0	1.5243	40.73	0.77302	-0.13099	-0.48689	1.1017	2.3781	0.8215	1.5833
42.0	1.4945	39.57	0.78140	-0.12800	-0.48467	1.1003	2.3213	0.8281	1.5515
43.0	1.4663	38.43	0.79016	-0.12497	-0.48228	1.0980	2.2672	0.8349	1.5213
44.0	1.4396	37.32	0.79935	-0.12190	-0.47973	1.0975	2.2154	0.8419	1.4926
45.0	1.4142	36.22	0.80898	-0.11878	-0.47699	1.0960	2.1660	0.8490	1.4654
46.0	1.3902	35.14	0.81907	-0.11562	-0.47406	1.0944	2.1187	0.8564	1.4396
47.0	1.3673	34.08	0.82967	-0.11242	-0.47093	1.0929	2.0733	0.8639	1.4150
48.0	1.3456	33.03	0.84079	-0.10919	-0.46758	1.0912	2.0299	0.8717	1.3916
49.0	1.3250	32.00	0.85247	-0.10593	-0.46399	1.0896	1.9881	0.8796	1.3693
50.0	1.3054	30.98	0.86475	-0.10263	-0.46016	1.0879	1.9480	0.8877	1.3481
51.0	1.2868	29.97	0.87767	-0.09930	-0.45606	1.0861	1.9095	0.8960	1.3279
52.0	1.2690	28.98	0.89128	-0.09595	-0.45169	1.0843	1.8724	0.9044	1.3087
53.0	1.2521	27.99	0.90563	-0.09258	-0.44702	1.0825	1.8366	0.9131	1.2903
54.0	1.2361	27.02	0.92076	-0.08918	-0.44203	1.0806	1.8021	0.9219	1.2728
55.0	1.2208	26.06	0.93685	-0.08577	-0.43670	1.0786	1.7689	0.9309	1.2561
56.0	1.2062	25.10	0.95367	-0.08235	-0.43102	1.0766	1.7368	0.9400	1.2402
57.0	1.1924	24.16	0.97159	-0.07892	-0.42497	1.0746	1.7057	0.9492	1.2250
58.0	1.1792	23.22	0.99059	-0.07548	-0.41851	1.0725	1.6757	0.9586	1.2104
59.0	1.1666	22.29	1.01078	-0.07204	-0.41163	1.0704	1.6467	0.9680	1.1966
60.0	1.1547	21.36	1.03227	-0.06860	-0.40430	1.0682	1.6185	0.9776	1.1834
θ	Ω_K	A_{MIN}	σ_0	σ_1	σ_3	Ω_1	Ω_2	Ω_3	Ω_4

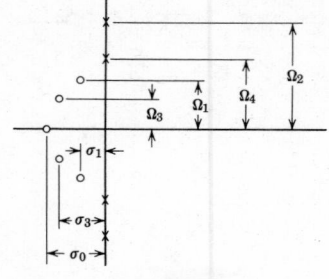

214

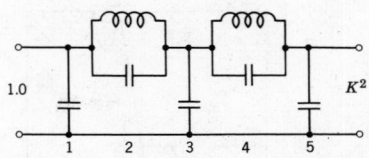

K²=1.0							K²=∞						
C_1	C_2	L_2	C_3	C_4	L_4	C_5	C_1	C_2	L_2	C_3	C_4	L_4	C_5
0.9732	0.0000	1.372	1.803	0.0000	1.372	0.9732	0.4866	0.0000	1.1388	1.4496	0.0000	1.5271	1.3109
0.97315	0.00008	1.37219	1.80298	0.00020	1.37200	0.97302	0.4865	0.0001	1.1387	1.4494	0.0002	1.5268	1.3107
0.97296	0.00031	1.37192	1.80240	0.00080	1.37117	0.97246	0.4863	0.0004	1.1383	1.4488	0.0007	1.5260	1.3104
0.97265	0.00069	1.37148	1.80144	0.00181	1.36978	0.97153	0.4858	0.0008	1.1377	1.4479	0.0016	1.5245	1.3097
0.97221	0.00123	1.37086	1.80009	0.00322	1.36784	0.97022	0.4852	0.0015	1.1368	1.4466	0.0029	1.5225	1.3088
0.97164	0.00192	1.37007	1.79835	0.00503	1.36535	0.96854	0.4845	0.0023	1.1356	1.4449	0.0045	1.5198	1.3076
0.97095	0.00277	1.36909	1.79624	0.00726	1.36230	0.96649	0.4835	0.0033	1.1343	1.4428	0.0065	1.5166	1.3062
0.97014	0.00377	1.36794	1.79374	0.00989	1.35870	0.96406	0.4824	0.0046	1.1326	1.4404	0.0089	1.5128	1.3045
0.96920	0.00493	1.36662	1.79086	0.01295	1.35455	0.96126	0.4811	0.0060	1.1307	1.4376	0.0116	1.5085	1.3025
0.96813	0.00624	1.36511	1.78759	0.01642	1.34985	0.95808	0.4797	0.0076	1.1286	1.4344	0.0147	1.5035	1.3003
0.96693	0.00772	1.36342	1.78395	0.02031	1.34460	0.95453	0.4780	0.0093	1.1262	1.4309	0.0182	1.4980	1.2978
0.96561	0.00935	1.36156	1.77993	0.02464	1.33880	0.95061	0.4762	0.0113	1.1235	1.4269	0.0221	1.4919	1.2951
0.96416	0.01114	1.35952	1.77553	0.02941	1.33245	0.94630	0.4742	0.0135	1.1206	1.4226	0.0264	1.4852	1.2921
0.96258	0.01310	1.35729	1.77076	0.03461	1.32555	0.94163	0.4721	0.0159	1.1175	1.4180	0.0310	1.4780	1.2889
0.96087	0.01522	1.35489	1.76561	0.04028	1.31811	0.93657	0.4697	0.0185	1.1140	1.4130	0.0361	1.4701	1.2854
0.95903	0.01750	1.35230	1.76009	0.04640	1.31012	0.93115	0.4672	0.0213	1.1104	1.4076	0.0416	1.4617	1.2816
0.95706	0.01996	1.34953	1.75421	0.05299	1.30160	0.92534	0.4645	0.0234	1.1064	1.4018	0.0475	1.4528	1.2776
0.95496	0.02258	1.34658	1.74795	0.06007	1.29253	0.91915	0.4616	0.0276	1.1022	1.3957	0.0538	1.4432	1.2733
0.95272	0.02537	1.34344	1.74134	0.06764	1.28292	0.91259	0.4585	0.0310	1.0977	1.3892	0.0606	1.4331	1.2688
0.95035	0.02833	1.34012	1.73436	0.07572	1.27277	0.90565	0.4552	0.0347	1.0930	1.3824	0.0678	1.4225	1.2640
0.94785	0.03148	1.33661	1.72702	0.08432	1.26209	0.89833	0.4513	0.0387	1.0880	1.3752	0.0754	1.4113	1.2590
0.94521	0.03480	1.33292	1.71933	0.09346	1.25088	0.89062	0.4481	0.0428	1.0828	1.3677	0.0835	1.3995	1.2537
0.94244	0.03830	1.32903	1.71129	0.10316	1.23913	0.88253	0.4442	0.0473	1.0772	1.3598	0.0921	1.3872	1.2482
0.93953	0.04199	1.32495	1.70290	0.11343	1.22686	0.87406	0.4402	0.0519	1.0714	1.3516	0.1013	1.3743	1.2424
0.93647	0.04586	1.32068	1.69416	0.12429	1.21406	0.86521	0.4359	0.0569	1.0653	1.3430	0.1109	1.3609	1.2364
0.93328	0.04993	1.31622	1.68509	0.13577	1.20074	0.85596	0.4314	0.0621	1.0590	1.3341	0.1210	1.3469	1.2301
0.92994	0.05420	1.31156	1.67568	0.14789	1.18689	0.84633	0.4268	0.0675	1.0524	1.3248	0.1317	1.3324	1.2236
0.92646	0.05866	1.30670	1.66594	0.16068	1.17253	0.83631	0.4219	0.0733	1.0455	1.3152	0.1430	1.3173	1.2169
0.92284	0.06333	1.30164	1.65587	0.17418	1.15766	0.82590	0.4168	0.0794	1.0383	1.3053	0.1549	1.3017	1.2099
0.91906	0.06821	1.29637	1.64548	0.18840	1.14228	0.81509	0.4114	0.0858	1.0308	1.2950	0.1674	1.2856	1.2027
0.91514	0.07330	1.29090	1.63478	0.20339	1.12639	0.80388	0.4059	0.0925	1.0231	1.2845	0.1805	1.2690	1.1952
0.91106	0.07862	1.28522	1.62377	0.21919	1.11000	0.79227	0.4001	0.0995	1.0150	1.2736	0.1944	1.2518	1.1875
0.90683	0.08416	1.27933	1.61246	0.23584	1.09311	0.78026	0.3941	0.1070	1.0067	1.2624	0.2089	1.2341	1.1796
0.90244	0.08993	1.27322	1.60084	0.25339	1.07573	0.76784	0.3878	0.1147	0.9980	1.2508	0.2242	1.2159	1.1714
0.89789	0.09594	1.26688	1.58895	0.27188	1.05785	0.75502	0.3813	0.1229	0.9891	1.2390	0.2402	1.1972	1.1630
0.89318	0.10220	1.26033	1.57677	0.29139	1.03950	0.74177	0.3745	0.1315	0.9798	1.2269	0.2571	1.1780	1.1544
0.88830	0.10871	1.25354	1.56431	0.31197	1.02066	0.72811	0.3674	0.1404	0.9703	1.2145	0.2749	1.1583	1.1456
0.88325	0.11549	1.24651	1.55159	0.33370	1.00135	0.71403	0.3601	0.1499	0.9604	1.2018	0.2936	1.1381	1.1365
0.87803	0.12254	1.23925	1.53862	0.35665	0.98158	0.69951	0.3525	0.1598	0.9502	1.1888	0.3133	1.1175	1.1272
0.87263	0.12987	1.23173	1.52540	0.38092	0.96134	0.68456	0.3447	0.1702	0.9397	1.1755	0.3340	1.0963	1.1177
0.86706	0.13749	1.22397	1.51195	0.40660	0.94065	0.66917	0.3365	0.1812	0.9288	1.1619	0.3559	1.0747	1.1080
0.86130	0.14542	1.21594	1.49827	0.43380	0.91952	0.65332	0.3280	0.1927	0.9176	1.1481	0.3790	1.0526	1.0980
0.85535	0.15367	1.20763	1.48438	0.46265	0.89794	0.63702	0.3192	0.2048	0.9061	1.1341	0.4033	1.0300	1.0879
0.84920	0.16225	1.19905	1.47030	0.49328	0.87594	0.62025	0.3101	0.2176	0.8942	1.1198	0.4291	1.0070	1.0776
0.84286	0.17119	1.19018	1.45603	0.52586	0.85352	0.60301	0.3006	0.2310	0.8820	1.1052	0.4564	0.9835	1.0670
0.83631	0.18048	1.18101	1.44159	0.56057	0.83069	0.58528	0.2907	0.2452	0.8694	1.0904	0.4852	0.9596	1.0563
0.82956	0.19016	1.17152	1.42699	0.59759	0.80746	0.56704	0.2805	0.2601	0.8564	1.0754	0.5159	0.9353	1.0453
0.82258	0.20025	1.16170	1.41227	0.63717	0.78385	0.54830	0.2699	0.2759	0.8430	1.0602	0.5485	0.9106	1.0342
0.81538	0.21076	1.15155	1.39742	0.67957	0.75986	0.52902	0.2589	0.2927	0.8293	1.0448	0.5832	0.8854	1.0229
0.80795	0.22173	1.14103	1.32849	0.72508	0.73551	0.50920	0.2474	0.3104	0.8151	1.0293	0.6202	0.8599	1.0114
0.80029	0.23317	1.13013	1.36748	0.77404	0.71083	0.48882	0.2355	0.3292	0.8005	1.0135	0.6598	0.8339	0.9998
0.79237	0.24514	1.11884	1.35242	0.82687	0.68582	0.46786	0.2231	0.3492	0.7855	0.9976	0.7022	0.8076	0.9880
0.78420	0.25765	1.10713	1.33735	0.88402	0.66050	0.44629	0.2102	0.3704	0.7700	0.9815	0.7477	0.7809	0.9760
0.77576	0.27075	1.09497	1.32228	0.94602	0.63490	0.42410	0.1968	0.3931	0.7541	0.9653	0.7967	0.7539	0.9639
0.76705	0.28448	1.08235	1.30726	1.01351	0.60905	0.40124	0.1828	0.4174	0.7377	0.9490	0.8496	0.7266	0.9516
0.75805	0.29891	1.06922	1.29232	1.08722	0.58296	0.37770	0.1682	0.4434	0.7208	0.9326	0.9068	0.6989	0.9393
0.74875	0.31408	1.05556	1.27750	1.16804	0.55666	0.35343	0.1530	0.4713	0.7034	0.9162	0.9690	0.6710	0.9268
0.73914	0.33006	1.04132	1.26284	1.25698	0.53019	0.32840	0.1371	0.5014	0.6855	0.8997	1.0368	0.6428	0.9142
0.72920	0.34694	1.02648	1.24841	1.35531	0.50358	0.30256	0.1204	0.5338	0.6671	0.8831	1.1111	0.6143	0.9016
0.71892	0.36479	1.01097	1.23424	1.46451	0.47688	0.27587	0.1030	0.5690	0.6481	0.8666	1.1926	0.5856	0.8888
0.70829	0.38374	0.99477	1.22041	1.58638	0.45013	0.24827	0.0847	0.6073	0.6285	0.8502	1.2826	0.5567	0.8761
L_1	L_2	C_2	L_3	L_4	C_4	L_5	L_1	L_2	C_2	L_3	L_4	C_4	L_5

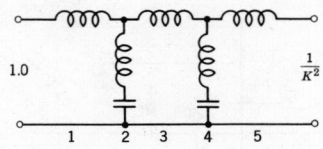

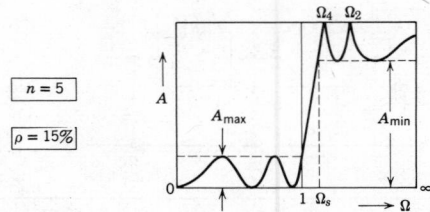

θ	Ω_K	A_MIN	σ_0	σ_1	σ_3	Ω_1	Ω_2	Ω_3	Ω_4
C	∞	∞	0.54025	−0.16695	−0.43707	1.0810	∞	0.6681	∞
1.0	57.2987	207.60	0.54030	−0.16691	−0.43706	1.0809	97.4775	0.6681	60.2470
2.0	28.6537	177.49	0.54046	−0.16682	−0.43705	1.0809	48.7389	0.6683	30.1274
3.0	19.1073	159.88	0.54072	−0.16666	−0.43702	1.0809	32.4927	0.6685	20.0893
4.0	14.3356	147.37	0.54110	−0.16644	−0.43699	1.0808	24.3697	0.6689	15.0716
5.0	11.4737	137.67	0.54158	−0.16615	−0.43694	1.0808	19.4959	0.6695	12.0620
6.0	9.5668	129.74	0.54216	−0.16581	−0.43689	1.0807	16.2468	0.6701	10.0565
7.0	8.2055	123.03	0.54286	−0.16540	−0.43682	1.0806	13.9260	0.6708	8.6247
8.0	7.1853	117.22	0.54367	−0.16493	−0.43674	1.0805	12.1854	0.6717	7.5516
9.0	6.3925	112.08	0.54458	−0.16439	−0.43665	1.0804	10.8316	0.6727	6.7175
10.0	5.7588	107.49	0.54561	−0.16380	−0.43655	1.0802	9.7486	0.6738	6.0507
11.0	5.2408	103.32	0.54675	−0.16314	−0.43643	1.0801	8.8625	0.6750	5.5057
12.0	4.8097	99.52	0.54800	−0.16242	−0.43630	1.0799	8.1241	0.6763	5.0520
13.0	4.4454	96.02	0.54937	−0.16164	−0.43615	1.0797	7.4993	0.6778	4.6684
14.0	4.1336	92.77	0.55085	−0.16080	−0.43599	1.0795	6.9638	0.6794	4.3401
15.0	3.8637	89.74	0.55246	−0.15898	−0.43581	1.0793	6.4997	0.6811	4.0559
16.0	3.6280	86.90	0.55418	−0.15893	−0.43562	1.0791	6.0936	0.6829	3.8076
17.0	3.4203	84.23	0.55603	−0.15790	−0.43450	1.0788	5.7353	0.6848	3.5888
18.0	3.2361	81.71	0.55800	−0.15681	−0.43517	1.0786	5.4168	0.6869	3.3946
19.0	3.0716	79.32	0.56009	−0.15567	−0.43491	1.0783	5.1318	0.6891	3.2212
20.0	2.9238	77.05	0.56232	−0.15446	−0.43463	1.0780	4.8753	0.6914	3.0654
21.0	2.7904	74.88	0.56468	−0.15320	−0.43432	1.0777	4.6433	0.6938	2.9246
22.0	2.6695	72.81	0.56718	−0.15187	−0.43399	1.0774	4.4323	0.6964	2.7970
23.0	2.5593	70.83	0.56981	−0.15049	−0.43363	1.0770	4.2397	0.6991	2.6807
24.0	2.4586	68.92	0.57259	−0.14905	−0.43324	1.0767	4.0631	0.7019	2.5743
25.0	2.3662	67.09	0.57551	−0.14755	−0.43282	1.0763	3.9007	0.7049	2.4767
26.0	2.2812	65.33	0.57858	−0.14599	−0.43236	1.0759	3.7507	0.7080	2.3868
27.0	2.2027	63.63	0.58181	−0.14438	−0.43186	1.0754	3.6119	0.7112	2.3038
28.0	2.1301	61.98	0.58519	−0.14271	−0.43133	1.0750	3.4829	0.7146	2.2270
29.0	2.0627	60.39	0.58874	−0.14099	−0.43075	1.0745	3.3629	0.7181	2.1556
30.0	2.0000	58.85	0.59246	−0.13921	−0.43012	1.0740	3.2508	0.7218	2.0892
31.0	1.9416	57.35	0.59635	−0.13738	−0.42945	1.0735	3.1460	0.7256	2.0274
32.0	1.8871	55.90	0.60042	−0.13550	−0.42873	1.0730	3.0476	0.7295	1.9695
33.0	1.8361	54.48	0.60468	−0.13356	−0.42795	1.0725	2.9553	0.7336	1.9154
34.0	1.7883	53.10	0.60913	−0.13157	−0.42711	1.0719	2.8683	0.7379	1.8646
35.0	1.7434	51.76	0.61378	−0.12953	−0.42621	1.0713	2.7864	0.7423	1.8170
36.0	1.7013	50.44	0.61865	−0.12744	−0.42524	1.0707	2.7089	0.7468	1.7722
37.0	1.6616	49.16	0.62373	−0.12530	−0.42419	1.0700	2.6356	0.7515	1.7299
38.0	1.6243	47.91	0.62903	−0.12311	−0.42307	1.0694	2.5662	0.7564	1.6901
39.0	1.5890	46.68	0.63457	−0.12087	−0.42187	1.0687	2.5003	0.7614	1.6525
40.0	1.5557	45.48	0.64036	−0.11859	−0.42058	1.0679	2.4377	0.7666	1.6170
41.0	1.5243	44.30	0.64641	−0.11626	−0.41919	1.0672	2.3781	0.7719	1.5833
42.0	1.4945	43.15	0.65273	−0.11389	−0.41771	1.0664	2.3213	0.7774	1.5515
43.0	1.4663	42.01	0.65933	−0.11147	−0.41612	1.0656	2.2672	0.7831	1.5213
44.0	1.4396	40.90	0.66623	−0.10902	−0.41441	1.0648	2.2154	0.7889	1.4926
45.0	1.4142	39.80	0.67344	−0.10652	−0.41258	1.0639	2.1660	0.7950	1.4654
46.0	1.3902	38.72	0.68099	−0.10398	−0.41063	1.0630	2.1187	0.8012	1.4396
47.0	1.3673	37.66	0.68888	−0.10140	−0.40583	1.0621	2.0733	0.8075	1.4150
48.0	1.3456	36.61	0.69714	−0.09879	−0.40628	1.0612	2.0299	0.8140	1.3916
49.0	1.3250	35.57	0.70579	−0.09615	−0.40388	1.0602	1.9881	0.8208	1.3693
50.0	1.3054	34.55	0.71485	−0.09347	−0.40131	1.0592	1.9480	0.8276	1.3481
51.0	1.2868	33.55	0.72436	−0.09075	−0.39856	1.0581	1.9095	0.8347	1.3279
52.0	1.2690	32.55	0.73433	−0.08801	−0.39562	1.0571	1.8724	0.8420	1.3087
53.0	1.2521	31.57	0.74481	−0.08524	−0.39248	1.0560	1.8366	0.8494	1.2903
54.0	1.2361	30.59	0.75581	−0.08245	−0.38912	1.0548	1.8021	0.8570	1.2829
55.0	1.2208	29.63	0.76739	−0.07963	−0.38552	1.0537	1.7689	0.8647	1.2561
56.0	1.2062	28.67	0.77959	−0.07679	−0.38168	1.0525	1.7368	0.8727	1.2402
57.0	1.1924	27.72	0.79245	−0.07392	−0.37757	1.0512	1.7057	0.8808	1.2250
58.0	1.1792	26.78	0.80602	−0.07105	−0.37319	1.0500	1.6757	0.8890	1.2104
59.0	1.1666	25.85	0.82036	−0.06815	−0.36850	1.0487	1.6467	0.8975	1.1966
60.0	1.1547	24.92	0.83554	−0.06525	−0.36350	1.0473	1.6185	0.9060	1.1834
θ	Ω_K	A_MIN	σ_0	σ_1	σ_3	Ω_1	Ω_2	Ω_3	Ω_4

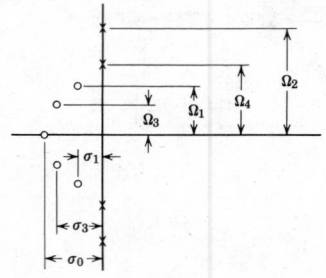

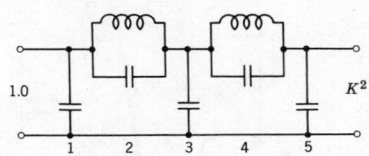

K²=1.0							K²=∞						
C_1	C_2	L_2	C_3	C_4	L_4	C_5	C_1	C_2	L_2	C_3	C_4	L_4	C_5
1.144	0.0000	1.372	1.972	0.0000	1.372	1.144	0.5720	0.0000	1.2474	1.5546	0.0000	1.5916	1.3749
1.14392	0.00008	1.37142	1.97200	0.00020	1.37124	1.14379	0.5719	0.0001	1.2473	1.5545	0.0002	1.5913	1.3748
1.14372	0.00031	1.37116	1.97139	0.00008	1.37047	1.14323	0.5717	0.0003	1.2470	1.5539	0.0007	1.5904	1.3744
1.14341	0.00069	1.37074	1.97036	0.00181	1.36919	1.14229	0.5713	0.0008	1.2464	1.5530	0.0016	1.5890	1.3737
1.14296	0.00123	1.37015	1.96892	0.00322	1.36739	1.14097	0.5707	0.0014	1.2456	1.5517	0.0028	1.5870	1.3728
1.14238	0.00192	1.36938	1.96707	0.00504	1.36507	1.13928	0.5700	0.0021	1.2445	1.5500	0.0043	1.5844	1.3716
1.14168	0.00277	1.36845	1.96481	0.00726	1.36224	1.13721	0.5692	0.0030	1.2432	1.5479	0.0063	1.5813	1.3701
1.14085	0.00377	1.36735	1.96214	0.00989	1.35890	1.13477	0.5682	0.0042	1.2416	1.5455	0.0085	1.5776	1.3684
1.13989	0.00493	1.36607	1.95906	0.01294	1.35505	1.13195	0.5670	0.0054	1.2398	1.5427	0.0111	1.5733	1.3664
1.13880	0.00625	1.36463	1.95558	0.01641	1.35068	1.12875	0.5657	0.0069	1.2378	1.5395	0.0141	1.5684	1.3641
1.13758	0.00772	1.36301	1.95169	0.02030	1.34581	1.12518	0.5642	0.0085	1.2356	1.5359	0.0175	1.5630	1.3615
1.13623	0.00935	1.36123	1.94739	0.02461	1.34042	1.12123	0.5625	0.0103	1.2331	1.5320	0.0212	1.5570	1.3587
1.13475	0.01115	1.35927	1.94269	0.02936	1.33453	1.11690	0.5607	0.0123	1.2303	1.5277	0.0253	1.5505	1.3557
1.13313	0.01310	1.35713	1.93759	0.03455	1.32813	1.11220	0.5587	0.0145	1.2273	1.5231	0.0297	1.5434	1.3523
1.13139	0.01522	1.35483	1.93208	0.04018	1.32122	1.10712	0.5566	0.0168	1.2241	1.5180	0.0346	1.5357	1.3487
1.12951	0.01750	1.35235	1.92618	0.04627	1.31381	1.10167	0.5543	0.0194	1.2206	1.5126	0.0398	1.5275	1.3448
1.12750	0.01995	1.34969	1.91988	0.05282	1.30589	1.09584	0.5518	0.0221	1.2169	1.5069	0.0454	1.5187	1.3407
1.12536	0.02257	1.34686	1.91318	0.05984	1.29747	1.08963	0.5492	0.0251	1.2129	1.5008	0.0514	1.5093	1.3363
1.12308	0.02536	1.34385	1.90609	0.06735	1.28856	1.08304	0.5464	0.0282	1.2087	1.4943	0.0579	1.4994	1.3317
1.12067	0.02832	1.34066	1.89860	0.07535	1.27914	1.07608	0.5434	0.0315	1.2043	1.4874	0.0647	1.4890	1.3268
1.11812	0.03146	1.33730	1.89073	0.08385	1.26923	1.06874	0.5402	0.0351	1.1995	1.4802	0.0720	1.4780	1.3216
1.11543	0.03478	1.33375	1.88248	0.09287	1.25882	1.06102	0.5369	0.0388	1.1946	1.4727	0.0797	1.4664	1.3162
1.11260	0.03827	1.33002	1.87383	0.10243	1.24792	1.05292	0.5334	0.0428	1.1894	1.4647	0.0879	1.4543	1.3105
1.10963	0.04195	1.32611	1.86481	0.11254	1.23653	1.04444	0.5297	0.0470	1.1839	1.4565	0.0965	1.4417	1.3045
1.10651	0.04582	1.32201	1.85541	0.12321	1.22465	1.03558	0.5259	0.0514	1.1781	1.4479	0.1056	1.4285	1.2983
1.10326	0.04988	1.31773	1.84564	0.13447	1.21229	1.02634	0.5218	0.0561	1.1722	1.4389	0.1152	1.4148	1.2919
1.09986	0.05413	1.31326	1.83550	0.14634	1.19945	1.01672	0.5176	0.0610	1.1659	1.4296	0.1253	1.4006	1.2852
1.09631	0.05858	1.30860	1.82498	0.15884	1.18613	1.00672	0.5132	0.0661	1.1594	1.4199	0.1360	1.3858	1.2782
1.09262	0.06323	1.30374	1.81411	0.17200	1.17233	0.99633	0.5086	0.0715	1.1526	1.4099	0.1471	1.3706	1.2710
1.08877	0.06809	1.29869	1.80288	0.18583	1.15805	0.98556	0.5038	0.0772	1.1455	1.3996	0.1589	1.3548	1.2636
1.08477	0.07316	1.29345	1.79129	0.20038	1.14331	0.97441	0.4988	0.0831	1.1382	1.3889	0.1712	1.3384	1.2559
1.08062	0.07845	1.28800	1.77935	0.21567	1.12810	0.96286	0.4936	0.0894	1.1306	1.3779	0.1841	1.3216	1.2480
1.07631	0.08396	1.28235	1.76706	0.23174	1.11243	0.95093	0.4881	0.0959	1.1227	1.3666	0.1977	1.3043	1.2398
1.07184	0.08970	1.27650	1.75444	0.24863	1.09630	0.93860	0.4825	0.1027	1.1146	1.3550	0.2119	1.2864	1.2314
1.06721	0.09567	1.27043	1.74148	0.26638	1.07971	0.92589	0.4767	0.1099	1.1061	1.3430	0.2268	1.2681	1.2227
1.06242	0.10189	1.26415	1.72819	0.28504	1.06267	0.91277	0.4706	0.1174	1.0974	1.3307	0.2425	1.2492	1.2138
1.05745	0.10836	1.25765	1.71457	0.30465	1.04518	0.89926	0.4643	0.1252	1.0883	1.3181	0.2589	1.2299	1.2047
1.05232	0.11508	1.25093	1.70064	0.32528	1.02726	0.88535	0.4578	0.1334	1.0790	1.3052	0.2761	1.2101	1.1953
1.04701	0.12207	1.24398	1.68640	0.34700	1.00889	0.87103	0.4510	0.1420	1.0693	1.2920	0.2942	1.1898	1.1857
1.04152	0.12934	1.23681	1.67186	0.36985	0.99010	0.85631	0.4440	0.1510	1.0594	1.2785	0.3132	1.1690	1.1759
1.03586	0.13689	1.22939	1.65702	0.39394	0.97088	0.84117	0.4367	0.1604	1.0491	1.2647	0.3332	1.1478	1.1658
1.03000	0.14473	1.22173	1.64189	0.41933	0.95124	0.82562	0.4292	0.1703	1.0385	1.2506	0.3542	1.1261	1.1555
1.02396	0.15289	1.21381	1.62649	0.44613	0.93118	0.80964	0.4214	0.1806	1.0276	1.2362	0.3763	1.1039	1.1450
1.01772	0.16137	1.20564	1.61081	0.47445	0.91072	0.79324	0.4133	0.1914	1.0164	1.2215	0.3996	1.0813	1.1343
1.01128	0.17018	1.19721	1.59488	0.50439	0.88985	0.77641	0.4049	0.2028	1.0048	1.2066	0.4241	1.0583	1.1233
1.00463	0.17934	1.18849	1.57869	0.53610	0.86860	0.75913	0.3962	0.2147	0.9928	1.1914	0.4500	1.0348	1.1122
0.99777	0.18888	1.17950	1.56227	0.56972	0.84696	0.74141	0.3872	0.2272	0.9805	1.1759	0.4773	1.0109	1.1008
0.99070	0.19879	1.17020	1.54562	0.60543	0.82495	0.72323	0.3779	0.2404	0.9678	1.1602	0.5062	0.9866	1.0892
0.98340	0.20911	1.16060	1.52876	0.64341	0.80256	0.70459	0.3683	0.2542	0.9548	1.1442	0.5368	0.9619	1.0774
0.97587	0.21987	1.15068	1.51170	0.68387	0.77983	0.68547	0.3583	0.2688	0.9413	1.1280	0.5693	0.9367	1.0654
0.96810	0.23107	1.14043	1.49445	0.72708	0.75675	0.66586	0.3479	0.2841	0.9275	1.1116	0.6038	0.9112	1.0532
0.96008	0.24275	1.12982	1.47704	0.77329	0.73333	0.64576	0.3372	0.3003	0.9132	1.0950	0.6406	0.8853	1.0408
0.95181	0.25495	1.11886	1.45947	0.82285	0.70960	0.62514	0.3261	0.3175	0.8985	1.0781	0.6797	0.8590	1.0283
0.94327	0.26769	1.10750	1.44178	0.87612	0.68556	0.60399	0.3145	0.3356	0.8834	1.0611	0.7216	0.8324	1.0155
0.93445	0.28101	1.09574	1.42397	0.93352	0.66124	0.58230	0.3025	0.3548	0.8678	1.0438	0.7664	0.8054	1.0026
0.92535	0.29495	1.08356	1.40607	0.99555	0.63664	0.56003	0.2900	0.3752	0.8518	1.0264	0.8146	0.7781	0.9894
0.91594	0.30957	1.07092	1.38811	1.06279	0.61178	0.53718	0.2770	0.3969	0.8352	1.0088	0.8664	0.7505	0.9762
0.90623	0.32492	1.05781	1.37012	1.13592	0.58670	0.51372	0.2635	0.4201	0.8182	0.9911	0.9224	0.7225	0.9627
0.89618	0.34105	1.04418	1.35212	1.21574	0.56140	0.48961	0.2494	0.4448	0.8006	0.9732	0.9830	0.6943	0.9491
0.88580	0.35805	1.03001	1.33416	1.30318	0.53592	0.46483	0.2347	0.4713	0.7824	0.9553	1.0489	0.6658	0.9354
0.87505	0.37599	1.01526	1.31625	1.39938	0.51028	0.43935	0.2194	0.4998	0.7637	0.9372	1.1208	0.6371	0.9215
L_1	L_2	C_2	L_3	L_4	C_4	L_5	L_1	L_2	C_2	L_3	L_4	C_4	L_5

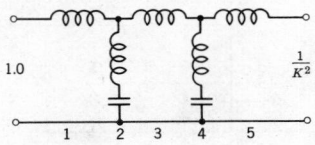

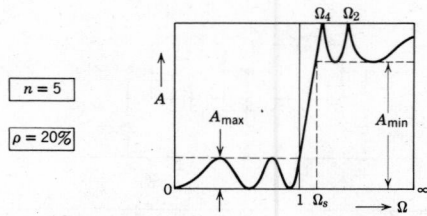

$n = 5$

$\rho = 20\%$

θ	Ω_K	A_{MIN}	σ_0	σ_1	σ_3	Ω_1	Ω_2	Ω_3	Ω_4
c	∞	∞	0.47472	-0.14670	-0.38406	1.0528	∞	0.6506	∞
1.0	57.2987	210.17	0.47476	-0.14667	-0.38405	1.0527	97.4775	0.6507	60.2470
2.0	28.6537	180.07	0.47489	-0.14659	-0.38404	1.0527	48.7389	0.6508	30.1274
3.0	19.1073	162.45	0.47511	-0.14646	-0.38402	1.0527	32.4927	0.6511	20.0893
4.0	14.3356	149.95	0.47542	-0.14628	-0.38399	1.0527	24.3697	0.6514	15.0716
5.0	11.4737	140.25	0.47582	-0.14605	-0.38396	1.0526	19.4959	0.6519	12.0620
6.0	9.5668	132.32	0.47631	-0.14577	-0.38392	1.0526	16.2468	0.6524	10.0565
7.0	8.2055	125.61	0.47689	-0.14543	-0.38387	1.0526	13.9260	0.6531	8.6247
8.0	7.1853	119.80	0.47757	-0.14505	-0.38381	1.0525	12.1854	0.6538	7.5516
9.0	6.3925	114.66	0.47833	-0.14461	-0.38375	1.0524	10.8316	0.6547	6.7175
10.0	5.7588	110.06	0.47918	-0.14412	-0.38367	1.0524	9.7486	0.6557	6.0507
11.0	5.2408	105.90	0.48013	-0.14358	-0.38359	1.0523	8.8625	0.6567	5.5057
12.0	4.8097	102.10	0.48117	-0.14299	-0.38349	1.0522	8.1241	0.6579	5.0520
13.0	4.4454	98.59	0.48231	-0.14235	-0.38338	1.0521	7.4993	0.6592	4.6684
14.0	4.1336	95.34	0.48355	-0.14166	-0.38327	1.0520	6.9638	0.6606	4.3401
15.0	3.8637	92.32	0.48488	-0.14092	-0.38314	1.0519	6.4997	0.6621	4.0559
16.0	3.6280	89.48	0.48631	-0.14013	-0.38299	1.0518	6.0936	0.6637	3.8076
17.0	3.4203	86.81	0.48785	-0.13929	-0.38283	1.0517	5.7353	0.6654	3.5888
18.0	3.2361	84.29	0.48949	-0.13840	-0.38266	1.0515	5.4168	0.6672	3.3946
19.0	3.0716	81.89	0.49123	-0.13746	-0.38247	1.0514	5.1318	0.6691	3.2212
20.0	2.9238	79.62	0.49308	-0.13647	-0.38227	1.0512	4.8753	0.6712	3.0654
21.0	2.7904	77.46	0.49503	-0.13543	-0.38204	1.0511	4.6433	0.6733	2.9246
22.0	2.6695	75.39	0.49710	-0.13434	-0.38180	1.0509	4.4323	0.6756	2.7970
23.0	2.5593	73.40	0.49929	-0.13320	-0.38153	1.0507	4.2397	0.6780	2.6807
24.0	2.4586	71.50	0.50159	-0.13201	-0.38125	1.0505	4.0631	0.6805	2.5743
25.0	2.3662	69.67	0.50401	-0.13078	-0.38093	1.0503	3.9007	0.6831	2.4767
26.0	2.2812	67.91	0.50655	-0.12950	-0.38060	1.0501	3.7507	0.6858	2.3868
27.0	2.2027	66.21	0.50922	-0.12817	-0.38023	1.0499	3.6119	0.6887	2.3038
28.0	2.1301	64.56	0.51201	-0.12679	-0.37983	1.0496	3.4829	0.6916	2.2270
29.0	2.0627	62.97	0.51494	-0.12536	-0.37941	1.0494	3.3629	0.6947	2.1556
30.0	2.0000	61.43	0.51801	-0.12389	-0.37895	1.0491	3.2508	0.6980	2.0892
31.0	1.9416	59.93	0.52122	-0.12238	-0.37845	1.0488	3.1460	0.7013	2.0274
32.0	1.8871	58.47	0.52458	-0.12081	-0.37791	1.0486	3.0476	0.7048	1.9695
33.0	1.8361	57.06	0.52808	-0.11920	-0.37734	1.0483	2.9553	0.7084	1.9154
34.0	1.7883	55.68	0.53175	-0.11755	-0.37672	1.0479	2.8683	0.7122	1.8646
35.0	1.7434	54.33	0.53557	-0.11585	-0.37605	1.0476	2.7864	0.7160	1.8170
36.0	1.7013	53.02	0.53956	-0.11411	-0.37533	1.0473	2.7089	0.7200	1.7722
37.0	1.6616	51.74	0.54373	-0.11233	-0.37456	1.0469	2.6356	0.7242	1.7299
38.0	1.6243	50.49	0.54808	-0.11050	-0.37372	1.0465	2.5662	0.7285	1.6901
39.0	1.5890	49.26	0.55261	-0.10863	-0.37283	1.0461	2.5003	0.7329	1.6525
40.0	1.5557	48.06	0.55734	-0.10672	-0.37187	1.0457	2.4377	0.7375	1.6170
41.0	1.5243	46.88	0.56228	-0.10477	-0.37085	1.0453	2.3781	0.7422	1.5833
42.0	1.4945	45.72	0.56743	-0.10278	-0.36974	1.0448	2.3213	0.7471	1.5515
43.0	1.4663	44.59	0.57281	-0.10075	-0.36856	1.0444	2.2672	0.7521	1.5213
44.0	1.4396	43.47	0.57842	-0.09868	-0.36829	1.0439	2.2154	0.7573	1.4926
45.0	1.4142	42.38	0.58428	-0.09658	-0.36593	1.0434	2.1660	0.7626	1.4654
46.0	1.3902	41.30	0.59039	-0.09444	-0.36448	1.0429	2.1187	0.7681	1.4396
47.0	1.3673	40.23	0.59678	-0.09226	-0.36291	1.0424	2.0733	0.7738	1.4150
48.0	1.3456	39.19	0.60345	-0.09004	-0.36124	1.0418	2.0299	0.7796	1.3916
49.0	1.3250	38.15	0.61043	-0.08780	-0.35945	1.0412	1.9881	0.7856	1.3693
50.0	1.3054	37.13	0.61773	-0.08552	-0.35753	1.0406	1.9480	0.7917	1.3481
51.0	1.2868	36.12	0.62537	-0.08321	-0.35548	1.0400	1.9095	0.7980	1.3279
52.0	1.2690	35.13	0.63336	-0.08087	-0.35329	1.0393	1.8724	0.8045	1.3087
53.0	1.2521	34.14	0.64174	-0.07850	-0.35094	1.0387	1.8366	0.8112	1.2903
54.0	1.2361	33.17	0.65053	-0.07610	-0.34842	1.0380	1.8021	0.8180	1.2728
55.0	1.2208	32.20	0.65975	-0.07368	-0.34573	1.0373	1.7689	0.8250	1.2561
56.0	1.2062	31.25	0.66943	-0.07123	-0.34285	1.0365	1.7368	0.8322	1.2402
57.0	1.1924	30.30	0.67962	-0.06876	-0.33977	1.0358	1.7057	0.8395	1.2250
58.0	1.1792	29.36	0.69033	-0.06627	-0.33647	1.0350	1.6757	0.8470	1.2104
59.0	1.1666	28.42	0.70163	-0.06376	-0.33295	1.0342	1.6467	0.8547	1.1966
60.0	1.1547	27.49	0.71354	-0.06124	-0.32917	1.0333	1.6185	0.8626	1.1834
θ	Ω_K	A_{MIN}	σ_0	σ_1	σ_3	Ω_1	Ω_2	Ω_3	Ω_4

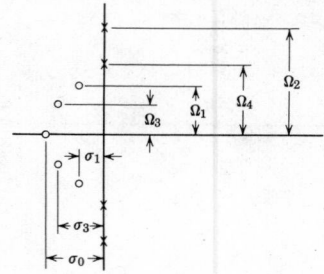

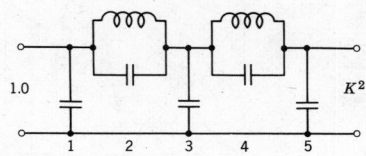

c₁	c₂	L₂	c₃	c₄	L₄	c₅	c₁	c₂	L₂	c₃	c₄	L₄	c₅
		K²=1.0							K²=∞				
1.302	0.0000	1.346	2.129	0.0000	1.346	1.302	0.6510	0.0000	1.3234	1.6362	0.0000	1.6265	1.4246
1.30183	0.00008	1.34548	2.12835	0.00020	1.34532	1.30170	0.6509	0.0001	1.3233	1.6360	0.0002	1.6263	1.4244
1.30163	0.00031	1.34523	2.12770	0.00082	1.34459	1.30112	0.6507	0.0003	1.3229	1.6355	0.0007	1.6254	1.4240
1.30130	0.00070	1.34483	2.12660	0.00184	1.34339	1.30016	0.6503	0.0007	1.3223	1.6345	0.0015	1.6240	1.4233
1.30084	0.00125	1.34426	2.12507	0.00328	1.34170	1.29881	0.6498	0.0013	1.3215	1.6332	0.0027	1.6220	1.4224
1.30024	0.00196	1.34353	2.12311	0.00513	1.33954	1.29708	0.6491	0.0020	1.3205	1.6315	0.0042	1.6195	1.4211
1.29951	0.00282	1.34264	2.12070	0.00740	1.33689	1.29496	0.6483	0.0029	1.3193	1.6294	0.0061	1.6164	1.4196
1.29865	0.00384	1.34159	2.11786	0.01008	1.33376	1.29246	0.6473	0.0039	1.3178	1.6270	0.0083	1.6128	1.4178
1.29766	0.00502	1.34037	2.11459	0.01318	1.33015	1.28957	0.6462	0.0051	1.3161	1.6241	0.0109	1.6085	1.4158
1.29653	0.00637	1.33899	2.11088	0.01671	1.32607	1.28630	0.6450	0.0065	1.3141	1.6209	0.0138	1.6038	1.4135
1.29527	0.00787	1.33744	2.10675	0.02067	1.32150	1.28264	0.6436	0.0080	1.3120	1.6174	0.0171	1.5984	1.4109
1.29387	0.00953	1.33573	2.10217	0.02506	1.31646	1.27859	0.6420	0.0097	1.3096	1.6134	0.0207	1.5926	1.4080
1.29234	0.01136	1.33386	2.09717	0.02989	1.31094	1.27417	0.6403	0.0116	1.3069	1.6091	0.0247	1.5861	1.4048
1.29067	0.01335	1.33182	2.09172	0.03516	1.30495	1.26936	0.6384	0.0136	1.3041	1.6044	0.0291	1.5791	1.4014
1.28887	0.01551	1.32961	2.08588	0.04089	1.29848	1.26416	0.6364	0.0159	1.3010	1.5993	0.0338	1.5716	1.3977
1.28693	0.01783	1.32724	2.07959	0.04707	1.29154	1.25858	0.6343	0.0182	1.2976	1.5939	0.0389	1.5635	1.3938
1.28485	0.02033	1.32470	2.07288	0.05371	1.28413	1.25261	0.6319	0.0208	1.2941	1.5881	0.0444	1.5548	1.3895
1.28263	0.02300	1.32199	2.06574	0.06084	1.27625	1.24627	0.6295	0.0236	1.2903	1.5819	0.0502	1.5456	1.3850
1.28027	0.02584	1.31911	2.05819	0.06844	1.26790	1.23953	0.6268	0.0265	1.2862	1.5754	0.0565	1.5359	1.3803
1.27778	0.02885	1.31607	2.05021	0.07655	1.25909	1.23241	0.6240	0.0296	1.2820	1.5685	0.0632	1.5256	1.3753
1.27514	0.03205	1.31285	2.04182	0.08515	1.24981	1.22491	0.6211	0.0329	1.2774	1.5612	0.0703	1.5148	1.3700
1.27236	0.03542	1.30945	2.03301	0.09428	1.24007	1.21703	0.6180	0.0364	1.2727	1.5536	0.0778	1.5035	1.3644
1.26943	0.03898	1.30589	2.02379	0.10393	1.22987	1.20876	0.6147	0.0402	1.2677	1.5456	0.0857	1.4916	1.3586
1.26636	0.04272	1.30215	2.01416	0.11414	1.21921	1.20010	0.6112	0.0441	1.2624	1.5373	0.0941	1.4791	1.3525
1.26314	0.04666	1.29823	2.00412	0.12490	1.20809	1.19107	0.6076	0.0482	1.2569	1.5286	0.1029	1.4662	1.3461
1.25978	0.05079	1.29413	1.99368	0.13625	1.19652	1.18164	0.6038	0.0525	1.2512	1.5195	0.1122	1.4527	1.3395
1.25262	0.05511	1.28985	1.98283	0.14819	1.18450	1.17183	0.5998	0.0571	1.2452	1.5101	0.1220	1.4387	1.3327
1.25259	0.05963	1.28540	1.97159	0.16075	1.17203	1.16164	0.5957	0.0619	1.2389	1.5004	0.1323	1.4242	1.3256
1.24877	0.06436	1.28075	1.95995	0.17396	1.15911	1.15106	0.5914	0.0669	1.2324	1.4903	0.1431	1.4092	1.3182
1.22480	0.06930	1.27592	1.94792	0.18783	1.14576	1.14010	0.5869	0.0721	1.2256	1.4798	0.1544	1.3936	1.3105
1.24067	0.07446	1.27091	1.93550	0.20239	1.13196	1.12874	0.5822	0.0777	1.2186	1.4690	0.1663	1.3776	1.3027
1.23638	0.07983	1.26570	1.92270	0.21768	1.11772	1.11700	0.5773	0.0834	1.2113	1.4579	0.1788	1.3610	1.2945
1.23192	0.08543	1.26030	1.90952	0.23371	1.10305	1.10487	0.5723	0.0894	1.2037	1.4464	0.1918	1.3439	1.2861
1.22731	0.09126	1.25470	1.89596	0.25054	1.08795	1.09235	0.5670	0.0957	1.1959	1.4346	0.2055	1.3264	1.2775
1.22252	0.09732	1.24890	1.88203	0.26819	1.07242	1.07944	0.5616	0.1023	1.1878	1.4225	0.2198	1.3083	1.2686
1.21757	0.10363	1.24290	1.86773	0.28671	1.05648	1.06614	0.5559	0.1092	1.1794	1.4100	0.2348	1.2898	1.2595
1.21244	0.11019	1.23669	1.85307	0.30614	1.04011	1.05244	0.5500	0.1164	1.1707	1.3972	0.2506	1.2707	1.2501
1.20714	0.11701	1.23028	1.83806	0.32654	1.02332	1.03835	0.5439	0.1239	1.1618	1.3841	0.2671	1.2512	1.2405
1.20166	0.12410	1.22364	1.82269	0.34795	1.00613	1.02386	0.5376	0.1318	1.1525	1.3707	0.2843	1.2313	1.2306
1.19600	0.13146	1.21679	1.80698	0.37044	0.98853	1.00897	0.5311	0.1400	1.1429	1.3570	0.3024	1.2108	1.2205
1.19015	0.13911	1.20971	1.79093	0.39408	0.97053	0.99368	0.5244	0.1485	1.1331	1.3429	0.3214	1.1899	1.2102
1.18411	0.14706	1.20241	1.77455	0.41894	0.95213	0.97798	0.5174	0.1575	1.1229	1.3286	0.3413	1.1686	1.1996
1.17787	0.15532	1.19486	1.75784	0.44510	0.93335	0.96187	0.5102	0.1668	1.1124	1.3139	0.3623	1.1467	1.1888
1.17144	0.16389	1.18708	1.74081	0.47265	0.91417	0.94535	0.5027	0.1766	1.1016	1.2989	0.3843	1.1245	1.1778
1.16480	0.17280	1.17904	1.72347	0.50170	0.89462	0.92841	0.4949	0.1868	1.0905	1.2837	0.4074	1.1018	1.1665
1.15794	0.18206	1.17075	1.70583	0.53236	0.87470	0.91105	0.4869	0.1975	1.0790	1.2681	0.4317	1.0787	1.1551
1.15088	0.19169	1.16219	1.68789	0.56476	0.85441	0.89326	0.4787	0.2087	1.0672	1.2523	0.4573	1.0551	1.1433
1.14359	0.20169	1.15336	1.66967	0.59903	0.83376	0.87504	0.4701	0.2205	1.0550	1.2362	0.4844	1.0311	1.1314
1.13607	0.21210	1.14425	1.65117	0.63534	0.81276	0.85638	0.4612	0.2328	1.0425	1.2198	0.5129	1.0068	1.1193
1.12831	0.22293	1.13484	1.63241	0.67386	0.79141	0.83727	0.4520	0.2457	1.0296	1.2032	0.5431	0.9820	1.1069
1.12031	0.23421	1.12513	1.61339	0.71481	0.76973	0.81771	0.4425	0.2593	1.0163	1.1863	0.5750	0.9568	1.0943
1.11206	0.24596	1.11509	1.59413	0.75841	0.74773	0.79768	0.4327	0.2736	1.0026	1.1691	0.6089	0.9313	1.0815
1.10354	0.25821	1.10473	1.57465	0.80492	0.72541	0.77717	0.4225	0.2886	0.9884	1.1517	0.6449	0.9053	1.0685
1.09476	0.27099	1.09401	1.55494	0.85465	0.70278	0.75619	0.4120	0.3044	0.9739	1.1341	0.6833	0.8790	1.0553
1.08569	0.28433	1.08293	1.53504	0.90794	0.67986	0.73470	0.4010	0.3211	0.9589	1.1162	0.7242	0.8524	1.0419
1.07633	0.29828	1.07147	1.51496	0.96518	0.65667	0.71270	0.3897	0.3388	0.9435	1.0981	0.7679	0.8254	1.0283
1.06666	0.31288	1.05960	1.49471	1.02684	0.63320	0.69016	0.3779	0.3574	0.9275	1.0798	0.8147	0.7981	1.0144
1.05668	0.32817	1.04731	1.47431	1.09344	0.60949	0.66709	0.3657	0.3772	0.9111	1.0613	0.8650	0.7705	1.0004
1.04636	0.34422	1.03456	1.45379	1.16561	0.58554	0.64344	0.3530	0.3983	0.8942	1.0426	0.9192	0.7425	0.9863
1.03570	0.36109	1.02134	1.43317	1.24407	0.56138	0.61920	0.3398	0.4207	0.8767	1.0237	0.9777	0.7143	0.9719
1.02467	0.37885	1.00760	1.41247	1.32969	0.53702	0.59435	0.3261	0.4446	0.8587	1.0046	1.0412	0.6858	0.9574
L₁	L₂	C₂	L₃	L₄	C₄	L₅	L₁	L₂	C₂	L₃	L₄	C₄	L₅

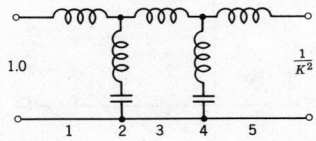

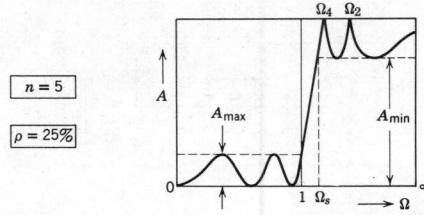

$n = 5$

$\rho = 25\%$

θ	Ω_K	A_{MIN}	σ_0	σ_1	σ_3	Ω_1	Ω_2	Ω_3	Ω_4
C	∞	∞	0.42450	−0.13118	−0.34343	1.0332	∞	0.6386	∞
1.0	57.2987	212.22	0.42454	−0.13115	−0.34342	1.0331	97.4775	0.6385	60.2470
2.0	28.6537	182.11	0.42465	−0.13109	−0.34341	1.0331	48.7389	0.6387	30.1274
3.0	19.1073	164.49	0.42484	−0.13098	−0.34340	1.0331	32.4927	0.6389	20.0893
4.0	14.3356	151.99	0.42511	−0.13082	−0.34338	1.0331	24.3697	0.6392	15.0716
5.0	11.4737	142.29	0.42545	−0.13063	−0.34335	1.0331	19.4959	0.6397	12.0620
6.0	9.5668	134.36	0.42587	−0.13039	−0.34332	1.0331	16.2468	0.6402	10.0565
7.0	8.2055	127.65	0.42637	−0.13010	−0.34329	1.0331	13.9260	0.6408	8.6247
8.0	7.1853	121.84	0.42695	−0.12978	−0.34324	1.0331	12.1854	0.6415	7.5516
9.0	6.3925	116.70	0.42761	−0.12941	−0.34319	1.0330	10.8316	0.6422	6.7175
10.0	5.7588	112.11	0.42834	−0.12899	−0.34313	1.0330	9.7486	0.6431	6.0507
11.0	5.2408	107.94	0.42916	−0.12854	−0.34307	1.0330	8.8625	0.6441	5.5057
12.0	4.8097	104.14	0.43006	−0.12804	−0.34299	1.0330	8.1241	0.6452	5.0520
13.0	4.4454	100.63	0.43104	−0.12749	−0.34291	1.0329	7.4993	0.6463	4.6684
14.0	4.1336	97.39	0.43210	−0.12690	−0.34282	1.0329	6.9638	0.6476	4.3401
15.0	3.8637	94.36	0.43325	−0.12627	−0.34272	1.0328	6.4997	0.6490	4.0559
16.0	3.6280	91.52	0.43448	−0.12560	−0.34261	1.0328	6.0936	0.6504	3.8076
17.0	3.4203	88.85	0.43580	−0.12489	−0.34248	1.0328	5.7353	0.6520	3.5888
18.0	3.2361	86.33	0.43720	−0.12413	−0.34235	1.0327	5.4168	0.6536	3.3946
19.0	3.0716	83.94	0.43870	−0.12333	−0.34220	1.0326	5.1318	0.6554	3.2212
20.0	2.9238	81.66	0.44029	−0.12248	−0.34204	1.0326	4.8753	0.6572	3.0654
21.0	2.7904	79.50	0.44197	−0.12160	−0.34187	1.0325	4.6433	0.6592	2.9246
22.0	2.6695	77.43	0.44375	−0.12067	−0.34168	1.0324	4.4323	0.6613	2.7970
23.0	2.5593	75.45	0.44562	−0.11970	−0.34147	1.0324	4.2397	0.6634	2.6807
24.0	2.4586	73.54	0.44759	−0.11869	−0.34125	1.0323	4.0631	0.6657	2.5743
25.0	2.3662	71.71	0.44967	−0.11763	−0.34101	1.0322	3.9007	0.6681	2.4767
26.0	2.2812	69.95	0.45185	−0.11654	−0.34074	1.0321	3.7507	0.6706	2.3868
27.0	2.2027	68.25	0.45413	−0.11540	−0.34046	1.0320	3.6119	0.6732	2.3038
28.0	2.1301	66.60	0.45653	−0.11422	−0.34015	1.0319	3.4829	0.6759	2.2270
29.0	2.0627	65.01	0.45904	−0.11301	−0.33981	1.0318	3.3629	0.6787	2.1556
30.0	2.0000	63.47	0.46166	−0.11175	−0.33945	1.0317	3.2508	0.6817	2.0892
31.0	1.9416	61.97	0.46440	−0.11045	−0.33906	1.0315	3.1460	0.6847	2.0274
32.0	1.8871	60.51	0.46727	−0.10911	−0.33864	1.0314	3.0476	0.6879	1.9695
33.0	1.8361	59.10	0.47027	−0.10773	−0.33819	1.0313	2.9553	0.6912	1.9154
34.0	1.7883	57.72	0.47339	−0.10631	−0.33770	1.0311	2.8683	0.6946	1.8646
35.0	1.7434	56.37	0.47666	−0.10485	−0.33718	1.0310	2.7864	0.6981	1.8170
36.0	1.7013	55.06	0.48006	−0.10336	−0.33662	1.0308	2.7089	0.7018	1.7722
37.0	1.6616	53.78	0.48361	−0.10182	−0.33601	1.0306	2.6356	0.7056	1.7299
38.0	1.6243	52.53	0.48731	−0.10025	−0.33535	1.0304	2.5662	0.7095	1.6901
39.0	1.5890	51.30	0.49117	−0.09864	−0.33465	1.0302	2.5003	0.7135	1.6525
40.0	1.5557	50.10	0.49519	−0.09699	−0.33390	1.0300	2.4377	0.7177	1.6170
41.0	1.5243	48.92	0.49939	−0.09531	−0.33309	1.0298	2.3781	0.7220	1.5833
42.0	1.4945	47.76	0.50376	−0.09359	−0.33222	1.0296	2.3213	0.7265	1.5515
43.0	1.4663	46.63	0.50832	−0.09183	−0.33129	1.0294	2.2672	0.7311	1.5213
44.0	1.4396	45.51	0.51307	−0.09004	−0.33029	1.0291	2.2154	0.7358	1.4926
45.0	1.4142	44.42	0.51803	−0.08822	−0.32922	1.0288	2.1660	0.7407	1.4654
46.0	1.3902	43.34	0.52320	−0.08636	−0.32807	1.0286	2.1187	0.7457	1.4396
47.0	1.3673	42.27	0.52859	−0.08446	−0.32684	1.0283	2.0733	0.5709	1.4150
48.0	1.3456	41.23	0.53422	−0.08254	−0.32551	1.0280	2.0299	0.7562	1.3916
49.0	1.3250	40.19	0.54010	−0.08058	−0.32410	1.0277	1.9881	0.7617	1.3693
50.0	1.3054	39.17	0.54624	−0.07860	−0.32258	1.0273	1.9480	0.7673	1.3481
51.0	1.2868	38.16	0.55266	−0.07658	−0.32096	1.0270	1.9095	0.7731	1.3279
52.0	1.2690	37.17	0.55937	−0.07453	−0.31922	1.0266	1.8724	0.7791	1.3087
53.0	1.2521	36.18	0.56639	−0.07246	−0.31736	1.0262	1.8366	0.7852	1.2903
54.0	1.2361	35.21	0.57374	−0.07036	−0.31537	1.0259	1.8021	0.7915	1.2728
55.0	1.2208	34.24	0.58144	−0.06823	−0.31323	1.0254	1.7689	0.7980	1.2561
56.0	1.2062	33.29	0.58952	−0.06608	−0.31095	1.0250	1.7368	0.8046	1.2402
57.0	1.1924	32.34	0.59799	−0.06390	−0.30850	1.0246	1.7057	0.8114	1.2250
58.0	1.1792	31.40	0.60690	−0.06170	−0.30588	1.0241	1.6757	0.8184	1.2104
59.0	1.1666	30.46	0.61626	−0.05948	−0.30308	1.0236	1.6467	0.8255	1.1966
60.0	1.1547	29.53	0.62612	−0.05725	−0.30008	1.0231	1.6185	0.8328	1.1834
θ	Ω_K	A_{MIN}	σ_0	σ_1	σ_3	Ω_1	Ω_2	Ω_3	Ω_4

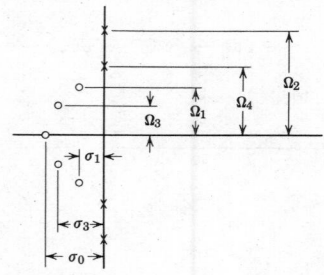

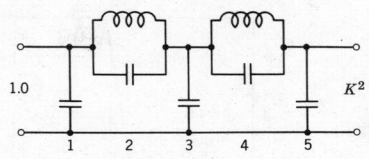

K²=1.0							K²=∞						
c_1	c_2	L_2	c_3	c_4	L_4	c_5	c_1	c_2	L_2	c_3	c_4	L_4	c_5
1.456	0.0000	1.307	2.283	0.0000	1.307	1.456	0.7280	0.0000	1.3772	1.7073	0.0000	1.6431	1.4689
1.45584	0.00008	1.30650	2.28310	0.00021	1.30635	1.45571	0.7279	0.0001	1.3771	1.7071	0.0002	1.6429	1.4687
1.45563	0.00032	1.30627	2.28240	0.00084	1.30567	1.45511	0.7277	0.0003	1.3768	1.7065	0.0007	1.6420	1.4683
1.45528	0.00073	1.30588	2.28124	0.00190	1.30454	1.45411	0.7273	0.0007	1.3762	1.7056	0.0015	1.6406	1.4676
1.45480	0.00129	1.30534	2.27962	0.00338	1.30295	1.45272	0.7268	0.0012	1.3754	1.7043	0.0027	1.6387	1.4666
1.45418	0.00202	1.30464	2.27753	0.00528	1.30092	1.45092	0.7262	0.0019	1.3745	1.7025	0.0042	1.6362	1.4654
1.45342	0.00291	1.30379	2.27498	0.00762	1.29843	1.44873	0.7254	0.0028	1.3732	1.7004	0.0061	1.6332	1.4638
1.45252	0.00396	1.30279	2.27197	0.01038	1.29549	1.44615	0.7245	0.0038	1.3718	1.6980	0.0082	1.6296	1.4620
1.45149	0.00517	1.30163	2.26849	0.01357	1.29210	1.44316	0.7234	0.0049	1.3702	1.6951	0.0108	1.6254	1.4599
1.45031	0.00655	1.30031	2.26455	0.01720	1.28826	1.43978	0.7222	0.0062	1.3683	1.6919	0.0137	1.6208	1.4575
1.44900	0.00810	1.29883	2.26016	0.02127	1.28398	1.43599	0.7209	0.0077	1.3662	1.6883	0.0169	1.6155	1.4549
1.44754	0.00981	1.29720	2.25530	0.02579	1.27924	1.43182	0.7194	0.0093	1.3639	1.6843	0.0205	1.6097	1.4519
1.44594	0.01170	1.29542	2.24999	0.03075	1.27406	1.42724	0.7177	0.0111	1.3613	1.6799	0.0244	1.6034	1.4487
1.44421	0.01375	1.29347	2.24422	0.03617	1.26843	1.42227	0.7159	0.0131	1.3586	1.6751	0.0287	1.5965	1.4452
1.44233	0.01597	1.29137	2.23799	0.04206	1.26235	1.41690	0.7140	0.0152	1.3556	1.6700	0.0334	1.5891	1.4414
1.44030	0.01836	1.28910	2.23131	0.04841	1.25583	1.41113	0.7119	0.0175	1.3524	1.6645	0.0384	1.5811	1.4374
1.43814	0.02093	1.28668	2.22417	0.05523	1.24887	1.40497	0.7097	0.0200	1.3489	1.6586	0.0439	1.5726	1.4331
1.43582	0.02368	1.28409	2.21659	0.06254	1.24147	1.39841	0.7073	0.0226	1.3453	1.6524	0.0497	1.5636	1.4285
1.43337	0.02660	1.28135	2.20855	0.07035	1.23363	1.39146	0.7048	0.0254	1.3413	1.6458	0.0558	1.5540	1.4236
1.43077	0.02970	1.27844	2.20007	0.07865	1.22535	1.38411	0.7021	0.0284	1.3372	1.6388	0.0624	1.5439	1.4184
1.42801	0.03299	1.27537	2.19114	0.08747	1.21663	1.37637	0.6993	0.0316	1.3329	1.6315	0.0694	1.5333	1.4130
1.42512	0.03646	1.27213	2.18177	0.09682	1.20748	1.36823	0.6963	0.0349	1.3283	1.6237	0.0768	1.5221	1.4073
1.42207	0.04012	1.26873	2.17196	0.10671	1.19790	1.35970	0.6931	0.0385	1.3234	1.6156	0.0846	1.5104	1.4014
1.41887	0.04397	1.26516	2.16171	0.11715	1.18788	1.35077	0.6898	0.0422	1.3184	1.6072	0.0929	1.4982	1.3952
1.41552	0.04802	1.26142	2.15102	0.12815	1.17744	1.34145	0.6864	0.0461	1.3130	1.5984	0.1016	1.4854	1.3887
1.41201	0.05226	1.25752	2.13990	0.13975	1.16658	1.33173	0.6827	0.0503	1.3075	1.5892	0.1107	1.4722	1.3819
1.40835	0.05671	1.25344	2.12834	0.15194	1.15528	1.32162	0.6789	0.0546	1.3017	1.5797	0.1204	1.4585	1.3749
1.40453	0.06136	1.24918	2.11636	0.16475	1.14357	1.31112	0.6750	0.0592	1.2957	1.5698	0.1305	1.4441	1.3676
1.40055	0.06623	1.24475	2.10395	0.17821	1.13144	1.30022	0.6708	0.0639	1.2894	1.5596	0.1411	1.4293	1.3601
1.39641	0.07130	1.24015	2.09112	0.19234	1.11889	1.28892	0.6665	0.0689	1.2828	1.5490	0.1522	1.4141	1.3522
1.39210	0.07660	1.23536	2.07787	0.20715	1.10593	1.27724	0.6621	0.0742	1.2760	1.5380	0.1638	1.3983	1.3442
1.38764	0.08212	1.23040	2.06421	0.22269	1.09256	1.26515	0.6574	0.0796	1.2690	1.5267	0.1761	1.3820	1.3358
1.38300	0.08787	1.22524	2.05013	0.23897	1.07878	1.25268	0.6526	0.0853	1.2617	1.5151	0.1888	1.3652	1.3273
1.37819	0.09386	1.21991	2.03564	0.25603	1.06460	1.23980	0.6476	0.0913	1.2541	1.5031	0.2022	1.3479	1.3184
1.37321	0.10009	1.21438	2.02075	0.27391	1.05001	1.22653	0.6423	0.0975	1.2463	1.4908	0.2162	1.3301	1.3093
1.36806	0.10657	1.20866	2.00546	0.29265	1.03503	1.21287	0.6370	0.1040	1.2382	1.4781	0.2309	1.3119	1.3000
1.36272	0.11330	1.20274	1.98978	0.31228	1.01966	1.19880	0.6314	0.1108	1.2298	1.4651	0.2462	1.2932	1.2904
1.35720	0.12030	1.19662	1.97370	0.33286	1.00389	1.18434	0.6256	0.1179	1.2211	1.4518	0.2623	1.2740	1.2805
1.35150	0.12757	1.19030	1.95723	0.35443	0.98774	1.16947	0.6196	0.1253	1.2122	1.4381	0.2791	1.2543	1.2704
1.34561	0.13513	1.18377	1.94038	0.37705	0.97121	1.15421	0.6134	0.1330	1.2030	1.4241	0.2967	1.2342	1.2601
1.33952	0.14298	1.17703	1.92316	0.40078	0.95429	1.13854	0.6069	0.1410	1.1934	1.4098	0.3151	1.2136	1.2495
1.33324	0.15112	1.17007	1.90557	0.42570	0.93701	1.12246	0.6003	0.1494	1.1836	1.3952	0.3345	1.1926	1.2387
1.32675	0.15959	1.16289	1.88761	0.45187	0.91935	1.10598	0.5934	0.1581	1.1735	1.3803	0.3547	1.1711	1.2276
1.32006	0.16837	1.15548	1.86929	0.47939	0.90133	1.08908	0.5863	0.1631	1.1631	1.3650	0.3760	1.1492	1.2163
1.31315	0.17750	1.14784	1.85061	0.50833	0.88296	1.07177	0.5790	0.1768	1.1523	1.3494	0.3983	1.1269	1.2047
1.30603	0.18698	1.13995	1.83159	0.53881	0.86423	1.05405	0.5714	0.1868	1.1412	1.3336	0.4217	1.1041	1.1929
1.29868	0.19683	1.13182	1.81224	0.57094	0.84515	1.03590	0.5636	0.1972	1.1298	1.3174	0.4464	1.0809	1.1809
1.29111	0.20707	1.12343	1.79255	0.60485	0.82573	1.01733	0.5555	0.2081	1.1180	1.3009	0.4724	1.0573	1.1687
1.28329	0.21771	1.11477	1.77253	0.64069	0.80597	0.99832	0.5471	0.2195	1.1059	1.2842	0.4997	1.0333	1.1562
1.27524	0.22878	1.10584	1.75220	0.67860	0.78589	0.97888	0.5384	0.2314	1.0934	1.2671	0.5286	1.0089	1.1435
1.26693	0.24030	1.09663	1.73156	0.71878	0.76549	0.95900	0.5295	0.2439	1.0806	1.2498	0.5591	0.9841	1.1306
1.25836	0.25229	1.08712	1.71063	0.76142	0.74477	0.93867	0.5202	0.2570	1.0673	1.2322	0.5914	0.9589	1.1174
1.24952	0.26478	1.07730	1.68941	0.80677	0.72374	0.91788	0.5106	0.2707	1.0537	1.2143	0.6256	0.9333	1.1040
1.24040	0.27781	1.06716	1.66791	0.85508	0.70243	0.89662	0.5007	0.2852	1.0396	1.1961	0.6619	0.9074	1.0904
1.23099	0.29139	1.05668	1.64615	0.90666	0.68082	0.87489	0.4904	0.3004	1.0251	1.1777	0.7005	0.8811	1.0766
1.22128	0.30559	1.04585	1.62414	0.96185	0.65894	0.85267	0.4798	0.3164	1.0102	1.1591	0.7417	0.8545	1.0626
1.21126	0.32042	1.03466	1.60189	1.02105	0.63679	0.82995	0.4688	0.3333	0.9948	1.1402	0.7857	0.8276	1.0483
1.20091	0.33595	1.02307	1.57942	1.08472	0.61439	0.80671	0.4574	0.3511	0.9789	1.1210	0.8328	0.8003	1.0339
1.19022	0.35222	1.01108	1.55674	1.15338	0.59175	0.78295	0.4455	0.3700	0.9625	1.1016	0.8833	0.7727	1.0192
1.17916	0.36930	0.99865	1.53387	1.22766	0.56888	0.75863	0.4332	0.3900	0.9456	1.0820	0.9377	0.7448	1.0044
1.16774	0.38725	0.98576	1.51083	1.30829	0.54581	0.73374	0.4202	0.4113	0.9282	1.0622	0.9965	0.7166	0.9893
L_1	L_2	c_2	L_3	L_4	c_4	L_5	L_1	L_2	c_2	L_3	L_4	c_4	L_5

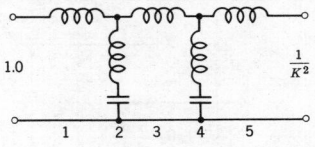

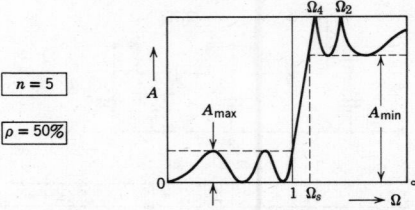

$n = 5$

$\rho = 50\%$

θ	Ω_K	A_{MIN}	σ_0	σ_1	σ_3	Ω_1	Ω_2	Ω_3	Ω_4
c	∞	∞	0.26645	-0.08234	-0.21556	0.98424	∞	0.6083	∞
1.0	57.2987	219.20	0.26646	-0.08232	-0.21556	0.98424	97.4775	0.6083	60.2470
2.0	28.6537	189.10	0.26653	-0.08228	-0.21555	0.9842	48.7389	0.60843	30.1274
3.0	19.1073	171.48	0.26664	-0.08223	-0.21555	0.9842	32.4927	0.60861	20.0893
4.0	14.3356	158.98	0.26679	-0.08214	-0.21554	0.9842	24.3697	0.60886	15.0716
5.0	11.4737	149.28	0.26699	-0.08203	-0.21553	0.9843	19.4959	0.60918	12.0620
6.0	9.5568	141.35	0.26723	-0.08190	-0.21552	0.9843	16.2468	0.60957	10.0565
7.0	8.2055	134.64	0.26751	-0.08175	-0.21550	0.9844	13.9260	0.61004	8.6247
8.0	7.1853	128.83	0.26784	-0.08157	-0.21549	0.9844	12.1854	0.61058	7.5516
9.0	6.3925	123.69	0.26821	-0.08137	-0.21547	0.9845	10.8316	0.61119	6.7175
10.0	5.7588	119.10	0.26863	-0.08114	-0.21544	0.9846	9.7486	0.61187	6.0507
11.0	5.2408	114.93	0.26910	-0.08089	-0.21542	0.9847	8.8625	0.61262	5.5057
12.0	4.8097	111.13	0.26961	-0.08062	-0.21539	0.9847	8.1241	0.61345	5.0520
13.0	4.4454	107.62	0.27016	-0.08032	-0.21536	0.9848	7.4993	0.61436	4.6684
14.0	4.1336	104.37	0.27077	-0.08000	-0.21532	0.9849	6.9638	0.61533	4.3401
15.0	3.8637	101.35	0.27142	-0.07965	-0.21528	0.9851	6.4997	0.61639	4.0559
16.0	3.6280	98.51	0.27212	-0.07928	-0.21524	0.9852	6.0936	0.61752	3.8076
17.0	3.4203	95.84	0.27286	-0.07889	-0.21519	0.9853	5.7353	0.61872	3.5888
18.0	3.2361	93.32	0.27366	-0.07847	-0.21514	0.9854	5.4168	0.62000	3.3946
19.0	3.0716	90.93	0.27451	-0.07803	-0.21508	0.9856	5.1318	0.62136	3.2212
20.0	2.9238	88.65	0.27541	-0.07757	-0.21501	0.9857	4.8753	0.62280	3.0654
21.0	2.7904	86.49	0.27636	-0.07708	-0.21494	0.9859	4.6433	0.62432	2.9246
22.0	2.6695	84.42	0.27736	-0.07657	-0.21486	0.9860	4.4323	0.62592	2.7970
23.0	2.5593	82.44	0.27842	-0.07603	-0.21478	0.9862	4.2397	0.62760	2.6807
24.0	2.4586	80.53	0.27953	-0.07547	-0.21469	0.9864	4.0631	0.62936	2.5743
25.0	2.3662	78.70	0.28070	-0.07489	-0.21459	0.9865	3.9007	0.63120	2.4767
26.0	2.2812	76.94	0.28193	-0.07428	-0.21448	0.9867	3.7507	0.63313	2.3868
27.0	2.2027	75.24	0.28321	-0.07365	-0.21436	0.9869	3.6119	0.63514	2.3038
28.0	2.1301	73.59	0.28456	-0.07300	-0.21423	0.9871	3.4829	0.63724	2.2270
29.0	2.0627	72.00	0.28597	-0.07232	-0.21409	0.9873	3.3629	0.63942	2.1556
30.0	2.0000	70.46	0.28744	-0.07162	-0.21394	0.9875	3.2508	0.64170	2.0892
31.0	1.9416	68.96	0.28897	-0.07089	-0.21377	0.9878	3.1460	0.64407	2.0274
32.0	1.8871	67.50	0.29058	-0.07014	-0.21360	0.9880	3.0476	0.64652	1.9695
33.0	1.8361	66.09	0.29225	-0.06937	-0.21341	0.9882	2.9553	0.64907	1.9154
34.0	1.7883	64.71	0.29399	-0.06857	-0.21320	0.9884	2.8683	0.65172	1.8646
35.0	1.7434	63.36	0.29581	-0.06775	-0.21297	0.9887	2.7864	0.65446	1.8170
36.0	1.7013	62.05	0.29770	-0.06691	-0.21273	0.9889	2.7089	0.65729	1.7722
37.0	1.6616	60.77	0.29967	-0.06604	-0.21247	0.9892	2.6356	0.66023	1.7299
38.0	1.6243	59.52	0.30172	-0.06515	-0.21219	0.9894	2.5662	0.66327	1.6901
39.0	1.5890	58.29	0.30386	-0.06424	-0.21189	0.9897	2.5003	0.66641	1.6525
40.0	1.5557	57.09	0.30608	-0.06330	-0.21156	0.9899	2.4377	0.66966	1.6170
41.0	1.5243	55.91	0.30839	-0.06234	-0.21121	0.9902	2.3781	0.67301	1.5833
42.0	1.4945	54.75	0.31079	-0.06136	-0.21084	0.9905	2.3213	0.67648	1.5515
43.0	1.4663	53.62	0.31330	-0.06036	-0.21043	0.9908	2.2672	0.68005	1.5213
44.0	1.4396	52.50	0.31590	-0.05933	-0.21000	0.9910	2.2154	0.68374	1.4926
45.0	1.4142	51.41	0.31860	-0.05828	-0.20953	0.9913	2.1660	0.68755	1.4654
46.0	1.3902	50.33	0.32142	-0.05721	-0.20902	0.9916	2.1187	0.69147	1.4396
47.0	1.3673	49.26	0.32435	-0.05611	-0.20848	0.9919	2.0733	0.69552	1.4150
48.0	1.3456	48.22	0.32740	-0.05499	-0.20790	0.9922	2.0299	0.69969	1.3916
49.0	1.3250	47.18	0.33058	-0.05386	-0.20728	0.9925	1.9881	0.70399	1.3693
50.0	1.3054	46.16	0.33389	-0.05270	-0.20661	0.9928	1.9480	0.70841	1.3481
51.0	1.2868	45.15	0.33734	-0.05151	-0.20590	0.9931	1.9095	0.71297	1.3279
52.0	1.2690	44.16	0.34093	-0.05031	-0.20513	0.9934	1.8724	0.71767	1.3087
53.0	1.2521	43.17	0.34468	-0.04809	-0.20430	0.9937	1.8366	0.72250	1.2903
54.0	1.2361	42.20	0.34859	-0.04784	-0.20342	0.9940	1.8021	0.72748	1.2728
55.0	1.2208	41.23	0.35267	-0.04658	-0.20247	0.9943	1.7689	0.73260	1.2561
56.0	1.2062	40.27	0.35693	-0.04529	-0.20145	0.9946	1.7368	0.73787	1.2402
57.0	1.1924	39.33	0.36138	-0.04399	-0.20036	0.9949	1.7057	0.74330	1.2250
58.0	1.1792	38.38	0.36604	-0.04267	-0.19918	0.9952	1.6757	0.74888	1.2104
59.0	1.1666	37.45	0.37091	-0.04133	-0.19792	0.9955	1.6467	0.75463	1.1966
60.0	1.1547	36.52	0.37602	-0.03997	-0.19657	0.9957	1.6185	0.76054	1.1834
θ	Ω_K	A_{MIN}	σ_0	σ_1	σ_3	Ω_1	Ω_2	Ω_3	Ω_4

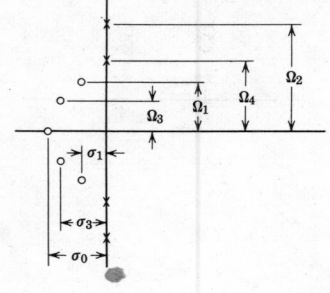

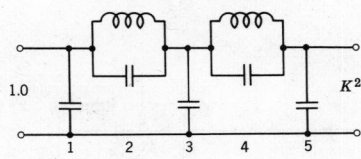

K²=1.0							K²=∞						
C_1	C_2	L_2	C_3	C_4	L_4	C_5	C_1	C_2	L_2	C_3	C_4	L_4	C_5
2.320	0.0000	1.035	3.205	0.0000	1.035	2.320	1.1598	0.0000	1.4315	2.0733	0.0000	1.5567	1.7240
2.31944	0.00010	1.03508	3.20440	0.00027	1.03498	2.31928	1.1597	0.0001	1.4314	2.0731	0.0002	1.5565	1.7238
2.31917	0.00041	1.03491	3.20345	0.00107	1.03448	2.31851	1.1595	0.0003	1.4311	2.0724	0.0007	1.5558	1.7233
2.31872	0.00092	1.03461	3.20186	0.00240	1.03366	2.31724	1.1592	0.0007	1.4306	2.0714	0.0016	1.5545	1.7225
2.31809	0.00163	1.03420	3.19964	0.00426	1.03250	2.31545	1.1587	0.0012	1.4300	2.0699	0.0028	1.5528	1.7214
2.31727	0.00255	1.03368	3.19678	0.00667	1.03102	2.31316	1.1581	0.0018	1.4291	2.0680	0.0044	1.5506	1.7199
2.31627	0.00367	1.03303	3.19329	0.00961	1.02921	2.31036	1.1573	0.0027	1.4281	2.0656	0.0064	1.5479	1.7182
2.31510	0.00500	1.03227	3.18916	0.01309	1.02707	2.30705	1.1564	0.0036	1.4269	2.0628	0.0087	1.5448	1.7161
2.31373	0.00653	1.03139	3.18441	0.01711	1.02461	2.30323	1.1554	0.0047	1.4255	2.0596	0.0114	1.5411	1.7136
2.31219	0.00827	1.03039	3.17902	0.02169	1.02181	2.29890	1.1543	0.0060	1.4239	2.0560	0.0144	1.5369	1.7109
2.31046	0.01022	1.02927	3.17300	0.02681	1.01869	2.29406	1.1530	0.0074	1.4222	2.0520	0.0178	1.5323	1.7078
2.30855	0.01238	1.02804	3.16635	0.03249	1.01524	2.28872	1.1515	0.0090	1.4202	2.0475	0.0216	1.5272	1.7045
2.30646	0.01476	1.02668	3.15907	0.03874	1.01147	2.28288	1.1499	0.0107	1.4181	2.0426	0.0258	1.5216	1.7008
2.30417	0.01734	1.02521	3.15116	0.04555	1.00737	2.27652	1.1482	0.0126	1.4158	2.0373	0.0303	1.5155	1.6967
2.30171	0.02015	1.02361	3.14262	0.05293	1.00295	2.26966	1.1464	0.0146	1.4132	2.0315	0.0352	1.5089	1.6924
2.29905	0.02316	1.02190	3.13346	0.06090	0.99821	2.26230	1.1444	0.0168	1.4105	2.0253	0.0405	1.5019	1.6877
2.29621	0.02640	1.02006	3.12367	0.06945	0.99314	2.25444	1.1422	0.0191	1.4076	2.0188	0.0462	1.4944	1.6828
2.29318	0.02986	1.01810	3.11327	0.07861	0.98775	2.24607	1.1400	0.0216	1.4045	2.0117	0.0522	1.4864	1.6775
2.28995	0.03354	1.01602	3.10224	0.08837	0.98205	2.23720	1.1375	0.0243	1.4012	2.0043	0.0587	1.4779	1.6719
2.28654	0.03745	1.01382	3.09059	0.09875	0.97602	2.22784	1.1350	0.0272	1.3977	1.9965	0.0656	1.4689	1.6660
2.28293	0.04159	1.01149	3.07832	0.10975	0.96968	2.21797	1.1323	0.0302	1.3941	1.9882	0.0729	1.4595	1.6597
2.27913	0.04597	1.00904	3.06544	0.12140	0.96302	2.20760	1.1294	0.0334	1.3902	1.9795	0.0806	1.4496	1.6532
2.27513	0.05058	1.00646	3.05194	0.13370	0.95605	2.19674	1.1264	0.0367	1.3861	1.9705	0.0888	1.4393	1.6463
2.27094	0.05542	1.00375	3.03784	0.14667	0.94876	2.18538	1.1232	0.0403	1.3818	1.9610	0.0974	1.4285	1.6392
2.26654	0.06052	1.00092	3.02312	0.16033	0.94117	2.17353	1.1199	0.0440	1.3774	1.9510	0.1065	1.4172	1.6317
2.26195	0.06586	0.99796	3.00780	0.17468	0.93326	2.16118	1.1164	0.0479	1.3727	1.9407	0.1160	1.4054	1.6239
2.25715	0.07145	0.99487	2.99187	0.18976	0.92504	2.14834	1.1128	0.0520	1.3678	1.9300	0.1260	1.3932	1.6158
2.25215	0.07730	0.99164	2.97533	0.20557	0.91652	2.13501	1.1091	0.0563	1.3627	1.9188	0.1365	1.3806	1.6074
2.24693	0.08341	0.98829	2.95820	0.22214	0.90770	2.12119	1.1051	0.0607	1.3574	1.9073	0.1474	1.3675	1.5987
2.24151	0.08979	0.98480	2.94047	0.23950	0.89857	2.10688	1.1010	0.0654	1.3519	1.8953	0.1589	1.3540	1.5897
2.23588	0.09644	0.98117	2.92215	0.25766	0.88914	2.09208	1.0968	0.0703	1.3461	1.8830	0.1710	1.3400	1.5804
2.23003	0.10338	0.97741	2.90323	0.27666	0.87941	2.07680	1.0924	0.0754	1.3402	1.8702	0.1835	1.3255	1.5708
2.22396	0.11059	0.97351	2.88373	0.29653	0.86939	2.06103	1.0878	0.0807	1.3340	1.8571	0.1967	1.3107	1.5609
2.21767	0.11810	0.96947	2.86363	0.31729	0.85907	2.04478	1.0830	0.0862	1.3276	1.8435	0.2104	1.2953	1.5507
2.21116	0.12592	0.96528	2.84296	0.33898	0.84846	2.02804	1.0781	0.0920	1.3210	1.8296	0.2248	1.2796	1.5403
2.20442	0.13404	0.96096	2.82170	0.36165	0.83756	2.01082	1.0730	0.0980	1.3142	1.8153	0.2397	1.2634	1.5295
2.19744	0.14247	0.95648	2.79987	0.38532	0.82637	1.99311	1.0677	0.1043	1.3071	1.8005	0.2554	1.2468	1.5184
2.19023	0.15124	0.95185	2.77747	0.41005	0.81490	1.97493	1.0623	0.1108	1.2998	1.7854	0.2717	1.2298	1.5070
2.18278	0.16034	0.94707	2.75449	0.43589	0.80314	1.95627	1.0566	0.1175	1.2923	1.7699	0.2888	1.2124	1.4954
2.17509	0.16979	0.94214	2.73095	0.46288	0.79111	1.93712	1.0508	0.1245	1.2845	1.7540	0.3066	1.1946	1.4834
2.16715	0.17959	0.93705	2.70684	0.49110	0.77880	1.91750	1.0447	0.1318	1.2764	1.7377	0.3251	1.1763	1.4712
2.15895	0.18977	0.93179	2.68218	0.52059	0.76621	1.89740	1.0385	0.1394	1.2682	1.7211	0.3446	1.1576	1.4587
2.15049	0.20033	0.92638	2.65695	0.55144	0.75336	1.87682	1.0321	0.1473	1.2596	1.7040	0.3649	1.1386	1.4459
2.14177	0.21129	0.92079	2.63118	0.58372	0.74024	1.85576	1.0255	0.1555	1.2508	1.6866	0.3861	1.1191	1.4328
2.13277	0.22266	0.91503	2.60486	0.61751	0.72685	1.83422	1.0186	0.1641	1.2417	1.6688	0.4083	1.0993	1.4194
2.12349	0.23446	0.90900	2.57799	0.65291	0.71320	1.81220	1.0116	0.1730	1.2324	1.6506	0.4315	1.0791	1.4058
2.11393	0.24671	0.90297	2.55058	0.69003	0.69929	1.78970	1.0043	0.1822	1.2227	1.6321	0.4559	1.0585	1.3919
2.10408	0.25943	0.89667	2.52264	0.72898	0.68513	1.76672	0.9968	0.1918	1.2128	1.6132	0.4814	1.0375	1.3777
2.09392	0.27264	0.89017	2.49416	0.76988	0.67072	1.74325	0.9890	0.2018	1.2025	1.5939	0.5082	1.0161	1.3632
2.08345	0.28636	0.88348	2.46515	0.81289	0.65606	1.71931	0.9810	0.2122	1.1920	1.5743	0.5363	0.9944	1.3485
2.07266	0.30062	0.87658	2.43563	0.85815	0.64116	1.69487	0.9728	0.2231	1.1812	1.5543	0.5659	0.9723	1.3334
2.06154	0.31544	0.86947	2.40558	0.90586	0.62602	1.66994	0.9643	0.2344	1.1700	1.5339	0.5970	0.9499	1.3181
2.05008	0.33086	0.86214	2.37501	0.95620	0.61064	1.64452	0.9555	0.2462	1.1585	1.5132	0.6298	0.9271	1.3026
2.03827	0.34691	0.85458	2.34394	1.00940	0.59504	1.61861	0.9464	0.2586	1.1466	1.4921	0.6644	0.9040	1.2867
2.02608	0.36362	0.84678	2.31236	1.06572	0.57921	1.59219	0.9371	0.2714	1.1344	1.4706	0.7010	0.8806	1.2706
2.01352	0.38104	0.83874	2.28028	1.12544	0.56316	1.56527	0.9274	0.2849	1.1218	1.4488	0.7397	0.8568	1.2543
2.00056	0.39921	0.83045	2.24770	1.18888	0.54690	1.53783	0.9174	0.2990	1.1088	1.4267	0.7808	0.8327	1.2376
1.98719	0.41819	0.82189	2.21464	1.25643	0.53042	1.50987	0.9071	0.3138	1.0954	1.4042	0.8245	0.8083	1.2207
1.97338	0.43801	0.81304	2.18109	1.32849	0.51375	1.48138	0.8965	0.3293	1.0816	1.3814	0.8710	0.7836	1.2035
1.95913	0.45876	0.80390	2.14705	1.40557	0.49688	1.45236	0.8854	0.3455	1.0673	1.3582	0.9206	0.7586	1.1861
1.94441	0.48050	0.79445	2.11254	1.48823	0.47981	1.42279	0.8740	0.3627	1.0526	1.3346	0.9737	0.7333	1.1683
L_1	L_2	C_2	L_3	L_4	C_4	L_5	L_1	L_2	C_2	L_3	L_4	C_4	L_5

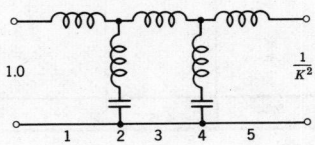

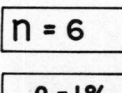

n = 6

ρ = 1%

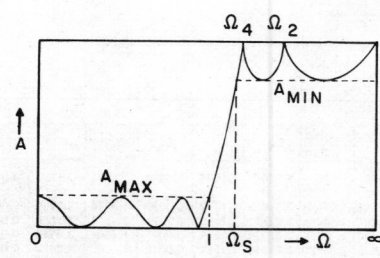

θ	Ω_S	A_MIN	σ_1	σ_3	σ_5	Ω_1	Ω_2	Ω_3	Ω_4	Ω_5
c	∞	∞	−0.2594325	−0.7087829	−0.9682154	1.3676454	∞	1.0011859	∞	0.3664595
11.0	5.422373	107.89	−0.2501021	−0.7001221	−0.9801741	1.3603378	7.922792	1.0131764	5.624706	0.3774122
12.0	4.975749	103.33	−0.2483461	−0.6984418	−0.9824704	1.3589464	7.263825	1.0154480	5.160811	0.3795395
13.0	4.598265	99.13	−0.2464438	−0.6966029	−0.9849749	1.3574331	6.706339	1.0179142	4.768680	0.3818685
14.0	4.275108	95.24	−0.2443967	−0.6946023	−0.9876897	1.3557978	6.228589	1.0205742	4.432936	0.3844036
15.0	3.995415	91.62	−0.2422067	−0.6924365	−0.9906171	1.3540403	5.814626	1.0234270	4.142305	0.3871497
16.0	3.751039	88.22	−0.2398756	−0.6901019	−0.9937598	1.3521603	5.452491	1.0264717	3.888329	0.3901119
17.0	3.535748	85.02	−0.2374054	−0.6875944	−0.9971203	1.3501578	5.133037	1.0297072	3.664543	0.3932961
18.0	3.344698	82.01	−0.2347984	−0.6849097	−1.0007018	1.3480324	4.849152	1.0331320	3.465915	0.3967084
19.0	3.174064	79.15	−0.2320567	−0.6820433	−1.0045072	1.3457840	4.595218	1.0367447	3.288476	0.4003555
20.0	3.020785	76.43	−0.2291828	−0.6789903	−1.0085400	1.3434124	4.366743	1.0405439	3.129050	0.4042447
21.0	2.882384	73.84	−0.2261793	−0.6757455	−1.0128037	1.3409174	4.160091	1.0445277	2.985065	0.4083839
22.0	2.756834	71.37	−0.2230489	−0.6723033	−1.0173022	1.3382989	3.972284	1.0486941	2.854418	0.4127815
23.0	2.642462	69.01	−0.2197945	−0.6686581	−1.0220393	1.3355567	3.800865	1.0530410	2.735370	0.4174467
24.0	2.537873	66.74	−0.2164191	−0.6648035	−1.0270195	1.3326907	3.643786	1.0575659	2.626475	0.4223892
25.0	2.441895	64.55	−0.2129258	−0.6607331	−1.0322471	1.3297007	3.499325	1.0622663	2.526516	0.4276195
26.0	2.353536	62.45	−0.2093180	−0.6564402	−1.0377269	1.3265868	3.366027	1.0671392	2.434463	0.4331490
27.0	2.271953	60.42	−0.2055991	−0.6519174	−1.0434640	1.3233488	3.242651	1.0721813	2.349441	0.4389897
28.0	2.196422	58.47	−0.2017728	−0.6471573	−1.0494636	1.3199868	3.128134	1.0773890	2.270699	0.4451547
29.0	2.126320	56.57	−0.1978427	−0.6421519	−1.0557314	1.3165008	3.021559	1.0827585	2.197588	0.4516580
30.0	2.061105	54.74	−0.1938129	−0.6368929	−1.0622731	1.3128910	2.922133	1.0882853	2.129549	0.4585144
31.0	2.000308	52.96	−0.1896873	−0.6313717	−1.0690951	1.3091576	2.829163	1.0939647	2.066092	0.4657402
32.0	1.943517	51.23	−0.1854703	−0.6255792	−1.0762038	1.3053006	2.742043	1.0997914	2.006790	0.4733525
33.0	1.890370	49.55	−0.1811662	−0.6195059	−1.0836062	1.3013206	2.660243	1.1057594	1.951268	0.4813699
34.0	1.840548	47.91	−0.1767795	−0.6131420	−1.0913095	1.2972177	2.583292	1.1118625	1.899195	0.4898121
35.0	1.793769	46.32	−0.1723150	−0.6064773	−1.0993214	1.2929926	2.510774	1.1180934	1.850277	0.4987006
36.0	1.749781	44.77	−0.1677774	−0.5995011	−1.1076501	1.2886457	2.442321	1.1244444	1.804254	0.5080581
37.0	1.708362	43.25	−0.1631719	−0.5922024	−1.1163040	1.2841777	2.377602	1.1309068	1.760893	0.5179093
38.0	1.669312	41.77	−0.1585035	−0.5845697	−1.1252924	1.2795895	2.316322	1.1374712	1.719987	0.5282807
39.0	1.632449	40.32	−0.1537776	−0.5765912	−1.1346249	1.2748818	2.258218	1.1441269	1.681350	0.5392007
40.0	1.597615	38.90	−0.1489997	−0.5682549	−1.1443119	1.2700557	2.203049	1.1508626	1.644814	0.5506998

θ	Ω_S	A_MIN	σ_1	σ_3	σ_5	Ω_1	Ω_2	Ω_3	Ω_4	Ω_5

224

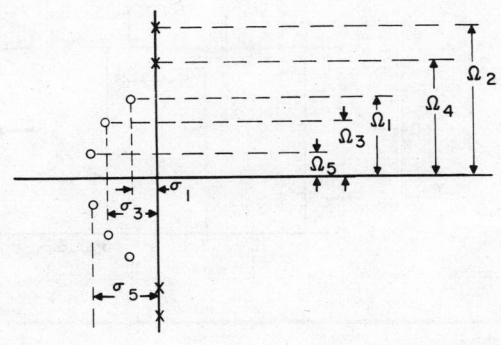

K²=0.9802							
C_1	C_2	L_2	C_3	C_4	L_4	C_5	L_6
0.5164	0.00000	1.130	1.378	0.00000	1.351	1.153	0.5062
0.5038	0.01434	1.111	1.354	0.02406	1.314	1.133	0.5078
0.5014	0.01711	1.108	1.350	0.02874	1.306	1.130	0.5081
0.4988	0.02015	1.104	1.345	0.03386	1.299	1.126	0.5084
0.4960	0.02344	1.099	1.340	0.03944	1.290	1.122	0.5087
0.4929	0.02701	1.095	1.334	0.04548	1.281	1.117	0.5091
0.4896	0.03085	1.090	1.328	0.05200	1.272	1.112	0.5095
0.4861	0.03498	1.085	1.322	0.05902	1.262	1.107	0.5100
0.4824	0.03939	1.080	1.315	0.06654	1.251	1.101	0.5104
0.4785	0.04410	1.074	1.308	0.07460	1.240	1.095	0.5109
0.4743	0.04911	1.068	1.300	0.08320	1.228	1.089	0.5114
0.4699	0.05443	1.062	1.292	0.09236	1.215	1.083	0.5119
0.4653	0.06008	1.055	1.284	0.1021	1.202	1.076	0.5125
0.4604	0.06605	1.048	1.276	0.1125	1.188	1.068	0.5131
0.4553	0.07238	1.041	1.267	0.1235	1.174	1.061	0.5137
0.4499	0.07905	1.033	1.258	0.1352	1.159	1.053	0.5143
0.4443	0.08610	1.025	1.248	0.1476	1.143	1.045	0.5149
0.4385	0.09352	1.017	1.239	0.1608	1.127	1.036	0.5156
0.4324	0.1013	1.008	1.228	0.1748	1.110	1.027	0.5163
0.4260	0.1096	0.9995	1.218	0.1896	1.092	1.018	0.5171
0.4194	0.1183	0.9903	1.207	0.2052	1.074	1.008	0.5178
0.4125	0.1274	0.9808	1.196	0.2219	1.056	0.9978	0.5186
0.4053	0.1370	0.9710	1.185	0.2396	1.036	0.9873	0.5194
0.3978	0.1471	0.9608	1.174	0.2583	1.017	0.9765	0.5202
0.3901	0.1577	0.9504	1.162	0.2783	0.9962	0.9653	0.5210
0.3820	0.1688	0.9396	1.150	0.2995	0.9752	0.9537	0.5219
0.3737	0.1806	0.9284	1.138	0.3221	0.9536	0.9417	0.5228
0.3651	0.1929	0.9170	1.126	0.3463	0.9314	0.9292	0.5237
0.3561	0.2059	0.9052	1.113	0.3720	0.9086	0.9164	0.5246
0.3468	0.2196	0.8931	1.101	0.3996	0.8853	0.9031	0.5256
0.3372	0.2340	0.8806	1.088	0.4291	0.8614	0.8893	0.5265
L_1	L_2	C_2	L_3	L_4	C_4	L_5	C_6

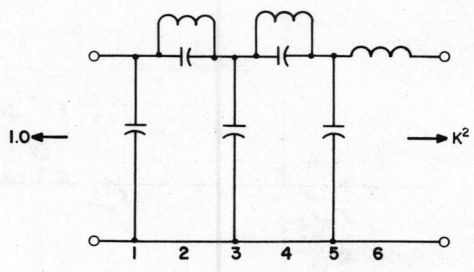

				$K^2 = \infty$				
θ	C_1	C_2	L_2	C_3	C_4	L_4	C_5	L_6
C	0.25821	0.000000	0.70882	1.02394	0.000000	1.21110	1.28556	1.09672
11.0	0.23611	0.02343	0.67998	0.99825	0.02690	1.17511	1.27467	1.10660
12.0	0.23182	0.02810	0.67448	0.99345	0.03214	1.16828	1.27261	1.10848
13.0	0.22712	0.03326	0.66849	0.98826	0.03788	1.16087	1.27038	1.11052
14.0	0.22200	0.03894	0.66202	0.98269	0.04414	1.15288	1.26798	1.11273
15.0	0.21646	0.04515	0.65506	0.97675	0.05093	1.14429	1.26540	1.11510
16.0	0.21047	0.05194	0.64762	0.97045	0.05827	1.13513	1.26265	1.11763
17.0	0.20404	0.05933	0.63968	0.96381	0.06617	1.12539	1.25974	1.12033
18.0	0.19714	0.06737	0.63126	0.95682	0.07466	1.11506	1.25665	1.12319
19.0	0.18975	0.07610	0.62234	0.94952	0.08375	1.10416	1.25340	1.12622
20.0	0.18187	0.08556	0.61293	0.94191	0.09347	1.09268	1.24999	1.12941
21.0	0.17347	0.09582	0.60303	0.93401	0.10385	1.08062	1.24642	1.13277
22.0	0.16453	0.10694	0.59263	0.92584	0.11492	1.06799	1.24269	1.13629
23.0	0.15502	0.11899	0.58174	0.91743	0.12671	1.05478	1.23880	1.13997
24.0	0.14492	0.13205	0.57036	0.90879	0.13925	1.04100	1.23476	1.14382
25.0	0.13421	0.14623	0.55848	0.89995	0.15259	1.02664	1.23056	1.14783
26.0	0.12284	0.16161	0.54611	0.89094	0.16678	1.01171	1.22622	1.15201
27.0	0.11078	0.17835	0.53326	0.88179	0.18186	0.99620	1.22174	1.15634
28.0	0.09799	0.19656	0.51991	0.87254	0.19788	0.98011	1.21712	1.16084
29.0	0.08443	0.21643	0.50608	0.86323	0.21492	0.96344	1.21237	1.16550
30.0	0.07005	0.23814	0.49178	0.85391	0.23305	0.94618	1.20750	1.17032
31.0	0.05477	0.26192	0.47700	0.84462	0.25235	0.92834	1.20251	1.17530
32.0	0.03856	0.28803	0.46176	0.83542	0.27290	0.90990	1.19740	1.18043
33.0	0.02131	0.31678	0.44606	0.82637	0.29482	0.89086	1.19220	1.18572
34.0	0.00297	0.34855	0.42992	0.81755	0.31823	0.87121	1.18691	1.19116
35.0	-0.01659	0.38377	0.41335	0.80904	0.34327	0.85094	1.18155	1.19675
36.0	-0.03745	0.42296	0.39636	0.80094	0.37009	0.83004	1.17613	1.20248
37.0	-0.05976	0.46677	0.37898	0.79334	0.39889	0.80849	1.17068	1.20834
38.0	-0.08365	0.51597	0.36123	0.78638	0.42990	0.78629	1.16522	1.21434
39.0	-0.10931	0.57150	0.34313	0.78021	0.46337	0.76341	1.15977	1.22045
40.0	-0.13694	0.63453	0.32471	0.77499	0.49960	0.73985	1.15438	1.22667
θ	L_1	L_2	C_2	L_3	L_4	C_4	L_5	C_6

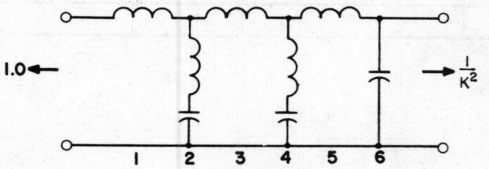

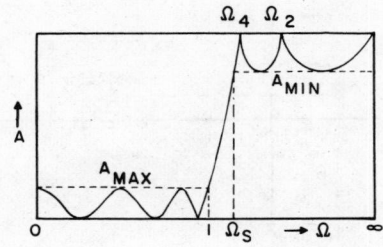

n = 6

P = 2%

θ	Ω_S	A_{MIN}	σ_1	σ_3	σ_5	Ω_1	Ω_2	Ω_3	Ω_4	Ω_5
c	∞	∞	−0.2187321	−0.5975873	−0.8163194	I.2646699	∞	0.9258027	∞	0.3388673
11.0	5.422373	113.91	−0.2122216	−0.5917062	−0.8249983	1.2600584	7.922792	0.9347674	5.624706	0.3467872
12.0	4.975749	109.35	−0.2109922	−0.5905677	−0.8266635	1.2591794	7.263825	0.9364699	5.160811	0.3483203
13.0	4.598265	105.16	−0.2096589	−0.5893226	−0.8284791	1.2582231	6.706339	0.9383198	4.768680	0.3499969
14.0	4.275108	101.27	−0.2082224	−0.5879690	−0.8304465	1.2571893	6.228589	0.9403169	4.432936	0.3518197
15.0	3.995415	97.64	−0.2066835	−0.5865048	−0.8325674	1.2560778	5.814626	0.9424610	4.142305	0.3537915
16.0	3.751039	94.24	−0.2050432	−0.5849278	−0.8348435	1.2548883	5.452491	0.9447518	3.888329	0.3559156
17.0	3.535748	91.05	−0.2033024	−0.5832354	−0.8372765	1.2536206	5.133037	0.9471891	3.664543	0.3581954
18.0	3.344698	88.03	−0.2014620	−0.5814252	−0.8398684	1.2522744	4.849152	0.9497724	3.465915	0.3606346
19.0	3.174064	85.17	−0.1995232	−0.5794942	−0.8426214	1.2508495	4.595218	0.9525013	3.288476	0.3632373
20.0	3.020785	82.45	−0.1974872	−0.5774395	−0.8455376	1.2493455	4.366743	0.9553754	3.129050	0.3660078
21.0	2.882384	79.87	−0.1953552	−0.5752578	−0.8486195	1.2477623	4.160091	0.9583942	2.985065	0.3689508
22.0	2.756834	77.39	−0.1931285	−0.5729458	−0.8518696	1.2460994	3.972284	0.9615570	2.854418	0.3720714
23.0	2.642462	75.03	−0.1908085	−0.5704999	−0.8552905	1.2443567	3.800865	0.9648632	2.735370	0.3753750
24.0	2.537873	72.76	−0.1883966	−0.5679163	−0.8588853	1.2425338	3.643786	0.9683120	2.626475	0.3788673
25.0	2.441895	70.58	−0.1858945	−0.5651909	−0.8626568	1.2406305	3.499325	0.9719026	2.526516	0.3825547
26.0	2.353536	68.47	−0.1833036	−0.5623194	−0.8666083	1.2386464	3.366027	0.9756338	2.434463	0.3864438
27.0	2.271953	66.45	−0.1806258	−0.5592974	−0.8707431	1.2365812	3.242651	0.9795046	2.349441	0.3905417
28.0	2.196422	64.49	−0.1778627	−0.5561201	−0.8750649	1.2344346	3.128134	0.9835139	2.270699	0.3948560
29.0	2.126320	62.59	−0.1750164	−0.5527826	−0.8795773	1.2322064	3.021559	0.9876601	2.197588	0.3993951
30.0	2.061105	60.76	−0.1720887	−0.5492795	−0.8842844	1.2298964	2.922133	0.9919416	2.129549	0.4041675
31.0	2.000308	58.98	−0.1690818	−0.5456054	−0.8891904	1.2275042	2.829163	0.9963568	2.066092	0.4091827
32.0	1.943517	57.25	−0.1659977	−0.5417545	−0.8942996	1.2250296	2.742043	1.0009037	2.006790	0.4144507
33.0	1.890370	55.57	−0.1628388	−0.5377207	−0.8996166	1.2224724	2.660243	1.0055799	1.951268	0.4199822
34.0	1.840548	53.93	−0.1596073	−0.5334977	−0.9051464	1.2198324	2.583292	1.0103833	1.899195	0.4257887
35.0	1.793769	52.34	−0.1563057	−0.5290787	−0.9108941	1.2171094	2.510774	1.0153108	1.850277	0.4318820
36.0	1.749781	50.79	−0.1529366	−0.5244568	−0.9168651	1.2143034	2.442321	1.0203596	1.804254	0.4382766
37.0	1.708362	49.27	−0.1495026	−0.5196247	−0.9230652	1.2114141	2.377602	1.0255261	1.760893	0.4449852
38.0	1.669312	47.79	−0.1460066	−0.5145746	−0.9295005	1.2084415	2.316322	1.0308067	1.719987	0.4520233
39.0	1.632449	46.34	−0.1424513	−0.5092987	−0.9361772	1.2053857	2.258218	1.0361971	1.681350	0.4594073
40.0	1.597615	44.92	−0.1388397	−0.5037886	−0.9431022	1.2022466	2.203049	1.0416925	1.644814	0.4671543

θ	Ω_S	A_{MIN}	σ_1	σ_3	σ_5	Ω_1	Ω_2	Ω_3	Ω_4	Ω_5

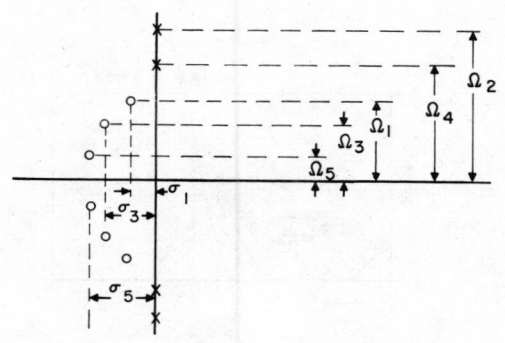

θ	c_1	c_2	L_2	c_3	c_4	L_4	c_5	L_6
				K²=0.9608				
c	0.6125	0.00000	1.240	1.505	0.00000	1.446	1.290	0.5885
11.0	0.6010	0.01304	1.222	1.481	0.02240	1.411	1.272	0.5898
12.0	0.5988	0.01555	1.219	1.477	0.02674	1.404	1.268	0.5901
13.0	0.5964	0.01830	1.215	1.472	0.03148	1.397	1.264	0.5904
14.0	0.5938	0.02129	1.211	1.467	0.03664	1.389	1.260	0.5907
15.0	0.5910	0.02451	1.207	1.461	0.04221	1.381	1.256	0.5910
16.0	0.5880	0.02798	1.202	1.455	0.04822	1.372	1.251	0.5913
17.0	0.5848	0.03170	1.197	1.449	0.05468	1.362	1.246	0.5917
18.0	0.5814	0.03567	1.192	1.442	0.06158	1.352	1.241	0.5921
19.0	0.5778	0.03990	1.187	1.435	0.06896	1.341	1.235	0.5925
20.0	0.5740	0.04440	1.181	1.427	0.07681	1.330	1.229	0.5929
21.0	0.5700	0.04917	1.175	1.419	0.08517	1.318	1.223	0.5934
22.0	0.5657	0.05422	1.169	1.411	0.09403	1.305	1.216	0.5938
23.0	0.5613	0.05956	1.162	1.402	0.1034	1.292	1.210	0.5943
24.0	0.5566	0.06520	1.155	1.393	0.1134	1.279	1.202	0.5948
25.0	0.5518	0.07114	1.148	1.384	0.1239	1.264	1.195	0.5954
26.0	0.5467	0.07739	1.140	1.374	0.1350	1.250	1.187	0.5959
27.0	0.5414	0.08396	1.133	1.364	0.1468	1.234	1.179	0.5965
28.0	0.5358	0.09088	1.125	1.354	0.1592	1.218	1.170	0.5971
29.0	0.5300	0.09814	1.116	1.343	0.1723	1.202	1.161	0.5977
30.0	0.5240	0.1058	1.107	1.332	0.1861	1.185	1.152	0.5984
31.0	0.5178	0.1137	1.098	1.321	0.2007	1.167	1.143	0.5990
32.0	0.5113	0.1221	1.089	1.309	0.2161	1.149	1.133	0.5997
33.0	0.5046	0.1309	1.079	1.297	0.2323	1.131	1.123	0.6004
34.0	0.4976	0.1401	1.069	1.285	0.2495	1.111	1.112	0.6011
35.0	0.4904	0.1498	1.059	1.273	0.2676	1.092	1.102	0.6019
36.0	0.4829	0.1599	1.048	1.260	0.2868	1.071	1.090	0.6027
37.0	0.4751	0.1705	1.037	1.247	0.3070	1.050	1.079	0.6034
38.0	0.4671	0.1816	1.026	1.234	0.3285	1.029	1.067	0.6042
39.0	0.4588	0.1933	1.015	1.220	0.3513	1.007	1.055	0.6051
40.0	0.4502	0.2055	1.003	1.207	0.3754	0.9845	1.042	0.6059

θ	L_1	L_2	c_2	L_3	L_4	c_4	L_5	c_6

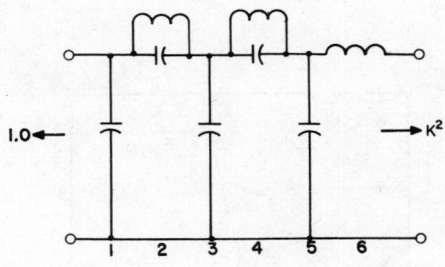

			$K^2 = \infty$				
C_1	C_2	L_2	C_3	C_4	L_4	C_5	L_6
0.30625	0.000000	0.81987	0.000000	1.15198	1.33406	1.39864	1.18584
0.28723	0.02006	0.79413	0.02432	1.12729	1.29945	1.38678	1.19350
0.28356	0.02401	0.78923	0.02904	1.12264	1.29289	1.38453	1.19496
0.27954	0.02836	0.78389	0.03420	1.11761	1.28576	1.38210	1.19654
0.27518	0.03313	0.77811	0.03982	1.11221	1.27808	1.37947	1.19825
0.27047	0.03832	0.77191	0.04590	1.10643	1.26983	1.37666	1.20009
0.26539	0.04395	0.76526	0.05245	1.10028	1.26102	1.37365	1.20205
0.25995	0.05006	0.75818	0.05949	1.09377	1.25165	1.37046	1.20414
0.25412	0.05665	0.75066	0.06704	1.08690	1.24173	1.36708	1.20636
0.24791	0.06376	0.74270	0.07510	1.07969	1.23125	1.36352	1.20871
0.24131	0.07142	0.73430	0.08370	1.07215	1.22022	1.35978	1.21118
0.23429	0.07965	0.72546	0.09285	1.06428	1.20864	1.35585	1.21378
0.22686	0.08849	0.71617	0.10258	1.05609	1.19651	1.35175	1.21651
0.21899	0.09799	0.70643	0.11290	1.04760	1.18384	1.34746	1.21936
0.21067	0.10818	0.69625	0.12383	1.03882	1.17062	1.34301	1.22234
0.20188	0.11911	0.68562	0.13542	1.02977	1.15685	1.33837	1.22545
0.19261	0.13085	0.67454	0.14768	1.02046	1.14255	1.33357	1.22869
0.18284	0.14344	0.66301	0.16065	1.01090	1.12770	1.32860	1.23205
0.17255	0.15698	0.65103	0.17436	1.00112	1.11232	1.32346	1.23553
0.16170	0.17152	0.63859	0.18886	0.99114	1.09639	1.31816	1.23915
0.15028	0.18717	0.62571	0.20419	0.98097	1.07993	1.31270	1.24289
0.13826	0.20402	0.61237	0.22039	0.97066	1.06293	1.30709	1.24675
0.12560	0.22219	0.59859	0.23753	0.96021	1.04539	1.30132	1.25074
0.11227	0.24181	0.58436	0.25566	0.94967	1.02732	1.29541	1.25485
0.09823	0.26304	0.56968	0.27485	0.93907	1.00870	1.28936	1.25908
0.08344	0.28605	0.55456	0.29518	0.92845	0.98955	1.28317	1.26343
0.06784	0.31103	0.53900	0.31674	0.91784	0.96985	1.27686	1.26790
0.05139	0.33823	0.52301	0.33962	0.90729	0.94960	1.27042	1.27249
0.03401	0.36792	0.50659	0.36394	0.89686	0.92880	1.26388	1.27719
0.01565	0.40040	0.48975	0.38982	0.88660	0.90745	1.25724	1.28201
-0.00378	0.43607	0.47250	0.41741	0.87658	0.88554	1.25052	1.28693
L_1	L_2	C_2	L_3	L_4	C_4	L_5	C_6

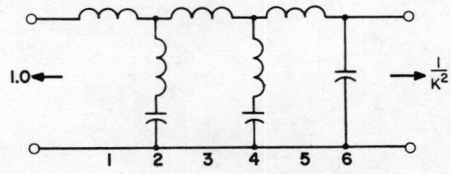

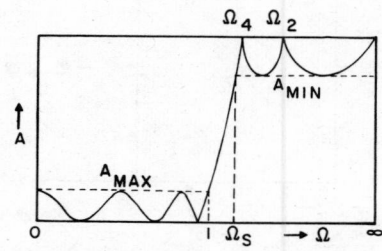

$n = 6$

$\rho = 3\%$

θ	Ω_S	A_{MIN}	σ_1	σ_3	σ_5	Ω_1	Ω_2	Ω_3	Ω_4	Ω_5
c	∞	∞	-0.1963077	-0.5363227	-0.7326305	1.2123366	∞	0.8874920	∞	0.3248446
11.0	5.422373	117.44	-0.1910491	-0.5316528	-0.7398178	1.2089007	7.922792	0.8951096	5.624706	0.3314285
12.0	4.975749	112.88	-0.1900544	-0.5307497	-0.7411963	1.2082454	7.263825	0.8965579	5.160811	0.3327009
13.0	4.598265	108.68	-0.1889752	-0.5297624	-0.7426990	1.2075324	6.706339	0.8981323	4.768680	0.3340915
14.0	4.275108	104.79	-0.1878117	-0.5286894	-0.7443272	1.2067614	6.228589	0.8998327	4.432936	0.3356026
15.0	3.995415	101.16	-0.1865645	-0.5275292	-0.7460821	1.2059323	5.814626	0.9016591	4.142305	0.3372361
16.0	3.751039	97.77	-0.1852341	-0.5262801	-0.7479651	1.2050448	5.452491	0.9036115	3.888329	0.3389946
17.0	3.535748	94.57	-0.1838212	-0.5249403	-0.7499776	1.2040987	5.133037	0.9056898	3.664543	0.3408805
18.0	3.344698	91.55	-0.1823264	-0.5235078	-0.7521211	1.2030937	4.849152	0.9078941	3.465915	0.3428968
19.0	3.174064	88.69	-0.1807503	-0.5219804	-0.7543974	1.2020296	4.595218	0.9102241	3.288476	0.3450463
20.0	3.020785	85.98	-0.1790936	-0.5203559	-0.7568082	1.2009062	4.366743	0.9126799	3.129050	0.3473325
21.0	2.882384	83.39	-0.1773572	-0.5186319	-0.7593554	1.1997232	4.160091	0.9152612	2.985065	0.3497587
22.0	2.756834	80.92	-0.1755419	-0.5168059	-0.7620410	1.1984801	3.972284	0.9179679	2.854418	0.3523288
23.0	2.642462	78.55	-0.1736484	-0.5148751	-0.7648671	1.1971769	3.800865	0.9207997	2.735370	0.3550469
24.0	2.537873	76.28	-0.1716778	-0.5128366	-0.7678361	1.1958132	3.643786	0.9237565	2.626475	0.3579171
25.0	2.441895	74.10	-0.1696310	-0.5106874	-0.7709504	1.1943886	3.499325	0.9268379	2.526516	0.3609442
26.0	2.353536	72.00	-0.1675091	-0.5084243	-0.7742124	1.1929028	3.366027	0.9300435	2.434463	0.3641331
27.0	2.271953	69.97	-0.1653131	-0.5060438	-0.7776249	1.1913556	3.242651	0.9333729	2.349441	0.3674891
28.0	2.196422	68.01	-0.1630441	-0.5035424	-0.7811907	1.1897466	3.128134	0.9368256	2.270699	0.3710179
29.0	2.126320	66.12	-0.1607034	-0.5009163	-0.7849128	1.1880755	3.021559	0.9404010	2.197588	0.3747256
30.0	2.061105	64.28	-0.1582922	-0.4981614	-0.7887943	1.1863419	2.922133	0.9440984	2.129549	0.3786185
31.0	2.000308	62.50	-0.1558118	-0.4952735	-0.7928387	1.1845455	2.829163	0.9479171	2.066092	0.3827037
32.0	1.943517	60.77	-0.1532637	-0.4922484	-0.7970493	1.1826860	2.742043	0.9518560	2.006790	0.3869884
33.0	1.890370	59.09	-0.1506492	-0.4890811	-0.8014300	1.1807632	2.660243	0.9559141	1.951268	0.3914805
34.0	1.840548	57.46	-0.1479700	-0.4857670	-0.8059845	1.1787765	2.583292	0.9600903	1.899195	0.3961886
35.0	1.793769	55.87	-0.1452276	-0.4823009	-0.8107169	1.1767259	2.510774	0.9643832	1.850277	0.4011215
36.0	1.749781	54.31	-0.1424237	-0.4786773	-0.8156317	1.1746109	2.442321	0.9687913	1.804254	0.4062889
37.0	1.708362	52.79	-0.1395600	-0.4748905	-0.8207333	1.1724313	2.377602	0.9733127	1.760893	0.4117011
38.0	1.669312	51.31	-0.1366384	-0.4709348	-0.8260266	1.1701869	2.316322	0.9779457	1.719987	0.4173691
39.0	1.632449	49.86	-0.1336609	-0.4668037	-0.8315166	1.1678773	2.258218	0.9826878	1.681350	0.4233046
40.0	1.597615	48.45	-0.1306293	-0.4624909	-0.8372087	1.1655024	2.203049	0.9875366	1.644814	0.4295204
41.0	1.564662	47.06	-0.1275458	-0.4579894	-0.8431084	1.1630620	2.150602	0.9924894	1.610227	0.4360299
42.0	1.533460	45.69	-0.1244126	-0.4532921	-0.8492219	1.1605559	2.100682	0.9975428	1.577454	0.4428477
43.0	1.503888	44.36	-0.1212320	-0.4483914	-0.8555554	1.1579840	2.053113	1.0026934	1.546369	0.4499893
44.0	1.475840	43.05	-0.1180063	-0.4432796	-0.8621156	1.1553461	2.007733	1.0079371	1.516862	0.4574715
45.0	1.449216	41.76	-0.1147382	-0.4379485	-0.8689096	1.1526423	1.964398	1.0132696	1.488829	0.4653124

θ	Ω_S	A_{MIN}	σ_1	σ_3	σ_5	Ω_1	Ω_2	Ω_3	Ω_4	Ω_5

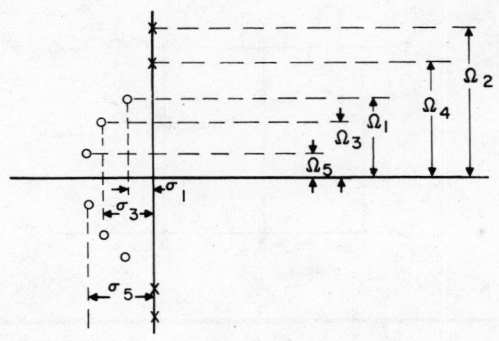

			K²=0.9417				
C_1	C_2	L_2	C_3	C_4	L_4	C_5	L_6
0.6825	0.00000	1.300	1.586	0.00000	1.494	1.380	0.6427
0.6714	0.01242	1.283	1.562	0.02166	1.459	1.362	0.6439
0.6693	0.01481	1.279	1.557	0.02584	1.453	1.359	0.6442
0.6670	0.01743	1.276	1.553	0.03041	1.446	1.355	0.6444
0.6645	0.02026	1.272	1.547	0.03538	1.438	1.351	0.6447
0.6619	0.02333	1.268	1.541	0.04075	1.430	1.346	0.6450
0.6590	0.02662	1.264	1.535	0.04654	1.421	1.342	0.6453
0.6559	0.03015	1.259	1.529	0.05274	1.412	1.337	0.6456
0.6527	0.03391	1.254	1.522	0.05937	1.402	1.332	0.6459
0.6492	0.03793	1.249	1.515	0.06644	1.392	1.326	0.6463
0.6456	0.04218	1.243	1.507	0.07397	1.381	1.320	0.6467
0.6418	0.04670	1.237	1.499	0.08196	1.369	1.314	0.6471
0.6377	0.05147	1.231	1.491	0.09043	1.357	1.308	0.6475
0.6335	0.05651	1.225	1.482	0.09940	1.345	1.301	0.6480
0.6290	0.06183	1.218	1.473	0.1089	1.331	1.294	0.6484
0.6244	0.06743	1.211	1.463	0.1189	1.318	1.286	0.6489
0.6195	0.07332	1.204	1.454	0.1295	1.303	1.279	0.6494
0.6145	0.07950	1.196	1.443	0.1406	1.289	1.271	0.6499
0.6092	0.08600	1.188	1.433	0.1523	1.273	1.262	0.6505
0.6037	0.09281	1.180	1.422	0.1647	1.257	1.254	0.6510
0.5980	0.09995	1.172	1.411	0.1777	1.241	1.245	0.6516
0.5920	0.1074	1.163	1.399	0.1914	1.224	1.236	0.6522
0.5859	0.1153	1.154	1.387	0.2058	1.206	1.226	0.6528
0.5795	0.1235	1.144	1.375	0.2210	1.188	1.216	0.6535
0.5728	0.1321	1.135	1.363	0.2370	1.170	1.206	0.6541
0.5660	0.1410	1.125	1.350	0.2539	1.151	1.196	0.6548
0.5588	0.1504	1.114	1.337	0.2716	1.131	1.185	0.6555
0.5515	0.1603	1.104	1.323	0.2903	1.111	1.174	0.6562
0.5439	0.1705	1.093	1.310	0.3101	1.090	1.162	0.6569
0.5360	0.1813	1.082	1.296	0.3309	1.069	1.150	0.6577
0.5278	0.1926	1.070	1.282	0.3529	1.047	1.138	0.6584
0.5194	0.6592	1.125	1.025	0.3762	1.267	1.058	0.2044
0.5107	0.6600	1.113	1.002	0.4009	1.252	1.046	0.2167
0.5018	0.6608	1.099	0.9791	0.4271	1.237	1.033	0.2296
0.4925	0.6616	1.086	0.9554	0.4549	1.222	1.020	0.2432
0.4829	0.6625	1.072	0.9312	0.4845	1.207	1.007	0.2574
L_1	L_2	C_2	L_3	L_4	C_4	L_5	C_6

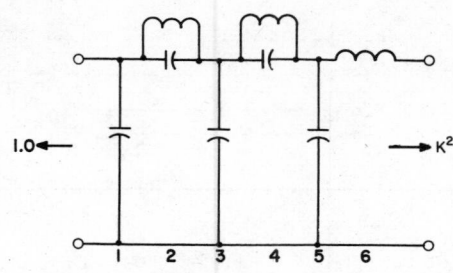

				$K^2=\infty$				
θ	C_1	C_2	L_2	C_3	C_4	L_4	C_5	L_6
C	0.34124	0.000000	0.89499	1.23295	0.000000	1.40751	1.46464	1.23672
11.0	0.32385	0.01829	0.87088	1.20869	0.02301	1.37353	1.45221	1.24320
12.0	0.32050	0.02188	0.86628	1.20411	0.02746	1.36709	1.44985	1.24443
13.0	0.31684	0.02582	0.86128	1.19916	0.03233	1.36010	1.44730	1.24577
14.0	0.31287	0.03012	0.85587	1.19382	0.03762	1.35255	1.44454	1.24722
15.0	0.30859	0.03479	0.85005	1.18811	0.04335	1.34445	1.44159	1.24878
16.0	0.30398	0.03986	0.84383	1.18203	0.04951	1.33581	1.43844	1.25044
17.0	0.29904	0.04533	0.83719	1.17558	0.05613	1.32661	1.43509	1.25221
18.0	0.29377	0.05123	0.83015	1.16877	0.06322	1.31687	1.43154	1.25408
19.0	0.28815	0.05756	0.82269	1.16161	0.07077	1.30659	1.42780	1.25607
20.0	0.28219	0.06436	0.81482	1.15411	0.07882	1.29576	1.42387	1.25816
21.0	0.27587	0.07164	0.80653	1.14626	0.08738	1.28440	1.41974	1.26036
22.0	0.26918	0.07944	0.79782	1.13808	0.09645	1.27250	1.41542	1.26267
23.0	0.26211	0.08777	0.78870	1.12958	0.10607	1.26006	1.41091	1.26508
24.0	0.25466	0.09667	0.77915	1.12076	0.11624	1.24709	1.40622	1.26760
25.0	0.24680	0.10617	0.76918	1.11164	0.12699	1.23359	1.40134	1.27023
26.0	0.23854	0.11632	0.75879	1.10222	0.13835	1.21956	1.39628	1.27297
27.0	0.22985	0.12715	0.74797	1.09253	0.15034	1.20500	1.39103	1.27581
28.0	0.22072	0.13872	0.73672	1.08257	0.16299	1.18991	1.38561	1.27876
29.0	0.21113	0.15107	0.72504	1.07235	0.17633	1.17431	1.38001	1.28181
30.0	0.20106	0.16427	0.71294	1.06190	0.19039	1.15818	1.37423	1.28498
31.0	0.19050	0.17838	0.70040	1.05123	0.20522	1.14154	1.36828	1.28825
32.0	0.17942	0.19347	0.68743	1.04036	0.22084	1.12437	1.36217	1.29162
33.0	0.16780	0.20964	0.67403	1.02930	0.23732	1.10669	1.35589	1.29510
34.0	0.15562	0.22698	0.66020	1.01809	0.25470	1.08849	1.34944	1.29868
35.0	0.14283	0.24558	0.64593	1.00674	0.27304	1.06978	1.34285	1.30237
36.0	0.12942	0.26559	0.63123	0.99528	0.29241	1.05055	1.33610	1.30616
37.0	0.11535	0.28713	0.61610	0.98374	0.31286	1.03081	1.32920	1.31005
38.0	0.10058	0.31036	0.60054	0.97216	0.33450	1.01055	1.32216	1.31404
39.0	0.08506	0.33546	0.58455	0.96058	0.35739	0.98978	1.31499	1.31813
40.0	0.06875	0.36266	0.56814	0.94902	0.38166	0.96848	1.30769	1.32232
41.0	0.05159	0.39218	0.55131	0.93753	0.40741	0.94667	1.30028	1.32659
42.0	0.03352	0.42431	0.53406	0.92618	0.43477	0.92433	1.29276	1.33096
43.0	0.01448	0.45939	0.51641	0.91500	0.46390	0.90147	1.28514	1.33542
44.0	-0.00562	0.49780	0.49835	0.90407	0.49497	0.87808	1.27745	1.33995
45.0	-0.02685	0.53999	0.47991	0.89345	0.52817	0.85415	1.26969	1.34456
θ	L_1	L_2	C_2	L_3	L_4	C_4	L_5	C_6

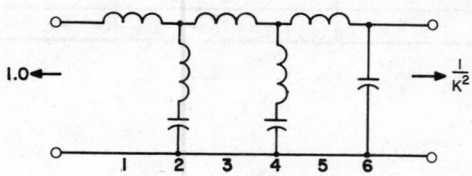

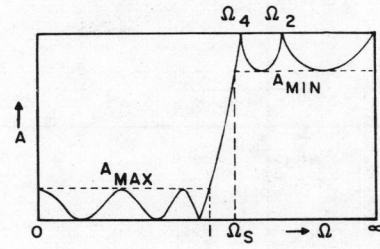

n = 6

ρ = 4%

θ	Ω_S	A_{MIN}	σ_1	σ_3	σ_5	Ω_1	Ω_2	Ω_3	Ω_4	Ω_5
C	∞	∞	-0.1809429	-0.4943453	-0.6752882	1.1785698	∞	0.8627730	∞	0.3157968
11.0	5.422373	119.94	-0.1764320	-0.4903895	-0.6815677	1.1758262	7.922792	0.8695865	5.624706	0.3215881
12.0	4.975749	115.38	-0.1755779	-0.4896250	-0.6827717	1.1753027	7.263825	0.8708829	5.160811	0.3227061
13.0	4.598265	111.18	-0.1746508	-0.4887893	-0.6840842	1.1747330	6.706339	0.8722924	4.768680	0.3239276
14.0	4.275108	107.29	-0.1736511	-0.4878815	-0.6855061	1.1741170	6.228589	0.8738152	4.432936	0.3252544
15.0	3.995415	103.66	-0.1725789	-0.4869000	-0.6870386	1.1734544	5.814626	0.8754513	4.142305	0.3266881
16.0	3.751039	100.27	-0.1714348	-0.4858436	-0.6886827	1.1727450	5.452491	0.8772008	3.888329	0.3282308
17.0	3.535748	97.07	-0.1702191	-0.4847108	-0.6904396	1.1719886	5.133037	0.8790638	3.664543	0.3298845
18.0	3.344698	94.05	-0.1689323	-0.4834999	-0.6923108	1.1711850	4.849152	0.8810404	3.465915	0.3316516
19.0	3.174064	91.20	-0.1675748	-0.4822093	-0.6942974	1.1703340	4.595218	0.8831307	3.288476	0.3335345
20.0	3.020785	88.48	-0.1661472	-0.4808370	-0.6964016	1.1694354	4.366743	0.8853347	3.129050	0.3355360
21.0	2.882384	85.89	-0.1646499	-0.4793811	-0.6986242	1.1684887	4.160091	0.8876524	2.985065	0.3376588
22.0	2.756834	83.42	-0.1630836	-0.4778396	-0.7009673	1.1674939	3.972284	0.8900840	2.854418	0.3399060
23.0	2.642462	81.05	-0.1614488	-0.4762101	-0.7034327	1.1664506	3.800865	0.8926294	2.735370	0.3422811
24.0	2.537873	78.78	-0.1597462	-0.4744903	-0.7060223	1.1653585	3.643786	0.8952885	2.626475	0.3447874
25.0	2.441895	76.60	-0.1579764	-0.4726778	-0.7087382	1.1642174	3.499325	0.8980614	2.526516	0.3474288
26.0	2.353536	74.50	-0.1561403	-0.4707699	-0.7115825	1.1630269	3.366027	0.9009480	2.434463	0.3502092
27.0	2.271953	72.47	-0.1542385	-0.4687637	-0.7145575	1.1617867	3.242651	0.9039481	2.349441	0.3531331
28.0	2.196422	70.51	-0.1522718	-0.4666563	-0.7176656	1.1604965	3.128134	0.9070617	2.270699	0.3562050
29.0	2.126320	68.62	-0.1502412	-0.4644447	-0.7209093	1.1591560	3.021559	0.9102884	2.197588	0.3594298
30.0	2.061105	66.78	-0.1481475	-0.4621255	-0.7242914	1.1577649	2.922133	0.9136280	2.129549	0.3628128
31.0	2.000308	65.00	-0.1459916	-0.4596952	-0.7278147	1.1563228	2.829163	0.9170803	2.066092	0.3663595
32.0	1.943517	63.28	-0.1437745	-0.4571502	-0.7314822	1.1548293	2.742043	0.9206448	2.006790	0.3700760
33.0	1.890370	61.60	-0.1414973	-0.4544867	-0.7352969	1.1532842	2.660243	0.9243209	1.951268	0.3739686
34.0	1.840548	59.96	-0.1391611	-0.4517006	-0.7392623	1.1516871	2.583292	0.9281082	1.899195	0.3780440
35.0	1.793769	58.37	-0.1367671	-0.4487877	-0.7433818	1.1500377	2.510774	0.9320059	1.850277	0.3823097
36.0	1.749781	56.81	-0.1343163	-0.4457435	-0.7476591	1.1483355	2.442321	0.9360131	1.804254	0.3867732
37.0	1.708362	55.30	-0.1318102	-0.4425632	-0.7520980	1.1465804	2.377602	0.9401291	1.760893	0.3914430
38.0	1.669312	53.81	-0.1292501	-0.4392421	-0.7567027	1.1447720	2.316322	0.9443526	1.719987	0.3963278
39.0	1.632449	52.37	-0.1266373	-0.4357747	-0.7614774	1.1429099	2.258218	0.9486824	1.681350	0.4014371
40.0	1.597615	50.95	-0.1239733	-0.4321558	-0.7664268	1.1409939	2.203049	0.9531170	1.644814	0.4067811
41.0	1.564662	49.56	-0.1212597	-0.4283795	-0.7715555	1.1390236	2.150602	0.9576548	1.610227	0.4123706
42.0	1.533460	48.20	-0.1184981	-0.4244399	-0.7768687	1.1369989	2.100682	0.9622939	1.577454	0.4182172
43.0	1.503888	46.86	-0.1156903	-0.4203307	-0.7823716	1.1349193	2.053113	0.9670321	1.546369	0.4243335
44.0	1.475840	45.55	-0.1128379	-0.4160451	-0.7880701	1.1327847	2.007733	0.9718669	1.516862	0.4307326
45.0	1.449216	44.26	-0.1099428	-0.4115763	-0.7939701	1.1305949	1.964398	0.9767954	1.488829	0.4374292
46.0	1.423927	42.99	-0.1070071	-0.4069170	-0.8000779	1.1283496	1.922972	0.9818146	1.462178	0.4444385
47.0	1.399891	41.75	-0.1040327	-0.4020595	-0.8064004	1.1260487	1.883335	0.9869209	1.436822	0.4517772
48.0	1.377032	40.52	-0.1010219	-0.3969959	-0.8129448	1.1236921	1.845375	0.9921100	1.412684	0.4594633
49.0	1.355282	39.31	-0.0979769	-0.3917177	-0.8197188	1.1212797	1.808987	0.9973775	1.389693	0.4675160
50.0	1.334577	38.12	-0.0949000	-0.3862161	-0.8267306	1.1188113	1.774078	1.0027182	1.367782	0.4759562
θ	Ω_S	A_{MIN}	σ_1	σ_3	σ_5	Ω_1	Ω_2	Ω_3	Ω_4	Ω_5

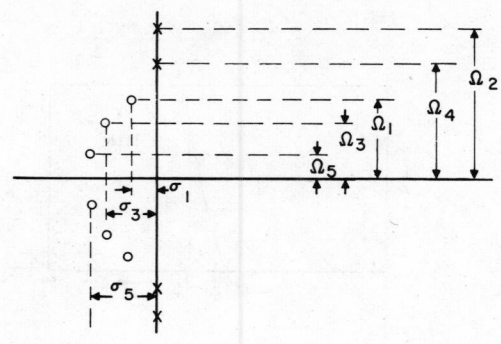

				$K^2 = 0.9231$				
θ	C_1	C_2	L_2	C_3	C_4	L_4	C_5	L_6
c	0.7404	0.00000	1.338	1.648	0.00000	1.521	1.450	0.6835
11.0	0.7297	0.01205	1.322	1.624	0.02125	1.488	1.432	0.6846
12.0	0.7276	0.01437	1.319	1.619	0.02535	1.481	1.428	0.6848
13.0	0.7254	0.01691	1.315	1.614	0.02982	1.474	1.425	0.6850
14.0	0.7229	0.01966	1.311	1.609	0.03469	1.467	1.421	0.6853
15.0	0.7203	0.02262	1.307	1.603	0.03994	1.459	1.416	0.6855
16.0	0.7175	0.02581	1.303	1.597	0.04560	1.451	1.412	0.6858
17.0	0.7146	0.02923	1.298	1.590	0.05166	1.441	1.407	0.6861
18.0	0.7114	0.03287	1.294	1.583	0.05814	1.432	1.402	0.6864
19.0	0.7081	0.03675	1.289	1.576	0.06505	1.422	1.396	0.6868
20.0	0.7045	0.04087	1.283	1.568	0.07239	1.411	1.390	0.6871
21.0	0.7008	0.04523	1.277	1.560	0.08018	1.400	1.384	0.6875
22.0	0.6968	0.04984	1.271	1.552	0.08843	1.388	1.378	0.6879
23.0	0.6927	0.05471	1.265	1.543	0.09716	1.376	1.371	0.6883
24.0	0.6884	0.05984	1.259	1.533	0.1064	1.363	1.364	0.6887
25.0	0.6839	0.06524	1.252	1.524	0.1161	1.349	1.357	0.6892
26.0	0.6792	0.07091	1.245	1.514	0.1264	1.335	1.349	0.6896
27.0	0.6742	0.07686	1.237	1.504	0.1371	1.321	1.341	0.6901
28.0	0.6691	0.08311	1.230	1.493	0.1485	1.306	1.333	0.6906
29.0	0.6638	0.08966	1.222	1.482	0.1605	1.290	1.325	0.6911
30.0	0.6582	0.09652	1.213	1.470	0.1730	1.274	1.316	0.6917
31.0	0.6524	0.1037	1.205	1.459	0.1862	1.258	1.307	0.6922
32.0	0.6464	0.1112	1.196	1.447	0.2001	1.241	1.297	0.6928
33.0	0.6402	0.1191	1.187	1.434	0.2147	1.223	1.287	0.6934
34.0	0.6338	0.1273	1.177	1.422	0.2301	1.205	1.277	0.6940
35.0	0.6271	0.1359	1.167	1.408	0.2462	1.187	1.267	0.6946
36.0	0.6202	0.1449	1.157	1.395	0.2631	1.167	1.256	0.6952
37.0	0.6131	0.1542	1.147	1.381	0.2810	1.148	1.245	0.6959
38.0	0.6057	0.1640	1.136	1.367	0.2998	1.128	1.234	0.6966
39.0	0.5981	0.1743	1.125	1.353	0.3196	1.107	1.222	0.6972
40.0	0.5902	0.1850	1.114	1.338	0.3404	1.086	1.210	0.6979
41.0	0.5821	0.1962	1.102	1.324	0.3624	1.064	1.198	0.6987
42.0	0.5737	0.2079	1.090	1.308	0.3857	1.042	1.185	0.6994
43.0	0.5650	0.2201	1.078	1.293	0.4102	1.019	1.172	0.7002
44.0	0.5561	0.2329	1.065	1.277	0.4362	0.9963	1.159	0.7009
45.0	0.5468	0.2463	1.052	1.261	0.4638	0.9728	1.145	0.7017
46.0	0.5373	0.2604	1.039	1.245	0.4930	0.9487	1.131	0.7025
47.0	0.5275	0.2751	1.025	1.229	0.5241	0.9242	1.117	0.7033
48.0	0.5173	0.2905	1.017	1.212	0.5573	0.8992	1.102	0.7041
49.0	0.5069	0.3067	0.9963	1.196	0.5926	0.8737	1.087	0.7050
50.0	0.4960	0.3238	0.9804	1.179	0.6304	0.8478	1.072	0.7058
θ	L_1	L_2	C_2	L_3	L_4	C_4	L_5	C_6

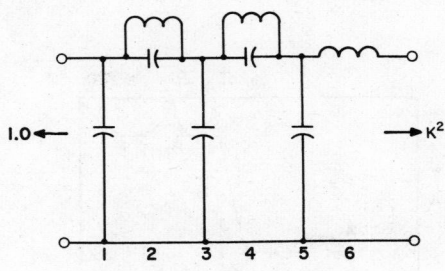

			K²= ∞				
C₁	C₂	L₂	C₃	C₄	L₄	C₅	L₆
0.37021	0.000000	0.95339	1.29300	0.000000	1.45959	1.51103	1.27171
0.35391	0.01712	0.93034	1.26898	0.02217	1.42600	1.49820	1.27741
0.35078	0.02047	0.92595	1.26444	0.02645	1.41964	1.49577	1.27849
0.34736	0.02414	0.92116	1.25953	0.03113	1.41272	1.49313	1.27967
0.34365	0.02814	0.91600	1.25423	0.03621	1.40526	1.49029	1.28094
0.33964	0.03249	0.91044	1.24856	0.04171	1.39725	1.48724	1.28231
0.33534	0.03719	0.90449	1.24252	0.04763	1.38870	1.48398	1.28377
0.33073	0.04226	0.89815	1.23610	0.05398	1.37961	1.48052	1.28533
0.32582	0.04771	0.89142	1.22933	0.06076	1.36998	1.47686	1.28698
0.32059	0.05355	0.88429	1.22219	0.06800	1.35981	1.47299	1.28872
0.31504	0.05981	0.87676	1.21470	0.07571	1.34911	1.46892	1.29056
0.30916	0.06650	0.86884	1.20687	0.08388	1.33787	1.46466	1.29249
0.30295	0.07365	0.86052	1.19869	0.09255	1.32611	1.46019	1.29452
0.29640	0.08126	0.85180	1.19018	0.10173	1.31381	1.45553	1.29664
0.28950	0.08938	0.84267	1.18134	0.11142	1.30099	1.45067	1.29886
0.28223	0.09802	0.83314	1.17218	0.12166	1.28764	1.44562	1.30117
0.27460	0.10722	0.82320	1.16272	0.13247	1.27377	1.44038	1.30358
0.26659	0.11700	0.81285	1.15295	0.14385	1.25938	1.43494	1.30608
0.25818	0.12741	0.80210	1.14289	0.15585	1.24448	1.42932	1.30867
0.24937	0.13848	0.79092	1.13255	0.16848	1.22906	1.42351	1.31136
0.24014	0.15027	0.77934	1.12194	0.18177	1.21312	1.41752	1.31414
0.23047	0.16282	0.76734	1.11108	0.19576	1.19668	1.41135	1.31701
0.22035	0.17618	0.75492	1.09998	0.21048	1.17973	1.40499	1.31998
0.20977	0.19042	0.74208	1.08865	0.22597	1.16227	1.39847	1.32304
0.19869	0.20560	0.72882	1.07711	0.24228	1.14431	1.39177	1.32619
0.18710	0.22182	0.71514	1.06538	0.25945	1.12584	1.38489	1.32943
0.17498	0.23914	0.70104	1.05348	0.27753	1.10687	1.37786	1.33277
0.16230	0.25768	0.68651	1.04144	0.29658	1.08741	1.37066	1.33619
0.14902	0.27753	0.67156	1.02927	0.31667	1.06744	1.36330	1.33971
0.13513	0.29884	0.65618	1.01700	0.33787	1.04698	1.35580	1.34331
0.12058	0.32174	0.64039	1.00466	0.36026	1.02602	1.34814	1.34700
0.10534	0.34640	0.62416	0.99228	0.38393	1.00456	1.34035	1.35077
0.08936	0.37301	0.60752	0.97989	0.40899	0.98260	1.33242	1.35463
0.07259	0.40178	0.59046	0.96754	0.43555	0.96015	1.32437	1.35856
0.05499	0.43296	0.57298	0.95527	0.46374	0.93719	1.31620	1.36258
0.03648	0.46685	0.55509	0.94313	0.49373	0.91373	1.30793	1.36666
0.01701	0.50378	0.53680	0.93117	0.52568	0.88977	1.29957	1.37082
-0.00350	0.54416	0.51811	0.91944	0.55979	0.86530	1.29113	1.37504
-0.02514	0.58845	0.49902	0.90802	0.59631	0.84031	1.28263	1.37931
-0.04801	0.63721	0.47956	0.89699	0.63548	0.81481	1.27410	1.38363
-0.07222	0.69112	0.45973	0.88643	0.67765	0.78879	1.26555	1.38798
L₁	L₂	C₂	L₃	L₄	C₄	L₅	C₆

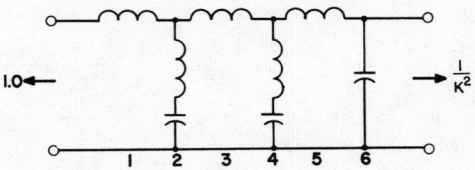

235

n = 6

ρ = 5%

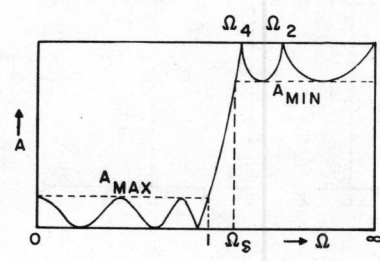

θ	Ω_S	A_{MIN}	σ_1	σ_3	σ_5	Ω_1	Ω_2	Ω_3	Ω_4	Ω_5
c	∞	∞	−0.1693091	−0.4625610	−0.6318701	1.1542411	∞	0.8449631	∞	0.3092780
11.0	5.422373	121.88	−0.1653096	−0.4590887	−0.6375188	1.1519649	7.922792	0.8512280	5.624706	0.3145308
12.0	4.975749	117.32	−0.1645518	−0.4584179	−0.6386017	1.1515305	7.263825	0.8524206	5.160811	0.3155441
13.0	4.598265	113.12	−0.1637290	−0.4576848	−0.6397820	1.1510578	6.706339	0.8537175	4.768680	0.3166509
14.0	4.275108	109.23	−0.1628415	−0.4568885	−0.6410607	1.1505464	6.228589	0.8551189	4.432936	0.3178528
15.0	3.995415	105.61	−0.1618895	−0.4560278	−0.6424387	1.1499963	5.814626	0.8566249	4.142305	0.3191511
16.0	3.751039	102.21	−0.1608733	−0.4551016	−0.6439170	1.1494074	5.452491	0.8582356	3.888329	0.3205477
17.0	3.535748	99.01	−0.1597931	−0.4541085	−0.6454967	1.1487793	5.133037	0.8599513	3.664543	0.3220443
18.0	3.344698	96.00	−0.1586494	−0.4530473	−0.6471789	1.1481119	4.849152	0.8617720	3.465915	0.3236430
19.0	3.174064	93.14	−0.1574423	−0.4519163	−0.6489648	1.1474051	4.595218	0.8636979	3.288476	0.3253457
20.0	3.020785	90.42	−0.1561724	−0.4507141	−0.6508559	1.1466585	4.366743	0.8657293	3.129050	0.3271549
21.0	2.882384	87.83	−0.1548400	−0.4494389	−0.6528535	1.1458719	4.160091	0.8678662	2.985065	0.3290730
22.0	2.756834	85.36	−0.1534455	−0.4480890	−0.6549592	1.1450451	3.972284	0.8701087	2.854418	0.3311027
23.0	2.642462	83.00	−0.1519894	−0.4466624	−0.6571745	1.1441778	3.800865	0.8724571	2.735370	0.3332466
24.0	2.537873	80.73	−0.1504721	−0.4451572	−0.6595012	1.1432698	3.643786	0.8749115	2.626475	0.3355080
25.0	2.441895	78.54	−0.1488942	−0.4435712	−0.6619411	1.1423208	3.499325	0.8774719	2.526516	0.3378900
26.0	2.353536	76.44	−0.1472561	−0.4419020	−0.6644960	1.1413304	3.366027	0.8801385	2.434463	0.3403961
27.0	2.271953	74.41	−0.1455586	−0.4401474	−0.6671680	1.1402985	3.242651	0.8829113	2.349441	0.3430299
28.0	2.196422	72.46	−0.1438020	−0.4383048	−0.6699592	1.1392246	3.128134	0.8857904	2.270699	0.3457954
29.0	2.126320	70.56	−0.1419872	−0.4363715	−0.6728718	1.1381086	3.021559	0.8887757	2.197588	0.3486968
30.0	2.061105	68.73	−0.1401148	−0.4343447	−0.6759083	1.1369500	2.922133	0.8918673	2.129549	0.3517386
31.0	2.000308	66.95	−0.1381854	−0.4322214	−0.6790711	1.1357486	2.829163	0.8950652	2.066092	0.3549255
32.0	1.943517	65.22	−0.1361999	−0.4299984	−0.6823629	1.1345040	2.742043	0.8983691	2.006790	0.3582626
33.0	1.890370	63.54	−0.1341590	−0.4276726	−0.6857864	1.1332159	2.660243	0.9017789	1.951268	0.3617553
34.0	1.840548	61.90	−0.1320635	−0.4252403	−0.6893446	1.1318839	2.583292	0.9052944	1.899195	0.3654094
35.0	1.793769	60.31	−0.1299144	−0.4226979	−0.6930405	1.1305078	2.510774	0.9089153	1.850277	0.3692312
36.0	1.749781	58.76	−0.1277126	−0.4200416	−0.6968774	1.1290871	2.442321	0.9126412	1.804254	0.3732271
37.0	1.708362	57.24	−0.1254590	−0.4172672	−0.7008587	1.1276215	2.377602	0.9164715	1.760893	0.3774042
38.0	1.669312	55.76	−0.1231546	−0.4143706	−0.7049880	1.1261107	2.316322	0.9204058	1.719987	0.3817700
39.0	1.632449	54.31	−0.1208007	−0.4113472	−0.7092691	1.1245544	2.258218	0.9244432	1.681350	0.3863325
40.0	1.597615	52.89	−0.1183982	−0.4081924	−0.7137060	1.1229522	2.203049	0.9285830	1.644814	0.3911003
41.0	1.564662	51.50	−0.1159484	−0.4049010	−0.7183029	1.1213037	2.150602	0.9328241	1.610227	0.3960826
42.0	1.533460	50.14	−0.1134525	−0.4014679	−0.7230644	1.1196087	2.100682	0.9371654	1.577454	0.4012892
43.0	1.503888	48.80	−0.1109120	−0.3978876	−0.7279950	1.1178669	2.053113	0.9416055	1.546369	0.4067305
44.0	1.475840	47.49	−0.1083281	−0.3941543	−0.7330998	1.1160778	2.007733	0.9461429	1.516862	0.4124179
45.0	1.449216	46.20	−0.1057023	−0.3902620	−0.7383842	1.1142413	1.964398	0.9507756	1.488829	0.4183635
46.0	1.423927	44.94	−0.1030363	−0.3862041	−0.7438536	1.1123570	1.922972	0.9555016	1.462178	0.4245803
47.0	1.399891	43.69	−0.1003316	−0.3819741	−0.7495140	1.1104248	1.883335	0.9603185	1.436822	0.4310823
48.0	1.377032	42.46	−0.0975900	−0.3775648	−0.7553718	1.1084443	1.845375	0.9652234	1.412684	0.4378844
49.0	1.355282	41.26	−0.0948132	−0.3729689	−0.7614337	1.1064154	1.808987	0.9702134	1.389693	0.4450030
50.0	1.334577	40.07	−0.0920031	−0.3681784	−0.7677068	1.1043378	1.774078	0.9752846	1.367782	0.4524555
θ	Ω_S	A_{MIN}	σ_1	σ_3	σ_5	Ω_1	Ω_2	Ω_3	Ω_4	Ω_5

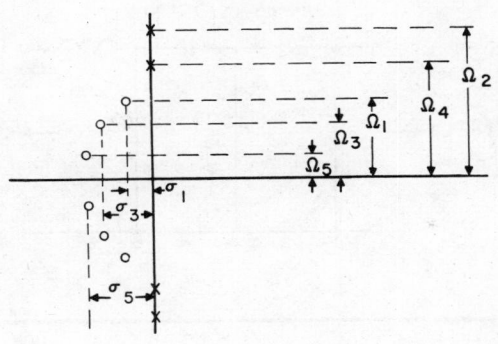

K²=0.9048							
c_1	c_2	L_2	c_3	c_4	L_4	c_5	L_6
0.7913	0.00000	1.365	1.700	0.00000	1.538	1.508	0.7159
0.7807	0.01182	1.348	1.675	0.02100	1.505	1.490	0.7170
0.7887	0.01409	1.345	1.670	0.02505	1.499	1.487	0.7172
0.7765	0.01657	1.342	1.665	0.02947	1.492	1.483	0.7174
0.7741	0.01926	1.338	1.660	0.03428	1.485	1.479	0.7176
0.7715	0.02217	1.334	1.654	0.03946	1.477	1.475	0.7179
0.7688	0.02529	1.330	1.648	0.04504	1.468	1.470	0.7181
0.7659	0.02863	1.326	1.641	0.05102	1.460	1.465	0.7184
0.7628	0.03220	1.321	1.634	0.05741	1.450	1.460	0.7187
0.7595	0.03599	1.316	1.627	0.06421	1.440	1.455	0.7190
0.7560	0.04002	1.311	1.619	0.07144	1.430	1.449	0.7194
0.7523	0.04428	1.305	1.611	0.07911	1.419	1.443	0.7197
0.7485	0.04878	1.299	1.602	0.08722	1.407	1.436	0.7201
0.7444	0.05354	1.293	1.593	0.09580	1.395	1.430	0.7205
0.7402	0.05854	1.287	1.584	0.1049	1.382	1.423	0.7209
0.7357	0.06381	1.280	1.574	0.1144	1.369	1.415	0.7213
0.7311	0.06934	1.273	1.564	0.1245	1.356	1.408	0.7217
0.7262	0.07515	1.266	1.554	0.1350	1.342	1.400	0.7221
0.7212	0.08123	1.258	1.543	0.1462	1.327	1.392	0.7226
0.7160	0.08761	1.250	1.532	0.1579	1.312	1.383	0.7231
0.7105	0.09429	1.242	0.520	0.1701	1.296	1.374	0.7236
0.7049	0.1013	1.234	1.508	0.1831	1.280	1.365	0.7241
0.6990	0.1086	1.225	1.496	0.1966	1.263	1.356	0.7246
0.6929	0.1162	1.216	1.483	0.2108	1.246	1.346	0.7252
0.6866	0.1242	1.207	1.470	0.2258	1.228	1.336	0.7258
0.6801	0.1325	1.197	1.457	0.2414	1.210	1.326	0.7263
0.6733	0.1412	1.187	1.443	0.2579	1.191	1.315	0.7269
0.6663	0.1503	1.177	1.429	0.2752	1.172	1.304	0.7275
0.6591	0.1598	1.166	1.415	0.2934	1.152	1.293	0.7282
0.6516	0.1697	1.155	1.400	0.3125	1.132	1.282	0.7288
0.6439	0.1801	1.144	1.386	0.3326	1.111	1.270	0.7295
0.6360	0.1909	1.133	1.370	0.3538	1.090	1.258	0.7301
0.6277	0.2022	1.121	1.355	0.3761	1.068	1.245	0.7308
0.6193	0.2139	1.109	1.339	0.3997	1.046	1.232	0.7315
0.6105	0.2263	1.096	1.323	0.4245	1.024	1.219	0.7323
0.6015	0.2392	1.084	1.307	0.4508	1.001	1.206	0.7330
0.5922	0.2526	1.070	1.290	0.4787	0.9771	1.192	0.7337
0.5826	0.2668	1.057	1.273	0.5082	0.9531	1.178	0.7345
0.5727	0.2815	1.043	1.256	0.5396	0.9286	1.163	0.7353
0.5625	0.2970	1.029	1.239	0.5730	0.9037	1.148	0.7360
0.5520	0.3133	1.014	1.221	0.6085	0.8784	1.133	0.7368
L_1	L_2	c_2	L_3	L_4	c_4	L_5	c_6

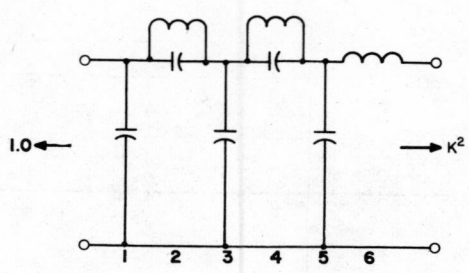

				$K^2 = \infty$				
θ	C_1	C_2	L_2	C_3	C_4	L_4	C_5	L_6
C	0.39565	0.000000	1.00174	1.34101	0.000000	1.49959	1.54668	1.29793
11.0	0.38015	0.01626	0.97947	1.31714	0.02156	1.46626	1.53354	1.30305
12.0	0.37718	0.01943	0.97522	1.31263	0.02572	1.45995	1.53106	1.30403
13.0	0.37393	0.02291	0.97060	1.30774	0.03026	1.45308	1.52835	1.30509
14.0	0.37040	0.02669	0.96561	1.30246	0.03520	1.44568	1.52544	1.30623
15.0	0.36661	0.03080	0.96024	1.29682	0.04054	1.43774	1.52232	1.30746
16.0	0.36253	0.03524	0.95450	1.29079	0.04628	1.42925	1.51898	1.30877
17.0	0.35816	0.04002	0.94837	1.28440	0.05243	1.42023	1.51543	1.31017
18.0	0.35351	0.04515	0.94187	1.27764	0.05901	1.41068	1.51168	1.31165
19.0	0.34856	0.05065	0.93498	1.27052	0.06602	1.40059	1.50772	1.31322
20.0	0.34331	0.05653	0.92771	1.26305	0.07348	1.38997	1.50355	1.31488
21.0	0.33775	0.06280	0.92006	1.25521	0.08139	1.37882	1.49917	1.31661
22.0	0.33189	0.06949	0.91202	1.24704	0.08977	1.36715	1.49460	1.31844
23.0	0.32571	0.07661	0.90359	1.23852	0.09864	1.35495	1.48982	1.32034
24.0	0.31920	0.08417	0.89478	1.22966	0.10800	1.34223	1.48483	1.32234
25.0	0.31236	0.09222	0.88557	1.22048	0.11788	1.32899	1.47965	1.32442
26.0	0.30518	0.10076	0.87596	1.21097	0.12829	1.31523	1.47427	1.32658
27.0	0.29764	0.10982	0.86596	1.20115	0.13925	1.30095	1.46869	1.32882
28.0	0.28975	0.11945	0.85556	1.19103	0.15079	1.28617	1.46292	1.33116
29.0	0.28149	0.12966	0.84477	1.18060	0.16293	1.27087	1.45696	1.33357
30.0	0.27284	0.14049	0.83357	1.16989	0.17569	1.25507	1.45080	1.33607
31.0	0.26380	0.15200	0.82196	1.15891	0.18911	1.23876	1.44446	1.33866
32.0	0.25435	0.16421	0.80995	1.14766	0.20321	1.22195	1.43793	1.34132
33.0	0.24448	0.17718	0.79753	1.13615	0.21803	1.20464	1.43122	1.34408
34.0	0.23417	0.19096	0.78471	1.12441	0.23360	1.18683	1.42432	1.34691
35.0	0.22340	0.20562	0.77146	1.11244	0.24997	1.16852	1.41725	1.34983
36.0	0.21215	0.22123	0.75781	1.10026	0.26718	1.14973	1.41000	1.35283
37.0	0.20041	0.23785	0.74374	1.08789	0.28529	1.13044	1.40258	1.35591
38.0	0.18815	0.25558	0.72925	1.07535	0.30435	1.11066	1.39499	1.35907
39.0	0.17535	0.27451	0.71434	1.06266	0.32441	1.09040	1.38724	1.36231
40.0	0.16197	0.29476	0.69902	1.04983	0.34556	1.06965	1.37933	1.36563
41.0	0.14799	0.31644	0.68327	1.03690	0.36787	1.04841	1.37126	1.36903
42.0	0.13338	0.33969	0.66710	1.02389	0.39142	1.02670	1.36304	1.37250
43.0	0.11810	0.36468	0.65052	1.01083	0.41632	1.00449	1.35469	1.37605
44.0	0.10210	0.39159	0.63351	0.99775	0.44267	0.98181	1.34619	1.37967
45.0	0.08534	0.42063	0.61608	0.98470	0.47060	0.95864	1.33757	1.38336
46.0	0.06778	0.45204	0.59824	0.97170	0.50026	0.93498	1.32884	1.38712
47.0	0.04935	0.48611	0.57998	0.95880	0.53180	0.91084	1.31999	1.39094
48.0	0.02999	0.52315	0.56131	0.94605	0.56542	0.88622	1.31106	1.39481
49.0	0.00962	0.56356	0.54224	0.93352	0.60132	0.86110	1.30204	1.39874
50.0	−0.01182	0.60778	0.52277	0.92125	0.63977	0.83550	1.29296	1.40271
θ	L_1	L_2	C_2	L_3	L_4	C_4	L_5	C_6

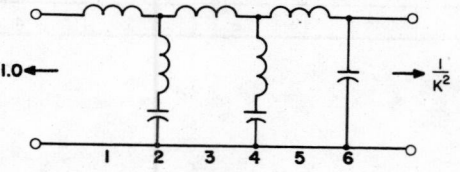

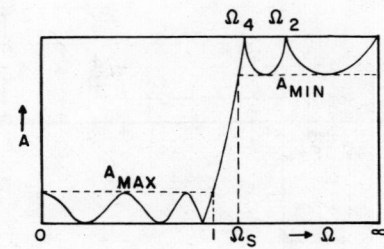

n = 6

ρ = 8%

θ	Ω_S	A_MIN	σ_1	σ_3	σ_5	Ω_1	Ω_2	Ω_3	Ω_4	Ω_5
c	∞	∞	-0.1455286	-0.3975916	-0.5431203	1.1081482	∞	0.8112208	∞	0.2969274
11.0	5.422373	125.98	-0.1424411	-0.3949684	-0.5476172	1.1066880	7.922792	0.8165150	5.624706	0.3012327
12.0	4.975749	121.42	-0.1418554	-0.3944621	-0.5484790	1.1064091	7.263825	0.8175238	5.160811	0.3020620
13.0	4.598265	117.22	-0.1412191	-0.3939090	-0.5494182	1.1061055	6.706339	0.8186211	4.768680	0.3029675
14.0	4.275108	113.33	-0.1405324	-0.3933082	-0.5504356	1.1057771	6.228589	0.8198072	4.432936	0.3039501
15.0	3.995415	109.71	-0.1397954	-0.3926592	-0.5515319	1.1054237	5.814626	0.8210823	4.142305	0.3050111
16.0	3.751039	106.31	-0.1390082	-0.3919609	-0.5527077	1.1050451	5.452491	0.8224467	3.888329	0.3061516
17.0	3.535748	103.11	-0.1381711	-0.3912125	-0.5539640	1.1046413	5.133037	0.8239006	3.664543	0.3073730
18.0	3.344698	100.10	-0.1372840	-0.3904130	-0.5553017	1.1042121	4.849152	0.8254443	3.465915	0.3086768
19.0	3.174064	97.24	-0.1363472	-0.3895613	-0.5567217	1.1037573	4.595218	0.8270780	3.288476	0.3100645
20.0	3.020785	94.52	-0.1353609	-0.3886564	-0.5582245	1.1032767	4.366743	0.8288020	3.129050	0.3115378
21.0	2.882384	91.93	-0.1343253	-0.3876969	-0.5598126	1.1027702	4.160091	0.8306167	2.985065	0.3130985
22.0	2.756834	89.46	-0.1332404	-0.3866817	-0.5614859	1.1022376	3.972284	0.8325223	2.854418	0.3147485
23.0	2.642462	87.10	-0.1321067	-0.3856092	-0.5632459	1.1016786	3.800865	0.8345192	2.735370	0.3164899
24.0	2.537873	84.82	-0.1309243	-0.3844782	-0.5650940	1.1010930	3.643786	0.8366076	2.626475	0.3183249
25.0	2.441895	82.64	-0.1296934	-0.3832869	-0.5670316	1.1004806	3.499325	0.8387879	2.526516	0.3202559
26.0	2.353536	80.54	-0.1284143	-0.3820339	-0.5690601	1.0998412	3.366027	0.8410605	2.434463	0.3222853
27.0	2.271953	78.51	-0.1270873	-0.3807174	-0.5711811	1.0991746	3.242651	0.8434255	2.349441	0.3244159
28.0	2.196422	76.56	-0.1257128	-0.3793354	-0.5733962	1.0984804	3.128134	0.8458835	2.270699	0.3266505
29.0	2.126320	74.66	-0.1242909	-0.3778862	-0.5757072	1.0977585	3.021559	0.8484347	2.197588	0.3289921
30.0	2.061105	72.83	-0.1228221	-0.3763677	-0.5781158	1.0970085	2.922133	0.8510794	2.129549	0.3314440
31.0	2.000308	71.05	-0.1213067	-0.3747776	-0.5806240	1.0962303	2.829163	0.8538179	2.066092	0.3340096
32.0	1.943517	69.32	-0.1197451	-0.3731138	-0.5832337	1.0954235	2.742043	0.8566505	2.006790	0.3366925
33.0	1.890370	67.64	-0.1181377	-0.3713738	-0.5859472	1.0945877	2.660243	0.8595776	1.951268	0.3394968
34.0	1.840548	66.00	-0.1164850	-0.3695551	-0.5887664	1.0937228	2.583292	0.8625993	1.899195	0.3424265
35.0	1.793769	64.41	-0.1147874	-0.3676550	-0.5916943	1.0928284	2.510774	0.8657159	1.850277	0.3454861
36.0	1.749781	62.85	-0.1130454	-0.3656707	-0.5947329	1.0919043	2.442321	0.8689276	1.804254	0.3486803
37.0	1.708362	61.34	-0.1112595	-0.3635993	-0.5978848	1.0909500	2.377602	0.8722345	1.760893	0.3520141
38.0	1.669312	59.86	-0.1094302	-0.3614376	-0.6011530	1.0899652	2.316322	0.8756369	1.719987	0.3554929
39.0	1.632449	58.41	-0.1075582	-0.3591824	-0.6045402	1.0889497	2.258218	0.8791348	1.681350	0.3591224
40.0	1.597615	56.99	-0.1056441	-0.3568301	-0.6080496	1.0879032	2.203049	0.8827281	1.644814	0.3629086
41.0	1.564662	55.60	-0.1036886	-0.3543771	-0.6116845	1.0868251	2.150602	0.8864169	1.610227	0.3668582
42.0	1.533460	54.24	-0.1016922	-0.3518195	-0.6154481	1.0857154	2.100682	0.8902010	1.577454	0.3709780
43.0	1.503888	52.90	-0.0996559	-0.3491533	-0.6193442	1.0845735	2.053113	0.8940802	1.546369	0.3752755
44.0	1.475840	51.59	-0.0975803	-0.3463741	-0.6233765	1.0833991	2.007733	0.8980542	1.516862	0.3797587
45.0	1.449216	50.30	-0.0954663	-0.3434775	-0.6275491	1.0821920	1.964398	0.9021226	1.488829	0.3844361
46.0	1.423927	49.04	-0.0933148	-0.3404586	-0.6318662	1.0809517	1.922972	0.9062848	1.462178	0.3893168
47.0	1.399891	47.79	-0.0911268	-0.3373125	-0.6363324	1.0796780	1.883335	0.9105401	1.436822	0.3944107
48.0	1.377032	46.56	-0.0889032	-0.3340338	-0.6409524	1.0783704	1.845375	0.9148876	1.412684	0.3997282
49.0	1.355282	45.36	-0.0866451	-0.3306168	-0.6457314	1.0770287	1.808987	0.9193262	1.389693	0.4052809
50.0	1.334577	44.16	-0.0843537	-0.3270557	-0.6506748	1.0756525	1.774078	0.9238547	1.367782	0.4110807
51.0	1.314859	42.99	-0.0820301	-0.3233442	-0.6557883	1.0742415	1.740561	0.9284714	1.346891	0.4171410
52.0	1.296076	41.83	-0.0796756	-0.3194757	-0.6610781	1.0727954	1.708356	0.9331745	1.326965	0.4234760
53.0	1.278176	40.68	-0.0772917	-0.3154432	-0.6665509	1.0713139	1.677386	0.9379618	1.307952	0.4301009
54.0	1.261116	39.55	-0.0748797	-0.3112392	-0.6722135	1.0697967	1.647585	0.9428307	1.289804	0.4370325
55.0	1.244853	38.43	-0.0724411	-0.3068561	-0.6780738	1.0682436	1.618888	0.9477783	1.272479	0.4442888
θ	Ω_S	A_MIN	σ_1	σ_3	σ_5	Ω_1	Ω_2	Ω_3	Ω_4	Ω_5

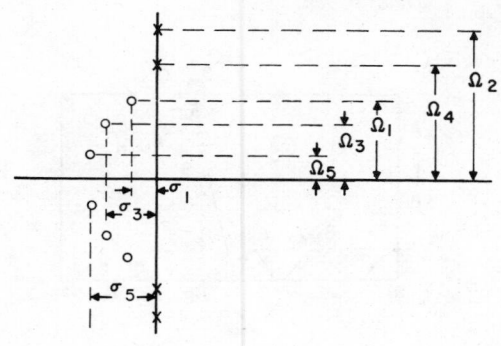

				K²=0.8519				
θ	C_1	C_2	L_2	C_3	C_4	L_4	C_5	L_6
C	0.9206	0.00000	1.405	1.824	0.00000	1.554	1.649	0.7842
11.0	0.9102	0.01147	1.389	1.799	0.02076	1.523	1.631	0.7851
12.0	0.9083	0.01367	1.386	1.794	0.02475	1.517	1.627	0.7853
13.0	0.9061	0.01608	1.383	1.789	0.02911	1.511	1.623	0.7854
14.0	0.9038	0.01869	1.379	1.783	0.03384	1.504	1.619	0.7856
15.0	0.9013	0.02150	1.376	1.777	0.03895	1.496	1.615	0.7859
16.0	0.8986	0.02452	1.372	1.771	0.04444	1.488	1.610	0.7861
17.0	0.8957	0.02776	1.367	1.764	0.05033	1.480	1.605	0.7863
18.0	0.8927	0.03121	1.363	1.757	0.05660	1.471	1.600	0.7866
19.0	0.8895	0.03487	1.358	1.749	0.06328	1.461	1.595	0.7868
20.0	0.8861	0.03876	1.353	1.741	0.07038	1.451	1.589	0.7871
21.0	0.8825	0.04288	1.348	1.733	0.07789	1.441	1.583	0.7874
22.0	0.8787	0.04722	1.342	1.724	0.08584	1.430	1.576	0.7877
23.0	0.8747	0.05180	1.336	1.714	0.09424	1.418	1.569	0.7880
24.0	0.8706	0.05663	1.330	1.705	0.1031	1.406	1.562	0.7884
25.0	0.8663	0.06169	1.324	1.695	0.1124	1.394	1.555	0.7887
26.0	0.8617	0.06701	1.317	1.684	0.1222	1.381	1.547	0.7891
27.0	0.8570	0.07259	1.310	1.673	0.1325	1.367	1.539	0.7895
28.0	0.8521	0.07844	1.303	1.662	0.1433	1.353	1.531	0.7899
29.0	0.8470	0.08455	1.295	1.650	0.1547	1.339	1.523	0.7903
30.0	0.8417	0.09095	1.288	1.639	0.1666	1.324	1.514	0.7907
31.0	0.8362	0.09764	1.280	1.626	0.1790	1.309	1.505	0.7912
32.0	0.8305	0.1046	1.271	1.613	0.1921	1.293	1.495	0.7916
33.0	0.8245	0.1119	1.263	1.600	0.2058	1.276	1.486	0.7921
34.0	0.8184	0.1195	1.254	1.587	0.2201	1.259	1.476	0.7926
35.0	0.8121	0.1275	1.245	1.573	0.2352	1.242	1.465	0.7931
36.0	0.8055	0.1357	1.235	1.558	0.2509	1.224	1.455	0.7936
37.0	0.7987	0.1444	1.225	1.544	0.2674	1.206	1.444	0.7941
38.0	0.7917	0.1534	1.215	1.529	0.2847	1.187	1.433	0.7946
39.0	0.7845	0.1628	1.205	1.514	0.3028	1.168	1.421	0.7952
40.0	0.7770	0.1725	1.194	1.498	0.3219	1.148	1.409	0.7957
41.0	0.7693	0.1827	1.183	1.482	0.3418	1.128	1.397	0.7963
42.0	0.7613	0.1934	1.172	1.466	0.3628	1.108	1.385	0.7969
43.0	0.7531	0.2045	1.160	1.449	0.3848	1.087	1.372	0.7975
44.0	0.7447	0.2160	1.148	1.432	0.4080	1.065	1.359	0.7981
45.0	0.7360	0.2281	1.136	1.415	0.4324	1.043	1.345	0.7987
46.0	0.7270	0.2407	1.123	1.397	0.4581	1.021	1.332	0.7994
47.0	0.7177	0.2539	1.111	1.379	0.4853	0.9981	1.318	0.8000
48.0	0.7082	0.2676	1.097	1.361	0.5140	0.9749	1.303	0.8007
49.0	0.6983	0.2820	1.084	1.342	0.5443	0.9513	1.289	0.8014
50.0	0.6882	0.2971	1.070	1.324	0.5765	0.9272	1.274	0.8021
51.0	0.6778	0.3128	1.055	1.305	0.6107	0.9027	1.258	0.8028
52.0	0.6670	0.3293	1.041	1.285	0.6470	0.8778	1.243	0.8035
53.0	0.6559	0.3466	1.025	1.266	0.6857	0.8525	1.227	0.8042
54.0	0.6445	0.3648	1.010	1.246	0.7271	0.8267	1.210	0.8049
55.0	0.6327	0.3839	0.9940	1.226	0.7714	0.8006	1.194	0.8056
θ	L_1	L_2	C_2	L_3	L_4	C_4	L_5	C_6

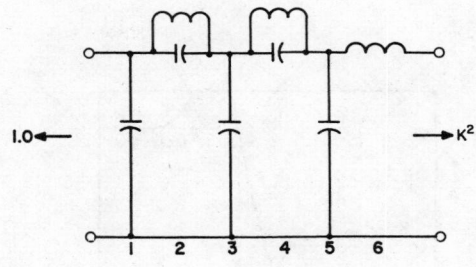

				$K^2 = \infty$			
C_1	C_2	L_2	C_3	C_4	L_4	C_5	L_6
0.46030	0.000000	1.11221	1.44625	0.000000	1.58088	1.62094	1.34927
0.44637	0.01460	1.09146	1.42257	0.02042	1.54805	1.60718	1.35327
0.44370	0.01743	1.08751	1.41809	0.02435	1.54182	1.60457	1.35404
0.44078	0.02053	1.08321	1.41323	0.02865	1.53506	1.60173	1.35486
0.43763	0.02390	1.07856	1.40798	0.03331	1.52777	1.59868	1.35576
0.43423	0.02755	1.07356	1.40236	0.03834	1.51994	1.59540	1.35672
0.43058	0.03149	1.06820	1.39636	0.04376	1.51158	1.59190	1.35774
0.42668	0.03572	1.06250	1.38999	0.04956	1.50269	1.58817	1.35884
0.42252	0.04026	1.05644	1.38324	0.05575	1.49327	1.58423	1.35999
0.41811	0.04510	1.05003	1.37613	0.06234	1.48333	1.58007	1.36122
0.41344	0.05027	1.04326	1.36865	0.06934	1.47287	1.57569	1.36251
0.40850	0.05577	1.03613	1.36081	0.07677	1.46188	1.57110	1.36387
0.40329	0.06161	1.02864	1.35261	0.08462	1.45037	1.56628	1.36529
0.39781	0.06781	1.02079	1.34406	0.09292	1.43835	1.56126	1.36678
0.39204	0.07438	1.01257	1.33516	0.10167	1.42581	1.55602	1.36834
0.38600	0.08134	1.00399	1.32591	0.11089	1.41276	1.55057	1.36996
0.37966	0.08870	0.99504	1.31632	0.12059	1.39920	1.54491	1.37165
0.37303	0.09648	0.98573	1.30640	0.13079	1.38513	1.53903	1.37341
0.36609	0.10470	0.97604	1.29614	0.14151	1.37056	1.53296	1.37523
0.35884	0.11339	0.96597	1.28556	0.15276	1.35548	1.52667	1.37712
0.35128	0.12256	0.95553	1.27466	0.16457	1.33991	1.52018	1.37907
0.34338	0.13225	0.94471	1.26345	0.17696	1.32384	1.51349	1.38109
0.33516	0.14247	0.93350	1.25193	0.18994	1.30728	1.50660	1.38318
0.32658	0.15327	0.92192	1.24012	0.20356	1.29023	1.49951	1.38533
0.31765	0.16468	0.90994	1.22803	0.21784	1.27269	1.49222	1.38754
0.30836	0.17673	0.89758	1.21565	0.23281	1.25467	1.48474	1.38982
0.29868	0.18947	0.88482	1.20300	0.24850	1.23616	1.47707	1.39217
0.28861	0.20294	0.87166	1.19010	0.26496	1.21718	1.46921	1.39458
0.27814	0.21720	0.85811	1.17694	0.28222	1.19772	1.46116	1.39705
0.26724	0.23230	0.84416	1.16356	0.30034	1.17779	1.45292	1.39959
0.25591	0.24830	0.82980	1.14995	0.31937	1.15739	1.44451	1.40218
0.24411	0.26528	0.81503	1.13613	0.33935	1.13652	1.43591	1.40485
0.23184	0.28331	0.79986	1.12212	0.36036	1.11518	1.42715	1.40757
0.21907	0.30249	0.78427	1.10794	0.38247	1.09339	1.41821	1.41035
0.20577	0.32291	0.76826	1.09360	0.40576	1.07113	1.40910	1.41320
0.19192	0.34468	0.75184	1.07912	0.43030	1.04842	1.39984	1.41610
0.17749	0.36793	0.73499	1.06453	0.45621	1.02525	1.39042	1.41905
0.16246	0.39281	0.71773	1.04985	0.48360	1.00163	1.38085	1.42207
0.14677	0.41948	0.70003	1.03510	0.51259	0.97756	1.37113	1.42513
0.13040	0.44813	0.68191	1.02032	0.54332	0.95303	1.36128	1.42825
0.11329	0.47896	0.66336	1.00553	0.57596	0.92805	1.35131	1.43140
0.09541	0.51224	0.64438	0.99078	0.61070	0.90263	1.34121	1.43461
0.07670	0.54825	0.62498	0.97610	0.64774	0.87676	1.33101	1.43785
0.05709	0.58732	0.60514	0.96154	0.68735	0.85044	1.32072	1.44112
0.03651	0.62985	0.58488	0.94714	0.72980	0.82367	1.31035	1.44441
0.01490	0.67630	0.56419	0.93296	0.77543	0.79645	1.29991	1.44772
L_1	L_2	C_2	L_3	L_4	C_4	L_5	C_6

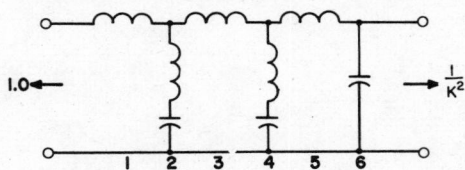

241

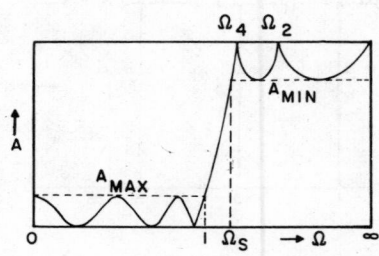

$n = 6$

$P = 10\%$

θ	Ω_S	A_MIN	σ_1	σ_3	σ_5	Ω_1	Ω_2	Ω_3	Ω_4	Ω_5
c	∞	∞	-0.1345398	-0.3675696	-0.5021094	1.0886352	∞	0.7969363	∞	0.2916989
11.0	5.422373	127.94	-0.1318187	-0.3652817	-0.5061306	1.0874937	7.922792	0.8018460	5.624706	0.2956309
12.0	4.975749	123.38	-0.1313022	-0.3648403	-0.5069011	1.0872756	7.263825	0.8027819	5.160811	0.2963878
13.0	4.598265	119.18	-0.1307410	-0.3643580	-0.5077409	1.0870381	6.706339	0.8038001	4.768680	0.2972141
14.0	4.275108	115.29	-0.1301352	-0.3638344	-0.5086504	1.0867811	6.228589	0.8049007	4.432936	0.2981106
15.0	3.995415	111.66	-0.1294850	-0.3632687	-0.5096304	1.0865046	5.814626	0.8060842	4.142305	0.2990783
16.0	3.751039	108.26	-0.1287903	-0.3626602	-0.5106815	1.0862083	5.452491	0.8073507	3.888329	0.3001183
17.0	3.535748	105.07	-0.1280512	-0.3620081	-0.5118045	1.0858922	5.133037	0.8087006	3.664543	0.3012318
18.0	3.344698	102.05	-0.1272679	-0.3613116	-0.5130001	1.0855561	4.849152	0.8101340	3.465915	0.3024200
19.0	3.174064	99.19	-0.1264405	-0.3605698	-0.5142691	1.0851999	4.595218	0.8116514	3.288476	0.3036843
20.0	3.020785	96.48	-0.1255690	-0.3597817	-0.5156126	1.0848235	4.366743	0.8132531	3.129050	0.3050261
21.0	2.882384	93.89	-0.1246537	-0.3589462	-0.5170313	1.0844266	4.160091	0.8149393	2.985065	0.3064471
22.0	2.756834	91.42	-0.1236946	-0.3580624	-0.5185264	1.0840091	3.972284	0.8167105	2.854418	0.3079487
23.0	2.642462	89.05	-0.1226918	-0.3571289	-0.5200989	1.0835708	3.800865	0.8185670	2.735370	0.3095330
24.0	2.537873	86.78	-0.1216456	-0.3561446	-0.5217500	1.0831116	3.643786	0.8205092	2.626475	0.3112017
25.0	2.441895	84.60	-0.1205561	-0.3551082	-0.5234808	1.0826312	3.499325	0.8225375	2.526516	0.3129570
26.0	2.353536	82.49	-0.1194234	-0.3540182	-0.5252927	1.0821294	3.366027	0.8246522	2.434463	0.3148009
27.0	2.271953	80.47	-0.1182479	-0.3528732	-0.5271871	1.0816061	3.242651	0.8268538	2.349441	0.3167358
28.0	2.196422	78.51	-0.1170295	-0.3516715	-0.5291653	1.0810609	3.128134	0.8291427	2.270699	0.3187642
29.0	2.126320	76.62	-0.1157687	-0.3504116	-0.5312289	1.0804938	3.021559	0.8315193	2.197588	0.3208886
30.0	2.061105	74.78	-0.1144655	-0.3490918	-0.5333794	1.0799044	2.922133	0.8339841	2.129549	0.3231119
31.0	2.000308	73.00	-0.1131204	-0.3477100	-0.5356186	1.0792925	2.829163	0.8365374	2.066092	0.3254371
32.0	1.943517	71.27	-0.1117334	-0.3462645	-0.5379483	1.0786578	2.742043	0.8391796	2.006790	0.3278673
33.0	1.890370	69.59	-0.1103050	-0.3447531	-0.5403702	1.0780001	2.660243	0.8419112	1.951268	0.3304058
34.0	1.840548	67.96	-0.1088354	-0.3431737	-0.5428864	1.0773191	2.583292	0.8447327	1.899195	0.3330563
35.0	1.793769	66.36	-0.1073249	-0.3415240	-0.5454990	1.0766146	2.510774	0.8476443	1.850277	0.3358225
36.0	1.749781	64.81	-0.1057738	-0.3398016	-0.5482101	1.0758862	2.442321	0.8506465	1.804254	0.3387085
37.0	1.708362	63.29	-0.1041827	-0.3380039	-0.5510221	1.0751337	2.377602	0.8537396	1.760893	0.3417186
38.0	1.669312	61.81	-0.1025517	-0.3361282	-0.5539373	1.0743568	2.316322	0.8569241	1.719987	0.3448574
39.0	1.632449	60.36	-0.1008815	-0.3341718	-0.5569584	1.0735551	2.258218	0.8602002	1.681350	0.3481298
40.0	1.597615	58.94	-0.0991723	-0.3321315	-0.5600881	1.0727284	2.203049	0.8635682	1.644814	0.3515410
41.0	1.564662	57.55	-0.0974247	-0.3300044	-0.5633291	1.0718763	2.150602	0.8670283	1.610227	0.3550965
42.0	1.533460	56.19	-0.0956392	-0.3277870	-0.5666845	1.0709985	2.100682	0.8705809	1.577454	0.3588024
43.0	1.503888	54.86	-0.0938163	-0.3254758	-0.5701574	1.0700948	2.053113	0.8742260	1.546369	0.3626650
44.0	1.475840	53.54	-0.0919567	-0.3230671	-0.5737513	1.0691647	2.007733	0.8779636	1.516862	0.3666911
45.0	1.449216	52.26	-0.0900608	-0.3205571	-0.5774697	1.0682079	1.964398	0.8817939	1.488829	0.3708879
46.0	1.423927	50.99	-0.0881295	-0.3179415	-0.5813162	1.0672241	1.922972	0.8857167	1.462178	0.3752632
47.0	1.399891	49.74	-0.0861633	-0.3152160	-0.5852950	1.0662130	1.883335	0.8897319	1.436822	0.3798254
48.0	1.377032	48.52	-0.0841631	-0.3123761	-0.5894101	1.0651742	1.845375	0.8938391	1.412684	0.3845855
49.0	1.355282	47.31	-0.0821296	-0.3094167	-0.5936661	1.0641073	1.808987	0.8980380	1.389693	0.3895470
50.0	1.334577	46.12	-0.0800638	-0.3063328	-0.5980677	1.0630121	1.774078	0.9023279	1.367782	0.3947264
51.0	1.314859	44.94	-0.0779664	-0.3031188	-0.6026199	1.0618882	1.740561	0.9067081	1.346891	0.4001327
52.0	1.296076	43.78	-0.0758386	-0.2997691	-0.6073282	1.0607352	1.708356	0.9111777	1.326965	0.4057781
53.0	1.278176	42.64	-0.0736813	-0.2962774	-0.6121983	1.0595528	1.677386	0.9157354	1.307952	0.4116755
54.0	1.261116	41.51	-0.0714956	-0.2926373	-0.6172363	1.0583407	1.647585	0.9203797	1.289804	0.4178390
55.0	1.244853	40.39	-0.0692829	-0.2888419	-0.6224489	1.0570986	1.618888	0.9251089	1.272479	0.4242838
56.0	1.229348	39.28	-0.0670443	-0.2848840	-0.6278431	1.0558261	1.591235	0.9299209	1.255935	0.4310265
57.0	1.214564	38.18	-0.0647812	-0.2807558	-0.6334266	1.0545230	1.564571	0.9348131	1.240135	0.4380850
58.0	1.200469	37.09	-0.0624952	-0.2764492	-0.6392076	1.0531890	1.538846	0.9397824	1.225044	0.4454788
59.0	1.187032	36.01	-0.0601878	-0.2719555	-0.6451950	1.0518238	1.514011	0.9448253	1.210630	0.4532294
60.0	1.174224	34.94	-0.0578607	-0.2672656	-0.6513985	1.0504271	1.490022	0.9499376	1.196863	0.4613599
θ	Ω_S	A_MIN	σ_1	σ_3	σ_5	Ω_1	Ω_2	Ω_3	Ω_4	Ω_5

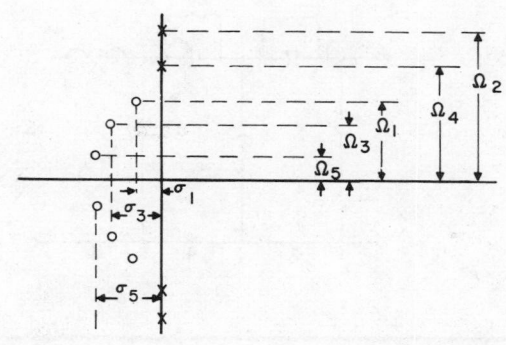

			K²=0.8182				
C_1	C_2	L_2	C_3	C_4	L_4	C_5	L_6
0.9958	0.00000	1.413	1.895	0.00000	1.550	1.727	0.8147
0.9855	0.01140	1.398	1.869	0.02079	1.520	1.709	0.8155
0.9835	0.01358	1.395	1.864	0.02479	1.514	1.705	0.8157
0.9813	0.01597	1.392	1.859	0.02916	1.508	1.701	0.8159
0.9790	0.01856	1.389	1.853	0.03390	1.501	1.697	0.8160
0.9765	0.02136	1.385	1.847	0.03901	1.494	1.693	0.8162
0.9738	0.02435	1.381	1.840	0.04450	1.486	1.688	0.8164
0.9710	0.02756	1.377	1.833	0.05038	1.478	1.683	0.8167
0.9679	0.03098	1.373	1.826	0.05666	1.469	1.678	0.8169
0.9647	0.03462	1.368	1.818	0.06334	1.460	1.672	0.8171
0.9613	0.03848	1.363	1.810	0.07042	1.450	1.666	0.8174
0.9578	0.04256	1.358	1.801	0.07793	1.440	1.660	0.8177
0.9540	0.04686	1.352	1.792	0.08586	1.429	1.654	0.8180
0.9501	0.05140	1.347	1.783	0.09424	1.418	1.647	0.8182
0.9459	0.05618	1.341	1.773	0.1031	1.407	1.640	0.8186
0.9416	0.06119	1.335	1.762	0.1123	1.394	1.632	0.8189
0.9371	0.06646	1.328	1.752	0.1221	1.382	1.625	0.8192
0.9324	0.07198	1.321	1.740	0.1324	1.369	1.617	0.8196
0.9275	0.07776	1.314	1.729	0.1431	1.355	1.608	0.8199
0.9224	0.08381	1.307	1.717	0.1544	1.341	1.600	0.8203
0.9171	0.09013	1.299	1.705	0.1662	1.327	1.591	0.8207
0.9116	0.09673	1.292	1.692	0.1786	1.312	1.582	0.8211
0.9059	0.1036	1.283	1.679	0.1916	1.296	1.572	0.8215
0.9000	0.1108	1.275	1.665	0.2052	1.280	1.562	0.8219
0.8939	0.1183	1.266	1.652	0.2194	1.264	1.552	0.8224
0.8876	0.1262	1.257	1.637	0.2342	1.247	1.542	0.8228
0.8811	0.1343	1.248	1.623	0.2498	1.230	1.531	0.8233
0.8743	0.1428	1.239	1.608	0.2661	1.212	1.520	0.8238
0.8674	0.1517	1.229	1.592	0.2832	1.194	1.509	0.8243
0.8602	0.1609	1.219	1.577	0.3010	1.175	1.497	0.8248
0.8527	0.1705	1.208	1.560	0.3198	1.156	1.485	0.8253
0.8451	0.1805	1.198	1.544	0.3394	1.146	1.473	0.8258
0.8372	0.1910	1.187	1.527	0.3600	1.116	1.460	0.8264
0.8290	0.2019	1.175	1.510	0.3816	1.096	1.448	0.8269
0.8206	0.2132	1.164	1.492	0.4042	1.075	1.435	0.8275
0.8120	0.2250	1.152	1.475	0.4281	1.054	1.421	0.8280
0.8031	0.2373	1.139	1.456	0.4532	1.032	1.407	0.8286
0.7939	0.2502	1.127	1.438	0.4796	1.010	1.393	0.8292
0.7845	0.2636	1.114	1.419	0.5075	0.9874	1.379	0.8298
0.7747	0.2777	1.101	1.400	0.5369	0.9645	1.364	0.8305
0.7647	0.2923	1.087	1.381	0.5680	0.9411	1.349	0.8311
0.7543	0.3077	1.073	1.361	0.6010	0.9173	1.334	0.8317
0.7437	0.3237	1.059	1.341	0.6359	0.8936	1.318	0.8324
0.7327	0.3405	1.044	1.321	0.6731	0.8685	1.302	0.8330
0.7214	0.3581	1.029	1.300	0.7127	0.8435	1.286	0.8337
0.7098	0.3766	1.013	1.279	0.7549	0.8181	1.269	0.8344
0.6978	0.3961	0.9972	1.258	0.8001	0.7923	1.252	0.8351
0.6854	0.4165	0.9809	1.237	0.8486	0.7662	1.235	0.8358
0.6726	0.4381	0.9641	1.215	0.9008	0.7397	1.217	0.8365
0.6595	0.4609	0.9468	1.194	0.9572	0.7128	1.199	0.8372
0.6458	0.4849	0.9291	1.172	1.018	0.6856	1.180	0.8379
L_1	L_2	C_2	L_3	L_4	C_4	L_5	C_6

243

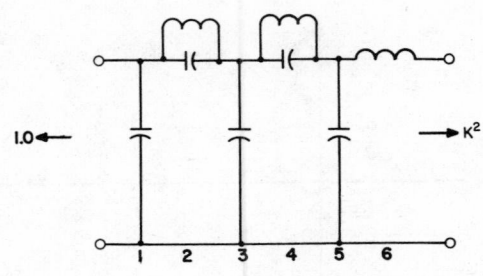

				$K^2 = \infty$				
θ	C_1	C_2	L_2	C_3	C_4	L_4	C_5	L_6
C	0.49790	0.000000	1.16832	1.49838	0.000000	1.61690	1.65621	1.37097
11.0	0.48464	0.01387	1.14825	1.47473	0.01995	1.58428	1.64214	1.37448
12.0	0.48210	0.01656	1.14442	1.47025	0.02379	1.57810	1.63947	1.37515
13.0	0.47934	0.01950	1.14026	1.46539	0.02798	1.57138	1.63658	1.37588
14.0	0.47634	0.02270	1.13576	1.46015	0.03253	1.56413	1.63345	1.37666
15.0	0.47311	0.02615	1.13092	1.45453	0.03745	1.55636	1.63010	1.37750
16.0	0.46964	0.02988	1.12574	1.44852	0.04273	1.54805	1.62652	1.37840
17.0	0.46594	0.03388	1.12022	1.44215	0.04838	1.53922	1.62271	1.37936
18.0	0.46200	0.03816	1.11436	1.43539	0.05441	1.52987	1.61868	1.38038
19.0	0.45782	0.04274	1.10816	1.42827	0.06084	1.51999	1.61443	1.38146
20.0	0.45339	0.04761	1.10160	1.42078	0.06766	1.50959	1.60995	1.38259
21.0	0.44871	0.05278	1.09471	1.41292	0.07488	1.49867	1.60525	1.38378
22.0	0.44378	0.05828	1.08746	1.40470	0.08252	1.48724	1.60033	1.38503
23.0	0.43860	0.06410	1.07986	1.39612	0.09059	1.47529	1.59519	1.38634
24.0	0.43315	0.07026	1.07192	1.38719	0.09910	1.46284	1.58983	1.38771
25.0	0.42744	0.07678	1.06361	1.37790	0.10805	1.44987	1.58425	1.38914
26.0	0.42145	0.08366	1.05496	1.36826	0.11747	1.43639	1.57845	1.39062
27.0	0.41520	0.09093	1.04594	1.35828	0.12736	1.42242	1.57244	1.39216
28.0	0.40866	0.09859	1.03656	1.34795	0.13775	1.40794	1.56622	1.39376
29.0	0.40183	0.10667	1.02682	1.33730	0.14865	1.39296	1.55979	1.39542
30.0	0.39471	0.11519	1.01671	1.32631	0.16008	1.37749	1.55314	1.39713
31.0	0.38729	0.12416	1.00624	1.31500	0.17206	1.36152	1.54629	1.39891
32.0	0.37957	0.13361	0.99540	1.30337	0.18461	1.34507	1.53923	1.40074
33.0	0.37152	0.14358	0.98418	1.29143	0.19775	1.32812	1.53197	1.40262
34.0	0.36316	0.15407	0.97259	1.27918	0.21152	1.31070	1.52450	1.40457
35.0	0.35445	0.16513	0.96061	1.26663	0.22594	1.29279	1.51683	1.40657
36.0	0.34541	0.17679	0.94826	1.25379	0.24104	1.27441	1.50896	1.40863
37.0	0.33601	0.18909	0.93552	1.24067	0.25686	1.25555	1.50089	1.41075
38.0	0.32624	0.20206	0.92239	1.22728	0.27343	1.23622	1.49263	1.41292
39.0	0.31610	0.21576	0.90887	1.21363	0.29080	1.21642	1.48418	1.41515
40.0	0.30556	0.23022	0.89495	1.19972	0.30901	1.19616	1.47554	1.41743
41.0	0.29462	0.24552	0.88064	1.18556	0.32811	1.17544	1.46671	1.41977
42.0	0.28325	0.26170	0.86592	1.17118	0.34816	1.15426	1.45770	1.42216
43.0	0.27145	0.27884	0.85080	1.15658	0.36922	1.13262	1.44850	1.42461
44.0	0.25918	0.29701	0.83526	1.14178	0.39136	1.11053	1.43913	1.42711
45.0	0.24643	0.31629	0.81932	1.12679	0.41466	1.08798	1.42959	1.42966
46.0	0.23318	0.33679	0.80295	1.11163	0.43919	1.06500	1.41988	1.43227
47.0	0.21940	0.35862	0.78617	1.09632	0.46506	1.04156	1.41000	1.43492
48.0	0.20506	0.38188	0.76896	1.08087	0.49237	1.01769	1.39996	1.43762
49.0	0.19014	0.40673	0.75132	1.06531	0.52125	0.99337	1.38977	1.44037
50.0	0.17460	0.43331	0.73325	1.04967	0.55184	0.96862	1.37943	1.44316
51.0	0.15840	0.46182	0.71475	1.03396	0.58429	0.94343	1.36895	1.44599
52.0	0.14151	0.49244	0.69581	1.01822	0.61877	0.91781	1.35834	1.44886
53.0	0.12387	0.52543	0.67642	1.00247	0.65550	0.89175	1.34761	1.45176
54.0	0.10544	0.56105	0.65660	0.98677	0.69471	0.86527	1.33676	1.45469
55.0	0.08616	0.59963	0.63633	0.97113	0.73667	0.83835	1.32582	1.45764
56.0	0.06596	0.64154	0.61562	0.95562	0.78170	0.81101	1.31479	1.46061
57.0	0.04477	0.68721	0.59446	0.94027	0.83017	0.78324	1.30369	1.46358
58.0	0.02251	0.73716	0.57286	0.92514	0.88252	0.75504	1.29254	1.46655
59.0	-0.00092	0.79202	0.55081	0.91031	0.93927	0.72642	1.28137	1.46950
60.0	-0.02563	0.85252	0.52833	0.89583	1.00103	0.69737	1.27020	1.47241
θ	L_1	L_2	C_2	L_3	L_4	C_4	L_5	C_6

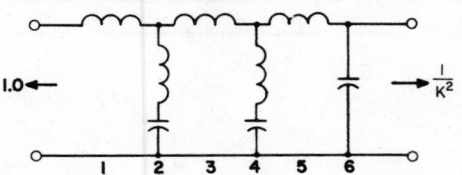

244

$n = 6$

$P = 15\%$

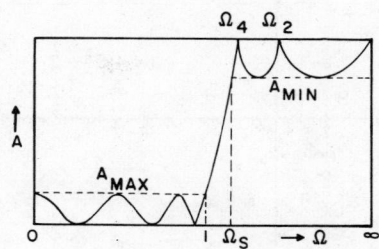

θ	Ω_S	A_{MIN}	σ_1	σ_3	σ_5	Ω_1	Ω_2	Ω_3	Ω_4	Ω_5
c	∞	∞	−0.1149705	−0.3141052	−0.4290757	1.0569384	∞	0.7737326	∞	0.2832058
11.0	5.422373	131.51	−0.1128261	−0.3123360	−0.4323275	1.0562814	7.922792	0.7780504	5.624706	0.2865655
12.0	4.975749	126.95	−0.1124187	−0.3119948	−0.4329504	1.0561557	7.263825	0.7788739	5.160811	0.2872116
13.0	4.598265	122.75	−0.1119759	−0.3116222	−0.4336293	1.0560188	6.706339	0.7797700	4.768680	0.2879167
14.0	4.275108	118.86	−0.1114978	−0.3112176	−0.4343645	1.0558706	6.228589	0.7807390	4.432936	0.2886814
15.0	3.995415	115.24	−0.1109844	−0.3107807	−0.4351566	1.0557111	5.814626	0.7817811	4.142305	0.2895065
16.0	3.751039	111.84	−0.1104357	−0.3103108	−0.4360061	1.0555400	5.452491	0.7828966	3.888329	0.2903929
17.0	3.535748	108.64	−0.1098517	−0.3098073	−0.4369136	1.0553574	5.133037	0.7840859	3.664543	0.2913415
18.0	3.344698	105.63	−0.1092325	−0.3092697	−0.4378797	1.0551632	4.849152	0.7853491	3.465915	0.2923532
19.0	3.174064	102.77	−0.1085781	−0.3086973	−0.4389051	1.0549572	4.595218	0.7866868	3.288476	0.2934291
20.0	3.020785	100.05	−0.1078886	−0.3080893	−0.4399904	1.0547393	4.366743	0.7880993	3.129050	0.2945704
21.0	2.882384	97.47	−0.1071639	−0.3074450	−0.4411364	1.0545094	4.160091	0.7895869	2.985065	0.2957783
22.0	2.756834	94.99	−0.1064042	−0.3067635	−0.4423439	1.0542674	3.972284	0.7911501	2.854418	0.2970540
23.0	2.642462	92.63	−0.1056094	−0.3060440	−0.4436138	1.0540132	3.800865	0.7927892	2.735370	0.2983991
24.0	2.537873	90.36	−0.1047797	−0.3052856	−0.4449470	1.0537466	3.643786	0.7945048	2.626475	0.2998148
25.0	2.441895	88.17	−0.1039151	−0.3044872	−0.4463444	1.0534675	3.499325	0.7962972	2.526516	0.3013030
26.0	2.353536	86.07	−0.1030157	−0.3036478	−0.4478070	1.0531757	3.366027	0.7981670	2.434463	0.3028651
27.0	2.271953	84.04	−0.1020815	−0.3027664	−0.4493359	1.0528710	3.242651	0.8001147	2.349441	0.3045031
28.0	2.196422	82.09	−0.1011126	−0.3018417	−0.4509322	1.0525534	3.128134	0.8021407	2.270699	0.3062188
29.0	2.126320	80.19	−0.1001091	−0.3008725	−0.4525971	1.0522226	3.021559	0.8042456	2.197588	0.3080142
30.0	2.061105	78.36	−0.0990712	−0.2998574	−0.4543320	1.0518784	2.922133	0.8064300	2.129549	0.3098916
31.0	2.000308	76.58	−0.0979988	−0.2987953	−0.4561380	1.0515207	2.829163	0.8086943	2.066092	0.3118531
32.0	1.943517	74.85	−0.0968921	−0.2976844	−0.4580166	1.0511493	2.742043	0.8110392	2.006790	0.3139013
33.0	1.890370	73.17	−0.0957513	−0.2965234	−0.4599693	1.0507639	2.660243	0.8134652	1.951268	0.3160388
34.0	1.840548	71.53	−0.0945765	−0.2953105	−0.4619977	1.0503645	2.583292	0.8159730	1.899195	0.3182682
35.0	1.793769	69.94	−0.0933678	−0.2940441	−0.4641033	1.0499506	2.510774	0.8185631	1.850277	0.3205925
36.0	1.749781	68.39	−0.0921253	−0.2927224	−0.4662878	1.0495222	2.442321	0.8212361	1.804254	0.3230148
37.0	1.708362	66.87	−0.0908493	−0.2913433	−0.4685532	1.0490790	2.377602	0.8239927	1.760893	0.3255384
38.0	1.669312	65.39	−0.0895400	−0.2899050	−0.4709013	1.0486208	2.316322	0.8268335	1.719987	0.3281668
39.0	1.632449	63.94	−0.0881974	−0.2884052	−0.4733341	1.0481473	2.258218	0.8297591	1.681350	0.3309038
40.0	1.597615	62.52	−0.0868220	−0.2868418	−0.4758538	1.0476582	2.203049	0.8327702	1.644814	0.3337533
41.0	1.564662	61.13	−0.0854138	−0.2852123	−0.4784625	1.0471534	2.150602	0.8358673	1.610227	0.3367196
42.0	1.533460	59.77	−0.0839731	−0.2835142	−0.4811627	1.0466325	2.100682	0.8390510	1.577454	0.3398072
43.0	1.503888	58.43	−0.0825002	−0.2817448	−0.4839569	1.0460952	2.053113	0.8423221	1.546369	0.3430209
44.0	1.475840	57.12	−0.0809955	−0.2799014	−0.4868477	1.0455414	2.007733	0.8456809	1.516862	0.3463658
45.0	1.449216	55.83	−0.0794592	−0.2779809	−0.4898378	1.0449707	1.964398	0.8491282	1.488829	0.3498475
46.0	1.423927	54.57	−0.0778917	−0.2759803	−0.4929302	1.0443827	1.922972	0.8526644	1.462178	0.3534718
47.0	1.399891	53.32	−0.0762933	−0.2738962	−0.4961281	1.0437773	1.883335	0.8562900	1.436822	0.3572451
48.0	1.377032	52.09	−0.0746645	−0.2717250	−0.4994348	1.0431541	1.845375	0.8600056	1.412684	0.3611741
49.0	1.355282	50.89	−0.0730058	−0.2694630	−0.5028537	1.0425128	1.808987	0.8638114	1.389693	0.3652659
50.0	1.334577	49.70	−0.0713175	−0.2671064	−0.5063885	1.0418530	1.774078	0.8677078	1.367782	0.3695285
51.0	1.314859	48.52	−0.0696003	−0.2646507	−0.5100433	1.0411745	1.740561	0.8716950	1.346891	0.3739701
52.0	1.296076	47.36	−0.0678546	−0.2620917	−0.5138223	1.0404768	1.708356	0.8757733	1.326965	0.3785998
53.0	1.278176	46.21	−0.0660811	−0.2594245	−0.5177298	1.0397598	1.677386	0.8799428	1.307952	0.3834273
54.0	1.261116	45.08	−0.0642805	−0.2566441	−0.5217708	1.0390229	1.647585	0.8842032	1.289804	0.3884629
55.0	1.244853	43.96	−0.0624534	−0.2537453	−0.5259504	1.0382660	1.618888	0.8885543	1.272479	0.3937182
56.0	1.229348	42.85	−0.0606007	−0.2507222	−0.5302740	1.0374886	1.591235	0.8929959	1.255935	0.3992053
57.0	1.214564	41.76	−0.0587231	−0.2475688	−0.5347476	1.0366904	1.564571	0.8975273	1.240135	0.4049375
58.0	1.200469	40.67	−0.0568216	−0.2442787	−0.5393775	1.0358710	1.538846	0.9021476	1.225044	0.4109294
59.0	1.187032	39.59	−0.0548972	−0.2408450	−0.5441706	1.0350301	1.514011	0.9068556	1.210630	0.4171968
60.0	1.174224	38.52	−0.0529510	−0.2372603	−0.5491342	1.0341673	1.490022	0.9116500	1.196863	0.4237567
θ	Ω_S	A_{MIN}	σ_1	σ_3	σ_5	Ω_1	Ω_2	Ω_3	Ω_4	Ω_5

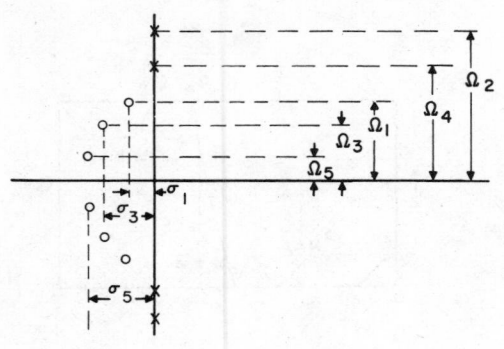

					K²=0.7391			
θ	C_1	C_2	L_2	C_3	C_4	L_4	C_5	L_6
C	1.165	0.00000	1.404	2.054	0.00000	1.518	1.900	0.8613
11.0	1.155	0.01146	1.390	2.026	0.02122	1.489	1.881	0.8620
12.0	1.153	0.01366	1.387	2.021	0.02530	1.484	1.877	0.7621
13.0	1.151	0.01606	1.385	2.015	0.02975	1.478	1.873	0.8622
14.0	1.148	0.01866	1.381	2.009	0.03458	1.472	1.869	0.8624
15.0	1.146	0.02146	1.378	2.003	0.03979	1.465	1.864	0.8625
16.0	1.143	0.02448	1.374	1.996	0.04538	1.458	1.860	0.8627
17.0	1.140	0.02770	1.370	1.989	0.05136	1.450	1.854	0.8629
18.0	1.137	0.03113	1.366	1.981	0.05775	1.442	1.849	0.8631
19.0	1.134	0.03477	1.362	1.972	0.06453	1.433	1.843	0.8633
20.0	1.130	0.03864	1.357	1.964	0.07173	1.424	1.837	0.8635
21.0	1.127	0.04273	1.352	1.954	0.07935	1.414	1.830	0.8637
22.0	1.123	0.04704	1.347	1.945	0.08741	1.404	1.824	0.8640
23.0	1.119	0.05158	1.342	1.935	0.09590	1.394	1.817	0.8642
24.0	1.115	0.05636	1.336	1.924	0.1048	1.383	1.809	0.8645
25.0	1.110	0.06138	1.330	1.913	0.1142	1.371	1.802	0.8647
26.0	1.106	0.06664	1.324	1.902	0.1241	1.360	1.794	0.8650
27.0	1.101	0.07216	1.318	1.890	0.1345	1.347	1.785	0.8653
28.0	1.096	0.07793	1.311	1.878	0.1453	1.334	1.777	0.8656
29.0	1.091	0.08396	1.305	1.865	0.1567	1.321	1.768	0.8659
30.0	1.086	0.09027	1.297	1.852	0.1686	1.308	1.758	0.8662
31.0	1.080	0.09685	1.290	1.839	0.1811	1.294	1.749	0.8666
32.0	1.074	0.1037	1.282	1.825	0.1941	1.279	1.739	0.8669
33.0	1.068	0.1109	1.274	1.811	0.2078	1.264	1.729	0.8673
34.0	1.062	0.1183	1.266	1.796	0.2220	1.249	1.718	0.8676
35.0	1.056	0.1261	1.258	1.781	0.2369	1.233	1.708	0.8680
36.0	1.049	0.1342	1.249	1.765	0.2525	1.217	1.697	0.8684
37.0	1.042	0.1426	1.240	1.749	0.2687	1.200	1.685	0.8688
38.0	1.035	0.1514	1.231	1.733	0.2857	1.183	1.674	0.8692
39.0	1.028	0.1605	1.221	1.716	0.3035	1.166	1.662	0.8696
40.0	1.021	0.1700	1.212	1.699	0.3221	1.148	1.649	0.8701
41.0	1.013	0.1799	1.202	1.682	0.3415	1.129	1.637	0.8705
42.0	1.005	0.1902	1.191	1.664	0.3619	1.111	1.624	0.8709
43.0	0.9967	0.2010	1.181	1.645	0.3832	1.091	1.611	0.8714
44.0	0.9883	0.2121	1.170	1.627	0.4055	1.072	1.597	0.8719
45.0	0.9796	0.2237	1.158	1.607	0.4289	1.052	1.583	0.8724
46.0	0.9706	0.2358	1.147	1.588	0.4534	1.032	1.569	0.8728
47.0	0.9614	0.2484	1.135	1.568	0.4792	1.011	1.555	0.8733
48.0	0.9519	0.2616	1.123	1.548	0.5063	0.9896	1.540	0.8739
49.0	0.9422	0.2753	1.110	1.527	0.5349	0.9681	1.525	0.8744
50.0	0.9321	0.2896	1.097	1.506	0.5649	0.9462	1.510	0.8749
51.0	0.9217	0.3045	1.084	1.485	0.5966	0.9239	1.494	0.8754
52.0	0.9111	0.3201	1.070	1.463	0.6302	0.9012	1.478	0.8760
53.0	0.9001	0.3364	1.057	1.442	0.6657	0.8781	1.462	0.8765
54.0	0.8888	0.3535	1.042	1.419	0.7033	0.8547	1.445	0.8771
55.0	0.8772	0.3713	1.028	1.397	0.7433	0.8309	1.428	0.8777
56.0	0.8652	0.3901	1.013	1.374	0.7858	0.8068	1.411	0.8782
57.0	0.8529	0.4097	0.9972	1.350	0.8312	0.7822	1.393	0.8788
58.0	0.8402	0.4304	0.9813	1.327	0.8798	0.7574	1.375	0.8794
59.0	0.8271	0.4521	0.9651	1.303	0.9319	0.7322	1.357	0.8800
60.0	0.8135	0.4751	0.9484	1.279	0.9879	0.7066	1.338	0.8806
θ	L_1	L_2	C_2	L_3	L_4	C_4	L_5	C_6

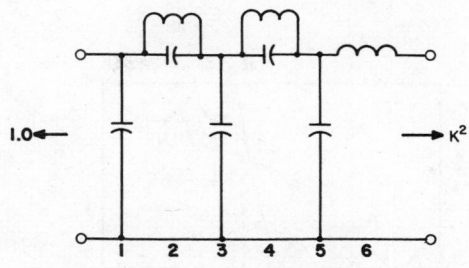

K²=∞							
C₁	C₂	L₂	C₃	C₄	L₄	C₅	L₆
0.58265	0.000000	1.27371	1.59836	0.000000	1.67421	1.72259	1.40314
0.57050	0.01270	1.25480	1.57462	0.01925	1.64202	1.70795	1.40584
0.56818	0.01515	1.25120	1.57013	0.02295	1.63592	1.70518	1.40635
0.56565	0.01783	1.24728	1.56524	0.02699	1.62929	1.70216	1.40691
0.56291	0.02074	1.24304	1.55997	0.03137	1.62214	1.69891	1.40751
0.55996	0.02388	1.23848	1.55432	0.03610	1.61446	1.69542	1.40816
0.55679	0.02727	1.23360	1.54828	0.04118	1.60626	1.69170	1.40885
0.55341	0.03090	1.22840	1.54186	0.04661	1.59755	1.68773	1.40959
0.54982	0.03478	1.22288	1.53507	0.05241	1.58831	1.68354	1.41037
0.54601	0.03891	1.21703	1.52789	0.05858	1.57856	1.67911	1.41120
0.54198	0.04331	1.21086	1.52034	0.06512	1.56830	1.67444	1.41207
0.53772	0.04798	1.20436	1.51241	0.07205	1.55752	1.66955	1.41299
0.53324	0.05292	1.19754	1.50412	0.07938	1.54624	1.66442	1.41395
0.52853	0.05815	1.19038	1.49545	0.08710	1.53444	1.65906	1.41495
0.52359	0.06367	1.18289	1.48642	0.09524	1.52214	1.65348	1.41600
0.51841	0.06950	1.17507	1.47702	0.10379	1.50934	1.64766	1.41710
0.51299	0.07564	1.16692	1.46727	0.11278	1.49604	1.64162	1.41824
0.50733	0.08210	1.15842	1.45715	0.12222	1.48224	1.63535	1.41942
0.50143	0.08890	1.14959	1.44668	0.13212	1.46794	1.62886	1.42065
0.49527	0.09605	1.14041	1.43585	0.14249	1.45315	1.62215	1.42192
0.48886	0.10356	1.13089	1.42468	0.15336	1.43788	1.61522	1.42324
0.48218	0.11145	1.12102	1.41316	0.16473	1.42211	1.60807	1.42460
0.47525	0.11973	1.11080	1.40130	0.17663	1.40586	1.60069	1.42601
0.46803	0.12843	1.10023	1.38911	0.18907	1.38913	1.59311	1.42746
0.46054	0.13756	1.08930	1.37658	0.20208	1.37193	1.58530	1.42896
0.45277	0.14715	1.07801	1.36373	0.21569	1.35424	1.57729	1.43049
0.44471	0.15721	1.06636	1.35055	0.22992	1.33609	1.56906	1.43208
0.43634	0.16778	1.05435	1.33706	0.24479	1.31747	1.56062	1.43370
0.42767	0.17888	1.04196	1.32326	0.26034	1.29839	1.55198	1.43537
0.41868	0.19053	1.02920	1.30915	0.27661	1.27884	1.54313	1.43709
0.40937	0.20278	1.01607	1.29474	0.29363	1.25884	1.53408	1.43884
0.39972	0.21566	1.00255	1.28005	0.31144	1.23838	1.52482	1.44064
0.38972	0.22921	0.98865	1.26507	0.33009	1.21747	1.51537	1.44248
0.37937	0.24348	0.97436	1.24981	0.34962	1.19612	1.50572	1.44437
0.36864	0.25850	0.95967	1.23429	0.37010	1.17432	1.49587	1.44629
0.35753	0.27435	0.94459	1.21851	0.39159	1.15208	1.48584	1.44826
0.34602	0.29107	0.92910	1.20248	0.41414	1.12940	1.47561	1.45027
0.33409	0.30873	0.91320	1.18621	0.43785	1.10629	1.46520	1.45231
0.32173	0.32741	0.89689	1.16973	0.46279	1.08275	1.45461	1.45440
0.30891	0.34719	0.88016	1.15303	0.48905	1.05878	1.44384	1.45652
0.29562	0.36817	0.86300	1.13613	0.51675	1.03439	1.43290	1.45868
0.28184	0.39044	0.84541	1.11904	0.54600	1.00959	1.42178	1.46087
0.26752	0.41413	0.82738	1.10179	0.57694	0.98436	1.41050	1.46310
0.25266	0.43938	0.80890	1.08439	0.60971	0.95872	1.39906	1.46536
0.23721	0.46633	0.78998	1.06687	0.64450	0.93267	1.38747	1.46764
0.22115	0.49515	0.77059	1.04923	0.68151	0.90621	1.37572	1.46996
0.20443	0.52606	0.75075	1.03151	0.72095	0.87935	1.36384	1.47229
0.18701	0.55928	0.73043	1.01373	0.76310	0.85208	1.35183	1.47464
0.16884	0.59508	0.70964	0.99592	0.80826	0.82441	1.33969	1.47701
0.14988	0.63376	0.68836	0.97811	0.85678	0.79635	1.32745	1.47938
0.13006	0.67570	0.66659	0.96034	0.90909	0.76790	1.31510	1.48175
L₁	L₂	C₂	L₃	L₄	C₄	L₅	C₆

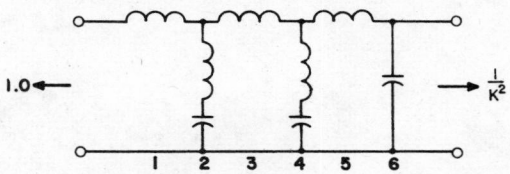

247

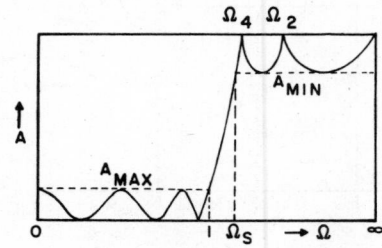

$n = 6$

$\rho = 20\%$

θ	Ω_S	A_{MIN}	σ_1	σ_3	σ_5	Ω_1	Ω_2	Ω_3	Ω_4	Ω_5
0	∞	∞	-0.1013110	-0.2767869	-0.3780979	1.0372901	∞	0.7593490	∞	0.2779410
11.0	5.422373	134.09	-0.0995180	-0.2753269	-0.3808650	1.0369134	7.922792	0.7633198	5.624706	0.2809668
12.0	4.975749	129.53	-0.0991771	-0.2750454	-0.3813950	1.0368412	7.263825	0.7640774	5.160811	0.2815483
13.0	4.598265	125.33	-0.0988066	-0.2747380	-0.3819725	1.0367626	6.706339	0.7649018	4.768680	0.2821827
14.0	4.275108	121.44	-0.0984064	-0.2744044	-0.3825980	1.0366773	6.228589	0.7657935	4.432936	0.2828706
15.0	3.995415	117.81	-0.0979766	-0.2740441	-0.3832719	1.0365854	5.814626	0.7667525	4.142305	0.2836127
16.0	3.751039	114.42	-0.0975172	-0.2736566	-0.3839945	1.0364869	5.452491	0.7677793	3.888329	0.2844097
17.0	3.535748	111.22	-0.0970281	-0.2732416	-0.3847665	1.0363816	5.133037	0.7688741	3.664543	0.2852623
18.0	3.344698	108.20	-0.0965094	-0.2727984	-0.3855882	1.0362695	4.849152	0.7700373	3.465915	0.2861714
19.0	3.174064	105.35	-0.0959610	-0.2723266	-0.3864602	1.0361505	4.595218	0.7712693	3.288476	0.2871379
20.0	3.020785	102.63	-0.0953830	-0.2718256	-0.3873832	1.0360245	4.366743	0.7725704	3.129050	0.2881628
21.0	2.882384	100.04	-0.0947754	-0.2712948	-0.3883577	1.0358914	4.160091	0.7739410	2.985065	0.2892470
22.0	2.756834	97.57	-0.0941381	-0.2707334	-0.3893845	1.0357512	3.972284	0.7753816	2.854418	0.2903917
23.0	2.642462	95.20	-0.0934712	-0.2701409	-0.3904642	1.0356037	3.800865	0.7768926	2.735370	0.2915981
24.0	2.537873	92.93	-0.0927748	-0.2695163	-0.3915976	1.0354488	3.643786	0.7784744	2.626475	0.2928674
25.0	2.441895	90.75	-0.0920487	-0.2688591	-0.3927855	1.0352864	3.499325	0.7801276	2.526516	0.2942009
26.0	2.353536	88.65	-0.0912931	-0.2681682	-0.3940287	1.0351164	3.366027	0.7818526	2.434463	0.2956001
27.0	2.271953	86.62	-0.0905080	-0.2674428	-0.3953282	1.0349387	3.242651	0.7836501	2.349441	0.2970665
28.0	2.196422	84.66	-0.0896933	-0.2666820	-0.3966848	1.0347532	3.128134	0.7855205	2.270699	0.2986016
29.0	2.126320	82.77	-0.0888491	-0.2658847	-0.3980997	1.0345596	3.021559	0.7874644	2.197588	0.3002073
30.0	2.061105	80.93	-0.0879755	-0.2650500	-0.3995737	1.0343580	2.922133	0.7894824	2.129549	0.3018853
31.0	2.000308	79.15	-0.0870725	-0.2641766	-0.4011081	1.0341481	2.829163	0.7915751	2.066092	0.3036375
32.0	1.943517	77.43	-0.0861401	-0.2632635	-0.4027039	1.0339297	2.742043	0.7937432	2.006790	0.3054660
33.0	1.890370	75.75	-0.0851784	-0.2623092	-0.4043626	1.0337028	2.660243	0.7959873	1.951268	0.3073729
34.0	1.840548	74.11	-0.0841874	-0.2613127	-0.4060852	1.0334672	2.583292	0.7983081	1.899195	0.3093606
35.0	1.793769	72.52	-0.0831672	-0.2602723	-0.4078732	1.0332226	2.510774	0.8007063	1.850277	0.3114315
36.0	1.749781	70.96	-0.0821178	-0.2591867	-0.4097281	1.0329690	2.442321	0.8031825	1.804254	0.3135882
37.0	1.708362	69.45	-0.0810395	-0.2580544	-0.4116514	1.0327061	2.377602	0.8057375	1.760893	0.3158335
38.0	1.669312	67.96	-0.0799321	-0.2568736	-0.4136446	1.0324338	2.316322	0.8083722	1.719987	0.3181702
39.0	1.632449	66.52	-0.0787959	-0.2556426	-0.4157094	1.0321518	2.258218	0.8110871	1.681350	0.3206016
40.0	1.597615	65.10	-0.0776309	-0.2543597	-0.4178477	1.0318600	2.203049	0.8138830	1.644814	0.3231309
41.0	1.564662	63.71	-0.0764372	-0.2530228	-0.4200613	1.0315581	2.150602	0.8167609	1.610227	0.3257616
42.0	1.533460	62.35	-0.0752151	-0.2516300	-0.4223521	1.0312459	2.100682	0.8197214	1.577454	0.3284975
43.0	1.503888	61.01	-0.0739646	-0.2501790	-0.4247223	1.0309232	2.053113	0.8227654	1.546369	0.3313427
44.0	1.475840	59.70	-0.0726859	-0.2486676	-0.4271741	1.0305898	2.007733	0.8258937	1.516862	0.3343013
45.0	1.449216	58.41	-0.0713792	-0.2470933	-0.4297098	1.0302454	1.964398	0.8291070	1.488829	0.3373780
46.0	1.423927	57.14	-0.0700446	-0.2454536	-0.4323318	1.0298898	1.922972	0.8324062	1.462178	0.3405776
47.0	1.399891	55.90	-0.0686825	-0.2437458	-0.4350428	1.0295227	1.883335	0.8357922	1.436822	0.3439053
48.0	1.377032	54.67	-0.0672930	-0.2419670	-0.4378455	1.0291439	1.845375	0.8392656	1.412684	0.3473667
49.0	1.355282	53.46	-0.0658764	-0.2401141	-0.4407429	1.0287530	1.808987	0.8428273	1.389693	0.3509677
50.0	1.334577	52.27	-0.0644331	-0.2381839	-0.4437381	1.0283499	1.774078	0.8464780	1.367782	0.3547149
51.0	1.314859	51.10	-0.0629632	-0.2361729	-0.4468343	1.0279341	1.740561	0.8502185	1.346891	0.3586149
52.0	1.296076	49.94	-0.0614672	-0.2340775	-0.4500351	1.0275055	1.708356	0.8540494	1.326965	0.3626753
53.0	1.278176	48.79	-0.0599455	-0.2318937	-0.4533443	1.0270637	1.677386	0.8579714	1.307952	0.3669041
54.0	1.261116	47.66	-0.0583984	-0.2296175	-0.4567658	1.0266084	1.647585	0.8619851	1.289804	0.3713090
55.0	1.244853	46.54	-0.0568265	-0.2272444	-0.4603039	1.0261392	1.618888	0.8660909	1.272479	0.3759016
56.0	1.229348	45.43	-0.0552302	-0.2247698	-0.4639631	1.0256560	1.591235	0.8702893	1.255935	0.3806897
57.0	1.214564	44.33	-0.0536101	-0.2221885	-0.4677485	1.0251582	1.564571	0.8745806	1.240135	0.3856848
58.0	1.200460	43.25	-0.0519668	-0.2194953	-0.4716652	1.0246455	1.538846	0.8789650	1.225044	0.3908990
59.0	1.187032	42.17	-0.0503010	-0.2166844	-0.4757190	1.0241177	1.514011	0.8834424	1.210630	0.3963449
60.0	1.174224	41.10	-0.0486135	-0.2137497	-0.4799161	1.0235744	1.490022	0.8880127	1.196863	0.4020366
θ	Ω_S	A_{MIN}	σ_1	σ_3	σ_5	Ω_1	Ω_2	Ω_3	Ω_4	Ω_5

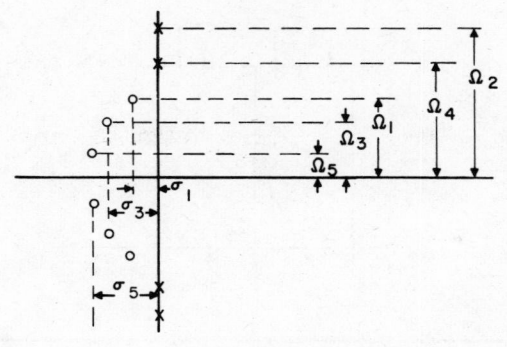

C_1	C_2	L_2	C_3	C_4	L_4	C_5	L_6
			$K^2 = 0.6667$				
1.322	0.00000	1.373	2.203	0.00000	1.469	2.059	0.8816
1.312	0.01172	1.359	2.174	0.02192	1.442	2.039	0.8822
1.310	0.01397	1.357	2.169	0.02613	1.437	2.035	0.8823
1.307	0.01642	1.354	2.163	0.03072	1.431	2.031	0.8824
1.305	0.01908	1.351	2.157	0.03570	1.425	2.027	0.8825
1.302	0.02194	1.348	2.150	0.04108	1.419	2.022	0.8827
1.299	0.02502	1.344	2.142	0.04684	1.412	2.017	0.8828
1.296	0.02831	1.341	2.135	0.05301	1.405	2.012	0.8830
1.293	0.03181	1.337	2.126	0.05959	1.397	2.006	0.8831
1.290	0.03553	1.333	2.118	0.06659	1.389	2.000	0.8833
1.286	0.03948	1.328	2.108	0.07400	1.380	1.993	0.8835
1.283	0.04365	1.324	2.099	0.08185	1.371	1.987	0.8837
1.279	0.04805	1.319	2.089	0.09013	1.362	1.979	0.8839
1.275	0.05268	1.314	2.078	0.09887	1.352	1.972	0.8841
1.270	0.05755	1.309	2.067	0.1081	1.341	1.964	0.8843
1.266	0.06267	1.303	2.055	0.1177	1.331	1.956	0.8845
1.261	0.06803	1.297	2.043	0.1279	1.320	1.948	0.8848
1.256	0.07365	1.291	2.031	0.1385	1.308	1.939	0.8850
1.251	0.07952	1.285	2.018	0.1497	1.296	1.930	0.8853
1.246	0.08566	1.279	2.005	0.1613	1.284	1.921	0.8855
1.240	0.09208	1.272	1.991	0.1735	1.271	1.911	0.8858
1.235	0.09877	1.265	1.977	0.1863	1.257	1.901	0.8861
1.229	0.1057	1.258	1.962	0.1996	1.244	1.891	0.8864
1.223	0.1130	1.250	1.947	0.2136	1.230	1.881	0.8867
1.216	0.1206	1.243	1.931	0.2281	1.215	1.870	0.8870
1.210	0.1285	1.235	1.915	0.2433	1.200	1.859	0.8873
1.203	0.1367	1.226	1.899	0.2592	1.185	1.847	0.8877
1.196	0.1452	1.218	1.882	0.2758	1.169	1.835	0.8880
1.189	0.1541	1.209	1.864	0.2931	1.153	1.823	0.8884
1.181	0.1634	1.200	1.847	0.3112	1.137	1.811	0.8887
1.174	0.1730	1.191	1.828	0.3301	1.120	1.798	0.8891
1.166	0.1830	1.181	1.810	0.3498	1.103	1.785	0.8895
1.158	0.1934	1.172	1.791	0.3704	1.085	1.771	0.8898
1.149	0.2043	1.161	1.771	0.3920	1.067	1.758	0.8902
1.141	0.2155	1.151	1.751	0.4145	1.049	1.744	0.8906
1.132	0.2272	1.140	1.731	0.4381	1.030	1.729	0.8910
1.123	0.2394	1.130	1.710	0.4628	1.011	1.715	0.8915
1.113	0.2521	1.118	1.689	0.4888	0.9910	1.700	0.8919
1.103	0.2653	1.107	1.668	0.5160	0.9711	1.684	0.8923
1.093	0.2791	1.095	1.646	0.5446	0.9508	1.669	0.8928
1.083	0.2935	1.083	1.623	0.5747	0.9302	1.653	0.8932
1.073	0.3084	1.070	1.600	0.6063	0.9092	1.637	0.8937
1.062	0.3241	1.057	1.577	0.6397	0.8878	1.620	0.8942
1.050	0.3404	1.044	1.554	0.6749	0.8661	1.603	0.8946
1.039	0.3574	1.031	1.530	0.7122	0.8440	1.586	0.8951
1.027	0.3752	1.017	1.506	0.7517	0.8216	1.568	0.8956
1.015	0.3939	1.003	1.481	0.7936	0.7989	1.551	0.8961
1.002	0.4135	0.9881	1.456	0.8382	0.7758	1.532	0.8966
0.9894	0.4340	0.9732	1.431	0.8857	0.7523	1.514	0.8971
0.9760	0.4556	0.9578	1.405	0.9365	0.7286	1.495	0.8976
0.9623	0.4783	0.9420	1.379	0.9909	0.7045	1.476	0.8981
L_1	L_2	C_2	L_3	L_4	C_4	L_5	C_6

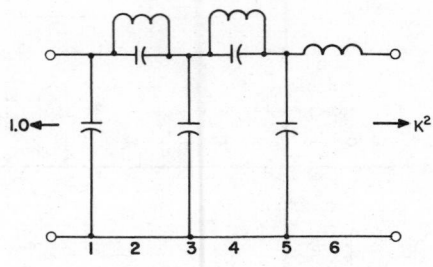

	$K^2 = \infty$							
θ	C_1	C_2	L_2	C_3	C_4	L_4	C_5	L_6
C	0.66120	0.000000	1.34717	1.67649	0.000000	1.70395	1.77516	1.41686
11.0	0.64973	0.01199	1.32907	1.65255	0.01890	1.67216	1.76008	1.41904
12.0	0.64753	0.01430	1.32562	1.64801	0.02253	1.66614	1.75722	1.41945
13.0	0.64515	0.01682	1.32187	1.64308	0.02650	1.65959	1.75411	1.41990
14.0	0.64256	0.01956	1.31781	1.63776	0.03079	1.65253	1.75076	1.42039
15.0	0.63978	0.02252	1.31345	1.63205	0.03543	1.64495	1.74716	1.42091
16.0	0.63679	0.02570	1.30878	1.62595	0.04041	1.63685	1.74332	1.42147
17.0	0.63361	0.02911	1.30380	1.61947	0.04573	1.62825	1.73924	1.42207
18.0	0.63022	0.03275	1.29852	1.61260	0.05141	1.61913	1.73491	1.42270
19.0	0.62664	0.03663	1.29292	1.60534	0.05745	1.60950	1.73034	1.42337
20.0	0.62284	0.04075	1.28702	1.59771	0.06386	1.59936	1.72553	1.42407
21.0	0.61883	0.04511	1.28080	1.58969	0.07064	1.58872	1.72048	1.42481
22.0	0.61462	0.04973	1.27427	1.58129	0.07780	1.57757	1.71519	1.42558
23.0	0.61019	0.05462	1.26742	1.57252	0.08535	1.56592	1.70967	1.42639
24.0	0.60555	0.05976	1.26025	1.56337	0.09330	1.55377	1.70390	1.42724
25.0	0.60069	0.C6519	1.25277	1.55385	0.10165	1.54113	1.69790	1.42813
26.0	0.59561	0.07089	1.24496	1.54396	0.11043	1.52799	1.69167	1.42904
27.0	0.59030	0.07689	1.23683	1.53370	0.11963	1.51435	1.68520	1.43000
28.0	0.58477	0.08320	1.22838	1.52307	0.12928	1.50023	1.67850	1.43099
29.0	0.57901	0.08981	1.21959	1.51208	0.13938	1.48562	1.67158	1.43202
30.0	0.57301	0.09675	1.21048	1.50073	0.14995	1.47053	1.66442	1.43308
31.0	0.56677	0.10402	1.20104	1.48902	0.16101	1.45495	1.65703	1.43418
32.0	0.56030	0.11165	1.19126	1.47696	0.17257	1.43890	1.64942	1.43532
33.0	0.55357	0.11963	1.18114	1.46454	0.18465	1.42237	1.64158	1.43649
34.0	0.54659	0.12800	1.17068	1.45177	0.19727	1.40537	1.63352	1.43769
35.0	0.53935	0.13677	1.15987	1.43866	0.21046	1.38790	1.62524	1.43894
36.0	0.53185	0.14594	1.14872	1.42521	0.22423	1.36997	1.61673	1.44021
37.0	0.52409	0.15555	1.13721	1.41142	0.23861	1.35157	1.60801	1.44153
38.0	0.51604	0.16562	1.12536	1.39730	0.25364	1.33271	1.59907	1.44287
39.0	0.50771	0.17616	1.11314	1.38285	0.26933	1.31340	1.58992	1.44426
40.0	0.49910	0.18721	1.10056	1.36808	0.28573	1.29363	1.58056	1.44568
41.0	0.49018	0.19880	1.08761	1.35299	0.30287	1.27342	1.57098	1.44713
42.0	0.48096	0.21094	1.07430	1.33758	0.32079	1.25276	1.56119	1.44862
43.0	0.47143	0.22368	1.06060	1.32187	0.33953	1.23166	1.55120	1.45014
44.0	0.46157	0.23705	1.04653	1.30585	0.35915	1.21012	1.54100	1.45169
45.0	0.45137	0.25109	1.03207	1.28955	0.37970	1.18814	1.53060	1.45328
46.0	0.44082	0.26585	1.01721	1.27295	0.40123	1.16574	1.52000	1.45490
47.0	0.42992	0.28138	1.00197	1.25607	0.42382	1.14291	1.50920	1.45656
48.0	0.41864	0.29773	0.98631	1.23892	0.44753	1.11965	1.49820	1.45824
49.0	0.40698	0.31495	0.97025	1.22151	0.47245	1.09598	1.48702	1.45996
50.0	0.39491	0.33313	0.95377	1.20384	0.49867	1.07189	1.47564	1.46171
51.0	0.38242	0.35232	0.93687	1.18593	0.52629	1.04739	1.46408	1.46349
52.0	0.36949	0.37262	0.91954	1.16778	0.55543	1.02248	1.45233	1.46529
53.0	0.35609	0.39413	0.90178	1.14941	0.58621	0.99716	1.44040	1.46713
54.0	0.34222	0.41693	0.88356	1.13083	0.61878	0.97144	1.42830	1.46898
55.0	0.32784	0.44117	0.86489	1.11206	0.65331	0.94533	1.41603	1.47087
56.0	0.31292	0.46696	0.84576	1.09311	0.68998	0.91882	1.40359	1.47277
57.0	0.29743	0.49448	0.82616	1.07399	0.72902	0.89192	1.39099	1.47470
58.0	0.28135	0.52388	0.80607	1.05472	0.77067	0.86463	1.37824	1.47664
59.0	0.26463	0.55539	0.78550	1.03533	0.81522	0.83696	1.36534	1.47859
60.0	0.24723	0.58923	0.76442	1.01584	0.86301	0.80890	1.35230	1.48054
θ	L_1	L_2	C_2	L_3	L_4	C_4	L_5	C_6

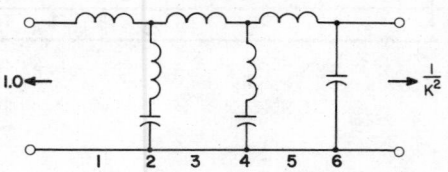

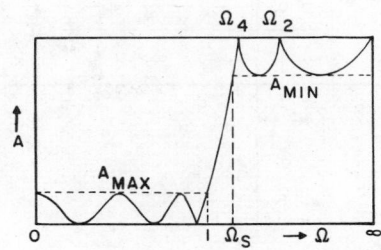

n = 6

ρ = 25%

θ	Ω_S	A_{MIN}	σ_1	σ_3	σ_5	Ω_1	Ω_2	Ω_3	Ω_4	Ω_5
c	∞	∞	-0.0907744	-0.2480003	-0.3387747	1.0236117	∞	0.7493358	∞	0.2742759
11.0	5.422373	136.13	-0.0892271	-0.2467529	-0.3411936	1.0234213	7.922792	0.7530737	5.624706	0.2770783
12.0	4.975749	131.57	-0.0889329	-0.2465125	-0.3416569	1.0233847	7.263825	0.7537870	5.160811	0.2776167
13.0	4.598265	127.37	-0.0886130	-0.2462500	-0.3421617	1.0233447	6.706339	0.7545634	4.768680	0.2782039
14.0	4.275108	123.48	-0.0882675	-0.2459650	-0.3427085	1.0233013	6.228589	0.7554031	4.432936	0.2788406
15.0	3.995415	119.86	-0.0878963	-0.2456573	-0.3432974	1.0232544	5.814626	0.7563064	4.142305	0.2795272
16.0	3.751039	116.46	-0.0874995	-0.2453264	-0.3439290	1.0232041	5.452491	0.7572736	3.888329	0.2802645
17.0	3.535748	113.26	-0.0870769	-0.2449721	-0.3446036	1.0231502	5.133037	0.7583050	3.664543	0.2810531
18.0	3.344698	110.25	-0.0866287	-0.2445938	-0.3453217	1.0230927	4.849152	0.7594009	3.465915	0.2818938
19.0	3.174064	107.39	-0.0861548	-0.2441910	-0.3460838	1.0230315	4.595218	0.7605618	3.288476	0.2827873
20.0	3.020785	104.67	-0.0856552	-0.2437634	-0.3468903	1.0229667	4.366743	0.7617880	3.129050	0.2837345
21.0	2.882384	102.08	-0.0851298	-0.2433104	-0.3477418	1.0228980	4.160091	0.7630799	2.985065	0.2847363
22.0	2.756834	99.61	-0.0845787	-0.2428314	-0.3486389	1.0228254	3.972284	0.7644380	2.854418	0.2857938
23.0	2.642462	97.24	-0.0840019	-0.2423258	-0.3495823	1.0227489	3.800865	0.7658627	2.735370	0.2869078
24.0	2.537873	94.97	-0.0833993	-0.2417930	-0.3505724	1.0226683	3.643786	0.7673544	2.626475	0.2880796
25.0	2.441895	92.79	-0.0827709	-0.2412323	-0.3516101	1.0225837	3.499325	0.7689138	2.526516	0.2893103
26.0	2.353536	90.69	-0.0821168	-0.2406431	-0.3526961	1.0224948	3.366027	0.7705412	2.434463	0.2906012
27.0	2.271953	88.66	-0.0814368	-0.2400245	-0.3538312	1.0224016	3.242651	0.7722373	2.349441	0.2919535
28.0	2.196422	86.70	-0.0807311	-0.2393759	-0.3550161	1.0223041	3.128134	0.7740027	2.270699	0.2933688
29.0	2.126320	84.81	-0.0799996	-0.2386962	-0.3562517	1.0222020	3.021559	0.7758379	2.197588	0.2948485
30.0	2.061105	82.98	-0.0792423	-0.2379847	-0.3575389	1.0220953	2.922133	0.7777435	2.129549	0.2963942
31.0	2.000308	81.20	-0.0784593	-0.2372404	-0.3588788	1.0219839	2.829163	0.7797202	2.066092	0.2980077
32.0	1.943517	79.47	-0.0776505	-0.2364622	-0.3602722	1.0218676	2.742043	0.7817687	2.006790	0.2996906
33.0	1.890370	77.79	-0.0768159	-0.2356493	-0.3617203	1.0217464	2.660243	0.7838896	1.951268	0.3014450
34.0	1.840548	76.15	-0.0759555	-0.2348003	-0.3632242	1.0216201	2.583292	0.7860837	1.899195	0.3032727
35.0	1.793769	74.56	-0.0750694	-0.2339142	-0.3647850	1.0214885	2.510774	0.7883516	1.850277	0.3051761
36.0	1.749781	73.00	-0.0741577	-0.2329898	-0.3664041	1.0213515	2.442321	0.7906943	1.804254	0.3071573
37.0	1.708362	71.49	-0.0732202	-0.2320256	-0.3680826	1.0212091	2.377602	0.7931123	1.760893	0.3092188
38.0	1.669312	70.01	-0.0722571	-0.2310204	-0.3698221	1.0210609	2.316322	0.7956066	1.719987	0.3113631
39.0	1.632449	68.56	-0.0712684	-0.2299727	-0.3716239	1.0209069	2.258218	0.7981780	1.681350	0.3135930
40.0	1.597615	67.14	-0.0702541	-0.2288808	-0.3734897	1.0207469	2.203049	0.8008273	1.644814	0.3159113
41.0	1.564662	65.75	-0.0692144	-0.2277433	-0.3754209	1.0205808	2.150602	0.8035554	1.610227	0.3183212
42.0	1.533460	64.39	-0.0681492	-0.2265582	-0.3774193	1.0204083	2.100682	0.8063633	1.577454	0.3208259
43.0	1.503888	63.05	-0.0670586	-0.2253240	-0.3794868	1.0202293	2.053113	0.8092517	1.546369	0.3234289
44.0	1.475840	61.74	-0.0659428	-0.2240385	-0.3816252	1.0200435	2.007733	0.8122217	1.516862	0.3261339
45.0	1.449216	60.45	-0.0648019	-0.2226997	-0.3838365	1.0198508	1.964398	0.8152741	1.488829	0.3289449
46.0	1.423927	59.19	-0.0636358	-0.2213055	-0.3861229	1.0196509	1.922972	0.8184101	1.462178	0.3318662
47.0	1.399891	57.94	-0.0624449	-0.2198535	-0.3884867	1.0194438	1.883335	0.8216305	1.436822	0.3349022
48.0	1.377032	56.71	-0.0612292	-0.2183414	-0.3909301	1.0192290	1.845375	0.8249363	1.412684	0.3380578
49.0	1.355282	55.50	-0.0599888	-0.2167665	-0.3934557	1.0190064	1.808987	0.8283285	1.389693	0.3413381
50.0	1.334577	54.31	-0.0587240	-0.2151261	-0.3960663	1.0187758	1.774078	0.8318082	1.367782	0.3447488
51.0	1.314859	53.14	-0.0574350	-0.2134172	-0.3987646	1.0185369	1.740561	0.8353762	1.346891	0.3482958
52.0	1.296076	51.98	-0.0561220	-0.2116367	-0.4015537	1.0182893	1.708356	0.8390337	1.326965	0.3519854
53.0	1.278176	50.83	-0.0547852	-0.2097814	-0.4044369	1.0180330	1.677386	0.8427815	1.307952	0.3558246
54.0	1.261116	49.70	-0.0534249	-0.2078477	-0.4074176	1.0177676	1.647585	0.8466206	1.289804	0.3598207
55.0	1.244853	48.58	-0.0520415	-0.2058317	-0.4104994	1.0174927	1.618888	0.8505520	1.272479	0.3639819
56.0	1.229348	47.47	-0.0506353	-0.2037296	-0.4136863	1.0172082	1.591235	0.8545765	1.255935	0.3683166
57.0	1.214564	46.37	-0.0492066	-0.2015371	-0.4169825	1.0169136	1.564571	0.8586949	1.240135	0.3728344
58.0	1.200469	45.29	-0.0477559	-0.1992494	-0.4203927	1.0166088	1.538846	0.8629080	1.225044	0.3775453
59.0	1.187032	44.21	-0.0462836	-0.1968618	-0.4239216	1.0162932	1.514011	0.8672164	1.210630	0.3824605
60.0	1.174224	43.14	-0.0447904	-0.1943690	-0.4275746	1.0159667	1.490022	0.8716207	1.196863	0.3875919
θ	Ω_S	A_{MIN}	σ_1	σ_3	σ_5	Ω_1	Ω_2	Ω_3	Ω_4	Ω_5

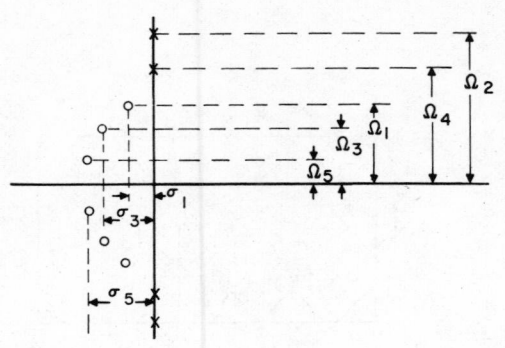

| θ | c₁ | c₂ | L₂ | c₃ | c₄ | L₄ | c₅ | L₆ |

θ	c_1	c_2	L_2	c_3	c_4	L_4	c_5	L_6
c	1.476	0.00000	1.330	2.353	0.00000	1.412	2.216	0.8855
11.0	1.465	0.01210	1.317	2.323	0.02260	1.387	2.195	0.8860
12.0	1.463	0.01442	1.315	2.317	0.02717	1.382	2.191	0.8861
13.0	1.460	0.01695	1.312	2.311	0.03194	1.377	2.187	0.8862
14.0	1.458	0.01969	1.309	2.304	0.03712	1.371	2.182	0.8863
15.0	1.455	0.02264	1.306	2.297	0.04270	1.365	2.177	0.8865
16.0	1.452	0.02582	1.303	2.289	0.04869	1.358	2.172	0.8866
17.0	1.449	0.02921	1.299	2.281	0.05510	1.352	2.166	0.8867
18.0	1.446	0.03282	1.296	2.272	0.06193	1.344	2.160	0.8869
19.0	1.442	0.03666	1.292	2.263	0.06919	1.336	2.154	0.8870
20.0	1.439	0.04073	1.288	2.253	0.07689	1.328	2.147	0.8872
21.0	1.435	0.04502	1.283	2.243	0.08503	1.320	2.140	0.8873
22.0	1.431	0.04956	1.279	2.232	0.09362	1.311	2.132	0.8875
23.0	1.426	0.05433	1.274	2.221	0.1027	1.302	2.125	0.8877
24.0	1.422	0.05935	1.269	2.209	0.1122	1.292	2.117	0.8879
25.0	1.417	0.06462	1.264	2.197	0.1222	1.282	2.108	0.8881
26.0	1.412	0.07014	1.258	2.184	0.1327	1.271	2.099	0.8883
27.0	1.407	0.07592	1.253	2.171	0.1438	1.260	2.090	0.8885
28.0	1.402	0.08197	1.247	2.158	0.1553	1.249	2.081	0.8887
29.0	1.397	0.08828	1.241	2.144	0.1674	1.237	2.071	0.8890
30.0	1.391	0.09488	1.234	2.129	0.1800	1.225	2.061	0.8892
31.0	1.385	0.1018	1.228	2.114	0.1932	1.213	2.051	0.8894
32.0	1.379	0.1089	1.221	2.098	0.2070	1.200	2.040	0.8897
33.0	1.373	0.1164	1.214	2.082	0.2214	1.186	2.029	0.8900
34.0	1.366	0.1242	1.207	2.066	0.2364	1.173	2.017	0.8902
35.0	1.359	0.1323	1.199	2.049	0.2521	1.159	2.006	0.8905
36.0	1.352	0.1407	1.191	2.031	0.2685	1.144	1.994	0.8908
37.0	1.345	0.1495	1.183	2.013	0.2855	1.129	1.981	0.8911
38.0	1.337	0.1586	1.175	1.995	0.3034	1.114	1.969	0.8914
39.0	1.330	0.1681	1.167	1.976	0.3220	1.099	1.956	0.8917
40.0	1.322	0.1780	1.158	1.957	0.3414	1.083	1.942	0.8920
41.0	1.314	0.1882	1.149	1.937	0.3616	1.066	1.928	0.8924
42.0	1.305	0.1989	1.139	1.917	0.3828	1.050	1.914	0.8927
43.0	1.296	0.2100	1.130	1.896	0.4049	1.033	1.900	0.8930
44.0	1.288	0.2215	1.120	1.875	0.4280	1.015	1.885	0.8934
45.0	1.278	0.2335	1.110	1.854	0.4522	0.9977	1.870	0.8937
46.0	1.269	0.2459	1.100	1.832	0.4775	0.9796	1.855	0.8941
47.0	1.259	0.2589	1.089	1.809	0.5040	0.9612	1.840	0.8945
48.0	1.249	0.2724	1.078	1.786	0.5317	0.9424	1.824	0.8949
49.0	1.239	0.2864	1.067	1.763	0.5608	0.9233	1.807	0.8952
50.0	1.228	0.3011	1.055	1.739	0.5914	0.9038	1.791	0.8956
51.0	1.217	0.3163	1.044	1.715	0.6236	0.8840	1.774	0.8960
52.0	1.206	0.3322	1.032	1.690	0.6574	0.8538	1.757	0.8964
53.0	1.194	0.3488	1.019	1.665	0.6931	0.8433	1.739	0.8969
54.0	1.183	0.3661	1.006	1.640	0.7308	0.8225	1.721	0.8973
55.0	1.170	0.3842	0.9932	1.614	0.7707	0.8014	1.703	0.8977
56.0	1.158	0.4031	0.9798	1.588	0.8129	0.7799	1.684	0.8981
57.0	1.145	0.4230	0.9660	1.561	0.8577	0.7581	1.665	0.8986
58.0	1.131	0.4438	0.9518	1.534	0.9053	0.7360	1.646	0.8990
59.0	1.118	0.4656	0.9373	1.506	0.9561	0.7136	1.627	0.8995
60.0	1.104	0.4885	0.9223	1.478	1.010	0.6909	1.607	0.8999
θ	L_1	L_2	c_2	L_3	L_4	c_4	L_5	c_6

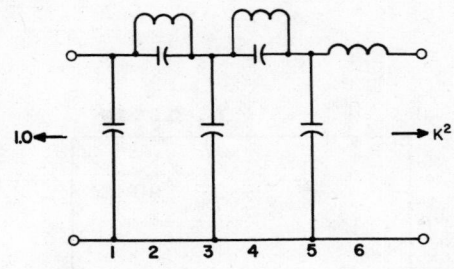

C₁	C₂	L₂	C₃	C₄	L₄	C₅	L₆
C_1	C_2	L_2	C_3	C_4	L_4	C_5	L_6

$K^2 = \infty$

C_1	C_2	L_2	C_3	C_4	L_4	C_5	L_6
0.73795	0.000000	1.39901	1.74513	0.000000	1.71631	1.82337	1.41887
0.72691	0.01153	1.38157	1.72089	0.01876	1.68497	1.80787	1.42067
0.72480	0.01375	1.37824	1.71629	0.02236	1.67903	1.80494	1.42102
0.72250	0.01618	1.37463	1.71130	0.02629	1.67257	1.80174	1.42139
0.72001	0.01881	1.37071	1.70591	0.03055	1.66561	1.79830	1.42179
0.71734	0.02164	1.36651	1.70012	0.03515	1.65813	1.79460	1.42223
0.71447	0.02470	1.36201	1.69394	0.04008	1.65015	1.79066	1.42269
0.71141	0.02796	1.35722	1.68737	0.04536	1.64166	1.78646	1.42318
0.70816	0.03145	1.35212	1.68040	0.05099	1.63267	1.78201	1.42371
0.70471	0.03516	1.34673	1.67305	0.05697	1.62317	1.77732	1.42426
0.70107	0.03911	1.34104	1.66530	0.06331	1.61318	1.77237	1.42484
0.69722	0.04328	1.33505	1.65717	0.07002	1.60268	1.76719	1.42546
0.69318	0.04770	1.32875	1.64865	0.07711	1.59168	1.76175	1.42610
0.68894	0.05235	1.32216	1.63975	0.08458	1.58020	1.75607	1.42677
0.68448	0.05726	1.31525	1.63047	0.09244	1.56821	1.75014	1.42747
0.67983	0.06243	1.30804	1.62080	0.10070	1.55574	1.74398	1.42821
0.67496	0.06787	1.30052	1.61075	0.10937	1.54278	1.73757	1.42897
0.66988	0.07357	1.29268	1.60033	0.11846	1.52933	1.73092	1.42976
0.66459	0.07956	1.28453	1.58952	0.12798	1.51540	1.72403	1.43058
0.65908	0.08583	1.27607	1.57835	0.13795	1.50099	1.71691	1.43143
0.65334	0.09241	1.26729	1.56680	0.14838	1.48610	1.70955	1.43232
0.64738	0.09930	1.25819	1.55489	0.15928	1.47074	1.70195	1.43323
0.64120	0.10651	1.24876	1.54261	0.17067	1.45490	1.69412	1.43417
0.63478	0.11405	1.23901	1.52996	0.18257	1.43860	1.68606	1.43514
0.62813	0.12193	1.22893	1.51695	0.19499	1.42182	1.67776	1.43614
0.62123	0.13018	1.21852	1.50359	0.20796	1.40459	1.66924	1.43717
0.61409	0.13881	1.20777	1.48986	0.22149	1.38689	1.66049	1.43823
0.60670	0.14782	1.19668	1.47579	0.23562	1.36874	1.65151	1.43932
0.59905	0.15725	1.18525	1.46136	0.25036	1.35014	1.64231	1.44043
0.59114	0.16711	1.17347	1.44660	0.26575	1.33108	1.63289	1.44158
0.58297	0.17741	1.16135	1.43148	0.28182	1.31158	1.62324	1.44276
0.57451	0.18820	1.14887	1.41604	0.29860	1.29163	1.61338	1.44396
0.56578	0.19948	1.13603	1.40025	0.31612	1.27125	1.60330	1.44520
0.55676	0.21128	1.12282	1.38414	0.33444	1.25042	1.59300	1.44646
0.54744	0.22364	1.10925	1.36770	0.35359	1.22917	1.58249	1.44775
0.53781	0.23660	1.09531	1.35094	0.37362	1.20749	1.57177	1.44907
0.52787	0.25017	1.08098	1.33387	0.39459	1.18538	1.56084	1.45041
0.51760	0.26441	1.06627	1.31649	0.41655	1.16285	1.54970	1.45178
0.50700	0.27936	1.05117	1.29880	0.43958	1.13990	1.53836	1.45318
0.49605	0.29506	1.03567	1.28082	0.46375	1.11654	1.52681	1.45461
0.48473	0.31157	1.01976	1.26254	0.48914	1.09277	1.51507	1.45606
0.47304	0.32895	1.00344	1.24398	0.51585	1.06859	1.50312	1.45754
0.46096	0.34726	0.98670	1.22515	0.54397	1.04401	1.49098	1.45904
0.44847	0.36658	0.96954	1.20605	0.57362	1.01903	1.47865	1.46057
0.43555	0.38699	0.95193	1.18669	0.60494	0.99366	1.46613	1.46212
0.42219	0.40858	0.93389	1.16708	0.63807	0.96790	1.45342	1.46369
0.40836	0.43145	0.91538	1.14723	0.67318	0.94174	1.44053	1.46527
0.39404	0.45572	0.89641	1.12716	0.71046	0.91521	1.42746	1.46688
0.37920	0.48154	0.87696	1.10687	0.75014	0.88829	1.41422	1.46850
0.36381	0.50903	0.85703	1.08638	0.79245	0.86100	1.40081	1.47014
0.34785	0.53839	0.83659	1.06571	0.83770	0.83334	1.38723	1.47178

L_1	L_2	C_2	L_3	L_4	C_4	L_5	C_6

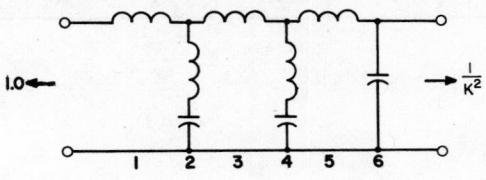

253

n = 6

P = 50%

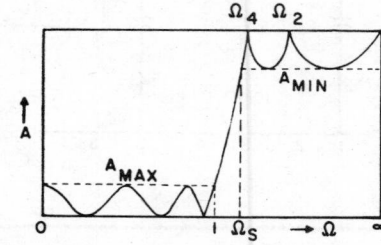

θ	Ω_S	A_MIN	σ_1	σ_3	σ_5	Ω_1	Ω_2	Ω_3	Ω_4	Ω_5
c	∞	∞	-0.0572662	-0.1564542	-0.2137204	0.9892872	∞	0.7242085	∞	0.2650787
11.0	5.422373	143.12	-0.0563819	-0.1557611	-0.2151531	0.9895326	7.922792	0.7273932	5.624706	0.2673530
12.0	4.975749	138.56	-0.0562136	-0.1556276	-0.2154274	0.9895792	7.263825	0.7280014	5.160811	0.2677894
13.0	4.598265	134.36	-0.0560305	-0.1554819	-0.2157263	0.9896298	6.706339	0.7286634	4.768680	0.2682652
14.0	4.275108	130.47	-0.0558327	-0.1553237	-0.2160500	0.9896844	6.228589	0.7293795	4.432936	0.2687808
15.0	3.995415	126.84	-0.0556201	-0.1551529	-0.2163987	0.9897429	5.814626	0.7301501	4.142305	0.2693366
16.0	3.751039	123.45	-0.0553928	-0.1549694	-0.2167725	0.9898055	5.452491	0.7309754	3.888329	0.2699331
17.0	3.535748	120.25	-0.0551506	-0.1547728	-0.2171718	0.9898720	5.133037	0.7318558	3.664543	0.2705708
18.0	3.344698	117.24	-0.0548936	-0.1545630	-0.2175968	0.9899425	4.849152	0.7327915	3.465915	0.2712502
19.0	3.174064	114.38	-0.0546217	-0.1543398	-0.2180478	0.9900169	4.595218	0.7337830	3.288476	0.2719718
20.0	3.020785	111.66	-0.0543349	-0.1541028	-0.2185250	0.9900953	4.366743	0.7348307	3.129050	0.2727363
21.0	2.882384	109.07	-0.0540332	-0.1538517	-0.2190287	0.9901775	4.160091	0.7359349	2.985065	0.2735443
22.0	2.756834	106.60	-0.0537165	-0.1535864	-0.2195594	0.9902637	3.972284	0.7370960	2.854418	0.2743965
23.0	2.642462	104.23	-0.0533848	-0.1533064	-0.2201173	0.9903537	3.800865	0.7383147	2.735370	0.2752936
24.0	2.537873	101.96	-0.0530381	-0.1530115	-0.2207029	0.9904476	3.643786	0.7395913	2.626475	0.2762365
25.0	2.441895	99.78	-0.0526763	-0.1527013	-0.2213165	0.9905453	3.499325	0.7409263	2.526516	0.2772259
26.0	2.353536	97.68	-0.0522994	-0.1523753	-0.2219585	0.9906468	3.366027	0.7423203	2.434463	0.2782627
27.0	2.271953	95.65	-0.0519074	-0.1520333	-0.2226294	0.9907521	3.242651	0.7437739	2.349441	0.2793479
28.0	2.196422	93.69	-0.0515001	-0.1516747	-0.2233296	0.9908611	3.128134	0.7452877	2.270699	0.2804825
29.0	2.126320	91.80	-0.0510777	-0.1512991	-0.2240597	0.9909739	3.021559	0.7468622	2.197588	0.2816675
30.0	2.061105	89.97	-0.0506400	-0.1509061	-0.2248202	0.9910904	2.922133	0.7484982	2.129549	0.2829041
31.0	2.000308	88.19	-0.0501869	-0.1504951	-0.2256116	0.9912105	2.829163	0.7501962	2.066092	0.2841933
32.0	1.943517	86.46	-0.0497186	-0.1500656	-0.2264345	0.9913343	2.742043	0.7519571	2.006790	0.2855366
33.0	1.890370	84.78	-0.0492348	-0.1496171	-0.2272895	0.9914617	2.660243	0.7537816	1.951268	0.2869352
34.0	1.840548	83.14	-0.0487356	-0.1491489	-0.2281772	0.9915926	2.583292	0.7556703	1.899195	0.2883904
35.0	1.793769	81.55	-0.0482209	-0.1486604	-0.2290984	0.9917270	2.510774	0.7576243	1.850277	0.2899039
36.0	1.749781	79.99	-0.0476907	-0.1481510	-0.2300537	0.9918650	2.442321	0.7596442	1.804254	0.2914772
37.0	1.708362	78.48	-0.0471450	-0.1476199	-0.2310439	0.9920063	2.377602	0.7617309	1.760893	0.2931118
38.0	1.669312	77.00	-0.0465837	-0.1470664	-0.2320699	0.9921510	2.316322	0.7638854	1.719987	0.2948098
39.0	1.632449	75.55	-0.0460067	-0.1464897	-0.2331323	0.9922990	2.258218	0.7661086	1.681350	0.2965728
40.0	1.597615	74.13	-0.0454140	-0.1458890	-0.2342322	0.9924503	2.203049	0.7684015	1.644814	0.2984028
41.0	1.564662	72.74	-0.0448057	-0.1452634	-0.2353704	0.9926048	2.150602	0.7707651	1.610227	0.3003021
42.0	1.533460	71.38	-0.0441816	-0.1446119	-0.2365479	0.9927625	2.100682	0.7732005	1.577454	0.3022728
43.0	1.503888	70.04	-0.0435417	-0.1439336	-0.2377658	0.9929232	2.053113	0.7757088	1.546369	0.3043174
44.0	1.475840	68.73	-0.0428860	-0.1432275	-0.2390251	0.9930869	2.007733	0.7782911	1.516862	0.3064382
45.0	1.449216	67.44	-0.0422144	-0.1424923	-0.2403270	0.9932535	1.964398	0.7809485	1.488829	0.3086381
46.0	1.423927	66.18	-0.0415270	-0.1417271	-0.2416728	0.9934230	1.922972	0.7836824	1.462178	0.3109199
47.0	1.399891	64.93	-0.0408237	-0.1409304	-0.2430637	0.9935952	1.883335	0.7864940	1.436822	0.3132867
48.0	1.377032	63.70	-0.0401045	-0.1401009	-0.2445011	0.9937700	1.845375	0.7893846	1.412684	0.3157416
49.0	1.355282	62.49	-0.0393694	-0.1392374	-0.2459864	0.9939475	1.808987	0.7923556	1.389693	0.3182883
50.0	1.334577	61.30	-0.0386184	-0.1383382	-0.2475212	0.9941273	1.774078	0.7954084	1.367782	0.3209303
51.0	1.314859	60.13	-0.0378515	-0.1374018	-0.2491072	0.9943095	1.740561	0.7985445	1.346891	0.3236718
52.0	1.296076	58.97	-0.0370686	-0.1364264	-0.2507460	0.9944939	1.708356	0.8017653	1.326965	0.3265168
53.0	1.278176	57.82	-0.0362698	-0.1354103	-0.2524396	0.9946804	1.677386	0.8050725	1.307952	0.3294702
54.0	1.261116	56.69	-0.0354552	-0.1343516	-0.2541898	0.9948688	1.647585	0.8084675	1.289804	0.3325366
55.0	1.244853	55.57	-0.0346248	-0.1332481	-0.2559990	0.9950590	1.618888	0.8119522	1.272479	0.3357215
56.0	1.229348	54.46	-0.0337785	-0.1320977	-0.2578692	0.9952508	1.591235	0.8155281	1.255935	0.3390305
57.0	1.214564	53.36	-0.0329166	-0.1308980	-0.2598029	0.9954440	1.564571	0.8191971	1.240135	0.3424698
58.0	1.200469	52.28	-0.0320390	-0.1296464	-0.2618028	0.9956386	1.538846	0.8229608	1.225044	0.3460461
59.0	1.187032	51.20	-0.0311458	-0.1283404	-0.2638717	0.9958341	1.514011	0.8268212	1.210630	0.3497665
60.0	1.174224	50.13	-0.0302373	-0.1269769	-0.2660125	0.9960305	1.490022	0.8307801	1.196863	0.3536390
θ	Ω_S	A_MIN	σ_1	σ_3	σ_5	Ω_1	Ω_2	Ω_3	Ω_4	Ω_5

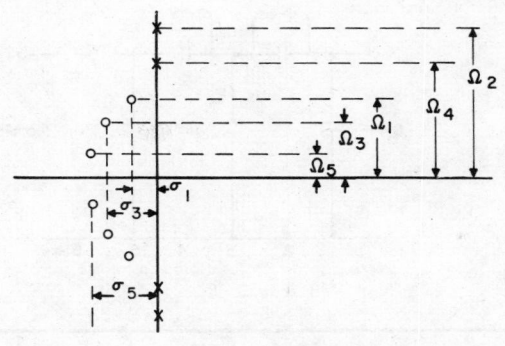

K²=0.3333							
c_1	c_2	L_2	c_3	c_4	L_4	c_5	L_6
2.340	0.00000	1.047	3.267	0.00000	1.089	3.140	0.7798
2.325	0.01536	1.037	3.226	0.02953	1.070	3.112	0.7801
2.322	0.01830	1.036	3.218	0.03520	1.067	3.107	0.7801
2.319	0.02151	1.034	3.210	0.04137	1.063	3.101	0.7802
2.316	0.02499	1.032	3.201	0.04807	1.059	3.095	0.7803
2.312	0.02874	1.029	3.191	0.05528	1.054	3.088	0.7803
2.309	0.03276	1.027	3.180	0.06302	1.049	3.081	0.7804
2.305	0.03705	1.024	3.169	0.07130	1.044	3.073	0.7805
2.301	0.04163	1.022	3.157	0.08012	1.039	3.065	0.7805
2.296	0.04649	1.019	3.145	0.08949	1.033	3.057	0.7806
2.291	0.05163	1.016	3.132	0.09941	1.027	3.048	0.7807
2.286	0.05707	1.012	3.118	0.1099	1.021	3.039	0.7808
2.281	0.06281	1.009	3.104	0.1210	1.015	3.029	0.7809
2.276	0.06884	1.006	3.088	0.1326	1.008	3.019	0.7810
2.270	0.07518	1.002	3.073	0.1449	1.000	3.008	0.7811
2.264	0.08183	0.9980	3.056	0.1578	0.9930	2.997	0.7812
2.258	0.08880	0.9940	3.039	0.1713	0.9852	2.985	0.7813
2.251	0.09609	0.9898	3.021	0.1854	0.9772	2.973	0.7814
2.244	0.1037	0.9854	3.003	0.2002	0.9688	2.961	0.7815
2.237	0.1117	0.9809	2.984	0.2156	0.9602	2.948	0.7817
2.230	0.1200	0.9762	2.964	0.2318	0.9513	2.934	0.7818
2.222	0.1286	0.9713	2.943	0.2487	0.9421	2.921	0.7819
2.214	0.1376	0.9663	2.922	0.2663	0.9326	2.906	0.7821
2.206	0.1470	0.9611	2.901	0.2846	0.9228	2.892	0.7822
2.198	0.1568	0.9557	2.878	0.3037	0.9128	2.877	0.7824
2.189	0.1670	0.9501	2.855	0.3237	0.9024	2.861	0.7825
2.180	0.1775	0.9444	2.832	0.3445	0.8918	2.845	0.7827
2.171	0.1885	0.9385	2.807	0.3661	0.8809	2.829	0.7828
2.161	0.1999	0.9324	2.782	0.3886	0.8698	2.812	0.7830
2.151	0.2118	0.9261	2.757	0.4121	0.8583	2.795	0.7832
2.141	0.2241	0.9196	2.730	0.4366	0.8466	2.777	0.7833
2.130	0.2368	0.9129	2.703	0.4621	0.8346	2.759	0.7835
2.120	0.2501	0.9060	2.676	0.4887	0.8224	2.741	0.7837
2.109	0.2635	0.8990	2.648	0.5164	0.8099	2.722	0.7839
2.097	0.2782	0.8917	2.619	0.5453	0.7971	2.703	0.7841
2.085	0.2931	0.8843	2.590	0.5754	0.7841	2.683	0.7842
2.073	0.3085	0.8766	2.559	0.6069	0.7708	2.663	0.7844
2.061	0.3245	0.8687	2.529	0.6397	0.7572	2.642	0.7846
2.048	0.3412	0.8607	2.497	0.6741	0.7434	2.621	0.7849
2.035	0.3585	0.8524	2.465	0.7100	0.7293	2.600	0.7851
2.021	0.3765	0.8438	2.433	0.7476	0.7150	2.578	0.7853
2.007	0.3953	0.8351	2.400	0.7870	0.7004	2.556	0.7855
1.993	0.4148	0.8261	2.366	0.8283	0.6856	2.534	0.7857
1.978	0.4351	0.8169	2.331	0.8717	0.6706	2.511	0.7859
1.963	0.4563	0.8075	2.296	0.9173	0.6553	2.487	0.7862
1.948	0.4783	0.7978	2.260	0.9654	0.6397	2.464	0.7864
1.932	0.5014	0.7878	2.224	1.016	0.6239	2.440	0.7866
1.915	0.5254	0.7776	2.187	1.070	0.6079	2.415	0.7869
1.898	0.5506	0.7671	2.149	1.126	0.5917	2.390	0.7871
1.881	0.5769	0.7563	2.111	1.186	0.5752	2.365	0.7874
1.863	0.6045	0.7453	2.072	1.250	0.5585	2.339	0.7876
L_1	L_2	c_2	L_3	L_4	c_4	L_5	c_6

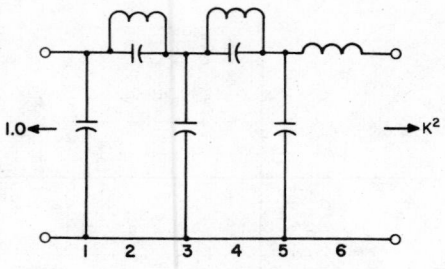

				K²=∞				
θ	C_1	C_2	L_2	C_3	C_4	L_4	C_5	L_6
C	1.16975	0.000000	1.44629	2.10623	0.000000	1.61389	2.11881	1.31311
11.0	1.15908	0.01113	1.43168	2.07899	0.01993	1.58598	2.10080	1.31393
12.0	1.15704	0.01326	1.42890	2.07382	0.02375	1.58068	2.09738	1.31408
13.0	1.15483	0.01559	1.42587	2.06820	0.02792	1.57493	2.09367	1.31425
14.0	1.15243	0.01812	1.42260	2.06213	0.03244	1.56873	2.08967	1.31443
15.0	1.14985	0.02084	1.41908	2.05562	0.03731	1.56207	2.08537	1.31463
16.0	1.14709	0.02377	1.41531	2.04866	0.04254	1.55495	2.08078	1.31484
17.0	1.14415	0.02689	1.41129	2.04126	0.04812	1.54739	2.07590	1.31506
18.0	1.14102	0.03022	1.40703	2.03341	0.05408	1.53938	2.07073	1.31530
19.0	1.13771	0.03377	1.40252	2.02512	0.06040	1.53091	2.06527	1.31555
20.0	1.13421	0.03752	1.39775	2.01639	0.06711	1.52200	2.05952	1.31582
21.0	1.13052	0.04149	1.39274	2.00721	0.07419	1.51265	2.05348	1.31610
22.0	1.12664	0.04568	1.38747	1.99760	0.08167	1.50285	2.04715	1.31639
23.0	1.12258	0.05009	1.38195	1.98755	0.08954	1.49261	2.04054	1.31669
24.0	1.11832	0.05473	1.37617	1.97705	0.09782	1.48193	2.03364	1.31701
25.0	1.11387	0.05960	1.37013	1.96612	0.10651	1.47081	2.02646	1.31734
26.0	1.10922	0.06471	1.36383	1.95475	0.11563	1.45925	2.01900	1.31769
27.0	1.10437	0.07007	1.35728	1.94295	0.12518	1.44726	2.01126	1.31805
28.0	1.09933	0.07567	1.35046	1.93071	0.13517	1.43484	2.00323	1.31842
29.0	1.09409	0.08153	1.34337	1.91804	0.14562	1.42199	1.99493	1.31881
30.0	1.08864	0.08766	1.33603	1.90494	0.15653	1.40871	1.98635	1.31921
31.0	1.08298	0.09405	1.32841	1.89141	0.16793	1.39501	1.97750	1.31962
32.0	1.07712	0.10072	1.32052	1.87745	0.17982	1.38088	1.96837	1.32005
33.0	1.07105	0.10767	1.31236	1.86306	0.19222	1.36634	1.95897	1.32049
34.0	1.06476	0.11492	1.30392	1.84824	0.20516	1.35137	1.94929	1.32095
35.0	1.05826	0.12247	1.29520	1.83300	0.21864	1.33600	1.93935	1.32141
36.0	1.05154	0.13034	1.28621	1.81733	0.23268	1.32021	1.92914	1.32189
37.0	1.04459	0.13853	1.27693	1.80125	0.24732	1.30401	1.91866	1.32239
38.0	1.03742	0.14706	1.26736	1.78474	0.26256	1.28740	1.90791	1.32290
39.0	1.03002	0.15594	1.25750	1.76781	0.27845	1.27039	1.89691	1.32342
40.0	1.02238	0.16518	1.24735	1.75047	0.29500	1.25298	1.88564	1.32395
41.0	1.01450	0.17480	1.23690	1.73271	0.31225	1.23517	1.87411	1.32450
42.0	1.00638	0.18481	1.22615	1.71453	0.33022	1.21697	1.86232	1.32506
43.0	0.99801	0.19524	1.21509	1.69594	0.34896	1.19837	1.85027	1.32563
44.0	0.98939	0.20609	1.20372	1.67694	0.36851	1.17939	1.83797	1.32622
45.0	0.98051	0.21740	1.19204	1.65754	0.38891	1.16002	1.82541	1.32682
46.0	0.97136	0.22917	1.18004	1.63772	0.41020	1.14027	1.81261	1.32743
47.0	0.96194	0.24144	1.16771	1.61750	0.43244	1.12014	1.79955	1.32806
48.0	0.95224	0.25423	1.15505	1.59688	0.45568	1.09963	1.78624	1.32870
49.0	0.94226	0.26757	1.14205	1.57585	0.48000	1.07876	1.77269	1.32935
50.0	0.93198	0.28150	1.12871	1.55442	0.50545	1.05751	1.75889	1.33001
51.0	0.92140	0.29603	1.11501	1.53260	0.53213	1.03590	1.74485	1.33068
52.0	0.91050	0.31122	1.10096	1.51037	0.56011	1.01392	1.73057	1.33137
53.0	0.89929	0.32710	1.08655	1.48776	0.58950	0.99159	1.71605	1.33207
54.0	0.88774	0.34372	1.07176	1.46475	0.62040	0.96890	1.70128	1.33277
55.0	0.87584	0.36113	1.05658	1.44135	0.65294	0.94586	1.68629	1.33349
56.0	0.86359	0.37938	1.04101	1.41756	0.68725	0.92247	1.67105	1.33422
57.0	0.85097	0.39854	1.02504	1.39339	0.72349	0.89873	1.65559	1.33496
58.0	0.83796	0.41866	1.00866	1.36883	0.76183	0.87465	1.63989	1.33571
59.0	0.82455	0.43984	0.99185	1.34389	0.80249	0.85023	1.62396	1.33647
60.0	0.81072	0.46216	0.97460	1.31857	0.84568	0.82548	1.60781	1.33723
θ	L_1	L_2	C_2	L_3	L_4	C_4	L_5	C_6

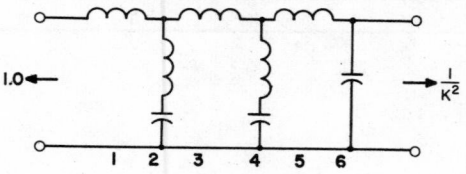

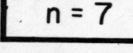

n = 7

ρ = 1%

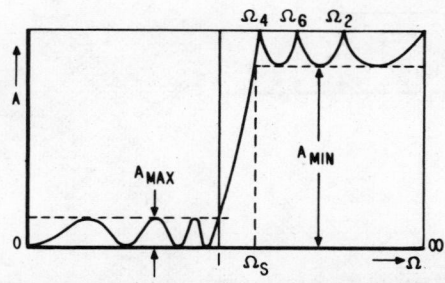

θ	Ω_S	A_{MIN}	σ_0	σ_1	σ_3	σ_5	Ω_1	Ω_2	Ω_3	Ω_4	Ω_5	Ω_6
C	∞	∞	0.831268	0.184975	0.518287	0.748947	1.267784	∞	1.016684	∞	0.564217	∞
11.0	5.2408	132.40	0.8444163	0.1785830	0.5094545	0.7530752	1.2618967	11.9886	1.0238421	5.3732	0.5766752	6.6793
12.0	4.8097	127.08	0.8469605	0.1773825	0.5077571	0.7538474	1.2607796	10.9867	1.0251852	4.9307	0.5790812	6.1257
13.0	4.4454	122.17	0.8497427	0.1760829	0.5059055	0.7546819	1.2595660	10.1387	1.0266389	4.5568	0.5817102	5.6575
14.0	4.1336	117.62	0.8527671	0.1746855	0.5038981	0.7555775	1.2582563	9.4116	1.0282010	4.2367	0.5845661	5.2564
15.0	3.8637	113.38	0.8560388	0.1731916	0.5017332	0.7565326	1.2568505	8.7812	1.0298699	3.9597	0.5876525	4.9090
16.0	3.6280	109.41	0.8595628	0.1716026	0.4994089	0.7575455	1.2553489	8.2295	1.0316437	3.7177	0.5909738	4.6052
17.0	3.4203	105.67	0.8633450	0.1699203	0.4969231	0.7586145	1.2537519	7.7424	1.0335200	3.5044	0.5945345	4.3374
18.0	3.2361	102.14	0.8673916	0.1681462	0.4942740	0.7597374	1.2520598	7.3093	1.0354964	3.3152	0.5983395	4.0995
19.0	3.0716	98.79	0.8717095	0.1661823	0.4914595	0.7609119	1.2502727	6.9216	1.0375707	3.1463	0.6023942	3.8868
20.0	2.9238	95.61	0.8763061	0.1643302	0.4884771	0.7621356	1.2483912	6.5725	1.0397396	2.9945	0.6067043	3.6955
21.0	2.7904	92.58	0.8811893	0.1622920	0.4853245	0.7634055	1.2464156	6.2564	1.0420007	2.8574	0.6112757	3.5226
22.0	2.6695	89.68	0.8863678	0.1601697	0.4819995	0.7647186	1.2443463	5.9690	1.0443503	2.7331	0.6161151	3.3656
23.0	2.5593	86.91	0.8918511	0.1579655	0.4784996	0.7660714	1.2421838	5.7063	1.0467852	2.6199	0.6212292	3.2224
24.0	2.4586	84.24	0.8976491	0.1556816	0.4748220	0.7674600	1.2399286	5.4654	1.0493019	2.5164	0.6266255	3.0912
25.0	2.3662	81.68	0.9037728	0.1533203	0.4709643	0.7688804	1.2375813	5.2436	1.0518964	2.4214	0.6323116	2.9707
26.0	2.2812	79.21	0.9102341	0.1508839	0.4669241	0.7703277	1.2351424	5.0388	1.0545645	2.3339	0.6382959	2.8596
27.0	2.2027	76.83	0.9170457	0.1483750	0.4626982	0.7717969	1.2326127	4.8489	1.0573018	2.2532	0.6445871	2.7568
28.0	2.1301	74.53	0.9242213	0.1457960	0.4582845	0.7732822	1.2299926	4.6725	1.0601037	2.1784	0.6511943	2.6615
29.0	2.0627	72.30	0.9317759	0.1431497	0.4536799	0.7747775	1.2272832	4.5080	1.0629652	2.1090	0.6581274	2.5729
30.0	2.0000	70.14	0.9397255	0.1404386	0.4488820	0.7762757	1.2244851	4.3544	1.0658809	2.0445	0.6653965	2.4903
31.0	1.9416	68.04	0.9480876	0.1376657	0.4438879	0.7777692	1.2215992	4.2106	1.0688456	1.9844	0.6730124	2.4132
32.0	1.8871	66.00	0.9568811	0.1348337	0.4386953	0.7792495	1.2186264	4.0756	1.0718529	1.9282	0.6809866	2.3410
33.0	1.8361	64.02	0.9661263	0.1319454	0.4333015	0.7807074	1.2155677	3.9486	1.0748969	1.8756	0.6893308	2.2733
34.0	1.7883	62.09	0.9758456	0.1290041	0.4277041	0.7821324	1.2124242	3.8290	1.0779708	1.8264	0.6980577	2.2096
35.0	1.7434	60.21	0.9860630	0.1260126	0.4219006	0.7835133	1.2091969	3.7161	1.0810679	1.7801	0.7071801	2.1497
36.0	1.7013	58.37	0.9968047	0.1229740	0.4158891	0.7848375	1.2058871	3.6093	1.0841809	1.7366	0.7167118	2.0933
37.0	1.6616	56.57	1.0080993	0.1198917	0.4096673	0.7860912	1.2024959	3.5081	1.0873020	1.6957	0.7266671	2.0400
38.0	1.6243	54.82	1.0199779	0.1167688	0.4032333	0.7872589	1.1990249	3.4121	1.0904233	1.6571	0.7370606	1.9896
39.0	1.5890	53.10	1.0324747	0.1136085	0.3965856	0.7883239	1.1954754	3.3209	1.0935362	1.6206	0.7479078	1.9418
40.0	1.5557	51.42	1.0456269	0.1104143	0.3897227	0.7892673	1.1918489	3.2341	1.0966318	1.5862	0.7592245	1.8966
41.0	1.5243	49.77	1.0594754	0.1071896	0.3826434	0.7900686	1.1881469	3.1513	1.0997012	1.5537	0.7710272	1.8537
42.0	1.4945	48.15	1.0740651	0.1039377	0.3753470	0.7907051	1.1843713	3.0724	1.1027342	1.5229	0.7833325	1.8129
43.0	1.4663	46.56	1.0894455	0.1006624	0.3678329	0.7911514	1.1805238	2.9970	1.1057212	1.4937	0.7961578	1.7740
44.0	1.4396	45.00	1.1056711	0.0973670	0.3601011	0.7913797	1.1766063	2.9248	1.1086514	1.4660	0.8095206	1.7371
45.0	1.4142	43.47	1.1228025	0.0940553	0.3521518	0.7913594	1.1726208	2.8557	1.1115138	1.4397	0.8234384	1.7019
46.0	1.3902	41.95	1.1409067	0.0907311	0.3439858	0.7910565	1.1685693	2.7895	1.1142972	1.4147	0.8379287	1.6683
47.0	1.3673	40.47	1.1600581	0.0873979	0.3356045	0.7904334	1.1644541	2.7259	1.1169898	1.3910	0.8530094	1.6362
48.0	1.3456	39.00	1.1803399	0.0840595	0.3270100	0.7894488	1.1602775	2.6648	1.1195793	1.3685	0.8686972	1.6056
49.0	1.3250	37.55	1.2018452	0.0807200	0.3182046	0.7880569	1.1560419	2.6060	1.1220530	1.3470	0.8850086	1.5763
50.0	1.3054	36.12	1.2246784	0.0773831	0.3091916	0.7862076	1.1517503	2.5494	1.1243974	1.3266	0.9019589	1.5482
51.0	1.2868	34.71	1.2489570	0.0740528	0.2999752	0.7838450	1.1474047	2.4948	1.1266000	1.3072	0.9195617	1.5213
52.0	1.2690	33.32	1.2748140	0.0707330	0.2905601	0.7809082	1.1430084	2.4422	1.1286459	1.2887	0.9378288	1.4956
53.0	1.2521	31.94	1.3023999	0.0674281	0.2809517	0.7773300	1.1385643	2.3913	1.1305212	1.2711	0.9567690	1.4709
54.0	1.2361	30.58	1.3318866	0.0641419	0.2711571	0.7730365	1.1340755	2.3422	1.1322110	1.2543	0.9763878	1.4473
55.0	1.2208	29.23	1.3634704	0.0608788	0.2611839	0.7679467	1.1295452	2.2946	1.1337005	1.2383	0.9966858	1.4245
56.0	1.2062	27.89	1.3973766	0.0576432	0.2510409	0.7619722	1.1249768	2.2485	1.1349741	1.2230	1.0176579	1.4027
57.0	1.1924	26.56	1.4338655	0.0544393	0.2407384	0.7550167	1.1203739	2.2039	1.1360161	1.2085	1.0392920	1.3817
58.0	1.1792	25.24	1.4732385	0.0512718	0.2302878	0.7469754	1.1157404	2.1606	1.1368101	1.1946	1.0615666	1.3616
59.0	1.1666	23.94	1.5158471	0.0481451	0.2197021	0.7377355	1.1110799	2.1185	1.1373400	1.1814	1.0844494	1.3422
60.0	1.1547	22.64	1.5621027	0.0450641	0.2089960	0.7271756	1.1063968	2.0776	1.1375889	1.1689	1.1078943	1.3235
θ	Ω_S	A_{MIN}	σ_0	σ_1	σ_3	σ_5	Ω_1	Ω_2	Ω_3	Ω_4	Ω_5	Ω_6

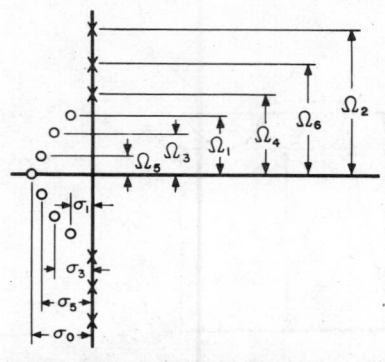

					$K^2 = 1.0$					
θ	C_1	C_2	L_2	C_3	C_4	L_4	C_5	C_6	L_6	C_7
C	0.5354	0.0000	1.179	1.464	0.0000	1.5000	1.464	0.00000	1.179	0.5354
11.0	0.53077	0.00593	1.17280	1.44295	0.02366	1.46366	1.43552	0.01947	1.15152	0.51747
12.0	0.52989	0.00707	1.17163	1.43903	0.02823	1.45676	1.43022	0.02325	1.14631	0.51404
13.0	0.52892	0.00831	1.17036	1.43478	0.03323	1.44927	1.42448	0.02739	1.14065	0.51030
14.0	0.52788	0.00966	1.16898	1.43019	0.03866	1.44118	1.41829	0.03190	1.13454	0.50626
15.0	0.52676	0.01111	1.16750	1.42526	0.04452	1.43251	1.41167	0.03679	1.12797	0.50190
16.0	0.52556	0.01266	1.16590	1.41999	0.05084	1.42325	1.40460	0.04206	1.12094	0.49723
17.0	0.52427	0.01433	1.16420	1.41439	0.05761	1.41341	1.39711	0.04774	1.11347	0.49225
18.0	0.52291	0.01610	1.16239	1.40845	0.06485	1.40299	1.38919	0.05382	1.10554	0.48694
19.0	0.52146	0.01799	1.16047	1.40217	0.07257	1.39198	1.38085	0.06033	1.09715	0.48130
20.0	0.51992	0.01998	1.15844	1.39556	0.08079	1.38040	1.37209	0.06728	1.08831	0.47534
21.0	0.51830	0.02209	1.15630	1.38862	0.08951	1.36825	1.36292	0.07469	1.07902	0.46905
22.0	0.51659	0.02432	1.15404	1.38134	0.09876	1.35553	1.35335	0.08256	1.06927	0.46242
23.0	0.51480	0.02667	1.15167	1.37373	0.10854	1.34224	1.34339	0.09093	1.05907	0.45544
24.0	0.51291	0.02913	1.14918	1.36578	0.11889	1.32838	1.33304	0.09982	1.04841	0.44812
25.0	0.51094	0.03172	1.14657	1.35750	0.12981	1.31396	1.32231	0.10924	1.03731	0.44045
26.0	0.50887	0.03443	1.14384	1.34890	0.14133	1.29898	1.31121	0.11922	1.02574	0.43242
27.0	0.50671	0.03728	1.14098	1.33996	0.15348	1.28345	1.29974	0.12980	1.01373	0.42402
28.0	0.50445	0.04025	1.13800	1.33069	0.16627	1.26737	1.28793	0.14099	1.00126	0.41525
29.0	0.50209	0.04336	1.13489	1.32109	0.17975	1.25073	1.27577	0.15284	0.98834	0.40609
30.0	0.49964	0.04660	1.13166	1.31116	0.19394	1.23355	1.26329	0.16538	0.97497	0.39655
31.0	0.49708	0.04999	1.12828	1.30090	0.20887	1.21583	1.25048	0.17866	0.96114	0.38661
32.0	0.49442	0.05352	1.12478	1.29031	0.22459	1.19757	1.23738	0.19271	0.94687	0.37626
33.0	0.49166	0.05721	1.12113	1.27939	0.24114	1.17878	1.22398	0.20759	0.93215	0.36549
34.0	0.48878	0.06104	1.11734	1.26815	0.25857	1.15945	1.21031	0.22336	0.91698	0.35429
35.0	0.48580	0.06504	1.11341	1.25658	0.27693	1.13959	1.19637	0.24007	0.90137	0.34265
36.0	0.48269	0.06920	1.10933	1.24468	0.29627	1.11921	1.18220	0.25778	0.88531	0.33055
37.0	0.47947	0.07353	1.10509	1.23245	0.31666	1.09830	1.16780	0.27659	0.86881	0.31798
38.0	0.47613	0.07804	1.10069	1.21990	0.33819	1.07688	1.15320	0.29656	0.85187	0.30492
39.0	0.47267	0.08272	1.09613	1.20702	0.36091	1.05493	1.13842	0.31780	0.83449	0.29136
40.0	0.46907	0.08760	1.09140	1.19381	0.38494	1.03247	1.12348	0.34041	0.81668	0.27727
41.0	0.46534	0.09268	1.08650	1.18028	0.41037	1.00950	1.10841	0.36450	0.79844	0.26264
42.0	0.46148	0.09796	1.08141	1.16643	0.43732	0.98602	1.09322	0.39022	0.77977	0.24743
43.0	0.45747	0.10346	1.07613	1.15225	0.46592	0.96204	1.07796	0.41770	0.76068	0.23164
44.0	0.45331	0.10918	1.07066	1.13774	0.49632	0.93755	1.06265	0.44713	0.74117	0.21522
45.0	0.44900	0.11514	1.06498	1.12292	0.52870	0.91255	1.04734	0.47869	0.72125	0.19814
46.0	0.44453	0.12135	1.05908	1.10777	0.56325	0.88706	1.03204	0.51261	0.70093	0.18038
47.0	0.43989	0.12781	1.05296	1.09231	0.60021	0.86106	1.01681	0.54914	0.68020	0.16190
48.0	0.43508	0.13456	1.04659	1.07653	0.63983	0.83457	1.00169	0.58857	0.65909	0.14264
49.0	0.43009	0.14159	1.03997	1.06043	0.68243	0.80759	0.98673	0.63125	0.63760	0.12258
50.0	0.42490	0.14894	1.03308	1.04402	0.72836	0.78012	0.97199	0.67756	0.61574	0.10164
51.0	0.41952	0.15661	1.02589	1.02731	0.77806	0.75216	0.95752	0.72797	0.59351	0.07978
52.0	0.41393	0.16464	1.01839	1.01029	0.83201	0.72373	0.94339	0.78301	0.57094	0.05693
53.0	0.40812	0.17305	1.01055	0.99297	0.89083	0.69482	0.92967	0.84334	0.54804	0.03301
54.0	0.40207	0.18186	1.00234	0.97537	0.95523	0.66544	0.91646	0.90970	0.52482	0.00794
55.0	0.39579	0.19113	0.99372	0.95749	1.02606	0.63562	0.90384	0.98300	0.50130	−0.01839
56.0	0.38924	0.20087	0.98465	0.93933	1.10439	0.60536	0.89193	1.06436	0.47751	−0.04607
57.0	0.38242	0.21114	0.97508	0.92091	1.19149	0.57468	0.88084	1.15509	0.45345	−0.07525
58.0	0.37532	0.22200	0.96495	0.90225	1.28895	0.54361	0.87073	1.25684	0.42918	−0.10606
59.0	0.36790	0.23352	0.95419	0.88336	1.39875	0.51219	0.86176	1.37164	0.40470	−0.13869
60.0	0.36017	0.24576	0.94272	0.86425	1.52338	0.48046	0.85413	1.50202	0.38006	−0.17332
θ	L_1	L_2	C_2	L_3	L_4	C_4	L_5	L_6	C_6	L_7

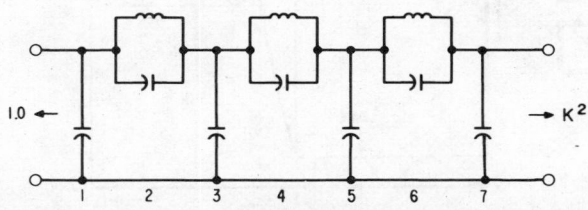

				$K^2 = \infty$					
C_1	C_2	L_2	C_3	C_4	L_4	C_5	C_6	L_6	C_7
0.267688	0.000000	0.736355	1.068843	0.000000	1.277167	1.393488	0.000000	1.424958	1.197893
0.2588	0.0096	0.7252	1.0431	0.0281	1.2346	1.3650	0.0160	1.4049	1.1905
0.2571	0.0115	0.7231	1.0381	0.0335	1.2265	1.3596	0.0190	1.4012	1.1891
0.2553	0.0135	0.7208	1.0328	0.0395	1.2177	1.3538	0.0224	1.3970	1.1875
0.2532	0.0157	0.7183	1.0270	0.0461	1.2082	1.3475	0.0260	1.3926	1.1859
0.2511	0.0181	0.7156	1.0208	0.0532	1.1980	1.3408	0.0299	1.3879	1.1841
0.2487	0.0207	0.7127	1.0142	0.0609	1.1871	1.3336	0.0341	1.3828	1.1822
0.2462	.0.0235	0.7096	1.0071	0.0693	1.1755	1.3260	0.0386	1.3774	1.1802
0.2436	0.0265	0.7063	0.9996	0.0782	1.1632	1.3179	0.0434	1.3717	1.1781
0.2407	0.0297	0.7028	0.9917	0.0878	1.1502	1.3094	0.0485	1.3657	1.1759
0.2377	0.0331	0.6992	0.9833	0.0981	1.1365	1.3005	0.0539	1.3594	1.1735
0.2345	0.0367	0.6953	0.9745	0.1091	1.1221	1.2912	0.0596	1.3528	1.1710
0.2311	0.0406	0.6912	0.9652	0.1209	1.1071	1.2815	0.0656	1.3459	1.1684
0.2276	0.0447	0.6869	0.9555	0.1335	1.0913	1.2714	0.0719	1.3387	1.1657
0.2238	0.0491	0.6825	0.9454	0.1469	1.0748	1.2608	0.0786	1.3311	1.1629
0.2199	0.0537	0.6778	0.9347	0.1613	1.0576	1.2499	0.0856	1.3233	1.1600
0.2157	0.0585	0.6729	0.9237	0.1766	1.0398	1.2386	0.0930	1.3152	1.1569
0.2114	0.0637	0.6677	0.9121	0.1929	1.0212	1.2270	0.1007	1.3069	1.1538
0.2068	0.0691	0.6624	0.9001	0.2103	1.0019	1.2150	0.1087	1.2982	1.1505
0.2020	0.0749	0.6569	0.8877	0.2289	0.9820	1.2027	0.1172	1.2892	1.1471
0.1970	0.0810	0.6511	0.8747	0.2488	0.9614	1.1900	0.1260	1.2800	1.1436
0.1918	0.0874	0.6451	0.8613	0.2702	0.9400	1.1770	0.1352	1.2705	1.1400
0.1863	0.0942	0.6389	0.8473	0.2930	0.9180	1.1638	0.1447	1.2608	1.1363
0.1806	0.1014	0.6324	0.8329	0.3175	0.8953	1.1503	0.1547	1.2508	1.1325
0.1746	0.1090	0.6257	0.8179	0.3438	0.8719	1.1365	0.1651	1.2405	1.1286
0.1684	0.1170	0.6188	0.8024	0.3722	0.8479	1.1225	0.1759	1.2299	1.1246
0.1618	0.1255	0.6116	0.7864	0.4028	0.8231	1.1083	0.1872	1.2191	1.1205
0.1550	0.1345	0.6042	0.7698	0.4360	0.7977	1.0940	0.1989	1.2081	1.1162
0.1479	0.1440	0.5965	0.7526	0.4720	0.7716	1.0795	0.2111	1.1968	1.1119
0.1404	0.1541	0.5885	0.7347	0.5111	0.7449	1.0649	0.2238	1.1852	1.1075
0.1326	0.1647	0.5803	0.7163	0.5539	0.7175	1.0502	0.2369	1.1734	1.1029
0.1245	0.1761	0.5719	0.6971	0.6009	0.6895	1.0356	0.2506	1.1613	1.0983
0.1160	0.1881	0.5631	0.6773	0.6525	0.6608	1.0210	0.2648	1.1490	1.0936
0.1070	0.2009	0.5541	0.6567	0.7098	0.6315	1.0066	0.2796	1.1364	1.0888
0.0977	0.2146	0.5448	0.6352	0.7734	0.6016	0.9924	0.2950	1.1235	1.0839
0.0880	0.2291	0.5351	0.6129	0.8447	0.5712	0.9784	0.3110	1.1102	1.0788
0.0777	0.2447	0.5252	0.5896	0.9249	0.5402	0.9649	0.3276	1.0967	1.0737
0.0670	0.2613	0.5150	0.5652	1.0160	0.5087	0.9518	0.3450	1.0828	1.0685
0.0557	0.2792	0.5044	0.5397	1.1201	0.4767	0.9395	0.3630	1.0686	1.0632
0.0439	0.2983	0.4936	0.5128	1.2402	0.4444	0.9280	0.3819	1.0538	1.0578
0.0314	0.3190	0.4823	0.4843	1.3802	0.4117	0.9176	0.4017	1.0386	1.0524
0.0183	0.3413	0.4708	0.4541	1.5452	0.3787	0.9086	0.4224	1.0228	1.0468
0.0044	0.3654	0.4588	0.4219	1.7421	0.3456	0.9013	0.4442	1.0063	1.0412
−0.0102	0.3916	0.4465	0.3872	1.9805	0.3125	0.8962	0.4673	0.9891	1.0354
−0.0257	0.4202	0.4339	0.3495	2.2736	0.2796	0.8938	0.4917	0.9709	1.0296
−0.0420	0.4514	0.4208	0.3082	2.6410	0.2469	0.8948	0.5178	0.9516	1.0238
−0.0594	0.4856	0.4073	0.2622	3.1114	0.2149	0.9004	0.5459	0.9310	1.0178
−0.0780	0.5234	0.3934	0.2104	3.7288	0.1836	0.9118	0.5763	0.9088	1.0119
−0.0977	0.5652	0.3790	0.1507	4.5637	0.1535	0.9311	0.6096	0.8848	1.0059
−0.1189	0.6118	0.3642	0.0804	5.7336	0.1250	0.9613	0.6466	0.8586	0.9999
−0.1415	0.6640	0.3489	0.0050	7.4461	0.0983	1.0068	0.6880	0.8298	0.9939
L_1	L_2	C_2	L_3	L_4	C_4	L_5	L_6	C_6	L_7

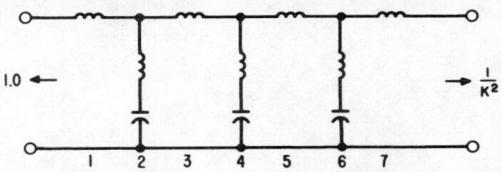

259

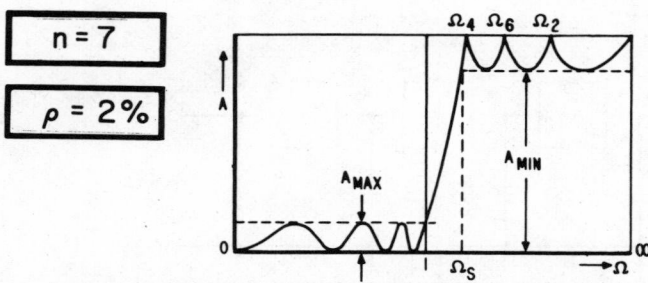

n = 7

ρ = 2%

θ	Ω_S	A_{MIN}	σ_O	σ_1	σ_3	σ_5	Ω_1	Ω_2	Ω_3	Ω_4	Ω_5	Ω_6
C	∞	∞	0.706358	0.157179	0.440407	0.636406	1.193616	∞	0.957206	∞	0.531209	∞
11.0	5.2408	138.42	0.7162432	0.1525073	0.4340404	0.6396525	1.1897613	11.9886	0.9630380	5.3732	0.5407826	6.6793
12.0	4.8097	133.10	0.7181520	0.1516269	0.4328174	0.6402629	1.1890288	10.9867	0.9641374	4.9307	0.5426279	6.1257
13.0	4.4454	128.19	0.7202378	0.1506728	0.4314835	0.6409237	1.1882327	10.1387	0.9653292	4.5568	0.5446431	5.6575
14.0	4.1336	123.64	0.7225034	0.1496456	0.4300376	0.6416342	1.1873730	9.4116	0.9666124	4.2367	0.5468307	5.2564
15.0	3.8637	119.40	0.7249521	0.1485462	0.4284781	0.6423937	1.1864499	8.7812	0.9679860	3.9597	0.5491932	4.9090
16.0	3.6280	115.43	0.7275872	0.1473753	0.4268041	0.6432014	1.1854633	8.2295	0.9694489	3.7177	0.5567335	4.6052
17.0	3.4203	111.69	0.7304125	0.1461337	0.4250140	0.6440560	1.1844134	7.7424	0.9710001	3.5044	0.5544547	4.3374
18.0	3.2361	108.16	0.7334320	0.1448224	0.4231063	0.6449567	1.1833001	7.3093	0.9726382	3.3152	0.5573601	4.0995
19.0	3.0716	104.81	0.7366503	0.1434424	0.4210794	0.6459021	1.1821236	6.9216	0.9743620	3.1463	0.5604534	3.8868
20.0	2.9238	101.63	0.7400720	0.1419947	0.4189318	0.6468909	1.1808840	6.5725	0.9761699	2.9945	0.5637383	3.6955
21.0	2.7904	98.60	0.7437023	0.1404804	0.4166617	0.6479216	1.1795813	6.2564	0.9780605	2.8574	0.5672190	3.5226
22.0	2.6695	95.70	0.7475467	0.1389005	0.4142672	0.6489925	1.1782157	5.9690	0.9800319	2.7331	0.5709000	3.3656
23.0	2.5593	92.93	0.7516113	0.1372564	0.4117464	0.6501017	1.1767874	5.7063	0.9820824	2.6199	0.5747857	3.2224
24.0	2.4586	90.26	0.7559025	0.1355491	0.4090976	0.6512473	1.1752964	5.4654	0.9842101	2.5164	0.5788813	3.0912
25.0	2.3662	87.70	0.7604272	0.1337801	0.4063186	0.6524270	1.1737429	5.2436	0.9864128	2.4214	0.5831919	2.9707
26.0	2.2812	85.23	0.7651929	0.1319507	0.4034073	0.6536383	1.1721272	5.0388	0.9886884	2.3339	0.5877232	2.8596
27.0	2.2027	82.85	0.7702076	0.1300623	0.4003616	0.6548786	1.1704494	4.8489	0.9910344	2.2532	0.5924812	2.7568
28.0	2.1301	80.55	0.7754798	0.1281163	0.3971793	0.6561446	1.1687098	4.6725	0.9934483	2.1784	0.5974720	2.6615
29.0	2.0627	78.32	0.7810190	0.1261142	0.3938583	0.6574333	1.1669085	4.5080	0.9959274	2.1090	0.6027025	2.5729
30.0	2.0000	76.16	0.7868351	0.1240577	0.3903960	0.6587409	1.1650460	4.3544	0.9984688	2.0445	0.6081794	2.4903
31.0	1.9416	74.06	0.7929388	0.1219482	0.3867903	0.6600633	1.1631225	4.2106	1.0010694	1.9844	0.6139104	2.4132
32.0	1.8871	72.02	0.7993418	0.1197875	0.3830390	0.6613962	1.1611384	4.0756	1.0037259	1.9282	0.6199031	2.3410
33.0	1.8361	70.04	0.8060564	0.1175773	0.3791394	0.6627345	1.1590938	3.9486	1.0064351	1.8756	0.6261658	2.2733
34.0	1.7883	68.11	0.8130961	0.1153194	0.3750893	0.6640731	1.1569894	3.8290	1.0091928	1.8264	0.6327072	2.2096
35.0	1.7434	66.23	0.8204755	0.1130155	0.3708862	0.6654057	1.1548256	3.7161	1.0119955	1.7801	0.6395365	2.1497
36.0	1.7013	64.39	0.8282103	0.1106676	0.3665278	0.6667259	1.1526027	3.6093	1.0148389	1.7366	0.6466631	2.0933
37.0	1.6616	62.60	0.8363174	0.1082776	0.3620118	0.6680265	1.1503214	3.5081	1.0177186	1.6957	0.6540973	2.0400
38.0	1.6243	60.84	0.8448153	0.1058474	0.3573358	0.6692993	1.1479822	3.4121	1.0206300	1.6571	0.6618495	1.9896
39.0	1.5890	59.12	0.8537238	0.1033792	0.3524973	0.6705357	1.1455856	3.3209	1.0235681	1.6206	0.6699307	1.9418
40.0	1.5557	57.44	0.8630647	0.1008749	0.3474945	0.6717257	1.1431323	3.2341	1.0265277	1.5862	0.6783528	1.8966
41.0	1.5243	55.79	0.8728615	0.0983367	0.3423249	0.6728587	1.1406230	3.1513	1.0295032	1.5537	0.6871278	1.8537
42.0	1.4945	54.17	0.8831399	0.0957668	0.3369866	0.6739228	1.1380586	3.0724	1.0324890	1.5229	0.6962683	1.8129
43.0	1.4663	52.58	0.8939278	0.0931675	0.3314776	0.6749047	1.1354396	2.9970	1.0354789	1.4937	0.7057875	1.7740
44.0	1.4396	51.02	0.9052560	0.0905411	0.3257961	0.6757900	1.1327671	2.9248	1.0384664	1.4660	0.7156993	1.7371
45.0	1.4142	49.49	0.9171579	0.0878898	0.3199404	0.6765627	1.1300421	2.8557	1.0414447	1.4397	0.7260178	1.7019
46.0	1.3902	47.98	0.9296704	0.0852163	0.3139091	0.6772047	1.1272652	2.7895	1.0444068	1.4147	0.7367578	1.6683
47.0	1.3673	46.49	0.9428341	0.0825229	0.3077008	0.6776967	1.1244379	2.7259	1.0473449	1.3910	0.7479345	1.6362
48.0	1.3456	45.02	0.9566936	0.0798121	0.3013145	0.6780167	1.1215611	2.6648	1.0502514	1.3685	0.7595637	1.6056
49.0	1.3250	43.57	0.9712984	0.0770867	0.2947496	0.6781407	1.1186359	2.6060	1.0531177	1.3470	0.7716613	1.5763
50.0	1.3054	42.14	0.9867033	0.0743492	0.2880053	0.6780419	1.1156639	2.5494	1.0559354	1.3266	0.7842435	1.5482
51.0	1.2868	40.73	1.0029691	0.0716024	0.2810817	0.6776909	1.1126463	2.4948	1.0586953	1.3072	0.7973268	1.5213
52.0	1.2690	39.34	1.0201639	0.0688491	0.2739790	0.6770548	1.1095846	2.4422	1.0613877	1.2887	0.8109278	1.4956
53.0	1.2521	37.96	1.0383636	0.0660922	0.2666980	0.6760972	1.1064804	2.3913	1.0640028	1.2711	0.8250627	1.4709
54.0	1.2361	36.60	1.0576535	0.0633348	0.2592396	0.6747781	1.1033353	2.3422	1.0665300	1.2543	0.8397475	1.4473
55.0	1.2208	35.24	1.0781296	0.0605797	0.2516056	0.6730526	1.1001512	2.2946	1.0689585	1.2383	0.8549974	1.4245
56.0	1.2062	33.90	1.0999006	0.0578300	0.2437983	0.6708713	1.0969298	2.2485	1.0712768	1.2230	0.8708266	1.4027
57.0	1.1924	32.58	1.1230896	0.0550893	0.2358204	0.6681796	1.0936733	2.2039	1.0734732	1.2085	0.8872477	1.3817
58.0	1.1792	31.26	1.1478372	0.0523606	0.2276758	0.6649167	1.0903838	2.1606	1.0755351	1.1946	0.9042711	1.3616
59.0	1.1666	29.95	1.1743043	0.0496474	0.2193689	0.6610156	1.0870635	2.1185	1.0774497	1.1814	0.9219048	1.3422
60.0	1.1547	28.64	1.2026757	0.0469533	0.2109048	0.6564020	1.0837149	2.0776	1.0792037	1.1689	0.9401526	1.3235
θ	Ω_S	A_{MIN}	σ_O	σ_1	σ_3	σ_5	Ω_1	Ω_2	Ω_3	Ω_4	Ω_5	Ω_6

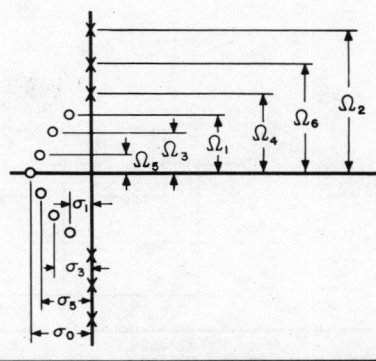

				K² = 1.0					
C_1	C_2	L_2	C_3	C_4	L_4	C_5	C_6	L_6	C_7
0.6301	0.00000	1.282	1.579	0.00000	1.575	1.579	0.00000	1.282	0.6301
0.62575	0.00545	1.27567	1.55827	0.02250	1.53919	1.55072	0.01785	1.25594	0.61354
0.62493	0.00650	1.27451	1.55433	0.02684	1.53247	1.54537	0.02130	1.25104	0.61038
0.62403	0.00764	1.27324	1.55004	0.03158	1.52518	1.53956	0.02508	1.24570	0.60695
0.62306	0.00888	1.27187	1.54542	0.03672	1.51731	1.53330	0.02919	1.23994	0.60323
0.62201	0.01021	1.27039	1.54046	0.04227	1.50887	1.52658	0.03363	1.23375	0.59923
0.62089	0.01164	1.26880	1.53515	0.04824	1.49986	1.51942	0.03842	1.22713	0.59494
0.61969	0.01317	1.26711	1.52951	0.05464	1.49029	1.51182	0.04357	1.22008	0.59036
0.61842	0.01479	1.26531	1.52354	0.06147	1.48014	1.50378	0.04907	1.21261	0.58549
0.61706	0.01652	1.26340	1.51722	0.06875	1.46944	1.49530	0.05495	1.20470	0.58033
0.61563	0.01835	1.26138	1.51057	0.07648	1.45817	1.48639	0.06120	1.19637	0.57488
0.61412	0.02029	1.25925	1.50358	0.08468	1.44635	1.47706	0.06786	1.18761	0.56913
0.61253	0.02233	1.25700	1.49626	0.09335	1.43397	1.46730	0.07492	1.17842	0.56307
0.61085	0.02448	1.25464	1.48860	0.10252	1.42105	1.45712	0.08240	1.16880	0.55671
0.60909	0.02674	1.25216	1.48062	0.11220	1.40757	1.44654	0.09031	1.15874	0.55004
0.60725	0.02911	1.24957	1.47229	0.12239	1.39355	1.43555	0.09868	1.14826	0.54306
0.60532	0.03159	1.24685	1.46364	0.13313	1.37899	1.42415	0.10752	1.13735	0.53577
0.60331	0.03419	1.24401	1.45465	0.14442	1.36389	1.41237	0.11685	1.12601	0.52815
0.60121	0.03691	1.24105	1.44534	0.15630	1.34826	1.40020	0.12670	1.11423	0.52020
0.59901	0.03975	1.23796	1.43569	0.16877	1.33209	1.38766	0.13707	1.10203	0.51193
0.59673	0.04271	1.23474	1.42572	0.18187	1.31540	1.37474	0.14801	1.08939	0.50331
0.59435	0.04581	1.23139	1.41541	0.19562	1.29819	1.36146	0.15954	1.07633	0.49436
0.59187	0.04903	1.22791	1.40478	0.21006	1.28046	1.34783	0.17169	1.06283	0.48505
0.58929	0.05239	1.22429	1.39383	0.22520	1.26221	1.33386	0.18449	1.04890	0.47539
0.58662	0.05588	1.22052	1.38254	0.24110	1.24346	1.31956	0.19798	1.03454	0.46536
0.58384	0.05952	1.21662	1.37093	0.25779	1.22419	1.30493	0.21220	1.01976	0.45496
0.58095	0.06331	1.21257	1.35900	0.27531	1.20443	1.29000	0.22719	1.00454	0.44418
0.57796	0.06725	1.20836	1.34674	0.29370	1.18416	1.27477	0.24300	0.98889	0.43301
0.57485	0.07134	1.20400	1.33415	0.31303	1.16340	1.25925	0.25969	0.97281	0.42144
0.57163	0.07560	1.19948	1.32125	0.33335	1.14215	1.24346	0.27732	0.95630	0.40945
0.56829	0.08002	1.19480	1.30801	0.35473	1.12042	1.22742	0.29595	0.93936	0.39704
0.56483	0.08462	1.18994	1.29446	0.37723	1.09820	1.21114	0.31566	0.92200	0.38420
0.56124	0.08941	1.18491	1.28059	0.40094	1.07550	1.19464	0.33652	0.90420	0.37090
0.55753	0.09438	1.17970	1.26639	0.42594	1.05233	1.17794	0.35863	0.88599	0.35713
0.55367	0.09955	1.17429	1.25187	0.45235	1.02868	1.16105	0.38208	0.86735	0.34288
0.54967	0.10492	1.16870	1.23703	0.48027	1.00457	1.14400	0.40700	0.84829	0.32812
0.54553	0.11051	1.16289	1.22187	0.50983	0.98000	1.12681	0.43352	0.82881	0.31284
0.54124	0.11633	1.15688	1.20640	0.54119	0.95497	1.10951	0.46176	0.80891	0.29701
0.53679	0.12239	1.15064	1.19060	0.57450	0.92948	1.09211	0.49191	0.78860	0.28061
0.53217	0.12870	1.14417	1.17448	0.60996	0.90354	1.07467	0.52415	0.76788	0.26362
0.52738	0.13527	1.13745	1.15805	0.64779	0.87715	1.05719	0.55869	0.74675	0.24599
0.52241	0.14212	1.13047	1.14131	0.68825	0.85031	1.03972	0.59577	0.72521	0.22770
0.51725	0.14928	1.12321	1.12425	0.73162	0.82304	1.02230	0.63567	0.70328	0.20871
0.51189	0.15674	1.11567	1.10688	0.77825	0.79533	1.04496	0.67873	0.68095	0.18898
0.50632	0.16455	1.10781	1.08921	0.82854	0.76719	0.98775	0.72531	0.65824	0.16846
0.50054	0.17272	1.09961	1.07123	0.88297	0.73862	0.97073	0.77585	0.63515	0.14710
0.49451	0.18128	1.09106	1.05295	0.94209	0.70964	0.95394	0.83087	0.61169	0.12485
0.48824	0.19026	1.08212	1.03438	1.00658	0.68025	0.93745	0.89099	0.58787	0.10162
0.48171	0.19969	1.07276	1.01551	1.07723	0.65046	0.92133	0.95692	0.56369	0.07736
0.47491	0.20963	1.06293	0.99637	1.15502	0.62028	0.90566	1.02953	0.53918	0.05197
0.46780	0.22011	1.05259	0.97696	1.24113	0.58973	0.89053	1.10988	0.51434	0.02536
L_1	L_2	C_2	L_3	L_4	C_4	L_5	L_6	C_6	L_7

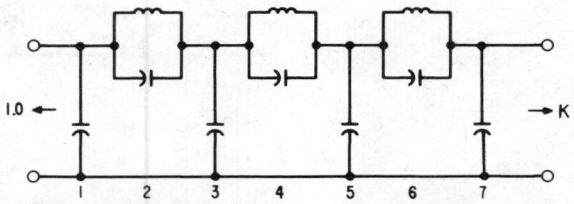

				$K^2 = \infty$						
θ	C_1	C_2	L_2	C_3	C_4	L_4	C_5	C_6	L_6	C_7
C	0.315026	0.0000	0.844825	1.191989	0.0000	1.393011	1.498631	0.0000	1.519144	1.272517
11.0	0.3073	0.0083	0.8348	1.1677	0.0256	1.3525	1.4704	0.0150	1.4986	1.2642
12.0	0.3058	0.0099	0.8328	1.1631	0.0306	1.3449	1.4650	0.0178	1.4947	1.2626
13.0	0.3042	0.0117	0.8307	1.1580	0.0360	1.3365	1.4592	0.0210	1.4905	1.2609
14.0	0.3025	0.0136	0.8285	1.1526	0.0420	1.3275	1.4530	0.0244	1.4860	1.2590
15.0	0.3006	0.0157	0.8260	1.1468	0.0484	1.3178	1.4463	0.0280	1.4811	1.2570
16.0	0.2986	0.0179	0.8234	1.1405	0.0553	1.3075	1.4391	0.0319	1.4759	1.2549
17.0	0.2964	0.0203	0.8206	1.1339	0.0628	1.2965	1.4315	0.0362	1.4703	1.2526
18.0	0.2941	0.0229	0.8176	1.1269	0.0708	1.2849	1.4235	0.0406	1.4644	1.2502
19.0	0.2916	0.0256	0.8145	1.1194	0.0794	1.2726	1.4150	0.0454	1.4582	1.2477
20.0	0.2890	0.0285	0.8112	1.1116	0.0885	1.2596	1.4061	0.0504	1.4517	1.2450
21.0	0.2863	0.0316	0.8077	1.1033	0.0983	1.2460	1.3968	0.0558	1.4448	1.2422
22.0	0.2834	0.0349	0.8040	1.0946	0.1087	1.2317	1.3871	0.0614	1.4377	1.2392
23.0	0.2803	0.0384	0.8001	1.0855	0.1197	1.2168	1.3769	0.0673	1.4302	1.2362
24.0	0.2771	0.0421	0.7960	1.0761	0.1315	1.2012	1.3663	0.0736	1.4224	1.2329
25.0	0.2737	0.0459	0.7918	1.0662	0.1439	1.1850	1.3553	0.0801	1.4142	1.2296
26.0	0.2701	0.0500	0.7873	1.0558	0.1572	1.1682	1.3439	0.0870	1.4058	1.2261
27.0	0.2664	0.0543	0.7827	1.0451	0.1712	1.1507	1.3322	0.0942	1.3970	1.2225
28.0	0.2625	0.0589	0.7778	1.0340	0.1861	1.1325	1.3200	0.1017	1.3880	1.2188
29.0	0.2584	0.0637	0.7728	1.0224	0.2019	1.1138	1.3075	0.1096	1.3786	1.2149
30.0	0.2542	0.0687	0.7676	1.0104	0.2186	1.0943	1.2946	0.1178	1.3689	1.2109
31.0	0.2497	0.0740	0.7621	0.9980	0.2364	1.0743	1.2813	0.1264	1.3589	1.2067
32.0	0.2451	0.0796	0.7565	0.9851	0.2553	1.0536	1.2677	0.1353	1.3487	1.2024
33.0	0.2402	0.0854	0.7506	0.9718	0.2754	1.0323	1.2538	0.1446	1.3381	1.1980
34.0	0.2352	0.0916	0.7445	0.9581	0.2967	1.0104	1.2395	0.1543	1.3272	1.1935
35.0	0.2299	0.0981	0.7382	0.9439	0.3195	0.9878	1.2249	0.1644	1.3160	1.1888
36.0	0.2244	0.1049	0.7316	0.9293	0.3438	0.9646	1.2101	0.1749	1.3046	1.1840
37.0	0.2187	0.1121	0.7249	0.9142	0.3697	0.9408	1.1949	0.1859	1.2928	1.1791
38.0	0.2127	0.1197	0.7179	0.8987	0.3974	0.9164	1.1795	0.1973	1.2808	1.1740
39.0	0.2065	0.1276	0.7106	0.8826	0.4271	0.8914	1.1639	0.2091	1.2684	1.1688
40.0	0.2000	0.1360	0.7031	0.8661	0.4591	0.8658	1.1480	0.2214	1.2558	1.1635
41.0	0.1933	0.1448	0.6954	0.8490	0.4934	0.8395	1.1320	0.2342	1.2429	1.1581
42.0	0.1862	0.1541	0.6874	0.8315	0.5306	0.8127	1.1158	0.2474	1.2297	1.1525
43.0	0.1789	0.1639	0.6791	0.8134	0.5707	0.7854	1.0994	0.2612	1.2162	1.1468
44.0	0.1713	0.1743	0.6706	0.7947	0.6144	0.7574	1.0830	0.2756	1.2025	1.1409
45.0	0.1633	0.1853	0.6618	0.7755	0.6619	0.7289	1.0665	0.2905	1.1884	1.1350
46.0	0.1550	0.1969	0.6527	0.7556	0.7140	0.6998	1.0499	0.3060	1.1740	1.1288
47.0	0.1463	0.2092	0.6433	0.7351	0.7712	0.6702	1.0335	0.3222	1.1594	1.1226
48.0	0.1372	0.2223	0.6336	0.7139	0.8343	0.6401	1.0171	0.3390	1.1444	1.1162
49.0	0.1277	0.2362	0.6235	0.6920	0.9043	0.6094	1.0008	0.3565	1.1290	1.1097
50.0	0.1178	0.2509	0.6132	0.6692	0.9825	0.5783	0.9848	0.3747	1.1133	1.1031
51.0	0.1074	0.2667	0.6025	0.6457	1.0702	0.5468	0.9691	0.3938	1.0973	1.0963
52.0	0.0966	0.2835	0.5914	0.6211	1.1695	0.5149	0.9539	0.4136	1.0808	1.0894
53.0	0.0851	0.3015	0.5800	0.5956	1.2826	0.4826	0.9392	0.4344	1.0639	1.0824
54.0	0.0731	0.3208	0.5683	0.5689	1.4125	0.4500	0.9253	0.4562	1.0465	1.0752
55.0	0.0605	0.3415	0.5561	0.5408	1.5633	0.4172	0.9123	0.4791	1.0285	1.0679
56.0	0.0473	0.3639	0.5435	0.5113	1.7402	0.3842	0.9003	0.5032	1.0100	1.0604
57.0	0.0333	0.3881	0.5304	0.4800	1.9500	0.3511	0.8899	0.5287	0.9907	1.0528
58.0	0.0185	0.4144	0.5170	0.4467	2.2023	0.3182	0.8811	0.5557	0.9706	1.0451
59.0	0.0029	0.4430	0.5030	0.4109	2.5105	0.2854	0.8747	0.5846	0.9496	1.0372
60.0	−0.0137	0.4742	0.4886	0.3721	2.8934	0.2530	0.8710	0.6155	0.9275	1.0292
θ	L_1	L_2	C_2	L_3	L_4	C_4	L_5	L_6	C_6	L_7

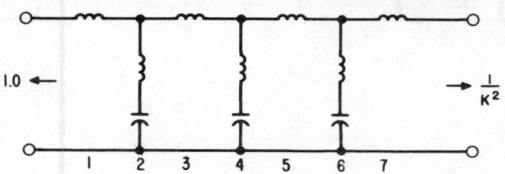

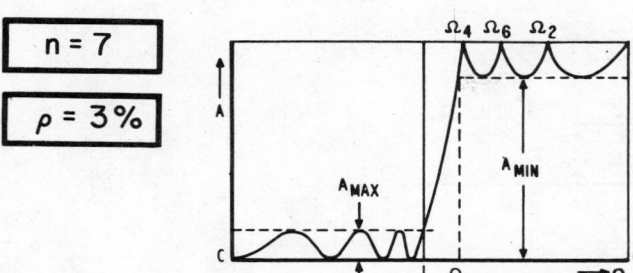

$n = 7$

$\rho = 3\%$

θ	Ω_S	A_{MIN}	σ_O	σ_1	σ_3	σ_5	Ω_1	Ω_2	Ω_3	Ω_4	Ω_5	Ω_6
C	∞	∞	0.636566	0.141649	0.396892	0.573526	1.155697	∞	0.926797	∞	0.514334	∞
11.0	5.2408	141.95	0.6449099	0.1377737	0.3916554	0.5763352	1.1527628	11.9886	0.9320122	5.3732	0.5225860	6.6793
12.0	4.8097	136.62	0.6465195	0.1370423	0.3906497	0.5768647	1.1522050	10.9867	0.9329976	4.9307	0.5241753	6.1257
13.0	4.4454	131.72	0.6482777	0.1362492	0.3895527	0.5774384	1.1515985	10.1387	0.9340666	4.5568	0.5259103	5.6575
14.0	4.1336	127.17	0.6501867	0.1353949	0.3883637	0.5780558	1.1509435	9.4116	0.9352184	4.2367	0.5277930	5.2564
15.0	3.8637	122.93	0.6522490	0.1344799	0.3870814	0.5787166	1.1502399	8.7812	0.9364527	3.9597	0.5298255	4.9090
16.0	3.6280	118.95	0.6544674	0.1335047	0.3857050	0.5794199	1.1494876	8.2295	0.9377686	3.7177	0.5320101	4.6052
17.0	3.4203	115.22	0.6568447	0.1324700	0.3842331	0.5801652	1.1486769	7.7424	0.9391654	3.5044	0.5344393	4.3374
18.0	3.2361	111.68	0.6593842	0.1313763	0.3726647	0.5809517	1.1478375	7.3093	0.9406422	3.3152	0.5368458	4.0995
19.0	3.0716	108.34	0.6620892	0.1302244	0.3809983	0.5817785	1.1469396	6.9216	0.9421983	3.1463	0.5395024	3.8868
20.0	2.9238	105.16	0.6649635	0.1290149	0.3792327	0.5826447	1.1459931	6.5725	0.9438325	2.9945	0.5423223	3.6955
21.0	2.7904	102.12	0.6680111	0.1277485	0.3773663	0.5835494	1.1449981	6.2564	0.9455440	2.8574	0.5453087	3.5226
22.0	2.6695	99.23	0.6712363	0.1264262	0.3753977	0.5844913	1.1439545	5.9690	0.9473314	2.7331	0.5484651	3.3656
23.0	2.5593	96.45	0.6746437	0.1250486	0.3733253	0.5854692	1.1428625	5.7063	0.9491937	2.6199	0.5517954	3.2224
24.0	2.4586	93.79	0.6782384	0.1236167	0.3711475	0.5864816	1.1417221	5.4654	0.9511297	2.5164	0.5553034	3.0912
25.0	2.3662	91.23	0.6820256	0.1221314	0.3688625	0.5875271	1.1405332	5.2436	0.9531377	2.4214	0.5589934	2.9707
26.0	2.2812	88.76	0.6860111	0.1205936	0.3664685	0.5886041	1.1392961	5.0388	0.9552164	2.3339	0.5628699	2.8596
27.0	2.2027	86.37	0.6902011	0.1190043	0.3639639	0.5897105	1.1380107	4.8489	0.9573642	2.2532	0.5669377	2.7568
28.0	2.1301	84.07	0.6946022	0.1173644	0.3613465	0.5908445	1.1366771	4.6725	0.9595793	2.1784	0.5712018	2.6615
29.0	2.0627	81.84	0.6992215	0.1156751	0.3586144	0.5920035	1.1352954	4.5080	0.9618600	2.1090	0.5756676	2.5729
30.0	2.0000	79.68	0.7040667	0.1139375	0.3557659	0.5931853	1.1338658	4.3544	0.9642043	2.0445	0.5803407	2.4903
31.0	1.9416	77.59	0.7091458	0.1121527	0.3527988	0.5943869	1.1323883	4.2106	0.9666100	1.9844	0.5852270	2.4132
32.0	1.8871	75.55	0.7144677	0.1103218	0.3497108	0.5956056	1.1308632	4.0756	0.9690748	1.9282	0.5903329	2.3410
33.0	1.8361	73.57	0.7200418	0.1084462	0.3464999	0.5968377	1.1292906	3.9486	0.9715966	1.8756	0.5956650	2.2733
34.0	1.7883	71.64	0.7258784	0.1065271	0.3431638	0.5980797	1.1276706	3.8290	0.9741726	1.8264	0.6012303	2.2096
35.0	1.7434	69,75	0.7319883	0.1045659	0.3397006	0.5993275	1.1260034	3.7161	0.9768003	1.7801	0.6070362	2.1497
36.0	1.7013	67.92	0.7383832	0.1025637	0.3361077	0.6005765	1.1242895	3.6093	0.9794766	1.7366	0.6130905	2.0933
37.0	1.6616	66.12	0.7450760	0.1005221	0.3323831	0.6018220	1.1225288	3.5081	0.9821986	1.6957	0.6194014	2.0400
38.0	1.6243	64.37	0.7520803	0.0974425	0.3285243	0.6030584	1.1207217	3.4121	0.9849630	1.6571	0.6259774	1.9896
39.0	1.5890	62.65	0.7594110	0.0963263	0.3245292	0.6042796	1.1187687	3.3209	0.9877662	1.6206	0.6328278	1.9418
40.0	1.5557	60.97	0.7670739	0.0941752	0.3203955	0.6054791	1.1169700	3.2341	0.9906047	1.5862	0.6399620	1.8966
41.0	1.5243	59.32	0.7751166	0.0919907	0.3161208	0.6066494	1.1150259	3.1513	0.9934745	1.5537	0.6473900	1.8537
42.0	1.4945	57.70	0.7835278	0.0897744	0.3117030	0.6077826	1.1130369	3.0724	0.9963714	1.5229	0.6551223	1.8129
43.0	1.4663	56.11	0.7923381	0.0875279	0.3071398	0.6088697	1.1110034	2.9970	0.9992911	1.4937	0.6631700	1.7740
44.0	1.4396	54.55	0.8015695	0.0852532	0.3024290	0.6099006	1.1089260	2.9248	1.0022288	1.4660	0.6715445	1.7371
45.0	1.4142	53.01	0.8112466	0.0829518	0.2975676	0.6108646	1.1068052	2.8557	1.0051798	1.4397	0.6802581	1.7019
46.0	1.3902	51.50	0.8213958	0.0806256	0.2925567	0.6117494	1.1046416	2.7895	1.0081386	1.4147	0.6893231	1.6683
47.0	1.3673	50.01	0.8320461	0.0782767	0.2873911	0.6125417	1.1024357	2.7259	1.0110998	1.3910	0.6987529	1.6362
48.0	1.3456	48.54	0.8432295	0.0759067	0.2820700	0.6132267	1.1001881	2.6648	1.0140575	1.3685	0.7085612	1.6056
49.0	1.3250	47.10	0.8549809	0.0735179	0.2765919	0.6137877	1.0978999	2.6060	1.0170056	1.3470	0.7187620	1.5763
50.0	1.3054	45.67	0.8673389	0.0711123	0.2709552	0.6142067	1.0955717	2.5494	1.0199374	1.3266	0.7293702	1.5482
51.0	1.2868	44.26	0.8803463	0.0686918	0.2651585	0.6144634	1.0932043	2.4948	1.0228461	1.3072	0.7404011	1.5213
52.0	1.2690	42.86	0.8940503	0.0662590	0.2592007	0.6145352	1.0907987	2.4422	1.0257243	1.2887	0.7518702	1.4956
53.0	1.2521	41.48	0.9085034	0.0638158	0.2530710	0.6143973	1.0883560	2.3913	1.0285644	1.2711	0.7637935	1.4709
54.0	1.2361	40.12	0.9237642	0.0613648	0.2467987	0.6140218	1.0858770	2.3422	1.0313582	1.2543	0.7761875	1.4473
55.0	1.2208	38.77	0.9398980	0.0589084	0.2403533	0.6133780	1.0833631	2.2946	1.0340972	1.2383	0.7890684	1.4245
56.0	1.2062	37.43	0.9569784	0.0564491	0.2337452	0.6124315	1.0808154	2.2485	1.0367725	1.2230	0.8024529	1.4027
57.0	1.1924	36.10	0.9750880	0.0539894	0.2269746	0.6111441	1.0782354	2.2039	1.0393745	1.2085	0.8163571	1.3817
58.0	1.1792	34.78	0.9943203	0.0515320	0.2200426	0.6094734	1.0756245	2.1606	1.0418931	1.1946	0.8307970	1.3616
59.0	1.1666	33.47	1.0147812	0.0490798	0.2129505	0.6073719	1.0729841	2.1185	1.0443178	1.1814	0.8457876	1.3422
60.0	1.1547	32.17	1.0365919	0.0466357	0.2057000	0.6047870	1.0703160	2.0776	1.0466376	1.1689	0.8613425	1.3235
θ	Ω_S	A_{MIN}	σ_O	σ_1	σ_3	σ_5	Ω_1	Ω_2	Ω_3	Ω_4	Ω_5	Ω_6

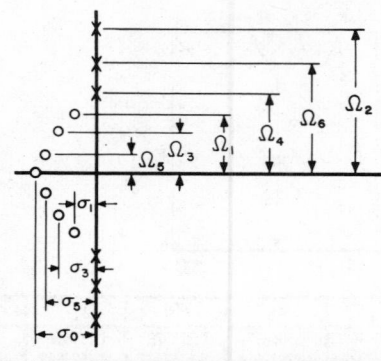

					K² = 1.0					
θ	C₁	C₂	L₂	C₃	C₄	L₄	C₅	C₆	L₆	C₇
C	0.6991	0.00000	1.338	1.653	0.00000	1.608	1.653	0.00000	1.338	0.6991
11.0	0.69496	0.00523	1.33147	1.63173	0.02201	1.57363	1.62408	0.01708	1.31263	0.68327
12.0	0.69417	0.00623	1.33031	1.62775	0.02625	1.56703	1.61866	0.02038	1.30789	0.68024
13.0	0.69330	0.00732	1.32905	1.62342	0.03077	1.55977	1.61278	0.02398	1.30274	0.67695
14.0	0.69236	0.00850	1.32768	1.61875	0.03589	1.55214	1.60644	0.02790	1.29718	0.67338
15.0	0.69134	0.00978	1.32621	1.61374	0.04131	1.54385	1.59964	0.03214	1.29120	0.66954
16.0	0.69026	0.01115	1.32463	1.60838	0.04714	1.53500	1.59238	0.03670	1.28482	0.66543
17.0	0.68910	0.01261	1.32295	1.60269	0.05337	1.52559	1.58468	0.04159	1.27802	0.66105
18.0	0.68786	0.01417	1.32115	1.59665	0.06003	1.51563	1.57652	0.04682	1.27080	0.65639
19.0	0.68655	0.01582	1.31925	1.59027	0.06712	1.50511	1.56792	0.05240	1.26317	0.65145
20.0	0.68517	0.01757	1.31724	1.58355	0.07464	1.49405	1.55888	0.05834	1.25513	0.64623
21.0	0.68370	0.01943	1.31512	1.57650	0.08262	1.48243	1.54940	0.06464	1.24667	0.64073
22.0	0.68216	0.02138	1.31288	1.56910	0.09105	1.47028	1.53948	0.07132	1.23779	0.63494
23.0	0.68054	0.02343	1.31053	1.56137	0.09995	1.45758	1.52914	0.07839	1.22850	0.62887
24.0	0.67884	0.02559	1.30806	1.55331	0.10935	1.44435	1.51837	0.08586	1.21880	0.62250
25.0	0.67706	0.02786	1.30548	1.54490	0.11923	1.43058	1.50719	0.09375	1.20867	0.61584
26.0	0.67519	0.03023	1.30277	1.53617	0.12962	1.41628	1.49558	0.10207	1.19814	0.60878
27.0	0.67324	0.03272	1.29995	1.52710	0.14055	1.40145	1.48357	0.11083	1.18718	0.60162
28.0	0.67121	0.03532	1.29700	1.51769	0.15203	1.38610	1.47116	0.12006	1.17581	0.59406
29.0	0.66909	0.03803	1.29392	1.50796	0.16407	1.37023	1.45834	0.12977	1.16402	0.58619
30.0	0.66688	0.04086	1.29072	1.49789	0.17671	1.35384	1.44514	0.13999	1.15181	0.57800
31.0	0.66457	0.04381	1.28739	1.48749	0.18995	1.33695	1.43156	0.15074	1.13918	0.56949
32.0	0.66218	0.04689	1.28392	1.47676	0.20383	1.31954	1.41760	0.16204	1.12614	0.56067
33.0	0.65969	0.05009	1.28032	1.46571	0.21838	1.30163	1.40327	0.17391	1.11267	0.55151
34.0	0.65710	0.05343	1.27658	1.45432	0.23363	1.28322	1.38858	0.18640	1.09879	0.54201
35.0	0.65442	0.05690	1.27269	1.44261	0.24961	1.26432	1.37354	0.19953	1.08449	0.53218
36.0	0.65163	0.06051	1.26866	1.43057	0.26635	1.24493	1.35716	0.21334	1.06976	0.52199
37.0	0.64874	0.06426	1.26449	1.41821	0.28390	1.22505	1.34245	0.22786	1.05462	0.51145
38.0	0.64574	0.06816	1.26015	1.40552	0.30231	1.20468	1.32641	0.24314	1.03905	0.50054
39.0	0.64263	0.07221	1.25566	1.39250	0.32161	1.18384	1.31007	0.25922	1.02307	0.48927
40.0	0.63940	0.07643	1.25101	1.37916	0.34188	1.16253	1.29343	0.27617	1.00666	0.47760
41.0	0.63606	0.08080	1.24619	1.36550	0.36316	1.14075	1.27651	0.29402	0.98983	0.46555
42.0	0.63260	0.08535	1.24120	1.35151	0.37552	1.11851	1.25931	0.31286	0.97259	0.45309
43.0	0.62901	0.09008	1.23602	1.33720	0.40904	1.09580	1.24186	0.33274	0.95492	0.44022
44.0	0.62529	0.09499	1.23067	1.32257	0.43381	1.07264	1.22416	0.35375	0.93683	0.42691
45.0	0.62144	0.10009	1.22512	1.30762	0.45991	1.04903	1.20624	0.37597	0.91832	0.41316
46.0	0.61744	0.10539	1.21937	1.29234	0.48746	1.02498	1.18811	0.39950	0.89939	0.39896
47.0	0.61331	0.11091	1.21342	1.27675	0.51657	1.00048	1.16979	0.42444	0.88004	0.38427
48.0	0.60902	0.11665	1.20725	1.26083	0.54737	0.97554	1.15131	0.45093	0.86027	0.36909
49.0	0.60457	0.12262	1.20085	1.24459	0.58002	0.95017	1.13268	0.47909	0.84009	0.35340
50.0	0.59996	0.12884	1.19422	1.22803	0.61469	0.92438	1.11392	0.50909	0.81949	0.33717
51.0	0.59518	0.13532	1.18733	1.21116	0.65158	0.89816	1.09506	0.54110	0.79848	0.32038
52.0	0.59021	0.14207	1.18019	1.19396	0.69092	0.87152	1.07614	0.57532	0.77705	0.30299
53.0	0.58506	0.14911	1.17277	1.17645	0.73297	0.84446	1.05717	0.61199	0.75521	0.28499
54.0	0.57971	0.15646	1.16506	1.15863	0.77803	0.81699	1.03720	0.65136	0.73297	0.26633
55.0	0.57415	0.16415	1.15704	1.14048	0.82647	0.78912	1.01925	0.69375	0.71032	0.24698
56.0	0.56837	0.17219	1.14868	1.12203	0.87867	0.76085	1.00038	0.73950	0.68727	0.22690
57.0	0.56235	0.18060	1.13997	1.10327	0.93517	0.73219	0.98161	0.78903	0.66383	0.20603
58.0	0.55609	0.18943	1.13088	1.08420	0.99652	0.70314	0.96300	0.84274	0.63999	0.18433
59.0	0.54956	0.19870	1.12138	1.06483	1.06341	0.67371	0.94460	0.90148	0.61577	0.16174
60.0	0.54275	0.20845	1.11142	1.04516	1.13668	0.64392	0.92647	0.96565	0.59117	0.13818
θ	L₁	L₂	C₂	L₃	L₄	C₄	L₅	L₆	C₆	L₇

264

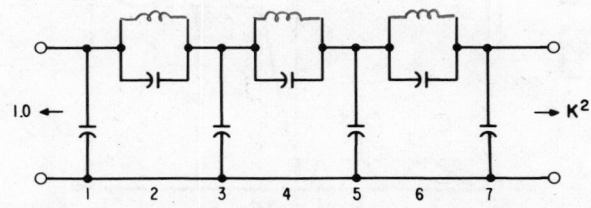

					K² = ∞				
C_1	C_2	L_2	C_3	C_4	L_4	C_5	C_6	L_6	C_7
0.349565	0.0000	0.918275	1.269836	0.0000	1.461975	1.559603	0.0000	1.572179	1.314598
0.3425	0.0077	0.9088	1.2463	0.0243	1.4226	1.5314	0.0144	1.5514	1.3057
0.3411	0.0091	0.9069	1.2418	0.0291	1.4151	1.5261	0.0172	1.5474	1.3041
0.3396	0.0107	0.9050	1.2369	0.0342	1.4070	1.5203	0.0202	1.5431	1.3022
0.3380	0.0125	0.9028	1.2316	0.0398	1.3982	1.5140	0.0235	1.5385	1.3003
0.3363	0.0144	0.9005	1.2260	0.0459	1.3878	1.5073	0.0271	1.5335	1.2981
0.3344	0.0164	0.8970	1.2199	0.0525	1.3787	1.5002	0.0309	1.5282	1.2959
0.3324	0.0186	0.8954	1.2135	0.0595	1.3680	1.4926	0.0349	1.5226	1.2935
0.3303	0.0210	0.8926	1.2067	0.0671	1.3567	1.4846	0.0392	1.5166	1.2909
0.3281	0.0235	0.8896	1.1995	0.0751	1.3448	1.4761	0.0438	1.5103	1.2882
0.3257	0.0261	0.8865	1.1919	0.0837	1.3322	1.4672	0.0487	1.5036	1.2854
0.3232	0.0289	0.8832	1.1839	0.0929	1.3189	1.4578	0.0538	1.4967	1.2824
0.3205	0.0319	0.8797	1.1755	0.1026	1.3051	1.4481	0.0593	1.4893	1.2792
0.3177	0.0351	0.8760	1.1667	0.1129	1.2906	1.4379	0.0650	1.4817	1.2760
0.3148	0.0384	0.8722	1.1576	0.1238	1.2755	1.4273	0.0710	1.4737	1.2725
0.3117	0.0419	0.8681	1.1480	0.1354	1.2597	1.4162	0.0773	1.4654	1.2690
0.3084	0.0456	0.8639	1.1380	0.1477	1.2433	1.4048	0.0839	1.4568	1.2652
0.3050	0.0495	0.8596	1.1277	0.1606	1.2264	1.3930	0.0909	1.4478	1.2614
0.3015	0.0536	0.8550	1.1169	0.1743	1.2078	1.3807	0.0981	1.4386	1.2574
0.2978	0.0579	0.8502	1.1058	0.1888	1.1906	1.3681	0.1057	1.4290	1.2532
0.2939	0.0624	0.7452	1.0942	0.2042	1.1717	1.3551	0.1136	1.4191	1.2489
0.2899	0.0671	0.8401	1.0823	0.2204	1.1523	1.3417	0.1219	1.4088	1.2445
0.2856	0.0721	0.8347	1.0699	0.2376	1.1323	1.3279	0.1305	1.3983	1.2399
0.2812	0.0773	0.8292	1.0572	0.2557	1.1116	1.3138	0.1395	1.3874	1.2352
0.2767	0.0828	0.8234	1.0440	0.2750	1.0904	1.2993	0.1488	1.3762	1.2303
0.2719	0.0886	0.8174	1.0304	0.2953	1.0685	1.2745	0.1586	1.3647	1.2253
0.2669	0.0946	0.8112	1.0164	0.3170	1.0461	1.2694	0.1687	1.3529	1.2202
0.2617	0.1010	0.8048	1.0020	0.3399	1.0231	1.2539	0.1792	1.3408	1.2149
0.2564	0.1076	0.7982	0.9871	0.3644	0.9995	1.2381	0.1902	1.3274	1.2094
0.2508	0.1146	0.7913	0.9719	0.3904	0.9753	1.2221	0.2016	1.3156	1.2038
0.2449	0.1219	0.7842	0.9561	0.4181	0.9506	1.2057	0.2134	1.3026	1.1981
0.2389	0.1296	0.7769	0.9400	0.4477	0.9252	1.1891	0.2257	1.2892	1.1922
0.2326	0.1377	0.7693	0.9233	0.4795	0.8993	1.1723	0.2385	1.2756	1.1862
0.2260	0.1462	0.7615	0.9062	0.5135	0.8729	1.1553	0.2518	1.2616	1.1800
0.2192	0.1552	0.7534	0.8887	0.5501	0.8459	1.1380	0.2657	1.2473	1.1737
0.2121	0.1646	0.7450	0.8706	0.5895	0.8184	1.1206	0.2801	1.2328	1.1672
0.2047	0.1745	0.7363	0.8520	0.6322	0.7903	1.1030	0.2950	1.2179	1.1606
0.1969	0.1850	0.7274	0.8329	0.6785	0.7617	1.0854	0.3106	1.2027	1.1538
0.1889	0.1961	0.7182	0.8132	0.7289	0.7326	1.0676	0.3268	1.1871	1.1469
0.1805	0.2078	0.7086	0.7930	0.7840	0.7030	1.0499	0.3436	1.1713	1.1398
0.1718	0.2202	0.6988	0.7721	0.8444	0.6729	1.0321	0.3612	1.1551	1.1326
0.1626	0.2333	0.6886	0.7507	0.9111	0.6423	1.0144	0.3795	1.1385	1.1252
0.1531	0.2473	0.6781	0.7285	0.9849	0.6114	0.9968	0.3986	1.1216	1.1177
0.1431	0.2621	0.6672	0.7056	1.0673	0.5799	0.9794	0.4185	1.1043	1.1100
0.1326	0.2779	0.6560	0.6819	1.1596	0.5482	0.9623	0.4394	1.0866	1.1021
0.1217	0.2948	0.6443	0.6574	1.2639	0.5160	0.9455	0.4612	1.0685	1.0941
0.1102	0.3128	0.6323	0.6319	1.3825	0.4836	0.9293	0.4841	1.0499	1.0859
0.0981	0.3321	0.6199	0.6053	1.5187	0.4509	0.9136	0.5082	1.0308	1.0775
0.0854	0.3529	0.6070	0.5776	1.6763	0.4180	0.8987	0.5335	1.0110	1.0690
0.0721	0.3754	0.5936	0.5485	1.8608	0.3850	0.8849	0.5603	0.9907	1.0603
0.0579	0.3996	0.5797	0.5179	2.0793	0.3520	0.8723	0.5887	0.9696	1.0514
L_1	L_2	C_2	L_3	L_4	C_4	L_5	L_6	C_6	L_7

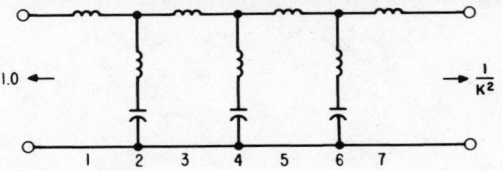

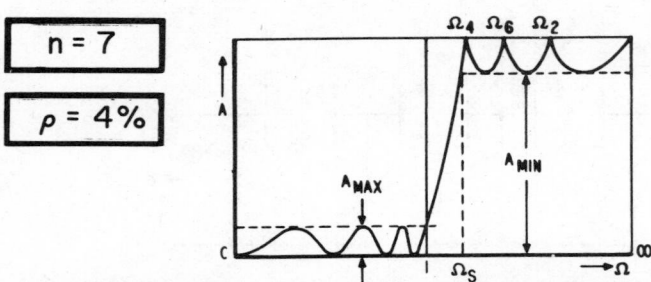

θ	Ω_S	A_{MIN}	σ_0	σ_1	σ_3	σ_5	Ω_1	Ω_2	Ω_3	Ω_4	Ω_5	Ω_6
C	∞	∞	0.588343	0.130919	0.366826	0.530079	1.131146	∞	0.907109	∞	0.503407	∞
11.0	5.2408	144.45	0.5957268	0.1275323	0.3622773	0.5326073	1.1287674	11.9886	0.9119462	5.3732	0.5108583	6.6793
12.0	4.8097	139.12	0.5971502	0.1268926	0.3614039	0.5330845	1.1283149	10.9867	0.9128615	4.9307	0.5122922	6.1257
13.0	4.4454	134.22	0.5987046	0.1261986	0.3604513	0.5336019	1.1278229	10.1387	0.9138548	4.5568	0.5138574	5.6575
14.0	4.1336	129.67	0.6003920	0.1254509	0.3594186	0.5341590	1.1272914	9.4116	0.9149260	4.2367	0.5155554	5.2564
15.0	3.8637	125.43	0.6022144	0.1246496	0.3583052	0.5347556	1.1267204	8.7812	0.9160742	3.9597	0.5173881	4.9090
16.0	3.6280	121.46	0.6041742	0.1237954	0.3571099	0.5353910	1.1261098	8.2295	0.9172992	3.7177	0.5193574	4.6052
17.0	3.4203	117.72	0.6062737	0.1228885	0.3558318	0.5360650	1.1254596	7.7424	0.9186004	3.5044	0.5214655	4.3374
18.0	3.2361	114.19	0.6085157	0.1219295	0.3544700	0.5367767	1.1247699	7.3093	0.9199771	3.3152	0.5237147	4.0995
19.0	3.0716	110.84	0.6109030	0.1209189	0.3530230	0.5375256	1.1240405	6.9216	0.9214288	3.1463	0.5261074	3.8868
20.0	2.9238	107.66	0.6134387	0.1198572	0.3514899	0.5383110	1.1232715	6.5725	0.9229547	2.9945	0.5286464	3.6955
21.0	2.7904	104.63	0.6161263	0.1187450	0.3498694	0.5391321	1.1224628	6.2564	0.9245542	2.8574	0.5313343	3.5226
22.0	2.6695	101.73	0.6189693	0.1175829	0.3481602	0.5399881	1.1216145	5.9690	0.9262264	2.7331	0.5341743	3.3656
23.0	2.5593	98.95	0.6219716	0.1163715	0.3463608	0.5408780	1.1207264	5.7063	0.9279704	2.6199	0.5371696	3.2224
24.0	2.4586	96.29	0.6251373	0.1151114	0.3444697	0.5418007	1.1197987	5.4654	0.9297852	2.5164	0.5403235	3.0912
25.0	2.3662	93.73	0.6284709	0.1138034	0.3424855	0.5427552	1.1188313	5.2436	0.9316699	2.4214	0.5436398	2.9707
26.0	2.2812	91.26	0.6319772	0.1124482	0.3404068	0.5437399	1.1178241	5.0388	0.9336234	2.3339	0.5471223	2.8596
27.0	2.2027	88.88	0.6356614	0.1110466	0.3382317	0.5447537	1.1167774	4.8489	0.9356444	2.2532	0.5507750	2.7568
28.0	2.1301	86.57	0.6395289	0.1095992	0.3359586	0.5457949	1.1156909	4.6725	0.9377319	2.1784	0.5546024	2.6615
29.0	2.0627	84.34	0.6435856	0.1081079	0.3335858	0.5468618	1.1145648	4.5080	0.9398843	2.1090	0.5586089	2.5729
30.0	2.0000	82.18	0.6478378	0.1065707	0.3311115	0.5479525	1.1133990	4.3544	0.9421001	2.0445	0.5627995	2.4903
31.0	1.9416	80.09	0.6522923	0.1049913	0.3285338	0.5490650	1.1121938	4.2106	0.9443779	1.9844	0.5671792	2.4132
32.0	1.8871	78.05	0.6569564	0.1033697	0.3258508	0.5501968	1.1109490	4.0756	0.9467161	1.9282	0.5717536	2.3410
33.0	1.8361	76.07	0.6618379	0.1017068	0.3230605	0.5513454	1.1096648	3.9486	0.9491127	1.8756	0.5765282	2.2733
34.0	1.7883	74.14	0.6669450	0.1000036	0.3201610	0.5525081	1.1083413	3.8290	0.9515661	1.8264	0.5815092	2.2096
35.0	1.7434	72.25	0.6722868	0.0982611	0.3171500	0.5536818	1.1069786	3.7161	0.9540738	1.7801	0.5867030	2.1497
36.0	1.7013	70.42	0.6778729	0.0964805	0.3140258	0.5548631	1.1055767	3.6093	0.9566340	1.7366	0.5921161	2.0933
37.0	1.6616	68.62	0.6837138	0.0946628	0.3107861	0.5560481	1.1041358	3.5081	0.9592443	1.6957	0.5977558	2.0400
38.0	1.6243	66.87	0.6898205	0.0928091	0.3074285	0.5572328	1.1026560	3.4121	0.9619022	1.6571	0.6036294	1.9896
39.0	1.5890	65.15	0.6962052	0.0909206	0.3039512	0.5584126	1.1011377	3.3209	0.9646049	1.6206	0.6097449	1.9418
40.0	1.5557	63.47	0.7028809	0.0889985	0.3003516	0.5595824	1.0995809	3.2341	0.9673499	1.5862	0.6161105	1.8966
41.0	1.5243	61.82	0.7098616	0.0870442	0.2966279	0.5607366	1.0979857	3.1513	0.9701339	1.5537	0.6227349	1.8537
42.0	1.4945	60.20	0.7171626	0.0850587	0.2927775	0.5618692	1.0963527	3.0724	0.9729538	1.5229	0.6296273	1.8129
43.0	1.4663	58.61	0.7248003	0.0830436	0.2887983	0.5629733	1.0946818	2.9970	0.9758063	1.4937	0.6367974	1.7740
44.0	1.4396	57.05	0.7327926	0.0810002	0.2846882	0.5640413	1.0929736	2.9248	0.9786875	1.4660	0.6442552	1.7371
45.0	1.4142	55.51	0.7411591	0.0789300	0.2804447	0.5650650	1.0912283	2.8557	0.9815937	1.4397	0.6520113	1.7019
46.0	1.3902	54.00	0.7499208	0.0768345	0.2760659	0.5660351	1.0894461	2.7895	0.9845210	1.4147	0.6600770	1.6683
47.0	1.3673	52.51	0.7591010	0.0747150	0.2715495	0.5669416	1.0876278	2.7259	0.9874646	1.3910	0.6684640	1.6362
48.0	1.3456	51.05	0.7687250	0.0725733	0.2668934	0.5677735	1.0857735	2.6648	0.9904202	1.3685	0.6771844	1.6056
49.0	1.3250	49.60	0.7788203	0.0704110	0.2620957	0.5685167	1.0838839	2.6060	0.9933827	1.3470	0.6862512	1.5763
50.0	1.3054	48.17	0.7894175	0.0682297	0.2571543	0.5691591	1.0819595	2.5494	0.9963469	1.3266	0.6956775	1.5482
51.0	1.2868	46.76	0.8005501	0.0660313	0.2520674	0.5696846	1.0800007	2.4948	0.9993073	1.3072	0.7054776	1.5213
52.0	1.2690	45.37	0.8122549	0.0638177	0.2468333	0.5700760	1.0780084	2.4422	1.0022579	1.2887	0.7156657	1.4956
53.0	1.2521	43.99	0.8245729	0.0615906	0.2414505	0.5703144	1.0759832	2.3913	1.0051928	1.2711	0.7262569	1.4709
54.0	1.2361	42.62	0.8375495	0.0593520	0.2359174	0.5703768	1.0739258	2.3422	1.0081047	1.2543	0.7372669	1.4473
55.0	1.2208	41.27	0.8512352	0.0571039	0.2302330	0.5702448	1.0718369	2.2946	1.0109871	1.2383	0.7487116	1.4245
56.0	1.2062	39.93	0.8656863	0.0548485	0.2243964	0.5698870	1.0697178	2.2485	1.0138325	1.2230	0.7606074	1.4027
57.0	1.1924	38.60	0.8809661	0.0525879	0.2184064	0.5692756	1.0675690	2.2039	1.0166328	1.2085	0.7729711	1.3817
58.0	1.1792	37.28	0.8971456	0.0503245	0.2122631	0.5683779	1.0653919	2.1606	1.0193798	1.1946	0.7858193	1.3616
59.0	1.1666	35.97	0.9143049	0.0480607	0.2059661	0.5671575	1.0631874	2.1185	1.0220644	1.1814	0.7991692	1.3422
60.0	1.1547	34.67	0.9325351	0.0457990	0.1995158	0.5655734	1.0609569	2.0776	1.0246774	1.1689	0.8130371	1.3235
θ	Ω_S	A_{MIN}	σ_0	σ_1	σ_3	σ_5	Ω_1	Ω_2	Ω_3	Ω_4	Ω_5	Ω_6

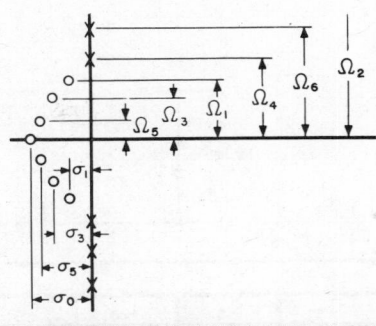

				$K^2 = 1.0$					
C_1	C_2	L_2	C_3	C_4	L_4	C_5	C_6	L_6	C_7
0.7564	0.00000	1.373	1.710	0.00000	1.626	1.710	0.0000	1.373	0.7564
0.75234	0.00509	1.36681	1.68833	0.02176	1.59151	1.68058	0.01662	1.34859	0.74095
0.75156	0.00607	1.36566	1.68431	0.02595	1.58500	1.67509	0.01983	1.34398	0.73800
0.75070	0.00713	1.36440	1.67993	0.03052	1.57794	1.66914	0.02333	1.33897	0.73479
0.74978	0.00828	1.36304	1.67521	0.03548	1.57033	1.66272	0.02714	1.33355	0.73131
0.74878	0.00952	1.36157	1.67014	0.04083	1.56216	1.65583	0.03125	1.32773	0.72757
0.74772	0.01086	1.36000	1.66473	0.04658	1.55344	1.64849	0.03568	1.32151	0.72356
0.74658	0.01228	1.35833	1.65897	0.05273	1.54417	1.64068	0.04043	1.31489	0.71929
0.74536	0.01380	1.35654	1.65287	0.05930	1.53435	1.63242	0.04550	1.30786	0.71475
0.74408	0.01541	1.35465	1.64642	0.06629	1.52399	1.62370	0.05090	1.30043	0.70994
0.74272	0.01711	1.35264	1.63963	0.07370	1.51308	1.61454	0.05665	1.29259	0.70485
0.74128	0.01892	1.35053	1.63250	0.08156	1.50164	1.60492	0.06275	1.28436	0.69950
0.73976	0.02082	1.34830	1.62503	0.08986	1.48966	1.59487	0.06920	1.27571	0.69387
0.73817	0.02282	1.34596	1.61722	0.09863	1.47715	1.58437	0.07603	1.26666	0.68796
0.73650	0.02492	1.34351	1.60906	0.10786	1.46411	1.57344	0.08324	1.25721	0.68177
0.73475	0.02712	1.34094	1.60057	0.11758	1.45054	1.56208	0.09084	1.24735	0.67529
0.73292	0.02943	1.33825	1.59174	0.12780	1.43645	1.55030	0.09885	1.23708	0.66853
0.73101	0.03185	1.33544	1.58258	0.13854	1.42185	1.53809	0.10729	1.22641	0.66148
0.72901	0.03437	1.33250	1.57308	0.14980	1.40672	1.52546	0.11616	1.21533	0.65414
0.72693	0.03701	1.32944	1.56324	0.16161	1.39109	1.51243	0.12548	1.20384	0.64650
0.72476	0.03977	1.32626	1.55307	0.17399	1.37494	1.49899	0.13528	1.19194	0.63856
0.72250	0.04264	1.32294	1.54256	0.18697	1.35830	1.48516	0.14557	1.17964	0.63031
0.72015	0.04563	1.31950	1.53172	0.20055	1.34115	1.47093	0.15637	1.16692	0.62176
0.71770	0.04874	1.31591	1.52055	0.21477	1.32351	1.45631	0.16772	1.15380	0.61289
0.71517	0.05198	1.31219	1.50905	0.22967	1.30538	1.44132	0.17962	1.14026	0.60371
0.71253	0.05535	1.30833	1.49722	0.24525	1.28676	1.42596	0.19212	1.12632	0.59420
0.70980	0.05885	1.30433	1.48506	0.26157	1.26766	1.41024	0.20524	1.11196	0.58436
0.70696	0.06250	1.30018	1.47258	0.27866	1.24808	1.39417	0.21902	1.09719	0.57418
0.70402	0.06628	1.29587	1.45976	0.29656	1.22803	1.37775	0.23349	1.08201	0.56366
0.70097	0.07022	1.29141	1.44662	0.31531	1.20751	1.36099	0.24869	1.06641	0.55279
0.69781	0.07430	1.28679	1.43315	0.33496	1.18653	1.34391	0.26467	1.05040	0.54156
0.69453	0.07855	1.28200	1.41935	0.35557	1.16508	1.32652	0.28147	1.03398	0.52996
0.69114	0.08295	1.27704	1.40523	0.37720	1.14319	1.30882	0.29915	1.01714	0.51798
0.68762	0.08753	1.27191	1.39078	0.39991	1.12084	1.29084	0.31777	0.99988	0.50562
0.68398	0.09229	1.26659	1.37601	0.42377	1.09805	1.27258	0.33740	0.98221	0.49286
0.68020	0.09723	1.26109	1.36091	0.44888	1.07482	1.25405	0.35810	0.96413	0.47969
0.67629	0.10237	1.25539	1.34549	0.47532	1.05116	1.23528	0.37996	0.94562	0.46610
0.67224	0.10771	1.24949	1.32975	0.50320	1.02706	1.21627	0.40307	0.92670	0.45207
0.66804	0.11326	1.24337	1.31368	0.53263	1.00254	1.19705	0.42753	0.90737	0.43759
0.66369	0.11903	1.23704	1.29729	0.56375	0.97760	1.17763	0.45345	0.88761	0.42263
0.65918	0.12504	1.23047	1.28057	0.59671	0.95224	1.15803	0.48095	0.86744	0.40719
0.65450	0.13130	1.22366	1.26353	0.63167	0.92647	1.13827	0.51020	0.84685	0.39125
0.64965	0.13782	1.21660	1.24617	0.66884	0.90029	1.11837	0.54133	0.82584	0.37477
0.64461	0.14461	1.20927	1.22849	0.70843	0.87372	1.09837	0.57455	0.80442	0.35773
0.63938	0.15170	1.20166	1.21049	0.75070	0.84674	1.07827	0.61007	0.78258	0.34012
0.63395	0.15910	1.19375	1.19217	0.79595	0.81938	1.05812	0.64812	0.76033	0.32189
0.62831	0.16683	1.18553	1.17352	0.84452	0.79163	1.03795	0.68898	0.73766	0.30301
0.62243	0.17493	1.17697	1.15456	0.89683	0.76350	1.01778	0.73299	0.71458	0.28345
0.61632	0.18340	1.16804	1.13528	0.95334	0.73499	0.99765	0.78051	0.69109	0.26316
0.60995	0.19230	1.15873	1.11568	1.01461	0.70612	0.97762	0.83199	0.66720	0.24210
0.60332	0.20164	1.14900	1.09578	1.08132	0.67689	0.95771	0.88794	0.64290	0.22021
L_1	L_2	C_2	L_3	L_4	C_4	L_5	L_6	C_6	L_7

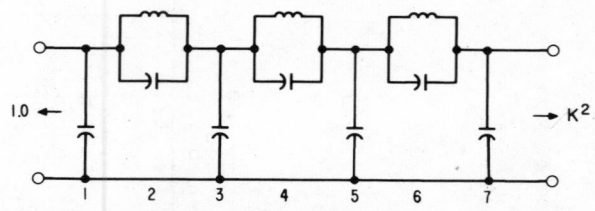

θ	C_1	C_2	L_2	C_3	C_4	L_4	C_5	C_6	L_6	C_7
				$K^2 = \infty$						
C	0.378216	0.0000	0.975415	1.327583	0.00000	1.510781	1.602333	0.0000	1.608317	1.343601
11.0	0.3715	0.0072	0.9663	1.3045	0.0235	1.4720	1.5742	0.0141	1.5873	1.3344
12.0	0.3703	0.0086	0.9645	1.3001	0.0281	1.4647	1.5688	0.0168	1.5833	1.3327
13.0	0.3689	0.0101	0.9626	1.2953	0.0331	1.4567	1.5630	0.0198	1.5790	1.3308
14.0	0.3674	0.0118	0.9606	1.2901	0.0385	1.4481	1.5568	0.0230	1.5743	1.3287
15.0	0.3657	0.0135	0.9583	1.2846	0.0443	1.4389	1.5501	0.0264	1.5693	1.3265
16.0	0.3640	0.0154	0.9560	1.2787	0.0506	1.4290	1.5429	0.0302	1.5639	1.3241
17.0	0.3621	0.0175	0.9534	1.2724	0.0574	1.4185	1.5353	0.0341	1.5582	1.3216
18.0	0.3601	0.0197	0.9507	1.2657	0.0646	1.4074	1.5273	0.0383	1.5522	1.3190
19.0	0.3580	0.0220	0.9479	1.2586	0.0724	1.3956	1.5188	0.0428	1.5458	1.3162
20.0	0.3558	0.0245	0.9449	1.2512	0.0806	1.3832	1.5099	0.0476	1.5390	1.3132
21.0	0.3534	0.0271	0.9417	1.2433	0.0894	1.3703	1.5005	0.0526	1.5320	1.3101
22.0	0.3509	0.0299	0.9383	1.2351	0.0987	1.3566	1.4907	0.0579	1.5246	1.3068
23.0	0.3483	0.0329	0.9348	1.2265	0.1085	1.3424	1.4805	0.0635	1.5168	1.3034
24.0	0.3455	0.0360	0.9311	1.2176	0.1190	1.3276	1.4699	0.0694	1.5087	1.2998
25.0	0.3426	0.0392	0.9272	1.2082	0.1300	1.3122	1.4588	0.0755	1.5003	1.2961
26.0	0.3396	0.0427	0.9232	1.1985	0.1416	1.2961	1.4473	0.0820	1.4916	1.2923
27.0	0.3364	0.0463	0.9190	1.1883	0.1540	1.2794	1.4354	0.0888	1.4825	1.2882
28.0	0.3331	0.0501	0.9145	1.1778	0.1670	1.2622	1.4231	0.0958	1.4731	1.2841
29.0	0.3296	0.0541	0.9100	1.1669	0.1807	1.2443	1.4104	0.1032	1.4633	1.2797
30.0	0.3260	0.0583	0.9052	1.1557	0.1952	1.2259	1.3973	0.1110	1.4532	1.2753
31.0	0.3222	0.0627	0.9002	1.1440	0.2104	1.2068	1.3839	0.1190	1.4428	1.2706
32.0	0.3183	0.0673	0.8951	1.1319	0.2266	1.1872	1.3700	0.1274	1.4321	1.2658
33.0	0.3142	0.0721	0.8897	1.1195	0.2436	1.1670	1.3558	0.1362	1.4210	1.2609
34.0	0.3099	0.0771	0.8842	1.1067	0.2616	1.1462	1.3412	0.1453	1.4097	1.2558
35.0	0.3055	0.0824	0.8784	1.0934	0.2806	1.1248	1.3262	0.1548	1.3979	1.2506
36.0	0.3008	0.0880	0.8725	1.0798	0.3007	1.1028	1.3109	0.1647	1.3859	1.2452
37.0	0.2960	0.0938	0.8663	1.0658	0.3219	1.0803	1.2952	0.1749	1.3736	1.2396
38.0	0.2910	0.0999	0.8599	1.0513	0.3445	1.0572	1.2793	0.1856	1.3609	1.2339
39.0	0.2858	0.1063	0.8533	1.0365	0.3684	1.0336	1.2630	0.1968	1.3479	1.2281
40.0	0.2804	0.1130	0.8465	1.0212	0.3938	1.0094	1.2464	0.2083	1.3346	1.2221
41.0	0.2784	0.1200	0.8394	1.0056	0.4208	0.9846	1.2295	0.2203	1.3209	1.2159
42.0	0.2689	0.1273	0.8321	0.9895	0.4495	0.9593	1.2123	0.2328	1.3070	1.2096
43.0	0.2629	0.1350	0.8245	0.9729	0.4802	0.9334	1.1949	0.2458	1.2927	1.2031
44.0	0.2565	0.1431	0.8167	0.9560	0.5130	0.9070	1.1772	0.2593	1.2781	1.1965
45.0	0.2500	0.1516	0.8086	0.9385	0.5482	0.8801	1.1593	0.2733	1.2632	1.1897
46.0	0.2431	0.1606	0.8003	0.9207	0.5859	0.8527	1.1411	0.2879	1.2479	1.1827
47.0	0.2360	0.1700	0.7917	0.9023	0.6266	0.8248	1.1229	0.3031	1.2324	1.1756
48.0	0.2286	0.1799	0.7828	0.8835	0.6706	0.7963	1.1044	0.3189	1.2165	1.1683
49.0	0.2209	0.1903	0.7736	0.8641	0.7182	0.7674	1.0858	0.3353	1.2002	1.1609
50.0	0.2129	0.2014	0.7641	0.8443	0.7700	0.7379	1.0671	0.3525	1.1837	1.1533
51.0	0.2045	0.2130	0.7543	0.8238	0.8265	0.7081	1.0484	0.3703	1.1668	1.1455
52.0	0.1958	0.2253	0.7441	0.8029	0.8885	0.6777	1.0297	0.3889	1.1495	1.1376
53.0	0.1867	0.2384	0.7336	0.7813	0.9567	0.6470	1.0109	0.4084	1.1318	1.1294
54.0	0.1771	0.2522	0.7227	0.7590	1.0322	0.6158	0.9923	0.4287	1.1138	1.1211
55.0	0.1672	0.2669	0.7115	0.7361	1.1163	0.5842	0.9738	0.4499	1.0954	1.1126
56.0	0.1568	0.2826	0.6999	0.7125	1.2104	0.5523	0.9556	0.4721	1.0765	1.1040
57.0	0.1458	0.2993	0.6878	0.6881	1.3165	0.5201	0.9376	0.4955	1.0572	1.0951
58.0	0.1344	0.3172	0.6753	0.6628	1.4370	0.4876	0.9200	0.5200	1.0373	1.0861
59.0	0.1223	0.3364	0.6624	0.6365	1.5750	0.4549	0.9030	0.5458	1.0170	1.0769
60.0	0.1097	0.3570	0.6490	0.6091	1.7344	0.4220	0.8866	0.5731	0.9961	1.0675
θ	L_1	L_2	C_2	L_3	L_4	C_4	L_5	L_6	C_6	L_7

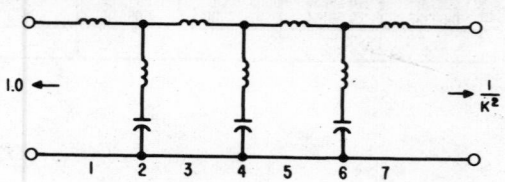

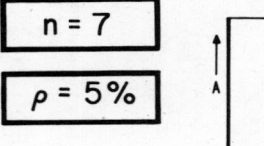

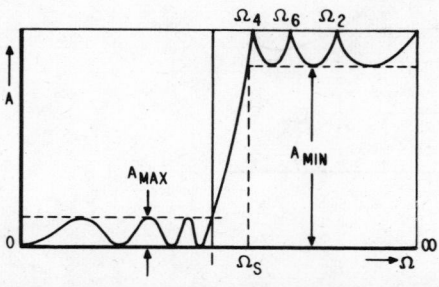

θ	Ω_S	A_{MIN}	σ_0	σ_1	σ_3	σ_5	Ω_1	Ω_2	Ω_3	Ω_4	Ω_5	Ω_6
C	∞	∞	0.551613	0.122745	0.343925	0.496986	1.113416	∞	0.892890	∞	0.495517	∞
11.0	5.2408	146.39	0.5583182	0.1197008	0.3398545	0.4993119	1.1114185	11.9886	0.8974649	5.3732	0.5024140	6.6793
12.0	4.8097	141.07	0.5596101	0.1191252	0.3390728	0.4997512	1.1110383	10.9867	0.8983315	4.9307	0.5037408	6.1257
13.0	4.4454	136.16	0.5610207	0.1185006	0.3382203	0.5002277	1.1106250	10.1387	0.8992722	4.5568	0.5051888	5.6575
14.0	4.1336	131.61	0.5625517	0.1178274	0.3372963	0.5007410	1.1101785	9.4116	0.9002868	4.2367	0.5067594	5.2564
15.0	3.8637	127.37	0.5642050	0.1171059	0.3362999	0.5012908	1.1096986	8.7812	0.9013750	3.9597	0.5084544	4.9090
16.0	3.6280	123.40	0.5659824	0.1163363	0.3352304	0.5018767	1.1091855	8.2295	0.9025364	3.7177	0.5102753	4.6052
17.0	3.4203	119.66	0.5678862	0.1155192	0.3340867	0.5024984	1.1086390	7.7424	0.9037707	3.5044	0.5122242	4.3374
18.0	3.2361	116.13	0.5699187	0.1146548	0.3328681	0.5031554	1.1080591	7.3093	0.9050774	3.3152	0.5143031	4.0995
19.0	3.0716	112.78	0.5720824	0.1137435	0.3315734	0.5038472	1.1074458	6.9216	0.9064557	3.1463	0.5165142	3.8868
20.0	2.9238	109.60	0.5743801	0.1127857	0.3302017	0.5045731	1.1067991	6.5725	0.9079056	2.9945	0.5188598	3.6955
21.0	2.7904	106.57	0.5768146	0.1117820	0.3287517	0.5053326	1.1061187	6.2564	0.9094263	2.8574	0.5213424	3.5226
22.0	2.6695	103.67	0.5793891	0.1107328	0.3272223	0.5061250	1.1054049	5.9690	0.9110171	2.7331	0.5239649	3.3656
23.0	2.5593	100.90	0.5821070	0.1096386	0.3256122	0.5069496	1.1046575	5.7063	0.9126775	2.6199	0.5267299	3.2224
24.0	2.4586	98.23	0.5849719	0.1084999	0.3239202	0.5078054	1.1038765	5.4654	0.9144066	2.5164	0.5296406	3.0912
25.0	2.3662	95.67	0.5879877	0.1073173	0.3221449	0.5086915	1.1030618	5.2436	0.9162037	2.4214	0.5327002	2.9707
26.0	2.2812	93.20	0.5911586	0.1060914	0.3202848	0.5096070	1.1022135	5.0388	0.9180680	2.3339	0.5359122	2.8596
27.0	2.2027	90.82	0.5944888	0.1048226	0.3183375	0.5105507	1.1013316	4.8489	0.9199986	2.2532	0.5392801	2.7568
28.0	2.1301	88.52	0.5979835	0.1035119	0.3163043	0.5115212	1.1004159	4.6725	0.9219944	2.1784	0.5428080	2.6615
29.0	2.0627	86.29	0.6016475	0.1021596	0.3141809	0.5125173	1.0994665	4.5080	0.9240545	2.1090	0.5464997	2.5729
30.0	2.0000	84.13	0.6054863	0.1007666	0.3119664	0.5135374	1.0984834	4.3544	0.9261778	2.0445	0.5503597	2.4903
31.0	1.9416	82.03	0.6095058	0.0993336	0.3096592	0.5145798	1.0974666	4.2106	0.9283629	1.9844	0.5543925	2.4132
32.0	1.8871	79.99	0.6137123	0.0978612	0.3072576	0.5156427	1.0964161	4.0756	0.9306087	1.9282	0.5586030	2.3410
33.0	1.8361	78.01	0.6181124	0.0963504	0.3047596	0.5167239	1.0953319	3.9486	0.9329137	1.8756	0.5629962	2.2733
34.0	1.7883	76.08	0.6227134	0.0948019	0.3021635	0.5178213	1.0942140	3.8290	0.9352764	1.8264	0.5675776	2.2096
35.0	1.7434	74.20	0.6275229	0.0932164	0.2994673	0.5189324	1.0930626	3.7161	0.9376950	1.7801	0.5723529	2.1497
36.0	1.7013	72.36	0.6325494	0.0915950	0.2966691	0.5200544	1.0918775	3.6093	0.9401683	1.7366	0.5773279	2.0933
37.0	1.6616	70.56	0.6378016	0.0899385	0.2937669	0.5211843	1.0906591	3.5081	0.9426939	1.6957	0.5825091	2.0400
38.0	1.6243	68.81	0.6432891	0.0882478	0.2907586	0.5223187	1.0894071	3.4121	0.9452702	1.6571	0.5879032	1.9896
39.0	1.5890	67.09	0.6490223	0.0865239	0.2876422	0.5234539	1.0881219	3.3209	0.9478948	1.6206	0.5935171	1.9418
40.0	1.5557	65.41	0.6550123	0.0847678	0.2844155	0.5245860	1.0868034	3.2341	0.9505657	1.5862	0.5993584	1.8966
41.0	1.5243	63.76	0.6612710	0.0829806	0.2810764	0.5257103	1.0854518	3.1513	0.9532804	1.5537	0.6054366	1.8537
42.0	1.4945	62.14	0.6678114	0.0811633	0.2776228	0.5268219	1.0840673	3.0724	0.9560363	1.5229	0.6117543	1.8129
43.0	1.4663	60.55	0.6746464	0.0793171	0.2740523	0.5279154	1.0826500	2.9970	0.9588306	1.4937	0.6183260	1.7740
44.0	1.4396	58.99	0.6817943	0.0774430	0.2703626	0.5289847	1.0812001	2.9248	0.9616604	1.4660	0.6251588	1.7371
45.0	1.4142	57.46	0.6892685	0.0755424	0.2665520	0.5300229	1.0797180	2.8557	0.9645225	1.4397	0.6322624	1.7019
46.0	1.3902	55.94	0.6970879	0.0736164	0.2626178	0.5310228	1.0782036	2.7895	0.9674136	1.4147	0.6396469	1.6683
47.0	1.3673	54.46	0.7052719	0.0716664	0.2585580	0.5319761	1.0766574	2.7259	0.9703302	1.3910	0.6473228	1.6362
48.0	1.3456	52.99	0.7138417	0.0696937	0.2543703	0.5328737	1.0750797	2.6648	0.9732683	1.3685	0.6553014	1.6056
49.0	1.3250	51.54	0.7228207	0.0676996	0.2500526	0.5337054	1.0734708	2.6060	0.9762240	1.3470	0.6635945	1.5763
50.0	1.3054	50.11	0.7322341	0.0656858	0.2456027	0.5344600	1.0718310	2.5494	0.9791929	1.3266	0.6722142	1.5482
51.0	1.2868	48.70	0.7421100	0.0636534	0.2410187	0.5351250	1.0701609	2.4948	0.9821703	1.3072	0.6811734	1.5213
52.0	1.2690	47.31	0.7524788	0.0616043	0.2362985	0.5356867	1.0684609	2.4422	0.9851514	1.2887	0.6904757	1.4956
53.0	1.2521	45.93	0.7633747	0.0595400	0.2314399	0.5361296	1.0667313	2.3913	0.9881308	1.2711	0.7001652	1.4709
54.0	1.2361	44.56	0.7748349	0.0574622	0.2264416	0.5364365	1.0649730	2.3422	0.9911029	1.2543	0.7102265	1.4473
55.0	1.2208	43.21	0.7869012	0.0553726	0.2213016	0.5365883	1.0631861	2.2946	0.9940621	1.2383	0.7206848	1.4245
56.0	1.2062	41.87	0.7996197	0.0532732	0.2160184	0.5365636	1.0613718	2.2485	0.9970016	1.2230	0.7315561	1.4027
57.0	1.1924	40.54	0.8130421	0.0511656	0.2105907	0.5363386	1.0595306	2.2039	0.9999184	1.2085	0.7428566	1.3817
58.0	1.1792	39.22	0.8272264	0.0490521	0.2050172	0.5358868	1.0576633	2.1606	1.0027947	1.1946	0.7546033	1.3616
59.0	1.1666	37.91	0.8422375	0.0469347	0.1992972	0.5351783	1.0557707	2.1185	1.0056335	1.1814	0.7668131	1.3422
60.0	1.1547	36.61	0.8581490	0.0448156	0.1934299	0.5341796	1.0538538	2.0776	1.0084229	1.1689	0.7795036	1.3235
θ	Ω_S	A_{MIN}	σ_0	σ_1	σ_3	σ_5	Ω_1	Ω_2	Ω_3	Ω_4	Ω_5	Ω_6

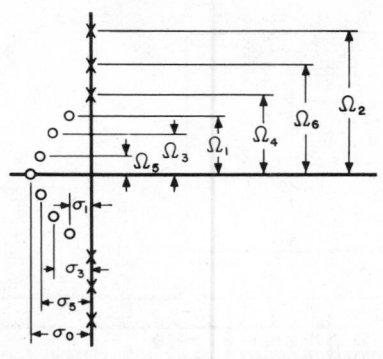

					K² = 1.0				
θ	C₁	C₂	L₂	C₃	C₄	L₄	C₅	C₆	L₆	C₇
C	0.8068	0.00000	1.397	1.757	0.00000	1.634	1.757	0.00000	1.397	0.8068
11.0	0.80275	0.00500	1.39056	1.73592	0.02164	1.60048	1.72807	0.01633	1.37283	0.79156
12.0	0.80197	0.00596	1.38941	1.73185	0.02570	1.59407	1.72252	0.01948	1.36832	0.78865
13.0	0.80113	0.00701	1.38816	1.72743	0.03034	1.58710	1.71649	0.02292	1.36342	0.78549
14.0	0.80021	0.00814	1.38680	1.72266	0.03527	1.57958	1.71000	0.02665	1.35812	0.78207
15.0	0.79923	0.00936	1.38535	1.71754	0.04058	1.57152	1.70303	0.03068	1.35243	0.77839
16.0	0.79817	0.01067	1.38378	1.71206	0.04629	1.56292	1.69559	0.03502	1.34634	0.77445
17.0	0.79704	0.01207	1.38211	1.70624	0.05241	1.55377	1.68769	0.03967	1.33986	0.77024
18.0	0.79584	0.01356	1.38034	1.70007	0.05893	1.54408	1.67932	0.04464	1.33298	0.76578
19.0	0.79457	0.01514	1.37845	1.69355	0.06586	1.53385	1.67050	0.04993	1.32571	0.76105
20.0	0.79322	0.01682	1.37646	1.68669	0.07322	1.52309	1.66121	0.05555	1.31805	0.75605
21.0	0.79180	0.01859	1.37436	1.67948	0.08101	1.51180	1.65147	0.06152	1.30998	0.75079
22.0	0.79030	0.02046	1.37214	1.67193	0.08925	1.49998	1.64128	0.06783	1.30153	0.74526
23.0	0.78873	0.02242	1.36982	1.66403	0.09793	1.48763	1.63064	0.07450	1.29267	0.73945
24.0	0.78707	0.02448	1.36738	1.65579	0.10709	1.47476	1.61956	0.08154	1.28341	0.73337
25.0	0.78534	0.02665	1.36482	1.64721	0.11671	1.46137	1.60804	0.08896	1.27377	0.72701
26.0	0.78353	0.02892	1.36214	1.63828	0.12683	1.44747	1.59608	0.09677	1.26373	0.72038
27.0	0.78164	0.03129	1.35935	1.62902	0.13745	1.43305	1.58370	0.10499	1.25328	0.71346
28.0	0.77966	0.03377	1.35643	1.61941	0.14860	1.41813	1.57088	0.11362	1.24244	0.70626
29.0	0.77760	0.03636	1.35339	1.60947	0.16028	1.40270	1.55765	0.12269	1.23119	0.69877
30.0	0.77545	0.03906	1.35022	1.59919	0.17251	1.38677	1.54400	0.13222	1.21955	0.69099
31.0	0.77322	0.04188	1.34693	1.58857	0.18532	1.37034	1.52994	0.14221	1.20750	0.68291
32.0	0.77089	0.04471	1.34350	1.57762	0.19873	1.35342	1.51547	0.15269	1.19506	0.67453
33.0	0.76847	0.04786	1.33994	1.56633	0.21276	1.33601	1.50061	0.16368	1.18221	0.66585
34.0	0.76597	0.05104	1.33624	1.55471	0.22744	1.31812	1.48536	0.17521	1.16896	0.65686
35.0	0.76336	0.05435	1.33241	1.54275	0.24280	1.29975	1.46972	0.18730	1.15531	0.64756
36.0	0.76066	0.05779	1.32843	1.53046	0.25887	1.28091	1.45370	0.19997	1.14125	0.63794
37.0	0.75785	0.06136	1.32430	1.51784	0.27568	1.26159	1.43732	0.21326	1.12678	0.62799
38.0	0.75494	0.06507	1.32003	1.50489	0.29327	1.24181	1.42057	0.22721	1.11191	0.61772
39.0	0.75193	0.06892	1.31559	1.49161	0.31168	1.22157	1.40347	0.24183	1.09664	0.60711
40.0	0.74880	0.07293	1.31100	1.47800	0.33096	1.20087	1.38603	0.25719	1.08095	0.59616
41.0	0.74557	0.07709	1.30625	1.46406	0.35116	1.17972	1.36825	0.27331	1.06486	0.58475
42.0	0.74221	0.08141	1.30133	1.44979	0.37233	1.15813	1.35015	0.29024	1.04836	0.57319
43.0	0.73874	0.08579	1.29624	1.43520	0.39454	1.13609	1.33174	0.30805	1.03145	0.56115
44.0	0.73514	0.09055	1.29096	1.42028	0.41785	1.11362	1.31302	0.32678	1.01413	0.54874
45.0	0.73141	0.09539	1.28550	1.40502	0.44234	1.09071	1.29401	0.34651	0.99639	0.53595
46.0	0.72755	0.10041	1.27975	1.38945	0.46810	1.06738	1.27473	0.36729	0.97825	0.52275
47.0	0.72355	0.10564	1.27399	1.37354	0.49521	1.04362	1.25518	0.38922	0.95968	0.50914
48.0	0.71941	0.11107	1.26793	1.35731	0.52380	1.01945	1.23538	0.41237	0.94071	0.49510
49.0	0.71511	0.11671	1.26165	1.34075	0.55396	0.99487	1.21535	0.43685	0.92132	0.48062
50.0	0.71066	0.12259	1.25515	1.32387	0.58585	0.96988	1.19511	0.46277	0.90151	0.46569
51.0	0.70604	0.12870	1.24840	1.30666	0.61962	0.94449	1.17466	0.49026	0.88129	0.45029
52.0	0.70126	0.13506	1.24141	1.28912	0.65544	0.91770	1.15403	0.51944	0.86065	0.43439
53.0	0.69629	0.14169	1.23416	1.27126	0.69350	0.89252	1.13325	0.55047	0.83959	0.41797
54.0	0.69114	0.14861	1.22664	1.25307	0.73404	0.86596	1.11232	0.58357	0.81812	0.40102
55.0	0.68578	0.15583	1.21882	1.23455	0.77733	0.83901	1.09129	0.61890	0.79622	0.38351
56.0	0.68022	0.16337	1.21070	1.21571	0.82366	0.81168	1.07016	0.65672	0.77391	0.36540
57.0	0.67443	0.17125	1.20225	1.19654	0.87338	0.78399	1.04898	0.69729	0.75117	0.34667
58.0	0.66841	0.17950	1.19346	1.17704	0.92692	0.75593	1.02776	0.74092	0.72802	0.32728
59.0	0.66215	0.18815	1.18429	1.15723	0.98476	0.72752	1.00656	0.78800	0.70445	0.30719
60.0	0.65561	0.19722	1.17472	1.13708	1.04748	0.69875	0.98540	0.83893	0.68046	0.28635
θ	L₁	L₂	C₂	L₃	L₄	C₄	L₅	L₆	C₆	L₇

270

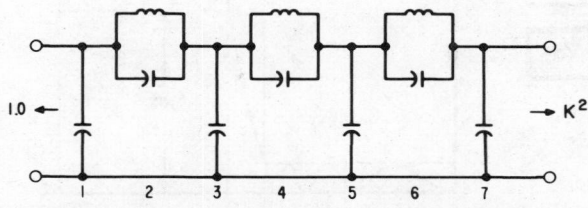

				K² = ∞					
C_1	C_2	L_2	C_3	C_4	L_4	C_5	C_6	L_6	C_7
0.403400	0.0000	1.022752	1.373772	0.0000	1.548193	1.635121	0.0000	1.635196	1.365613
0.3970	0.0069	1.0139	1.3510	0.0229	1.5099	1.6069	0.0139	1.6140	1.3562
0.3958	0.0082	1.0122	1.3466	0.0274	1.5027	1.6016	0.0166	1.6100	1.3544
0.3945	0.0096	1.0103	1.3419	0.0322	1.4948	1.5958	0.0195	1.6056	1.3524
0.3931	0.0112	1.0083	1.3368	0.0375	1.4863	1.5896	0.0226	1.6009	1.3503
0.3915	0.0129	1.0062	1.3314	0.0432	1.4772	1.5828	0.0260	1.5959	1.3480
0.3898	0.0147	1.0039	1.3255	0.0493	1.4674	1.5757	0.0296	1.5904	1.3456
0.3881	0.0167	1.0014	1.3193	0.0559	1.4571	1.5681	0.0335	1.5847	1.3430
0.3862	0.0187	0.9988	1.3127	0.0629	1.4461	1.5600	0.0377	1.5786	1.3403
0.3841	0.0210	0.9960	1.3057	0.0704	1.4345	1.5515	0.0421	1.5721	1.3374
0.3820	0.0233	0.9931	1.2984	0.0784	1.4223	1.5426	0.0468	1.5654	1.3344
0.3798	0.0258	0.9900	1.2907	0.0869	1.4095	1.5332	0.0517	1.5582	1.3312
0.3774	0.0284	0.9867	1.2826	0.0959	1.3961	1.5234	0.0569	1.5507	1.3278
0.3749	0.0312	0.9833	1.2741	0.1054	1.3820	1.5132	0.0624	1.5429	1.3243
0.3723	0.0342	0.9797	1.2653	0.1155	1.3674	1.5025	0.0682	1.5348	1.3206
0.3695	0.0373	0.9760	1.2561	0.1261	1.3522	1.4914	0.0742	1.5263	1.3168
0.3666	0.0405	0.9720	1.2465	0.1374	1.3363	1.4799	0.0806	1.5174	1.3128
0.3636	0.0439	0.9679	1.2365	0.1492	1.3199	1.4680	0.0872	1.5082	1.3086
0.3605	0.0475	0.9637	1.2262	0.1617	1.3029	1.4556	0.0942	1.4987	1.3043
0.3572	0.0513	0.9592	1.2155	0.1749	1.2853	1.4429	0.1015	1.4879	1.2999
0.3537	0.0552	0.9546	1.2044	0.1888	1.2671	1.4297	0.1090	1.4787	1.2953
0.3501	0.0594	0.9498	1.1929	0.2034	1.2484	1.4162	0.1170	1.4682	1.2905
0.3464	0.0637	0.9448	1.1811	0.2188	1.2290	1.4022	0.1252	1.4573	1.2856
0.3425	0.0683	0.9396	1.1688	0.2351	1.2091	1.3879	0.1338	1.4461	1.2805
0.3385	0.0730	0.9342	1.1562	0.2522	1.1886	1.3732	0.1428	1.4346	1.2752
0.3342	0.0780	0.9286	1.1432	0.2703	1.1675	1.3581	0.1521	1.4227	1.2698
0.3299	0.0832	0.9228	1.1298	0.2894	1.1459	1.3427	0.1618	1.4105	1.2642
0.3253	0.0886	0.9168	1.1161	0.3095	1.1237	1.3269	0.1719	1.3980	1.2585
0.3206	0.0943	0.9106	1.1019	0.3308	1.1010	1.3108	0.1824	1.3851	1.2526
0.3157	0.1003	0.9042	1.0874	0.3533	1.0777	1.2943	0.1933	1.3719	1.2466
0.3106	0.1065	0.8975	1.0724	0.3771	1.0539	1.2776	0.2047	1.3584	1.2404
0.3053	0.1131	0.8907	1.0571	0.4024	1.0295	1.2605	0.2164	1.3446	1.2340
0.2997	0.1199	0.8836	1.0413	0.4292	1.0046	1.2431	0.2287	1.3304	1.2275
0.2940	0.1271	0.8762	1.0252	0.4578	0.9792	1.2254	0.2415	1.3159	1.2208
0.2881	0.1346	0.8686	1.0086	0.4881	0.9532	1.2074	0.2547	1.3011	1.2139
0.2819	0.1425	0.8608	0.9916	0.5206	0.9268	1.1892	0.2685	1.2859	1.2069
0.2755	0.1507	0.8527	0.9742	0.5553	0.8998	1.1707	0.2828	1.2704	1.1997
0.2678	0.1594	0.8443	0.9563	0.5925	0.8723	1.1520	0.2977	1.2546	1.1923
0.2617	0.1685	0.8357	0.9380	0.6324	0.8443	1.1331	0.3132	1.2384	1.1848
0.2546	0.1781	0.8267	0.9192	0.6755	0.8159	1.1140	0.3294	1.2219	1.1770
0.2471	0.1882	0.8175	0.7999	0.7220	0.7869	1.0948	0.3462	1.2050	1.1691
0.2392	0.1989	0.8079	0.8802	0.7725	0.7576	1.0754	0.3637	1.1878	1.1611
0.2311	0.2101	0.7980	0.8599	0.8275	0.7277	1.0559	0.3820	1.1703	1.1528
0.2226	0.2220	0.7878	0.8391	0.8875	0.6974	1.0364	0.4011	1.1523	1.1444
0.2137	0.2345	0.7772	0.8178	0.9534	0.6667	1.0169	0.4210	1.1340	1.1358
0.2044	0.2479	0.7663	0.7959	1.0260	0.6357	0.9973	0.4418	1.1153	1.1270
0.1947	0.2620	0.7550	0.7733	1.1065	0.6042	0.9779	0.4636	1.0962	1.1180
0.1846	0.2770	0.7432	0.7501	1.1963	0.5724	0.9586	0.4865	1.0766	1.1088
0.1740	0.2930	0.7311	0.7261	1.2970	0.5403	0.9395	0.5105	1.0566	1.0994
0.1628	0.3101	0.7185	0.7014	1.4107	0.5079	0.9207	0.5358	1.0361	1.0898
0.1511	0.3285	0.7054	0.6758	1.5402	0.4752	0.9023	0.5624	1.0151	1.0800
L_1	L_2	C_2	L_3	L_4	C_4	L_5	L_6	C_6	L_7

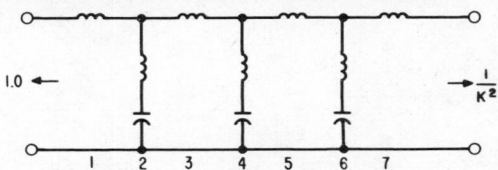

271

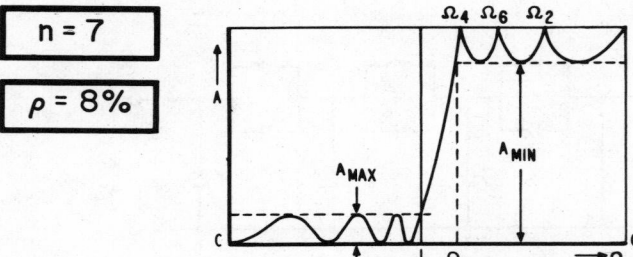

θ	Ω_S	A_{MIN}	σ_0	σ_1	σ_3	σ_5	Ω_1	Ω_2	Ω_3	Ω_4	Ω_5	Ω_6
C	∞	∞	0.475964	0.105912	0.296758	0.428828	1.079726	∞	0.865873	∞	0.480523	∞
11.0	5.2408	150.49	0.4814003	0.1034935	0.2935563	0.4307623	1.0784090	11.9886	0.8699723	5.3732	0.4864262	6.6793
12.0	4.8097	145.16	0.4824469	0.1030358	0.2929414	0.4311283	1.0781582	10.9867	0.8707501	4.9307	0.4875608	6.1257
13.0	4.4454	140.26	0.4835894	0.1025388	0.2922710	0.4315254	1.0778854	10.1387	0.8715951	4.5568	0.4887986	5.6575
14.0	4.1336	135.71	0.4848289	0.1020028	0.2915442	0.4319536	1.0775905	9.4116	0.8725071	4.2367	0.4901409	5.2564
15.0	3.8637	131.47	0.4861670	0.1014280	0.2907607	0.4324125	1.0772736	8.7812	0.8734860	3.9597	0.4915888	4.9090
16.0	3.6280	127.50	0.4876051	0.1008148	0.2899196	0.4329021	1.0769345	8.2295	0.8745315	3.7177	0.4931438	4.6052
17.0	3.4203	123.76	0.4891447	0.1001631	0.2890203	0.4334220	1.0765733	7.7424	0.8756436	3.5044	0.4948075	4.3374
18.0	3.2361	120.23	0.4907877	0.0994732	0.2880621	0.4339719	1.0761899	7.3093	0.8768219	3.3152	0.4965814	4.0995
19.0	3.0716	116.88	0.4925360	0.0987455	0.2870441	0.4345516	1.0757842	6.9216	0.8780663	3.1463	0.4984672	3.8868
20.0	2.9238	113.70	0.4943916	0.0979802	0.2859654	0.4351607	1.0753561	6.5725	0.8793765	2.9945	0.5004668	3.6955
21.0	2.7904	110.67	0.4963567	0.0971773	0.2848254	0.4357988	1.0749057	6.2564	0.8807522	2.8574	0.5025824	3.5226
22.0	2.6695	107.77	0.4980337	0.0963376	0.2836229	0.4364655	1.0744328	5.9690	0.8821932	2.7331	0.5048158	3.3656
23.0	2.5593	104.99	0.5006251	0.0954609	0.2823570	0.4371602	1.0739374	5.7063	0.8836991	2.6199	0.5071695	3.2224
24.0	2.4586	102.33	0.5029335	0.0945478	0.2810266	0.4378826	1.0734195	5.4654	0.8852695	2.5164	0.5096458	3.0912
25.0	2.3662	99.77	0.5053620	0.0935986	0.2796307	0.4386320	1.0728790	5.2436	0.8869040	2.4214	0.5122473	2.9707
26.0	2.2812	97.30	0.5079135	0.0926137	0.2781680	0.4394078	1.0723158	5.0388	0.8886023	2.3339	0.5149767	2.8596
27.0	2.2027	94.92	0.5105915	0.0915934	0.2766375	0.4402092	1.0717298	4.8489	0.8903639	2.2532	0.5178369	2.7568
28.0	2.1301	92.62	0.5133993	0.0905381	0.2750379	0.4410355	1.0711211	4.6725	0.8921881	2.1784	0.5208309	2.6615
29.0	2.0627	90.39	0.5163409	0.0894481	0.2733678	0.4418858	1.0704895	4.5080	0.8940746	2.1090	0.5239620	2.5729
30.0	2.0000	88.23	0.5194202	0.0883240	0.2716260	0.4427591	1.0698350	4.3544	0.8960227	2.0445	0.5272336	2.4903
31.0	1.9416	86.13	0.5226416	0.0871662	0.2698111	0.4436545	1.0691576	4.2106	0.8980316	1.9844	0.5306493	2.4132
32.0	1.8871	84.09	0.5260097	0.0859752	0.2679214	0.4445707	1.0684571	4.0756	0.9001008	1.9282	0.5342129	2.3410
33.0	1.8361	82.11	0.5295294	0.0847515	0.2659557	0.4455064	1.0677336	3.9486	0.9022294	1.8756	0.5379285	2.2733
34.0	1.7883	80.18	0.5332059	0.0834954	0.2639123	0.4464603	1.0669871	3.8290	0.9044167	1.8264	0.5418004	2.2096
35.0	1.7434	78.30	0.5370450	0.0822077	0.2617895	0.4474307	1.0662173	3.7161	0.9066618	1.7801	0.5458329	2.1497
36.0	1.7013	76.46	0.5410526	0.0808888	0.2595859	0.4484159	1.0654244	3.6093	0.9089635	1.7366	0.5500309	2.0933
37.0	1.6616	74.66	0.5452352	0.0795394	0.2572995	0.4494141	1.0646084	3.5081	0.9113210	1.6957	0.5543994	2.0400
38.0	1.6243	72.91	0.5495998	0.0781599	0.2549288	0.4504231	1.0637691	3.4121	0.9137330	1.6571	0.5589437	1.9896
39.0	1.5890	71.19	0.5541538	0.0767511	0.2524718	0.4514406	1.0629066	3.3209	0.9161984	1.6206	0.5636694	1.9418
40.0	1.5557	69.51	0.5589053	0.0753136	0.2499267	0.4524640	1.0620210	3.2341	0.9187158	1.5862	0.5685825	1.8966
41.0	1.5243	67.86	0.5638627	0.0738482	0.2472917	0.4534904	1.0611120	3.1513	0.9212838	1.5537	0.5736890	1.8537
42.0	1.4945	66.24	0.5690353	0.0723554	0.2445647	0.4545169	1.0601799	3.0724	0.9239008	1.5229	0.5789957	1.8129
43.0	1.4663	64.65	0.5744333	0.0708361	0.2417437	0.4555397	1.0592246	2.9970	0.9265651	1.4937	0.5845095	1.7740
44.0	1.4396	63.09	0.5800672	0.0692909	0.2388268	0.4565552	1.0582461	2.9248	0.9292751	1.4660	0.5902377	1.7371
45.0	1.4142	61.55	0.5859487	0.0677209	0.2358117	0.4575591	1.0572445	2.8557	0.9320288	1.4397	0.5961881	1.7019
46.0	1.3902	60.04	0.5920905	0.0661266	0.2326966	0.4585466	1.0562199	2.7895	0.9348241	1.4147	0.6023688	1.6683
47.0	1.3673	58.55	0.5985062	0.0645091	0.2294790	0.4595126	1.0551723	2.7259	0.9376586	1.3910	0.6087885	1.6362
48.0	1.3456	57.09	0.6052107	0.0628693	0.2261569	0.4604512	1.0541020	2.6648	0.9405302	1.3685	0.6154563	1.6056
49.0	1.3250	55.64	0.6122200	0.0612081	0.2227281	0.4613561	1.0530088	2.6060	0.9434362	1.3470	0.6223818	1.5763
50.0	1.3054	54.21	0.6195518	0.0595265	0.2191904	0.4622201	1.0518931	2.5494	0.9463737	1.3266	0.6295752	1.5482
51.0	1.2868	52.80	0.6272252	0.0578255	0.2155413	0.4630353	1.0507550	2.4948	0.9493398	1.3072	0.6370472	1.5213
52.0	1.2690	51.41	0.6352614	0.0561062	0.2117790	0.4637928	1.0495945	2.4422	0.9523314	1.2887	0.6448091	1.4956
53.0	1.2521	50.03	0.6436834	0.0543698	0.2079008	0.4644830	1.0484121	2.3913	0.9553447	1.2711	0.6528728	1.4709
54.0	1.2361	48.66	0.6525167	0.0526175	0.2039047	0.4650947	1.0472078	2.3422	0.9583764	1.2543	0.6612509	1.4473
55.0	1.2208	47.31	0.6617892	0.0508505	0.1997886	0.4656159	1.0459821	2.2946	0.9614222	1.2383	0.6699567	1.4245
56.0	1.2062	45.97	0.6715319	0.0490702	0.1955500	0.4660328	1.0447351	2.2485	0.9644779	1.2230	0.6790041	1.4027
57.0	1.1924	44.64	0.6817792	0.0472778	0.1911870	0.4663303	1.0434672	2.2039	0.9675389	1.2085	0.6884078	1.3817
58.0	1.1792	43.32	0.6925694	0.0454750	0.1866974	0.4664912	1.0421790	2.1606	0.9706002	1.1946	0.6981832	1.3616
59.0	1.1666	42.01	0.7039452	0.0436632	0.1820794	0.4664963	1.0408706	2.1185	0.9736563	1.1814	0.7083463	1.3422
60.0	1.1547	40.71	0.7159545	0.0418439	0.1773310	0.4663242	1.0395427	2.0776	0.9767017	1.1689	0.7189142	1.3235
θ	Ω_S	A_{MIN}	σ_0	σ_1	σ_3	σ_5	Ω_1	Ω_2	Ω_3	Ω_4	Ω_5	Ω_6

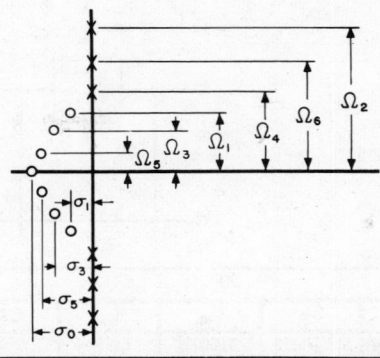

				$K^2 = 1.0$					
C_1	C_2	L_2	C_3	C_4	L_4	C_5	C_6	L_6	C_7
0.9350	0.00000	1.431	1.874	0.00000	1.634	1.874	0.00000	1.431	0.9350
0.93101	0.00488	1.42497	1.85222	0.02164	1.60084	1.84408	0.01592	1.40835	0.92010
0.93024	0.00582	1.42384	1.84802	0.02579	1.59464	1.83834	0.01898	1.40407	0.91725
0.92940	0.00684	1.42261	1.84345	0.03033	1.58792	1.83210	0.02233	1.39942	0.91415
0.92849	0.00794	1.42128	1.83852	0.03524	1.58067	1.82538	0.02596	1.39439	0.91081
0.92752	0.00913	1.41985	1.83322	0.04055	1.57289	1.81816	0.02988	1.38899	0.90720
0.92647	0.01041	1.41832	1.82757	0.04624	1.56458	1.81046	0.03409	1.38321	0.90335
0.92535	0.01178	1.41668	1.82155	0.05234	1.55575	1.80227	0.03860	1.37706	0.89924
0.92416	0.01323	1.41494	1.81518	0.05884	1.54640	1.79360	0.04342	1.37054	0.89487
0.92289	0.01477	1.41309	1.80844	0.06575	1.53653	1.78445	0.04854	1.36364	0.89025
0.92156	0.01640	1.41113	1.80135	0.07307	1.52615	1.77482	0.05399	1.35636	0.88537
0.92015	0.01813	1.40907	1.79390	0.08083	1.51525	1.76472	0.05975	1.34871	0.88023
0.91866	0.01995	1.40690	1.78610	0.08902	1.50384	1.75414	0.06585	1.34069	0.87483
0.91710	0.02186	1.40461	1.77794	0.09765	1.49193	1.74310	0.07229	1.33228	0.86916
0.91546	0.02387	1.40222	1.76942	0.10674	1.47951	1.73159	0.07907	1.32350	0.86324
0.91374	0.02598	1.39971	1.76056	0.11630	1.46659	1.71962	0.08621	1.31434	0.85704
0.91195	0.02819	1.39709	1.75133	0.12633	1.45317	1.70719	0.09372	1.30481	0.85058
0.91007	0.03050	1.39434	1.74176	0.13686	1.43926	1.69431	0.10161	1.29489	0.84384
0.90811	0.03292	1.39148	1.73184	0.14789	1.42486	1.68097	0.10989	1.28460	0.83683
-0.90607	0.03544	1.38850	1.72157	0.15945	1.40997	1.66719	0.11858	1.27392	0.82955
0.90394	0.03807	1.38540	1.71095	0.17154	1.39460	1.65297	0.12768	1.26287	0.82198
0.90173	0.04081	1.38217	1.69998	0.18419	1.37875	1.63831	0.13722	1.25143	0.81414
0.89942	0.04366	1.37881	1.68867	0.19742	1.36243	1.62322	0.14720	1.23961	0.80601
0.89703	0.04663	1.37533	1.67701	0.21124	1.34563	1.60770	0.15766	1.22740	0.79759
0.89455	0.04972	1.37171	1.66501	0.22569	1.32837	1.59176	0.16860	1.21482	0.78888
0.89197	0.05294	1.36795	1.65266	0.24078	1.31065	1.57541	0.18005	1.20184	0.77988
0.88929	0.05628	1.36405	1.63997	0.25655	1.29247	1.55864	0.19202	1.18849	0.77058
0.88651	0.05975	1.36002	1.62694	0.27303	1.27384	1.54147	0.20456	1.17474	0.76097
0.88364	0.06335	1.35583	1.61357	0.29024	1.25476	1.52390	0.21767	1.16061	0.75105
0.88065	0.06709	1.35150	1.59986	0.30823	1.23524	1.50594	0.23140	1.14608	0.74082
0.87756	0.07098	1.34701	1.58581	0.32704	1.21527	1.48760	0.24577	1.13117	0.73027
0.87436	0.07501	1.34236	1.57142	0.34670	1.19488	1.46888	0.26081	1.11586	0.71940
0.87105	0.07920	1.33756	1.55669	0.36728	1.17405	1.44980	0.27658	1.10016	0.70819
0.86762	0.08355	1.33258	1.54162	0.38882	1.15281	1.43035	0.29310	1.08407	0.69664
0.86406	0.08806	1.32743	1.52622	0.41138	1.13114	1.41056	0.31042	1.06758	0.68475
0.86038	0.09275	1.32210	1.51048	0.43502	1.10906	1.39042	0.32860	1.05069	0.67250
0.85657	0.09761	1.31658	1.49440	0.45983	1.08658	1.36995	0.34769	1.03341	0.65990
0.85262	0.10267	1.31087	1.47799	0.48587	1.06369	1.34916	0.36775	1.01572	0.64691
0.84853	0.10792	1.30496	1.46124	0.51325	1.04040	1.32806	0.38885	0.99763	0.63355
0.84430	0.11337	1.29885	1.44415	0.54206	1.01672	1.30666	0.41106	0.97913	0.61979
0.83991	0.11904	1.29251	1.42672	0.57242	0.99265	1.28497	0.43448	0.96023	0.60563
0.83537	0.12494	1.28595	1.40896	0.60444	0.96820	1.26301	0.45919	0.94092	0.59105
0.83065	0.13108	1.27916	1.39086	0.63829	0.94338	1.24079	0.48530	0.92120	0.57603
0.82576	0.13747	1.27212	1.37242	0.67412	0.91818	1.21833	0.51293	0.90106	0.56057
0.82069	0.14412	1.26481	1.35364	0.71212	0.89262	1.19564	0.54222	0.88051	0.54463
0.81543	0.15107	1.25724	1.33452	0.75249	0.86670	1.17273	0.57331	0.85954	0.52821
0.80996	0.15831	1.24937	1.31506	0.79549	0.84042	1.14963	0.60638	0.83815	0.51129
0.80428	0.16587	1.24121	1.29526	0.84140	0.81380	1.12636	0.64162	0.81634	0.49383
0.79837	0.17378	1.23271	1.27511	0.89053	0.78683	1.10293	0.67927	0.79410	0.47581
0.79223	0.18206	1.22388	1.25462	0.94327	0.75952	1.07938	0.71957	0.77143	0.45721
0.78583	0.19074	1.21467	1.23378	1.00006	0.73188	1.05571	0.76284	0.74833	0.43800
L_1	L_2	C_2	L_3	L_4	C_4	L_5	L_6	C_6	L_7

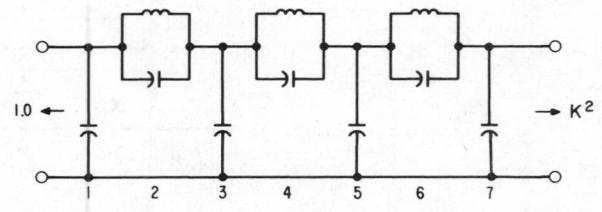

					$K^2 = \infty$					
θ	C_1	C_2	L_2	C_3	C_4	L_4	C_5	C_6	L_6	C_7
C	0.467517	0.0000	1.130992	1.475112	0.0000	1.623996	1.703432	0.0000	1.687138	1.410984
11.0	0.4618	0.0062	1.1226	1.4528	0.0218	1.5867	1.6752	0.0135	1.6657	1.4010
12.0	0.4607	0.0074	1.1211	1.4486	0.0260	1.5796	1.6698	0.0160	1.6616	1.3991
13.0	0.4595	0.0087	1.1193	1.4440	0.0306	1.5719	1.6640	0.0189	1.6572	1.3971
14.0	0.4582	0.0101	1.1174	1.4390	0.0356	1.5636	1.6577	0.0219	1.6524	1.3948
15.0	0.4568	0.0116	1.1154	1.4337	0.0410	1.5548	1.6510	0.0252	1.6472	1.3924
16.0	0.4553	0.0133	1.1132	1.4280	0.0468	1.5453	1.6438	0.0287	1.6417	1.3899
17.0	0.4537	0.0150	1.1109	1.4220	0.0530	1.5352	1.6362	0.0325	1.6359	1.3872
18.0	0.4520	0.0169	1.1085	1.4155	0.0597	1.5245	1.6281	0.0365	1.6297	1.3843
19.0	0.4501	0.0189	1.1059	1.4087	0.0668	1.5132	1.6196	0.0408	1.6231	1.3812
20.0	0.4482	0.0210	1.1031	1.4016	0.0743	1.5013	1.6106	0.0453	1.6162	1.3780
21.0	0.4462	0.0232	1.1002	1.3941	0.0823	1.4888	1.6012	0.0501	1.6090	1.3746
22.0	0.4441	0.0256	1.0971	1.3862	0.0907	1.4757	1.5913	0.0551	1.6014	1.3711
23.0	0.4418	0.0281	1.0939	1.3779	0.0996	1.4621	1.5810	0.0604	1.5934	1.3673
24.0	0.4394	0.0307	1.0905	1.3693	0.1091	1.4478	1.5702	0.0660	1.5851	1.3634
25.0	0.4370	0.0335	1.0870	1.3604	0.1190	1.4330	1.5591	0.0719	1.5765	1.3594
26.0	0.4344	0.0364	1.0833	1.3510	0.1295	1.4176	1.5475	0.0780	1.5675	1.3552
27.0	0.4317	0.0394	1.0794	1.3413	0.1405	1.4016	1.5354	0.0844	1.5581	1.3508
28.0	0.4288	0.0426	1.0754	1.3313	0.1521	1.3851	1.5230	0.0912	1.5484	1.3462
29.0	0.4259	0.0459	1.0712	1.3209	0.1643	1.3679	1.5101	0.0982	1.5383	1.3415
30.0	0.4228	0.0494	1.0669	1.3101	0.1772	1.3503	1.4968	0.1055	1.5279	1.3366
31.0	0.4196	0.0531	1.0623	1.2989	0.1907	1.3320	1.4832	0.1132	1.5172	1.3316
32.0	0.4162	0.0569	1.0576	1.2874	0.2048	1.3132	1.4691	0.1212	1.5061	1.3263
33.0	0.4128	0.0609	1.0527	1.2756	0.2197	1.2939	1.4546	0.1295	1.4946	1.3209
34.0	0.4091	0.0651	1.0476	1.2634	0.2353	1.2740	1.4397	0.1381	1.4828	1.3154
35.0	0.4054	0.0695	1.0424	1.2508	0.2518	1.2535	1.4244	0.1471	1.4707	1.3096
36.0	0.4015	0.0740	1.0369	1.2378	0.2690	1.2325	1.4087	0.1565	1.4582	1.3037
37.0	0.3974	0.0788	1.0313	1.2245	0.2872	1.2110	1.3927	0.1663	1.4454	1.2976
38.0	0.3932	0.0838	1.0254	1.2108	0.3063	1.1890	1.3763	0.1764	1.4322	1.2914
39.0	0.3888	0.0890	1.0194	1.1967	0.3264	1.1664	1.3595	0.1869	1.4187	1.2849
40.0	0.3843	0.0944	1.0131	1.1823	0.3476	1.1433	1.3424	0.1979	1.4048	1.2783
41.0	0.3796	0.1000	1.0067	1.1675	0.3700	1.1197	1.3249	0.2093	1.3906	1.2716
42.0	0.3747	0.1059	1.0000	1.1524	0.3936	1.0955	1.3071	0.2211	1.3760	1.2646
43.0	0.3696	0.1121	0.9930	1.1368	0.4186	1.0709	1.2889	0.2335	1.3611	1.2575
44.0	0.3644	0.1186	0.9859	1.1209	0.4450	1.0457	1.2704	0.2463	1.3458	1.2501
45.0	0.3589	0.1253	0.9785	1.1046	0.4730	1.0201	1.2517	0.2596	1.3302	1.2426
46.0	0.3533	0.1324	0.9709	1.0879	0.5027	0.9940	1.2326	0.2734	1.3142	1.2350
47.0	0.3474	0.1398	0.9630	1.0708	0.5342	0.9674	1.2132	0.2878	1.2978	1.2271
48.0	0.3413	0.1475	0.9548	1.0533	0.5679	0.9403	1.1935	0.3028	1.2812	1.2190
49.0	0.3349	0.1556	0.9464	1.0354	0.6038	0.9128	1.1736	0.3184	1.2641	1.2108
50.0	0.3284	0.1641	0.9376	1.0172	0.6422	0.8848	1.1534	0.3346	1.2467	1.2024
51.0	0.3215	0.1730	0.9286	0.9984	0.6834	0.8563	1.1330	0.3516	1.2289	1.1937
52.0	0.3144	0.1824	0.9193	0.9793	0.7277	0.8274	1.1124	0.3692	1.2108	1.1849
53.0	0.3070	0.1922	0.9096	0.9597	0.7755	0.7981	1.0916	0.3877	1.1923	1.1759
54.0	0.2993	0.2026	0.8996	0.9397	0.8272	0.7684	1.0706	0.4069	1.1734	1.1666
55.0	0.2913	0.2136	0.8893	0.9192	0.8834	0.7383	1.0495	0.4270	1.1541	1.1572
56.0	0.2829	0.2251	0.8786	0.8982	0.9446	0.7077	1.0283	0.4480	1.1344	1.1476
57.0	0.2741	0.2373	0.8675	0.8767	1.0117	0.6768	1.0070	0.4701	1.1142	1.1377
58.0	0.2650	0.2503	0.8560	0.8547	1.0854	0.6456	0.9856	0.4932	1.0937	1.1276
59.0	0.2555	0.2640	0.8441	0.8321	1.1669	0.6140	0.9642	0.5175	1.0727	1.1173
60.0	0.2455	0.2786	0.8317	0.8090	1.2575	0.5821	0.9429	0.5430	1.0513	1.1067
θ	L_1	L_2	C_2	L_3	L_4	C_4	L_5	L_6	C_6	L_7

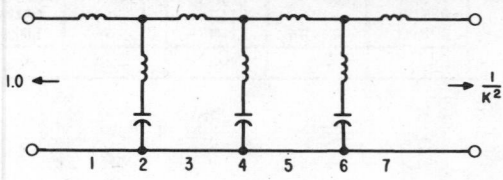

274

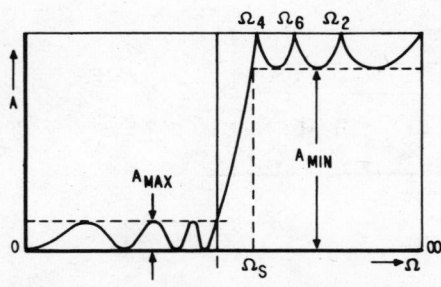

n = 7
ρ = 10%

θ	Ω_S	A_{MIN}	σ_0	σ_1	σ_3	σ_5	Ω_1	Ω_2	Ω_3	Ω_4	Ω_5	Ω_6
C	∞	∞	0.440754	0.098077	0.274805	0.397105	1.065424	∞	0.854404	∞	0.474158	∞
11.0	5.2408	152.45	0.4456544	0.0959179	0.2719594	0.3988682	1.0643790	11.9886	0.8583104	5.3732	0.4796616	6.6793
12.0	4.8097	147.12	0.4465974	0.0955090	0.2714131	0.3992021	1.0641798	10.9867	0.8590523	4.9307	0.4807189	6.1257
13.0	4.4454	142.21	0.4476267	0.0950650	0.2708172	0.3995644	1.0639631	10.1387	0.8598584	4.5568	0.4818723	5.6575
14.0	4.1336	137.66	9.4487434	0.0945860	0.2701714	0.3999552	1.0637288	9.4116	0.7607286	4.2367	0.4831227	5.2564
15.0	3.8637	133.42	0.4499485	0.0940723	0.2694751	0.4003742	1.0634770	8.7812	0.8616629	3.9597	0.4844716	4.9090
16.0	3.6280	129.45	0.4512436	0.0935240	0.2687276	0.4008212	1.0632076	8.2295	0.8626612	3.7177	0.4859200	4.6052
17.0	3.4203	125.71	0.4526299	0.0929412	0.2679285	0.4012962	1.0629205	7.7424	0.8637234	3.5044	0.4874692	4.3374
18.0	3.2361	122.18	0.4541090	0.0923240	0.2670769	0.4017988	1.0626156	7.3093	0.8648492	3.3152	0.4891207	4.0995
19.0	3.0716	118.84	0.4556827	0.0916729	0.2661723	0.4023288	1.0622930	6.9216	0.8660387	3.1463	0.4908762	3.8868
20.0	2.9238	115.65	0.4573525	0.0909877	0.2652139	0.4028860	1.0619525	6.5725	0.7672916	2.9945	0.4927373	3.6955
21.0	2.7904	112.62	0.4591205	0.0902690	0.2642008	0.4034700	1.0615942	6.2564	0.8686078	2.8574	0.4947058	3.5226
22.0	2.6695	109.72	0.4609887	0.0895168	0.2631323	0.4040805	1.0612179	5.9690	0.8699871	2.7331	0.4967836	3.3656
23.0	2.5593	106.95	0.4629594	0.0897314	0.2620074	0.4047171	1.0608335	5.7063	0.8714292	2.6199	0.4989727	3.2224
24.0	2.4586	104.28	0.4650348	0.0879129	0.2608253	0.4053795	1.0604111	5.4654	0.8729340	2.5164	0.5012752	3.0912
25.0	2.3662	101.72	0.4672175	0.0870617	0.2595849	0.4060671	1.0599806	5.2436	0.8745012	2.4214	0.5036937	2.9707
26.0	2.2812	99.25	0.4695101	0.0861782	0.2582851	0.4067795	1.0595318	5.0388	0.8761305	2.3339	0.5062303	2.8596
27.0	2.2027	96.87	0.4719156	0.0852624	0.2569251	0.4075160	1.0590647	4.8489	0.8778216	2.2532	0.5088878	2.7568
28.0	2.1301	94.57	0.4744371	0.0843149	0.2555036	0.4082762	1.0585795	4.6725	0.8795741	2.1784	0.5116688	2.6615
29.0	2.0627	92.34	0.4770777	0.0833358	0.2540194	0.4090592	1.0580757	4.5080	0.8813876	2.1090	0.5145763	2.5729
30.0	2.0000	90.18	0.4798410	0.0823255	0.2524714	0.4098643	1.0575534	4.3544	0.8832619	2.0445	0.5176134	2.4903
31.0	1.9416	88.08	0.4827308	0.0812844	0.2508583	0.4106908	1.0570126	4.2106	0.8851964	1.9844	0.5207832	2.4132
32.0	1.8871	86.05	0.4857510	0.0802129	0.2491788	0.4115376	1.0564533	4.0756	0.8871906	1.9282	0.5240893	2.3410
33.0	1.8361	84.06	0.4889058	0.0791113	0.2474315	0.4124037	1.0558753	3.9486	0.8892439	1.8756	0.5275352	2.2733
34.0	1.7883	82.13	0.4921998	0.0779800	0.2456150	0.4132880	1.0552785	3.8290	0.8913559	1.8264	0.5311248	2.2096
35.0	1.7434	80.25	0.4956379	0.0768194	0.2437279	0.4141893	1.0546630	3.7161	0.8935258	1.7801	0.5348621	2.1497
36.0	1.7013	78.41	0.4992253	0.0756301	0.2417685	0.4151062	1.0540287	3.6093	0.8957529	1.7366	0.5387514	2.0933
37.0	1.6616	76.62	0.5029676	0.0744125	0.2397354	0.4160371	1.0533755	3.5081	0.8980366	1.6957	0.5427972	2.0400
38.0	1.6243	74.86	0.5068706	0.0731669	0.2376269	0.4169804	1.0527034	3.4121	0.9003759	1.6571	0.5470044	1.9896
39.0	1.5890	73.15	0.5109409	0.0718940	0.2354415	0.4179342	1.0520122	3.3209	0.9027699	1.6206	0.5513778	1.9418
40.0	1.5557	71.46	0.5151851	0.0705944	0.2331773	0.4188964	1.0513021	3.2341	0.9052177	1.5862	0.5559229	1.8966
41.0	1.5243	69.81	0.5196108	0.0692683	0.2308325	0.4198648	1.0505730	3.1513	0.9077182	1.5537	0.5606452	1.8537
42.0	1.4945	68.20	0.5242258	0.0679166	0.2284053	0.4208368	1.0498249	3.0724	0.9102703	1.5229	0.5655508	1.8129
43.0	1.4663	66.61	0.5290386	0.0665397	0.2258941	0.4218097	1.0490576	2.9970	0.9128726	1.4937	0.5706457	1.7740
44.0	1.4396	65.04	0.5340584	0.0651384	0.2232966	0.4227804	1.0482712	2.9248	0.9155240	1.4660	0.5759368	1.7371
45.0	1.4142	63.51	0.5392951	0.0637133	0.2206110	0.4237455	1.0474658	2.8557	0.9182229	1.4397	0.5814309	1.7019
46.0	1.3902	62.00	0.5447594	0.0622650	0.2178354	0.4247011	1.0466413	2.7895	0.9209675	1.4147	0.5871355	1.6683
47.0	1.3673	60.51	0.5504628	0.0607943	0.2149676	0.4256432	1.0457978	2.7259	0.9237564	1.3910	0.5930584	1.6362
48.0	1.3456	59.04	0.5564180	0.0593018	0.2120054	0.4265670	1.0449352	2.6648	0.9265875	1.3685	0.5992079	1.6056
49.0	1.3250	57.59	0.5626386	0.0577886	0.2089468	0.4274675	1.0440537	2.6060	0.9294589	1.3470	0.6055926	1.5763
50.0	1.3054	56.17	0.5691393	0.0562553	0.2057897	0.4283387	1.0431533	2.5494	0.9323685	1.3266	0.6122219	1.5482
51.0	1.2868	54.76	0.5759364	0.0547028	0.2025317	0.4291745	1.0422341	2.4948	0.9353137	1.3072	0.6191055	1.5213
52.0	1.2690	53.36	0.5830476	0.0531320	0.1991707	0.4299677	1.0412961	2.4422	0.9382922	1.2887	0.6262539	1.4956
53.0	1.2521	51.98	0.5904921	0.0515438	0.1957043	0.4307105	1.0403396	2.3913	0.9413011	1.2711	0.6336778	1.4709
54.0	1.2361	50.62	0.5982912	0.0499393	0.1921303	0.4313939	1.0393645	2.3422	0.9443375	1.2543	0.6413889	1.4473
55.0	1.2208	49.26	0.6064682	0.0483195	0.1884465	0.4320083	1.0383713	2.2946	0.9473980	1.2383	0.6493995	1.4245
56.0	1.2062	47.92	0.6150490	0.0466856	0.1846504	0.4325425	1.0373599	2.2485	0.9504793	1.2230	0.6577224	1.4027
57.0	1.1924	46.59	0.6240620	0.0450386	0.1807397	0.4329842	1.0363306	2.2039	0.9535776	1.2085	0.6663715	1.3817
58.0	1.1792	45.27	0.6335389	0.0433799	0.1767124	0.4333198	1.0352836	2.1606	0.9566889	1.1946	0.6753610	1.3616
59.0	1.1666	43.96	0.6435150	0.0417106	0.1725661	0.4335336	1.0342194	2.1185	0.9598085	1.1814	0.6847063	1.3422
60.0	1.1547	42.66	0.6540295	0.0400323	0.1682986	0.4336083	i.0331381	2.0776	0.9629320	1.1689	0.6944236	1.3235
θ	Ω_S	A_{MIN}	σ_0	σ_1	σ_3	σ_5	Ω_1	Ω_2	Ω_3	Ω_4	Ω_5	Ω_6

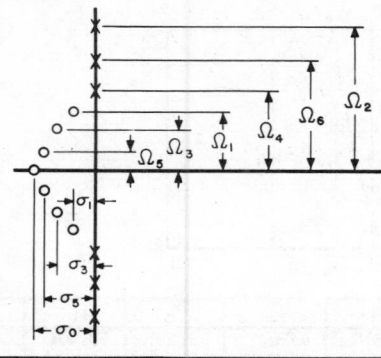

					K² = 1.0					
θ	C_1	C_2	L_2	C_3	C_4	L_4	C_5	C_6	L_6	C_7
C	1.010	0.00000	1.437	1.94i	0.00000	1.622	1.941	0.00000	1.437	1.010
11.0	1.00569	0.00486	1.43101	1.91876	0.02179	1.58963	1.91043	0.01574	1.41497	0.99482
12.0	1.00491	0.00579	1.42990	1.91446	0.02597	1.58357	1.90456	0.01889	1.41081	0.99198
13.0	1.00407	0.00681	1.42869	1.90979	0.03054	1.57700	1.89818	0.02222	1.40629	0.98889
14.0	1.00316	0.00791	1.42737	1.90475	0.03549	1.56991	1.89130	0.02583	1.40141	0.98555
15.0	1.00218	0.00909	1.42596	1.89934	0.04082	1.56230	1.88392	0.02972	1.39616	0.98196
16.0	1.00112	0.01037	1.42445	1.89356	0.04655	1.55417	1.87604	0.03391	1.39056	0.97811
17.0	1.00000	0.01172	1.42284	1.88742	0.05268	1.54554	1.86767	0.03839	1.38458	0.97402
18.0	0.99880	0.01317	1.42112	1.88090	0.05922	1.53639	1.85880	0.04317	1.37825	0.96966
19.0	0.99753	0.01471	1.41929	1.87402	0.06617	1.52674	1.84943	0.04826	1.37155	0.96505
20.0	0.99619	0.01633	1.41737	1.86677	0.07353	1.51659	1.83958	0.05366	1.36449	0.96019
21.0	0.99477	0.01805	1.41533	1.85916	0.08133	1.50593	1.82924	0.05938	1.35706	0.95507
22.0	0.99328	0.01986	1.41319	1.75119	0.08956	1.49477	1.81842	0.06543	1.34927	0.94969
23.0	0.99171	0.02177	1.41094	1.84285	0.09823	1.48312	1.80711	0.07181	1.34111	0.94404
24.0	0.99007	0.02377	1.40858	1.83415	0.10736	1.47097	1.79533	0.07853	1.33258	0.93814
25.0	0.98834	0.02587	1.40611	1.82508	0.11696	1.45834	1.78307	0.08560	1.32369	0.93197
26.0	0.98654	0.02806	1.40352	1.81566	0.12703	1.44522	1.77033	0.09304	1.31443	0.92553
27.0	0.98465	0.03036	1.40082	1.80588	0.13759	1.43161	1.75713	0.10084	1.30480	0.91883
28.0	0.98269	0.03276	1.39800	1.79574	0.14866	1.41753	1.74347	0.10903	1.29480	0.91185
29.0	0.98064	0.03527	1.39507	1.78525	0.16024	1.40297	1.72935	0.11761	1.28443	0.90460
30.0	0.97850	0.03789	1.39201	1.77440	0.17237	1.38793	1.71476	0.12660	1.27370	0.89708
31.0	0.97628	0.04061	1.38883	1.76319	0.18504	1.37243	1.69973	0.13600	1.26259	0.88928
32.0	0.97397	0.04345	1.38552	1.75163	0.19829	1.35647	1.68425	0.14585	1.25111	0.88120
33.0	0.97157	0.04641	1.38209	1.73972	0.21212	1.34005	1.66833	0.15615	1.23926	0.87284
34.0	0.96907	0.04948	1.37852	1.72746	0.22658	1.32317	1.65197	0.16692	1.22703	0.86419
35.0	0.96648	0.05267	1.37482	1.71484	0.24267	1.30583	1.63517	0.17818	1.21443	0.85525
36.0	0.96380	0.05599	1.37099	1.70188	0.25743	1.28806	1.61795	0.18995	1.20146	0.84601
37.0	0.96101	0.05944	1.36701	1.68856	0.27389	1.26984	1.60031	0.20226	1.18810	0.83648
38.0	0.95813	0.06302	1.36289	1.67490	0.29107	1.25118	1.58225	0.21512	1.17437	0.82664
39.0	0.95513	0.06674	1.35863	1.66089	0.30902	1.23209	1.56378	0.22857	1.16026	0.81650
40.0	0.95204	0.07060	1.35421	1.64654	0.32777	1.21257	1.54491	0.24264	1.14577	0.80604
41.0	0.94883	0.07461	1.34964	1.63184	0.34736	1.19262	1.52565	0.25735	1.13090	0.79527
42.0	0.94550	0.07877	1.34491	1.61679	0.36784	1.17226	1.50599	0.27274	1.11564	0.78418
43.0	0.94206	0.08309	1.34001	1.60140	0.38926	1.15149	1.48595	0.28885	1.10000	0.77275
44.0	0.93850	0.08757	1.33495	1.58566	0.41168	1.13030	1.46554	0.30573	1.08397	0.76099
45.0	0.93481	0.09222	1.32971	1.56958	0.43516	1.10872	1.44477	0.32341	1.06755	0.74889
46.0	0.93099	0.09705	1.32428	1.55316	0.45976	1.08673	1.42364	0.34195	1.05075	0.73644
47.0	0.92704	0.10206	1.31867	1.53639	0.48557	1.06436	1.40216	0.36140	1.03355	0.72362
48.0	0.92294	0.10727	1.31287	1.51928	0.51266	1.04159	1.38034	0.38183	1.01595	0.71044
49.0	0.91870	0.11268	1.30686	1.50183	0.54114	1.01844	1.35820	0.40331	0.99796	0.69688
50.0	0.91431	0.11830	1.30064	1.48403	0.57111	0.99492	1.33573	0.42590	0.97957	0.68293
51.0	0.90976	0.12414	1.29419	1.46588	0.60269	0.97103	1.31297	0.44970	0.96078	0.66858
52.0	0.90504	0.13022	1.28752	1.44739	0.63601	0.94676	1.28991	0.47479	0.94158	0.65382
53.0	0.90015	0.13655	1.28061	1.42855	0.67123	0.92214	1.26657	0.50130	0.92198	0.63863
54.0	0.89508	0.14315	1.27345	1.40937	0.70851	0.89716	1.24296	0.52932	0.90196	0.62299
55.0	0.88981	0.15002	1.26602	1.38983	0.74806	0.87184	1.21910	0.55900	0.88153	0.60689
56.0	0.88435	0.15718	1.25831	1.36995	0.79009	0.84617	1.19500	0.59050	0.86069	0.59032
57.0	0.87867	0.16467	1.25031	1.34971	0.83487	0.82016	1.17068	0.62397	0.83943	0.57325
58.0	0.87276	0.17248	1.24199	1.32912	0.88269	0.79381	1.14616	0.65963	0.81774	0.55565
59.0	0.86662	0.18066	1.23334	1.30817	0.93390	0.76714	1.12145	0.69769	0.79563	0.53751
60.0	0.86023	0.18923	1.22434	1.28687	0.98889	0.74015	1.09658	0.73842	0.77308	0.51879
θ	L_1	L_2	C_2	L_3	L_4	C_4	L_5	L_6	C_6	L_7

276

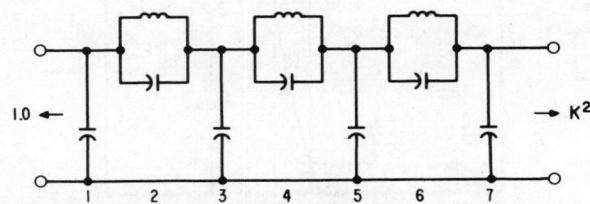

				$K^2 = \infty$					
C_1	C_2	L_2	C_3	C_4	L_4	C_5	C_6	L_6	C_7
0.504865	0.0000	1.185996	1.525398	0.0000	1.657407	1.735974	0.0000	1.708627	1.432522
0.4994	0.0059	1.1779	1.5033	0.0214	1.6205	1.7076	0.0133	1.6870	1.4223
0.4983	0.0070	1.1763	1.4991	0.0255	1.6135	1.7023	0.0158	1.6829	1.4204
0.4972	0.0083	1.1746	1.4946	0.0300	1.6059	1.6964	0.0186	1.6785	1.4183
0.4960	0.0096	1.1728	1.4896	0.0349	1.5977	1.6902	0.0216	1.6737	1.4160
0.4946	0.0111	1.1708	1.4844	0.0401	1.5890	1.6834	0.0249	1.6675	1.4135
0.4932	0.0126	1.1687	1.4787	0.0458	1.5796	1.6762	0.0284	1.6630	1.4109
0.4917	0.0143	1.1665	1.4727	0.0519	1.5696	1.6685	0.0321	1.6571	1.4081
0.4900	0.0161	1.1641	1.4663	0.0584	1.5590	1.6604	0.0360	1.6509	1.4052
0.4883	0.0180	1.1616	1.4596	0.0653	1.5479	1.6519	0.0403	1.6443	1.4020
0.4865	0.0200	1.1589	1.4525	0.0726	1.5361	1.6428	0.0447	1.6373	1.3987
0.4845	0.0221	1.1560	1.4451	0.0804	1.5238	1.6334	0.0494	1.6300	1.3953
0.4825	0.0243	1.1531	1.4373	0.0886	1.5109	1.6235	0.0544	1.6224	1.3916
0.4804	0.0267	1.1499	1.4291	0.0973	1.4974	1.6131	0.0597	1.6144	1.3878
0.4781	0.0292	1.1466	1.4206	0.1065	1.4833	1.6024	0.0652	1.6060	1.3838
0.4758	0.0318	1.1432	1.4117	0.1161	1.4687	1.5911	0.0709	1.5973	1.3797
0.4733	0.0346	1.1396	1.4025	0.1263	1.4534	1.5795	0.0770	1.5882	1.3754
0.4707	0.0374	1.1359	1.3929	0.1370	1.4377	1.5674	0.0833	1.5788	1.3709
0.4680	0.0405	1.1320	1.3829	0.1483	1.4213	1.5549	0.0900	1.5690	1.3662
0.4652	0.0436	1.1279	1.3726	0.1601	1.4044	1.5420	0.0969	1.5589	1.3613
0.4623	0.0469	1.1236	1.3620	0.1725	1.3870	1.5286	0.1041	1.5484	1.3563
0.4593	0.0504	1.1192	1.3509	0.1855	1.3689	1.5148	0.1117	1.5375	1.3511
0.4561	0.0540	1.1147	1.3396	0.1992	1.3504	1.5007	0.1195	1.5263	1.3458
0.4528	0.0578	1.1099	1.3278	0.2135	1.3313	1.4861	0.1277	1.5148	1.3402
0.4493	0.0617	1.1050	1.3157	0.2286	1.3116	1.4711	0.1363	1.5029	1.3345
0.4458	0.0658	1.0998	1.3033	0.2444	1.2915	1.4557	0.1452	1.4907	1.3286
0.4421	0.0701	1.0945	1.2905	0.2609	1.2708	1.4399	0.1544	1.4780	1.3226
0.4382	0.0746	1.0891	1.2773	0.2783	1.2495	1.4238	0.1640	1.4651	1.3163
0.4342	0.0793	1.0834	1.2638	0.2966	1.2278	1.4072	0.1740	1.4518	1.3099
0.4301	0.0842	1.0775	1.2499	0.3158	1.2055	1.3903	0.1844	1.4381	1.3033
0.4258	0.0892	1.0714	1.2357	0.3360	1.1827	1.3730	0.1952	1.4241	1.2965
0.4213	0.0945	1.0651	1.2211	0.3573	1.1594	1.3554	0.2064	1.4097	1.2895
0.4167	0.1001	1.0586	1.2061	0.3797	1.1356	1.3374	0.2181	1.3950	1.2824
0.4119	0.1058	1.0519	1.1908	0.4033	1.1113	1.3190	0.2303	1.3799	1.2751
0.4070	0.1119	1.0449	1.1751	0.4283	1.0865	1.3003	0.2429	1.3644	1.2675
0.4018	0.1182	1.0377	1.1590	0.4546	1.0613	1.2813	0.2560	1.3486	1.2598
0.3965	0.1247	1.0303	1.1426	0.4825	1.0355	1.2620	0.2697	1.3325	1.2519
0.3909	0.1316	1.0226	1.1258	0.5121	1.0093	1.2423	0.2838	1.3159	1.2439
0.3852	0.1388	1.0147	1.1086	0.5434	0.9826	1.2223	0.2986	1.2990	1.2356
0.3792	0.1463	1.0064	1.0910	0.5768	0.9555	1.2021	0.3140	1.2817	1.2271
0.3730	0.1542	0.9980	1.0731	0.6124	0.9279	1.1816	0.3300	1.2642	1.2184
0.3666	0.1624	0.9892	1.0547	0.6504	0.8998	1.1608	0.3467	1.2461	1.2095
0.3599	0.1711	0.9801	1.0360	0.6910	0.8714	1.1397	0.3641	1.2278	1.2005
0.3529	0.1801	0.9707	1.0168	0.7347	0.8425	1.1184	0.3823	1.2090	1.1912
0.3457	0.1897	0.9610	0.9972	0.7817	0.8132	1.0969	0.4013	1.1898	1.1817
0.3382	0.1997	0.9509	0.9772	0.8324	0.7835	1.0752	0.4211	1.1703	1.1719
0.3303	0.2103	0.9405	0.9567	0.8874	0.7534	1.0534	0.4418	1.1503	1.1620
0.3222	0.2214	0.9297	0.9358	0.9472	0.7229	1.0313	0.4635	1.1300	1.1518
0.3137	0.2332	0.9185	0.9144	1.0125	0.6921	1.0092	0.4863	1.1092	1.1414
0.3048	0.2457	0.9069	0.8925	1.0840	0.6609	0.9869	0.5102	1.0879	1.1308
0.2955	0.2589	0.8949	0.8700	1.1629	0.6294	0.9646	0.5354	1.0662	1.1199
L_1	L_2	C_2	L_3	L_4	C_4	L_5	L_6	C_6	L_7

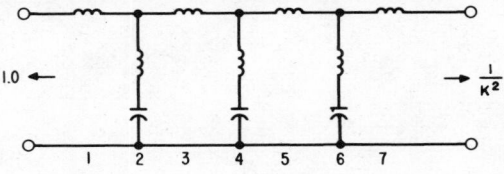

n = 7

ρ = 15%

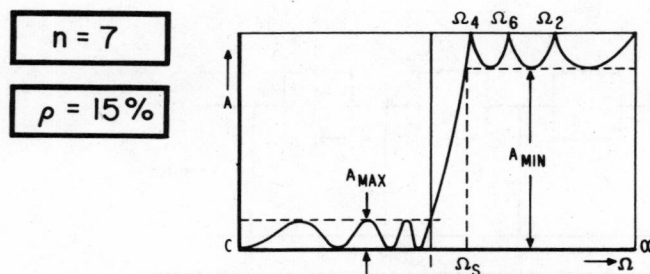

θ	Ω_S	A_MIN	σ_0	σ_1	σ_3	σ_5	Ω_1	Ω_2	Ω_3	Ω_4	Ω_5	Ω_6
C	∞	∞	0.377675	0.084040	0.235476	0.340273	1.042142	∞	0.835733	∞	0.463797	∞
11.0	5.2408	156.02	0.3816903	0.0823008	0.2332010	0.3417450	1.0415170	11.9886	0.8393370	5.3732	0.4686768	6.6793
12.0	4.8097	150.70	0.3824627	0.0819710	0.2327642	0.3420241	1.0413979	10.9867	0.8400222	4.9307	0.469138	6.1257
13.0	4.4454	145.79	0.3833055	0.0816127	0.2322879	0.3423271	1.0412682	10.1387	0.8407670	4.5568	0.4706357	5.6575
14.0	4.1336	141.24	0.3842196	0.0812262	0.2317717	0.3426540	1.0411279	9.4116	0.8415716	4.2367	0.4717434	5.2564
15.0	3.8637	137.00	0.3852060	0.0808115	0.2312151	0.3430047	1.0409770	8.7812	0.8424357	3.9597	0.4729379	4.9090
16.0	3.6280	133.03	0.3862657	0.0803686	0.2306177	0.3433791	1.0408156	8.2295	0.8433594	3.7177	0.4742202	4.6052
17.0	3.4203	129.29	0.3873997	0.0798976	0.2299790	0.3437771	1.0406434	7.7424	0.8443429	3.5044	0.4755915	4.3374
18.0	3.2361	125.76	0.3886093	0.0793988	0.2292984	0.3441985	1.0404605	7.3093	0.8453859	3.3152	0.4770528	4.0995
19.0	3.0716	122.41	0.3898958	0.0788722	0.2285753	0.3446431	1.0402669	6.9216	0.8464885	3.1463	0.4786057	3.8868
20.0	2.9238	119.23	0.3912606	0.0783179	0.2278093	0.3451109	1.0400623	6.5725	0.8476508	2.9945	0.4802513	3.6955
21.0	2.7904	116.20	0.3927051	0.0777359	0.2269997	0.3456016	1.0398469	6.2564	0.8488726	2.8574	0.4819914	3.5226
22.0	2.6695	113.30	0.3942308	0.0771266	0.2261458	0.3461151	1.0396206	5.9690	0.8501540	2.7331	0.4838273	3.3656
23.0	2.5593	110.53	0.3958396	0.0764899	0.2252468	0.3466510	1.0393832	5.7063	0.8514949	2.6199	0.4857609	3.2224
24.0	2.4586	107.86	0.3975333	0.0758263	0.2243020	0.3472091	1.0391349	5.4654	0.8528952	2.5164	0.4877939	3.0912
25.0	2.3662	105.30	0.3993137	0.0751354	0.2233106	0.3477891	1.0388753	5.2436	0.8543549	2.4214	0.4899282	2.9707
26.0	2.2812	102.83	0.4011830	0.0744180	0.2222719	0.3483908	1.0386046	5.0388	0.8558739	2.3339	0.4921659	2.8596
27.0	2.2027	100.45	0.4031434	0.0736738	0.2211849	0.3490136	1.0383226	4.8489	0.8574522	2.2532	0.4945091	2.7568
28.0	2.1301	98.15	0.4051971	0.0729032	0.2200487	0.3496573	1.0380294	4.6725	0.8590895	2.1784	0.4969602	2.6615
29.0	2.0627	95.92	0.4073469	0.0721063	0.2188625	0.3503214	1.0377247	4.5080	0.8607858	2.1090	0.4995214	2.5729
30.0	2.0000	93.76	0.4095953	0.0712834	0.2176251	0.3510055	1.0374086	4.3544	0.8625410	2.0445	0.5021953	2.4903
31.0	1.9416	91.66	0.4119451	0.0704347	0.2163356	0.3517088	1.0370809	4.2106	0.8643550	1.9844	0.5049847	2.4132
32.0	1.8871	89.62	0.4143996	0.0695605	0.2149929	0.3524309	1.0367416	4.0756	0.8662274	1.9282	0.5078924	2.3410
33.0	1.8361	87.64	0.4169618	0.0686608	0.2135958	0.3531711	1.0363906	3.9486	0.8681580	1.8756	0.5109215	2.2733
34.0	1.7883	85.71	0.4196353	0.0677362	0.2121433	0.3539287	1.0360280	3.8290	0.8701467	1.8264	0.5140750	2.2096
35.0	1.7434	83.83	0.4224237	0.0667866	0.2106340	0.3547028	1.0356534	3.7161	0.8721932	1.7801	0.5173563	2.1497
36.0	1.7013	81.99	0.4253311	0.0658126	0.2090667	0.3554925	1.0352670	3.6093	0.8742970	1.7366	0.5207691	2.0933
37.0	1.6616	80.20	0.4283616	0.0648144	0.2074402	0.3562968	1.0348686	3.5081	0.8764582	1.6957	0.5243170	2.0400
38.0	1.6243	78.44	0.4315197	0.0637922	0.2057529	0.3571146	1.0344582	3.4121	0.8786759	1.6571	0.5280040	1.9896
39.0	1.5890	76.72	0.4348104	0.0627463	0.2040037	0.3579447	1.0340357	3.3209	0.8809499	1.6206	0.5318343	1.9418
40.0	1.5557	75.04	0.4382387	0.0616773	0.2021910	0.3587856	1.0336010	3.2341	0.8832799	1.5862	0.5358123	1.8966
41.0	1.5243	73.39	0.4418102	0.0605853	0.2003132	0.3596361	1.0331541	3.1513	0.8856649	1.5537	0.5399427	1.8537
42.0	1.4945	71.77	0.4455308	0.0594708	0.1983688	0.3604942	1.0326949	3.0724	0.8881047	1.5229	0.5442305	1.8129
43.0	1.4663	70.18	0.4494069	0.0583342	0.1963563	0.3613582	1.0322233	2.9970	0.8905985	1.4937	0.5486808	1.7740
44.0	1.4396	68.62	0.4534454	0.0571760	0.1942738	0.3622261	1.0317393	2.9248	0.8931456	1.4660	0.5532992	1.7371
45.0	1.4142	67.09	0.4576536	0.0559963	0.1921200	0.3630956	1.0312428	2.8557	0.8957452	1.4397	0.5580915	1.7019
46.0	1.3902	65.57	0.4620395	0.0547959	0.1898926	0.3639641	1.0307338	2.7895	0.8983964	1.4147	0.5630639	1.6683
47.0	1.3673	64.09	0.4666117	0.0535752	0.1875900	0.3648289	1.0302123	2.7259	0.9010983	1.3910	0.5682230	1.6362
48.0	1.3456	62.62	0.4713794	0.0523347	0.1852105	0.3656868	1.0296781	2.6648	0.9038497	1.3685	0.5735757	1.6056
49.0	1.3250	61.17	0.4763528	0.0510749	0.1827519	0.3665345	1.0291313	2.6060	0.9066495	1.3470	0.5791293	1.5763
50.0	1.3054	59.74	0.4815427	0.0497964	0.1802123	0.3673681	1.0285719	2.5494	0.9094963	1.3266	0.5848917	1.5482
51.0	1.2868	58.33	0.4869609	0.0484997	0.1775897	0.3681833	1.0279997	2.4948	0.9123889	1.3072	0.5908711	1.5213
52.0	1.2690	56.94	0.4926204	0.0471856	0.1748820	0.3689754	1.0274150	2.4422	0.9153256	1.2887	0.5970763	1.4956
53.0	1.2521	55.56	0.4985352	0.0458547	0.1720869	0.3697392	1.0268175	2.3913	0.9183047	1.2711	0.6035166	1.4709
54.0	1.2361	54.19	0.5047207	0.0445077	0.1692024	0.3704686	1.0262073	2.3422	0.9213244	1.2543	0.6102019	1.4473
55.0	1.2208	52.84	0.5111937	0.0431453	0.1662262	0.3711572	1.0255844	2.2946	0.9243827	1.2383	0.6171427	1.4245
56.0	1.2062	51.50	0.5179729	0.0417683	0.1631560	0.3717974	1.0249490	2.2485	0.9274774	1.2230	0.6243503	1.4027
57.0	1.1924	50.17	0.5250786	0.0403777	0.1599892	0.3723812	1.0243010	2.2039	0.9306059	1.2085	0.6318365	1.3817
58.0	1.1792	48.85	0.5325335	0.0389742	0.1567239	0.3728991	1.0236405	2.1606	0.9337657	1.1946	0.6396140	1.3616
59.0	1.1666	47.54	0.5403624	0.0375588	0.1533575	0.3733408	1.0229677	2.1185	0.9369537	1.1814	0.6476962	1.3422
60.0	1.1547	46.24	0.5485933	0.0361325	0.1498874	0.3736944	1.0222826	2.0776	0.9401669	1.1689	0.6560976	1.3235
θ	Ω_S	A_MIN	σ_0	σ_1	σ_3	σ_5	Ω_1	Ω_2	Ω_3	Ω_4	Ω_5	Ω_6

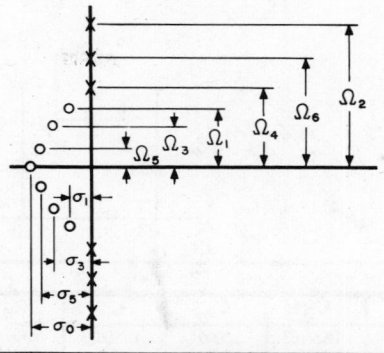

				K² = 1.0					
C_1	C_2	L_2	C_3	C_4	L_4	C_5	C_6	L_6	C_7
1.178	0.00000	1.423	2.094	0.00000	1.574	2.094	0.00000	1.423	1.178
1.17422	0.00491	1.41769	2.07019	0.02243	1.54408	2.06138	0.01598	1.40281	1.16326
1.17343	0.00585	1.41662	2.06566	0.02674	1.53834	2.05518	0.01905	1.39892	1.16038
1.17256	0.00687	1.41545	2.06073	0.03143	1.53211	2.04844	0.02240	1.39469	1.15725
1.17162	0.00798	1.41419	2.05541	0.03652	1.52539	2.04117	0.02604	1.39012	1.15386
1.17062	0.00918	1.41284	2.04970	0.04201	1.51818	2.03337	0.02996	1.38521	1.15022
1.16953	0.01046	1.41138	2.04360	0.04790	1.51048	2.02505	0.03417	1.37996	1.14633
1.16838	0.01183	1.40983	2.03711	0.05420	1.50230	2.01619	0.03868	1.37437	1.14218
1.16715	0.01329	1.40818	2.03024	0.06092	1.49363	2.00682	0.04348	1.36844	1.13777
1.16585	0.01484	1.40643	2.02298	0.06805	1.48449	1.99692	0.04859	1.36217	1.13311
1.16447	0.01648	1.40458	2.01533	0.07561	1.47486	1.98650	0.05402	1.35555	1.12818
1.16301	0.01821	1.40262	2.00730	0.08361	1.46477	1.97556	0.05976	1.34860	1.12300
1.16148	0.02004	1.40057	1.99888	0.09206	1.45419	1.96411	0.06582	1.34130	1.11756
1.15987	0.02196	1.39840	1.99008	0.10095	1.44315	1.95214	0.07221	1.33366	1.11185
1.15818	0.02398	1.39614	1.98090	0.11031	1.43164	1.93967	0.07894	1.32568	1.10588
1.15641	0.02609	1.29276	1.97134	0.12014	1.41967	1.92669	0.08602	1.31735	1.09965
1.15456	0.02831	1.39128	1.96139	0.13046	1.40724	1.91320	0.09345	1.30868	1.09315
1.15263	0.03063	1.38868	1.95107	0.14127	1.39434	1.89922	0.10124	1.29966	1.08638
1.15061	0.03305	1.38598	1.94037	0.15259	1.38100	1.88474	0.10941	1.29030	1.07934
1.14851	0.03558	1.38316	1.92930	0.16444	1.36720	1.86976	0.11796	1.28059	1.07203
1.14631	0.03821	1.38023	1.91785	0.17682	1.35296	1.85430	0.12691	1.27053	1.06444
1.14403	0.04096	1.37717	1.90603	0.18976	1.33827	1.83835	0.13627	1.26013	1.05658
1.14166	0.04382	1.37400	1.89383	0.20328	1.32314	1.82192	0.14605	1.24937	1.04844
1.13920	0.04679	1.37070	1.88126	0.21739	1.30758	1.80501	0.15627	1.23827	1.04001
1.13665	0.04988	1.36729	1.86832	0.23212	1.29159	1.78763	0.16695	1.22682	1.03131
1.13399	0.05310	1.36374	1.85501	0.24748	1.27517	1.76978	0.17809	1.21501	1.02231
1.13124	0.05644	1.36006	1.84133	0.26352	1.25832	1.75147	0.18973	1.20286	1.01303
1.12839	0.05991	1.35625	1.82729	0.28024	1.24106	1.73270	0.20188	1.19034	1.00345
1.12543	0.06352	1.35230	1.81287	0.29769	1.22338	1.71347	0.21455	1.17748	0.99357
1.12237	0.06726	1.34821	1.79810	0.31589	1.20529	1.69380	0.22779	1.16425	0.98339
1.11920	0.07114	1.34398	1.78295	0.33489	1.18680	1.67368	0.24160	1.15067	0.97291
1.11591	0.07517	1.33960	1.76744	0.35471	1.16790	1.65313	0.25603	1.13673	0.96212
1.11251	0.07935	1.33507	1.75157	0.37541	1.14861	1.63215	0.27109	1.12242	0.95101
1.10899	0.08369	1.33039	1.73534	0.39704	1.12893	1.61074	0.28683	1.10776	0.93959
1.10534	0.08819	1.32554	1.71874	0.41964	1.10886	1.58891	0.30328	1.09272	0.92783
1.10157	0.09286	1.32052	1.70178	0.44327	1.08842	1.56667	0.32048	1.07732	0.91575
1.09767	0.09771	1.31534	1.68445	0.46800	1.06759	1.54403	0.33847	1.06156	0.90333
1.09363	0.10274	1.30997	1.66677	0.49390	1.04640	1.52099	0.35730	1.04542	0.89056
1.08945	0.10796	1.30442	1.64872	0.52104	1.02484	1.49757	0.37703	1.02890	0.87744
1.08511	0.11338	1.29868	1.63031	0.54952	1.00291	1.47376	0.39771	1.01201	0.86396
1.08063	0.11902	1.29274	1.61153	0.57943	0.98064	1.44958	0.41940	0.99474	0.85010
1.07599	0.12488	1.28659	1.59239	0.61088	0.95801	1.42504	0.44219	0.97709	0.83587
1.07117	0.13097	1.28022	1.57288	0.64398	0.93504	1.40014	0.46614	0.95905	0.82125
1.06619	0.13730	1.27363	1.55301	0.67889	0.91173	1.37491	0.49136	0.94063	0.80623
1.06102	0.14390	1.26680	1.53277	0.71575	0.88808	1.34933	0.51793	0.92181	0.79079
1.05565	0.15077	1.25972	1.51216	0.75475	0.86411	1.32344	0.54596	0.90259	0.77492
1.05009	0.15793	1.25239	1.49118	0.79607	0.83981	1.29723	0.57560	0.88297	0.75861
1.04431	0.16540	1.24477	1.46982	0.83995	0.81520	1.27073	0.60696	0.86295	0.74184
1.03830	0.17320	1.23687	1.44809	0.88664	0.79028	1.24394	0.64023	0.84252	0.72459
1.03206	0.18135	1.22865	1.42597	0.93646	0.76504	1.21688	0.67557	0.82167	0.70683
1.02556	0.18989	1.22011	1.40347	0.98974	0.73951	1.18957	0.71321	0.80041	0.68856
L_1	L_2	C_2	L_3	L_4	C_4	L_5	L_6	C_6	L_7

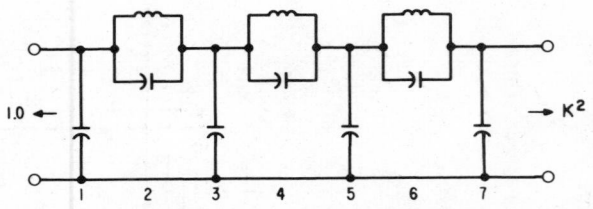

					$K^2 = \infty$				
θ	C_1	C_2	L_2	C_3	C_4	L_4	C_5	C_6	L_6	C_7
C	0.589187	0.0000	1.289299	1.622103	0.0000	1.710023	1.797779	0.0000	1.739136	1.473860
11.0	0.5841	0.0054	1.2816	1.6002	0.0207	1.6739	1.7692	0.0131	1.7174	1.4632
12.0	0.5832	0.0065	1.2801	1.5961	0.0247	1.6671	1.7638	0.0156	1.7133	1.4612
13.0	0.5821	0.0076	1.2785	1.5916	0.0290	1.6597	1.7579	0.0183	1.7088	1.4590
14.0	0.5810	0.0088	1.2768	1.5867	0.0337	1.6517	1.7516	0.0212	1.7040	1.4566
15.0	0.5798	0.0102	1.2749	1.5815	0.0388	1.6431	1.7448	0.0244	1.6988	1.4541
16.0	0.5785	0.0116	1.2729	1.5759	0.0443	1.6339	1.7375	0.0278	1.6932	1.4513
17.0	0.5771	0.0131	1.2708	1.5700	0.0501	1.6241	1.7298	0.0315	1.6873	1.4484
18.0	0.5756	0.0148	1.2685	1.5637	0.0564	1.6138	1.7216	0.0354	1.6810	1.4453
19.0	0.5740	0.0165	1.2661	1.5570	0.0630	1.6029	1.7129	0.0395	1.6744	1.4421
20.0	0.5723	0.0183	1.2635	1.5500	0.0701	1.5914	1.7038	0.0439	1.6674	1.4386
21.0	0.5705	0.0203	1.2608	1.5426	0.0775	1.5793	1.6943	0.0485	1.6600	1.4350
22.0	0.5687	0.0223	1.2580	1.5349	0.0854	1.5667	1.6843	0.0534	1.6523	1.4312
23.0	0.5667	0.0245	1.2550	1.5268	0.0938	1.5535	1.6738	0.0586	1.6442	1.4272
24.0	0.5646	0.0267	1.2519	1.5184	0.1026	1.5398	1.6629	0.0640	1.6358	1.4231
25.0	0.5625	0.0291	1.2486	1.5096	0.1118	1.5254	1.6516	0.0696	1.6270	1.4187
26.0	0.5602	0.0316	1.2452	1.5005	0.1215	1.5106	1.6398	0.0756	1.6179	1.4142
27.0	0.5579	0.0343	1.2417	1.4910	0.1317	1.4952	1.6276	0.0818	1.6083	1.4095
28.0	0.5554	0.0370	1.2379	1.4812	0.1425	1.4792	1.6149	0.0883	1.5985	1.4046
29.0	0.5528	0.0399	1.2341	1.4710	0.1537	1.4627	1.6018	0.0951	1.5882	1.3996
30.0	0.5502	0.0429	1.2300	1.4605	0.1655	1.4456	1.5883	0.1022	1.5777	1.3943
31.0	0.5474	0.0460	1.2259	1.4496	0.1778	1.4281	1.5744	0.1096	1.5667	1.3889
32.0	0.5445	0.0493	1.2215	1.4384	0.1908	1.4099	1.5600	0.1173	1.5554	1.3833
33.0	0.5415	0.0527	1.2170	1.4268	0.2043	1.3913	1.5452	0.1254	1.5437	1.3775
34.0	0.5383	0.0563	1.2123	1.4149	0.2185	1.3721	1.5301	0.1337	1.5317	1.3715
35.0	0.5351	0.0600	1.2074	1.4026	0.2333	1.3524	1.5144	0.1424	1.5193	1.3654
36.0	0.5317	0.0638	1.2024	1.3900	0.2489	1.3322	1.4984	0.1515	1.5066	1.3590
37.0	0.5282	0.0679	1.1972	1.3770	0.2652	1.3115	1.4820	0.1609	1.4935	1.3525
38.0	0.5246	0.0721	1.1918	1.3637	0.2823	1.2903	1.4652	0.1707	1.4800	1.3458
39.0	0.5208	0.0764	1.1862	1.3501	0.3001	1.2686	1.4480	0.1809	1.4661	1.3389
40.0	0.5169	0.0810	1.1804	1.3360	0.3189	1.2463	1.4304	0.1915	1.4519	1.3318
41.0	0.5128	0.0857	1.1744	1.3217	0.3386	1.2236	1.4124	0.2025	1.4374	1.3245
42.0	0.5086	0.0907	1.1682	1.3070	0.3592	1.2004	1.3941	0.2139	1.4224	1.3170
43.0	0.5043	0.0958	1.1618	1.2919	0.3809	1.1768	1.3753	0.2258	1.4071	1.3093
44.0	0.4998	0.1012	1.1552	1.2765	0.4037	1.1526	1.3563	0.2382	1.3914	1.3014
45.0	0.4951	0.1068	1.1484	1.2608	0.4277	1.1280	1.3368	0.2510	1.3754	1.2933
46.0	0.4903	0.1126	1.1413	1.2446	0.4530	1.1029	1.3170	0.2644	1.3590	1.2850
47.0	0.4853	0.1187	1.1340	1.2282	0.4797	1.0774	1.2968	0.2783	1.3422	1.2766
48.0	0.4801	0.1250	1.1265	1.2113	0.5079	1.0514	1.2764	0.2928	1.3250	1.2679
49.0	0.4747	0.1316	1.1186	1.1941	0.5377	1.0250	1.2555	0.3078	1.3074	1.2590
50.0	0.4691	0.1385	1.1106	1.1766	0.5693	0.9981	1.2344	0.3235	1.2895	1.2499
51.0	0.4633	0.1458	1.1022	1.1587	0.6028	0.9708	1.2129	0.3399	1.2712	1.2406
52.0	0.4573	0.1533	1.0936	1.1401	0.6385	0.9431	1.1912	0.3569	1.2525	1.2310
53.0	0.4511	0.1612	1.0847	1.1217	0.6765	0.9150	1.1691	0.3747	1.2333	1.2213
54.0	0.4446	0.1695	1.0754	1.1026	0.7171	0.8865	1.1468	0.3933	1.2138	1.2113
55.0	0.4379	0.1782	1.0659	1.0832	0.7605	0.8575	1.1242	0.4127	1.1939	1.2011
56.0	0.4309	0.1873	1.0560	1.0633	0.8072	0.8282	1.1013	0.4331	1.1736	1.1906
57.0	0.4236	0.1969	1.0457	1.0431	0.8575	0.7986	1.0782	0.4543	1.1528	1.1799
58.0	0.4160	0.2070	1.0351	1.0224	0.9118	0.7685	1.0549	0.4767	1.1317	1.1690
59.0	0.4081	0.2176	1.0240	1.0013	0.9706	0.7381	1.0314	0.5001	1.1100	1.1578
60.0	0.3998	0.2288	1.0125	0.9798	1.0347	0.7074	1.0076	0.5247	1.0880	1.1463
θ	L_1	L_2	C_2	L_3	L_4	C_4	L_5	L_6	C_6	L_7

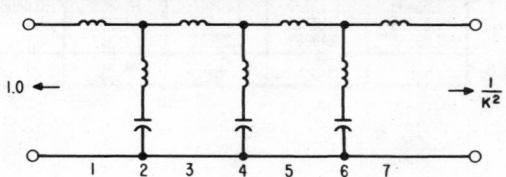

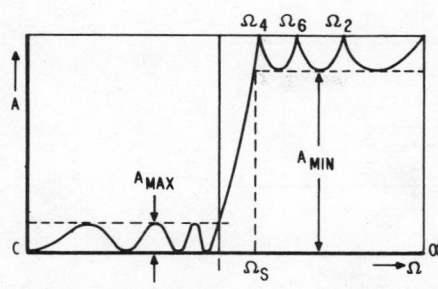

n = 7

ρ = 20%

θ	Ω_S	A_MIN	σ_0	σ_1	σ_3	σ_5	Ω_1	Ω_2	Ω_3	Ω_4	Ω_5	Ω_6
C	∞	∞	0.333376	0.074183	0.207856	0.300361	1.027677	∞	0.824133	∞	0.457359	∞
11.0	5.2408	158.60	0.3368216	0.0727069	0.2059359	0.3016396	1.0273006	11.9886	0.8275565	5.3732	0.4618693	6.6793
12.0	4.8097	153.27	0.3374842	0.0724269	0.2055673	0.3018821	1.0272286	10.9867	0.8282078	4.9307	0.4627349	6.1257
13.0	4.4454	148.37	0.3382070	0.0721227	0.2051653	0.3021455	1.0271503	10.1387	0.8289160	4.5568	0.4636788	5.6575
14.0	4.1336	143.82	0.3389910	0.0717944	0.2047297	0.3024298	1.0270655	9.4116	0.8296810	4.2367	0.4647018	5.2564
15.0	3.8637	139.58	0.3398368	0.0714421	0.2042600	0.3027349	1.0269742	8.7812	0.8305030	3.9597	0.4658048	4.9090
16.0	3.6280	135.61	0.3407454	0.0710657	0.2037557	0.3030606	1.0268765	8.2295	0.8313820	3.7177	0.4669886	4.6052
17.0	3.4203	131.87	0.3417175	0.0706655	0.2032167	0.3034070	1.0267722	7.7424	0.8323181	3.5044	0.4682543	4.3374
18.0	3.2361	128.34	0.3427542	0.0702414	0.2026424	0.3037739	1.0266614	7.3093	0.8333112	3.3152	0.4696030	4.0995
19.0	3.0716	124.99	0.3438566	0.0697935	0.2020323	0.3041612	1.0265439	6.9216	0.8343616	3.1463	0.4710358	3.8868
20.0	2.9238	121.81	0.3450259	0.0693219	0.2013859	0.3045688	1.0264198	6.5725	0.8354691	2.9945	0.4725538	3.6955
21.0	2.7904	118.78	0.3462631	0.0688267	0.2007027	0.3049966	1.0262889	6.2564	0.8366340	2.8574	0.4741586	3.5226
22.0	2.6695	115.88	0.3475698	0.0683080	0.1999820	0.3054444	1.0261514	5.9690	0.8378561	2.7331	0.4758515	3.3656
23.0	2.5593	113.10	0.3489472	0.0677658	0.1992234	0.3059121	1.0260069	5.7063	0.8391356	2.6199	0.4776339	3.2224
24.0	2.4586	110.44	0.3503969	0.0672003	0.1984262	0.3063994	1.0258556	5.4654	0.8404725	2.5164	0.4795074	3.0912
25.0	2.3662	107.88	0.3519204	0.0666116	0.1975896	0.3069062	1.0256974	5.2436	0.8418669	2.4214	0.4814738	2.9707
26.0	2.2812	105.41	0.3535196	0.0659999	0.1967130	0.3074323	1.0255322	5.0388	0.8433188	2.3339	0.4835348	2.8596
27.0	2.2027	103.03	0.3551962	0.0653651	0.1957958	0.3079773	1.0253599	4.8489	0.8448282	2.2532	0.4856924	2.7568
28.0	2.1301	100.72	0.3569521	0.0647075	0.1948370	0.3085410	1.0251805	4.6725	0.8463952	2.1784	0.4879486	2.6615
29.0	2.0627	98.49	0.3587896	0.0640272	0.1938359	0.3091231	1.0249939	4.5080	0.8480196	2.1090	0.4903054	2.5729
30.0	2.0000	96.33	0.3607106	0.0633243	0.1927916	0.3097231	1.0248001	4.3544	0.8497017	2.0445	0.4927652	2.4903
31.0	1.9416	94.24	0.3627177	0.0625990	0.1917033	0.3103408	1.0245990	4.2106	0.8514412	1.9844	0.4953303	2.4132
32.0	1.8871	92.20	0.3648133	0.0618514	0.1905700	0.3109757	1.0243905	4.0756	0.8532384	1.9282	0.4980032	2.3410
33.0	1.8361	90.22	0.3670001	0.0610817	0.1893908	0.3116272	1.0241745	3.9486	0.8550929	1.8756	0.5007866	2.2733
34.0	1.7883	88.29	0.3692809	0.0602902	0.1881647	0.3122949	1.0239511	3.8290	0.8570049	1.8264	0.5036833	2.2096
35.0	1.7434	86.40	0.3716588	0.0594768	0.1868906	0.3129782	1.0237200	3.7161	0.8589743	1.7801	0.5066963	2.1497
36.0	1.7013	84.57	0.3741369	0.0586420	0.1855674	0.3136763	1.0234812	3.6093	0.8610009	1.7366	0.5098287	2.0933
37.0	1.6616	82.77	0.3767189	0.0577859	0.1841940	0.3143886	1.0232348	3.5081	0.8630846	1.6957	0.5130839	2.0400
38.0	1.6243	81.02	0.3794082	0.0569087	0.1827692	0.3151142	1.0229804	3.4121	0.8652252	1.6571	0.5164652	1.9896
39.0	1.5890	79.30	0.3822090	0.0560106	0.1812918	0.3158523	1.0227183	3.3209	0.8674227	1.6206	0.5199765	1.9418
40.0	1.5557	77.62	0.3851253	0.0550918	0.1797607	0.3166017	1.0224481	3.2341	0.8696767	1.5862	0.5236216	1.8966
41.0	1.5243	75.97	0.3881618	0.0541527	0.1781742	0.3173615	1.0221698	3.1513	0.8719871	1.5537	0.5274046	1.8537
42.0	1.4945	74.35	0.3913231	0.0531935	0.1765312	0.3181303	1.0218835	3.0724	0.8743535	1.5229	0.5313300	1.8129
43.0	1.4663	72.76	0.3946146	0.0522146	0.1748303	0.3189068	1.0215890	2.9970	0.8767757	1.4937	0.5354023	1.7740
44.0	1.4396	71.20	0.3980417	0.0512161	0.1730699	0.3196895	1.0212862	2.9248	0.8792531	1.4660	0.5396264	1.7371
45.0	1.4142	69.66	0.4016104	0.0501984	0.1712486	0.3204767	1.0209751	2.8557	0.8817855	1.4397	0.5440075	1.7019
46.0	1.3902	68.15	0.4053272	0.0491619	0.1693647	0.3212666	1.0206555	2.7895	0.8843722	1.4147	0.5485510	1.6683
47.0	1.3673	66.66	0.4091989	0.0481069	0.1674166	0.3220570	1.0203275	2.7259	0.8870128	1.3910	0.5532629	1.6362
48.0	1.3456	65.20	0.4132331	0.0470339	0.1654026	0.3228456	1.0199910	2.6648	0.8897067	1.3685	0.5581492	1.6056
49.0	1.3250	63.75	0.4174377	0.0459431	0.1633209	0.3236300	1.0196458	2.6060	0.8924530	1.3470	0.5632164	1.5763
50.0	1.3054	62.62	0.4218217	0.0448351	0.1611699	0.3244071	1.0192920	2.5494	0.8952511	1.3266	0.5684715	1.5482
51.0	1.2868	60.91	0.4263943	0.0437101	0.1589477	0.3251739	1.0189295	2.4948	0.8981000	1.3072	0.5739218	1.5213
52.0	1.2690	59.52	0.4311660	0.0425689	0.1566522	0.3259269	1.0185581	2.4422	0.9009989	1.2887	0.5795753	1.4956
53.0	1.2521	58.14	0.4361480	0.0414118	0.1542814	0.3266620	1.0181780	2.3913	0.9039465	1.2711	0.5854401	1.4709
54.0	1.2361	56.77	0.4413524	0.0402394	0.1518335	0.3273748	1.0177889	2.3422	0.9069416	1.2543	0.5915251	1.4473
55.0	1.2208	55.42	0.4467927	0.0390523	0.1493062	0.3280605	1.0173910	2.2946	0.9099829	1.2383	0.5978398	1.4245
56.0	1.2062	54.08	0.4524835	0.0378510	0.1466974	0.3287134	1.0169842	2.2485	0.9130690	1.2230	0.6043942	1.4027
57.0	1.1924	52.75	0.4584410	0.0366363	0.1440048	0.3293274	1.0165683	2.2039	0.9161982	1.2085	0.6111991	1.3817
58.0	1.1792	51.43	0.4646830	0.0354088	0.1412261	0.3298956	1.0161435	2.1606	0.9193686	1.1946	0.6182658	1.3616
59.0	1.1666	50.12	0.4712290	0.0341693	0.1383591	0.3304101	1.0157097	2.1185	0.9225781	1.1814	0.6256067	1.3422
60.0	1.1547	48.81	0.4781008	0.0329186	0.1354013	0.3308621	1.0152669	2.0776	0.9258247	1.1689	0.6332348	1.3235
θ	Ω_S	A_MIN	σ_0	σ_1	σ_3	σ_5	Ω_1	Ω_2	Ω_3	Ω_4	Ω_5	Ω_6

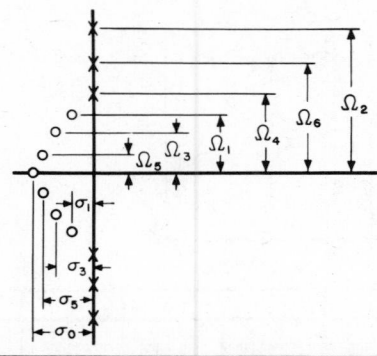

θ	C₁	C₂	L₂	C₃	C₄	L₄	C₅	C₆	L₆	C₇
					K² = 1.0					
C	1.335	0.00000	1.389	2.240	0.00000	1.515	2.240	0.00000	1.389	1.335
11.0	1.33064	0.00503	1.38316	2.21490	0.02330	1.48669	2.20558	0.01637	1.36926	1.31941
12.0	1.32982	0.00599	1.38214	2.21011	0.02777	1.48125	2.19903	0.01952	1.36559	1.31645
13.0	1.32892	0.00704	1.38102	2.20491	0.03264	1.47534	2.19192	0.02295	1.36161	1.31323
14.0	1.32794	0.00818	1.37982	2.19929	0.03792	1.46897	2.18424	0.02667	1.35731	1.30975
15.0	1.32690	0.00941	1.37852	2.19327	0.04362	1.46213	2.17601	0.03068	1.35269	1.30600
16.0	1.32577	0.01072	1.37712	2.18683	0.04973	1.45484	2.16721	0.03499	1.34775	1.30200
17.0	1.32457	0.01213	1.37564	2.17999	0.05627	1.44708	2.15786	0.03959	1.34249	1.29774
18.0	1.32330	0.01362	1.37406	2.17273	0.06323	1.43886	2.14796	0.04451	1.33691	1.29321
19.0	1.32194	0.01521	1.37238	2.16507	0.07063	1.43019	2.13750	0.04973	1.33100	1.28841
20.0	1.32051	0.01689	1.37061	2.15700	0.07848	1.42107	2.12649	0.05527	1.32478	1.28336
21.0	1.31900	0.01866	1.36874	2.14852	0.08677	1.41149	2.11493	0.06113	1.31823	1.27803
22.0	1.31741	0.02054	1.36677	2.13964	0.09552	1.40147	2.10283	0.06732	1.31137	1.27244
23.0	1.31574	0.02250	1.36470	2.13035	0.10474	1.39100	2.09018	0.07384	1.30418	1.26658
24.0	1.31398	0.02457	1.36253	2.12066	0.11443	1.38009	2.07699	0.08071	1.29666	1.26045
25.0	1.31215	0.02674	1.36026	2.11057	0.12461	1.36874	2.06327	0.08792	1.28882	1.25405
26.0	1.31022	0.02901	1.35788	2.10008	0.13529	1.35695	2.04901	0.09549	1.28066	1.24738
27.0	1.30822	0.03138	1.35540	2.08919	0.14648	1.34473	2.03422	0.10343	1.27218	1.24044
28.0	1.30612	0.03386	1.35281	2.07790	0.15820	1.33207	2.01890	0.11174	1.26336	1.23322
29.0	1.30394	0.03645	1.35012	2.06621	0.17045	1.31899	2.00305	0.12044	1.25423	1.22572
30.0	1.30167	0.03914	1.34731	2.05413	0.18325	1.30549	1.98669	0.12954	1.24476	1.21794
31.0	1.29930	0.04196	1.34439	2.04165	0.19663	1.29156	1.96980	0.13905	1.23497	1.20988
32.0	1.29684	0.04488	1.34136	2.02878	0.21059	1.27722	1.95241	0.14898	1.22485	1.20154
33.0	1.29429	0.04793	1.33821	2.01552	0.22516	1.26247	1.93450	0.15935	1.21440	1.19291
34.0	1.29164	0.05109	1.33494	2.00187	0.24036	1.24730	1.91609	0.17017	1.20362	1.18399
35.0	1.28889	0.05438	1.33155	1.98782	0.25621	1.23173	1.89717	0.18146	1.19250	1.17479
36.0	1.28603	0.05780	1.32803	1.97339	0.27274	1.21576	1.87776	0.19323	1.18106	1.16529
37.0	1.28307	0.06135	1.32439	1.95857	0.28998	1.19939	1.85786	0.20551	1.16928	1.15549
38.0	1.28001	0.06504	1.32062	1.94336	0.30794	1.18263	1.83747	0.21832	1.15716	1.14539
39.0	1.27683	0.06887	1.31671	1.92777	0.32668	1.16548	1.81659	0.23168	1.14471	1.13499
40.0	1.27355	0.07284	1.31267	1.91179	0.34622	1.14795	1.79524	0.24560	1.13192	1.12428
41.0	1.27014	0.07696	1.30849	1.89542	0.36660	1.13003	1.77342	0.26013	1.11879	1.11326
42.0	1.26662	0.08123	1.30416	1.87867	0.38787	1.11174	1.75113	0.27529	l.10532	1.10192
43.0	1.26297	0.08566	1.29969	1.86154	0.41006	1.09308	1.72837	0.29110	1.09151	1.09026
44.0	1.25920	0.09026	1.29506	1.84403	0.43324	1.07406	1.70517	0.30761	1.07735	1.07828
45.0	1.25529	0.09504	1.29027	1.82614	0.45746	1.05467	1.6815⊦	0.32484	1.06285	1.06596
46.0	1.25125	0.09999	1.28532	1.80786	0.48277	1.03493	1.65741	0.34285	1.04799	1.05331
47.0	1.24707	0.10513	1.28020	1.78920	0.50926	1.01484	1.63287	0.36167	1.03278	1.04032
48.0	1.24274	0.11046	1.27491	1.77015	0.53699	0.99439	1.60791	0.38135	1.01722	1.02697
49.0	1.23826	0.11600	1.26943	1.75073	0.56606	0.97361	1.58252	0.40196	1.00131	1.01327
50.0	1.23362	0.12175	1.26377	1.73092	0.59655	0.95250	1.55672	0.42354	0.98503	0.99920
51.0	1.22882	0.12772	1.25791	1.71072	0.62857	0.93105	1.53051	0.44616	0.96839	0.98475
52.0	1.22385	0.13394	1.25184	1.69014	0.66223	0.90927	1.50390	0.46990	0.95138	0.96992
53.0	1.21869	0.14040	1.24556	1.66917	0.69768	0.88718	1.47690	0.49484	0.93401	0.95470
54.0	1.21335	0.14712	1.23906	1.64782	0.73505	0.86477	1.44952	0.52106	0.91626	0.93907
55.0	1.20781	0.15412	1.23233	1.62607	0.77452	0.84205	1.42177	0.54868	0.89813	0.92302
56.0	1.20207	0.16141	1.22534	1.60392	0.81628	0.81902	1.39365	0.57779	0.87962	0.90654
57.0	1.19610	0.16902	1.21810	1.58138	0.86054	0.79570	1.36518	0.60854	0.86072	0.88961
58.0	1.18991	0.17696	1.21058	1.55844	0.90754	0.77208	1.33637	0.64106	0.84143	0.87222
59.0	1.18347	0.18525	1.20278	1.53510	0.95758	0.74817	1.30723	0.67552	0.82174	0.85435
60.0	1.17677	0.19393	1.19467	1.51134	1.01098	0.72398	1.27776	0.71211	0.80165	0.83597
θ	L₁	L₂	C₂	L₃	L₄	C₄	L₅	L₆	C₆	L₇

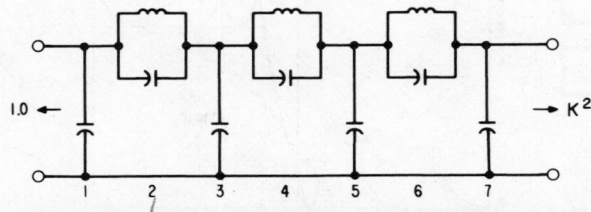

				$K^2 = \infty$					
C_1	C_2	L_2	C_3	C_4	L_4	C_5	C_6	L_6	C_7
0.667478	0.0000	1.361207	1.698055	0.0000	1.736469	1.847593	0.0000	1.750003	1.508166
0.6627	0.0051	1.3538	1.6762	0.0204	1.7010	1.8187	0.0130	1.7283	1.4972
0.6618	0.0061	1.3524	1.6721	0.0243	1.6943	1.8133	0.0155	1.7242	1.4951
0.6608	0.0072	1.3508	1.6675	0.0285	1.6870	1.8073	0.0182	1.7197	1.4928
0.6597	0.0084	1.3491	1.6627	0.0332	1.6792	1.8009	0.0211	1.7149	1.4904
0.6586	0.0096	1.3473	1.6575	0.0382	1.6707	1.7940	0.0243	1.7097	1.4877
0.6573	0.0110	1.3454	1.6519	0.0435	1.6617	1.7867	0.0277	1.7041	1.4849
0.6560	0.0124	1.3434	1.6459	0.0493	1.6521	1.7788	0.0313	1.6982	1.4819
0.6546	0.0140	1.3412	1.6397	0.0554	1.6420	1.7706	0.0352	1.6919	1.4787
0.6531	0.0156	1.3388	1.6330	0.0619	1.6313	1.7618	0.0393	1.6853	1.4754
0.6515	0.0173	1.3364	1.6260	0.0688	1.6200	1.7526	0.0436	1.6783	1.4718
0.6498	0.0192	1.3338	1.6186	0.0762	1.6082	1.7429	0.0482	1.6709	1.4681
0.6481	0.0211	1.3311	1.6109	0.0839	1.5958	1.7328	0.0531	1.6632	1.4641
0.6462	0.0231	1.3282	1.6029	0.0920	1.5829	1.7223	0.0582	1.6551	1.4600
0.6443	0.0253	1.3252	1.5945	0.1006	1.5694	1.7112	0.0636	1.6467	1.4557
0.6422	0.0275	1.3221	1.5857	0.1097	1.5553	1.6997	0.0692	1.6379	1.4513
0.6401	0.0299	1.3188	1.5766	0.1192	1.5407	1.6878	0.0751	1.6288	1.4466
0.6379	0.0323	1.3153	1.5671	0.1291	1.5256	1.6754	0.0813	1.6193	1.4418
0.6355	0.0349	1.3118	1.5573	0.1396	1.5100	1.6626	0.0877	1.6094	1.4367
0.6331	0.0376	1.3080	1.5472	0.1505	1.4938	1.6494	0.0945	1.5991	1.4315
0.6306	0.0404	1.3041	1.5366	0.1620	1.4771	1.6357	0.1015	1.5885	1.4261
0.6279	0.0434	1.3001	1.5258	0.1740	1.4598	1.6216	0.1088	1.5776	1.4205
0.6252	0.0465	1.2959	1.5146	0.1865	1.4420	1.6070	0.1165	1.5663	1.4147
0.6224	0.0497	1.2916	1.5030	0.1997	1.4238	1.5920	0.1245	1.5546	1.4087
0.6194	0.0530	1.2871	1.4911	0.2134	1.4050	1.5766	0.1328	1.5425	1.4025
0.6164	0.0565	1.2824	1.4789	0.2277	1.3857	1.5608	0.1414	1.5301	1.3962
0.6132	0.0601	1.2775	1.4663	0.2428	1.3659	1.5445	0.1504	1.5173	1.3896
0.6099	0.0639	1.2725	1.4534	0.2585	1.3456	1.5279	0.1598	1.5042	1.3829
0.6065	0.0678	1.2673	1.4401	0.2749	1.3248	1.5108	0.1695	1.4907	1.3759
0.6029	0.0719	1.2619	1.4265	0.2921	1.3035	1.4933	0.1796	1.4768	1.3688
0.5992	0.0761	1.2563	1.4126	0.3101	1.2817	1.4755	0.1901	1.4625	1.3614
0.5954	0.0805	1.2506	1.3983	0.3289	1.2594	1.4572	0.2010	1.4479	1.3539
0.5915	0.0851	1.2446	1.3836	0.3487	1.2367	1.4385	0.2123	1.4329	1.3461
0.5874	0.0899	1.2385	1.3686	0.3694	1.2135	1.4195	0.2241	1.4176	1.3382
0.5832	0.0949	1.2321	1.3533	0.3911	1.1899	1.4001	0.2364	1.4018	1.3300
0.5788	0.1001	1.2255	1.3376	0.4139	1.1658	1.3802	0.2492	1.3857	1.3217
0.5743	0.1055	1.2187	1.3216	0.4378	1.1412	1.3601	0.2624	1.3692	1.3131
0.5696	0.1111	1.2117	1.3052	0.4630	1.1162	1.3395	0.2762	1.3523	1.3043
0.5647	0.1169	1.2044	1.2885	0.4895	1.0908	1.3186	0.2906	1.3351	1.2953
0.5597	0.1230	1.1969	1.2714	0.5175	1.0649	1.2973	0.3055	1.3174	1.2861
0.5545	0.1294	1.1891	1.2540	0.5471	1.0386	1.2757	0.3211	1.2994	1.2767
0.5490	0.1360	1.1811	1.2362	0.5783	1.0119	1.2538	0.3373	1.2810	1.2671
0.5434	0.1430	1.1728	1.2180	0.6115	0.9848	1.2315	0.3542	1.2621	1.2572
0.5376	0.1502	1.1642	1.1995	0.6466	0.9573	1.2089	0.3719	1.2429	1.2471
0.5316	0.1578	1.1553	1.1806	0.6840	0.9293	1.1860	0.3903	1.2233	1.2367
0.5253	0.1657	1.1461	1.1614	0.7238	0.9010	1.1628	0.4096	1.2032	1.2262
0.5188	0.1740	1.1366	1.1418	0.7664	0.8723	1.1392	0.4297	1.1827	1.2153
0.5120	0.1827	1.1267	1.1217	0.8120	0.8433	1.1154	0.4508	1.1619	1.2043
0.5049	0.1919	1.1164	1.1013	0.8609	0.8139	1.0913	0.4729	1.1405	1.1929
0.4976	0.2015	1.1058	1.0805	0.9137	0.7841	1.0670	0.4962	1.1188	1.1813
0.4899	0.2116	1.0948	1.0593	0.9707	0.7540	1.0424	0.5206	1.0965	1.1695
L_1	L_2	C_2	L_3	L_4	C_4	L_5	L_6	C_6	L_7

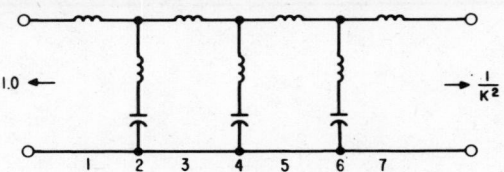

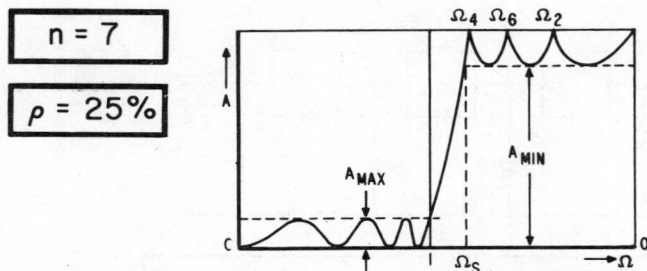

n = 7

ρ = 25%

θ	Ω_S	A_{MIN}	σ_0	σ_1	σ_3	σ_5	Ω_1	Ω_2	Ω_3	Ω_4	Ω_5	Ω_6
C	∞	∞	0.299064	0.066548	0.186464	0.269448	1.017593	∞	0.816046	∞	0.452871	∞
11.0	5.2408	160.64	0.3020947	0.0652607	0.1847953	0.2705817	1.0173832	11.9886	0.8193465	5.3732	0.4571309	6.6793
12.0	4.8097	155.31	0.3026772	0.0650164	0.1844750	0.2707970	1.0173431	10.9867	0.8199747	4.9307	0.4579482	6.1257
13.0	4.4454	150.41	0.3033127	0.0647509	0.1841258	0.2710308	1.0172993	10.1387	0.8206579	4.5568	0.4588393	5.6575
14.0	4.1336	145.86	0.3040018	0.0644644	0.1837473	0.2712831	1.0172519	9.4116	0.8213961	4.2367	0.4598051	5.2564
15.0	3.8637	141.62	0.3047452	0.0641568	0.1833393	0.2715539	1.0172009	8.7812	0.8221894	3.9597	0.4608461	4.9090
16.0	3.6280	137.65	0.3055437	0.0638283	0.1829013	0.2718432	1.0171461	8.2295	0.8230379	3.7177	0.4619634	4.6052
17.0	3.4203	133.91	0.3063979	0.0634788	0.1824331	0.2721509	1.0170876	7.7424	0.8239417	3.5044	0.4631578	4.3374
18.0	3.2361	130.38	0.3073088	0.0631084	0.1819341	0.2724768	1.0170254	7.3093	0.8249008	3.3152	0.4644303	4.0995
19.0	3.0716	127.03	0.3082773	0.0627171	0.1814042	0.2728210	1.0169593	6.9216	0.8259154	3.1463	0.4657819	3.8868
20.0	2.9238	123.85	0.3093044	0.0623051	0.1808427	0.2731834	1.0168893	6.5725	0.8269856	2.9945	0.4672139	3.6955
21.0	2.7904	120.82	0.3103911	0.0618723	0.1802492	0.2735637	1.0168155	6.2564	0.8281114	2.8574	0.4687273	3.5226
22.0	2.6695	117.92	0.3115385	0.0614189	0.1796233	0.2739621	1.0167378	5.9690	0.8292930	2.7331	0.4703235	3.3656
23.0	2.5593	115.14	0.3127479	0.0609449	0.1789643	0.2743783	1.0166561	5.7063	0.8305304	2.6199	0.4720039	3.2224
24.0	2.4586	112.48	0.3140205	0.0604503	0.1782717	0.2748121	1.0165703	5.4654	0.8318239	2.5164	0.4737698	3.0912
25.0	2.3662	109.92	0.3153577	0.0599353	0.1775451	0.2752634	1.0164804	5.2436	0.8331734	2.4214	0.4756230	2.9707
26.0	2.2812	107.45	0.3167610	0.0593999	0.1767837	0.2757321	1.0163864	5.0388	0.8345792	2.3339	0.4775648	2.8596
27.0	2.2027	105.07	0.3182320	0.0588443	0.1759869	0.2762179	1.0162883	4.8489	0.8360411	2.2532	0.4795973	2.7568
28.0	2.1301	102.76	0.3197723	0.0582685	0.1751540	0.2767206	1.0161858	4.6725	0.8375596	2.1784	0.4817220	2.6615
29.0	2.0627	100.54	0.3213836	0.0576726	0.1742844	0.2772401	1.0160792	4.5080	0.8391344	2.1090	0.4839412	2.5729
30.0	2.0000	98.38	0.3230680	0.0570567	0.1733773	0.2777759	1.0159680	4.3544	0.8407659	2.0445	0.4862566	2.4903
31.0	1.9416	96.28	0.3248273	0.0564209	0.1724319	0.2783278	1.0158525	4.2106	0.8424540	1.9844	0.4886706	2.4132
32.0	1.8871	94.24	0.3266638	0.0557654	0.1714473	0.2788955	1.0157324	4.0756	0.8441989	1.9282	0.4911855	2.3410
33.0	1.8361	92.26	0.3285796	0.0550902	0.1704229	0.2794786	1.0156079	3.9486	0.8460005	1.8756	0.4937036	2.2733
34.0	1.7883	90.33	0.3305773	0.0543956	0.1693576	0.2800766	1.0154786	3.8290	0.8478590	1.8264	0.4965275	2.2096
35.0	1.7434	88.45	0.3326594	0.0536816	0.1682506	0.2806891	1.0153447	3.7161	0.8497745	1.7801	0.4993600	2.1497
36.0	1.7013	86.61	0.3348286	0.0529485	0.1671007	0.2813157	1.0152060	3.6093	0.8517469	1.7366	0.5023039	2.0933
37.0	1.6616	84.81	0.3370879	0.0521962	0.1659074	0.2819556	1.0150625	3.5081	0.8537763	1.6957	0.5053622	2.0400
38.0	1.6243	83.06	0.3394405	0.0514250	0.1646692	0.2826083	1.0149140	3.4121	0.8558626	1.6571	0.5085382	1.9896
39.0	1.5890	81.34	0.3418896	0.0506351	0.1633852	0.2832730	1.0147606	3.3209	0.8580059	1.6206	0.5118351	1.9418
40.0	1.5557	79.66	0.3444389	0.0498268	0.1620542	0.2839490	1.0146021	3.2341	0.8602061	1.5862	0.5152566	1.8966
41.0	1.5243	78.01	0.3470921	0.0490000	0.1606752	0.2846354	1.0144385	3.1513	0.8624632	1.5537	0.5188064	1.8537
42.0	1.4945	76.39	0.3498534	0.0481551	0.1592468	0.2853313	1.0142697	3.0724	0.8647770	1.5229	0.5224886	1.8129
43.0	1.4663	74.80	0.3527271	0.0472924	0.1577677	0.2860355	1.0140956	2.9970	0.8671474	1.4937	0.5263072	1.7740
44.0	1.4396	73.24	0.3557179	0.0464119	0.1562368	0.2867468	1.0139161	2.9248	0.8695743	1.4660	0.5302669	1.7371
45.0	1.4142	71.70	0.3588308	0.0455141	0.1546525	0.2874640	1.0137311	2.8557	0.8720575	1.4397	0.5343723	1.7019
46.0	1.3902	70.19	0.3620713	0.0445990	0.1530136	0.2881855	1.0135406	2.7895	0.8745967	1.4147	0.5386284	1.6683
47.0	1.3673	68.70	0.3654452	0.0436601	0.1513185	0.2889098	1.0133445	2.7259	0.8771917	1.3910	0.5430405	1.6362
48.0	1.3456	67.24	0.3689588	0.0427185	0.1495656	0.2896349	1.0131428	2.6648	0.8798421	1.3685	0.5476142	1.6056
49.0	1.3250	65.79	0.3726188	0.0417537	0.1477535	0.2903589	1.0129352	2.6060	0.8825474	1.3470	0.5523556	1.5763
50.0	1.3054	64.36	0.3764326	0.0407730	0.1458804	0.2910796	1.0127218	2.5494	0.8853073	1.3266	0.5572709	1.5482
51.0	1.2868	62.95	0.3804082	0.0397767	0.1439446	0.2917943	1.0125026	2.4948	0.8881213	1.3072	0.5623670	1.5213
52.0	1.2690	61.56	0.3845540	0.0387650	0.1419446	0.2925003	1.0122773	2.4422	0.8909886	1.2887	0.5676509	1.4956
53.0	1.2521	60.18	0.3888796	0.0377387	0.1398783	0.2931945	1.0120460	2.3913	0.8939086	1.2711	0.5731303	1.4709
54.0	1.2361	58.81	0.3933951	0.0366979	0.1377439	0.2938733	1.0118085	2.3422	0.8968806	1.2543	0.5788133	1.4473
55.0	1.2208	57.46	0.3981117	0.0356433	0.1355394	0.2945329	1.0115648	2.2946	0.8999035	1.2383	0.5847087	1.4245
56.0	1.2062	56.12	0.4030415	0.0345751	0.1332629	0.2951688	1.0113148	2.2485	0.9029764	1.2230	0.5908256	1.4027
57.0	1.1924	54.79	0.4081980	0.0334941	0.1309122	0.2957762	1.0110586	2.2039	0.9060971	1.2085	0.5971740	1.3817
58.0	1.1792	53.47	0.4145958	0.0324008	0.1284850	0.2963495	1.0107959	2.1606	0.9092675	1.1946	0.6037644	1.3616
59.0	1.1666	52.16	0.4192512	0.0312957	0.1259794	0.2968825	1.0105268	2.1185	0.9124829	1.1814	0.6106083	1.3422
60.0	1.1547	50.86	0.4251822	0.0301795	0.1233928	0.2973682	1.0102511	2.0776	0.9157429	1.1689	0.6177176	1.3235
θ	Ω_S	A_{MIN}	σ_0	σ_1	σ_3	σ_5	Ω_1	Ω_2	Ω_3	Ω_4	Ω_5	Ω_6

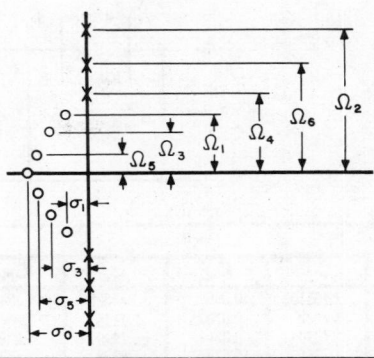

K² = 1.0									
C₁	C₂	L₂	C₃	C₄	L₄	C₅	C₆	L₆	C₇
1.488	0.00000	1.343	2.388	0.00000	1.451	2.388	0.00000	1.343	1.488
1.48360	0.00520	1.33783	2.36123	0.02432	1.42409	2.35139	0.01692	1.32479	1.47199
1.48274	0.00620	1.33685	2.35617	0.02899	1.41894	2.34447	0.02017	1.32134	1.46892
1.48180	0.00728	1.33579	2.35068	0.03407	1.41334	2.33695	0.02371	1.31758	1.46558
1.48078	0.00846	1.33464	2.34476	0.03959	1.40730	2.32885	0.02755	1.31353	1.46198
1.47969	0.00973	1.33340	2.33839	0.04553	1.40082	2.32015	0.03170	1.30917	1.45810
1.47851	0.01108	1.33207	2.33160	0.0519i	1.39391	2.31075	0.03615	1.30451	1.45395
1.47726	0.01254	1.33065	2.32437	0.05872	1.38656	2.30097	0.04090	1.29955	1.44953
1.47593	0.01408	1.32914	2.31671	0.06599	1.37877	2.29051	0.04597	1.29429	1.44483
1.47451	0.01572	1.32754	2.30862	0.07371	1.37056	2.27945	0.05136	1.28873	1.43987
1.47302	0.01746	1.32585	2.30009	0.08189	1.36191	2.26782	0.05708	1.28286	1.43463
1.47144	0.01929	1.32407	2.29114	0.09053	1.35284	2.25560	0.06312	1.27669	1.42911
1.46978	0.02123	1.32219	2.28176	0.09965	1.34334	2.24281	0.06950	1.27021	1.42332
1.46803	0.02326	1.32021	2.27196	0.10926	1.33342	2.22944	0.07622	1.26343	1.41725
1.46620	0.02540	1.31814	2.26173	0.11936	1.32308	2.21550	0.08330	1.25635	1.41091
1.46428	0.02764	1.31597	2.25107	0.12997	1.31232	2.20099	0.09073	1.24896	1.40428
1.46227	0.02998	1.31371	2.23999	0.14109	1.30115	2.18591	0.09852	1.24127	1.39737
1.46017	0.03243	1.31134	2.22849	0.15275	1.28956	2.17027	0.10669	1.23326	1.39018
1.45799	0.03500	1.30887	2.21657	0.16494	1.27757	2.15407	0.11524	1.22496	1.38271
1.45571	0.03767	1.30630	2.20023	0.17770	1.26518	2.13731	0.12419	1.21634	1.37495
1.45333	0.04046	1.30362	2.19148	0.19102	1.25238	2.11999	0.13354	1.20742	1.36690
1.45086	0.04336	1.30084	2.17830	0.20494	1.23918	2.10213	0.14332	1.19818	1.35857
1.44829	0.04638	1.29794	2.16471	0.21946	1.22559	2.08372	0.15352	1.18864	1.34994
1.44563	0.04953	1.29494	2.15071	0.23461	1.21161	2.06477	0.16416	1.17879	1.34102
1.44286	0.05280	1.29182	2.13629	0.25041	1.19723	2.04527	0.17526	1.16862	1.33180
1.43999	0.05620	1.28859	2.12146	0.26678	1.18248	2.02525	0.18684	1.15814	1.32229
1.43701	0.05973	1.28524	2.10622	0.28405	1.16734	2.00469	0.19891	1.14735	1.31247
1.43392	0.06339	1.28177	2.09057	0.30195	1.15183	1.98361	0.21149	1.13624	1.30235
1.43072	0.06720	1.27817	2.07451	0.32060	1.13595	1.96201	0.22460	1.12482	1.29192
1.42741	0.07115	1.27445	2.05805	0.34004	1.11969	1.93989	0.23826	1.11308	1.28118
1.42398	0.07525	1.27060	2.04118	0.36030	1.10308	1.91725	0.25250	1.10101	1.27013
1.42043	0.07950	1.26661	2.02390	0.38143	1.08610	1.89411	0.26734	1.08863	1.25876
1.41676	0.08391	1.26249	2.00622	0.40346	1.06877	1.87047	0.28281	1.07593	1.24706
1.41295	0.08849	1.25823	1.98813	0.42645	1.05108	1.84634	0.29894	1.06290	1.23504
1.40902	0.09323	1.25382	1.96964	0.45044	1.03305	1.82171	0.31576	1.04954	1.22269
1.40495	0.09816	1.24926	1.95074	0.47549	1.01468	1.79660	0.33330	1.03586	1.21000
1.40074	0.10326	1.24455	1.93145	0.50166	0.99597	1.77101	0.35162	1.02185	1.19697
1.39638	0.10856	1.23968	1.91174	0.52902	0.97693	1.74494	0.37074	1.00750	1.18360
1.39187	0.11406	1.23464	1.89164	0.55765	0.95756	1.7184i	0.39073	0.99282	1.16986
1.38720	0.11977	1.22943	1.87113	0.58764	0.93786	1.69142	0.41162	0.97780	1.15577
1.38237	0.12570	1.22404	1.85021	0.61906	0.91785	1.66398	0.43348	0.96244	1.14130
1.37737	0.13186	1.21846	1.82889	0.65204	0.89753	1.63609	0.45637	0.94674	1.12646
1.37219	0.13826	1.21269	1.80716	0.68669	0.87689	1.60776	0.48035	0.93069	1.11123
1.36683	0.14492	1.20672	1.78502	0.72313	0.85596	1.57900	0.50551	0.91428	1.09561
1.36127	0.15184	1.20054	1.76247	0.76151	0.83472	1.54981	0.53194	0.89753	1.07957
1.35551	0.15905	1.19413	1.73950	0.80200	0.81319	1.52021	0.55972	0.88041	1.06312
1.34953	0.16656	1.18750	1.71612	0.84479	0.79137	1.49020	0.58897	0.86293	1.04624
1.34333	0.17439	1.18061	1.69232	0.89009	0.76927	1.45979	0.61980	0.84508	1.02891
1.33688	0.18255	1.17347	1.66809	0.93714	0.74690	1.42899	0.65236	0.82686	1.01113
1.33019	0.19109	1.16606	1.64343	0.98921	0.72424	1.39782	0.68679	0.80825	0.99286
1.32323	0.20001	1.15836	1.61834	1.04364	0.70132	1.36627	0.72328	0.78926	0.97410
L₁	L₂	C₂	L₃	L₄	C₄	L₅	L₆	C₆	L₇

285

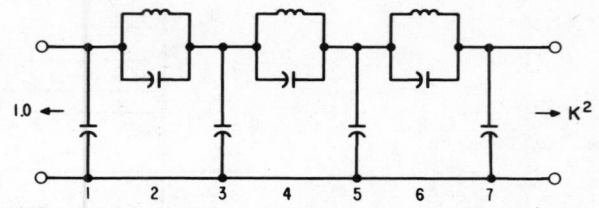

					$K^2 = \infty$				
θ	C_1	C_2	L_2	C_3	C_4	L_4	C_5	C_6	L_6	C_7
C	0.744057	0.0000	1.411805	1.765133	0.0000	1.746286	1.894103	0.0000	1.748136	1.541281
11.0	0.7395	0.0050	1.4046	1.7432	0.0202	1.7115	1.8649	0.0130	1.7266	1.5300
12.0	0.7386	0.0059	1.4032	1.7390	0.0241	1.7049	1.8593	0.0155	1.7225	1.5278
13.0	0.7376	0.0069	1.4017	1.7345	0.0284	1.6978	1.8533	0.0182	1.7180	1.5255
14.0	0.7366	0.0081	1.4001	1.7296	0.0330	1.6900	1.8468	0.0211	1.7132	1.5230
15.0	0.7355	0.0093	1.3984	1.7243	0.0379	1.6817	1.8398	0.0243	1.7080	1.5203
16.0	0.7343	0.0106	1.3965	1.7187	0.0433	1.6729	1.8324	0.0277	1.7025	1.5174
17.0	0.7330	0.0120	1.3945	1.7127	0.0489	1.6635	1.8244	0.0313	1.6966	1.5143
18.0	0.7316	0.0134	1.3924	1.7064	0.0550	1.6535	1.8160	0.0352	1.6904	1.5110
19.0	0.7302	0.0150	1.3902	1.6997	0.0615	1.6430	1.8072	0.0393	1.6838	1.5076
20.0	0.7287	0.0167	1.3878	1.6927	0.0683	1.6320	1.7978	0.0437	1.6768	1.5039
21.0	0.7270	0.0184	1.3853	1.6853	0.0756	1.6204	1.7880	0.0483	1.6695	1.5001
22.0	0.7253	0.0203	1.3826	1.6775	0.0832	1.6082	1.7778	0.0531	1.6619	1.4960
23.0	0.7236	0.0223	1.3799	1.6694	0.0913	1.5955	1.7671	0.0582	1.6538	1.4918
24.0	0.7217	0.0243	1.3769	1.6610	0.0998	1.5823	1.7559	0.0636	1.6454	1.4874
25.0	0.7197	0.0265	1.3739	1.6522	0.1087	1.5685	1.7442	0.0692	1.6367	1.4828
26.0	0.7177	0.0287	1.3707	1.6430	0.1181	1.5542	1.7321	0.0751	1.6276	1.4780
27.0	0.7155	0.0311	1.3674	1.6335	0.1280	1.5393	1.7196	0.0813	1.6181	1.4730
28.0	0.7133	0.0336	1.3639	1.6237	0.1383	1.5240	1.7066	0.0878	1.6083	1.4678
29.0	0.7110	0.0362	1.3603	1.6134	0.1491	1.5081	1.6931	0.0945	1.5981	1.4625
30.0	0.7085	0.0389	1.3566	1.6029	0.1604	1.4917	1.6792	0.1016	1.5876	1.4569
31.0	0.7060	0.0417	1.3526	1.5920	0.1722	1.4747	1.6649	0.1089	1.5767	1.4511
32.0	0.7034	0.0446	1.3486	1.5807	0.1846	1.4573	1.6501	0.1166	1.5654	1.4452
33.0	0.7006	0.0477	1.3444	1.5691	0.1975	1.4394	1.6349	0.1245	1.5537	1.4390
34.0	0.6978	0.0509	1.3400	1.5572	0.2110	1.4209	1.6193	0.1328	1.5417	1.4327
35.0	0.6948	0.0542	1.3355	1.5449	0.2251	1.4020	1.6032	0.1415	1.5294	1.4261
36.0	0.6918	0.0577	1.3308	1.5323	0.2398	1.3826	1.5867	0.1505	1.5167	1.4194
37.0	0.6886	0.0613	1.3259	1.5193	0.2552	1.3627	1.5698	0.1598	1.5036	1.4124
38.0	0.6853	0.0650	1.3209	1.5060	0.2713	1.3423	1.5524	0.1695	1.4901	1.4053
39.0	0.6819	0.0689	1.3157	1.4923	0.2881	1.3214	1.5347	0.1796	1.4763	1.3979
40.0	0.6784	0.0730	1.3103	1.4773	0.3057	1.3000	1.5165	0.1901	1.4621	1.3904
41.0	0.6748	0.0772	1.3047	1.4640	0.3241	1.2782	1.4979	0.2011	1.4475	1.3826
42.0	0.6710	0.0816	1.2979	1.4493	0.3433	1.2560	1.4789	0.2124	1.4325	1.3747
43.0	0.6671	0.0861	1.2930	1.4342	0.3635	1.2332	1.4596	0.2242	1.4172	1.3665
44.0	0.6630	0.0908	1.2868	1.4188	0.3846	1.2100	1.4398	0.2365	1.4015	1.3581
45.0	0.6588	0.0958	1.2804	1.4031	0.4067	1.1864	1.4196	0.2492	1.3855	1.3495
46.0	0.6545	0.1009	1.2738	1.3871	0.4299	1.1623	1.3990	0.2625	1.3690	1.3407
47.0	0.6500	0.1062	1.2670	1.3706	0.4542	1.1378	1.3781	0.2762	1.3522	1.3316
48.0	0.6453	0.1118	1.2600	1.3539	0.4798	1.1129	1.3568	0.2906	1.3349	1.3224
49.0	0.6405	0.1175	1.2527	1.3368	0.5068	1.0875	1.3351	0.3055	1.3173	1.3129
50.0	0.6355	0.1236	1.2452	1.3193	0.5352	1.0618	1.3131	0.3211	1.2993	1.3032
51.0	0.6304	0.1298	1.2374	1.3015	0.5651	1.0356	1.2907	0.3373	1.2809	1.2932
52.0	0.6250	0.1364	1.2294	1.2833	0.5968	1.0090	1.2679	0.3542	1.2621	1.2831
53.0	0.6194	0.1432	1.2211	1.2648	0.6303	0.9820	1.2448	0.3719	1.2429	1.2727
54.0	0.6137	0.1503	1.2125	1.2459	0.6658	0.9547	1.2213	0.3903	1.2233	1.2620
55.0	0.6077	0.1578	1.2036	1.2267	0.7036	0.9270	1.1976	0.4095	1.2033	1.2511
56.0	0.6015	0.1656	1.1943	1.2070	0.7438	0.8988	1.1735	0.4297	1.1728	1.2400
57.0	0.5950	0.1738	1.1848	1.1870	0.7867	0.8704	1.1490	0.4508	1.1619	1.2285
58.0	0.5883	0.1823	1.1749	1.1667	0.8326	0.8415	1.1243	0.4729	1.1406	1.2169
59.0	0.5813	0.1913	1.1646	1.1459	0.8819	0.8124	1.0993	0.4961	1.1189	1.2049
60.0	0.5741	0.2008	1.1539	1.1247	0.9349	0.7829	1.0740	0.5205	1.0967	1.1927
θ	L_1	L_2	C_2	L_3	L_4	C_4	L_5	L_6	C_6	L_7

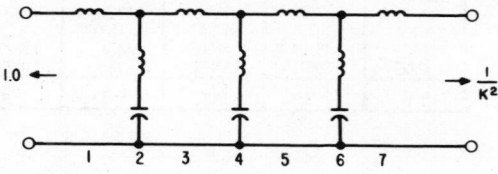

286

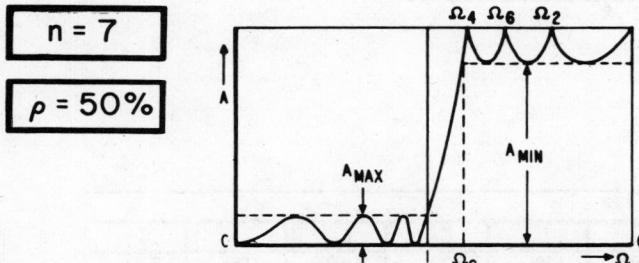

n = 7

ρ = 50%

θ	Ω_S	A_MIN	σ_O	σ_1	σ_3	σ_5	Ω_1	Ω_2	Ω_3	Ω_4	Ω_5	Ω_6
C	∞	∞	0.189249	0.042112	0.117995	0.170507	0.992233	∞	0.795709	∞	0.441585	∞
11.0	5.2408	167.63	0.1910710	0.0413547	0.1170239	0.1712046	0.9924223	11.9886	0.7987116	5.3732	0.4452420	6.6793
12.0	4.8097	162.30	0.1914211	0.0412108	0.1168376	0.1713371	0.9924581	10.9867	0.7992838	4.9307	0.4459431	6.1257
13.0	4.4454	157.40	0.1918029	0.0410546	0.1166345	0.1714811	0.9924971	10.1387	0.7999063	4.5568	0.4467073	5.6575
14.0	4.1336	152.85	0.1922169	0.0408858	0.1164143	0.1716366	0.9925390	9.4116	0.8005792	4.2367	0.4475353	5.2564
15.0	3.8637	148.61	0.1926634	0.0407046	0.1161769	0.1718036	0.9925840	8.7812	0.8013027	3.9597	0.4484277	4.9090
16.0	3.6280	144.64	0.1931428	0.0405108	0.1159222	0.1719820	0.9926319	8.2295	0.8020770	3.7177	0.4493850	4.6052
17.0	3.4203	140.90	0.1936556	0.0403047	0.1156498	0.1721718	0.9926829	7.7424	0.8029021	3.5044	0.4504081	4.3374
18.0	3.2361	137.37	0.1942023	0.0400862	0.1153596	0.1723730	0.9927369	7.3093	0.8037782	3.3152	0.4514976	4.0995
19.0	3.0716	134.02	0.1947833	0.0398551	0.1150513	0.1725857	0.9927938	6.9216	0.8047057	3.1463	0.4526545	3.8868
20.0	2.9238	130.84	0.1953992	0.0396117	0.1147247	0.1728097	0.9928536	6.5725	0.8056845	2.9945	0.4538796	3.6955
21.0	2.7904	127.81	0.1960507	0.0393559	0.1143795	0.1730450	0.9929164	6.2564	0.8067150	2.8574	0.4551738	3.5226
22.0	2.6695	124.91	0.1967383	0.0390878	0.1140154	0.1732916	0.9929820	5.9690	0.8077973	2.7331	0.4565382	3.3656
23.0	2.5593	122.13	0.1974627	0.0388072	0.1136321	0.1735495	0.9930505	5.7063	0.8089318	2.6199	0.4579739	3.2224
24.0	2.4586	119.47	0.1982248	0.0385144	0.1132293	0.1738186	0.9931218	5.4654	0.8101185	2.5164	0.4594820	3.0912
25.0	2.3662	116.91	0.1990251	0.0382091	0.1128067	0.1740988	0.9931958	5.2436	0.8113578	2.4214	0.4610636	2.9707
26.0	2.2812	114.44	0.1998646	0.0378917	0.1123639	0.1743900	0.9932727	5.0388	0.8126499	2.3339	0.4627201	2.8596
27.0	2.2027	112.06	0.2007442	0.0375619	0.1119004	0.1746923	0.9933523	4.8489	0.8139950	2.2532	0.4644528	2.7568
28.0	2.1301	109.75	0.2016647	0.0372199	0.1114159	0.1750055	0.9934346	4.6725	0.8153935	2.1784	0.4662632	2.6615
29.0	2.0627	107.53	0.2026272	0.0368657	0.1109101	0.1753296	0.9935196	4.5080	0.8168456	2.1090	0.4681528	2.5729
30.0	2.0000	105.37	0.2036327	0.0364992	0.1103824	0.1756643	0.9936072	4.3544	0.8183516	2.0445	0.4701232	2.4903
31.0	1.9416	103.27	0.2046823	0.0361207	0.1098324	0.1760096	0.9936974	4.2106	0.8199117	1.9844	0.4721760	2.4132
32.0	1.8871	101.23	0.2057773	0.0357299	0.1092596	0.1763654	0.9937900	4.0756	0.8215264	1.9282	0.4743132	2.3410
33.0	1.8361	99.25	0.2069189	0.0353272	0.1086636	0.1767315	0.9938853	3.9486	0.8231958	1.8756	0.4765365	2.2733
34.0	1.7883	97.32	0.2081084	0.0349123	0.1080436	0.1771077	0.9939829	3.8290	0.8249203	1.8264	0.4788480	2.2096
35.0	1.7434	95.44	0.2093472	0.0344854	0.1073994	0.1774939	0.9940830	3.7161	0.8267002	1.7801	0.4812498	2.1497
36.0	1.7013	93.60	0.2106370	0.0340466	0.1067302	0.1778898	0.9941854	3.6093	0.8285359	1.7366	0.4837442	2.0933
37.0	1.6616	91.80	0.2119793	0.0335959	0.1060354	0.1782952	0.9942902	3.5081	0.8304276	1.6957	0.4863333	2.0400
38.0	1.6243	90.05	0.2133758	0.0331333	0.1053145	0.1787097	0.9943972	3.4121	0.8323757	1.6571	0.4890199	1.9896
39.0	1.5890	88.33	0.2148284	0.0326590	0.1045668	0.1791332	0.9945064	3.3209	0.8343806	1.6206	0.4918064	1.9418
40.0	1.5557	86.65	0.2163390	0.0321729	0.1037914	0.1795653	0.9946178	3.2341	0.8364424	1.5862	0.4946956	1.8966
41.0	1.5243	85.00	0.2179097	0.0316751	0.1029879	0.1800056	0.9947312	3.1513	0.8385617	1.5537	0.4976906	1.8537
42.0	1.4945	83.38	0.2195428	0.0311659	0.1021554	0.1804537	0.9948467	3.0724	0.8407386	1.5229	0.5007943	1.8129
43.0	1.4663	81.79	0.2212407	0.0306450	0.1012930	0.1809091	0.9949641	2.9970	0.8429737	1.4937	0.5040101	1.7740
44.0	1.4396	80.23	0.2230058	0.0301128	0.1004002	0.1813712	0.9950834	2.9248	0.8452670	1.4660	0.5073414	1.7371
45.0	1.4142	78.69	0.2248409	0.0295692	0.0994758	0.1818396	0.9952045	2.8557	0.8476191	1.4397	0.5107919	1.7019
46.0	1.3902	77.18	0.2267490	0.0290144	0.0985192	0.1823136	0.9953273	2.7895	0.8500302	1.4147	0.5143655	1.6683
47.0	1.3673	75.69	0.2287331	0.0284485	0.0975292	0.1827924	0.9954519	2.7259	0.8525006	1.3910	0.5180664	1.6362
48.0	1.3456	74.23	0.2307967	0.0278716	0.0965051	0.1832753	0.9955780	2.6648	0.8550306	1.3685	0.5218988	1.6056
49.0	1.3250	72.78	0.2329434	0.0272838	0.0954456	0.1837613	0.9957055	2.6060	0.8576206	1.3470	0.5258675	1.5763
50.0	1.3054	71.35	0.2351771	0.0266853	0.0943500	0.1842495	0.9958345	2.5494	0.8602707	1.3266	0.5299774	1.5482
51.0	1.2868	69.94	0.2375020	0.0260762	0.0932168	0.1847387	0.9959648	2.4948	0.8629811	1.3072	0.5342337	1.5213
52.0	1.2690	68.55	0.2399227	0.0254567	0.0920452	0.1852277	0.9960964	2.4422	0.8657522	1.2887	0.5386422	1.4956
53.0	1.2521	67.17	0.2424442	0.0248269	0.0908338	0.1857150	0.9962290	2.3913	0.8685841	1.2711	0.5432087	1.4709
54.0	1.2361	65.80	0.2450718	0.0241872	0.0895813	0.1861992	0.9963627	2.3422	0.8714770	1.2543	0.5479396	1.4473
55.0	1.2208	64.45	0.2478113	0.0235374	0.0882864	0.1866784	0.9964973	2.2946	0.8744309	1.2383	0.5528417	1.4245
56.0	1.2062	63.11	0.2506692	0.0228781	0.0869479	0.1871507	0.9966326	2.2485	0.8774460	1.2230	0.5579225	1.4027
57.0	1.1924	61.78	0.2536523	0.0222093	0.0855643	0.1876139	0.9967686	2.2039	0.8805224	1.2085	0.5631896	1.3817
58.0	1.1792	60.46	0.2567683	0.0215314	0.0841341	0.1880654	0.9969051	2.1606	0.8836600	1.1946	0.5686514	1.3616
59.0	1.1666	59.15	0.2600257	0.0208446	0.0826556	0.1885025	0.9970420	2.1185	0.8868586	1.1814	0.5743171	1.3422
60.0	1.1547	57.84	0.2634335	0.0201493	0.0811273	0.1889219	0.9971791	2.0776	0.8901183	1.1689	0.5801964	1.3235
θ	Ω_S	A_MIN	σ_O	σ_1	σ_3	σ_5	Ω_1	Ω_2	Ω_3	Ω_4	Ω_5	Ω_6

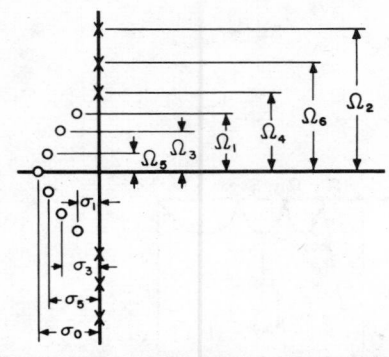

					$K^2 = 1.0$				
θ	C_1	C_2	L_2	C_3	C_4	L_4	C_5	C_6	L_6	C_7
C	2.352	0.00000	1.053	3.297	0.00000	1.108	3.297	0.00000	1.053	2.352
11.0	2.34567	0.00663	1.04932	3.26121	0.03184	1.08799	3.24791	0.02156	1.03987	2.33087
12.0	2.34453	0.00790	1.04858	3.25439	0.03794	1.08415	3.23857	0.02569	1.03734	2.32692
13.0	2.34329	0.00928	1.04777	3.24699	0.04459	1.07999	3.22843	0.03020	1.03458	2.32262
14.0	2.34194	0.01078	1.04689	3.23900	0.05180	1.07550	3.21749	0.03508	1.03161	2.31798
15.0	2.34050	0.01240	1.04595	3.23042	0.05957	1.07069	3.20574	0.04035	1.02841	2.31299
16.0	2.33895	0.01413	1.04495	3.22125	0.06790	1.06554	3.19319	0.04600	1.02499	2.30765
17.0	2.33729	0.01598	1.04387	3.21151	0.07681	1.06008	3.17985	0.05204	1.02135	2.30197
18.0	2.33553	0.01795	1.04273	3.20118	0.08630	1.05429	3.16571	0.05848	1.01749	2.29594
19.0	2.33367	0.02004	1.04151	3.19027	0.09638	1.04818	3.15079	0.06532	1.01340	2.28955
20.0	2.33169	0.02225	1.04023	3.17877	0.10705	1.04175	3.13507	0.07256	1.00909	2.28282
21.0	2.32961	0.02459	1.03888	3.16671	0.11833	1.03500	3.11856	0.08022	1.00456	2.27574
22.0	2.32742	0.02705	1.03745	3.15406	0.13023	1.02794	3.10127	0.08830	0.99981	2.26830
23.0	2.32511	0.02964	1.03596	3.14084	0.14275	1.02056	3.08320	0.09680	0.99484	2.26051
24.0	2.32270	0.03236	1.03439	3.12704	0.15592	1.01287	3.06434	0.10575	0.98964	2.25237
25.0	2.32017	0.03522	1.03275	3.11267	0.16974	1.00487	3.04472	0.11513	0.98421	2.24387
26.0	2.31752	0.03820	1.03103	3.09774	0.18422	0.99656	3.02432	0.12497	0.97856	2.23501
27.0	2.31476	0.04132	1.02924	3.08223	0.19938	0.98795	3.00315	0.13527	0.97269	2.22579
28.0	2.31187	0.04458	1.02737	3.06616	0.21524	0.97903	2.98122	0.14605	0.96659	2.21622
29.0	2.30887	0.04799	1.02542	3.04952	0.23182	0.96981	2.95853	0.15731	0.96027	2.20628
30.0	2.30574	0.05153	1.02339	3.03231	0.24912	0.96029	2.93508	0.16907	0.95372	2.19598
31.0	2.30249	0.05523	1.02129	3.01355	0.26719	0.95048	2.91088	0.18134	0.94694	2.18531
32.0	2.29911	0.05907	1.01910	2.99623	0.28642	0.94037	2.88593	0.19414	0.93993	2.17428
33.0	2.29560	0.06308	1.01682	2.97734	0.30566	0.92997	2.86023	0.20747	0.93269	2.16287
34.0	2.29196	0.06723	1.01447	2.95790	0.32612	0.91928	2.83379	0.22137	0.92523	2.15110
35.0	2.28819	0.07156	1.01202	2.93791	0.34744	0.90831	2.80662	0.23584	0.91754	2.13895
36.0	2.28427	0.07604	1.00949	2.91736	0.36964	0.89705	2.77871	0.25090	0.90961	2.12642
37.0	2.28021	0.08070	1.00686	2.89626	0.39276	0.88551	2.75008	0.26657	0.90145	2.11351
38.0	2.27601	0.08554	1.00415	2.87461	0.41683	0.87370	2.72072	0.28288	0.89306	2.10022
39.0	2.27166	0.09056	1.00134	2.85241	0.44189	0.86161	2.69064	0.29985	0.88444	2.08654
40.0	2.26716	0.09576	0.99843	2.82966	0.46799	0.84925	2.65986	0.31751	0.87558	2.07247
41.0	2.26250	0.10116	0.99542	2.80637	0.49517	0.83662	2.62836	0.33588	0.86648	2.05801
42.0	2.25768	0.10676	0.99231	2.78253	0.52348	0.82373	2.59616	0.35499	0.85715	2.04315
43.0	2.25269	0.11256	0.98909	2.75814	0.55298	0.81058	2.56327	0.37488	0.84758	2.02789
44.0	2.24754	0.11858	0.98577	2.73321	0.58372	0.79717	2.52968	0.39558	0.83776	2.01222
45.0	2.24220	0.12483	0.98233	2.70773	0.61578	0.78350	2.49541	0.41712	0.82771	1.99614
46.0	2.23669	0.13130	0.97878	2.68171	0.64923	0.76958	2.46045	0.43956	0.81741	1.97964
47.0	2.23098	0.13802	0.97511	2.65515	0.68414	0.75542	2.42482	0.46294	0.80686	1.96272
48.0	2.22508	0.14498	0.97132	2.62804	0.72062	0.74101	2.38852	0.48730	0.79607	1.94537
49.0	2.21898	0.15221	0.96739	2.60038	0.75874	0.72636	2.35155	0.51270	0.78502	1.92758
50.0	2.21267	0.15972	0.96334	2.57218	0.79863	0.71148	2.31392	0.53920	0.77373	1.90934
51.0	2.20614	0.16751	0.95915	2.54343	0.84041	0.69636	2.27565	0.56688	0.76218	1.89066
52.0	2.19938	0.17560	0.95481	2.51413	0.88420	0.68101	2.23672	0.59579	0.75036	1.87151
53.0	2.19239	0.18401	0.95032	2.48428	0.93016	0.66544	2.19716	0.62601	0.73829	1.85189
54.0	2.18514	0.19276	0.94568	2.45386	0.97846	0.64965	2.15695	0.65765	0.72596	1.83179
55.0	2.17763	0.20186	0.94088	2.42289	1.02928	0.63363	2.11612	0.69080	0.71335	1.81120
56.0	2.16986	0.21133	0.93590	2.39135	1.08283	0.61741	2.07467	0.72556	0.70048	1.79010
57.0	2.16179	0.22120	0.93074	2.35925	1.13936	0.60097	2.03260	0.76206	0.68732	1.76849
58.0	2.15342	0.23149	0.92540	2.32656	1.19914	0.58433	1.98992	0.80044	0.67389	1.74634
59.0	2.14473	0.24223	0.91985	2.29329	1.26247	0.56748	1.94663	0.84084	0.66017	1.72364
60.0	2.13570	0.25345	0.91409	2.25943	1.32971	0.55044	1.90275	0.88346	0.64616	1.70037
θ	L_1	L_2	C_2	L_3	L_4	C_4	L_5	L_6	C_6	L_7

288

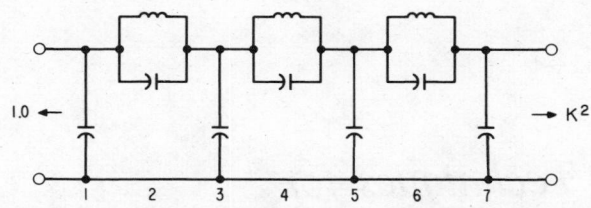

				$K^2 = \infty$					
C_1	C_2	L_2	C_3	C_4	L_4	C_5	C_6	L_6	C_7
1.175812	0.0000	1.454830	2.122869	0.0000	1.635410	2.190128	0.0000	1.607506	1.767247
1.1714	0.0048	1.4487	2.0986	0.0216	1.6048	2.1573	0.0141	1.5879	1.7541
1.1705	0.0057	1.4476	2.0940	0.0257	1.5990	2.1511	0.0168	1.5842	1.7516
1.1696	0.0067	1.4463	2.0891	0.0302	1.5927	2.1443	0.0198	1.5802	1.7489
1.1686	0.0078	1.4449	2.0837	0.0351	1.5859	2.1370	0.0230	1.5758	1.7460
1.1675	0.0090	1.4434	2.0779	0.0404	1.5786	2.1292	0.0264	1.5711	1.7428
1.1663	0.0102	1.4418	2.0717	0.0461	1.5708	2.1208	0.0301	1.5661	1.7395
1.1651	0.0116	1.4402	2.0651	0.0521	1.5625	2.1119	0.0341	1.5607	1.7359
1.1638	0.0130	1.4383	2.0581	0.0586	1.5538	2.1024	0.0383	1.5551	1.7321
1.1624	0.0145	1.4364	2.0508	0.0654	1.5445	2.0925	0.0427	1.5491	1.7281
1.1609	0.0161	1.4344	2.0430	0.0727	1.5348	2.0820	0.0475	1.5427	1.7238
1.1593	0.0178	1.4323	2.0349	0.0803	1.5246	2.0710	0.0525	1.5361	1.7193
1.1577	0.0196	1.4300	2.0264	0.0884	1.5139	2.0594	0.0577	1.5291	1.7147
1.1560	0.0215	1.4277	2.0174	0.0970	1.5027	2.0474	0.0633	1.5218	1.7097
1.1541	0.0235	1.4252	2.0081	0.1059	1.4911	2.0348	0.0691	1.5142	1.7046
1.1522	0.0256	1.4226	1.9984	0.1153	1.4789	2.0217	0.0752	1.5062	1.6993
1.1503	0.0277	1.4199	1.9883	0.1252	1.4664	2.0081	0.0816	1.4979	1.6937
1.1482	0.0300	1.4171	1.9779	0.1355	1.4533	1.9940	0.0883	1.4893	1.6879
1.1460	0.0324	1.4142	1.9670	0.1464	1.4398	1.9793	0.0954	1.4804	1.6819
1.1438	0.0349	1.4111	1.9558	0.1577	1.4258	1.9642	0.1027	1.4711	1.6756
1.1414	0.0375	1.4079	1.9442	0.1695	1.4114	1.9485	0.1103	1.4615	1.6691
1.1390	0.0402	1.4046	1.9322	0.1818	1.3965	1.9324	0.1183	1.4516	1.6624
1.1365	0.0430	1.4012	1.9198	0.1947	1.3812	1.9157	0.1266	1.4413	1.6555
1.1338	0.0459	1.3976	1.9070	0.2082	1.3655	1.8986	0.1353	1.4307	1.6483
1.1311	0.0489	1.3939	1.8939	0.2222	1.3493	1.8809	0.1443	1.4197	1.6409
1.1283	0.0521	1.3900	1.8804	0.2368	1.3326	1.8628	0.1536	1.4085	1.6333
1.1253	0.0554	1.3860	1.8665	0.2520	1.3156	1.8442	0.1634	1.3968	1.6254
1.1223	0.0588	1.3819	1.8523	0.2679	1.2981	1.8250	0.1735	1.3849	1.6173
1.1191	0.0623	1.3777	1.8376	0.2845	1.2802	1.8054	0.1841	1.3726	1.6090
1.1159	0.0660	1.3732	1.8226	0.3017	1.2618	1.7854	0.1950	1.3600	1.6004
1.1125	0.0699	1.3687	1.8073	0.3197	1.2431	1.7648	0.2064	1.3470	1.5916
1.1090	0.0738	1.3639	1.7915	0.3385	1.2239	1.7438	0.2182	1.3337	1.5826
1.1054	0.0779	1.3590	1.7754	0.3580	1.2044	1.7223	0.2305	1.3200	1.5733
1.1016	0.0822	1.3540	1.7589	0.3784	1.1844	1.7003	0.2433	1.3060	1.5638
1.0978	0.0867	1.3488	1.7420	0.3997	1.1641	1.6779	0.2566	1.2917	1.5540
1.0938	0.0913	1.3434	1.7248	0.4220	1.1434	1.6550	0.2704	1.2769	1.5439
1.0896	0.0961	1.3378	1.7072	0.4452	1.1222	1.6317	0.2847	1.2619	1.5337
1.0853	0.1010	1.3320	1.6892	0.4695	1.1007	1.6079	0.2997	1.2465	1.5231
1.0809	0.1062	1.3261	1.6709	0.4949	1.0789	1.5836	0.3152	1.2307	1.5123
1.0763	0.1116	1.3199	1.6522	0.5216	1.0566	1.5589	0.3314	1.2145	1.5012
1.0716	0.1171	1.3135	1.6331	0.5495	1.0341	1.5338	0.3482	1.1980	1.4899
1.0667	0.1229	1.3069	1.6136	0.5788	1.0111	1.5082	0.3658	1.1811	1.4783
1.0616	0.1290	1.3001	1.5937	0.6096	0.9878	1.4822	0.3841	1.1639	1.4664
1.0563	0.1352	1.2931	1.5735	0.6420	0.9642	1.4558	0.4032	1.1462	1.4543
1.0509	0.1418	1.2858	1.5529	0.6761	0.9402	1.4289	0.4232	1.1282	1.4418
1.0452	0.1486	1.2783	1.5319	0.7121	0.9159	1.4017	0.4440	1.1098	1.4291
1.0394	0.1557	1.2705	1.5105	0.7501	0.8912	1.3740	0.4658	1.0910	1.4161
1.0333	0.1631	1.2624	1.4887	0.7904	0.8663	1.3459	0.4887	1.0718	1.4027
1.0270	0.1708	1.2540	1.4665	0.8331	0.8410	1.3173	0.5127	1.0522	1.3891
1.0204	0.1789	1.2453	1.4440	0.8786	0.8155	1.2884	0.5378	1.0322	1.3751
1.0136	0.1874	1.2363	1.4210	0.9270	0.7896	1.2591	0.5643	1.0117	1.3608
L_1	L_2	C_2	L_3	L_4	C_4	L_5	L_6	C_6	L_7

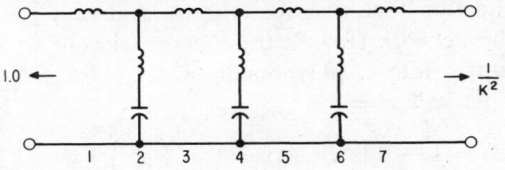

6

Design Techniques for Polynomial Filters

6.1 INTRODUCTION TO TABLES OF NORMALIZED ELEMENT VALUES

A majority of modern high-frequency filters can be designed with the aid of tables of normalized parameters as developed in the previous chapter. This is especially true where polynomial filters are concerned. There are many types of polynomial filters and they are indeed adaptable for tabulation.

Polynomial Filters

The amplitude response of Cauer parameter (CC) filters are characterized by peaks of attenuation in the stopband. The Butterworth, Chebyshev, and other filters to be considered here have no attenuation peaks at finite frequencies. These filters have become known as *all-pole* (transfer function) filters, since their voltage ratio

$$W(s) = \frac{V_2(s)}{V_1(s)}$$

is an all-pole function, with zeros only at infinity. To the filter designer, however, the term *pole* is often linked with "pole of attenuation," which is exactly opposite to the type of amplitude response expressed by the foregoing ratio. Pole-zero diagrams for transmission function, as we know, include only zeros and no poles which constitute the denominator polynomial in the filtering function. The term *polynomial filters* will be used here to describe these filters since the rational function that describes the transmission property of the network (for all these cases without poles) degenerates into a polynomial in *s*. In the Butterworth filter with $n = 3$

$$H(s)H(-s) = 1 + D(s)D(-s) = 1 + \left(\frac{s}{j\omega_c}\right)^6$$

Therefore,

$$D = \pm\left(\frac{s}{\omega_c}\right)^3$$

From now on we assume *s* a normalized parameter (relative to ω_c).

$$D = \frac{F}{P}$$

$$F = \pm s^3$$

and

$$P = 1.$$

Taking a negative sign for *s*,

$$D = \frac{-s^3}{1}.$$

The transmission function is

$$H = \frac{E}{P} = (s + 1)(s^2 + s + 1)$$

which is polynomial in *s*.

The polynomial filter is physically simpler than the CC filter, since the lowpass model in Fig. 6.1 consists only of inductors in the series arm and capacitors in the shunt arm. The transformation to the highpass, bandpass, and bandstop schematics, as well as the impedance and frequency scaling technique is performed in the same way as for the CC filter.

Principles of Tabulation

Section 6.9 presents the normalized element values for polynomial filters of a variety of response shapes. Included is data describing the Butterworth, Chebyshev, maximally flat delay, linear phase with equiripple

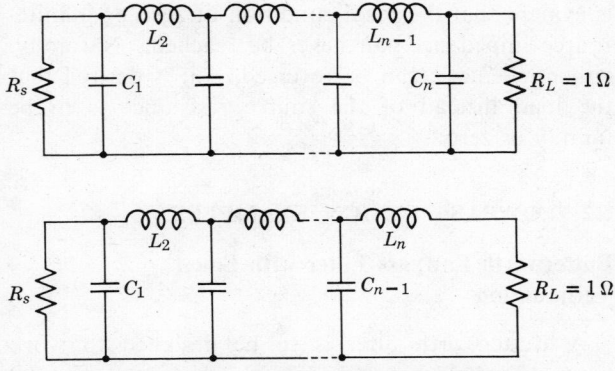

Fig. 6.1. Odd- and even-order polynomial filters.

error, Gaussian, modified Gaussian, and Legendre types. The tabulation is, of course, divided into groups according to the above types of responses, and within each of the catalogs, the normalized element values are given for filters up to the tenth order.

In order to limit the size of the tables which give element values corresponding to particular approximation, some restrictions must be employed. For example, only a limited number of load to source resistances can be used, and only a finite number of physical quality factors of components can be considered.

Even if the load-to-source resistance ratio and element Q are known, there are still a number of different ladder networks which will give the same

transmission properties. These different networks are characterized by the difference in the location of the reflection coefficient singularities. (See, for example, the tables of CC filters of Chapter 5 where one of the parameters is the reflection coefficient.) It can be seen that if all of the foregoing factors are included, just one simple filter would yield quite a large table.

An extremely useful set of tables can be produced having one limitation imposed: All the reflection zeros must be in the left-half plane. This gives a network with a small input capacitor for some types of filters, and hence a larger gain bandwidth product is possible.

Most filters are either designed with only one end loaded, or equally terminated so that there is a maximum power transfer. For the latter type, at the frequency where the magnitude characteristic is at its largest, there will be an equal amount of loss in the load and in the source. Nevertheless, in order to compliment the two cases, the tables list element values calculated for a variety of ratios of source-to-load impedances. The single-terminated filters are used where it is inconvenient or wasteful of power to use the double-terminated filter. On the other hand, it has been observed that filters designed on the basis of maximum power transfer are less critical to align. In Fig. 6.2 a single-terminated basic circuit is changed into several circuits each with the same frequency transmission properties.

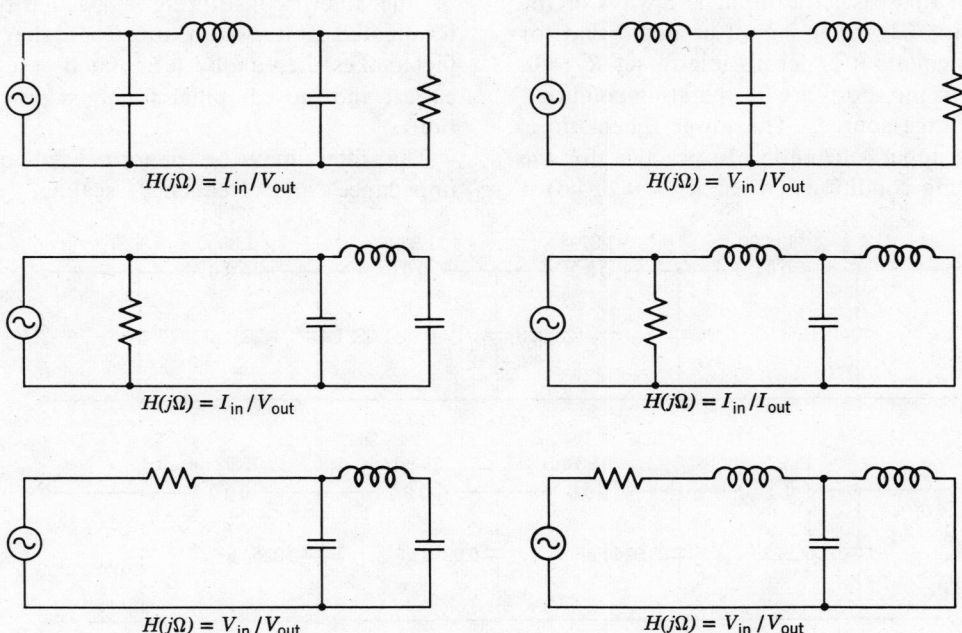

Fig. 6.2. Transformation of single-terminated filter.

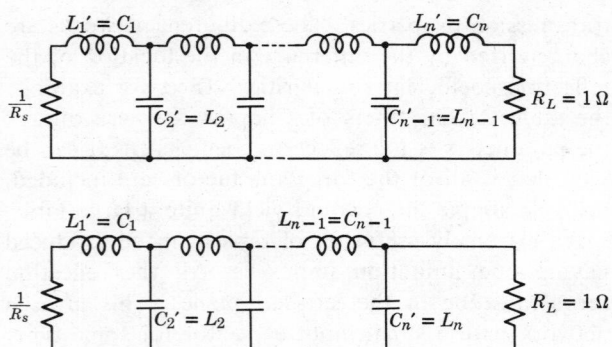

Fig. 6.3. Odd- and even-order polynomial filters, dual schematics of Fig. 6.1.

The element values listed in the tables correspond to the components of the basic lowpass model shown in Fig. 6.1. All element values have been calculated to give a unity 3-dB bandwidth.

The tables set the load, or output resistance as equal to unity and allow the input, or source resistance (R_s) to vary from one to infinity. In the case of n even, R_s increases from unity, and for n odd, R_s decreases. When using the tables, you will note that the schematic at the top of the page corresponds to the top column headings. This schematic always has a π-type input, and can be used, if desired, with open-circuit input resistance corresponding to an ideal current source. On the other hand, if the lower schematic is used, the column headings at the bottom of the tables are to be used. In this case, the input is always of the T-type as in Fig. 6.3 and it should be noted that for this type of schematic it is permissible to set $R_s = 0$, corresponding to the short-circuit operating condition of an ideal voltage source. The lower schematic is the dual of the upper schematic. In practice, for the extreme operating conditions (when $R_s = 0$ or ∞) it

is evident that the ideal condition of zero or infinite-source impedance can never be reached. Naturally, either schematic can be reversed, if it is desired that the load instead of the source resistance is to be infinity or zero.

6.2 LOWPASS DESIGN EXAMPLES

Butterworth Lowpass Filter with Equal Termination

A Butterworth filter is to be designed to work between equal source and load resistances of 1000 ohms. The filter must pass frequencies between 0 and 10 kc, the maximum attenuation at 10 kc being 3-dB. Not less then 30-dB attenuation must be provided at 15 kc and above.

From the nomogram by Kawakami (Fig. 5.1), it can be determined that the filter must be between the eighth and ninth order. The ninth order must be chosen. The desired information is given on page 314 of the tables, with $n = 9$, and $R_s = 1$ the normalized element values can then be read from the table. We are using the schematic at the top of the page, having a π input, since this schematic contains four inductors and five capacitors. Its dual contains five inductors and four capacitors.

In Fig. 6.4a, the lowpass network is shown with normalized element values. It should be noted that the element values are symmetrical about the center of the filter, constituting a pecularity of equally terminated Butterworth and Chebyshev filters. This fact makes the equally terminated case certainly the easiest and most popular for most practical applications.

The filter may be denormalized by the usual impedance and frequency scaling. A reference

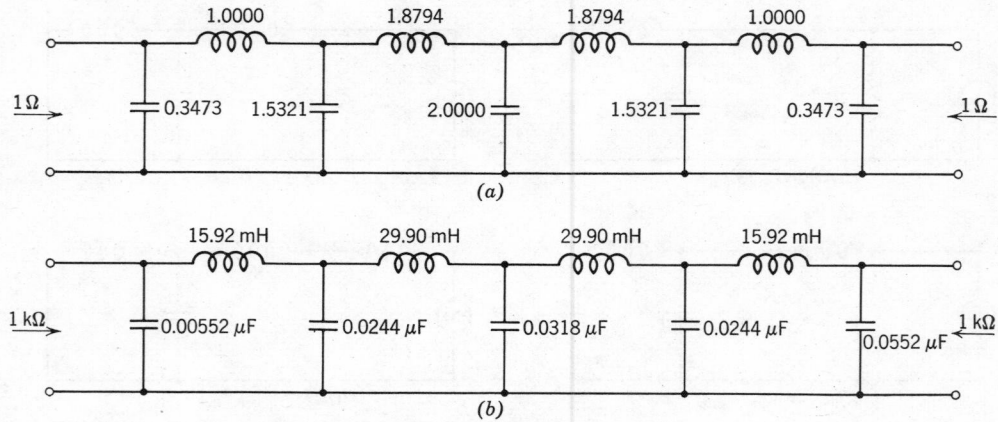

Fig. 6.4. Butterworth lowpass filter.

inductor is defined as

$$L_r = \frac{R_r}{\omega_r}$$

Likewise, a reference capacitor is defined as

$$C_r = \frac{1}{R_r \omega_r}.$$

With the known values of $R_r = 1000$ ohms and $\omega_r = 2\pi\, 10{,}000$ rad,

$$L_r = \frac{10^3}{2\pi 10^4} = 15.92\ \text{mH},$$

and

$$C_r = \frac{1}{10^3 2\pi 10^4} = 0.01592\ \mu\text{F}.$$

Multiplying the reference values by the normalized element values of Fig. 6.4*a*, denormalizes the filter, and gives the actual element values of Fig. 6.4*b*.

We can conclude from this example that the evaluation of circuit elements from the given filter requirements presents no difficulty, and the entire procedure can be accomplished with the aid of a slide rule.

Butterworth Lowpass Filter with Unequal Termination

A Butterworth lowpass filter must operate between a source resistance of 50 ohms and a load resistance of 1000 ohms. The passband is from 0 to 50 kc, the maximum allowable attenuation at 50 kc being 0.2 dB. At 100 kc, the filter must provide no less than a 30-dB discrimination. The ratio of the source to load impedance is

$$\frac{R_s}{R_L} = \frac{50}{1000} = 0.05$$

and is such that the extreme case of impedance transformation is to be applied. That is, it can be assumed that $R_s = 0$. The lower schematic (*T*-input) is acceptable for this purpose. Here, the lower line is to be used, with $1/R_s = $ INF.

The complexity of the network can be again determined by use of a nomogram. Noting that the maximum passband attenuation is 0.2 dB, we find that a seventh-order filter will provide more than 30 dB at the specified stopband frequency. If the maximum attenuation in the passband were allowed to be 3 dB as in the previous example, a fifth-order filter would be sufficient to provide the necessary attenuation.

The normalized element values are obtained from the catalog, with $n = 7$ and $1/R_s = $ INF.

$$L_1 = 1.5576$$
$$C_2 = 1.7988$$
$$L_3 = 1.6588$$
$$C_4 = 1.3972$$
$$L_5 = 1.0550$$
$$C_6 = 0.6560$$
$$L_7 = 0.2225$$

The entire structure is not symmetrical and tends to be more difficult to align. The remainder of the design procedure—that is, denormalizing, is similar to the first example.

Chebyshev Filter Design

On pages 315–322 of the catalog, data for the lowpass models of Chebyshev filters are presented in three groups. The first group (pages 315–317) are for filters with a passband ripple of 0.01 dB, or approximately 4.5% reflection coefficient ρ. Element value information for ripples below 0.01 dB can be found in the upper rows of the catalogs of Chapter 5. The second group (pages 318–320) and the third group (pages 321–322) are element values for Chebyshev filters with 0.1- and 0.5-dB passband ripple, corresponding to approximately 14% and 35% reflection coefficient respectively. The element value information and schematics are presented in the same manner as Butterworth filters.

The Chebyshev response, from many points of view, is more advantageous than the Butterworth response. In order to appreciate the difference we will consider a Chebyshev design in the same way as we did for the Butterworth filter. In the case of the Chebyshev filter there is an accurately definable ripple bandwidth which is, evidently, narrower than a 3-dB bandwidth. Filter specifications can be formulated in terms of ripple bandwidth and correlated with 3-dB data in the tables by Fig. 6.5, giving values of $\Omega_{3\,\text{dB}}/\Omega_{\text{ripple}}$.

Chebyshev Lowpass Filter Having Specifications Similar to Butterworth with Equal Termination

A lowpass filter is allowed to exhibit an equal ripple amplitude behavior in the passband of 0.1 dB. The attenuation at 10 kc must be no more than 3 dB. No less than 30-dB attenuation is required at 15 kc.

It can be shown that such a filter will be of sixth order (six elements). The Butterworth filter design discussed previously required the use of nine elements. Economy of elements is not the only advantage of this

Ratio of $\Omega_{3\,dB}/\Omega_{ripple}$

Ripple n	0.001 dB	0.005 dB	0.01 dB	0.05 dB	0.10 dB	0.25 dB	0.50 dB	1.00 dB
2	5.7834930	3.9027831	3.3036192	2.2685899	1.9432194	1.5981413	1.3897437	1.2176261
3	2.6427081	2.0740079	1.8771819	1.5120983	1.3889948	1.2528880	1.1674852	1.0948680
4	1.8416695	1.5656920	1.4669048	1.2783955	1.2130992	1.1397678	1.0931019	1.0530019
5	1.5155888	1.3510908	1.2912179	1.1753684	1.1347180	1.0887238	1.0592591	1.0338146
6	1.3495755	1.2397596	1.1994127	1.1207360	1.0929306	1.0613406	1.0410296	1.0234422
7	1.2531352	1.1743735	1.1452685	1.0882424	1.0680005	1.0449460	1.0300900	1.0172051
8	1.1919877	1.1326279	1.1106090	1.0673321	1.0519266	1.0343519	1.0230107	1.0131638
9	1.1507149	1.1043196	1.0870644	1.0530771	1.0409547	1.0271099	1.0181668	1.0103963
10	1.1215143	1.0842257	1.0703312	1.0429210	1.0331307	1.0219402	1.0147066	1.0084182

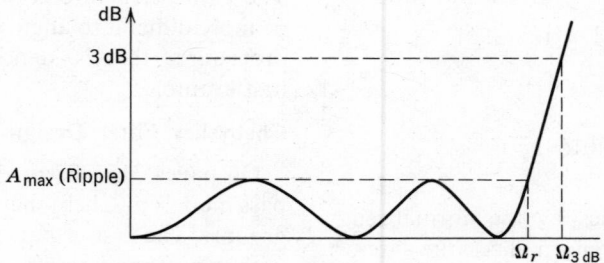

Fig. 6.5.   Ratios of $\Omega_{3\,dB}/\Omega_{ripple}$ for Chebyshev filters.

design over the Butterworth design. It can be shown that the usable bandwidth in the Chebyshev filter is wider and the response curve is sharper.

Unlike Butterworth filters, Chebyshev filters are distinctively different, so far as even and odd orders are involved. In the case of the odd-order filters ($n = 3, 5, \ldots,$) when symmetrical, they can be realized with equal input and output impedances without any complication. This is reflected in the table in which the R_s column gives the ratio 1.000. For the even-order filter, however, the ratio is never unity, and is always higher depending on the amplitude of the passband ripples. Therefore, even-order Chebyshev filters always have an impedance transforming ratio within its schematic.

Another peculiarity of even-order filters is that they do not provide a maximum transmission at zero frequency; the attenuation is finite, even in the purely reactive network, and is equal to the maximum value of passband ripple. The finite attenuation at zero frequency is particularly undesirable if pulsed- or step-function information is to be transmitted where a maximum final value of the signal is important. In order to overcome this inconvenience, a small amount of stopband attenuation can be sacrificed to force the maximum transmission to occur at zero frequency.

Chapter 5 explained the method of avoidance of the transforming action of the even-order Chebyshev polynomial filter and how to force, at the same time,

the attenuation at zero frequency to be equal to zero. In Table 5.1 the even-order schematic and passband response near zero frequency are shown.

Second Example of Chebyshev Filter

A lowpass filter is to be developed and is to have a passband from 0 to 3200 cycles. At 4100 cycles an attenuation of 40 dB required. The passband ripple is allowed to be 0.5 dB. The frequency, 3200 cycles, is the half-dB point.

With the nomogram for Chebyshev filters, Fig. 5.2, with $A_{max} = 0.5$ and $A_{min} = 40$, and the ratio $\Omega_s/\Omega_c = 1.28$, the filter will be of the eighth order and will exhibit a 41-dB loss at 4100 cycles. The filter is antimetric. According to the table on page 322 for $n = 8$, the minimum ratio of input to output impedance in this filter will be 1.9841. Denormalization of element values is performed in the same manner as before. If one desires to modify the filter so that the input and output impedances are equal, and that zero dB will be obtained at zero frequency, the modified Chebyshev filter can be used.

In Fig. 6.6, the normalized response curves for even-order filters are shown. The curves marked b, belong to the regular Chebyshev realization, and those marked c are the modified Chebyshev realization having equal termination. To appreciate the loss of attenuation outside of the passband as a result of forcing the attenuation to be zero at $\Omega = 0$, we

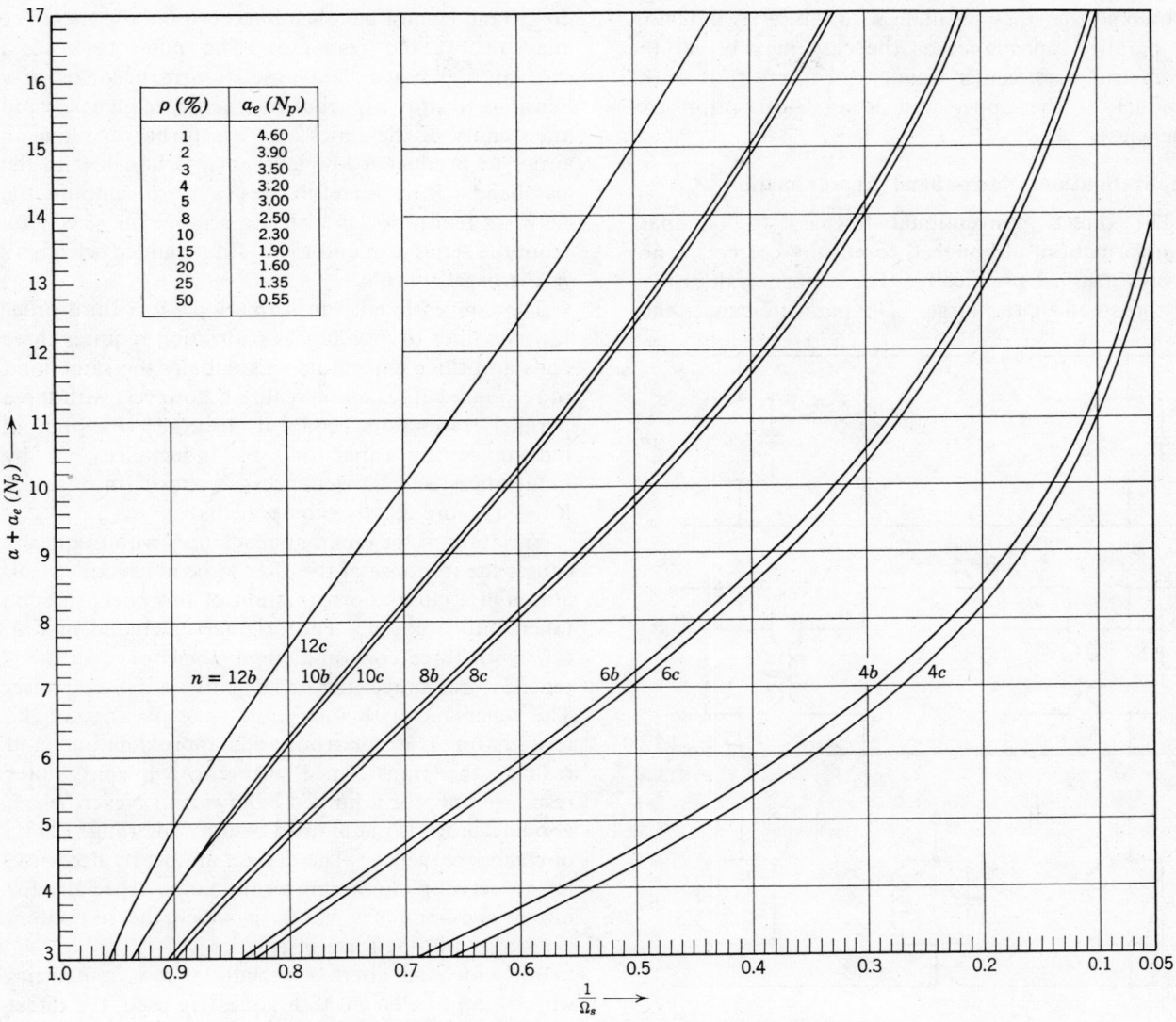

ρ (%)	a_e (N_p)
1	4.60
2	3.90
3	3.50
4	3.20
5	3.00
8	2.50
10	2.30
15	1.90
20	1.60
25	1.35
50	0.55

Fig. 6.6. Normalized response curves of even-order Chebyshev filters.

compare the attenuation in this example with the previous example of the sixth-order filter. At $\Omega = 1.5$ the unmodified Chebyshev response produced 33.8 dB.

In the corresponding modified filter $6c$ the attenuation will be 32.8 dB at the same frequency. The comparison shows that the modified filter provides less attenuation at any given stopband frequency, but when the filter is complicated (of higher order) the difference is negligible, and both filters can be considered as equal.

6.3 BANDPASS FILTER DESIGN

A bandpass filter can be obtained from its lowpass prototype which will result in exact geometric symmetry regardless of the bandwidth. In many low-frequency applications this realization may be attractive since it has the minimum number of reactive components. Three transmission zeros require three pairs of lumped reactances and therefore the resulting network is canonic.

To design a bandpass filter using the above technique, a lowpass prototype filter is first determined (see similar examples in Chapter 5). The impedance level and the 3-dB cutoff of the lowpass prototype are the same as the impedance level and over-all 3-dB bandwidth of the desired bandpass filter. The next step is to place a capacitance in series with all inductors and an inductor in parallel with all capacitances. The values of these added components are chosen so that they resonate with the series inductor, or parallel capacitance as the case may be, at the geometric mean center frequency (square root of the product of the upper and lower 3-dB cutoff frequencies).

Realization and Narrowband Approximation

The direct, conventional lowpass to bandpass transformation, although theoretically correct, is not always justified practically. The element values may be too small or too large. The parasitic capacitance

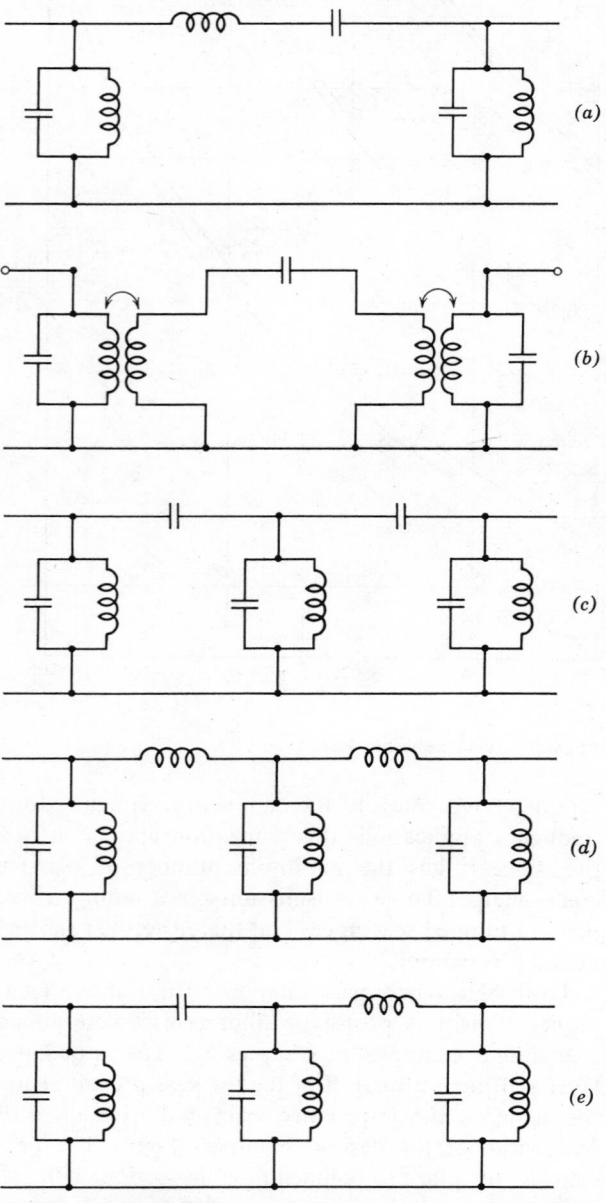

Fig. 6.7. Third-order bandpass filter (*a*), and its narrowband approximations.

to ground cannot be taken into account and therefore may distort the response. The node between a capacitor and a coil in a series arm becomes very sensitive to stray capacitance at some frequencies, and the quality of the series arm has to be very high in order to produce a low level of insertion loss in the passband. It is therefore desirable to simplify the network realization in order to remove the selectivity from the series arm and to substitute added selectivity in the parallel arms.

The conventional transformation for a third-order lowpass filter to a bandpass realization requires three coils and three capacitors. Essentially the same kind of response curve can be realized, however, with three parallel resonators coupled together by mutual inductance, by capacitors, by inductances, or by inductances and capacitances as shown in Fig. 6.7 [(*n* − 1) more reactive components].

Variation of the coupling reactances with frequency causes the response of the filter to be nonsymmetrical, although a close approximation of this effect may be taken into account. The original schematic in Fig. 6.7*a* with three coils and three capacitors will theoretically guarantee the ideal polynomial response. The schematic with only coils or capacitors in the series arm is a narrowband approximation and realizes the transformed theoretical lowpass filter response only for a limited bandwidth. Nevertheless, good accuracy is maintained over a wide range (20%) of center frequency. The design, almost by necessity, must use coupling circuit terminology, especially for microwave-frequency filters in which the resonators are realized in the form of cavities.

For wideband filters, especially at low frequencies where lumped-element techniques are used, the direct transformation is useful. It does not create too much of a problem except that the number of bulky and expensive coils (especially at very low frequencies) will complicate the realization.

6.4 CONCEPT OF COUPLING

Neither image-parameter theory nor synthesis theory requires a specific concept of coupling in order to design or to explain the physical operation of filters. So-called "half" or "full" sections can be connected together if they have appropriate characteristic impedances. In a discussion of the polynomial filter the term *coupling* loses its sense, because filter components are a result of the basic design procedure. Coupling in modern filters refers only to the parasitic effects of one component on another. The electronics engineer sometimes uses the term to describe a particular response as being over-coupled, under-coupled, critically or optimally coupled, and transitionally coupled, but even these descriptive terms are being replaced by Chebyshev or Butterworth responses, since they, in reality, describe a shape of an amplitude response.

The term coupling was reintroduced in network synthesis theory by Milton Dishal. There is no physical (magnetic or electrostatic) coupling between coil and capacitor except in the sense of connection, but the value of a coupling term has an effect on the bandwidth similar to that of the familiar magnetic coupling coefficient. Filter design in terms of coupling has become very popular with electronic engineers especially when narrow bandpass filters are concerned. The transmission property of polynomial filters in terms of decrements, relative bandwidths and coupling may be symbolically expressed by

$$H(j\Omega) = \frac{I_{in}}{V_{out}} = \frac{E(d + \Omega, k)}{B}$$

where Ω is a frequency parameter, k coupling coefficient, and d a decrement.

Normalized Circuit Parameters

In the previous examples of design all reactive element values were related to a terminating resistance (or source resistance) or to a completely arbitrary normalizing resistance R_r. The reduced impedance level resulted in simplified calculations. Another normalizing parameter was frequency; the frequency of the 3-dB down point was normalized to 1 rad/sec.

The results were tabulated for various types of amplitude responses in the form of normalized element values given in pages 312–340 in Section 6.9.

Another form of normalization results when the reactive component of each element is related to the reactive part of the immediately preceding element, and to a definite bandwidth defined in the same manner as in the first method. Here, the numerical results are not normalized element values, but normalized coefficients of coupling. Structures in terms of coupling include not only bandpass configurations where coupling is used in its original physical sense but also in lowpass, highpass, and bandstop structures.

The resistive component of each element related to its reactive part is called a normalized decrement d or, when inverted, normalized quality factor q.

The first type of normalization has been extensively used throughout the text. In the following treatment, this second form of normalization will be employed, and familiarity with coefficient of coupling k and normalized q is desirable.

Definitions

With reference to Fig. 6.8a the relations between k and q and the circuit elements can be defined. When the first arm is a shunt arm and the last arm is shunt or series, the coupling coefficient gives the ratio

$$k_{12} = \frac{\Omega_{12}}{\Omega_{3\,dB}}$$

where Ω_{12} is the resonant frequency of two adjacent

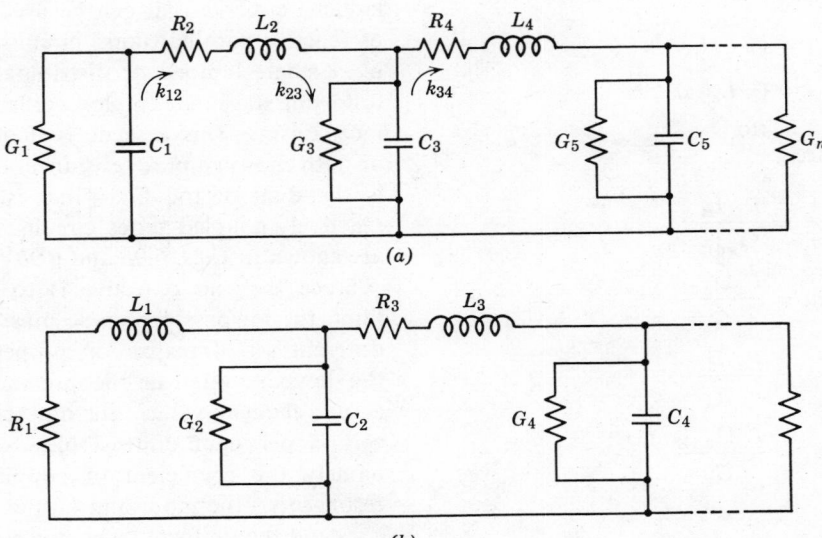

Fig. 6.8. Lowpass filters used for defining k and q values.

elements, and $\Omega_{3\,\mathrm{dB}}$ is the over-all 3-dB down frequency of the filter.

$$\frac{\Omega_{12}}{\Omega_{3\,\mathrm{dB}}} = k_{12}; \quad \frac{\Omega_{23}}{\Omega_{3\,\mathrm{dB}}} = k_{23}; \quad \frac{\Omega_{34}}{\Omega_{3\,\mathrm{dB}}} = k_{34}$$

where

$$\Omega_{12} = (\sqrt{C_1 L_2})^{-1}; \quad \Omega_{23} = (\sqrt{L_2 C_3})^{-1};$$
$$\Omega_{34} = (\sqrt{C_3 L_4})^{-1}$$

The expression for resonant frequency shows that it is equal to $\Omega_{3\,\mathrm{dB}}$ times the coupling coefficient k_{ik}; $\Omega_{i,k} = \Omega_{3\,\mathrm{dB}} k_{i,k}$.

The expressions for normalized quality factors of the circuits are

$$\frac{1}{q_1} = \frac{G_1/C_1}{\Omega_{3\,\mathrm{dB}}}$$

$$q_2 = \frac{\Omega_{3\,\mathrm{dB}} L_2}{R_2}$$

$$q_3 = \frac{\Omega_{3\,\mathrm{dB}} C_3}{G_3}$$

$$q_4 = \frac{\Omega_{3\,\mathrm{dB}} L_4}{R_4}$$

When the lowpass corresponds to Fig. 6.8*b* having a series arm at one end,

$$k_{12} = \frac{\Omega_{12}}{\Omega_{3\,\mathrm{dB}}}$$

where

$$\Omega_{12} = (\sqrt{L_1 C_1})^{-1}$$

The remaining resonant frequencies which enter in the expression for k's will be

$$\Omega_{23} = (\sqrt{C_2 L_3})^{-1}$$
$$\Omega_{34} = (\sqrt{L_3 C_4})^{-1}$$
$$\text{etc.}$$

The normalized q's are

$$\frac{1}{q_1} = \frac{R_1/L_1}{\Omega_{3\,\mathrm{dB}}}$$

$$q_2 = \frac{C_2 \Omega_{3\,\mathrm{dB}}}{G_2}$$

$$q_3 = \frac{L_3 \Omega_{3\,\mathrm{dB}}}{R_3}$$

$$q_4 = \frac{C_4 \Omega_{3\,\mathrm{dB}}}{G_4}$$

$$\cdot$$
$$\cdot$$
$$\cdot$$

To design a bandpass filter based on the foregoing lowpass model, the total required 3-dB down bandwidth should be inserted in the above expressions for k's and q's.

Every inductance in the series branch of the ladder has to be supplemented by a series capacitor and to every capacitor in a shunt circuit an inductor should be connected in parallel. The configuration will constitute a bandpass filter which includes only resonant circuits (tuned to f_m, the midband frequency) instead of single reactances as in a lowpass structure.

The input circuit of the filter is expressed by the reciprocal of q. The value $1/q$ is given as a ratio of 3-dB level bandwidth of a single element resulting from the resistive load and losses associated with reactance, to the required over-all 3-dB bandwidth of the filter. Thus in Fig. 6.8*a* the value of $1/R_1 C_1$ expresses the 3-dB radian bandwidth of C_1 and the conductance G_1 that must be put across capacitor C_1. If C_1 and G_1 are properly chosen, the bandwidth of these elements at their 3-dB point will be $1/q_1$ times the required over-all 3-dB bandwidth of the filter.

In most high-frequency designs, especially the microwave type, the use of the normalized value of k may simplify the adjustment procedure since the numerical value of k could be directly applied to the tuning of the actual filter.

6.5   COUPLED RESONATORS

There are basically two types of coupled resonator ladder networks that can be used. The first consists of reactive parallel (tank) circuits which are coupled by a single lumped or distributed reactance. Every following stage may employ a different type of coupling mechanism. This cascade is then terminated at one or both ends in a pure resistance. The second network is the dual of the first—that is, it is made up of reactively coupled series circuits. These two ladders are shown in Figs. 6.9*a* and 6.9*b* respectively.

Since the end objective is to obtain a bandpass filter, the lowpass prototype must be designed with a different set of realization properties from those of the lowpass filter in the previous discussion. The actual element values are not tabulated, but rather certain pertinant dimensionless ratios in particular; namely the coefficient of coupling k between the resonators; the input and output circuits normalized q's; and the uniform resonator q_0 due to finite losses. These quantities are defined for a lowpass prototype

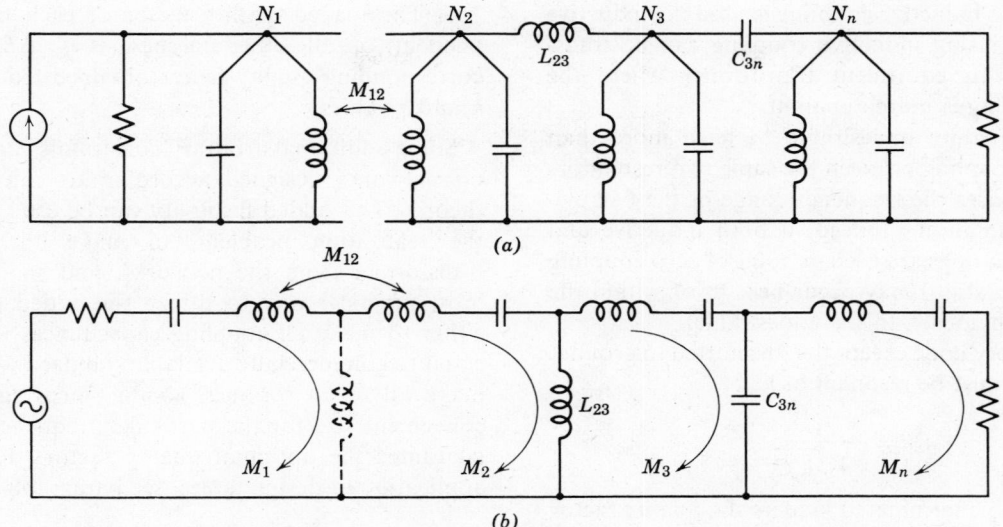

Fig. 6.9. (*a*) Basic nodal circuit with mutual, inductive, and capacitive coupling. (*b*) Basic mesh circuit with mutual, inductive, and capacitive coupling.

circuit as follows:

$k_{i,k}$ = the ratio of the series-resonant frequency of the ith and kth reactive elements to the 3-dB cutoff frequency.

q_i = the (Q) quality of the ith reactive element influenced by the source and load resistance, if present, and is in series or parallel with it.

From a standpoint of realizability it would be desirable if (1) the two end loaded q's and the coefficients of coupling were equal, and (2) the prototype filter was designed on a maximum power transfer basis so that the alignment will not be so difficult.

Nodal Circuit Design

Knowing the bandwidth Δf, the center frequency f_m, and the uniform resonator quality factor Q_0 due to parasitic losses, the normalized resonator quality factor can be determined from the formula

$$q_0 = \frac{\Delta f}{f_m} Q_0$$

If the degree n (also the number of resonators) and the type of polynomial approximation are known, then with the above value of Q, a set of normalized coupling coefficients and loaded q's can be found from tables for double-loaded networks. If the exact normalized resonator q is not in the tables, interpolation can be used. From the normalized loaded q's and k's the corresponding unnormalized quantities

can be obtained from the formulas

$$K_{i,k} = k_{i,k} \frac{\Delta f}{f_m}$$

$$Q_i = q_i \frac{f_m}{\Delta f}$$

These quantities are directly related to the bandpass circuit elements. However, because the number of circuit elements is greater than the minimum number, the element values are not uniquely defined.

With regard to the nodal circuit of Fig. 6.9a the following relationships must hold:

1. When capacitive coupling is used between the ith and kth node, the coupling capacitance $C_{i,k}$ must be equal to

$$C_{i,k} = K_{i,k}\sqrt{C_i C_k}$$

where $K_{i,k}$ is the unnormalized coupling coefficient determined previously, and C_i is the total shunt (nodal) capacitance of ith node when all other nodes are shorted to ground.

2. If inductive coupling is used between the ith and kth node, the coupling inductance $L_{i,k}$ is given by

$$L_{i,k} = \frac{\sqrt{L_i L_k}}{K_{i,k}}$$

where L_i is the total shunt inductance of the ith node when all other nodes are shorted to ground.

3. If mutual inductive coupling is used the inductive π formed by using inductive coupling can be transformed into an equivalent transformer where the mutual coupling is more apparent.

4. It is generally undesirable to have more than one type of coupling between the same two resonators since this reduces the frequency range of the narrowband approximation. Indeed, if both inductive and capacitive couplings are used, a point of zero coupling (infinite attenuation) may occur next to or within the passband. Obviously, this is undesirable.

5. With every node except the ith shorted to ground, the ith node must be resonant at f_m

$$f_m = \frac{1}{2\pi\sqrt{L_i C_i}}$$

6. For double terminated ladders the quality factor Q of the two end nodes with all other nodes shorted to ground and the source and load resistance included in the evaluation of the quality factor is equal to $\frac{f_m}{\Delta f}$ times the corresponding quality factors q_1 and q_n obtained from the tables of k and q values. A similar statement holds for tables and ladders terminated at one end only.

Mesh Circuit Design

A dual set of relationships are required to make a bandpass filter from the mesh circuit (dual of the nodal circuit) of Fig. 6.9*b*. Briefly, these relationships are as follows:

1. The value of the coupling capacitance $C_{i,k}$ is

$$C_{i,k} = \frac{\sqrt{C_i C_k}}{K_{i,k}}$$

where $K_{i,k}$ is the unnormalized coupling coefficient, C_i is the total series capacitive of the ith mesh when all other meshes are open circuited.

2. If inductive coupling is used, the inductance $L_{i,k}$ is

$$L_{i,k} = K_{i,k}\sqrt{L_i L_k}$$

where L_i is the total series capacitance of mesh i when all other meshes are open.

3. Regarding multiple or mutual coupling, the same discussion applies to mesh circuits as that used previously for nodal circuits.

4. Each mesh with all others open circuited must resonate at f_m.

$$f_m = \frac{1}{2\pi\sqrt{L_i C_i}}$$

5. The loaded quality factor of each mesh, independent of all other meshes, is $f_m/\Delta f$ times the corresponding quality factor obtained from tables of k and q values.

Some additional arbitrary constraints can be placed on networks designed according to coupled-circuit theory. This added flexibility can be used to simplify the realization problem, to make an impedance transformer from the networks, and so forth. The most practical way to utilize the added freedom is either to make all coupling capacitances identical or equal to commercially available standard values, or to make all nodal (or mesh) inductances equal and a convenient size for the particular frequency range to guarantee the optimum quality factor. For specific application of design tables, see paragraph 6.7.

6.6 SECOND-ORDER BANDPASS FILTER

The design of multiresonator filters generally follows one of two different methods. In the synchronously tuned method, all the resonant circuits are tuned to the desired center frequency at each node. The symmetrically detuned method (staggered tuning) has one or more resonant circuits detuned symmetrically about the center frequency as shown in Fig. 6.10. Strong coupling is equivalent to strong resonant detuning.

Both methods may arrive at the same amplitude-frequency response, but the absolute value of the output at center frequency is generally greater for the synchronously tuned filter. This is the reason for realizing only synchronously tuned filters in practice, especially when the network's transmission function is higher than the second order.

To understand the physical concept of coupling and selectivity, we should concentrate our attention on simple bandpass filters in the form of a double-tuned circuit, such as that generally used between amplifier stages and similar circuits.

A second-order bandpass filter can produce the same amplitude-frequency response as a two-stage

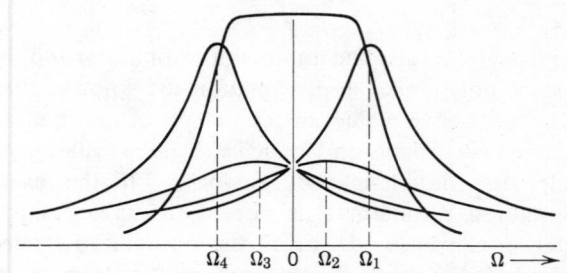

Fig. 6.10. Partial contribution for maximally flat design.

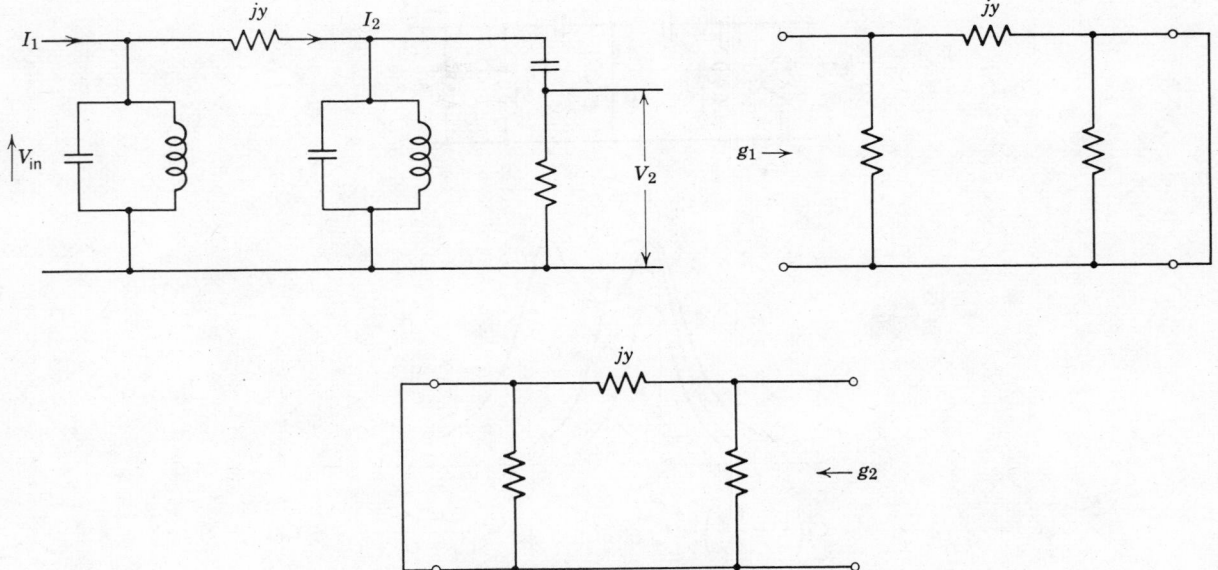

Fig. 6.11. General schematic of filter having two resonant circuits.

tuned amplifier with a single resonant circuit in each stage. Moreover, the response of several stages of single-tuned circuits in cascade can be replaced by a single filter with an equal number of resonant circuits, which may be coupled by either capacitive or inductive reactances. The method of coupling may alternate from each resonant circuit to the next as required by the situation.

Two basic assumptions will be used here. First, we will assume that coupling reactance is independent of frequency; the change in reactance is so small that, within the limits of the frequencies in which the filter is operating, the change can be completely disregarded. Second, we will assume that each resonant circuit is coupled only with the circuit adjacent to it.

The general schematic of the filter with two anti-resonant circuits is shown in Fig. 6.11. It can be shown that five different ways of coupling produce essentially the same response and that all of these may be represented by the schematics of Fig. 6.11.

Figures 6.12 and 6.13 illustrate methods of capacitively coupling two resonant circuits. In Figs. 6.14 and 6.15 methods of inductively coupling the same circuits are shown. The three capacitors in Fig. 6.12 can be transformed into the T configuration of Fig. 6.13, by using the well-known delta-star transformation. A similar transformation may be used for the inductive coupling networks in Figs. 6.14 and 6.15. Formulas to facilitate such transformations are shown in Fig. 6.16.

A filter with magnetic coupling between the first

and second resonant circuits is shown in Fig. 6.17. From transformer theory, we know that two different values of the coupling coefficient can result depending on the sign for mutual inductance M, which can be positive or negative (depending upon the polarity of the secondary winding). The value of y in Fig. 6.11 can therefore be either positive or negative. Very often, instead of pure inductive, pure capacitive, or

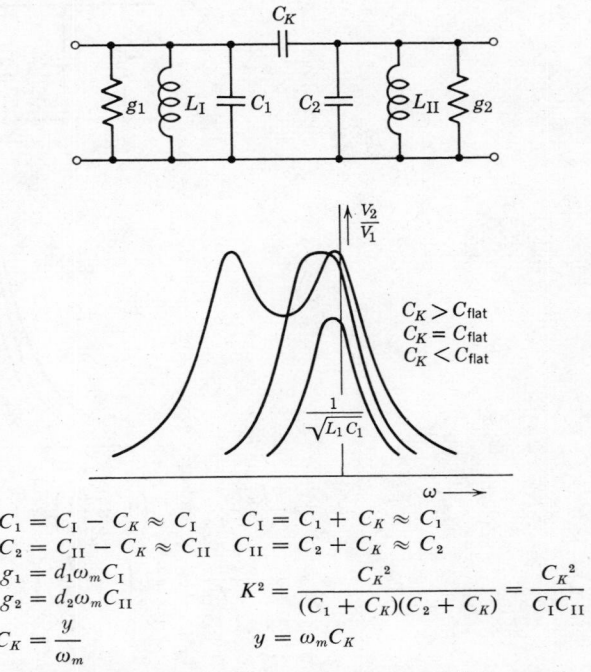

$$C_1 = C_I - C_K \approx C_I \quad C_I = C_1 + C_K \approx C_1$$
$$C_2 = C_{II} - C_K \approx C_{II} \quad C_{II} = C_2 + C_K \approx C_2$$
$$g_1 = d_1\omega_m C_I$$
$$g_2 = d_2\omega_m C_{II}$$
$$C_K = \frac{y}{\omega_m} \quad K^2 = \frac{C_K^2}{(C_1 + C_K)(C_2 + C_K)} = \frac{C_K^2}{C_I C_{II}}$$
$$y = \omega_m C_K$$

Fig. 6.12. Second-order filter with series capacitive coupling.

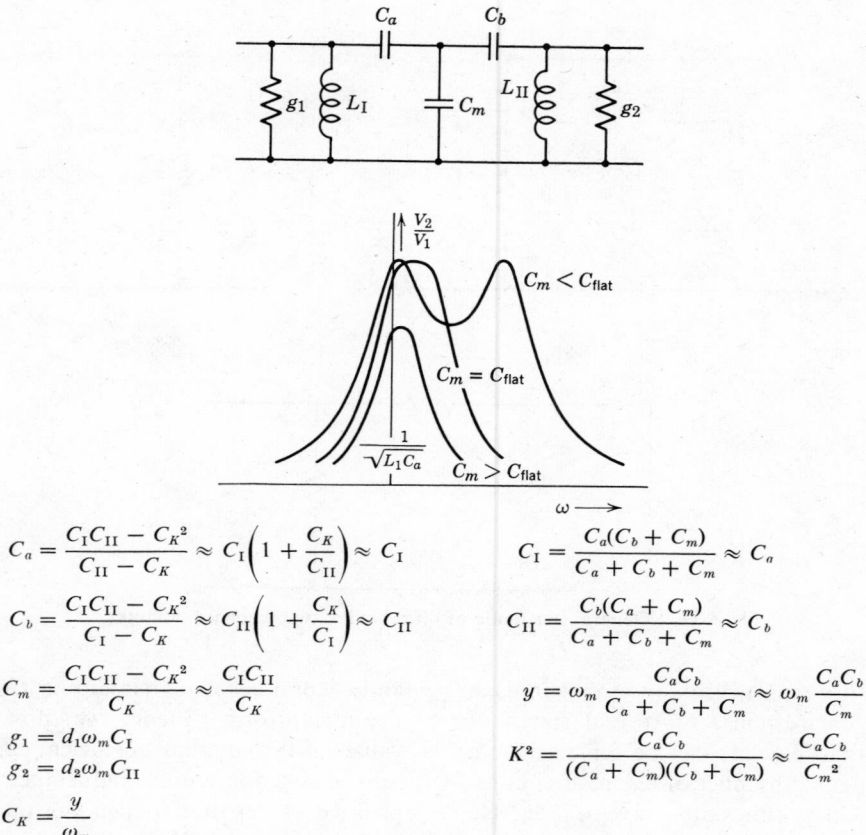

$$C_a = \frac{C_I C_{II} - C_K{}^2}{C_{II} - C_K} \approx C_I \left(1 + \frac{C_K}{C_{II}}\right) \approx C_I \qquad C_I = \frac{C_a(C_b + C_m)}{C_a + C_b + C_m} \approx C_a$$

$$C_b = \frac{C_I C_{II} - C_K{}^2}{C_I - C_K} \approx C_{II} \left(1 + \frac{C_K}{C_I}\right) \approx C_{II} \qquad C_{II} = \frac{C_b(C_a + C_m)}{C_a + C_b + C_m} \approx C_b$$

$$C_m = \frac{C_I C_{II} - C_K{}^2}{C_K} \approx \frac{C_I C_{II}}{C_K} \qquad y = \omega_m \frac{C_a C_b}{C_a + C_b + C_m} \approx \omega_m \frac{C_a C_b}{C_m}$$

$$g_1 = d_1 \omega_m C_I$$
$$g_2 = d_2 \omega_m C_{II} \qquad K^2 = \frac{C_a C_b}{(C_a + C_m)(C_b + C_m)} \approx \frac{C_a C_b}{C_m{}^2}$$

$$C_K = \frac{y}{\omega_m}$$

Fig. 6.13. Second-order filter with shunt capacitive coupling.

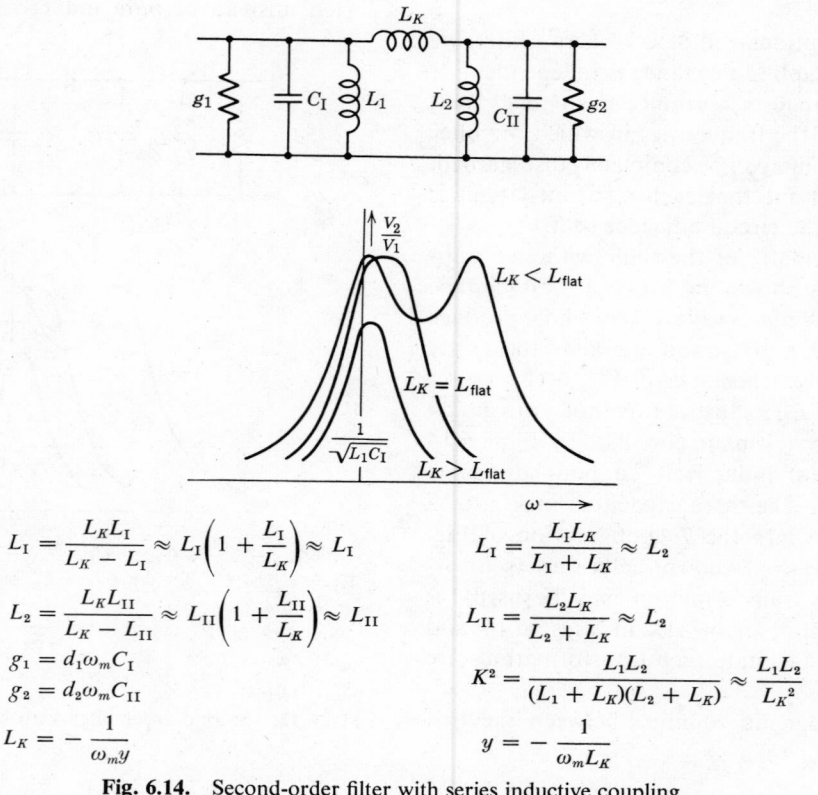

$$L_I = \frac{L_K L_I}{L_K - L_I} \approx L_I \left(1 + \frac{L_I}{L_K}\right) \approx L_I \qquad L_I = \frac{L_I L_K}{L_I + L_K} \approx L_2$$

$$L_2 = \frac{L_K L_{II}}{L_K - L_{II}} \approx L_{II} \left(1 + \frac{L_{II}}{L_K}\right) \approx L_{II} \qquad L_{II} = \frac{L_2 L_K}{L_2 + L_K} \approx L_2$$

$$g_1 = d_1 \omega_m C_I$$
$$g_2 = d_2 \omega_m C_{II} \qquad K^2 = \frac{L_1 L_2}{(L_1 + L_K)(L_2 + L_K)} \approx \frac{L_1 L_2}{L_K{}^2}$$

$$L_K = -\frac{1}{\omega_m y} \qquad y = -\frac{1}{\omega_m L_K}$$

Fig. 6.14. Second-order filter with series inductive coupling.

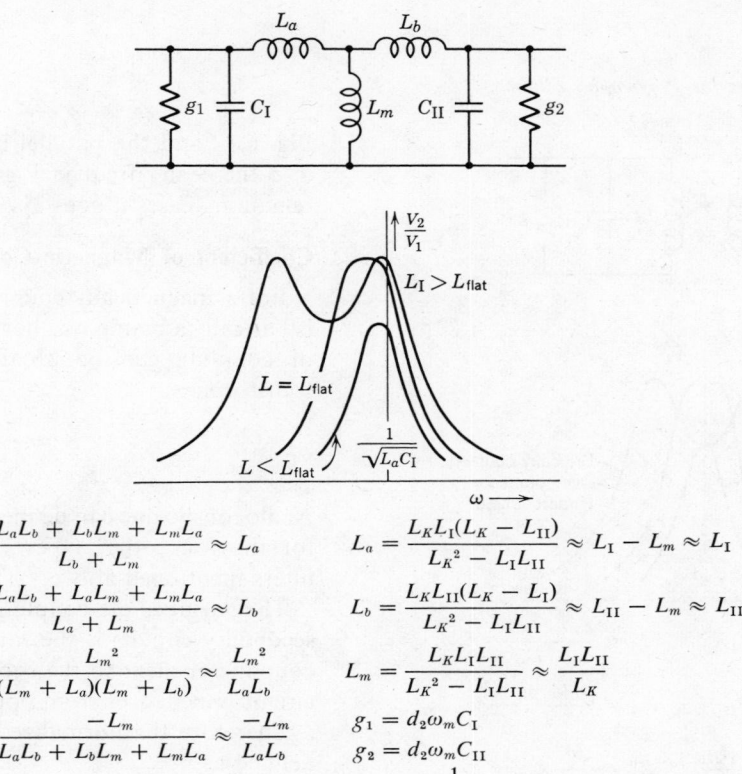

$$L_\mathrm{I} = \frac{L_a L_b + L_b L_m + L_m L_a}{L_b + L_m} \approx L_a \qquad L_a = \frac{L_K L_\mathrm{I}(L_K - L_\mathrm{II})}{L_K{}^2 - L_\mathrm{I} L_\mathrm{II}} \approx L_\mathrm{I} - L_m \approx L_\mathrm{I}$$

$$L_\mathrm{II} = \frac{L_a L_b + L_a L_m + L_m L_a}{L_a + L_m} \approx L_b \qquad L_b = \frac{L_K L_\mathrm{II}(L_K - L_\mathrm{I})}{L_K{}^2 - L_\mathrm{I} L_\mathrm{II}} \approx L_\mathrm{II} - L_m \approx L_\mathrm{II}$$

$$K^2 = \frac{L_m{}^2}{(L_m + L_a)(L_m + L_b)} \approx \frac{L_m{}^2}{L_a L_b} \qquad L_m = \frac{L_K L_\mathrm{I} L_\mathrm{II}}{L_K{}^2 - L_\mathrm{I} L_\mathrm{II}} \approx \frac{L_\mathrm{I} L_\mathrm{II}}{L_K}$$

$$\omega_m y = \frac{-L_m}{L_a L_b + L_b L_m + L_m L_a} \approx \frac{-L_m}{L_a L_b} \qquad g_1 = d_2 \omega_m C_\mathrm{I}$$
$$g_2 = d_2 \omega_m C_\mathrm{II}$$
$$L_K = -\frac{1}{\omega_m y}$$

Fig. 6.15. Second-order filter with shunt inductive coupling.

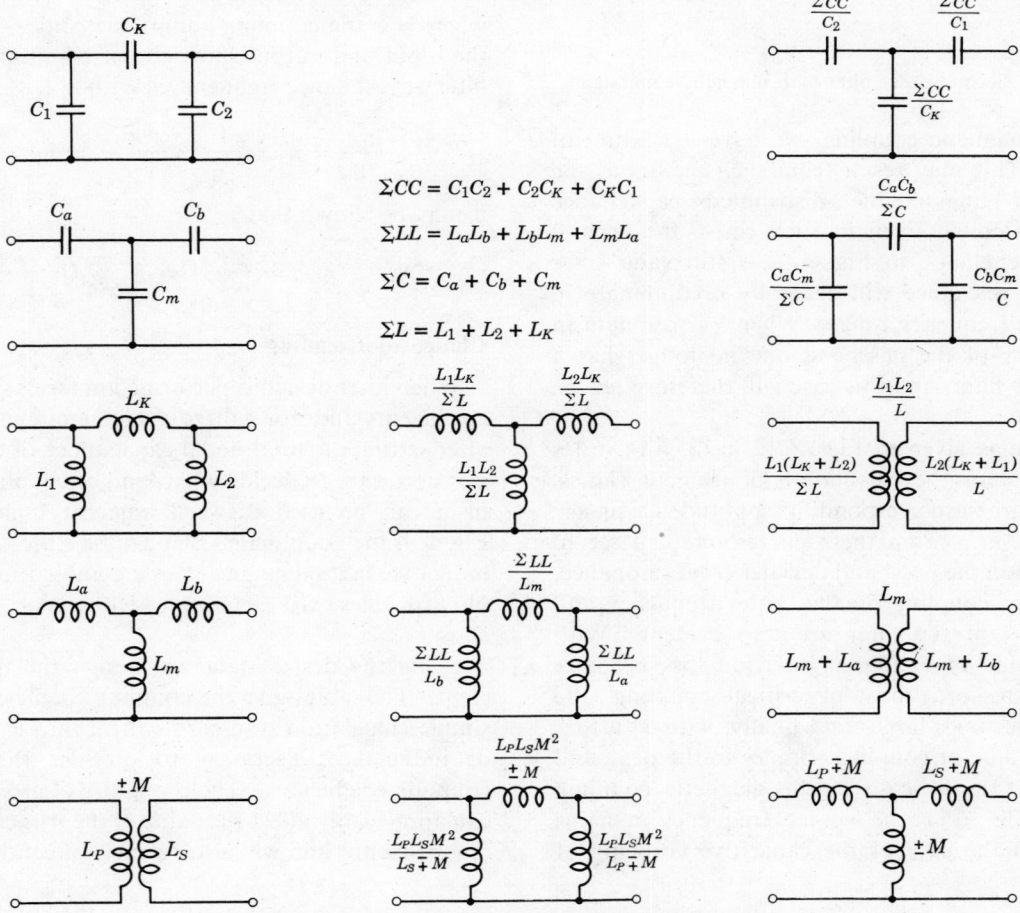

$$\Sigma CC = C_1 C_2 + C_2 C_K + C_K C_1$$

$$\Sigma LL = L_a L_b + L_b L_m + L_m L_a$$

$$\Sigma C = C_a + C_b + C_m$$

$$\Sigma L = L_1 + L_2 + L_K$$

Fig. 6.16. Equivalent schematics and related element values.

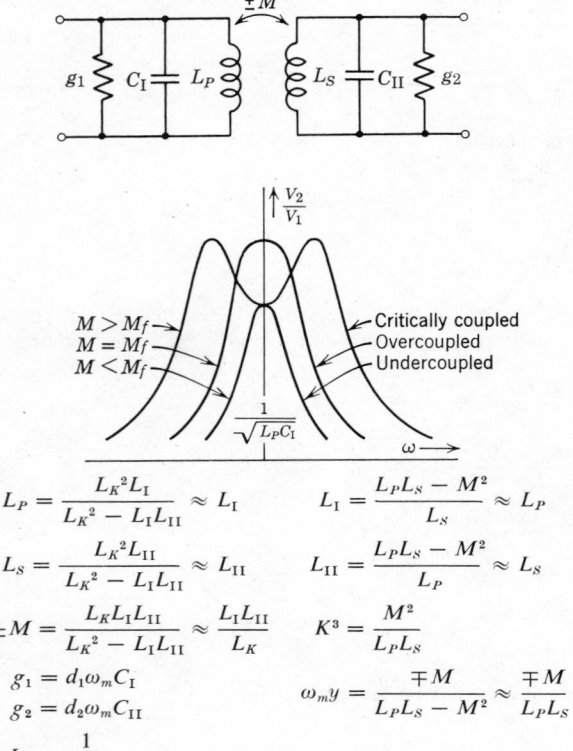

$$L_P = \frac{L_K{}^2 L_I}{L_K{}^2 - L_I L_{II}} \approx L_I \qquad L_I = \frac{L_P L_S - M^2}{L_S} \approx L_P$$

$$L_S = \frac{L_K{}^2 L_{II}}{L_K{}^2 - L_I L_{II}} \approx L_{II} \qquad L_{II} = \frac{L_P L_S - M^2}{L_P} \approx L_S$$

$$\pm M = \frac{L_K L_I L_{II}}{L_K{}^2 - L_I L_{II}} \approx \frac{L_I L_{II}}{L_K} \qquad K^3 = \frac{M^2}{L_P L_S}$$

$$g_1 = d_1 \omega_m C_I$$
$$g_2 = d_2 \omega_m C_{II} \qquad\qquad \omega_m y = \frac{\mp M}{L_P L_S - M^2} \approx \frac{\mp M}{L_P L_S}$$

$$\pm L_K = \frac{1}{\omega_m y}$$

Fig. 6.17. Second-order filter with magnetic coupling.

even pure magnetic coupling, we have a mixture of couplings. This may result from such factors as the presence of unavoidable distributed capacitance between the coils. Even in such cases, the general equivalent schematic in Fig. 6.11 is still valid since one type of reactance will generally predominate in each specific frequency range. When y is resonant in the proximity of the passband one no longer has a second-order filter, and this case will therefore not be considered here.

The formulas given in Figs. 6.12, 6.13, 6.14, 6.15, and 6.17 are for the calculation of element values. Also shown are the corresponding amplitude-frequency response curves. From these curves one can see in which direction the passband deviates from resonance. The effects of coupling on the center frequency and insertion loss of the filter are also evident. With increased value of coupling, insertion loss becomes lower. In the proximity of critical coupling, the bandwidth becomes larger and finally, with a further increase in value of coupling, ripples in the passband will appear. In the case of the magnetic coupling shown in Fig. 6.17 the center frequency remains constant. In the case of the capacitive coupling of

Fig. 6.12 and the parallel inductive coupling of Fig. 6.15 the center frequency goes down, and in the two remaining cases it goes up.

Coefficient of Magnetic Coupling

For a magnetically coupled resonant circuit, which is basically a bandpass filter the value of the coefficient of coupling can be obtained from the theory of transformers:

$$K = \frac{M}{\sqrt{L_p L_s}}$$

Analogously, one can define the coefficient of coupling for the four other types of second-order bandpass filters mentioned above.

The degree of coupling between primary and secondary circuits is the ratio of the reactance of the coupling element to the geometric mean of the open-circuit impedances from opposite terminals.

Therefore the normalized coupling k, will now be defined as

$$k = \frac{y}{\sqrt{g_1 g_2}}$$

where y is the coupling admittance and g_1 and g_2 are the input and output short-circuit admittances of the filter at resonance frequency. With

$$\frac{g_1}{\omega_m C_1} = d_1 = \frac{1}{Q_1} \; ; \quad \frac{g_2}{\omega_m C_{11}} = d_2 = \frac{1}{Q_2}$$

it may be shown that

$$k = \frac{K}{\sqrt{d_1 d_2}} = K \sqrt{Q_1 Q_2}$$

Choice of Coupling

When filter design is performed in terms of coupling coefficients and normalized q of the circuit elements, after setting the total nodal capacitance of each node, it is necessary to decide what kind of coupling mechanism can be used between adjacent nodes. As in Fig. 6.9, the coupling chosen can be either capacitive, inductive, mutual magnetic, or a combination of these. No firm rules exist dictating which type of coupling to chose.

Using the design data, we design the network in steps. The tables give the coupling coefficients, and a simple calculation transforms them into capacitances or inductances necessary to produce the required coupling coefficients. The proximity of ground planes can appreciably affect the value of the true capacitance of a capacitor and will also strongly affect the value of self-inductance of an inductor or the mutual inductance of a pair of inductors. Therefore, particularly when small element values are called for, the actual value of coupling capacitors and inductors to be used in the filter cannot always be determined, since the actual values of these components could change because of proximity effects. Because of this, the adjustment procedure given in Section 9.5 is recommended. For setting the coefficients of coupling while tuning the filter, the procedure requires the measurement of bandwidth between two sharply defined response peaks with the filter components in the exact position that they will occupy in the final embodient. It is possible to satisfy each coefficient within 1% of its required value. Once the correct value has been obtained, Q-meter or physical configuration measurement can be used to duplicate these correct values.

In the following example, the evaluation of coupling parameters and the application of the different types of physical equivalents will be demonstrated. Coupling in narrow bandpass filters does not contribute to the amount of discrimination obtained, and therefore does not change the order of polynomial. The selectivity of the filter of given bandwidth is only a function of n.

The minimum number of resonators required to satisfy a certain requirement may be obtained from conventional nomographs or curves. After the number of resonators and the coupling mechanism is determined the components can be found. To obtain the first resonator for the narrowband type of schematic of Fig. 6.7*b, c, d,* and *e,* the Q factor at the first node is selected. This means that the source resistance has determined the first node inductance and capacitance. For a very small relative bandwidth, the ratio of coupling to shunt elements is approximately reciprocal to relative bandwidth and is very small. Physically, this means that capacitor coupling is not always realizable and one must sometimes employ a mutual inductive coupling. (See second-order bandpass filters). The remainder of the circuit elements are rigidly determined by a step-by-step design procedure with known values of loaded q and k. To obtain the results predicted by theory, the unloaded Q of each element must be greater than a certain minimum.

In narrow bandpass filters with coupling in the form of single reactances, the mid-frequency reactance of coil or capacitance for internal resonators may be chosen absolutely independently of the response requirements. This property constitutes the greatest advantage of narrowband approximation and appropriate realizations.

The choice of the impedance level can be based on the desire to obtain maximum unloaded Q and some mechanical and economical considerations.

6.7 DESIGN WITH TABLES OF PREDISTORTED k AND q PARAMETERS

When each of the filter elements has the same finite Q, then the real part of each root (see Chapter 2) can be predistorted by an amount equal to the normalized unloaded decrement of each element.

It is a difficult problem to synthesize a predistorted network if all of the reactive elements do not have the same Q. This means that the excellent Q of capacitances must be degraded with a parallel resistance in order to make their Q the same as that of the lossy inductances. If only moderate accuracy is needed in the network's frequency characteristics it is sometimes feasible to use a value of Q which is twice as high as that of the inductances and then in the final network the capacitances are left as they are—that is, their Q is not reduced artificially.

It should be noted that from the definitions of Q

$$Q = \omega_{3\,\mathrm{dB}} \frac{L}{R_L}$$

$$Q = \omega_{3\,\mathrm{dB}} \frac{C}{R_C}$$

for an inductance and capacitance, that Q is invariant with both frequency and impedance scaling. This last statement makes the use of finite Q tables feasible.

The values of resistance that are placed in series with the inductances and in parallel with the capacitance can be determined from the formulas

$$R_L = \frac{\omega_{3\,\mathrm{dB}}L}{Q}$$

$$R_C = \frac{Q}{\omega_{3\,\mathrm{dB}}C}$$

Predistortion Parameter d_0

By definition $R_c C \omega_{3\,\mathrm{dB}} = q_0$

$$\frac{L\omega_{3\,\mathrm{dB}}}{R_L} = q_0$$

for lowpass reactances and

$$q_0 = \frac{Q_0}{a}$$

for bandpass resonators where a here corresponds to the bandwidth at the 3-dB points.

The resulting modified or predistorted root values are then used in the n-factors that are involved in polynomial expression for voltage transmission function. In the case of lossless element design, the conventional tables with normalized circuit element values can be used. In practice these lossless element design data are mainly applicable to lowpass and wide bandpass filters because the assumption of infinite normalized unloaded Q is almost satisfied for this case.

In the case of narrow bandpass filters the normalized unloaded resonator q_0 is equal to unloaded Q_0 of a resonator divided by normalizing 3-dB bandwidth of the filter.

$$q_0 = \frac{Q_0}{a}$$

This value is seldom greater than ten, and unfortunately in many actual cases the normalized quality factor is as low as two or three. Thus the assumption that the unloaded Q is infinite cannot be made. When roots of the approximate polynomial are predistorted by $1/q_0$, the design data can be an important addition to the lossless design data.

In addition to the element values, the tables in terms of k and q also include the fixed insertion loss which is the price paid for using uniform predistortion techniques.

Introduction to the Tables of 3-dB Down k and q Values

Section 6.10 presents, in tabular form, k and q values for polynomial filters just as Section 6.9 provides the normalized element values for these same filters. The tables are divided into groups according to the type of response (Butterworth, Chebyshev, and so forth) and k and q values are given for filters up to the eight order.

The tables, for each type and order of filter given, tabulates the k and q values for various values of q_0, starting from q_0 equal to infinity (the lossless case) and then decreasing to a certain finite value. The lossless case, in practice, is used where the quality factor of the components of the filter is such that an insertion loss of 1 dB or less is expected. In many cases of filter design, however, the available elements will be far from ideal, and the tabulated values of predistorted k and q values will here be extremely valuable. Note that for each case of finite q_0, the expected insertion loss of the filter is given. Linear interpolation between values of q_0 is usually accurate for most practical design problems.

In many of the tables, two or more solutions are given for a particular requirement. Choice of which solution to pick is usually dictated by the comparative values of q_1 to q_n and to the range of k values within each solution.

6.8 DESIGN EXAMPLES USING TABLES OF k AND q VALUES

Example of Lowpass Butterworth Filter

A fifth-order Butterworth filter is to be designed having its 3-dB cutoff at 1.5 Mc. At this frequency it is not difficult to obtain inductors having a Q of 100.

The normalized unloaded q_0 for the inductors of the lowpass ladder is

$$q_0 = \frac{L\omega_{3\,dB}}{R_L}$$

where R_L is the resistance of the coil. It can be proven that with $Q = 100$ the values of k's and q's have negligible difference from the k's and q's for the ideal unloaded $Q = \infty$. Therefore the assumption of infinite normalized unloaded q_0 can be made. A similar assumption can be made with wide bandpass filters when bandwidth parameter a is small and q_0 is large.

From the Butterworth table, Section 6.10 pages 341–343, for $n = 5$ and $q_0 = $ INF, the following values are read:

$$q_1 = 0.618$$
$$q_5 = 0.618$$
$$k_{12} = 1.0$$
$$k_{23} = 0.5559$$
$$k_{34} = 0.5559$$
$$k_{45} = 1.0$$

The corresponding circuit is shown in Fig. 6.18. Since $q_1 = q_5$, the network will be terminated equally on the input and output sides. Also, since $k_{12} = k_{45}$, and $k_{23} = k_{34}$, the network is symmetrical.

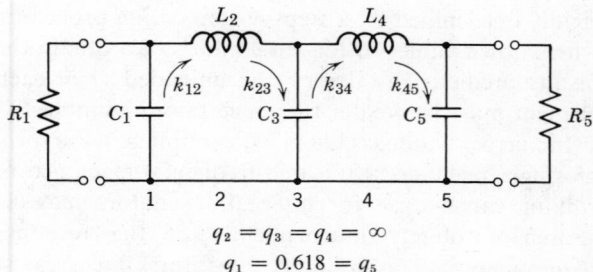

$$q_2 = q_3 = q_4 = \infty$$
$$q_1 = 0.618 = q_5$$

Fig. 6.18. Schematic of example of Butterworth lowpass filter designed by k and q values.

To absorb the shunt capacitance associated with the signal source, the lowpass ladder with a shunt capacitance for its first element will be used. The lowpass design with the table is accomplished by starting with R_1 and working to the right, element by element.

$$q_1 = C_1 R_1 \omega_{3\,dB} = 0.618$$

The time constant,

$$C_1 R_1 = T = \frac{q}{\omega_{3\,dB}}$$

or the radian frequency

$$\frac{1}{C_1 R_1} = \frac{1}{q_1} \omega_{3\,dB} = \frac{1}{0.618} 2\pi \times 1.5 \times 10^6$$

$1/C_1 R_1$ is the radian frequency of the $R_1 C_1$ combination, and is the 3-dB down radian frequency of the filter.

Assuming in this example, that the source (and load) resistances are to be 75 ohms,

$$\frac{1}{R_1 C_1} = 2\pi \times 2.423 \times 10^6$$

$$C_1 = \frac{10^6}{2\pi \times 2.423 \times 75} = \frac{10^4}{2\pi \times 2.42 \times 0.75} = 876 \text{ pF}$$

The 3-dB down frequency of the *RC* combination is 2.423 Mc, and can be used to test the correctness of the values of R_1 and C_1 in the finished filter. For the lowpass ladder, the coupling coefficient between C_1 and L_2 is

$$k_{12} = (\omega_{3\,dB} \sqrt{C_1 L_2})^{-1}$$

Therefore

$$(\sqrt{C_1 L_2})^{-1} = k_{12} \omega_{3\,dB} = 1 \times 2\pi \times 1.5 \times 10^6$$

$$(C_1 L_2)^{-1} = (2\pi \times 1.5 \times 10^6)^2$$

$$(L_2)^{-1} = (2\pi \times 1.5 \times 10^6)^2 \times 876 \cdot 10^{-12}$$

$$L_2 = (876 \times 39.5 \times 2.25)^{-1}$$

$$= (0.077800 \times 10^6)^{-1} = 12.8 \ \mu H$$

The expression $1/\sqrt{C_1 L_2}$ is of course, the resonant radian frequency of the $C_1 L_2$ combination. Both element values are known and the resonant frequency is 1.5 Mc. Evidently it can be used for checking the correctness of the inductance value by measuring the resonant frequency of C_1 and L_2.

The following element of the filter is the capacitor C_3. Its value can be found from the following expression for coupling coefficient:

$$(\sqrt{L_2 C_3})^{-1} = k_{23} \omega_{3\,dB} = 0.5559 \times 2\pi \times 1.5 \times 10^6$$

$$= 2\pi \times 0.833 \times 10^6$$

$$(C_3)^{-1} = (2\pi \times 0.833 \times 10^6)^2 12.8 \times 10^{-6}$$

$$C_3 = \frac{876 \times (1.5)^2}{(0.833)^2} = \frac{1970}{0.694} = 2840 \text{ pF}$$

The resonant frequency for $L_2 C_3$ combination is 0.833×10^6 cps. A similar technique can be used to evaluate the rest of the circuit parameters

$$(\sqrt{C_3 L_4})^{-1} = k_{34} \omega_{3\,dB} = 0.5559 \times 2\pi \times 1.5 \times 10^6$$

which are

$$L_4 = L_2 = 12.8 \ \mu H$$

$$C_5 = C_1 = 876 \text{ pF}$$

Example of Predistorted Design of a Linear-Phase Bandpass Filter

To illustrate the predistorted design procedure a linear-phase bandpass filter will be designed to the following specifications:

1. The phase ripple is 0.5°.
2. The order of complexity $n = 5$.
3. The center frequency $f_m = 20$ Mc.
4. The 3-dB bandwidth $\Delta f = 1.0$ Mc.
5. All nodal inductances are chosen to be 2.5 μH.
6. Capacitive coupling will be used.
7. The quality factor, Q_0 of each resonator is equal to 120.
8. The filter is terminated at both ends.

SOLUTION:
   a. Normalized Q factor of resonators.
      The normalized quality factor of the resonators due to losses is found:

$$q_0 = \frac{\Delta f}{f_m} Q_0 = \frac{1.0}{20} 120 = 6.0$$

   b. Tabulated values.
      Using the above q_0 the couplings and normalized loaded q's can be found from the table for $n = 5$, $q_0 = 6.159$ (page 362 of predistorted designs). Two solutions are given for this design in the table. In most cases, when more than one solution is possible, it is best to chose the one in which q_1

and q_n are of the same order of magnitude.

$$q_1 = 0.7793$$
$$k_{12} = 1.6491$$
$$k_{23} = 0.9616$$
$$k_{34} = 0.5742$$
$$k_{45} = 1.0233$$
$$q_5 = 0.7284$$

c. Denormalized values.

The tabulated parameters are denormalized by the factor $f_m/\Delta f$ or its reciprocal to obtain

$$Q_1 = \frac{f_m}{\Delta f} q_1 = \frac{20}{1.0} 0.7793 = 15.586$$

$$K_{12} = \frac{\Delta f}{f_m} k_{12} = \frac{60}{20} 1.6491 = 0.082455$$

$$K_{23} = 0.04808$$
$$K_{34} = 0.02871$$
$$K_{45} = 0.051165$$
$$Q_5 = 14.568$$

d. The nodal capacitance.

With known value of nodal inductance and realizing that it and the nodal capacitance resonates at f_m, the nodal capacitance is easily determined as

$$C_N = [(2\pi)^2 f_m^2 L]^{-1} = 25.3 \text{ pF}$$

e. The coupling capacitances.

The coupling capacitances, where C_I, C_II, C_III, and so forth are all equal to the nodal capacitance C_n, are now found as follows:

$$C_{12} = K_{12}\sqrt{C_1 C_2} = 2.09 \text{ pF}$$
$$C_{23} = 1.22 \text{ pF}$$
$$C_{34} = 0.726 \text{ pF}$$
$$C_{45} = 1.29 \text{ pF}$$

f. Calculation of tuning capacitances.

Knowing the value of all coupling capacitors and the nodal capacitance, the values of the tuning capacitances may be calculated as follows:

$$C_1 = C_n - C_{12} = 23.2 \text{ pF}$$
$$C_2 = C_n - C_{12} - C_{23} = 22.0 \text{ pF}$$
$$C_3 = C_n - C_{23} - C_{34} = 23.4 \text{ pF}$$
$$C_4 = C_n - C_{34} - C_{45} = 23.3 \text{ pF}$$
$$C_5 = C_n - C_{45} = 24.0 \text{ pF}$$

g. Required load.

The required load and source resistances are found from Q_1, Q_5, and Q_0. First the total resistance, including the parasitic loss resistance is found

$$R_1 = \omega_m L Q_1 = 4.89 \text{ k}\Omega$$
$$R_5 = 4.57 \text{ k}\Omega$$

The parasitic resistance due to the finite Q_0 is

$$R_p = \omega_m L Q_0 = 37.7 \text{ k}\Omega$$

Hence, the load and source resistances are

$$R_L = \frac{R_1 R_p}{R_p - R_1} = 5.62 \text{ k}\Omega$$
$$R_s = 5.20 \text{ k}\Omega$$

h. The completed structure.

The completed filter schematic is shown in Fig. 6.19. The insertion loss of the network will be approximately 4 dB as stated in the design tabulation. The required load and source resistance of this filter is too high to match 50- or 75-ohm cables or loads. If this were required, impedance scaling as well as some sort of impedance transformation performed at both ends of the filter would be necessary.

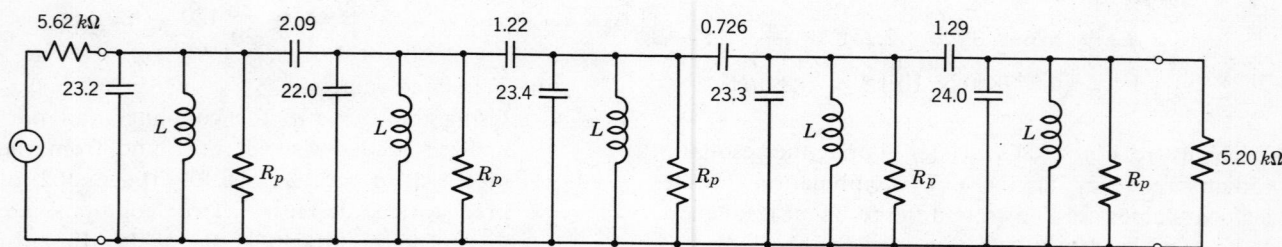

Fig. 6.19. Completed five-resonator linear-phase filter. All capacitances are in pF, all inductances L are 2.5 μH, and all loss resistances R_p are 37.7 kΩ.

i. Tuning.

The final step is to build and align the filter. It is appropriate to note that the coefficients of coupling and quality factors Q are simply related to physically observable phenomenon. This means that the k's and q's can be set typically to within 1 or 2% of the desired values.

It should also be noted that a more linear phase, or constant group delay filter, can be obtained for wider relative bandwidths if the amplitude response can be made to have a symmetry which is closer to arithmetic than geometric. If geometrical symmetry is present, the high side of a bandpass filter has a slower rate of attenuation than does the low side (this assumes a linear frequency scale). The attenuation rates on the high and low side can, however, be made more nearly equal if inductive coupling is used for nodal circuits and capacitive coupling for mesh circuits. This gives a better arithmetic symmetry and hence, a more linear phase will result.

Example of Butterworth Bandpass Filter Design

Let us assume that the following filter specification is given:

1. Input and output impedance = 50 ohms (voltage source).
2. Passband = maximally flat (no ripples).
3. $\text{bw}_{1\,dB} = 8.4$ Mc.
4. $\text{bw}_{40\,dB} = 35$ Mc.
5. Center frequency $f_m = \sqrt{f_1 f_2} = 100$ Mc.

SOLUTION:

1. Response form factor,
$$\text{bw}_{40\,dB}/\text{bw}_{1\,dB} = 35/8.4 = 4.17.$$

2. From Fig. 5.1, using $A_{max} = 1$ dB, $A_{min} = 40$ dB, and at $\Omega = 4.17$, the required number of resonators falls between three and four. Therefore we must choose $n = 4$.

3. With $\text{bw}_{1\,dB}$ known, and by using the curve of the attenuation characteristics for Butterworth filters in Chapter 3, the value of $\text{bw}_{3\,dB}$ can be easily obtained:
$$\frac{\text{bw}_{1\,dB}}{\text{bw}_{3\,dB}} = 0.84$$

Since $\text{bw}_{1\,dB} = 8.4$ Mc, then $\text{bw}_{3\,dB} = 10.0$ Mc. From this same curve, the stopband rejection can be checked. At $\text{bw}_{40\,dB}/\text{bw}_{3\,dB} = 35/10 = 3.5$, for $n = 4$, the attenuation will be approximately 42 dB.

4. The k and q values are found in the table for the Butterworth response, page 341. The infinite Q case is normally used if it is possible to obtain elements with high enough Q factor such that the insertion loss of the filter will be 1 dB or less. In order to obtain a 1 dB insertion loss, the value of q_0 must be approximately 26, as read from the table. Then the minimum unloaded Q of the internal resonators must be
$$Q_{min} = q_0 \frac{f_m}{\text{bw}_{3\,dB}} = 26 \frac{100}{10} = 260$$

(For additional explanation of Q_{min} see Chapter 9.)

Assuming that the desired value, or a greater one, can be achieved the following parameters are read from the table:
$$q_1 = q_4 = 0.7654$$
$$k_{12} = k_{34} = 0.8409$$
$$k_{23} = 0.5412$$

The insertion loss of the filter will be less than 1 dB, and the response shape will be satisfied providing the minimum unloaded Q of the resonators is 260.

5. The next decision to be made concerns the type of coupling between the resonators. The required Q_1 of the first resonator may be obtained by choosing the nodal inductance and capacitance (in the case of capacitive coupling) of such a value that the generator resistance produces the desired Q_1. The alternative method is to use a transforming circuit to couple a nonresonant generator to the first node. For most applications this technique is advisable because it allows one to choose values of L and C that are easily realized in practice. A reasonable value of inductance for a filter designed to operate at 100 Mc is about 0.075 μH. This value will be used for all coils which, by the same token, will require a nodal capacity of 33.8 pF in all cases. The capacitive coupling elements and shunt capacitors can be calculated in the following steps:
$$C_{12} = k_{12} \frac{\Delta f_{3\,dB}}{f_0} \sqrt{C_I C_{II}}$$

where the C's with subscript Roman numerals designate nodal capacitors.
$$C_{12} = 0.84 \frac{10}{100} 33.8 = 2.48 \text{ pF}$$

The first nodal capacitor was set at 33.8 so $C_a = C_I - C_{12} = 31.32$.
$$C_{23} = k_{23} \frac{f_{3\,dB}}{f_0} \sqrt{C_{II} C_{III}} = 1.83 \text{ pF}$$

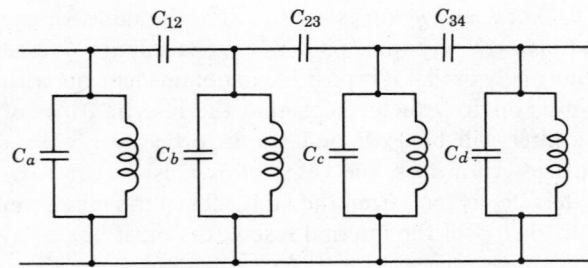

Fig. 6.20.   Four-pole bandpass filter.

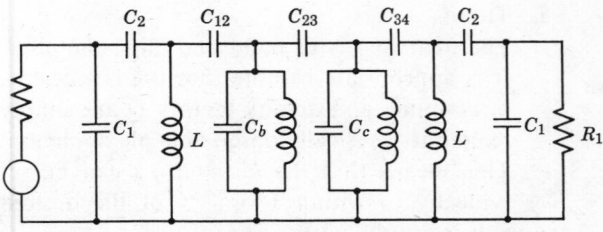

Fig. 6.22.   The final circuit of completed design.

where $\sqrt{C_1 C_{11}} = \sqrt{C_{11}C_{111}} = \cdots = C_I = 33.8$ pF.

$$C_b = 33.8 - 1.83 - 2.48 = 39.49 \text{ pF.}$$
$$C_{34} = C_{12} = 2.48 \text{ pF.}$$
$$C_c = C_b = 29.49 \text{ pF.}$$
$$C_d = C_a = 31.32 \text{ pF.}$$

6. The circuit for the above calculated filter is shown in Fig. 6.20. The design is considered to be complete except that the generator impedance must be transformed to produce the required quality factor in the first resonator. The same transformation must

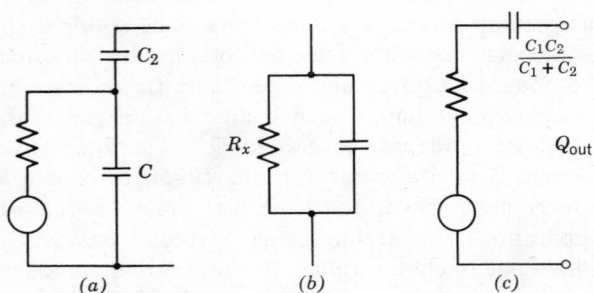

Fig. 6.21.   Transformation of impedance.

be performed with the last resonator also. The most practical transformation for the above case is shown in Fig. 6.21, because the shunt capacitor may be used to absorb the distributed capacitance usually associated with the input and output circuits. The transforming circuit can be calculated in the following fashion:

$$\frac{C_1}{C_2} = \frac{R_x}{R_1}$$
$$\frac{C_1 C_2}{C_1 + C_2} = C_a$$

where R_x is the transformed value of R_1 required to produce the specified Q. The account for incidental coil dissipation (neglecting capacitive losses) can be estimated in the following way:

$$\frac{1}{R} = \frac{1}{R_x} + \frac{1}{Q_1 X_1}$$

In this equation R is the value of nodal shunt resistance required to obtain the specific Q_{out}. The transformation is similar to the input transformation. The final circuit of the completed filter is shown in Fig. 6.22.

6.9   TABLES OF LOWPASS ELEMENT VALUES

σ = the value of the real part of the complex pole pair location

Ω_p = the value of the imaginary part of the complex pole pair location

Ω_z = the value of the imaginary part of the complex zero pair location (the value of the real part of the complex zero pair location is always 0)

$$c = \sqrt{\sigma^2 + \Omega_p^2}$$

$$R_3 = \frac{b + 1}{4.5}$$

$$C_3 = \frac{C_1}{1.5b}$$

$$a = \frac{2\sigma}{c}$$

$$R_4 = 4.5R_3$$

$$C_4 = \frac{C_3}{2}$$

$$b = \frac{\Omega_z}{c^2}$$

$$C_1 = \frac{4.5b}{(b + 1)c}$$

$$K = \frac{(2.5 - a)(b + 1)}{1.5b}$$

$$R_1 = \frac{b + 1}{3b}$$

$$C_2 = \frac{C_1}{4.5}$$

$$\text{Section gain} = \frac{2.5 - a}{1.5}$$

$$R_2 = 2R_1$$

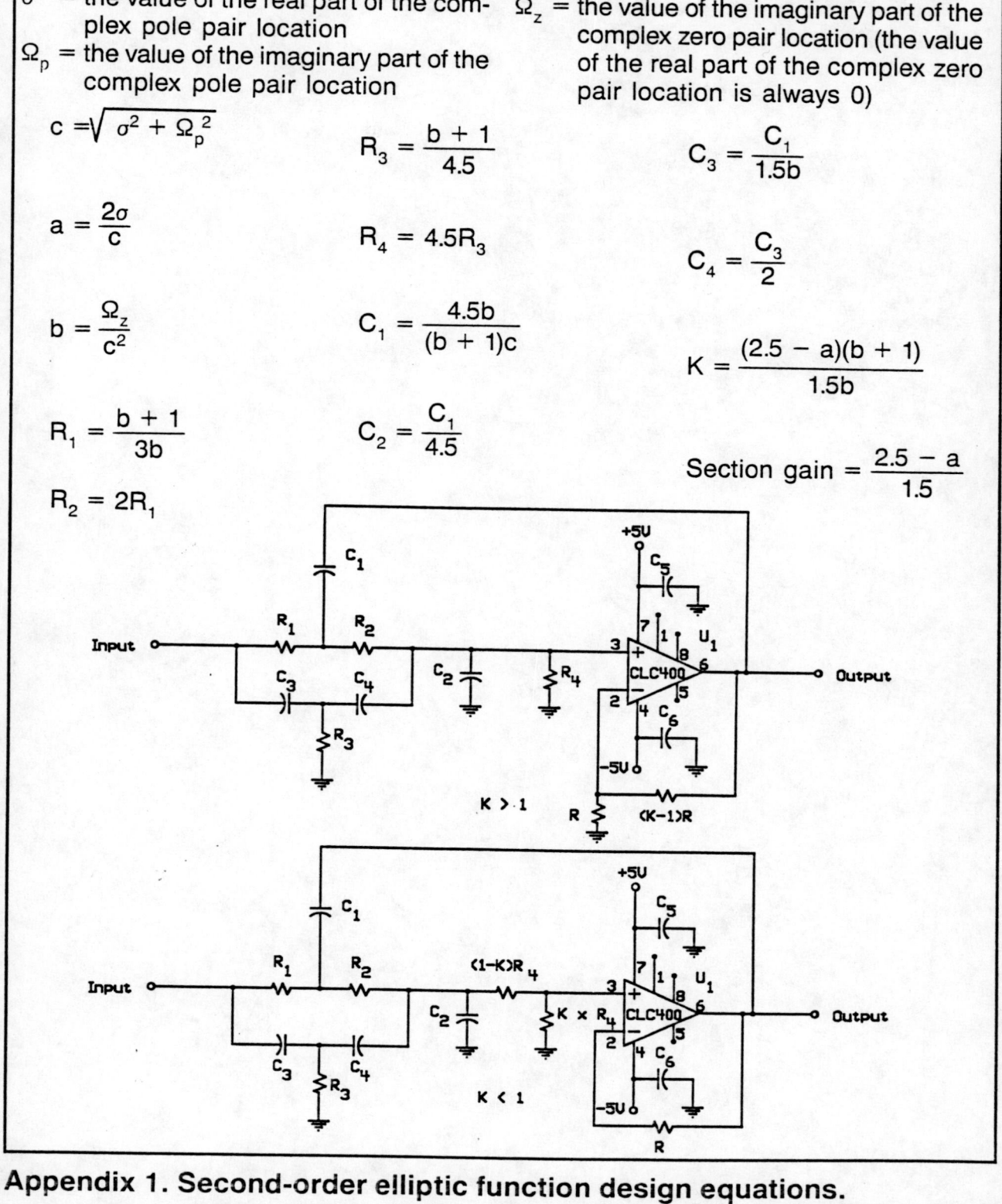

Appendix 1. Second-order elliptic function design equations.

σ = the value of the real part of the complex pole pair location

Ω_p = the value of the imaginary part of the complex pole pair location

Ω_z = the value of the imaginary part of the complex zero pair location (the value of the real part of the complex zero pair location is always 0)

$$c = \sqrt{\sigma^2 + \Omega_p^2}$$

$$R_3 = \frac{b + 1}{4.5}$$

$$C_3 = \frac{C_1}{1.5b}$$

$$a = \frac{2\sigma}{c}$$

$$R_4 = 4.5R_3$$

$$C_4 = \frac{C_3}{2}$$

$$b = \frac{\Omega_z}{c^2}$$

$$C_1 = \frac{4.5b}{(b + 1)c}$$

$$K = \frac{(2.5 - a)(b + 1)}{1.5b}$$

$$R_1 = \frac{b + 1}{3b}$$

$$C_2 = \frac{C_1}{4.5}$$

$$\text{Section gain} = \frac{2.5 - a}{1.5}$$

$$R_2 = 2R_1$$

Appendix 1. Second-order elliptic function design equations.

σ = the value of the real part of the complex pole pair location

Ω_p = the value of the imaginary part of the complex pole pair location

Ω_z = the value of the imaginary part of the complex zero pair location (the value of the real part of the complex zero pair location is always 0)

$$c = \sqrt{\sigma^2 + \Omega_p^2}$$

$$R_3 = \frac{b + 1}{4.5}$$

$$C_3 = \frac{C_1}{1.5b}$$

$$a = \frac{2\sigma}{c}$$

$$R_4 = 4.5R_3$$

$$C_4 = \frac{C_3}{2}$$

$$b = \frac{\Omega_z}{c^2}$$

$$C_1 = \frac{4.5b}{(b + 1)c}$$

$$K = \frac{(2.5 - a)(b + 1)}{1.5b}$$

$$R_1 = \frac{b + 1}{3b}$$

$$C_2 = \frac{C_1}{4.5}$$

$$R_2 = 2R_1$$

$$\text{Section gain} = \frac{2.5 - a}{1.5}$$

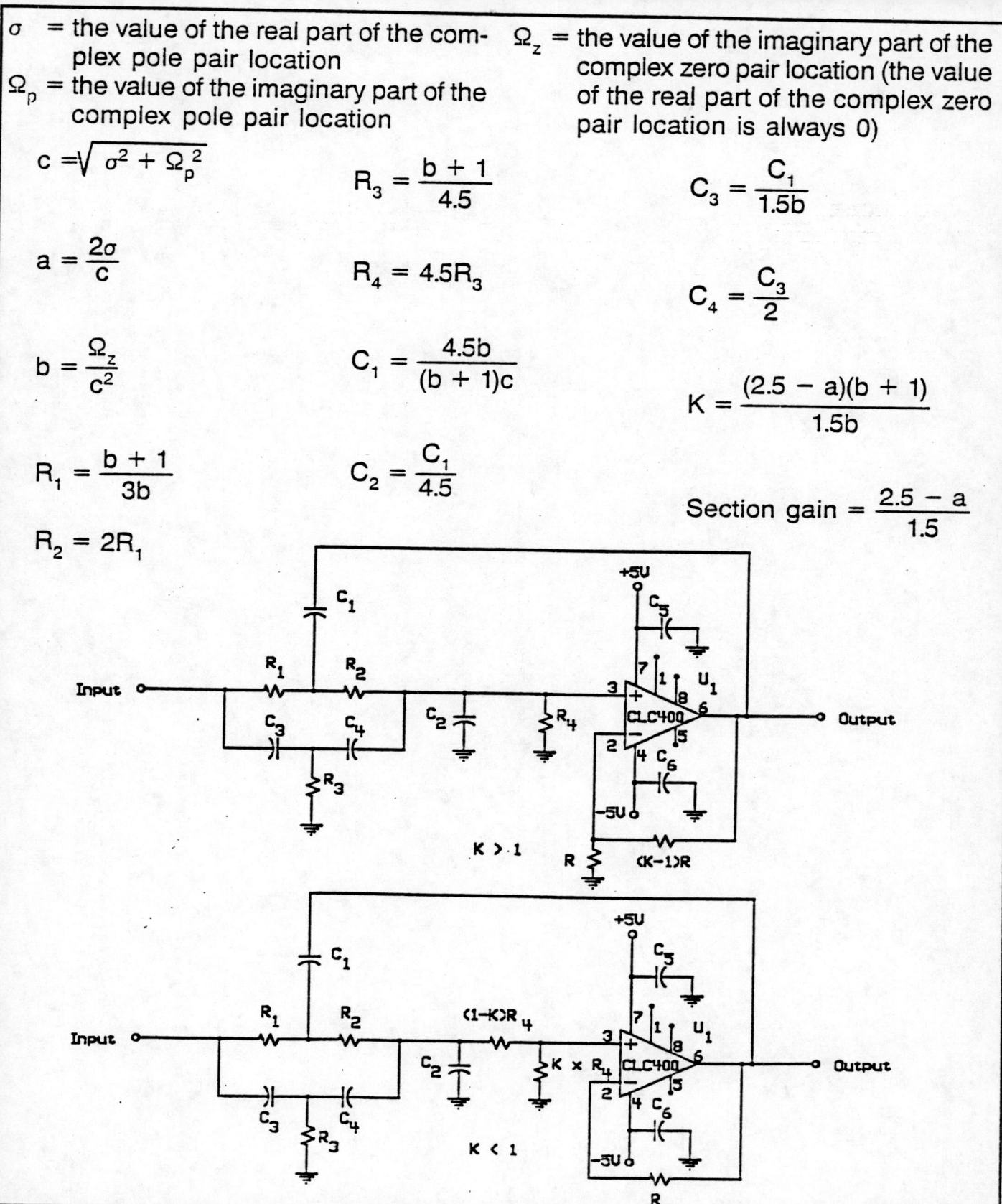

Appendix 1. Second-order elliptic function design equations.

6.10   TABLES OF 3-dB DOWN k AND q VALUES   *page*

n	$1/R_s$	L_1	C_2	L_3	C_4
3	0.7000	0.9192			
	0.6000	1.0225	0.9650	2.7024	
	0.5000	1.1811	0.7789	3.2612	
	0.4000	1.4254	0.6042	4.0642	
	0.3000	1.8380	0.4396	5.3634	
	0.2000	2.6687	0.2842	7.9102	
	0.1000	5.1672	0.1377	15.4554	
	INF.	1.5000	1.3333	0.5000	
4	1.0000	0.7654	1.8478	1.8478	0.7654
	1.1111	0.4657	1.5924	1.7439	1.4690
	1.2500	0.3882	1.6946	1.5110	1.8109
	1.4286	0.3251	1.8618	1.2913	2.1752
	1.6667	0.2690	2.1029	1.0824	2.6131
	2.0000	0.2175	2.4524	0.8826	3.1868
	2.5000	0.1692	2.9858	0.6911	4.0094
	3.3333	0.1237	3.8826	0.5072	5.3381
	5.0000	0.0804	5.6835	0.3307	7.9397
	10.0000	0.0392	11.0942	0.1616	15.6421
	INF.	1.5307	1.5772	1.0824	0.3827

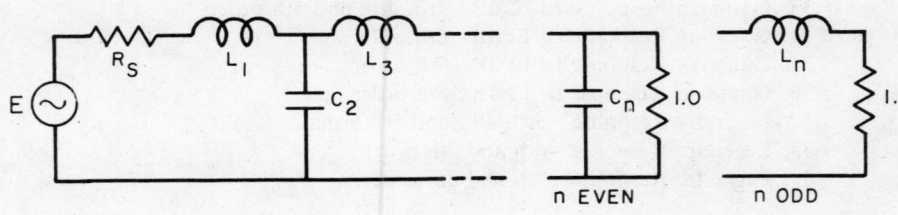

LOW PASS ELEMENT VALUES

BUTTERWORTH RESPONSE

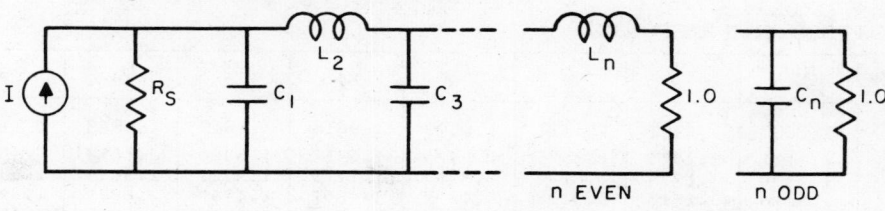

n	R_s	C_1	L_2	C_3	L_4	C_5	L_6	C_7
5	1.0000	0.6180	1.6180	2.0000	1.6180	0.6180		
	0.9000	0.4416	1.0265	1.9095	1.7562	1.3887		
	0.8000	0.4698	0.8660	2.0605	1.5443	1.7380		
	0.7000	0.5173	0.7313	2.2849	1.3326	2.1083		
	0.6000	0.5860	0.6094	2.5998	1.1255	2.5524		
	0.5000	0.6857	0.4955	3.0510	0.9237	3.1331		
	0.4000	0.8378	0.3877	3.7357	0.7274	3.9648		
	0.3000	1.0937	0.2848	4.8835	0.5367	5.3073		
	0.2000	1.6077	0.1861	7.1849	0.3518	7.9345		
	0.1000	3.1522	0.0912	14.0945	0.1727	15.7103		
	INF.	1.5451	1.6944	1.3820	0.8944	0.3090		
6	1.0000	0.5176	1.4142	1.9319	1.9319	1.4142	0.5176	
	1.1111	0.2890	1.0403	1.3217	2.0539	1.7443	1.3347	
	1.2500	0.2445	1.1163	1.1257	2.2389	1.5498	1.6881	
	1.4286	0.2072	1.2363	0.9567	2.4991	1.3464	2.0618	
	1.6667	0.1732	1.4071	0.8011	2.8580	1.1431	2.5092	
	2.0000	0.1412	1.6531	0.6542	3.3687	0.9423	3.0938	
	2.5000	0.1108	2.0275	0.5139	4.1408	0.7450	3.9305	
	3.3333	0.0816	2.6559	0.3788	5.4325	0.5517	5.2804	
	5.0000	0.0535	3.9170	0.2484	8.0201	0.3628	7.9216	
	10.0000	0.0263	7.7053	0.1222	15.7855	0.1788	15.7375	
	INF.	1.5529	1.7593	1.5529	1.2016	0.7579	0.2588	
7	1.0000	0.4450	1.2470	1.8019	2.0000	1.8019	1.2470	0.4450
	0.9000	0.2985	0.7111	1.4043	1.4891	2.1249	1.7268	1.2961
	0.8000	0.3215	0.6057	1.5174	1.2777	2.3338	1.5461	1.6520
	0.7000	0.3571	0.5154	1.6883	1.0910	2.6177	1.3498	2.0277
	0.6000	0.4075	0.4322	1.9284	0.9170	3.0050	1.1503	2.4771
	0.5000	0.4799	0.3536	2.2726	0.7512	3.5532	0.9513	3.0640
	0.4000	0.5899	0.2782	2.7950	0.5917	4.3799	0.7542	3.9037
	0.3000	0.7745	0.2055	3.6706	0.4373	5.7612	0.5600	5.2583
	0.2000	1.1448	0.1350	5.4267	0.2874	8.5263	0.3692	7.9079
	0.1000	2.2571	0.0665	10.7004	0.1417	16.8222	0.1823	15.7480
	INF.	1.5576	1.7988	1.6588	1.3972	1.0550	0.6560	0.2225
n	$1/R_s$	L_1	C_2	L_3	C_4	L_5	C_6	L_7

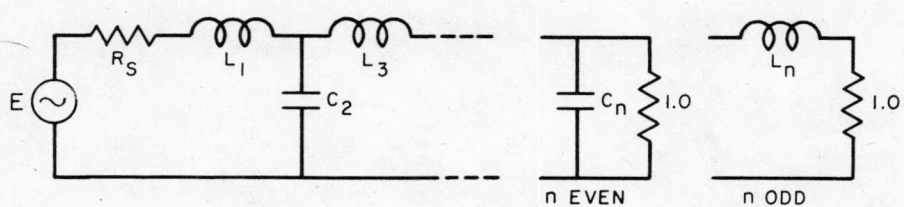

313

LOW PASS ELEMENT VALUES

BUTTERWORTH RESPONSE

n	R_s	C_1	L_2	C_3	L_4	C_5	L_6	C_7	L_8	C_9	L_{10}
8	1.0000	0.3902	1.1111	1.6629	1.9616	1.9616	1.6629	1.1111	0.3902		
	1.1111	0.2075	0.7575	0.9925	1.6362	1.5900	2.1612	1.7092	1.2671		
	1.2500	0.1774	0.8199	0.8499	1.7779	1.3721	2.3874	1.5393	1.6246		
	1.4286	0.1513	0.9138	0.7257	1.9852	1.1760	2.6879	1.3490	2.0017		
	1.6667	0.1272	1.0455	0.6102	2.2740	0.9912	3.0945	1.1530	2.4524		
	2.0000	0.1042	1.2341	0.5003	2.6863	0.8139	3.6678	0.9558	3.0408		
	2.5000	0.0822	1.5201	0.3945	3.3106	0.6424	4.5308	0.7594	3.8825		
	3.3333	0.0608	1.9995	0.2919	4.3563	0.4757	5.9714	0.5650	5.2400		
	5.0000	0.0400	2.9608	0.1921	6.4523	0.3133	8.8538	0.3732	7.8952		
	10.0000	0.0198	5.8479	0.0949	12.7455	0.1547	17.4999	0.1846	15.7510		
	INF.	1.5607	1.8246	1.7287	1.5283	1.2588	0.9371	0.5776	0.1951		
9	1.0000	0.3473	1.0000	1.5321	1.8794	2.0000	1.8794	1.5321	1.0000	0.3473	
	0.9000	0.2242	0.5388	1.0835	1.1859	1.7905	1.6538	2.1796	1.6930	1.2447	
	0.8000	0.2434	0.4623	1.1777	1.0200	1.9542	1.4336	2.4189	1.5318	1.6033	
	0.7000	0.2719	0.3954	1.3162	0.8734	2.1885	1.2323	2.7314	1.3464	1.9812	
	0.6000	0.3117	0.3330	1.5092	0.7361	2.5124	1.0410	3.1516	1.1533	2.4328	
	0.5000	0.3685	0.2735	1.7846	0.6046	2.9734	0.8565	3.7426	0.9579	3.0223	
	0.4000	0.4545	0.2159	2.2019	0.4775	3.6706	0.6771	4.6310	0.7624	3.8654	
	0.3000	0.5987	0.1600	2.9006	0.3539	4.8373	0.5022	6.1128	0.5680	5.2249	
	0.2000	0.8878	0.1054	4.3014	0.2333	7.1750	0.3312	9.0766	0.3757	7.8838	
	0.1000	1.7558	0.0521	8.5074	0.1153	14.1930	0.1638	17.9654	0.1862	15.7504	
	INF.	1.5628	1.8424	1.7772	1.6202	1.4037	1.1408	0.8414	0.5155	0.1736	
10	1.0000	0.3129	0.9080	1.4142	1.7820	1.9754	1.9754	1.7820	1.4142	0.9080	0.3129
	1.1111	0.1614	0.5924	0.7853	1.3202	1.3230	1.8968	1.6956	2.1883	1.6785	1.2267
	1.2500	0.1388	0.6452	0.6762	1.4400	1.1420	2.0779	1.4754	2.4377	1.5245	1.5861
	1.4286	0.1190	0.7222	0.5797	1.6130	0.9802	2.3324	1.2712	2.7592	1.3431	1.9646
	1.6667	0.1004	0.8292	0.4891	1.8528	0.8275	2.6825	1.0758	3.1895	1.1526	2.4169
	2.0000	0.0825	0.9818	0.4021	2.1943	0.6808	3.1795	0.8864	3.7934	0.9588	3.0072
	2.5000	0.0652	1.2127	0.3179	2.7108	0.5384	3.9302	0.7018	4.7002	0.7641	3.8512
	3.3333	0.0484	1.5992	0.2358	3.5754	0.3995	5.1858	0.5211	6.2118	0.5700	5.2122
	5.0000	0.0319	2.3740	0.1556	5.3082	0.2636	7.7010	0.3440	9.2343	0.3775	7.8738
	10.0000	0.0158	4.7005	0.0770	10.5104	0.1305	15.2505	0.1704	18.2981	0.1872	15.7481
	INF.	1.5643	1.8552	1.8121	1.6869	1.5100	1.2921	1.0406	0.7626	0.4654	0.1564
n	$1/R_s$	L_1	C_2	L_3	C_4	L_5	C_6	L_7	C_8	L_9	C_{10}

LOW PASS ELEMENT VALUES

CHEBYSHEV RESPONSE

RIPPLE = 0.01 db

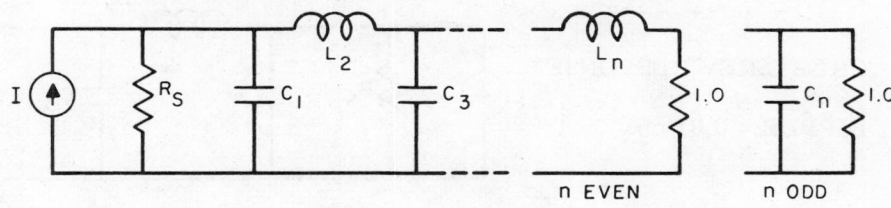

n	R_s	C_1	L_2	C_3	L_4
	1.1007	1.3472	1.4829		
	1.1111	1.2472	1.5947		
	1.2500	0.9434	1.9974		
	1.4286	0.7591	2.3442		
	1.6667	0.6091	2.7490		
2	2.0000	0.4791	3.2772		
	2.5000	0.3634	4.0328		
	3.3333	0.2590	5.2546		
	5.0000	0.1642	7.6498		
	10.0000	0.0781	14.7492		
	INF.	1.4118	0.7415		
	1.0000	1.1811	1.8214	1.1811	
	0.9000	1.0917	1.6597	1.4802	
	0.8000	1.0969	1.4431	1.8057	
	0.7000	1.1600	1.2283	2.1653	
	0.6000	1.2737	1.0236	2.5984	
3	0.5000	1.4521	0.8294	3.1644	
	0.4000	1.7340	0.6452	3.9742	
	0.3000	2.2164	0.4704	5.2800	
	0.2000	3.1934	0.3047	7.8338	
	0.1000	6.1411	0.1479	15.3899	
	INF.	1.5012	1.4330	0.5905	
	1.1007	0.9500	1.9382	1.7608	1.0457
	1.1111	0.8539	1.9460	1.7439	1.1647
	1.2500	0.6182	2.0749	1.5417	1.6170
	1.4286	0.4948	2.2787	1.3336	2.0083
	1.6667	0.3983	2.5709	1.1277	2.4611
4	2.0000	0.3156	2.9943	0.9260	3.0448
	2.5000	0.2418	3.6406	0.7293	3.8746
	3.3333	0.1744	4.7274	0.5379	5.2085
	5.0000	0.1121	6.9102	0.3523	7.8126
	10.0000	0.0541	13.4690	0.1729	15.5100
	INF.	1.5287	1.6939	1.3122	0.5229
n	$1/R_s$	L_1	C_2	L_3	C_4

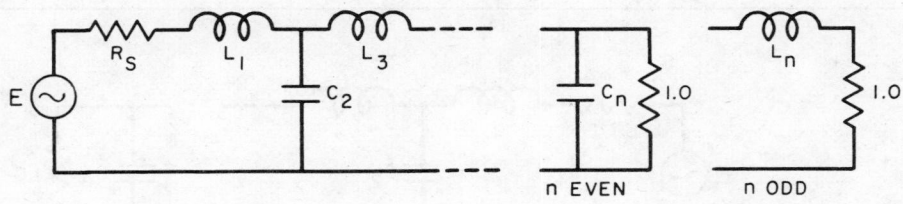

315

LOW PASS ELEMENT VALUES

CHEBYSHEV RESPONSE

RIPPLE = 0.01 db

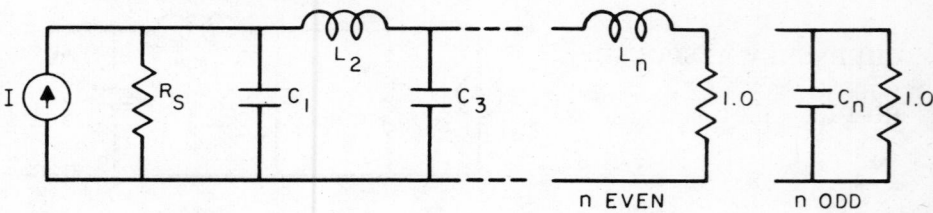

n EVEN   n ODD

n	R_s	C_1	L_2	C_3	L_4	C_5	L_6	C_7
5	1.0000	0.9766	1.6849	2.0366	1.6849	0.9766		
	0.9000	0.8798	1.4558	2.1738	1.6412	1.2739		
	0.8000	0.8769	1.2350	2.3785	1.4991	1.6066		
	0.7000	0.9263	1.0398	2.6582	1.3228	1.9772		
	0.6000	1.0191	0.8626	3.0408	1.1345	2.4244		
	0.5000	1.1658	0.6985	3.5835	0.9421	3.0092		
	0.4000	1.3983	0.5442	4.4027	0.7491	3.8453		
	0.3000	1.7966	0.3982	5.7721	0.5573	5.1925		
	0.2000	2.6039	0.2592	8.5140	0.3679	7.8257		
	0.1000	5.0406	0.1266	16.7406	0.1819	15.6126		
	INF.	1.5466	1.7950	1.6449	1.2365	0.4883		
6	1.1007	0.8514	1.7956	1.8411	2.0266	1.6312	0.9372	
	1.1111	0.7597	1.7817	1.7752	2.0941	1.6380	1.0533	
	1.2500	0.5445	1.8637	1.4886	2.4025	1.5067	1.5041	
	1.4286	0.4355	2.0383	1.2655	2.7346	1.3318	1.8987	
	1.6667	0.3509	2.2978	1.0607	3.1671	1.1451	2.3568	
	2.0000	0.2786	2.6781	0.8671	3.7683	0.9536	2.9483	
	2.5000	0.2139	3.2614	0.6816	4.6673	0.7606	3.7899	
	3.3333	0.1547	4.2448	0.5028	6.1631	0.5676	5.1430	
	5.0000	0.0997	6.2227	0.3299	9.1507	0.3760	7.7852	
	10.0000	0.0483	12.1707	0.1623	18.1048	0.1865	15.5950	
	INF.	1.5510	1.8471	1.7897	1.5976	1.1904	0.4686	
7	1.0000	0.9127	1.5947	2.0021	1.8704	2.0021	1.5947	0.9127
	0.9000	0.8157	1.3619	2.0886	1.7217	2.2017	1.5805	1.2060
	0.8000	0.8111	1.1504	2.2618	1.5252	2.4647	1.4644	1.5380
	0.7000	0.8567	0.9673	2.5158	1.3234	2.8018	1.3066	1.9096
	0.6000	0.9430	0.8025	2.8720	1.1237	3.2496	1.1310	2.3592
	0.5000	1.0799	0.6502	3.3822	0.9276	3.8750	0.9468	2.9478
	0.4000	1.2971	0.5072	4.1563	0.7350	4.8115	0.7584	3.7900
	0.3000	1.6692	0.3716	5.4540	0.5459	6.3703	0.5682	5.1476
	0.2000	2.4235	0.2423	8.0565	0.3604	9.4844	0.3776	7.8019
	0.1000	4.7006	0.1186	15.8718	0.1784	18.8179	0.1879	15.6523
	INF.	1.5593	1.8671	1.8657	1.7651	1.5633	1.1610	0.4564
n	$1/R_s$	L_1	C_2	L_3	C_4	L_5	C_6	L_7

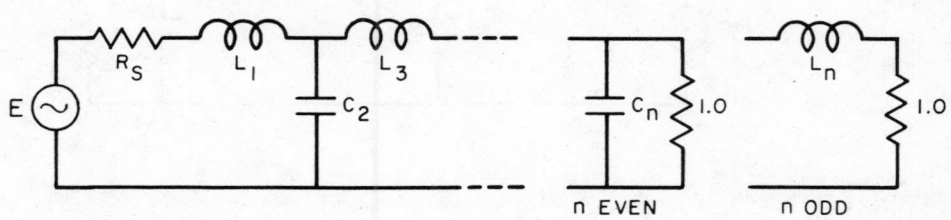

n EVEN   n ODD

LOW PASS ELEMENT VALUES

CHEBYSHEV RESPONSE   RIPPLE = 0.01 db

n	R_s	C_1	L_2	C_3	L_4	C_5	L_6	C_7	L_8	C_9	L_{10}
8	1.1007	0.8145	1.7275	1.7984	2.0579	1.8695	1.9796	1.5694	0.8966		
	1.1111	0.7248	1.7081	1.7239	2.1019	1.8259	2.0595	1.5827	1.0111		
	1.2500	0.5176	1.7772	1.4315	2.3601	1.5855	2.4101	1.4754	1.4597		
	1.4286	0.4138	1.9422	1.2141	2.6686	1.3723	2.7734	1.3142	1.8544		
	1.6667	0.3336	2.1896	1.0169	3.0808	1.1660	3.2393	1.1369	2.3136		
	2.0000	0.2650	2.5533	0.8313	3.6598	0.9639	3.8820	0.9518	2.9073		
	2.5000	0.2036	3.1118	0.6537	4.5303	0.7653	4.8393	0.7627	3.7524		
	3.3333	0.1474	4.0539	0.4826	5.9828	0.5697	6.4287	0.5718	5.1118		
	5.0000	0.0951	5.9495	0.3170	8.8889	0.3770	9.6002	0.3804	7.7668		
	10.0000	0.0462	11.6509	0.1562	17.6067	0.1870	19.1009	0.1895	15.6158		
	INF.	1.5588	1.8848	1.8988	1.8556	1.7433	1.5391	1.1412	0.4483		
9	1.0000	0.8854	1.5513	1.9614	1.8616	2.0717	1.8616	1.9614	1.5513	0.8854	
	0.9000	0.7886	1.3192	2.0330	1.6941	2.2249	1.7402	2.1774	1.5478	1.1764	
	0.8000	0.7834	1.1127	2.1959	1.4930	2.4614	1.5603	2.4565	1.4423	1.5076	
	0.7000	0.8273	0.9353	2.4404	1.2924	2.7808	1.3662	2.8093	1.2927	1.8793	
	0.6000	0.9109	0.7761	2.7852	1.0962	3.2140	1.1688	3.2747	1.1233	2.3295	
	0.5000	1.0436	0.6290	3.2805	0.9045	3.8249	0.9710	3.9223	0.9436	2.9193	
	0.4000	1.2542	0.4910	4.0329	0.7167	4.7444	0.7739	4.8900	0.7582	3.7637	
	0.3000	1.6151	0.3599	5.2951	0.5325	6.2792	0.5780	6.4989	0.5697	5.1254	
	0.2000	2.3468	0.2349	7.8274	0.3518	9.3504	0.3835	9.7114	0.3797	7.7882	
	0.1000	4.5556	0.1150	15.4334	0.1743	18.5641	0.1908	19.3382	0.1895	15.6645	
	INF.	1.5646	1.8884	1.9242	1.8977	1.8425	1.7261	1.5217	1.1273	0.4427	
10	1.1007	0.7970	1.6930	1.7690	2.0395	1.8827	2.0724	1.8529	1.9472	1.5380	0.8773
	1.1111	0.7083	1.6714	1.6921	2.0763	1.8281	2.1308	1.8167	2.0310	1.5541	0.9910
	1.2500	0.5049	1.7353	1.4005	2.3184	1.5706	2.4371	1.5953	2.3952	1.4574	1.4381
	1.4286	0.4037	1.8958	1.1871	2.6178	1.3552	2.7830	1.3895	2.7685	1.3027	1.8327
	1.6667	0.3255	2.1375	0.9942	3.0205	1.1497	3.2370	1.1863	3.2448	1.1300	2.2923
	2.0000	0.2586	2.4932	0.8128	3.5878	0.9497	3.8698	0.9849	3.9004	0.9484	2.8867
	2.5000	0.1988	3.0398	0.6394	4.4418	0.7538	4.8173	0.7849	4.8757	0.7617	3.7333
	3.3333	0.1440	3.9619	0.4723	5.8678	0.5612	6.3951	0.5863	6.4939	0.5722	5.0955
	5.0000	0.0930	5.8175	0.3103	8.7220	0.3715	9.5486	0.3893	9.7217	0.3814	7.7563
	10.0000	0.0451	11.3993	0.1530	17.2866	0.1844	19.0046	0.1938	19.3905	0.1904	15.6234
	INF.	1.5625	1.8978	1.9323	1.9288	1.8907	1.8309	1.7128	1.5088	1.1173	0.4386
n	$1/R_s$	L_1	C_2	L_3	C_4	L_5	C_6	L_7	C_8	L_9	C_{10}

LOW PASS ELEMENT VALUES

CHEBYSHEV RESPONSE

RIPPLE = 0.1 db

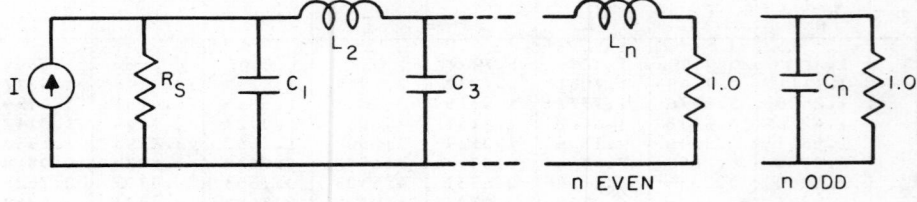

n	R_S	C_1	L_2	C_3	L_4
2	1.3554	1.2087	1.6382		
	1.4286	0.9771	1.9824		
	1.6667	0.7326	2.4585		
	2.0000	0.5597	3.0538		
	2.5000	0.4169	3.8265		
	3.3333	0.2933	5.0502		
	5.0000	0.1841	7.4257		
	10.0000	0.0868	14.4332		
	INF.	1.3911	0.8191		
3	1.0000	1.4328	1.5937	1.4328	
	0.9000	1.4258	1.4935	1.6219	
	0.8000	1.4511	1.3557	1.8711	
	0.7000	1.5210	1.1927	2.1901	
	0.6000	1.6475	1.0174	2.6026	
	0.5000	1.8530	0.8383	3.1594	
	0.4000	2.1857	0.6603	3.9675	
	0.3000	2.7630	0.4860	5.2788	
	0.2000	3.9418	0.3172	7.8503	
	0.1000	7.5121	0.1549	15.4656	
	INF.	1.5133	1.5090	0.7164	
4	1.3554	0.9924	2.1476	1.5845	1.3451
	1.4286	0.7789	2.3480	1.4292	1.7001
	1.6667	0.5764	2.7304	1.1851	2.2425
	2.0000	0.4398	3.2269	0.9672	2.8563
	2.5000	0.3288	3.9605	0.7599	3.6976
	3.3333	0.2329	5.1777	0.5602	5.0301
	5.0000	0.1475	7.6072	0.3670	7.6143
	10.0000	0.0704	14.8873	0.1802	15.2297
	INF.	1.5107	1.7682	1.4550	0.6725
n	$1/R_S$	L_1	C_2	L_3	C_4

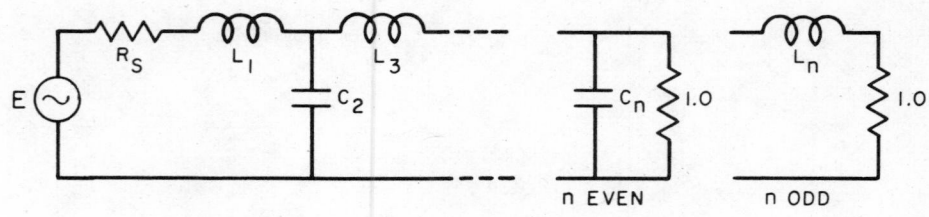

318

LOW PASS ELEMENT VALUES

CHEBYSHEV RESPONSE

RIPPLE = 0.1 db

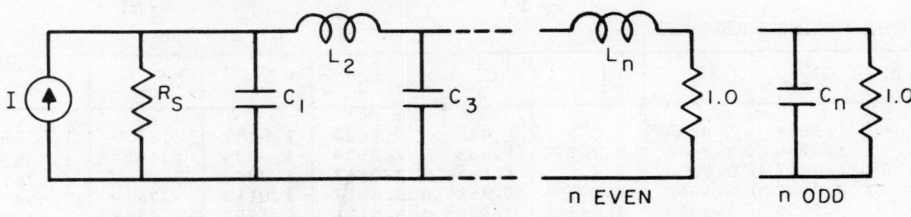

n	R_s	C_1	L_2	C_3	L_4	C_5	L_6	C_7
5	1.0000	1.3013	1.5559	2.2411	1.5559	1.3013		
	0.9000	1.2845	1.4329	2.3794	1.4878	1.4883		
	0.8000	1.2998	1.2824	2.5819	1.3815	1.7384		
	0.7000	1.3580	1.1170	2.8679	1.2437	2.0621		
	0.6000	1.4694	0.9469	3.2688	1.0846	2.4835		
	0.5000	1.6535	0.7777	3.8446	0.9126	3.0548		
	0.4000	1.9538	0.6119	4.7193	0.7333	3.8861		
	0.3000	2.4765	0.4509	6.1861	0.5503	5.2373		
	0.2000	3.5457	0.2950	9.1272	0.3659	7.8890		
	0.1000	6.7870	0.1447	17.9569	0.1820	15.7447		
	INF.	1.5613	1.8069	1.7659	1.4173	0.6507		
6	1.3554	0.9419	2.0797	1.6581	2.2473	1.5344	1.2767	
	1.4286	0.7347	2.2492	1.4537	2.5437	1.4051	1.6293	
	1.6667	0.5422	2.6003	1.1830	3.0641	1.1850	2.1739	
	2.0000	0.4137	3.0679	0.9575	3.7119	0.9794	2.7936	
	2.5000	0.3095	3.7652	0.7492	4.6512	0.7781	3.6453	
	3.3333	0.2195	4.9266	0.5514	6.1947	0.5795	4.9962	
	5.0000	0.1393	7.2500	0.3613	9.2605	0.3835	7.6184	
	10.0000	0.0666	14.2200	0.1777	18.4267	0.1901	15.3495	
	INF.	1.5339	1.8838	1.8306	1.7485	1.3937	0.6383	
7	1.0000	1.2615	1.5196	2.2392	1.6804	2.2392	1.5196	1.2615
	0.9000	1.2422	1.3946	2.3613	1.5784	2.3966	1.4593	1.4472
	0.8000	1.2550	1.2449	2.5481	1.4430	2.6242	1.3619	1.6967
	0.7000	1.3100	1.0826	2.8192	1.2833	2.9422	1.2326	2.0207
	0.6000	1.4170	0.9169	3.2052	1.1092	3.3841	1.0807	2.4437
	0.5000	1.5948	0.7529	3.7642	0.9276	4.0150	0.9142	3.0182
	0.4000	1.8853	0.5926	4.6179	0.7423	4.9702	0.7384	3.8552
	0.3000	2.3917	0.4369	6.0535	0.5557	6.5685	0.5569	5.2167
	0.2000	3.4278	0.2862	8.9371	0.3692	9.7697	0.3723	7.8901
	0.1000	6.5695	0.1405	17.6031	0.1838	19.3760	0.1862	15.8127
	INF.	1.5748	1.8577	1.9210	1.8270	1.7340	1.3786	0.6307
n	$1/R_s$	L_1	C_2	L_3	C_4	L_5	C_6	L_7

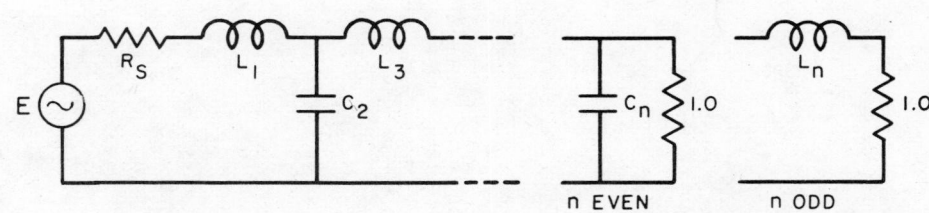

319

LOW PASS ELEMENT VALUES

CHEBYSHEV RESPONSE

n	R_s	C_1	L_2	C_3	L_4	C_5	L_6	C_7	L_8	C_9	L_{10}.
8	1.3554	0.9234	2.0454	1.6453	2.2826	1.6841	2.2300	1.5091	1.2515		
	1.4286	0.7186	2.2054	1.4350	2.5554	1.4974	2.5422	1.3882	1.6029		
	1.6667	0.5298	2.5459	1.1644	3.0567	1.2367	3.0869	1.1769	2.1477		
	2.0000	0.4042	3.0029	0.9415	3.6917	1.0118	3.7619	0.9767	2.7690		
	2.5000	0.3025	3.6859	0.7365	4.6191	0.7990	4.7388	0.7787	3.6240		
	3.3333	0.2147	4.8250	0.5421	6.1483	0.5930	6.3423	0.5820	4.9811		
	5.0000	0.1364	7.1050	0.3554	9.1917	0.3917	9.5260	0.3863	7.6164		
	10.0000	0.0652	13.9469	0.1749	18.3007	0.1942	19.0437	0.1922	15.3880		
	INF.	1.5422	1.9106	1.9008	1.9252	1.8200	1.7231	1.3683	0.6258		
9	1.0000	1.2446	1.5017	2.2220	1.6829	2.2957	1.6829	2.2220	1.5017	1.2446	
	0.9000	1.2244	1.3765	2.3388	1.5756	2.4400	1.5870	2.3835	1.4444	1.4297	
	0.8000	1.2361	1.2276	2.5201	1.4365	2.6561	1.4572	2.6168	1.3505	1.6788	
	0.7000	1.2898	1.0670	2.7856	1.2751	2.9647	1.3019	2.9422	1.2248	2.0029	
	0.6000	1.3950	0.9035	3.1653	1.1008	3.3992	1.1304	3.3937	1.0761	2.4264	
	0.5000	1.5701	0.7419	3.7166	0.9198	4.0244	0.9494	4.0377	0.9121	3.0020	
	0.4000	1.8566	0.5840	4.5594	0.7359	4.9750	0.7630	5.0118	0.7382	3.8412	
	0.3000	2.3560	0.4307	5.9781	0.5509	6.5700	0.5736	6.6413	0.5579	5.2068	
	0.2000	3.3781	0.2822	8.8291	0.3661	9.7699	0.3827	9.9047	0.3737	7.8891	
	0.1000	6.4777	0.1386	17.3994	0.1823	19.3816	0.1912	19.6976	0.1873	15.8393	
	INF.	1.5804	1.8727	1.9584	1.9094	1.9229	1.8136	1.7150	1.3611	0.6223	
10	1.3554	0.9146	2.0279	1.6346	2.2777	1.6963	2.2991	1.6805	2.2155	1.4962	1.2397
	1.4286	0.7110	2.1837	1.4231	2.5425	1.5002	2.5915	1.5000	2.5322	1.3789	1.5903
	1.6667	0.5240	2.5194	1.1536	3.0362	1.2349	3.1229	1.2444	3.0839	1.1717	2.1351
	2.0000	0.3998	2.9713	0.9326	3.6647	1.0089	3.7923	1.0214	3.7669	0.9741	2.7572
	2.5000	0.2993	3.6476	0.7295	4.5843	0.7962	4.7673	0.8090	4.7547	0.7779	3.6136
	3.3333	0.2124	4.7758	0.5370	6.1022	0.5907	6.3734	0.6020	6.3758	0.5822	4.9735
	5.0000	0.1350	7.0347	0.3522	9.1248	0.3902	9.5681	0.3987	9.5942	0.3871	7.6148
	10.0000	0.0646	13.8141	0.1734	18.1739	0.1935	19.1282	0.1981	19.2158	0.1929	15.4052
	INF.	1.5460	1.9201	1.9216	1.9700	1.9102	1.9194	1.8083	1.7090	1.3559	0.6198
n	$1/R_s$	L_1	C_2	L_3	C_4	L_5	C_6	L_7	C_8	L_9	C_{10}

LOW PASS ELEMENT VALUES

CHEBYSHEV RESPONSE   RIPPLE = 0.5 db

n	R_s	C_1	L_2	C_3	L_4	C_5	L_6
2	1.9841	0.9827	1.9497				
	2.0000	0.9086	2.1030				
	2.5000	0.5635	3.1647				
	3.3333	0.3754	4.4111				
	5.0000	0.2282	6.6995				
	10.0000	0.1052	13.3221				
	INF.	1.3067	0.9748				
3	1.0000	1.8636	1.2804	1.8636			
	0.9000	1.9175	1.2086	2.0255			
	0.8000	1.9965	1.1203	2.2368			
	0.7000	2.1135	1.0149	2.5172			
	0.6000	2.2889	0.8937	2.8984			
	0.5000	2.5571	0.7592	3.4360			
	0.4000	2.9854	0.6146	4.2416			
	0.3000	3.7292	0.4633	5.5762			
	0.2000	5.2543	0.3087	8.2251			
	0.1000	9.8899	0.1534	16.1177			
	INF.	1.5720	1.5179	0.9318			
4	1.9841	0.9202	2.5864	1.3036	1.8258		
	2.0000	0.8452	2.7198	1.2383	1.9849		
	2.5000	0.5162	3.7659	0.8693	3.1205		
	3.3333	0.3440	5.1196	0.6208	4.4790		
	5.0000	0.2100	7.7076	0.3996	6.9874		
	10.0000	0.0975	15.3520	0.1940	14.2616		
	INF.	1.4361	1.8888	1.5211	0.9129		
5	1.0000	1.8068	1.3025	2.6914	1.3025	1.8068	
	0.9000	1.8540	1.2220	2.8478	1.2379	1.9701	
	0.8000	1.9257	1.1261	3.0599	1.1569	2.1845	
	0.7000	2.0347	1.0150	3.3525	1.0582	2.4704	
	0.6000	2.2006	0.8901	3.7651	0.9420	2.8609	
	0.5000	2.4571	0.7537	4.3672	0.8098	3.4137	
	0.4000	2.8692	0.6091	5.2960	0.6640	4.2447	
	0.3000	3.5877	0.4590	6.8714	0.5075	5.6245	
	0.2000	5.0639	0.3060	10.0537	0.3430	8.3674	
	0.1000	9.5560	0.1525	19.6465	0.1731	16.5474	
	INF.	1.6299	1.7400	1.9217	1.5138	0.9034	
6	1.9841	0.9053	2.5774	1.3675	2.7133	1.2991	1.7961
	2.0000	0.8303	2.7042	1.2912	2.8721	1.2372	1.9557
	2.5000	0.5056	3.7219	0.8900	4.1092	0.8808	3.1025
	3.3333	0.3370	5.0554	0.6323	5.6994	0.6348	4.4810
	5.0000	0.2059	7.6145	0.4063	8.7319	0.4121	7.0310
	10.0000	0.0958	15.1862	0.1974	17.6806	0.2017	14.4328
	INF.	1.4618	1.9799	1.7803	1.9253	1.5077	0.8981
n	$1/R_s$	L_1	C_2	L_3	C_4	L_5	C_6

CHEBYSHEV RESPONSE   RIPPLE = 0.5 db

n	R_s	C_1	L_2	C_3	L_4	C_5	L_6	C_7	L_8	C_9	L_{10}
7	1.0000	1.7896	1.2961	2.7177	1.3848	2.7177	1.2961	1.7896			
	0.9000	1.8348	1.2146	2.8691	1.3080	2.8829	1.2335	1.9531			
	0.8000	1.9045	1.1182	3.0761	1.2149	3.1071	1.1546	2.1681			
	0.7000	2.0112	1.0070	3.3638	1.1050	3.4163	1.0582	2.4554			
	0.6000	2.1744	0.8824	3.7717	0.9786	3.8524	0.9441	2.8481			
	0.5000	2.4275	0.7470	4.3695	0.8377	4.4886	0.8137	3.4050			
	0.4000	2.8348	0.6035	5.2947	0.6846	5.4698	0.6690	4.2428			
	0.3000	3.5456	0.4548	6.8674	0.5221	7.1341	0.5129	5.6350			
	0.2000	5.0070	0.3034	10.0491	0.3524	10.4959	0.3478	8.4041			
	0.1000	9.4555	0.1513	19.6486	0.1778	20.6314	0.1761	16.6654			
	INF.	1.6464	1.7772	2.0306	1.7892	1.9239	1.5034	0.8948			
8	1.9841	0.8998	2.5670	1.3697	2.7585	1.3903	2.7175	1.2938	1.7852		
	2.0000	0.8249	2.6916	1.2919	2.9134	1.3160	2.8800	1.2331	1.9449		
	2.5000	0.5017	3.6988	0.8878	4.1404	0.9184	4.1470	0.8815	3.0953		
	3.3333	0.3344	5.0234	0.6304	5.7323	0.6577	5.7761	0.6370	4.4807		
	5.0000	0.2044	7.5682	0.4052	8.7771	0.4257	8.8833	0.4146	7.0453		
	10.0000	0.0951	15.1014	0.1969	17.7747	0.2081	18.0544	0.2035	14.4924		
	INF.	1.4710	2.0022	1.8248	2.0440	1.7911	1.9218	1.5003	0.8926		
9	1.0000	1.7822	1.2921	2.7162	1.3922	2.7734	1.3922	2.7162	1.2921	1.7822	
	0.9000	1.8267	1.2103	2.8658	1.3135	2.9353	1.3165	2.8834	1.2302	1.9458	
	0.8000	1.8955	1.1139	3.0709	1.2189	3.1565	1.2246	3.1102	1.1523	2.1611	
	0.7000	2.0013	1.0028	3.3565	1.1075	3.4635	1.1157	3.4232	1.0568	2.4489	
	0.6000	2.1634	0.8786	3.7621	0.9801	3.8985	0.9900	3.8647	0.9436	2.8426	
	0.5000	2.4150	0.7436	4.3573	0.8385	4.5355	0.8493	4.5087	0.8140	3.4010	
	0.4000	2.8203	0.6008	5.2792	0.6850	5.5207	0.6957	5.5023	0.6700	4.2416	
	0.3000	3.5279	0.4528	6.8474	0.5223	7.1951	0.5318	7.1876	0.5142	5.6390	
	0.2000	4.9830	0.3021	10.0212	0.3526	10.5818	0.3600	10.5925	0.3491	8.4189	
	0.1000	9.4131	0.1507	19.5995	0.1779	20.8006	0.1822	20.8588	0.1770	16.7140	
	INF.	1.6533	1.7890	2.0570	1.8383	2.0481	1.7910	1.9199	1.4981	0.8911	
10	1.9841	0.8972	2.5610	1.3683	2.7631	1.4009	2.7795	1.3927	2.7148	1.2908	1.7801
	2.0000	0.8223	2.6845	1.2901	2.9166	1.3246	2.9390	1.3191	2.8783	1.2306	1.9398
	2.5000	0.4999	3.6868	0.8858	4.1383	0.9216	4.2020	0.9238	4.1540	0.8812	3.0919
	3.3333	0.3332	5.0071	0.6289	5.7274	0.6594	5.8399	0.6631	5.7948	0.6376	4.4804
	5.0000	0.2037	7.5446	0.4042	8.7695	0.4266	8.9727	0.4300	8.9249	0.4154	7.0518
	10.0000	0.0948	15.0578	0.1965	17.7624	0.2086	18.2313	0.2107	18.1644	0.2041	14.5199
	INF.	1.4753	2.0107	1.8386	2.0733	1.8432	2.0494	1.7904	1.9183	1.4965	0.8900
n	$1/R_s$	L_1	C_2	L_3	C_4	L_5	C_6	L_7	C_8	L_9	C_{10}

LOW PASS ELEMENT VALUES

MAXIMALLY FLAT DELAY

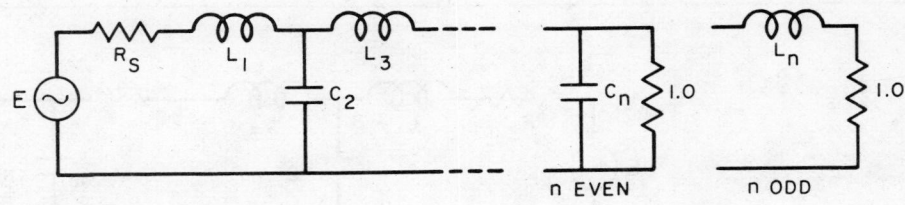

n	R_s	C_1	L_2	C_3	L_4
2	1.0000	0.5755	2.1478		
	1.1111	0.5084	2.3097		
	1.2500	0.4433	2.5096		
	1.4286	0.3801	2.7638		
	1.6667	0.3191	3.0993		
	2.0000	0.2601	3.5649		
	2.5000	0.2032	4.2577		
	3.3333	0.1486	5.4050		
	5.0000	0.0965	7.6876		
	10.0000	0.0469	14.5097		
	INF.	1.3617	0.4539		
3	1.0000	0.3374	0.9705	2.2034	
	0.9000	0.3708	0.8650	2.3745	
	0.8000	0.4124	0.7609	2.5867	
	0.7000	0.4657	0.6584	2.8575	
	0.6000	0.5365	0.5576	3.2159	
	0.5000	0.6353	0.4587	3.7144	
	0.4000	0.7829	0.3618	4.4573	
	0.3000	1.0283	0.2673	5.6888	
	0.2000	1.5176	0.1752	8.1403	
	0.1000	2.9825	0.0860	15.4697	
	INF.	1.4631	0.8427	0.2926	
4	1.0000	0.2334	0.6725	1.0815	2.2404
	1.1111	0.2085	0.7423	0.9670	2.4143
	1.2500	0.1839	0.8292	0.8534	2.6304
	1.4286	0.1596	0.9406	0.7410	2.9066
	1.6667	0.1356	1.0886	0.6299	3.2727
	2.0000	0.1120	1.2952	0.5202	3.7824
	2.5000	0.0887	1.6040	0.4120	4.5430
	3.3333	0.0658	2.1174	0.3056	5.8048
	5.0000	0.0434	3.1416	0.2013	8.3185
	10.0000	0.0214	6.2086	0.0993	15.8372
	INF.	1.5012	0.9781	0.6127	0.2114
n	$1/R_s$	L_1	C_2	L_3	C_4

LOW PASS ELEMENT VALUES

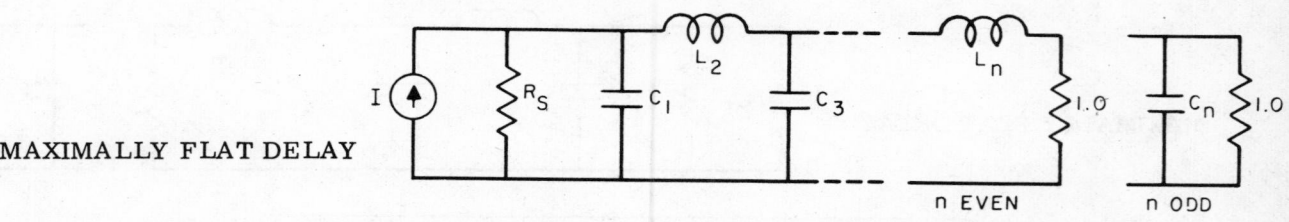

MAXIMALLY FLAT DELAY

n	R_S	C_1	L_2	C_3	L_4	C_5	L_6	C_7
5	1.0000	0.1743	0.5072	0.8040	1.1110	2.2582		
	0.9000	0.1926	0.4542	0.8894	0.9945	2.4328		
	0.8000	0.2154	0.4016	0.9959	0.8789	2.6497		
	0.7000	0.2447	0.3494	1.1323	0.7642	2.9272		
	0.6000	0.2836	0.2977	1.3138	0.6506	3.2952		
	0.5000	0.3380	0.2465	1.5672	0.5382	3.8077		
	0.4000	0.4194	0.1958	1.9464	0.4270	4.5731		
	0.3000	0.5548	0.1457	2.5768	0.3174	5.8433		
	0.2000	0.8251	0.0964	3.8352	0.2095	8.3747		
	0.1000	1.6349	0.0478	7.6043	0.1036	15.9487		
	INF.	1.5125	1.0232	0.7531	0.4729	0.1618		
6	1.0000	0.1365	0.4002	0.6392	0.8538	1.1126	2.2645	
	1.1111	0.1223	0.4429	0.5732	0.9456	0.9964	2.4388	
	1.2500	0.1082	0.4961	0.5076	1.0600	0.8810	2.6554	
	1.4286	0.0943	0.5644	0.4424	1.2069	0.7665	2.9325	
	1.6667	0.0804	0.6553	0.3775	1.4022	0.6530	3.3001	
	2.0000	0.0666	0.7824	0.3131	1.6752	0.5405	3.8122	
	2.5000	0.0530	0.9725	0.2492	2.0837	0.4292	4.5770	
	3.3333	0.0395	1.2890	0.1859	2.7633	0.3193	5.8467	
	5.0000	0.0261	1.9209	0.1232	4.1204	0.2110	8.3775	
	10.0000	0.0130	3.8146	0.0612	8.1860	0.1045	15.9506	
	INF.	1.5124	1.0329	0.8125	0.6072	0.3785	0.1287	
7	1.0000	0.1106	0.3259	0.5249	0.7020	0.8690	1.1052	2.2659
	0.9000	0.1224	0.2923	0.5815	0.6302	0.9630	0.9899	2.4396
	0.8000	0.1372	0.2589	0.6521	0.5586	1.0803	0.8754	2.6556
	0.7000	0.1562	0.2257	0.7428	0.4873	1.2308	0.7618	2.9319
	0.6000	0.1815	0.1927	0.8634	0.4163	1.4312	0.6491	3.2984
	0.5000	0.2168	0.1599	1.0321	0.3457	1.7111	0.5374	3.8090
	0.4000	0.2698	0.1274	1.2847	0.2755	2.1304	0.4269	4.5718
	0.3000	0.3579	0.0951	1.7051	0.2058	2.8280	0.3177	5.8380
	0.2000	0.5338	0.0630	2.5448	0.1365	4.2214	0.2100	8.3623
	0.1000	1.0612	0.0313	5.0616	0.0679	8.3967	0.1040	15.9166
	INF.	1.5087	1.0293	0.8345	0.6752	0.5031	0.3113	0.1054
n	$1/R_S$	L_1	C_2	L_3	C_4	L_5	C_6	L_7

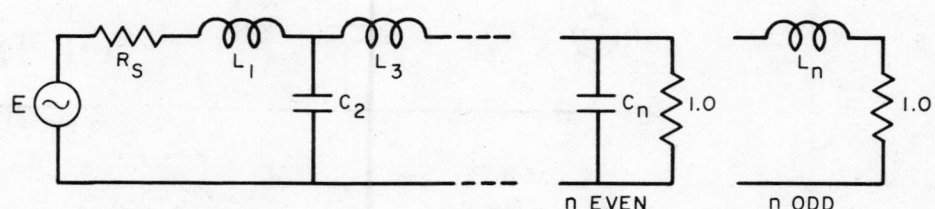

324

LOW PASS ELEMENT VALUES

MAXIMALLY FLAT DELAY

n	R_s	C_1	L_2	C_3	L_4	C_5	L_6	C_7	L_8	C_9	L_{10}
8	1.0000	0.0919	0.2719	0.4409	0.5936	0.7303	0.8695	1.0956	2.2656		
	1.1111	0.0825	0.3013	0.3958	0.6580	0.6559	0.9639	0.9813	2.4388		
	1.2500	0.0731	0.3380	0.3509	0.7385	0.5817	1.0816	0.8678	2.6541		
	1.4286	0.0637	0.3850	0.3061	0.8418	0.5078	1.2328	0.7552	2.9295		
	1.6667	0.0545	0.4477	0.2616	0.9794	0.4342	1.4340	0.6435	3.2949		
	2.0000	0.0452	0.5354	0.2173	1.1718	0.3608	1.7153	0.5329	3.8041		
	2.5000	0.0360	0.6667	0.1732	1.4599	0.2878	2.1367	0.4233	4.5645		
	3.3333	0.0269	0.8852	0.1294	1.9396	0.2151	2.8380	0.3151	5.8271		
	5.0000	0.0179	1.3218	0.0859	2.8981	0.1429	4.2389	0.2083	8.3441		
	10.0000	0.0089	2.6307	0.0427	5.7710	0.0711	8.4376	0.1032	15.8768		
	INF.	1.5044	1.0214	0.8392	0.7081	0.5743	0.4253	0.2616	0.0883		
9	1.0000	0.0780	0.2313	0.3770	0.5108	0.6306	0.7407	0.8639	1.0863	2.2649	
	0.9000	0.0864	0.2077	0.4180	0.4588	0.6994	0.6655	0.9578	0.9730	2.4376	
	0.8000	0.0970	0.1841	0.4691	0.4069	0.7854	0.5905	1.0750	0.8604	2.6524	
	0.7000	0.1105	0.1607	0.5348	0.3553	0.8957	0.5157	1.2255	0.7488	2.9271	
	0.6000	0.1286	0.1373	0.6222	0.3038	1.0427	0.4411	1.4258	0.6380	3.2915	
	0.5000	0.1538	0.1141	0.7445	0.2525	1.2483	0.3667	1.7059	0.5283	3.7993	
	0.4000	0.1916	0.0910	0.9278	0.2014	1.5563	0.2926	2.1256	0.4197	4.5578	
	0.3000	0.2545	0.0680	1.2329	0.1506	2.0692	0.2189	2.8241	0.3124	5.8171	
	0.2000	0.3803	0.0452	1.8426	0.1000	3.0941	0.1455	4.2196	0.2065	8.3276	
	0.1000	0.7573	0.0225	3.6704	0.0498	6.1666	0.0725	8.4023	0.1023	15.8408	
	INF.	1.5006	1.0127	0.8361	0.7220	0.6142	0.4963	0.3654	0.2238	0.0754	
10	1.0000	0.0672	0.1998	0.3270	0.4454	0.5528	0.6493	0.7420	0.8561	1.0781	2.2641
	1.1111	0.0604	0.2216	0.2937	0.4941	0.4967	0.7205	0.6668	0.9492	0.9656	2.4365
	1.2500	0.0536	0.2488	0.2606	0.5548	0.4408	0.8093	0.5918	1.0654	0.8539	2.6508
	1.4286	0.0467	0.2836	0.2275	0.6327	0.3850	0.9233	0.5170	1.2147	0.7430	2.9249
	1.6667	0.0400	0.3301	0.1945	0.7366	0.3294	1.0753	0.4423	1.4134	0.6331	3.2885
	2.0000	0.0332	0.3951	0.1617	0.8818	0.2739	1.2879	0.3678	1.6913	0.5242	3.7953
	2.5000	0.0265	0.4924	0.1290	1.0995	0.2186	1.6064	0.2936	2.1076	0.4164	4.5521
	3.3333	0.0198	0.6546	0.0965	1.4620	0.1635	2.1369	0.2197	2.8007	0.3099	5.8087
	5.0000	0.0132	0.9786	0.0641	2.1864	0.1087	3.1971	0.1461	4.1854	0.2049	8.3137
	10.0000	0.0066	1.9499	0.0319	4.3583	0.0542	6.3759	0.0728	8.3359	0.1015	15.8108
	INF.	1.4973	1.0045	0.8297	0.7258	0.6355	0.5401	0.4342	0.3182	0.1942	0.0653
n	$1/R_s$	L_1	C_2	L_3	C_4	L_5	C_6	L_7	C_8	L_9	C_{10}

LOW PASS ELEMENT VALUES

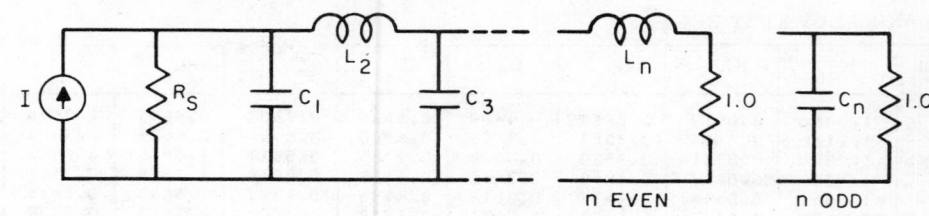

LINEAR PHASE WITH
EQUIRIPPLE ERROR

PHASE ERROR = 0.05°

n	R_s	C_1	L_2	C_3	L_4
2	1.0000	0.6480	2.1085		
	1.1111	0.5703	2.2760		
	1.2500	0.4955	2.4817		
	1.4286	0.4235	2.7422		
	1.6667	0.3544	3.0848		
	2.0000	0.2880	3.5589		
	2.5000	0.2244	4.2630		
	3.3333	0.1637	5.4270		
	5.0000	0.1059	7.7400		
	10.0000	0.0513	14.6480		
	INF.	1.3783	0.4957		
3	1.0000	0.4328	1.0427	2.2542	
	0.9000	0.4745	0.9330	2.4258	
	0.8000	0.5262	0.8238	2.6400	
	0.7000	0.5925	0.7153	2.9146	
	0.6000	0.6805	0.6078	3.2795	
	0.5000	0.8032	0.5015	3.7884	
	0.4000	0.9865	0.3967	4.5487	
	0.3000	1.2910	0.2938	5.8106	
	0.2000	1.8983	0.1931	8.3253	
	0.1000	3.7161	0.0950	15.8472	
	INF.	1.5018	0.9328	0.3631	
4	1.0000	0.3363	0.7963	1.1428	2.2459
	1.1111	0.2993	0.8810	1.0212	2.4241
	1.2500	0.2631	0.9865	0.9012	2.6445
	1.4286	0.2275	1.1216	0.7826	2.9254
	1.6667	0.1926	1.3009	0.6657	3.2970
	2.0000	0.1584	1.5509	0.5502	3.8138
	2.5000	0.1250	1.9244	0.4364	4.5844
	3.3333	0.0923	2.5448	0.3242	5.8626
	5.0000	0.0606	3.7818	0.2139	8.4091
	10.0000	0.0298	7.4845	0.1058	16.0266
	INF.	1.5211	1.0444	0.7395	0.2925
n	$1/R_s$	L_1	C_2	L_3	C_4

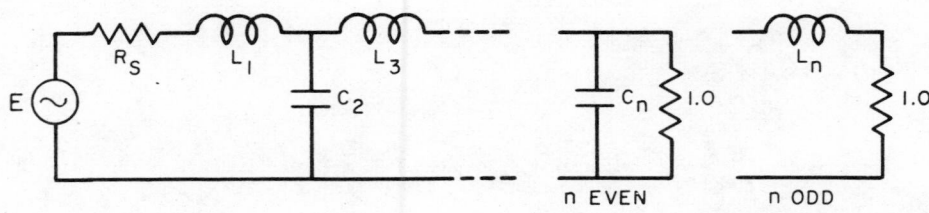

LOW PASS ELEMENT VALUES

LINEAR PHASE WITH
EQUIRIPPLE ERROR

PHASE ERROR = 0.05°

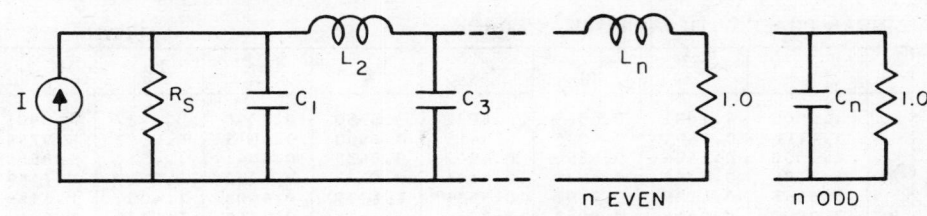

n	R_s	C_1	L_2	C_3	L_4	C_5	L_6	C_7
5	1.0000	0.2751	0.6541	0.8892	1.1034	2.2873		
	0.9000	0.3031	0.5868	0.9841	0.9904	2.4589		
	0.8000	0.3380	0.5197	1.1026	0.8774	2.6733		
	0.7000	0.3827	0.4529	1.2548	0.7648	2.9484		
	0.6000	0.4420	0.3865	1.4575	0.6526	3.3144		
	0.5000	0.5248	0.3204	1.7408	0.5410	3.8254		
	0.4000	0.6486	0.2549	2.1651	0.4302	4.5896		
	0.3000	0.8544	0.1899	2.8713	0.3205	5.8595		
	0.2000	1.2649	0.1257	4.2817	0.2120	8.3922		
	0.1000	2.4940	0.0624	8.5082	0.1051	15.9739		
	INF.	1.5144	1.0407	0.8447	0.6177	0.2456		
6	1.0000	0.2374	0.5662	0.7578	0.8760	1.1163	2.2448	
	1.1111	0.2120	0.6272	0.6799	0.9726	0.9977	2.4214	
	1.2500	0.1870	0.7032	0.6023	1.0931	0.8807	2.6396	
	1.4286	0.1622	0.8008	0.5253	1.2475	0.7652	2.9174	
	1.6667	0.1378	0.9306	0.4487	1.4530	0.6512	3.2849	
	2.0000	0.1138	1.1118	0.3725	1.7401	0.5387	3.7958	
	2.5000	0.0901	1.3830	0.2969	2.1698	0.4277	4.5579	
	3.3333	0.0669	1.8340	0.2217	2.8849	0.3182	5.8220	
	5.0000	0.0441	2.7343	0.1472	4.3129	0.2103	8.3408	
	10.0000	0.0218	5.4312	0.0732	8.5924	0.1041	15.8769	
	INF.	1.5050	1.0306	0.8554	0.7283	0.5389	0.2147	
7	1.0000	0.2085	0.4999	0.6653	0.7521	0.8749	1.0671	2.2845
	0.9000	0.2302	0.4488	0.7374	0.6768	0.9687	0.9580	2.4538
	0.8000	0.2573	0.3978	0.8274	0.6013	1.0861	0.8489	2.6655
	0.7000	0.2919	0.3470	0.9431	0.5258	1.2369	0.7400	2.9375
	0.6000	0.3380	0.2964	1.0972	0.4503	1.4381	0.6314	3.2996
	0.5000	0.4023	0.2461	1.3127	0.3749	1.7196	0.5235	3.8051
	0.4000	0.4986	0.1960	1.6356	0.2995	2.1416	0.4163	4.5613
	0.3000	0.6585	0.1463	2.1734	0.2242	2.8445	0.3101	5.8180
	0.2000	0.9778	0.0970	3.2480	0.1492	4.2496	0.2052	8.3246
	0.1000	1.9340	0.0482	6.4698	0.0744	8.4623	0.1017	15.8281
	INF.	1.4988	1.0071	0.8422	0.7421	0.6441	0.4791	0.1911
n	$1/R_s$	L_1	C_2	L_3	C_4	L_5	C_6	L_7

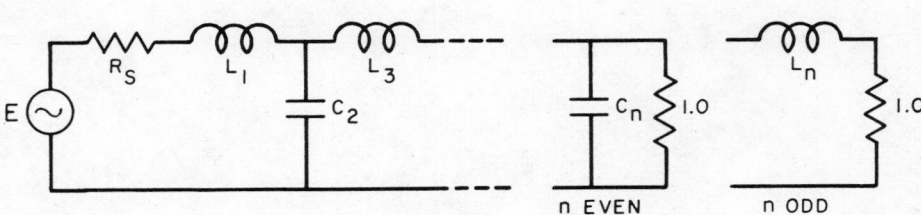

327

LOW PASS ELEMENT VALUES

LINEAR PHASE WITH EQUIRIPPLE ERROR

PHASE ERROR = 0.05°

n	R_s	C_1	L_2	C_3	L_4	C_5	L_6	C_7	L_8	C_9	L_{10}
8	1.0000	0.1891	0.4543	0.6031	0.6750	0.7590	0.8427	1.0901	2.2415		
	1.1111	0.1691	0.5035	0.5415	0.7500	0.6813	0.9362	0.9735	2.4176		
	1.2500	0.1494	0.5650	0.4802	0.8435	0.6041	1.0527	0.8588	2.6349		
	1.4286	0.1298	0.6438	0.4191	0.9637	0.5272	1.2019	0.7459	2.9113		
	1.6667	0.1105	0.7487	0.3583	1.1237	0.4508	1.4004	0.6345	3.2767		
	2.0000	0.0914	0.8953	0.2978	1.3475	0.3748	1.6776	0.5247	3.7846		
	2.5000	0.0725	1.1148	0.2376	1.6827	0.2991	2.0927	0.4164	4.5418		
	3.3333	0.0539	1.4801	0.1776	2.2411	0.2237	2.7833	0.3096	5.7978		
	5.0000	0.0356	2.2095	0.1180	3.3568	0.1488	4.1627	0.2046	8.3004		
	10.0000	0.0176	4.3954	0.0588	6.7021	0.0742	8.2969	0.1013	15.7878		
	INF.	1.4953	1.0018	0.8264	0.7396	0.6688	0.5858	0.4369	0.1743		
9	1.0000	0.1718	0.4146	0.5498	0.6132	0.6774	0.7252	0.8450	1.0447	2.2834	
	0.9000	0.1900	0.3724	0.6097	0.5519	0.7513	0.6529	0.9352	0.9382	2.4512	
	0.8000	0.2125	0.3302	0.6846	0.4905	0.8436	0.5805	1.0481	0.8314	2.6613	
	0.7000	0.2415	0.2882	0.7807	0.4291	0.9624	0.5079	1.1933	0.7247	2.9315	
	0.6000	0.2800	0.2463	0.9088	0.3676	1.1207	0.4352	1.3870	0.6184	3.2914	
	0.5000	0.3337	0.2046	1.0880	0.3062	1.3424	0.3624	1.6581	0.5125	3.7941	
	0.4000	0.4141	0.1631	1.3565	0.2448	1.6749	0.2897	2.0647	0.4075	4.5462	
	0.3000	0.5478	0.1219	1.8038	0.1834	2.2289	0.2170	2.7420	0.3035	5.7960	
	0.2000	0.8148	0.0809	2.6977	0.1222	3.3369	0.1445	4.0960	0.2007	8.2890	
	0.1000	1.6146	0.0403	5.3782	0.0610	6.6602	0.0721	8.1556	0.0995	15.7520	
	INF.	1.4907	0.9845	0.8116	0.7197	0.6646	0.6089	0.5359	0.4003	0.1598	
10	1.0000	0.1601	0.3867	0.5125	0.5702	0.6243	0.6557	0.7319	0.8178	1.0767	2.2387
	1.1111	0.1433	0.4288	0.4604	0.6336	0.5609	0.7290	0.6567	0.9089	0.9608	2.4151
	1.2500	0.1267	0.4812	0.4084	0.7127	0.4977	0.8205	0.5820	1.0221	0.8471	2.6323
	1.4286	0.1102	0.5486	0.3567	0.8143	0.4348	0.9380	0.5079	1.1672	0.7354	2.9082
	1.6667	0.0939	0.6383	0.3051	0.9498	0.3721	1.0944	0.4342	1.3600	0.6254	3.2727
	2.0000	0.0778	0.7637	0.2537	1.1392	0.3096	1.3131	0.3609	1.6291	0.5170	3.7791
	2.5000	0.0618	0.9515	0.2024	1.4232	0.2473	1.6408	0.2880	2.0320	0.4102	4.5340
	3.3333	0.0460	1.2641	0.1515	1.8961	0.1852	2.1866	0.2154	2.7022	0.3049	5.7860
	5.0000	0.0304	1.8885	0.1007	2.8416	0.1232	3.2775	0.1433	4.0406	0.2014	8.2806
	10.0000	0.0151	3.7600	0.0502	5.6766	0.0615	6.5485	0.0714	8.0520	0.0997	15.7441
	INF.	1.4905	0.9858	0.8018	0.7123	0.6540	0.6141	0.5669	0.5003	0.3741	0.1494
n	$1/R_s$	L_1	C_2	L_3	C_4	L_5	C_6	L_7	C_8	L_9	C_{10}

LOW PASS ELEMENT VALUES

LINEAR PHASE WITH
EQUIRIPPLE ERROR

PHASE ERROR = 0.5°

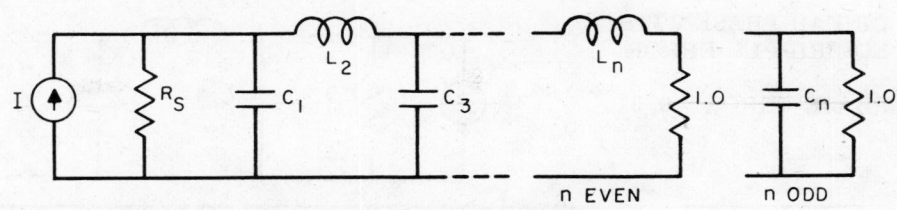

n	R_s	C_1	L_2	C_3	L_4
2	1.0000	0.8245	1.9800		
	1.1111	0.7166	2.1640		
	1.2500	0.6160	2.3850		
	1.4286	0.5216	2.6603		
	1.6667	0.4327	3.0181		
	2.0000	0.3489	3.5088		
	2.5000	0.2700	4.2329		
	3.3333	0.1956	5.4242		
	5.0000	0.1258	7.7842		
	10.0000	0.0606	14.8185		
	INF.	1.4022	0.5821		
3	1.0000	0.5534	1.0218	2.4250	
	0.9000	0.6059	0.9213	2.5929	
	0.8000	0.6710	0.8197	2.8046	
	0.7000	0.7540	0.7173	3.0787	
	0.6000	0.8639	0.6141	3.4462	
	0.5000	1.0168	0.5105	3.9625	
	0.4000	1.2448	0.4068	4.7385	
	0.3000	1.6231	0.3034	6.0326	
	0.2000	2.3771	0.2008	8.6197	
	0.1000	4.6332	0.0994	16.3738	
	INF.	1.5495	0.9820	0.4506	
4	1.0000	0.4526	0.7967	1.2669	2.0504
	1.1111	0.3996	0.8889	1.1137	2.2502
	1.2500	0.3486	1.0028	0.9699	2.4866
	1.4286	0.2995	1.1481	0.8333	2.7788
	1.6667	0.2521	1.3405	0.7024	3.1576
	2.0000	0.2062	1.6083	0.5762	3.6769
	2.5000	0.1618	2.0079	0.4542	4.4438
	3.3333	0.1190	2.6708	0.3358	5.7080
	5.0000	0.0777	3.9916	0.2207	8.2168
	10.0000	0.0380	7.9426	0.1088	15.7068
	INF.	1.4944	1.0715	0.7889	0.3708
n	$1/R_s$	L_1	C_2	L_3	C_4

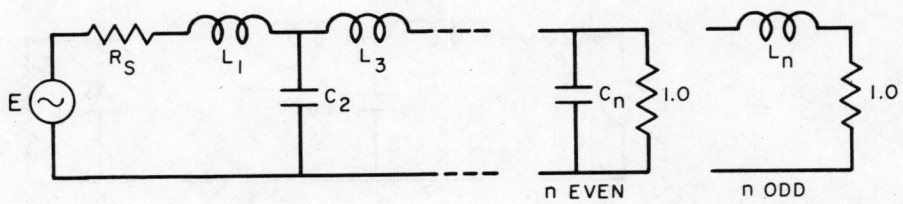

329

LOW PASS ELEMENT VALUES

LINEAR PHASE WITH
EQUIRIPPLE ERROR

PHASE ERROR = 0.5°

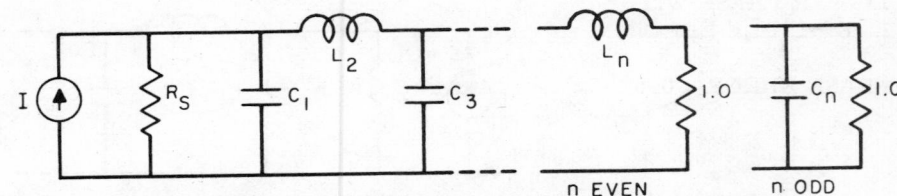

n	R_s	C_1	L_2	C_3	L_4	C_5	L_6	C_7
5	1.0000	0.3658	0.6768	0.9513	1.0113	2.4446		
	0.9000	0.4027	0.6099	1.0486	0.9157	2.6062		
	0.8000	0.4485	0.5427	1.1700	0.8182	2.8114		
	0.7000	0.5069	0.4752	1.3260	0.7189	3.0787		
	0.6000	0.5843	0.4074	1.5341	0.6181	3.4387		
	0.5000	0.6921	0.3395	1.8253	0.5160	3.9462		
	0.4000	0.8530	0.2714	2.2623	0.4130	4.7108		
	0.3000	1.1201	0.2033	2.9908	0.3094	5.9881		
	0.2000	1.6524	0.1352	4.4478	0.2057	8.5444		
	0.1000	3.2454	0.0674	8.8185	0.1024	16.2117		
	INF.	1.5327	1.0180	0.8740	0.6709	0.3182		
6	1.0000	0.3313	0.5984	0.8390	0.7964	1.2734	2.0111	
	1.1111	0.2934	0.6667	0.7446	0.8985	1.1050	2.2282	
	1.2500	0.2571	0.7515	0.6542	1.0223	0.9549	2.4742	
	1.4286	0.2219	0.8600	0.5666	1.1787	0.8164	2.7718	
	1.6667	0.1876	1.0040	0.4812	1.3848	0.6859	3.1529	
	2.0000	0.1541	1.2051	0.3976	1.6709	0.5615	3.6720	
	2.5000	0.1216	1.5058	0.3155	2.0972	0.4420	4.4362	
	3.3333	0.0898	2.0058	0.2347	2.8044	0.3266	5.6935	
	5.0000	0.0589	3.0038	0.1553	4.2137	0.2146	8.1871	
	10.0000	0.0290	5.9928	0.0771	8.4320	0.1058	15.6296	
	INF.	1.4849	1.0430	0.8427	0.7651	0.5972	0.2844	
7	1.0000	0.2826	0.5332	0.7142	0.6988	0.9219	0.9600	2.4404
	0.9000	0.3118	0.4802	0.7896	0.6322	1.0137	0.8718	2.5953
	0.8000	0.3481	0.4271	0.8836	0.5649	1.1287	0.7809	2.7936
	0.7000	0.3945	0.3739	1.0043	0.4967	1.2768	0.6875	3.0535
	0.6000	0.4560	0.3206	1.1650	0.4277	1.4750	0.5919	3.4051
	0.5000	0.5416	0.2671	1.3899	0.3580	1.7531	0.4947	3.9025
	0.4000	0.6695	0.2136	1.7271	0.2874	2.1714	0.3961	4.6534
	0.3000	0.8819	0.1601	2.2890	0.2163	2.8700	0.2969	5.9091
	0.2000	1.3054	0.1066	3.4127	0.1445	4.2690	0.1974	8.4236
	0.1000	2.5731	0.0532	6.7835	0.0724	8.4691	0.0983	15.9666
	INF.	1.5079	0.9763	0.8402	0.7248	0.6741	0.5305	0.2532
n	$1/R_s$	L_1	C_2	L_3	C_4	L_5	C_6	L_7

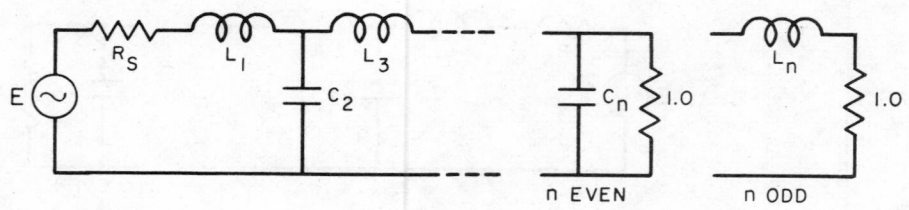

LOW PASS ELEMENT VALUES

LINEAR PHASE WITH EQUIRIPPLE ERROR   PHASE ERROR = 0.5°

n	R_s	C_1	L_2	C_3	L_4	C_5	L_6	C_7	L_8	C_9	L_{10}
	1.0000	0.2718	0.4999	0.6800	0.6312	0.8498	0.7447	1.3174	1.9626		
	1.1111	0.2408	0.5567	0.6045	0.7116	0.7452	0.8529	1.1169	2.2146		
	1.2500	0.2114	0.6271	0.5324	0.8086	0.6506	0.9780	0.9551	2.4766		
	1.4286	0.1828	0.7173	0.4622	0.9315	0.5612	1.1331	0.8117	2.7837		
	1.6667	0.1549	0.8373	0.3934	1.0939	0.4753	1.3355	0.6795	3.1715		
8	2.0000	0.1276	1.0049	0.3256	1.3201	0.3920	1.6148	0.5550	3.6960		
	2.5000	0.1009	1.2559	0.2589	1.6580	0.3107	2.0297	0.4362	4.4654		
	3.3333	0.0747	1.6734	0.1930	2.2194	0.2311	2.7164	0.3220	5.7294		
	5.0000	0.0492	2.5074	0.1279	3.3400	0.1530	4.0835	0.2114	8.2345		
	10.0000	0.0242	5.0066	0.0636	6.6971	0.0760	8.1733	0.1042	15.7101		
	INF.	1.4915	1.0265	0.8169	0.7548	0.6709	0.6318	0.4995	0.2387		
	1.0000	0.2347	0.4493	0.5914	0.5747	0.7027	0.6552	0.8944	0.9255	2.4332	
	0.9000	0.2594	0.4045	0.6547	0.5193	0.7754	0.5943	0.9809	0.8427	2.5822	
	0.8000	0.2900	0.3597	0.7336	0.4635	0.8662	0.5322	1.0895	0.7566	2.7745	
	0.7000	0.3291	0.3148	0.8348	0.4073	0.9829	0.4690	1.2299	0.6673	3.0283	
	0.6000	0.3810	0.2699	0.9695	0.3505	1.1388	0.4046	1.4183	0.5753	3.3734	
9	0.5000	0.4533	0.2249	1.1580	0.2932	1.3572	0.3392	1.6834	0.4812	3.8629	
	0.4000	0.5613	0.1799	1.4405	0.2355	1.6854	0.2727	2.0828	0.3855	4.6032	
	0.3000	0.7407	0.1348	1.9111	0.1772	2.2331	0.2054	2.7508	0.2889	5.8424	
	0.2000	1.0986	0.0898	2.8522	0.1185	3.3299	0.1373	4.0895	0.1921	8.3246	
	0.1000	2.1702	0.0448	5.6749	0.0594	6.6230	0.0688	8.1099	0.0956	15.7718	
	INF.	1.4888	0.9495	0.8044	0.6892	0.6589	0.5952	0.5645	0.4475	0.2141	
	1.0000	0.2359	0.4369	0.5887	0.5428	0.7034	0.5827	0.8720	0.6869	1.4317	1.8431
	1.1111	0.2081	0.4866	0.5218	0.6141	0.6141	0.6729	0.7394	0.8187	1.1397	2.1907
	1.2500	0.1827	0.5480	0.4601	0.6972	0.5376	0.7708	0.6394	0.9483	0.9616	2.4734
	1.4286	0.1582	0.6267	0.3999	0.8024	0.4651	0.8922	0.5487	1.1042	0.8122	2.7907
	1.6667	0.1343	0.7314	0.3407	0.9416	0.3948	1.0514	0.4631	1.3052	0.6777	3.1847
10	2.0000	0.1108	0.8777	0.2823	1.1356	0.3263	1.2719	0.3811	1.5809	0.5525	3.7138
	2.5000	0.0877	1.0969	0.2247	1.4258	0.2591	1.6003	0.3017	1.9888	0.4338	4.4876
	3.3333	0.0651	1.4619	0.1676	1.9085	0.1931	2.1451	0.2242	2.6628	0.3200	5.7573
	5.0000	0.0429	2.1910	0.1112	2.8724	0.1280	3.2311	0.1483	4.0033	0.2100	8.2726
	10.0000	0.0212	4.3764	0.0553	5.7612	0.0636	6.4830	0.0736	8.0118	0.1035	15.7776
	INF.	1.4973	1.0192	0.8005	0.7312	0.6498	0.6331	0.5775	0.5501	0.4369	0.2091

n	$1/R_s$	L_1	C_2	L_3	C_4	L_5	C_6	L_7	C_8	L_9	C_{10}

LOW PASS ELEMENT VALUES

GAUSSIAN RESPONSE

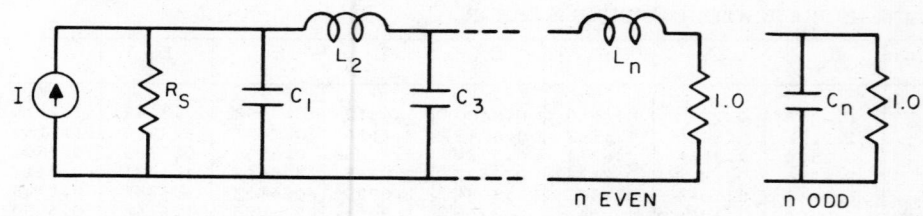

n	R_s	C_1	L_2	C_3	L_4
2	1.0000	0.4738	2.1850		
	1.1111	0.4204	2.3394		
	1.2500	0.3681	2.5310		
	1.4286	0.3170	2.7757		
	1.6667	0.2672	3.0998		
	2.0000	0.2187	3.5509		
	2.5000	0.1716	4.2240		
	3.3333	0.1260	5.3408		
	5.0000	0.0821	7.5660		
	10.0000	0.0400	14.2232		
	INF.	1.3294	0.3894		
3	1.0000	0.2624	0.8167	2.2262	
	0.9000	0.2892	0.7284	2.3904	
	0.8000	0.3226	0.6413	2.5943	
	0.7000	0.3654	0.5554	2.8548	
	0.6000	0.4222	0.4708	3.2002	
	0.5000	0.5014	0.3877	3.6812	
	0.4000	0.6199	0.3061	4.3990	
	0.3000	0.8167	0.2264	5.5901	
	0.2000	1.2093	0.1486	7.9635	
	0.1000	2.3844	0.0730	15.0642	
	INF.	1.4179	0.7167	0.2347	
4	1.0000	0.1772	0.5302	0.9321	2.2450
	1.1111	0.1586	0.5859	0.8330	2.4126
	1.2500	0.1401	0.6554	0.7349	2.6209
	1.4286	0.1219	0.7444	0.6379	2.8871
	1.6667	0.1038	0.8627	0.5420	3.2403
	2.0000	0.0859	1.0279	0.4475	3.7322
	2.5000	0.0682	1.2750	0.3543	4.4666
	3.3333	0.0507	1.6857	0.2628	5.6856
	5.0000	0.0335	2.5055	0.1731	8.1152
	10.0000	0.0166	4.9605	0.0854	15.3849
	INF.	1.4518	0.8406	0.4905	0.1642
n	$1/R_s$	L_1	C_2	L_3	C_4

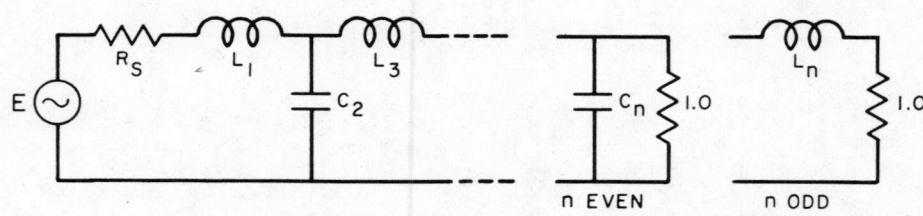

LOW PASS ELEMENT VALUES

GAUSSIAN RESPONSE

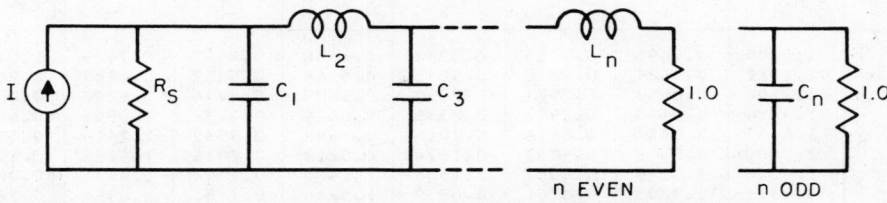

n	R_s	C_1	L_2	C_3	L_4	C_5	L_6	C_7
	1.0000	0.1312	0.3896	0.6485	0.9782	2.2533		
	0.9000	0.1451	0.3493	0.7177	0.8750	2.4222		
	0.8000	0.1626	0.3092	0.8039	0.7727	2.6322		
	0.7000	0.1850	0.2693	0.9144	0.6713	2.9007		
	0.6000	0.2148	0.2297	1.0615	0.5711	3.2569		
5	0.5000	0.2564	0.1905	1.2668	0.4720	3.7532		
	0.4000	0.3187	0.1515	1.5742	0.3742	4.4943		
	0.3000	0.4223	0.1130	2.0854	0.2779	5.7245		
	0.2000	0.6293	0.0748	3.1060	0.1833	8.1767		
	0.1000	1.2495	0.0371	6.1633	0.0906	15.5150		
	INF.	1.4655	0.8934	0.6109	0.3684	0.1239		
	1.0000	0.1026	0.3045	0.5004	0.7050	0.9982	2.2568	
	1.1111	0.0920	0.3373	0.4490	0.7807	0.8932	2.4263	
	1.2500	0.0815	0.3782	0.3979	0.8752	0.7891	2.6370	
	1.4286	0.0711	0.4307	0.3469	0.9963	0.6859	2.9064	
	1.6667	0.0607	0.5006	0.2963	1.1576	0.5838	3.2638	
6	2.0000	0.0504	0.5983	0.2459	1.3829	0.4828	3.7619	
	2.5000	0.0401	0.7447	0.1959	1.7202	0.3830	4.5057	
	3.3333	0.0300	0.9883	0.1462	2.2813	0.2846	5.7405	
	5.0000	0.0199	1.4748	0.0970	3.4019	0.1879	8.2022	
	10.0000	0.0099	2.9331	0.0482	6.7594	0.0929	15.5689	
	INF.	1.4713	0.9174	0.6710	0.4792	0.2915	0.0981	
	1.0000	0.0833	0.2473	0.4055	0.5606	0.7333	1.0073	2.2583
	0.9000	0.0923	0.2221	0.4495	0.5033	0.8123	0.9016	2.4280
	0.8000	0.1035	0.1969	0.5044	0.4463	0.9110	0.7966	2.6390
	0.7000	0.1180	0.1718	0.5749	0.3894	1.0376	0.6926	2.9088
	0.6000	0.1372	0.1468	0.6687	0.3328	1.2062	0.5896	3.2667
7	0.5000	0.1641	0.1219	0.7999	0.2764	1.4417	0.4877	3.7655
	0.4000	0.2044	0.0972	0.9965	0.2204	1.7944	0.3871	4.5105
	0.3000	0.2715	0.0726	1.3237	0.1646	2.3813	0.2878	5.7473
	0.2000	0.4056	0.0482	1.9775	0.1093	3.5534	0.1900	8.2129
	0.1000	0.8073	0.0240	3.9372	0.0544	7.0659	0.0940	15.5917
	INF.	1.4737	0.9286	0.7020	0.5412	0.3918	0.2387	0.0803
n	$1/R_s$	L_1	C_2	L_3	C_4	L_5	C_6	L_7

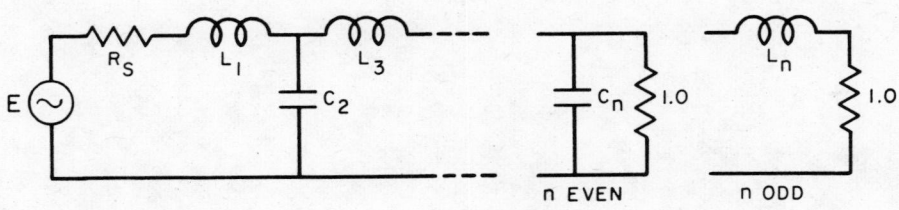

LOW PASS ELEMENT VALUES

GAUSSIAN RESPONSE

n	R_s	C_1	L_2	C_3	L_4	C_5	L_6	C_7	L_8	C_9	L_{10}
8	1.0000	0.0695	0.2065	0.3388	0.4658	0.5942	0.7479	1.0116	2.2590		
	1.1111	0.0624	0.2290	0.3043	0.5166	0.5337	0.8287	0.9055	2.4287		
	1.2500	0.0553	0.2571	0.2699	0.5800	0.4734	0.9296	0.8002	2.6398		
	1.4286	0.0483	0.2931	0.2356	0.6614	0.4132	1.0591	0.6958	2.9098		
	1.6667	0.0413	0.3411	0.2015	0.7698	0.3533	1.2314	0.5924	3.2680		
	2.0000	0.0343	0.4082	0.1674	0.9213	0.2936	1.4723	0.4901	3.7671		
	2.5000	0.0274	0.5087	0.1336	1.1485	0.2342	1.8331	0.3890	4.5126		
	3.3333	0.0205	0.6761	0.0999	1.5266	0.1751	2.4336	0.2892	5.7502		
	5.0000	0.0136	1.0106	0.0664	2.2823	0.1163	3.6330	0.1910	8.2176		
	10.0000	0.0068	2.0135	0.0331	4.5478	0.0579	7.2275	0.0945	15.6017		
	INF.	1.4748	0.9341	0.7185	0.5766	0.4529	0.3292	0.2005	0.0674		
9	1.0000	0.0591	0.1761	0.2892	0.3973	0.5025	0.6134	0.7556	1.0137	2.2593	
	0.9000	0.0656	0.1582	0.3207	0.3570	0.5574	0.5510	0.8373	0.9074	2.4291	
	0.8000	0.0737	0.1403	0.3602	0.3168	0.6260	0.4889	0.9394	0.8019	2.6402	
	0.7000	0.0840	0.1225	0.4108	0.2767	0.7141	0.4269	1.0704	0.6974	2.9103	
	0.6000	0.0978	0.1048	0.4783	0.2367	0.8314	0.3651	1.2448	0.5938	3.2686	
	0.5000	0.1171	0.0871	0.5727	0.1968	0.9956	0.3035	1.4886	0.4913	3.7678	
	0.4000	0.1460	0.0695	0.7141	0.1571	1.2416	0.2421	1.8538	0.3899	4.5135	
	0.3000	0.1941	0.0520	0.9496	0.1175	1.6512	0.1811	2.4615	0.2900	5.7515	
	0.2000	0.2902	0.0346	1.4203	0.0781	2.4698	0.1203	3.6755	0.1915	8.2197	
	0.1000	0.5785	0.0172	2.8317	0.0389	4.9241	0.0600	7.3141	0.0948	15.6062	
	INF.	1.4753	0.9367	0.7273	0.5972	0.4907	0.3880	0.2822	0.1716	0.0576	
10	1.0000	0.0512	0.1525	0.2509	0.3451	0.4353	0.5250	0.6244	0.7597	1.0147	2.2594
	1.1111	0.0460	0.1692	0.2255	0.3829	0.3912	0.5825	0.5610	0.8419	0.9083	2.4292
	1.2500	0.0408	0.1901	0.2001	0.4301	0.3472	0.6543	0.4978	0.9446	0.8028	2.6404
	1.4286	0.0357	0.2168	0.1748	0.4907	0.3033	0.7466	0.4347	1.0765	0.6981	2.9105
	1.6667	0.0305	0.2525	0.1496	0.5715	0.2596	0.8695	0.3719	1.2520	0.5944	3.2688
	2.0000	0.0254	0.3024	0.1244	0.6845	0.2159	1.0414	0.3092	1.4973	0.4918	3.7681
	2.5000	0.0202	0.3771	0.0993	0.8539	0.1724	1.2991	0.2468	1.8648	0.3904	4.5139
	3.3333	0.0151	0.5017	0.0743	1.1360	0.1290	1.7283	0.1846	2.4765	0.2903	5.7521
	5.0000	0.0101	0.7505	0.0494	1.6999	0.0858	2.5861	0.1227	3.6985	0.1918	8.2206
	10.0000	0.0050	1.4968	0.0246	3.3907	0.0428	5.1581	0.0611	7.3609	0.0949	15.6082
	INF.	1.4755	0.9381	0.7321	0.6092	0.5142	0.4267	0.3381	0.2456	0.1492	0.0501
n	$1/R_s$	L_1	C_2	L_3	C_4	L_5	C_6	L_7	C_8	L_9	C_{10}

LOW PASS ELEMENT VALUES

GAUSSIAN RESPONSE TO 6 db

n	R_s	C_1	L_2	C_3	L_4	C_5	L_6	C_7
3	1.0000	0.4042	0.8955	2.3380				
	0.9000	0.4440	0.8038	2.5027				
	0.8000	0.4935	0.7121	2.7088				
	0.7000	0.5568	0.6205	2.9739				
	0.6000	0.6407	0.5292	3.3275				
	0.5000	0.7575	0.4384	3.8223				
	0.4000	0.9319	0.3482	4.5635				
	0.3000	1.2213	0.2590	5.7972				
	0.2000	1.7980	0.1709	8.2605				
	0.1000	3.5236	0.0845	15.6391				
	INF.	1.4742	0.8328	0.3446				
4	1.0000	0.4198	0.7832	1.1598	2.1427			
	1.1111	0.3720	0.8717	1.0279	2.3286			
	1.2500	0.3256	0.9816	0.9010	2.5539			
	1.4286	0.2804	1.1220	0.7781	2.8367			
	1.6667	0.2365	1.3083	0.6587	3.2069			
	2.0000	0.1938	1.5678	0.5424	3.7179			
	2.5000	0.1524	1.9552	0.4289	4.4761			
	3.3333	0.1122	2.5982	0.3180	5.7296			
	5.0000	0.0733	3.8797	0.2095	8.2221			
	10.0000	0.0359	7.7138	0.1035	15.6717			
	INF.	1.4871	1.0222	0.7656	0.3510			
5	1.0000	0.4544	0.8457	1.0924	1.0774	2.4138		
	0.9000	0.4991	0.7622	1.2046	0.9769	2.5746		
	0.8000	0.5543	0.6781	1.3452	0.8739	2.7797		
	0.7000	0.6247	0.5936	1.5263	0.7687	3.0475		
	0.6000	0.7179	0.5086	1.7683	0.6615	3.4087		
	0.5000	0.8476	0.4233	2.1077	0.5527	3.9183		
	0.4000	1.0411	0.3379	2.6176	0.4428	4.6863		
	0.3000	1.3621	0.2526	3.4680	0.3321	5.9690		
	0.2000	2.0019	0.1677	5.1693	0.2212	8.5360		
	0.1000	3.9166	0.0833	10.2723	0.1103	16.2354		
	INF.	1.5392	1.0993	1.0203	0.8269	0.3824		
6	1.0000	0.5041	0.9032	1.2159	1.0433	1.4212	2.0917	
	1.1111	0.4427	1.0079	1.0739	1.1892	1.2274	2.3324	
	1.2500	0.3853	1.1364	0.9415	1.3611	1.0620	2.5935	
	1.4286	0.3306	1.2999	0.8145	1.5753	0.9111	2.9053	
	1.6667	0.2779	1.5162	0.6914	1.8557	0.7692	3.3032	
	2.0000	0.2271	1.8169	0.5713	2.2433	0.6333	3.8456	
	2.5000	0.1780	2.2654	0.4534	2.8200	0.5016	4.6459	
	3.3333	0.1308	3.0091	0.3376	3.7758	0.3730	5.9662	
	5.0000	0.0853	4.4902	0.2235	5.6803	0.2468	8.5904	
	10.0000	0.0416	8.9199	0.1109	11.3810	0.1225	16.4352	
	INF.	1.5664	1.2166	1.1389	1.1010	0.8844	0.4062	
7	1.0000	0.4918	0.9232	1.2146	1.1224	1.3154	1.1407	2.5039
	0.9000	0.5403	0.8318	1.3393	1.0196	1.4426	1.0434	2.6575
	0.8000	0.6001	0.7399	1.4950	0.9141	1.6040	0.9401	2.8593
	0.7000	0.6760	0.6474	1.6952	0.8061	1.8144	0.8317	3.1285
	0.6000	0.7763	0.5545	1.9626	0.6956	2.0986	0.7190	3.4967
	0.5000	0.9157	0.4613	2.3373	0.5829	2.5004	0.6029	4.0203
	0.4000	1.1236	0.3681	2.9002	0.4684	3.1072	0.4844	4.8129
	0.3000	1.4685	0.2750	3.8389	0.3524	4.1232	0.3644	6.1397
	0.2000	2.1560	0.1823	5.7166	0.2354	6.1604	0.2433	8.7977
	0.1000	4.2137	0.0905	11.3483	0.1178	12.2787	0.1217	16.7743
	INF.	1.5950	1.2166	1.2240	1.1784	1.1260	0.8975	0.4110
n	$1/R_s$	L_1	C_2	L_3	C_4	L_5	C_6	L_7

GAUSSIAN RESPONSE TO 6 db

n	R_s	C_1	L_2	C_3	L_4	C_5	L_6	C_7	L_8	C_9	L_{10}
	1.0502	0.5031	0.9699	1.2319	1.1324	1.4262	1.0449	1.6000	1.9285		
	1.1111	0.4586	1.0338	1.1286	1.2497	1.2635	1.2099	1.3372	2.2286		
	1.2500	0.3964	1.1670	0.9831	1.4404	1.0842	1.4259	1.1197	2.5453		
	1.4286	0.3392	1.3351	0.8487	1.6698	0.9299	1.6706	0.9502	2.8771		
8	1.6667	0.2848	1.5571	0.7195	1.9674	0.7863	1.9808	0.7989	3.2846		
	2.0000	0.2325	1.8656	0.5939	2.3776	0.6487	2.4039	0.6569	3.8326		
	2.5000	0.1822	2.3255	0.4710	2.9870	0.5151	3.0295	0.5204	4.6374		
	3.3333	0.1337	3.0879	0.3504	3.9962	0.3840	4.0640	0.3874	5.9636		
	5.0000	0.0872	4.6062	0.2318	6.0063	0.2547	6.1243	0.2566	8.5995		
	10.0000	0.0425	9.1467	0.1150	12.0217	0.1268	12.2919	0.1276	16.4808		
	INF.	1.5739	1.2698	1.2325	1.2633	1.2017	1.1404	0.9066	0.4148		
	1.0000	0.4979	0.9367	1.2371	1.1589	1.3845	1.1670	1.3983	1.1422	2.5277	
	0.9000	0.5475	0.8439	1.3648	1.0517	1.5194	1.0673	1.5233	1.0527	2.6698	
	0.8000	0.6083	0.7505	1.5238	0.9424	1.6894	0.9625	1.6850	0.9540	2.8635	
	0.7000	0.6854	0.6567	1.7278	0.8306	1.9103	0.8527	1.8996	0.8472	3.1279	
9	0.6000	0.7870	0.5624	1.9998	0.7165	2.2081	0.7383	2.1929	0.7342	3.4938	
	0.5000	0.9280	0.4679	2.3811	0.6002	2.6288	0.6202	2.6105	0.6166	4.0174	
	0.4000	1.1383	0.3732	2.9537	0.4820	3.2641	0.4993	3.2438	0.4960	4.8118	
	0.3000	1.4873	0.2788	3.9087	0.3625	4.3274	0.3763	4.3062	0.3734	6.1429	
	0.2000	2.1829	0.1848	5.8191	0.2421	6.4590	0.2518	6.4381	0.2496	8.8109	
	0.1000	4.2652	0.0918	11.5490	0.1211	12.8601	0.1262	12.8438	0.1250	16.8186	
	INF.	1.6014	1.2508	1.2817	1.2644	1.2805	1.2103	1.1456	0.9096	0.4160	
	1.1372	0.4682	1.0839	1.1516	1.2991	1.3293	1.2748	1.4216	1.1730	1.5040	2.1225
	1.1372	0.4682	1.0839	1.1516	1.2991	1.3293	1.2748	1.4216	1.1730	1.5040	2.1225
	1.2500	0.4087	1.1987	1.0148	1.4855	1.1389	1.5155	1.1705	1.4593	1.1798	2.5537
	1.4286	0.3489	1.3718	0.8744	1.7253	0.9733	1.7813	0.9908	1.7344	0.9878	2.9155
10	1.6667	0.2928	1.6000	0.7409	2.0334	0.8219	2.1124	0.8338	2.0664	0.8275	3.3380
	2.0000	0.2389	1.9169	0.6114	2.4574	0.6776	2.5622	0.6868	2.5129	0.6799	3.8995
	2.5000	0.1872	2.3893	0.4848	3.0868	0.5377	3.2264	0.5451	3.1699	0.5387	4.7218
	3.3333	0.1373	3.1723	0.3606	4.1290	0.4007	4.3241	0.4065	4.2549	0.4011	6.0762
	5.0000	0.0895	4.7317	0.2385	6.2048	0.2657	6.5094	0.2698	6.4154	0.2659	8.7681
	10.0000	0.0437	9.3953	0.1183	12.4165	0.1322	13.0503	0.1345	12.8837	0.1323	16.8178
	INF.	1.6077	1.3178	1.2927	1.3406	1.3070	1.3160	1.2409	1.1733	0.9311	0.4257
n	$1/R_s$	L_1	C_2	L_3	C_4	L_5	C_6	L_7	C_8	L_9	C_{10}

LOW PASS ELEMENT VALUES

GAUSSIAN RESPONSE TO 12 db

n	R_s	C_1	L_2	C_3	L_4	C_5	L_6	C_7
3	1.0000	0.4152	0.9050	2.3452				
	0.9000	0.4560	0.8126	2.5101				
	0.8000	0.5067	0.7202	2.7166				
	0.7000	0.5715	0.6278	2.9825				
	0.6000	0.6573	0.5356	3.3372				
	0.5000	0.7769	0.4438	3.8336				
	0.4000	0.9554	0.3526	4.5775				
	0.3000	1.2517	0.2623	5.8157				
	0.2000	1.8420	0.1732	8.2884				
	0.1000	3.6083	0.0856	15.6955				
	INF.	1.4800	0.8440	0.3527				
4	1.0000	0.3097	0.6545	1.0598	2.1518			
	1.1111	0.2757	0.7262	0.9418	2.3289			
	1.2500	0.2423	0.8156	0.8268	2.5459			
	1.4286	0.2096	0.9300	0.7146	2.8203			
	1.6667	0.1775	1.0821	0.6050	3.1814			
	2.0000	0.1461	1.2944	0.4980	3.6812			
	2.5000	0.1153	1.6118	0.3934	4.4241			
	3.3333	0.0853	2.1393	0.2913	5.6532			
	5.0000	0.0560	3.1917	0.1916	8.0979			
	10.0000	0.0276	6.3425	0.0944	15.4048			
	INF.	1.4585	0.9300	0.6294	0.2707			
5	1.0000	0.2909	0.5837	0.8112	0.9660	2.3745		
	0.9000	0.3207	0.5253	0.8961	0.8707	2.5377		
	0.8000	0.3577	0.4667	1.0019	0.7746	2.7433		
	0.7000	0.4051	0.4081	1.1379	0.6777	3.0092		
	0.6000	0.4680	0.3495	1.3192	0.5804	3.3650		
	0.5000	0.5556	0.2908	1.5727	0.4827	3.8642		
	0.4000	0.6865	0.2322	1.9528	0.3850	4.6138		
	0.3000	0.9038	0.1738	2.5859	0.2875	5.8631		
	0.2000	1.3372	0.1155	3.8515	0.1907	8.3597		
	0.1000	2.6347	0.0575	7.6464	0.0947	15.8420		
	INF.	1.4953	0.9388	0.7587	0.5724	0.2592		
6	1.0000	0.3164	0.6070	0.7962	0.7880	1.1448	2.1154	
	1.1111	0.2813	0.6750	0.7108	0.8826	1.0087	2.3076	
	1.2500	0.2470	0.7597	0.6273	0.9994	0.8804	2.5365	
	1.4286	0.2135	0.8681	0.5452	1.1481	0.7580	2.8209	
	1.6667	0.1807	1.0123	0.4644	1.3451	0.6402	3.1908	
	2.0000	0.1487	1.2136	0.3847	1.6194	0.5263	3.6993	
	2.5000	0.1174	1.5148	0.3060	2.0292	0.4157	4.4522	
	3.3333	0.0868	2.0154	0.2282	2.7098	0.3080	5.6952	
	5.0000	0.0570	3.0146	0.1513	4.0679	0.2028	8.1654	
	10.0000	0.0280	6.0071	0.0753	8.1355	0.1002	15.5460	
	INF.	1.4732	0.9894	0.8129	0.7484	0.5979	0.2752	
7	1.0000	0.3207	0.6267	0.8091	0.7753	0.9241	0.9649	2.3829
	0.9000	0.3534	0.5641	0.8946	0.7016	1.0176	0.8750	2.5374
	0.8000	0.3940	0.5015	1.0015	0.6270	1.1350	0.7824	2.7351
	0.7000	0.4458	0.4387	1.1388	0.5513	1.2867	0.6876	2.9937
	0.6000	0.5146	0.3758	1.3218	0.4747	1.4899	0.5910	3.3428
	0.5000	0.6102	0.3128	1.5781	0.3972	1.7755	0.4931	3.8355
	0.4000	0.7531	0.2498	1.9626	0.3189	2.2054	0.3943	4.5779
	0.3000	0.9902	0.1869	2.6034	0.2399	2.9236	0.2951	5.8175
	0.2000	1.4630	0.1243	3.8851	0.1603	4.3624	0.1961	8.2974
	0.1000	2.8780	0.0619	7.7299	0.0803	8.6825	0.0976	15.7326
	INF.	1.4861	0.9693	0.8643	0.8040	0.7689	0.6157	0.2826
n	$1/R_s$	L_1	C_2	L_3	C_4	L_5	C_6	L_7

LOW PASS ELEMENT VALUES

GAUSSIAN RESPONSE TO 12 db

n	R_s	C_1	L_2	C_3	L_4	C_5	L_6	C_7	L_8	C_9	L_{10}
8	1.0000	0.3449	0.6565	0.8686	0.8028	0.9701	0.8182	1.2503	2.0612		
	1.1111	0.3053	0.7304	0.7729	0.9044	0.8550	0.9339	1.0753	2.2930		
	1.2500	0.2674	0.8221	0.6810	1.0276	0.7489	1.0694	0.9272	2.5445		
	1.4286	0.2308	0.9394	0.5913	1.1837	0.6480	1.2378	0.7929	2.8444		
	1.6667	0.1952	1.0953	0.5034	1.3898	0.5504	1.4579	0.6673	3.2267		
	2.0000	0.1604	1.3128	0.4168	1.6764	0.4553	1.7620	0.5476	3.7473		
	2.5000	0.1265	1.6381	0.3314	2.1045	0.3621	2.2143	0.4323	4.5145		
	3.3333	0.0934	2.1788	0.2471	2.8155	0.2702	2.9640	0.3203	5.7789		
	5.0000	0.0613	3.2575	0.1638	4.2343	0.1794	4.4583	0.2111	8.2899		
	10.0000	0.0301	6.4874	0.0814	8.4843	0.0894	8.9330	0.1044	15.7917		
	INF.	1.4974	1.0324	0.8943	0.8908	0.8494	0.8098	0.6452	0.2955		
9	1.0000	0.3318	0.6500	0.8467	0.8167	0.9426	0.8239	0.9857	0.9630	2.4140	
	0.9000	0.3657	0.5852	0.9363	0.7389	1.0390	0.7492	1.0803	0.8785	2.5608	
	0.8000	0.4078	0.5201	1.0480	0.6602	1.1599	0.6725	1.2003	0.7894	2.7524	
	0.7000	0.4614	0.4550	1.1914	0.5806	1.3160	0.5936	1.3568	0.6963	3.0070	
	0.6000	0.5324	0.3897	1.3825	0.4999	1.5251	0.5127	1.5681	0.6001	3.3538	
	0.5000	0.6312	0.3243	1.6500	0.4183	1.8192	0.4301	1.8667	0.5015	3.8460	
	0.4000	0.7787	0.2590	2.0512	0.3358	2.2620	0.3460	2.3177	0.4016	4.5897	
	0.3000	1.0234	0.1938	2.7200	0.2526	3.0021	0.2607	3.0729	0.3009	5.8332	
	0.2000	1.5113	0.1288	4.0574	0.1687	4.4850	0.1744	4.5875	0.2000	8.3219	
	0.1000	2.9716	0.0641	8.0692	0.0845	8.9384	0.0875	9.1379	0.0996	15.7849	
	INF.	1.4917	0.9908	0.9105	0.8770	0.8910	0.8457	0.8022	0.6376	0.2917	
10	1.0139	0.3500	0.6698	0.8817	0.8148	1.0183	0.7949	1.0929	0.7508	1.4303	1.8322
	1.1111	0.3092	0.7364	0.7856	0.9200	0.8864	0.9293	0.9147	0.9187	1.1138	2.2110
	1.2500	0.2701	0.8290	0.6907	1.0477	0.7734	1.0708	0.7890	1.0712	0.9404	2.4914
	1.4286	0.2328	0.9474	0.5992	1.2075	0.6681	1.2428	0.6777	1.2503	0.7978	2.8009
	1.6667	0.1968	1.1046	0.5099	1.4179	0.5671	1.4663	0.5734	1.4791	0.6690	3.1853
	2.0000	0.1616	1.3239	0.4221	1.7103	0.4689	1.7746	0.4734	1.7920	0.5483	3.7034
	2.5000	0.1274	1.6517	0.3355	2.1467	0.3728	2.2329	0.3761	2.2552	0.4326	4.4640
	3.3333	0.0941	2.1966	0.2501	2.8714	0.2781	2.9923	0.2806	3.0215	0.3206	5.7162
	5.0000	0.0617	3.2837	0.1658	4.3171	0.1845	4.5061	0.1863	4.5478	0.2114	8.2022
	10.0000	0.0303	6.5385	0.0824	8.6473	0.0919	9.0397	0.0929	9.1181	0.1046	15.6293
	INF.	1.4826	1.0350	0.9134	0.9263	0.9061	0.9159	0.8654	0.8190	0.6502	0.2973
n	$1/R_s$	L_1	C_2	L_3	C_4	L_5	C_6	L_7	C_8	L_9	C_{10}

LOW PASS ELEMENT VALUES

LEGENDRE RESPONSE

n	R_s	C_1	L_2	C_3	L_4	C_5	L_6	C_7
3	1.0000	1.1737	1.3538	2.1801				
	0.9000	1.2634	1.2481	2.3191				
	0.8000	1.3705	1.1335	2.5087				
	0.7000	1.5043	1.0095	2.7700				
	0.6000	1.6805	0.8766	3.1353				
	0.5000	1.9276	0.7363	3.6613				
	0.4000	2.3009	0.5906	4.4609				
	0.3000	2.9285	0.4420	5.7978				
	0.2000	4.1926	0.2928	8.4664				
	0.1000	8.0007	0.1448	16.4429				
	INF.	1.5909	1.4270	0.7629				
4	1.0000	1.0826	1.4769	1.9584	1.5645			
	1.1111	0.8572	1.7431	1.5976	1.9497			
	1.2500	0.7192	1.9825	1.3930	2.2200			
	1.4286	0.5991	2.2677	1.2091	2.5350			
	1.6667	0.4907	2.6313	1.0324	2.9401			
	2.0000	0.3915	3.1262	0.8581	3.4982			
	2.5000	0.3004	3.8558	0.6844	4.3266			
	3.3333	0.2163	5.0593	0.5111	5.6949			
	5.0000	0.1385	7.4522	0.3386	8.4094			
	10.0000	0.0666	14.6083	0.1679	16.5000			
	INF.	1.6120	1.6616	1.4292	0.6399			
5	1.0000	0.9512	1.4780	2.0673	1.5395	1.9990		
	0.9000	1.0280	1.3432	2.2401	1.4367	2.1246		
	0.8000	1.1170	1.2022	2.4619	1.3162	2.3125		
	0.7000	1.2270	1.0548	2.7562	1.1790	2.5825		
	0.6000	1.3726	0.9025	3.1587	1.0281	2.9643		
	0.5000	1.5787	0.7477	3.7309	0.8672	3.5132		
	0.4000	1.8921	0.5926	4.5960	0.6994	4.3434		
	0.3000	2.4209	0.4391	6.0424	0.5269	5.7258		
	0.2000	3.4878	0.2886	8.9373	0.3519	8.4778		
	0.1000	6.7041	0.1420	17.6200	0.1757	16.6914		
	INF.	1.6372	1.7509	1.7358	1.3945	0.6445		
6	1.0000	0.9160	1.4852	1.9857	1.7442	1.9040	1.5763	
	1.1111	0.7493	1.6903	1.6710	2.0862	1.5976	1.9047	
	1.2500	0.6344	1.9031	1.4574	2.3724	1.4181	2.1500	
	1.4286	0.5310	2.1625	1.2599	2.7152	1.2479	2.4521	
	1.6667	0.4361	2.4970	1.0698	3.1600	1.0768	2.8536	
	2.0000	0.3488	2.9561	0.8838	3.7756	0.9025	3.4151	
	2.5000	0.2683	3.6372	0.7008	4.6940	0.7250	4.2535	
	3.3333	0.1937	4.7662	0.5207	6.2197	0.5450	5.6409	
	5.0000	0.1245	7.0180	0.3437	9.2642	0.3635	8.3942	
	10.0000	0.0600	13.7648	0.1700	18.3835	0.1815	16.6001	
	INF.	1.6348	1.8088	1.8223	1.6795	1.3486	0.5793	
7	1.0000	0.8394	1.4770	1.9394	1.7270	2.1506	1.5895	1.8640
	0.9000	0.9082	1.3344	2.1177	1.5865	2.3194	1.4918	1.9822
	0.8000	0.9861	1.1863	2.3416	1.4328	2.5500	1.3690	2.1725
	0.7000	1.0825	1.0333	2.6345	1.2668	2.8662	1.2256	2.4498
	0.6000	1.2115	0.8780	3.0307	1.0922	3.3033	1.0679	2.8401
	0.5000	1.3952	0.7230	3.5898	0.9126	3.9263	0.9004	3.3970
	0.4000	1.6759	0.5702	4.4314	0.7302	4.8683	0.7263	4.2345
	0.3000	2.1505	0.4209	5.8357	0.5468	6.4434	0.5477	5.6239
	0.2000	3.1083	0.2758	8.6447	0.3634	9.5964	0.3663	8.3835
	0.1000	5.9964	0.1355	17.0696	0.1809	19.0546	0.1833	16.6105
	INF.	1.6391	1.8312	1.8911	1.7988	1.6845	1.3290	0.5787
n	$1/R_s$	L_1	C_2	L_3	C_4	L_5	C_6	L_7

LOW PASS ELEMENT VALUES

LEGENDRE RESPONSE

n	R_s	C_1	L_2	C_3	L_4	C_5	L_6	C_7	L_8	C_9	L_{10}
8	1.0000	0.8205	1.4688	1.9115	1.7672	2.0515	1.8411	1.8501	1.5564		
	1.1111	0.6826	1.6451	1.6429	2.0600	1.7361	2.1816	1.5863	1.8418		
	1.2500	0.5801	1.8419	1.4373	2.3360	1.5286	2.4637	1.4226	2.0731		
	1.4286	0.4863	2.0838	1.2428	2.6725	1.3332	2.8094	1.2593	2.3705		
	1.6667	0.3998	2.3977	1.0542	3.1112	1.1408	3.2660	1.0906	2.7720		
	2.0000	0.3202	2.8310	0.8696	3.7193	0.9492	3.9045	0.9163	3.3353		
	2.5000	0.2466	3.4767	0.6885	4.6264	0.7578	4.8621	0.7375	4.1761		
	3.3333	0.1784	4.5504	0.5108	6.1337	0.5669	6.4572	0.5554	5.5658		
	5.0000	0.1149	6.6961	0.3367	9.1422	0.3767	9.6449	0.3711	8.3208		
	10.0000	0.0555	13.1315	0.1664	18.1571	0.1876	19.2009	0.1856	16.5270		
	INF.	1.6345	1.8542	1.9102	1.8673	1.8019	1.6437	1.2869	0.5372		
9	1.0000	0.7695	1.4555	1.8674	1.7755	2.0662	1.7816	2.1585	1.6134	1.7645	
	0.9000	0.8328	1.3104	2.0462	1.6202	2.2488	1.6453	2.3201	1.5192	1.8791	
	0.8000	0.9031	1.1597	2.2682	1.4544	2.4896	1.4886	2.5533	1.3937	2.0738	
	0.7000	0.9906	1.0054	2.5565	1.2792	2.8129	1.3167	2.8778	1.2463	2.3575	
	0.6000	1.1088	0.8507	2.9443	1.0981	3.2539	1.1359	3.3265	1.0848	2.7535	
	0.5000	1.2783	0.6981	3.4897	0.9143	3.8777	0.9501	3.9649	0.9141	3.3147	
	0.4000	1.5379	0.5491	4.3092	0.7295	4.8171	0.7615	4.9292	0.7372	4.1546	
	0.3000	1.9771	0.4045	5.6760	0.5450	6.3852	0.5714	6.5409	0.5560	5.5439	
	0.2000	2.8640	0.2648	8.4097	0.3616	9.5222	0.3807	9.7668	0.3720	8.2985	
	0.1000	5.5379	0.1299	16.6100	0.1798	18.9317	0.1901	19.4438	0.1863	16.5032	
	INF.	1.6341	1.8625	1.9349	1.8961	1.8815	1.7860	1.6397	1.2740	0.5358	
10	1.0000	0.7575	1.4454	1.8537	1.7839	2.0327	1.8453	2.0409	1.8953	1.8122	1.5286
	1.1111	0.6367	1.6053	1.6118	2.0469	1.7480	2.1553	1.7403	2.2189	1.5801	1.7832
	1.2500	0.5419	1.7907	1.4125	2.3146	1.5395	2.4377	1.5419	2.4925	1.4255	2.0076
	1.4286	0.4544	2.0194	1.2214	2.6444	1.3409	2.7857	1.3504	2.8365	1.2654	2.3046
	1.6667	0.3738	2.3177	1.0353	3.0760	1.1455	3.2444	1.1589	3.2963	1.0974	2.7076
	2.0000	0.2996	2.7314	0.8532	3.6751	0.9515	3.8843	0.9665	3.9424	0.9227	3.2726
	2.5000	0.2310	3.3500	0.6749	4.5694	0.7585	4.8425	0.7733	4.9133	0.7432	4.1142
	3.3333	0.1673	4.3811	0.5004	6.0559	0.5667	6.4374	0.5796	6.5322	0.5600	5.5027
	5.0000	0.1079	6.4443	0.3297	9.0242	0.3762	9.6244	0.3860	9.7691	0.3745	8.2524
	10.0000	0.0523	12.6366	0.1628	17.9209	0.1872	19.1790	0.1927	19.4749	0.1875	16.4374
	INF.	1.6298	1.8741	1.9405	1.9223	1.9102	1.8629	1.7785	1.6082	1.2386	0.5065
n	$1/R_s$	L_1	C_2	L_3	C_4	L_5	C_6	L_7	C_8	L_9	C_{10}

3-db DOWN k AND q VALUES

BUTTERWORTH RESPONSE

n	q_0	I. L.	q_1	q_n	k_{12}	k_{23}	k_{34}	k_{45}
2	INF.	0.000	1.4142	1.4142	0.7071			
	14.142	0.915	1.4142	1.4142	0.7071			
	7.071	1.938	1.4142	1.4142	0.7071			
	4.714	3.098	1.4142	1.4142	0.7071			
	3.536	4.437	1.4142	1.4142	0.7071			
	2.828	6.021	1.4142	1.4142	0.7071			
	2.357	7.959	1.4142	1.4142	0.7071			
	2.020	10.458	1.4142	1.4142	0.7071			
3	INF.	0.000	1.0000	1.0000	0.7071	0.7071		
	20.000	0.958	0.8041	1.4156	0.7687	0.6582		
	10.000	2.052	0.8007	1.5359	0.7388	0.6716		
	6.667	3.300	0.8087	1.6301	0.7005	0.6879		
	5.000	4.742	0.8226	1.7115	0.6567	0.7060		
	4.000	6.443	0.8406	1.7844	0.6077	0.7256		
	3.333	8.512	0.8625	1.8497	0.5524	0.7470		
	2.857	11.157	0.8884	1.9068	0.4883	0.7706		
4	INF.	0.000	0.7654	0.7654	0.8409	0.5412	0.8409	
	26.131	1.002	0.5376	1.4782	1.0927	0.5668	0.6670	
	13.066	2.162	0.5355	1.6875	1.0745	0.5546	0.6805	
	8.710	3.489	0.5417	1.8605	1.0411	0.5373	0.6992	
	6.533	5.020	0.5521	2.0170	1.0004	0.5161	0.7207	
	5.226	6.822	0.5656	2.1621	0.9547	0.4906	0.7444	
	4.355	9.003	0.5819	2.2961	0.9051	0.4592	0.7706	
	3.733	11.772	0.6012	2.4159	0.8518	0.4192	0.7998	
5	INF.	0.000	0.6180	0.6180	1.0000	0.5559	0.5559	1.0000
	32.361	1.045	0.4001	1.5527	1.4542	0.6946	0.5285	0.6750
	32.361	1.045	0.5662	0.7261	1.0947	0.5636	0.5800	0.8106
	16.180	2.263	0.3990	1.8372	1.4414	0.6886	0.5200	0.6874
	16.180	2.263	0.5777	0.7577	1.0711	0.5408	0.6160	0.7452
	10.787	3.657	0.4036	2.0825	1.4088	0.6750	0.5080	0.7066
	10.787	3.657	0.5927	0.7869	1.0408	0.5144	0.6520	0.6860
	8.090	5.265	0.4111	2.3118	1.3670	0.6576	0.4927	0.7290
	8.090	5.265	0.6100	0.8157	1.0075	0.4844	0.6887	0.6278
	6.472	7.151	0.4206	2.5307	1.3195	0.6374	0.4732	0.7542
	6.472	7.151	0.6293	0.8449	0.9722	0.4501	0.7267	0.5681
	5.393	9.425	0.4321	2.7375	1.2675	0.6148	0.4479	0.7821
	5.393	9.425	0.6508	0.8748	0.9355	0.4103	0.7663	0.5048

3-db DOWN k AND q VALUES

BUTTERWORTH RESPONSE

n	q_0	I. L.	q_1	q_n	k_{12}	k_{23}	k_{34}	k_{45}	k_{56}	k_{67}
6	INF.	0.000	0.5176	0.5176	1.1688	0.6050	0.5176	0.6050	1.1688	
	38.637	1.084	0.3179	1.6276	1.8246	0.8492	0.5986	0.5162	0.6803	
	38.637	1.084	0.6091	0.4721	1.0792	0.5589	0.5705	0.5779	1.2068	
	19.319	2.354	0.3173	1.9792	1.8142	0.8457	0.5955	0.5089	0.6916	
	19.319	2.354	0.6334	0.4813	1.0553	0.5272	0.6111	0.5503	1.1587	
	12.879	3.808	0.3208	2.2924	1.7811	0.8326	0.5873	0.4991	0.7108	
	12.879	3.808	0.6573	0.4922	1.0299	0.4936	0.6503	0.5233	1.1101	
	9.659	5.481	0.3264	2.5936	1.7375	0.8151	0.5762	0.4862	0.7336	
	9.659	5.481	0.6819	0.5043	1.0039	0.4571	0.6896	0.4950	1.0610	
	7.727	7.439	0.3334	2.8876	1.6873	0.7948	0.5631	0.4693	0.7594	
	7.727	7.439	0.7076	0.5174	0.9776	0.4171	0.7297	0.4637	1.0113	
	6.440	9.791	0.3416	3.1707	1.6324	0.7722	0.5479	0.4466	0.7882	
	6.440	9.791	0.7348	0.5315	0.9510	0.3724	0.7713	0.4276	0.9610	
7	INF.	0.000	0.4450	0.4450	1.3424	0.6671	0.5268	0.5268	0.6671	1.3424
	44.940	1.121	0.2636	1.6995	2.1977	1.0104	0.6940	0.5579	0.5118	0.6835
	44.940	1.121	0.3360	0.7110	1.7554	0.8400	0.6101	0.5029	0.5938	0.8089
	44.940	1.121	0.4272	0.4897	1.3980	0.7143	0.5146	0.5740	0.5745	1.1986
	44.940	1.121	0.6568	0.3496	1.0599	0.5533	0.5702	0.5299	0.7351	1.6355
	22.470	2.437	0.2633	2.1126	2.1883	1.0076	0.6929	0.5558	0.5050	0.6942
	22.470	2.437	0.3391	0.7562	1.7324	0.8328	0.6092	0.4872	0.6313	0.7251
	22.470	2.437	0.4366	0.5047	1.3622	0.7090	0.4901	0.6171	0.5397	1.1487
	22.470	2.437	0.6939	0.3533	1.0377	0.5146	0.6111	0.5151	0.7117	1.6005
	14.980	3.943	0.2661	2.4907	2.1541	0.9941	0.6854	0.5499	0.4962	0.7132
	14.980	3.943	0.3449	0.7933	1.6958	0.8191	0.6043	0.4688	0.6691	0.6559
	14.980	3.943	0.4475	0.5193	1.3225	0.7006	0.4636	0.6586	0.5071	1.1027
	14.980	3.943	0.7270	0.3591	1.0173	0.4756	0.6499	0.4999	0.6873	1.5584
	11.235	5.674	0.2703	2.8626	2.1087	0.9757	0.6748	0.5418	0.4846	0.7362
	11.235	5.674	0.3521	0.8271	1.6530	0.8023	0.5974	0.4471	0.7075	0.5914
	11.235	5.674	0.4594	0.5341	1.2810	0.6908	0.4342	0.6999	0.4737	1.0585
	11.235	5.674	0.7593	0.3660	0.9976	0.4347	0.6885	0.4828	0.6622	1.5129

BUTTERWORTH RESPONSE

n	q_0	I.L.	q_1	q_n	k_{12}	k_{23}	k_{34}	k_{45}	k_{56}	k_{67}	k_{78}
7	8.988	7.694	0.2756	3.2330	2.0563	0.9542	0.6622	0.5318	0.4691	0.7622	
	8.988	7.694	0.3605	0.8593	1.6063	0.7836	0.5893	0.4212	0.7470	0.5278	
	8.988	7.694	0.4723	0.5493	1.2383	0.6801	0.4010	0.7418	0.4378	1.0151	
	8.988	7.694	0.7920	0.3738	0.9784	0.3910	0.7278	0.4623	0.6362	1.4654	
	7.490	10.113	0.2818	3.5954	1.9988	0.9304	0.6480	0.5201	0.4477	0.7915	
	7.490	10.113	0.3699	0.8903	1.5568	0.7633	0.5800	0.3898	0.7879	0.4623	
	7.490	10.113	0.4863	0.5650	1.1945	0.6688	0.3630	0.7849	0.3976	0.9723	
	7.490	10.113	0.8255	0.3824	0.9595	0.3436	0.7687	0.4365	0.6092	1.4164	
8	INF.	0.000	0.3902	0.3902	1.5187	0.7357	0.5537	0.5098	0.5537	0.7357	1.5187
	51.258	1.155	0.2251	1.7676	2.5714	1.1739	0.7958	0.6243	0.5367	0.5105	0.6854
	51.258	1.155	0.3584	0.4508	1.6793	0.8252	0.6208	0.4937	0.5790	0.5783	1.2471
	51.258	1.155	0.3258	0.5156	1.7864	0.8475	0.6164	0.4983	0.5787	0.5691	1.1813
	51.258	1.155	0.7059	0.2784	1.0408	0.5475	0.5709	0.5151	0.6144	0.9230	2.0623
	25.629	2.513	0.2250	2.2379	2.5623	1.1712	0.7952	0.6242	0.5351	0.5039	0.6955
	25.629	2.513	0.3650	0.4646	1.6485	0.8148	0.6234	0.4697	0.6230	0.5402	1.1842
	25.629	2.513	0.3302	0.5366	1.7524	0.8369	0.6134	0.4783	0.6229	0.5282	1.1325
	25.629	2.513	0.7561	0.2802	1.0213	0.5035	0.6102	0.5082	0.5978	0.9066	2.0352
	17.086	4.065	0.2272	2.6783	2.5271	1.1570	0.7874	0.6193	0.5303	0.4957	0.7145
	17.086	4.065	0.3732	0.4773	1.6108	0.8001	0.6227	0.4436	0.6652	0.5065	1.1299
	17.086	4.065	0.3363	0.5552	1.7107	0.8226	0.6075	0.4564	0.6652	0.4908	1.0906
	17.086	4.065	0.7988	0.2839	1.0054	0.4606	0.6471	0.5003	0.5806	0.8865	1.9960
	12.815	5.848	0.2306	3.1201	2.4800	1.1376	0.7762	0.6120	0.5237	0.4848	0.7375
	12.815	5.848	0.3823	0.4898	1.5699	0.7834	0.6206	0.4147	0.7070	0.4730	1.0794
	12.815	5.848	0.3433	0.5731	1.6654	0.8066	0.6003	0.4316	0.7072	0.4532	1.0517
	12.815	5.848	0.8393	0.2885	0.9910	0.4165	0.6837	0.4903	0.5624	0.8647	1.9513
	10.252	7.923	0.2347	3.5677	2.4257	1.1149	0.7630	0.6031	0.5155	0.4700	0.7637
	10.252	7.923	0.3922	0.5023	1.5271	0.7652	0.6177	0.3822	0.7493	0.4374	1.0310
	10.252	7.923	0.3511	0.5908	1.6179	0.7894	0.5921	0.4030	0.7497	0.4137	1.0146
	10.252	7.923	0.8793	0.2938	0.9775	0.3704	0.7210	0.4768	0.5432	0.8418	1.9034
	8.543	10.401	0.2396	4.0118	2.3661	1.0899	0.7481	0.5929	0.5056	0.4493	0.7934
	8.543	10.401	0.4030	0.5149	1.4828	0.7459	0.6140	0.3452	0.7924	0.3976	0.9841
	8.543	10.401	0.3597	0.6086	1.5687	0.7714	0.5831	0.3691	0.7930	0.3705	0.9789
	8.543	10.401	0.9195	0.2998	0.9643	0.3214	0.7602	0.4577	0.5226	0.8180	1.8531

CHEBYSHEV RESPONSE   RIPPLE = 0.01 db

n	q_0	I. L.	q_1	q_n	k_{12}	k_{23}	k_{34}	k_{45}
2	INF.	0.000	1.4829	1.4829	0.7075			
	14.829	0.915	1.4829	1.4829	0.7075			
	7.415	1.938	1.4829	1.4829	0.7075			
	4.943	3.098	1.4829	1.4829	0.7075			
	3.707	4.437	1.4829	1.4829	0.7075			
	2.966	6.021	1.4829	1.4829	0.7075			
	2.472	7.959	1.4829	1.4829	0.7075			
	2.118	10.458	1.4829	1.4829	0.7075			
3	INF.	0.000	1.1811	1.1811	0.6818	0.6818		
	23.622	0.937	0.9796	1.5869	0.7027	0.6630		
	11.811	2.006	0.9649	1.7472	0.6785	0.6759		
	7.874	3.227	0.9686	1.8730	0.6465	0.6917		
	5.905	4.642	0.9810	1.9817	0.6091	0.7093		
	4.724	6.313	0.9994	2.0787	0.5667	0.7283		
	3.937	8.351	1.0230	2.1653	0.5180	0.7491		
	3.375	10.963	1.0518	2.2408	0.4607	0.7721		
4	INF.	0.000	1.0457	1.0457	0.7369	0.5413	0.7369	
	35.703	0.949	0.7764	1.7588	0.8469	0.5368	0.6717	
	17.851	2.046	0.7588	2.0720	0.8418	0.5261	0.6837	
	11.901	3.304	0.7593	2.3392	0.8227	0.5121	0.7011	
	8.926	4.763	0.7681	2.5867	0.7966	0.4948	0.7211	
	7.141	6.486	0.7825	2.8209	0.7659	0.4736	0.7436	
	5.950	8.584	0.8015	3.0408	0.7315	0.4468	0.7685	
	5.100	11.262	0.8252	3.2409	0.6934	0.4115	0.7968	
5	INF.	0.000	0.9766	0.9766	0.7796	0.5398	0.5398	0.7796
	51.135	0.957	0.6792	1.9345	0.9585	0.5688	0.5188	0.6755
	51.135	0.957	0.9157	1.1146	0.8320	0.5060	0.5943	0.6691
	25.567	2.069	0.6608	2.3970	0.9659	0.5660	0.5106	0.6858
	25.567	2.069	0.9273	1.1733	0.8259	0.4806	0.6328	0.6190
	17.045	3.349	0.6591	2.8196	0.9553	0.5582	0.5009	0.7027
	17.045	3.349	0.9472	1.2252	0.8135	0.4541	0.6692	0.5741
	12.784	4.833	0.6648	3.2353	0.9361	0.5476	0.4887	0.7231
	12.784	4.833	0.9717	1.2752	0.7980	0.4256	0.7052	0.5298
	10.227	6.587	0.6754	3.6507	0.9117	0.5348	0.4730	0.7462
	10.227	6.587	1.0001	1.3251	0.7806	0.3940	0.7418	0.4837
	8.522	8.719	0.6898	4.0602	0.8833	0.5199	0.4520	0.7725
	8.522	8.719	1.0321	1.3755	0.7616	0.3581	0.7794	0.4338
	7.305	11.436	0.7080	4.4468	0.8514	0.5025	0.4223	0.8030
	7.305	11.436	1.0680	1.4266	0.7412	0.3161	0.8186	0.3774

CHEBYSHEV RESPONSE   RIPPLE = 0.01 db

n	q_0	I. L.	q_1	q_n	k_{12}	k_{23}	k_{34}	k_{45}	k_{56}	k_{67}
6	INF.	0.000	0.9372	0.9372	0.8088	0.5500	0.5177	0.5500	0.8088	
	69.951	0.962	0.6250	2.0966	1.0375	6030	0.5214	0.5189	0.6750	
	69.951	0.962	1.0775	0.8704	0.8228	0.4863	0.5739	0.5413	0.7732	
	34.975	2.083	0.6062	2.7015	1.0532	0.6051	0.5180	0.5120	0.6836	
	34.975	2.083	1.1288	0.8819	0.8217	0.4480	0.6130	0.5287	0.7404	
	23.317	3.376	0.6031	3.2860	1.0485	0.6011	0.5125	0.5045	0.6998	
	23.317	3.376	1.1766	0.8987	0.8173	0.4118	0.6494	0.5150	0.7086	
	17.488	4.877	0.6067	3.8912	1.0343	0.5939	0.5053	0.4950	0.7197	
	17.488	4.877	1.2246	0.9185	0.8113	0.3754	0.6853	0.4989	0.6766	
	13.990	6.649	0.6145	4.5267	1.0143	0.5846	0.4965	0.4821	0.7428	
	13.990	6.649	1.2739	0.9405	0.8042	0.3374	0.7221	0.4787	0.6443	
	11.658	8.803	0.6258	5.1830	0.9901	0.5735	0.4859	0.4642	0.7694	
	11.658	8.803	1.3255	0.9647	0.7961	0.2970	0.7609	0.4525	0.6112	
	9.993	11.543	0.6401	5.8295	0.9624	0.5609	0.4731	0.4376	0.8010	
	9.993	11.543	1.3799	0.9909	^.7869	0.2529	0.8029	0.4164	0.5772	
7	INF.	0.000	0.9127	0.9127	0.8289	0.5597	0.5168	0.5168	0.5597	0.8289
	92.164	0.965	0.5914	2.2411	1.0940	0.6295	0.5355	0.5092	0.5219	0.6723
	92.164	0.965	0.7299	1.3038	0.9698	0.5838	0.5452	0.4687	0.6075	0.6323
	92.164	0.965	0.8898	0.9870	0.8526	0.5839	0.4613	0.5932	0.5124	0.7644
	92.164	0.965	1.2453	0.7496	0.8150	0.4793	0.5497	0.5409	0.5420	0.8742
	46.082	2.092	0.5724	2.9791	1.1156	0.6348	0.5354	0.5055	0.5162	0.6795
	46.082	2.092	0.7224	1.4318	0.9747	0.5807	0.5503	0.4449	0.6492	0.5728
	46.082	2.092	0.9049	1.0228	0.8420	0.5863	0.4300	0.6385	0.4880	0.7352
	46.082	2.092	1.3471	0.7460	0.8203	0.4330	0.5798	0.5469	0.5244	0.8594
	30.721	3.394	0.5682	3.7253	1.1152	0.6330	0.5325	0.5008	0.5102	0.6948
	30.721	3.394	0.7264	1.5339	0.9675	0.5740	0.5526	0.4222	0.6880	0.5237
	30.721	3.394	0.9247	1.0559	0.8278	0.5862	0.4001	0.6795	0.4646	0.7096
	30.721	3.394	1.4327	0.7516	0.8233	0.3907	0.6085	0.5496	0.5072	0.8399
	23.041	4.905	0.5702	4.5328	1.1049	0.6278	0.5279	0.4949	0.5024	0.7140
	23.041	4.905	0.7359	1.6251	0.9550	0.5653	0.5535	0.3985	0.7258	0.4774
	23.041	4.905	0.9473	1.0886	0.8117	0.5851	0.3695	0.7192	0.4396	0.6854
	23.041	4.905	1.5131	0.7614	0.8250	0.3494	0.6377	0.5492	0.4894	0.8181

3-db DOWN k AND q VALUES

CHEBYSHEV RESPONSE   RIPPLE = 0.01 db

n	q_0	I. L.	q_1	q_n	k_{12}	k_{23}	k_{34}	k_{45}	k_{56}	k_{67}	k_{78}
7	18.433	6.690	0.5762	5.4187	1.0885	0.6204	0.5221	0.4877	0.4917	0.7367	
	18.433	6.690	0.7490	1.7099	0.9392	0.5553	0.5533	0.3726	0.7636	0.4306	
	18.433	6.690	0.9723	1.1216	0.7945	0.5832	0.3371	0.7586	0.4116	0.6619	
	18.433	6.690	1.5919	0.7741	0.8255	0.3075	0.6692	0.5447	0.4703	0.7947	
	15.361	8.857	0.5852	6.3749	1.0676	0.6114	0.5151	0.4787	0.4760	0.7632	
	15.361	8.857	0.7650	1.7901	0.9209	0.5442	0.5520	0.3432	0.8020	0.3810	
	15.361	8.857	0.9997	1.1554	0.7762	0.5809	0.3016	0.7985	0.3786	0.6386	
	15.361	8.857	1.6708	0.7891	0.8250	0.2641	0.7047	0.5338	0.4493	0.7701	
8	INF.	0.000	0.8966	0.8966	0.8430	0.5673	0.5198	0.5098	0.5198	0.5673	0.8430
	117.783	0.967	0.5690	2.3686	1.1355	0.6495	0.5485	0.5144	0.5063	0.5249	0.6686
	117.783	0.967	0.7790	1.1159	0.9016	0.6046	0.5102	0.4777	0.5985	0.4855	0.7656
	117.783	0.967	0.8454	1.0031	0.9259	0.5587	0.5724	0.4473	0.5663	0.5541	0.7023
	117.783	0.967	1.4106	0.6799	0.8066	0.4813	0.5267	0.5430	0.5196	0.5640	0.9551
	58.891	2.099	0.5497	3.2290	1.1614	0.6570	0.5504	0.5133	0.5026	0.5201	0.6745
	58.891	2.099	0.7790	1.1833	0.8944	0.6076	0.5043	0.4556	0.6460	0.4492	0.7438
	58.891	2.099	0.8539	1.0442	0.9269	0.5473	0.5897	0.4099	0.6058	0.5448	0.6546
	58.891	2.099	1.5729	0.6698	0.8180	0.4313	0.5443	0.5595	0.5106	0.5523	0.9534
	39.261	3.406	0.5447	4.1324	1.1645	0.6568	0.5490	0.5106	0.4984	0.5153	0.6889
	39.261	3.406	0.7871	1.2386	0.8812	0.6075	0.4981	0.4331	0.6887	0.4168	0.7261
	39.261	3.406	0.8685	1.0794	0.9206	0.5352	0.6022	0.3761	0.6430	0.5336	0.6158
	39.261	3.406	1.7062	0.6703	0.8273	0.3859	0.5623	0.5725	0.5006	0.5402	0.9424
	29.446	4.924	0.5456	5.1473	1.1572	0.6529	0.5457	0.5070	0.4933	0.5090	0.7074
	29.446	4.924	0.7992	1.2893	0.8652	0.6059	0.4918	0.4086	0.7295	0.3846	0.7100
	29.446	4.924	0.8864	1.1126	0.9112	0.5226	0.6123	0.3429	0.6803	0.5190	0.5806
	29.446	4.924	1.8288	0.6756	0.8352	0.3416	0.5819	0.5830	0.4890	0.5274	0.9272
	23.557	6.717	0.5502	6.3053	1.1437	0.6469	0.5413	0.5024	0.4868	0.4999	0.7294
	23.557	6.717	0.8140	1.3378	0.8474	0.6035	0.4853	0.3809	0.7696	0.3510	0.6947
	23.557	6.717	0.9069	1.1450	0.8998	0.5095	0.6207	0.3090	0.7192	0.4993	0.5473
	23.557	6.717	1.9465	0.6839	0.8418	0.2971	0.6046	0.5906	0.4751	0.5141	0.9093
	19.630	8.893	0.5575	7.6074	1.1256	0.6392	0.5360	0.4970	0.4787	0.4863	0.7557
	19.630	8.893	0.8311	1.3852	0.8282	0.6004	0.4787	0.3486	0.8098	0.3144	0.6798
	19.630	8.893	0.9298	1.1770	0.8870	0.4959	0.6278	0.2733	0.7609	0.4720	0.5151
	19.630	8.893	2.0619	0.6944	0.8470	0.2511	0.6327	0.5936	0.4578	0.4999	0.8895

3-db DOWN k AND q VALUES

CHEBYSHEV RESPONSE   RIPPLE = 0.1 db

n	q_0	I. L.	q_1	q_n	k_{12}	k_{23}	k_{34}	k_{45}
2	INF.	0.000	1.6382	1.6382	0.7106			
	16.382	0.915	1.6382	1.6382	0.7106			
	8.191	1.938	1.6382	1.6382	0.7106			
	5.461	3.098	1.6382	1.6382	0.7106			
	4.096	4.437	1.6382	1.6382	0.7106			
	3.276	6.021	1.6382	1.6382	0.7106			
	2.730	7.959	1.6382	1.6382	0.7106			
	2.340	10.458	1.6382	1.6382	0.7106			
3	INF.	0.000	1.4328	1.4328	0.6618	0.6618		
	28.657	0.926	1.2184	1.8511	0.6539	0.6667		
	14.328	1.977	1.1922	2.0522	0.6311	0.6807		
	9.552	3.177	1.1907	2.2158	0.6029	0.6966		
	7.164	4.567	1.2013	2.3595	0.5703	0.7138		
	5.731	6.212	1.2202	2.4887	0.5329	0.7325		
	4.776	8.221	1.2461	2.6045	0.4897	0.7527		
	4.094	10.801	1.2789	2.7052	0.4380	0.7751		
4	INF.	0.000	1.3451	1.3451	0.6850	0.5421	0.6850	
	45.924	0.932	1.0380	2.0836	0.7279	0.5313	0.6687	
	22.962	1.996	1.0028	2.4837	0.7234	0.5200	0.6826	
	15.308	3.217	0.9958	2.8404	0.7091	0.5068	0.7002	
	11.481	4.633	1.0017	3.1803	0.6895	0.4910	0.7200	
	9.185	6.309	1.0161	3.5089	0.6660	0.4715	0.7419	
	7.654	8.354	1.0373	3.8226	0.6392	0.4466	0.7665	
	6.561	10.974	1.0652	4.1105	0.6091	0.4132	0.7945	
5	INF.	0.000	1.3013	1.3013	0.7028	0.5355	0.5355	0.7028
	68.137	0.935	0.9470	2.2887	0.7849	0.5382	0.5234	0.6667
	68.137	0.935	1.2368	1.4612	0.7419	0.4830	0.6031	0.6203
	34.069	2.007	0.9082	2.8755	0.7922	0.5332	0.5156	0.6790
	34.069	2.007	1.2482	1.5439	0.7416	0.4546	0.6428	0.5770
	22.712	3.238	0.8977	3.4377	0.7865	0.5258	0.5067	0.6962
	22.712	3.238	1.2715	1.6172	0.7356	0.4274	0.6792	0.5377
	17.034	4.668	0.8997	4.0108	0.7742	0.5165	0.4957	0.7161
	17.034	4.668	1.3018	1.6875	0.7267	0.3991	0.7147	0.4985
	13.627	6.360	0.9095	4.6017	0.7576	0.5054	0.4814	0.7387
	13.627	6.360	1.3375	1.7571	0.7159	0.3686	0.7502	0.4572
	11.356	8.425	0.9255	5.2012	0.7377	0.4924	0.4620	0.7645
	11.356	8.425	1.3786	1.8270	0.7035	0.3346	0.7866	0.4119
	9.734	11.068	0.9472	5.7821	0.7148	0.4771	0.4340	0.7950
	9.734	11.068	1.4251	1.8974	0.6896	0.2952	0.8244	0.3598

CHEBYSHEV RESPONSE   RIPPLE = 0.1 db

n	q_0	I. L.	q_1	q_n	k_{12}	k_{23}	k_{34}	k_{45}	k_{56}	k_{67}
	INF.	0.000	1.2767	1.2767	0.7145	0.5385	0.5180	0.5385	0.7145	
	95.292	0.936	0.8941	2.4617	0.8244	0.5530	0.5134	0.5252	0.6625	
	95.292	0.936	1.1990	1.4481	0.6600	0.5526	0.5646	0.4651	0.7524	
	47.646	2.013	0.8534	3.2171	0.8394	0.5527	0.5084	0.5197	0.6730	
	47.646	2.013	1.2102	1.5236	0.6287	0.5473	0.6009	0.4237	0.7602	
	31.764	3.250	0.8406	3.9826	0.8393	0.5490	0.5026	0.5135	0.6891	
6	31.764	3.250	1.5923	1.2306	0.7632	0.3864	0.6350	0.5390	0.5997	
	23.823	4.688	0.8399	4.8071	0.8318	0.5433	0.4956	0.5054	0.7083	
	23.823	4.688	1.6602	1.2557	0.7636	0.3499	0.6692	0.5272	0.5713	
	19.058	6.391	0.8464	5.7065	0.8195	0.5360	0.4873	0.4943	0.7306	
	19.058	6.391	1.7296	1.2844	0.7624	0.3127	0.7049	0.5107	0.5428	
	15.882	8.467	0.8585	6.6715	0.8038	0.5275	0.4772	0.4784	0.7567	
	15.882	8.467	1.8017	1.3163	0.7598	0.2738	0.7435	0.4871	0.5137	
	13.613	11.122	0.8755	7.6585	0.7850	0.5175	0.4647	0.4541	0.7881	
	13.613	11.122	1.8774	1.3512	0.7557	0.2319	0.7866	0.4525	0.4836	
	INF.	0.000	1.2615	1.2615	0.7223	0.5421	0.5155	0.5155	0.5421	0.7223
	127.384	0.938	0.8602	2.6064	0.8524	0.5655	0.5170	0.5089	0.5276	0.6577
	127.384	0.938	1.0409	1.7081	0.8047	0.5301	0.5465	0.4609	0.6086	0.5978
	127.384	0.938	1.2395	1.3525	0.7397	0.5608	0.4453	0.6004	0.5059	0.6758
	127.384	0.938	1.6528	1.0625	0.7556	0.4674	0.5266	0.5551	0.5291	0.7110
	63.692	2.017	0.8183	3.5111	0.8726	0.5682	0.5149	0.5045	0.5237	0.6663
	63.692	2.017	1.0201	1.8989	0.8128	0.5239	0.5541	0.4339	0.6517	0.5468
	63.692	2.017	1.2580	1.4045	0.7349	0.5643	0.4109	0.6461	0.4844	0.6519
	63.692	2.017	1.8079	1.0484	0.7711	0.4200	0.5482	0.5700	0.5124	0.6955
7	42.461	3.258	0.8039	4.4691	0.8765	0.5667	0.5115	0.4996	0.5193	0.6812
	42.461	3.258	1.0195	2.0544	0.8113	0.5164	0.5586	0.4097	0.6907	0.5032
	42.461	3.258	1.2835	1.4524	0.7266	0.5658	0.3797	0.6866	0.4631	0.6310
	42.461	3.258	1.9375	1.0509	0.7817	0.3772	0.5707	0.5801	0.4958	0.6784
	31.846	4.701	0.8012	5.5491	0.8723	0.5630	0.5072	0.4938	0.5135	0.6995
	31.846	4.701	1.0281	2.1941	0.8053	0.5081	0.5615	0.3855	0.7284	0.4611
	31.846	4.701	1.3133	1.4993	0.7166	0.5663	0.3487	0.7253	0.4398	0.6113
	31.846	4.701	2.0581	1.0610	0.7898	0.3358	0.5950	0.5864	0.4783	0.6601

CHEBYSHEV RESPONSE

RIPPLE = 0.1 db

n	q_0	I. L.	q_1	q_n	k_{12}	k_{23}	k_{34}	k_{45}	k_{56}	k_{67}	k_{78}
7	25.477	6.410	0.8053	6.7852	0.8632	0.5578	0.5019	0.4867	0.5049	0.7211	
	25.477	6.410	1.0428	2.3243	0.7964	0.4989	0.5630	0.3600	0.7657	0.4176	
	25.477	6.410	1.3467	1.5465	0.7054	0.5662	0.3167	0.7636	0.4132	0.5921	
	25.477	6.410	2.1754	1.0759	0.7959	0.2941	0.6227	0.5885	0.4591	0.6411	
	21.231	8.493	0.8145	8.1811	0.8505	0.5514	0.4957	0.4778	0.4918	0.7467	
	21.231	8.493	1.0624	2.4469	0.7854	0.4889	0.5633	0.3317	0.8035	0.3706	
	21.231	8.493	1.3837	1.5944	0.6931	0.5656	0.2824	0.8024	0.3813	0.5730	
	21.231	8.493	2.2921	1.0945	0.8003	0.2510	0.6560	0.5841	0.4374	0.6213	
8	INF.	0.000	1.2515	1.2515	0.7276	0.5451	0.5160	0.5100	0.5160	0.5451	0.7276
	164.415	0.938	0.8368	2.7279	0.8730	0.5752	0.5222	0.5077	0.5088	0.5293	0.6530
	164.415	0.938	1.1103	1.5132	0.7480	0.5812	0.4842	0.4816	0.6023	0.4724	0.6943
	164.415	0.938	0.9813	1.8433	0.7552	0.5281	0.5312	0.5437	0.5013	0.4780	0.7531
	164.415	0.938	1.3798	1.1950	0.6140	0.5732	0.5426	0.4523	0.5703	0.5050	0.7971
	82.207	2.019	0.7941	3.7636	0.8970	0.5798	0.5219	0.5051	0.5050	0.5265	0.6600
	82.207	2.019	1.1030	1.6171	0.7431	0.5875	0.4742	0.4592	0.6507	0.4365	0.6805
	82.207	2.019	0.9556	2.0894	0.7517	0.5137	0.5278	0.5655	0.5076	0.4297	0.7747
	82.207	2.019	1.4414	1.2029	0.5678	0.5762	0.5752	0.4125	0.5935	0.4882	0.8050
	54.805	3.264	0.7785	4.8997	0.9037	0.5799	0.5201	0.5020	0.5009	0.5236	0.6736
	54.805	3.264	1.1098	1.7021	0.7341	0.5906	0.4658	0.4365	0.6934	0.4045	0.6688
	54.805	3.264	0.9500	2.2940	0.7429	0.5004	0.5215	0.5839	0.5170	0.3851	0.7917
	54.805	3.264	1.4943	1.2205	0.5303	0.5738	0.6080	0.3768	0.6105	0.4736	0.8055
	41.104	4.710	0.7743	6.2294	0.9021	0.5776	0.5172	0.4982	0.4959	0.5194	0.6908
	41.104	4.710	1.1235	1.7794	0.7231	0.5922	0.4581	0.4119	0.7337	0.3731	0.6577
	41.104	4.710	1.2433	1.5437	0.8027	0.4596	0.6244	0.3421	0.6427	0.5663	0.4963
	41.104	4.710	2.4825	0.9531	0.8059	0.3411	0.5292	0.6001	0.5125	0.4870	0.7314
	32.883	6.422	0.7766	7.8157	0.8954	0.5737	0.5136	0.4938	0.4897	0.5128	0.7114
	32.883	6.422	1.1418	1.8527	0.7106	0.5928	0.4507	0.3842	0.7732	0.3403	0.6470
	32.883	6.422	1.2703	1.5915	0.7978	0.4459	0.6360	0.3070	0.6805	0.5523	0.4640
	32.883	6.422	2.6635	0.9614	0.8179	0.2963	0.5454	0.6145	0.5002	0.4733	0.7181
	27.402	8.510	0.7838	9.6910	0.8850	0.5685	0.5092	0.4885	0.4817	0.5022	0.7364
	27.402	8.510	1.1638	1.9237	0.6971	0.5927	0.4436	0.3520	0.8126	0.3047	0.6361
	27.402	8.510	1.3008	1.6384	0.7914	0.4324	0.6456	0.2704	0.7232	0.5291	0.4326
	27.402	8.510	2.8402	0.9737	0.8278	0.2498	0.5679	0.6258	0.4833	0.4589	0.7035

CHEBYSHEV RESPONSE RIPPLE = 0.5 db

n	q_0	I. L.	q_1	q_n	k_{12}	k_{23}	k_{34}	k_{45}
2	INF.	0.000	1.9497	1.9497	0.7225			
	19.497	0.915	1.9497	1.9497	0.7225			
	9.748	1.938	1.9497	1.9497	0.7225			
	6.499	3.098	1.9497	1.9497	0.7225			
	4.874	4.437	1.9497	1.9497	0.7225			
	3.899	6.021	1.9497	1.9497	0.7225			
	3.249	7.959	1.9497	1.9497	0.7225			
	2.785	10.458	1.9497	1.9497	0.7225			
3	INF.	0.000	1.8636	1.8636	0.6474	0.6474		
	37.273	0.920	1.6161	2.3388	0.6182	0.6718		
	18.636	1.958	1.5765	2.5962	0.5943	0.6884		
	12.424	3.140	1.5685	2.8158	0.5680	0.7051		
	9.318	4.508	1.5771	3.0140	0.5387	0.7224		
	7.455	6.128	1.5974	3.1949	0.5055	0.7407		
	6.212	8.109	1.6275	3.3584	0.4667	0.7606		
	5.325	10.658	1.6674	3.5011	0.4197	0.7825		
4	INF.	0.000	1.8258	1.8258	0.6482	0.5446	0.6482	
	62.337	0.923	1.4557	2.6571	0.6477	0.5338	0.6630	
	31.169	1.967	1.3966	3.1724	0.6387	0.5225	0.6804	
	20.779	3.159	1.3780	3.6570	0.6256	0.5098	0.6991	
	15.584	4.541	1.3789	4.1352	0.6093	0.4948	0.7192	
	12.467	6.177	1.3927	4.6098	0.5904	0.4763	0.7411	
	10.390	8.178	1.4169	5.0723	0.5690	0.4525	0.7656	
	8.905	10.748	1.4511	5.5035	0.5447	0.4202	0.7938	
5	INF.	0.000	1.8068	1.8068	0.6519	0.5341	0.5341	0.6519
	94.608	0.924	1.3689	2.9013	0.6742	0.5259	0.5298	0.6547
	94.608	0.924	1.7348	2.0051	0.6836	0.4673	0.6116	0.5873
	47.304	1.971	1.2999	3.6468	0.6751	0.5185	0.5236	0.6704
	47.304	1.971	1.7482	2.1211	0.6866	0.4357	0.6525	0.5500
	31.536	3.168	1.2746	4.4007	0.6697	0.5104	0.5160	0.6884
	31.536	3.168	1.7775	2.2263	0.6848	0.4071	0.6889	0.5151
	23.652	4.557	1.2694	5.1999	0.6605	0.5012	0.5062	0.7084
	23.652	4.557	1.8166	2.3279	0.6804	0.3788	0.7239	0.4796
	18.922	6.202	1.2769	6.0524	0.6484	0.4907	0.4932	0.7309
	18.922	6.202	1.8637	2.4284	0.6742	0.3491	0.7587	0.4415
	15.768	8.212	1.2943	6.9443	0.6338	0.4785	0.4751	0.7567
	15.768	8.212	1.9186	2.5290	0.6664	0.3166	0.7941	0.3990
	13.515	10.795	1.3205	7.8328	0.6171	0.4640	0.4484	0.7874
	13.515	10.795	1.9814	2.6299	0.6569	0.2794	0.8307	0.3496

CHEBYSHEV RESPONSE   RIPPLE = 0.5 db

n	q_0	I. L.	q_1	q_n	k_{12}	k_{23}	k_{34}	k_{45}	k_{56}	k_{67}
6	INF.	0.000	1.7962	1.7962	0.6547	0.5326	0.5191	0.5326	0.6547	
	134.067	0.925	1.3159	3.0894	0.6931	0.5300	0.5133	0.5306	0.6474	
	134.067	0.925	1.7077	2.0078	0.5905	0.5675	0.5535	0.4539	0.7047	
	67.033	1.974	1.2411	4.0312	0.7004	0.5263	0.5077	0.5274	0.6607	
	67.033	1.974	1.7177	2.1203	0.5585	0.5705	0.5848	0.4104	0.7193	
	44.689	3.174	1.2115	5.0368	0.6999	0.5216	0.5015	0.5229	0.6772	
	44.689	3.174	1.7423	2.2222	0.5303	0.5678	0.6158	0.3720	0.7279	
	33.517	4.566	1.2020	6.1652	0.6948	0.5160	0.4944	0.5166	0.6962	
	33.517	4.566	1.7746	2.3221	0.5033	0.5604	0.6480	0.3351	0.7334	
	26.813	6.216	1.2049	7.4439	0.6867	0.5095	0.4861	0.5073	0.7181	
	26.813	6.216	1.8127	2.4234	0.4765	0.5472	0.6828	0.2981	0.7366	
	22.344	8.232	1.2169	8.8696	0.6760	0.5020	0.4759	0.4934	0.7437	
	22.344	8.232	1.8556	2.5279	0.4493	0.5260	0.7216	0.2599	0.7380	
	19.152	10.821	1.2371	10.3842	0.6631	0.4933	0.4632	0.4713	0.7752	
	19.152	10.821	1.9032	2.6371	0.4213	0.4925	0.7665	0.2191	0.7375	
7	INF.	0.000	1.7896	1.7896	0.6566	0.5328	0.5155	0.5155	0.5328	0.6566
	180.707	0.925	1.2804	3.2369	0.7067	0.5349	0.5113	0.5118	0.5312	0.6411
	180.707	0.925	1.5180	2.3193	0.7057	0.4972	0.5530	0.4579	0.6070	0.5745
	180.707	0.925	1.7684	1.9068	0.6710	0.5482	0.4342	0.6065	0.5037	0.6211
	180.707	0.925	2.2561	1.5464	0.7124	0.4640	0.5080	0.5618	0.5333	0.6129
	90.354	1.975	1.2019	4.3436	0.7183	0.5338	0.5075	0.5076	0.5298	0.6522
	90.354	1.975	1.4780	2.5930	0.7134	0.4857	0.5636	0.4290	0.6507	0.5319
	90.354	1.975	1.7933	1.9817	0.6694	0.5521	0.3967	0.6524	0.4846	0.6013
	90.354	1.975	2.4863	1.5151	0.7357	0.4178	0.5197	0.5835	0.5198	0.5930
	60.236	3.177	1.1693	5.5753	0.7211	0.5313	0.5034	0.5028	0.5275	0.6671
	60.236	3.177	1.4689	2.8263	0.7143	0.4760	0.5703	0.4037	0.6899	0.4932
	60.236	3.177	1.8277	2.0515	0.6650	0.5543	0.3642	0.6925	0.4649	0.5837
	60.236	3.177	2.6832	1.5108	0.7525	0.3753	0.5351	0.5997	0.5045	0.5751
	45.177	4.572	1.1568	7.0210	0.7189	0.5277	0.4988	0.4972	0.5235	0.6848
	45.177	4.572	1.4750	3.0402	0.7119	0.4667	0.5748	0.3793	0.7276	0.4542
	45.177	4.572	1.8683	2.1201	0.6589	0.5558	0.3328	0.7304	0.4428	0.5670
	45.177	4.572	2.8674	1.5194	0.7658	0.3338	0.5540	0.6121	0.4873	0.5577

CHEBYSHEV RESPONSE RIPPLE = 0.5 db

n	q_0	I. L.	q_1	q_n	k_{12}	k_{23}	k_{34}	k_{45}	k_{56}	k_{67}	k_{78}
7	36.141	6.225	1.1562	8.7438	0.7134	0.5232	0.4935	0.4902	0.5171	0.7055	
	36.141	6.225	1.4909	3.2415	0.7072	0.4573	0.5776	0.3541	0.7649	0.4129	
	36.141	6.225	1.9143	2.1889	0.6516	0.5567	0.3012	0.7679	0.4170	0.5506	
	36.141	6.225	3.0465	1.5361	0.7764	0.2917	0.5776	0.6203	0.4676	0.5404	
	30.118	8.244	1.1645	10.7756	0.7053	0.5180	0.4875	0.4813	0.5066	0.7303	
	30.118	8.244	1.5147	3.4318	0.7008	0.4476	0.5787	0.3266	0.8028	0.3674	
	30.118	8.244	1.9654	2.2585	0.6432	0.5573	0.2679	0.8059	0.3856	0.5341	
	30.118	8.244	3.2237	1.5592	0.7848	0.2482	0.6081	0.6225	0.4445	0.5229	
8	INF.	0.000	1.7852	1.7852	0.6580	0.5333	0.5145	0.5106	0.5145	0.5333	0.6580
	234.528	0.925	1.2553	3.3549	0.7167	0.5392	0.5123	0.5072	0.5120	0.5313	0.6359
	234.528	0.925	1.6140	2.1046	0.6577	0.5724	0.4647	0.4880	0.6034	0.4662	0.6481
	234.528	0.925	1.4479	2.4748	0.6363	0.5191	0.5433	0.5371	0.4856	0.4800	0.7121
	234.528	0.925	1.9471	1.7207	0.5632	0.5872	0.5211	0.4658	0.5667	0.4729	0.7188
	117.264	1.976	1.1741	4.6001	0.7315	0.5398	0.5100	0.5039	0.5090	0.5310	0.6450
	117.264	1.976	1.5960	2.2595	0.6521	0.5821	0.4494	0.4668	0.6520	0.4308	0.6410
	117.264	1.976	1.3964	2.8328	0.6265	0.5037	0.5474	0.5590	0.4819	0.4361	0.7411
	117.264	1.976	2.0369	1.7295	0.5187	0.6019	0.5441	0.4277	0.5937	0.4492	0.7310
	78.176	3.179	1.1393	6.0309	0.7366	0.5387	0.5073	0.5004	0.5054	0.5301	0.6584
	78.176	3.179	1.5999	2.3894	0.6447	0.5879	0.4380	0.4445	0.6946	0.3991	0.6340
	78.176	3.179	1.3786	3.1427	0.6162	0.4891	0.5463	0.5793	0.4834	0.3934	0.7641
	78.176	3.179	2.1157	1.7516	0.4816	0.6083	0.5707	0.3917	0.6145	0.4307	0.7362
	58.632	4.576	1.1245	7.7719	0.7366	0.5364	0.5042	0.4965	0.5009	0.5279	0.6746
	58.632	4.576	1.6148	2.5081	0.6362	0.5919	0.4283	0.4201	0.7347	0.3679	0.6268
	58.632	4.576	1.3761	3.4334	0.6053	0.4744	0.5412	0.5987	0.4888	0.3501	0.7836
	58.632	4.576	2.1897	1.7816	0.4477	0.6081	0.6013	0.3559	0.6318	0.4144	0.7379
	46.906	6.230	1.1214	9.9355	0.7330	0.5332	0.5006	0.4919	0.4950	0.5237	0.6940
	46.906	6.230	1.6372	2.6206	0.6268	0.5948	0.4196	0.3926	0.7739	0.3354	0.6193
	46.906	6.230	1.3828	3.7148	0.5938	0.4593	0.5317	0.6177	0.4989	0.3050	0.8003
	46.906	6.230	2.2612	1.8177	0.4151	0.6004	0.6371	0.3190	0.6464	0.3994	0.7374
	39.088	8.252	1.1267	12.6150	0.7267	0.5293	0.4965	0.4864	0.4872	0.5160	0.7177
	39.088	8.252	1.6657	2.7289	0.6166	0.5967	0.4117	0.3604	0.8130	0.3002	0.6113
	39.088	8.252	1.3963	3.9903	0.5817	0.4433	0.5162	0.6355	0.5158	0.2573	0.8145
	39.088	8.252	2.3310	1.8595	0.3831	0.5824	0.6800	0.2801	0.6586	0.3852	0.7352

3-db DOWN k AND q VALUES

MAXIMALLY FLAT DELAY

n	q_0	I. L.	q_1	q_n	k_{12}	k_{23}	k_{34}	k_{45}
2	INF.	0.000	0.5755	2.1478	0.8995			
	9.078	1.335	0.6100	1.7737	0.8329			
	4.539	2.734	0.6523	1.4920	0.7685			
	3.026	4.187	0.7093	1.2606	0.7068			
	2.269	5.680	0.8138	1.0263	0.6486			
	1.816	7.270	0.9078	0.9078	0.6360			
	1.513	9.208	0.9078	0.9078	0.6360			
	1.297	11.707	0.9078	0.9078	0.6360			
3	INF.	0.000	0.3374	2.2034	1.7475	0.6838		
	9.547	1.627	0.3625	1.8051	1.5994	0.6580		
	4.774	3.311	0.3921	1.5211	1.4537	0.6359		
	3.182	5.054	0.4275	1.3085	1.3119	0.6185		
	2.387	6.853	0.4706	1.1444	1.1766	0.6076		
	1.909	8.709	0.5236	1.0164	1.0512	0.6056		
	1.591	10.631	0.5893	0.9157	0.9375	0.6176		
4	INF.	0.000	0.2334	2.2404	2.5239	1.1725	0.6424	
	INF.	0.000	0.3672	0.4982	1.8260	0.6935	0.8880	
	10.048	1.855	0.2508	1.8384	2.3161	1.1086	0.6292	
	10.048	1.855	0.3891	0.5098	1.7120	0.6794	0.8512	
	5.024	3.764	0.2711	1.5536	2.1095	1.0481	0.6189	
	5.024	3.764	0.4139	0.5218	1.6009	0.6691	0.8155	
	3.349	5.728	0.2952	1.3411	1.9046	0.9922	0.6122	
	3.349	5.728	0.4420	0.5344	1.4932	0.6635	0.7809	
	2.512	7.744	0.3242	1.1761	1.7022	0.9420	0.6099	
	2.512	7.744	0.4740	0.5480	1.3891	0.6644	0.7474	
	2.010	9.813	0.3601	1.0435	1.5029	0.8994	0.6133	
	2.010	9.813	0.5104	0.5631	1.2878	0.6742	0.7157	
	1.675	11.930	0.4058	0.9330	1.3069	0.8666	0.6243	
	1.675	11.930	0.5514	0.5805	1.1860	0.6973	0.6880	
5	INF.	0.000	0.1743	2.2582	3.3629	1.5659	1.0580	0.6313
	INF.	0.000	0.3938	0.2747	1.9094	0.7477	0.6500	1.9872
	10.442	2.045	0.1868	1.8595	3.1024	1.4712	1.0204	0.6239
	10.442	2.045	0.4141	0.2876	1.8152	0.7433	0.6306	1.8593
	5.221	4.143	0.2012	1.5767	2.8425	1.3788	0.9864	0.6189
	5.221	4.143	0.4366	0.3017	1.7229	0.7432	0.6127	1.7326

3-db DOWN k AND q VALUES

MAXIMALLY FLAT DELAY

n	q_0	I. L.	q_1	q_n	k_{12}	k_{23}	k_{34}	k_{45}	k_{56}
	3.481	6.291	0.2181	1.3656	2.5833	1.2893	0.9566	0.6167	
	3.481	6.291	0.4618	0.3173	1.6322	0.7484	0.5967	1.6072	
5	2.610	8.491	0.2382	1.2019	2.3253	1.2034	0.9322	0.6179	
	2.610	8.491	0.4899	0.3346	1.5425	0.7604	0.5833	1.4831	
	2.088	10.741	0.2625	1.0711	2.0685	1.1217	0.9146	0.6233	
	2.088	10.741	0.5213	0.3540	1.4520	0.7814	0.5737	1.3610	
	INF.	0.000	0.1365	2.2645	4.2787	1.9773	1.3537	1.0260	0.6300
	INF.	0.000	0.1803	0.4499	3.3375	1.6154	1.1753	0.6434	0.9302
	INF.	0.000	0.2385	0.2798	2.5687	1.3884	0.6861	0.6788	2.0706
	INF.	0.000	0.4145	0.1867	1.9999	0.8105	0.6006	1.2525	3.0377
	10.745	2.210	0.1457	1.8714	3.9686	1.8556	1.2940	1.0019	0.6257
	10.745	2.210	0.1921	0.4566	3.1054	1.5310	1.1474	0.6409	0.9017
	10.745	2.210	0.2525	0.2911	2.4015	1.3478	0.6833	0.6631	1.9581
	10.745	2.210	0.4338	0.1964	1.9180	0.8139	0.5887	1.1872	2.8455
	5.373	4.470	0.1564	1.5917	3.6587	1.7355	1.2370	0.9811	0.6233
	5.373	4.470	0.2056	0.4632	2.8744	1.4481	1.1230	0.6412	0.8739
	5.373	4.470	0.2682	0.3034	2.2348	1.3106	0.6834	0.6486	1.8471
	5.373	4.470	0.4551	0.2072	1.8370	0.8220	0.5785	1.1234	2.6544
6	3.582	6.780	0.1687	1.3826	3.3493	1.6171	1.1831	0.9643	0.6233
	3.582	6.780	0.2211	0.4698	2.6448	1.3668	1.1026	0.6451	0.8469
	3.582	6.780	0.2862	0.3168	2.0686	1.2773	0.6870	0.6357	1.7380
	3.582	6.780	0.4787	0.2192	1.7561	0.8360	0.5707	1.0612	2.4644
	2.686	9.139	0.1832	1.2201	3.0403	1.5008	1.1332	0.9524	0.6259
	2.686	9.139	0.2392	0.4765	2.4168	1.2871	1.0869	0.6533	0.8208
	2.686	9.139	0.3067	0.3313	1.9025	1.2485	0.6949	0.6247	1.6310
	2.686	9.139	0.5048	0.2327	1.6743	0.8572	0.5659	1.0010	2.2757
	2.149	11.546	0.2004	1.0902	2.7318	1.3868	1.0878	0.9467	0.6317
	2.149	11.546	0.2606	0.4832	2.1908	1.2089	1.0768	0.6676	0.7957
	2.149	11.546	0.3305	0.3471	1.7353	1.2249	0.7087	0.6162	1.5265
	2.149	11.546	0.5338	0.2479	1.5893	0.8878	0.5656	0.9433	2.0884

MAXIMALLY FLAT DELAY

n	q_0	I. L.	q_1	q_n	k_{12}	k_{23}	k_{34}	k_{45}	k_{56}	k_{67}
	INF.	0.000	0.1106	2.2659	5.2682	2.4179	1.6474	1.2803	1.0204	0.6319
	INF.	0.000	0.1865	0.2424	3.3252	1.6591	1.2931	0.6953	0.6737	2.2028
	INF.	0.000	0.1664	0.2876	3.5280	1.7078	1.2521	0.6537	0.7142	2.1500
	INF.	0.000	0.4301	0.1396	2.0942	0.8750	0.5867	1.0730	1.7892	4.1096
	10.991	2.357	0.1177	1.8788	4.9104	2.2716	1.5679	1.2390	1.0044	0.6294
	10.991	2.357	0.1979	0.2513	3.1151	1.5810	1.2726	0.6996	0.6571	2.0850
	10.991	2.357	0.1762	0.2979	3.2960	1.6292	1.2287	0.6508	0.7019	2.0502
	10.991	2.357	0.4489	0.1470	2.0205	0.8851	0.5794	1.0282	1.6935	3.8616
	5.495	4.762	0.1258	1.6024	4.5526	2.1264	1.4903	1.2006	0.9915	0.6285
	5.495	4.762	0.2109	0.2608	2.9062	1.5035	1.2552	0.7077	0.6416	1.9683
	5.495	4.762	0.1873	0.3090	3.0640	1.5526	1.2089	0.6498	0.6909	1.9520
7	5.495	4.762	0.4696	0.1551	1.9466	0.9002	0.5741	0.9849	1.5991	3.6147
	3.664	7.215	0.1351	1.3952	4.1951	1.9822	1.4148	1.1655	0.9822	0.6295
	3.664	7.215	0.2257	0.2709	2.6989	1.4265	1.2409	0.7207	0.6272	1.8528
	3.664	7.215	0.1998	0.3210	2.8319	1.4781	1.1936	0.6509	0.6815	1.8556
	3.664	7.215	0.4923	0.1642	1.8717	0.9215	0.5712	0.9435	1.5061	3.3688
	2.748	9.715	0.1459	1.2339	3.8377	1.8393	1.3418	1.1344	0.9774	0.6326
	2.748	9.715	0.2428	0.2818	2.4933	1.3498	1.2301	0.7398	0.6145	1.7383
	2.748	9.715	0.2141	0.3338	2.5994	1.4057	1.1837	0.6545	0.6737	1.7612
	2.748	9.715	0.5174	0.1744	1.7943	0.9506	0.5717	0.9041	1.4148	3.1240
	2.198	12.262	0.1586	1.1049	3.4805	1.6976	1.2715	1.1082	0.9782	0.6382
	2.198	12.262	0.2628	0.2936	2.2897	1.2727	1.2226	0.7672	0.6039	1.6250
	2.198	12.262	0.2307	0.3477	2.3657	1.3353	1.1808	0.6616	0.6682	1.6691
	2.198	12.262	0.5452	0.1860	1.7120	0.9897	0.5769	0.8674	1.3255	2.8804

3-db DOWN k AND q VALUES

MAXIMALLY FLAT DELAY

n	q_0	I. L.	q_1	q_n	k_{12}	k_{23}	k_{34}	k_{45}	k_{56}	k_{67}	k_{78}
	INF.	0.000	0.0919	2.2656	6.3258	2.8881	1.9547	1.5189	1.2550	1.0246	0.6347
	INF.	0.000	0.1124	0.4123	5.2298	2.4495	1.7089	1.3508	1.1312	0.6408	0.9742
	INF.	0.000	0.1391	0.2418	4.2836	2.0791	1.4838	1.2462	0.6780	0.6973	2.2856
	INF.	0.000	0.1922	0.1634	3.3296	1.6987	1.4074	0.7575	0.6136	1.3565	3.4255
	INF.	0.000	0.1259	0.2957	4.6152	2.1309	1.5112	1.1853	0.6339	0.7525	2.2277
	INF.	0.000	0.1679	0.1863	3.5556	1.7616	1.3592	0.6968	0.6498	1.4195	3.1366
	INF.	0.000	0.2355	0.1413	2.6540	1.6002	0.7776	0.6085	1.1597	1.8453	4.1157
	INF.	0.000	0.4420	0.1104	2.1899	0.9390	0.5850	1.0006	1.4723	2.3156	5.2219
	11.200	2.489	0.0975	1.8842	5.9216	2.7188	1.8573	1.4616	1.2253	1.0138	0.6333
	11.200	2.489	0.1192	0.4169	4.8959	2.3142	1.6364	1.3128	1.1222	0.6417	0.9493
	11.200	2.489	0.1474	0.2499	4.0137	1.9766	1.4331	1.2361	0.6818	0.6825	2.1762
	11.200	2.489	0.2034	0.1704	3.1367	1.6246	1.3923	0.7691	0.6027	1.2931	3.2425
	11.200	2.489	0.1332	0.3052	4.3253	2.0181	1.4629	1.1695	0.6307	0.7436	2.1384
	11.200	2.489	0.1772	0.1944	3.3376	1.6876	1.3420	0.6980	0.6410	1.3675	3.8868
	11.200	2.489	0.2476	0.1482	2.5022	1.5729	0.7831	0.6019	1.1206	1.7580	3.8868
	11.200	2.489	0.4606	0.1161	2.1216	0.9547	0.5810	0.9670	1.4066	2.1941	4.9228
8	5.600	5.025	0.1038	1.6109	5.5175	2.5502	1.7613	1.4064	1.1986	1.0059	0.6332
	5.600	5.025	0.1269	0.4215	4.5621	2.1797	1.5655	1.2769	1.1165	0.6448	0.9249
	5.600	5.025	0.1567	0.2584	3.7441	1.8753	1.3834	1.2295	0.6885	0.6685	2.0681
	5.600	5.025	0.2161	0.1779	2.9451	1.5503	1.3795	0.7854	0.5931	1.2307	3.0608
	5.600	5.025	0.1412	0.3155	4.0355	1.9061	1.4177	1.1569	0.6287	0.7361	2.0506
	5.600	5.025	0.1876	0.2032	3.1198	1.6149	1.3286	0.7015	0.6334	1.3172	2.8092
	5.600	5.025	0.2610	0.1558	2.3498	1.5486	0.7919	0.5966	1.0832	1.6719	3.6590
	5.600	5.025	0.4809	0.1224	2.0524	0.9759	0.5789	0.9350	1.3422	2.0737	4.6245
	3.733	7.607	0.1111	1.4054	5.1134	2.3823	1.6667	1.3536	1.1753	1.0013	0.6346
	3.733	7.607	0.1357	0.4260	4.2285	2.0461	1.4965	1.2432	1.1149	0.6504	0.9010
	3.733	7.607	0.1674	0.2674	3.4747	1.7754	1.3347	1.2271	0.6988	0.6554	1.9614
	3.733	7.607	0.2305	0.1860	2.7549	1.4756	1.3688	0.8078	0.5853	1.1694	2.8802
	3.733	7.607	0.1503	0.3265	2.9020	1.7948	1.3759	1.1483	0.6279	0.7302	1.9643
	3.733	7.607	0.1994	0.2128	2.9020	1.5432	1.3195	0.7078	0.6271	1.2687	2.6471
	3.733	7.607	0.2760	0.1642	2.1963	1.5277	0.8049	0.5931	1.0476	1.5871	3.4322
	3.733	7.607	0.5033	0.1294	1.9812	1.0038	0.5795	0.9047	1.2795	1.9546	4.3272
	2.800	10.236	0.1194	1.2452	4.7093	2.2152	1.5736	1.3036	1.1561	1.0007	0.6378
	2.800	10.236	0.1459	0.4304	3.8954	1.9133	1.4294	1.2120	1.1182	0.6591	0.8776
	2.800	10.236	0.1607	0.3382	3.4557	1.6840	1.3382	1.1445	0.6286	0.7261	1.8796
	2.800	10.236	0.2127	0.2234	2.6838	1.4724	1.3158	0.7176	0.6224	1.2223	2.4862
	2.800	10.236	0.1796	0.2769	3.2058	1.6769	1.2867	1.2296	0.7137	0.6436	1.8561
	2.800	10.236	0.2470	0.1949	2.5665	1.4002	1.3597	0.8381	0.5797	1.1093	2.7009
	2.800	10.236	0.2928	0.1735	2.0407	1.5105	0.8233	0.5916	1.0142	1.5038	3.2067
	2.800	10.236	0.5279	0.1373	1.9064	1.0399	0.5836	0.8766	1.2185	1.8368	4.0308

3-db DOWN k AND q VALUES

LINEAR PHASE WITH EQUIRIPPLE ERROR   PHASE ERROR = 0.05

n	q_0	I. L.	q_1	q_n	k_{12}	k_{23}	k_{34}
2	INF.	0.000	0.6480	2.1085	0.8555		
	9.914	1.230	0.6928	1.7421	0.7970		
	4.957	2.502	0.7526	1.4522	0.7407		
	3.305	3.802	0.8544	1.1807	0.6873		
	2.478	5.155	0.9914	0.9914	0.6680		
	1.983	6.738	0.9914	0.9914	0.6680		
	1.652	8.676	0.9914	0.9914	0.6680		
	1.416	11.175	0.9914	0.9914	0.6680		
3	INF.	0.000	0.4328	2.2542	1.4886	0.6523	
	11.708	1.405	0.4661	1.9113	1.3820	0.6371	
	5.854	2.856	0.5047	1.6618	1.2814	0.6251	
	3.903	4.357	0.5494	1.4753	1.1883	0.6170	
	2.927	5.915	0.6015	1.3337	1.1040	0.6137	
	2.342	7.546	0.6620	1.2246	1.0289	0.6169	
	1.951	9.275	0.7337	1.1380	0.9605	0.6304	
	1.673	11.148	0.8236	1.0616	0.8897	0.6648	
4	INF.	0.000	0.3363	2.2459	1.9325	1.0483	0.6242
	INF.	0.000	0.4934	0.7182	1.6320	0.7181	0.7391
	13.427	1.477	0.3640	1.9126	1.7921	1.0121	0.6182
	13.427	1.477	0.5189	0.7446	1.5625	0.7227	0.7091
	6.713	2.981	0.3973	1.6544	1.6553	0.9784	0.6138
	6.713	2.981	0.5457	0.7761	1.4945	0.7335	0.6798
	4.476	4.510	0.4388	1.4420	1.5230	0.9467	0.6109
	4.476	4.510	0.5725	0.8159	1.4270	0.7531	0.6520
	3.357	6.058	0.4943	1.2494	1.3965	0.9142	0.6095
	3.357	6.058	0.5954	0.8744	1.3574	0.7873	0.6271
	2.685	7.618	0.5917	1.0159	1.2815	0.8569	0.6139
	2.238	9.255	0.6541	1.0035	1.1964	0.8562	0.6347
	1.918	11.024	0.7313	0.9911	1.1074	0.8673	0.6630

3-db DOWN k AND q VALUES

LINEAR PHASE WITH EQUIRIPPLE ERROR   PHASE ERROR = 0.05°

n	q_0	I. L.	q_1	q_n	k_{12}	k_{23}	k_{34}	k_{45}	k_{56}
	INF.	0.000	0.2751	2.2873	2.3573	1.3112	1.0096	0.6295	
	INF.	0.000	0.5470	0.4456	1.7988	0.8484	0.5839	1.3717	
	14.856	1.537	0.2969	1.9936	2.1945	1.2625	0.9959	0.6279	
	14.856	1.537	0.5725	0.4709	1.7425	0.8610	0.5764	1.2908	
	7.428	3.104	0.3222	1.7705	2.0362	1.2168	0.9855	0.6277	
	7.428	3.104	0.6009	0.4990	1.6858	0.8769	0.5710	1.2134	
	4.952	4.703	0.3521	1.5975	1.8834	1.1744	0.9788	0.6292	
5	4.952	4.703	0.6328	0.5302	1.6281	0.8965	0.5681	1.1405	
	3.714	6.337	0.3875	1.4619	1.7370	1.1356	0.9767	0.6329	
	3.714	6.337	0.6696	0.5647	1.5684	0.9201	0.5685	1.0731	
	2.971	8.013	0.4302	1.3553	1.5976	1.1006	0.9803	0.6393	
	2.971	8.013	0.7127	0.6026	1.5055	0.9487	0.5729	1.0126	
	2.476	9.742	0.4822	1.2714	1.4647	1.0695	0.9912	0.6498	
	2.476	9.742	0.7646	0.6440	1.4369	0.9841	0.5831	0.9599	
	2.122	11.541	0.5468	1.2051	1.3351	1.0419	1.0129	0.6671	
	2.122	11.541	0.8285	0.6889	1.3564	1.0302	0.6030	0.9163	
	INF.	0.000	0.2374	2.2448	2.7277	1.5267	1.2274	1.0113	0.6317
	INF.	0.000	0.2952	0.7870	2.4035	1.3859	1.2352	0.7356	0.7124
	INF.	0.000	0.3967	0.4679	1.9929	1.3787	0.6832	0.6609	1.5532
	INF.	0.000	0.5898	0.3376	1.9556	1.0009	0.5777	0.9428	1.8719
	16.255	1.559	0.2563	1.9598	2.5426	1.4670	1.2059	1.0039	0.6305
6	16.255	1.559	0.3146	0.8107	2.2776	1.3432	1.2293	0.7531	0.6905
	16.255	1.559	0.4244	0.4865	1.9025	1.3665	0.6792	0.6614	1.4961
	16.255	1.559	0.6124	0.3599	1.9041	1.0277	0.5781	0.9103	1.7563
	8.128	3.138	0.2787	1.7309	2.3606	1.4097	1.1875	0.9977	0.6298
	8.128	3.138	0.3363	0.8387	2.1550	1.3026	1.2237	0.7763	0.6694
	8.128	3.138	0.4571	0.5056	1.8146	1.3562	0.6750	0.6651	1.4412
	8.128	3.138	0.6356	0.3857	1.8492	1.0613	0.5810	0.8790	1.6432

LINEAR PHASE WITH EQUIRIPPLE ERROR   PHASE ERROR = 0.05°

n	q_0	I. L.	q_1	q_n	k_{12}	k_{23}	k_{34}	k_{45}	k_{56}	k_{67}
6	5.418	4.733	0.3059	1.5383	2.1820	1.3545	1.1730	0.9921	0.6292	
	5.418	4.733	0.3604	0.8734	2.0357	1.2653	1.2174	0.8070	0.6498	
	5.418	4.733	0.4967	0.5246	1.7300	1.3476	0.6705	0.6735	1.3884	
	5.418	4.733	0.6586	0.4165	1.7884	1.1042	0.5875	0.8482	1.5328	
	4.064	6.343	0.3401	1.3644	2.0071	1.3006	1.1642	0.9855	0.6282	
	4.064	6.343	0.3864	0.9211	1.9196	1.2333	1.2082	0.8489	0.6329	
	4.064	6.343	0.5474	0.5415	1.6504	1.3385	0.6648	0.6895	1.3380	
	4.064	6.343	0.6784	0.4547	1.7172	1.1609	0.5998	0.8164	1.4253	
	3.251	7.962	0.3890	1.1671	1.8329	1.2411	1.1679	0.9665	0.6247	
	3.251	7.962	0.4092	1.0163	1.8088	1.2148	1.1889	0.9159	0.6219	
	3.251	7.962	0.6258	0.5466	1.5858	1.3146	0.6521	0.7259	1.2931	
	3.251	7.962	0.6799	0.5111	1.6204	1.2495	0.6263	0.7727	1.3178	
	2.709	9.609	0.4443	1.0744	1.6848	1.1880	1.1810	0.9577	0.6357	
	2.709	9.609	0.7054	0.5669	1.5250	1.3053	0.6698	0.7410	1.2383	
	2.322	11.335	0.5011	1.0641	1.5509	1.1483	1.1913	0.9808	0.6530	
	2.322	11.335	0.7697	0.6112	1.4350	1.3354	0.7135	0.7358	1.1783	
7	INF.	0.000	0.2085	2.2845	3.0973	1.7340	1.4137	1.2328	1.0350	0.6405
	INF.	0.000	0.3157	0.4841	2.4902	1.4418	1.4463	0.9271	0.5792	1.2602
	INF.	0.000	0.3107	0.4963	2.3076	1.5738	1.1950	0.6106	0.7660	1.7390
	INF.	0.000	0.6285	0.2745	2.1234	1.1457	0.5988	0.8928	1.2355	2.3237
	17.461	1.591	0.2245	2.0311	2.8961	1.6663	1.3861	1.2218	1.0333	0.6408
	17.461	1.591	0.3355	0.5085	2.3839	1.3999	1.4387	0.9576	0.5773	1.1937
	17.461	1.591	0.3329	0.5146	2.1841	1.5550	1.1846	0.6061	0.7699	1.6911
	17.461	1.591	0.6537	0.2926	2.0756	1.1750	0.6033	0.8779	1.1919	2.1858
	8.730	3.203	0.2430	1.8324	2.6990	1.6012	1.3614	1.2128	1.0334	0.6419
	8.730	3.203	0.3581	0.5352	2.2791	1.3590	1.4298	0.9923	0.5777	1.1298
	8.730	3.203	0.3585	0.5344	2.0636	1.5399	1.1761	0.6023	0.7751	1.6437
	8.730	3.203	0.6817	0.3131	2.0249	1.2082	0.6100	0.8649	1.1503	2.0517
	5.820	4.839	0.2647	1.6745	2.5068	1.5390	1.3401	1.2057	1.0357	0.6440
	5.820	4.839	0.3842	0.5645	2.1758	1.3193	1.4193	1.0315	0.5810	1.0694
	5.820	4.839	0.3881	0.5562	1.9460	1.5286	1.1703	0.5995	0.7813	1.5968
	5.820	4.839	0.7131	0.3364	1.9706	1.2455	0.6194	0.8544	1.1110	1.9223

LINEAR PHASE WITH EQUIRIPPLE ERROR   PHASE ERROR = 0.05°

n	q_0	I. L.	q_1	q_n	k_{12}	k_{23}	k_{34}	k_{45}	k_{56}	k_{67}	k_{78}
7	4.365	6.502	0.2905	1.5481	2.3204	1.4797	1.3227	1.2011	1.0406	0.6473	
	4.365	6.502	0.4146	0.5964	2.0734	1.2812	1.4069	1.0752	0.5876	1.0133	
	4.365	6.502	0.4227	0.5805	1.8312	1.5213	1.1685	0.5982	0.7886	1.5503	
	4.365	6.502	0.7491	0.3632	1.9112	1.2877	0.6323	0.8467	1.0740	1.7985	
	3.492	8.195	0.3215	1.4469	2.1402	1.4230	1.3099	1.1996	1.0488	0.6521	
	3.492	8.195	0.4509	0.6312	1.9714	1.2450	1.3927	1.1238	0.5980	0.9621	
	3.492	8.195	0.4636	0.6080	1.7179	1.5180	1.1728	0.5989	0.7966	1.5044	
	3.492	8.195	0.7912	0.3940	1.8448	1.3356	0.6500	0.8424	1.0396	1.6815	
	2.910	9.926	0.3592	1.3663	1.9663	1.3684	1.3023	1.2023	1.0612	0.6589	
	2.910	9.926	0.4950	0.6687	1.8684	1.2108	1.3766	1.1782	0.6134	0.9167	
	2.910	9.926	0.5122	0.6397	1.6035	1.5182	1.1868	0.6026	0.8057	1.4592	
	2.910	9.926	0.8413	0.4298	1.7675	1.3913	0.6749	0.8421	1.0079	1.5722	
	2.494	11.705	0.4063	1.3025	1.7970	1.3140	1.3001	1.2122	1.0800	0.6685	
	2.494	11.705	0.5498	0.7090	1.7618	1.1781	1.3589	1.2414	0.6353	0.8777	
	2.494	11.705	0.5711	0.6765	1.4821	1.5197	1.2173	0.6117	0.8162	1.4150	
	2.494	11.705	0.9022	0.4715	1.6713	1.4583	0.7118	0.8469	0.9796	1.4713	
8	INF.	0.000	0.1891	2.2415	3.4121	1.9104	1.5673	1.3971	1.2504	1.0433	0.6397
	INF.	0.000	0.2189	0.8566	3.1090	1.8034	1.5506	1.3164	1.2832	0.7803	0.6915
	INF.	0.000	0.2701	0.4918	2.7445	1.6959	1.4260	1.4459	0.7572	0.6358	1.4435
	INF.	0.000	0.3354	0.3630	2.5854	1.4842	1.6021	1.1594	0.5958	0.8675	1.7334
	INF.	0.000	0.2614	0.5236	2.6287	1.7049	1.4627	1.0380	0.5830	0.8942	1.9082
	INF.	0.000	0.3221	0.3800	2.3993	1.6415	1.4216	0.7387	0.6255	1.2344	2.0052
	INF.	0.000	0.4467	0.2860	2.1377	1.7137	0.7562	0.6672	1.1818	1.3372	2.4016
	INF.	0.000	0.6591	0.2371	2.2644	1.2996	0.6187	0.8859	1.1297	1.4686	2.7026
	18.724	1.597	0.2036	1.9886	3.1933	1.8350	1.5344	1.3819	1.2460	1.0413	0.6391
	18.724	1.597	0.2337	0.8803	2.9409	1.7427	1.5300	1.3047	1.2868	0.8024	0.6739
	18.724	1.597	0.2897	0.5093	2.6101	1.6488	1.4050	1.4589	0.7629	0.6367	1.3947
	18.724	1.597	0.3548	0.3851	2.4942	1.4487	1.5815	1.2088	0.6053	0.8402	1.6341
	18.724	1.597	0.2809	0.5392	2.4770	1.6735	1.4552	1.0203	0.5801	0.9092	1.8659
	18.724	1.597	0.3416	0.4020	2.2950	1.6218	1.4231	0.7481	0.6197	1.2159	1.9278
	18.724	1.597	0.4762	0.3016	2.0687	1.7145	0.7524	0.6778	1.0939	1.2996	2.2917
	18.724	1.597	0.6818	0.2533	2.2138	1.3416	0.6261	0.8759	1.1074	1.4153	2.5423
	9.362	3.209	0.2206	1.7794	2.9774	1.7617	1.5042	1.3683	1.2436	1.0394	0.6385
	9.362	3.209	0.2504	0.9084	2.7765	1.6837	1.5111	1.2955	1.2897	0.8294	0.6572
	9.362	3.209	0.3128	0.5270	2.4789	1.6018	1.3846	1.4744	0.7688	0.6401	1.3477
	9.362	3.209	0.3762	0.4105	2.4036	1.4169	1.5546	1.2648	0.6194	0.8139	1.5366

LINEAR PHASE WITH EQUIRIPPLE ERROR   PHASE ERROR = 0.05°

n	q_0	I. L.	q_1	q_n	k_{12}	k_{23}	k_{34}	k_{45}	k_{56}	k_{67}	k_{78}
	9.362	3.209	0.3037	0.5550	2.3277	1.6461	1.4503	1.0014	0.5780	0.9283	1.8227
	9.362	3.209	0.3631	0.4272	2.1918	1.6048	1.4268	0.7610	0.6136	1.1975	1.8532
	9.362	3.209	0.5109	0.3187	2.0010	1.7172	0.7486	0.6919	1.0864	1.2638	2.1851
	9.362	3.209	0.7050	0.2720	2.1568	1.3913	0.6361	0.8663	1.0869	1.3636	2.3848
	6.241	4.834	0.2410	1.5989	2.7646	1.6904	1.4776	1.3561	1.2438	1.0371	0.6374
	6.241	4.834	0.2692	0.9434	2.6162	1.6268	1.4935	1.2903	1.2910	0.8628	0.6421
	6.241	4.834	0.3404	0.5444	2.3513	1.5542	1.3655	1.4922	0.7742	0.6472	1.3028
	6.241	4.834	0.3995	0.4403	2.3125	1.3910	1.5192	1.3282	0.6397	0.7882	1.4410
	6.241	4.834	0.3311	0.5702	2.1807	1.6227	1.4482	0.9806	0.5769	0.9533	1.7778
	6.241	4.834	0.3867	0.4570	2.0891	1.5909	1.4329	0.7791	0.6071	1.1781	1.7821
	6.241	4.834	0.5528	0.3372	1.9355	1.7210	0.7450	0.7111	1.0786	1.2305	2.0824
	6.241	4.834	0.7275	0.2941	2.0905	1.4515	0.6500	0.8566	1.0687	1.3133	2.2301
	4.681	6.469	0.2663	1.4321	2.5547	1.6203	1.4556	1.3447	1.2477	1.0326	0.6353
	4.681	6.469	0.2902	0.9918	2.4606	1.5733	1.4760	1.2914	1.2891	0.9053	0.6296
	4.681	6.469	0.3750	0.5596	2.2285	1.5035	1.3498	1.5115	0.7774	0.6602	1.2606
	4.681	6.469	0.4243	0.4769	2.2190	1.3753	1.4713	1.4004	0.6701	0.7618	1.3473
8	4.681	6.469	0.3651	0.5830	2.0361	1.6029	1.4498	0.9558	0.5774	0.9882	1.7294
	4.681	6.469	0.4117	0.4938	1.9857	1.5809	1.4422	0.8060	0.5999	1.1549	1.7159
	4.681	6.469	0.6060	0.3567	1.8741	1.7237	0.7413	0.7380	1.0693	1.2014	1.9843
	4.681	6.469	0.7462	0.3212	2.0083	1.5273	0.6703	0.8453	1.0536	1.2633	2.0784
	3.745	8.111	0.3007	1.2372	2.3429	1.5462	1.4427	1.3285	1.2608	1.0149	0.6293
	3.745	8.111	0.3110	1.0892	2.3147	1.5284	1.4538	1.3069	1.2783	0.9691	0.6234
	3.745	8.111	0.4241	0.5633	2.1136	1.4361	1.3498	1.5251	0.7672	0.6895	1.2254
	3.745	8.111	0.4448	0.5305	2.1178	1.3857	1.3969	1.4889	0.7246	0.7260	1.2511
	3.745	8.111	0.4133	0.5834	1.8920	1.5845	1.4567	0.9133	0.5821	1.0502	1.6716
	3.745	8.111	0.4330	0.5483	1.8804	1.5767	1.4554	0.8579	0.5903	1.1119	1.6602
	3.745	8.111	0.6885	0.3730	1.8316	1.7090	0.7324	0.7836	1.0543	1.1838	1.8964
	3.745	8.111	0.7448	0.3583	1.8887	1.6385	0.7058	0.8229	1.0452	1.2074	1.9249
	3.121	9.774	0.3401	1.1463	2.1555	1.4774	1.4368	1.3177	1.2801	1.0060	0.6361
	3.121	9.774	0.4776	0.5817	2.0107	1.3792	1.3449	1.5453	0.7905	0.7017	1.1797
	3.121	9.774	0.4660	0.5999	1.7625	1.5677	1.4778	0.8860	0.5992	1.0815	1.6159
	3.121	9.774	0.7697	0.3978	1.7763	1.7141	0.7566	0.8092	1.0439	1.1589	1.7975
	2.675	11.496	0.3828	1.1364	1.9858	1.4160	1.4291	1.3234	1.2975	1.0224	0.6486
	2.675	11.496	0.5309	0.6215	1.9026	1.3492	1.3150	1.5864	0.8473	0.6978	1.1267
	2.675	11.496	0.5186	0.6393	1.6331	1.5507	1.5152	0.8930	0.6156	1.0828	1.5679
	2.675	11.496	0.8337	0.4361	1.6732	1.7624	0.8102	0.8179	1.0417	1.1241	1.6914

3-db DOWN k AND q VALUES

LINEAR PHASE WITH EQUIRIPPLE ERROR   PHASE ERROR = 0.5°

n	q_0	I. L.	q_1	q_n	k_{12}	k_{23}	k_{34}	k_{45}
2	INF.	0.000	0.8245	1.9800	0.7827			
	11.642	1.056	0.9142	1.6023	0.7365			
	5.821	2.124	1.1642	1.1642	0.6981			
	3.881	3.284	1.1642	1.1642	0.6981			
	2.910	4.622	1.1642	1.1642	0.6981			
	2.328	6.206	1.1642	1.1642	0.6981			
	1.940	8.144	1.1642	1.1642	0.6981			
	1.663	10.643	1.1642	1.1642	0.6981			
3	INF.	0.000	0.5534	2.4250	1.3299	0.6353		
	14.350	1.232	0.5948	2.1350	1.2599	0.6266		
	7.175	2.511	0.6416	1.9181	1.1958	0.6203		
	4.783	3.848	0.6946	1.7522	1.1379	0.6169		
	3.588	5.258	0.7551	1.6227	1.0860	0.6170		
	2.870	6.767	0.8246	1.5191	1.0392	0.6222		
	2.392	8.413	0.9061	1.4333	0.9951	0.6355		
	2.050	10.262	1.0054	1.3569	0.9476	0.6645		
4	INF.	0.000	0.4526	2.0504	1.6652	0.9953	0.6204	
	INF.	0.000	0.5808	1.0256	1.5752	0.7974	0.6559	
	16.565	1.202	0.5019	1.7129	1.5651	0.9587	0.6160	
	16.565	1.202	0.5949	1.1171	1.5227	0.8289	0.6332	
	8.282	2.410	0.5854	1.3383	1.4713	0.8919	0.6181	
	5.522	3.670	0.6329	1.3251	1.4079	0.8814	0.6294	
	4.141	5.011	0.6884	1.3133	1.3479	0.8746	0.6409	
	3.313	6.450	0.7542	1.3032	1.2905	0.8728	0.6532	
	2.761	8.017	0.8331	1.2951	1.2336	0.8781	0.6680	
	2.366	9.760	0.9299	1.2880	1.1717	0.8943	0.6890	
	2.071	11.764	1.0542	1.2781	1.0867	0.9298	0.7302	
5	INF.	0.000	0.3658	2.4446	2.0098	1.2462	1.0195	0.6360
	INF.	0.000	0.6637	0.6111	1.7785	0.9188	0.5761	1.1621
	18.478	1.295	0.3954	2.2150	1.8996	1.2142	1.0158	0.6368
	18.478	1.295	0.6970	0.6469	1.7355	0.9314	0.5739	1.1115
	9.239	2.623	0.4297	2.0367	1.7949	1.1842	1.0149	0.6388
	9.239	2.623	0.7351	0.6859	1.6924	0.9455	0.5732	1.0652
	6.159	3.989	0.4698	1.8966	1.6958	1.1565	1.0169	0.6423
	6.159	3.989	0.7793	0.7284	1.6491	0.9616	0.5742	1.0233

3-db DOWN k AND q VALUES

LINEAR PHASE WITH EQUIRIPPLE ERROR   **PHASE ERROR = 0.5°**

n	q_0	I.L.	q_1	q_n	k_{12}	k_{23}	k_{34}	k_{45}	k_{56}
5	4.620	5.405	0.5172	1.7854	1.6024	1.1314	1.0222	0.6476	
	4.620	5.405	0.8313	0.7748	1.6049	0.9804	0.5772	0.9860	
	3.696	6.886	0.5741	1.6966	1.5141	1.1094	1.0313	0.6552	
	3.696	6.886	0.8930	0.8255	1.5589	1.0035	0.5831	0.9534	
	3.080	8.454	0.6437	1.6250	1.4292	1.0913	1.0455	0.6659	
	3.080	8.454	0.9673	0.8810	1.5085	1.0333	0.5932	0.9258	
	2.640	10.148	0.7309	1.5665	1.3435	1.0773	1.0676	0.6819	
	2.640	10.148	1.0580	0.9422	1.4478	1.0747	0.6109	0.9035	
6	INF.	0.000	0.3313	2.0111	2.2460	1.4113	1.2234	0.9930	0.6249
	INF.	0.000	0.3715	1.2130	2.1610	1.3416	1.2591	0.8592	0.6310
	INF.	0.000	0.5524	0.5862	1.8741	1.3552	0.6409	0.7210	1.4294
	INF.	0.000	0.6743	0.4919	1.9221	1.1645	0.5925	0.8491	1.5075
	20.436	1.245	0.3683	1.6533	2.1085	1.3585	1.2223	0.9662	0.6201
	20.436	1.245	0.3864	1.3657	2.0807	1.3294	1.2401	0.9127	0.6194
	20.436	1.245	0.6181	0.5874	1.8319	1.3144	0.6251	0.7549	1.3956
	20.436	1.245	0.6709	0.5466	1.8580	1.2374	0.6054	0.8061	1.4209
	10.218	2.506	0.4084	1.4797	1.9916	1.3107	1.2274	0.9443	0.6264
	10.218	2.506	0.6842	0.6013	1.7981	1.2819	0.6289	0.7703	1.3542
	6.812	3.815	0.4452	1.4654	1.8933	1.2782	1.2269	0.9486	0.6361
	6.812	3.815	0.7309	0.6408	1.7522	1.2880	0.6464	0.7597	1.3045
	5.109	5.181	0.4890	1.4536	1.7993	1.2478	1.2285	0.9556	0.6456
	5.109	5.181	0.7850	0.6853	1.7049	1.2982	0.6651	0.7508	1.2585
	4.087	6.617	0.5419	1.4447	1.7092	1.2197	1.2326	0.9660	0.6551
	4.087	6.617	0.8487	0.7358	1.6543	1.3139	0.6859	0.7440	1.2164
	3.406	8.141	0.6072	1.4393	1.6216	1.1944	1.2402	0.9810	0.6650
	3.406	8.141	0.9248	0.7934	1.5972	1.3371	0.7105	0.7399	1.1786
	2.919	9.784	0.6895	1.4376	1.5329	1.1715	1.2530	1.0038	0.6766
	2.919	9.784	1.0176	0.8597	1.5269	1.3714	0.7433	0.7398	1.1451
	2.555	11.597	0.7966	1.4391	1.4342	1.1469	1.2748	1.0427	0.6929
	2.555	11.597	1.1335	0.9363	1.4261	1.4227	0.7958	0.7469	1.1163

3-db DOWN k AND q VALUES

LINEAR PHASE WITH EQUIRIPPLE ERROR   PHASE ERROR = 0.5°

n	q_0	I. L.	q_1	q_n	k_{12}	k_{23}	k_{34}	k_{45}	k_{56}	k_{67}
	INF.	0.000	0.2826	2.4404	2.5763	1.6204	1.4155	1.2459	1.0630	0.6533
	INF.	0.000	0.4008	0.6878	2.3244	1.3935	1.4998	1.0789	0.5903	1.0449
	INF.	0.000	0.4208	0.6359	2.0793	1.6187	1.1573	0.5971	0.8223	1.6654
	INF.	0.000	0.7507	0.3821	2.1536	1.2700	0.6266	0.8909	1.1368	1.9003
	21.750	1.319	0.3053	2.2520	2.4380	1.5759	1.4030	1.2389	1.0658	0.6556
	21.750	1.319	0.4276	0.7243	2.2476	1.3622	1.4899	1.1098	0.5945	1.0048
	21.750	1.319	0.4524	0.6628	2.0000	1.6121	1.1501	0.5950	0.8267	1.6298
	21.750	1.319	0.7851	0.4089	2.1145	1.2949	0.6335	0.8861	1.1084	1.8090
	10.875	2.665	0.3318	2.1023	2.3050	1.5332	1.3931	1.2330	1.0701	0.6587
	10.875	2.665	0.4587	0.7638	2.1723	1.3320	1.4791	1.1424	0.6005	0.9681
	10.875	2.665	0.4887	0.6930	1.9236	1.6080	1.1453	0.5937	0.8312	1.5950
	10.875	2.665	0.8247	0.4393	2.0740	1.3215	0.6421	0.8833	1.0814	1.7226
	7.250	4.041	0.3630	1.9827	2.1778	1.4926	1.3862	1.2285	1.0760	0.6629
	7.250	4.041	0.4953	0.8063	2.0986	1.3032	1.4674	1.1768	0.6082	0.9349
	7.250	4.041	0.5307	0.7274	1.8496	1.6065	1.1437	0.5933	0.8358	1.5611
	7.250	4.041	0.8706	0.4738	2.0314	1.3504	0.6528	0.8824	1.0558	1.6414
7	5.438	5.458	0.4001	1.8869	2.0564	1.4541	1.3825	1.2257	1.0837	0.6683
	5.438	5.458	0.5388	0.8522	2.0262	1.2763	1.4550	1.2131	0.6180	0.9052
	5.438	5.458	0.5797	0.7667	1.7769	1.6080	1.1463	0.5941	0.8406	1.5284
	5.438	5.458	0.9246	0.5134	1.9857	1.3827	0.6661	0.8835	1.0318	1.5659
	4.350	6.926	0.4450	1.8101	1.9404	1.4177	1.3824	1.2257	1.0932	0.6749
	4.350	6.926	0.5915	0.9018	1.9545	1.2520	1.4424	1.2519	0.6301	0.8793
	4.350	6.926	0.6378	0.8120	1.7036	1.6128	1.1547	0.5964	0.8459	1.4970
	4.350	6.926	0.9886	0.5593	1.9350	1.4201	0.6829	0.8866	1.0097	1.4962
	3.625	8.465	0.5006	1.7487	1.8286	1.3831	1.3863	1.2297	1.1053	0.6830
	3.625	8.465	0.6565	0.9556	1.8821	1.2305	1.4300	1.2942	0.6452	0.8570
	3.625	8.465	0.7077	0.8645	1.6262	1.6214	1.1720	0.6009	0.8521	1.4671
	3.625	8.465	1.0657	0.6130	1.8754	1.4656	0.7053	0.8919	0.9899	1.4327
	3.107	10.103	0.5709	1.7000	1.7179	1.3491	1.3949	1.2410	1.1216	0.6929
	3.107	10.103	0.7386	1.0143	1.8055	1.2123	1.4185	1.3433	0.6647	0.8385
	3.107	10.103	0.7937	0.9260	1.5379	1.6335	1.2045	0.6095	0.8601	1.4385
	3.107	10.103	1.1598	0.6768	1.7988	1.5245	0.7372	0.9001	0.9731	1.3752
	2.719	11.883	0.6631	1.6616	1.6012	1.3102	1.4083	1.2672	1.1477	0.7063
	2.719	11.883	0.8454	1.0787	1.7162	1.1943	1.4075	1.4086	0.6932	0.8244
	2.719	11.883	0.9022	0.9985	1.4233	1.6449	1.2683	0.6268	0.8713	1.4111
	2.719	11.883	1.2768	0.7538	1.6850	1.6068	0.7900	0.9133	0.9605	1.3237

3-db DOWN k AND q VALUES

LINEAR PHASE WITH EQUIRIPPLE ERROR　　　　　　　　　　　　　　PHASE ERROR = 0.5°

n	q_0	I. L.	q_1	q_n	k_{12}	k_{23}	k_{34}	k_{45}	k_{56}	k_{67}	k_{78}
	INF.	0.000	0.2718	1.9626	2.7128	1.7151	1.5264	1.3654	1.2571	1.0096	0.6219
	INF.	0.000	0.2874	1.4108	2.6547	1.6811	1.5399	1.3257	1.2853	0.9248	0.6161
	INF.	0.000	0.3830	0.6338	2.4179	1.5572	1.4431	1.4972	0.7217	0.6841	1.2854
	INF.	0.000	0.4147	0.5627	2.4102	1.4694	1.5269	1.4087	0.6590	0.7554	1.3347
	INF.	0.000	0.3824	0.6354	2.2101	1.6910	1.4188	0.9068	0.5726	1.0406	1.8029
	INF.	0.000	0.4140	0.5640	2.1832	1.6641	1.4118	0.8113	0.5872	1.1585	1.7885
	INF.	0.000	0.6466	0.3785	2.1123	1.6686	0.6795	0.7695	1.0998	1.2471	2.0505
	INF.	0.000	0.7422	0.3520	2.1940	1.5174	0.6448	0.8448	1.0863	1.2906	2.1062
	23.902	1.267	0.3011	1.6231	2.5507	1.6529	1.5197	1.3394	1.2700	0.9783	0.6189
	23.902	1.267	0.4259	0.6290	2.3373	1.4857	1.4633	1.4823	0.7063	0.7144	1.2611
	23.902	1.267	0.4267	0.6274	2.1116	1.6692	1.4122	0.8553	0.5819	1.1015	1.7557
	23.902	1.267	0.7299	0.3895	2.1096	1.6134	0.6699	0.8117	1.0847	1.2392	1.9875
	11.951	2.563	0.3264	1.6066	2.4228	1.6082	1.5100	1.3319	1.2733	0.9835	0.6273
	11.951	2.563	0.4579	0.6655	2.2644	1.4561	1.4468	1.5033	0.7330	0.7064	1.2168
	11.951	2.563	0.4603	0.6606	2.0305	1.6612	1.4132	0.8471	0.5921	1.0975	1.7199
	11.951	2.563	0.7738	0.4177	2.0690	1.6273	0.6877	0.8119	1.0800	1.2091	1.9017
8	7.967	3.898	0.3562	1.5927	2.2995	1.5648	1.5028	1.3261	1.2774	0.9901	0.6354
	7.967	3.898	0.4953	0.7060	2.1932	1.4286	1.4288	1.5261	0.7616	0.6997	1.1754
	7.967	3.898	0.4994	0.6978	1.9513	1.6554	1.4170	0.8401	0.6019	1.0947	1.6853
	7.967	3.898	0.8238	0.4501	2.0258	1.6450	0.7066	0.8134	1.0765	1.1801	1.8201
	5.976	5.279	0.3918	1.5820	2.1810	1.5229	1.4983	1.3223	1.2824	0.9984	0.6433
	5.976	5.279	0.5395	0.7513	2.1230	1.4038	1.4094	1.5504	0.7923	0.6943	1.1371
	5.976	5.279	0.5455	0.7399	1.8733	1.6515	1.4247	0.8347	0.6115	1.0933	1.6517
	5.976	5.279	0.8817	0.4877	1.9786	1.6675	0.7275	0.8164	1.0742	1.1524	1.7434
	4.780	6.716	0.4350	1.5750	2.0671	1.4824	1.4966	1.3216	1.2885	1.0086	0.6510
	4.780	6.716	0.5928	0.8021	2.0532	1.3823	1.3888	1.5766	0.8256	0.6906	1.1023
	4.780	6.716	0.6007	0.7882	1.7947	1.6495	1.4380	0.8319	0.6209	1.0935	1.6193
	4.780	6.716	0.9496	0.5318	1.9250	1.6963	0.7514	0.8214	1.0731	1.1262	1.6718
	3.984	8.226	0.4886	1.5723	1.9568	1.4430	1.4978	1.3252	1.2966	1.0210	0.6588
	3.984	8.226	0.6584	0.8593	1.9818	1.3646	1.3675	1.6055	0.8625	0.6888	1.0710
	3.984	8.226	0.6676	0.8442	1.7127	1.6485	1.4602	0.8331	0.6305	1.0956	1.5878
	3.984	8.226	1.0306	0.5840	1.8606	1.7338	0.7809	0.8288	1.0731	1.1019	1.6057
	3.415	9.834	0.5567	1.5743	1.8475	1.4034	1.5019	1.3359	1.3089	1.0367	0.6669
	3.415	9.834	0.7411	0.9242	1.9052	1.3509	1.3455	1.6392	0.9055	0.6895	1.0435
	3.415	9.834	0.7505	0.9099	1.6212	1.6458	1.4976	0.8413	0.6407	1.1001	1.5570
	3.415	9.834	1.1292	0.6469	1.7765	1.7842	0.8212	0.9394	1.0745	1.0803	1.5453

3-db DOWN k AND q VALUES

GAUSSIAN RESPONSE

n	q_0	I. L.	q_1	q_n	k_{12}	k_{23}	k_{34}	k_{45}
2	INF.	0.000	0.4738	2.1850	0.9828			
	7.787	1.537	0.4976	1.7902	0.8996			
	3.894	3.189	0.5248	1.5091	0.8180			
	2.596	4.964	0.5566	1.2958	0.7385			
	1.947	6.864	0.5959	1.1236	0.6620			
	1.557	8.878	0.6500	0.9710	0.5896			
	1.298	10.969	0.7787	0.7787	0.5319			
	1.112	13.468	0.7787	0.7787	0.5319			
3	INF.	-0.000	0.2624	2.2262	2.1603	0.7416		
	7.275	2.032	0.2813	1.7628	1.9646	0.6967		
	3.638	4.187	0.3032	1.4546	1.7713	0.6544		
	2.425	6.474	0.3292	1.2348	1.5820	0.6155		
	1.819	8.901	0.3602	1.0704	1.3989	0.5813		
	1.455	11.478	0.3977	0.9445	1.2268	0.5532		
4	INF.	0.000	0.1772	2.2450	3.2627	1.4225	0.6913	
	INF.	0.000	0.2747	0.4083	2.2792	0.7553	0.9896	
	7.053	2.454	0.1910	1.7484	2.9689	1.3232	0.6590	
	7.053	2.454	0.2913	0.4213	2.1131	0.7226	0.9268	
	3.526	5.036	0.2074	1.4285	2.6773	1.2276	0.6304	
	3.526	5.036	0.3102	0.4349	1.9513	0.6933	0.8650	
	2.351	7.755	0.2268	1.2052	2.3887	1.1371	0.6042	
	2.351	7.755	0.3318	0.4493	1.7951	0.6687	0.8040	
	1.763	10.617	0.2504	1.0403	2.1050	1.0532	0.5815	
	1.763	10.617	0.3565	0.4650	1.6462	0.6503	0.7436	
	1.411	13.629	0.2796	0.9134	1.8286	0.9783	0.5629	
	1.411	13.629	0.3847	0.4826	1.5063	0.6411	0.6833	
5	INF.	0.000	0.1312	2.2533	4.4236	1.9894	1.2555	0.6736
	INF.	0.000	0.2848	0.2194	2.4001	0.8055	0.7221	2.4514
	6.917	2.827	0.1417	1.7377	4.0345	1.8428	1.1872	0.6489
	6.917	2.827	0.2999	0.2326	2.2564	0.7829	0.6836	2.2488
	3.459	5.787	0.1540	1.4117	3.6471	1.6998	1.1229	0.6263
	3.459	5.787	0.3169	0.2472	2.1168	0.7646	0.6464	2.0483
	2.306	8.887	0.1688	1.1869	3.2623	1.5615	1.0635	0.6063
	2.306	8.887	0.3360	0.2638	1.9820	0.7521	0.6108	1.8500
	1.729	12.132	0.1867	1.0226	2.8811	1.4296	1.0104	0.5889
	1.729	12.132	0.3577	0.2827	1.8527	0.7472	0.5774	1.6540

3-db DOWN k AND q VALUES

GAUSSIAN RESPONSE

n	q_0	I. L.	q_1	q_n	k_{12}	k_{23}	k_{34}	k_{45}	k_{56}	k_{67}
6	INF.	0.000	0.1026	2.2568	5.6583	2.5620	1.6836	1.1921	0.6663	
	INF.	0.000	0.1336	0.3694	4.4308	2.0630	1.3810	0.6835	1.0373	
	INF.	0.000	0.1753	0.2228	3.4119	1.6683	0.7307	0.7520	2.5458	
	INF.	0.000	0.2908	0.1481	2.5283	0.8608	0.6600	1.5303	3.8038	
	6.820	3.165	0.1108	1.7290	5.1748	2.3697	1.5829	1.1401	0.6458	
	6.820	3.165	0.1436	0.3785	4.0699	1.9288	1.3245	0.6650	0.9813	
	6.820	3.165	0.1871	0.2348	3.1536	1.5936	0.7111	0.7179	2.3589	
	6.820	3.165	0.3048	0.1581	2.4013	0.8468	0.6309	1.4197	3.4918	
	3.410	6.467	0.1204	1.3993	4.6927	2.1805	1.4863	1.0919	0.6272	
	3.410	6.467	0.1553	0.3879	3.7110	1.7977	1.2727	0.6493	0.9261	
	3.410	6.467	0.2007	0.2479	2.8972	1.5238	0.6940	0.6853	2.1754	
	3.410	6.467	0.3205	0.1696	2.2774	0.8380	0.6035	1.3115	3.1827	
	2.273	9.910	0.1319	1.1739	4.2126	1.9955	1.3950	1.0483	0.6105	
	2.273	9.910	0.1691	0.3974	3.3554	1.6705	1.2269	0.6370	0.8715	
	2.273	9.910	0.2164	0.2624	2.6438	1.4601	0.6800	0.6546	1.9956	
	2.273	9.910	0.3382	0.1827	2.1571	0.8357	0.5782	1.2060	2.8767	
7	INF.	0.000	0.0833	2.2583	6.9676	3.1576	2.0974	1.5597	1.1636	0.6630
	INF.	0.000	0.1238	0.2286	4.7111	2.1899	1.4600	0.6914	0.7866	2.6295
	INF.	0.000	0.1362	0.1958	4.4462	2.1314	1.5070	0.7257	0.7472	2.6996
	INF.	0.000	0.2930	0.1106	2.6644	0.9164	0.6367	1.2941	2.2184	5.1648
	6.746	3.477	0.0899	1.7218	6.3900	2.9204	1.9650	1.4832	1.1217	0.6453
	6.746	3.477	0.1458	0.2062	4.1096	2.0064	1.4614	0.7142	0.7112	2.4951
	6.746	3.477	0.1327	0.2395	4.3369	2.0599	1.4059	0.6729	0.7570	2.4592
	6.746	3.477	0.3064	0.1184	2.5494	0.9101	0.6135	1.2131	2.0560	4.7531
	3.373	7.091	0.0976	1.3897	5.8137	2.6860	1.8364	1.4111	1.0832	0.6292
	3.373	7.091	0.1570	0.2175	3.7749	1.8836	1.4211	0.7067	0.6762	2.2930
	3.373	7.091	0.1431	0.2513	3.9637	1.9341	1.3560	0.6557	0.7293	2.2929
	3.373	7.091	0.3214	0.1273	2.4366	0.9094	0.5921	1.1348	1.8964	4.3448
	2.249	10.851	0.1067	1.1639	5.2389	2.4550	1.7128	1.3442	1.0488	0.6146
	2.249	10.851	0.1701	0.2300	3.4435	1.7633	1.3869	0.7041	0.6422	2.0932
	2.249	10.851	0.1552	0.2643	3.5924	1.8141	1.3112	0.6400	0.7039	2.1308
	2.249	10.851	0.3384	0.1375	2.3261	0.9160	0.5729	1.0595	1.7400	3.9398

3-db DOWN k AND q VALUES

GAUSSIAN RESPONSE

n	q_0	I. L.	q_1	q_n	k_{12}	k_{23}	k_{34}	k_{45}	k_{56}	k_{67}	k_{78}
8	INF.	0.000	0.0695	2.2590	8.3490	3.7806	2.5173	1.9006	1.5001	1.1497	0.6615
	INF.	0.000	0.0840	0.3404	6.9571	3.2074	2.1751	1.6464	1.2740	0.6679	1.0818
	INF.	0.000	0.1027	0.1958	5.7435	2.7079	1.8543	1.4158	0.7021	0.7716	2.7893
	INF.	0.000	0.1382	0.1316	4.4806	2.1983	1.6307	0.7746	0.6734	1.6513	4.2385
	INF.	0.000	0.0946	0.2346	6.1340	2.7916	1.8802	1.3530	0.6677	0.8223	2.7081
	INF.	0.000	0.1237	0.1480	4.7811	2.2742	1.5750	0.7272	0.7089	1.7065	3.9067
	INF.	0.000	0.1691	0.1120	3.6169	1.9117	0.8066	0.6599	1.3853	2.2754	5.1607
	INF.	0.000	0.2926	0.0876	2.8076	0.9715	0.6274	1.1913	1.8036	2.8842	6.5640
	6.686	3.766	0.0749	1.7159	7.6776	3.4992	2.3538	1.7998	1.4388	1.1145	0.6457
	6.686	3.766	0.0904	0.3477	6.4057	2.9832	2.0519	1.5762	1.2408	0.6545	1.0295
	6.686	3.766	0.1102	0.2055	5.3005	2.5376	1.7682	1.3818	0.6908	0.7380	2.5935
	6.686	3.766	0.1475	0.1397	4.1641	2.0800	1.5951	0.7704	0.6453	1.5366	3.9172
	6.686	3.766	0.1015	0.2442	5.6467	2.6033	1.7909	1.3078	0.6499	0.7973	2.5546
	6.686	3.766	0.1324	0.1565	4.4181	2.1500	1.5289	0.7128	0.6848	1.6120	3.6297
	6.686	3.766	0.1797	0.1194	3.3706	1.8545	0.7954	0.6386	1.3116	2.1242	4.7747
	6.686	3.766	0.3058	0.0937	2.7006	0.9719	0.6082	1.1266	1.6873	2.6760	6.0582
	3.343	7.673	0.0812	1.3821	7.0072	3.2201	2.1938	1.7032	1.3818	1.0824	0.6313
	3.343	7.673	0.0978	0.3549	5.8551	2.7616	1.9324	1.5099	1.2121	0.6431	0.9777
	3.343	7.673	0.1189	0.2159	4.8580	2.3702	1.6853	1.3529	0.6822	0.7053	2.4008
	3.343	7.673	0.1582	0.1487	3.8491	1.9628	1.5648	0.7713	0.6185	1.4239	3.5995
	3.343	7.673	0.1096	0.2548	5.1612	2.4184	1.7069	1.2656	0.6329	0.7747	2.4049
	3.343	7.673	0.1423	0.1662	4.0569	2.0300	1.4875	0.7002	0.6623	1.5208	3.3552
	3.343	7.673	0.1918	0.1277	3.1256	1.8024	0.7870	0.6189	1.2413	1.9760	4.3916
	3.343	7.673	0.3206	0.1007	2.5950	0.9784	0.5909	1.0646	1.5739	2.4707	5.5556
	2.229	11.725	0.0886	1.1561	6.3381	2.9439	2.0381	1.6119	1.3299	1.0537	0.6181
	2.229	11.725	0.1065	0.3622	5.3059	2.5432	1.8177	1.4481	1.1885	0.6343	0.9264
	2.229	11.725	0.1291	0.2273	4.4166	2.2064	1.6062	1.3302	0.6769	0.6738	2.2115
	2.229	11.725	0.1709	0.1589	3.5370	1.8465	1.5404	0.7790	0.5932	1.3134	3.2854
	2.229	11.725	0.1191	0.2665	4.6782	2.2381	1.6299	1.2267	0.6165	0.7549	2.2589
	2.229	11.725	0.1539	0.1770	3.6985	1.9151	1.4518	0.6896	0.6414	1.4335	3.0832
	2.229	11.725	0.2057	0.1372	2.8826	1.7567	0.7817	0.6011	1.1747	1.8308	4.0116
	2.229	11.725	0.3373	0.1089	2.4909	0.9929	0.5760	1.0058	1.4639	2.2684	5.0560

3-db DOWN k AND q VALUES

GAUSSIAN RESPONSE TO 6-db

n	q_0	I. L.	q_1	q_n	k_{12}	k_{23}	k_{34}	k_{45}
3	INF.	0.000	0.4042	2.3380	1.6622	0.6911		
	10.393	1.556	0.4345	1.9824	1.5556	0.6694		
	5.197	3.192	0.4691	1.7310	1.4574	0.6496		
	3.464	4.927	0.5084	1.5473	1.3690	0.6319		
	2.598	6.788	0.5531	1.4101	1.2915	0.6165		
	2.079	8.825	0.6042	1.3057	1.2257	0.6037		
	1.732	11.126	0.6630	1.2249	1.1717	0.5938		
	1.485	13.864	0.7315	1.1610	1.1292	0.5877		
4	INF.	0.000	0.4198	2.1427	1.7441	1.0493	0.6344	
	INF.	0.000	0.5700	0.9137	1.6232	0.7984	0.6816	
	15.863	1.241	0.4591	1.8365	1.6391	1.0207	0.6284	
	15.863	1.241	0.5919	0.9679	1.5685	0.8200	0.6567	
	7.931	2.493	0.5112	1.5611	1.5395	0.9878	0.6226	
	7.931	2.493	0.6083	1.0496	1.5120	0.8556	0.6339	
	5.288	3.754	0.6000	1.2438	1.4512	0.9245	0.6213	
	3.966	5.072	0.6520	1.2334	1.3864	0.9187	0.6335	
	3.173	6.482	0.7135	1.2242	1.3240	0.9182	0.6467	
	2.644	8.011	0.7872	1.2165	1.2616	0.9252	0.6626	
	2.266	9.700	0.8777	1.2093	1.1931	0.9438	0.6857	
	1.983	11.622	0.9941	1.1986	1.0985	0.9828	0.7311	
5	INF.	0.000	0.4544	2.4138	1.6130	1.0404	0.9218	0.6201
	INF.	0.000	0.8907	0.6702	1.4178	0.8637	0.5526	1.0457
	28.095	0.915	0.4847	2.2480	1.5347	1.0193	0.9247	0.6255
	28.095	0.915	0.9305	0.6977	1.3822	0.8795	0.5556	1.0127
	14.047	1.844	0.5189	2.1092	1.4579	0.9991	0.9300	0.6323
	14.047	1.844	0.9753	0.7268	1.3446	0.8973	0.5604	0.9820
	9.365	2.786	0.5579	1.9920	1.3824	0.9794	0.9380	0.6411
	9.365	2.786	1.0263	0.7575	1.3042	0.9176	0.5675	0.9538
	7.024	3.746	0.6027	1.8921	1.3073	0.9599	0.9495	0.6525
	7.024	3.746	1.0850	0.7899	1.2595	0.9411	0.5778	0.9285
	5.619	4.725	0.6549	1.8054	1.2315	0.9397	0.9657	0.6681
	5.619	4.725	1.1538	0.8236	1.2080	0.9688	0.5929	0.9069

3-db DOWN k AND q VALUES

GAUSSIAN RESPONSE TO 6-db

n	q_0	I. L.	q_1	q_n	k_{12}	k_{23}	k_{34}	k_{45}	k_{56}
5	4.682	5.728	0.7167	1.7277	1.1519	0.9160	0.9886	0.6912	
	4.682	5.728	1.2364	0.8581	1.1443	1.0019	0.6165	0.8904	
	4.014	6.760	0.7928	1.6501	1.0615	0.8803	1.0227	0.7323	
	4.014	6.760	1.3419	0.8912	1.0546	1.0419	0.6581	0.8837	
	3.512	8.360	0.8313	1.7933	0.9225	0.7759	1.0288	0.9490	
	3.512	8.360	0.9558	1.3999	0.9393	0.7246	1.1460	0.7841	
6	INF.	0.000	0.5041	2.0917	1.4820	0.9542	0.8879	0.8212	0.5800
	INF.	0.000	0.5674	1.4297	1.4259	0.9118	0.9237	0.7539	0.5794
	INF.	0.000	0.8834	0.7519	1.1722	1.0290	0.5952	0.6045	1.0939
	INF.	0.000	1.0982	0.6446	1.2072	0.9523	0.5502	0.6743	1.1490
	45.379	0.687	0.5427	1.8827	1.4131	0.9280	0.8892	0.8211	0.5838
	45.379	0.687	0.5834	1.5158	1.3837	0.9026	0.9125	0.7824	0.5810
	45.379	0.687	0.9585	0.7517	1.1407	1.0235	0.5954	0.6200	1.0801
	45.379	0.687	1.0932	0.6854	1.1650	0.9806	0.5676	0.6605	1.1093
	22.689	1.376	0.5941	1.6602	1.3433	0.8967	0.8973	0.8153	0.5897
	22.689	1.376	1.0678	0.7413	1.1154	1.0113	0.5957	0.6408	1.0697
	15.126	2.074	0.6297	1.6408	1.2886	0.8763	0.8958	0.8283	0.6021
	15.126	2.074	1.1193	0.7669	1.0771	1.0174	0.6163	0.6410	1.0470
	11.345	2.783	0.6701	1.6209	1.2330	0.8546	0.8939	0.8447	0.6166
	11.345	2.783	1.1771	0.7938	1.0338	1.0239	0.6408	0.6433	1.0259
	9.076	3.503	0.7164	1.5994	1.1753	0.8300	0.8908	0.8667	0.6349
	9.076	3.503	1.2435	0.8217	0.9827	1.0296	0.6716	0.6487	1.0072
	7.563	4.236	0.7709	1.5729	1.1131	0.7991	0.8834	0.8984	0.6614
	7.563	4.236	1.3230	0.8496	0.9179	1.0315	0.7137	0.6604	0.9927
	6.483	4.983	0.8430	1.5184	1.0332	0.7439	0.8512	0.9607	0.7238
	6.483	4.983	1.4401	0.8692	0.8092	1.0107	0.7911	0.6954	0.9944
	5.672	8.345	0.7020	3.0098	1.0296	0.7377	0.7333	0.9431	0.9615
	5.672	8.345	1.1733	1.1057	1.0415	0.4975	1.0575	0.8795	0.6591

3-db DOWN k AND q VALUES

GAUSSIAN RESPONSE TO 6-DB

n	q_0	I. L.	q_1	q_n	k_{12}	k_{23}	k_{34}	k_{45}	k_{56}	k_{67}
	INF.	0.000	0.4918	2.5039	1.4841	0.9443	0.8565	0.8230	0.8164	0.5917
	INF.	0.000	0.7297	0.9413	1.3234	0.8230	0.8872	0.8686	0.5419	0.7836
	INF.	0.000	0.7364	0.9304	1.1297	0.9546	0.8841	0.5339	0.6333	1.1040
	INF.	0.000	1.4389	0.5754	1.1230	0.9555	0.5700	0.6510	0.7772	1.2647
	65.716	0.577	0.5148	2.4136	1.4325	0.9253	0.8492	0.8223	0.8255	0.6006
	65.716	0.577	0.7597	0.9612	1.2915	0.8115	0.8753	0.8851	0.5542	0.7742
	65.716	0.577	0.7681	0.9481	1.0931	0.9468	0.8927	0.5411	0.6376	1.0915
	65.716	0.577	1.4795	0.5949	1.0940	0.9673	0.5838	0.6537	0.7695	1.2342
	32.858	1.157	0.5401	2.3306	1.3809	0.9058	0.8416	0.8218	0.8361	0.6107
	32.858	1.157	0.7927	0.9812	1.2588	0.7998	0.8620	0.9024	0.5686	0.7661
	32.858	1.157	0.8025	0.9666	1.0552	0.9376	0.9033	0.5498	0.6423	1.0793
	32.858	1.157	1.5242	0.6156	1.0621	0.9790	0.5999	0.6577	0.7625	1.2046
	21.905	1.741	0.5679	2.2533	1.3292	0.8855	0.8333	0.8215	0.8488	0.6227
	21.905	1.741	0.8293	1.0013	1.2249	0.7875	0.8467	0.9206	0.5859	0.7596
	21.905	1.741	0.8401	0.9859	1.0156	0.9262	0.9162	0.5608	0.6478	1.0675
	21.905	1.741	1.5739	0.6373	1.0263	0.9903	0.6191	0.6632	0.7564	1.1762
7	16.429	2.328	0.5989	2.1802	1.2768	0.8640	0.8237	0.8209	0.8646	0.6375
	16.429	2.328	0.8817	1.0058	0.9733	0.9111	0.9321	0.5751	0.6543	1.0565
	16.429	2.328	0.8703	1.0210	1.1893	0.7744	0.8285	0.9401	0.6073	0.7552
	16.429	2.328	1.6301	0.6601	0.9849	1.0004	0.6426	0.6711	0.7516	1.1492
	13.143	2.919	0.6337	2.1081	1.2230	0.8402	0.8112	0.8192	0.8854	0.6572
	13.143	2.919	0.9283	1.0255	0.9266	0.8890	0.9519	0.5953	0.6628	1.0468
	13.143	2.919	0.9170	1.0396	1.1510	0.7592	0.8053	0.9612	0.6354	0.7542
	13.143	2.919	1.6957	0.6837	0.9350	1.0076	0.6725	0.6827	0.7487	1.1243
	10.953	3.514	0.6742	2.0285	1.1653	0.8109	0.7913	0.8130	0.9167	0.6887
	10.953	3.514	0.9830	1.0428	0.8707	0.8512	0.9772	0.6282	0.6759	1.0410
	10.953	3.514	0.9728	1.0545	1.1070	0.7391	0.7705	0.9845	0.6784	0.7607
	10.953	3.514	1.7795	0.7070	0.8684	1.0067	0.7146	0.7027	0.7505	1.1037
	9.388	4.791	0.6562	2.6576	1.1313	0.7846	0.7331	0.7700	0.9460	0.8452
	9.388	4.791	0.7545	1.7398	1.0986	0.7420	0.7729	0.7293	1.0463	0.6820
	9.388	4.791	0.9780	1.1393	1.0341	0.7535	0.6154	1.0678	0.7292	0.7894
	9.388	4.791	1.2136	0.9292	1.0878	0.5993	0.7767	0.9966	0.7306	0.8054
	8.214	8.459	0.6277	4.3301	1.1366	0.7781	0.7122	0.7248	0.9157	0.9691
	8.214	8.459	0.7778	1.8570	1.0866	0.7043	0.7900	0.6096	1.1961	0.5288
	8.214	8.459	1.0060	1.2047	1.0025	0.7724	0.4764	1.2164	0.6094	0.7555
	8.214	8.459	1.4567	0.8790	1.1331	0.4171	0.9035	0.9958	0.6425	0.7847

3-db DOWN k AND q VALUES

GAUSSIAN RESPONSE TO 6-db

n	q_0	I. L.	q_1	q_n	k_{12}	k_{23}	k_{34}	k_{45}	k_{56}	k_{67}	k_{78}
	INF.	0.003	0.5284	1.9285	1.4315	0.9148	0.8467	0.7869	0.8192	0.7734	0.5693
	INF.	0.003	0.7862	0.8779	1.2579	0.8534	0.7503	0.9548	0.6713	0.5623	0.8920
	INF.	0.003	0.7376	0.9477	1.1237	0.8664	0.9218	0.7075	0.5012	0.7522	1.0888
	INF.	0.003	1.3602	0.5967	0.9906	1.0245	0.6263	0.6068	0.7151	0.7993	1.2740
	89.114	0.487	0.5496	1.9075	1.3906	0.8985	0.8394	0.7833	0.8214	0.7838	0.5787
	89.114	0.487	0.8154	0.8950	1.2311	0.8456	0.7384	0.9550	0.6921	0.5674	0.8819
	89.114	0.487	0.7642	0.9661	1.0930	0.8547	0.9261	0.7161	0.5088	0.7532	1.0774
	89.114	0.487	1.3975	0.6142	0.9651	1.0256	0.6433	0.6129	0.7156	0.7920	1.2490
	44.557	0.975	0.5727	1.8861	1.3495	0.8818	0.8318	0.7793	0.8238	0.7958	0.5892
	44.557	0.975	0.8471	0.9126	1.2034	0.8377	0.7256	0.9541	0.7150	0.5741	0.8727
	44.557	0.975	0.7929	0.9851	1.0615	0.8412	0.9306	0.7268	0.5173	0.7546	1.0662
	44.557	0.975	1.4376	0.6326	0.9372	1.0256	0.6625	0.6204	0.7167	0.7851	1.2245
	29.705	1.465	0.5979	1.8639	1.3080	0.8644	0.8233	0.7748	0.8264	0.8099	0.6012
	29.705	1.465	0.8240	1.0046	1.0287	0.8252	0.9350	0.7404	0.5270	0.7565	1.0554
	29.705	1.465	0.8818	0.9303	1.1746	0.8296	0.7113	0.9513	0.7406	0.5828	0.8645
	29.705	1.465	1.4811	0.6519	0.9060	1.0237	0.6842	0.6296	0.7187	0.7787	1.2007
8	22.279	1.959	0.6257	1.8400	1.2656	0.8458	0.8134	0.7691	0.8289	0.8275	0.6156
	22.279	1.959	0.8581	1.0243	0.9940	0.8053	0.9386	0.7584	0.5386	0.7593	1.0453
	22.279	1.959	0.9202	0.9479	1.1440	0.8210	0.6950	0.9459	0.7701	0.5946	0.8579
	22.279	1.959	1.5293	0.6721	0.8702	1.0185	0.7096	0.6416	0.7222	0.7733	1.1779
	17.823	2.457	0.6568	1.8122	1.2217	0.8254	0.8008	0.7606	0.8308	0.8510	0.6347
	17.823	2.457	0.8961	1.0433	0.9560	0.7785	0.9396	0.7836	0.5540	0.7636	1.0366
	17.823	2.457	0.9636	0.9646	1.1108	0.8111	0.6749	0.9355	0.8058	0.6119	0.8540
	17.823	2.457	1.5846	0.5929	0.8273	1.0070	0.7404	0.6582	0.7281	0.7695	1.1568
	14.852	2.958	0.6933	1.7708	1.1732	0.7998	0.7810	0.7433	0.8287	0.8904	0.6674
	14.852	2.958	0.9416	1.0581	0.9099	0.7347	0.9303	0.8262	0.5795	0.7718	1.0328
	14.852	2.958	1.0164	0.9774	1.0710	0.7972	0.6443	0.9113	0.8563	0.6438	0.8575
	14.852	2.958	1.6570	0.7125	0.7673	0.9788	0.7818	0.6861	0.7408	0.7702	1.1403
	12.731	4.872	0.6059	3.4555	1.2000	0.8106	0.7408	0.7219	0.7648	0.9224	0.8423
	12.731	4.872	0.8735	1.2580	1.0167	0.8467	0.6058	0.7261	1.0709	0.5783	0.8684
	12.731	4.872	0.8538	1.3013	1.1256	0.6916	0.8466	0.5852	0.9510	0.8841	0.6321
	12.731	4.872	1.5021	0.7850	1.1270	0.5779	0.6776	0.9600	0.8173	0.6684	0.9177
	11.139	8.556	0.5903	5.6102	1.2029	0.8058	0.7308	0.7021	0.7248	0.9042	0.9481
	11.139	8.556	0.8705	1.3820	0.9896	0.8561	0.5863	0.6068	1.2117	0.4528	0.8593
	11.139	8.556	0.9085	1.2958	1.1196	0.6382	0.8991	0.4485	1.0897	0.8111	0.5513
	11.139	8.556	1.8008	0.7593	1.1840	0.4007	0.7491	1.0144	0.7465	0.6274	0.9124

3-db DOWN k AND q VALUES

GAUSSIAN RESPONSE TO 12-db

n	q_0	I. L.	q_1	q_n	k_{12}	k_{23}	k_{34}	k_{45}
3	INF.	0.000	0.4152	2.3452	1.6314	0.6864		
	10.684	1.524	0.4463	1.9958	1.5287	0.6660		
	5.342	3.125	0.4818	1.7477	1.4342	0.6476		
	3.561	4.820	0.5220	1.5656	1.3492	0.6313		
	2.671	6.636	0.5679	1.4291	1.2747	0.6175		
	2.137	8.619	0.6202	1.3250	1.2113	0.6063		
	1.781	10.852	0.6804	1.2441	1.1590	0.5986		
	1.526	13.493	0.7506	1.1798	1.1170	0.5954		
4	INF.	0.000	0.3097	2.1518	2.2212	1.2007	0.6622	
	INF.	0.000	0.4188	0.7657	1.9888	0.8329	0.7396	
	10.879	1.685	0.3396	1.7670	2.0635	1.1480	0.6469	
	10.879	1.685	0.4374	0.8171	1.9126	0.8470	0.6979	
	5.439	3.400	0.3788	1.4566	1.9134	1.0884	0.6320	
	5.439	3.400	0.4534	0.8924	1.8388	0.8788	0.6577	
	3.626	5.135	0.4482	1.0974	1.7709	0.9732	0.6235	
	2.720	6.978	0.4901	1.0891	1.6836	0.9449	0.6346	
	2.176	9.034	0.5403	1.0825	1.6072	0.9191	0.6440	
	1.813	11.397	0.6012	1.0782	1.5434	0.8963	0.6513	
5	INF.	0.000	0.2909	2.3745	2.4267	1.4532	1.1296	0.6603
	INF.	0.000	0.5344	0.5032	2.0850	0.9760	0.6052	1.3325
	13.981	1.583	0.3159	2.0901	2.2764	1.4095	1.1184	0.6550
	13.981	1.583	0.5620	0.5363	2.0323	0.9874	0.5968	1.2586
	6.990	3.214	0.3451	1.8793	2.1342	1.3689	1.1106	0.6508
	6.990	3.214	0.5936	0.5731	1.9804	1.0003	0.5899	1.1904
	4.660	4.904	0.3797	1.7199	2.0012	1.3313	1.1067	0.6478
	4.660	4.904	0.6305	0.6139	1.9297	1.0147	0.5845	1.1288
	3.495	6.670	0.4211	1.5981	1.8781	1.2973	1.1070	0.6463
	3.495	6.670	0.6742	0.6592	1.8802	1.0308	0.5809	1.0743
	2.796	8.539	0.4715	1.5041	1.7655	1.2674	1.1120	0.6466
	2.796	8.539	0.7267	0.7094	1.8318	1.0497	0.5791	1.0272
	2.330	10.556	0.5341	1.4312	1.6625	1.2426	1.1224	0.6490
	2.330	10.556	0.7908	0.7653	1.7837	1.0733	0.5798	0.9876

3-db DOWN k AND q VALUES

GAUSSIAN RESPONSE TO 12-db

n	q_0	I. L.	q_1	q_n	k_{12}	k_{23}	k_{34}	k_{45}	k_{56}
	INF.	0.000	0.3164	2.1154	2.2817	1.4384	1.2625	1.0528	0.6426
	INF.	0.000	0.3716	1.0618	2.1520	1.3505	1.3132	0.8609	0.6550
	INF.	0.000	0.5427	0.5585	1.8394	1.4416	0.6857	0.7069	1.4676
	INF.	0.000	0.7280	0.4426	1.9169	1.1881	0.6051	0.8844	1.5958
	22.327	1.125	0.3424	1.8739	2.1568	1.3993	1.2583	1.0458	0.6393
	22.327	1.125	0.3908	1.1173	2.0691	1.3284	1.3069	0.8933	0.6416
	22.327	1.125	0.5869	0.5714	1.7858	1.4334	0.6786	0.7196	1.4318
	22.327	1.125	0.7449	0.4736	1.8648	1.2332	0.6124	0.8591	1.5191
	11.164	2.255	0.3755	1.6341	2.0329	1.3578	1.2602	1.0327	0.6346
	11.164	2.255	0.4090	1.2048	1.9885	1.3124	1.2959	0.9380	0.6309
	11.164	2.255	0.6464	0.5788	1.7399	1.4180	0.6682	0.7420	1.3982
	11.164	2.255	0.7525	0.5139	1.8006	1.2943	0.6256	0.8281	1.4431
	7.442	3.390	0.4222	1.3749	1.9112	1.3069	1.2770	0.9996	0.6321
	7.442	3.390	0.7394	0.5737	1.7156	1.3776	0.6567	0.7796	1.3693
6	5.582	4.557	0.4580	1.3641	1.8155	1.2773	1.2826	1.0088	0.6415
	5.582	4.557	0.7896	0.6061	1.6631	1.3933	0.6787	0.7705	1.3232
	4.465	5.768	0.5002	1.3548	1.7209	1.2481	1.2915	1.0222	0.6511
	4.465	5.768	0.8476	0.6420	1.6048	1.4148	0.7044	0.7631	1.2801
	3.721	7.031	0.5507	1.3473	1.6260	1.2182	1.3046	1.0421	0.6616
	3.721	7.031	0.9157	0.6821	1.5364	1.4436	0.7363	0.7581	1.2406
	3.190	8.359	0.6121	1.3416	1.5275	1.1843	1.3229	1.0737	0.6746
	3.190	8.359	0.9969	0.7268	1.4496	1.4818	0.7804	0.7570	1.2050
	2.791	9.770	0.6888	1.3370	1.4177	1.1362	1.3465	1.1308	0.6950
	2.791	9.770	1.0962	0.7766	1.3247	1.5302	0.8525	0.7639	1.1748
	2.481	11.290	0.7890	1.3275	1.2741	1.0266	1.3491	1.2777	0.7529
	2.481	11.290	1.2257	0.8300	1.0924	1.5633	1.0233	0.8024	1.1578

3-db DOWN k AND q VALUES

GAUSSIAN RESPONSE TO 12-db

n	q_0	I. L.	q_1	q_n	k_{12}	k_{23}	k_{34}	k_{45}	k_{56}	k_{67}
	INF.	0.000	0.3207	2.3829	2.2308	1.4044	1.2626	1.1814	1.0590	0.6595
	INF.	0.000	0.4794	0.6886	1.9727	1.2043	1.3588	1.0945	0.5981	1.0149
	INF.	0.000	0.4917	0.6647	1.7083	1.4402	1.1811	0.6106	0.7807	1.5407
	INF.	0.000	0.9987	0.3942	1.7733	1.2774	0.6413	0.8410	1.0667	1.7974
	27.537	1.079	0.3439	2.2284	2.1131	1.3644	1.2517	1.1828	1.0690	0.6640
	27.537	1.079	0.5095	0.7173	1.9054	1.1777	1.3472	1.1307	0.6071	0.9825
	27.537	1.079	0.5260	0.6871	1.6306	1.4354	1.1902	0.6127	0.7855	1.5119
	27.537	1.079	1.0460	0.4166	1.7254	1.3109	0.6552	0.8385	1.0446	1.7224
	13.769	2.173	0.3705	2.1000	1.9974	1.3244	1.2422	1.1863	1.0818	0.6697
	13.769	2.173	0.5441	0.7479	1.8374	1.1512	1.3338	1.1705	0.6187	0.9524
	13.769	2.173	0.5650	0.7117	1.5515	1.4311	1.2043	0.6165	0.7906	1.4834
	13.769	2.173	1.1001	0.4413	1.6714	1.3481	0.6729	0.8379	1.0237	1.6505
	9.179	3.282	0.4013	1.9927	1.8834	1.2838	1.2336	1.1925	1.0984	0.6768
	9.179	3.282	0.5842	0.7802	1.7679	1.1243	1.3181	1.2150	0.6339	0.9249
7	9.179	3.282	0.6097	0.7389	1.4693	1.4260	1.2256	0.6229	0.7963	1.4554
	9.179	3.282	1.1626	0.4687	1.6088	1.3899	0.6960	0.8395	1.0040	1.5817
	6.884	4.411	0.4374	1.9027	1.7702	1.2412	1.2251	1.2027	1.1208	0.6862
	6.884	4.411	0.6313	0.8143	1.6956	1.0960	1.2987	1.2663	0.6542	0.9004
	6.884	4.411	0.6616	0.7690	1.3813	1.4176	1.2578	0.6334	0.8028	1.4281
	6.884	4.411	1.2360	0.4993	1.5330	1.4372	0.7276	0.8440	0.9860	1.5166
	5.507	5.563	0.4801	1.8265	1.6564	1.1937	1.2137	1.2181	1.1536	0.6997
	5.507	5.563	0.6876	0.8503	1.6184	1.0637	1.2720	1.3289	0.6833	0.8797
	5.507	5.563	0.7224	0.8025	1.2825	1.3997	1.3074	0.6519	0.8108	1.4018
	5.507	5.563	1.3234	0.5334	1.4352	1.4907	0.7743	0.8533	0.9701	1.4555
	4.590	6.743	0.5317	1.7601	1.5385	1.1344	1.1899	1.2399	1.2097	0.7226
	4.590	6.743	0.7563	0.8877	1.5318	1.0206	1.2259	1.4130	0.7322	0.8648
	4.590	6.743	0.7952	0.8394	1.1630	1.3537	1.3886	0.6892	0.8225	1.3777
	4.590	6.743	1.4305	0.5715	1.2958	1.5472	0.8530	0.8727	0.9581	1.3992
	3.934	7.961	0.5971	1.6879	1.4052	1.0386	1.1022	1.2428	1.3565	0.7943
	3.934	7.961	0.8454	0.9222	1.4213	0.9389	1.0858	1.5480	0.8690	0.8724
	3.934	7.961	0.8878	0.8766	0.9906	1.1855	1.5377	0.8096	0.8519	1.3632
	3.934	7.961	1.5760	0.6125	1.0414	1.5616	1.0389	0.9368	0.9619	1.3545

GAUSSIAN RESPONSE TO 12-db

n	q_0	I. L.	q_1	q_n	k_{12}	k_{23}	k_{34}	k_{45}	k_{56}	k_{67}	k_{78}
	INF.	0.000	0.3449	2.0612	2.1014	1.3243	1.1976	1.1332	1.1224	0.9887	0.6229
	INF.	0.000	0.3736	1.4137	2.0343	1.2885	1.2157	1.0976	1.1635	0.9042	0.6115
	INF.	0.000	0.5108	0.7010	1.8146	1.2287	1.0824	1.3551	0.7861	0.6323	1.1397
	INF.	0.000	0.5762	0.6065	1.8108	1.1351	1.1638	1.3132	0.6956	0.6978	1.2003
	INF.	0.000	0.4893	0.7459	1.6267	1.2739	1.2807	0.9155	0.5581	0.9114	1.5133
	INF.	0.000	0.5490	0.6399	1.5859	1.2622	1.2960	0.8075	0.5692	1.0232	1.5042
	INF.	0.000	0.9073	0.4382	1.4364	1.4879	0.7350	0.6997	0.9540	1.0657	1.7468
	INF.	0.000	1.1363	0.3993	1.5295	1.3873	0.6744	0.7699	0.9408	1.1096	1.8132
	40.441	0.835	0.3711	1.8470	1.9983	1.2844	1.1873	1.1243	1.1357	0.9887	0.6223
	40.441	0.835	0.5921	0.6464	1.7595	1.1406	1.1188	1.3499	0.7372	0.6843	1.1556
	40.441	0.835	0.5287	0.7437	1.5512	1.2566	1.2945	0.9033	0.5619	0.9412	1.4862
	40.441	0.835	1.1291	0.4255	1.4594	1.4355	0.7037	0.7623	0.9382	1.0831	1.7404
	20.220	1.673	0.4032	1.6462	1.8961	1.2413	1.1807	1.1120	1.1564	0.9836	0.6239
	20.220	1.673	0.6054	0.6965	1.7040	1.1510	1.0705	1.3794	0.7955	0.6682	1.1130
	20.220	1.673	0.5783	0.7361	1.4761	1.2362	1.3149	0.8851	0.5711	0.9746	1.4568
	20.220	1.673	1.1101	0.4573	1.3777	1.4823	0.7470	0.7504	0.9401	1.0539	1.6692
8	13.480	2.521	0.4298	1.6318	1.8113	1.2067	1.1677	1.1089	1.1692	1.0020	0.6342
	13.480	2.521	0.6435	0.7219	1.6493	1.1328	1.0447	1.3959	0.8368	0.6705	1.0877
	13.480	2.521	0.6144	0.7624	1.4088	1.2138	1.3348	0.9001	0.5817	0.9749	1.4336
	13.480	2.521	1.1694	0.4798	1.3183	1.4997	0.7833	0.7578	0.9394	1.0372	1.6167
	10.110	3.383	0.4602	1.6181	1.7262	1.1701	1.1526	1.1053	1.1853	1.0259	0.6459
	10.110	3.383	0.6870	0.7490	1.5921	1.1139	1.0138	1.4121	0.8870	0.6758	1.0640
	10.110	3.383	0.6553	0.7906	1.3376	1.1837	1.3591	0.9234	0.5940	0.9764	1.4107
	10.110	3.383	1.2363	0.5045	1.2470	1.5166	0.8298	0.7687	0.9402	1.0213	1.5660
	8.088	4.260	0.4952	1.6048	1.6401	1.1297	1.1323	1.0989	1.2065	1.0602	0.6607
	8.088	4.260	0.7372	0.7776	1.5308	1.0931	0.9736	1.4262	0.9526	0.6863	1.0426
	8.088	4.260	0.7020	0.8210	1.2600	1.1389	1.3876	0.9624	0.6098	0.9793	1.3885
	8.088	4.260	1.3130	0.5317	1.1573	1.5285	0.8938	0.7861	0.9432	1.0067	1.5173
	6.740	5.153	0.5361	1.5901	1.5508	1.0818	1.0994	1.0808	1.2346	1.1191	0.6838
	6.740	5.153	0.7963	0.8075	1.4624	1.0665	0.9127	1.4298	1.0517	0.7093	1.0254
	6.740	5.153	0.7564	0.8531	1.1714	1.0619	1.4144	1.0364	0.6351	0.9851	1.3680
	6.740	5.153	1.4035	0.5613	1.0353	1.5203	0.9928	0.8187	0.9515	0.9948	1.4717
	5.777	6.342	0.5437	1.9749	1.4606	1.0348	0.9810	0.9714	1.1695	1.3671	0.8693
	5.777	6.342	0.6375	1.2867	1.4413	0.9567	1.0501	0.9009	1.2972	1.2941	0.7420
	5.777	6.342	0.7869	0.9303	1.3330	1.0778	0.7341	1.1986	1.4228	0.7967	1.0463
	5.777	6.342	1.0000	0.7431	1.4247	0.9010	0.7803	1.3662	1.2447	0.8448	1.0529

3-db DOWN k AND q VALUES

LEGENDRE RESPONSE

n	q_0	I. L.	q_1	q_n	k_{12}	k_{23}	k_{34}	k_{45}
3	INF.	-0.000	1.1737	2.1801	0.7933	0.5821		
	28.970	0.711	1.3002	1.9720	0.7502	0.6017		
	14.485	1.435	1.5592	1.6658	0.6787	0.6539		
	9.657	2.592	1.4090	2.0103	0.6306	0.6952		
	7.242	3.952	1.3829	2.2247	0.5988	0.7166		
	5.794	5.569	1.3856	2.4017	0.5627	0.7385		
	4.828	7.551	1.4041	2.5550	0.5206	0.7620		
	4.139	10.104	1.4350	2.6867	0.4695	0.7879		
4	INF.	0.000	1.0828	1.5642	0.7908	0.5880	0.5713	
	43.161	0.617	1.1876	1.4827	0.7356	0.5868	0.6031	
	21.581	1.421	1.0622	1.8913	0.6969	0.5691	0.6650	
	14.387	2.626	0.9930	2.3994	0.7013	0.5480	0.6880	
	10.790	4.030	0.9807	2.7950	0.6875	0.5291	0.7118	
	8.632	5.696	0.9859	3.1557	0.6660	0.5079	0.7370	
	7.194	7.733	1.0015	3.4926	0.6392	0.4818	0.7647	
	6.166	10.347	1.0253	3.7996	0.6079	0.4473	0.7960	
5	INF.	0.000	0.9512	1.9990	0.8434	0.5721	0.5605	0.5700
	INF.	0.000	1.1709	1.4337	0.6947	0.4839	0.5764	0.7477
	65.110	0.537	1.0494	1.8099	0.7861	0.5525	0.5627	0.6060
	65.110	0.537	1.1847	1.5121	0.7022	0.5050	0.5745	0.7052
	32.555	1.368	0.9908	2.2216	0.7638	0.5365	0.5484	0.6633
	32.555	1.368	1.2264	1.5529	0.7333	0.4953	0.6131	0.6183
	21.703	2.581	0.9363	2.8958	0.7734	0.5309	0.5359	0.6825
	21.703	2.581	1.2565	1.6196	0.7329	0.4590	0.6612	0.5704
	16.277	3.996	0.9234	3.5175	0.7672	0.5220	0.5240	0.7049
	16.277	3.996	1.2885	1.6917	0.7265	0.4262	0.7033	0.5267
	13.022	5.677	0.9261	4.1423	0.7535	0.5108	0.5095	0.7298
	13.022	5.677	1.3255	1.7645	0.7170	0.3925	0.7439	0.4822
	10.852	7.731	0.9384	4.7743	0.7349	0.4973	0.4904	0.7580
	10.852	7.731	1.3678	1.8382	0.7051	0.3558	0.7847	0.4340
	9.301	10.365	0.9583	5.3914	0.7125	0.4811	0.4628	0.7911
	9.301	10.365	1.4158	1.9131	0.6913	0.3137	0.8266	0.3792

3-db DOWN k AND q VALUES

LEGENDRE RESPONSE

n	q_0	I. L.	q_1	q_n	k_{12}	k_{23}	k_{34}	k_{45}	k_{56}	k_{67}
6	INF.	0.000	0.9160	1.5763	0.8574	0.5823	0.5373	0.5487	0.5772	
	INF.	0.000	1.2732	1.0631	0.7271	0.6070	0.4925	0.5097	0.7426	
	86.811	0.482	1.0042	1.4615	0.7991	0.5630	0.5274	0.5595	0.6193	
	86.811	0.482	1.0859	1.3173	0.7405	0.5274	0.5090	0.5894	0.6924	
	43.406	1.375	0.8214	2.4006	0.8474	0.5611	0.5160	0.5468	0.6563	
	43.406	1.375	1.1062	1.3699	0.6625	0.5627	0.5775	0.4737	0.7502	
	28.937	2.604	0.7819	3.2375	0.8667	0.5599	0.5104	0.5371	0.6741	
	28.937	2.604	1.4611	1.1069	0.7577	0.4248	0.6226	0.5504	0.6321	
	21.703	4.039	0.7719	4.0612	0.8660	0.5549	0.5031	0.5280	0.6957	
	21.703	4.039	1.5331	1.1242	0.7601	0.3823	0.6631	0.5360	0.6013	
	17.362	5.740	0.7733	4.9367	0.8563	0.5478	0.4943	0.5168	0.7201	
	17.362	5.740	1.6020	1.1474	0.7597	0.3409	0.7034	0.5174	0.5704	
	14.469	7.818	0.7817	5.8710	0.8410	0.5388	0.4834	0.5010	0.7483	
	14.469	7.818	1.6718	1.1744	0.7572	0.2984	0.7456	0.4921	0.5389	
	12.402	10.478	0.7954	6.8308	0.8215	0.5282	0.4699	0.4771	0.7820	
	12.402	10.478	1.7442	1.2045	0.7528	0.2529	0.7917	0.4560	0.5066	
7	INF.	0.000	0.8394	1.8640	0.8981	0.5909	0.5464	0.5189	0.5409	0.5810
	INF.	0.000	1.0632	1.2702	0.7624	0.5781	0.5640	0.4746	0.5096	0.7306
	INF.	0.000	0.9288	1.5356	0.8132	0.5335	0.5002	0.4986	0.5924	0.7044
	INF.	0.000	1.2109	1.1086	0.6562	0.4957	0.5554	0.5275	0.5617	0.8181
	115.998	0.434	0.9188	1.6767	0.8384	0.5675	0.5360	0.5148	0.5503	0.6246
	115.998	0.434	1.1099	1.2758	0.7329	0.5570	0.5587	0.4933	0.5260	0.7306
	115.998	0.434	0.9649	1.5422	0.7987	0.5408	0.5169	0.5099	0.5742	0.6783
	115.998	0.434	1.1779	1.1964	0.6854	0.5216	0.5538	0.5139	0.5493	0.7717
	57.999	1.361	0.7938	2.6189	0.8859	0.5712	0.5226	0.5130	0.5425	0.6522
	57.999	1.361	0.9548	1.6825	0.8263	0.5391	0.5469	0.4712	0.6206	0.5884
	57.999	1.361	1.1818	1.2571	0.7476	0.5611	0.4520	0.6145	0.5088	0.6786
	57.999	1.361	1.5781	0.9921	0.7586	0.4675	0.5366	0.5692	0.5258	0.7260
	38.666	2.596	0.7565	3.6135	0.9075	0.5742	0.5193	0.5076	0.5364	0.6672
	38.666	2.596	0.9441	1.8543	0.8322	0.5315	0.5544	0.4402	0.6697	0.5324
	38.666	2.596	1.1971	1.3103	0.7408	0.5661	0.4125	0.6657	0.4825	0.6530
	38.666	2.596	1.7458	0.9749	0.7744	0.4131	0.5650	0.5815	0.5060	0.7128

3-db DOWN k AND q VALUES

LEGENDRE RESPONSE

n	q_0	I. L.	q_1	q_n	k_{12}	k_{23}	k_{34}	k_{45}	k_{56}	k_{67}	k_{78}
7	28.999	4.037	0.7458	4.6602	0.9099	0.5722	0.5148	0.5014	0.5302	0.6870	
	28.999	4.037	0.9481	1.9974	0.8285	0.5230	0.5583	0.4126	0.7134	0.4836	
	28.999	4.037	1.2215	1.3572	0.7308	0.5680	0.3770	0.7107	0.4565	0.6302	
	28.999	4.037	1.8777	0.9776	0.7851	0.3650	0.5932	0.5886	0.4865	0.6951	
	23.200	5.747	0.7457	5.8366	0.9035	0.5676	0.5093	0.4938	0.5220	0.7102	
	23.200	5.747	0.9597	2.1263	0.8201	0.5135	0.5602	0.3847	0.7552	0.4356	
	23.200	5.747	1.2509	1.4027	0.7189	0.5684	0.3417	0.7538	0.4278	0.6086	
	23.200	5.747	1.9981	0.9883	0.7928	0.3185	0.6238	0.5910	0.4658	0.6755	
	19.333	7.834	0.7521	7.1642	0.8915	0.5611	0.5027	0.4844	0.5096	0.7375	
	19.333	7.834	0.9766	2.2458	0.8086	0.5030	0.5605	0.3544	0.7969	0.3851	
	19.333	7.834	1.2842	1.4481	0.7057	0.5679	0.3045	0.7967	0.3941	0.5874	
	19.333	7.834	2.1143	1.0037	0.7982	0.2715	0.6594	0.5867	0.4430	0.6547	
8	INF.	0.000	0.8205	1.5564	0.9109	0.5968	0.5441	0.5252	0.5145	0.5418	0.5893
	INF.	0.000	0.9924	1.1714	0.7864	0.5687	0.5597	0.5215	0.4612	0.5465	0.7403
	INF.	0.000	1.0568	1.0927	0.8140	0.5871	0.5049	0.5521	0.5195	0.4982	0.7039
	INF.	0.000	1.3603	0.8879	0.6912	0.6034	0.5128	0.4966	0.5031	0.5512	0.8428
	145.049	0.398	0.8954	1.4221	0.8510	0.5736	0.5308	0.5207	0.5126	0.5563	0.6379
	145.049	0.398	1.0337	1.1729	0.7594	0.5513	0.5452	0.5286	0.4828	0.5560	0.7391
	145.049	0.398	0.9262	1.3509	0.8228	0.5545	0.5150	0.5124	0.5154	0.5795	0.6764
	145.049	0.398	1.0750	1.1239	0.7267	0.5250	0.5306	0.5378	0.4978	0.5730	0.7696
	72.525	1.366	0.7060	2.7608	0.9564	0.5940	0.5272	0.5138	0.5116	0.5421	0.6469
	72.525	1.366	0.9568	1.3633	0.7838	0.5883	0.4932	0.4811	0.6180	0.4742	0.6986
	72.525	1.366	0.8298	1.7436	0.8164	0.5313	0.5296	0.5625	0.5082	0.4739	0.7597
	72.525	1.366	1.1993	1.0584	0.6236	0.5675	0.5698	0.4498	0.5715	0.5185	0.8218
	48.350	2.610	0.6756	3.8909	0.9824	0.5992	0.5271	0.5099	0.5072	0.5374	0.6605
	48.350	2.610	0.9519	1.4562	0.7782	0.5925	0.4864	0.4521	0.6719	0.4345	0.6821
	48.350	2.610	1.0719	1.2433	0.8265	0.5020	0.5935	0.4053	0.6125	0.5605	0.5774
	48.350	2.610	1.9872	0.8103	0.7817	0.4178	0.5221	0.5840	0.5206	0.5165	0.8165
	36.262	4.061	0.6663	5.1290	0.9876	0.5987	0.5248	0.5056	0.5019	0.5328	0.6791
	36.262	4.061	0.9596	1.5295	0.7673	0.5940	0.4796	0.4243	0.7188	0.3988	0.6678
	36.262	4.061	1.0900	1.2845	0.8244	0.4876	0.6090	0.3662	0.6532	0.5488	0.5387
	36.262	4.061	2.1808	0.8082	0.7985	0.3673	0.5380	0.6014	0.5093	0.5022	0.8075
	24.175	7.881	0.6701	8.2813	0.9732	0.5902	0.5166	0.4950	0.4868	0.5163	0.7277
	24.175	7.881	0.9902	1.6577	0.7388	0.5927	0.4662	0.3609	0.8063	0.3244	0.6413
	24.175	7.881	1.1378	1.3619	0.8114	0.4597	0.6307	0.2896	0.7405	0.5046	0.4700
	24.175	7.881	2.5225	0.8219	0.8228	0.2676	0.5828	0.6261	0.4767	0.4729	0.7791

7

Filter Characteristics
in the Time Domain

7.1 INTRODUCTION TO TRANSIENT CHARACTERISTICS

All previous discussions have been limited to filter behavior in the frequency domain, so important in communication technology. No less important is the behavior of filters in the time domain and the consideration of their response to input signals given as time functions. In pulse and telemetering techniques as well as in modern communications, the attenuation, phase, and group delay each have their counterpart in the time domain. The correlation between frequency and time domain parameters will be described by the Fourier or Laplace transform, since the circuits under consideration are linear and time invariant with lumped components.

It is worthwhile, first, to make a general comparison between these two representations. The stationary or steady-state consideration shows that for every sinusoidal signal on the input side, there is a corresponding sinusoidal signal of the same frequency on the output side. Therefore the transmission process in a linear system can be characterized by: (1) an attenuation or change in amplitude, and (2) a change in phase of these otherwise equivalent signals. Distortion of the signal can be easily explained and described by the frequency dependence of both of these parameters. An important parameter in the frequency domain is the bandwidth Δf.

Design considerations under dynamic conditions put the accent on the form of the signal. Distortion or deformation of the signal will be described by time parameters. One of these is the rise time τ_r, the time required for an impulse to rise from 10% to 90% of its final value. A long-known fundamental law of electrical communication states the relationship between these parameters in filters as follows:

$$\Delta f_{(3\,\text{dB})}\tau_r \approx 0.3 \rightarrow 0.45 \qquad (7.1.1)$$

A second dynamic parameter is the pulse width τ_m. Using the concept of the average pulse length τ_m and considering the case of unit input impulse signals

$$\tau_m \approx \tau_r \qquad (7.1.2)$$

In Fig. 7.1 the foregoing values are shown as related to a lowpass filter. Since the networks we deal with are linear, calculations can be simplified by applying the method of superposition: It is possible to represent a complex signal as a sum of component signals, as, for example, Fourier synthesis which uses as the basic elements sinusoidal oscillations of different frequencies. Practically, it is sufficient to know the transient response for such fundamental types of signals as delta functions (unit impulse), step functions, and brisk sinusoidal swings.

7.2 TIME AND FREQUENCY DOMAINS

The network may be considered as an operator $W(s)$ which is a function of the complex frequency $s = \sigma + j\omega$. Also, the time-varying driving function $f_1(t)$ can be transformed into a corresponding function of frequency $F_1(s)$ which can be operated upon by $W(s)$ to give the response

$$F_2(s) = W(s)F_1(s) \qquad (7.2.1)$$

A second transformation is then required to find the response as a function of time, $f_2(t)$. This method, then, involves the transformation of the resulting response from the frequency domain back to the time domain. In the earlier method of circuit analysis, a time-varying current was calculated as the response of the network to a time-varying applied voltage by solving the network differential equation. In the case

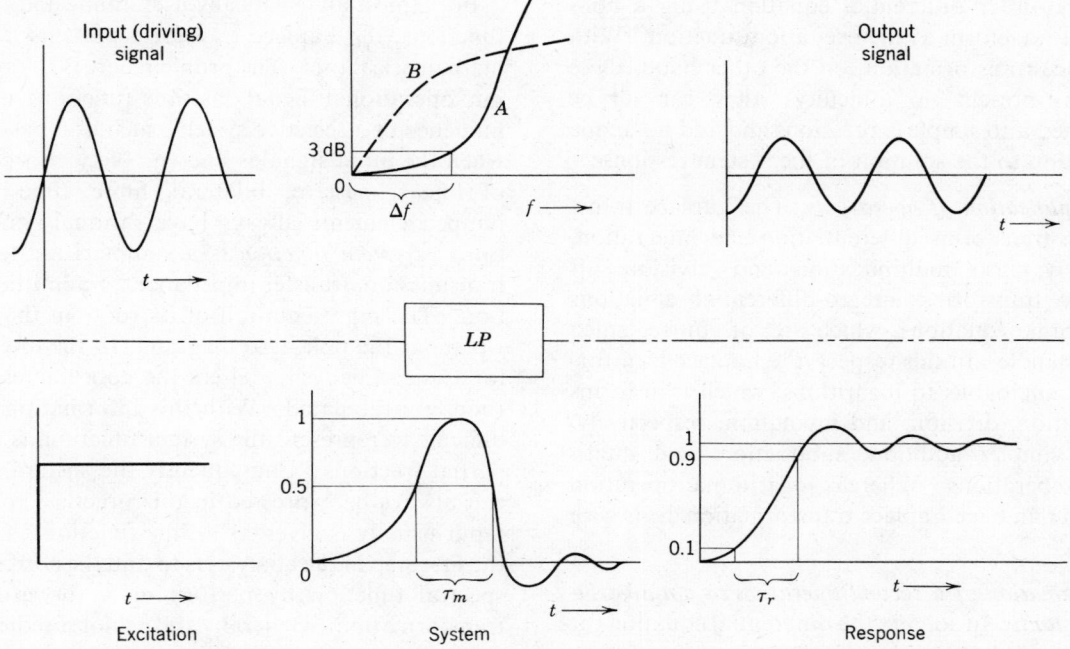

Fig. 7.1. Comparison of a lowpass system operating under steady-state and dynamic conditions.

of a multimesh network this could lead to differential equations of high order.

One fundamental limitation of the operational method is that the network must be linear—that is the principle of superposition holds and the magnitude of the response must be directly proportional to the magnitude of the driving force. If this linear relationship does not hold in a time-invariant network, modulation will result and additional frequency components which were not present in $f_1(t)$ will be present in $f_2(t)$.

Before the direct and inverse transformation between the time and frequency domains are considered, a few statements concerning $W(s)$ are in order. $W(s)$ is the same function that has been discussed earlier in Chapter 5; it may give the relationship between voltage and current at a single pair of terminals, or it may relate voltage or current at one pair of terminals to voltage at another pair of terminals or to current in another mesh of the network. This function, therefore, may have the dimension of impedance or admittance if it relates voltage to current, or it may be dimensionless when it relates voltage to voltage or current to current. It has been customary to call $W(s)$ the system function, or the transfer function.

From the previous expression, another definition of the transfer function can be formulated: The transfer function of a linear system is the ratio of the Laplace transforms of the output and input time

functions $f_2(t)$ and $f_1(t)$.

$$W(s) = \frac{F_2(s)}{F_1(s)} \qquad (7.2.2)$$

As a basis for calculation the one-sided Laplace transform in the following form is used

$$F(s) = \int_0^\infty f(t)e^{-st}\,dt \qquad (7.2.3)$$

and

$$f(t) = \frac{1}{2\pi j}\int_{\sigma_0-j\infty}^{\sigma_0+j\infty} F(s)e^{st}\,ds \quad \text{for } t > 0. \quad (7.2.4)$$

Advantages of the method of the Laplace transformation are:

1. Simplification of functions. The Laplace transformation transforms the frequently occurring exponential and transcendental functions and their combinations into simple, easier-to-handle algebraic functions. Of particular importance is its ability to transform nonsinusoidal periodic functions and functions with discontinuities or discontinuous derivatives into simple expressions. For example, it is quite difficult to determine the response of a system by the classical approach to excitation functions of types different than sinusoidal, no matter how simple the system is itself. The reader can easily satisfy himself that this is true by trying to find the solution of a simple first-order differential equation using a non-sinusoidal waveform as the excitation function. With the Laplace transformation, on the other hand, these waveforms present no difficulty; they can all be transformed into simple expressions and add no undue complication to the solution of the system response.

2. Simplification of operations. The Laplace transformation transforms differentiation and integration, respectively, into multiplication and division. It essentially transforms integro-differential equations into algebraic equations, which are, of course, much easier to handle. In this respect, the Laplace transformation is analogous to logarithms, which transforms multiplication, division, and involution, respectively, into the simpler addition, subtraction, and multiplication operations. Whereas logarithmic operation deals with numbers, Laplace transformation deals with functions.

3. Elimination of a second operation to obtain arbitrary constants. In solving the differential equation for a given system by the method of Laplace transforms, a complete solution is obtained; no separate constant-determining operation is necessary. In the classical method the determination of constants can be a tedious job, especially when there are many—that is, when the order of the differential equation is high.

4. Effective use of step and impulse responses. Briefly, if we know or have determined the response of a system to step or impulse excitation, then we can determine quite simply the response of the system to any other type of excitation function. This enables us to describe, or *label*, systems or networks in terms of their step or impulse responses. The method of Laplace transforms makes effective use of this property.

The foregoing points should become clear as we gradually expound the techniques of Laplace transformation.

For almost all technically useful time and frequency functions, the Laplace transform presents a valuable mathematical tool. The problem here is to obtain output operational signals as time functions under the influence of a selective system, such as a passive filter, when the input signal is known. Networks composed of linear, passive, bilateral, finite, time-invariant, lumped elements always have rational system functions. *System function* is a summarizing term for a transmission, transfer impedance, or admittance function. The representation of its roots in the complex s-plane is the pole-zero diagram. In the filter catalog for Cauer-Chebyshev filters the coordinates of these roots are tabulated. With this information it is not difficult to represent the system function as a sum of partial fractions. Consequently the spectral function can always be expressed in this practical form. The input usually is given as a time function. Therefore, the first step in the analysis is to find the corresponding spectral function by performing a forward (direct) transformation. Generally this is not a tedious task.

For several chosen, but typical, signal forms in Fig. 7.2 the quantative results are collected. The time functions are in the first row and the corresponding spectral functions are in the second row. The spectral function of the response signal can be expressed by Eq. 7.2.1, where

$F_1(s)$ is a spectral function of the input signal.

$W(s)$ is the transfer function of the system.

$F_2(s)$ is the spectral function of response signal.

The remainder of the problem is the translation of $F_2(s)$ into the time function $f_2(t)$. If, on the input side, the function is a unit impulse (delta function), the output time function will be

$$f_i(t) = L^{-1}[1 \cdot W(s)] \qquad (7.2.5)$$

Driving with a step function yields at the output

$$f_s(t) = L^{-1}\left[\frac{1}{s} W(s)\right] \qquad (7.2.6)$$

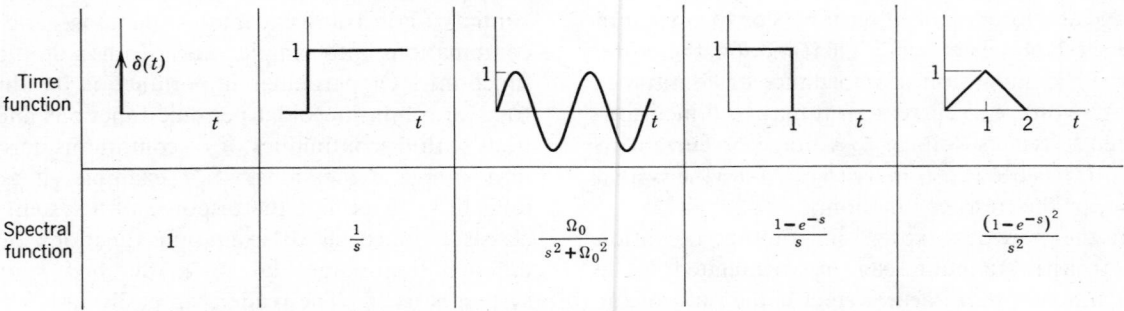

Fig. 7.2. Time and spectral representation of important normalized signal forms.

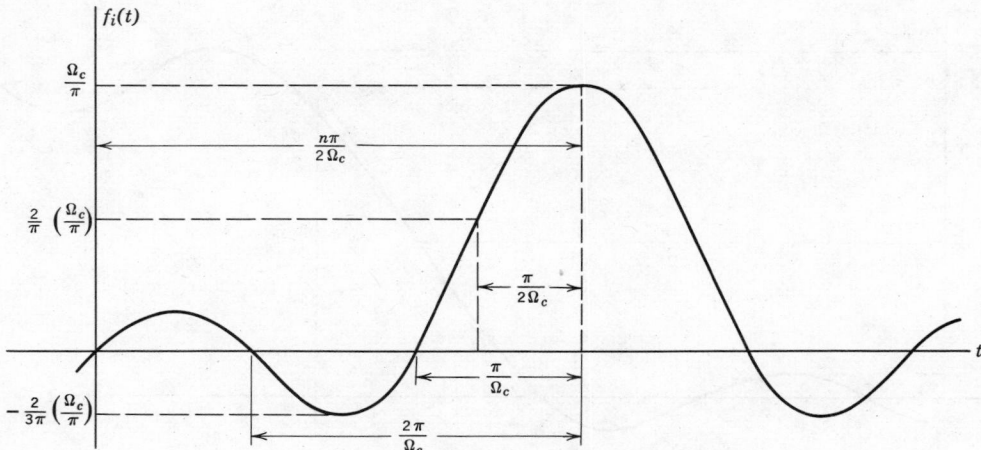

Fig. 7.3. Impulse response of ideal filter.

The problem usually is to choose the parameters of the rational function in such a way that the behavior of the system is optimal; for example, the rise time and overshoot are minimal for a given filter complexity.

7.3 INFORMATION CONTAINED IN THE IMPULSE RESPONSE

From the performance of the ideal system driven by the chosen signal, the conclusion can be drawn as to how the actual circuit will perform. Therefore, these ideal systems will serve as a first approximation to the actual circuit conditions. The most appropriate mathematical tool under the foregoing conditions, the simple time-frequency transformation, will be used.

The ideal lowpass filter transfer function has a magnitude of unity with linearly decreasing phase. For the ideal case the group delay is a constant (see Fig. 3.1*b*). Such a filter has the transfer function:

$$|W(j\Omega)| = 1 \quad \text{for } -\Omega_c \leq \Omega \leq \Omega_c$$
$$= 0 \quad \text{for all other } \Omega$$

$$b = \text{angle of } W(j\Omega) = -\frac{n\pi}{2}\frac{\Omega}{\Omega_c} \quad (7.3.1)$$

The unit impulse response $f_i(t)$ of the ideal filter is easily determined from the Fourier transform.

$$f_i(t) = \frac{1}{2\pi}\int_{-\Omega_c}^{\Omega_c} W(j\Omega) \cdot e^{j\Omega t}\, d\Omega$$

$$= \frac{1}{2\pi}\int_{-\Omega_c}^{\Omega_c} \exp\left[j\Omega\left(t - \frac{n\pi}{2\Omega_c}\right)\right] d\Omega \quad (7.3.2)$$

$$= \frac{\Omega_c}{\pi}\frac{\sin\left[\Omega_c t - (n\pi/2)\right]}{\left[\Omega_c t - (n\pi/2)\right]}$$

In Fig. 7.3 the impulse response of the ideal filter is shown.

From a purely mathematical point of view n must be infinite in order not to have an output before the impulse is applied to the input (anticipatory phenomenon). It can be seen that the pulse width $2\pi/\Omega_c$ as shown in Fig. 7.3 is inversely related to the bandwidth in the frequency domain. In other words, the wider the pulse width, the narrower the transmitting system (or filter). Conversely, if the transmitting system is very wide, the pulse width is very narrow and nearly equal to the incident pulse in width. This is more or less an obvious conclusion since the Fourier transform of the unit impulse contains all frequencies. On the other hand, the peak impulse amplitude is Ω_c/π which is proportional to the bandwidth. The larger the bandwidth the higher the amplitude of the impulse response in the time domain. The product of the pulse width and the amplitude of the impulse response is a constant, namely two.

7.4 STEP RESPONSE

To obtain the step response $f_s(t)$ one must integrate the impulse response so that

$$f_s(t) = \int_{-\infty}^{t} f_i(t)\, dt = \frac{\Omega_c}{\pi}\int_{-\infty}^{t} \frac{\sin\left(\Omega_c t - n\pi/2\right)}{\left(\Omega_c t - n\pi/2\right)}\, dt \quad (7.4.1)$$

In Fig. 7.4 the step response $f_s(t)$ is shown. With reference to the previous response for the impulse function, we notice that the time of the peak of the impulse corresponds to the time when the step response function reaches the halfway point. This time is equal

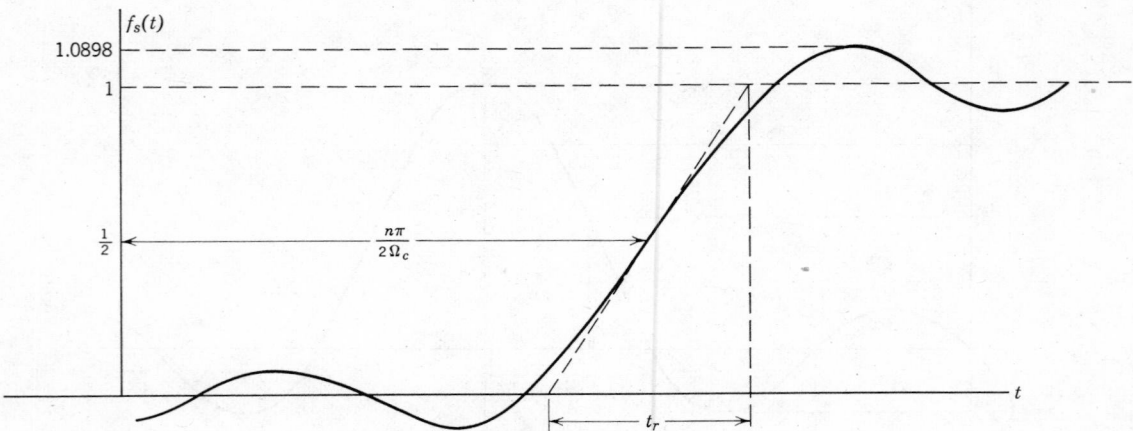

Fig. 7.4. Step response of ideal filter.

to $n\pi/2\Omega_c$. This point on the curve could be used to formulate the definition for the theoretical rise time. The rise time is that time required to go from zero to unity with a slope equal to that at the halfway point. (The usual definition is different. See Section 7.1.) Numerically the slope is equal to the peak amplitude of the impulse response, so that $t_r = \pi/\Omega_c$. Evidently the product of the bandwidth of the lowpass model and the rise time is a constant, namely π. Two conclusions could be made from visual analysis of the step response curve:

1. The filter has a prolonged oscillation, called ringing of pseudo period $2\pi/\Omega_c$.

2. Approximately 9% overshoot is present in the step response which makes the first overshoot higher than the original unit step.

Both effects, especially the ringing, are undesirable from a pulse transmission point of view and are present even though we have assumed ideal delay characteristics. The so-called ideal response which assumes rectangular amplitude and linear phase inside and outside the passband is not ideal in the time domain because of these two undesirable phenomena.

Therefore phase correction, sometimes used to supplement a chain of filters, will not improve the time-domain response of the so-called ideal filter, because the phase is already linear.

7.5 IMPULSE RESPONSE OF AN IDEAL GAUSSIAN FILTER

The Fourier transform of an idealized Gaussian characteristic having a constant in the exponential chosen such that $|W(j\Omega_c)| = \frac{1}{2}$ will provide the following impulse response

$$f_i(t) = \frac{\Omega_c}{2\sqrt{\pi \ln 2}} \exp - \frac{(\Omega_c t - n\pi/2)^2}{4 \ln 2} \quad (7.5.1)$$

A curve of this (impulse) response is shown in Fig. 7.5. The remarkable features of this impulse response are evident from the curve.

1. The impulse response is identical in shape to the absolute magnitude of the frequency response.

2. No overshoot or ringing is exhibited.

The peak of the impulse is within 7% of the peak of the ideal square-shape frequency response. Both could

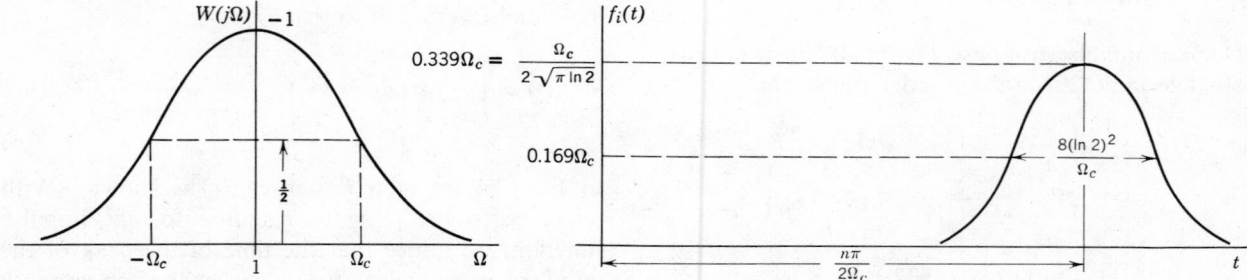

Fig. 7.5. Frequency and impulse response of a Gaussian filter.

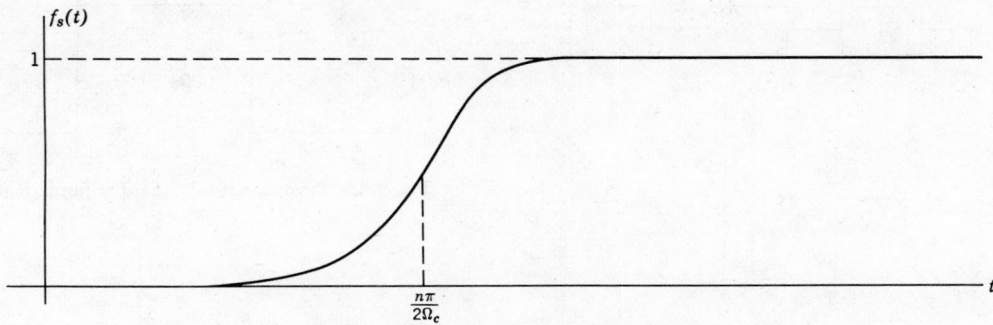

Fig. 7.6. Step response of a Gaussian filter.

have been made identical by the proper choice of bandwidth.

As in the previous case the step response is obtainable by integrating $f_i(t)$.

$$f_s(t) = \frac{\Omega_c}{2\sqrt{\pi \ln 2}} \int_{-\infty}^{t} \exp - \frac{(\Omega_c t - n\pi/2)^2}{4 \ln 2} \, dt \quad (7.5.2)$$

The practical procedure requires the use of tables because the step function expression is not integrable. The theoretical rise time from Fig. 7.6 is easily obtained to be

$$t_r = \frac{2\sqrt{\pi \ln 2}}{\Omega_c} = \frac{2.95}{\Omega_c} \quad (7.5.3)$$

which is quite close to the previous result for the ideal filter.

The Gaussian frequency response is very advantageous from the transient point of view. From the frequency domain point of view the Gaussian filter is usually not sufficiently selective.

7.6 RESIDUE DETERMINATION

Let us assume that $F_2(s)$ in (7.2.1) is a rational function and therefore the Heaviside operational method can be employed (instead of 7.2.3 and 7.2.4). For a better understanding of the calculations, first the residue determination will be explained and later the inverse transformation will be shown.

The first step consists of representing the spectral function as a sum of partial fractions.

Given is the rational function

$$F(s) = \frac{N(s)}{D(s)} = C \frac{a_0 + a_1 s + \cdots + s^m}{b_0 + b_1 s + \cdots + s^n} \quad (7.6.1a)$$

$$= C \frac{(s - s_{01})(s - s_{02}) \cdots (s - s_{0m})}{(s - s_{1\infty})(s - s_{2\infty}) \cdots (s - s_{n\infty})} \quad (7.6.1b)$$

with $n > m$, and n and m as integers. From the practical standpoint, it is preferable to work with the expression in (7.6.1b). The roots of the system are available and can be taken from the catalog to find the expression in the form

$$F(s) = \frac{\text{Res}_1}{(s - s_{1\infty})} + \frac{\text{Res}_2}{(s - s_{2\infty})} + \cdots + \frac{\text{Res}_n}{(s - s_{n\infty})} \quad (7.6.2)$$

This introductory description will be limited to the simple poles in (7.6.1). The problem now is to determine the complex values of Res_i.

The best known methods of solution are:

1. The coefficient comparison method. The solution is accomplished by equating (7.6.1b) with (7.6.2), and then equating coefficients of like powers of s.

2. The remainder function method. The residue Res_i can be regarded as a remainder function of $F(s)$ when the linear factor $(s - s_{i\infty})$ is removed from (7.6.1b) and s is replaced by $s_{i\infty}$.

3. The graphical method. When the function is very complicated the graphical method can be of great value for practical engineering work. In the pole-zero diagram (in the complex frequency plane) the $|\text{Res}_i|$ can be described (within a constant) by the straight lines from the reference pole s_i to all other roots. The angle of Res_i corresponds to the sum of angles with appropriate signs.

The time function which corresponds to $F(s)$ is then the sum of the time functions corresponding to each of the terms in (7.6.2). See Fig. 7.7.

7.7 NUMERICAL EXAMPLE

GIVEN: Input $v_1(t)$ (unit impulse).

SYSTEM: CC03 10 20 (see Fig. 7.8).

TO FIND: Impulse response $v_i(t)$.

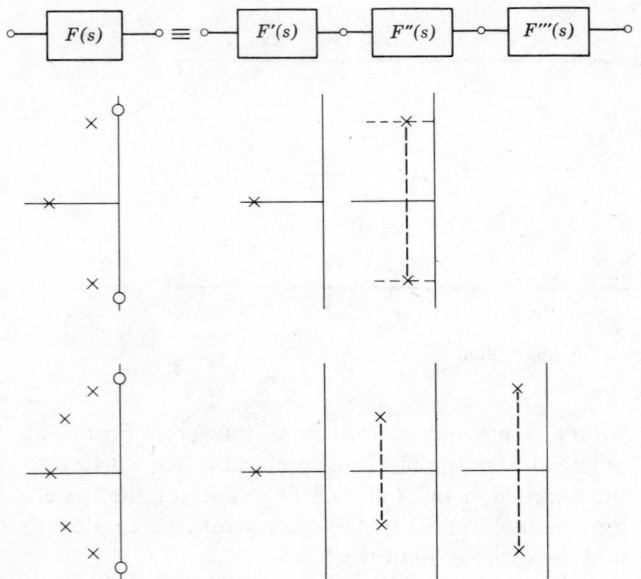

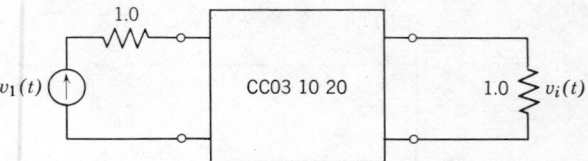

Fig. 7.8. Transmission system for numerical example.

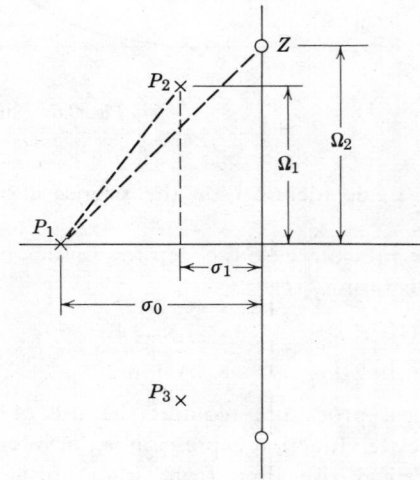

Fig. 7.7. Division of transmission system into partial systems. Contributions of complex conjugate transfer poles are combined together.

Fig. 7.9. Determination of residue with reference point P_1.

SOLUTION: According to Fig. 7.2 a unit impulse has a constant spectral function of value 1 (normalized representation and therefore, without dimensions).

The pole-zero data of transfer function

$$W(s) = \frac{2V_i(s)}{V_1(s)} \tag{7.7.1}$$

are found from the catalog of **Chapter 5**:

$$\left.\begin{array}{l} \sigma_0 = 1.263 \\ \sigma_1 = 0.5153 \\ \Omega_1 = \pm 1.341 \end{array}\right\} \begin{array}{l} \text{The coordinates of the transfer} \\ \text{poles (or transmission zeros).} \end{array}$$

Zero transfer (pole of attenuation) occurs at

$$\Omega_2 = \pm 3.35$$

With this information the pole-zero diagram for the output function can be drawn. This will be a geometric representation of the spectral function that corresponds to the output time function. In this example the spectral function of the input signal given by the value 1 has no poles and zeros.

Determination of Residue

From the known form (with $C = 1$ for simplicity)

$$V_i(s) = \frac{(s - s_{01})(s - s_{02})}{(s - s_{1\infty})(s - s_{2\infty})(s - s_{3\infty})} = \frac{(s - j3.35)(s + j3.35)}{(s + 1.263)(s + 0.5153 - j1.341)(s + 0.5153 + j1.341)} \tag{7.7.2}$$

the equivalent expression can be developed.

$$V_i(s) = \frac{\text{Res}_1}{s - s_{1\infty}} + \frac{\text{Res}_2}{s - s_{2\infty}} + \frac{\text{Res}_3}{s - s_{3\infty}} \tag{7.7.3}$$

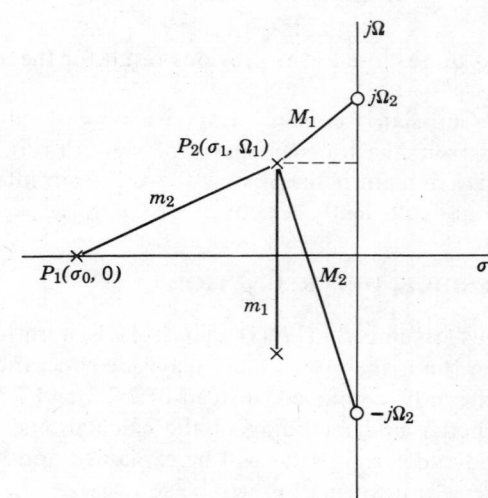

Fig. 7.10. Residue determination with reference point P_2.

The problem is reduced to the evaluation of the residues in (7.7.3). This will be done by methods 2 and 3 of Section 7.6.

With a known reference point such as P_1 (with coordinates $\sigma_0 + j0$). (See Fig. 7.9.)

$$\text{Res}_1 = \frac{(s - j\Omega_2)(s + j\Omega_2)}{(s + \sigma_1 - j\Omega_1)(s + \sigma_1 + j\Omega_1)}\bigg|_{s=-\sigma_0} \quad (7.7.4)$$

At $s = -1.263$

$$\text{Res}_1 = \frac{(-1.263 - j3.35)(-1.263 + j3.35)}{(-1.263 + 0.5153 - j1.341)(-1.263 + 0.5153 + j1.341)} = 5.44 + j0.0$$

or by the graphical method:

$$|\text{Res}_1| = \frac{\sigma_0^2 + \Omega_2^2}{(\sigma_0 - \sigma_1)^2 + \Omega_1^2} = 5.44 \quad (7.7.5)$$

$\measuredangle \text{Res}_1 = 0$

With reference point P_2 (coordinates $\sigma_1 + j\Omega_1$). (See Fig. 7.10.)

$$M_1 = \sqrt{\sigma_1^2 + (\Omega_2 - \Omega_1)^2} \quad (7.7.6)$$

$$M_2 = \sqrt{(\Omega_2 + \Omega_1)^2 + \sigma_1^2} \quad (7.7.7)$$

$$m_1 = 2\Omega_1 \quad (7.7.8)$$

$$m_2 = \sqrt{\Omega_1^2 + (\sigma_0 - \sigma_1)^2} \quad (7.7.9)$$

$$|\text{Res}_2| = \frac{m_1 m_2}{M_1 M_2} \quad (7.7.10)$$

$$= \frac{\sqrt{\sigma_1^2 + (\Omega_2 - \Omega_1)^2}}{\sqrt{(\sigma_0 - \sigma_1)^2 + \Omega_1^2}} \frac{\sqrt{\sigma_1^2 + (\Omega_2 + \Omega_1)^2}}{2\Omega_1}$$

At $s = -0.5153 + j1.341$

$|\text{Res}_2| = 2.375$

$\measuredangle |\text{Res}_2| = -159.1°$

Alternatively,

$$\text{Res}_2 = \frac{[-0.5153 + j(1.341 - 3.35)][-0.5153 + j(1.341 + 3.35)]}{(-0.5153 + 1.263 + j1.341)j(2.682)}$$

$$\text{Res}_2 = -2.22 - j0.847$$

With reference point P_3 (coordinates $\sigma_1 - j\Omega_1$)

$|\text{Res}_3| = |\text{Res}_2| = 2.375$

$\measuredangle \text{Res}_3 = - \measuredangle \text{Res}_2 = 159.1°$

or

$\text{Res}_3 = -2.22 + j0.847$

Therefore for $V_i(s)$:

$$V_i(s) = \frac{5.44}{s + 1.263} + \frac{2.22 + j0.847}{s + 0.5153 - j1.341}$$

$$- \frac{2.22 - j0.847}{s + 0.5153 + j1.341} \quad (7.7.11)$$

For the two complex conjugate components in (7.7.11) the following equation can be applied.

$$\frac{a + jb}{s + \sigma - j\Omega} + \frac{a - jb}{s + \sigma + j\Omega} = 2\frac{as + a\sigma - b\Omega}{s^2 + 2\sigma s + \Omega^2 + \sigma^2}$$

As a result of this recombination the expression for $V_i(s)$ will be simplified.

$$V_i(s) = \frac{5.44}{s + 1.263} + \frac{-4.44s}{s^2 + 1.031s + 2.065}$$

The inverse transformation to the time domain can be achieved with the aid of the expression

$$\frac{\text{Res}_i}{s - s_{i\infty}} \rightarrow \text{Res}_i\, e^{s_{i\infty}t} \quad (7.7.12)$$

Therefore, with $159.1° = 2.777$ rad

$$v_i(t) = \text{Res}_1 e^{s_1 \infty t} + \text{Res}_2 e^{s_2 \infty t} + \text{Res}_3 e^{s_3 \infty t}$$

$$= 5.44 e^{-1.263t} + 2.375 e^{-0.5153t} e^{j(1.341t - 2.777)}$$

$$= 5.44 e^{-1.263t} + 4.75 e^{-0.5153t} \cos(1.341t - 2.777)$$

$$(7.7.13)$$

which describes the composite impulse response in the form of an exponential decay plus a damped oscillation.

Using different values of t, $v_i(t)$ can be calculated and the response curve plotted as in Fig. 7.11. When C equals unity the curve starts out horizontally and equal to one at $t = 0$. The validity of the general approximation introduced in Eq. 7.1.1 and 7.1.2 is shown by the foregoing example.

In normalized form

$$t_m = \frac{0.3 \rightarrow 0.45}{f_{3\,\text{dB}}} = \frac{0.3 \rightarrow 0.45}{\Omega_{3\,\text{dB}}} 2\pi = 1.7 \rightarrow 2.6$$

with $\Omega_{3\,\text{dB}} = 1.1$ (estimated value, since only the bandwidth for $\rho = 10\%$ is known exactly). This

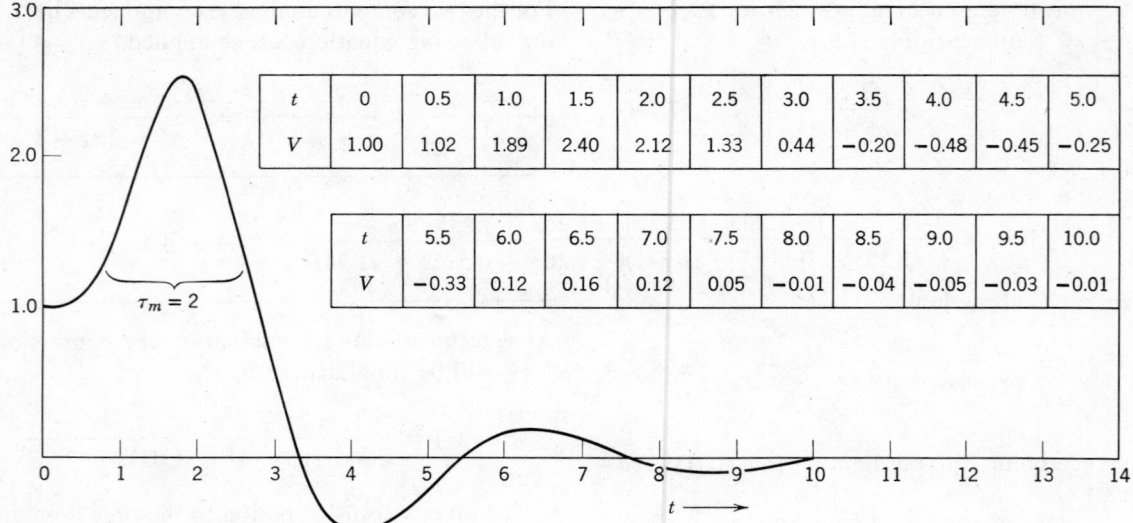

t	0	0.5	1.0	1.5	2.0	2.5	3.0	3.5	4.0	4.5	5.0
V	1.00	1.02	1.89	2.40	2.12	1.33	0.44	−0.20	−0.48	−0.45	−0.25

t	5.5	6.0	6.5	7.0	7.5	8.0	8.5	9.0	9.5	10.0
V	−0.33	0.12	0.16	0.12	0.05	−0.01	−0.04	−0.05	−0.03	−0.01

Fig. 7.11. Calculated impulse response.

range is in complete agreement with the example shown in Fig. 7.11.

7.8 PRACTICAL STEPS IN THE INVERSE TRANSFORMATION

Practical steps in the calculation procedure for the response function can be devised when the system is known and the time function of the input signal is given. With known pole-zero information the determination of the residues and the inverse transformation can be performed in one compact step in a trigonometric formulation. When certain familiarity with the solutions to simple problems is attained one can, for a given rational (spectral) function represented in the pole-zero diagram, write immediately the corresponding time function.

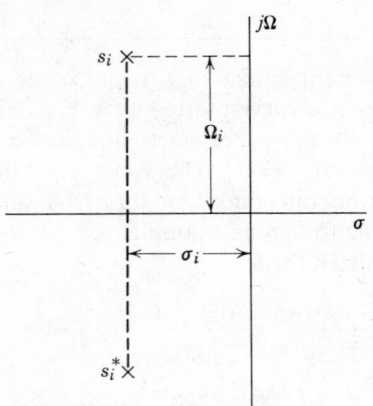

Fig. 7.12. Location of transfer pole pair.

In actual networks only the real poles and the conjugate-complex transfer poles (zeros of transmission) are possible. It is sensible, therefore, to investigate which time function belongs to such fundamental expressions. The complete answer is given in Eq. 7.7.12. For the real pole, the expression in Eq. 7.7.12 is the only one possible. For the conjugate complex pole the corresponding time function exists in a trigonometric form.

In Fig. 7.12 the pole-zero diagram of a transfer pole pair is given.

The transfer function corresponding to this diagram is:

$$W(s) = [(s + \sigma_i + j\Omega_i)(s + \sigma_i - j\Omega_i)]^{-1}$$

$$= \frac{K_1}{(s + \sigma_i + j\Omega_i)} + \frac{K_1{}^*}{(s + \sigma_i - j\Omega_i)} \quad (7.8.1)$$

or

$$[(s + \sigma_i - j\Omega_i)]^{-1} = K_1 + \frac{K_1{}^*(s + \sigma_i + j\Omega_i)}{s + \sigma_i - j\Omega_i}$$

At $s = -\sigma_i - j\Omega_i$

$$K_1 = (-\sigma_i - j\Omega_i + \sigma_i - j\Omega_i)^{-1} = (-2j\Omega_i)^{-1}$$

and

$$K_1{}^* = (-\sigma_i + j\Omega_i + \sigma_i + j\Omega_i)^{-1} = (2j\Omega_i)^{-1}$$

$$\text{at } s = -\sigma_i + j\Omega_i$$

Therefore in the time domain the time function for

the complex conjugate poles $f_c(t)$ is

$$f_c(t) = \frac{1}{2j\Omega_i} e^{(-\sigma_i + j\Omega_i)t} + \frac{1}{-2j\Omega_i} e^{(-\sigma_i - j\Omega_i)t}$$

$$= \frac{1}{\Omega_i} e^{-\sigma_i t} \left(\frac{e^{j\Omega_i t} - e^{-j\Omega_i t}}{2j} \right)$$

$$= \frac{1}{\Omega_i} e^{-\sigma_i t} \sin \Omega_i t \qquad (7.8.2)$$

The same result may be obtained using Eq. 7.7.12:

$$f_c(t) = |\text{Res}_i| e^{-\sigma_i t} [e^{j(\Omega_i t + \measuredangle \text{Res}_i)} + e^{-j(\Omega_i t + \measuredangle \text{Res}_i)}] \qquad (7.8.3)$$

with

$$\text{Res}_i = \frac{1}{2\Omega_i} e^{j270°} = \frac{1}{\Omega_i} \frac{1}{2j} \qquad (7.8.4)$$

$$\text{Res}_i{}^* = \frac{1}{2\Omega_i} e^{-j270°} = \frac{1}{\Omega_i} \frac{1}{-2j} \qquad (7.8.5)$$

or

$$f_c(t) = \frac{1}{\Omega_i} e^{-\sigma_i t} \frac{e^{j\Omega_i t} - e^{-j\Omega_i t}}{2j} = \frac{1}{\Omega_i} e^{-\sigma_i t} \sin \Omega_i t \quad (7.8.6)$$

Therefore, the pair of transfer poles with their spectral function

$$F(s) = [(s + \sigma_i)^2 + \Omega_i^2]^{-1} = (s^2 + 2\sigma_i s + \sigma_i^2 + \Omega_i^2)^{-1} \qquad (7.8.7)$$

has a corresponding time function $f(t)$ which is a sinusoidal decay with attenuation decrement σ_i. The governing factor in amplitude decay is the factor σ_i, and the frequency of free oscillation is Ω_i. The normalized amplitude is $1/\Omega_i$ and its phase angle is zero. This is indeed the simplest result.

7.9 INVERSE TRANSFORM OF RATIONAL SPECTRAL FUNCTIONS

The result just obtained can be used with advantage for the inverse transformations of spectral functions containing only simple transfer poles. This does not present any serious limitation, since the optimum system functions under consideration always contain simple poles. Naturally, the technique can be extended to cover the case of transfer functions with multiple poles. This related development will not be discussed here.

Using the combination of rules explained in the previous text, the following steps for obtaining the solution should be followed. Determine the residue for each pole and immediately transform each into its corresponding time function. The complex-conjugate transfer pole pairs should be considered as an operational unit and the calculations therefore will be minimized. The time function for the total system is the arithmetical sum of the partial time functions. With experience, the time function can be written without much difficulty.

When the reference pole is real, Fig. 7.13 permits us to write the following expression (with reference line as indicated).

$$f_R(t) = \frac{\prod\limits_{u=1}^{m} \overline{P_i Z_u}}{\prod\limits_{k=1}^{n} \overline{P_i P_k}} e^{-\sigma_i t} e^{j \measuredangle \text{Res}_i} \qquad (7.9.1)$$

with

$$\measuredangle \text{Res}_i = \sum_{u=1}^{m} \measuredangle \overline{P_i Z_u} - \sum_{k=1}^{n} \measuredangle \overline{P_i P_k} \qquad (7.9.2)$$

where m = number of zeros and n = number of poles (without P_i). The resulting phase angle either has the value 0° or 180° which corresponds to the plus or to the minus sign.

On the other hand, when a conjugate complex pole is considered as a reference pole (as shown in Fig. 7.10)

$$f_c(t) = \frac{\prod\limits_{u=1}^{m} \overline{P_i Z_u}}{\Omega_i \prod\limits_{k=1}^{n} P_i P_k} e^{-\sigma_i t} \sin (\Omega_i t + \measuredangle \text{Res}_i) \quad (7.9.3)$$

with

$$\measuredangle \text{Res}_i = \sum_{u=1}^{m} \measuredangle (\overline{P_i Z_u}) - \sum_{k=1}^{n} \measuredangle (\overline{P_i P_k}) \quad (7.9.4)$$

where m = number of zeros and n = number of poles (without P_i) of the transfer function. The complete time function is therefore

$$f(t) = \sum_{1}^{\gamma} f_R(t) + \sum_{1}^{\beta} f_c(t) \qquad (7.9.5)$$

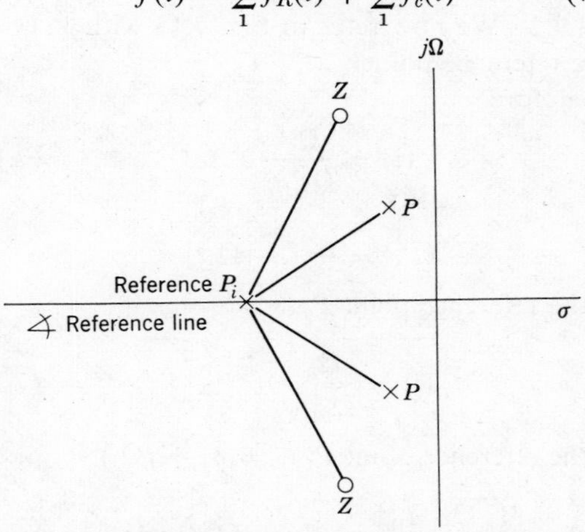

Fig. 7.13. Pole-zero diagram with real reference pole.

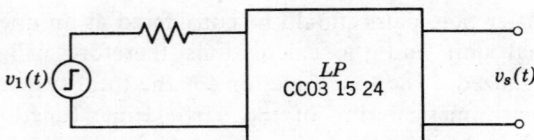

Fig. 7.14. Transmission system for numerical example.

where γ = number of real poles and β = number of conjugate-complex pole pairs.

7.10 NUMERICAL EXAMPLE

GIVEN: The input signal is a step function.

SYSTEM: LP CC03 15 24 (see Fig. 7.14).

PROBLEM: Find the response $v_s(t)$.

SOLUTION: According to Fig. 7.2, a step function has a spectral function $1/s$ (normalized representation, therefore dimensionless).

From the catalog of pole-zero data the effective

$$v_s(t) = 1 - e^{-\sigma_0 t} - \frac{\sigma_0}{\Omega_1(\sigma_0^{\ 2} + \Omega_2^{\ 2})} \left\{ \frac{[\sigma_1^{\ 2} + (\Omega_2 - \Omega_1)^2][\sigma_1^{\ 2} + (\Omega_2 + \Omega_1)^2][(\sigma_0 - \sigma_1)^2 + \Omega_1^{\ 2}]}{\sigma_1^{\ 2} + \Omega_1^{\ 2}} \right\}^{1/2} \cdot e^{-\sigma_1 t} \sin \Omega_1 t$$

$$(7.10.6)$$

transfer function has the following parameters.

$$\sigma_0 = 1.062$$
$$\sigma_1 = 0.4195$$
$$\Omega_1 = \pm 1.217$$
$$\Omega_2 = \pm 2.808$$

The superposition of the above pole-zero data, and the pole of the input spectral function is shown in Fig. 7.15, which is the geometric interpretation of Eq. 7.9.2. We now refer to Fig. 7.15 with P_s $(s = 0)$ as the reference point.

Therefore

$$f_s(t) = \frac{\Omega_2^{\ 2}}{\Omega_{s1}\sigma_0} e^{-0t} e^{j0°} \tag{7.10.1}$$

with

$$\Omega_{s1}^{\ 2} = (\sigma_1^{\ 2} + \Omega_1^{\ 2})$$

For the reference point P_0 $(s = -\sigma_0)$

$$f_R(t) = \frac{\Omega_{02}^{\ 2}}{\Omega_{01}^{\ 2}\sigma_0} e^{-\sigma_0 t} e^{-j\pi} \tag{7.10.2}$$

For the reference point P_1 $(s = \sigma_1 + j\Omega_1)$

$$f_c(t) = \frac{1}{\Omega_1} \frac{\Omega_{12}\Omega_{12*}}{\Omega_{1s}\Omega_{10}} e^{-\sigma_1 t} \sin(\Omega_1 t + 180°) \tag{7.10.3}$$

where

$$\beta_{12} + \beta_{12*} + \beta_{10} + \beta_{1s} = 180°$$

According to Eq. 7.9.5

$$v_s(t) = f_s(t) + f_R(t) + f_c(t)$$

However, for a unit step input the final value ($t = \infty$) at the output is always unity. Therefore $v_s(t)$ must be unnormalized. This is equivalent to multiplying $v_s(t)$ by $\sigma_0 \Omega_{s1}/\Omega_2^{\ 2}$. The final result is

$$v_s(t) = 1 - \frac{\Omega_{02}^{\ 2}\Omega_{s1}^{\ 2}}{\Omega_{01}^{\ 2}\Omega_2^{\ 2}} e^{-\sigma_0 t}$$
$$- \frac{\Omega_{12}\Omega_{12*}\sigma_0\Omega_{1s}}{\Omega_1\Omega_{10}\Omega_2^{\ 2}} e^{-\sigma_1 t} \sin \Omega_1 t \tag{7.10.4}$$

The pole-zero distribution obeys the following law (which is an important checking equation)

$$\Omega_2^{\ 2} = \frac{\sigma_0(\sigma_1^{\ 2} + \Omega_1^{\ 2})}{\sigma_0 - 2\sigma_1} \quad \text{(see Eq. 5.1.8)} \tag{7.10.5}$$

Poles and zeros cannot be chosen freely in the complex frequency plane. For the case under consideration the following simplification can be made:

Thus the step function response is expressed by the known (tabulated) parameters of the transfer function. It can be seen that at $t = 0$ the total response equals zero. With the values given in the catalog Eq. 7.10.6 reduces to:

$$v_s(t) = 1 - e^{1.062t} - 0.692 e^{-0.4195t} \sin 1.217t$$

It is seen that $v_s(t)$ consists of a constant component, a decreasing exponential, and a damped sine oscillation which has zero phase.

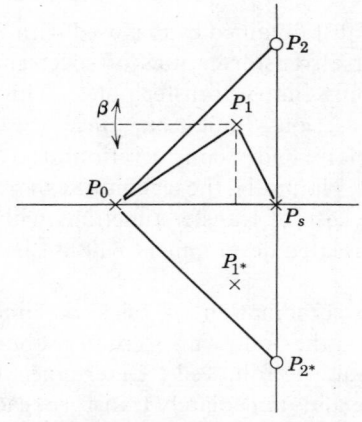

Fig. 7.15. Pole-zero diagram for design example of Fig. 7.14.

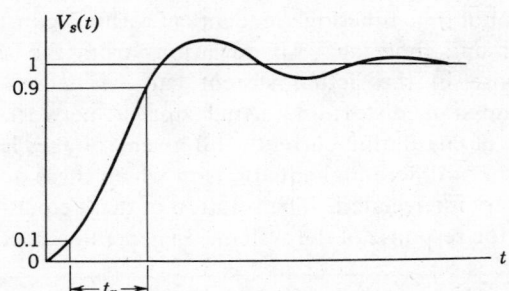

Fig. 7.16. Calculated step response of filter of Fig. 7.14.

In Fig. 7.16 the calculated step response of the above filter is shown. With the normalization chosen, it is evident that $v_s(\infty) = 1$. The derivative of $v_s(t)$ at $t = 0$ is finite. The normalized rise time is $t_r = 2$ which, once again, is in the limit given by the approximate expression in the beginning of this treatment as:

$$t_r = \frac{0.3 \to 0.45}{f_{3\,dB}} = 1.7 \to 2.6$$

or

$$t_r = \frac{0.3 \to 0.45}{1.1} \cdot 6.28$$

for $\Omega_c = 1.1$ (the approximate value, since only the bandwidth for $\rho = 15\%$ is known exactly). The rise time given by this expression is in good agreement with the calculated (or measured) value. The overshoot amounts to 12% of the final value.

The impulse response and step response of the system are important characteristics of the transmission system. The impulse response gives insight into the "fine structure" of the corresponding step response. Practical problems arise when the impulse response of the system is to be measured. The first is to put enough energy into the system. The second is that it may not be permissible to excite the system instantaneously. Often it is enough to know only a characteristic value of the system, such as the rise time, pulse width, etc., instead of a complete description of the system's properties.

If the normalization is such that step function response $f_s(\infty) = 1$, then the "time area" of the corresponding impulse response will be

$$\int_0^\infty f_i(t)\, dt = 1 \tag{7.10.7}$$

where $f_i(t)$ is the impulse response.

7.11   ESTIMATION THEORY

In Fig. 7.17 both the impulse and step responses are shown. In classical system theory the concept of approximating the impulse response by an equivalent rectangle whose width is the average rise time τ_r and whose area is unity (see Eq. 7.10.7) is well known. The equivalent rectangle is drawn with dashed lines in Fig. 7.17 and provides mathematical simplifications when only an estimate of the system behavior is needed. A second important parameter is the average time delay which can be defined as the time center of gravity of the impulse response, and is

$$t_0 = \frac{\displaystyle\int_0^\infty t f_i(t)\, dt}{\displaystyle\int_0^\infty f_i(t)\, dt} \tag{7.11.1}$$

This expression may be simplified further if the normalization is accomplished according to (7.10.7). The denominator in (7.11.1) will then be equal to unity. The step response in estimation theory is understood to be the shaded area shown in Fig. 7.17. It is important to note that in general, the position of the first, and most important, maximum in the response $f_i(t)$ does not coincide with t_0. Nevertheless, the value of t_0 can be easily determined from the roots of the transfer function (see "pole-zero" discussion, Section 5.1). For the example of Section 7.10, the transfer function is given by:

$$W(s) = B \frac{s^2 + \Omega_2{}^2}{s^3 + (\sigma_0 + 2\sigma_1)s^2 + (2\sigma_0\sigma_1 + \sigma_1{}^2 + \Omega_1{}^2)s + \sigma_0(\sigma_1{}^2 + \Omega_1{}^2)} \tag{7.11.2}$$

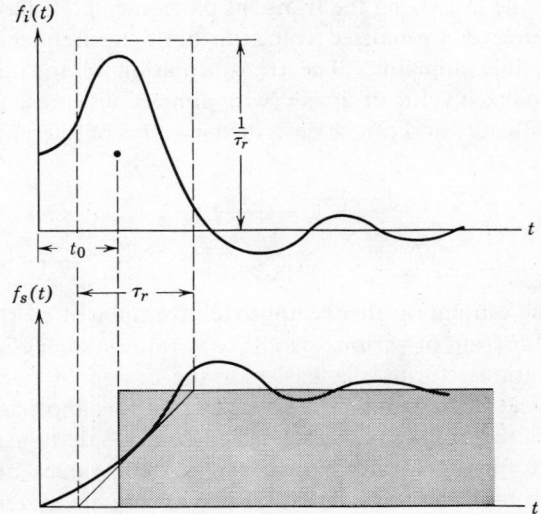

Fig. 7.17. Relationship between impulse and step response.

with

$$B = \frac{\sigma_0(\sigma_1{}^2 + \Omega_1{}^2)}{\Omega_2{}^2} \qquad (7.11.3)$$

Since $W(0) = 1$ when $s = 0$

$$\int_0^\infty f_i(t)\, dt = 1 \quad \text{or} f_s(\infty) = 1 \quad \text{(see Fig. 7.13)} \quad (7.11.4)$$

$W(s)$ can also be expressed as

$$W(s) = \frac{(1/\Omega_2{}^2)s^2 + 1}{[\sigma_0(\sigma_1{}^2 + \Omega_1{}^2)]^{-1}s^3 + \{[(\sigma_0 + 2\sigma_1)s^2]/\sigma_0(\sigma_1{}^2 + \Omega_1{}^2)\} + [(2\sigma_0\sigma_1 + \sigma_1 + \Omega_1{}^2)/\sigma_0(\sigma_1{}^2 + \Omega_1{}^2)]s + 1} \qquad (7.11.5)$$

Referring to Eq. 7.2.5 it is seen that

$$W(s) = \int_0^\infty f_i(t)e^{-st}\, dt$$

Then

$$\frac{dW(s)}{ds} = -\int_0^\infty t f_i(t)e^{-st}\, dt \qquad (7.11.6)$$

At $s = 0$, t_0 can be found from 7.11.5 and 7.11.6:

$$t_0 = -\left.\frac{dW(s)}{ds}\right|_{s=0} = \frac{2\sigma_0\sigma_1 + \sigma_1{}^2 + \Omega_1{}^2}{\sigma_0(\sigma_1{}^2 + \Omega_1{}^2)} \qquad (7.11.7)$$

This value is identical with the group delay for the steady-state condition at $\Omega = 0$ expressed by (see Eq. 5.3.3)

$$t_g(\Omega) = \frac{\sigma_0}{\sigma_0{}^2 + \Omega^2} + \frac{\sigma_1}{\sigma_1{}^2 + (\Omega - \Omega_1)^2}$$
$$+ \frac{\sigma_1}{\sigma_1{}^2 + (\Omega + \Omega_1)^2} \qquad (7.11.8)$$

While discussing the transient phenomena, we have introduced normalized values in both the frequency and time domains. The transformation to the unnormalized value of time τ (with dimensions) presents no difficulty and can be described in terms of reference values as:

$$\tau_{\text{ref}} = \frac{1}{\omega_{\text{ref}}} \qquad (7.11.9)$$

$$\tau = t\tau_{\text{ref}} \qquad (7.11.10)$$

The estimation theory approach formulated by the introduction of various signals to various system configurations, forms the basis for the design of communication systems. The advantage of this approach lies in the simplicity of calculations, since only typical input functions and only a few system values are involved. One ideal model of a transmission system is shown in Fig. 7.18. Here the output time function

and input time function are identical within a constant factor and time lag. All deviations from the ideal response in the actual system can be defined as transmission distortion. Analyzing a network in terms of circulating currents and given voltages leads to a set of differential equations in which these quantities are interrelated. The solution of these equations gives the response of the system. In general, especially for a complex system, solving these differential equations by classical methods is a tedious task. The use of the operational method reduces the tedium considerably. The problem is reduced to solving a set of simultaneous linear equations. The estimation theory approach permits one to eliminate difficulties of calculation from the start, and results in a catalog of output functions. The use of an operational method permits one to express the output function as an integral. The difficulties now are reduced to evaluating these explicit expressions—that is, to finding the values of the integrals.

7.12 TRANSIENT RESPONSE IN HIGHPASS AND BANDPASS FILTERS

When the linear system is known by its stationary transfer function

$$W(j\Omega) = |W(j\Omega)|\, e^{-jb} \qquad (7.12.1)$$

then the output (time) function $f_2(t)$ is given by

$$f_2(t) = \frac{1}{2\pi}\int_0^\infty |F_1(j\Omega)| \cdot |W(j\Omega)|$$
$$\times \sin\left[\Omega t + \angle F_1(j\Omega) - b(\Omega)\right] d\Omega \qquad (7.12.2)$$

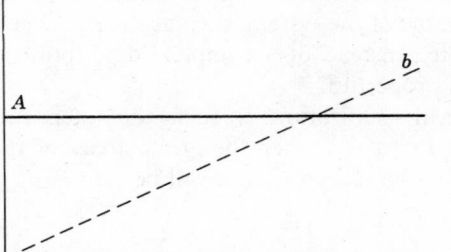

Fig. 7.18. Distortion-free transmission system.

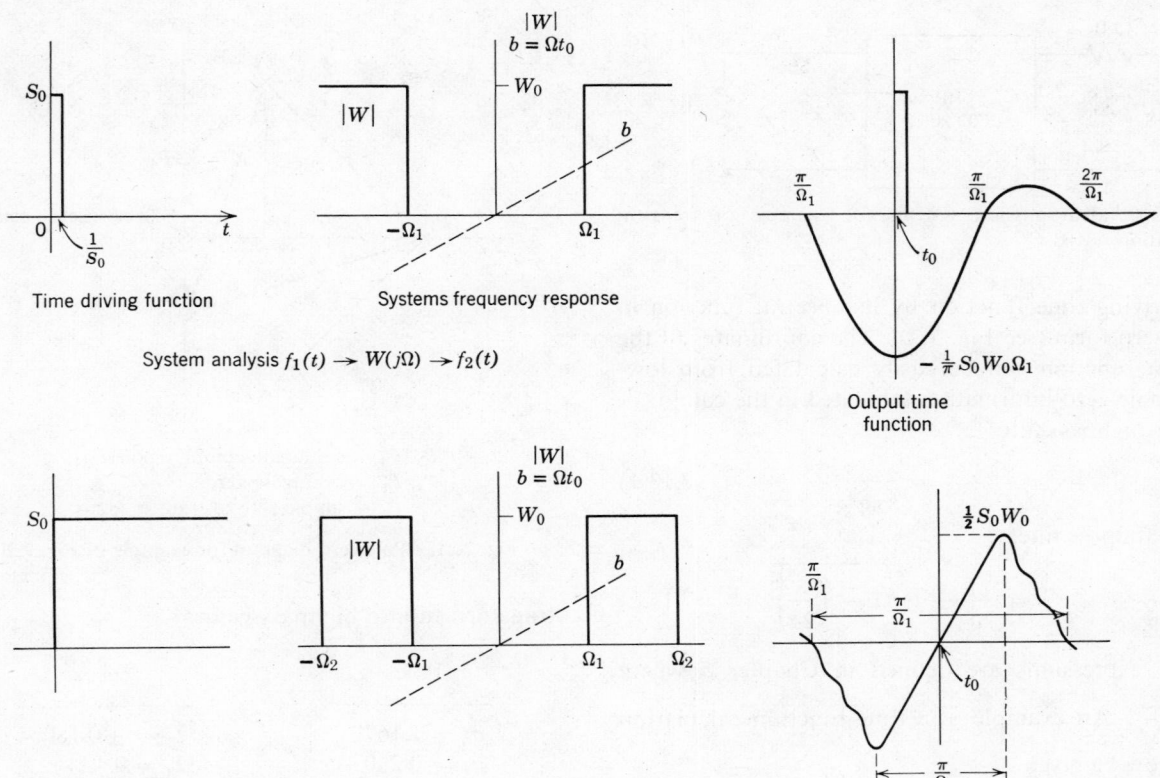

Fig. 7.19. Performance of the ideal system.

for a given input time function

$$f_1(t) = \frac{1}{2\pi} \int_0^\infty |F_1(j\Omega)| \sin\left[\Omega t + \angle F_1(j\Omega)\right] d\Omega$$

$$(7.12.3)$$

The solution to Eq. 7.12.2 consists in evaluating the integral. Parameters of the integrand

$$[F_1(j\Omega),\ \angle F_1(j\Omega),\ |W(j\Omega)|,\ b(\Omega)]$$

are always relatively easy to measure, and thus the ideal system and input can be described. As an example, consider a very narrow pulse of amplitude S_0, and width $1/S_0$, or a step function of amplitude S_0. Let

W_0 = value of the transfer function in the passband.
t_0 = time delay of the system.
Ω_1 = lower cutoff frequency.
Ω_2 = upper cutoff frequency.

The evaluation of Eq. 7.12.2 gives the output time function in terms of

$$\frac{\sin x}{x}$$

or

$$Si(x) = \int_0^x \frac{\sin y}{y}\, dy \qquad (7.12.4)$$

Two cases of interest are the pulse response of a highpass system and the step response of a broadband system. The output time functions for these two systems are shown on the right side of Fig. 7.19. The pulse response is

$$S_0 W_0 \frac{\sin x}{x}$$

and the maximum value occurs at t_0. The step response is the superposition of two $Si(x)$ functions.

7.13 THE EXACT CALCULATION OF TRANSIENT PHENOMENA FOR HIGHPASS SYSTEMS

The estimation theory approach is not sufficiently accurate when a nonideal network is involved. Therefore for exact calculation, as before, the Laplace transform is used. The graphical method which has been explained previously simplifies the exact design procedure. For this case it is necessary to represent

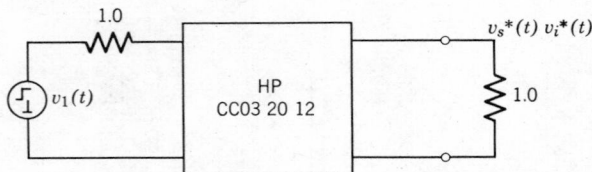

Fig. 7.20. Transmission systems for example of transient calculation.

the driving time function by its spectral function in pole-zero form (see Fig. 7.2). The coordinates of the system function can be easily calculated from low-pass pole-zero information tabulated in the catalog.

For highpass filters:

$$s_{\mathrm{HP}} = \frac{1}{s_{\mathrm{LP}}} \qquad (7.13.1)$$

For bandpass filters:

$$s_{\mathrm{B}} = \frac{s_{\mathrm{LP}}}{2a} \pm j\sqrt{1 - \left(\frac{s}{2a}\right)^2} \qquad (7.13.2)$$

These expressions are defined in Chapter 5 where $a = \dfrac{f_m}{\Delta f}$. An example of a time function calculation will now be given.

GIVEN: $v_1(t) = $ a unit impulse or step function.

SYSTEM: Highpass filter CC03 20 12. (See Fig. 7.20.)

TO FIND: Output signal $v_2(t)$ (which is the impulse response or the step response).

SOLUTION: According to Fig. 7.2, an impulse has a spectral function equal to unity, and a step function has a spectral function $1/s$. Both expressions are normalized and therefore without dimension. Both functions can pass from one to the other with aid of integration or differentiation. The step response will be calculated first. From the filter catalog of Chapter 5 the lowpass pole-zero normalized values for the transfer function are:

Poles	Zeros
$\sigma_0 = 0.85673$	$\Omega_2 = \pm 5.53857$
$\sigma_1 = 0.40807$	
$\Omega_1 = \pm 1.13438$	

One additional zero occurs at infinity. With the aid of Eq. 7.13.1 and Fig. 5.26 the above values can be

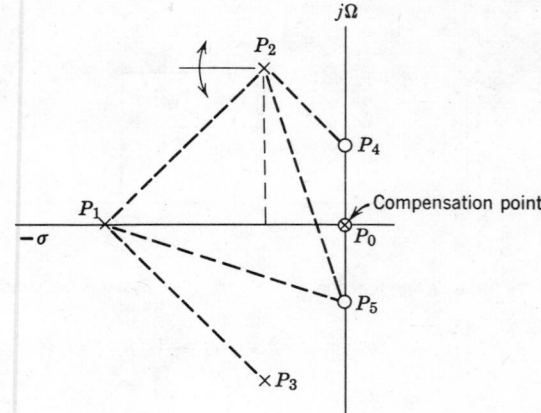

P_2, P_3 — conjugate complex poles
P_4, P_5 — conjugate zeros
P_0 — simple zero of transfer factor

Fig. 7.21. Pole-zero diagram for example of Fig. 7.20.

transformed into highpass values

Poles	Zeros
$\sigma_0 = 1.1672$	$\Omega_2 = \pm 0.18055$
$\sigma_1 = 0.28078$	$\Omega_3 = 0$
$\Omega_1 = 0.78053$	

The superposition of the pole of the transformed input function and the poles and zeros of the system function results in the diagram which is shown in Fig. 7.21. It can be noted that the pole of the input signal cancels the zero of the transfer function at $s = 0$. Consider the reference point P_1 ($s = -\sigma_0$). The component of the time function due to this pole is expressed by

$$f_R(t) = \frac{\Omega_2{}^2 + \sigma_0{}^2}{(\sigma_0 - \sigma_1)^2 + \Omega_1{}^2}\, e^{-\sigma_0 t}$$

and for P_2 ($s = -\sigma_1 + j\Omega_1$)

$$f_c(t) = \frac{\sqrt{\sigma_1{}^2 + (\Omega_1 - \Omega_2)^2}\,\sqrt{\sigma_1{}^2 + (\Omega_1 + \Omega_2)^2}}{\Omega_1\sqrt{(\sigma_0 - \sigma_1)^2 + \Omega_1{}^2}}$$
$$\times\, e^{-\sigma_1 t}\sin\left(\Omega_1 t + 180°\right)$$

Note, as a result of the symmetry condition for the reactive network (similar to the lowpass case) the resulting residue angle will always be 180°. In other words the free oscillation will be free of phase.

The total response to the step function will be

$$v_s(t) = e^{-\sigma_0 t} - \frac{\sqrt{\sigma_1{}^2 + (\Omega_1 - \Omega_2)^2}\,\sqrt{\sigma_1{}^2 + (\Omega_1 + \Omega_2)^2}\,\sqrt{(\sigma_0 - \sigma_1)^2 + \Omega_1{}^2}}{\Omega_1(\Omega_2{}^2 + \sigma_0{}^2)}\, e^{-\sigma_1 t}\sin\Omega_1 t$$

Multiplication by $[(\sigma_0 - \sigma_1)^2 + \Omega_1^2]/[\Omega_2^2 + \sigma_0^2]$ will normalize the amplitude so that $v_s(0) = 1$. Substitution of the known pole-zero data for the highpass filter yields

$$v_s(t) = e^{-1.1672t} - 0.71138 e^{-0.28078t} \sin 0.78053t$$

This expression shows that the step response is composed of two distinct parts: (1) exponential decay and (2) exponentially damped phase free oscillation.

Since this expression can be written in the general form

$$v_s(t) = e^{-\sigma_0 t} - M e^{-\sigma_1 t} \sin \Omega_1 t$$

the impulse response can be obtained directly by differentation of the preceding expression

$$v_i(t) = v_s{}'(t) = -\sigma_0 e^{-\sigma_0 t} + M\sqrt{\sigma_1^2 + \Omega_1^2}$$
$$\times e^{-\sigma_1 t} \sin (\Omega_1 t - \theta)$$

where

$$\theta = \sin^{-1} \frac{\Omega_1}{\sqrt{\sigma_1^2 + \Omega_1^2}}$$

For this example substitution of the numerical values gives

$$v_i(t) = -1.1672 e^{-1.1672t} + 0.5967 e^{-0.28078}$$
$$\times \sin (0.78053t - 1.2255)$$

Both the step and the impulse responses are shown in Fig. 7.22. The area of the equivalent rectangle that represents the impulse response assumes the value one.

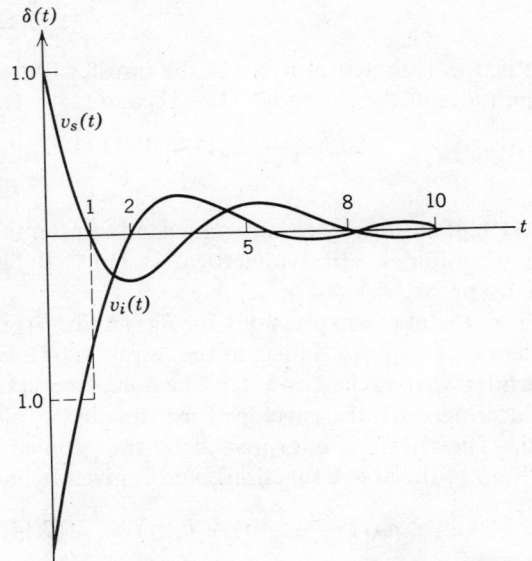

Fig. 7.22. Transient and impulse response for example of Fig. 7.20.

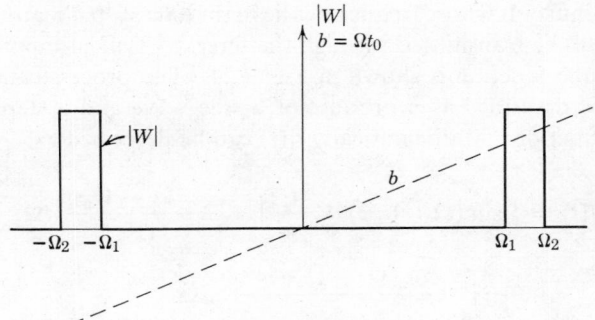

Fig. 7.23. Idealized narrowband system.

The validity of the estimation theory in calculating the distance between zeros of the step or frequency response is given for this example in normalized form:

$$t_1 = \frac{\pi}{\Omega_1} \approx 2.1$$

with

$$\Omega_1 \approx \Omega_{6\,\mathrm{dB}} \approx 1.5$$

The number 2.1 corresponds to the first zero of the impulse response in Fig. 7.22. The functions $v_s{}^*(t)$ and $v_i{}^*(t)$ in Fig. 7.20 are identical with the normalized functions $v_i(t)$ and $v_s(t)$ given by the above expressions within a factor $\frac{1}{2}$. This reduction is a simple result of voltage division by the loading resistances.

The interrelation between the steady state and dynamic properties that becomes evident in the foregoing example is important for many practical problems. The calculation of dynamic responses for broadband bandpass filter systems is in many respects similar to the highpass system.

7.14 ESTIMATE OF TRANSIENT RESPONSES IN NARROWBAND FILTERS

The transient response of narrowband networks can be discussed in essentially the same way as the previous examples. For a better understanding of the different approaches, we will discuss the estimation theory and the exact solutions.

From the previous discussion it is known that the classical estimation theory enables us to determine the response of idealized systems to typical input time functions. In Fig. 7.23 the ideal narrowband system without phase distortion is shown. The transmission band has an absolute value

$$\Delta\Omega = \Omega_2 - \Omega_1 \tag{7.14.1}$$

and relative bandwidth

$$\frac{1}{a} = \frac{\Delta\Omega}{\Omega_m} = \frac{\Omega_2 - \Omega_1}{\sqrt{\Omega_1 \Omega_2}} \tag{7.14.2}$$

Sinusoids whose frequencies lie in the narrow passband will be transmitted through the filter. A typical input time function is shown in Fig. 7.24. The process can be described as a product of a sine wave and a step function. Mathematically $f_1(t)$ can be described as

$$f_1(t) = \tfrac{1}{2} \sin (\Omega_0 t + \psi) + \frac{1}{2\pi} \int_0^\infty \frac{\cos (\Omega - \Omega_0)t}{\Omega} d\Omega$$

$$- \frac{1}{2\pi} \int_0^\infty \frac{\cos (\Omega + \Omega_0)t + \psi}{\Omega} d\Omega$$

$$= \tfrac{1}{2} \sin (\Omega_0 t + \psi)$$

$$+ \frac{1}{\pi} \int_0^\infty \frac{\Omega \sin \psi \sin \Omega t - \Omega_0 \cos \psi \cos \Omega t}{\Omega^2 - \Omega_0{}^2} d\Omega$$

$$(7.14.3)$$

In contrast to the constant spectral function which corresponds to an impulse and the $1/s$ spectral function which corresponds to a unit step, the spectral function of the sinusoid given in Eq. 7.14.3 has a pole at Ω_0 (see Fig. 7.25). The output of a narrowband filter, when the input is $f_1(t)$ is approximated closely by

$$f_2(t) = \tfrac{1}{2} |W(j\Omega_0)| \sin [\Omega_0 t + \psi - b\Omega_0]$$

$$+ \frac{1}{2\pi} \int_0^\infty |W(j\Omega - j\Omega_0)|$$

$$\times \frac{\cos [(\Omega - \Omega_0)t - \psi - b(\Omega - \Omega_0)]}{\Omega} d\Omega$$

$$- \frac{1}{2\pi} \int_0^\infty |W(j\Omega - j\Omega_0)|$$

$$\times \frac{\cos [(\Omega + \Omega_0)t + \psi - b(\Omega - \Omega_0)]}{\Omega} d\Omega$$

$$(7.14.4)$$

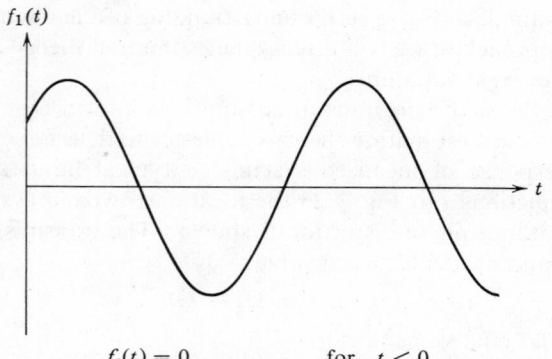

$$f_1(t) = 0 \qquad \text{for} \quad t < 0$$
$$f_1(t) = \sin (\Omega_0 t + \psi) \quad \text{for} \quad t > 0$$

Fig. 7.24.   Typical input time function.

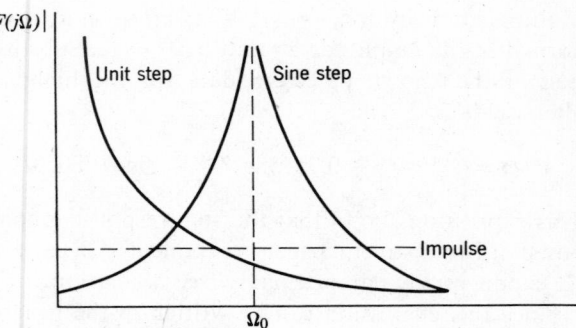

Fig. 7.25.   Amplitude spectrum of a sinusoidal signal shown in Fig. 7.24.

The response function can be written in a slightly different form:

$$f_2(t) = h_1(t) \sin [\Omega_0(t - t_0) + \psi]$$
$$+ h_2(t) \cos [\Omega_0(t - t_0) + \psi] \quad (7.14.5)$$

where

$$h_1(t) = \tfrac{1}{2} |W(j\Omega_0)| + \frac{1}{\pi} \int_0^\infty |W_1(j\Omega)|$$

$$\times \frac{\sin \Omega(t - t_0)}{\Omega} d\Omega \quad (7.14.6)$$

$$h_2(t) = \frac{1}{\pi} \int_0^\infty |W_2(j\Omega)| \frac{\cos \Omega(t - t_0)}{\Omega} d\Omega \quad (7.14.7)$$

with

$$b = \Omega t_0$$

$|W_1(j\Omega)|$ and $|W_2(j\Omega)|$ are defined as follows:

$$|W_1(j\Omega)| = \tfrac{1}{2}[|Wj(\Omega - \Omega_0)| + |Wj(\Omega + \Omega_0)|] \quad (7.14.8)$$

that is, the arithmetical mean of the transfer function magnitudes of the sidebands $\Omega - \Omega_0$ and $\Omega_0 + \Omega$.

$$|W_2(j\Omega)| = \tfrac{1}{2}[|Wj(\Omega - \Omega_0)| - |Wj(\Omega + \Omega_0)|] \quad (7.14.9)$$

that is, half the difference of the sideband transfer function magnitudes. In symmetrical systems $|W_2(j\Omega)|$ will disappear.

From the above expressions for $h_1(t)$ and $h_2(t)$ it is evident that they are similar to the response of a low-pass filter when excited by a step function. In practical engineering work the envelope time function is often used. The envelope is expressed by the geometrical addition of the above functions, and is given by $h_0(t)$.

$$h_0(t) = \sqrt{h_1{}^2(t) + h_2{}^2(t)} \quad (7.14.10)$$

For the special case of a symmetrical narrowband system without phase distortion as shown in Fig. 7.23

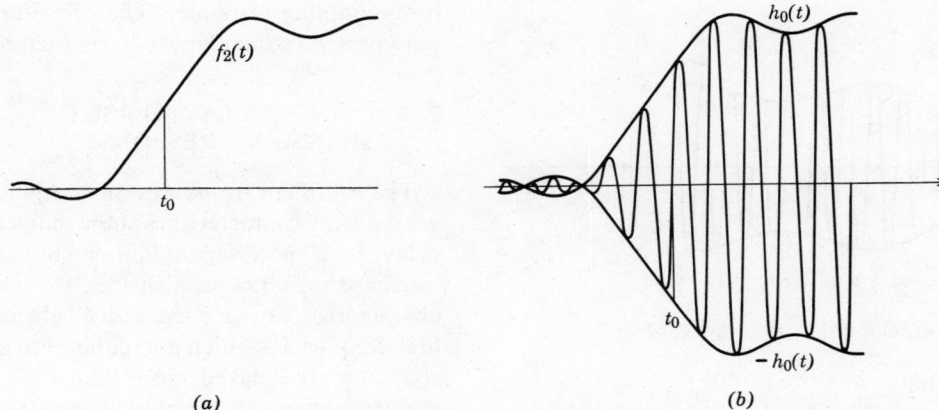

Fig. 7.26. Transient functions (*a*) ideal lowpass system and (*b*) ideal narrowband bandpass system.

and $\Delta\Omega$ and $1/a$ as previously defined, the integration from $2\Omega_0 - \Delta\Omega$ up to $2\Omega_0 + \Delta\Omega$ with $\Omega \approx 2\Omega_0$ will result in:

$$f_2(t) = W_0\left\{\tfrac{1}{2} + \frac{1}{\pi} Si\left[\frac{\Delta\Omega(t - t_0)}{2}\right]\right\}$$
$$\times \sin\left[\Omega_0(t - t_0) + \psi\right] + \frac{W_0}{2\pi}\frac{1}{a}$$
$$\times Si\left[\frac{\Delta\Omega(t - t_0)}{2}\right]\cos\left[\Omega_0(t - t_0) + \psi\right]$$
$$\approx W_0\left\{\tfrac{1}{2} + \frac{1}{\pi} Si\left[\frac{\Delta\Omega(t - t_0)}{2}\right]\right\}$$
$$\times \sin\left[\Omega_0(t - t_0) + \psi\right] \qquad (7.14.11)$$

The envelope of the ac response of symmetrical narrowband systems without phase distortion is practically identical with the transient response of the lowpass filters. The phase of the ac signal at the instant of application to the filter has no influence on the shape of the envelope. In Fig. 7.26 the corresponding curves are shown. The earlier discussion relating to rise time, overshoot, etc., of lowpass systems can be directly applied to symmetrical bandpass systems.

7.15 THE EXACT TRANSIENT CALCULATION IN NARROWBAND SYSTEMS

By means of the lowpass–highpass analogy, mentioned in previous paragraphs, transient calculations can be made in the same manner as for lowpass filters. The necessary roots of the system function can be taken from the catalog. Let us consider the following example:

GIVEN: Driving function $v_0(t)$—sudden application of a sine-wave.

SYSTEM: Bandpass filter with bandwidth $1/a = 5\%$ and $f_m = 100$ kc. This is a result of a frequency transformation from CC03 25 30, $K^2 = \infty$ (see Fig. 7.27).

TO FIND: $v_s(t)$ or more precisely the shape of the envelope of the output wave (since the frequency of oscillation is already known).

SOLUTION: The shape of the response, as has been explained, is identical with the corresponding lowpass step response.

For a symmetrical reactance network the expression given in (7.10.6) is valid.

$$v_s(t) = \pm h_0(t) = 1 - e^{-\sigma_0 t} - \frac{\sigma_0}{\Omega_1(\sigma_0{}^2 + \Omega_2{}^2)}$$
$$\times \left\{\frac{[\sigma_1{}^2 + (\Omega_2 - \Omega_1)^2][\sigma_1{}^2 + (\Omega_2 + \Omega_1)^2][(\sigma_0 - \sigma_1)^2 + \Omega_1]}{\sigma_1{}^2 + \Omega_1{}^2}\right\}^{\tfrac{1}{2}} e^{-\sigma_1 t}\sin\ \Omega_1 t$$

The data from the catalog is

Poles	Zeros
$\sigma_0 = 0.8312$	$\Omega_2 = \pm 2.2701$
$\sigma_1 = 0.3113$	
$\Omega_1 = 1.094$	

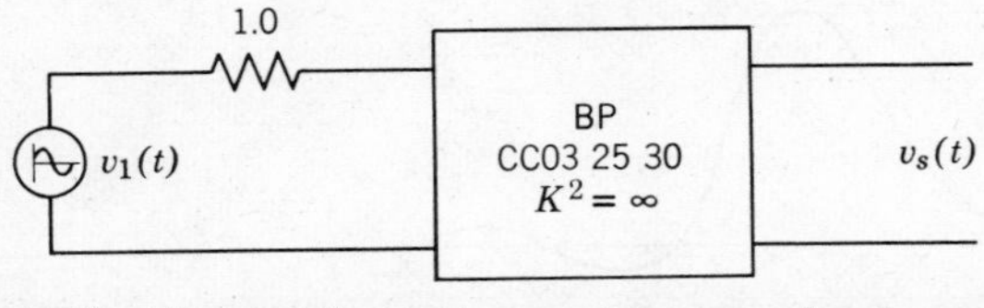

Fig. 7.27. Transmission system used in narrowband example.

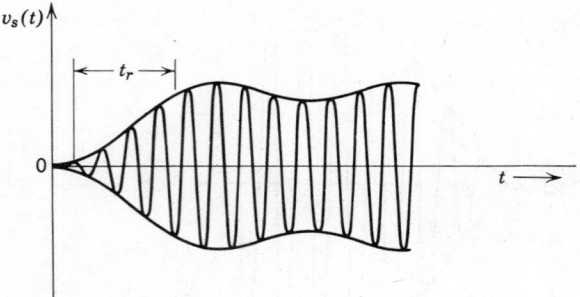

Fig. 7.28.   Envelope of output wave for Fig. 7.27.

and it follows that

$$\pm h_0(t) = 1 - e^{-0.8312t} - 0.5691\, e^{-0.3113t} \sin 1.094t$$

Evaluation of this expression for different values of t results in the curve shown in Fig. 7.28.

The validity of the approximate relation for rise time can be checked.

$$t_r = \frac{2\pi}{\Delta\Omega} \approx 2.4$$

where

$$\Delta\Omega \approx \Delta\Omega_{6\,\text{dB}} \approx 2.6$$

which is in good agreement with the value obtained from the exact expression.

The rise time is proportional to the absolute value of bandwidth. For the previous example the rise time can be determined from:

$$\tau_r = \frac{2\pi}{\omega_m}\, a = \frac{a}{f_m} = \frac{1}{\Delta f} = 0.2ms$$

The discussion of the transient characteristics has shown that when the tabulated roots of a filtering system are known, the impulse or step response may be easily determined for any real network. The approximate method provides an alternate approach.

Caution: The substitution of s by $1/s$ does not necessarily yield a highpass filter having transient characteristics comparable to the lowpass model. For example, the transformation of a lowpass Bessel filter does not yield a highpass filter with low overshoots

in the impulse response. The substitution of s by $1/s$ only preserves the amplitude characteristics.

7.16   GROUP DELAY VERSUS TRANSIENT RESPONSE

The transient responses in conjunction with the group-delay characteristics show that a constant group delay is a necessary requirement for good pulse transmission. Let us consider the increasing delay characteristics associated with Chebyshev filters. The high frequencies which are required to give a fast pulse rise time are delayed more than the low frequencies and thus give an amplitude peak at a later time (Fig. 7.29). Thus we would expect the rise time to increase as the high frequencies become delayed more and more with respect to the low frequencies. This is indeed the case.

For a filter with drooping group-delay characteristics, such as the synchronous or Gaussian magnitude filters, the high frequencies are delayed less than the low frequencies, thus the impulse response will come to a peak more rapidly and die off rather slowly because the low frequency components arrive at a later time as in Fig. 7.30. Both cases result in non-symmetrical pulse responses.

A constant group delay gives rise to a very desirable symmetrical pulse as shown in Fig. 7.31. The transient response is shown with a 1-sec group-delay normalization. The equal ripple delay filters which closely approximate this type of delay response in the frequency domain have an impulse response that is practically symmetrical about $t = 1$. (This is especially true for values of $n > 4$.)

7.17   COMPUTER DETERMINATION OF FILTER IMPULSE RESPONSE

The impulse response of a filter is the voltage-time function measured at the load when a unit impulse is the input signal. A unit impulse and its constituent frequency spectrum is shown in Fig. 7.32. Its duration

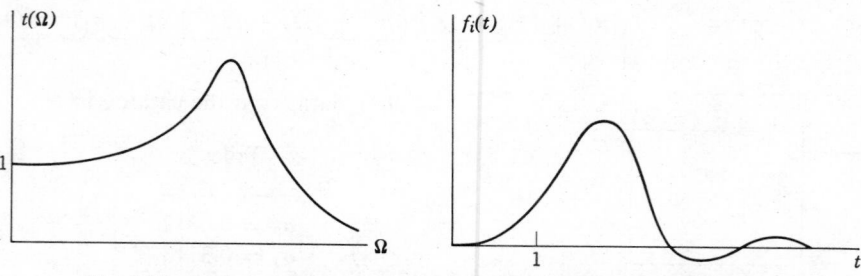

Fig. 7.29.   Delay-transient correspondence for an increasing delay filter.

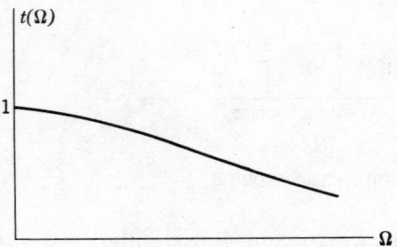

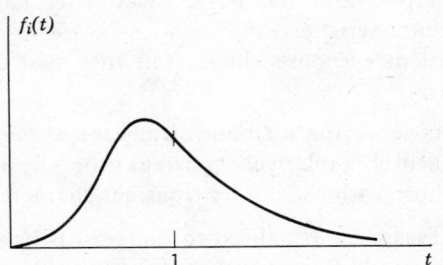

Fig. 7.30. Delay-transient correspondence for a decreasing delay filter.

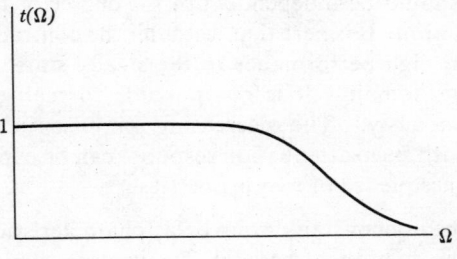

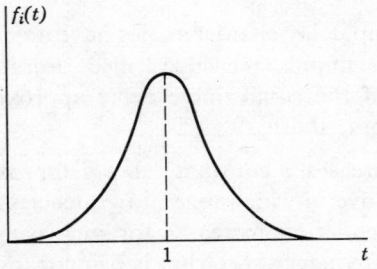

Fig. 7.31. Delay-transient correspondence for a constant group delay filter.

in time is infinitesimally small and its amplitude is such that the area under this time-amplitude function is unity. The frequency components which synthesize this function are constant with respect to frequency extending to infinity and they are infinitesimally spaced along the frequency axis.

To obtain the impulse response of a filter one must know the analytic expression of the transmission function. Knowing this, the inverse Laplace transform can be obtained which yields the impulse response. Only in simple cases is this function known by the engineer, owing to the effect of losses and tolerances of the elements of the filter.

A more realistic case is that in which only the amplitude and the phase responses of the filter are known from a computer loop equation program. This, however, is enough information to find the impulse response of the filter. A loop equation program prints out the attenuation A, and the phase angle ϕ, at a finite number of frequencies. Each point represents

$$A \cos (2\pi ft - \phi) \qquad (7.17.1)$$

A good approximation to applying an impulse to a filter is to have, as an input, a frequency spectrum consisting of equal amplitude sine waves spaced uniformly in frequency. Since it is not necessary to include any frequency components which might be attenuated by greater than, say, 80 dB by the filter, none of these frequency components will be considered. Such a frequency spectrum is shown in Fig. 7.33. Thus it is possible to generate an approximate impulse spectrum which is only one sense is an approximation—that is, it does not extend to infinite frequencies. The fact that the frequency components are finitely spaced along the frequency axis merely means that the impulse repeats itself in time.

If this approximate impulse spectrum was applied to a filter, the output frequency response would be exactly the phase and amplitude printout obtained from the computer loop equation program, as shown in Fig. 7.34. The summation, in time, of the cosine waves from Eq. 7.17.1, by the theorem of superposition, would result in the impulse response of the

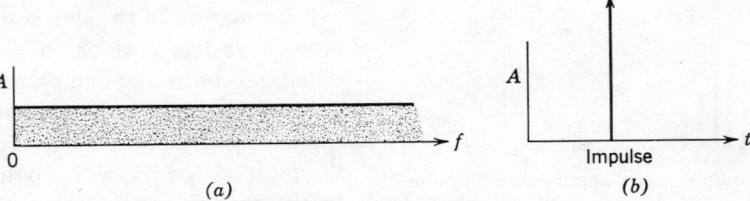

Fig. 7.32. (*a*) Frequency spectrum of (*b*) true impulse.

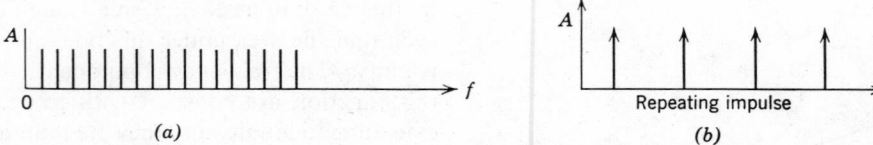

Fig. 7.33. (*a*) Frequency spectrum of (*b*) approximation to true impulse.

filter. Thus at any one time τ_A the computer performs the addition.

$$\sum_{i=f0}^{i=f_{80\ \mathrm{dB}}} A_i \cos\left(2\pi f_i t_A - \phi_i\right) \qquad (7.17.2)$$

One may choose as many times *t* as are necessary to obtain a smooth curve for the impulse response. At each time t_A the computer takes the summation of Eq. 7.17.2. Thus for 200 points from the loop equations frequency-amplitude response and for 300 time inputs, 60,000 numbers found from Eq. 7.17.1 are manipulated.

7.18   TRANSIENT RESPONSE CURVES

If a filter network is to be used for pulse applications, it is desirable that the transient portion of the response die out rapidly. This, in effect, means that the transient response should have little or no overshoot. One cause of the overshoot is the lack of all the frequency components which make up the pulse. Hence, as the bandwidth is widened, the pulse reproduction improves. However, Gibbs phenomena are always present. For a given bandwidth there exists one pulse width such that the overshoot is a minimum.

Butterworth filters, characterized by their maximally flat magnitude response, also have a fast step function response. But overshoot increases with increasing order, and this is undesirable. Bessel filters, which have maximally flat group delay characteristics, have negligible overshoot in the transient response, particularly for the higher order filters. However, the rise time is not so good as in the Butterworth case. There exist filters which are compromises between these two. One such compromise is the transitional Butterworth-Bessel class of filters which can be made to have a response characteristic lying between those

of the Butterworth and Bessel filters by varying a parameter in the transfer function. One can never obtain both the Butterworth maximally flat magnitude characteristic and the Bessel maximally flat group delay characteristic.

Transient response shapes fall into two different categories:

1. Those having a smooth transient response and, consequently, a relatively constant group delay.
2. Those with oscillatory transient characteristics.

The curves are normalized to unity 3-dB bandwidth. Also, it should be noted that the area under the unit impulse response curves is a constant. The peak of the impulse response should be as high as possible, and the height should be independent of the degree of the approximation. It is here that we suffer the contradiction between high performance in the steady state and in the time domain. It is not possible to realize both simultaneously. The decreasing impulse peak for filters with steep attenuation response can be explained as a consequence of two properties:

1. The nonconstant group delay characteristics, and
2. The fact that the high frequencies, which are necessary for large peaks, are attenuated much more in high rejection type filters.

That the group delay characteristics have an effect on the peak of the impulse is easily verified. For instance, in the case of the equal-ripple delay approximation for 0.5°, we note that:

1. As *n* increases, a constant delay is approximated more closely over a wide range of frequencies
2. The attenuation increases for increasing *n*, so that a small frequency spectrum is being transmitted.
3. The peak of the impulse response increases slightly for increasing *n*.

Examination of the step responses shows that some filters have delays which are two or more times greater than filters of similar complexity, but of different types. The reason for this phenomenon is again obtained by examining the group delay curves. The area under these curves is $n\pi/2$, which bounds the available delay. In the case of sharp attenuation type filters, almost

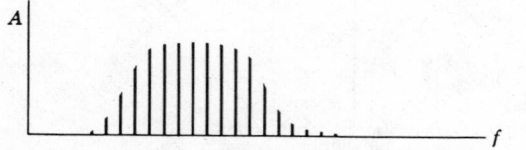

Fig. 7.34. Bandpass filter output when approximate impulse is the input.

all of the delay appears in the passband. This is not the case for the Gaussian type filters. For example, in the case of $n = 8$, only about 20% of the total delay area occurs within the passband. But in a Chebyshev filter with 2 dB ripples, practically all of the delay is within the passband. By the Hilbert transformation, we know that the attenuation of a minimum-phase network uniquely specifies the delay. Therefore, two filters with similar attenuation will have similar delay. A practical estimate is obtained by comparing the attenuation of the filter in question with the curves given in Chapter 3.

The rise time for a step input may be defined in one of two ways:

1. The time required to go from 10% to 90% of the final value. This definition was given in Section 7.1.

2. The time required to go from zero to the final value at the maximum slope which is present in the vicinity of the 50% value. This definition may be encountered in some other text.

The time delay of a step passing through a filter, is arbitrarily taken as the time difference between the application of this step and the moment the output reaches 50% of its final value. This time is also, approximately, the time at which the impulse reaches a peak.

To answer the question as to whether or not the time domain characteristics of a given filter are satisfactory, the transient response information is presented in Curves 1 through 24:

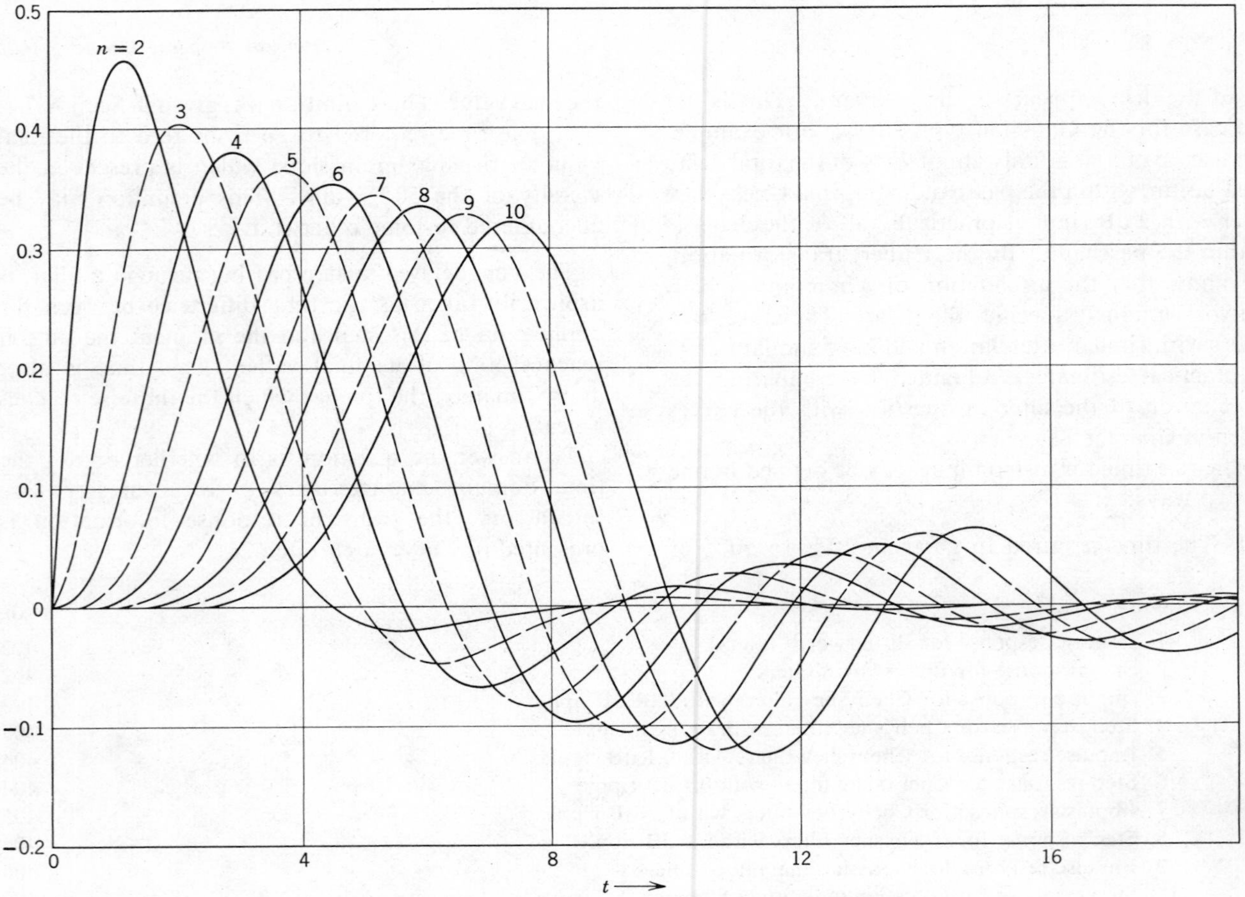

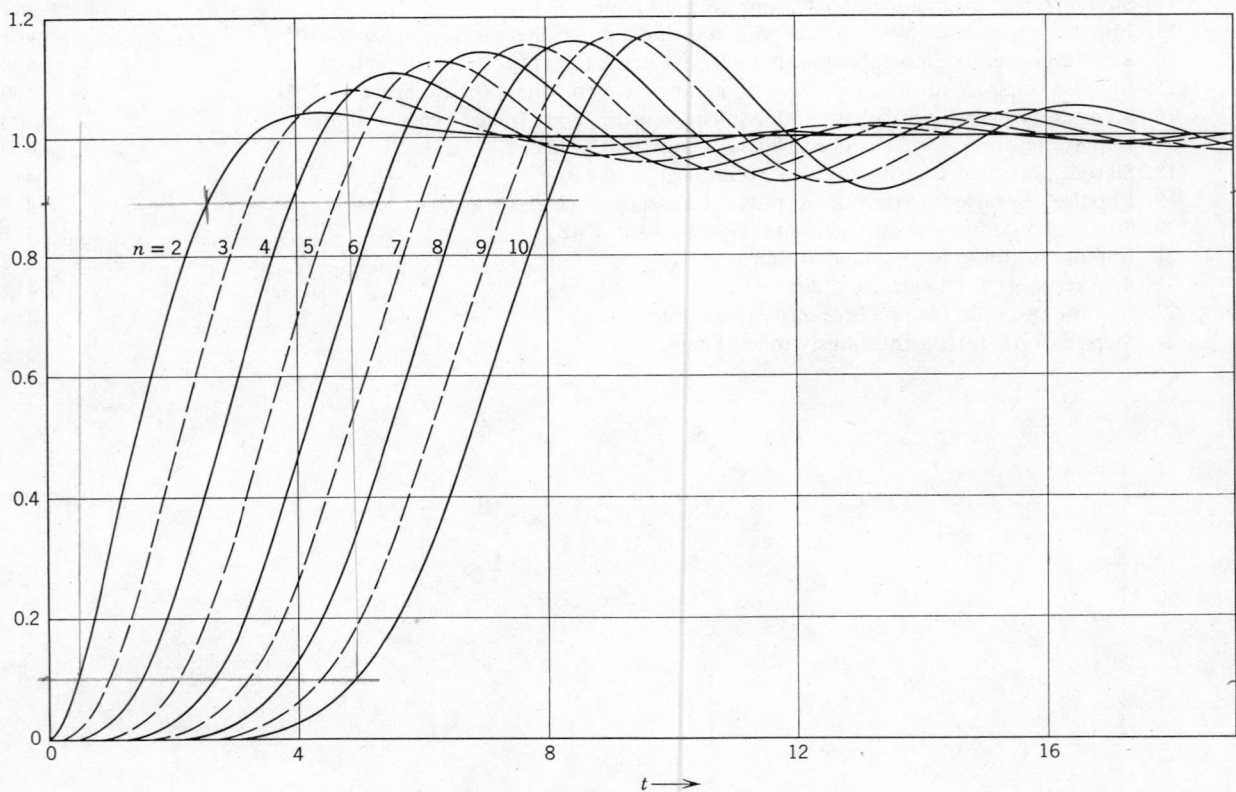

Curve 1. Impulse response for Butterworth filters.

Curve 2. Step response for Butterworth filters.

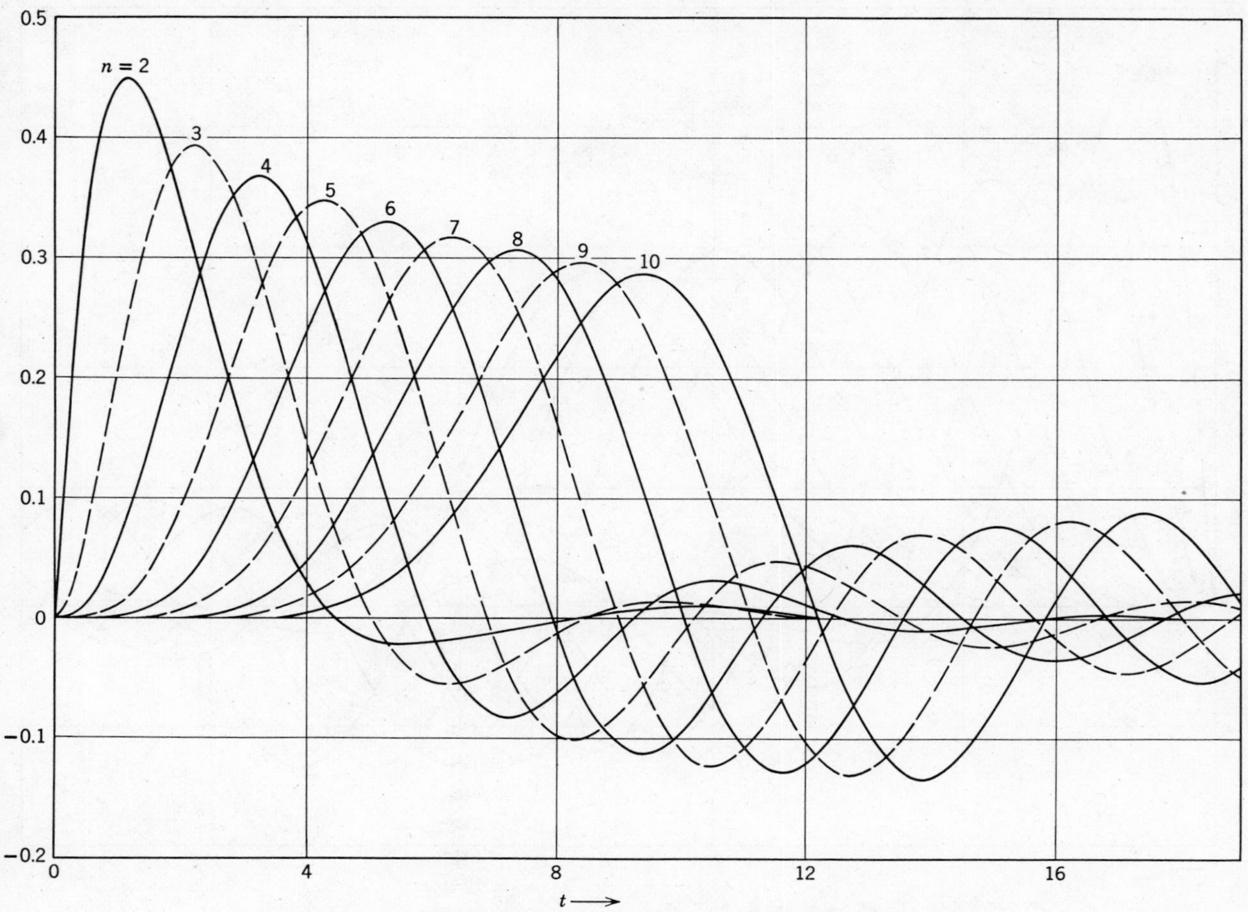

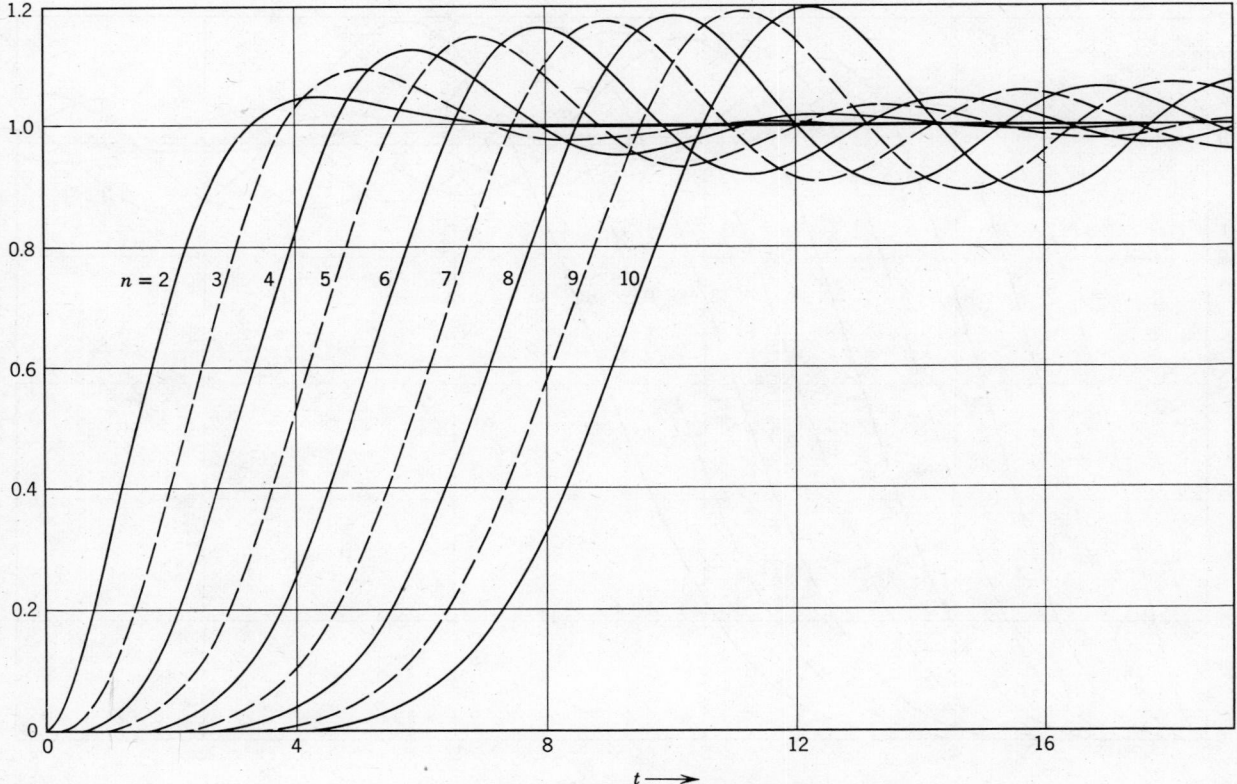

Curve 3. Impulse response for Chebyshev filters with 0.01 dB ripple.

Curve 4. Step response for Chebyshev filters with 0.01 dB ripple.

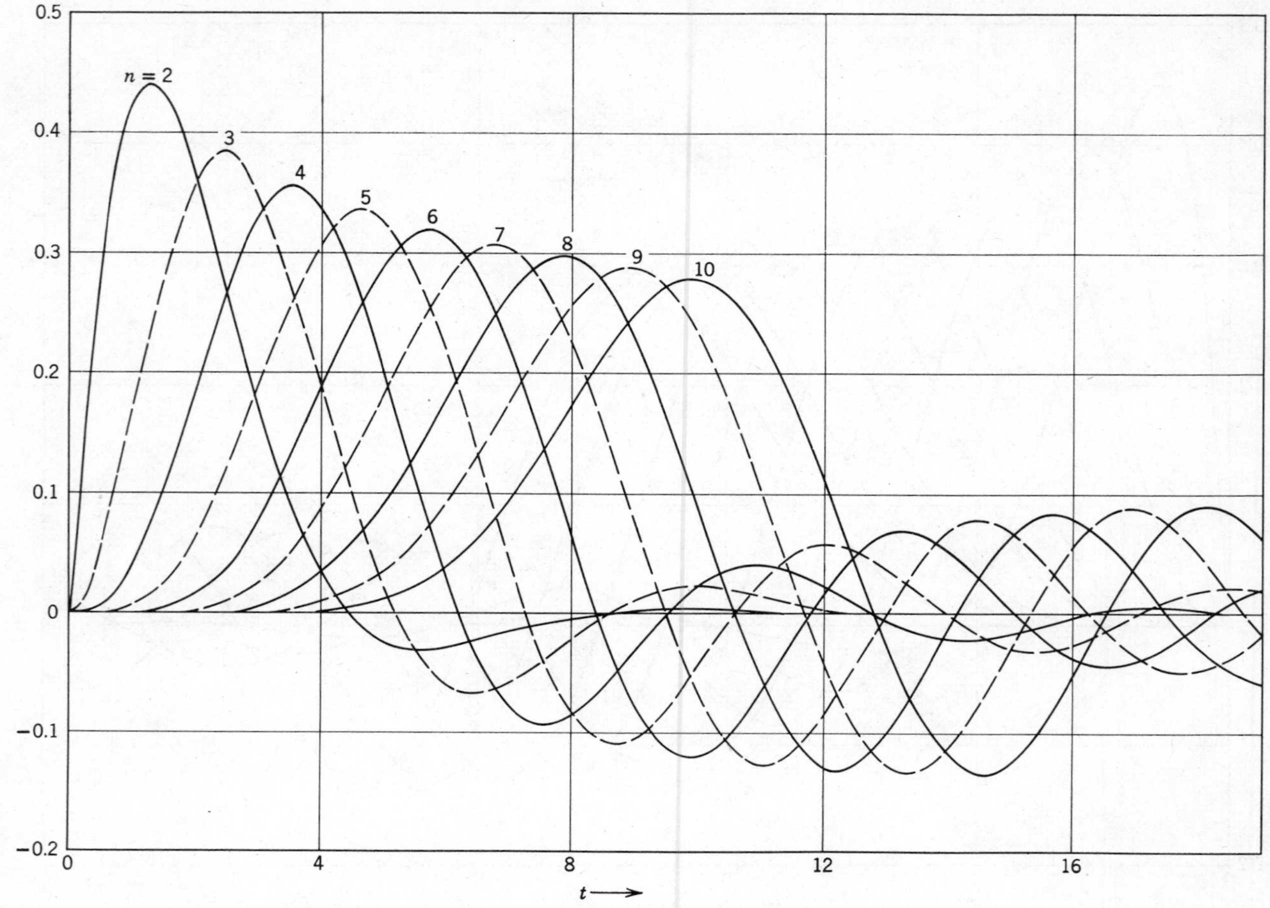

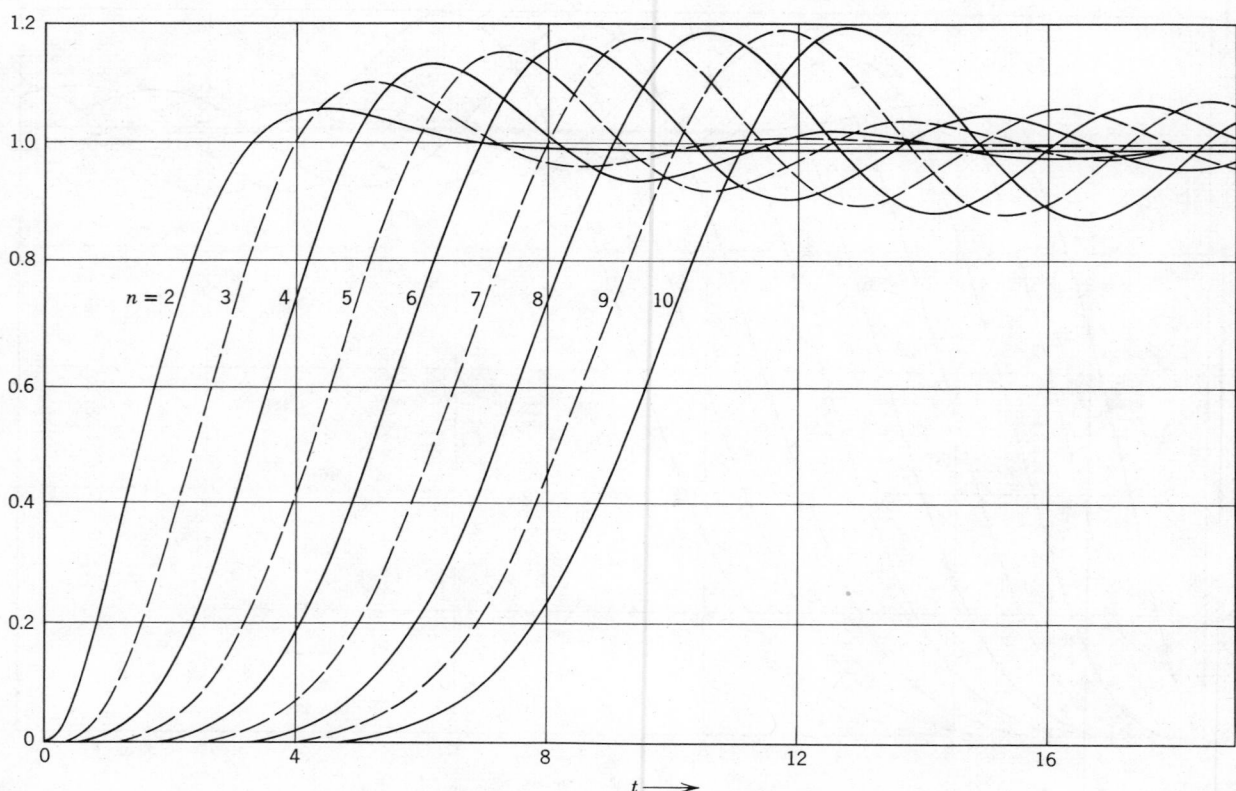

Curve 5. Impulse response for Chebyshev filters with 0.1 dB ripple.

Curve 6. Step response for Chebyshev filters with 0.1 dB ripple.

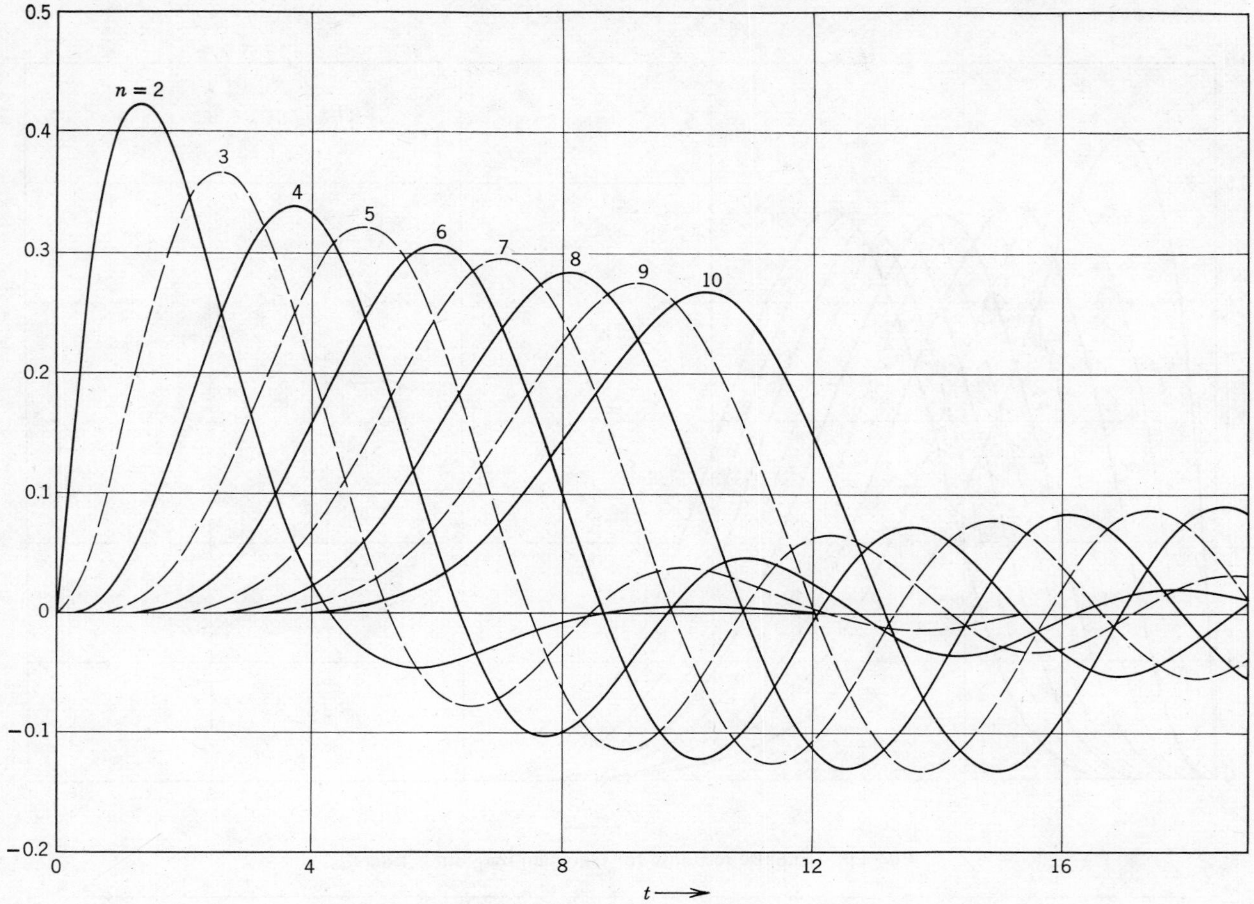

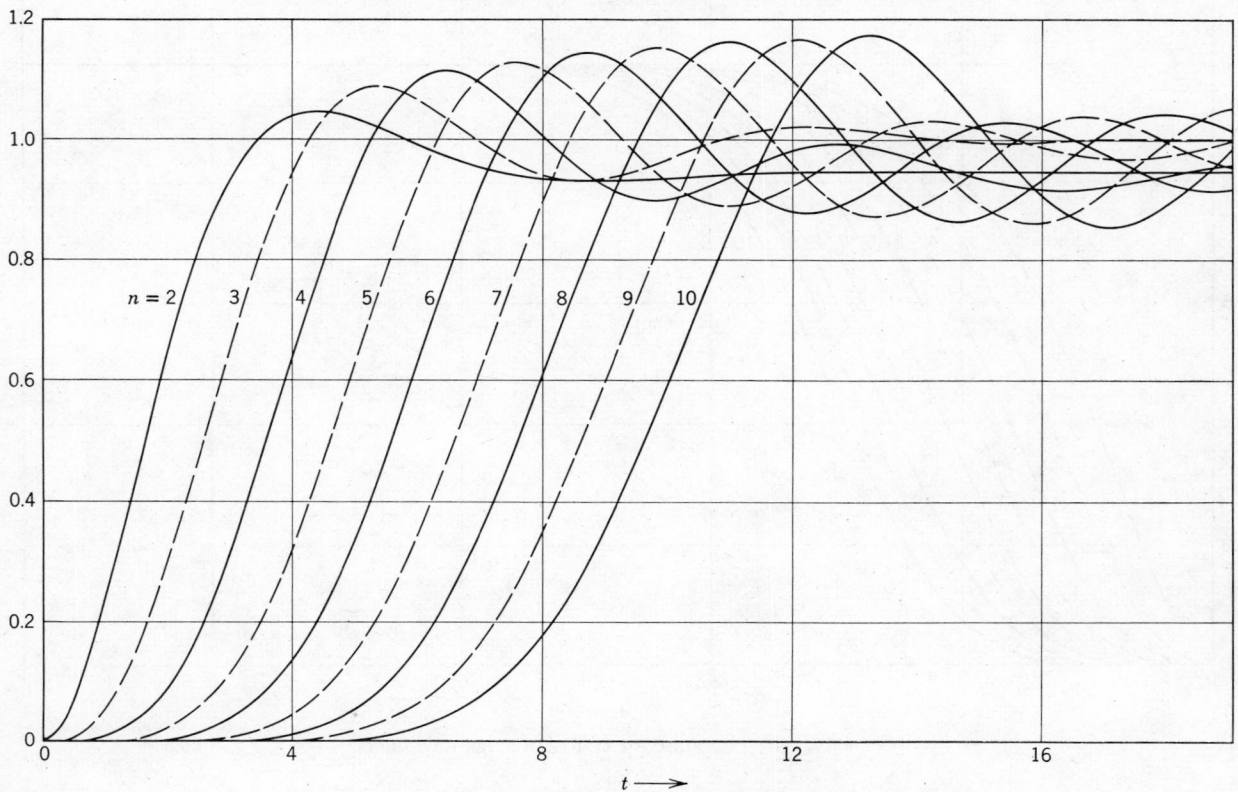

Curve 7. Impulse response for Chebyshev filters with 0.5 dB ripple.

Curve 8. Step response for Chebyshev filters with 0.5 dB ripple.

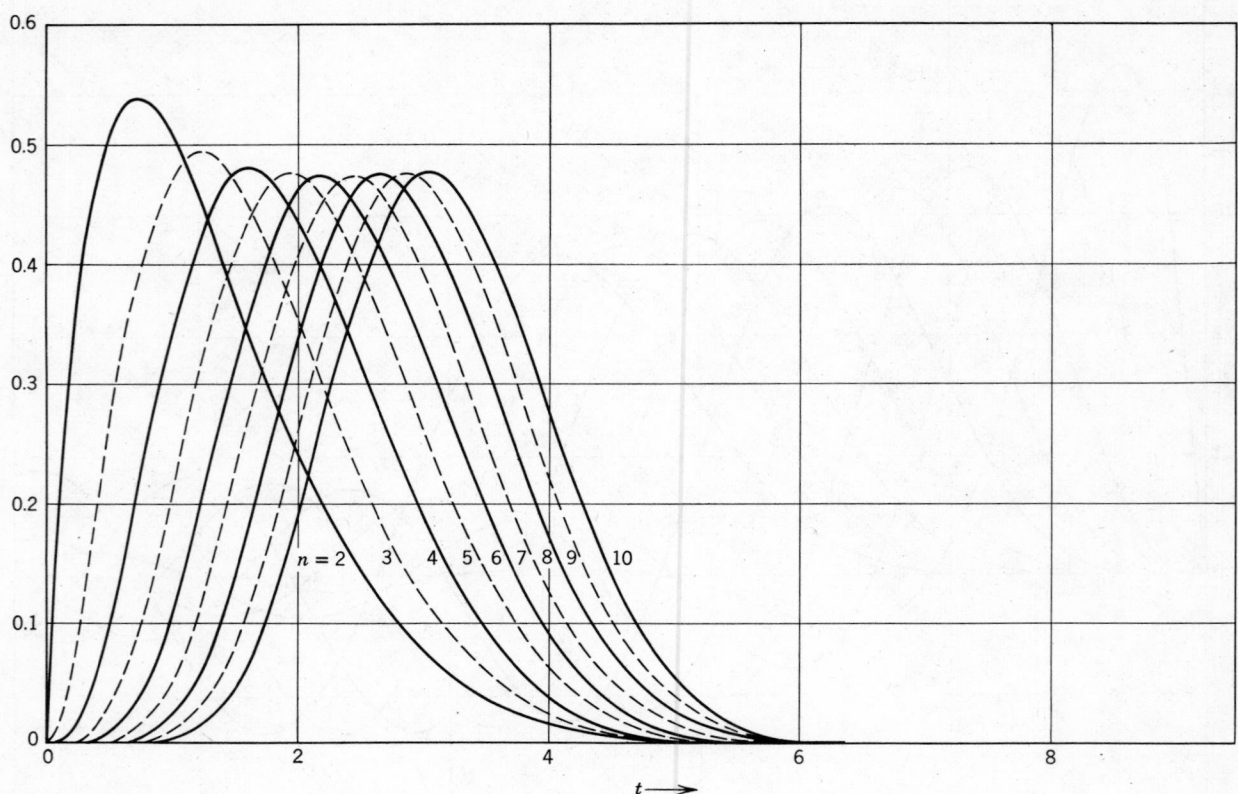

Curve 9. Impulse response for Gaussian magnitude filters.

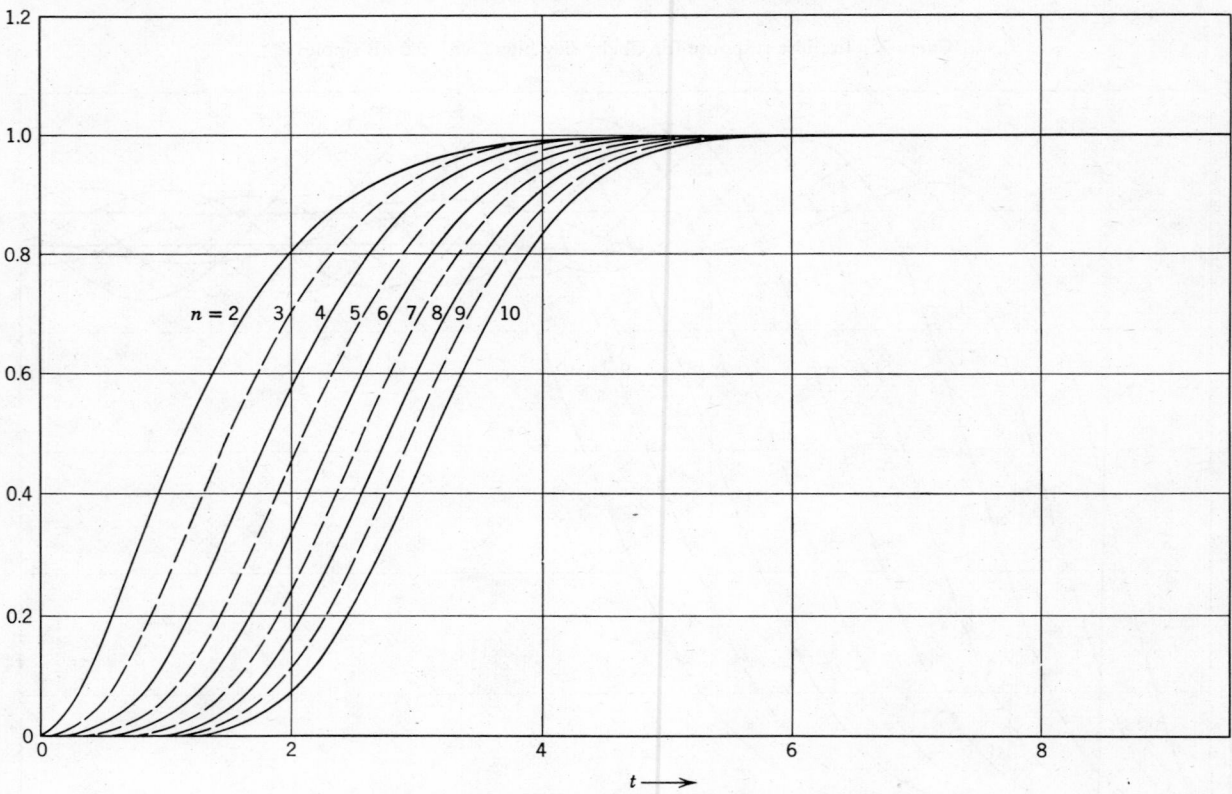

Curve 10. Step response for Gaussian magnitude filters.

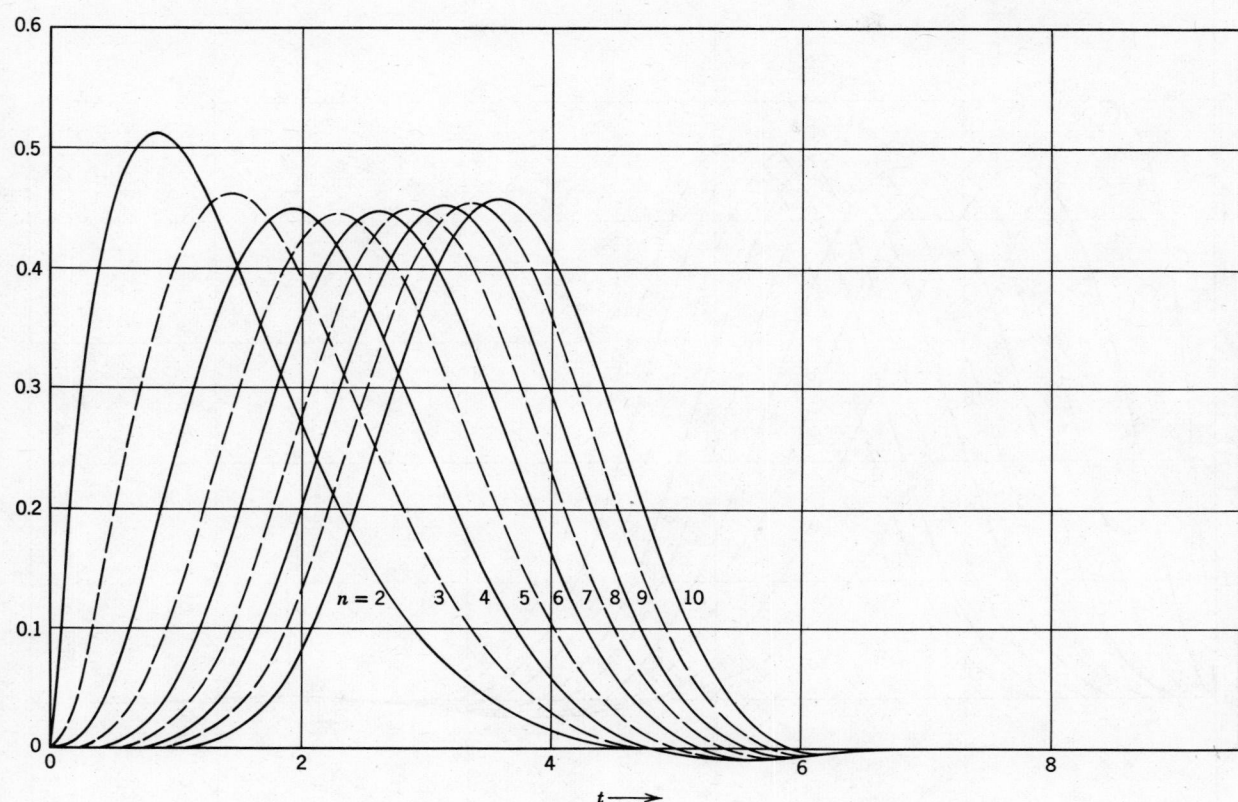

Curve 11. Impulse response for maximally flat delay (Bessel) filters.

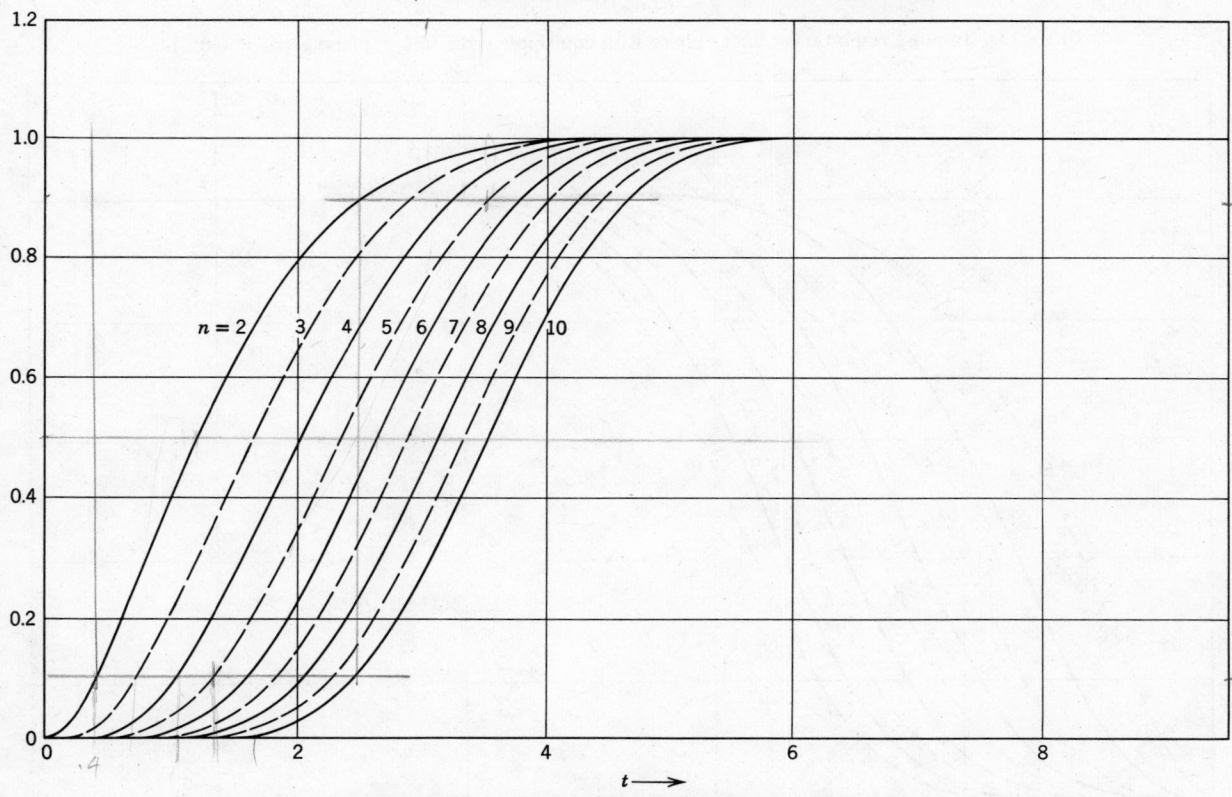

Curve 12. Step response for maximally flat delay (Bessel) filters.

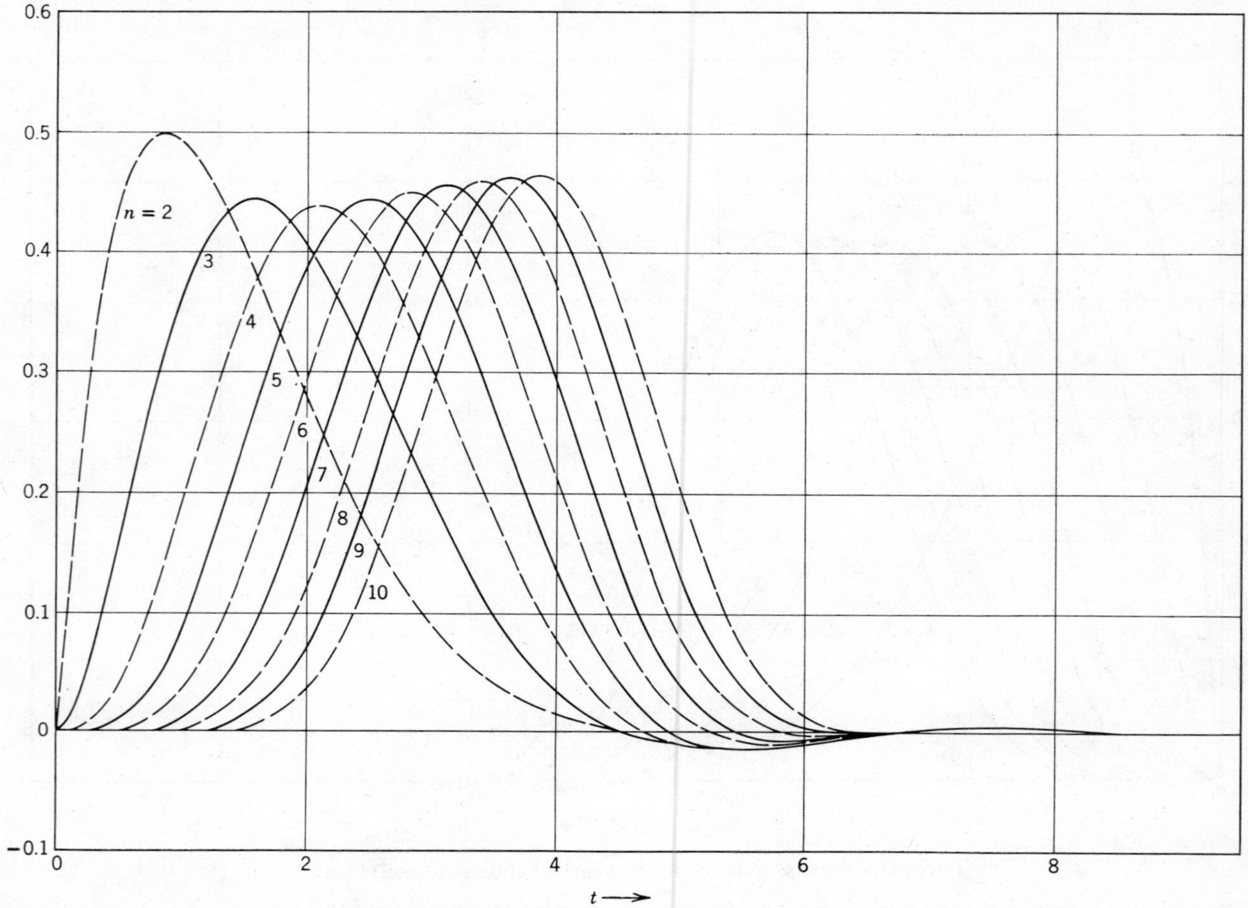

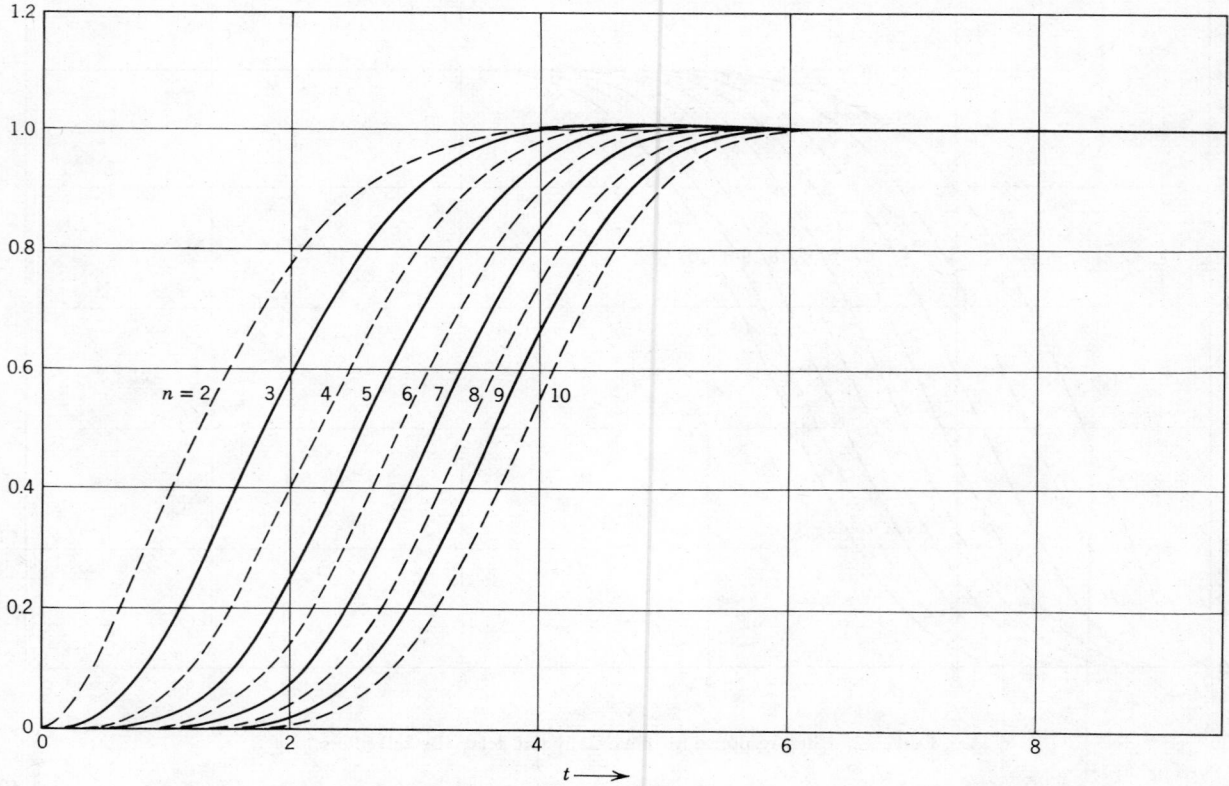

Curve 13. Impulse response for linear phase with equiripple error filters (phase error = 0.05°).

Curve 14. Step response for linear phase with equiripple error filters (phase error = 0.05°).

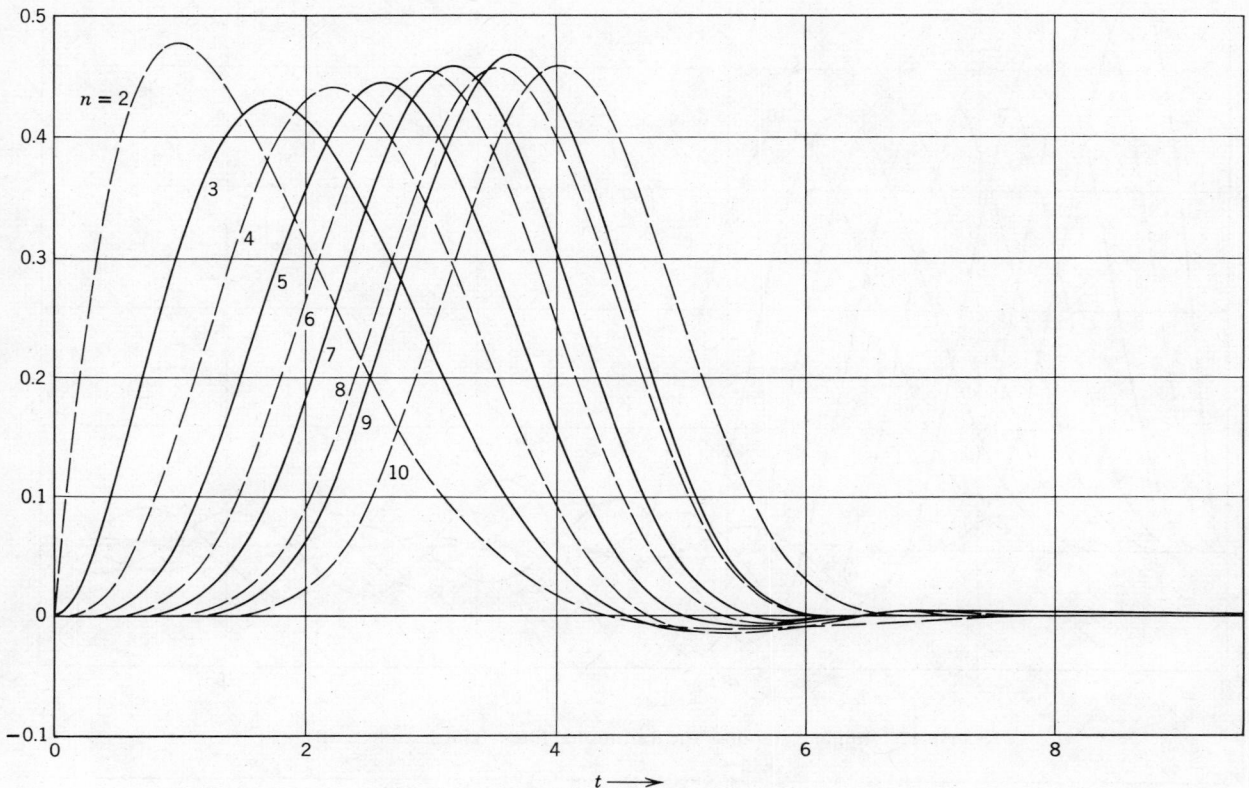

Curve 15. Impulse response for linear phase with equiripple error filters (phase error = 0.5°).

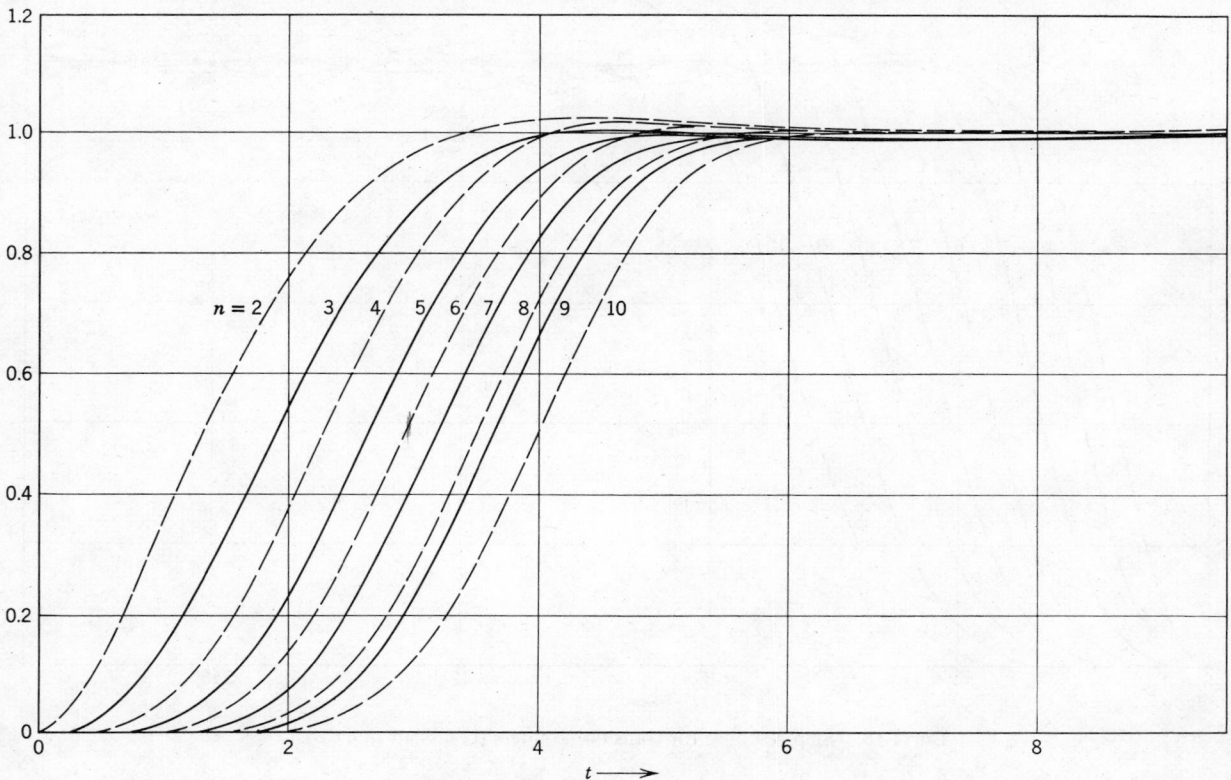

Curve 16. Step response for linear phase with equiripple error filters (phase error = 0.5°).

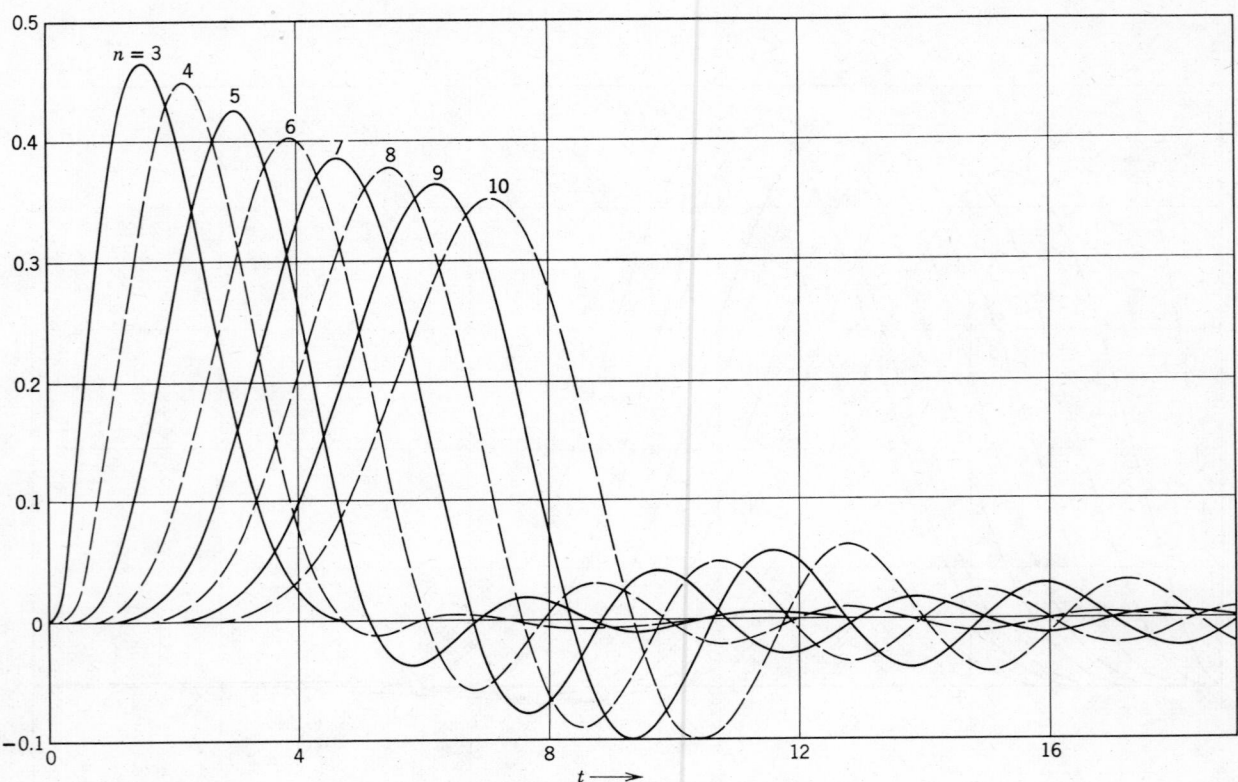

Curve 17. Impulse response for transitional filters (Gaussian to 6 dB).

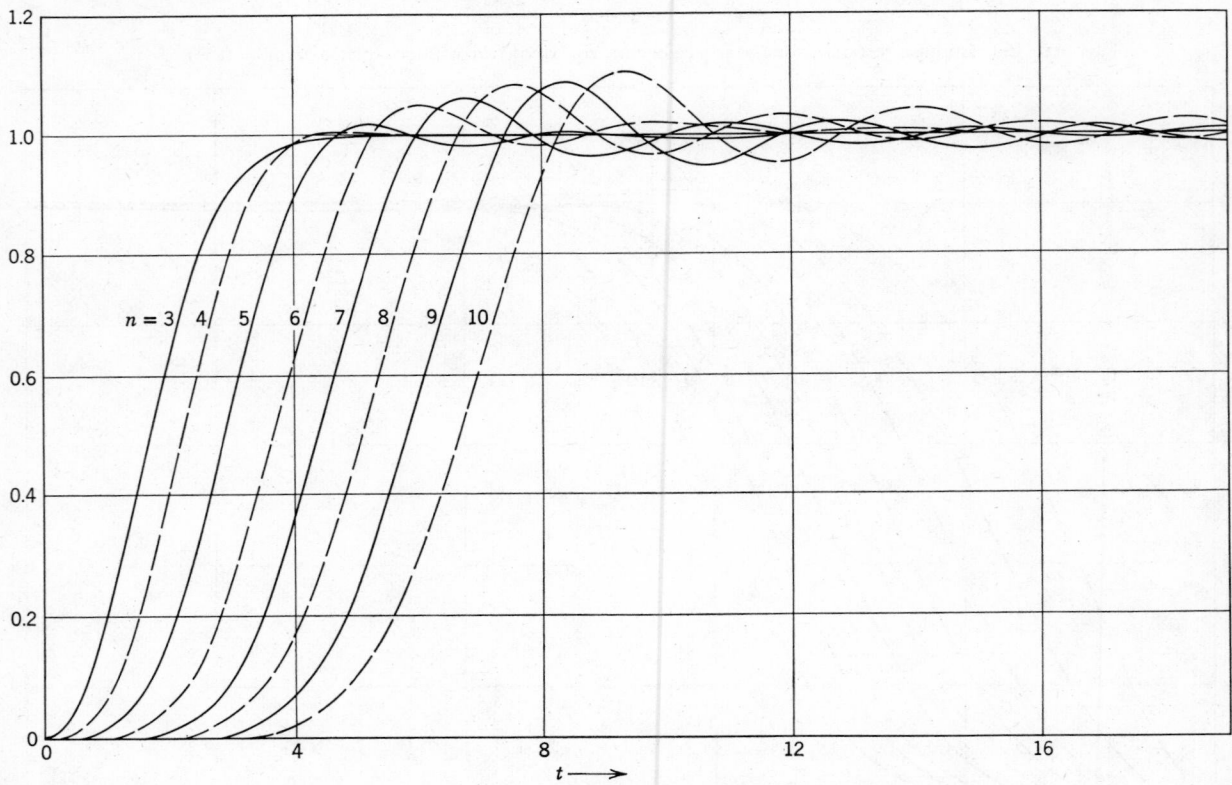

Curve 18. Step response for transitional filters (Gaussian to 6 dB).

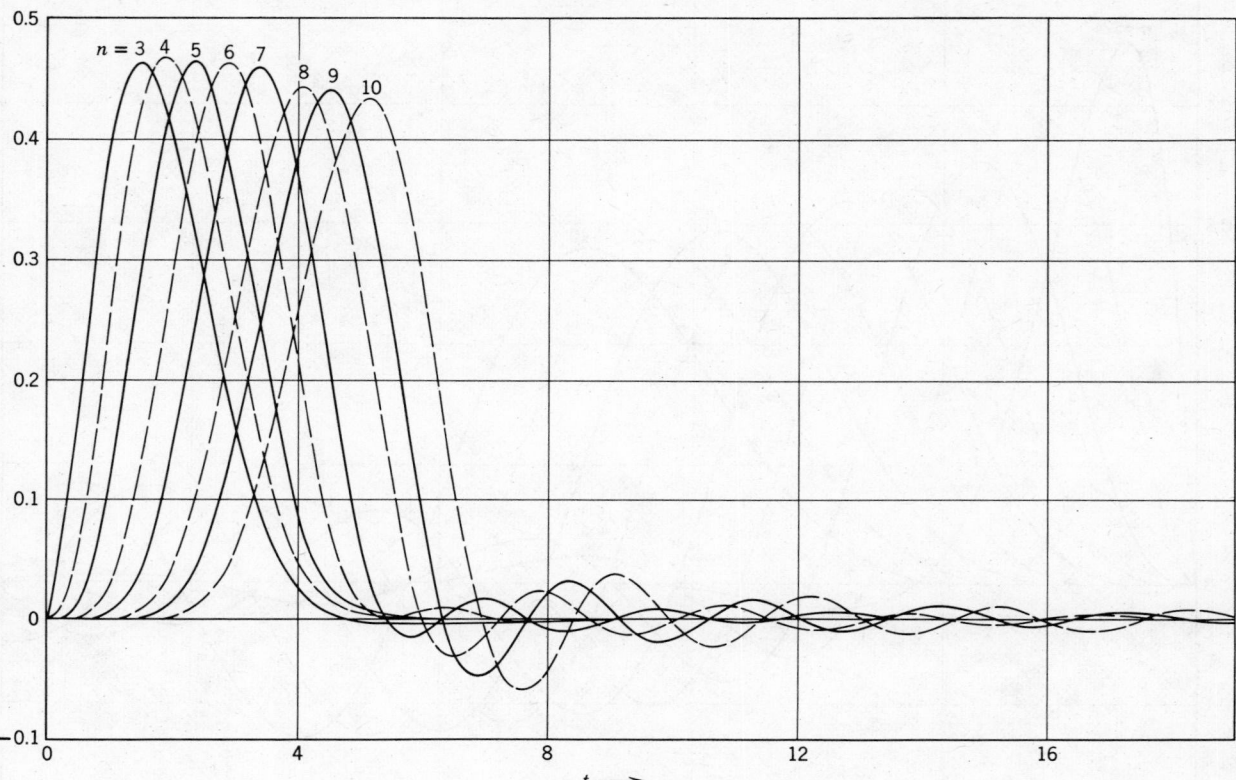

Curve 19. Impulse response for transitional filters (Gaussian to 12 dB).

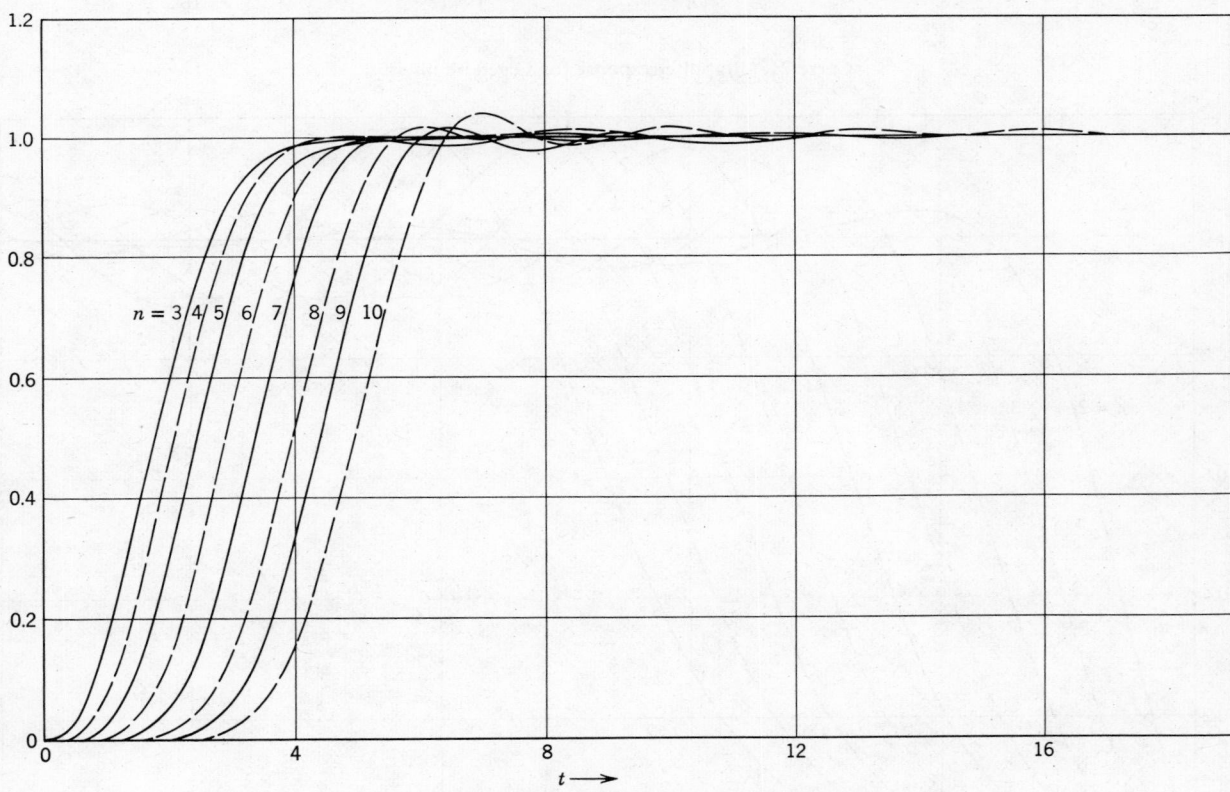

Curve 20. Step response for transitional filters (Gaussian to 12 dB).

411

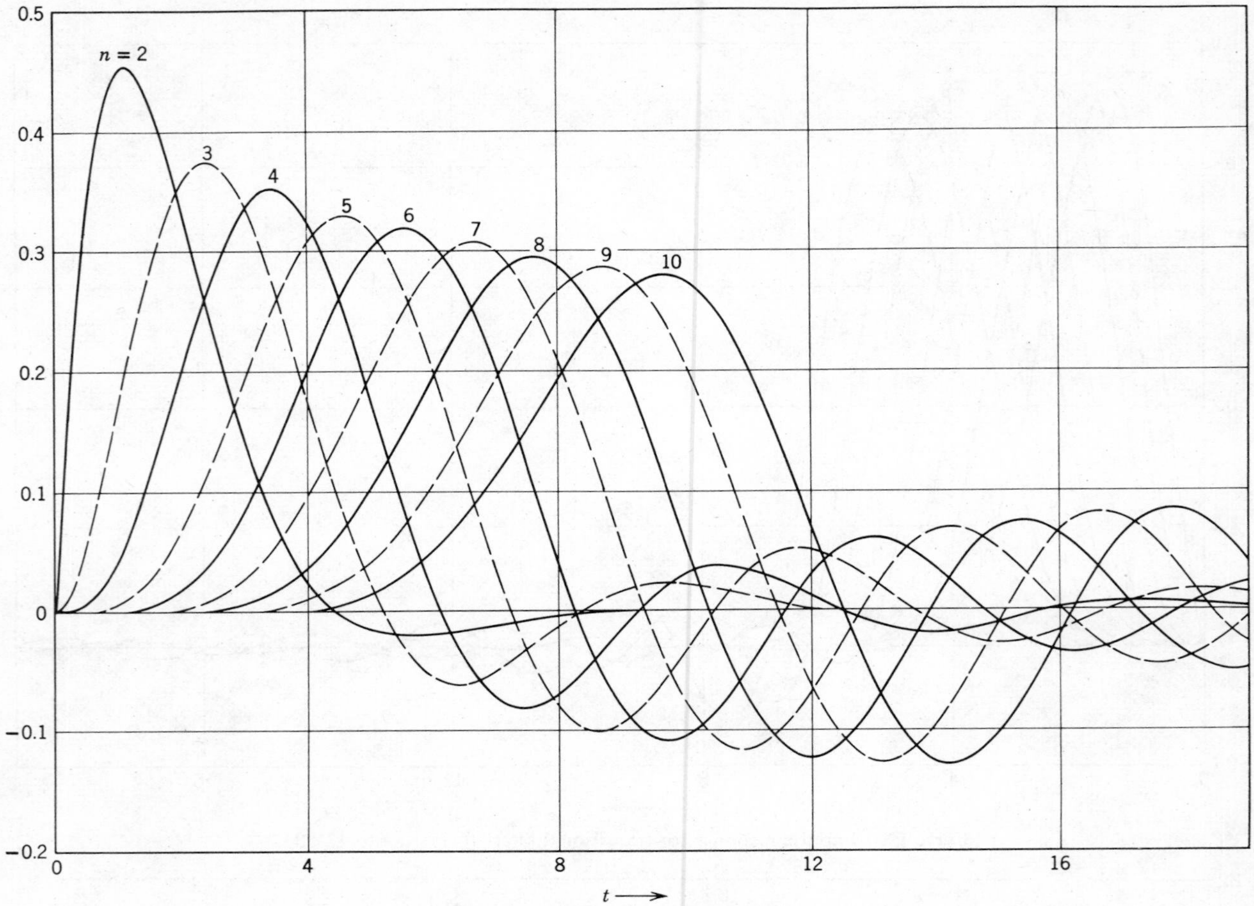

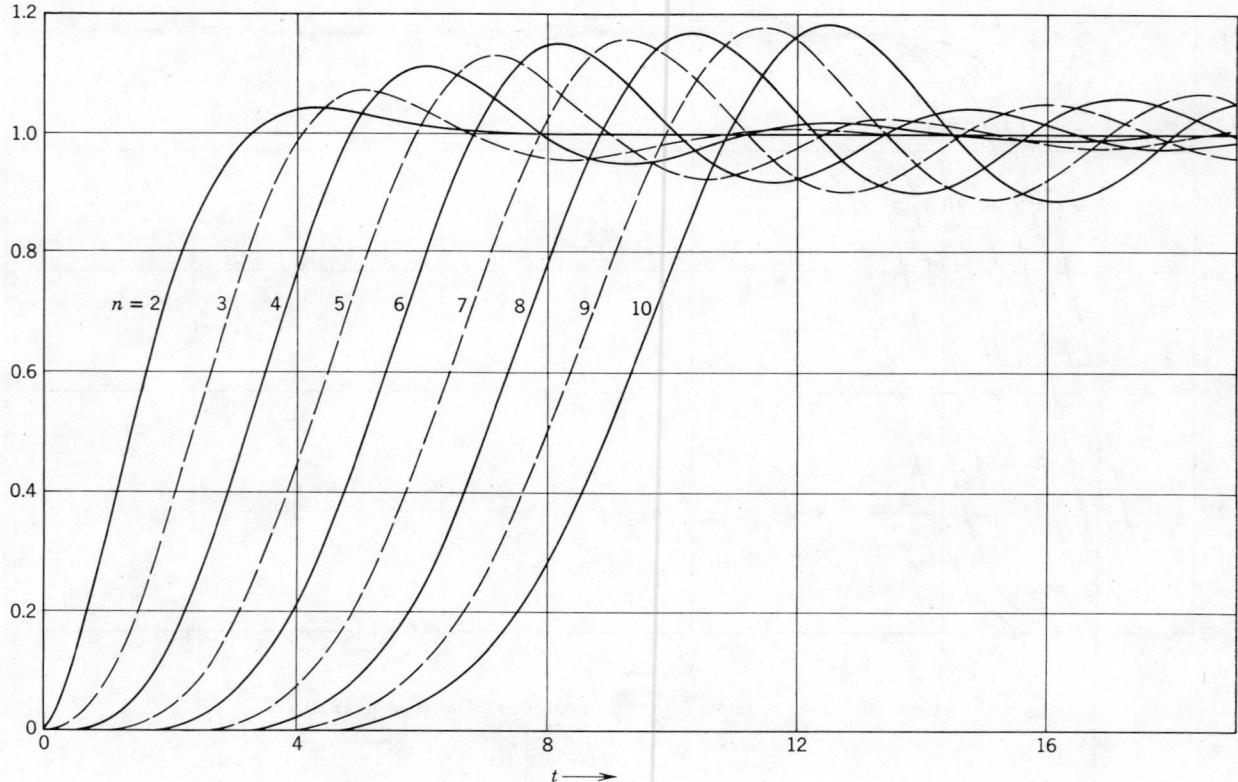

Curve 21. Impulse response for Legendre filters.

Curve 22. Step response for Legendre filters.

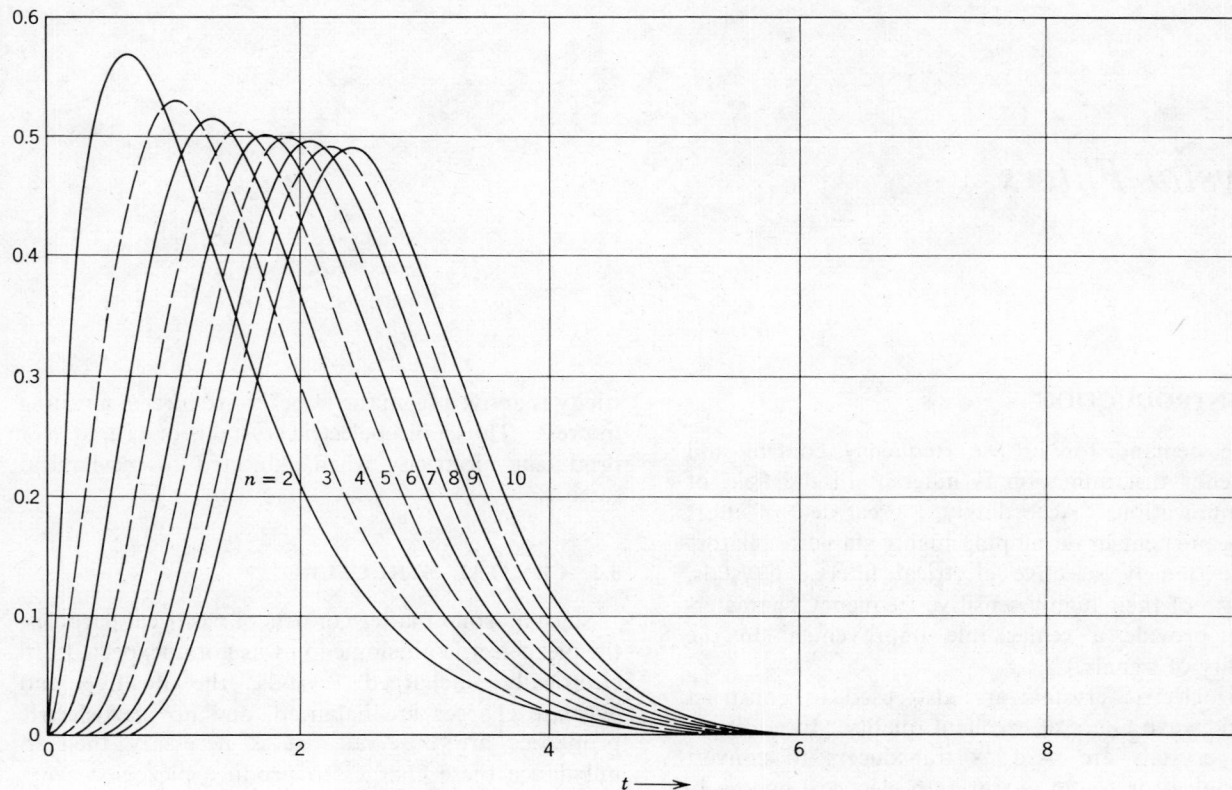

Curve 23. Impulse response for synchronously tuned filters.

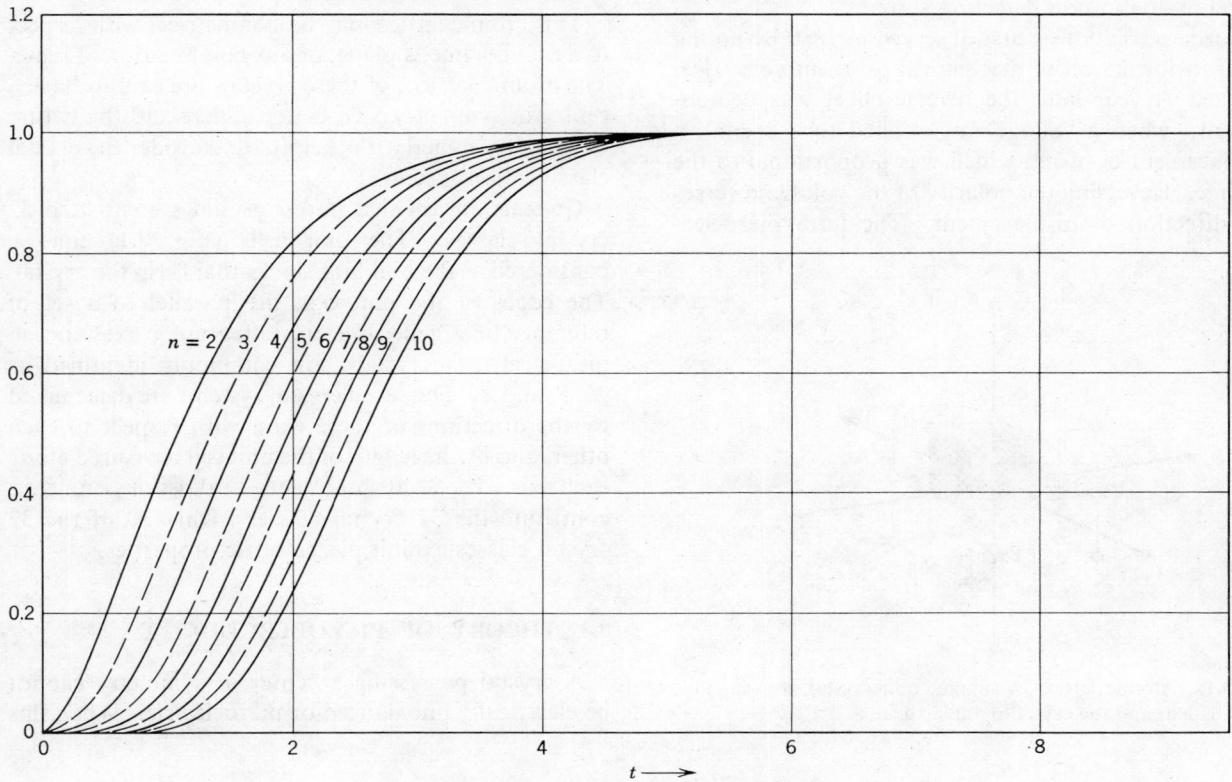

Curve 24. Step response for synchronously tuned filters.

8

Crystal Filters

8.1 INTRODUCTION

The demand for precise frequency control and frequency discrimination is inherent in the field of communications. Accordingly, a great deal of effort has been spent in developing highly stable oscillators and extremely selective electrical filters. Crystals, because of their highly sensitive frequency characteristics, provide a remarkable improvement in the stability of signals.

Piezoelectric crystals are also used to construct electric wave filters of excellent quality. In addition, these crystals are used as transducers to convert mechanical or sound energy into electrical energy in such devices as microphones, phonographs, and in sound and vibration-detection systems.

Piezoelectricity was first observed in 1880 by noting the transformation of mechanical force into electrical charge. A year later the reverse effect was demonstrated; when a voltage was applied to a crystal, a displacement occurred which was proportional to the voltage. Reversing the polarity of the voltage reverses the direction of displacement. The term piezoelec-

tricity is derived from the Greek word *piezein* meaning to press. Thus a piezoelectric crystal is one capable of producing electricity when subjected to mechanical force.

8.2 CRYSTAL STRUCTURE

Since all solid matter consists of electrical particles, the piezoelectric phenomenon was not unexpected. In electrically uncharged crystals, the positive and negative charges are balanced, and no piezoelectric properties are observed. It is necessary then to unbalance these charges to produce piezoelectricity. Only crystals possessing certain types of atomic symmetry can become electrically unbalanced.

This atomic lattice may be symmetrical with respect to a point, a line, a plane, or any combination of these. The atomic lattices of these crystals are said to have a center of symmetry. To better understand the nature of crystal symmetry it is helpfnl to consider the crystal classification system.

Crystals are divided into 7 crystal systems and 32 crystal classes. The unit cells (Fig. 8.1) can be considered as the building blocks that form the crystal. The edges of the unit cells are parallel to a set of reference lines called the crystallographic axes and, in piezoelectricity crystals, are commonly identified as X, Y, and Z. The seven crystal systems are determined by the directions of these axes, with respect to each other, and by the length of the unit cell measured along each axis. The 32 arrangements, or types of symmetry, constitute the 32 crystal classes. Only 20 of the 32 crystal classes exhibit piezoelectric properties.

8.3 THEORY OF PIEZOELECTRICITY

A crystal possessing a center of symmetry cannot be electrically unbalanced or piezoelectric. When this

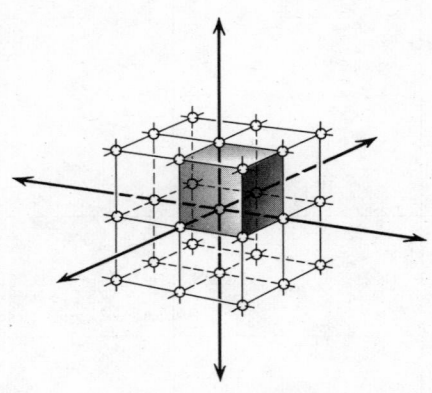

Fig. 8.1. Atomic lattice of simple cubic crystal, showing unit cell (shaded) and the crystallographic axes.

414

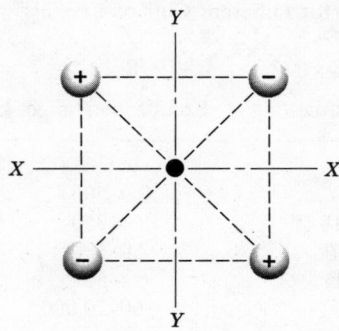

Fig. 8.2. A molecular structure that has no charge separation when deformed. A crystal with this type of molecular symmetry has no piezoelectric property.

type of crystal is subjected to pressure, the same displacement of positive and negative charges occurs in any direction. A distribution of charges having a center of symmetry is shown in Fig. 8.2.

An example of the distribution of charges for a crystal without a center of symmetry is shown in Fig. 8.3a. Note that the lines connecting like charges form an equilateral triangle and that the geometric centers of the two triangles coincide. As long as the centers coincide the charges remain neutral. If however, a so-called longitudinal stress is applied to this crystal in the direction of the X-axis, a displacement of the charges occurs as shown in Fig. 8.3b. In this example, the center of each set of like charges has shifted in opposite directions along the X-axis, thus creating a dipole.

If the stress is applied in the direction of the Y-axis, as shown in Fig. 8.3c, the charge separation still occurs along the X-axis, but is of opposite polarity. For this

reason the X-axis is called the electrical axis and the Y-axis is called the mechanical axis. (These names are slightly misleading because under certain types of stress a displacement of charge centers will occur along the Y-axis.) Perpendicular to these axes is the Z-axis. Because of the optical techniques used to locate this axis in a raw crystal, it is called the optical axis. No piezoelectric effect is associated with the optical axis.

8.4 PROPERTIES OF PIEZOELECTRIC QUARTZ CRYSTALS

The most commonly used piezoelectric crystal is quartz. Quartz is silicon dioxide (SiO_2) and crystallizes in the trigonal trapezohedral class of the trigonal system.

The three lateral axes are always equal and intersect each other at 60° angles. The fourth vertical axis may be shorter or longer than the lateral axis. A drawing of an ideal quartz crystal is shown in Fig. 8.4. In nature, crystals of such perfect symmetry are seldom found and usually only the top formations and parts of the prism faces are visible. The different faces associated with quartz crystals are customarily designated by the lower-case letters *m*, *r*, *s*, *x*, and *z*.

In its original form, a raw quartz crystal is not usable in electronic circuits. The quartz crystals must first be formed into various sizes and shapes, called plates or blanks to give them particular piezoelectric properties. By slicing the raw quartz at various angles with respect to its axes it is possible to obtain a variety of blanks having different temperature characteristics. Certain blanks have become standard and are classified into two groups; the X-Group and the Y-Group. The thickness dimension of X-Group blanks is parallel

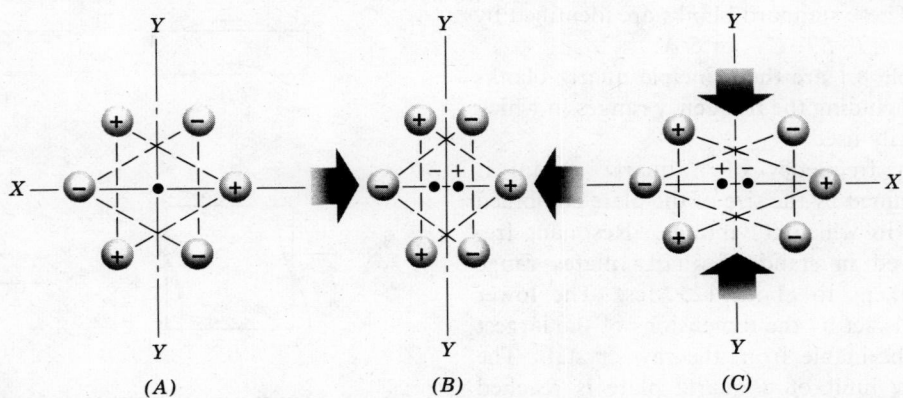

Fig. 8.3. *A*, model of one molecule of a quartz crystal; when longitudinal force is applied, the effect is shown in *B*. If a shearing force is applied, the effect is shown in *C*.

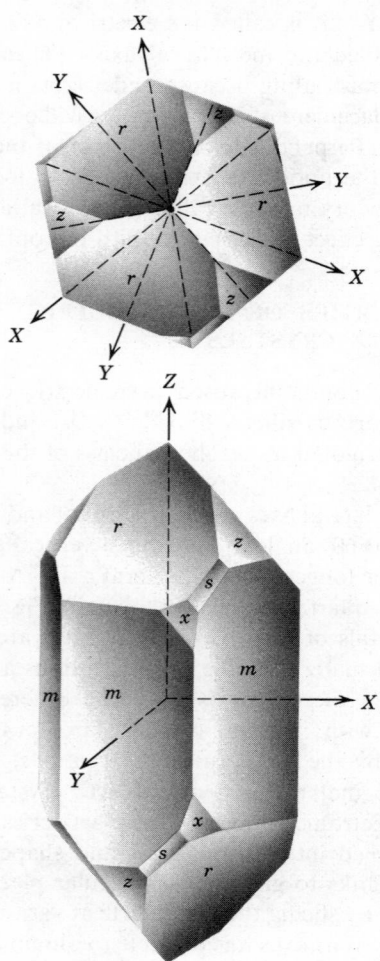

Fig. 8.4. Ideal quartz crystal, showing the directions of the crystallographic axes. The faces of the crystal are identified by the lower-case letters.

to the *X*-axis of the raw crystal from which it was cut. In *Y*-Group blanks the thickness dimension is parallel to the *Y*-axis. These standard blanks are identified by symbols such as *AT*, *BT*, *CT*, or 5°*X*.

Listed in Table 8.1 are the principle quartz blanks of two groups including the frequency ranges in which they are ordinarily used.

The resonant frequency of a quartz crystal is generally determined by the size of the plate combined with the mode in which it vibrates. Resonant frequencies achieved in standard quartz plates range from about 400 cps to about 125 Mcs. The lower frequency limit is set by the dimensions of the largest usable plates obtainable from the raw crystal. The upper frequency limit of a quartz plate is reached when its size becomes so small that it is difficult to handle and likely to shatter when put into use.

Table 8.1 Normal Range of Application for Different Cuts of Crystals

X-Group	
Name	Frequency Range, kc
X	40–20,000
5°*X*	0.9–500
−18°*X*	60–350
MT	50–100
NT	4–50
V	60–20,000

Y-Group	
Name	Frequency Range, kc
Y	1000–20,000
AT	500–100,000
BT	1000–75,000
CT	300–1100
DT	60–500
ET	600–1800
FT	150–1500
GT	100–550

The three basic modes of vibration associated with quartz crystals are the flexure mode, the extensional (or longitudinal) mode, and the shear mode. Flexure motion is a bending or bowing motion, whereas extensional motion consists of a displacement along the length of the plate and away from the center. The more complicated shear motion involves sliding two parallel planes of a quartz plate in opposite directions with both planes remaining parallel. Each type of vibration can occur in a fundamental mode or in an overtone (harmonic) mode. The fundamental vibration of each mode is illustrated in Fig. 8.5. The

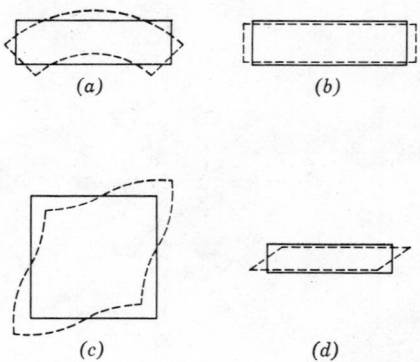

Fig. 8.5. Fundamental vibrational modes associated with quartz crystal plates. (*a*) Flexure, (*b*) extensional (or longitudinal), (*c*) face shear, and (*d*) thickness shear.

practical frequency ranges achieved in vibrating quartz plates in the three modes, including overtones, are:

Flexure mode	0.4–100 kc
Extensional mode	40–15,000 kc
Shear mode	100–125,000 kc

Quartz plates exhibit an extremely high quality factor, or Q, which guarantees highly stable and efficient performance.

Q-Factor

The Q of a circuit is defined as the ratio of reactance to resistance. At the resonant frequency of a circuit, the capacitive and inductive reactances are equal and opposite. Only the resistance of the circuit remains to oppose the flow of current. When this resistance is high, the Q will be low, and more power must be supplied to sustain oscillation. This added power contributes to instability and drift. Thus the higher the Q, the more stable the oscillations and the less the resistive losses.

Most of the losses in electrical-resonant circuits are caused by the high resistance of coils. As a result, the Q of these circuits is comparatively low, ranging from about 10–400. The losses of a crystal are in its internal dissipation, mechanical mounting, and the damping of its motion by the surrounding air. The sum of these losses is very small when compared to an electrical circuit. Because of this, the Q of a crystal is comparatively high and may range from ten thousand to over one million.

Equivalent Circuit and Applications

Crystals possess two resonant frequencies: a series-resonant frequency and a parallel-resonant (or anti-resonant) frequency. The series-resonant frequency is determined by the equivalent motional inductance L_s and the capacitance C_s as shown in Fig. 8.6a. The parallel-resonant frequency is determined essentially by the parallel capacitance C_p. Figure 8.6b shows the absolute value of impedance versus frequency of a typical quartz plate. The impedance of a crystal is lowest at its series-resonant frequency and highest at its parallel-resonant frequency. In Fig. 8.6c the values of practical L_s are plotted against the frequency of application.

Either the series-resonant or the parallel-resonant characteristics of a crystal may be used in an oscillator circuit. Both resonances are also used in filter circuits.

Under certain conditions it may be desirable to adjust the resonant frequency of a crystal. Slight adjustments can be made by adding a variable capacitor or inductor in series with the crystal in the series-resonant mode and in parallel with the crystal in the parallel-resonant mode. The amount of adjustment can only be very small, but it does provide a means of offsetting frequency drift.

Crystal filters are widely used in narrowband applications, such as intercepting the pilot frequency in a carrier system, or separating the sidebands in a single-sideband radio system, or comb filters used in doppler type of radars. Because of extremely high Q and stability, such networks exhibit superior frequency characteristics and, therefore, are used advantageously as carrier channel filters to multiplex voice-frequency

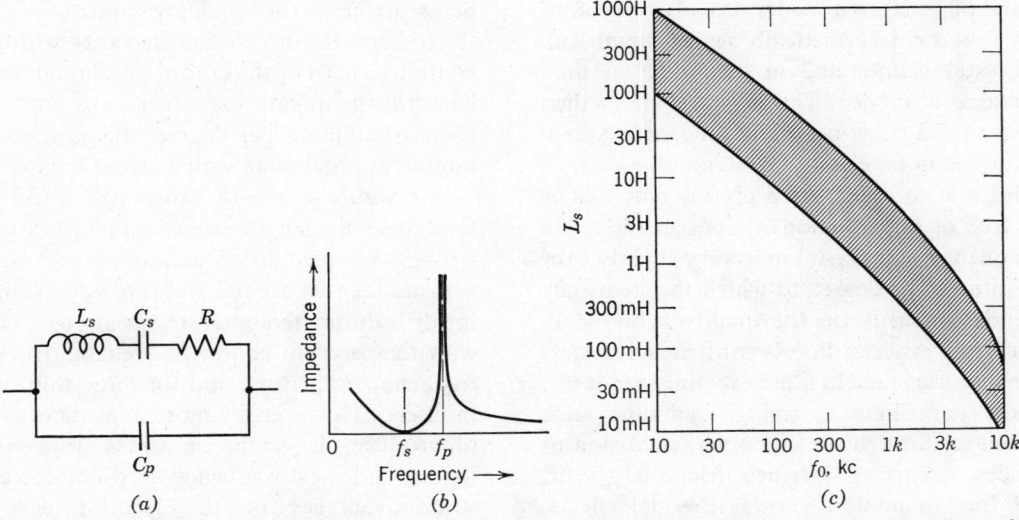

Fig. 8.6. Parameters of quartz crystal: (a) equivalent circuit (b) magnitude of motional inductance for different frequencies.

signals. The sharp cutoff feature permits closer spacing of carrier channels, thereby providing more channels within a given frequency range.

Manufacturing of Quartz Crystal

There are two types of raw quartz crystals, natural and cultured. Natural quartz is mined primarily in Brazil. Cultured (or synthetic) quartz is grown in a process that is analogous to growing cultured pearls. Cultured quartz crystals are usually better formed than natural quartz, resulting in a higher yield of quality plates per crystal.

Making a quartz crystal unit begins by carefully inspecting the natural or cultured crystal for imperfections and then by locating its optical (Z) axis. A face is ground on the raw crystal parallel to the optical axis. Next, the electrical (X) axis is located by an Xray process. The mechanical (Y) axis is then known to be perpendicular to the X- and Z-axes.

After locating the axes, the raw crystal is sawed into sections and then into thin wafers along the Z-axis so that the length of the plate is parallel to either the Y- or X-axes. Each wafer is then diced into small sections called blanks. These blanks are ground to the desired dimensions using a precision grinding process, known as lapping, and then inspected, cleaned, and etched in chemical solutions. After cleaning and etching, a spot of silver is applied to two opposite sides of each blank where the wire leads are to be attached. These same sides are then gold plated to form the electrodes required to electrically connect the crystal to the leads. Next, leads are soldered to the silver spots. The crystal can now be frequency calibrated and placed into a holder, usually a vacuum-sealed glass tube or a hermetically-sealed metal can. After the crystal is mounted in its holder, a final frequency check is made. The crystal unit is then allowed to age for a short period of time, after which it is ready for use in an electronic circuit.

It is impossible to construct a crystal unit that is completely free of imperfections. Consequently, the resonant frequency of a crystal may vary slightly over a period of time. The degree to which the frequency varies depends primarily on the quality achieved in the manufacturing process. Poorly constructed crystals age faster than those made to more exacting standards. Many factors contribute to aging, including such things as leakage through the container, corrosion of the electrodes, material fatigue, frictional wear, presence of foreign matter, various thermal effects, and surface erosion of the crystals. If, however, a crystal is carefully constructed it will have a long, stable life before deteriorating.

The *X*-Cut Crystals

The X-cut was the original quartz plate investigated by Curie. This cut was also the first to be used as a transducer of ultrasonic waves, the control element of radio-frequency oscillators and as a filter component. However, because of its comparatively large coefficient of temperature, the X-cut plate is now rarely used for frequency control. As a transducer of electrical to mechanical vibrations, the X-cut is still widely used to produce ultrasonic waves in gases, liquids, and solids.

The X-cut is used in frequency range from 40–20,000 kc in the fundamental mode, and at lower frequencies when coupled as a transducer for generating vibrations in liquids and solids.

The temperature coefficient is -20 to -25 parts per million per degree centigrade. (That is, for each degree of increase or decrease in temperature, the frequency respectively decreases or increases 20 to 25 cycles for each megacycle of the initial frequency—a rise in temperature of 10°C would thus cause the frequency of a 5000 kc crystal to drop 1000 to 1250 cps.)

Advantages of X-cut thickness-extension mode are its mechanical stability, the economy of cut, an efficiency of conversion of electrical to mechanical energy. The disadvantages are the large temperature coefficient, a tendency to jump from one mode to another, and the difficulty of clamping the crystal in a fixed position without greatly damping the normal vibration. These difficulties prevent this element from being preferred for oscillator control.

The temperature coefficient varies with the ratio of width to length of the crystal. Although for long, thin bars the temperature coefficient is only about two parts per million per degree, this is greater than the minimum obtainable with 5° X-cut bars.

For width to length ratios from 0.35 to 1.0, the fundamental length-extensional vibration is not strongly coupled to other modes, and hence the resonance is easily excited and is of good stability except for drift during temperature variations. This element, with temperature control, is reliable for use in low-frequency oscillators, and for long, thin bars, for use in filters. However, its most important application is to produce ultrasonic vibrations, when the ratio of highest to lowest frequency need not exceed 1.1.

Disadvantages of the length-extensional mode include its inefficiency as a transducer since the

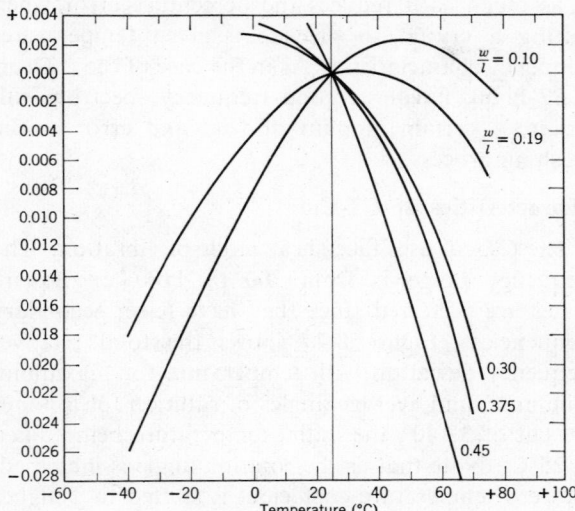

Fig. 8.7a. Temperature-frequency characteristics of *X*-cut.

electromechanical coupling is low. Strong coupling with a flexural mode makes the crystal useless at *w/l* ratios between 0.2 and 0.3 and a weak coupling with a shear mode causes the frequency to decrease as the *w/l* ratio approaches 1.0. This coupling to other modes interferes with the frequency response of the element when used in filters, unless the *w/l* ratio is 0.1 or less.

An advantage to the length-extensional mode is that it can be used in temperature controlled environments for low-frequency and narrow-bandpass filters. With the thickness dimension ground for a particular high frequency, the same crystal unit may be used to generate either of two widely separate frequencies.

The 5° X-Cuts. The 5°*X*-cut orientation provides a zero temperature coefficient for the lengthwise vibrations of long, thin *X*-bars. Thus this cut is preferred over the nonrotated *X*-cut for use in low-frequency filters and control devices. Its length-extensional, length-width-flexural, and duplex length-thickness-flexural modes are defined as the elements *E*, *H*, and *J* respectively; the last named element, *J* providing the lowest frequencies. However, the 5°*X*-elements are also coupled to the other modes, so that for *w/l* ratios much greater than 0.1 the frequency spectrum is little improved over that of the length-extensional mode of the *X*-cut. Furthermore, as the *w/l* ratio increases, so does the temperature coefficient. For these reasons the 5°*X*-elements are especially advantageous only when the *w/l* ratio is 0.1 or less. These long, thin bars are used commercially as filters, and

are particularly adaptable for use in telephone carrier systems.

Characteristics of *AT*-Cut Crystals

This popular type of crystal cut is used in frequency ranges 500–1000 kc (special cuts); 1000–15,000 kc (fundamental vibration); 10,000–100,000 kc (overtone modes). The best known property of this crystal is its low temperature coefficient as shown in Fig. 8.7b. This curve gives the total relative frequency deviation for the normal maximum, minimum, and average angles of this element; the temperature coefficient at each point on a curve is the slope at that point in parts per hundred. At $\phi = 35°15'$, the temperature coefficient will vanish at 45°C changing from negative to positive as the temperature increases. The curves demonstrate the excellent temperature-frequency characteristics. This element is preferred for high-frequency oscillator control wherever wide variations of temperature are encountered.

Because of its superior temperature-frequency and piezoelectric characteristics as compared with the *BT*-characteristics, the *AT*-element may well be preferred for filters and the control of frequencies in the VHF range.

Because of the large thickness dimensions that would be required, the *AT*-element is generally not economical for the generation of frequencies below 1000 kc, although special circular cuts have been used at frequencies as low as 500 kc. In its normal high-frequency range, the most troublesome problems are to find the proper length and width dimensions that will permit the desired frequency to be widely separated from other modes. It has been found that the unwanted modes are somewhat restricted by giving the plate a convex contour, and also by the use of circular plates. The convex contour is possible for all but the very thin plates that are used at frequencies above 15,000 kc. A 1000 kc crystal may have a contour of 3- to 5-microns.

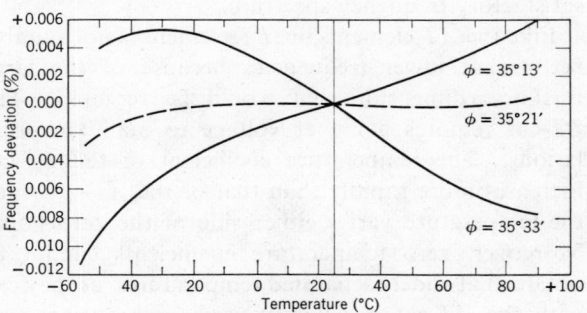

Fig. 8.7b. Temperature-frequency characteristics of *AT*-cut.

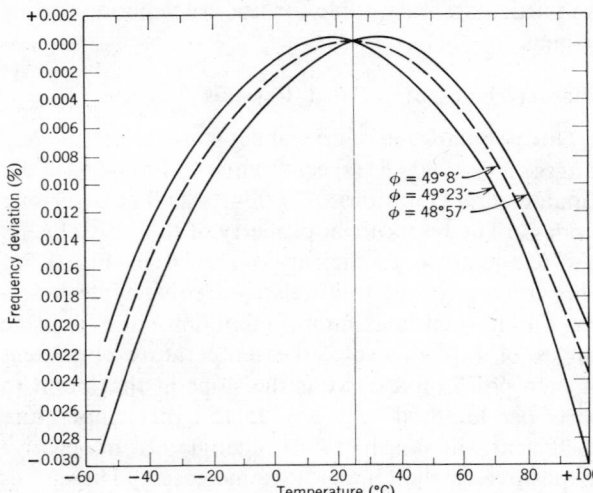

Fig. 8.7c. Temperature-frequency characteristics of *BT*-cut.

Characteristics of *BT*-Crystals

The *BT*-cut makes use of the length-thickness-shear mode, or length-width-shear mode, and has a frequency range of application from 1000 to 20,000 kc for the fundamental vibration. Using the overtone mode, the frequency range is from 15,000 to 75,000 kc. The temperature coefficient in Fig. 8.7c shows the total relative-frequency deviation for the normal maximum, minimum, and average angles of this element. The temperature coefficient in parts per hundred per degree centigrade at each point on the curve is the slope at that point. Zero coefficients are obtained at 20°C and 75°C when ϕ is −49°16′ and −47°22′ respectively.

The temperature-frequency characteristics make this element useful for high-frequency oscillator control where the temperature is not expected to vary too widely from the mean value. It is particularly applicable for use in radio equipment which is to operate at the high end of the high-frequency spectrum. Most high-frequency crystal oscillators employ either the *BT*- or the *AT*-cut. The *BT*-element exhibits a satisfactory frequency spectrum.

Like the *AT*-element, the *BT*-element is not suitable for use at lower frequencies because of the large thickness dimension that would be required. The *BT*-cut requires a higher voltage to maintain oscillations. The temperature coefficient of the *BT*-cut increases more rapidly than that of the *AT*-cut when the temperature varies either side of the zero point. Moreover, zero temperature coefficients cannot be obtained at widely separated temperatures, as they can with the *AT*-cut by slightly varying the orientation angle. This limitation, however, becomes an advantage

in as much as it reduces the percentage error when cutting a crystal to provide a given temperature-frequency characteristic. As in the case of the *AT*-cut, a *BT*-blank having a good frequency spectrum will require a certain amount of trial and error in the finishing process.

Characteristics of *CT*-Cut

The *CT*-cut uses face-shear mode of vibration. The frequency range is from 300 to 1100 kc. Square plates are preferred since they have fewer secondary frequencies. Figure 8.7d shows the total relative-frequency deviation with temperature for maximum, minimum, and average angles of rotation for a nominal cut of 37°40′, the initial temperature being taken at 25°C. Note that as the rotation angle is increased, the zero temperature coefficient is shifted to a higher temperature. For square plates, zero coefficients can be obtained at higher temperatures (50°C to 200°C) by rotation angles from 38°20′ to 41°50′ respectively.

The *CT*-cut is essentially a *BT*-cut rotated approximately 90° so that the face shear of the *C*-element corresponds to the thickness shear of the *B*-element. This orientation provides a zero-temperature-coefficient shear mode for generating low frequencies, without requiring a crystal of large thickness dimension.

The *C*-element is widely used both for low-frequency oscillator control and in filters, and does not require constant temperature control under normal operating conditions. One of its principal applications has been as the control element in frequency-modulated oscillators. But the crystal has certain disadvantages. Because of its larger frequency constant, the *CT*-element must be cut with larger face dimensions than some other elements to provide the same frequency of vibration. Thus for the generation of very low frequencies the smaller *DT*-cut is more economical to use. Care must be taken to prevent flexure modes from being strongly coupled to the face-shear mode.

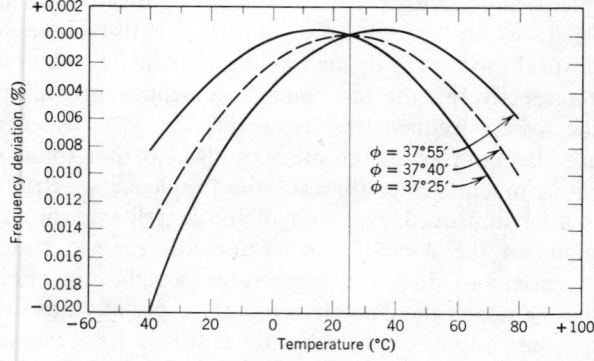

Fig. 8.7d. Temperature-frequency characteristics of *CT*-cut.

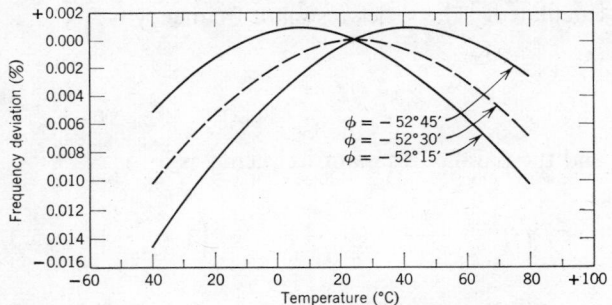

Fig. 8.7e. Temperature-frequency characteristics of *DT*-cut.

Characteristics of *DT*-Cut

The *DT*-cut makes use of the face-shear mode. The frequency range of application is from 60–500 kc. Square plates are preferred since they have fewer secondary frequencies. The temperature coefficient as shown in Fig. 8.7e demonstrates the total relative-frequency deviation with temperature for maximum, minimum, and average angles of rotation for a nominal cutoff $-52°30'$, where the initial temperature is taken at 25°C. Note, as the rotation angle is increased, the zero temperature coefficient is shifted to a higher temperature. The upper limit for a zero coefficient is approximately 200°C, when $\phi = 54°$.

Conclusion

Quartz crystals enjoy a very useful position in the communications industry and, from all observations, their usefulness will continue for sometime. In the past, several possible substitutes for quartz crystals were developed, but none were too successful. Prominent among these were ammonium dihydrogen phosphate (*ADP*), and dipotassium tartrate (*DKT*). These so-called synthetic crystals were developed because of a shortage of the large size raw quartz crystals required for making filter crystal plates. When the process of growing cultured quartz crystals developed, these substitutes were no longer needed.

Recently, a disc-shaped ceramic device with piezo-electric properties has been developed and used in filter circuits. One advantage of the ceramic disc is its characteristic bandwidth, which is about ten times larger than the bandwidth of quartz crystals. However, its Q is limited to below 2,000 and its frequency stability is about 0.1% variation (as opposed to about 0.01% for quartz crystals). It is doubtful, therefore, that these devices will compete seriously with quartz crystals in filter applications, except in certain special cases.

8.5 CLASSIFICATION OF CRYSTAL FILTERS

The great majority of crystal filters are bandpass filters. As such they can be classified in the following manner:

1. Narrowband filters.
2. Intermediate-bandpass filters.
3. Wideband filters.
4. Extra-large-bandwidth filters.

The first category of narrowband crystal filters can be realized as either a ladder or a bridge circuit as shown in Figs. 8.8 and 8.9. Narrowband crystal filters consist basically of only two kinds of components: crystals and capacitors. Both forms of realization have advantages and disadvantages which are a result of the physical nature of the quartz element and the properties of the schematics.

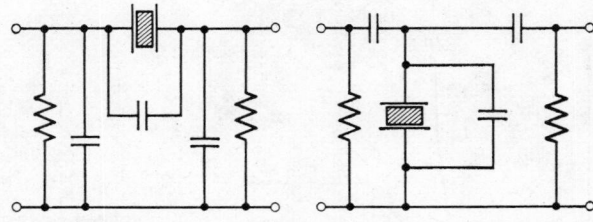

Fig. 8.8. Simplest ladder filters with crystal elements.

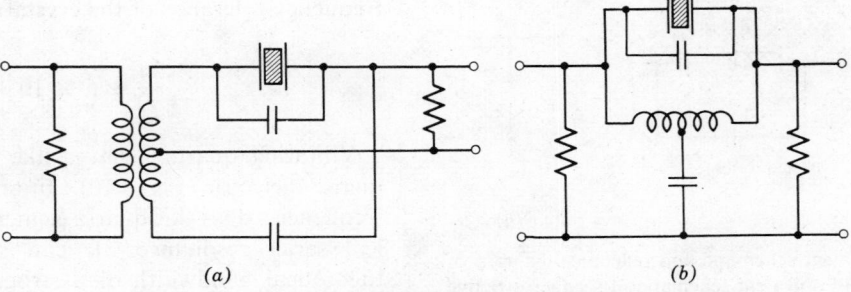

(a)   (b)

Fig. 8.9. Simplest bridge filters with crystal elements.

Crystal Elements

The equivalent electric circuit of the piezoelectric resonator is the dual to the resonator based on the magnetostrictive effect. In order to incorporate an electromechanical resonator into a network, the equivalent electrical properties, frequency range, electrical quality, and technologically achievable ratios of inductances and capacitances must be known. The duality of resonators permits the extension of the relations developed for crystals to schematics using ferrite resonators based on the magnetostrictive effect. Figure 8.10 shows both of the above mentioned electromechanical resonators.

It is worthwhile to review some of the fundamental properties of quartz crystals as reactive components for filters. The reactance as a function of the frequency can be expressed by the following formula:

$$X = \frac{1 + L_s C_s s^2}{s(s^2 L_s C_s C_p + C_s + C_p)} \qquad (8.5.1)$$

where $s = j\omega$

L_s = motional inductance of the quartz

C_s = motional (dynamic) capacitance of the quartz

C_p = parallel (static) capacitance across the quartz

The element values can be normalized in the usual fashion for unity impedance and reference frequency, ω_r. From Eq. 8.5.1, two important frequencies can be

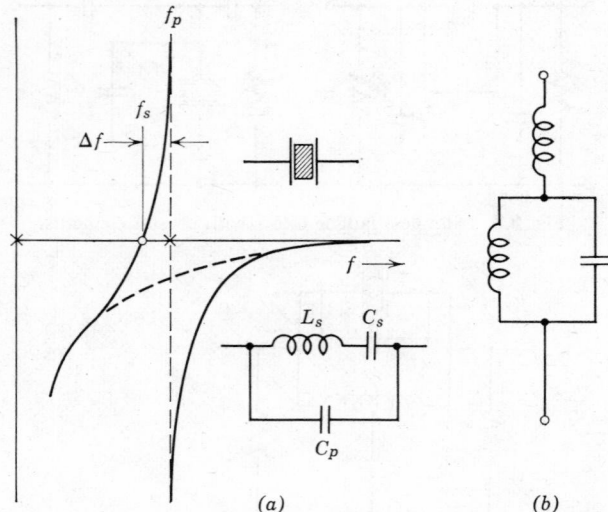

Fig. 8.10. (*a*) Equivalent schematic and reactance diagram of crystal resonator. (*b*) Equivalent schematic of magnetostrictive resonator.

calculated. The series-resonant frequency is

$$\omega_s{}^2 = \frac{1}{L_s C_s} \qquad (8.5.2)$$

and the parallel-resonant frequency is

$$\omega_p{}^2 = \frac{1}{L_s C_s}\left(1 + \frac{C_s}{C_p}\right) = \omega_s{}^2\left(1 + \frac{C_s}{C_p}\right) \qquad (8.5.3)$$

Figure 8.10 shows the reactance diagram of the crystal, drawn from Eqs. 8.5.2 and 8.5.3.

The distance between the two resonances is defined as,

$$\Delta f = f_p - f_s \qquad (8.5.4)$$

and therefore from

$$\frac{C_s}{C_p} = \frac{f_p{}^2 - f_s{}^2}{f_s{}^2} \qquad (8.5.5)$$

the approximate ratio of parallel to series capacitance of the crystal is given by

$$\frac{C_p}{C_s} \approx \frac{f_s}{2\,\Delta f} \qquad (8.5.6)$$

The obtainable bandwidth of the filter, as is easy to recognize, is dependent on this capacitance ratio. For the ratio $C_p/C_s = 140$, $\Delta f = f_s/280$.

Equation 8.5.6 is indeed a very important criterion for the realization of crystal filters. The ratio of C_p/C_s is dependent on the type of cut and the frequency range. Low-frequency crystals can be made with C_p/C_s ratio as low as 125.

The quality factor Q, of quartz usually lies within 10,000 to 150,000 dependent upon the method of capsulation (in atmosphere, an inert gas, or vacuum). The temperature coefficient of crystals varies with different types of cut, and is between 0.5–6 parts per million per degree centigrade. The relative long-term stability (relative to nominal value) is of the order of 10^{-5}. The achievable exactitude, $\Delta f/f_s$ (the setting frequency tolerance of the crystal) is

$$\frac{\Delta f}{f_s} \approx 5 \times 10^{-6} \qquad (8.5.7)$$

Vibrating quartz behaves like a capacitor with a quartz dielectric (Fig. 8.10). In only a narrow band of frequencies does this quartz element exhibit a mechanical series resonance. It can be shown that the theoretical bandwidth of a structure with spreading coils is approximately 14% of center frequency. The condition for that ultimate limit requires that the ratio between the static to dynamic capacitance in the crystal is equal to 139. This ratio can be reached only by longitudinally vibrating quartz used in the low-frequency domain up to 300 kc.

Practically, the bandwidth maximum must be reduced to about 7%–10%, since the wiring in the filter will provide a supplementary parasitic capacitance across the crystal.

The equivalent circuit of the crystal, such as given in Fig. 8.10, is a good approximation only near resonance. On the high side of resonance, and sometimes on the low side, so-called spurious responses occur which give random passbands of a few cycles bandwidth. Such occurrences are not predicted by the simple equivalent circuit of Fig. 8.10.

From audio frequencies to around 30 Mc, usually first overtone crystals are used. For higher frequency applications higher overtone crystals such as third and fifth may be employed. Their equivalent schematic includes only three reactances and is valid only for a very narrow frequency band about the specific overtone involved. Consequently the parameters of the same crystal at different overtones will be different. Regardless of the center frequency (series-resonant frequency) the equivalent motional capacitance of the crystal at the fundamental mode is in the order of 0.02 pF.

Inspite of the good stability of a crystal with both age and temperature, for extremely-narrow-bandwidth filters and when a wide variation of the environmental conditions is predicted, preaging and a temperature-controlled environment (ovens) may be necessary.

8.6 BRIDGE FILTERS

A full lattice (see also Sections 1.4 and 2.13) is one of the most general bridge network structures, but is not suitable for practical filter application because of component duplication and a lack of common ground between input and output. If, however, transformers are allowed as components, equivalent circuits overcome the faults of the full lattice.

In bridge circuits such as in Fig. 8.9a it is possible to balance the effect of the parasitic, or shunt, capacitance inherent in the crystals, and thus obtain a frequency-symmetrical response curve. Therefore as a filter network, the bridge is a very convenient structure. Their arms may be very complicated and consist of several crystals with lumped inductances and capacitors.

Differential transformers do not add to the selec-

tivity of the bridge circuit. However, in the case of the bridged-T schematic of Fig. 8.9b, the center-tapped inductance is required as a part of the filter and its presence will increase the selectivity.

The lattice and semilattice, as shown in Fig. 8.11a, are known to be equivalent without any restrictions. Both arms in the semilattice section are the same as Z_a and Z_b in the full lattice provided that the source and load impedances are half of the lattice impedance. Two repetitive arms of the lattice are eliminated, but the complementary component introduced in the semilattice is a differential transformer with a turns ratio of $1:1:1$, or center-taped inductance with a turns ratio of $1:-1$.

Figure 8.11b indicates the possibility of extracting a common capacitance which is included in the design for band limitation. Of course, C_d in the lattice does not include the interelectrode capacitance in the crystal (which naturally cannot be extracted from the crystal element). The presence of the supplementary capacitor across the crystal is an indication that the bandwidth is narrower than a certain maximum which has been discussed already. In Fig. 8.11c an example of a common inductance is shown for band-widening.

In Fig. 8.11d a bridged-T equivalent of the lattice with a spreading coil is shown. The crystal must be shunted by inductance. Many applications of bridge-T equivalent circuits can be found in lowpass and especially band-reject designs. In order to realize this section, the impedance ratio of the branches must be $1:4$ if no transformer is used in the circuit.

In the bridge structure, the graphic representation of both reactive arms as a function of frequency contributes to a closer understanding of the physical picture of frequency discrimination. The transmission function of a lattice network may be represented as

$$H(s) = \frac{[1 + Z_a(s)][1 + Z_b(s)]}{Z_b(s) - Z_a(s)}$$

The effective attenuation of that lattice is given by

$$a = \ln \left| \frac{V_0}{2V_2} \right| = \ln |H(s)| \ \text{Np}$$

Z_a and Z_b are the reactances of arms in the lattice structure. From the foregoing expression, we can see that when $Z_a = 1/Z_b$, zeros appear. When $Z_a = Z_b$ the poles of transmission will appear and no energy will flow from input to output. This expression for the transmission functions provides a good means for the physical interpretation of the filtering process.

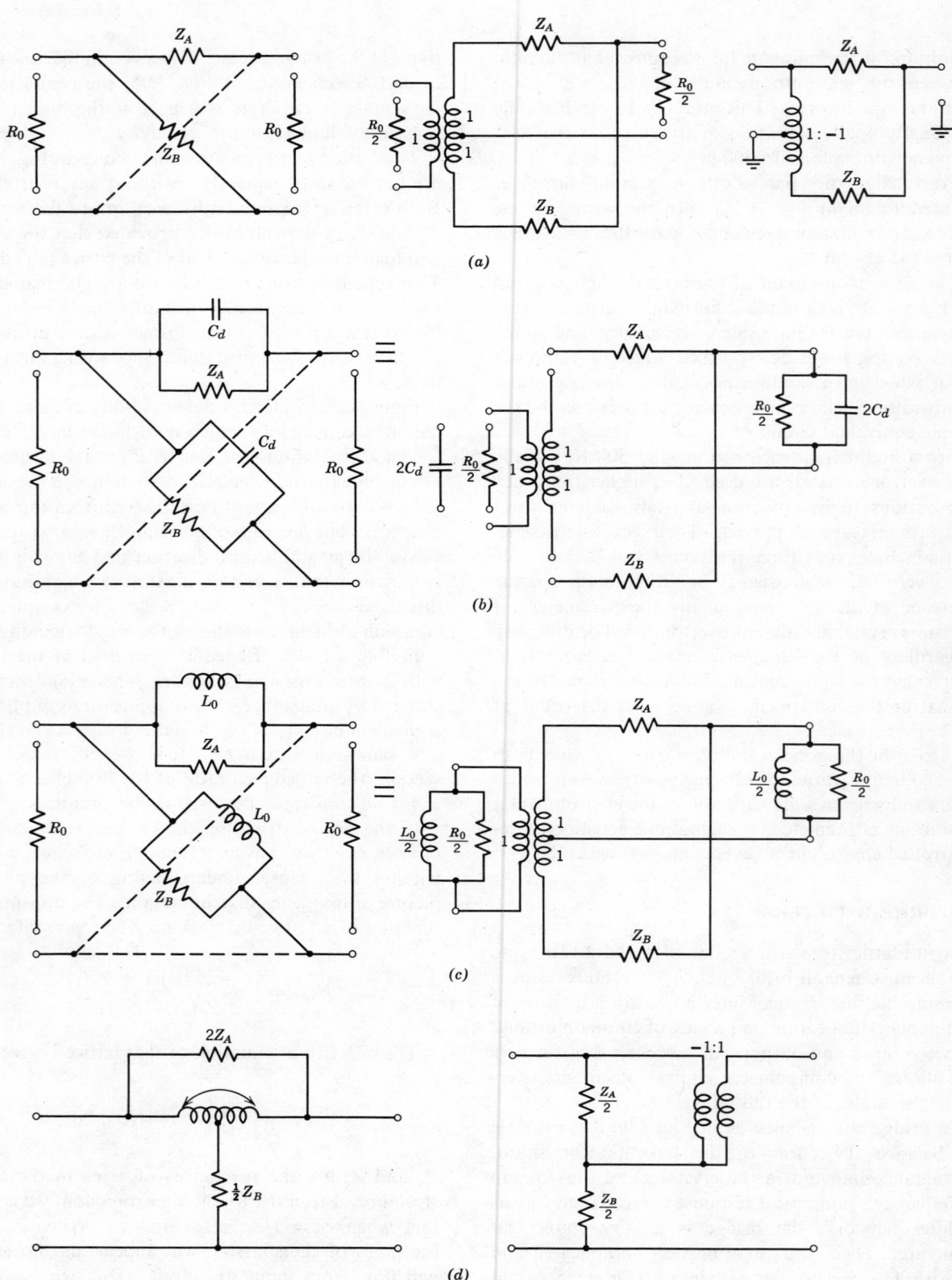

Fig. 8.11. (*a*) Lattice and equivalent semilattice. (*b*) Extracting common capacitance. (*c*) Extracting common inductance. (*d*) Bridge T equivalent and Cauer equivalent.

The attenuation curve of the network is dependent on reactances in both arms, and the balancing effect between the reactances. For filters having a requirement of less than 40-dB attenuation, the symmetrical differential transformer provides stability to quite an acceptable degree of perfection.

It is easy to recognize the shape of the attenuation response. The lattice (bridge) structure shown in Fig. 8.11a allows the simplest form of loss compensation in the filter. That compensation will result in the raising of the insertion loss in the passband although the magnitude of the discrimination curve will be only slightly damaged. All the equivalent networks in Fig. 8.11 have only two arms instead of four.

The ideal transformers required by semilattice structures are not entirely realizable but they can be approximated sufficiently well for most applications. Nevertheless, the realization of transformers presents a serious limitation in obtaining bandwidth wider than a particular maximum.

The method of filter design based on a single-section lattice unit suffers severe limitations such as the amount of stopband attenuation that can be obtained; no matter how many crystals are involved in one filter section, the ultimate attenuation cannot be higher than 40–50 dB. For small bandwidths (below 0.4%) such limitations may be overcome by capacitive coupling several two-crystal semilattice filters. In this arrangement the parasitic shunt capacitance of the crystal forms part of the capacitive coupling network. This phenonemon is well known and removal of the common reactance from within the lattice or semilattice (such as illustrated in Figs. 8.11b and 8.11c) is a popular design tool. Though extremely useful, this approach is restricted to narrow bandwidths and to responses derivable from polynomial lowpass filters without peaks of attenuation in the stopband.

A simple crystal filter in the form of a semilattice which includes a differential transformer or auto-transformer is an extremely practical building block.

8.7 LIMITATION OF BRIDGE CRYSTAL FILTERS

The distance Δf, between the series-resonant frequency of the quartz and its parallel resonance in Fig. 8.10a is very small and that physical fact governs the limitation of the bandwidth in crystal filters. In Fig. 8.12, the bridge and conventional ladder schematics are shown. Crystal filters without inductive components (pure crystal filters) can be realized as a ladder or bridge, but the maximum realizable bandwidth will be a fraction of 1% of the center frequency. The lower limit for bandwidth is of no practical consequence and depends only upon the quality factor of the elements involved and crystal frequency stability. With two crystals as shown in Fig. 8.12b, the bandwidth is doubled in comparison with 8.12a.

When the bandwidth of the filter has to be more than 0.3% of center frequency the design must include spreading coils. Presence of inductive components increases the bandwidth, and removes the narrowband limitation. In Fig. 8.12c a typical wideband crystal filter with a series spreading coil is shown. The filter with spreading coils in series or in parallel to the crystal are subdivided into two distinctive categories; intermediate band and wideband. Wideband crystal filters are those where the arm reactances produce poles and zeros responsible for attenuation and are all located in the passband of the filter. Poles and zeros responsible for impedance characteristics of the filter are located outside the passband. In intermediate-bandpass filters some of the resonant frequencies that are normally responsible for the attenuation are moved from the passband to outside the band and therefore will not contribute to attenuation, but rather to the impedance characteristic in the passband of the filter. The schematics of intermediate and wideband crystal filters are usually the same, but the number of transmission zeros are different. For example, two crystals and coils in a differential-bridge realization will be equivalent to one, two, or three zeros of the transmission characteristic in intermediate band filters, but in a wideband realization the same number of elements will provide four transmission zeros. In Fig. 8.13 the reactance diagram of wideband and intermediate band crystal filters for a specific example of image-parameter design are shown.

By using the inductive elements, a certain amount of instability will be incorporated. Since inductive components are quite poor and unreliable when compared to quartz resonators, good filters with band-spreading inductive elements are difficult to realize, especially when more than 55 dB of channel separation is to be achieved as in high-quality telephone-channel filters. The solution to this problem could be in the form of two separate filters having their individual attenuation requirements reduced. When connected in tandem, these two separate filters will provide enough channel separation. Nevertheless, this solution is not the ultimate one and the inconvenience of this type of network is very well known. The bridge requires impedance balance and is therefore extremely sensitive to changes in element values (see

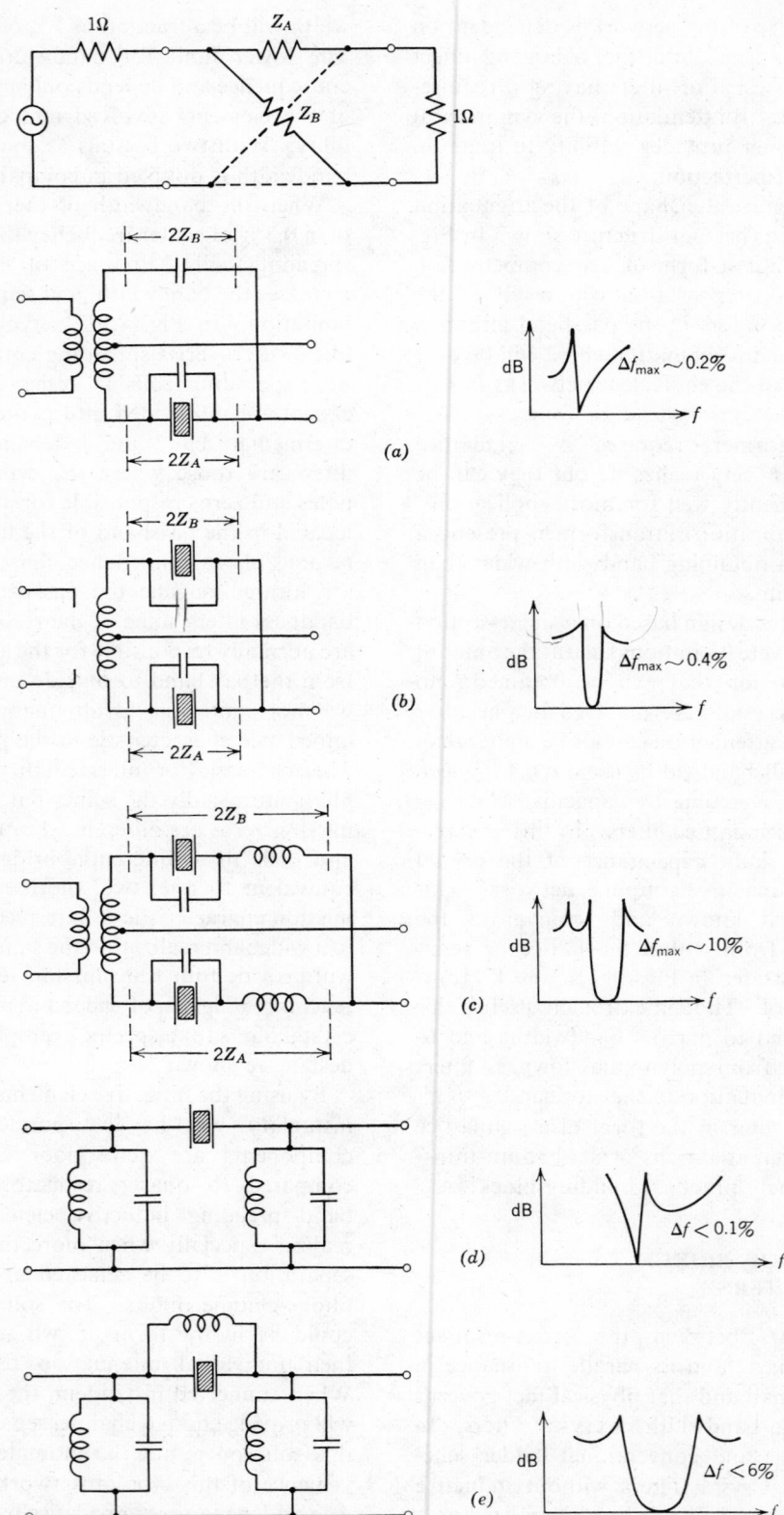

Fig. 8.12. Crystal filter configurations and their limitations.

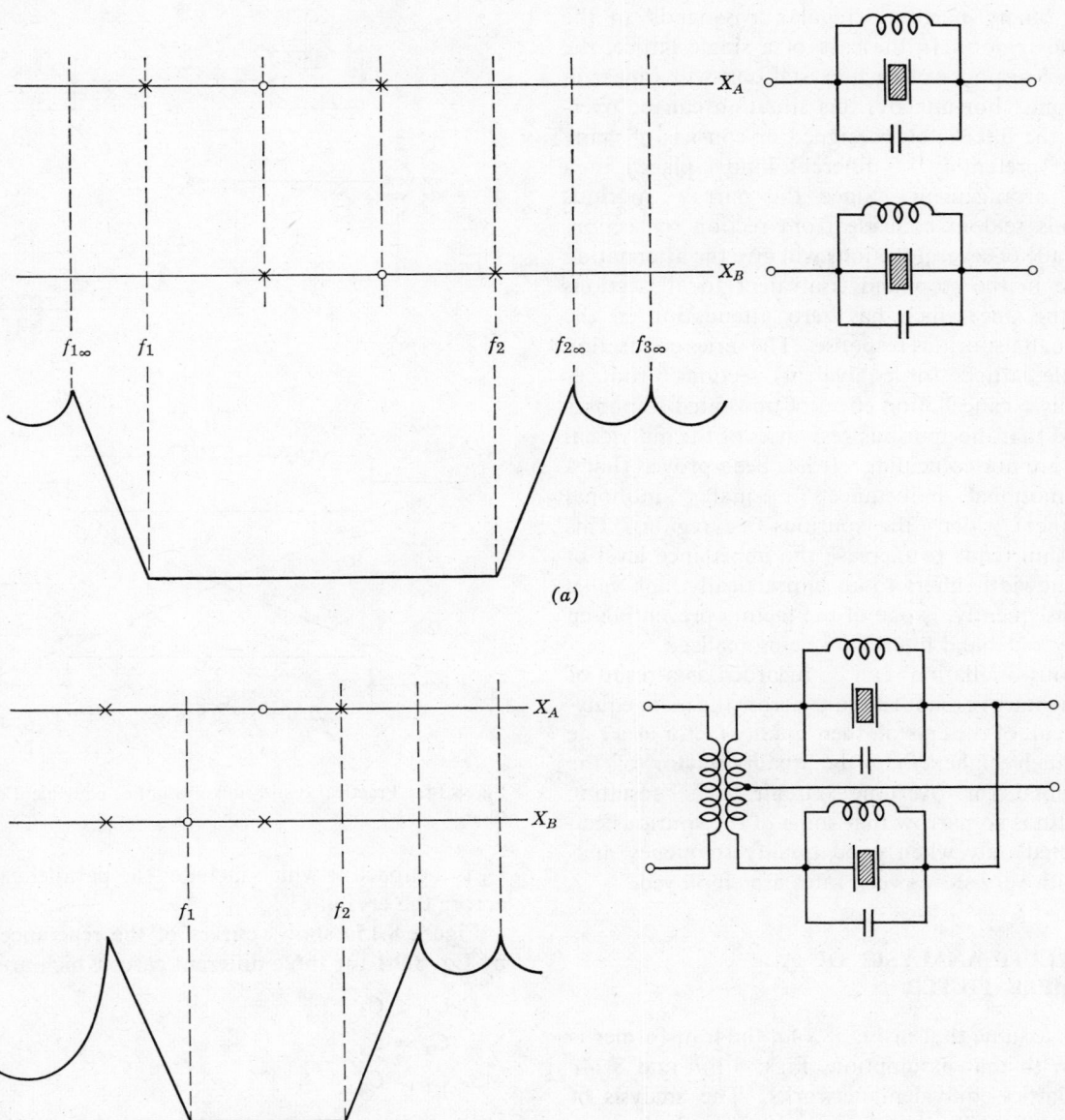

Fig. 8.13. (a) Wideband crystal filter. (b) Intermediate-band crystal filter.

Section 2.14). The attenuation poles in the stopband of that filter may be displaced with even a small change in one element. Because of the difficulty in obtaining sufficient stability, bridge networks are not used for *LC* filters.

The inductive elements common to both arms of the bridge, also a common capacitance, can usually be removed from the bridge and merged with load resistance on both sides. The common inductance on the transformer side of the bridge can be incorporated with the transformer itself, making it a finite inductive component.

8.8 SPURIOUS RESPONSE

As stated previously the equivalent circuit of the crystal resonator is valid only near the primary resonance. At frequencies somewhat removed from the main resonance other vibration modes are in existence but are not predicted by the equivalent circuit of the crystal.

Such modes are referred to as spurious modes. These responses are undesirable no matter where they occur. If they are in the passband, the behavior of the filter will be severely distorted. When they occur outside the passband, their presence will be noted by one or many narrow irregular passbands in the stopband region. In the case of a single lattice, the unwanted response of each crystal unit will appear in the output. Fortunately, this situation can be overcome if the filter is overdesigned or consists of many sections (preferably of different kinds) placed in a cascade arrangement. Since the narrow spurious passbands seldom coincide from section to section, the cascade of several sections will give the attenuation response in the stopband equivalent to all sections minus the one which has zero attenuation in the narrow band spurious response. The series connection of simple lattices (or equivalent) sections produces essentially a cancellation effect of unwanted responses provided that the spurious responses of the individual crystals are not coinciding. It has been proved that a larger motional inductance (a smaller motional capacitance) widens the spurious-free region. This relationship tends to increase the impedance level of wide-bandwidth filters to an impractically high value and, consequently, is one of the factors preventing an arbitrary wideband filter from being realized.

Spurious oscillations can be regarded as a result of supplementary resonant circuits across the main equivalent circuit of the crystal, their quality factor as a rule being much higher than the quality factor of the fundamental (or overtone) circuit. The resulting bandwidth is so narrow that some of the spurious can be detected only when good quality frequency analyzers with very slow sweep rates are employed.

8.9   CIRCUIT ANALYSIS OF A SIMPLE FILTER

Let us assume that in Fig. 8.14a, the transformer is ideal. With that assumption, Figs. 8.14b and 8.14c will be lattice-equivalent networks. The analysis of this schematic will be made assuming that all elements are ideal. The impedances for the reactive arms of the lattice are as follows: (see notation in Fig. 8.12)

$$Z_A = \frac{1}{j\omega C_1} \frac{\omega_1{}^2 - \omega^2}{\omega_2{}^2 - \omega^2} \qquad (8.9.1)$$

$$Z_B = \frac{1}{j\omega C_2}$$

where

$$\omega_1 = \frac{1}{\sqrt{L_s C_s}}$$

$$\omega_2 = \omega_1 \sqrt{1 + \frac{C_s}{C_1}} \qquad (8.9.2)$$

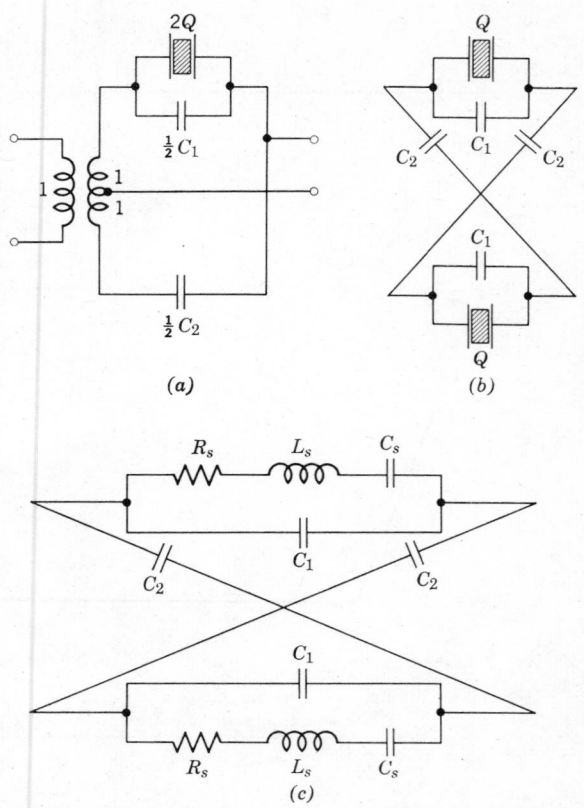

Fig. 8.14. Practical realization and lattice equivalent of simple filter.

C_1 is a capacitor which includes the parallel capacitor across the crystal C_p.

Figure 8.15a shows curves of the reactances given by Eq. 8.9.1 for three different cases which are:

1. $C_1 < C_2$
2. $C_1 = C_2$
3. $C_1 > C_2$

From these curves we can see that the reactances of the arms have opposite signs between f_1 and f_2. This range is the theoretical passband. For all frequencies outside of the passband Z_A and Z_B have the same signs. The network in all three cases is a bandpass filter with its theoretical stopband below f_1 and above f_2. The relative bandwidth is the ratio of absolute bandwidth

$$\Delta f = f_2 - f_1 \qquad (8.9.3)$$

to the arithmetic midband frequency,

$$f_a = \tfrac{1}{2}(f_1 + f_2) \qquad (8.9.4)$$

and is expressed by

$$\frac{1}{a} = \frac{\Delta f}{f_a} \approx \frac{C_s}{2C} \qquad (8.9.4a)$$

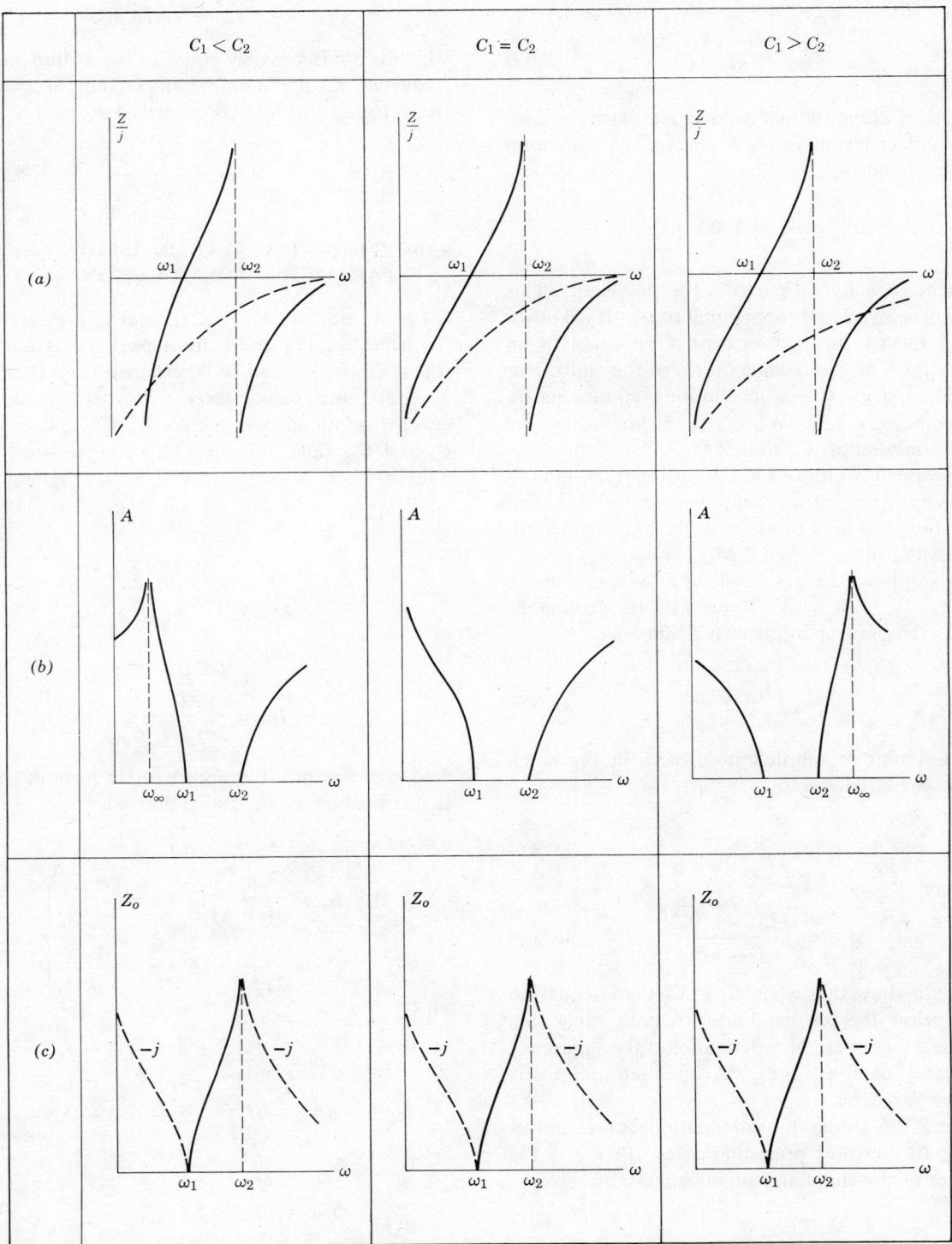

Fig. 8.15. (*a*) Branch reactances, (*b*) amplitude responses, and (*c*) characteristic impedances of Fig. 8.14.

The widest passband can be obtained when no complementary capacitor is used across the crystal:

$$\Delta f_{\max} = \frac{C_s}{2C_p} f_a \qquad (8.9.4b)$$

In the case of a longitudinal X-cut crystal with $C_p/C_s = 140$ and center frequency $f_a = 280$ kc, the maximum achievable bandwidth will be:

$$\Delta f = \frac{f_a}{280} = 1000 \text{ cps}$$

Additional capacity C_d narrows the passband. This narrowing is inversely proportional to C_d. It is known that piezoelectric resonators cannot be cut with an arbitrary ratio of static capacity C_p to the equivalent motional capacity C_s. Some limiting minimums do exist (see Section 8.5). When the thickness mode is used, the minimum is about 200. Correspondingly, the relative bandwidth of the filter using longitudinal-mode crystals theoretically cannot surpass 0.35%, and when using thickness mode crystals, the bandwidth theoretically cannot exceed 0.25%.

In the stopband there is a peak of attenuation at the frequency f_∞. At this frequency, the reactance $Z_1 = Z_2$. The peak parameter is defined by:

$$m = \sqrt{\frac{f_1^2 - f_\infty^2}{f_2^2 - f_\infty^2}}. \qquad (8.9.5)$$

which is similar to the definition of m in Fig. 1.25. Equation 8.9.5 reduces to

$$m = \sqrt{\frac{C_1}{C_2}} \qquad (8.9.6)$$

From here

$$f_\infty = \sqrt{\frac{f_1^2 - m^2 f_2^2}{1 - m^2}} \qquad (8.9.7)$$

It is easy to show that when $C_1 < C_2$ the attenuation peak is below the passband. When both capacitors are equal, the peak of attenuation takes place at $f = \infty$, and when $C_1 > C_2$ the attenuation peak is above the passband.

Figure 8.15b shows the attenuation curves corresponding to the three preceding cases. In Fig. 8.15c the curves of the characteristic impedance are given.

$$Z_0 = \sqrt{Z_A Z_B} = \frac{1}{j\omega\sqrt{C_1 C_2}} \sqrt{\frac{f_1^2 - f^2}{f_2^2 - f^2}} \qquad (8.9.8)$$

In all three cases the impedance is the same.

Assuming in Eq. 8.9.8 that

$$f_m = \sqrt{f_1 f_2}$$

which is approximately equal to the arithmetic center frequency f_a, the nominal impedance at the center frequency of this filter is expressed as

$$R_0 = \frac{1}{\omega_2 \sqrt{C_1 C_2}} \qquad (8.9.9)$$

8.10 ELEMENT VALUES IN IMAGE-PARAMETER FORMULATION

Let us assume that R_0, f_1, and f_2 are given. The frequency of the attenuation peak f_∞ is also given (the coefficient m can be determined from these data). The unknown values are f_s (series resonance of the crystal), motional parameters L_s, C_s, and capacitors C_1 and C_2. The following set of expressions can be written

$$\omega_1^2 = \frac{1}{L_s C_s} = 4\pi^2 f_1^2$$

$$f_2 = f_1 \sqrt{1 + \frac{C_s}{C_1}}$$

$$m = \sqrt{\frac{C_1}{C_2}} \qquad (8.10.1)$$

$$R_0 = \frac{1}{2\pi f_2 \sqrt{C_1 C_2}}$$

And consequently the approximate formulas for the element values can easily be derived:

$$f_s = f_1 \qquad (8.10.2)$$

$$L_s = \frac{R_0}{4\pi m \Delta f}, \quad C_s = \frac{m \Delta f}{\pi f_a^2 R_0} \left.\begin{array}{c} \\ \\ \\ \end{array}\right\}$$

$$C_1 = \frac{1}{2\pi f_a R_0} \quad C_2 = \frac{1}{2\pi m f_a R_0} \qquad (8.10.3)$$

where

$$m = \sqrt{\frac{\Omega_\infty + 1}{\Omega_\infty - 1}} = \sqrt{\frac{f_1 - f_\infty}{f_2 - f_\infty}} \qquad (8.10.4)$$

$$\Omega_\infty = \frac{\Delta f_\infty}{\Delta f/2} \qquad (8.10.5)$$

$$\Delta f = f_2 - f_1 \qquad (8.10.6)$$

$$\Delta f_\infty = (f_\infty - f_a) \qquad (8.10.7)$$

Calculations made according to Eq. 8.10.3 give, in comparison with the exact formulas, a relative error no greater than $\Delta f/f_a$ which is usually a fraction of 1%.

8.11 LADDER FILTERS

The voltage ratio of a doubly terminated lattice shown in Fig. 8.16a is found to be

$$\frac{V_2}{V_1} = \underbrace{\frac{R(Z_B - Z_A)}{(2R + Z_A)(2R + Z_B)}}_{\text{in terms of reactances}}$$

$$= \underbrace{\frac{Z_{12}R}{(R + Z_{11} - Z_{12})(R + Z_{11} + Z_{12})}}_{\text{in terms of matrix coefficients}}$$

The input impedance is correspondingly

$$Z_{\text{in}} = \frac{(2R + Z_A)(2R + Z_B)}{Z_A + Z_B + 4R}$$

$$= \frac{(R + Z_{11} - Z_{12})(R + Z_{11} + Z_{12})}{R + Z_{11}}$$

where the ideal value of Z parameters in terms of semilattice reactances Z_A and Z_B as shown in Fig. 8.16a is

$$Z_{11} = \frac{Z_A + Z_B}{4} = Z_{22}$$

$$Z_{12} = \frac{Z_B - Z_A}{4}$$

As far as analysis is concerned, the lattice structure is by far the most convenient. Consider initially the single-crystal lattice (crystal in Z_A only) structure of Fig. 8.16a. It is seen that the lattice has the same transmission property as a one-resonator ladder terminated in a capacitance at each end (π-section). Since the crystal filter is inherently a narrow bandwidth device, the parallel RC terminations can be transformed at the center frequency to a series RC circuit as shown in Fig. 8.16c.

In addition to bridge circuits, a ladder equivalent circuit composed of piezoelectric resonators and capacitors can find application in the design of narrowband filters especially for high frequencies. See Fig. 8.8.

A method shown in Section 8.5 permits the removal of parts common to both impedances Z_A and Z_B from the inside of the lattice, placing them at the input and output ends of the lattice without affecting either its input impedance conditions or transmission characteristics. The capacitor across the crystal, $2C_1$ of Fig. 8.16a, may be equal to the capacitor in the lattice arm ($2C_1$) and therefore can be removed from the inside completely.

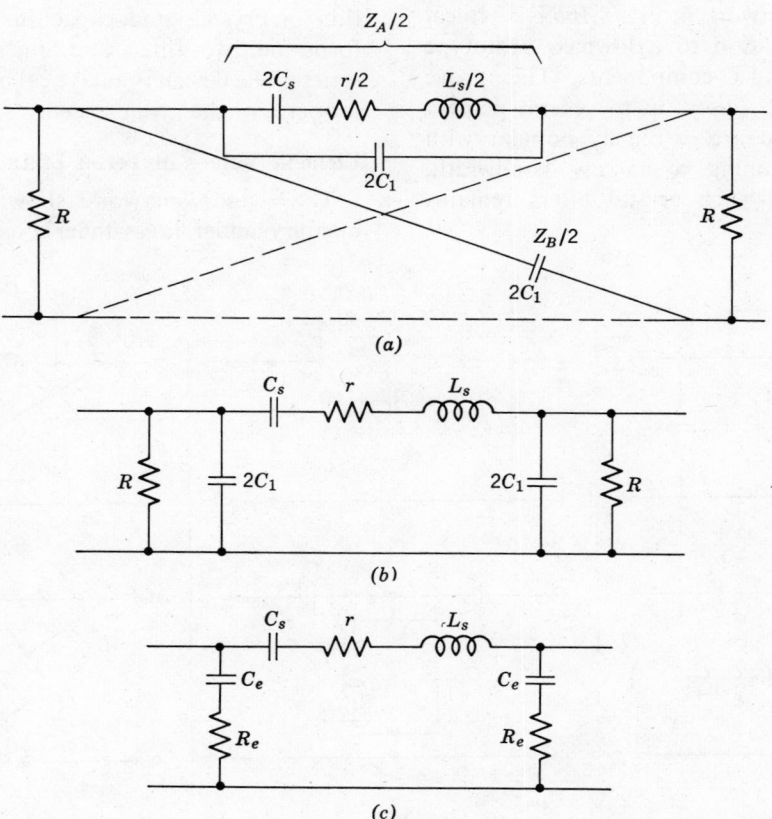

Fig. 8.16. Simple crystal lattice and its narrowband approximations for balanced-out parallel capacitance.

In Fig. 8.16*b* the removal is illustrated for the special case of a simple one-crystal lattice. The practical example of simple-lattice decomposition for the more general case $C_1 \neq C_2$ will be analyzed later in terms of image parameters.

It is noted that for the single-crystal bridge filter as shown in Fig. 8.16 the shunt capacitance can be balanced with a capacitance of the identical value in the opposite arm of the bridge. The amplitude response of this filter is therefore monotonic as has been demonstrated before in Fig. 8.15 for the case $C_1 = C_2$. These balance conditions are necessary in order to transform the lattice network into a minimum-phase ladder. Moreover, they make the transmission property of the lattice obtainable without special frequency transformation from the ordinary lowpass polynomial models tabulated in Chapter 6. The elementary filter structures with $C_1 \neq C_2$ are closer to Cauer parameter models, since they include poles and zeros of transmission. The filter in Fig. 8.16 is equivalent to one transmission-zero network.

In Figs. 8.16*b* and *c* equivalent ladder-narrowband structures are shown. In this form they can be treated in the same way as the coupled resonators in terms of k and q values. The network in Fig. 8.16*b* is a typical narrowband approximation to a lowpass prototype with conventional L and C components. They prove to be adequate for LC filters up to several percent relative bandwidth and are extremely popular with filter designers. Assuming a narrow bandwidth, the response curve of such crystal filters remains symmetrical.

Referring to Fig. 8.16*c* the following should be noted. Since the complete loop must resonate at center frequency, the resonant frequency of the crystal f_s must be lower than center frequency f_m of the filter to compensate for the pulling effect by the complex loads. The amount of crystal shifting is determined from the expression

$$f_m{}^2 = \frac{1}{2\pi L_s} \frac{C_s + C_e/2}{C_s C_e/2} = f_s{}^2 \left(1 + \frac{2C_s}{C_e}\right)$$

where C_e is the equivalent series capacitance of the loading circuit. The value of the crystal motional capacitance C_s is much smaller than C_e of the complex load and therefore the foregoing expression can be substituted by the following approximate formula, always valid in crystal filter design

$$f_m = f_s \left(1 + \frac{C_s}{C_e}\right) = f_s + \delta f$$

where δf is the amount of pulling.

For narrowband filters the main design unit is a semilattice, but the lowpass model is a ladder network. In the following discussion, the analysis of simple networks is undertaken and the formulas for the design of simple ladder filters are given. The combination of crystals and capacitors under consideration form one, two, three, and four sections, connected in series. The design is analytical in nature, based on the property of the given structure.

Element Values in Terms of Image Parameters

The π- and T-networks shown in 8.17*a* are the elementary ladder filters under consideration. Equivalent

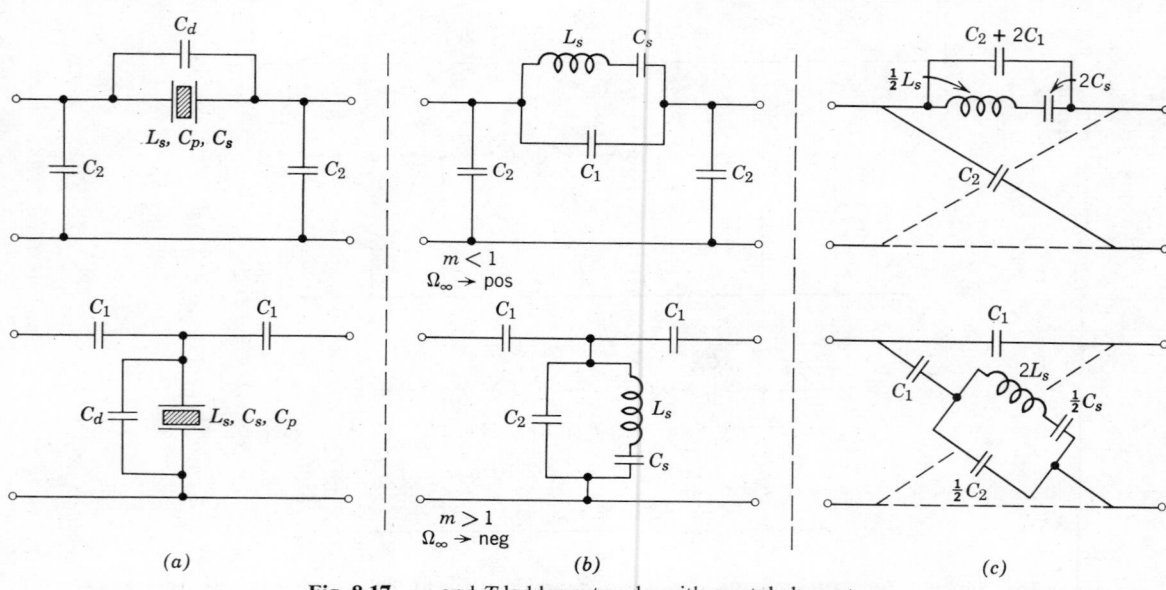

Fig. 8.17. π and T ladder networks with crystal elements.

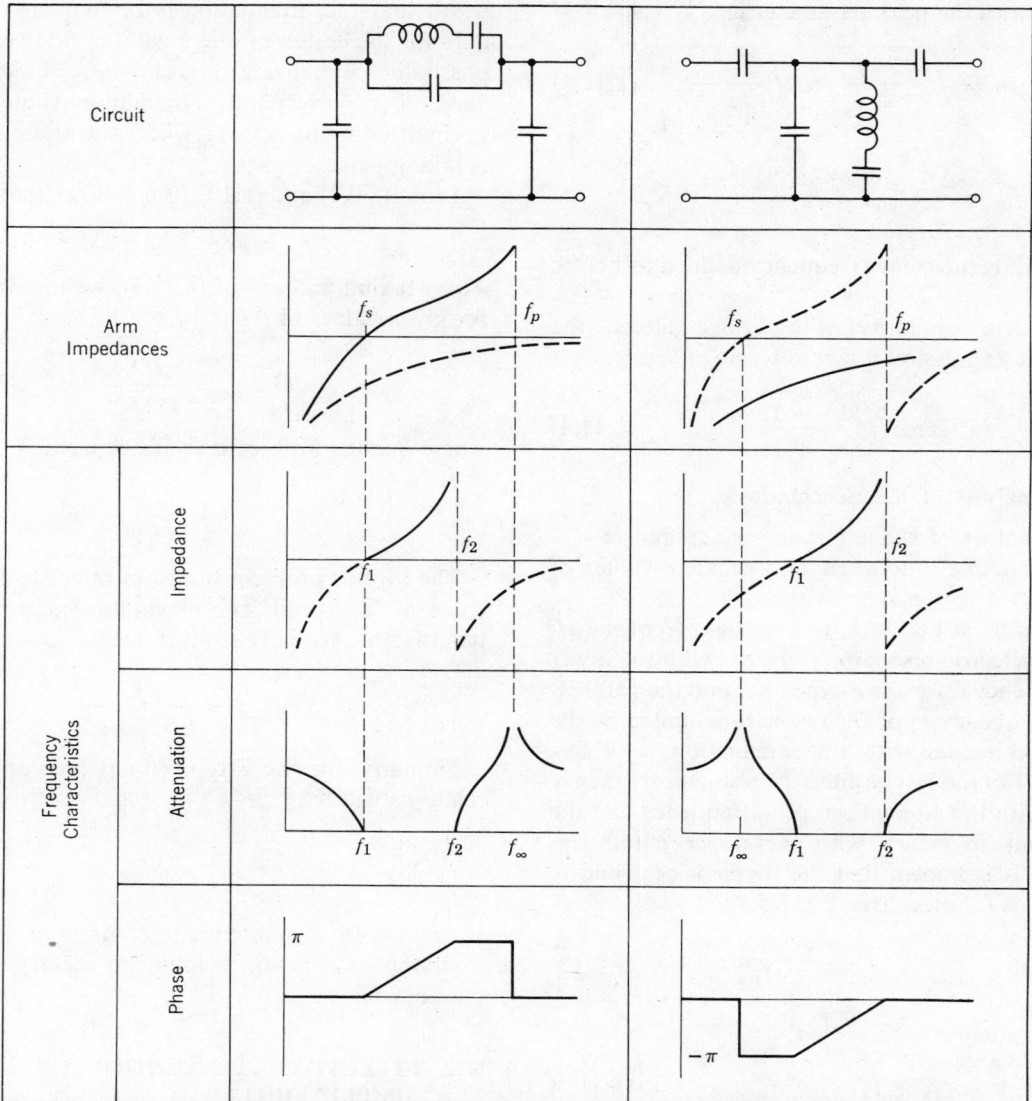

Fig. 8.18. Impedance, attenuation, and phase of simple crystal ladder sections.

circuits for both networks are given in 8.17b. The characteristic parameters of both circuits are shown in Fig. 8.18. The peak of attenuation of the π-type circuit is located in the upper attenuation band and for the T-type circuit, in the lower attenuation band.

The characteristic impedance of both circuits has the same form; the impedance is resistive in the passband and varies from zero to infinity (first-order), and it is capacitive in the stopband.

In narrowband filters for which the ratio of the passband width to the center frequency is less than approximately 0.05, simple approximate formulas for element values can be used:

For a π-type network

$$\left. \begin{array}{ll} C_1 \approx \dfrac{1 - m^2}{4\pi m f_a R_0} & C_2 \approx \dfrac{m}{2\pi f_a R_0} \\[3mm] C_s \approx \dfrac{\Delta f}{2\pi m f_a R_0}, & L_s \approx \dfrac{m R_0}{2\pi\,\Delta f} \end{array} \right\} \quad (8.11.1)$$

For a T-type network,

$$\left. \begin{array}{ll} C_1 \approx \dfrac{m}{2\pi f_a R_0}, & C_2 \approx \dfrac{m}{\pi (m^2 - 1) f_a R_0} \\[3mm] C_s \approx \dfrac{2m^3\,\Delta f}{\pi (m^2 - 1)^2 f_a R_0}, & L_s \approx \dfrac{(m^2 - 1)^2 R_0}{(8\pi m^3\,\Delta f\,)} \end{array} \right\} \quad (8.11.2)$$

In both cases the peak parameter is:

$$m = \sqrt{\frac{f_2{}^2 - f_\infty{}^2}{f_1{}^2 - f_\infty{}^2}} \approx \sqrt{\frac{f_2 - f_\infty}{f_1 - f_\infty}} \qquad (8.11.3)$$

where

$$f_\infty = \frac{\Delta f}{2}\Omega_\infty + f_a$$

with the arithmetic center frequency defined as before in Eq. 8.9.4.

In complete similarity with bridge filters, the approximate expression for m in terms of Ω_∞ is:

$$m \approx \sqrt{\frac{\Omega_\infty - 1}{\Omega_\infty + 1}} \qquad (8.11.4)$$

Physical Analysis of the Schematics

Positive values of shape parameter Ω_∞ and $m < 1$ correspond to the π-network, and negative values of Ω_∞ and $m > 1$ to the T-network.

As is seen from Fig. 8.18, the resonant frequency f_s of the piezoelectric resonator coincides with the lower cutoff frequency f_1 for the π-type filter and the parallel-resonant frequency f_p of the resonator shunted by the capacitor, coincides with the attenuation peak frequency f_∞. For the T-type filter, the resonant frequency coincides with the attenuation peak frequency and the anti-resonant frequency with the upper cutoff frequency f_2. It is known that the frequencies f_s and f_p are related as follows: (see Eq. 8.5.5)

$$f_p{}^2 - f_s{}^2 = \frac{C_s}{C_1} f_s{}^2$$

or approximately

$$f_p - f_s \approx \frac{C_s}{2C_1} f_s \qquad (8.11.5)$$

where C_1 is the total capacitance consisting of the static capacitance C_p of the resonator and the capacitance of the additional capacitor C_d as shown in Fig. 8.17.

The distance on the frequency scale between the series- and parallel-resonant frequencies of the crystal resonator corresponds to the distance between the lower cutoff frequency and the attenuation peak in the π-type filter. The greatest spacing between the cutoff frequency and the attenuation peak can be obtained in the case when capacitance C_1 (Fig. 8.17b) equals the static capacitance C_p of the resonator—that is, if the additional capacitance is not connected in parallel to the resonator ($C_d = 0$). Consequently, it is impossible to select at the same time both the passband

width and the attenuation-peak frequency as initial data for the design of filters without first verifying the possibility of realizing the resonator. To every given value of Δf corresponds a definite maximum value of normalized frequency Ω_∞ where the attenuation peak is taking place.

From (8.9.4) and (8.10.10) it follows that

$$f_\infty - f_1 = \tfrac{1}{2}\Delta f + \Delta f_\infty$$

Thus, taking account of (8.11.5), we obtain for Ω_{max} for given values of Δf and f_s

$$\Omega_{max\infty} = \frac{f_s}{k\,\Delta f} - 1 \qquad (8.11.6)$$

where k is the ratio of crystal's capacitors

$$k = \frac{C_p}{C_s} \qquad (8.11.7)$$

The inverse problem is also of interest; for a given value of Ω_∞ to find the maximum value of the pass bandwidth. We have from (8.11.6)

$$\Delta f_{max} = \frac{f_s}{k(\Omega_\infty + 1)}$$

Similarly, for the T-type circuit the corresponding relations will be:

$$\Omega_{max} = 1 - \frac{f_s}{k\,\Delta f}, \qquad \Delta f_{max} = \frac{f_s}{k(1 - \Omega_\infty)}$$

In Fig. 8.19 the attenuation curve of one simple π-section is shown, helping to define the above parameters.

8.12 EFFECTIVE ATTENUATION OF SIMPLE FILTERS

It is possible to compute the effective attenuation of ladder filters having a comparatively small number of elements by using the formulas for bridge circuits.

The lattice equivalents of the π- and T-type circuits are presented in Fig. 8.17c. As seen from the figures, these equivalents differ only in the relative location of the two-terminal networks forming the arms of the bridge circuit. The attenuation of such bridges is described by the same expression.

Figure 8.19 shows the attenuation that can be realized per simple section (a simple section has 180° of phase shift over the passband), having attenuation peaks placed at various locations above the upper cutoff frequency. The abscissa is expressed in multiples of half-bandwidths. These curves are valid at any

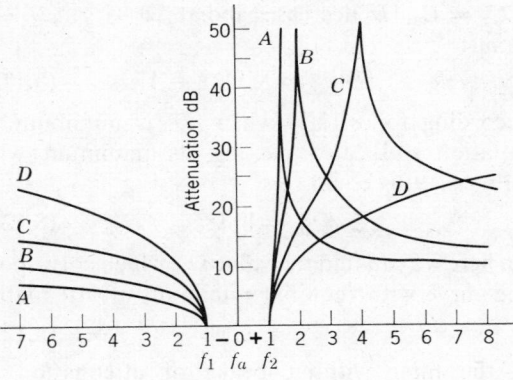
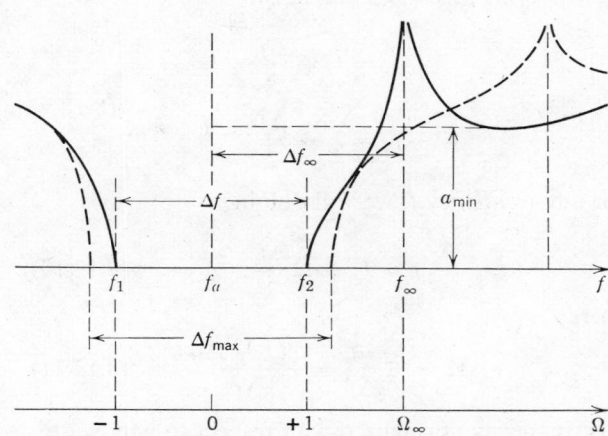

Fig. 8.19.   Attenuation in a simple section.

center frequency and any percent of bandwidth within reasonable limits. The curves A, B, and C show the resultant attenuation for sections having attenuation peaks to the right of the passband. It is apparent that attenuation is increased near the passband and drops fast at frequencies on the other side of the peak. To obtain the maximum attenuation near the edge of the band, peak-type sections such as A, B, and C offer sharper response cutoffs than the infinity-type such as D.

To secure peaks of attenuation near the passband, and achieve a flat response in the passband, it is necessary to have elements with sufficient quality factor. In general, if the percent bandwidth decreases, the Q of the respective elements must increase.

The derivation of the expressions for the effective attenuation of a π-type filter follows. Evidently, T-type networks can also be calculated by means of these formulas.

Effective Attenuation of Single-Section Lattice Filters

The expression to compute the effective attenuation of the single-section lattice circuit as shown in Figs. 8.12 and 8.17 is (according to Eq. 2.1.4)

$$a = \ln \sqrt{1 + D^2} = \tfrac{1}{2} \ln (1 + D^2) \quad (8.12.1)$$

where D is the discrimination or filtering function

$$D = \frac{r + X_A X_B / r}{X_A - X_B} \quad (8.12.2)$$

where

$$X_A = \frac{Z_A}{jR}, \quad X_B = \frac{Z_B}{jR}, \quad r = \frac{R}{R_0} \quad (8.12.3)$$

Z_1 and Z_2 are the impedances of the arms of the lattice circuit, R the load resistance of the filter. With Eqs. 8.9.1, 8.9.6, 8.9.9, and 8.11.3;

$$X_A = \frac{f_2}{mf} \frac{f_1^2 - f^2}{f_2^2 - f^2}$$

$$X_B = \frac{mf_2}{f} \quad (8.12.4)$$

By substituting in the formula for the calculation of the filtering function D

$$D = \frac{mr}{1 - m^2} \frac{(f^2 - f_{01}^2)(f^2 - f_{02}^2)}{f_2 f (f^2 - f_\infty^2)} \quad (8.12.5)$$

where f_{01} and f_{02} are the roots of the equation defining the positions of the zeros in the passband and f_∞ is a pole of attenuation. In the particular case when $C_1 = C_2$ or $m = 1$ the attenuation pole vanishes and Eq. 8.12.5 is reduced to

$$D = r \frac{(f^2 - f_{01}^2)(f^2 - f_{02}^2)}{f f_2 (f_2^2 - f_1^2)} \quad (8.12.6)$$

At $r = 1$ the zeros will merge together and will be equal to the geometric center frequency

$$f_{01} = f_{02} = \sqrt{f_1 f_2} \quad (8.12.7)$$

Consequently Eq. 8.12.6 is reduced to

$$D = \frac{(f^2 - f_1 f_2)^2}{f f_2 (f_2^2 - f_1^2)} \quad (8.12.8)$$

Calculation of effective attenuation with the aid of this formula requires very high accuracy. Therefore the approximate equations will be of certain interest. According to Eqs. 8.9.3 and 8.9.4

$$f_1 = f_a - \frac{\Delta f}{2}$$

$$f_2 = f_a + \frac{\Delta f}{2} \qquad (8.12.9)$$

For any frequency f, we will obtain

$$f = f_a + \Omega \frac{\Delta f}{2}, \qquad (8.12.10)$$

where

$$\Omega = \frac{f - f_a}{\frac{1}{2}\Delta f} = \frac{2(f - f_a)}{\Delta f} \qquad (8.12.11)$$

Ω is frequency normalized with respect to bandwidth. In the center of the band $\Omega = 0$, at f_1, $\Omega = -1$, and at f_2, $\Omega = 1$ (see Fig. 8.19). For frequencies not too far from f_a

$$f_1 + f \approx f_2 + f \approx 2f_a, \frac{f}{f_2} \approx 1$$

With these assumptions and simplifications

$$X_A = \frac{1}{m}\frac{\Omega + 1}{\Omega - 1}$$

$$X_B = m \qquad (8.12.12)$$

Substituting this expression in (8.12.2) the following approximate expression for D can be derived.

$$D = \frac{1}{2}\left(r + \frac{1}{r}\right)\frac{\Omega - \Omega_0}{\Omega - \Omega_\infty}\sqrt{\Omega_\infty{}^2 - 1} \quad (8.12.13)$$

where

$$\Omega_0 = \frac{r^2 - 1}{r^2 + 1} \qquad (8.12.14)$$

It is evident, when $\Omega = \Omega_0$, the effective attenuation is at its minimum, equal to zero. At $r = 1$, we obtain a very simple equation

$$D = \frac{\Omega}{\Omega - \Omega_\infty}\sqrt{\Omega_\infty{}^2 - 1} \qquad (8.12.15)$$

In this case the attenuation is passing its minimum value, equal to zero at $\Omega = 0$ (in other words, in the center of the passband). The expression for D (8.12.15) permits us to conclude that the selectivity of a simple bandpass filter is dependent only upon the passband width (included in Ω) and the position of the attenuation peak relative to the passband (Ω_∞). Selectivity does not depend on the mean frequency of the filter.

From (8.12.13) and (8.12.15) it can be seen that at $\Omega = \Omega_\infty$, $|D| = \infty$. By increasing $|\Omega|$ in the filter

with $C_1 \neq C_2$, $|D|$ decreases and at $|\Omega| \to \infty$, $|D| \to$ to the limit:

$$|D|_{\min} = \sqrt{\Omega_\infty{}^2 - 1} \qquad (8.12.16)$$

According to (8.12.1) when $|D|$ is minimum, the attenuation a is also passing its minimum which (see Fig. 8.19) is equal to

$$a_{\min} = \ln \Omega_\infty \qquad (8.12.17)$$

From here we can find the value Ω_∞ which corresponds to the curve with required minimum of attenuation:

$$\Omega_\infty = e^{a_{\min}} \qquad (8.12.18)$$

In the filter without peaks of attenuation Eq. 8.12.13 becomes

$$D = \frac{1}{2}\left(r + \frac{1}{r}\right)(\Omega - \Omega_0), \qquad (8.12.19)$$

and correspondingly when $r = 1$,

$$D = \Omega \qquad (8.12.20)$$

and

$$a = \ln \sqrt{1 + \Omega^2} \qquad (8.12.21)$$

Equation 8.12.21 describes the effective attenuation where $C_1 = C_2$ and with $r = 1$, which is symmetrical relative to the ordinate. To illustrate this, Fig. 8.20 shows the effective attenuation of a single-section calculated for the values of Ω_∞

$$-7, -20, -\infty$$

and for r

$$r = \tfrac{1}{2}, 2$$

From the foregoing formulas it can be seen that the curves remain unchanged if the values r, Ω_∞, Ω, and $-\Omega$ are substituted simultaneously by values $1/r$, $-\Omega_\infty$, $-\Omega$, and Ω. These substitutes are shown in brackets in the figures.

The curve $r = \tfrac{1}{2}$ is shifted along the abscissa to the right, and for $r = 2$ is shifted left by the value $\Omega = 0.6$. A comparison of these responses with one for $r = 1$ would show that the deviation r from unity in any direction leads to narrowing of the passband and to increasing of the stopband attenuation.

Design Steps for a One-Section Elementary Filter

The following design steps are recommended for one-section filters:

1. Assume a bandwidth Δf, and a midfrequency f_a of the theoretical passband, but keep in mind that $\Delta f/f_a$ cannot exceed certain values indicated in the analysis.

2. Choose according to the curves of Fig. 8.20, values of Ω_∞ and r satisfying the necessary effective-attenuation response.

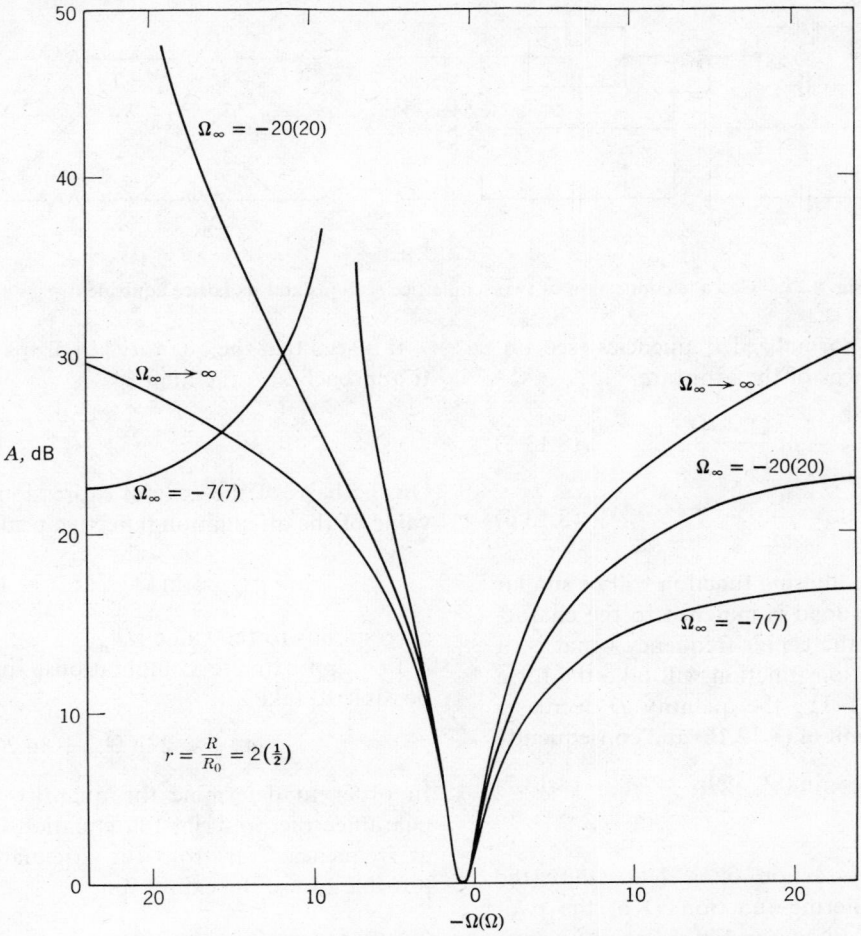

Fig. 8.20. Magnitude responses of simple filter with mismatched load.

3. Knowing the load resistance R and value r, find the nominal impedance of the filter according to the relation

$$R_0 = \frac{R}{r}$$

4. Using the approximate formulas, calculate all element values.

8.13 EFFECTIVE ATTENUATION OF LADDER NETWORKS

The following expressions can be written for the impedance of the arms of a lattice circuit which is equivalent to a π-circuit (Fig. 8.17c):

$$X_A = \frac{1}{j2\pi(2C_1 + C_2)} \frac{f_1^2 - f^2}{f(f_2^2 - f^2)} \tag{8.13.1}$$

$$X_B = \frac{1}{j2\pi C_2} \frac{1}{f} \tag{8.13.2}$$

where

$$f_1^2 = \frac{1}{4\pi^2 L_s C_s} \; ; \; f_2^2 = \frac{1}{4\pi L_s C_s}\left(1 + \frac{2C_s}{2C_1 + C_2}\right)$$

Using the last formulas, the normalized reactances of the arms X_A and X_B can be obtained:

$$X_A = -m\frac{f_2}{f}\frac{f_1^2 - f^2}{f_2^2 - f^2}$$

$$X_B = -\frac{1}{m}\frac{f_2}{f} \tag{8.13.3}$$

Substituting X_A and X_B into Eq. 8.12.2, the expression for D becomes

$$D = \frac{m}{f_2} \frac{rf^2(f_2^2 - f^2) - (1/r)f_2^2(f_1^2 - f^2)}{f(f^2 - f_\infty^2)} \tag{8.13.4}$$

The approximate formulas, by means of which the computation can be carried out with a slide rule, can

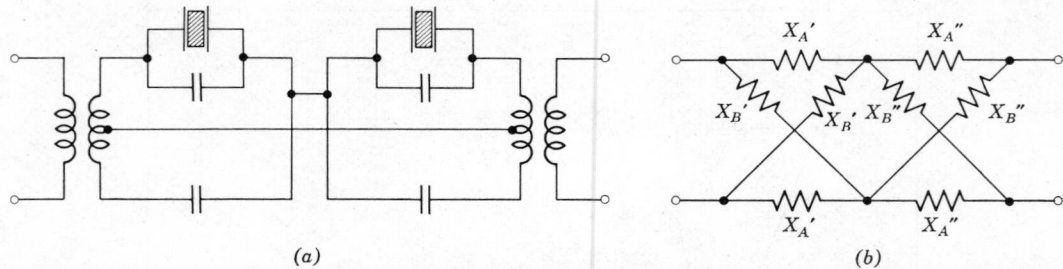

Fig. 8.21. Cascade connection of two semilattice sections and its lattice equivalent.

be given in terms of normalized frequencies (see Eq. 8.12.11). The reactances of the arms are

$$X_A \approx -m \frac{\Omega + 1}{\Omega - 1} \qquad (8.13.5)$$

$$X_B \approx -\frac{1}{m} \qquad (8.13.6)$$

The expression for the filtering function will be similar to Eq. 8.12.13. If the load is matched to the characteristic impedance at the center frequency—that is, if $r = 1$, the discrimination function will take the form of (8.12.15). For $\Omega > \Omega_\infty$ the quantity D decreases and approaches the limit of (8.12.16) and consequently

$$a_{\min} \approx \ln (\Omega_\infty) \text{ Np} \qquad (8.13.7)$$

Two-Section Filter

To compute a two-section filter, let us use the expression for the filtering function D of the two-section bridge filter as shown in Fig. 8.21.

$$D = \frac{(r + X_A'X_B'/r)(X_A' + X_B' + X_A'' + X_B'')}{(X_A' - X_B')(X_A'' - X_B'')}$$
$$(8.13.8)$$

The quantity of primes indicate the number of the section, and X_A and X_B are determined from (8.13.5) and (8.13.6).

Let us consider in more detail the case of identical sections, which is of great practical interest.

In this case the expression for D is

$$D = 2 \frac{(r + X_A X_B/r)(X_A + X_B)}{(X_A - X_B)^2} \qquad (8.13.9)$$

or

$$D \approx \sqrt{\Omega_\infty^2 - 1} \frac{[1/r - r + (1/r + r)\Omega](\Omega\Omega_\infty - 1)}{(\Omega - \Omega_\infty)^2}$$
$$(8.13.10)$$

For $r = 1$

$$D \approx \sqrt{\Omega_\infty^2 - 1} \frac{2\Omega(\Omega\Omega_\infty - 1)}{(\Omega - \Omega_\infty)^2} \qquad (8.13.11)$$

It is seen that the quantity $|D|$ drops when $\Omega > \Omega_\infty$. It approaches to the limit

$$|D|_{\min} \approx 2\Omega_\infty \sqrt{\Omega_\infty^2 - 1} \qquad (8.13.12)$$

On the basis of the general expression, the minimum value of the attenuation (effective peak)

$$a_{\min} \approx \ln \Omega_\infty \sqrt{\Omega_\infty^2 - 1}$$

corresponds to the value $|D|_{\min}$.

For approximate computations, for $\Omega_\infty \geq 3$ it is possible to take

$$a_{\min} \approx 2 \ln \Omega_\infty + \ln 2 \qquad (8.13.13)$$

In order to determine the quantity Ω_∞ which will guarantee the prescribed attenuation in the stopband at frequencies far from the attenuation peak, it is possible to use the expression

$$\Omega_\infty \approx \exp \frac{a - \ln 2}{2} \qquad (8.13.14)$$

Three-Section Filter

A three-section filter can also be represented as a series connection of two lattice networks; one of them can be equivalent to a two-section π-type circuit as shown in Fig. 8.22.

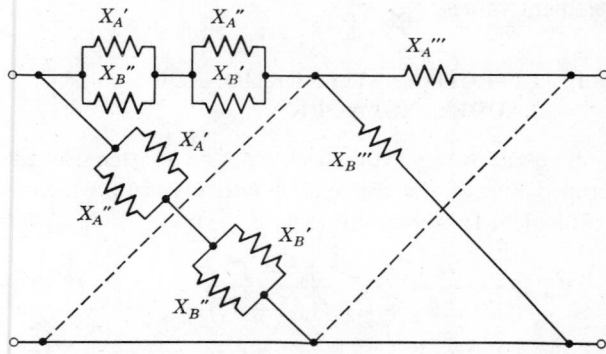

Fig. 8.22. Cascade connection of three lattice sections.

The expressions for the impedances X_A and X_B of the arms of the lattice equivalent to the series connection of two lattices are (see notation in Figs. 8.21 and 8.22)

$$X_A = \frac{X_A'X_B''}{X_A' + X_B''} + \frac{X_A''X_B'}{X_A'' + X_B'}$$
$$X_B = \frac{X_A'X_A''}{X_A' + X_A''} + \frac{X_B'X_B''}{X_B' + X_B''} \quad (8.13.15)$$

Substituting the expressions for the arm impedances into (8.13.8), the filtering function will be, in the case of identical sections,

$$D \approx \frac{1}{2}\sqrt{\Omega_\infty^2 - 1}\left[\frac{1}{r} - r + \left(\frac{1}{r} + r\right)\Omega\right]$$
$$\times \frac{4(\Omega\Omega_\infty - 1)^2 - (\Omega_\infty - \Omega)^2}{(\Omega_\infty - \Omega)^3} \quad (8.13.16)$$

For $r = 1$

$$D \approx \sqrt{\Omega_\infty^2 - 1}\,\frac{\Omega[4(\Omega\Omega_\infty - 1)^2 - (\Omega_\infty - \Omega)^2]}{(\Omega_\infty - \Omega)^3}$$

For $\Omega > \Omega_\infty$ the above expression for $|D|$ approaches

$$|D|_{\min} \approx 4\Omega_\infty^2\sqrt{\Omega_\infty^2 - 1}$$

For $\Omega_\infty \geq 3$ the estimate can be accomplished with the formula

$$a_{\min} \approx 3\ln\Omega_\infty + 2\ln 2 \quad (8.13.17)$$

from which

$$\Omega_\infty = \exp\frac{(a - 2\ln 2)}{3}$$

The computed characteristics of the effective attenuation of a three-section filter are given in Fig. 8.23 for different values of Ω_∞.

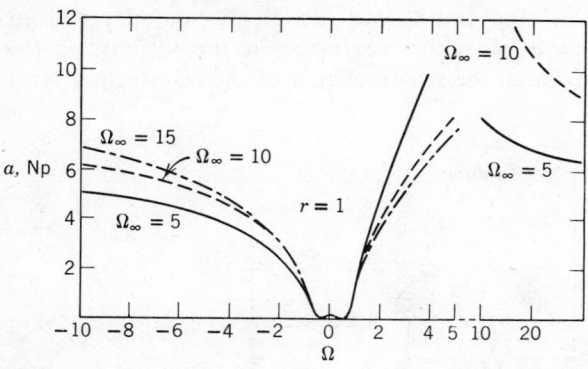

Fig. 8.23. Computed characteristics of effective attenuation of three simple sections.

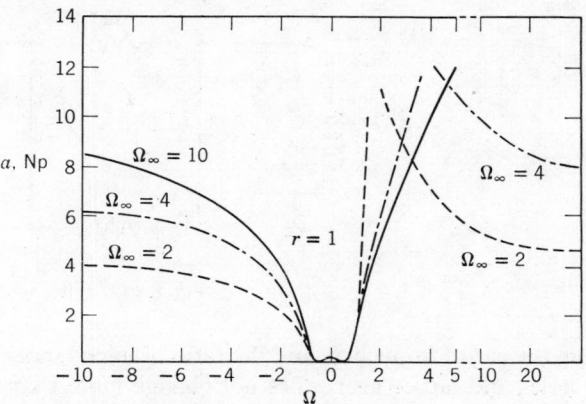

Fig. 8.24. Computed characteristics of effective attenuation of four-simple-section filter.

Four-Section Filter

Let us limit the discussion to the case of practical importance, when all four sections are identical. The series connection of four sections can be transformed into a two-section circuit consisting of two lattices. Here, we can use for the arm impedances values pertinent to the two-section filter treated previously.

The expression for D as a function of Ω is

$$D \approx 2\sqrt{\Omega_\infty^2 - 1}\left[\frac{1}{r} - r + \left(r + \frac{1}{r}\right)\Omega\right](\Omega\Omega_\infty - 1)$$
$$\times 2\frac{(\Omega\Omega_\infty - 1)^2 - (\Omega - \Omega_\infty)^2}{(\Omega - \Omega_\infty)^4}$$

For $r = 1$

$$D \approx 4\sqrt{\Omega_\infty^2 - 1}\,(\Omega\Omega_\infty - 1)\Omega$$
$$\times \frac{2(\Omega\Omega_\infty - 1)^2 - (\Omega - \Omega_\infty)^2}{(\Omega - \Omega_\infty)^4}$$

The minimum value of the attenuation at $\Omega_\infty \geq 3$

$$a_{\min} \approx 4\ln\Omega_\infty + 3\ln 2 \text{ Np}$$

from which

$$\Omega_\infty \approx \exp\frac{a - 3\ln 2}{4}$$

Computed characteristics of a four-section filter are given in Fig. 8.24, for $r = 1$ and different values of Ω_∞. Comparatively large nonuniformities of the attenuation appear in the passband for small values of Ω_∞.

8.14 LADDER VERSUS BRIDGE FILTERS

The requirements for the crystal units used in ladder and bridge filters may be very different especially if the sections include several crystals (semilattice as shown in Fig. 8.12b). The attenuation peaks, in the case of bridge filters, can be located at any normalized frequency from infinity to one-half the bandwidth

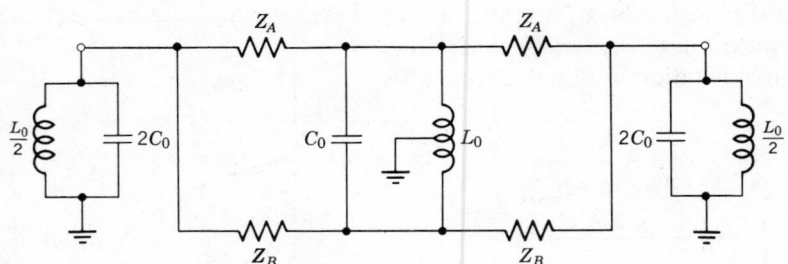

Fig. 8.25. Filter with two reversed semilattices.

from the cutoff frequency, and the ratio of impedances in series and lattice arms does not change more than 40%.

In the case of ladder filters the picture is drastically different. To achieve nearly equal impedances for the series and shunt arm crystals in one single section requires the placement of attenuation peaks as close as 0.2 part of the bandwidth from the cutoff. This would indicate that for all practical purposes (except very simple crystal filters) the lattice design is the only one having a large latitude for the position of attenuation peaks, and therefore provides the necessary flexibility for filter design.

Any practically encountered filter response can be physically realized with a lattice or with one of its equivalents in terms of peak parameters. The limitations are on the maximum obtainable bandwidth. After a certain limit, different for the various crystal cuts, a combination of LC elements and crystals can be used in a special ladder form which practically removes the upper bandwidth limitation.

The lower limit for bridge filters is a result of losses, mutual inductances, and distributed capacitances associated with auxiliary inductive components such as transformer and coils.

Figure 8.25 shows a transformation of two wideband lattices into a configuration that minimizes the number of transformers. This is essentially a bridge circuit having one center-tapped coil with both halves of the coil tightly coupled to each other. Figure 8.26 shows

the two semilattices of Fig. 8.25 reversed. It should be noted that not only does the differential transformer or center-tapped coil provide a ground connection to the filter, but that the inductance is calculated to provide a band-spreading effect. If no spreading coil is used, an "ideal transformer" with very high inductance is placed across the line to obtain high impedance which eliminates shunting effects to the filter. For small percentage bandwidths, Fig. 8.27 shows another approximate equivalence.

8.15 PRACTICAL DIFFERENTIAL TRANS-FORMER FOR CRYSTAL FILTERS

The most characteristic component of bridge-type filters is the differential transformer or its equivalent center-tapped coil. The property of this component has bearing on the general performance of the crystal filter.

Only with an ideal transformer are the schematics shown in Fig. 8.11a equivalent. Practical transformers, especially at high frequencies, are never ideal but only an approximation. The ideal transformer implies an infinitely large inductance. Large inductors are not always practical, since the windings are always associated with a distributed capacitance. More wire on the coil form means more distributed capacitance. When the transformer design is not supposed to present capacitive reactance in the vicinity of the passband, the self resonance of the transformer must

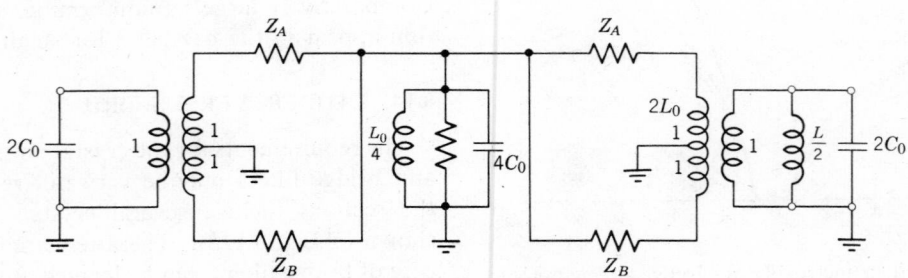

Fig. 8.26. Filter with two semilattices.

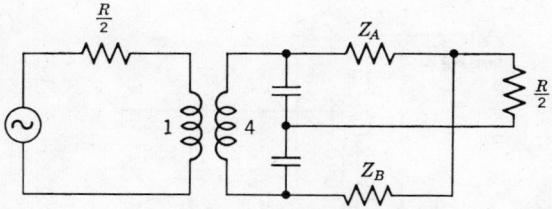

Fig. 8.27. Narrowband equivalent.

be above the frequency range occupied by the filter. Often it is advantageous to use a supplementary tunable capacitor to resonate the transformer at a desired frequency. However, it can easily be shown that this tuning method is unfavorable to the attenuation response of the filter. To obtain a wider passband, the transformer should not be frequency selective.

Further deviation of the transformer from ideal is exhibited by nonideal coupling between primary and secondary and between both parts of the secondary windings. A final deviation from the ideal transformer is the presence of losses in the coils and in the coupling.

The Symmetry of Differential Transformers

The appropriate distribution of differential windings on the coil form improves the symmetry of the transformer, but this technique can diminish the amount of coupling between the two halves of the secondary winding. Bifilar winding leads to better symmetry, but increases the capacitance between the windings and thus increases the losses. Certain improvements in symmetry can be reached by using two capacitors or a differential capacitor (see Figs. 1.19 and 8.27). This type of adjustment is useful, when the distributed coil capacitance puts the self resonance below the passband of the filter or when the capacitor, being a common part of the reactances of the arms, is to provide adjustment of the filter elements. Symmetry is especially important to maintain the designed attenuation in the stopband where the balance of the bridge is of paramount value.

Realization of crystal filters with an asymmetrical response automatically calls for different levels of arm impedances even though they are in the same frequency range. Two specific types of design practices are used to bring the crystal's motional parameters to somewhat similar values. First, nonsymmetric transformers can be used in the semilattice to reduce the impedance difference between Z_A and Z_B. Second, a series capacitor can be used to modify the crystal impedance without actually changing the arm impedance (see section 1b of Chapter 10).

Equivalent Representation of Differential Transformers

It is possible to represent a physical transformer as a combination of inductances, capacitances, resistances, and one ideal transformer. This type of representation permits a visual picture as to how parasitic reactive elements incorporated in the transformer, being resonant with an opposite type reactance, yield an over-all effect of an ideal transformer for a narrowband of frequencies. In Fig. 8.28 the differential transformer with number of turns n_1, n_2, and n_3 is shown and Z_1, Z_2, and Z_3 signify the respective inductances with losses.

Between different windings appear mutual inductances which are Z_{12}, Z_{13}, and Z_{23}. With this notation, the expression for elements of the equivalent schematics will be as follows:

$$Z_A = \frac{1}{K} Z_{13} \qquad (8.15.1)$$

$$Z_B = \frac{1}{K} Z_{23} \qquad (8.15.2)$$

$$Z_{33} = -Z_{12} \qquad (8.15.3)$$

$$Z_{01} = Z_1 + Z_{12} - \frac{1}{K} Z_{13} \qquad (8.15.4)$$

$$Z_{02} = Z_2 + Z_{12} - \frac{1}{K} Z_{23} \qquad (8.15.5)$$

$$Z_{03} = \frac{1}{K^2} Z_3 - \frac{1}{K} Z_{13} - \frac{1}{K} Z_{23} \qquad (8.15.6)$$

The foregoing equations permit the calculation of any unsymmetric differential transformer and to express nonsymmetry as a parameter of such a transformer. The equivalent circuit of the transformer alone can be supplemented by two capacitors and resistors between the points a, b, and 0 and one capacitor between points a and b of Fig. 8.29 which usually exist in practical transformer realizations. Finally two reactance arms can be added as the proper filtering components. The

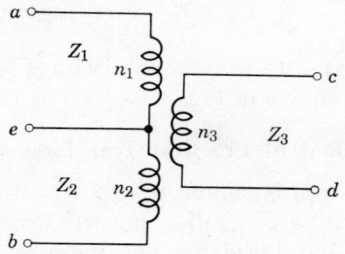

Fig. 8.28. Nomenclature used for differential transformer.

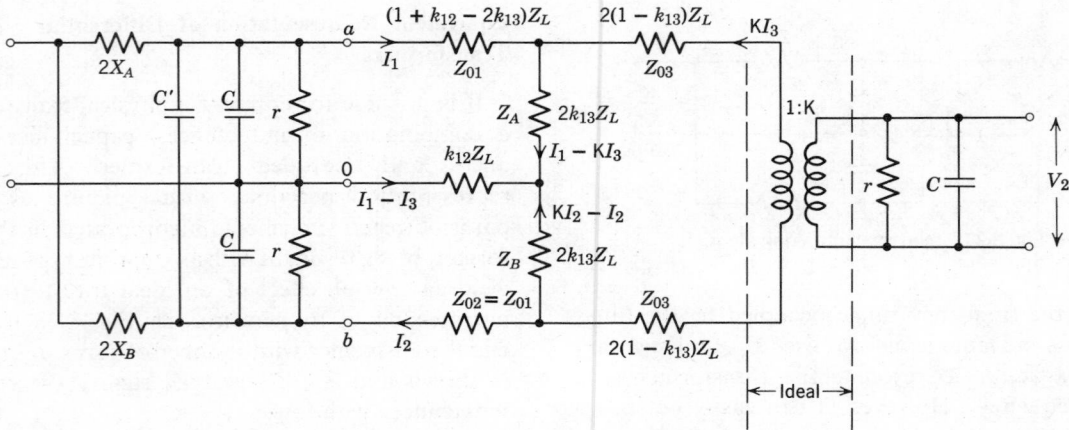

Fig. 8.29.   Equivalent differential transformer in terms of coupling k.

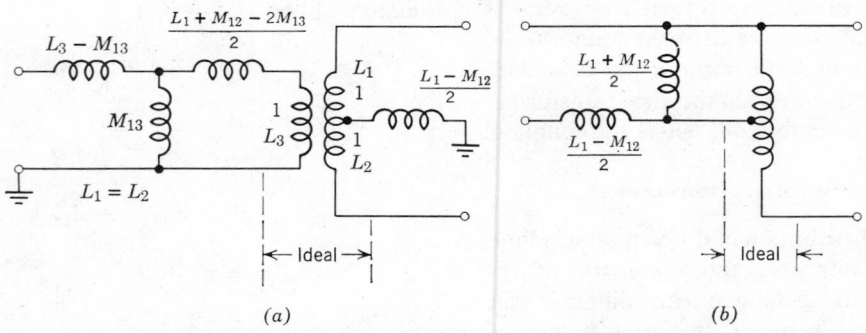

(a)   (b)

Fig. 8.30.   Approximate equivalent representation of (a) ideal differential transformer, (b) ideal center-tapped coil.

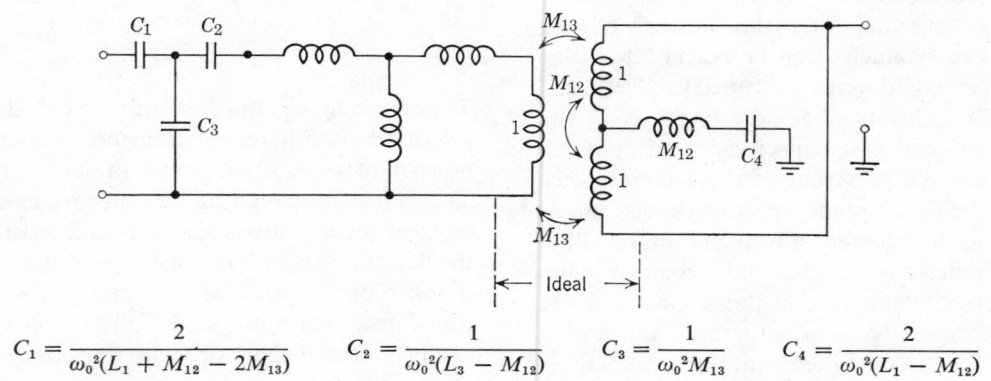

$$C_1 = \frac{2}{\omega_0{}^2(L_1 + M_{12} - 2M_{13})} \qquad C_2 = \frac{1}{\omega_0{}^2(L_3 - M_{12})} \qquad C_3 = \frac{1}{\omega_0{}^2 M_{13}} \qquad C_4 = \frac{2}{\omega_0{}^2(L_1 - M_{12})}$$

Fig. 8.31.   Use of capacitive pad to minimize effect of leakage and mutual inductance.

equivalent transformer circuit with the preceding additions is shown in Fig. 8.29.

Approximations of Practical Transformers

The first approximation to a practical transformer is the neglecting of its dissymmetric losses and stray capacitance and considering only the effect of nonunity coupling and finite inductances as shown on the right

part of the Fig. 8.29. These simplifications permit the differential transformer and auto-transformer used in Fig. 8.11a to be represented by the equivalent circuits in Figs. 8.30a and 8.30b respectively.

For a narrow band of frequencies the capacitive pads shown in Fig. 8.31 and Fig. 8.32 can be used to minimize the effect of leakage and mutual inductance. They effectively introduce the opposite effect

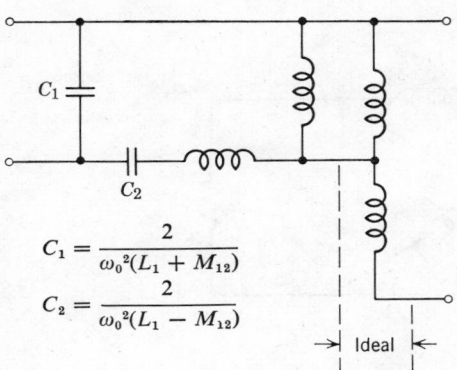

$$C_1 = \frac{2}{\omega_0^2(L_1 + M_{12})}$$

$$C_2 = \frac{2}{\omega_0^2(L_1 - M_{12})}$$

Fig. 8.32. Use of capacitive pad for the case of center-tapped coil.

introduced by inductances and consequently produce an exact single frequency equivalence to an ideal transformer at resonant frequency. As an approximation this structure is valid only for a limited frequency band, unless the coupling coefficient is close to unity.

Another approximation, generally adequate for most bifilar wound transformers (and those with a coefficient of coupling very close to unity), is to assume that the stray (distributed) capacitance represented by C in Fig. 8.29 can be lumped into a single capacitance in parallel with the finite mutual inductance, and that the transformer losses can be shown as single-parallel resistance at the input terminals. This kind of approximation is shown in Fig. 8.33.

8.16 DESIGN OF NARROWBAND FILTERS WITH THE AID OF LOWPASS MODEL

Semilattice filter circuits can be obtained by circuit transformations from a ladder network which has been designed to give the desired frequency-domain characteristic.

For the resulting single-crystal lattice circuit the shunt capacitance of the crystal is balanced by an equal capacitance in the opposite arm. For the two-crystal lattice this balance is achieved by making sure that the shunt capacitances C_1 of both crystals are equal. These balance conditions are necessary for such circuits to be equivalent to a simple ladder, and also to make it possible to derive the transmission properties of the lattice from a polynomial prototype filter.

Therefore, our concern is the design of a ladder network which can be easily converted into semilattice circuits. Since the ladder-to-lattice transformation requires one or more transformers whose exact

placement is somewhat arbitrary, several possibilities will be illustrated. From a practical point of view, a simpler circuit is obtained when all crystals have equal motional capacitances and equal loss resistances.

Ladder Equivalent for One- and Two-Crystal Semilattice

In Section 8.11, the lattice-to-ladder transformation for a one-crystal filter has been demonstrated. The method, essentially consists of removal of the common parts from the balanced lattice. The equivalent ladder network in Fig. 8.16*b* is an exact equivalent to Fig. 8.16*c* at the center frequency. The analysis of a two-crystal section is similar to the single-crystal section. Each of the two crystals is assumed to have the same motional capacitance C_s, and parallel (parasitic) capacitance C_p. The lattice shown in Fig. 8.34*a* decomposes into the equivalent inductively coupled ladder shown in Fig. 8.34*c*. According to coupled-ladder theory, each mesh must be resonant at the center frequency. The only way this can occur for the lattice structure is to have it equally terminated at both ends.

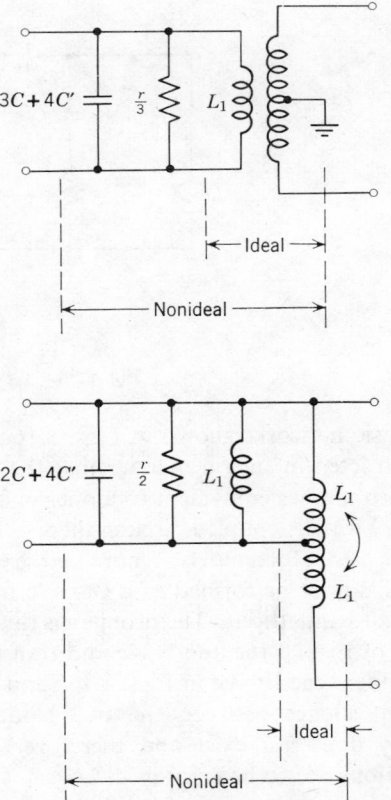

Fig. 8.33. Transformer approximation with lumped capacitances and resistances.

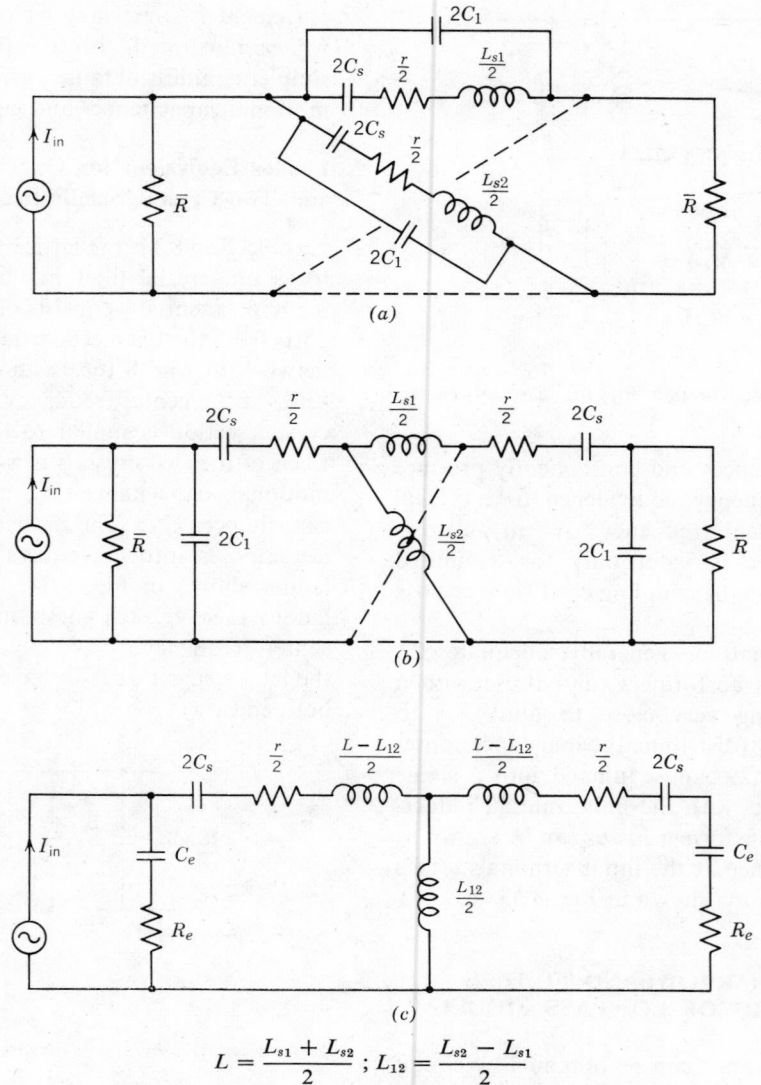

$$L = \frac{L_{s1} + L_{s2}}{2} \; ; \; L_{12} = \frac{L_{s2} - L_{s1}}{2}$$

Fig. 8.34. Crystal lattice and its equivalent ladder network.

The basic networks shown in Figs. 8.16c and 8.34c are constructed in such a manner that they represent one or two meshes equivalent to a one- or two-crystal filter for which coupled-circuit theory is directly applicable. Unfortunately, more complex ladder networks cannot be formed as a cascade of one- and two-crystal semilattices. The problem is thus to design a ladder of Figs. 8.16c and 8.34c and transform them to the ladder type shown in Figs. 8.16b and 8.34b. An exact equivalence between the two ladders at any frequency does not exist and therefore several approximations are to be employed.

1. The parallel combination of R and C at frequency f_0 (see Fig. 8.35a) is equivalent to the series combination of the same kind of elements, but having different values, namely C_e and R_e. This transformation also permits a negative capacitor to be absorbed by a positive capacitor existing in the adjacent circuitry.

2. The π-network of Fig. 8.35b with R and C_a in the shunt arm is approximately equivalent to the T-network with the $R_1 C_2$ combination in the shunt arm. In the network equivalent to two single-crystal semilattices, the resistance in series with the capacitor in Fig. 8.36a may be displaced to the series arm as shown in Fig. 8.36b.

Figure 8.37c represents the ladder equivalent to a fifth-order crystal filter of Fig. 8.37a or b. The attenuation response will be symmetrical, since it is derived from a lowpass prototype filter provided that each mesh in Fig. 8.37d, including the resonant

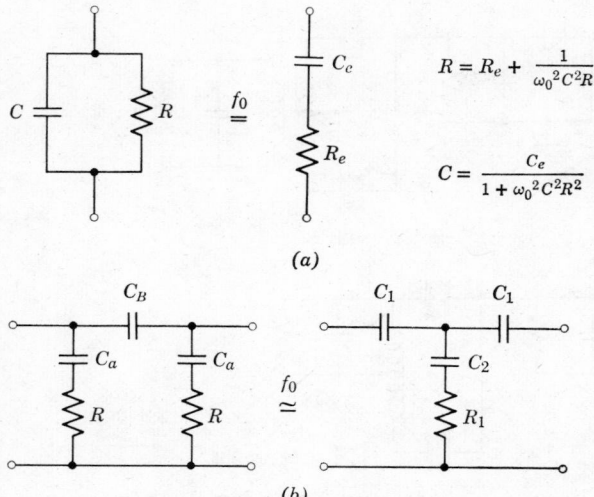

Fig. 8.35. Useful transformations for narrowband filters.

circuits and coupling mechanism, is resonant at the center frequency, and the mutual coupling element between each mesh is purely reactive.

In the typical crystal-filter realization as shown in Figs. 8.37a and b, lossy differential transformers or auto-transformers explain the presence of certain amounts of resistance in some of the coupling networks.

The Design Formulas for Narrow-Bandpass Filters

To begin the design of a crystal filter, the transmission requirements of the filter, as well as certain other parameters and constants must be known:

1. The bandwidth of the filter, for example the 3-dB bandwidth Δf_3.
2. The center frequency of the passband

$$f_a = \frac{f_1 + f_2}{2} \approx f_m = \sqrt{f_1 f_2}$$

3. The realizable motional inductance of the crystal L_s.
4. The Q of the crystal resonator Q_x.
5. The loss resistance of the transformer r.

In addition the normalized values of coupling coefficients (k's and q's) for the desired response characteristic must be available (see Section 6.10). From the given quantities the following intermediate design parameters can be evaluated:

$$q_x = \frac{\Delta f_3}{f_a} Q_x = \frac{Q_x}{a} \tag{8.16.1}$$

$$\bar{C} = \frac{1}{2\pi^2 \Delta f_3 f_a L_s} \tag{8.16.2}$$

$$\bar{R} = \pi \Delta f_3 L_s \tag{8.16.3}$$

The remaining design equations differ, depending on the number of crystals used in the filter.

One-crystal filter

$$R_1 = R_2 = 2\bar{R}\left(1 - \frac{1}{q_x}\right)$$

$$C_{\text{in}} = C_{\text{out}} = \frac{\bar{C}}{2(1 - 1/q_x)}$$

$$f_s = f_a - \frac{\Delta f_3}{2}(1 - 1/q_x)$$

The prototype and actual semilattice realization is shown in Fig. 8.38. Single and unequally terminated designs are also possible but are infrequently used.

Two-crystal filter. To design a two-crystal semilattice filter equivalent to Fig. 8.34 the following additional data must be known:

1. The shunt to motional capacitance ratio

$$C_p/C_s = k.$$

2. The motional capacitance C_s, assumed to be the same for both crystals.

Coupled-ladder theory gives the following design relations, referring to Fig. 8.34c:

$$Q = Q_1 = Q_2 = aq = \sqrt{\frac{L_s}{C_s}} \frac{1}{r + 2R_e} \tag{8.16.4}$$

and

$$\frac{k_{12}}{a} = \frac{L_{12}}{L} \tag{8.16.5}$$

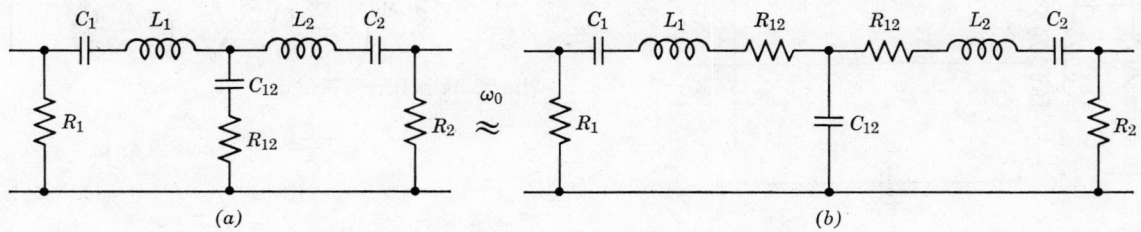

Fig. 8.36. Example of redistribution of losses in coupled networks.

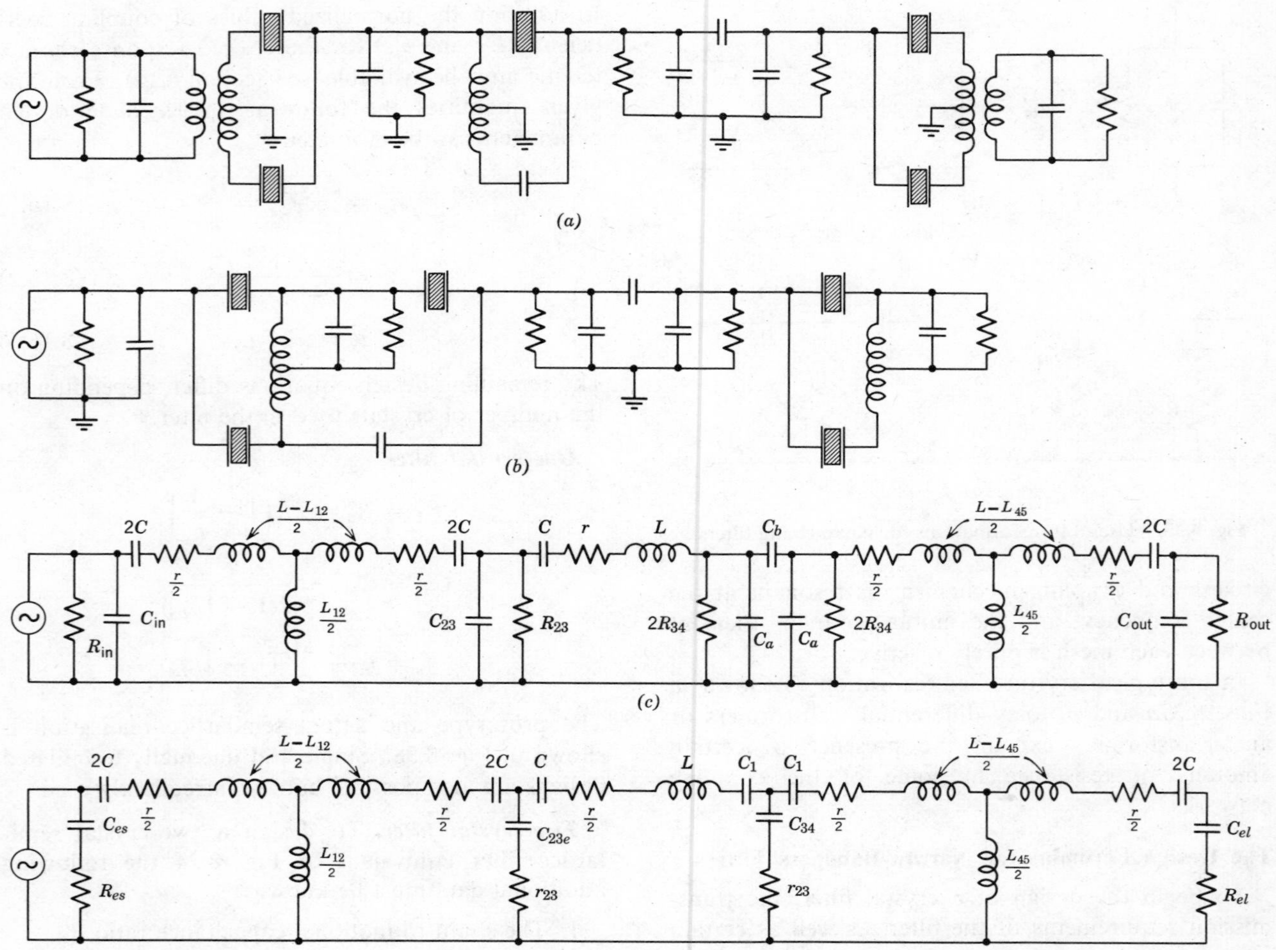

Fig. 8.37. Fifth-order crystal filter in form of (*a* and *b*) cascaded semilattices, and (*c* and *d*) ladder equivalents with five coupled meshes, as shown in *d*.

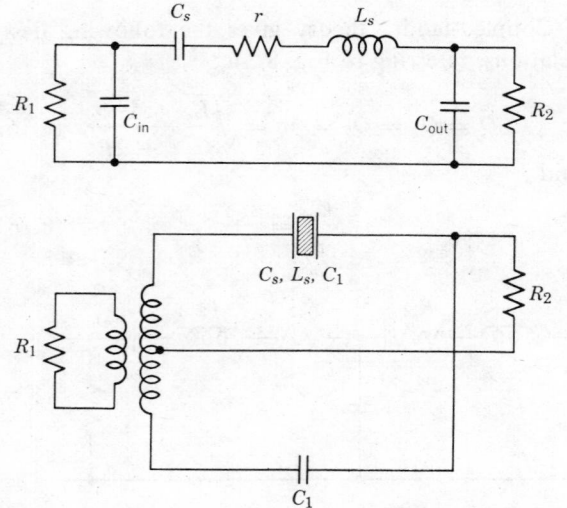

Fig. 8.38. Prototype and semilattice one crystal bandpass filter.

where q and k_{12} values are normalized quality factor and coupling coefficient respectively. The exact value of k_{12} and q are determined from the normalized lowpass prototype whose bandpass response should be duplicated. Equation (8.16.4) is similar to the one derived for the single-crystal section. Thus by analogy, the solution for the load resistance $\bar{R}$ is

$$\bar{R} = \frac{1}{a} Z \left[\frac{1}{\left(\frac{1}{q} - \frac{a}{Q} \right)} \pm \sqrt{ \frac{1}{\left(\frac{a}{q} - \frac{a^2}{Q} \right)^2} - 1 } \right]$$

where, as before

$$a = \frac{2C_1}{C_s} \quad \text{(See 8.9.4a)}$$

$$Z = \sqrt{ \frac{L_s}{C_s} }$$

The center frequency of the crystals using L_s and C_s is

$$\omega_s{}^2 = \frac{1}{L_s C_s}$$

It is also noted that the crystals in each of the meshes of Fig. 8.34c is pulled by the same amount as for the single-crystal section—that is,

$$\delta f = \frac{f_a}{a\left\{1 + \left[(1/a\bar{R})\sqrt{\frac{L}{C}}\right]^2\right\}} = \frac{f_a}{a + (1/a)(Z/\bar{R})^2}$$

The center frequencies of the crystals f_{s1} and f_{s2} can now be determined from the coefficient of coupling as follows:

$$f_{s1,\,s2} = \frac{1}{2\pi\sqrt{C_s(L \pm L_{12})}} \approx f_a\left(1 \mp \frac{L_{12}}{2L}\right) \quad (8.16.6)$$

Substituting (8.16.5)

$$f_{s1,\,s2} = f_m\left(1 \mp \frac{1}{a}\frac{k_{12}}{2}\right) \approx f_a \mp \Delta f_3\frac{k_{12}}{2} \quad (f_a = f_m)$$

However,

$$f_{s0} = f_a - \delta f, \quad \text{thus } f_{s1,\,s2} = f_a \mp \Delta f_3\frac{k_{12}}{2} - \delta f$$
$$(8.16.7)$$

In Table 8.2 a set of design equations for the filter with two crystals is given. Similar sets of design information are presented in the Tables 8.3 and 8.4 for third- and fourth-order narrowband filters respectively.

Design Equations for Narrowband Filters with $n = 5$

By inspection, the normalized Q's and coupling coefficients, in terms of the element values of the circuit shown in Fig. 8.37 are:

$$q_1 = \frac{1}{a}\frac{\omega_m L}{r_x + 2R_e} \quad (8.16.8)$$

$$q_m = q_2 = q_3 = q_4 = \frac{1}{a}\frac{\omega_m L}{r_x + 2r_{23}} \quad (8.16.9)$$

$$q_5 = \frac{1}{a}\frac{\omega_m L_s}{r_x + 2R_e} \quad (8.16.10)$$

$$k_{12} = a\frac{L_{12}}{L} \quad (8.16.11)$$

$$k_{23} = a\frac{\sqrt{2}\,C_s}{C_{23e} + 2C_s} \approx a\frac{\sqrt{2}\,C_s}{C_{23e}} \quad \text{(since } C_s \lll C_{23}\text{)}$$
$$(8.16.12)$$

$$k_{34} = a\frac{\sqrt{2}\,C}{C_{34e} + 2C_s} \approx a\frac{\sqrt{2}\,C_s}{C_{34s}} \quad (8.16.13)$$

$$k_{45} = a\frac{L_{45}}{L} \quad \text{where } a = \frac{f_a}{\Delta f_3} \quad (8.16.14)$$

As before, these values are given in terms of 3-dB bandwidth, center frequency, and crystal parameters. If the assumption that $C_{23} \leq C_{34}$ is made then

$$C_{es} = C_{el} = C_{23e} = \frac{C_1 C_{34e}}{C_1 + C_{34e}} \quad (8.16.15)$$

to guarantee that each mesh is resonant at the same frequency. In terms of coupling coefficients, the preceding equality will be satisfied if $k_{23} \geq k_{34}$.

The filter is always designed around several known parameters. For the narrowband filter it is assumed that besides Δf, f_m, Q_x and L, also the transformer loss resistance R_{23} is known

$$q_{23} = \frac{R_{23}}{2\pi\,\Delta f_3 L} \quad (8.16.16)$$

Before a proper set of normalized k's and q's can be obtained, the normalized resonator quality q_m must be found. The first step towards this is to determine an approximate expression for r_{23}.

From the tables of k and q values, the value of k_{23} can be found for the infinite Q case. Then using the equation for k_{23}, the value of C_{23e} can be calculated.

$$C_{23e} = \frac{\bar{C}}{\sqrt{2}\,k_{23}} \quad (8.16.17)$$

The transformer-loss resistance R_{23} shown in Fig. 8.37c is in parallel with C_{23} and is known; thus, from R_{23} and C_{23e} we must find r_{23}, the equivalent-series resistance, and the equivalent-parallel capacitance C_{23}. This is accomplished by using the equivalent approximations in Fig. 8.35. If R and C_e in this figure are given, the value of $C = C_{23}$ will be

$$C_{23} = C_{23e}\left[0.5 + 0.5\sqrt{\frac{1 - 2k_{23}{}^2}{q_{23}{}^2}}\right] \quad (8.16.18)$$

If $q_{23}{}^2 \gg 2k_{23}{}^2$ then $C_{23} = C_{23e}$ and

$$r_{23} = \frac{R_{23}}{1 + 2q_{23}{}^2/k_{23}{}^2} \approx \frac{R_{23}k_{23}{}^2}{2q_{23}{}^2} \quad (8.16.19)$$

The normalized q_m of Eq. 8.16.8 is now easily obtained with $q_x = Q/a$ and the tables of predistorted values can then be used to obtain a better value of k_{23}. Then, with this k_{23}, a new q_m can be calculated. With a complete set of k and q values, the design can proceed.

Table 8.2 Design Information for Two-Crystal Semilattice

Schematic	

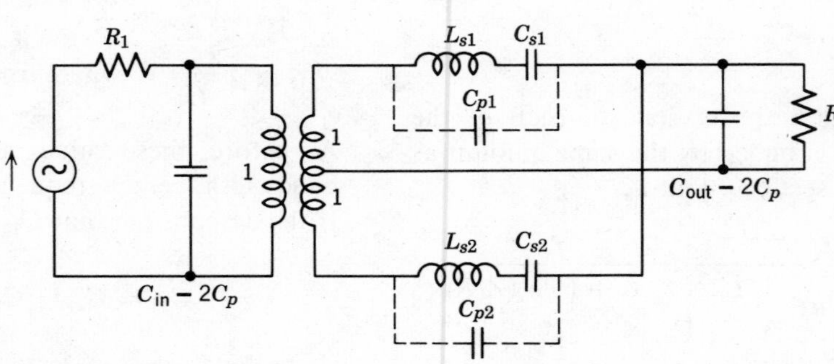

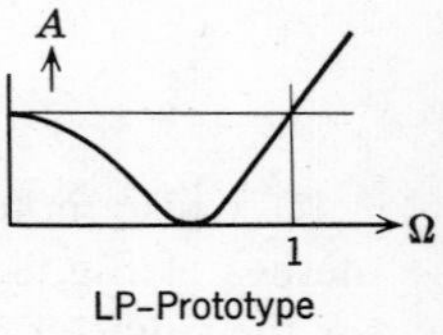

Given values	$L = L_{s1} = L_{s2}$ $\qquad C_{p1} = C_{p2}$ Q_x (crystal) $\qquad\qquad q_1 = q_2$ $\qquad\qquad$ From tables of 3-dB down Δf_{3dB} $\qquad\qquad\qquad k_{12}$ $\qquad\qquad\qquad$ k and q values
Resonant frequency of crystals	$f_{s1} = f_m - \dfrac{\Delta f}{2}\left(\dfrac{1}{q_1} - \dfrac{1}{q_x} + k_{12}\right)$ $f_{s2} = f_m - \dfrac{\Delta f}{2}\left(\dfrac{1}{q_1} - \dfrac{1}{q_x} - k_{12}\right)$
Source and load	with $\bar{R} = \pi L \Delta f$ and $\bar{C} = \dfrac{1}{2\pi^2\, \Delta f f_m L_s}$ (1) $\quad R_1 = R_2 = 2\bar{R}\left(\dfrac{1}{q_1} - \dfrac{1}{q_x}\right)$ (2) $\quad C_{\text{in}} = C_{\text{out}} = \dfrac{\bar{C}}{2\left(\dfrac{1}{q_1} - \dfrac{1}{q_x}\right)}$ $\qquad$ $C_{\text{in}} - 2C_p = C_{T\,\text{in}}$ $\quad$ (total capacitance) $\qquad\qquad\qquad\qquad\qquad\qquad\qquad$ $C_{\text{out}} - 2C_p = C_{T\,\text{out}}$

The solution of (8.16.8) for R_e will result in

$$R_e = \bar{R}\left(\frac{1}{q_1} - \frac{1}{q_x}\right)$$

The following steps must be followed in order to find the element values of the network:

1. Load resistances

$$R_{\text{in,out}} = \bar{R}\,\frac{2k_{23}{}^2 + \left[(1/q_{1,5}) - \dfrac{1}{q_x}\right]^2}{(1/q_{1,5}) - \dfrac{1}{q_x}}$$

2. Coupling resistance

$$2R_{34} = R_{23}\left(1 + \frac{k_{34}{}^2}{k_{23}{}^2}\right)$$

3. Input and output capacitances

$$C_{\text{in,out}} = \bar{C}\,\frac{k_{23}\sqrt{2}}{2k_{23}{}^2 + \left[(1/q_{1,5}) - \dfrac{1}{q_x}\right]^2}$$

4. Coupling capacitor C_{23}

$$C_{23} = \bar{C}/k_{23}\sqrt{2}$$

5. Capacitor of the coupling network

$$C_a = \bar{C}\,\frac{1/\sqrt{2}}{(k_{23} + k_{34})}$$

$$C_b = \bar{C}\,\frac{k_{34}/\sqrt{2}}{k_{23}{}^2 - k_{34}{}^2}$$

Table 8.3   Design Information for Three-Crystal Semilattice.

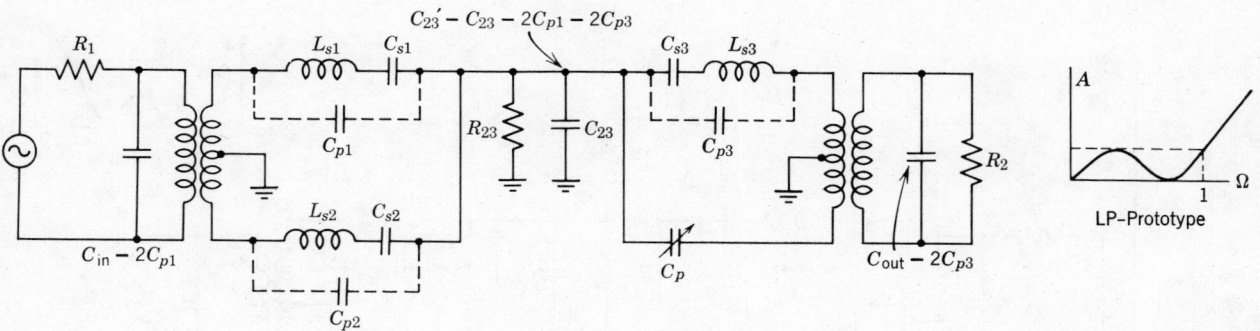

Given values	$L = L_{s1} = L_{s2} = L_{s3}$ $C_{p1} = C_{p2}$ $C_{p3} = C_{p4}$ Q_x q_1 k_{12} R_{23} (equivalent losses of the coupling capacitor) $\Delta f_{3\mathrm{dB}}$ q_3 k_{23}
Resonant frequency of crystals	with $q_{23} = \dfrac{R_{23}}{2\pi\,\Delta f}$ $f_{s3} = f_m - \dfrac{\Delta f}{2}(\sqrt{2}\,k_{23})$ $f_{s1} = f_m - \dfrac{\Delta f}{2}(\sqrt{2}k_{23} + k_{12})$ $f_{s1} = f_m - \dfrac{\Delta f}{2}(\sqrt{2}\,k_{23} - k_{12})$
Source and load	with $\bar{R} = \pi L\,\Delta f$; $\bar{C} = [2\pi^2\,\Delta f f_m L_s]^{-1}$ $R = \bar{R}\,\dfrac{2k_{23}{}^2 + (1/q_1 - 1/q_x)^2}{(1/q_1 - 1/q_x)}$ $R_2 = \bar{R}\,\dfrac{k_{23}{}^2 + 2(1/q_3 - 1/q_x - k_{23}{}^2/2q_{23})^2}{1/q_3 - 1/q_x(k_{23}{}^2/2q_{23})}$ $C_{\mathrm{in}} = \bar{C}\,\dfrac{\sqrt{2}\,k_{23}}{2k_{23}{}^2 + (1/q - 1/q_x)^2}$ $C_{\mathrm{out}} = \bar{C}\,\dfrac{k_{23}/\sqrt{2}}{k_{23}{}^2 + 2(1/q_3 - 1/q_x - k_{23}{}^2/2q_{23})^2}$ $C_{23} = \bar{C}/\sqrt{2}(k_{23})$

6. Series-resonant frequencies of the crystals

$$f_{s1,s2} = f_m - \frac{\Delta f}{2}(k_{23}\sqrt{2} \pm k_{12})$$

$$f_{s3} = f_m - \frac{\Delta f}{\sqrt{2}}\,k_{23}$$

$$f_{s4,s5} = f_m - \frac{\Delta f}{2}(k_{23}\sqrt{2} \pm k_{45})$$

It has been assumed $q_{23}{}^2 \gg 2k_{23}{}^2$.

Although transformer losses have been taken into account in the computations, the shunt mutual inductance has not been considered. The shunt equivalent-transformer inductance can be compensated with the single-frequency approximation by using a capacitance that will make the total coupling susceptance equal the designed value at the center frequency as shown in Figs. 8.37a and b.

Example 1

Let us consider the following simple-filter specification. The primary function of the required filter

Table 8.4 Design Information for Four-Crystal Semilattice.

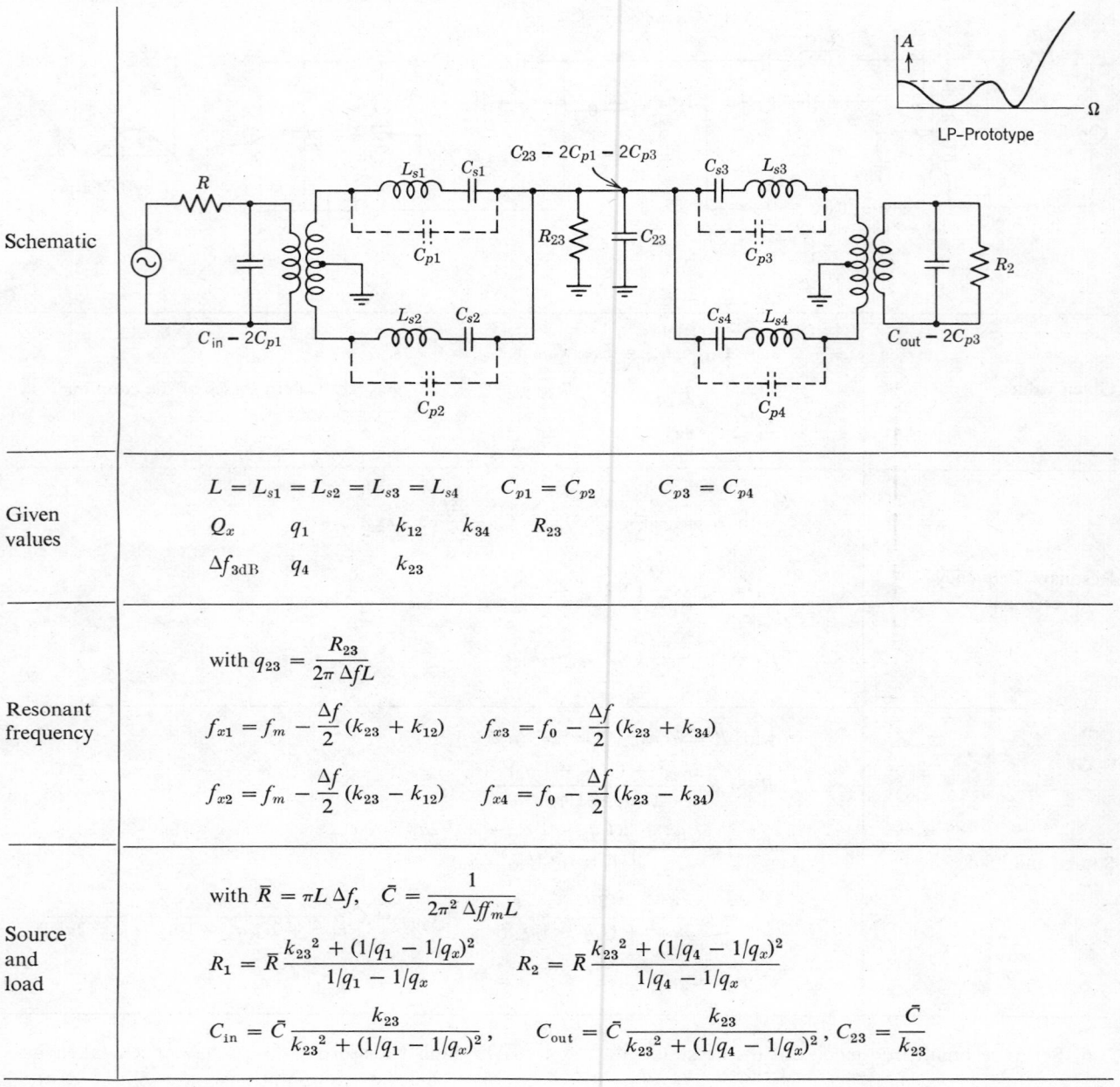

Schematic	(diagram) LP-Prototype

| Given values | $L = L_{s1} = L_{s2} = L_{s3} = L_{s4}$ $C_{p1} = C_{p2}$ $C_{p3} = C_{p4}$

 Q_x q_1 k_{12} k_{34} R_{23}

 $\Delta f_{3\text{dB}}$ q_4 k_{23} |

| Resonant frequency | with $q_{23} = \dfrac{R_{23}}{2\pi\,\Delta f L}$

 $f_{x1} = f_m - \dfrac{\Delta f}{2}(k_{23} + k_{12})$ $f_{x3} = f_0 - \dfrac{\Delta f}{2}(k_{23} + k_{34})$

 $f_{x2} = f_m - \dfrac{\Delta f}{2}(k_{23} - k_{12})$ $f_{x4} = f_0 - \dfrac{\Delta f}{2}(k_{23} - k_{34})$ |

| Source and load | with $\bar{R} = \pi L\,\Delta f,\quad \bar{C} = \dfrac{1}{2\pi^2\,\Delta f f_m L}$

 $R_1 = \bar{R}\,\dfrac{k_{23}{}^2 + (1/q_1 - 1/q_x)^2}{1/q_1 - 1/q_x}$ $R_2 = \bar{R}\,\dfrac{k_{23}{}^2 + (1/q_4 - 1/q_x)^2}{1/q_4 - 1/q_x}$

 $C_{\text{in}} = \bar{C}\,\dfrac{k_{23}}{k_{23}{}^2 + (1/q_1 - 1/q_x)^2},\quad C_{\text{out}} = \bar{C}\,\dfrac{k_{23}}{k_{23}{}^2 + (1/q_4 - 1/q_x)^2},\quad C_{23} = \dfrac{\bar{C}}{k_{23}}$ |

is to establish a certain noise bandwidth at 30 Mc. The minimum 3-dB bandwidth is 25 kc and the attenuation requirement is such that one crystal will be sufficient to satisfy the system's performance.

From practical considerations (aging and temperature shifting) the bandwidth used will be increased slightly, to $\Delta f_3 = 28$ kc. The available motional inductance of the crystal is $L_s = 10$ mH, and the unloaded quality factor of the crystal is $Q_x = 20,000$.

1. Determine auxiliary parameters

(a) $$\bar{C} = \frac{1}{2\pi^2\,\Delta f_3 f_m L_s} = 6.03 \text{ pF}$$

(b) $\bar{R} = \pi\,\Delta f_3 L_s = 879.65$ ohms. The normalized quality of the crystal is

$$q_x = \frac{Q_x}{a} = 18.66$$

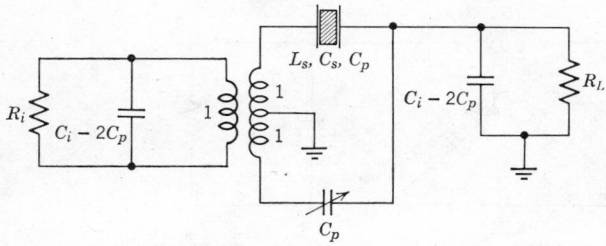

Fig. 8.39. Realization of one-crystal filter of example 1.

2. Calculate the resistor across the input and output terminals

$$R_{in} = R_{out} = 2\bar{R}\left(1 - \frac{1}{q_x}\right) = 1665 \text{ ohms}$$

3. Calculate the input and output capacitor of the equivalent semilattice

$$C_{in} = C_{out} = \frac{\bar{C}}{2(1 - 1/q_x)} = 3.18 \text{ pF}$$

The values of these capacitors is to be reduced by the amount of common capacitance which appears across the crystal (C_p) and which is balanced by the bridge schematic by tuning the branch with the trimmer. The numerical value of capacitance to subtract is twice that of C_p.

4. Calculate the resonant frequency of the crystal

$$f_{s1} = f_m \frac{\Delta f}{2}(1 - 1/q_x)$$

$$= 30 \times 10^6 - 13 \times 25 \times 10^3 = 29.9868 \text{ Mc.}$$

5. For completeness, calculate the motional capacitance of the crystal

$$C_{s1} = \frac{1}{4\pi^2 f_{s1}^2 L_s} = 0.00281696 \text{ pF}$$

6. The schematic to be used is shown in Fig. 8.39.

Example 2

A narrowband filter is to be designed to perform according to the following specifications.

1. Center frequency is 19.557 Mc.
2. 3-dB bandwidth is 31 kc.
3. Phase linearity across the band within 1° (±0.5° deviation).
4. Total phase shift across 3-dB bandwidth = 240° minimum.

For a linear phase with equiripple error (0.5° ripples) with $n = 4$, the normalized delay at the center frequency is 2.28 sec (obtained from the appropriate curve in Chapter 3) which corresponds to

$$\frac{db}{d\omega} = t_g = 2.28$$

or approximately, between the center frequency and the cutoff

$$\Delta b = 2.28 \times 57.3 = 130°$$

According to the appropriate table of 3-dB down k and q values in Chapter 6,

$$q_1 = 0.581 \qquad k_{12} = 1.575$$
$$q_4 = 1.03 \qquad k_{23} = 0.797$$
$$k_{34} = 0.656$$

The source resistance is

$$R_{in} = \frac{\bar{R}[k_{23}^2 + (1/q_1 - 1/q_x)^2]}{1/q_1 - 1/q_x}$$

The formula for R_{out} is identical, except q_4 is to be substituted for q_1.

1. The auxiliary resistance value is calculated, since the available motional inductance $L_s = 10$ mH, and $\Delta f = 31$ kc

$$\bar{R} = \pi L_s \Delta f_3 = 973.9 \text{ ohms}$$

Auxiliary capacitance $\bar{C}$ is

$$\bar{C} = [62 \times 10^3 \times \pi^2 \times 19.557 \times 10^6 \times 10 \times 10^{-3}]^{-1}$$
$$= 8.356 \text{ pF}$$

2. Therefore the input resistance (neglecting the contribution by the crystal) is:

$$R_{in} = \frac{973.9[(0.797)^2 + (1/0.581)^2]}{1/0.581} = 2035.45 \text{ ohms}$$

and the output resistance is

$$R_{out} = \frac{973.9[0.6352 + (1/1.06)]}{1/1.03} = 1583 \text{ ohms}$$

3. The input capacitor of the network is

$$C_{in} = \bar{C}\frac{k_{23}}{k_{23}^2 + (1/q_1 - 1/q_x)^2}$$
$$C_{in} = 8.356\frac{0.797}{(0.6352 + 2.9629)} = 1.85 \text{ pF}$$

and the output capacitor is

$$C_{out} = \frac{8.356 \times 0.797}{0.6352 + 0.9434} = 4.2187 \text{ pF}$$

4. The coupling capacitor is

$$C_{23} = \frac{C}{k_{23}} = \frac{8.356}{0.797} = 10.484 \text{ pF}$$

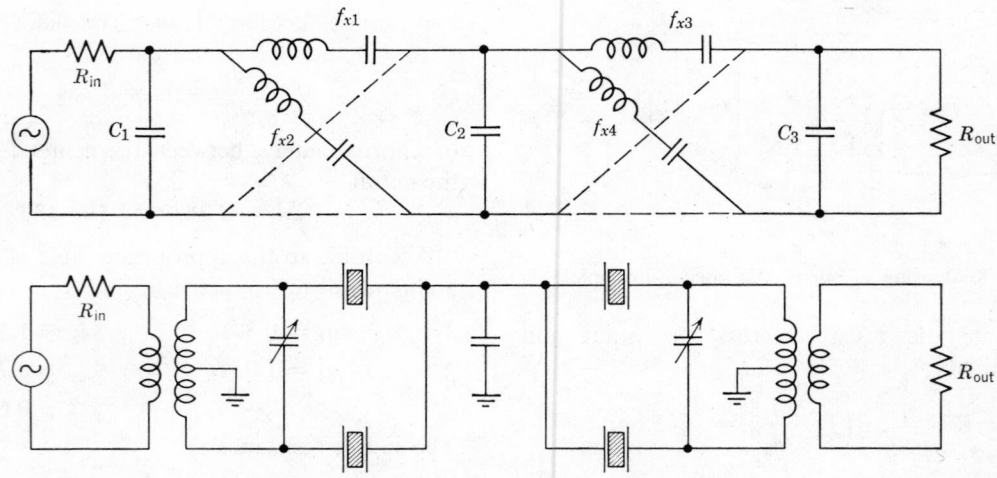

Fig. 8.40. Lattice equivalent and realization of four-crystal filter of Example 2.

5. The resonant frequencies of the crystals are

$$f_{s1} = 19.557 - 15.5(0.797 + 1.575) = 19.5202 \text{ Mc}$$
$$f_{s2} = 19.557 + 15.5(0.778) = 19.569 \text{ Mc}$$
$$f_{s3} = f_0 - 15.5(0.797 + 0.656) = f_0 - 15.5(1.453)$$
$$= 19.5345 \text{ Mc}$$
$$f_{s4} = 19.557 - 15.5(0.797 - 0.656) = 19.5548 \text{ Mc}$$

As we can see, the calculation of the filter is simple even in the case of the filter with four crystals. In Fig. 8.40 the schematic of the filter is given. The final values of elements are

$$L_s = 10 \cdot 10^{-3} \text{ H}$$
$$C_1 = 0.9255 \text{ pF}$$
$$C_2 = 5.242 \text{ pF}$$
$$C_3 = 2.1094 \text{ pF}$$
$$R_{\text{in}} = 4070 \text{ ohms}$$
$$R_{\text{out}} = 3166 \text{ ohms}$$

Figure 8.41 is the measured response curve of the filter.

Limitations of the Method Based on Lowpass Polynomial Model

As the required bandwidth is increased, transformers with higher Q or higher inductance, or both, are needed. Considerable passband distortion results when inductive components of insufficient quality factors are employed. The reason for this is that the current through the lossy mutual elements (transformer losses) is not the same for frequencies that are an equal distance above and below the center frequency. Therefore the uniform predistortion techniques relied

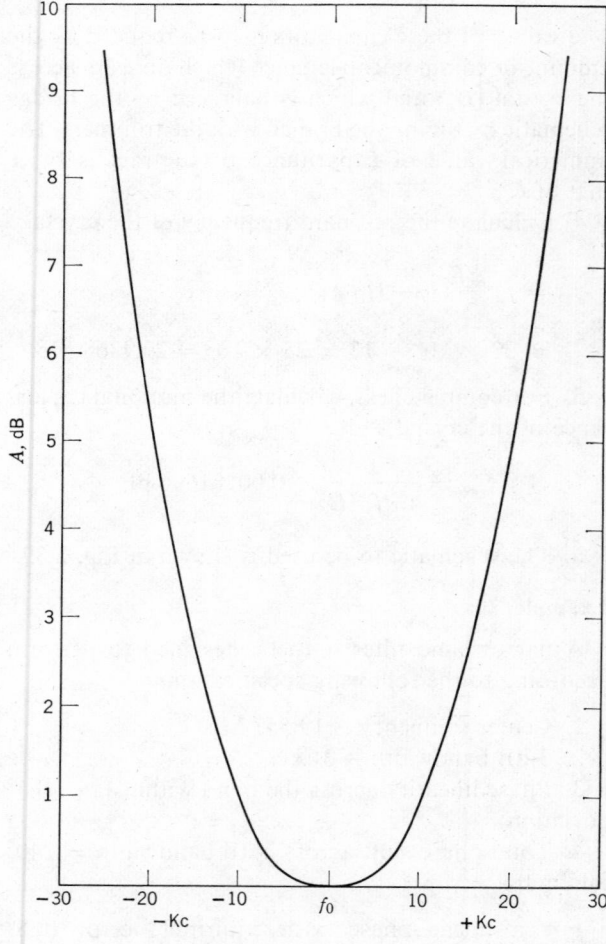

Fig. 8.41. Frequency response of $n = 4$ linear phase filter of Example 2.

on for design data, are only adequate in the center of the passband.

The effects of transformer losses can be reduced by choosing a set of coupling coefficients that give a minimum value of k_{23}. This acts to bypass the loss resistance with a larger lossless capacitance.

Finite transformer inductance is another limiting factor. If transformer losses are sufficiently low, as the bandwidth begins to approach about one percent, the transmission zeros created by a finite inductance are close enough to the dominant crystal zeros to influence the response in the passband and thus invalidate the design procedure.

Extremely small-percentage-bandwidth filters encounter difficulties, not from nonideal transformers, but from a lack of sufficient crystal Q and stability. Crystal stability can be increased by the use of a temperature controlled environment, but ultimately, a lower limit is placed on the bandwidth that can be achieved.

Although the foregoing examples recommend the use of predistorted q's and k's, it is possible to obtain a simplified circuit with identical crystals, realized when the coupling coefficients (k's) and the end q's are symmetrical. If some response degradation is allowed, symmetrical q's and k's can be used with finite-Q elements, but the response characteristic will not be entirely predictable.

8.17 SYNTHESIS OF LADDER SINGLE SIDEBAND FILTERS

As symmetrical bandpass filters are now almost exclusively designed by network tabulations, so asymmetrical SSB bandpass filters with attenuation peaks can also be designed with tables.

The structural appearance of the filters will be identical to those previously analyzed, but the entire network will reflect the property of a lowpass model which will be given in tables of normalized design parameters (such as 3-dB down k and q values).

Dishal, who introduced the characterization of filters by node (or loop) resonances, normalized decrements (d's), and normalized coefficients of coupling (k) has extended this characterization to include crystal ladder filters of the SSB type. The schematic of Fig. 8.42 is by far the most realizable and desirable schematic. The greatest use for this network is not single sideband applications, since it can be applied to any system where a narrow, stable bandwidth is required and a slight dissymmetry is not upsetting. Let us make a qualitative description of

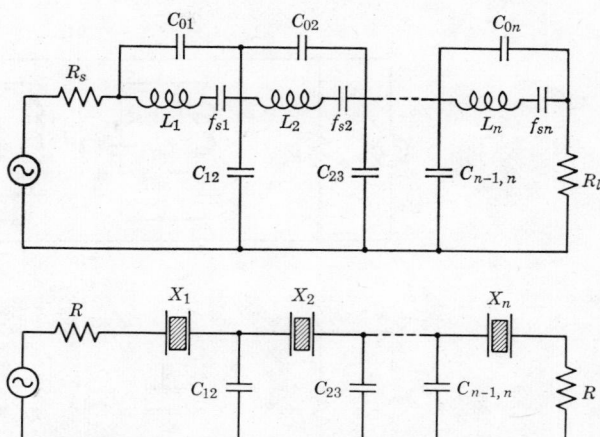

Fig. 8.42. Ladder capacitor-crystal lower sideband filter.

the two main variations of crystal ladder filters. In the series-crystal circuit

$$R \cong L_s \Delta\omega_{3\,\mathrm{dB}}$$

which in practical narrowband designs using realistic crystals is moderate (≈ 300 ohms at 10 kc bandwidth with $L = 6$ mH). The shunt crystal circuit shown in Fig. 8.43 on the other hand, is a high-impedance circuit with $C_p/C_s = 200$.

$$\frac{1}{G} = R \cong \frac{1}{\Delta\omega_{3\,\mathrm{dB}}C_p} \cong \frac{1}{\Delta\omega_{3\,\mathrm{dB}}C_s \times 200}$$

In this circuit, for the same data given above, R would be more than 1 MΩ.

The circuit of Fig. 8.43 depends on the parallel resonance to set the passband thus requiring the use of trimmers in parallel with the crystal. The series crystal circuit of Fig. 8.42 sets the passband with the crystal series resonance, and for most applications the crystal manufacturer can set the frequency so close to requirements that no trimmers are needed in the circuit. Only in a very narrow bandwidth and where the bandwidth must be accurately set is there any need of tuneable elements. Almost symmetrical bandpass requirements can be met with bandwidths up to 0.1%, but beyond this percentage, up to the limit of about 0.4%, dissymmetry is noted. The parallel-resonant frequency in Fig. 8.42 produces the peak of attenuation expected from the filter. Thus if all the peaks are not precisely coincident owing to variations of parallel capacity, there is no degradation in the performance of the filter. The same cannot be said of the parallel crystal circuit of Fig. 8.43.

The series crystal configuration can be designed by using the tables presented in this section for filters

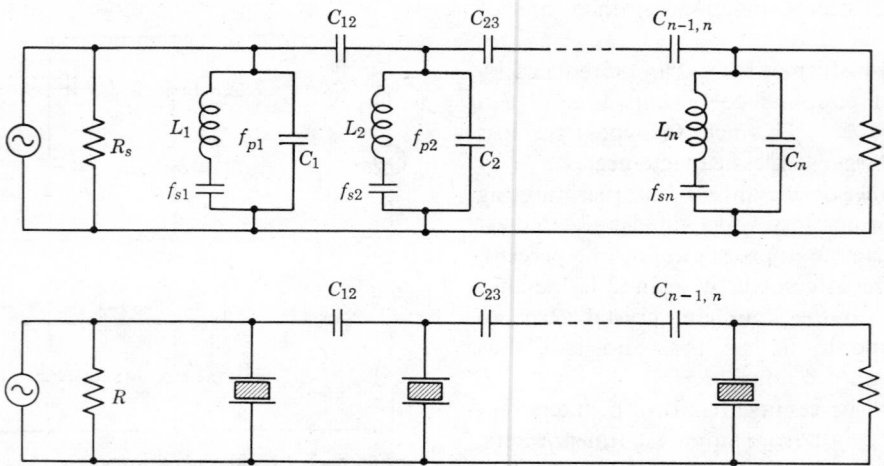

Fig. 8.43.   Ladder capacitor-crystal upper sideband filter.

with from two to five crystals. The ladder configuration has an advantage over the common half-lattice realization when it comes to ultimate rejection, since far from the passband the filters of Figs. 8.42 and 8.43 are merely ladders of capacitors, acting like a voltage divider. The equivalent half-lattice is dependent upon the balance of its transformer for its ultimate rejection. The problem of spurious responses in the crystals is less bothersome in the ladder schematic since each crystal tends to reject the spurious of the others.

The tables and curves to follow have element values for the Butterworth response, and for the Chebyshev response from 0.01 dB to 1.0 dB ripple. In addition, it is possible to make filters displaying a quasi-Gaussian shape which will have a linear phase shift over the passband and low time-transient response.

Frequency Transformation for Nonsymmetrical Filters

In Chapter 5 we discussed the technique of frequency transformation for resonant circuits which reflects symmetrical attenuation responses. Every reactance in the lowpass prototype transforms to a resonator in the bandpass circuit.

The corresponding equations for the series and parallel circuits of the bandpass schematic (see Section 5.5) are:

$$x_{si} = j \frac{L_{si}}{R_r} \omega_m \left(\frac{\omega}{\omega_m} - \frac{\omega_m}{\omega} \right) = j \frac{L_{si}}{R_r} \omega_m x$$

$$x_{pi} = \frac{1}{jC_{pi}R_r\omega_m \left(\dfrac{\omega}{\omega_m} - \dfrac{\omega_m}{\omega} \right)} = \frac{1}{jC_{pi}R_r\omega_m x}$$

Filter characteristics of the SSB type are never symmetrical. One side (the slow cutoff side) is of

little importance and the attenuation response is monotonic. The opposite side of passband (the sharp cutoff side) has to satisfy a certain attenuation requirement in a restricted, relatively narrow region of frequency in order to suppress the unwanted band of modulation products. Therefore the usual frequency transformation of lowpass to bandpass is not applicable. A similar difficulty was encountered in the case of coil-saving *LC* filters, where bandpass resonators consisted of three reactances (Section 4.16).

In Fig. 8.44 a basic-ladder narrowband filter is shown. The problem is to modify the ladder in such a way that no inductors will be needed for actual realization, and the blocks in shunt arms will accommodate crystal resonators.

In complete similarity to a conventional nodal network each node is resonated to the same frequency f_0 (with all other nodes short circuited). It is important to realize that the blocks shown as part of the coupling mechanism are simply impedance converters, the negative elements of which may be absorbed (no physical additions are required to realize the filter network).

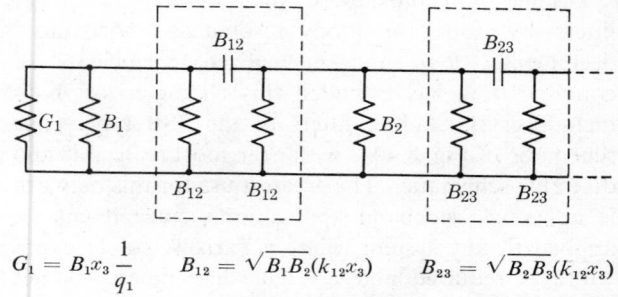

$$G_1 = B_1 x_3 \frac{1}{q_1} \qquad B_{12} = \sqrt{B_1 B_2}(k_{12}x_3) \qquad B_{23} = \sqrt{B_2 B_3}(k_{12}x_3)$$

Fig. 8.44.   Basic-ladder narrowband filter.

The Generalized Frequency Variable

The admittance of the shunt arms of Fig. 8.44 can be written in the form

$$Y = jB_i x \qquad (8.17.1)$$

where x is a general frequency variable, and B_i is the susceptance of the shunt arm. For the admittance in this form the regular design data which has been used for frequency symmetrical bandpass filters remains equally valid.

For example, let us consider the case where the admittance of the shunt arm is

$$Y = j\omega_0 C\left(\frac{\omega}{\omega_0} - \frac{\omega_0}{\omega}\right) = j\omega_0 C x_1 \qquad (8.17.2)$$

where ω_0 is the resonant frequency of the parallel circuit (see Fig. 8.45) and

$$B = \omega_0 C$$

For the narrow bandwidth case considered where ω/ω_0 is between 0.95 and 1.05 (less than 10% of bandwidth) the frequency variable x_1 of Eq. 8.17.2 approaches the linear bandwidth variable

$$\left(\frac{\omega}{\omega_0} - \frac{\omega_0}{\omega}\right) \approx \frac{2(f - f_0)}{f_0} = F$$

Thus for the parallel circuit in Fig. 8.45 to be used in the filter of Fig. 8.44, the symmetrical generalized variable x in the terms of frequency is

$$x_1 = \frac{2(f - f_0)}{f_0} = F$$

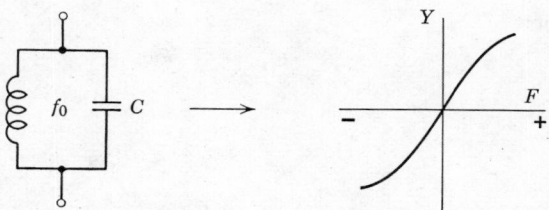

Fig. 8.45. Parallel-resonant circuit and its admittance diagram.

When a crystal is used in Fig. 8.44 instead of a simple parallel circuit, the frequency variable x is more complex, since the equivalent circuit of the crystal consists of three elements and consequently exhibits two resonances.

For such an element (shown in Fig. 8.46) and with the realistic approximation the $C_0/C_1 \gg 1$, the admittance of the arm may be expressed by

$$Y = j\omega_0 C_0 \frac{x_1}{1 + (C_0/C_1)(\omega_0/\omega)x_1}$$

Once again for the narrowband case ($\omega/\omega_0 = 0.95 - 1.05$) where $x_1 \cong F$, the linear variable may be introduced

$$Y = j\omega_0 C_0 \frac{F}{1 + (C_1/C_s)F}$$

The bandwidth variable F as we know from the previous discussion in Section 8.10 is a very convenient substitute for actual frequency.

The variable x and bandwidth variable are related by

$$x = \frac{F}{1 + kF}$$

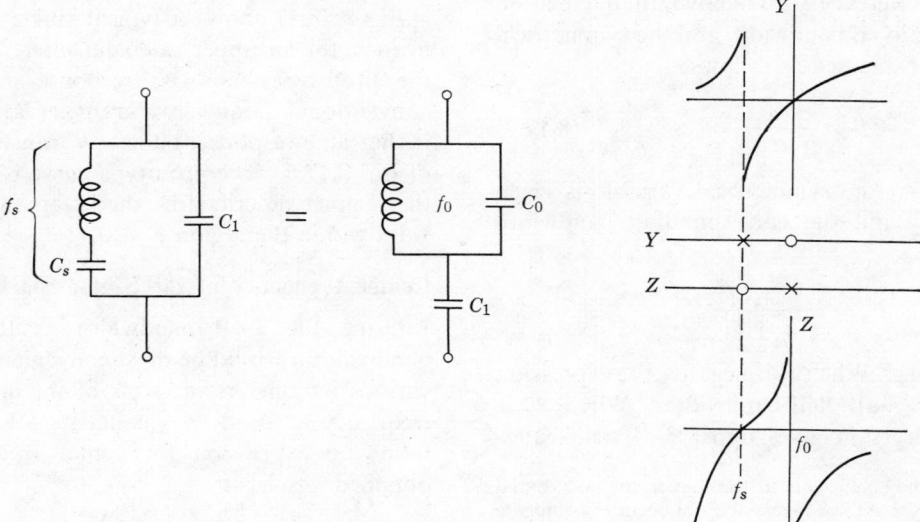

Fig. 8.46. Equivalent crystal resonator and its admittance and impedance diagram.

where

$$k = \frac{C_1}{C_s} \gg 1$$

$$C_s = \frac{C_0}{k^2}$$

$$f_s = f_0 - \frac{1}{2}\frac{f_0}{k}$$

The expression for transmission function in terms of symmetrical variable x can be written as

$$\frac{\text{Input}}{\text{Output}} = \frac{(jx)^n + A_1(jx)^{n-1} + A_2(jx)^{n-2} + \cdots A_n}{K}$$

where A_1, A_2, A_n are numbers derived from various combinations of normalized q's and k's.

If the expression for x is substituted in the expression for the transmission function the resulting equation has a numerator which is a polynomial in terms of (jF) of nth order having complex coefficients.[1] The resulting denominator K will be not a constant but $(1 + kF)^n$ with n equal to the number of crystals in the filter.

If the numerator described the passband behavior in terms of complex zeros in the left half of jF plane, the numerator shows that there are n coincident real frequency-transmission poles which are responsible for the poles of attenuation.

Normalized Transformations

Using the symmetrical variable x belonging to the Chebyshev or Gaussian family (or any other possible type of response shapes) and tabulated parameters normalized to a reference 3-dB bandwidth, the relation between the fractional bandwidth and the symmetrical variable will be expressed as follows:

$$F = \frac{x}{1 - kx} \qquad (8.17.3)$$

At the 3-dB point the symmetrical variable is given the subscript x_3 and the corresponding bandwidth variable will be:

$$F_3 = \frac{x_3}{1 - 1/\Omega_\infty}$$

where $kx_3 = 1/\Omega_\infty$. When x_3 is negative the expression gives the lower 3-dB half-bandwidth. When x_3 is positive the expression gives upper 3-dB half-band-

[1] The fact that the coefficients of the numerator polynomial are complex, means that these n-zeros do not occur in conjugate complex pairs; each zero has both a real and an imaginary coordinate which is different for any other zeros.

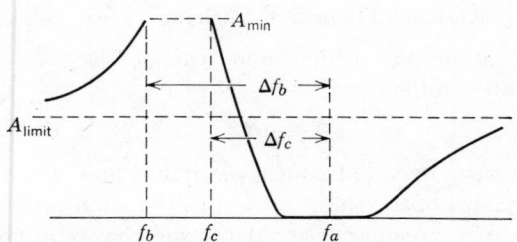

Fig. 8.47. Upper single sideband filter specification.

width. The sum of two magnitudes expresses the total 3-dB bandwidth

$$\frac{\Delta f_3}{f_0} = \frac{1}{a} = \frac{x_3}{1 - \dfrac{1}{\Omega_\infty^2}} \qquad (8.17.4)$$

In every expression involving the total bandwidth x_3 is a positive (absolute) value.

$$x_3 = \frac{\Delta f_3}{f_0} - \frac{\Delta f_3}{f_0}\left|\frac{1}{\Omega_\infty^2}\right| = \frac{1}{a}\left(1 - \frac{1}{\Omega_\infty^2}\right)$$

The normalized transformation will be expressed by

$$\frac{F}{\Delta f_3/f_0} = \frac{1 - \dfrac{1}{\Omega_\infty^2}}{\dfrac{x_3}{x} - \dfrac{1}{\Omega_\infty}} = \frac{1 - \dfrac{1}{\Omega_\infty^2}}{\dfrac{1}{\Omega} - \dfrac{1}{\Omega_\infty}} = Fa \qquad (8.17.5)$$

where only x can take either positive or negative values. Ω_∞ is an attenuation-peak parameter. With these expressions every symmetrical attenuation curve can be transformed into a nonsymmetrical response with one attenuation peak at Ω_∞. When this is done, the resulting passbands are no longer linear phase or Gaussian.

Figure 8.47 shows a typical single sideband specification for an upper sideband filter. Figure 8.48a is the prototype bandpass response in terms of the conventional frequency parameter Ω. Figure 8.48b is the same response, bilinearly transformed with aid of Eq. 8.17.5. The prototype curve is transformed in the manner described by the letters in the circle, and valid within the region $F < 0.1$.

Center Frequency of the Single Sideband Filter

Using the 3-dB bandwidth as the normalizing bandwidth it would be most convenient to express the various parameters in terms of the arithmetic center frequency f_a. With x/x_3 equal to -1, in (8.17.5) the following expression for center frequency will be obtained

$$f_a = f_0 + \frac{\Delta f_3}{2}\frac{1}{\Omega_\infty}$$

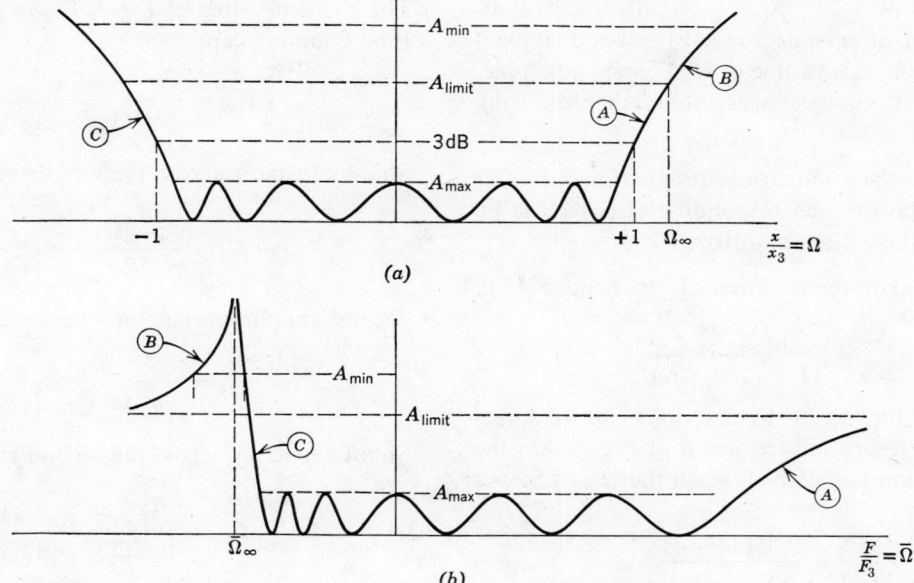

Fig. 8.48. Bandpass response (*a*) in terms of normalized frequency and (*b*) bilinearly transformed response into SSB shape.

The transformation formula itself in terms of f_a is

$$\bar{\Omega} = \frac{2(f - f_a)}{\Delta f_3} = \frac{\Omega \Omega_\infty - 1}{\Omega_\infty - \Omega} \qquad (8.17.6)$$

(Here, $\bar{\Omega}$ is the same as Ω in Eq. 8.12.11.) For the different regions of Fig. 8.48 the magnitude of Eq. 8.17.6 is given in Table 8.5. From Fig. 8.48*a* we see that the peak of attenuation is obtained when the symmetrical variable Ω equals infinity; inserting this value in (8.17.6) results in

$$\bar{\Omega}_\infty = \Omega_\infty = \frac{f_\infty - f_a}{\Delta f_3/2}$$

which is the same as (8.10.5) for Ω_∞.

Thus the parameter Ω_∞ sets directly the distance between the frequency of the attenuation peak and the center of the passband for a given 3-dB bandwidth.

In the upper sideband filter, Fig. 8.43, the series-resonant frequencies of the crystals coincide with the frequency of infinite attenuation; therefore the required crystal frequency is

$$f_s = f_a - \frac{\Delta f_3}{2} |\Omega_\infty|$$

For a given number of crystals (*n*) and a given value of passband ripple there is an additional characteristic of the attenuation, $A_{\min}$, which is immediately set when the parameter Ω_∞ is chosen. The ultimate minimum of attenuation (A_{limit} in Fig. 8.48*b*) is reached at a frequency relatively close to the passband. As Fig. 8.48*a* shows, this minimum attenuation occurs at the transition point between Region *A* and Region *B* at $\Omega = \Omega_\infty$. The value of minimum attenuation is obtained from calculations when all other values such as Ω_∞, the number of crystal and the passband ripples are known.

Design Procedure

Given parameters (see Fig. 8.47) are: (1) Required attenuation $A_{\min}$ at a specified frequency f_c and (2)

Table 8.5 Relationship between Ω and $\bar{\Omega}$

Range	Ω	$\bar{\Omega}$	Magnitude of $\bar{\Omega}$
A	$0 < \Omega < \Omega_\infty$	$-\left\|\dfrac{1}{\Omega_\infty}\right\| < \bar{\Omega} < +\infty$	$\dfrac{\|\Omega_\infty\| \, \|\Omega\| - 1}{\|\Omega_\infty\| - \|\Omega\|}$
B	$\Omega_\infty < \Omega < +\infty$	$-\infty < \bar{\Omega} < -\|\Omega_\infty\|$	$\dfrac{\|\Omega_\infty\| \, \|\Omega\| - 1}{\|\Omega\| - \|\Omega_\infty\|}$
C	$-\infty < \Omega < 0$	$-\|\Omega_\infty\| < \bar{\Omega} < -\left\|\dfrac{1}{\Omega_\infty}\right\|$	$\dfrac{\|\Omega_\infty\| \, \|\Omega\| + 1}{\|\Omega_\infty\| + \|\Omega\|}$

Required ratio of Δf_s to $\Delta f_3/2$. The assumed values are (3) Number of crystals and (4) Passband ripple.

To arrive at the exact value of Ω_∞, passband ripple, and the correct number of crystals, the following procedure may be used:

Step 1. From the symmetrical attenuation curves of Chapter 3, obtain the corresponding Ω (which is Ω_s in Chapter 5 when A_{max} is 3-dB).

Step 2. Calculate the required Ω_∞ with Eq. 8.17.6.

$$\Omega_\infty = \frac{(2\Delta f_s/\Delta f_3)\Omega - 1}{\Omega - (2\Delta f_s/\Delta f_3)}$$

Step 3. Use Eq. 8.17.6 to determine at what frequency on the return hill (region B of Fig. 8.48b) the specified rejection is obtained, when the Ω_∞ of Step 2 is used.

$$\frac{\Delta f_b}{\Delta f_3/2} = \frac{|\Omega_\infty|\, \Omega_s - 1}{\Omega_s - |\Omega_\infty|}$$

If the stopband between f_s and f_b is too narrow, then a larger ripple in the passband or a larger number of crystals must be chosen, and Steps 1 through 3 repeated, until the required reject bandwidth is obtained.

Upper Sideband Networks with Identical Crystals

It has already been pointed out that all the series-resonant crystal frequencies are identical. All have the same motional capacitance C_s (or the same motional inductance L_s).

In order to have the same required motional capacitance, the values B_1, B_2, B_3 in Fig. 8.44 must be the same for each node (since k sets the location of the attenuation-peak frequency, which is identical for each node).

The capacitance ratio as a function of Ω_∞ is

$$\frac{C_1}{C_s} = \frac{a}{\left[|\Omega_\infty| - \left| \dfrac{1}{\Omega_\infty} \right| \right]}$$

The required loading resistor is

$$R_1 = 2\pi q_1 (\Omega_\infty^2 - 1)\,\Delta f$$

where q_1 is given in the design tables. The frequency of infinite attenuation is given by

$$f_\infty = f_a - \frac{\Delta f}{2}\Omega_\infty$$

where

$$\Omega_\infty = \frac{f_\infty - f_a}{\Delta f_3/2}$$

The element values of Fig. 8.43 are given by:
First coupling capacitor

$$C_{12} = \frac{k_{12}}{4\pi^2(\Omega_\infty^2 - 1)\,\Delta f_3 f_a}$$

Shunt capacitor across the first crystal

$$C_1 = \frac{\Omega_\infty - k_{12}}{4\pi^2(\Omega_\infty^2 - 1)\,\Delta f_3 f_a}$$

Second coupling capacitor

$$C_{23} = \frac{k_{23}}{4\pi^2(\Omega_\infty^2 - 1)\,\Delta f_3 f_a}$$

Shunt capacitor across the second crystal

$$C_2 = \frac{\Omega_\infty - k_{12} - k_{23}}{4\pi^2(\Omega_\infty^2 - 1)\,\Delta f_3 f_a}$$

Shunt capacitor across the third crystal

$$C_3 = \frac{\Omega_\infty - k_{23} - k_{34}}{4\pi^2(\Omega_\infty^2 - 1)\,\Delta f_3 f_a}$$

Parallel-resonant frequency of first crystal

$$f_{p1} = f_a - \frac{\Delta f_3}{2}\frac{1 - k_{12}\Omega_\infty}{\Omega_\infty - k_{12}}$$

Parallel-resonant frequency of second crystal

$$f_{p2} = f_a - \frac{\Delta f_3}{2}\frac{1 - (k_{12} + k_{23})\Omega_\infty}{\Omega_\infty - k_{12} - k_{23}}$$

Parallel-resonant frequency of third crystal

$$f_{p3} = f_a - \frac{\Delta f_3}{2}\frac{1 - (k_{23} + k_{34})\Omega_\infty}{\Omega_\infty - k_{23} - k_{34}}$$

In the network of Fig. 8.44 the equivalent circuit of Fig. 8.46 will be placed. Inside each block there appears a capacitance (C_1) shunting the series-resonant circuit, and from the definition of k the value of C_1 is obtained

$$C_1 = \frac{C_0}{k} = kC_s$$

It is important to comprehend that C_1 is not the initial minimum C_p and is not the shunt capacitance to be used in the realized network. As can be seen from Fig. 8.44 this capacitance $C_1(C_2, C_3)$ inside each box is shunted by two negative capacitors (shown as B_{12} and B_{23} in the case of the midsection), one from the impedance converter preceding it, and one from the converter following it. Over the small-percentage bandwidth involved in these filters, the entire coupling network will appear as capacitor C_{12}.

The sum of the three capacitors that touch each node, for example, $C_{12} + C_2 + C_{23}$ for node 2, is a constant. For every node:

$$\frac{C_1}{C_s} = k = \frac{a}{\left| \Omega_\infty \right| - \left| \dfrac{1}{\Omega_\infty} \right|}$$

This fact can also be used as a convenient double check of the numerical design work.

The parallel-resonant frequency (f_0) of each node, with every other node short-circuited, must be the same; equations for f_{p1}, f_{p2} and f_{p3} give the required value of parallel-resonant frequency to make the nodal frequency the same.

In the case of AT-cut crystals, a minimum possible value of C_p/C_s is approximately 200 in the 5-Mc region. In the network of Fig. 8.43 the lowest value of C_1 is C_p. The following equation can be used for determining the maximum obtainable bandwidth:

$$\left(\frac{1}{a} \right)_{\max} = \left(\frac{C_s}{C_p} \right) \frac{\left| \Omega_\infty \right| - (k_{12} + k_{23})}{\left| \Omega_\infty \right|^2 - 1}$$

Lower Sideband Network

For the lower sideband ladder filter shown in Fig. 8.42 all attenuation data of the previous discussion are

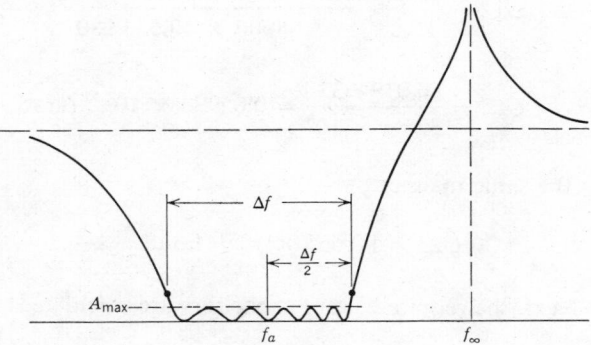

$A_{\max}$ = peak to valley ripple amplitude.
f_a = filter center frequency.
Δf = filter bandwidth between the 3-dB down points.
f_∞ = frequency of infinite attenuation.
$\Omega_\infty = \dfrac{f_\infty - f_a}{\Delta f/2}$, parameter telling the location of the infinite-attenuation point in units of filter half-band-widths.

Fig. 8.49. Lower single sideband response.

directly applicable if the A, B, and C regions are used to identify the three frequency regions involved. Therefore rather than redraw Fig. 8.48, it is only necessary to properly identify the frequency regions involved in Fig. 8.49.

The following charts and tables are computer results for the design of crystal-ladder filters containing from two to five crystals. All crystals have the same inductance, normalized to unity, and the ladder is made up of extremely simple components. Also, it is physically symmetrical about the center—that is, the same values of components are seen from either end. Consequently, only half the element values are given in the tabular print out.

The following defines the nomenclature used in the tables:

Ω_∞ = independent variable to be chosen by the designer. It is the ratio of the frequency difference from the arithmetic center of the passband to the infinite-attenuation point $(f_\infty - f_a)$ to one-half the 3-dB bandwidth Δf.

$\dfrac{2}{\Delta f}(\Delta F_{m,n})$ = constant which is related to the center frequency of the mth and nth crystal. Multiply this normalized value by one-half the 3-dB bandwidth $(\Delta f/2)$ and add the result to the center frequency f_a to obtain the series-resonant frequency of the crystal f_s.

$\dfrac{R}{\Delta f}$ = normalized load resistance. To un-normalize multiply by the bandwidth Δf.

$\dfrac{\Delta f(C0m)}{f_a(Cm)}$ = normalized capacitance ratio of the mth crystal. To denormalize multiply by $f_a/\Delta f$. (This is used only to check realizability of a particular design.)

$\Delta f f_a(Cmn)$ = normalized coupling capacitor between the mth and nth crystal. To denormalize divide by the product of the bandwidth Δf and the center frequency f_a.

$\Delta f f_a(C0m)$ = normalized crystal shunt capacitance. To denormalize, proceed as above.

Example of Design

A filter is required to have a center frequency of 40.654280 Mc, a 3-dB bandwidth of 8000 cycles, and an ultimate attenuation of 100 dB is required far from the passband. Assuming that the filter uses five crystals, with zero-dB peak ripple in the passband,

and that the attenuation pole (infinite-attenuation frequency) is located ten half-bandwidths from the center frequency, then

$$N = 5, \quad A_{\max} = 0.0 \text{ dB}, \quad \Omega_\infty = \frac{f_\infty - f_a}{\Delta f/2} = 10$$

All the design information for this filter is located in Table 8.13. From the design request:

$$\Delta f = 8000 \text{ cycles}, \quad f_a = 40.654280 \text{ Mc}$$

First, the resonant frequencies of the individual crystals will be found. $\Delta F_{1,5}$, $\Delta F_{2,4}$ etc gives the frequency deviation of the crystal series resonance with respect to the center frequency f_a. Therefore all ΔF's calculated must be added to the required center frequency in order to yield the final required crystal frequency.

$$\frac{2}{\Delta f}(\Delta F_{1,5}) = -1.062849$$

$$\Delta F_{1,5} = -1.062849\left(\frac{\Delta f}{2}\right) = -1.062849\left(\frac{8000}{2}\right)$$

$$\Delta F_{1,5} = -1.062849 \times 4000 = -4252 \text{ cps}$$

Thus the first and fifth crystals have series-resonant frequencies of $40.654280 + (\Delta F_{1,5})$,

$$f_{s1,5} = 40,654.280 - 4.252 = 40,650.028 \text{ kc}$$

The remaining series crystal frequencies are calculated in the same manner

$$f_{s2,4} = 40.647770 \text{ Mc}, \quad f_{s3} = 40.649510 \text{ Mc}$$

The terminating resistance R of the filter is calculated algebraically as are all element values obtained from these tables.

$$\frac{R}{\Delta f} = 12.568003$$

$$R = 12.568003 \times \Delta_f$$

$$R = 12.568003 \times 8000$$

$$R = 100,500 \text{ ohms}$$

At this point it should be recalled that the inductance of all crystals in these designs is 1 H. Of course, this

is not a limitation since the final schematic can be impedance scaled to any practical inductance value. This value of practical inductance is only learned through experience and close technical consultation with the crystal manufacturer. In this example, an inductance of 4 mH is easily realizable and therefore impedance scaling will be necessary.

The next item in the tables is a check on crystal realizability. The ratio of the parallel to the series capacity can be a minimum of 200:1 in the case of AT-cut crystals. Only the smallest ratio need be calculated as a check.

$$\frac{\Delta f}{f_a}\left(\frac{C02}{C2}\right) = 0.085996$$

$$\frac{C02}{C2} = 0.085996\,\frac{40,654,280}{8000}$$

$$= 0.085996 \times 5.082 \times 10^3$$

$$\frac{C02}{C2} = 435$$

This figure is easily obtained by adding a small capacitor in parallel with the crystal.

The next elements to be calculated are the coupling capacitors:

$$\Delta f f_a(C12) = 0.0194933$$

$$C12 = \frac{0.0194933}{\Delta f f_a} = \frac{0.0194933}{8000 \times 40,654,280}$$

$$C12 = \frac{0.0194933}{325.6 \times 10^9} = 0.05990 \times 10^{-12} \text{ farad}$$

In the same manner

$$C23 = 0.1065 \times 10^{-12} \text{ farad}$$

Next, the required shunt capacitors are calculated:

$$\Delta f f_a(C01) = 0.0022897$$

$$C01 = \frac{0.0022897}{325.6 \times 10^9} = 0.007042 \times 10^{-12} \text{ farad}$$

$$C02 = 0.006694 \times 10^{-12} \text{ farad}$$

$$C03 = 0.006963 \times 10^{-12} \text{ farad}$$

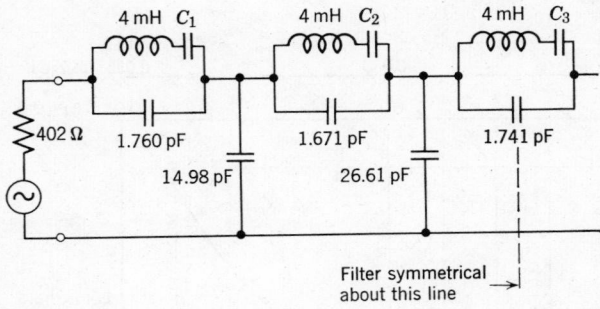

Fig. 8.50. Schematic of five-crystal filter showing element values calculated in text example.

Next, it is necessary to impedance scale all of the foregoing element values such that the crystal inductance is 4.0 mH. All capacitors are to be multiplied by 250, and all inductors and resistors are to be divided by 250. The resulting schematic is shown in Fig. 8.50. If desired, a computer-loop-equation response can be run for the schematic, after calculating

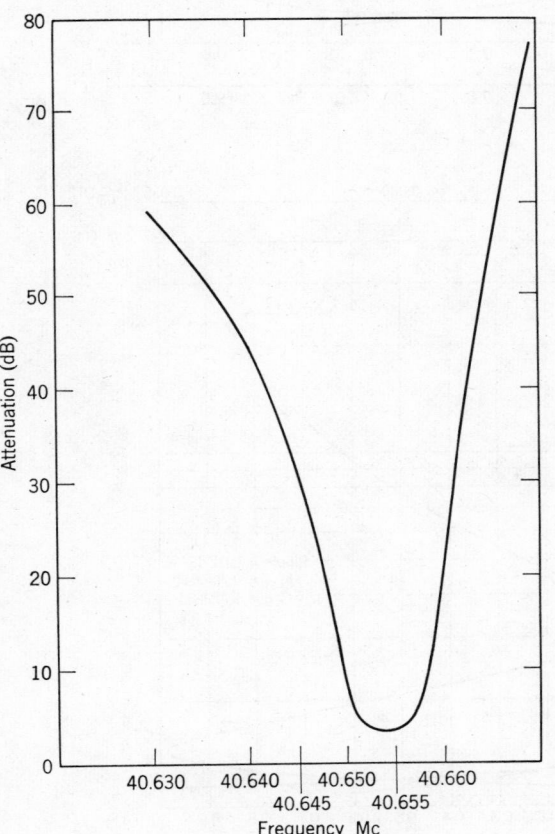

Fig. 8.51. Computed response of five-crystal filter of text example.

the values of the series crystal capacitances from

$$C1,5 = [4\pi^2(f_{s1,5})^2 \times 4 \times 10^{-3}]^{-1}$$
$$C2,4 = [4\pi^2(f_{s2,4})^2 \times 4 \times 10^{-3}]^{-1}, \text{ etc.}$$

Figure 8.51 is the calculated response of the filter assuming a crystal Q of 40,000 and that the coupling capacitors $Q = 2000$. Owing to losses, the 3-dB bandwidth is slightly narrower than the required bandwidth. Inspection of the response shows it to be 7.1 kc instead of the calculated value of 8.0 kc. The slope of the attenuation skirt is such that the ultimate attenuation will be 100 dB.

Design Curves and Tables

The following design information is presented in the form of curves and tables:

Crystal	Ladder	Parameters
n	Ripple, dB	Chart
	0.00	8.1
	0.01	8.2
	0.05	8.3
2	0.10	8.4
	0.25	8.5
	0.50	8.6
	1.00	8.7
	0.00	8.8
	0.01	8.9
	0.05	8.10
3	0.10	8.11
	0.25	8.12
	0.50	8.13
	1.00	8.14
		Table
	0.00	8.6
	0.01	8.7
	0.05	8.8
4	0.10	8.9
	0.25	8.10
	0.50	8.11
	1.00	8.12
	0.00	8.13
	0.01	8.14
	0.05	8.15
5	0.10	8.16
	0.25	8.17
	0.50	8.18
	1.00	8.19

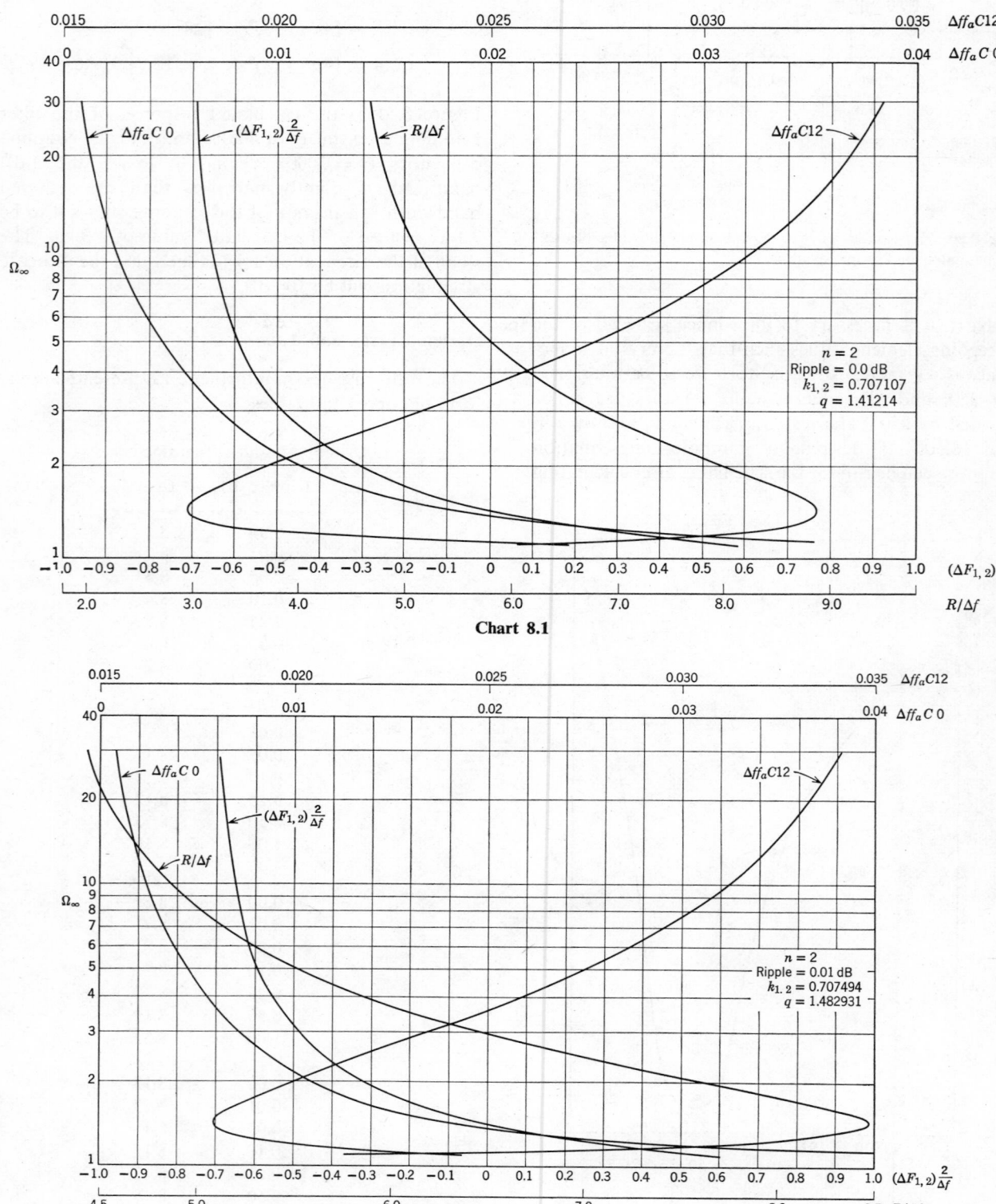

Chart 8.1

$n = 2$
Ripple = 0.0 dB
$k_{1,2} = 0.707107$
$q = 1.41214$

Chart 8.2

$n = 2$
Ripple = 0.01 dB
$k_{1,2} = 0.707494$
$q = 1.482931$

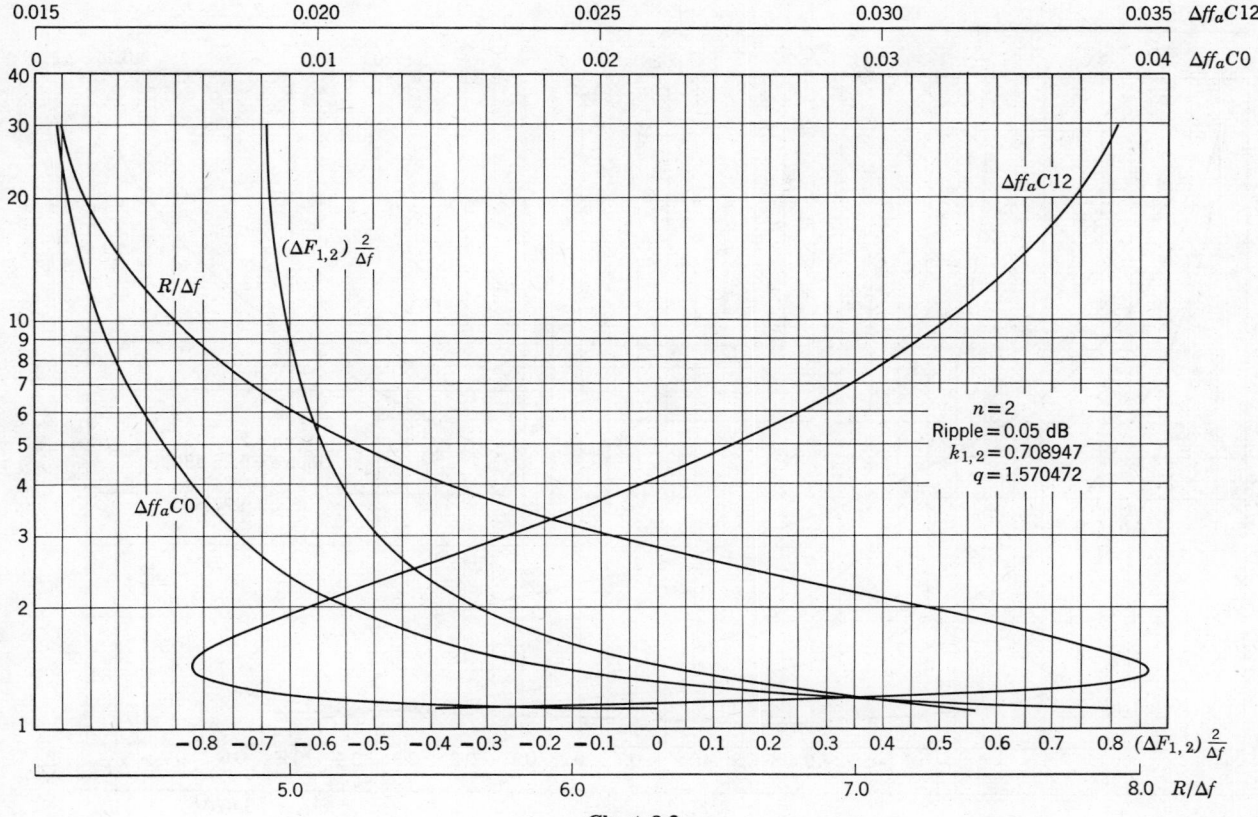

Chart 8.3

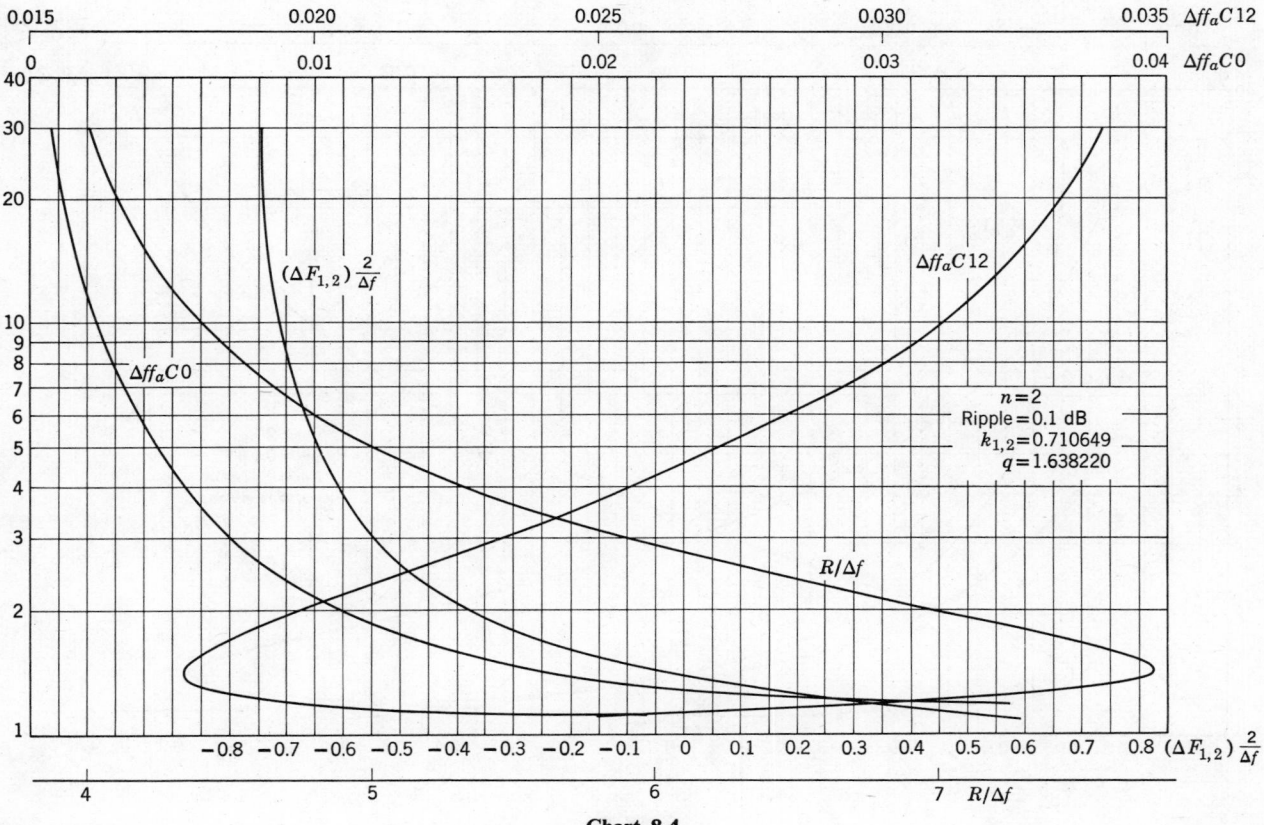

Chart 8.4

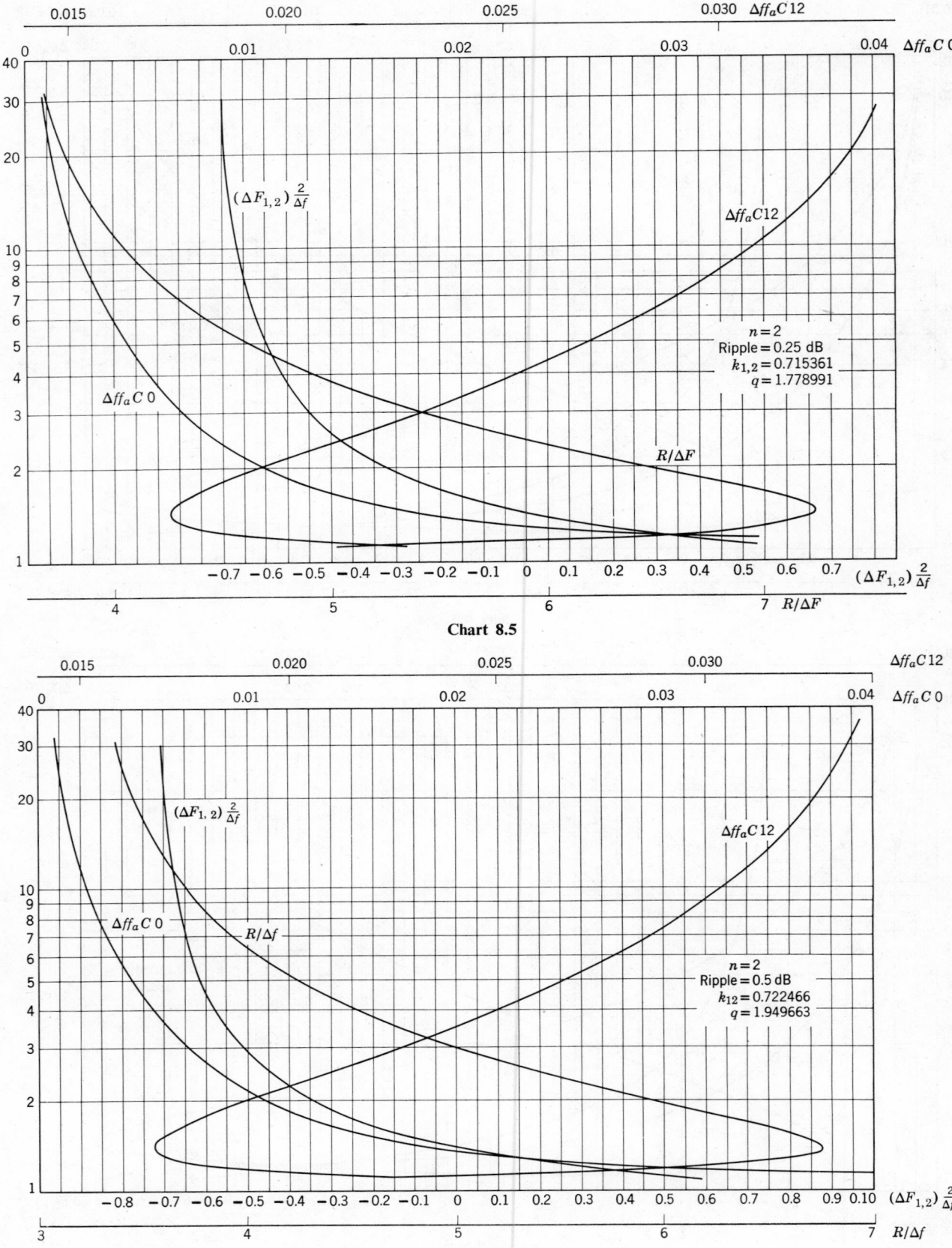

Chart 8.5

n = 2
Ripple = 0.25 dB
$k_{1,2} = 0.715361$
$q = 1.778991$

$(\Delta F_{1,2}) \frac{2}{\Delta f}$

$\Delta ff_a C\,12$

$\Delta ff_a C\,0$

$R/\Delta F$

Chart 8.6

n = 2
Ripple = 0.5 dB
$k_{12} = 0.722466$
$q = 1.949663$

$(\Delta F_{1,2}) \frac{2}{\Delta f}$

$\Delta ff_a C\,12$

$\Delta ff_a C\,0$

$R/\Delta f$

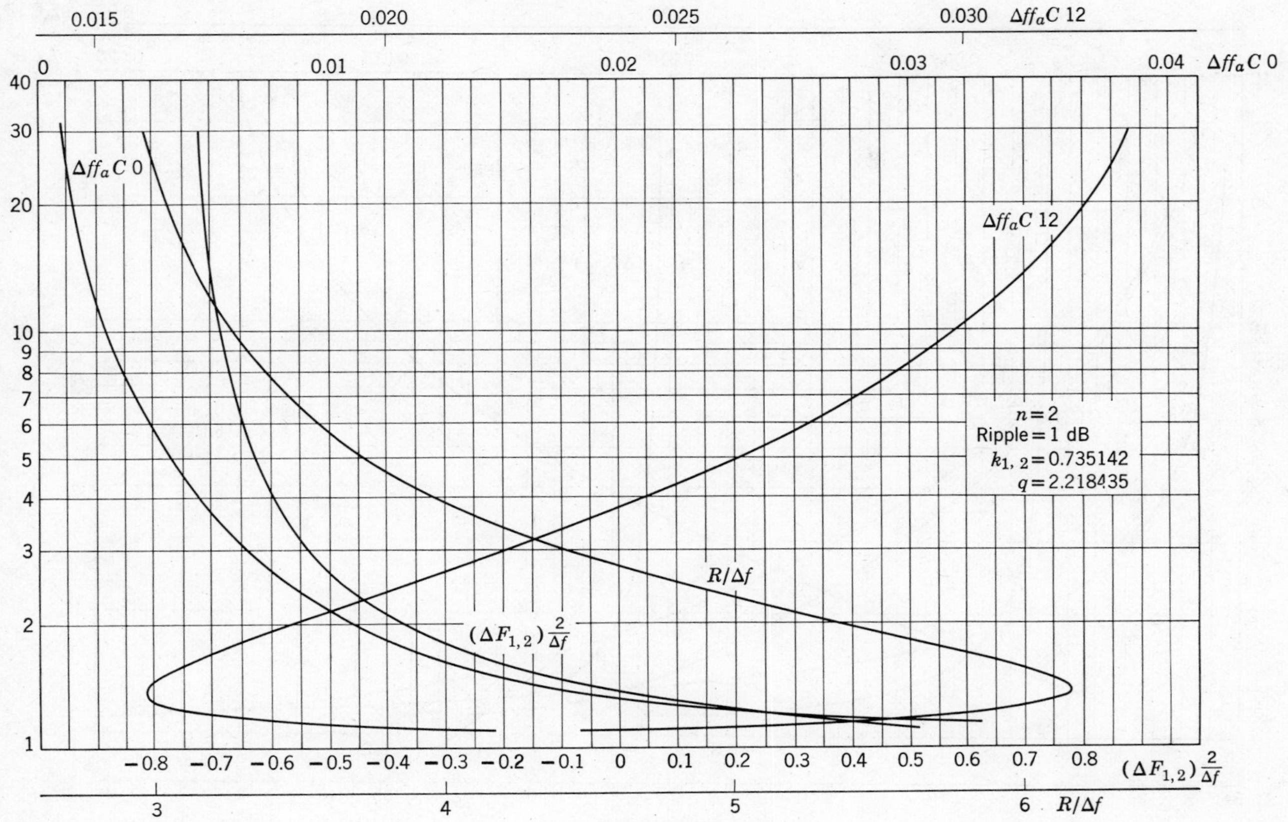

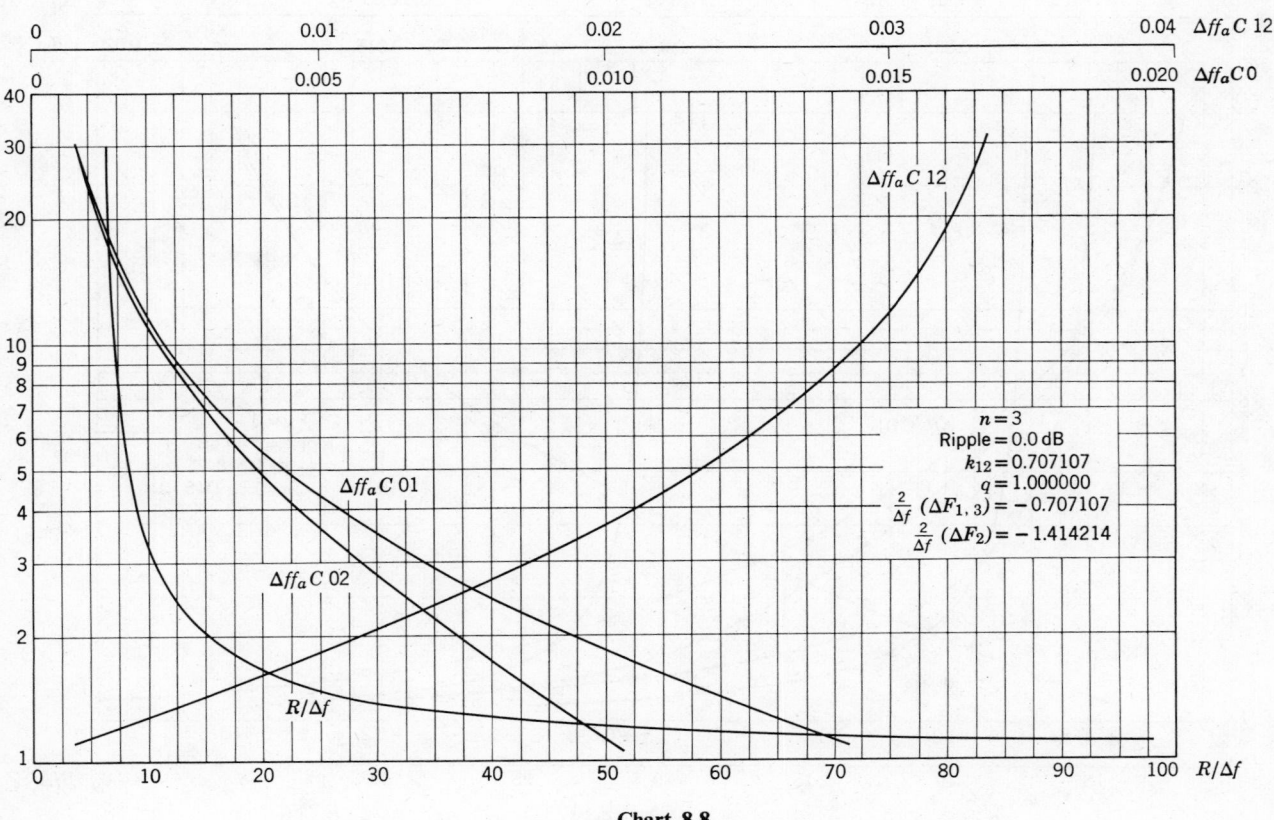

0.015 0.020 0.025 0.030 $\Delta ff_a C$ 12
0 0.01 0.02 0.03 0.04 $\Delta ff_a C$ 0
40 30 20 10 9 8 7 6 5 4 3 2 1
$\Delta ff_a C$ 0
$\Delta ff_a C$ 12
$n=2$
Ripple = 1 dB
$k_{1,2}=0.735142$
$q=2.218435$
$R/\Delta f$
$(\Delta F_{1,2})\frac{2}{\Delta f}$
−0.8 −0.7 −0.6 −0.5 −0.4 −0.3 −0.2 −0.1 0 0.1 0.2 0.3 0.4 0.5 0.6 0.7 0.8 $(\Delta F_{1,2})\frac{2}{\Delta f}$
3 4 5 6 $R/\Delta f$

Chart 8.7

0 0.01 0.02 0.03 0.04 $\Delta ff_a C$ 12
0 0.005 0.010 0.015 0.020 $\Delta ff_a C$ 0
40 30 20 10 9 8 7 6 5 4 3 2 1
$\Delta ff_a C$ 12
$n=3$
Ripple = 0.0 dB
$k_{12}=0.707107$
$q=1.000000$
$\frac{2}{\Delta f}(\Delta F_{1,3})=-0.707107$
$\frac{2}{\Delta f}(\Delta F_2)=-1.414214$
$\Delta ff_a C$ 01
$\Delta ff_a C$ 02
$R/\Delta f$
0 10 20 30 40 50 60 70 80 90 100 $R/\Delta f$

Chart 8.8

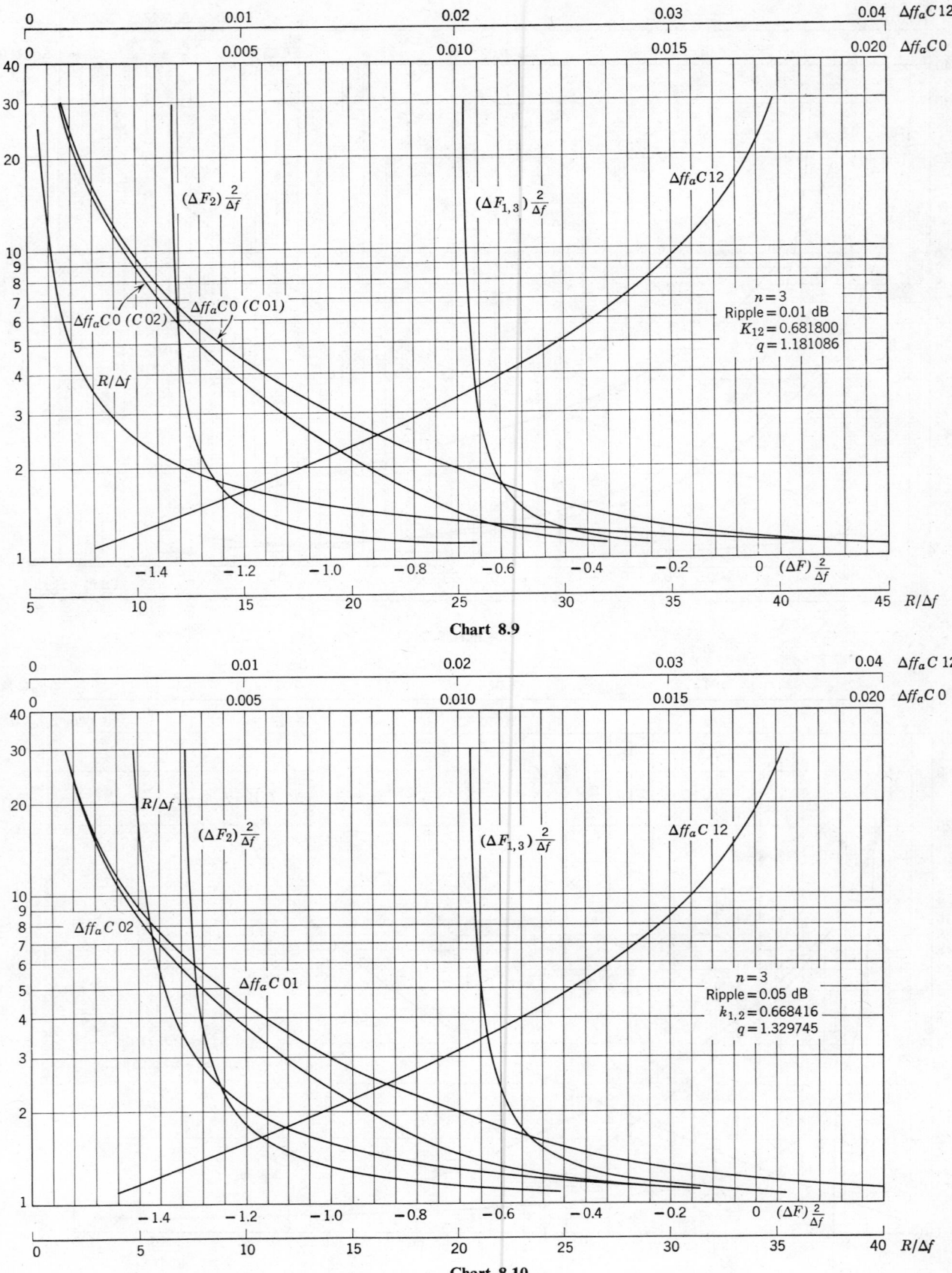

Chart 8.9

Chart 8.10

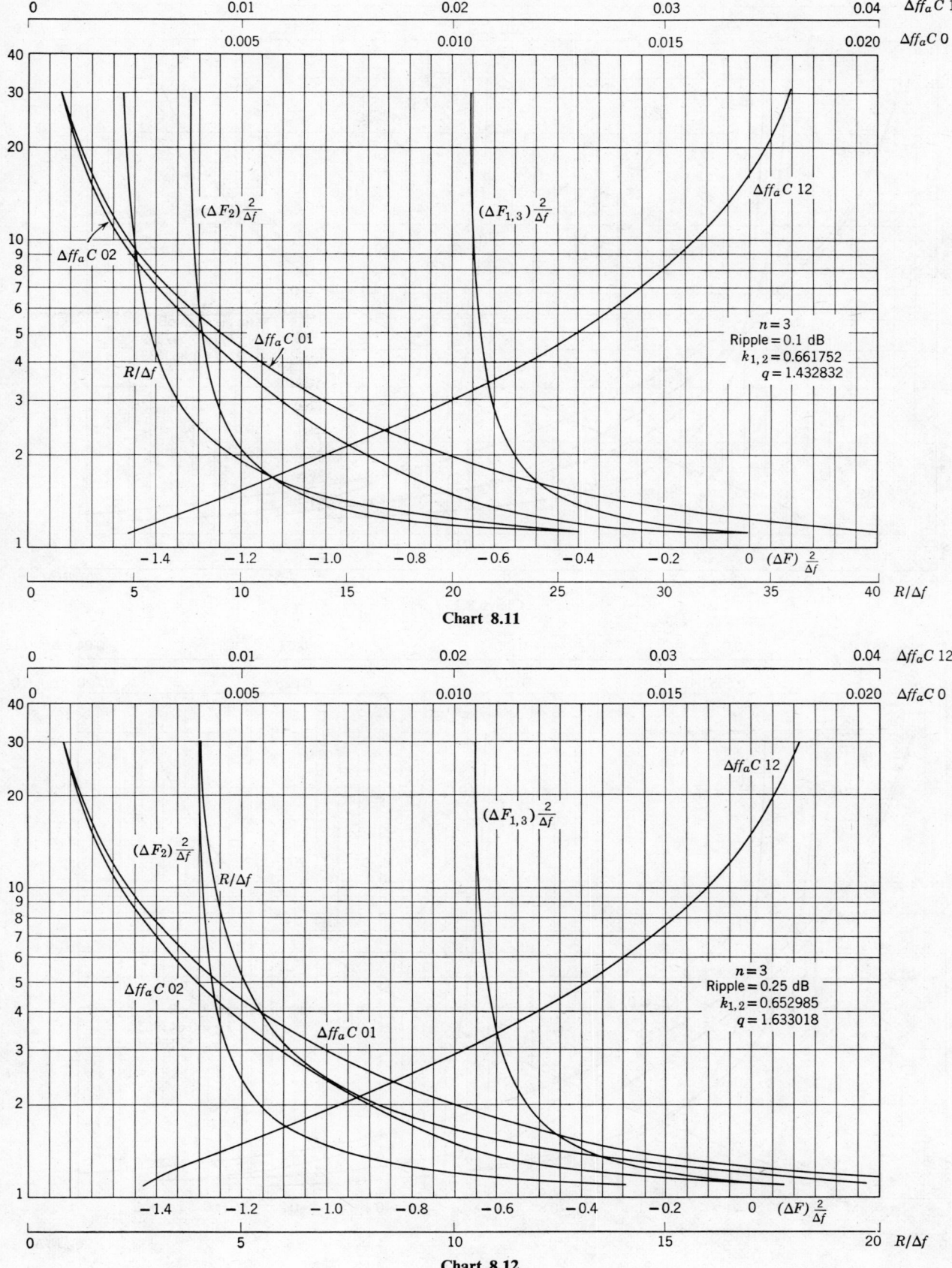

Chart 8.11

Chart 8.12

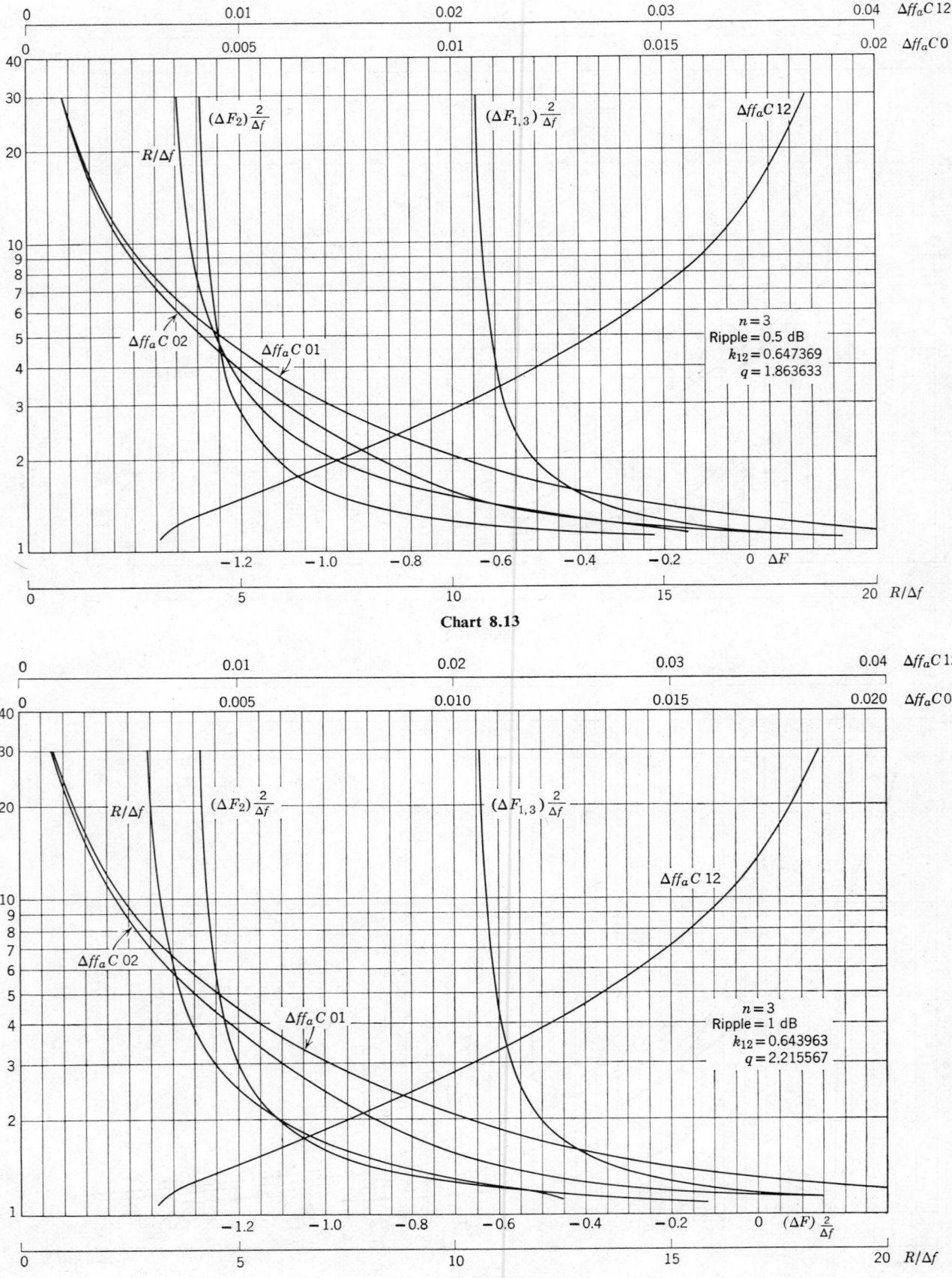

Chart 8.13

$n = 3$
Ripple $= 0.5$ dB
$k_{12} = 0.647369$
$q = 1.863633$

Chart 8.14

$n = 3$
Ripple $= 1$ dB
$k_{12} = 0.643963$
$q = 2.215567$

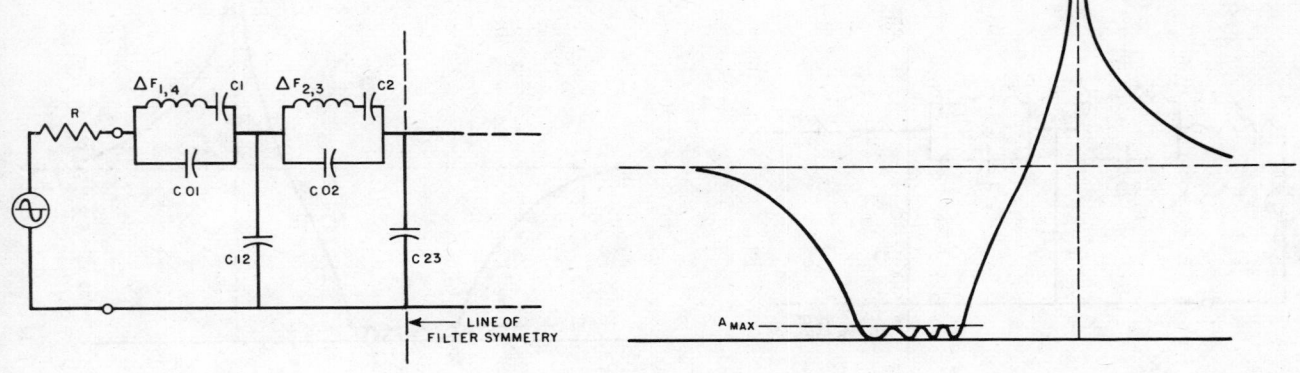

XTL LADDER PARAMETERS FOR N=4 ,L=1

A_{MAX} = .0000   K12= .840896   K23= .541196   Q= .765367

Ω_∞	$\frac{2}{\Delta f}(\Delta F_{1,4})$	$\frac{2}{\Delta f}(\Delta F_{2,3})$	$R/\Delta f$	$\Delta f(C0I)/f_0(C1)$	$\Delta f(C02)/f_0(C2)$	$\Delta f f_0(C12)$	$\Delta f f_0(C23)$	$\Delta f f_0(C0I)$	$\Delta f f_0(C02)$
1.100	4.280362	5.510054	395.406600	-.314430	-.226755	.0004510	.0005054	-.0079646	-.0057438
1.150	68.109105	93.105732	114129.992000	-.014934	-.010875	.0000016	.0000018	-.0003783	-.0002755
1.200	-6.706601	-9.559850	1166.372200	.126477	.092938	.0001558	.0001779	.0032037	.0023541
1.250	-3.622642	-5.324298	346.510770	.205227	.152107	.0005289	.0006091	.0051985	.0038529
1.300	-2.651457	-3.986278	185.770060	.253071	.189097	.0009947	.0011548	.0064104	.0047899
1.350	-2.176389	-3.333246	124.117981	.283576	.213527	.0015002	.0017552	.0071831	.0054087
1.400	-1.894933	-2.944062	92.839372	.303496	.230199	.0020203	.0023810	.0076877	.0058310
1.450	-1.708956	-2.686046	74.305137	.316560	.241777	.0025418	.0030164	.0080186	.0061243
1.500	-1.577046	-2.502358	62.182505	.324987	.249853	.0030574	.0036523	.0082320	.0063288
1.600	-1.402628	-2.258032	47.444818	.333042	.259199	.0040565	.0049054	.0084360	.0065656
1.800	-1.217391	-1.995121	33.367631	.331412	.263496	.0058923	.0072792	.0083948	.0066744
2.000	-1.121058	-1.855546	26.655850	.320404	.259367	.0075099	.0094457	.0081159	.0065698
2.250	-1.051192	-1.751899	22.022087	.302921	.249881	.0092631	.0118726	.0076731	.0063296
2.500	-1.008059	-1.686184	19.243526	.285058	.238681	.0107689	.0140219	.0072206	.0060509
3.000	-.958074	-1.607252	16.076372	.252648	.217049	.0132148	.0176397	.0063997	.0054979
3.500	-.930296	-1.561309	14.322644	.225719	.197577	.0151131	.0205548	.0057175	.0050047
4.000	-.912780	-1.531143	13.209109	.203551	.180794	.0166283	.0229481	.0051560	.0045796
4.500	-.900799	-1.509773	12.439289	.185158	.166396	.0178654	.0249460	.0046901	.0042149
5.000	-.892126	-1.493819	11.875257	.169718	.153993	.0188946	.0266377	.0042990	.0039007
6.000	-.880467	-1.471556	11.103965	.145339	.133841	.0205086	.0293448	.0036815	.0033902
7.000	-.873034	-1.456737	10.601157	.127016	.118249	.0217167	.0314139	.0032173	.0029953
8.000	-.867904	-1.446151	10.247352	.112766	.105863	.0226549	.0330458	.0028564	.0026815
10.000	-.861315	-1.432021	9.782274	.092070	.087474	.0240175	.0354548	.0023322	.0022157
12.000	-.857280	-1.423010	9.490135	.077777	.074499	.0249594	.0371468	.0019701	.0018871
16.000	-.852614	-1.412169	9.143334	.059338	.057431	.0261769	.0393659	.0015030	.0014547
20.000	-.850006	-1.405874	8.944369	.047962	.046716	.0269297	.0407561	.0012149	.0011833
25.000	-.848018	-1.400944	8.789823	.038688	.037877	.0275445	.0419017	.0009800	.0009594
30.000	-.846741	-1.397708	8.688986	.032418	.031849	.0279608	.0426825	.0008212	.0008068

Table 8.6

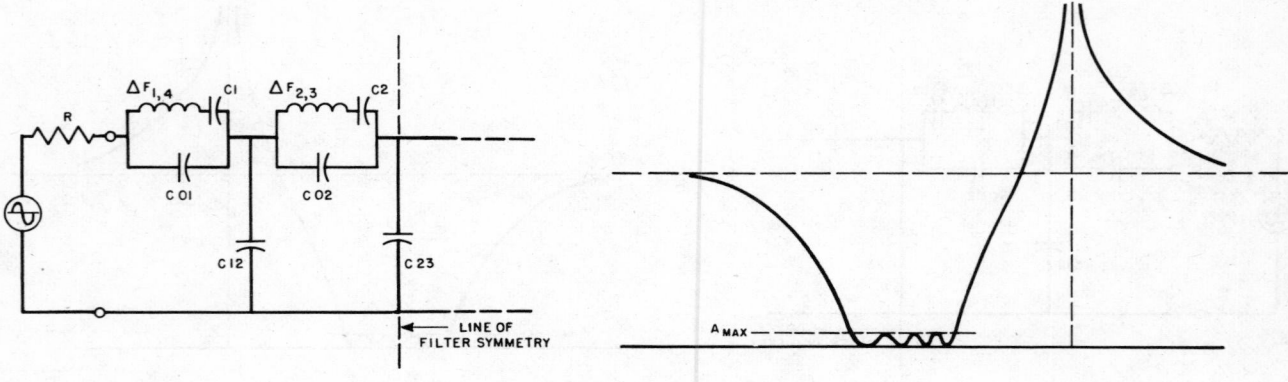

LINE OF
FILTER SYMMETRY

A MAX

XTL LADDER PARAMETERS FOR N=4 ,L=1

A_{MAX} = .0100   K12= .736952   K23= .541310 ω= 1.045709

Ω_∞	$\frac{2}{\Delta f}(\Delta F_{1,4})$	$\frac{2}{\Delta f}(\Delta F_{2,3})$	$R/\Delta f$	$\Delta f(C01)/f_a(C1)$	$\Delta f(C02)/f_a(C2)$	$\Delta ff_a(C12)$	$\Delta ff_a(C23)$	$\Delta ff_a(C01)$	$\Delta ff_a(C02)$
1.100	-2.707504	-4.298254	414.792140	.262639	.185245	.0003512	.0003372	.0066527	.0046923
1.150	-1.615994	-2.728666	142.541640	.361534	.257821	.0010332	.0010031	.0091578	.0065307
1.200	-1.282856	-2.245849	84.182018	.402762	.290204	.0017677	.0017340	.0102021	.0073510
1.250	-1.122404	-2.010732	60.120662	.421513	.306680	.0024993	.0024756	.0106771	.0077683
1.300	-1.028514	-1.871261	47.214761	.429458	.315332	.0032117	.0032105	.0108783	.0079875
1.350	-.967199	-1.778720	39.224701	.431556	.319619	.0038995	.0039318	.0109314	.0080961
1.400	-.924219	-1.712694	33.810534	.430252	.321265	.0045610	.0046365	.0108984	.0081377
1.450	-.892565	-1.663118	29.907070	.426882	.321221	.0051963	.0053233	.0108131	.0081366
1.500	-.868385	-1.624458	26.962703	.422229	.320056	.0058061	.0059918	.0106952	.0081071
1.600	-.834124	-1.567909	22.820741	.410825	.315666	.0069536	.0072740	.0104063	.0079959
1.800	-.795139	-1.499060	18.065178	.385336	.303117	.0089929	.0096308	.0097607	.0076780
2.000	-.774283	-1.458250	15.415201	.360454	.289164	.0107477	.0117382	.0091304	.0073246
2.250	-.759408	-1.425500	13.394851	.332291	.272072	.0126240	.0140720	.0084170	.0068917
2.500	-.750652	-1.403413	12.093461	.307631	.256186	.0142215	.0161237	.0077924	.0064893
3.000	-.741438	-1.375201	10.513715	.267277	.228561	.0167979	.0195564	.0067702	.0057895
3.500	-.737122	-1.357743	9.588670	.236009	.205657	.0187866	.0223089	.0059782	.0052144
4.000	-.734896	-1.345779	8.980462	.211198	.187063	.0203690	.0245620	.0053497	.0047384
4.500	-.733691	-1.337027	8.549771	.191070	.171320	.0216587	.0264388	.0048399	.0043396
5.000	-.733030	-1.330328	8.228605	.174428	.157970	.0227301	.0280255	.0044183	.0040014
6.000	-.732504	-1.320719	7.781336	.148533	.136599	.0244084	.0305601	.0037624	.0034601
7.000	-.732446	-1.314139	7.484479	.129325	.120277	.0256630	.0324938	.0032758	.0030467
8.000	-.732568	-1.309341	7.272980	.114514	.107419	.0266367	.0340171	.0029007	.0027210
10.000	-.732963	-1.302800	6.991543	.093171	.088474	.0280498	.0362625	.0023600	.0022411
12.000	-.733366	-1.298543	6.812703	.078534	.075196	.0290262	.0378375	.0019893	.0019047
16.000	-.734022	-1.293327	6.598261	.059758	.057826	.0302874	.0399004	.0015137	.0014647
20.000	-.734495	-1.290248	6.474154	.048229	.046970	.0310670	.0411913	.0012216	.0011898
25.000	-.734915	-1.287810	6.377198	.038858	.038040	.0317036	.0422542	.0009843	.0009636
30.000	-.735215	-1.286197	6.313669	.032536	.031963	.0321345	.0429782	.0008241	.0008096

Table 8.7

470

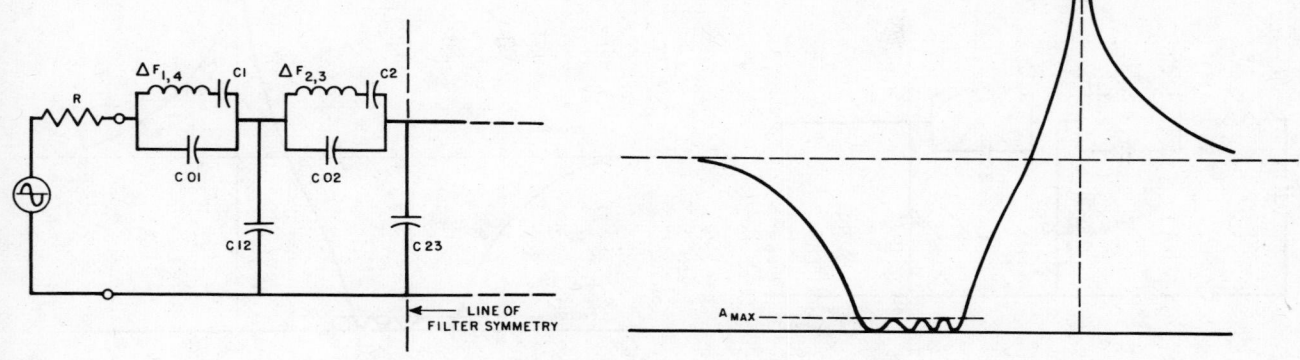

A_{MAX} = .0500   K12= .701463   K23= .541685   G= 1.225691

Ω_∞	$\frac{2}{\Delta f}(\Delta F_{1,4})$	$\frac{2}{\Delta f}(\Delta F_{2,3})$	$R/\Delta f$	$\Delta f(CO1)/f_a(C1)$	$\Delta f(CO2)/f_a(C2)$	$\Delta f f_a(C12)$	$\Delta f f_a(C23)$	$\Delta f f_a(CO1)$	$\Delta f f_a(CO2)$
1.100	-1.066747	-1.998414	114.602941	.461521	.322746	.0011296	.0010229	.0116905	.0081752
1.150	-.885553	-1.727397	65.861861	.491267	.347536	.0019883	.0018215	.0124439	.0088032
1.200	-.808207	-1.608202	46.985426	.497957	.356100	.0028174	.0026091	.0126134	.0090201
1.250	-.766046	-1.540679	37.040503	.496021	.358336	.0036103	.0033775	.0125643	.0090768
1.300	-.739952	-1.496923	30.916438	.490208	.357536	.0043670	.0041246	.0124171	.0090565
1.350	-.722501	-1.466081	26.770234	.482509	.355103	.0050890	.0048500	.0122221	.0089949
1.400	-.710209	-1.443049	23.778176	.473887	.351735	.0057783	.0055539	.0120037	.0089096
1.450	-.701229	-1.425111	21.517589	.464850	.347813	.0064368	.0062368	.0117748	.0088102
1.500	-.694491	-1.410684	19.749507	.455687	.343562	.0070667	.0068994	.0115427	.0087025
1.600	-.685322	-1.388776	17.162038	.437575	.334585	.0082474	.0081664	.0110839	.0084751
1.800	-.676089	-1.360483	14.030829	.403863	.316407	.0103363	.0104866	.0102300	.0080147
2.000	-.672352	-1.342593	12.202952	.374202	.299169	.0121277	.0125559	.0094787	.0075780
2.250	-.670858	-1.327427	10.765305	.342365	.279531	.0140394	.0148438	.0086722	.0070806
2.500	-.670905	-1.316684	9.817610	.315367	.262008	.0156648	.0168531	.0079883	.0066367
3.000	-.672614	-1.302215	8.642899	.272286	.232438	.0182835	.0202115	.0068971	.0058877
3.500	-.674841	-1.292760	7.941929	.239530	.208648	.0203031	.0229021	.0060674	.0052851
4.000	-.676987	-1.286022	7.475492	.213813	.189178	.0219095	.0251031	.0054159	.0047919
4.500	-.678915	-1.280948	7.142424	.193091	.172982	.0232182	.0269356	.0048910	.0043817
5.000	-.680613	-1.276976	6.892517	.176037	.159312	.0243053	.0284842	.0044591	.0040354
6.000	-.683405	-1.271137	6.542237	.149624	.137530	.0260078	.0309569	.0037900	.0034837
7.000	-.685574	-1.267039	6.308268	.130114	.120962	.0272804	.0328424	.0032958	.0030640
8.000	-.687294	-1.263997	6.140831	.115111	.107945	.0282679	.0343271	.0029158	.0027343
10.000	-.689831	-1.259776	5.917050	.093547	.088812	.0297009	.0365148	.0023696	.0022496
12.000	-.691607	-1.256983	5.774254	.078792	.075432	.0306910	.0380488	.0019958	.0019107
16.000	-.693918	-1.253507	5.602413	.059902	.057959	.0319698	.0400571	.0015173	.0014681
20.000	-.695354	-1.251430	5.502646	.048320	.047056	.0327603	.0413133	.0012240	.0011919
25.000	-.696529	-1.249770	5.424542	.038916	.038096	.0334057	.0423475	.0009857	.0009650
30.000	-.697326	-1.248665	5.373288	.032576	.032001	.0338426	.0430517	.0008252	.0008106

Table 8.8

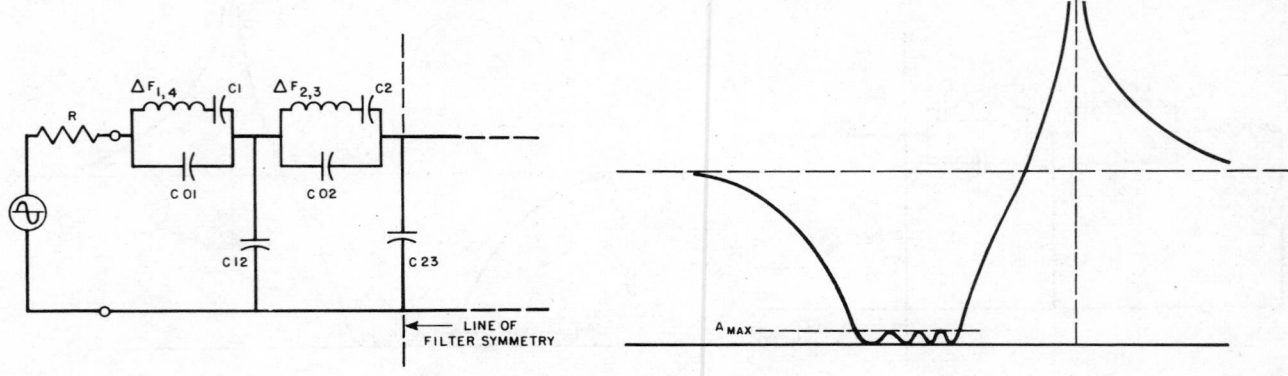

XTL LADDER PARAMETERS FOR N=4 ,L=1

A_{MAX} = .1000   K12= .684979   K23= .542090   Q= 1.345069

Ω_∞	$\frac{2}{\Delta f}(\Delta F_{1,4})$	$\frac{2}{\Delta f}(\Delta F_{2,3})$	$R/\Delta f$	$\Delta f(CO1)/f_a(C1)$	$\Delta f(CO2)/f_a(C2)$	$\Delta f f_a(C12)$	$\Delta f f_a(C23)$	$\Delta f f_a(CO1)$	$\Delta f f_a(CO2)$
1.100	-.706223	-1.494014	72.570257	.553641	.385503	.0016574	.0014583	.0140239	.0097649
1.150	-.663647	-1.424084	47.644294	.551375	.388488	.0025546	.0022743	.0139665	.0098405
1.200	-.644771	-1.389459	36.129961	.542073	.386181	.0034062	.0030662	.0137309	.0097821
1.250	-.634815	-1.368356	29.501921	.530556	.381919	.0042149	.0038338	.0134391	.0096741
1.300	-.629123	-1.353890	25.194501	.518370	.376805	.0049839	.0045777	.0131305	.0095446
1.350	-.625764	-1.343194	22.170196	.506133	.371306	.0057160	.0052987	.0128205	.0094053
1.400	-.623801	-1.334859	19.929639	.494120	.365650	.0064140	.0059975	.0125162	.0092620
1.450	-.622722	-1.328108	18.202812	.482457	.359957	.0070803	.0066750	.0122208	.0091178
1.500	-.622227	-1.322481	16.830956	.471203	.354298	.0077170	.0073319	.0119357	.0089745
1.600	-.622311	-1.313516	14.788371	.449982	.343228	.0089097	.0085873	.0113982	.0086941
1.800	-.624497	-1.300962	12.258304	.412457	.322481	.0110177	.0108849	.0104477	.0081685
2.000	-.627567	-1.292268	10.750327	.380580	.303742	.0128243	.0129330	.0096402	.0076939
2.250	-.631524	-1.284311	9.547427	.347039	.282941	.0147513	.0151968	.0087906	.0071670
2.500	-.635221	-1.278296	8.746057	.318957	.264670	.0163892	.0171845	.0080793	.0067042
3.000	-.641537	-1.269635	7.743096	.274609	.234212	.0190273	.0205058	.0069559	.0059327
3.500	-.646563	-1.263604	7.139357	.241164	.209925	.0210616	.0231659	.0061087	.0053175
4.000	-.650597	-1.259120	6.735368	.215026	.190146	.0226794	.0253416	.0054467	.0048165
4.500	-.653888	-1.255641	6.445763	.194028	.173743	.0239974	.0271526	.0049148	.0044010
5.000	-.656615	-1.252855	6.227836	.176784	.159927	.0250922	.0286829	.0044780	.0040510
6.000	-.660864	-1.248664	5.921452	.150131	.137956	.0268066	.0311258	.0038029	.0034945
7.000	-.664014	-1.245654	5.716189	.130480	.121276	.0280881	.0329883	.0033051	.0030720
8.000	-.666441	-1.243385	5.568986	.115388	.108185	.0290825	.0344546	.0029228	.0027404
10.000	-.669929	-1.240188	5.371840	.093721	.088966	.0305255	.0366148	.0023740	.0022535
12.000	-.672315	-1.238041	5.245792	.078912	.075540	.0315224	.0381291	.0019989	.0019134
16.000	-.675365	-1.235337	5.093845	.059969	.058020	.0328101	.0401115	.0015190	.0014697
20.000	-.677232	-1.233703	5.005495	.048362	.047095	.0336061	.0413514	.0012250	.0011929
25.000	-.678746	-1.232389	4.936262	.038943	.038121	.0342559	.0423718	.0009864	.0009656
30.000	-.679764	-1.231510	4.890796	.032595	.032019	.0346959	.0430668	.0008256	.0008110

Table 8.9

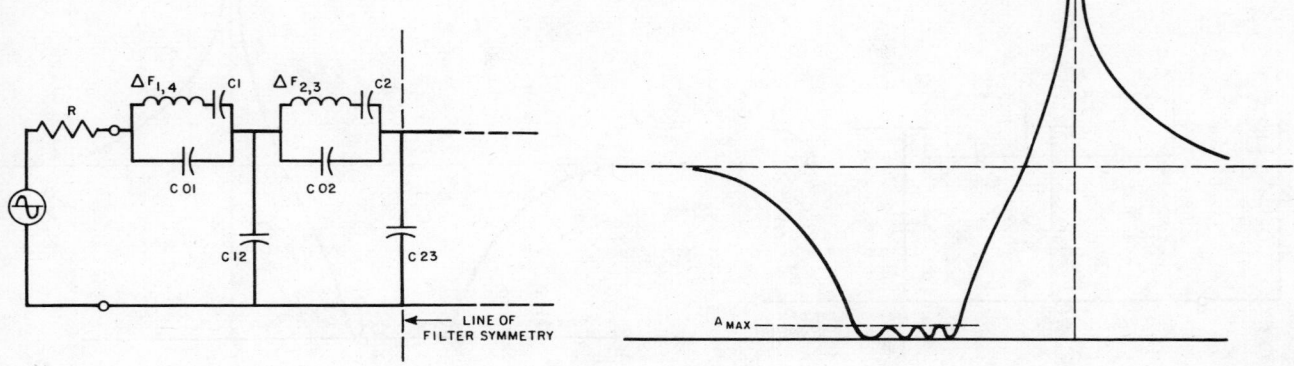

XTL LADDER PARAMETERS FOR N=4 ,L=1

A_{MAX} = .2500   K12= .663344   K23= .543136   Q= 1.570832

Ω_∞	$\frac{2}{\Delta f}(\Delta F_{1,4})$	$\frac{2}{\Delta f}(\Delta F_{2,3})$	$R/\Delta f$	$\Delta f(CO1)/f_0(C1)$	$\Delta f(CO2)/f_0(C2)$	$\Delta ff_0(C12)$	$\Delta ff_0(C23)$	$\Delta ff_0(CO1)$	$\Delta ff_0(CO2)$
1.100	-.385255	-1.046367	42.017743	.673285	.465904	.0025155	.0021259	.0170545	.0118015
1.150	-.438571	-1.117837	31.299237	.629497	.440949	.0034183	.0029244	.0159453	.0111694
1.200	-.468215	-1.154521	25.298887	.599443	.424715	.0042776	.0037015	.0151641	.0107582
1.250	-.487648	-1.176433	21.470957	.575490	.412128	.0050944	.0044557	.0145773	.0104393
1.300	-.501705	-1.190738	18.817787	.555030	.401487	.0058714	.0051871	.0140591	.0101698
1.350	-.512553	-1.200632	16.870648	.536897	.392060	.0066112	.0058962	.0135998	.0099310
1.400	-.521314	-1.207755	15.380082	.520477	.383472	.0073166	.0065837	.0131838	.0097134
1.450	-.528626	-1.213034	14.203623	.505401	.375512	.0079899	.0072503	.0128020	.0095118
1.500	-.534884	-1.217029	13.250107	.491429	.368049	.0086333	.0078968	.0124480	.0093228
1.600	-.545177	-1.222492	11.799185	.466162	.354297	.0098385	.0091325	.0118080	.0089744
1.800	-.560312	-1.227931	9.948116	.423673	.330258	.0119683	.0113943	.0107318	.0083655
2.000	-.571297	-1.229989	8.815226	.388909	.309599	.0137933	.0134107	.0098512	.0078422
2.250	-.581702	-1.230589	7.894493	.353145	.287308	.0157396	.0156394	.0089453	.0072776
2.500	-.589775	-1.230246	7.273518	.323648	.268079	.0173939	.0175961	.0081981	.0067905
3.000	-.601677	-1.228623	6.485894	.277648	.236484	.0200580	.0208653	.0070329	.0059902
3.500	-.610137	-1.226776	6.006345	.243301	.211561	.0221122	.0234830	.0061629	.0053589
4.000	-.616504	-1.225062	5.683100	.216614	.191385	.0237459	.0256236	.0054869	.0048478
4.500	-.621485	-1.223549	5.450185	.195256	.174717	.0250767	.0274051	.0049459	.0044256
5.000	-.625496	-1.222231	5.274249	.177762	.160714	.0261822	.0289102	.0045028	.0040709
6.000	-.631567	-1.220083	5.025907	.150794	.138503	.0279133	.0313123	.0038197	.0035083
7.000	-.635949	-1.218426	4.858867	.130959	.121678	.0292073	.0331433	.0033172	.0030821
8.000	-.639265	-1.217119	4.738744	.115751	.108494	.0302113	.0345845	.0029320	.0027482
10.000	-.643954	-1.215198	4.577422	.093950	.089165	.0316684	.0367074	.0023798	.0022586
12.000	-.647111	-1.213859	4.474009	.079069	.075678	.0326751	.0381952	.0020029	.0019169
16.000	-.651098	-1.212120	4.349063	.060056	.058099	.0339754	.0401424	.0015212	.0014717
20.000	-.653511	-1.211043	4.276269	.048418	.047145	.0347791	.0413600	.0012264	.0011942
25.000	-.655454	-1.210163	4.219150	.038978	.038153	.0354353	.0423621	.0009873	.0009664
30.000	-.656756	-1.209567	4.181604	.032619	.032041	.0358796	.0430443	.0008263	.0008116

Table 8.10

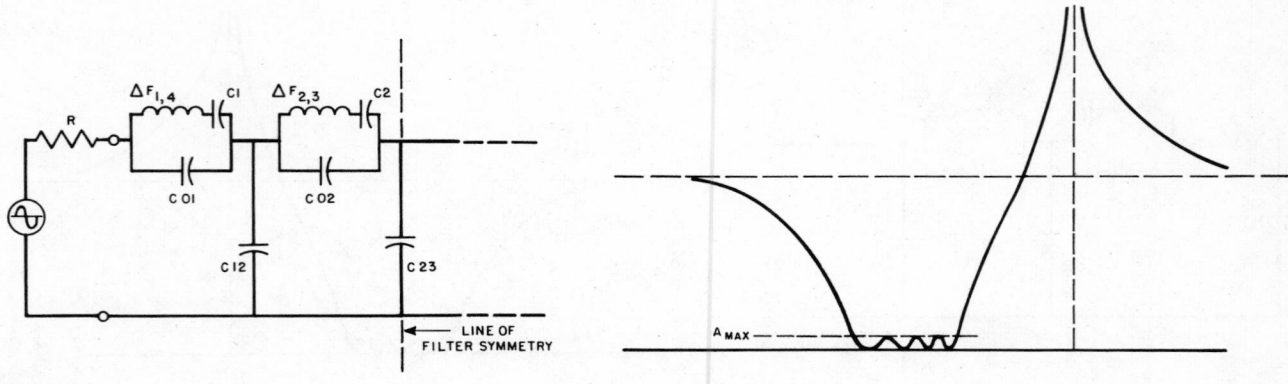

XTL LADDER PARAMETERS FOR N=4 ,L=1

A_{MAX} = .5000   K12= .648187   K23= .544604   Q= 1.825814

Ω_∞	$\frac{2}{\Delta f}(\Delta F_{1,4})$	$\frac{2}{\Delta f}(\Delta F_{2,3})$	$R/\Delta f$	$\Delta f(C01)/f_a(C1)$	$\Delta f(C02)/f_a(C2)$	$\Delta f f_a(C12)$	$\Delta f f_a(C23)$	$\Delta f f_a(C01)$	$\Delta f f_a(C02)$
1.100	-.224771	-.824210	28.759781	.754847	.519694	.0032193	.0026380	.0191205	.0131640
1.150	-.314470	-.950690	22.885179	.682841	.476034	.0040966	.0033991	.0172966	.0120581
1.200	-.365741	-1.019867	19.173899	.638675	.450477	.0049470	.0041530	.0161778	.0114107
1.250	-.399463	-1.063112	16.645084	.606258	.432318	.0057613	.0048898	.0153567	.0109507
1.300	-.423663	-1.092460	14.817635	.580160	.417980	.0065387	.0056069	.0146956	.0105876
1.350	-.442090	-1.113519	13.437140	.558008	.405923	.0072805	.0063035	.0141345	.0102822
1.400	-.456733	-1.129249	12.358079	.538580	.395374	.0079886	.0069799	.0136424	.0100149
1.450	-.468749	-1.141360	11.491614	.521173	.385898	.0086651	.0076363	.0132015	.0097749
1.500	-.478857	-1.150909	10.780576	.505342	.377229	.0093120	.0082733	.0128005	.0095553
1.600	-.495085	-1.164845	9.682825	.477308	.361684	.0105243	.0094917	.0120903	.0091616
1.800	-.517951	-1.181103	8.254369	.431415	.335446	.0126680	.0117234	.0109279	.0084970
2.000	-.533780	-1.189759	7.364442	.394667	.313503	.0145056	.0137140	.0099970	.0079411
2.250	-.548199	-1.195683	6.632649	.357373	.290218	.0164657	.0159148	.0090524	.0073513
2.500	-.559033	-1.198905	6.133837	.326901	.270350	.0181318	.0178473	.0082805	.0068441
3.000	-.574505	-1.201744	5.496234	.279759	.237996	.0208154	.0210761	.0070864	.0060285
3.500	-.585185	-1.202572	5.104981	.244787	.212650	.0228847	.0236615	.0062005	.0053865
4.000	-.593070	-1.202636	4.839920	.217719	.192210	.0245304	.0257753	.0055149	.0048687
4.500	-.599159	-1.202389	4.648251	.196111	.175365	.0258711	.0275345	.0049675	.0044420
5.000	-.604014	-1.202018	4.503087	.178444	.161238	.0269848	.0290205	.0045200	.0040842
6.000	-.611286	-1.201191	4.297611	.151256	.138866	.0287288	.0313920	.0038314	.0035175
7.000	-.616486	-1.200412	4.159023	.131294	.121945	.0300325	.0331994	.0033257	.0030889
8.000	-.620393	-1.199729	4.059166	.116004	.108699	.0310440	.0346219	.0029384	.0027534
10.000	-.625881	-1.198638	3.924004	.094110	.089297	.0325121	.0367168	.0023838	.0022619
12.000	-.629555	-1.197825	3.838516	.079179	.075770	.0335263	.0381849	.0020056	.0019193
16.000	-.634168	-1.196715	3.734094	.060117	.058151	.0348364	.0401060	.0015228	.0014730
20.000	-.636949	-1.196001	3.673172	.048457	.047179	.0356462	.0413071	.0012274	.0011951
25.000	-.639181	-1.195404	3.625325	.039003	.038175	.0363074	.0422955	.0009880	.0009670
30.000	-.640673	-1.194993	3.593852	.032636	.032056	.0367550	.0429685	.0008267	.0008120

Table 8.11

474

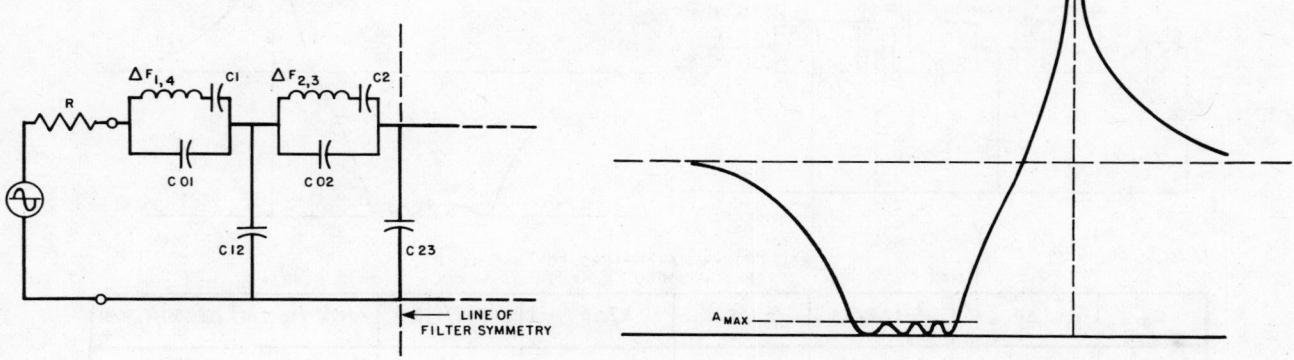

XTL LADDER PARAMETERS FOR N=4 , L=1

A_{MAX} = 1.0000   K12= .635329   K23= .547055 ω= 2.210305

Ω_∞	$\frac{2}{\Delta f}(\Delta F_{1,4})$	$\frac{2}{\Delta f}(\Delta F_{2,3})$	R/Δf	Δf(CO1)/f_a(C1)	Δf(CO2)/f_a(C2)	Δff_a(C12)	Δff_a(C23)	Δff_a(CO1)	Δff_a(CO2)
1.100	-.119656	-.681159	20.136452	.819903	.561432	.0038541	.0030649	.0207684	.0142212
1.150	-.228263	-.837164	16.744113	.725551	.503230	.0046947	.0037816	.0183784	.0127470
1.200	-.292113	-.925727	14.383956	.670191	.470427	.0055308	.0045087	.0169761	.0119161
1.250	-.334670	-.982446	12.690616	.631046	.447939	.0063393	.0052260	.0159846	.0113464
1.300	-.365391	-1.021645	11.426440	.600460	.430729	.0071151	.0059274	.0152098	.0109105
1.350	-.388826	-1.050197	10.449699	.575101	.416632	.0078573	.0066108	.0145675	.0105534
1.400	-.407437	-1.071909	9.673478	.553270	.404562	.0085671	.0072753	.0140145	.0102477
1.450	-.422676	-1.088655	9.042207	.533995	.393909	.0092460	.0079210	.0135263	.0099778
1.500	-.435457	-1.102093	8.518919	.516674	.384306	.0098957	.0085482	.0130875	.0097346
1.600	-.455866	-1.122034	7.701611	.486413	.367372	.0111142	.0097487	.0123210	.0093057
1.800	-.484295	-1.146082	6.621920	.437772	.339434	.0132707	.0119500	.0110689	.0085980
2.000	-.503678	-1.159555	5.939684	.399412	.316500	.0151202	.0139149	.0101172	.0080170
2.250	-.521086	-1.169401	5.373216	.360869	.292449	.0170937	.0160880	.0091409	.0074078
2.500	-.534000	-1.175264	4.984248	.329598	.272089	.0187714	.0179966	.0083488	.0068921
3.000	-.552194	-1.181428	4.483642	.281516	.239153	.0214739	.0211860	.0071309	.0060578
3.500	-.564583	-1.184255	4.174526	.246028	.213481	.0235580	.0237400	.0062320	.0054075
4.000	-.573649	-1.185652	3.964258	.218644	.192840	.0252155	.0258282	.0055383	.0048847
4.500	-.580601	-1.186356	3.811774	.196827	.175660	.0265658	.0275659	.0049857	.0044546
5.000	-.586117	-1.186697	3.696039	.179015	.161637	.0276875	.0290337	.0045345	.0040943
6.000	-.594337	-1.186764	3.531647	.151645	.139143	.0294442	.0313761	.0038412	.0035245
7.000	-.600182	-1.186745	3.420853	.131576	.122149	.0307573	.0331610	.0033329	.0030941
8.000	-.604559	-1.186533	3.340752	.116217	.108855	.0317762	.0345658	.0029438	.0027573
10.000	-.610683	-1.186667	3.232602	.094245	.089397	.0332549	.0366344	.0023872	.0022645
12.000	-.614769	-1.185650	3.163571	.079272	.075840	.0342765	.0380840	.0020080	.0019211
16.000	-.619883	-1.185015	3.079239	.060169	.058190	.0355962	.0399806	.0015241	.0014740
20.000	-.622959	-1.184575	3.030098	.048490	.047204	.0364119	.0411664	.0012283	.0011957
25.000	-.625423	-1.184191	2.991473	.039024	.038191	.0370780	.0421420	.0009885	.0009674
30.000	-.627068	-1.183919	2.966052	.032651	.032068	.0375289	.0428063	.0008271	.0008123

Table 8.12

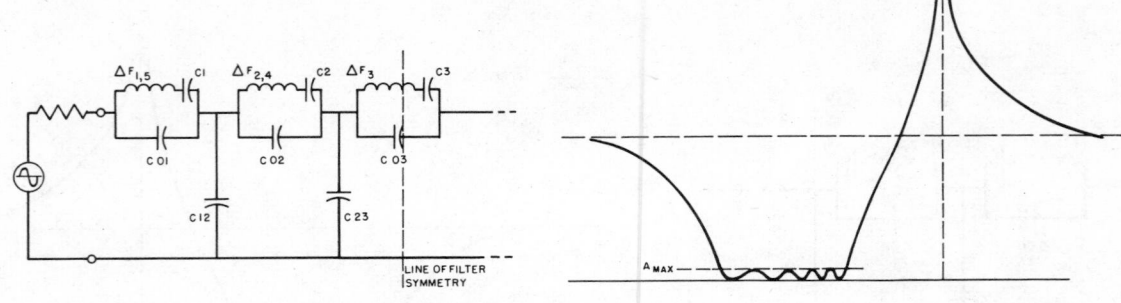

XTL LADDER PARAMETERS FOR N=5 ,L=1

A_{MAX} = 0.0000   K12= 1.000000   K23= 0.555893   Q= 0.618034

Ω_∞	$\frac{2}{\Delta f}(\Delta F_{1,5})$	$\frac{2}{\Delta f}(\Delta F_{2,4})$	$\frac{2}{\Delta f}(\Delta F_3)$	$R/\Delta f$	$\Delta f(C01)/f_a(C.1)$	$\Delta f(C02)/f_a(C2)$	$\Delta f(C03)/f_a(C3)$
30.000	-1.019253	-1.577602	-1.136918	10.881043	0.032238	0.031668	0.032116
26.000	-1.022352	-1.581146	-1.140936	10.997907	0.037006	0.036257	0.036845
22.000	-1.026640	-1.586073	-1.146485	11.160445	0.043428	0.042398	0.043203
20.000	-1.029467	-1.589334	-1.150136	11.268112	0.047552	0.046319	0.047281
18.000	-1.032966	-1.593383	-1.154647	11.401919	0.052540	0.051038	0.052207
16.000	-1.037409	-1.598546	-1.160363	11.572694	0.058694	0.056823	0.058274
14.000	-1.043239	-1.605353	-1.167844	11.798197	0.066475	0.064081	0.065929
12.000	-1.051228	-1.614735	-1.178056	12.109724	0.076621	0.073450	0.075884
10.000	-1.062849	-1.628487	-1.192839	12.568003	0.090393	0.085996	0.089343
8.000	-1.081321	-1.650566	-1.216170	13.308374	0.110116	0.103621	0.108505
7.000	-1.095356	-1.667493	-1.233771	13.880279	0.123528	0.115374	0.121451
6.000	-1.115286	-1.691714	-1.258590	14.705648	0.140542	0.130010	0.137768
5.000	-1.145836	-1.729179	-1.296281	15.999934	0.162712	0.148607	0.158824
4.500	-1.168171	-1.756771	-1.323598	16.967690	0.176424	0.159827	0.171715
4.000	-1.198661	-1.794645	-1.360601	18.317211	0.192357	0.172573	0.186546
3.500	-1.242797	-1.849791	-1.413664	20.327509	0.210846	0.186923	0.203514
3.000	-1.312441	-1.937322	-1.496415	23.633270	0.231887	0.202539	0.222399
2.500	-1.438825	-2.097063	-1.644380	30.042884	0.253883	0.217530	0.241291
2.250	-1.550568	-2.238777	-1.773582	36.146903	0.263119	0.222778	0.248535
2.000	-1.739259	-2.478519	-1.989577	47.382445	0.267433	0.223288	0.250663
1.800	-2.019235	-2.834623	-2.307060	66.202169	0.261833	0.215767	0.243483
1.600	-2.641568	-3.626508	-3.006724	117.245380	0.235762	0.191332	0.217074
1.500	-3.362945	-4.544418	-3.813399	192.334084	0.205637	0.165442	0.188203
1.450	-4.027942	-5.390515	-4.555302	276.709301	0.182550	0.146188	0.166520
1.400	-5.212906	-6.898070	-5.875482	463.106571	0.151219	0.120510	0.137448
1.350	-7.918118	-10.339461	-8.886056	1061.731598	0.107897	0.085547	0.097694
1.300	-20.293788	-26.081921	-22.648330	6870.306396	0.046310	0.036520	0.041757
1.250	23.536551	29.670679	26.078203	8976.989380	-0.044870	-0.035186	-0.040277
1.200	6.509182	8.011019	7.143663	651.283577	-0.188353	-0.146821	-0.168246
1.150	3.437211	4.102793	3.724238	164.911182	-0.437214	-0.338662	-0.388465
1.100	2.151718	2.466890	2.290626	53.548442	-0.950825	-0.731588	-0.839894

Ω_∞	Δff_a (C12)	Δff_a (C23)	Δff_a (C01)	Δff_a (C02)	Δff_a (C03)
30.000	0.0232482	0.0416633	0.0008166	0.0008022	0.0008135
26.000	0.0229408	0.0410881	0.0009374	0.0009184	0.0009333
22.000	0.0225269	0.0403140	0.0011000	0.0010740	0.0010943
20.000	0.0222610	0.0398171	0.0012045	0.0011733	0.0011976
18.000	0.0219395	0.0392164	0.0013309	0.0012928	0.0013224
16.000	0.0215427	0.0384757	0.0014867	0.0014393	0.0014761
14.000	0.0210407	0.0375393	0.0016838	0.0016232	0.0016700
12.000	0.0203852	0.0363182	0.0019408	0.0018605	0.0019222
10.000	0.0194933	0.0346594	0.0022897	0.0021783	0.0022631
8.000	0.0182087	0.0322765	0.0027893	0.0026247	0.0027485
7.000	0.0173281	0.0306477	0.0031290	0.0029224	0.0030764
6.000	0.0161992	0.0285655	0.0035600	0.0032932	0.0034897
5.000	0.0146997	0.0258116	0.0041215	0.0037642	0.0040231
4.500	0.0137492	0.0240734	0.0044689	0.0040485	0.0043496
4.000	0.0126128	0.0220039	0.0048725	0.0043713	0.0047253
3.500	0.0112311	0.0195011	0.0053408	0.0047348	0.0051551
3.000	0.0095173	0.0164203	0.0058738	0.0051304	0.0056334
2.500	0.0073443	0.0125565	0.0064309	0.0055101	0.0061120
2.250	0.0060319	0.0102495	0.0066649	0.0056430	0.0062955
2.000	0.0045378	0.0076508	0.0067741	0.0056560	0.0063491
1.800	0.0032055	0.0053623	0.0066323	0.0054654	0.0061675
1.600	0.0017825	0.0029524	0.0059719	0.0048465	0.0054985
1.500	0.0010772	0.0017735	0.0052088	0.0041907	0.0047672
1.450	0.0007453	0.0012229	0.0046241	0.0037030	0.0042180
1.400	0.0004431	0.0007246	0.0038304	0.0030526	0.0034816
1.350	0.0001923	0.0003132	0.0027331	0.0021669	0.0024746
1.300	0.0000296	0.0000479	0.0011730	0.0009251	0.0010577
1.250	0.0000225	0.0000363	-0.0011366	-0.0008913	-0.0010202
1.200	0.0003082	0.0004953	-0.0047710	-0.0037190	-0.0042617
1.150	0.0012096	0.0019333	-0.0110748	-0.0085784	-0.0098399
1.100	0.0037002	0.0058798	-0.0240847	-0.0185313	-0.0212748

Table 8.13

476

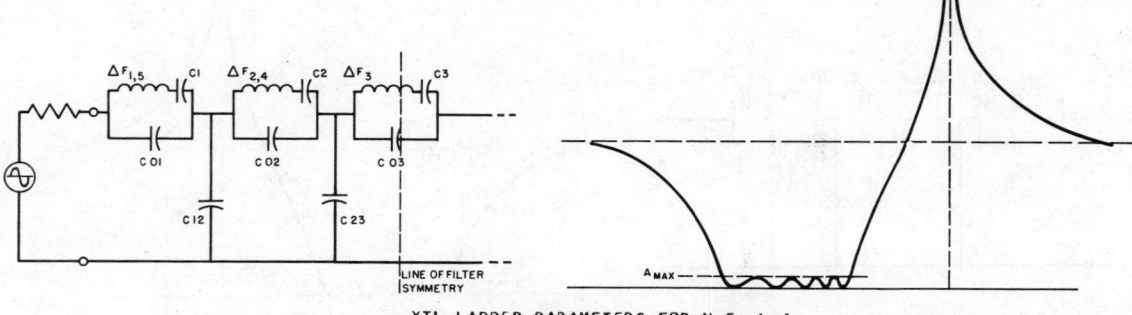

LINE OF FILTER
SYMMETRY

A MAX

XTL LADDER PARAMETERS FOR N=5 , L=1
A MAX = 0.0100   K12= 0.779566   K23= 0.539822   Q= 0.976589

Ω_∞	$\frac{2}{\Delta f}(\Delta F_{1,5})$	$\frac{2}{\Delta f}(\Delta F_{2,4})$	$\frac{2}{\Delta f}(\Delta F_3)$	$R/\Delta f$	$\Delta f(C01)/f_a(C\,1)$	$\Delta f(C02)/f_a(C\,2)$	$\Delta f(C03)/f_a(C\,3)$
30.000	-0.780690	-1.326027	-1.094033	6.780555	0.032488	0.031922	0.032161
26.000	-0.780895	-1.327102	-1.096291	6.836198	0.037340	0.036594	0.036905
22.000	-0.781189	-1.328591	-1.099390	6.913117	0.043896	0.042866	0.043291
20.000	-0.731390	-1.329574	-1.101418	6.963772	0.048120	0.046883	0.047390
18.000	-0.781645	-1.330791	-1.103910	7.026402	0.053243	0.051731	0.052345
16.000	-0.781981	-1.332339	-1.107048	7.105825	0.059588	0.057696	0.058455
14.000	-0.782439	-1.334371	-1.111120	7.209843	0.067648	0.065213	0.066176
12.000	-0.783099	-1.337159	-1.116616	7.351984	0.078228	0.074978	0.076239
10.000	-0.784122	-1.341215	-1.124448	7.557922	0.092729	0.088174	0.089892
8.000	-0.785889	-1.347658	-1.136518	7.883134	0.113819	0.106979	0.109451
7.000	-0.787336	-1.352543	-1.145396	8.128392	0.128414	0.119724	0.122769
6.000	-0.789529	-1.359452	-1.157598	8.473821	0.147286	0.135880	0.139712
5.000	-0.793160	-1.369959	-1.175458	8.996795	0.172617	0.156987	0.161931
4.500	-0.795988	-1.377564	-1.187928	9.374136	0.188822	0.170139	0.175811
4.000	-0.800040	-1.387826	-1.204234	9.882495	0.208332	0.185604	0.192151
3.500	-0.806216	-1.402418	-1.226534	10.604912	0.232222	0.203981	0.211572
3.000	-0.816466	-1.424773	-1.259060	11.713885	0.262022	0.226000	0.234794
2.500	-0.835837	-1.463224	-1.311549	13.636983	0.299775	0.252320	0.262361
2.250	-0.853111	-1.495035	-1.352580	15.249979	0.322257	0.267020	0.277579
2.000	-0.881588	-1.544508	-1.413404	17.807807	0.347031	0.282127	0.292963
1.800	-0.921026	-1.609264	-1.489048	21.265974	0.367508	0.293307	0.304006
1.600	-0.995312	-1.725571	-1.619021	27.779388	0.385310	0.300700	0.310653
1.500	-1.062116	-1.826437	-1.727579	33.787470	0.390302	0.300622	0.309830
1.450	-1.110334	-1.898000	-1.803206	38.254511	0.390574	0.298686	0.307389
1.400	-1.174921	-1.992774	-1.902135	44.434931	0.388361	0.294744	0.302834
1.350	-1.265463	-2.124213	-2.037714	53.509309	0.382341	0.287835	0.295184
1.300	-1.400747	-2.318647	-2.236039	68.012173	0.370268	0.276346	0.282802
1.250	-1.623236	-2.635557	-2.556015	94.425222	0.348040	0.257363	0.262742
1.200	-2.053448	-3.243680	-3.164662	154.775842	0.307366	0.225039	0.229113
1.150	-3.221382	-4.884888	-4.796115	381.220066	0.228761	0.165703	0.168177
1.100	-17.591752	-25.005553	-24.713266	10704.067383	0.053500	0.038306	0.038740

Ω_∞	Δff_a (C12)	Δff_a (C23)	Δff_a (C0I)	Δff_a (C02)	Δff_a (C03)	
30.000	0.0302945	0.0433079	0.0008229	0.0008086	0.0008146	
26.000	0.0299690	0.0427750	0.0009458	0.0009269	0.0009348	
22.000	0.0295304	0.0420579	0.0011119	0.0010858	0.0010966	
20.000	0.0292485	0.0415976	0.0012189	0.0011876	0.0012004	
18.000	0.0289073	0.0410412	0.0013487	0.0013104	0.0013259	
16.000	0.0284857	0.0403550	0.0015094	0.0014614	0.0014807	
14.000	0.0279518	0.0394877	0.0017135	0.0016519	0.0016763	
12.000	0.0272536	0.0383566	0.0019815	0.0018992	0.0019312	
10.000	0.0263014	0.0368202	0.0023489	0.0022335	0.0022770	
8.000	0.0249253	0.0346136	0.0028831	0.0027098	0.0027724	
7.000	0.0239785	0.0331055	0.0032528	0.0030326	0.0031098	
6.000	0.0227599	0.0311778	0.0037308	0.0034419	0.0035389	
5.000	0.0211323	0.0286283	0.0043724	0.0039765	0.0041018	
4.500	0.0200944	0.0270190	0.0047829	0.0043097	0.0044533	
4.000	0.0188461	0.0251021	0.0052771	0.0047014	0.0048672	
3.500	0.0173155	0.0227819	0.0058823	0.0051669	0.0053592	
3.000	0.0153931	0.0199193	0.0066371	0.0057247	0.0059474	
2.500	0.0129031	0.0163079	0.0075934	0.0063913	0.0066457	
2.250	0.0113587	0.0141291	0.0081629	0.0067637	0.0070312	
2.000	0.0095438	0.0116350	0.0087904	0.0071464	0.0074208	
1.800	0.0078456	0.0093722	0.0093091	0.0074296	0.0077006	
1.600	0.0058730	0.0068379	0.0097600	0.0076168	0.0078689	
1.500	0.0047656	0.0054632	0.0098865	0.0076148	0.0078481	
1.450	0.0041791	0.0047498	0.0098934	0.0075658	0.0077863	
1.400	0.0035706	0.0040208	0.0098373	0.0074660	0.0076709	
1.350	0.0029412	0.0032792	0.0096848	0.0072909	0.0074771	
1.300	0.0022941	0.0025303	0.0093790	0.0069999	0.0071635	
1.250	0.0016371	0.0017848	0.0088159	0.0065191	0.0066553	
1.200	0.0009889	0.0010645	0.0077857	0.0057003	0.0058035	
1.150	0.0003972	0.0004217	0.0057946	0.0041973	0.0042600	
1.100	0.0000140	0.0000146	0.0013552	0.0009703	0.0009813	

Table 8.14

477

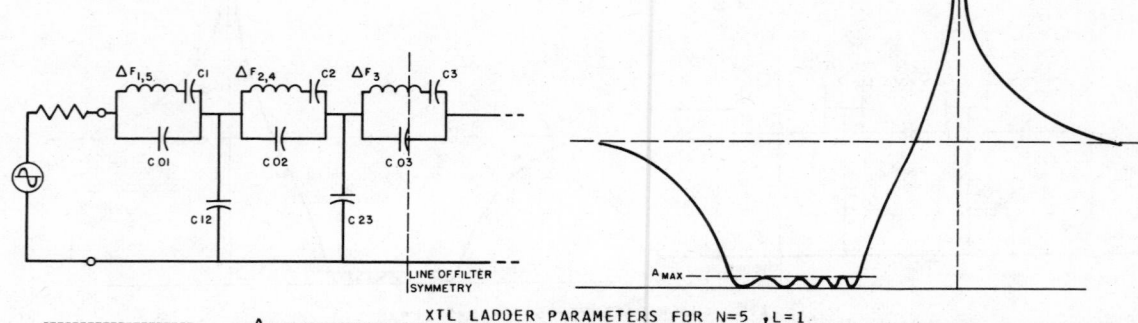

XTL LADDER PARAMETERS FOR N=5 , L=1

A_{MAX} = 0.0500   K12= 0.726257   K23= 0.536688   Q= 1.173513

Ω_∞	$\frac{2}{\Delta f}(\Delta F_{1,5})$	$\frac{2}{\Delta f}(\Delta F_{2,4})$	$\frac{2}{\Delta f}(\Delta F_3)$	$R/\Delta f$	$\Delta f(C01)/f_a(C1)$	$\Delta f(C02)/f_a(C2)$	$\Delta f(C03)/f_a(C3)$
30.000	-0.723585	-1.266545	-1.085358	5.621807	0.032548	0.031983	0.032169
26.000	-0.723190	-1.267126	-1.087226	5.664541	0.037421	0.036674	0.036918
22.000	-0.722659	-1.267931	-1.089783	5.723518	0.044009	0.042978	0.043309
20.000	-0.722318	-1.268462	-1.091452	5.762297	0.048257	0.047018	0.047413
18.000	-0.721907	-1.269120	-1.093500	5.810179	0.053413	0.051897	0.052374
16.000	-0.721402	-1.269955	-1.096072	5.870797	0.059804	0.057904	0.058493
14.000	-0.720765	-1.271051	-1.099398	5.950014	0.067931	0.065483	0.066228
12.000	-0.719940	-1.272553	-1.103869	6.057954	0.078617	0.075343	0.076313
10.000	-0.718835	-1.274734	-1.110202	6.213723	0.093294	0.088694	0.090007
8.000	-0.717294	-1.278190	-1.119873	6.458248	0.114715	0.107780	0.109651
7.000	-0.716284	-1.280803	-1.126916	6.641514	0.129596	0.120761	0.123048
6.000	-0.715068	-1.284489	-1.136499	6.898024	0.148919	0.137278	0.140125
5.000	-0.713642	-1.290075	-1.150322	7.282593	0.175020	0.158981	0.162593
4.500	-0.712890	-1.294102	-1.159833	7.558199	0.191832	0.172589	0.176684
4.000	-0.712203	-1.299517	-1.172093	7.925900	0.212215	0.188696	0.193345
3.500	-0.711785	-1.307178	-1.188545	8.442512	0.237429	0.208022	0.213286
3.000	-0.712193	-1.318828	-1.211908	9.222804	0.269383	0.231544	0.237422
2.500	-0.715133	-1.338633	-1.248118	10.542184	0.311029	0.260509	0.266801
2.250	-0.719115	-1.354798	-1.275245	11.618570	0.336801	0.277408	0.283668
2.000	-0.727214	-1.379558	-1.313767	13.274219	0.366675	0.295897	0.301771
1.800	-0.740147	-1.411350	-1.359436	15.422744	0.393678	0.311396	0.316512
1.600	-0.766882	-1.466422	-1.432139	19.227391	0.422497	0.326113	0.329800
1.500	-0.791857	-1.512412	-1.488966	22.498670	0.436327	0.331960	0.334564
1.450	-0.809973	-1.544065	-1.526728	24.803882	0.442483	0.333994	0.335939
1.400	-0.834091	-1.584794	-1.574146	27.836996	0.447609	0.335031	0.336231
1.350	-0.867325	-1.639147	-1.635909	32.004763	0.450994	0.334544	0.334906
1.300	-0.915302	-1.715314	-1.720434	38.081031	0.451406	0.331640	0.331078
1.250	-0.989323	-1.829700	-1.844539	47.731238	0.446564	0.324707	0.323150
1.200	-1.115850	-2.020651	-2.047472	65.261934	0.431807	0.310496	0.307932
1.150	-1.374617	-2.403577	-2.447236	105.816473	0.396100	0.281407	0.277991
1.100	-2.167631	-3.560050	-3.638167	272.231720	0.306032	0.214590	0.211052

Ω_∞	$\Delta f f_a$ (C12)	$\Delta f f_a$ (C23)	$\Delta f f_a$ (C01)	$\Delta f f_a$ (C02)	$\Delta f f_a$ (C03)	
30.000	0.0326406	0.0436559	0.0008245	0.0008101	0.0008149	
26.000	0.0323092	0.0431339	0.0009479	0.0009290	0.0009351	
22.000	0.0318625	0.0424315	0.0011148	0.0010886	0.0010970	
20.000	0.0315754	0.0419806	0.0012224	0.0011910	0.0012010	
18.000	0.0312277	0.0414356	0.0013530	0.0013146	0.0013266	
16.000	0.0307983	0.0407635	0.0015148	0.0014667	0.0014816	
14.000	0.0302542	0.0399139	0.0017207	0.0016587	0.0016776	
12.000	0.0295424	0.0388062	0.0019914	0.0019085	0.0019330	
10.000	0.0285714	0.0373014	0.0023632	0.0022466	0.0022799	
8.000	0.0271672	0.0351404	0.0029058	0.0027301	0.0027775	
7.000	0.0262006	0.0336637	0.0032827	0.0030589	0.0031168	
6.000	0.0249556	0.0317762	0.0037722	0.0034773	0.0035494	
5.000	0.0232912	0.0292803	0.0044333	0.0040270	0.0041185	
4.500	0.0222288	0.0277051	0.0048592	0.0043717	0.0044754	
4.000	0.0209499	0.0258290	0.0053755	0.0047797	0.0048975	
3.500	0.0193797	0.0235582	0.0060141	0.0052693	0.0054026	
3.000	0.0174038	0.0207570	0.0068235	0.0058651	0.0060140	
2.500	0.0148366	0.0172222	0.0078785	0.0065988	0.0067581	
2.250	0.0132384	0.0150883	0.0085313	0.0070268	0.0071854	
2.000	0.0113525	0.0126432	0.0092880	0.0074952	0.0076440	
1.800	0.0095775	0.0104200	0.0099720	0.0078877	0.0080173	
1.600	0.0074966	0.0079188	0.0107020	0.0082605	0.0083539	
1.500	0.0063148	0.0065523	0.0110523	0.0084086	0.0084746	
1.450	0.0056828	0.0058384	0.0112082	0.0084602	0.0085094	
1.400	0.0050212	0.0051040	0.0113381	0.0084864	0.0085168	
1.350	0.0043282	0.0043494	0.0114238	0.0084741	0.0084833	
1.300	0.0036027	0.0035757	0.0114342	0.0084005	0.0083863	
1.250	0.0028448	0.0027857	0.0113116	0.0082249	0.0081855	
1.200	0.0020575	0.0019856	0.0109378	0.0078650	0.0078000	
1.150	0.0012538	0.0011907	0.0100333	0.0071281	0.0070416	
1.100	0.0004810	0.0004489	0.0077519	0.0054356	0.0053460	

Table 8.15

478

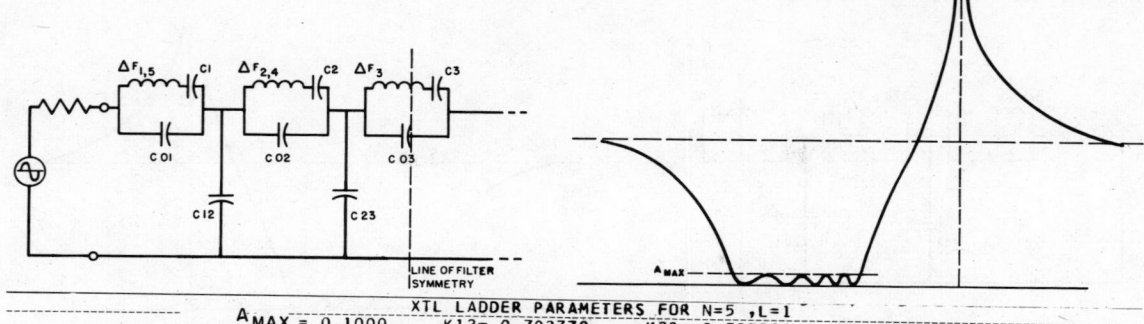

LINE OF FILTER SYMMETRY

A_{MAX}

		XTL LADDER PARAMETERS FOR N=5 , L=1					
	$A_{MAX} = 0.1000$	K12= 0.702770		K23= 0.535520		Q= 1.301309	

Ω_∞	$\frac{2}{\Delta f}(\Delta F_{1,5})$	$\frac{2}{\Delta f}(\Delta F_{2,4})$	$\frac{2}{\Delta f}(\Delta F_3)$	$R/\Delta f$	$\Delta f(C0I)/f_a(C\,1)$	$\Delta f(C02)/f_a(C\,2)$	$\Delta f(C03)/f_a(C\,3)$
30.000	-0.698500	-1.240634	-1.081999	5.061438	0.032575	0.032010	0.032173
26.000	-0.697853	-1.241012	-1.083701	5.098568	0.037456	0.036709	0.036923
22.000	-0.696975	-1.241536	-1.086028	5.149774	0.044059	0.043026	0.043316
20.000	-0.696408	-1.241881	-1.087545	5.183420	0.048318	0.047077	0.047421
18.000	-0.695718	-1.242309	-1.089405	5.224937	0.053488	0.051969	0.052385
16.000	-0.694860	-1.242851	-1.091737	5.277457	0.059899	0.057995	0.058508
14.000	-0.693766	-1.243563	-1.094748	5.346024	0.068056	0.065601	0.066248
12.000	-0.692322	-1.244537	-1.098786	5.439331	0.078788	0.075503	0.076343
10.000	-0.690332	-1.245952	-1.104487	5.573738	0.093542	0.088921	0.090054
8.000	-0.687419	-1.248192	-1.113148	5.784160	0.115109	0.108129	0.109732
7.000	-0.685396	-1.249883	-1.119422	5.941424	0.130117	0.121214	0.123161
6.000	-0.682779	-1.252266	-1.127911	6.160920	0.149638	0.137888	0.140294
5.000	-0.679288	-1.255872	-1.140058	6.488972	0.176078	0.159850	0.162865
4.500	-0.677084	-1.258467	-1.148344	6.722628	0.193159	0.173657	0.177043
4.000	-0.674486	-1.261951	-1.158940	7.033569	0.213927	0.190044	0.193838
3.500	-0.671433	-1.266870	-1.173003	7.468223	0.239726	0.209781	0.213995
3.000	-0.667957	-1.274328	-1.192659	8.120033	0.272631	0.233955	0.238512
2.540	-0.664540	-1.286943	-1.222387	9.210036	0.316002	0.264065	0.268645
2.250	-0.663491	-1.297184	-1.244079	10.088658	0.343231	0.281914	0.286198
2.000	-0.664059	-1.312772	-1.274056	11.422619	0.375367	0.301862	0.305432
1.800	-0.667506	-1.332618	-1.308422	13.124045	0.405268	0.319222	0.321707
1.600	-0.678004	-1.366555	-1.360812	16.061412	0.438981	0.337091	0.337745
1.500	-0.689434	-1.394475	-1.400066	18.516250	0.456739	0.345486	0.344820
1.450	-0.698182	-1.413474	-1.425445	20.209846	0.465510	0.349226	0.347772
1.400	-0.710162	-1.437664	-1.456609	22.395434	0.473897	0.352402	0.350065
1.350	-0.727011	-1.469505	-1.496151	25.324502	0.481461	0.354672	0.351352
1.300	-0.751617	-1.513299	-1.548582	29.453909	0.487420	0.355455	0.351052
1.250	-0.789562	-1.577315	-1.622534	35.706774	0.490301	0.353693	0.348125
1.200	-0.853162	-1.679738	-1.736929	46.258659	0.487054	0.347254	0.340492
1.150	-0.975902	-1.869887	-1.942975	67.663746	0.470389	0.331138	0.323313
1.100	-1.291655	-2.344682	-2.444903	131.515526	0.418121	0.290303	0.282095

Ω_∞	Δff_a (CI2)	Δff_a (C23)	Δff_a (C0I)	Δff_a (C02)	Δff_a (C03)	
30.000	0.0337870	0.0437921	0.0008251	0.0008108	0.0008150	
26.000	0.0334527	0.0432750	0.0009488	0.0009299	0.0009353	
22.000	0.0330021	0.0425792	0.0011160	0.0010899	0.0010972	
20.000	0.0327124	0.0421326	0.0012239	0.0011925	0.0012012	
18.000	0.0323616	0.0415927	0.0013507	0.0013164	0.0013269	
16.000	0.0319283	0.0409270	0.0015173	0.0014690	0.0014820	
14.000	0.0313792	0.0400854	0.0017239	0.0016617	0.0016781	
12.000	0.0306609	0.0389882	0.0019957	0.0019125	0.0019338	
10.000	0.0296807	0.0374977	0.0023695	0.0022524	0.0022811	
8.000	0.0282631	0.0353574	0.0029157	0.0027389	0.0027795	
7.000	0.0272869	0.0338948	0.0032959	0.0030704	0.0031197	
6.000	0.0260294	0.0320255	0.0037904	0.0034927	0.0035537	
5.000	0.0243476	0.0295539	0.0044601	0.0040490	0.0041254	
4.500	0.0232737	0.0279941	0.0048928	0.0043988	0.0044846	
4.000	0.0219805	0.0261366	0.0054188	0.0048139	0.0049100	
3.500	0.0203920	0.0238884	0.0060723	0.0053138	0.0054206	
3.000	0.0183918	0.0211153	0.0069058	0.0059261	0.0060416	
2.500	0.0157902	0.0176162	0.0080044	0.0066888	0.0068049	
2.250	0.0141685	0.0155039	0.0086941	0.0071410	0.0072495	
2.000	0.0122522	0.0130830	0.0095082	0.0076463	0.0077367	
1.800	0.0104450	0.0108809	0.0102655	0.0080860	0.0081489	
1.600	0.0083204	0.0084009	0.0111195	0.0085386	0.0085552	
1.500	0.0071094	0.0070436	0.0115693	0.0087513	0.0087344	
1.450	0.0064601	0.0063335	0.0117915	0.0088460	0.0088092	
1.400	0.0057786	0.0056018	0.0120040	0.0089265	0.0088673	
1.350	0.0050623	0.0048481	0.0121956	0.0089839	0.0088998	
1.300	0.0043089	0.0040726	0.0123465	0.0090038	0.0088922	
1.250	0.0035159	0.0032760	0.0124195	0.0089591	0.0088181	
1.200	0.0026823	0.0024608	0.0123372	0.0087960	0.0086248	
1.150	0.0018106	0.0016332	0.0119151	0.0083878	0.0081896	
1.100	0.0009188	0.0008134	0.0105911	0.0073534	0.0071456	

Table 8.16

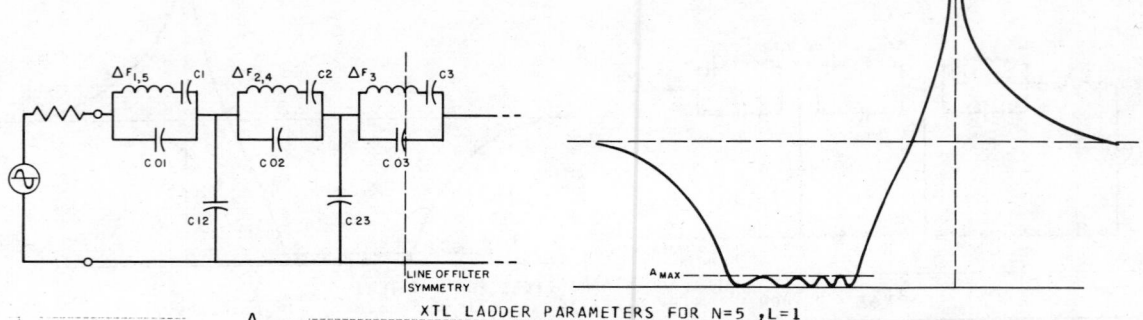

XTL LADDER PARAMETERS FOR N=5 ,L=1

$A_{MAX} = 0.2500$ K12= 0.672719   K23= 0.534400   Q= 1.539917

Ω_∞	$\frac{2}{\Delta f}$ ($\Delta F_{1,5}$)	$\frac{2}{\Delta f}$ ($\Delta F_{2,4}$)	$\frac{2}{\Delta f}$ (ΔF_3)	$R/\Delta f$	$\Delta f(C01)/f_a$ (C 1)	$\Delta f(C02)/f_a$ (C 2)	$\Delta f(C03)/f_a$ (C 3)
30.000	-0.666473	-1.207943	-1.078502	4.268256	0.032609	0.032043	0.032177
26.000	-0.665515	-1.208076	-1.080000	4.298121	0.037502	0.036754	0.036928
22.000	-0.664211	-1.208259	-1.082046	4.339269	0.044122	0.043088	0.043324
20.000	-0.663364	-1.208380	-1.083376	4.366281	0.048395	0.047151	0.047431
18.000	-0.662330	-1.208530	-1.085005	4.399585	0.053584	0.052060	0.052397
16.000	-0.661040	-1.208720	-1.087043	4.441672	0.060020	0.058110	0.058524
14.000	-0.659383	-1.208969	-1.089668	4.496549	0.068216	0.065751	0.066271
12.000	-0.657180	-1.209310	-1.093177	4.571096	0.079007	0.075704	0.076376
10.000	-0.654105	-1.209805	-1.098105	4.678227	0.093861	0.089208	0.090105
8.000	-0.649517	-1.210587	-1.105535	4.845356	0.115613	0.108571	0.109823
7.000	-0.646259	-1.211177	-1.110873	4.969805	0.130783	0.121785	0.123291
6.000	-0.641943	-1.212007	-1.118033	5.142860	0.150558	0.138658	0.140488
5.000	-0.635956	-1.213260	-1.128147	5.400157	0.177432	0.160946	0.163181
4.500	-0.632006	-1.214160	-1.134955	5.582465	0.194856	0.175004	0.177464
4.000	-0.627119	-1.215366	-1.143547	5.823883	0.216117	0.191741	0.194418
3.500	-0.620931	-1.217065	-1.154743	6.159150	0.242664	0.211996	0.214835
3.000	-0.612875	-1.219630	-1.169981	6.657305	0.276788	0.236988	0.239809
2.500	-0.602067	-1.223944	-1.192061	7.478692	0.322366	0.268533	0.270851
2.250	-0.595257	-1.227422	-1.207409	8.130780	0.351462	0.287569	0.289234
2.000	-0.587342	-1.232678	-1.227544	9.104770	0.386497	0.309341	0.309833
1.800	-0.580331	-1.239301	-1.249115	10.320704	0.420110	0.329023	0.327964
1.600	-0.573474	-1.250458	-1.279107	12.355682	0.460093	0.350821	0.347330
1.500	-0.570904	-1.259480	-1.299556	13.998845	0.482881	0.362387	0.357199
1.450	-0.570204	-1.265540	-1.311980	15.104084	0.494999	0.368251	0.362059
1.400	-0.570207	-1.273165	-1.326486	16.498134	0.507561	0.374088	0.366772
1.350	-0.571335	-1.283049	-1.343855	18.312721	0.520471	0.379788	0.371215
1.300	-0.574329	-1.296369	-1.365385	20.774226	0.533524	0.385153	0.375180
1.250	-0.580575	-1.315288	-1.393425	24.307199	0.546276	0.389820	0.378297
1.200	-0.592983	-1.344270	-1.432764	29.811379	0.557730	0.393040	0.379829
1.150	-0.618745	-1.394237	-1.495028	39.580694	0.565372	0.393045	0.378068
1.100	-0.681508	-1.500876	-1.618016	61.665032	0.561322	0.384486	0.367915

Ω_∞	$\Delta f f_a$ (C12)	$\Delta f f_a$ (C23)	$\Delta f f_a$ (C01)	$\Delta f f_a$ (C02)	$\Delta f f_a$ (C03)	
30.000	0.0353702	0.0439347	0.0008260	0.0008117	0.0008150	
26.000	0.0350318	0.0434241	0.0009499	0.0009310	0.0009354	
22.000	0.0345757	0.0427370	0.0011176	0.0010914	0.0010974	
20.000	0.0342824	0.0422959	0.0012259	0.0011944	0.0012014	
18.000	0.0339273	0.0417628	0.0013573	0.0013187	0.0013272	
16.000	0.0334885	0.0411053	0.0015203	0.0014719	0.0014824	
14.000	0.0329326	0.0402744	0.0017279	0.0016655	0.0016787	
12.000	0.0322052	0.0391908	0.0020013	0.0019176	0.0019346	
10.000	0.0312124	0.0377191	0.0023775	0.0022597	0.0022824	
8.000	0.0297761	0.0356058	0.0029285	0.0027501	0.0027819	
7.000	0.0287868	0.0341619	0.0033128	0.0030849	0.0031230	
6.000	0.0275121	0.0323166	0.0038137	0.0035122	0.0035586	
5.000	0.0258066	0.0298769	0.0044944	0.0040768	0.0041334	
4.500	0.0247171	0.0283375	0.0049357	0.0044329	0.0044952	
4.000	0.0234047	0.0265044	0.0054743	0.0048569	0.0049247	
3.500	0.0217917	0.0242861	0.0061467	0.0053699	0.0054418	
3.000	0.0197592	0.0215504	0.0070111	0.0060030	0.0060744	
2.500	0.0171124	0.0180993	0.0081656	0.0068020	0.0068607	
2.250	0.0154604	0.0160162	0.0089026	0.0072842	0.0073264	
2.000	0.0135055	0.0136289	0.0097901	0.0078357	0.0078482	
1.800	0.0116585	0.0114571	0.0106415	0.0083342	0.0083074	
1.600	0.0094812	0.0090100	0.0116543	0.0088864	0.0087980	
1.500	0.0082362	0.0076695	0.0122315	0.0091794	0.0090480	
1.450	0.0075672	0.0069675	0.0125385	0.0093273	0.0091711	
1.400	0.0068634	0.0062433	0.0128567	0.0094758	0.0092905	
1.350	0.0061218	0.0054964	0.0131837	0.0096201	0.0094030	
1.300	0.0053388	0.0047260	0.0135143	0.0097560	0.0095034	
1.250	0.0045103	0.0039318	0.0138373	0.0098743	0.0095824	
1.200	0.0036318	0.0031135	0.0141275	0.0099558	0.0096212	
1.150	0.0026984	0.0022715	0.0143211	0.0099559	0.0095766	
1.100	0.0017065	0.0014081	0.0142185	0.0097391	0.0093194	

Table 8.17

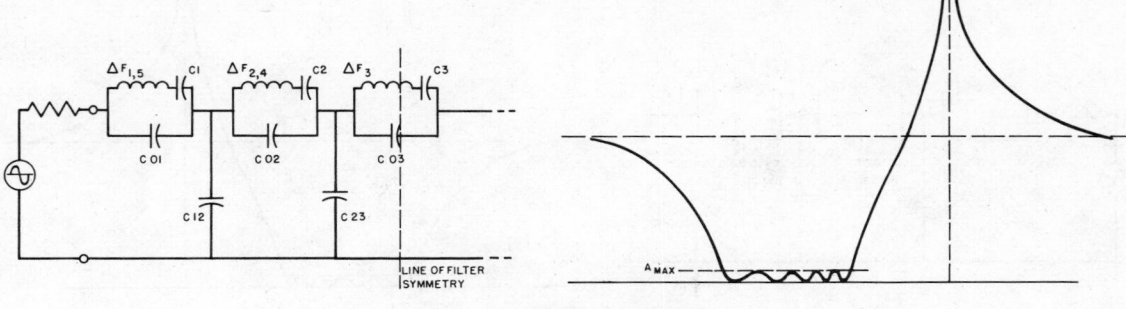

XTL LADDER PARAMETERS FOR N=5 ,L=1

A MAX = 0.5000   K12= 0.651855   K23= 0.534101   Q= 1.806852

Ω_∞	$\frac{2}{\Delta f}(\Delta F_{1,5})$	$\frac{2}{\Delta f}(\Delta F_{2,4})$	$\frac{2}{\Delta f}(\Delta F_3)$	$R/\Delta f$	$\Delta f(C01)/f_a(C1)$	$\Delta f(C02)/f_a(C2)$	$\Delta f(C03)/f_a(C3)$
30.000	-0.644287	-1.185798	-1.077091	3.632424	0.032633	0.032066	0.032178
26.000	-0.643122	-1.185772	-1.078458	3.656989	0.037533	0.036784	0.036930
22.000	-0.641533	-1.185737	-1.080321	3.690810	0.044167	0.043130	0.043327
20.000	-0.640501	-1.185714	-1.081532	3.712998	0.048448	0.047202	0.047435
18.000	-0.639238	-1.185685	-1.083011	3.740339	0.053650	0.052122	0.052403
16.000	-0.637659	-1.185649	-1.084859	3.774865	0.060105	0.058188	0.058531
14.000	-0.635629	-1.185601	-1.087235	3.819842	0.068326	0.065852	0.066281
12.000	-0.632921	-1.185536	-1.090402	3.880867	0.079158	0.075841	0.076392
10.000	-0.629128	-1.185441	-1.094832	3.968417	0.094081	0.089402	0.090132
8.000	-0.623435	-1.185292	-1.101469	4.104660	0.115963	0.108870	0.109872
7.000	-0.619364	-1.185179	-1.106204	4.205846	0.131245	0.122172	0.123362
6.000	-0.613932	-1.185020	-1.112509	4.346188	0.151196	0.139178	0.140597
5.000	-0.606316	-1.184782	-1.121320	4.554084	0.178370	0.161687	0.163363
4.500	-0.601231	-1.184610	-1.127181	4.700851	0.196031	0.175914	0.177709
4.000	-0.594866	-1.184381	-1.134492	4.894537	0.217634	0.192887	0.194761
3.500	-0.586665	-1.184059	-1.143864	5.162293	0.244698	0.213490	0.215338
3.000	-0.575694	-1.183573	-1.156301	5.557608	0.279666	0.239030	0.240599
2.500	-0.560250	-1.182759	-1.173573	6.203140	0.326771	0.271536	0.272215
2.250	-0.549882	-1.182106	-1.184971	6.710321	0.357158	0.291366	0.291123
2.000	-0.536815	-1.181124	-1.199042	7.459568	0.394195	0.314354	0.312594
1.800	-0.523588	-1.179894	-1.212850	8.381608	0.430369	0.335582	0.331912
1.600	-0.506718	-1.177841	-1.229561	9.893399	0.474672	0.359992	0.353412
1.500	-0.496329	-1.176199	-1.239149	11.086593	0.500919	0.373664	0.365077
1.450	-0.490474	-1.175105	-1.244237	11.876660	0.515338	0.380937	0.371163
1.400	-0.484074	-1.173739	-1.249481	12.858253	0.530765	0.388540	0.377432
1.350	-0.477012	-1.171984	-1.254806	14.112510	0.547342	0.396513	0.383906
1.300	-0.469116	-1.169645	-1.260063	15.773225	0.565254	0.404916	0.390615
1.250	-0.460103	-1.166380	-1.264946	18.079200	0.584760	0.413842	0.397623
1.200	-0.449462	-1.161498	-1.268798	21.502496	0.606258	0.423460	0.405055
1.150	-0.436076	-1.153408	-1.270005	27.125360	0.630487	0.434139	0.413222
1.100	-0.416784	-1.137403	-1.263636	38.096528	0.659290	0.446947	0.423077

Ω_∞	$\Delta f f_a$ (C12)	$\Delta f f_a$ (C23)	$\Delta f f_a$ (C01)	$\Delta f f_a$ (C02)	$\Delta f f_a$ (C03)	
30.000	0.0365547	0.0439926	0.0008266	0.0008122	0.0008151	
26.000	0.0362132	0.0434865	0.0009507	0.0009317	0.0009354	
22.000	0.0357528	0.0428056	0.0011188	0.0010925	0.0010975	
20.000	0.0354567	0.0423686	0.0012272	0.0011956	0.0012015	
18.000	0.0350983	0.0418403	0.0013590	0.0013203	0.0013274	
16.000	0.0346554	0.0411888	0.0015225	0.0014739	0.0014826	
14.000	0.0340942	0.0403654	0.0017307	0.0016680	0.0016789	
12.000	0.0333598	0.0392918	0.0020051	0.0019211	0.0019350	
10.000	0.0323574	0.0378336	0.0023831	0.0022646	0.0022831	
8.000	0.0309070	0.0357399	0.0029374	0.0027577	0.0027831	
7.000	0.0299077	0.0343093	0.0033245	0.0030947	0.0031248	
6.000	0.0286200	0.0324813	0.0038298	0.0035254	0.0035614	
5.000	0.0268967	0.0300648	0.0045182	0.0040956	0.0041380	
4.500	0.0257955	0.0285401	0.0049665	0.0044559	0.0045014	
4.000	0.0244687	0.0267247	0.0055127	0.0048859	0.0049334	
3.500	0.0228376	0.0245283	0.0061983	0.0054078	0.0054546	
3.000	0.0207812	0.0218199	0.0070840	0.0060547	0.0060944	
2.500	0.0181017	0.0184041	0.0082772	0.0068781	0.0068953	
2.250	0.0164279	0.0163428	0.0090469	0.0073804	0.0073742	
2.000	0.0144458	0.0139810	0.0099851	0.0079627	0.0079181	
1.800	0.0125712	0.0118328	0.0109014	0.0085004	0.0084074	
1.600	0.0103586	0.0094127	0.0120236	0.0091187	0.0089520	
1.500	0.0090918	0.0080871	0.0126884	0.0094650	0.0092475	
1.450	0.0084103	0.0073929	0.0130537	0.0096492	0.0094017	
1.400	0.0076930	0.0066767	0.0134444	0.0098418	0.0095605	
1.350	0.0069365	0.0059379	0.0138643	0.0100438	0.0097244	
1.300	0.0061369	0.0051758	0.0143181	0.0102567	0.0098944	
1.250	0.0052896	0.0043898	0.0148121	0.0104827	0.0100719	
1.200	0.0043895	0.0035793	0.0153567	0.0107264	0.0102602	
1.150	0.0034302	0.0027438	0.0159704	0.0109969	0.0104670	
1.100	0.0024046	0.0018833	0.0167000	0.0113213	0.0107167	

Table 8.18

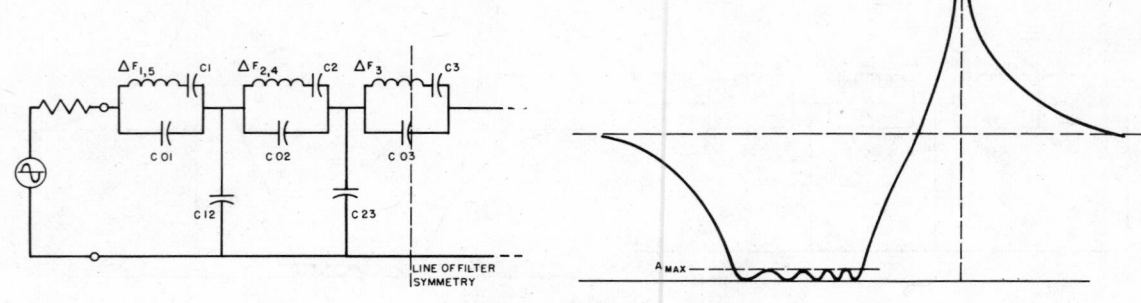

A_MAX = 1.0000   XTL LADDER PARAMETERS FOR N=5 , L=1   K12= 0.633777   K23= 0.534560   Q= 2.207072

Ω_∞	$\frac{2}{\Delta f}(\Delta F_{1,5})$	$\frac{2}{\Delta f}(\Delta F_{2,4})$	$\frac{2}{\Delta f}(\Delta F_3)$	$R/\Delta f$	$\Delta f(C01)/f_a(C1)$	$\Delta f(C02)/f_a(C2)$	$\Delta f(C03)/f_a(C3)$
30.000	-0.625107	-1.167407	-1.077389	2.970017	0.032653	0.032085	0.032178
26.000	-0.623769	-1.167258	-1.078655	2.989501	0.037560	0.036809	0.036929
22.000	-0.621943	-1.167051	-1.080379	3.016311	0.044205	0.043165	0.043327
20.000	-0.620756	-1.166914	-1.081498	3.033889	0.048495	0.047244	0.047435
18.000	-0.619303	-1.166746	-1.082863	3.055538	0.053708	0.052174	0.052403
16.000	-0.617485	-1.166532	-1.084567	3.082860	0.060178	0.058253	0.058532
14.000	-0.615145	-1.166251	-1.086752	3.118423	0.068422	0.065936	0.066283
12.000	-0.612019	-1.165868	-1.089657	3.166625	0.079289	0.075954	0.076396
10.000	-0.607632	-1.165312	-1.093705	3.235677	0.094272	0.089563	0.090141
8.000	-0.601026	-1.164436	-1.099734	3.342900	0.116265	0.109117	0.109893
7.000	-0.596288	-1.163776	-1.104007	3.422355	0.131643	0.122492	0.123396
6.000	-0.589942	-1.162849	-1.109657	3.532308	0.151746	0.139609	0.140654
5.000	-0.580998	-1.161452	-1.117465	3.694672	0.179179	0.162299	0.163466
4.500	-0.574992	-1.160451	-1.122599	3.808934	0.197045	0.176664	0.177854
4.000	-0.567432	-1.159113	-1.128927	3.959282	0.218941	0.193832	0.194973
3.500	-0.557614	-1.157235	-1.136897	4.166318	0.246450	0.214720	0.215661
3.000	-0.544324	-1.154410	-1.147187	4.470336	0.282141	0.240708	0.241127
2.500	-0.525237	-1.149694	-1.160783	4.962757	0.330553	0.273996	0.273166
2.250	-0.512114	-1.145919	-1.169175	5.346300	0.362042	0.294471	0.292468
2.000	-0.495114	-1.140265	-1.178659	5.907764	0.400783	0.318444	0.314598
1.800	-0.477227	-1.133219	-1.186645	6.590645	0.439131	0.340922	0.334824
1.600	-0.453036	-1.121550	-1.193539	7.691868	0.487084	0.367438	0.357969
1.500	-0.437074	-1.112291	-1.195234	8.545666	0.516242	0.382806	0.371025
1.450	-0.427611	-1.106159	-1.195061	9.103239	0.532592	0.391212	0.378063
1.400	-0.416798	-1.098534	-1.193740	9.788261	0.550419	0.400234	0.385544
1.350	-0.404198	-1.088818	-1.190657	10.650863	0.570061	0.410035	0.393599
1.300	-0.389114	-1.075990	-1.184782	11.771515	0.592026	0.420877	0.402450
1.250	-0.370351	-1.058285	-1.174266	13.287988	0.617150	0.433222	0.412496
1.200	-0.345640	-1.032279	-1.155477	15.457075	0.646981	0.447973	0.424542
1.150	-0.310006	-0.990352	-1.120253	18.816666	0.684929	0.467213	0.440480
1.100	-0.249836	-0.911462	-1.045902	24.700511	0.740831	0.497151	0.466004

Ω_∞	$\Delta ff_a(C12)$	$\Delta ff_a(C23)$	$\Delta ff_a(C01)$	$\Delta ff_a(C02)$	$\Delta ff_a(C03)$
30.000	0.0376431	0.0439803	0.0008271	0.0008127	0.0008151
26.000	0.0372986	0.0434785	0.0009514	0.0009324	0.0009354
22.000	0.0368341	0.0428033	0.0011197	0.0010934	0.0010975
20.000	0.0365354	0.0423700	0.0012284	0.0011967	0.0012015
18.000	0.0361738	0.0418461	0.0013604	0.0013216	0.0013274
16.000	0.0357269	0.0412000	0.0015243	0.0014756	0.0014826
14.000	0.0351606	0.0403836	0.0017332	0.0016702	0.0016790
12.000	0.0344196	0.0393190	0.0020084	0.0019239	0.0019351
10.000	0.0334079	0.0378732	0.0023879	0.0022687	0.0022833
8.000	0.0319439	0.0357973	0.0029450	0.0027640	0.0027836
7.000	0.0309352	0.0343791	0.0033346	0.0031028	0.0031257
6.000	0.0296350	0.0325670	0.0038438	0.0035363	0.0035628
5.000	0.0278946	0.0301718	0.0045387	0.0041111	0.0041407
4.500	0.0267823	0.0286607	0.0049912	0.0044750	0.0045051
4.000	0.0254418	0.0268617	0.0055459	0.0049098	0.0049387
3.500	0.0237934	0.0246854	0.0062427	0.0054389	0.0054628
3.000	0.0217145	0.0220025	0.0071467	0.0060972	0.0061078
2.500	0.0190041	0.0186197	0.0083730	0.0069404	0.0069194
2.250	0.0173100	0.0165790	0.0091706	0.0074590	0.0074083
2.000	0.0153027	0.0142415	0.0101520	0.0080663	0.0079689
1.800	0.0134030	0.0121161	0.0111233	0.0086357	0.0084812
1.600	0.0111587	0.0097229	0.0123380	0.0093073	0.0090675
1.500	0.0098729	0.0084127	0.0130766	0.0096966	0.0093982
1.450	0.0091810	0.0077268	0.0134907	0.0099095	0.0095764
1.400	0.0084524	0.0070194	0.0139423	0.0101380	0.0097659
1.350	0.0076839	0.0062901	0.0144398	0.0103863	0.0099700
1.300	0.0068715	0.0055381	0.0149962	0.0106609	0.0101942
1.250	0.0060107	0.0047632	0.0156326	0.0109736	0.0104486
1.200	0.0050968	0.0039652	0.0163882	0.0113473	0.0107538
1.150	0.0041247	0.0031450	0.0173495	0.0118346	0.0111575
1.100	0.0030912	0.0023054	0.0187655	0.0125930	0.0118040

Table 8.19

482

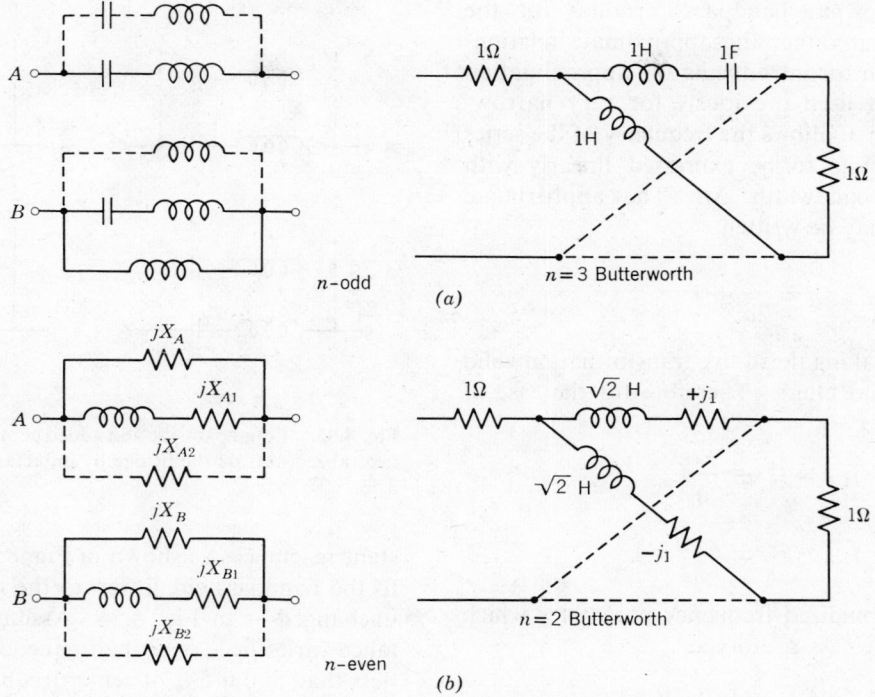

n-odd

$n=3$ Butterworth

(a)

n-even

$n=2$ Butterworth

(b)

Fig. 8.52. Examples of odd and even intermediate-band crystal filters.

8.18 THE SYNTHESIS OF INTERMEDIATE BANDPASS FILTERS

The cascade synthesis technique for narrow-bandwidth crystal filters, because of the bandwidth limitation, must be supplemented by other design methods which remove these limitations. The synthesis technique for $RLCX$ networks developed by D. S. Humpherys permits the preparation of tables of normalized crystal filter parameters for intermediate bandwidth filters (less than 1%).

Prototype Lowpass Filters

Though a ladder network is, in general, the preferred filter configuration it is not adaptable to crystal filters of this category, since it is unable to incorporate the quartz-crystal resonator when the bandwidth is extended beyond the limit formulated in the section on single sideband type of filters and it cannot effectively absorb the shunt capacitance of the crystal except in a few special designs. Bridge-type networks are generally used for filters broader than 0.2 to 0.3% bandwidth.

If the transmission function being synthesized is odd as in Fig. 8.52a, the model with conventional equivalent circuits in the arm is used. On the other hand, if n is even as shown in Fig. 8.52b the arms reactances

are unconventional. They will include reactive constants with frequency independent phase. Physically these constants are unrealizable. However, when a lowpass-to-bandpass transformation is performed, they become realizable, being equivalent to a negative capacitance and negative inductance to an excellent degree of approximation.

The arms of the lowpass prototype may contain one or more series LC combinations; by using the conventional lowpass-bandpass transformation

$$s_B = \frac{s}{2a} \pm j\sqrt{1 - \left(\frac{s}{2a}\right)^2}$$

each of these series circuits as shown in Fig. 8.53 transforms into other combinations of the series-resonant circuits. Other LC combinations cannot be incorporated in the multicrystal filter.

For this filter category it is found convenient not to

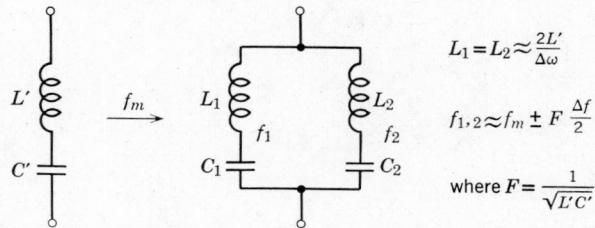

$$L_1 = L_2 \approx \frac{2L'}{\Delta\omega}$$

$$f_{1,2} \approx f_m \pm F\frac{\Delta f}{2}$$

where $F = \dfrac{1}{\sqrt{L'C'}}$

Fig. 8.53. Lowpass-bandpass reactance transformation.

use the exact lowpass-bandpass formulas for the element values but rather the approximate relationships. The reason for introducing the approximation is similar to that used previously for very narrow-bandwidth filters; it allows the frequency of the series LC-resonant circuits to be expressed linearly with respect to the bandwidth Δf. The approximate transformation may be written

$$s_B = \frac{s}{2a} \pm j$$

when $s/2a \ll 1$ making the above transformation valid for all narrowband filters. Therefore for the case of crystal frequencies

$$f_{1,2} = f_m \pm \frac{F}{2a} f_m$$

$$f_{1,2} = f_m \pm F \frac{\Delta f}{2}$$

where F is a normalized frequency parameter which belongs to the lowpass prototype

$$F = \frac{1}{\sqrt{L'C'}}$$

The above formula reflects the lowpass geometry into the bandpass plane and shows how far the crystal frequency is shifted from the center of the band $(f_m = f_a)$.

The error introduced by the approximation is negligible for bandwidths below 10%. Since the crystal filters discussed here are less than 1%, the deviation from the exact value is insignificant.

Arms of Bandpass Filters

For n-even, the lowpass prototype arms contain one or more series combinations of inductances and con-

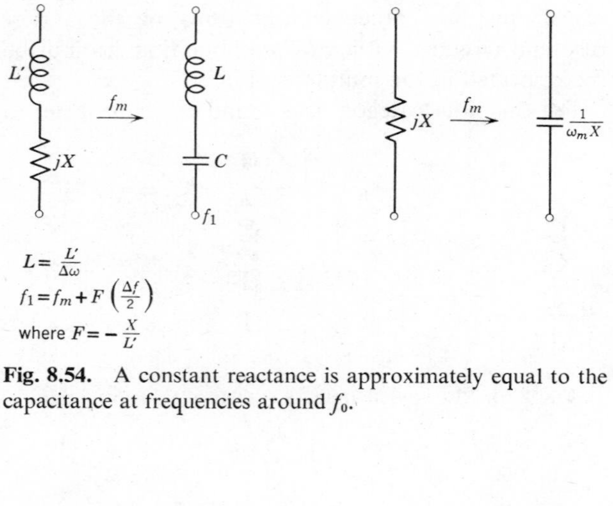

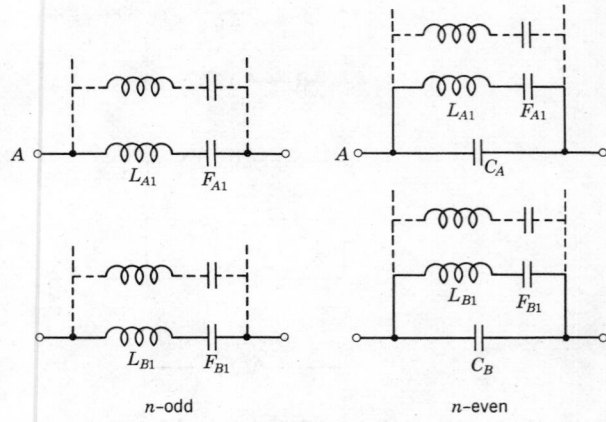

Fig. 8.55. Equal parasitic and added common capacities can be neutralized outside the bridge by inductances as shown in Fig. 8.56.

stant reactances as shown in Fig. 8.52. Transforming to the bandpass circuit leaves the constant reactance unchanged as in Fig. 8.54. Assuming that the reactance varies little over the frequency range of interest (less than about 5% of center frequency), the narrowband realization of constant reactance is a capacitance (or inductance).

Utilizing the lowpass-bandpass transformation, the lattice arms of bandpass filter schematics are shown in Fig. 8.55. After impedance scaling to make the smallest inductance of the crystal have a value of unity, the series-resonant circuits can be uniquely determined knowing only the inductance ratios and their series-resonant frequencies. Thus to simplify the design work, normalized resistances, capacitances, inductive relations, and center frequencies of an n-crystal filter section are derived from the parameters of the lowpass prototype as illustrated in Table 8.20 for $n = 3$. The Chebyshev family of filters (Butterworth, Chebyshev-polynomial, and elliptic function type filters) can be designed from this table.

Bandwidth Limitation

The schematic of bandpass filters with n-odd does not have shunt capacitances. When n is even one of the parallel capacitances in the arms is negative. Herein lies the reason for a bandwidth limitation making filters with bandwidth greater than 1% unrealizable. It would appear that such networks are not adequate for crystal filter application. Shunt capacitance which, in reality, is always present as part of the crystal equivalent schematic, can be considered as being in parallel with the load and source resistance. At one end of the semilattice an inductance is present.

$$L = \frac{L'}{\Delta\omega}$$
$$f_1 = f_m + F\left(\frac{\Delta f}{2}\right)$$
where $F = -\frac{X}{L'}$

Fig. 8.54. A constant reactance is approximately equal to the capacitance at frequencies around f_0.

Table 8.20 Intermediate Band Crystal Filters for $n = 3$

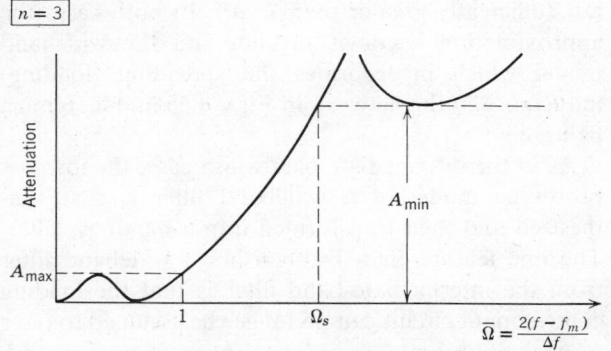

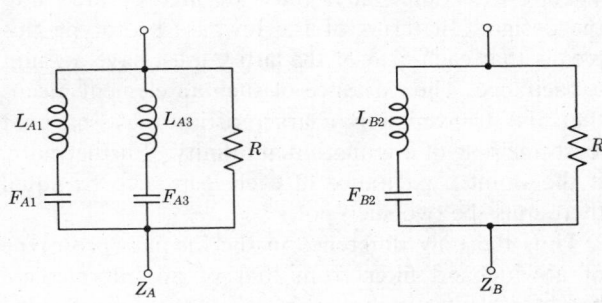

$$\Delta F_{A3} = -\Delta F_{A1} \qquad \Delta F_{B2} = 0$$

$$L_{A3} = L_{A1} \qquad L_{B2} = 1$$

Ω_s	A_{min}	L_{A1}	ΔF_{A1}	$R/\Delta f$
maximally flat		$A_{max} = 0.00$		Butterworth

Ω_s	A_{min}	L_{A1}	ΔF_{A1}	$R/\Delta f$
∞	∞	2.0000	1.0000	6.2832

$A_{max} = 0.01$

Ω_s	A_{min}	L_{A1}	ΔF_{A1}	$R/\Delta f$
∞	∞	2.0000	1.8100	9.9863
5.0	39.38	2.2141	1.7910	10.356
4.0	33.42	2.3520	1.7797	10.581
3.5	29.81	2.4807	1.7697	10.782
3.0	25.59	2.7028	1.7537	11.112
2.5	20.51	3.1488	1.7253	11.721

$A_{max} = 0.05$

Ω_s	A_{min}	L_{A1}	ΔF_{A1}	$R/\Delta f$
∞	∞	2.0000	1.4294	7.1448
5.0	46.39	2.1302	1.4235	7.3147
4.0	40.43	2.2106	1.4200	7.4158
3.5	36.81	2.2836	1.4169	7.5053
3.0	32.59	2.4054	1.4119	7.6498
2.5	27.49	2.6359	1.4029	7.9089
2.25	24.46	2.8420	1.3955	8.1267
2.00	20.99	3.1836	1.3841	8.4626

Ω_s	A_{min}	L_{A1}	ΔF_{A1}	$R/\Delta f$
		$A_{max} = 0.10$		
∞	∞	2.0000	1.2999	6.0910
5.0	49.43	2.1068	1.2969	6.2130
4.0	43.46	2.1722	1.2951	6.2853
3.5	39.85	2.2310	1.2935	6.3490
3.0	35.62	2.3280	1.2910	6.4513
2.5	30.52	2.5088	1.2863	6.6335
2.25	27.49	2.6674	1.2824	6.7851
2.00	24.01	2.9246	1.2764	7.0169
1.80	2.79	3.2743	1.2687	7.3079

$A_{max} = 0.25$

∞	∞	2.0000	1.1570	4.8206
5.00	53.48	2.0840	1.1564	4.8994
4.00	47.52	2.1348	1.1561	4.9459
3.50	43.90	2.1802	1.1557	4.9866
3.00	39.68	2.2543	1.1552	5.0517
2.50	34.57	2.3902	1.1541	5.1668
2.25	31.54	2.5072	1.1531	5.2617
2.00	28.05	2.6929	1.1515	5.4055
1.80	24.82	2.9384	1.1493	5.5837
1.60	21.02	3.3726	1.1454	5.8714

$A_{max} = 0.50$

∞	∞	2.0000	1.0689	3.9361
5.0	56.62	2.0714	1.0694	3.9925
4.0	50.65	2.1143	1.0697	4.0255
3.5	47.04	2.1525	1.0699	4.0544
3.0	42.82	2.2145	1.0703	4.1006
2.5	37.71	2.3271	1.0709	4.1818
2.25	34.67	2.4230	1.0712	4.2484
2.00	31.19	2.5736	1.0717	4.3487
1.80	27.95	2.7695	1.0720	4.4722
1.60	24.13	3.1085	1.0722	4.6697
1.50	21.92	3.3872	1.0720	4.8189

$A_{max} = 1.00$

∞	∞	2.0000	0.9971	3.1050
5.0	59.89	2.0620	0.9984	3.1446
4.0	53.92	2.0990	0.9991	3.1678
3.5	50.31	2.1318	0.9997	3.1881
3.0	46.08	2.1850	1.0007	3.2204
2.5	40.97	2.2808	1.0023	3.2770
2.25	37.94	2.3619	1.0035	3.3232
2.00	34.45	2.4879	1.0053	3.3927
1.80	31.21	2.6500	1.0072	3.4776
1.60	27.39	2.9259	1.0100	3.6124
1.50	25.18	3.1488	1.0118	3.7134
1.40	22.67	3.14840	1.0139	3.8542

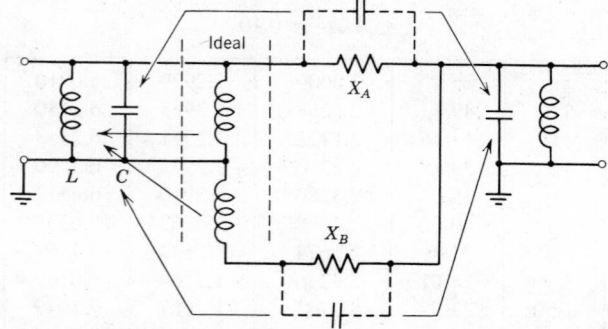

Fig. 8.56. Intermediate-bandpass filter with compensating inductances.

This inductance is part of the nonideal transformer. Both reactances (capacitance and inductance) can be made to resonate and therefore appear not to be present.

Another loading inductance of value identical to the effective-transformer inductance may be placed at the other end of the filter. Its reactance is tuned out with the effective crystal capacitance in the same manner as the inductance at the transformer side. In Fig. 8.56 the example of an intermediate filter is shown. If the inductance value is high enough (larger than L_s/a^2 where L_s is the smallest crystal inductance), the Q factor of the parallel circuit is still low. By themselves the bandwidths of these parallel circuits are much larger than those of the designed crystal filter. Consequently, the source and load circuits are almost entirely resistive with only a small reactive component. Thus the lowpass model does not include spreading coils to change the property of the lattice arm and does not add more to the complexity of original polynomial. Loading coils simply nullify the unavoidable narrowing effects of the parasitic capacitor across the filter arms.

The shunt capacitance of the crystal, supplemented by the distributed capacitance of the coil, places an upper limit on the value of loading inductance. If the winding is bifilar (for better coupling), the capacitance between the windings is especially high. Therefore, if a large inductance value is required the accent necessarily has to be put on the permeability of core material but not on the number of turns.

If the desired bandwidth is increased to above 1 or 2% of center frequency, the impedance level of the filter increases so that the source and load *LC* parallel circuits are not loaded enough to have bandwidths greater than the required five times the bandwidth of the filter. As a result of the excessive bandwidth, the necessary inequality condition may also be violated—

that is, the inductance of the tank circuit will become not sufficiently greater than L_s/a^2. In both cases the approximation becomes invalid, and the wideband model which incorporates the spreading (loading) inductance in the network in Figs. 8.55 and 8.56 must be used.

As in the intermediate bandwidth case, the lowpass prototype model of a wideband filter is first synthesized and then transformed into a bandpass filter. The one feature that distinguishes a wideband filter from the intermediate-band filter is that the loading coil resonant circuit can no longer be assumed to have a negligibly high impedance over the passband. Therefore its impedance must be incorporated into the design. In terms of the lowpass prototype this means that each arm of the lattice must have a shunt capacitance. The existance of such an element means that the transmission characteristics must have at least one pole of attenuation at infinity. Furthermore, if the shunt capacitance in each arm is to be equal there must be two such poles.

Thus the only difference in the lowpass prototype of a wideband filter from that of an intermediate bandwidth filter is that a shunt capacitance is added to each arm of the lattice in Fig. 8.55. The arm impedances, on the other hand, are found in exactly the same manner as in the intermediate-bandwidth case except the word admittance is substituted for impedance.

Numerical Example

The following specifications for a filter are required:

1. Center frequency-100 kc.
2. Maximum attenuation in the passband $A_{max} = 0.5$ dB.
3. Bandwidth-1000 cps.
4. Minimum attenuation in the stopband $A_{min} = 40$ dB at $f_m \pm 650$ cps.

SOLUTION:

The type of crystal to be used is *X*-cut. One section is required since only a 40-dB suppression in the stopband is needed.

$$a = \frac{f_m}{\Delta f} = \frac{100}{1} = 100$$

$$\Omega_s = 1.3$$

The nomograph for *CC* filters (Fig. 5.3) provides the following data. A filter with $n = 6$ guarantees more than 45 dB, even with lower ripples (0.25 dB) and a higher selectivity parameter $\Omega_s = 1.25$. The element values can be calculated

In A arm	In B arm
$L_{A2} = 1.687$	$L_{A1} = L_{B1} = 1.0$
$L_{A3} = 4.719$	$L_{B2} = L_{A2}$
$\Delta F_{A1} = 0.3997$	$L_{A3} = L_{B3}$
$\Delta F_{A2} = -0.9286$	$\Delta F_{B1} = -\Delta F_{A1}$
$R/\Delta f = 2.393$	$\Delta F_{B2} = -\Delta F_{A2}$
	$\Delta F_{B3} = -\Delta F_{A3}$

Denormalizing

$$R = \Delta f \times 2.393 = 2.393 \text{ k}\Omega$$

$$L_{A1} = 1.0 \text{ H}$$

$$L_{A2} = 1.687 \text{ H}$$

$$L_{A3} = 4.719 \text{ H}$$

$$f_{A1} = f_m + \frac{\Delta f}{2} \times \Delta F_{A1} = 100,200 \text{ cps}$$

$$f_{A2} = f_m - 500 \times 0.9286 = 99,536 \text{ cps}$$

$$f_{A3} = 100,000 + 500 \times 1.061 = 100,530 \text{ cps}$$

$$f_{B1} = f_m - 500 \times 0.3997 = 99,800 \text{ cps}$$

$$f_{B2} = f_m + \frac{\Delta f}{2} \times 0.9286 = 100,464 \text{ cps}$$

$$f_{B3} = f_m - \frac{\Delta f}{2} \times \Delta F_{B3} = 99,470 \text{ cps}$$

Assuming the lowest crystals motional inductance value equal 80 H, and the transformer with total inductance of 40 mH, the lattice structure can be impedance scaled and realized in the semilattice structure of Fig. 8.57.

The loading circuit which includes a tuning capacitor, inductance, and resistance has a bandwidth of 6.6 kc. This value of bandwidth does not distort the

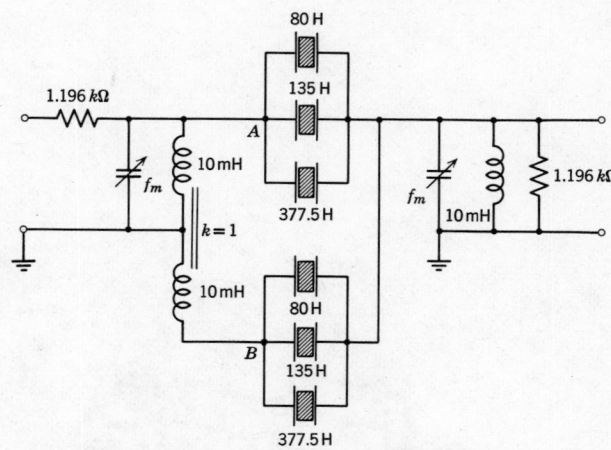

Fig. 8.57. Example of intermediate-bandpass filter design, one section.

passband characteristic of the filter with its own bandwidth of 1000 cps. The only influence of the load observed is the reduction of the ripples in the proximity of the edges.

The inductance inequality is satisfied since the inductance of the transformer (40 mH) is much larger than $L_s/a^2 = 8$ mH.

Example of Practical Synthesis for Single Sideband Application

Figure 8.58 shows another example of an intermediate-bandpass filter synthesized to reflect lowpass pole-zero geometry. The filter is realized in two cascaded semilattices with five crystals each. Inductances involved are not strictly part of the filter but rather parts of the load introduced to neutralize parasitic capacitances across the crystals. One crystal in each arm is connected in series with capacitors C_{s1}

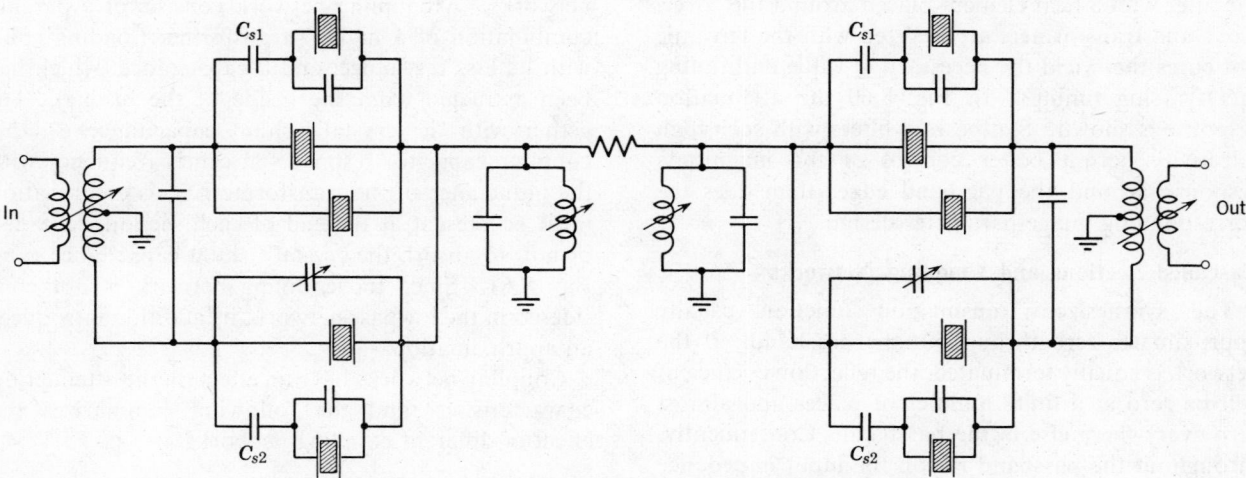

Fig. 8.58. Single sideband two-section filter synthesized with ten crystals.

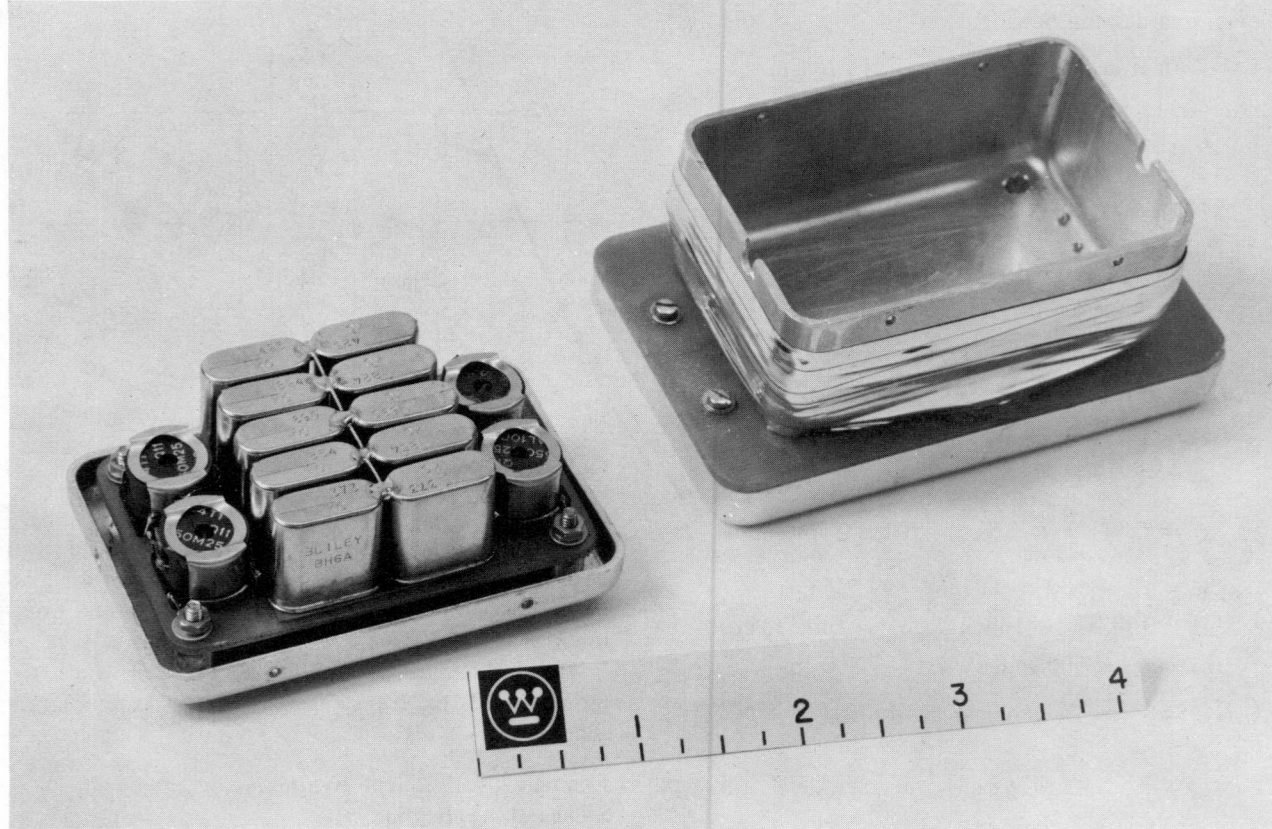

Fig. 8.59.   Practical realization of filter shown in Fig. 8.58.

and C_{s2} in order to bring those crystal impedances to a realizable value of the same order as the rest of the crystals.

For the physical network operating in unstable environmental conditions, a certain amount of temperature control is required. Figure 8.59 shows the filter with a heat element placed around the cover. Coils and transformers are realized with the ferramic pot cores that yield the necessary Q value and tuning facility (slug tuning). In Fig. 8.60 the attenuation response is shown. Synthesized filters with such high selectivity permit better control of the magnitude response around the passband edges than does the corresponding image-parameter design.

Cascaded Sections and Coupling Networks

The synthesized transmission function usually approximates zero throughout the passband. If the network is equally terminated, the reflection coefficient will be zero at a finite number of places and almost zero everywhere else in the passband. Consequently, throughout the passband region the input impedance of the network must be very nearly a pure resistance,

equal in value to the source resistance. When that condition is satisfied it is possible to cascade several sections together without increasing the passband ripple appreciably. The many benefits of cascading, mentioned previously, can be realized. Cascading bandpass crystal filters requires the use of coupling networks. A coupling network consists of a parallel combination of a high Q transformer (loading coil), with its loss resistance, and a capacitance (which has been extracted from the inside of the bridge). Together with the crystal's shunt capacitance C_p this coupling capacitor resonates at center frequency with the inductance of the transformer. This combination must be present at the end of each section, cascaded or not, to absorb the crystal's shunt capacitance as in Fig. 8.61. Since the coupling network is not considered in the lowpass network, its addition introduces an approximation.

Coupling networks have an effect on the attenuation characteristics, and the following summarizes the effect of different coupling networks:

1. The transition region from the passband to the

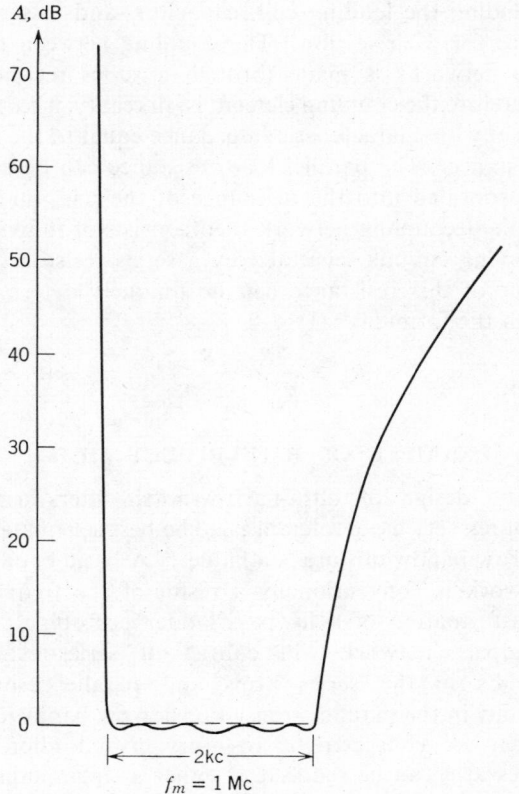

Fig. 8.60. Response of filter shown in Fig. 8.58.

stopband is essentially unaffected by the choice of coupling networks as long as the loading coils inductance is much larger than the product of the lowest motional inductance and the square of the relative bandwidth ($L_s/a^2 \ll L$).

2. The passband ripple increases with the number of sections cascaded, but the amount of losses introduced by coupling networks is normally small.

3. The passband ripple is reduced slightly by the addition of a lossless coupling network. From this point of view it is desirable to make the parallel resistance of the coupling networks much greater than

the load resistance. Otherwise an impedance mismatch occurs between sections, giving a sizable increase in passband ripple. On the other hand, a lossless coupling network will introduce undesirable passbands, one on each side of the main response. The exact location of these spurious bands is a function of the inductance of the transformer. A larger inductance moves the spurious bands further from the fundamental response. In most practical cases it is impossible to observe spurious bands, since the coupling loss dampens the spurious response to an almost insignificant level. Nevertheless the stopband attenuation can be decreased appreciably in certain frequency ranges. This necessitates some overdesigning. When large coupling inductances are used, the stopband attenuation is essentially the sum of the attenuation of the individual sections in exactly the same fashion as image-parameter-theory design.

4. Cascading can increase the ripple bandwidth considerably if the sections are designed with either low ripple content or a small number of crystals per section.

5. The analysis of cascaded identical sections, as shown in Fig. 8.61, reveals that when the ripples in individual sections are very small, the total ripple of the cascaded network is smaller than the sum of the ripple content of the individual sections. The reverse situation exists when the ripples of the individual sections are very large, for example 1 dB. Here the ripples of the bandpass filter will be greater than the sum of the individual section's contribution. The obvious conclusion is that this design technique is more appropriate to Chebyshev filters with a low level of ripples in the passband.

6. Sections having small values of ripples have a gradual change in amount of reflection in the proximity of the band limit. When cascaded, moderate reflections can occur at the beginning of the transition region. These reflections reinforce each other at one frequency to give an extremely narrow spurious

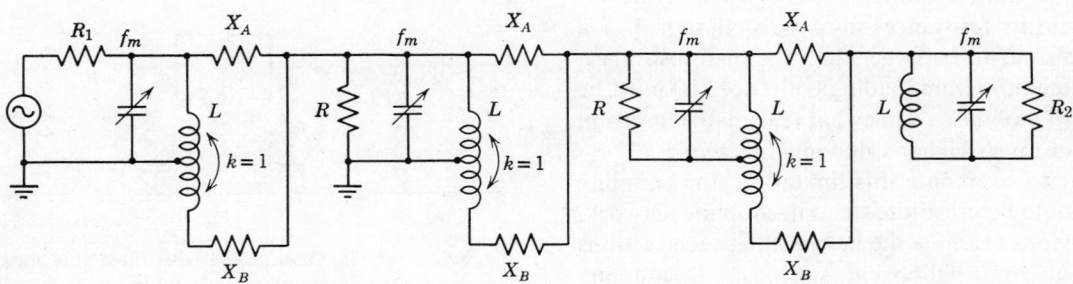

Fig. 8.61. Example of cascaded synthesis for filters below 1% bandwidth.

passband (for lossless elements). In a physical realizable model with finite Q and practical tolerances, this spurious response is dampened almost completely, but its presence in the calculated response indicates that the transition region is not as narrow as it could be. This is why the filter, in most cases, must be overdesigned.

Filter with Nonidentical Sections

Nonidentical sections are designed in essentially the same way as a cascade of identical sections. Each section is designed individually and one section is chosen as the reference. The load resistances of the remaining sections must be scaled so that they are the same as the reference section. After this operation, the sections can be connected in cascade. It must be emphasized that the main difference between these sections is the shape factor, which implies that the points of maximum attenuation are in different positions.

The study of response characteristics of several filters designed in this manner shows that the response is almost identical to those mentioned before with identical sections. Filters containing identical sections have obvious advantage of simplicity, although nonidentical sections give better spurious control.

Design with Lossy Coupling

When the relative bandwidth is increased beyond approximately 1%, the previously mentioned inductance inequality condition is difficult to satisfy. Moreover, the transformer losses begin to approach the value of load resistance, and the motional inductance of the crystal has to be increased to keep the passband spurious free. A bandwidth is eventually reached, where the methods of cascaded filter design discussed in previous sections cannot be employed. In the range of bandwidths between 1 and 1.5%, the inductance of the transformer is approximately equal to the square of the relative bandwidth times the inductance of the crystal. The meaning of this expression is that the inductance is too small, and yet the parallel loss resistances must be maintained at a high value so as to be larger than the load resistance. The absolute minimum loading coil Q of 100 must be available. To obtain a somewhat reasonable insertion loss, a Q of much higher value must be used.

In order to overcome this limitation, the coupling network could be substituted by a decoupling network. In the previous example the network between sections was shared by neighboring sections. Decoupling sections between two filters consist of shunt arms

including the loading coil, capacitor, and loss resistance for each section. The coupling between these two networks is made through a series resistance. Therefore the coupling element is, in reality, a resistive π-pad with characteristic impedance equal to the load resistance. The parallel load resistance can be easily incorporated into the resistance of the pad, and the whole decoupling network then consists of individual coupling circuits separated by a series resistor. The value of this resistance can be uniquely determined from the formula

$$R_s = \frac{2R_{\text{Loss}}R_{\text{Load}}}{R_{\text{Loss}}^2 - R_{\text{Load}}^2}$$

8.19 EXAMPLE OF BAND-REJECT FILTER

The design of ultra-narrow-notch filters usually requires very close tolerances. The best schematic for narrow bandwidth use is a ladder. A basic bandpass network is conventionally a result of the frequency transformation of a lowpass ladder prototype. The bandpass network will consist of series-resonant circuits in the series arms and parallel-resonant circuits in the parallel arms. For narrow bandwidths, which are characteristic to every crystal filter, the series arm can be reduced to either a single inductor or a single capacitor without appreciably affecting the response characteristics. This technique is already known and used in modern synthesis of LC filters. The basic band-reject schematic may be derived from a highpass filter in ladder form, and consequently will consist of series-resonant circuits in the parallel arms and parallel-resonant circuits in the series arms as illustrated in Fig. 8.62a and b. Using an argument similar to the one used for bandpass filters with narrow bandwidths, the series arms will reduce to a single capacitor or single inductor which will be assumed a constant reactance in a narrow band of frequencies. The narrowband approximation is given in Fig. 8.62c.

From the preceding, one might consider a possible

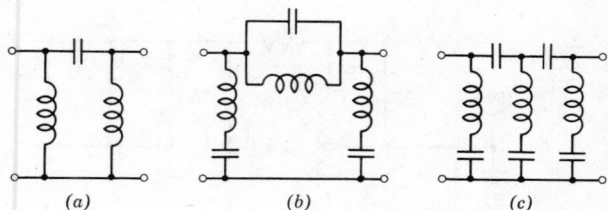

(a)   (b)   (c)

Fig. 8.62. (*a*) Highpass model with three transmission zeros transformed to (*b*) third-order band-reject filter. (*c*) narrowband equivalent of band-reject filter.

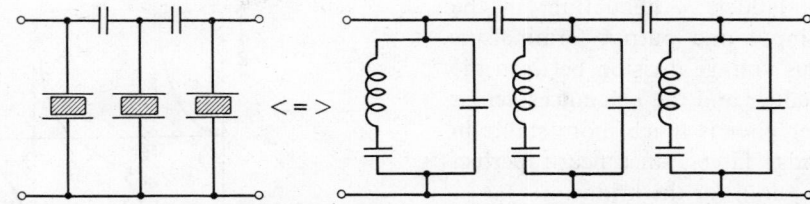

Fig. 8.63. Crystal ladder approximating circuit of Fig. 8.62*c*, and its equivalent schematic.

crystal notch configuration, as shown in Fig. 8.63, if the supplementary crystal capacitance across the resonant circuit can be neutralized. The schematic of this figure, of course, violates the narrowband notch schematic of Fig. 8.62*c*. The behavior of the schematic of Fig. 8.63 may be made to duplicate the desired notch configuration by adding a shunting inductor across the parallel arms and resonating the parallel combination at the series resonant frequency of the shunt arms. The final schematic is shown in Fig. 8.64. The bandwidth of the parallel-resonant circuit is much larger than the bandwidth of the series circuit. The effect of coil imperfection (Q) is negligible.

The final schematic has as its shunt arms a combination of the narrowband bandpass and narrowband band-reject filter, the series arms being the same. One might therefore expect both a bandpass and notch characteristic from this same network by an appropriate choice of element values. This approximation to a bandstop filter results in making the attenuation far above and far below the stopband not equal to zero. Experimentation does varify these results. The synthesis procedure for this network permits optimum performance and allows the filter to be derived from a highpass model which can be of a high order.

8.20 LADDER FILTERS WITH LARGE BANDWIDTH

A relatively wideband filter can be designed as a combination of *LC* components and crystal resonators in a ladder realization. For example, the input and output circuit may be a conventional *LC* tank circuit, but the coupling between them in series arm will be a crystal shunted by spreading coil as shown in Fig.

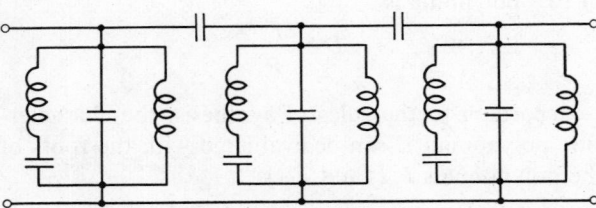

Fig. 8.64. Final band-reject ladder filter.

8.65. The maximum bandwidth for this type of filter is approximately 6% of the center frequency, the widest filter possible with *X*-cut low-frequency crystals.

In ladder schematics, in general, the attenuation poles are produced by anti-resonant circuits in the series arms or by traps (series-resonant circuits) in the parallel arms. The position of the attenuation poles depends only on the sections where such a circuit is involved. Matching between sections of a complex ladder or between filter and load is simple because the attenuation poles can be prescribed independently in every section without interference to the rest of the filter.

In dealing with extremely sharp filters another general parameter, that of selectivity, must be added to the terminology. The numerical definition of selectivity is

$$S = \frac{f_c}{f_\infty - f_c} \qquad (8.20.1)$$

where f_c is the cutoff frequency and f_∞ is the frequency of the closest pole outside of the passband. Design experience indicates that when S is less than 60, the filter can be realized with coils and capacitors (*LC* filters), but if it is greater than 60, the filter by necessity is a crystal filter.

The ultimate solution for high attenuation and stability and a wide passband is achieved only with a ladder schematic or with a schematic where only a small part of whole composite network is a bridge,

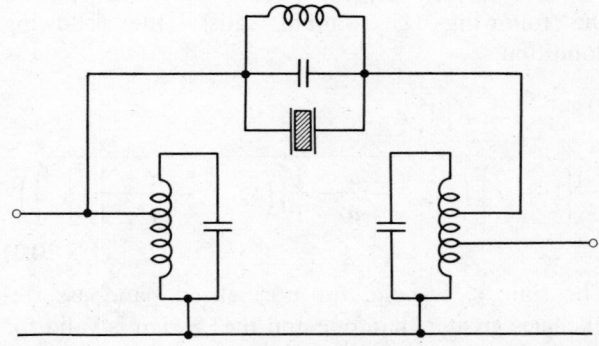

Fig. 8.65. Simplest configuration of ladder filter with *LC*-crystal components.

the majority being a ladder. Attenuation, in the ladder, between the input and output terminals is provided by continuous voltage division between the reactances along the ladder and the reactances across the ladder. The ladder filter is much more stable in contrast with pure bridge filters, since nearly perfect electrical balance is required for the bridge.

Crystal elements can be either inserted in the series arm or in the shunt arm of the ladder. In 1956, Poschenrieder devised the method of inserting the piezoelectric resonator in the parallel arm of a conventional ladder schematic. The design procedure was based on the image-parameter theory. This method of development is analytical in nature.

Synthesized *LC* Filters with Quartz Elements

The method of synthesis permits us to prescribe the property of the network and, with the aid of the mathematical theory of effective parameters, to translate these properties into the elements of the networks. As we know, the function D is the starting point of the synthesis of a network. The relationship between the discrimination and transmission function is expressed by

$$H(s)H(-s) = 1 + D(s)D(-s) \qquad (8.20.2)$$
$$|H|^2 = 1 + |D|^2$$

Besides the usual necessary conditions for the filtering function, the following supplementary requirements are to be satisfied:

1. At frequency $\Omega = 0$, a minimum of one double pole must exist. Theoretically, a simple pole would be sufficient, but from the point of view of realization a multiple pole at $\Omega = 0$ permits more structural variations. From the practical approach, this type of flexibility leads to optimum realization of the network components.

2. When the frequency Ω_1 in Fig. 8.66 is realized as the crystal pole $\Omega_{1\infty}$, the distance between $\Omega_{1\infty}$ and the following $\Omega_{2\infty}$ should satisfy the following condition

$$\Omega_{2\infty}^2 \underset{\geq}{\lessgtr} \Omega_{1\infty}^2 \left(1 + \right.$$

$$\left. \frac{C_s}{C_p} \left\{ 1 + \sqrt{\left[\left(\Omega_{1\infty}^2 - \frac{a+1}{a-1}\right)\left(\Omega_{1\infty}^2 - \frac{a-1}{a+1}\right)\right]^{-1}} \right\} \right)$$
$$(8.20.3)$$

The sign $\leq$ is valid for normalized bandpass frequencies greater than one and the $\geq$ sign is valid for normalized frequencies less than one. The relative bandwidth of the network is known to be $1/a$, and the

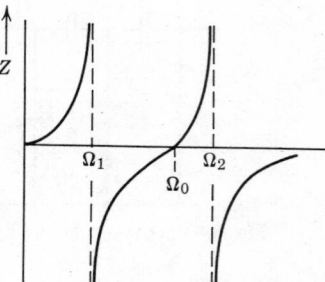

Fig. 8.66. Reactance diagram of the series arm of Fig. 8.65.

value of C_s/C_p is the given ratio of the dynamic to static capacitance of the quartz crystal. The approximating expression (8.20.3) yields the limit for the $\Omega_{2\infty}$ pole position provided that the choice of frequency $\Omega_{1\infty}$ is made, the bandwidth coefficient a is known, and the technically realizable ratio C_s/C_p of the crystal is given. Because only the sharpest pole must be realized with the crystal, all the rest of the critical frequencies except $\Omega_{2\infty}$ are still free for choice.

Realization

In bandpass filters with *LC* and piezoelectric crystals, the schematic usually requires one capacitance and one inductance in parallel with the crystal. The reactance diagram for this combination of elements contain one zero and two poles at finite frequencies, as shown in Fig. 8.66. At zero and infinite frequency simple zeros exist. The expression for reactance can be written in the following form

$$Z = A \frac{s(s^2 + \Omega_0^2)}{(s^2 + \Omega_1^2)(s^2 + \Omega_2^2)}$$

where A is a constant. With the conditions 1 and 2 above, the discrimination function will be uniquely defined as

$$D = B \frac{\displaystyle\prod_1^v s^2 + \Omega_{2v}^2}{s^x \displaystyle\prod_v^{v-2} s^2 + \Omega_{2v-3}^2} = \frac{F(s)}{P(s)}$$

with $x = 2, 3, 4$, according to the number of poles at $s = 0$. The filtering function D, is defined as the ratio of two polynomials

$$D = \frac{F}{P}$$

According to the rules of synthesis, the characteristic polynomial E can be evaluated with the roots of the polynomials $P(s)$ and $F(s)$

$$E(s)E(-s) = P(s)P(-s) + F(s)F(-s)$$

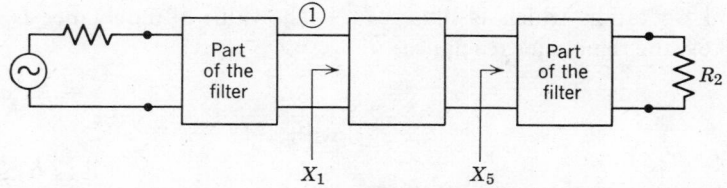

Fig. 8.67. Complex filter network subdivided in order to introduce a crystal.

The transmission function will be

$$H(s) = \frac{E(s)}{P(s)} = \frac{E_e + E_o}{P(s)}$$

where $E = E_e + E_0$ and E_e stands for the even and E_0 for the odd part of the function.

The chain matrix for $P(s)$ even is

$$|M_e| = \frac{1}{P} \begin{vmatrix} E_e + F_e & E_o + F_o \\ E_o - F_o & E_e - F_e \end{vmatrix}$$

and for $P(s)$ odd is

$$|M_o| = \frac{1}{P} \begin{vmatrix} E_o + F_o & E_e + F_e \\ E_e - F_e & E_o - F_o \end{vmatrix}$$

From these expressions and from knowing the general structure of the network (symmetric or antimetric) one can proceed with the removal of the open-circuit or short-circuit impedances.

The following discussion considers the realization with crystals inserted across the line. This method of realization which has been described in Chapter 4, allows more generalizations. Usually the removal is accomplished only with the aid of preliminary partial removals at $s = 0$ or $s = \infty$. The generalizations call for a partial removal at finite frequencies, which instead of a coil or a capacitor the partially removed elements would consist of a resonant circuit which constitutes a part of an attenuation pole as mentioned in Section 4.19, suggesting that the full removal of this pole will be performed later in the circuit.

Sequence of Removals

In Fig. 8.67 the parts of a complicated filtering network are shown. The left part of the entire network is developed in the regular manner by the sequential removal of reactive elements up to point 1. The remaining part of the network is developed with the reactance function:

$$X_1(s) = \frac{N_1 \text{ (numerator)}}{D_1 \text{ (denominator)}}$$

The function is prepared for the step of partial removal of the parallel circuit at $\Omega_{2\infty}$, as shown in Fig. 8.68. (The poles at zero or infinity are already partially removed so that the remainder function suggests the partial removal of the pole at the finite frequency.)

To produce the equivalent schematic of a quartz crystal across the line, the parallel-resonant circuit at $\Omega_{2\infty}$ must be removed in such a manner that the remainder function is able to produce the pole of attenuation $\Omega_{1\infty}$. The remainder function after partial removal of the resonant circuit at $\Omega_{2\infty}$ is expressed by

$$X_2 = X_1 - \frac{1}{C_3} \frac{s}{(s^2 + \Omega_{2\infty}^2)} = \frac{N_1 - sD_1(1/C_3)}{D_1(s^2 + \Omega_{2\infty}^2)}$$

The remainder reactance $X_2 = 0$ for $s^2 = -\Omega_{1\infty}^2$. Therefore,

$$N_1 - \frac{1}{C_3} sD_1 \bigg|_{s^2 = -\Omega_{1\infty}^2} = 0$$

$$C_3 = \frac{sD_1}{N_1} \bigg|_{s^2 = -\Omega_{1\infty}^2} \quad \text{and} \quad L_3 = \frac{1}{C_3 \Omega_{2\infty}^2}$$

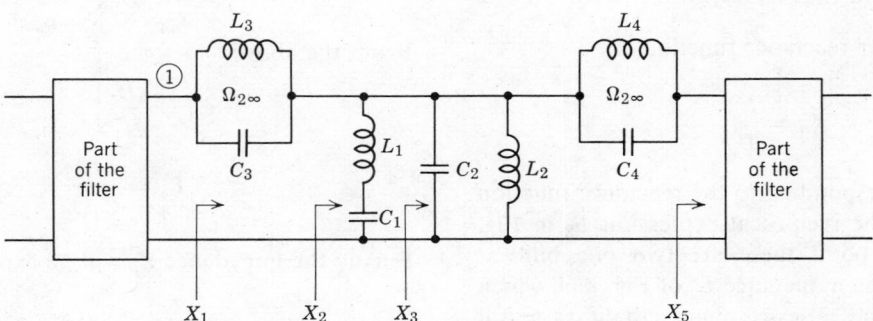

Fig. 8.68. Part of the filter used to illustrate the removal technique.

After this partial removal operation which is illustrated on line X_2 of Fig. 8.69, the remainder reactance will be

$$X_2 = \frac{N_2}{D_2}$$

Since the pole at $\Omega_{2\infty}$ is only partially removed,

$$N_2 = \frac{N_1 - sD_1(1/C_3)}{s^2 + \Omega_{2\infty}{}^2}$$

Because the parallel circuit is removed in a special manner so that at $\Omega_{1\infty}$ the reactance $X_2 = 0$, the numerator N_2 must include the factor $(s^2 + \Omega_{1\infty}{}^2)$. Therefore it can be constructed as

$$N_3 = \frac{N_2}{s^2 + \Omega_{1\infty}{}^2}$$

The remainder admittance (susceptance) can be made

$$B_3 = \frac{1}{X_2} - \frac{1}{L_1}\frac{s}{(s^2 + \Omega_{1\infty}{}^2)} = \frac{D_2 - \frac{1}{L_1}sN_3}{N_3(s^2 + \Omega_{1\infty}{}^2)}$$

In the numerator of the reactance equation, the factor $s^2 + \Omega_{1\infty}{}^2$ must exist and so the following equation is valid:

$$D_2 - \frac{1}{L_1}sN_3 \bigg|_{s^2=-\Omega_{1\infty}{}^2} = 0$$

From which the elements of the series-resonant circuit can be obtained

$$L_1 = \frac{sN_3}{D_2}\bigg|_{s^2=-\Omega_{1\infty}{}^2} \quad \text{and} \quad C_1 = \frac{1}{L_1\Omega_{1\infty}{}^2}$$

For the remainder function X_3, the denominator is

$$D_3 = \frac{D_2 - (1/L_1)sN_3}{s^2 + \Omega_{2\infty}{}^2}$$

Then the remainder reactance function is

$$X_3 = \frac{N_3}{D_3}$$

The situation corresponding to the reactance function X_3 is shown for the reciprocal expression B_3 in Fig. 8.69. From this point there are two possibilities: either to remove the inductance L_2 of Fig. 8.68 which is in parallel to the series-resonant circuit as a full removal or as partial removal at $\Omega = 0$.

The value of inductance L_2 is

$$\frac{1}{L_2} = K\frac{sD_3}{N_3}\bigg|_{s\to 0}$$

with

$$0 \le K \le 1$$

The remainder reactance is obtainable from

$$D_4 = D_3 - \frac{N_3}{sL_2}$$

$$X_4 = \frac{N_4}{D_4} \quad \text{with} \quad N_3 = N_4.$$

The following step can be, for example, the partial removal at infinite frequency, so that the remainder impedance has a pole at $\Omega_{2\infty}$. This step serves as a preparatory step for a full removal of a parallel circuit in the series arm of the network.

$$\frac{1}{X_4} - sC_2 \bigg|_{s^2=-\Omega_{2\infty}{}^2} = 0$$

or

$$C_2 = \frac{D_4}{sN_4}\bigg|_{s^2=-\Omega_{2\infty}{}^2}$$

In the case when the following step is a full removal of a parallel resonant circuit, the denominator must consist of

$$D_5 = \frac{(D_4 - sC_2N_4)}{(s^2 + \Omega_{2\infty}{}^2)}$$

and the remainder reactance function will be of the following form

$$X_5 = X_4 - \frac{1}{C_2}\frac{s}{(s^2 + \Omega_{2\infty}{}^2)} = \frac{N_4 - sD_5(1/C_4)}{D_5(s^2 + \Omega_{2\infty}{}^2)}$$

Since the numerator N_5 has the factor $(s^2 + \Omega_{2\infty}{}^2)$ the following has to be valid:

$$N_4 - \frac{1}{C_4}sD_5 \bigg|_{s^2=-\Omega_{2\infty}{}^2} = 0$$

From the above,

$$C_4 = \frac{sD_5}{N_4}\bigg|_{s^2=-\Omega_{2\infty}{}^2}$$

$$L_4 = \frac{1}{C_4\Omega_{2\infty}{}^2}$$

Finally the impedance Z_5 will be expressed by

$$Z_5 = \frac{N_5}{D_5}$$

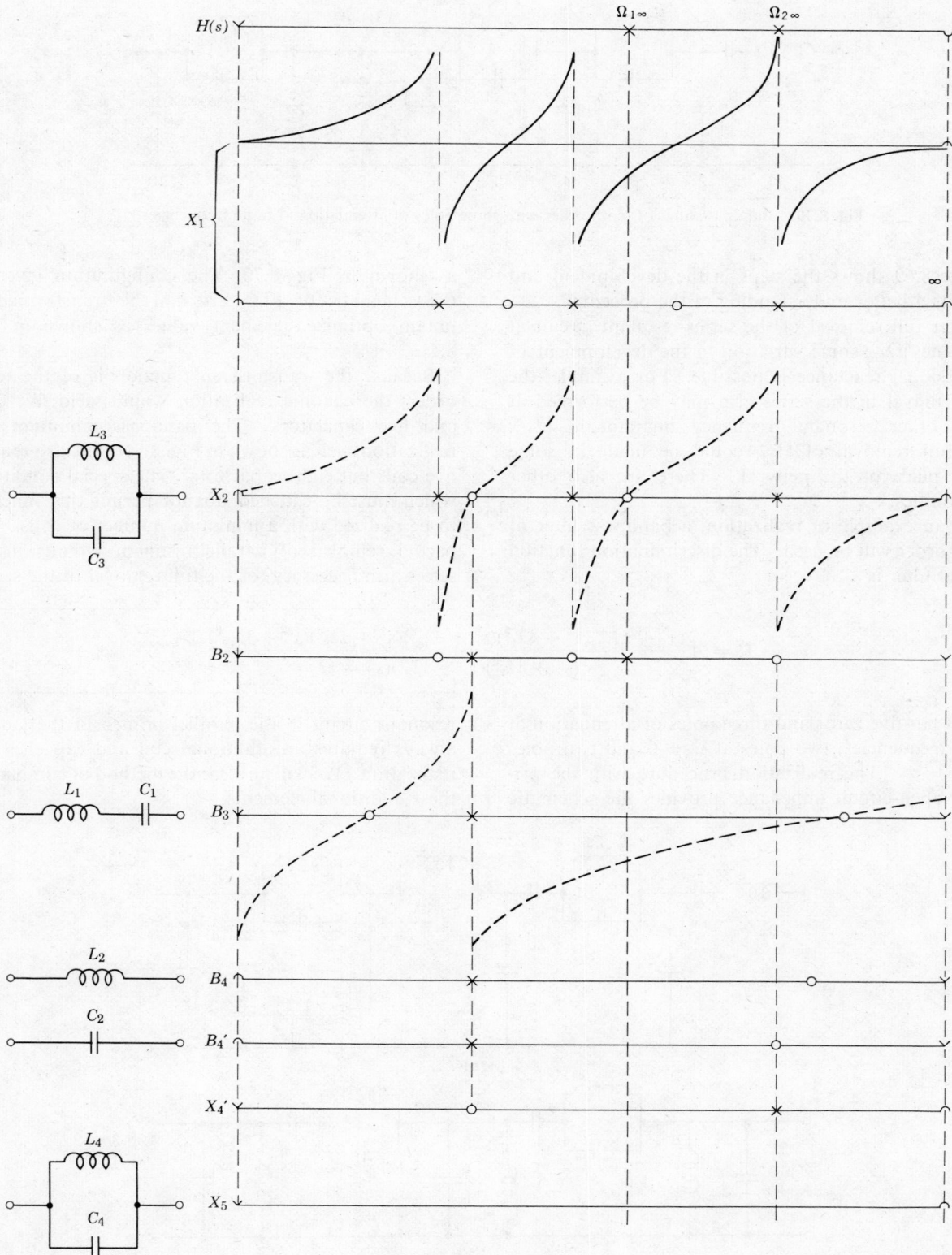

Fig. 8.69. Sequence of partial and full removals.

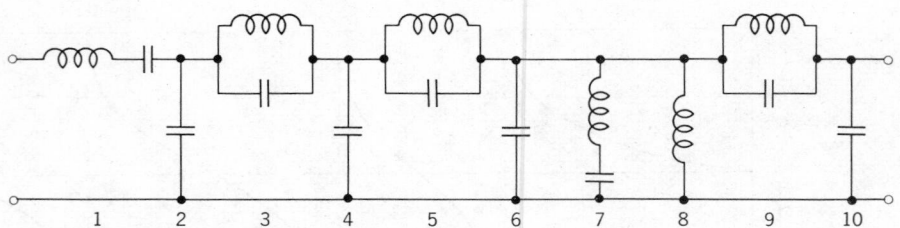

Fig. 8.70.   Bandpass filter of tenth order with three poles of attenuation at finite frequencies.

Figure 8.69 shows the steps in the development and permits a better understanding of the procedure.

After full removal of the series-resonant circuit at frequency $\Omega_{1\infty}$ some variation in the development of the residual reactance is possible. For example, the full removal in the series arm may be performed at some other resonant frequency but not at $\Omega_{2\infty}$. The full removal of $\Omega_{2\infty}$ could be made in some other place of the network. There are also other possibilities.

As an example of realization, a bandpass filter of tenth order will be used. The discrimination function of that filter is

$$D = A \frac{(s^2 + \Omega_2{}^2)(s^2 + \Omega_4{}^2)(s^2 + \Omega_6{}^2)(s^2 + \Omega_8{}^2)(s^2 + \Omega_{10}{}^2)}{s^2(s^2 + \Omega_1{}^2)(s^2 + \Omega_3{}^2)(s^2 + \Omega_5{}^2)}$$

which has five zeros and three poles of attenuation at finite frequencies, two poles at $\Omega = 0$ and two poles at $\Omega = \infty$. The realization procedure with the primary open-circuit impedance provides the schematic

as shown in Fig. 8.70. The configuration given in the schematic of Fig. 8.70 can be transformed to obtain optimized element values as shown in Fig. 8.71a.

Because the transmission function is of the tenth order, the canonic realization would yield five coils and five capacitors. The bandpass minimum coil realization such as shown in Fig. 8.71b will also require five coils but eight capacitors. The special conditions which must be satisfied do not permit this function to be realized with a minimum number of coils. The partial removal of parallel-resonant circuits in the series arm necessary for the full removal of the series-resonant circuit in the parallel branch of the ladder, always requires an additional coil and capacitor for realization. We will now see the method of eliminating these additional elements.

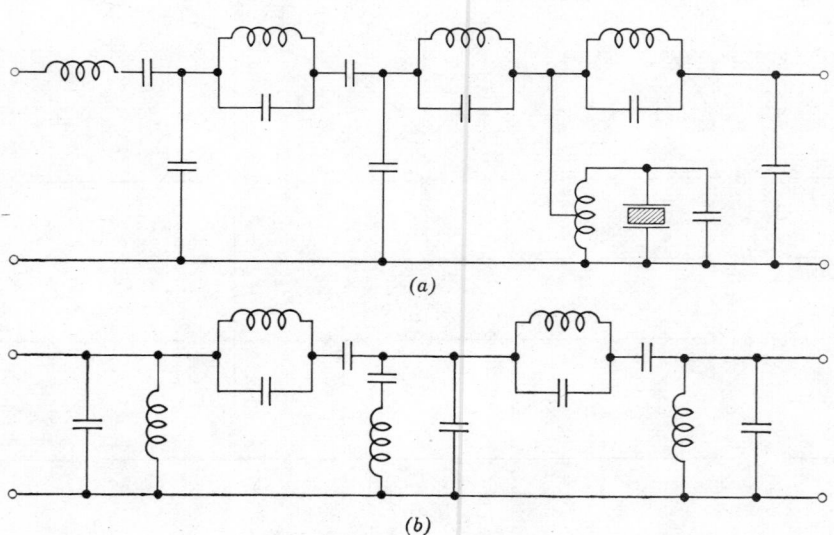

(a)

(b)

Fig. 8.71.   Crystal filter (a) and minimum coil filter (b) of equal order.

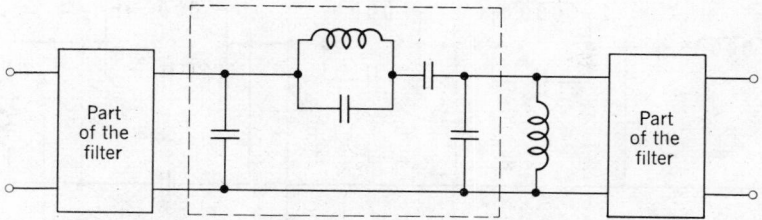

Fig. 8.72. *LC* filter with sharp attenuation peak in series arm.

The Use of Crystals in the Series Arm of the Ladder Filter

Once more we must begin with the discrimination function *D*. The conditions 1 and 2 must be fulfilled. From the short-circuit or open-circuit impedance, the circuit elements and schematic are developed. When the schematic is developed, there is no equivalent crystal configuration. But it is important to make sure that the sharpest pole is a parallel circuit in the series branch of the ladder. A series capacitor must come before or after this circuit. The parallel branches of the filter should have capacitors on both sides of the above mentioned combination and also an inductance across the line after the section, as shown in Fig. 8.72.

The dipole in the series branch can be transformed into an equivalent network with a series-resonant circuit having a capacitor across it. But the ratio C_p/C_s in this circuit may happen to be unrealizable with a crystal. The necessary transformation to accommodate the crystal in the circuit is shown in Fig. 8.73. First the network is made symmetrical and then it is transformed into a lattice.

From comparison of matrix coefficients it follows

$$C_0 = C_3'.$$

$$C_p = \frac{1}{C_1 + C_2}(2C_1C_2 + C_1C_3' + C_2C_3')$$

$$C_s = \frac{2C_2^2}{C_1 + C_2}$$

$$L_s = \frac{L_1}{2}\left(\frac{C_1 + C_2}{C_2}\right)^2$$

The last expression indicates that L_s is considerably larger than L_1, generally because capacitor C_1 is much larger than C_2. Consequently, C_s is very small. The values in this schematic suggest the possibility of inserting crystal resonators.

The transformation permits the conversion of a part of the ladder into a semilattice. Therefore the attenuation between the input and output is not entirely provided by the voltage division action of the ladder schematic but partly by a compensation effect between the arms of the bridge circuit.

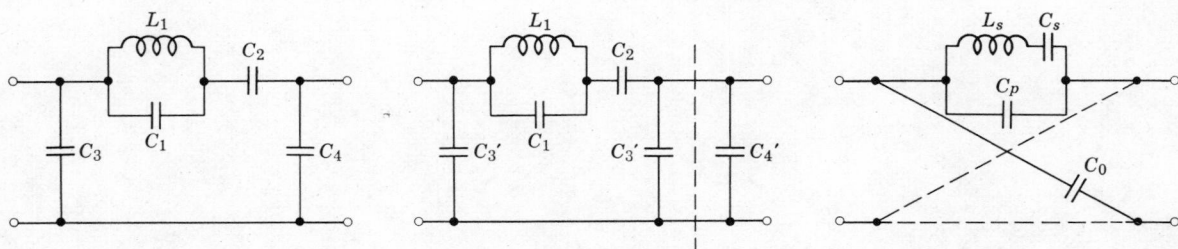

Fig. 8.73. Equivalent circuits used to substitute a crystal for the *LC* circuit of Fig. 8.72.

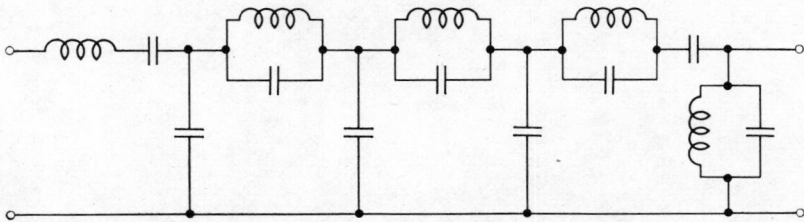

Fig. 8.74. *LC* equivalent schematic of tenth-order filter which has been prepared for crystal insertion.

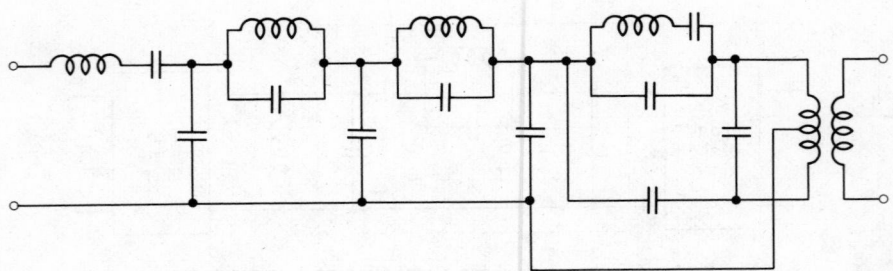

Fig. 8.75. Practical realization of the network of Fig. 8.74 with one crystal in semilattice circuit.

As long as only a small part of the total network is being transformed into a bridge configuration, the network does not exhibit the disadvantages of a pure bridge. In most cases it is sufficient to transform only one or a maximum of two attenuation poles on the sharpest side of the filter for crystal use.

The requirement for symmetry and exactitude in a partial bridge network is evidently very much lower, because the largest part of the attenuation is still provided by the ladder part of the complete network.

As an example of the transformation, once more a bandpass filter of the tenth order is used. To satisfy the attenuation requirements, the filter is to have the schematic as shown in Fig. 8.74. With the aid of the relations for network transformation, the extreme right side of the filter is modified as shown in Fig. 8.75.

9

Helical Filters

9.1 INTRODUCTION

In the VHF range, high-selectivity filters cannot be realized by conventional techniques. All filters require certain minimum values for the quality factor of their resonant circuits. At frequencies above 30 Mc, high-quality lumped elements, inductances, and capacitances do not exist. The piezoelectric crystal resonator, although of high quality, is not very flexible in realization of filters with wide bandwidths. Finally, the coaxial resonator cannot be used in the VHF range because of its large size. However, in the UHF range the cavity resonator (shown in Fig. 9.1) with tuning adjustment and coupling loops is very useful.

The toroidal inductor is not generally used at frequencies above 30 Mc because of its high distributed capacitance. The single-layer solenoid can be used to a degree, but the maximum Q realizable is about 200. Also, in filter applications the coils must be isolated from each other, requiring the use of shields, which, when placed in close proximity to the coils, reduces their Q. The quality factor can be improved by increasing the size of the solenoid, but the resulting filter is unproportionally large in comparison with the size of other components used in modern circuitry.

The piezoelectric crystal, although of high Q and small size, becomes impractical at higher frequencies. It is practically impossible to construct crystals whose fundamental frequency is above approximately 35 Mc since their physical dimensions become so small that good quality crystals cannot be produced. Harmonic crystals are generally utilized at frequencies above 30 Mc. The crystal is a high-quality device, but it has certain performance weaknesses, such as an unpredictable amount of spurious modes above the fundamental frequency (or any harmonic frequency), which discourages its use except for very low-percentage bandwidths (below 1 %). These shortcomings reduce its value for use in filter construction. Any attempt to create wideband crystal filters meets unsurpassable difficulties because of the necessity to include lossy spreading coils, which reduce the obtainable bandwidth from its maximum value. A crystal filter is basically a very-narrow-bandpass element.

For a long time, the filter-design engineer has looked for a new type of resonator which could be used to produce selective, lowloss filters in the VHF domain. Coaxial lines with helical inner conductors have been used in traveling-wave tubes, parametric amplifiers, high-Q resonators, high characteristic-impedance transmission lines, low-frequency antenna designs, and in impedance-matching techniques for frequencies as low as 300 kc. The use of helical resonators for filtering in the VHF–UHF range will be discussed in this chapter.

9.2 HELICAL RESONATORS

Helical resonators of practical size and form factor and with high Q (of the order of 1000) can be constructed for the VHF and UHF ranges. Basically they resemble

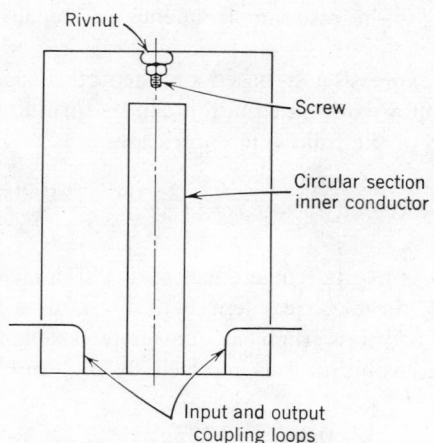

Fig. 9.1. Coaxial cavity for use in UHF filtering.

a coaxial quarter-wave resonator, except that the inner conductor is in the form of a single-layer solenoid, or helix. The helix is enclosed in a highly conductive shield of either circular or square cross-section. One lead of the helical winding is connected directly to the shield and the other end is open circuited.

As an example of space saving and a superior form factor, consider a coaxial resonator at 54 Mc with an unloaded Q of 550. The coaxial resonator would be 4.5 ft long by 0.7 in. in diameter. The same quality helical resonator would be 2 in. long by 1.5 in. in diameter.

Figure 9.2 is a sketch of the resonator with a circular cross-section. With these notations, the following set of equations can be given:

$$L = 0.025n^2d^2[1 - d/D^2] \; \mu\text{H per axial inch} \quad (9.2.1)$$

where

L = the equivalent inductance of the resonator in μH per axial inch.

d = the mean diameter of the turns in inches.

D = the inside diameter of the shield in inches.

$n = 1/\tau$ = turns per inch, where τ is the pitch of the winding in inches. $\quad (9.2.2)$

Empirically for an air dielectric,

$$C = \frac{0.75}{\log_{10}(D/d)} \; \mu\mu\text{F per axial inch} \quad (9.2.3)$$

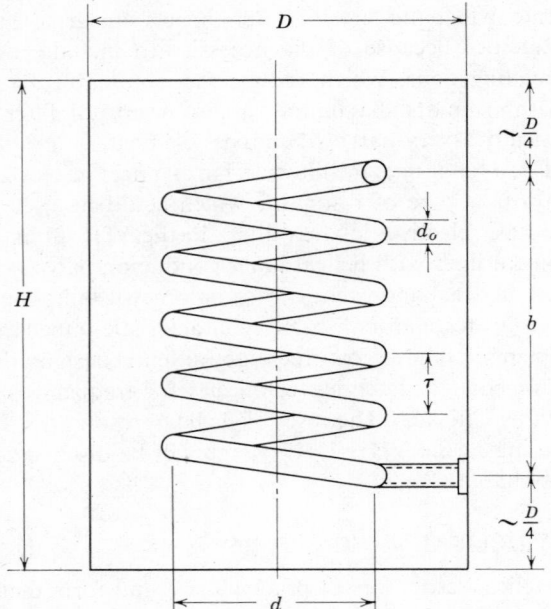

Fig. 9.2. Helical resonator with circular cross section.

This equation is valid only for the following condition,

$$\frac{b}{d} = 1.5 \quad (9.2.4)$$

where b is the axial length of the coil in inches.

These equations and all those below are accurate for the resonator when it is realized between the following limits:

$$1.0 < \frac{b}{d} < 4.0$$

$$0.45 < \frac{d}{D} < 0.6$$

$$0.4 < \frac{d_0}{\tau} < 0.6 \text{ at } \frac{b}{d} = 1.5$$

$$0.5 < \frac{d_0}{\tau} < 0.7 \text{ at } \frac{b}{d} = 4.0$$

$$\tau < \frac{d}{2}$$

where d_0 is the diameter of the conductor in inches.

The axial length of the coil is approximately equivalent to a quarter wavelength. This actual length is much shorter than the free-space length, which is given by the expression,

$$\frac{\lambda}{4} = \frac{c}{4f_0} 10^{-6} \quad (9.2.5)$$

where c is the speed of light in free-space and f_0 is the operating frequency in megacycles per second. The actual length of the coil in inches can be expressed by the following equation:

$$b = \frac{250}{f_0\sqrt{LC}} \quad (9.2.6)$$

where f_0 is the resonant frequency in megacycles per second.

This expression is based on theoretical considerations, but a working equation can be formulated with the help of the following expression.

$$\text{wave velocity } v = f_0\lambda = \frac{2\pi \text{ rad}}{2\pi\sqrt{LC}} = \frac{1000}{\sqrt{LC}} \quad (9.2.7)$$

Because of the fringe effect and self capacitance of the coil, the electrical length of the coil is approximately 6% less than a quarter wavelength. The empirical value of b is reduced by 6% and is given below.

$$b = \frac{0.94\lambda}{4} = \frac{0.235v}{f_0} = \frac{235}{f_0\sqrt{LC}} \quad (9.2.8)$$

The number of turns per inch is obtained by substituting Eqs. 9.2.1 and 9.2.3 into Eq. 9.2.8.

$$\frac{1}{\tau} = n = \frac{1720}{f_0 bd}\left[\frac{\log_{10}(D/d)}{1 - (d/D)^2}\right]^{\frac{1}{2}} \text{ turns per inch} \quad (9.2.9)$$

The total number of turns N is given by

$$N = nb = \frac{1720}{f_0 D(d/D)}\left[\frac{\log_{10}(D/d)}{1 - (d/D)^2}\right]^{\frac{1}{2}} \text{ turns} \quad (9.2.10)$$

The characteristic impedance of the resonator is expressed by

$$Z_0 = 1000\sqrt{\frac{L}{C}} = 183nd\left[\left(1 - \frac{d}{D}\right)^2 \log_{10}\frac{D}{d}\right]^{\frac{1}{2}} \text{ ohms} \quad (9.2.11)$$

$$\frac{d}{D} = 0.55 \text{ and } \frac{b}{d} = 1.5, \quad \text{then } N = \frac{1900}{f_0 D} \text{ turns} \quad (9.2.12)$$

and

$$Z_0 = \frac{98000}{f_0 D} \text{ ohms} \quad (9.2.13)$$

If the shield is of square cross section, the following equations are applicable:

$$S = \text{length of one side of the square} \approx D/1.2 \quad (9.2.14)$$

$$Q = 60S\sqrt{f_0} \quad (9.2.15)$$

$$N = \frac{1600}{f_0 S} \quad (9.2.16)$$

$$n = \frac{1}{\tau} = \frac{1600}{S^2 f_0} \quad (9.2.17)$$

$$Z_0 = \frac{81500}{f_0 S} \quad (9.2.18)$$

$$d = 0.66S \text{ for } \frac{d}{D} = 0.55 \quad (9.2.19)$$

$$b = S \text{ for } \frac{b}{d} = 1.5 \quad (9.2.20)$$

$$H = 1.6S \quad (9.2.21)$$

Figure 9.3 shows a nomogram constructed from Eqs. 9.2.14 to 9.2.21. This nomogram is to be used for helical resonators in shields of a square cross section, the resonator that physically lends itself best to filter design.

Quality Factor

The helical resonator has solved the problem of high-quality resonators in the VHF range. In a reasonable volume, they provide a tuned circuit whose Q is higher than the normal lumped circuit. Possible causes of dissipation in the resonator are losses in the conductor, in the shield, and in the dielectric. The Q of a resonator is defined by

$$Q = 2\pi f \frac{\text{energy stored}}{\text{power dissipated}} \quad (9.2.22)$$

Losses in the helical resonator include the actual loss in the helix, a copper loss as influenced by the skin and the proximity effects. There is also an additional loss owing to currents in the shield. The resistance of the coil can be expressed as

$$R_c = \frac{n\pi d\phi\sqrt{f}}{12,000d_0}$$

or,

$$R_c = \frac{0.083}{1000}\frac{\phi}{nd_0}n^2\pi\sqrt{f} \text{ ohms per axial inch} \quad (9.2.23)$$

An additional resistance due to the shield is given by

$$R_s = \frac{9.37n^2 b^2 (d/2)^4\sqrt{1.724f}}{b[D^2(b+d)/8]^{\frac{1}{3}}}$$
$$\times \sqrt{\frac{\rho_s}{\rho_{cu}}} \times 10^{-4} \text{ ohms per axial inch} \quad (9.2.24)$$

The unloaded Q of a resonant line is given by

$$Q_u = \frac{\beta}{2\alpha} \quad (9.2.25)$$

If R_c and R_s are assumed in series, the Q of the resonant line is expressed as

$$Q_u = \frac{2\pi f_0 L}{R_s + R_c} \quad (9.2.26)$$

In this form, the dielectric losses are neglected. For a resonator with a copper coil and copper shield, Eqs. 9.2.23 and 9.2.24 can be substituted into Eq. 9.2.26 and the final expression for the unloaded Q is

$$Q_u = 220\frac{(d/D) - (d/D)^3}{1.5 + (d/D)^3}D\sqrt{f_0} \quad (9.2.27)$$
$$Q_u \approx 50D\sqrt{f_0} = 60S\sqrt{f_0} \quad (9.2.28)$$

The simplified equation is accurate to $\pm 10\%$ and is derived with three practical limitations; when

$$0.45 < \frac{d}{D} < 0.6, \quad \frac{b}{d} > 1.0 \text{ and } d_0 > 5\delta$$

where δ is the skin depth.

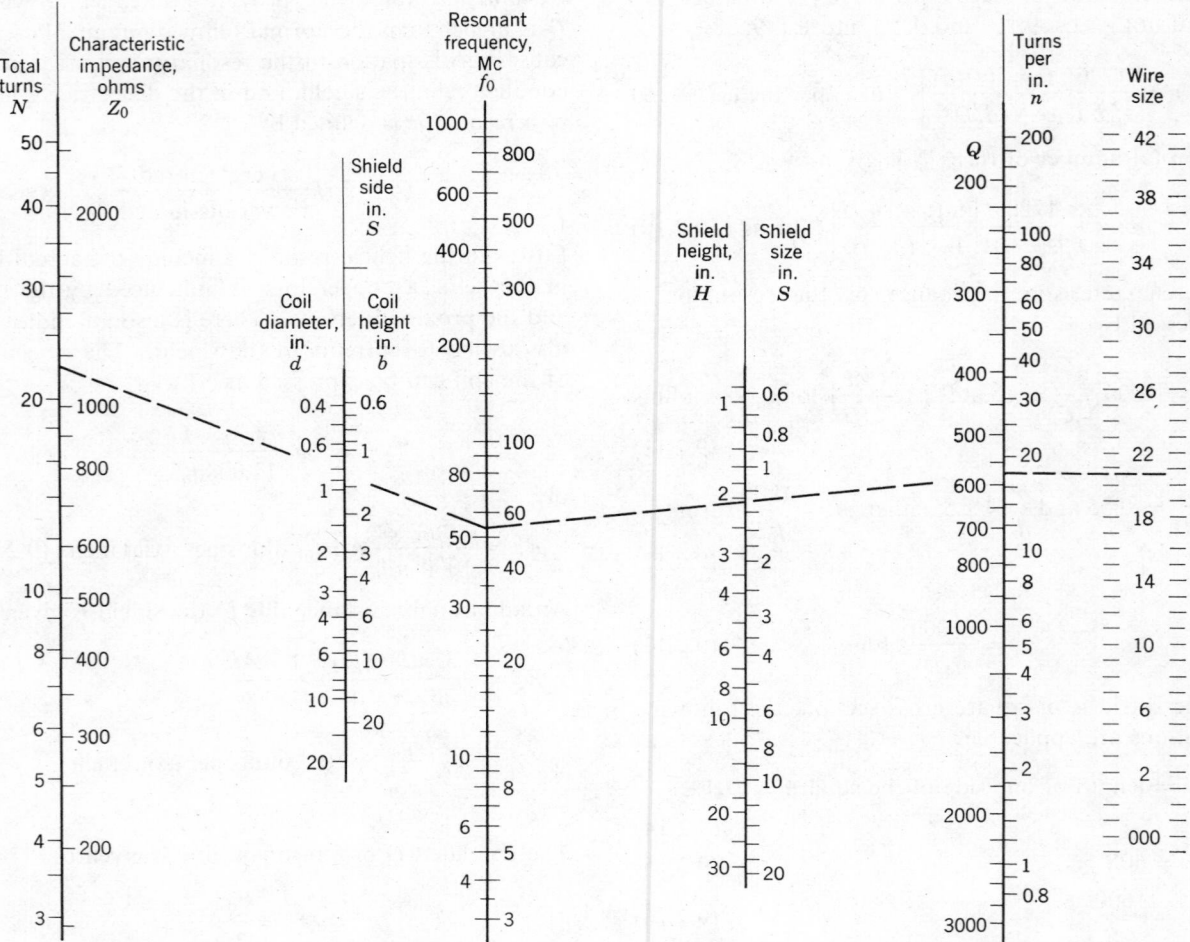

Fig. 9.3.   Nomogram for helical resonators in shields of square cross section.

For copper conductors,

$$\delta = \frac{2.60 \times 10^{-3}}{\sqrt{f_0}} \text{ inch} \qquad (9.2.29)$$

To show how important the volume of the resonator is, the Q_u as a function of volume is

$$Q_u = 50 \sqrt[3]{\text{Vol}} \sqrt{f_0}, \qquad (9.2.30)$$

when

$$0.4 < \frac{d}{D} < 0.6 \text{ and } 1 < \frac{b}{d} < 3.$$

Figure 9.4 illustrates how rapidly the unloaded Q decreases as a function of d/D and how important it is to keep this ratio between the specified limits.

If the condition $d_0 > 5\delta$ is not fulfilled, the Q of the resonator will be lower than that predicted by Eq.

9.2.28. For $d_0/\tau = 0.5$, the wire diameter d_0 is given by

$$d_0 = \frac{dDf_0}{3,800} \text{ inches} \qquad (9.2.31)$$

In a given space, d_0 can only be increased if the number of turns is reduced. This means that the inductance must be reduced. Thus for a fixed frequency, the capacitance must be increased.

Measurement of Resonator Q

The problem of finding the unloaded Q of the resonator is not easy. Many methods have been proposed but most have been inconvenient or impractical. However, the unloaded Q can be estimated quite accurately from the loaded Q and the insertion loss. This relation between the insertion loss and Q

when generator and load impedances are equal, is

$$L_{dB} = 20 \log \frac{U}{U-1} \qquad (9.2.32)$$

where

$$U = \frac{Q_{unloaded}}{Q_{minimum}}$$

In this case, Q_{min} is the loaded Q determined from the relation

$$Q_{min} = \frac{f_0}{BW_{3dB}}$$

where BW_{3dB} is the measured 3-dB bandwidth. When loop coupling is used into and out of the resonator, the insertion loss, and hence Q_{min}, will be a function of the coupling between loops and the losses in the loop circuits. It is desirable to use very loose coupling in order that the effect of coupling between loops may

be neglected. Figure 9.5 gives a plot of Eq. 9.2.32. The insertion loss is measured by the substitution method when all coaxial cables are as short as possible. The value of unloaded Q is evaluated by multiplying the value of Q_{min} by the value of U which corresponds to the measured insertion loss. If the insertion loss of the resonator is greater than 25 dB, the correction factor for the unloaded Q will be 1.05 or less. At this condition, Q_{min} will only be in error of Q_{unl} by 5% and it is self evident that Q_{unl} could be calculated from

$$\frac{f_0}{BW_{3dB}}$$

Physical Construction of Resonator

To obtain the predicted unloaded Q, several points for construction of the resonator should be remembered. The shield can be cylindrical, rectangular, or any other shape, but for simplicity of calculations, only the shields of circular and square cross sections have been considered. Any seams in the shield parallel to the coil axis should have good physical and electrical connection. Dip-brazed cans have been used extensively. If the coil end is run to the bottom cover of the shield, the cover must be solidly connected to the shield to reduce the losses and ensure the high-quality factor of the resonator. This connection may be done best by soldering, but in actual practice, the use of screws every few inches is permissible.

The length of the shield must be extended beyond the coil on each side by approximately one-quarter of the shield diameter, or for a square shield, by approximately 0.3 times the side of the square. If the coil were carried to the bottom of the resonator can without having this clearance, the lower few turns would be ineffective for storage of energy but would still contribute loss. The clearance at the top of the resonator is to reduce capacitive loading. Actually, the resonator could be built without top and bottom covers, since they have little effect upon the frequency and Q. However, the external field is greatly reduced by the use of these covers.

At the open end of the helix a high voltage exists. The coil should end smoothly, without sharp edges, and should not turn into or out of the helix. The coil form supporting the helix must be made of a low-loss material; otherwise, the quality factor of the resonator will be degraded seriously. Polystyrene rods have been used for this application in many helical filter designs. This material is known for its low-dissipation factor, making it electrically suitable. Two mechanical properties of polystyrene, however, limit its usefulness.

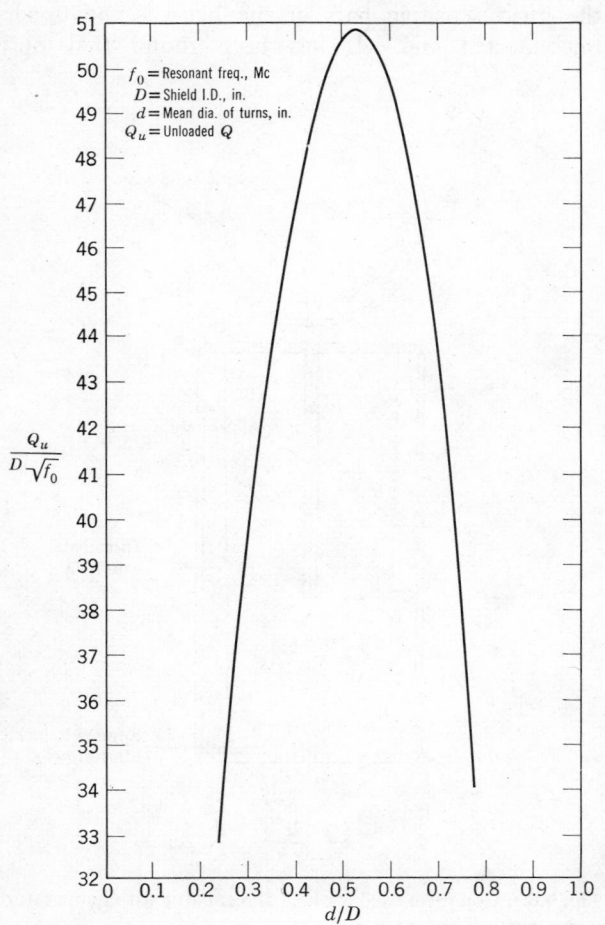

f_0 = Resonant freq., Mc
D = Shield I.D., in.
d = Mean dia. of turns, in.
Q_u = Unloaded Q

Fig. 9.4. Unloaded Q of the helical resonator.

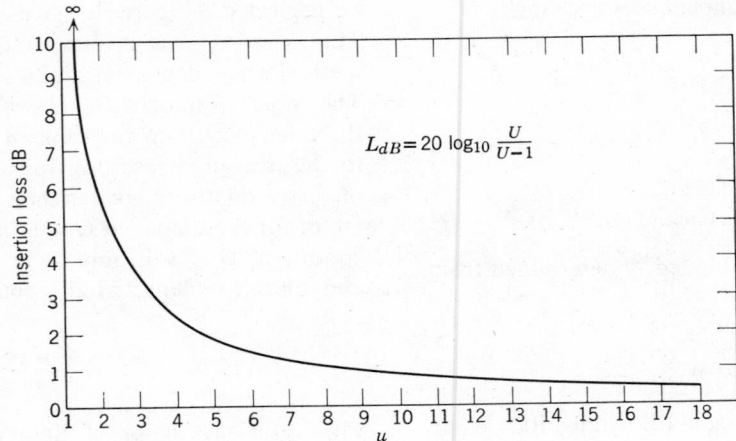

$$L_{dB} = 20 \log_{10} \frac{U}{U-1}$$

Fig. 9.5.   General curve for insertion loss.

First, the softening temperature of polystyrene is relatively low, approximately 82°C. At this temperature distortion of the material occurs. Second, the coefficient of expansion of polystyrene is rather high, and special provisions must be made to achieve a filter which is temperature stable.

Temperature Compensation in Helical Filters

In variable temperature conditions, most of the components used for selective devices and coupling networks, exhibit appreciable change in losses as well as electrical values. For example, the quality factor of a coil decreases as the temperature increases, and the value of inductance usually increases with the temperature increase. Capacitors, such as the silver mica type also exhibit a similar effect, with a positive temperature coefficient.

Experience proves that temperature compensating devices only solve part of the problem, that of stabilizing resonant frequencies, since their presence in the circuit may deteriorate the quality factor of the original network. This is especially true in high-quality, high-frequency filters, where the midband insertion loss is increased due to their use. Since their loss factor is temperature dependent, filters employing temperature-compensating capacitors will still exhibit a varying insertion loss with temperature.

In the helical resonator filter, the polystyrene form on which the coil is wound expands with increasing temperature. This expansion is greater than the expansion of the wire itself, and also greater than the cavity expansion. The expansion of the form tends to force the coil to increase in diameter, and lowers the center frequency of the filter. By using the coil form

of Fig. 9.6, this problem is almost entirely eliminated. Here, four slits running down the form absorb the expansion of the material, and the perimeter of the form will not change. Experiments have shown that the most sensitive part of the helix is the upper, unconnected end. It has been found that only

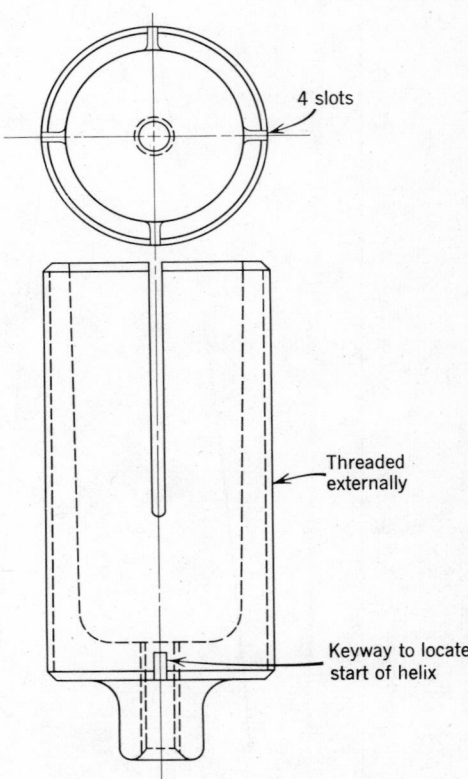

Fig. 9.6.   Coil form used for helical resonator with temperature compensating feature.

approximately 60% of the total length of the coil form need be slotted to absorb the variation.

When insertion loss variation is critical, the change can be completely cancelled by use of resistive pads with temperature-sensitive resistors. The pads may be of the L or π variety using one or more sensistors, the temperature-sensitive element, along with regular carbon resistors.

9.3 FILTER WITH HELICAL RESONATORS

To develop a filter, it is first necessary to consider the required attenuation requirements, paying particular attention to the filters relative bandwidth, and its relationship to the minimum quality factor of the resonators necessary to realize the design.

For a given set of specifications, the value of the unloaded Q must exceed a certain Q_{min} for that filter to be realizable. Figure 9.7 shows the relationship between the required q_{min} [$Q_{min} = q_{min}(f_0/\Delta f)$] for a Butterworth and three different Chebyshev filters.

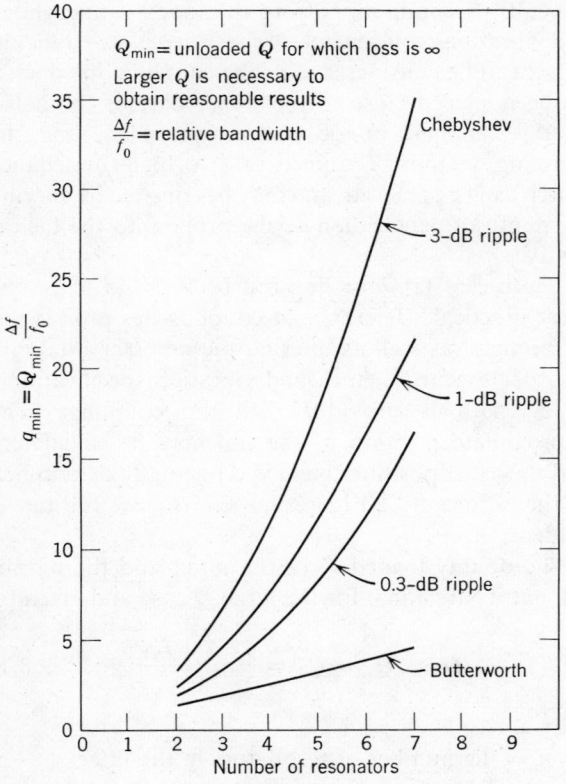

Fig. 9.7. Relative minimum unloaded Q for Butterworth and Chebyshev filters.

Example 1

A seven-pole filter (seven resonators) possessing a Butterworth response requires a q_{min} of 4.6. For a Chebyshev response with a 1-dB ripple in the passband, $q_{min} = 21.9$. If Q_{unl} were equal to Q_{min}, the insertion loss would be infinite. It is self evident that the unloaded Q must exceed Q_{min}.

For this example, assuming a 1% bandwidth, Q_{min} equals 460 for the Butterworth filter and 2190 for the Chebyshev. It must be remembered that if components whose unloaded Q is barely equal to Q_{min} are used, the response can be achieved, provided certain predistorted values of coupling coefficients are used, but an extremely large value of insertion loss will result.

When the unloaded Q is greater than Q_{min}, the loss of the filter does not primarily depend on the number of sections, but is exclusively controlled by the ratio U, given in Eq. 9.2.32. Once the minimum value of unloaded Q is obtained, and the quality factor of available components is determined, the loss in the filter is almost completely defined and varies very little with the shape of the filter, the number of sections, the bandwidth, etc.

Fubini and Guillemin give a curve which shows the minimum insertion loss at midband of Butterworth filters plotted as a function of the ratio U. The following two conclusions can be made:

(1) For moderate losses, the curves are very close to each other.

(2) The curves for one- and two-section filters are exactly the same and are expressed by Eq. 9.2.32.

From the previous Example 1, of a seven-pole Butterworth filter, assume the available unloaded Q of each section is 3000. The value of U can be computed as follows:

$$U = \frac{Q_{unl}}{Q_{min}} = \frac{3000}{460} = 6.52$$

From Eq. 9.2.32 or Fig. 9.5,

$$L = 20 \log \frac{6.52}{5.52} = 20 \times 0.072 = 1.44 \text{ dB}$$

As mentioned before, this equation is only valid for one and two sections. For the example with seven resonators, a correction factor must be used. Figure 9.8 plots this correction factor and shows that the loss is always greater as the number of sections is increased.

Since the number of sections is seven, the correction factor is 1.27 and the actual insertion loss at midband

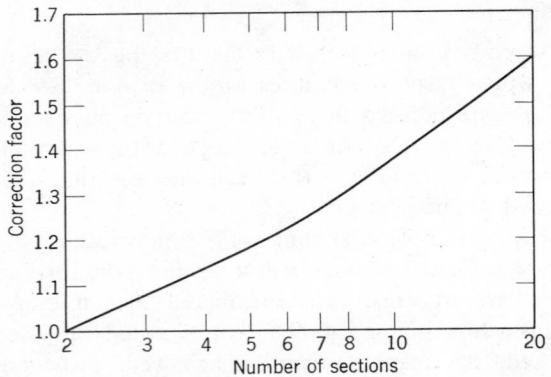

Fig. 9.8. Correction factor for insertion loss for more than two sections.

will be
$$L = 1.44 \times 1.27 = 1.83 \text{ dB}$$

It should be stated that this method of insertion loss is approximate, due to the empirical correction factor used. An exact method of calculation will be given later.

For a realizable Chebyshev filter, q_{min} is always higher than that of a Butterworth filter and may be found from Fig. 9.7. For the same unloaded Q and bandwidth, the insertion loss of a Chebyshev filter will be several times higher than that of a Butterworth filter.

Example 2

If the available unloaded Q is 5000 and the relative bandwidth is 1%, the insertion loss for a four-pole Butterworth filter will be 0.53 dB. For a Chebyshev filter with four poles (3-dB ripple) the expected insertion loss will be 2.67 dB. If the Q factor is only 3000, the former will result in 0.898 dB loss and the latter will exhibit 4.95 dB insertion loss. If the Q is 2000 the corresponding insertion losses will be 1.38 dB and 8.8 dB respectively.

Filter Construction

As previously stated, the only reason for using helical resonators is to reduce the size of the filter and to provide a low insertion loss in the passband. The design of a filter for a Butterworth and Chebyshev response is straightforward, and the evaluation of their circuit elements is well known. More interesting, however, are the equivalent schematics and the mechanical realization of the filter.

Even after carefully calculating the number of turns, and all dimensions of the resonator, it is very possible that the resonant frequency may be in error by as much as 10%. This must be adjusted without any

distortion to the other dimensions so that the predicted Q will be obtained. This adjustment is made by using a tuning screw at the top of the helix. In the equivalent schematic the screw, because it is connected to ground, has the effect of providing capacitive loading for the helix.

Coupling into the Filter

The problem to be considered here is most important, namely, coupling into and out of the helical resonator filter. There are three methods of achieving this coupling: loop, probe, and tap coupling.

For loop coupling, a loop of approximately one turn is placed around the first helix. The loop is usually positioned slightly below the helix, in a plane perpendicular to the helix axis. The distance between the coupling loop and helix can be adjusted until a proper match is achieved between the filter and its load. The use of the loop generally yields relatively low impedance coupling and features a dc short to ground.

Loop coupling, although rather simple to use, has disadvantages. First, positioning of the loop so that the desired coupling requirements are satisfied is difficult. Second, supporting the loop so that shock and vibration requirements are satisfied is also difficult.

Next to be considered is probe coupling, in which a probe is placed close to the upper part of the helix. In this manner, no dc path is provided, and the coupling is mostly capacitive. A high impedance match can be achieved, and may be adjusted by varying the depth of penetration of the probe into the helical cavity.

The use of tap coupling has been found to be the most practical. This type of coupling has production advantages, as well as offering the necessary stability in order to satisfy shock and vibration specifications. A dc path is provided with tap coupling. The approximate position of the tap may be calculated, and the exact position then experimentally determined in the laboratory. Figure 9.9 shows how the tap is made.

The doubly loaded Q of the input and the output resonators are equal for the large Q case and given by

$$Q_d = Q_1 = Q_n = \frac{1}{2} q_1 \frac{f_0}{\text{BW}_{3\,\text{dB}}} \qquad (9.3.1)$$

where

n = the number of resonators in the filter

Q_d = the doubly loaded Q

q_1 = normalized quality factor given in tables of 3-dB down k and q values in Chapter 6.

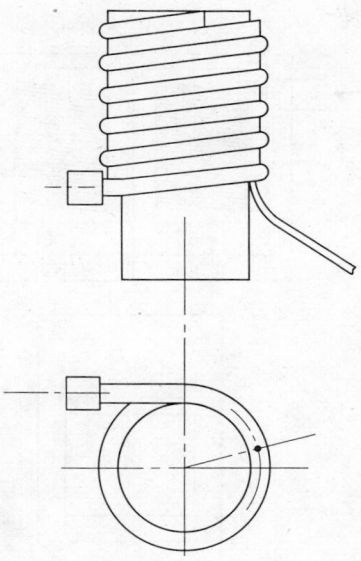

Fig. 9.9. Tap coupling.

For the Butterworth response, the list below may be referred to:

No. of Resonators	q_1
2	1.414
3	1.000
4	0.766
5	0.618
6	0.518
7	0.445

Applying transmission line theory, the following equation may be obtained.

$$\frac{R_b}{Z_0} = \frac{\pi}{4}\left(\frac{1}{Q_d} - \frac{1}{Q_u}\right) \qquad (9.3.2)$$

which is divided equally between the generator and the second resonator when tapping the input coil. When tapping the output coil, R_b/Z_0 is divided equally between the load and the $(n-1)$ resonator.

Also from resonant line theory,

$$\sin \theta = \sqrt{\frac{R_b}{2Z_0}\frac{R_{\text{tap}}}{Z_0}} \qquad (9.3.3)$$

where θ is the electrical angle from the voltage standing wave minimum point (here, the helix ground). The tap is then placed $N\theta/90°$ turns from the grounded end of the helix.

Example 3

A helical filter exhibiting Butterworth characteristics is to be placed between the source impedance of 50 ohms and a load whose impedance is 1000 ohms. It is a four-section filter with a 2.5% bandwidth. Each resonator has 71.5 turns and has an unloaded Q of 300. The characteristic impedance of each resonator is 3630 ohms. Using Eqs. 9.3.1, 9.3.2 and 9.3.3:

$$Q_d = \frac{1}{2} 0.766 \frac{1}{0.025} = 15.32$$

$$\frac{R_b}{Z_0} = \frac{\pi}{4}\left(\frac{1}{15.32} - \frac{1}{300}\right) = 0.04865$$

For the 50-ohm source,

$$\sin \theta = \sqrt{0.02433\frac{50}{3630}} = 0.0183$$

$$\theta = 1.05°$$

$$\text{tap} = \frac{71.5 \times 1.05°}{90°} = 0.83 \text{ turns from grounded end}$$

For the 1000-ohm load,

$$\sin \theta = 0.0819$$
$$\theta = 4.70°$$

Tap = 3.73 turns from the grounded end.

For all three types of end coupling (that is, loop, probe, and tap), fine adjustment of the coupling from the outside of the filter is difficult. To add some flexibility to the filter, a form of fine coupling adjustment is often desirable. Coupling adjustment may be achieved by connecting a variable capacitor between the ground end of the helix and ground as in Fig. 9.10. Tap coupling is calculated as above for the approximate location of the tap. If the variable capacitor is large, the coupling will be unaffected, as if no capacitor were used, and the ground end of the resonator grounded. As the capacitance is decreased, the coupling into the resonator is decreased. This continues until no power is coupled in, which occurs when

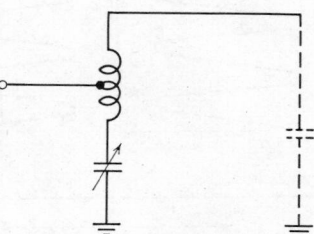

Fig. 9.10. Fine adjustment of tap coupling.

that portion of the helix below the tap and the capacitor is at series resonance. As the series capacitor is decreased further, coupling begins to increase. The fundamental frequency of the helix will shift slightly because of the supplementary capacitance, but the shift can be eliminated by the conventional method of tuning the helix.

This modification is especially useful for limited production, where the tap cannot be preset exactly and is of particular advantage when the source and load impedances are not known precisely. A dc short does not exist in this configuration.

Coupling Between Resonators

The coupling of helical resonators is considered the most complicated problem in the realization of a specific filter design. The problem arises because of the difficulty encountered in the mathematical analysis of the coupling.

Figure 9.11 defines the dimension h to be that part of one helix exposed to the adjacent helix. The shield may be either open at the bottom or the top of the resonator. The shield is made of the same material as the can and is solidly connected to the sides and top of the can (dip-brazed, or soldered). The dimension h determines the amount of coupling between resonators. The problem of accurately determining the dimension h for helical filters of all practical sizes and frequencies have not yet been solved. However, measurements taken on a coupling test block (Fig. 9.12) have yielded data that may be extended for use at frequencies in the 30-Mc range, with $S \approx 1.25$.

The test block is actually a two resonator filter, featuring a replaceable shield. The parameters necessary to calculate the coupling was measured,

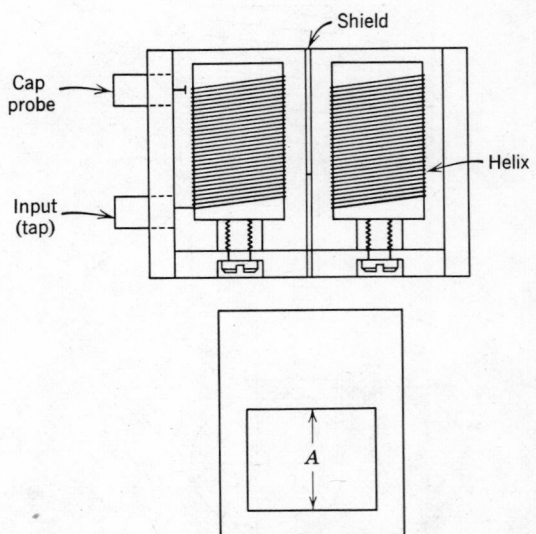

Fig. 9.12. Coupling test block.

and the results were plotted (Fig. 9.13). With this test block, the shield thickness was $\frac{1}{32}$-inch. Most filters are made with $\frac{1}{16}$-inch thick shields. To obtain h for $\frac{1}{16}$-inch material, multiply the value of h obtained from the curve by 1.075.

The curve of Fig. 9.13 has been reduced to

$$K \times 10^{-3} = 0.071 \left(\frac{h}{d}\right)^{1.91} \tag{9.3.4}$$

Use of this design method will yield results accurate to within approximately 6%.

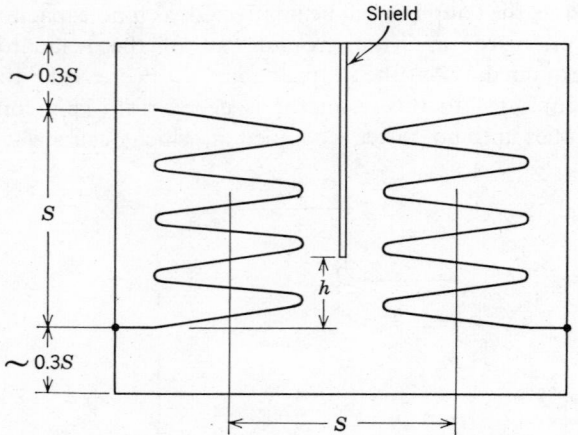

Fig. 9.11. Position of coupling shield between resonators.

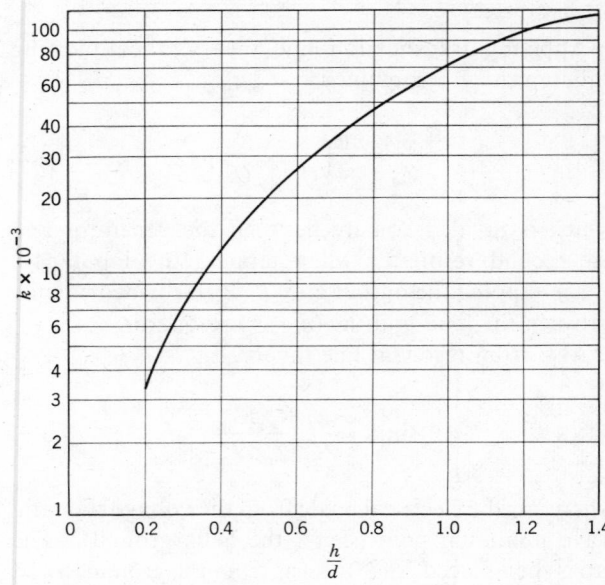

Fig. 9.13. Coupling between resonators as a function of h/d.

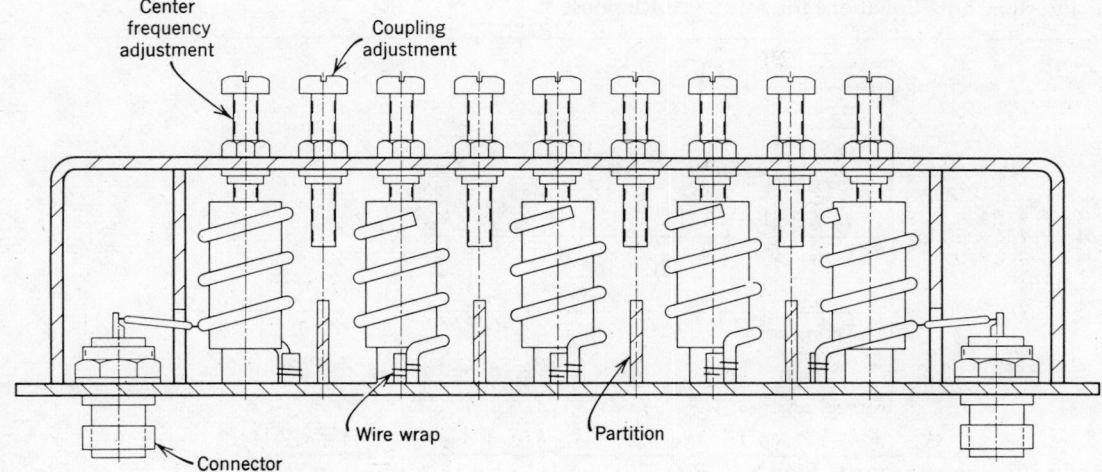

Fig. 9.14. 540-Mc bandpass helical filter with variable coupling feature.

Openings at the upper part of the resonator yield mostly capacitive coupling, and more attenuation is obtained on the lower side of the passband. Openings in the lower part of the partitions provide mostly inductive coupling, and the filter will exhibit more attenuation on the high-frequency side of the response curve.

When using capacitive coupling, a fine coupling adjustment can be obtained by using a screw inserted through the top of the can in line with the shield. Figure 9.14 shows a 540-Mc filter using this arrangement.

Determination of the Number of Resonators

Given any design problem, the first step is to determine the number of resonators necessary to fulfill the rejection requirement. The attenuation curves of Chapter 3 may be reviewed, or for Butterworth or Chebyshev filters, the nomographs of Figs. 5.1 and 5.2 may be used.

Example 4

A bandpass filter is required with the following specifications:

$$f_0 = 30 \text{ Mc}$$

$$\text{BW}_{3 \text{ dB}} = 1 \text{ Mc}$$

$$\text{BW}_{50 \text{ dB}} = 4 \text{ Mc}$$

It is desired that the response shape be Butterworth.

Using the attenuation curve for Butterworth filters in Chapter 3, and keeping in mind

$$\frac{\text{BW}}{\text{BW}_{3 \text{ dB}}} = \frac{4}{1} = 4$$

it can be seen by using four resonators, 48 dB is obtained at the frequency where 50-dB rejection is required. Going to five resonators, 60 dB is obtained.

Determination of Required Unloaded Q

To determine the size of the resonators required for the filter, we must first find the value of unloaded Q necessary to fulfill the insertion loss requirement. The method given earlier, in Examples 1 and 2, is only approximate. An exact method will be given here, making use of the quantities:

$$Q_u = \text{the unloaded } Q \text{ required}$$

$$Q_L = \text{the loaded } Q = f_0/\text{BW}_{3 \text{ dB}}$$

$$Q_0 = Q_u/Q_L$$

$$q_1, q_n = \text{the normalized } Q \text{ of the first and}$$
$$\text{last resonators}$$

$$k_{12} \cdots k_{(n-1)(n)} = \text{the normalized coupling coefficients}$$

The equations for insertion loss given in Table 9.1 are exact for any type response. The equations of Table 9.2 are for very low-loss Butterworth responses. For low-insertion losses, the higher order terms can be neglected.

Example 5

What unloaded Q is needed for a four-pole Butterworth filter with a 1 dB insertion loss, and a 3-dB bandwidth of 15 Mc at a center frequency of 500 Mc?

SOLUTION

$$IL = 20 \log \left[\frac{2.62}{Q_0^3} + \frac{3.41}{Q_0^2} + \frac{2.62}{Q_0} + 1 \right]$$

for Q_0, neglecting the Q_0^3 term yields

$$Q_0 = 22.7$$

Table 9.1   Insertion Loss Equations for Any Type Response

$n = 2$ $IL = 20 \log \left[\dfrac{\sqrt{q_n/q_1}}{k_{12}} \left(\dfrac{1}{Q_0} + \dfrac{1}{q_2} \right) \right]$

$n = 3$ $IL = 20 \log \left[\dfrac{\sqrt{q_n/q_1}}{k_{12}k_{23}} \left(\dfrac{1}{Q_0^2} + \dfrac{1}{q_n Q_0} + k_{23}^2 \right) \right]$

$n = 4$ $IL = 20 \log \left\{ \dfrac{\sqrt{q_n/q_1}}{k_{12}k_{23}k_{34}} \left[\dfrac{1}{Q_0^3} + \dfrac{1}{q_n Q_0^2} + \dfrac{(k_{23}^2 + k_{34}^2)}{Q_0} + \dfrac{k_{23}^2}{q_n} \right] \right\}$

$n = 5$ $IL = 20 \log \left\{ \dfrac{\sqrt{q_n/q_1}}{k_{12}k_{23}k_{34}k_{45}} \left[\dfrac{1}{Q_0^4} + \dfrac{1}{q_n Q_0^3} + \dfrac{(k_{23}^2 + k_{34}^2 + k_{45}^2)}{Q_0^2} + \dfrac{(k_{23}^2 + k_{34}^2)}{q_n Q_0} + k_{23}^2 k_{45}^2 \right] \right\}$

$n = 6$ $IL = 20 \log \left\{ \dfrac{\sqrt{q_n/q_1}}{k_{12}k_{23}k_{34}k_{45}k_{56}} \left[\dfrac{1}{Q_0^5} + \dfrac{1}{q_n Q_0^4} + \dfrac{(k_{23}^2 + k_{34}^2 + k_{45}^2 + k_{56}^2)}{Q_0^3} \right. \right.$

$\left. \left. + \dfrac{(k_{23}^2 + k_{34}^2 + k_{45}^2)}{q_n Q_0^2} + \dfrac{(k_{23}^2 k_{56}^2 + k_{23}^2 k_{45}^2 + k_{34}^2 k_{56}^2)}{Q_0} + \dfrac{k_{23}^2 k_{45}^2}{q_n} \right] \right\}$

$n = 7$ $IL = 20 \log \left\{ \dfrac{\sqrt{q_n/q_1}}{k_{12}k_{23}k_{34}k_{45}k_{56}k_{67}} \left[\dfrac{1}{Q_0^6} + \dfrac{1}{q_n Q_0^5} + \dfrac{(k_{23}^2 + k_{34}^2 + k_{45}^2 + k_{56}^2 + k_{67}^2)}{Q_0^4} \right. \right.$

$+ \dfrac{(k_{23}^2 + k_{34}^2 + k_{45}^2 + k_{56}^2)}{q_n Q_0^3} + \dfrac{(k_{23}^2 k_{56}^2 + k_{23}^2 k_{67}^2 + k_{23}^2 k_{45}^2 + k_{34}^2 k_{56}^2 + k_{34}^2 k_{67}^2 + k_{45}^2 k_{67}^2)}{Q_0^2}$

$\left. \left. + \dfrac{(k_{23}^2 k_{45}^2 + k_{23}^2 k_{56}^2 + k_{34}^2 k_{56}^2)}{q_n Q_0} + k_{23}^2 k_{45}^2 k_{67}^2 \right] \right\}$

Then

$$Q_u = Q_L Q_0 = \frac{f_0}{BW_{3\,dB}} Q_0$$

$$Q_u = 756$$

Predistorted k and q Values

The values of k and q given for the infinite q case in Chapter 6 can be used provided

$$Q_u \geq \left(\frac{f_0}{BW_{3\,dB}} \right) q_{min} = Q_{min}$$

As mentioned before, if $Q_u = Q_{min}$, an infinite midband insertion loss will result. If Q_u is only slightly larger than Q_{min}, the resulting filter will have a high value of insertion loss, and the 3-dB bandwidth of the resulting filter will be narrower than the design bandwidth. Only when

$$Q_u \approx 10 Q_{min}$$

or using Dishal's notation

$$Q_u \approx \left(\frac{f_0}{BW_{3\,dB}} \right) q_{2,3,\ldots,(n-1)}$$

Table 9.2   Insertion Loss Equations for Low-Loss Butterworth Responses

$n = 2$ $IL = 20 \log \left(\dfrac{1.414}{Q_0} + 1 \right)$

$n = 3$ $IL = 20 \log \left(\dfrac{2}{Q_0^2} + \dfrac{2}{Q_0} + 1 \right)$

$n = 4$ $IL = 20 \log \left(\dfrac{2.62}{Q_0^3} + \dfrac{3.41}{Q_0^2} + \dfrac{2.62}{Q_0} + 1 \right)$

$n = 5$ $IL = 20 \log \left(\dfrac{3.24}{Q_0^4} + \dfrac{5.23}{Q_0^3} + \dfrac{5.23}{Q_0^2} + \dfrac{3.24}{Q_0} + 1 \right)$

$n = 6$ $IL = 20 \log \left(\dfrac{3.84}{Q_0^5} + \dfrac{7.42}{Q_0^4} + \dfrac{9.11}{Q_0^3} + \dfrac{7.43}{Q_0^2} + \dfrac{3.84}{Q_0} + 1 \right)$

$n = 7$ $IL = 20 \log \left(\dfrac{4.46}{Q_0^6} + \dfrac{10.0}{Q_0^5} + \dfrac{14.5}{Q_0^4} + \dfrac{14.6}{Q_0^3} + \dfrac{10.0}{Q_0^2} + \dfrac{4.46}{Q_0} + 1 \right)$

will the resulting bandwidth be equal to the design bandwidth.

For the low Q case, when $Q_u < 10Q_{min}$, the predistorted k and q values given should be used in order to realize the design bandwidth.

This can be summarized as follows: If $Q_u \geq 10Q_{min}$, use the infinite Q values of k and q. If $Q_u < 10Q_{min}$, use the predistorted values. It should be mentioned, for the high-Q case, an insertion loss of 1 dB or less will result. For the low-Q case, the insertion loss of the network is given in the tables, having been defined as

$$IL = 10 \log \frac{P_{out\,max}}{P_{out}} = 10 \log \frac{R_2}{4R_1} \left| \frac{V_{in}}{V_{out}} \right|^2 \quad (9.3.5)$$

where

$$P_{out\,max} = \frac{V_{in}^2}{4R_1}$$

It should be noticed that for most of the predistorted cases, the values of q_1 and q_n are different, and the input and output resonators should be tapped in different points, for the helical filter to be loaded properly.

Example 6

A filter is required with the following specifications:

$f_0 = 30$ Mc

$BW_{3\,dB} = 0.9$ Mc

$BW_{50\,dB} = 4.5$ Mc

$Z_{in} = Z_{out} = 50$ ohms

Insertion loss: 3 dB maximum

Maximum dimensions of filter: $6\frac{3}{8}$ in. by $1\frac{3}{4}$ in. by $2\frac{3}{4}$ in.

Response: Butterworth

SOLUTION

1. Determine the number of resonators. Using Fig. 5.1, with $A_{max} = 3$ dB, $A_{min} = 50$ dB, and $\Omega = 4.5/0.9 = 5.0$, we see that between three and four resonators will be required for realization. We must use four resonators.

We could also have used the curve of the Butterworth attenuation characteristics of Chapter 3, to find that a three-pole would only provide 42 dB at the 50-dB point, but the four-pole network satisfies the requirement with 56 dB.

2. Calculate the Q that can be obtained. Since we are using four resonators, five metal thicknesses must be subtracted from the maximum length. For $\frac{1}{16}$ in.

walls, this results in approximately 6 in. remaining for four resonators. Each resonator will have

$$S = 6/4 = 1.5 \text{ in.}$$

Then,

$$Q_u = 60S\sqrt{f_0} = 492.9$$

3. Determine if the high-Q case can be realized. For a four-pole Butterworth, $q_{min} = 2.6$

$$Q_{min} = 2.6 \frac{30}{0.90} = 86.6$$

Since $Q_u < 10Q_{min}$, we cannot use the high-Q case and obtain the exact bandwidth.

4. Determine predistorted k and q values and insertion loss. Using the table of 3-dB down k and q values for the Butterworth response, with $n = 4$, and calculating

$$q_0 = \left(\frac{BW_{3\,dB}}{f_0} \right) Q = \frac{0.9}{30} 490 = 14.7$$

We could use the k and q values associated with $q_0 = 13.066$, the closest value to the calculated q_0. It is also possible to plot the values given in the table, as done in Fig. 9.15. Here, the abscissa is

$$d_0 = \frac{1}{q_0}$$

For our example $d_0 = 0.068$, and using the curves of Fig. 9.15 we find

$q_1 = 0.533$

$q_4 = 1.642$

$k_{12} = 1.076$

$k_{23} = 0.554$

$k_{34} = 0.680$

$IL = 1.9$ dB, satisfying the insertion-loss requirement.

5. Computer run (optional). To check the resulting filter, element values can be determined and the filter run on a digital computer to check bandwidth, rejection, and insertion loss. The lumped representation of the helical filter is shown in Fig. 9.16. Figure 9.17 is the computed response.

At the 50-dB rejection points, on the low side 59 dB is achieved, and on the high side the attenuation is 52 dB. The 3-dB bandwidth is 0.9 Mc, as required.

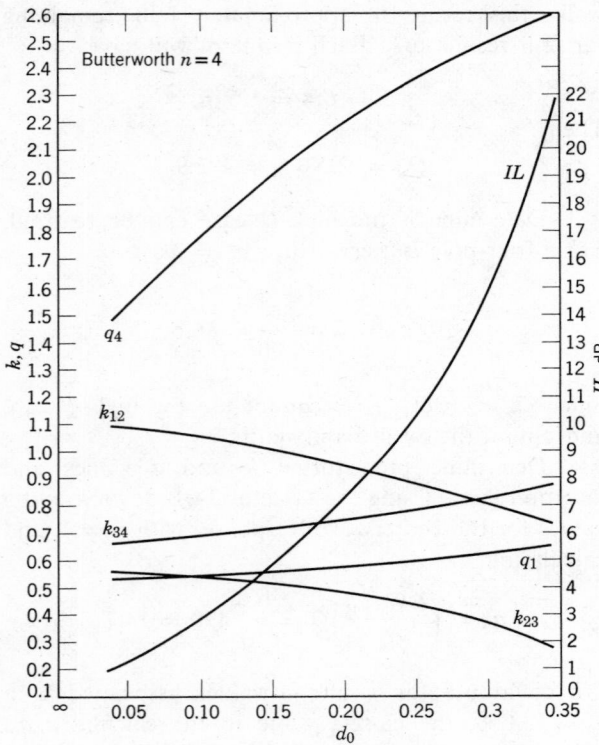

Fig. 9.15. 3-dB down k and q values for Butterworth filter with $n = 4$.

The computed insertion gain is 3.14 dB. The voltage gain due to the impedance step-up is

$$\frac{V_{\text{out}}}{V_{\text{in}}} = \sqrt{\frac{Z_{\text{out}}}{Z_{\text{in}}}} = 1.755$$

This corresponds to 4.88 dB gain. Since the filter's actual insertion gain is 3.14 dB, the insertion loss not considering the impedance step-up is 1.74 dB. This checks closely with the 1.9-dB insertion loss obtained previously, and satisfies the specification of the filter.

6. Compute the resonator dimensions. The necessary design quantities are obtained from Eqs. 9.2.16 through 9.2.21.

$$N = \frac{1600}{f_0 S} = 35.5 \text{ turns}$$

$$n = \frac{1600}{S^2 f_0} = 23.7 \text{ turns per in.}$$

$$Z_0 = \frac{81,500}{f_0 S} = 1811.1 \text{ ohms}$$

$$d = 0.66S = 0.99 \text{ in.}$$

$$b = S = 1.5 \text{ in.}$$

$$H = 1.6S = 2.4 \text{ in.}$$

$$d_0 = \frac{1}{2n} = 0.0210 \text{ in., which corresponds to No. 24 copper wire.}$$

7. Compute the shield heights. The dimensions of the coupling shields will now be calculated. We have previously found

$$k_{12} = 1.076$$
$$k_{23} = 0.554$$
$$k_{34} = 0.680$$

Now

$$K = k\frac{\text{BW}_{3\,\text{dB}}}{f_0}$$

$$K = 30 \times 10^{-3}k$$

and

$$K_{12} = 32.3 \times 10^{-3}$$
$$K_{23} = 16.6 \times 10^{-3}$$
$$K_{34} = 20.4 \times 10^{-3}$$

From Fig. 9.13,

$$\frac{h_{12}}{d} = 0.66$$

$$\frac{h_{23}}{d} = 0.46$$

$$\frac{h_{34}}{d} = 0.52$$

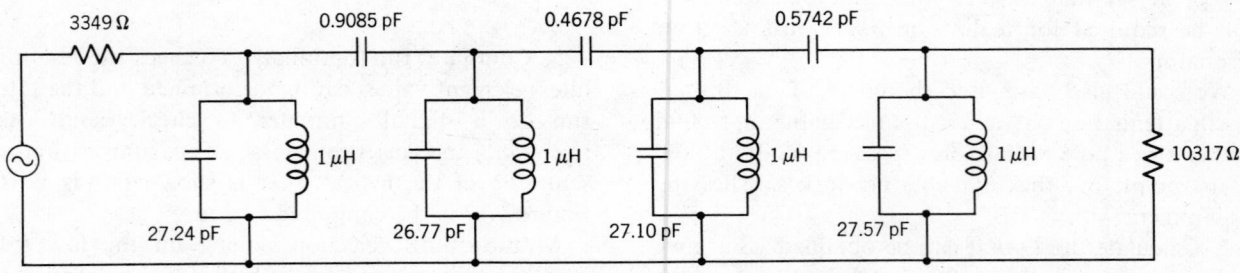

Fig. 9.16. Lumped representation of helical filter of Example 6.

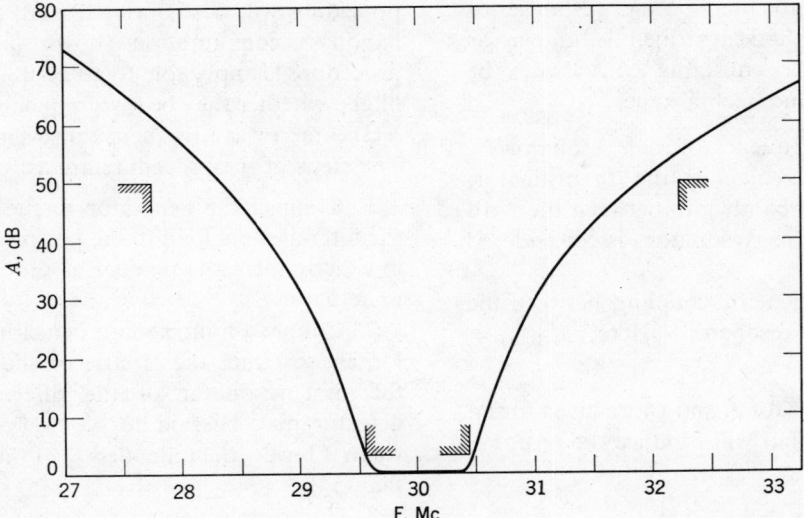

Fig. 9.17.   Computed response of network of Fig. 9.16.

We have calculated $d = 0.99$ inches.   Therefore,

$$h_{12} = 0.653 \text{ in.}$$

$$h_{23} = 0.455 \text{ in.}$$

$$h_{34} = 0.515 \text{ in.}$$

We will use $\frac{1}{16}$ in. thick material for the shields.   As mentioned earlier, the correction factor is 1.075.   So,

$$h_{12} = 0.702 \text{ in.}$$

$$h_{23} = 0.489 \text{ in.}$$

$$h_{34} = 0.554 \text{ in.}$$

This is the opening between adjacent helices, and may be at the top or bottom of the resonator.

8. Calculate the input and output tap.   We must determine the position of the input and output tap to match a 50-ohm source and load.   We have obtained and calculated the following information.

$$q_1 = 0.533$$

$$q_4 = 1.642$$

$$Q_u = 490$$

$$Z_0 = 1811.1 \text{ ohms}$$

$$N = 35.5 \text{ turns}$$

Using Eqs. 9.3.1, 9.3.2, and 9.3.3 for the input coil,

$$Q_d = \tfrac{1}{2}q_1 \frac{f_0}{\text{BW}_{3\,\text{dB}}} = 8.9$$

$$\frac{R_b}{Z_0} = \frac{\pi}{4}\left(\frac{1}{Q_d} - \frac{1}{Q_u}\right) = 0.0866$$

$$\sin\theta = \sqrt{\frac{R_b}{2Z_0}\frac{R_{\text{tap}}}{Z_0}} = 0.0346$$

$$\theta = 1.98°$$

$$\text{tap} = \frac{N\theta}{90°} = 0.78 \text{ turn from ground.}$$

For the output coil the resulting tap position is approximately 0.44 turn from ground.

9. Determine the final outside dimensions of the can.   Finally, the dimensions of the can, considering $\frac{1}{16}$ in. thick metal thickness is

$$\text{Length} = 4S + (5 \times \tfrac{1}{16}) = 6.313 \text{ in.}$$

$$\text{Width} = S + (2 \times \tfrac{1}{16}) = 1.625 \text{ in.}$$

$$\text{Height} = H + (2 \times \tfrac{1}{16}) = 2.525 \text{ in.}$$

10. Tuning the filter. The necessary design information has now been obtained.   Tuning of the filter is accomplished by use of a sweep generator, to be described in Section 9.4.

9.4   ALIGNMENT OF HELICAL FILTERS

When a filter is designed and constructed to the best of the engineering ability, the next very important problem is to adjust to the specific response or "tuning up". The method described is known as Dishal's method. Filter constants can always be reduced to only three fundamental types:

1. f_0—the resonant frequency of each resonator
2. $d_i = 1/Q_i$—the decrement of the ith resonator, defined as the fractional bandwidth between the 3-dB down points, when the resonator is considered separately
3. $K_{i(i+1)}$—the coefficient of coupling between the ith and the $(i + 1)$th resonator. Here, $K_{i(i+1)} = (\Delta f/f_0)k_{i(i+1)}$.

All of the required values for K and Q are given for an n-resonant circuit filter that will produce the response

$$\left(\frac{V_p}{V}\right)^2 = 1 + \left(\frac{\Delta f}{\Delta f_{3\,dB}}\right)^{2n}$$

where V_p is the voltage output at the peak of the response curve in the passband.

Most selective circuit designs incorporate a trimming adjustment for setting the resonant frequency of each resonator, such as a tuning screw for the helical resonator filter. The coefficient of coupling between adjacent resonators may be variable, and a method is given for easily adjusting the exact desired value.

In this tuning method, the filter is completely assembled, and attention is concentrated on amplitude phenomena occurring in the first resonator of the filter at the desired resonant frequency. The alignment procedure will be described using the four-resonator bandpass configuration shown in Fig. 9.18. The procedure is applicable to all coupled-resonant circuit filters, whether they be low frequency, high-frequency, VHF, microwave-frequency or even waveguide filters. The steps of the procedure are as follows:

1. Connect the generator to the first resonator of the filter and the load to the last resonator of the filter in exactly the same manner as they will be connected in actual use.

2. Couple a nonresonant detector directly and very loosely to either the electric or the magnetic field of the first resonator of the filter. A nonresonant detector may be said to be "very loosely" coupled when it lowers the unloaded Q of the resonator by less than 5%.

3. Completely detune all resonators. A resonator is sufficiently detuned when its resonant frequency is at least 10 passband widths away from passband midfrequency.

4. Set the generator frequency to the desired mid-frequency of the filter.

5. Tune resonator 1 for maximum output indication on the detector. Lock the tuning adjustment.

6. Tune resonator 2 for minimum output indication on the detector. Lock the tuning adjustment.

7. Tune resonator 3 for maximum output and lock the tuning adjustment.

8. Tune resonator 4 for minimum output and lock the tuning adjustment.

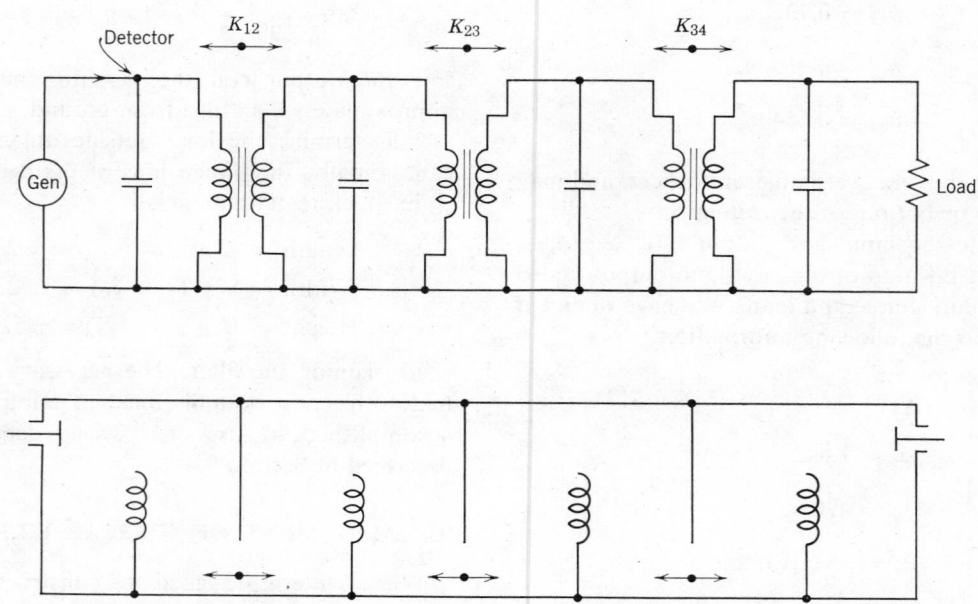

Fig. 9.18. Four-pole conventional helical network used to illustrate alignment procedure.

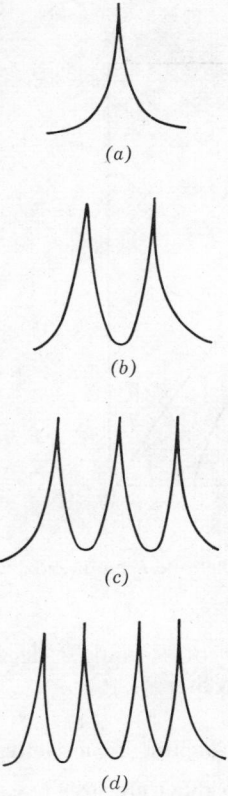

Fig. 9.19. Reproductions of oscillograms of the amplitude frequency phenomenon occurring in resonator 1 as alignment steps are performed.

The alignment of the filter is now complete. Figure 9.19 shows the amplitude-frequency phenomena occurring in resonator 1 as the alignment steps are performed. They may be observed on an oscilloscope, as the filter is swept. It should be realized that since the alignment adjustments depend exclusively on the amplitude of the response at f_0, a sweep-frequency generator is not required, and all adjustments can be made with a single-frequency input, f_0.

Figure 9.19a is the oscillogram produced when resonator 2 is detuned or short circuited, and the input resonator 1 is adjusted for a maximum signal at f_0. Figure 9.19b occurs when the second resonator is tuned for minimum amplitude at f_0. The third resonator is detuned. Figures 9.19c and 9.19d show the continuation of the procedure as detected in resonator 1.

It can be seen that when the ith resonator is tuned, there will be i peaks and $i - 1$ valleys produced in resonator 1. It should be remembered that if it is impossible to completely detune all resonators, a single device may be used to short circuit the resonator immediately following the one being tuned. This short must be effective at the frequency involved.

Measurement of Coupling

The fundamental procedure is based upon the consideration that every pair of adjacent resonators is a double-tuned—that is, a two-pole circuit (with all the other resonators completely detuned). In a double-tuned circuit with Q_A and Q_B equal to infinity, the fractional bandwidth between primary response peaks is exactly equal to the coefficient of coupling between resonators A and B. This direct relationship makes this phenomena an excellent one to use as the basis of an experimental procedure for adjusting of coefficient coupling to a desired value.

When the unloaded Q of the resonators are very high, but Q is not infinity, the curve of Fig. 9.20 supplies a way of finding the exact coupling between adjacent resonators. To determine the amount of coupling between adjacent resonators, the following steps must be taken:

1. Designate as A the lower Q resonator and B the higher Q resonator.
2. Couple a nonresonant signal generator directly and very loosely to either the electric or magnetic field of resonator A.
3. Couple a nonresonant detector directly and loosely to either the electric or magnetic field of resonator A.
4. Completely detune all the resonators in the filter.
5. Tune resonator A for maximum output from the detector. Record the signal generator input and the detector output.
6. Tune resonator B for minimum output from the detector (as in alignment procedure in the previous section). Increase the signal generator input to produce the same output obtained in Step 5 (see Fig. 9.21).
7. The ratio of the signal generator input in Step 6 to that in Step 5 is the abscissa in Fig. 9.20. From the ordinate of this graph, read the ratio of coupling K between resonators A and B to the percentage bandwidth $\Delta f_p/f_0$ between the amplitude peaks that are now present across resonator A.
8. Carefully measure the bandwidth Δf_p between the response peaks of resonator A.
9. The exact coefficient of coupling is equal to the fractional bandwidth between these peaks time the ordinate value obtained in Step 7.

For the low-Q case, in a double-tuned circuit with Q_A and Q_B only two or three times the fractional bandwidth, the coefficient of coupling required in the

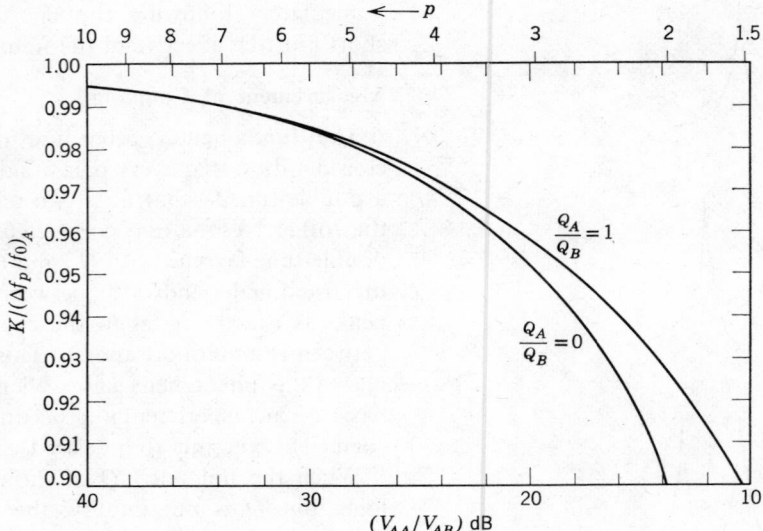

Fig. 9.20. Method of finding exact coefficient of coupling K between two resonators, when the K required is much greater than the unloaded decrements.

network to produce desirable response shapes are not much greater than the unloaded decrements. The resulting peak to valley ratio will not be very great when the previous procedure is used, and the peaks will not be sharply defined. For this case, the coefficient of coupling can be found in terms of the measured Q of the resonators being used.

In order to determine the amount of coupling between adjacent resonators for the low-Q case, the following steps must be taken:

(Repeat steps 1 through 6, as for the high Q case.)

7. The ratio of signal generator input of Step 6 to that in Step 5 is the abscissa of the graph of Fig. 9.22. From the ordinate of this graph read the ratio of the coefficient of the coupling K_{AB} to the geometric mean of the decrement of the resonators A and B.

8. Carefully measure the Q of each resonator A and B by measuring the 3-dB bandwidth. See Eq. 9.4.7 and related discussion.

9. The exact coefficient of coupling is equal to the

geometric mean of A and B decrement times the ordinate value in Step 7.

Adjustment of Coupling in a Helical Filter

It is often desirable to be able to set or to check each coefficient of coupling of a filter without going through the procedure of converting into a double-tuned circuit. This can be accomplished by measurements made entirely in the input resonator.

There are, in practice, two cases which must be considered. In the first case, the unloaded Q's of the resonators being used are very much greater than the fractional midfrequency $f_0/\mathrm{BW}_{3\,\mathrm{dB}}$ being used—that is, the unloaded individual Q's are essentially infinite. In the second case, the unloaded Q's of the resonators are only four, or five, or fewer times $f_0/\mathrm{BW}_{3\,\mathrm{dB}}$.

For the first case above, which is the most common case in helical filtering, the K's can be approximately measured, in consecutive order, by measuring the

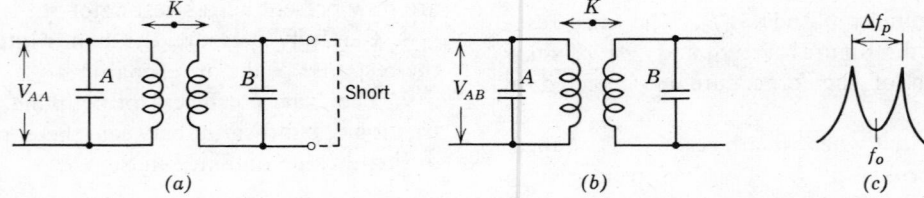

Fig. 9.21. Procedure for obtaining coefficient of coupling K between two resonators when the K required is much greater than the unloaded decrement. (*a*) Resonate circuit A for maximum V_{AA} (B shorted). (*b*) Resonate circuit B for minimum V_{AB} (short removed). (*c*) Bandwidth between primary peaks of V_{AB}.

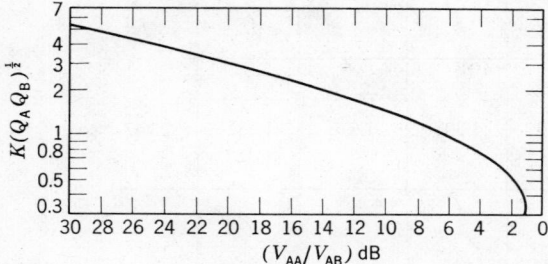

Fig. 9.22. Method of obtaining exact coefficient of coupling K between two resonators when the K required is not much greater than the unloaded decrements.

bandwidth between the various response peaks appearing in resonator 1, as each of the following resonators is resonated in consecutive order. If the unloaded Q's are approximately 100 times $f_0/\text{BW}_{3\,\text{dB}}$, then this procedure gives exact answers. It should be remembered that there will be i response peaks

occurring in the input resonator, when the ith resonator is correctly tuned.

The first step is to be sure that the input resonator is correctly loaded by adjusting the position of the tap. The 3-dB bandwidth of the response of the first resonator is given by

$$\Delta_{3\,\text{dB}} = \frac{\text{BW}_{3\,\text{dB}}}{q_1} \qquad (9.4.1)$$

when the following resonator is shorted. Here, the value of q_1 is obtained from the tables of 3 dB down k and q values of Chapter 6, and $\text{BW}_{3\,\text{dB}}$ is the desired filter bandwidth.

Next, in tuning the completed filter, we must calculate the frequencies of the peaks obtained as the filter is tuned resonator by resonator. This is done by using Eqs. 9.4.2 through 9.4.5. The values of k, are obtained from the tables of Chapter 6, and $\text{BW}_{3\,\text{dB}}$ is again the desired filter 3-dB bandwidth.

The resulting Δf_p is the distance between amplitude response peaks occurring in the first resonator, and by adding to and subtracting from f_0 the quantities $\Delta f_p/2$, the absolute frequencies of the peaks are obtained.

In setting the first resonator,

$$\Delta f_p = 0 \qquad (9.4.2)$$

Here, one peak occurring at f_0 is obtained and the 3-dB bandwidth is given by Eq. 9.4.1. For the second resonator,

$$\Delta f_p = k_{12}\text{BW}_{3\,\text{dB}} \qquad (9.4.3)$$

For the third resonator,

$$\left(\frac{\Delta f_p}{\text{BW}_{3\,\text{dB}}}\right)^3 - (k_{12}{}^2 + k_{23}{}^2)\frac{\Delta f_p}{\text{BW}_{3\,\text{dB}}} = 0$$

yields two solutions:

$$\Delta f_p = 0$$

and

$$\Delta f_p = \sqrt{k_{12}{}^2 + k_{23}{}^2}\,\text{BW}_{3\,\text{dB}} \qquad (9.4.4)$$

This indicates that three peaks are obtained, one at f_0, and a lower and higher peak separated by the distance Δf_p.

For the fourth resonator,

$$\left(\frac{\Delta f_p}{\text{BW}_{3\,\text{dB}}}\right)^4 - (k_{12}{}^2 + k_{23}{}^2 + k_{34}{}^2)$$
$$\times \left(\frac{\Delta f_p}{\text{BW}_{3\,\text{dB}}}\right)^2 + (k_{12}k_{34})^2 = 0$$

again yields two solutions:

$$\Delta f_p = \text{BW}_{3\,\text{dB}}\left(\frac{(k_{12}{}^2 + k_{23}{}^2 + k_{34}{}^2) \pm \sqrt{(k_{12}{}^2 + k_{23}{}^2 + k_{34}{}^2)^2 - 4(k_{12}k_{34})^2}}{2}\right)^{1/2} \qquad (9.4.5)$$

Here, four peaks are obtained; the larger value of Δf_p is the distance between the outer pair, the smaller value of Δf_p is the distance between the inner pair.

With Eqs. 9.4.2 through 9.4.5, filters up to and including seven poles may be completely tuned. This is done by first working from input to output and then turning the filter around, working from output to input. Figure 9.23 lists values of normalized $\Delta f_p/\text{BW}_{3\,\text{dB}}$ for the Butterworth response shape for 2-, 3-, 4-, 5-, 6-, and 7-pole filters, as well as values of $1/q_1 = \Delta_{3\,\text{dB}}/\text{BW}_{3\,\text{dB}}$ for the purpose of adjusting the input and output tap.

For the second case described—that is, where the unloaded Q is only four or five times $f_0/\text{BW}_{3\,\text{dB}}$, the coefficients of coupling should be set or measured in consecutive order as follows: Accurately measure the Q of each resonator in the filter, then precede step by step, through the alignment procedure, accurately measuring and recording the magnitudes of the maxima and minima produced.

The ratio of the detector output obtained when resonator 1 is alone resonated to that obtained when resonator i is resonated is given by

$$\left(\frac{V_{1,1}}{V_{1,i}}\right) = \left[1 + \cfrac{p_{12}{}^2}{1 + \cfrac{p_{23}{}^2}{1 + \cfrac{\cdots}{1 + p_{(i-1)i}{}^2}}}\right] \qquad (9.4.6)$$

Number of resonators			2	3	4	5	6	7
Tuning resonator no. 1, and setting tap.	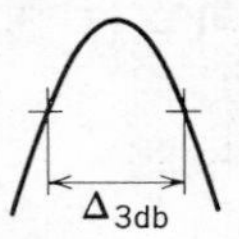 Δ_{3db}	$\dfrac{\Delta_{3db}}{BW_{3db}} =$	0.707	1.000	1.305	1.618	1.931	2.247
Tuning resonator no. 2, and setting K_{12}	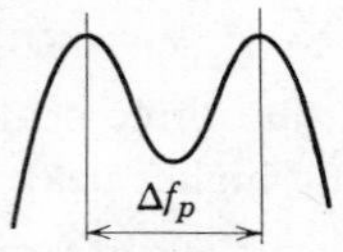 Δf_p	$\dfrac{\Delta f_p}{BW_{3db}} =$	0.707	0.707	0.840	1.000	1.170	1.340
Tuning resonator no. 3, and setting K_{23}	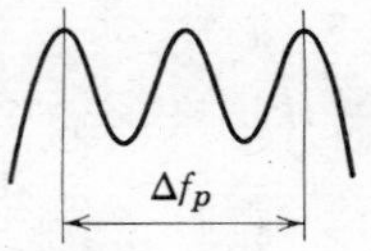 Δf_p	$\dfrac{\Delta f_p}{BW_{3db}} =$		1.000	1.000	1.144	1.318	1.498
Tuning resonator no. 4, and setting K_{34}	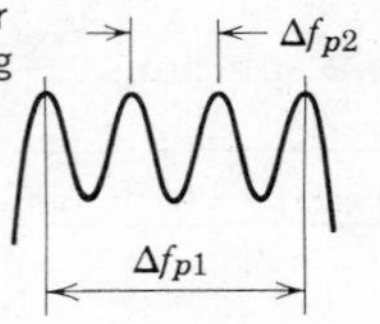 Δf_{p2}, Δf_{p1}	$\dfrac{\Delta f_{p1}}{BW_{3db}} =$			1.154	1.182	1.342	1.518
		$\dfrac{\Delta f_{p2}}{BW_{3db}} =$			0.612	0.470	0.452	0.466

Fig. 9.23. Normalized Δf_p and Δ_{3dB} for Butterworth filters.

where $p_{12}^2 = K_{12}^2 Q_1 Q_2$, and so on, and

$$K_{12} = \frac{\Delta f}{f_0} k_{12}$$

Since we know the desired value of K, and have measured each Q, we can calculate the required $V_{1,1}/V_{1,i}$ ratio from Eq. 9.4.6, and compare to the measured value. The most trustworthy method of making accurate loaded or unloaded Q measurements on a resonator that is part of the filter chain seems to be as follows:

1. Completely assemble the filter.
2. Completely detune all resonators except the one to be measured. Obviously complete detuning of the resonator on each side of the one being measured should be satisfactory.
3. A nonresonant signal generator is coupled directly and very loosely to either the electrical or magnetic field of the resonator.
4. A nonresonant detector is coupled very loosely and preferably to the field opposite to that being used for the generator. In other words, make sure that there is negligible direct coupling between generator and detector.
5. Using an unmodulated wave, or an amplitude-modulated wave checked for negligible frequency modulation from the signal generator, measure the frequency difference Δf_β between the points that are V_p/V_β down from the peak response. The resonator Q is given by

$$Q = \frac{f_0}{\Delta f_\beta} \sqrt{\left(\frac{V_p}{V_\beta}\right)^2 - 1} \qquad (9.4.7)$$

When high Q's are to be measured, the setup must be capable of measuring very small-percentage bandwidths.

Sweep Display Setups

Figure 9.24 illustrates a typical sweep display setup for viewing the transmission characteristics of a VHF filter. To view the impedance characteristics of the filter, the block diagram of Fig. 9.25 can be used. Finally, Fig. 9.26 shows a setup for tuning the filter by means of a small probe placed in the first resonant cavity. The signal from this probe must be approximately 20-dB down, in order not to load down the first resonator.

9.5 EXAMPLES OF HELICAL FILTERING

The filter of Fig. 9.14 is a 540-Mc helical filter which is tuned in order to provide an equiripple group

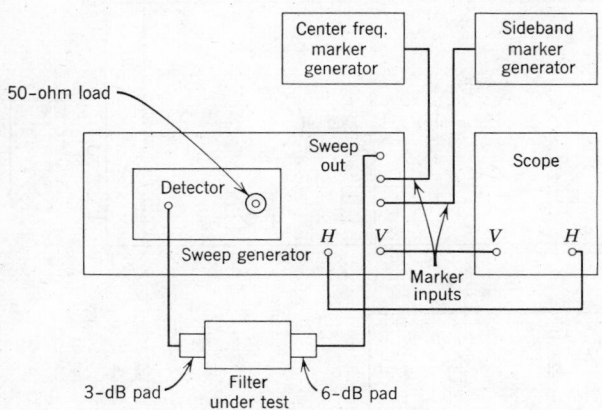

Fig. 9.24. Typical sweep display setup of transmission characteristics of a filter at VHF.

delay. The can of Fig. 9.27 is used for two filters, with center frequencies at 28 and 32 Mc. The length is approximately 4 inches. The filters have 3-dB bandwidths of 1.5 Mc and an insertion loss of 1.4 dB was obtained. The coils have approximately 65 turns. Tap coupling was used for the input and output coils, providing an impedance of 75 ohms. One unusual specification for these filters was that at 30 Mc, 29 dB $\pm$ 0.5 dB was to be provided for both filters.

When a high value of attenuation is needed at frequencies near $3f_0$, the helical filter must be followed by a lowpass *LC* filter, since helical filters have a secondary or spurious response at this frequency. Figure 9.28 shows a VHF helical filter–lowpass combination. A minimum attenuation of 60-dB is achieved at frequencies up to 6.6 times f_0.

The design shown in Fig. 9.29 is actually two filters in one case. The helical section is a 30-Mc bandpass

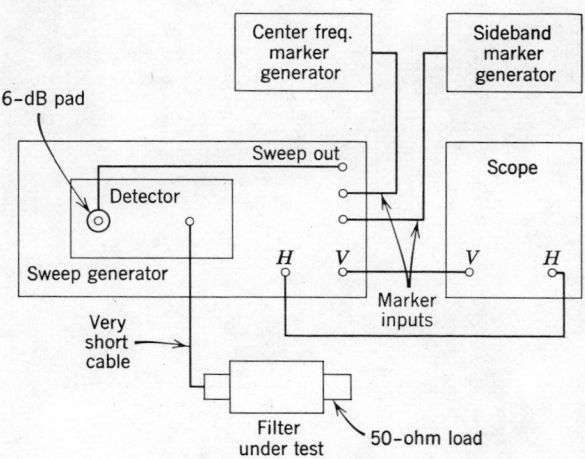

Fig. 9.25. Typical sweep display setup of impedance characteristic of a filter at VHF.

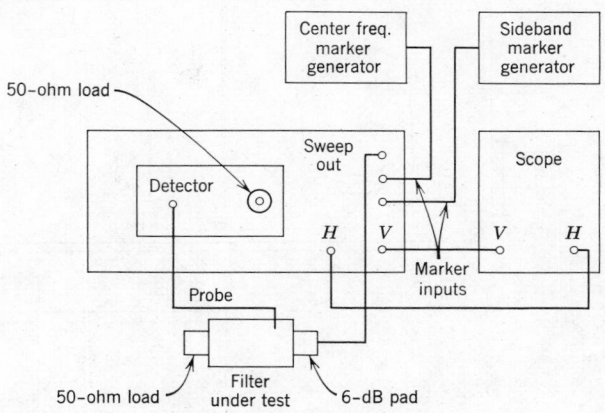

Fig. 9.26. Typical sweep display setup of amplitude phenomena in first cavity for VHF filters.

filter with a 3-dB bandwidth of 330 kc, and the filter mounted on the vertical printed circuit board is a 1 Mc wide *LC* filter at the same center frequency. Both filters are tuned to provide a Gaussian response. This design is unusual because of the small physical size (1 $\times$ 1.5 $\times$ 2 inches), made possible with the square layout of the helical resonators.

Use of Helical Filters in Parametric Multipliers

For frequency multipliers, a varactor diode is usually inserted between bandpass filters. The first filter is designed for the fundamental signal to be multiplied, and the output filter is designed for the specific harmonic in which the fundamental frequency is transformed by the parametric diode. The entire network from input to output is still considered as a filter, because the nonlinear diode is absorbed by the adjacent elements in such a fashion that the low-frequency filter presents an appropriate load, and to the harmonic filter, the diode represents a nonlinear capacitor providing a source of harmonic frequencies with high impedance.

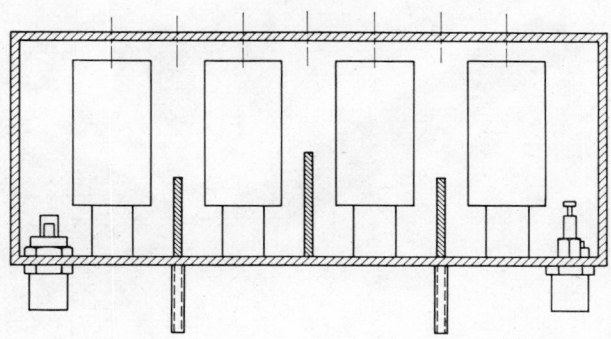

Fig. 9.27. Four resonator 28- and 32-Mc filter.

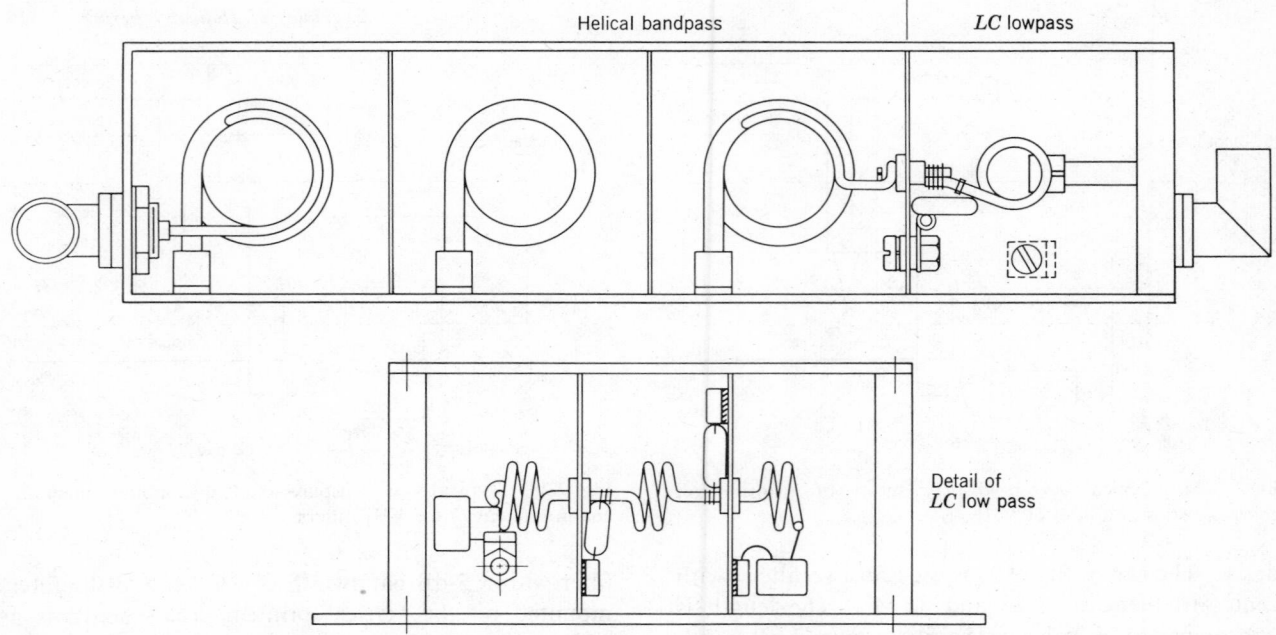

Helical bandpass   *LC* lowpass

Detail of
LC low pass

Fig. 9.28. Helical bandpass and *LC* lowpass filter combination.

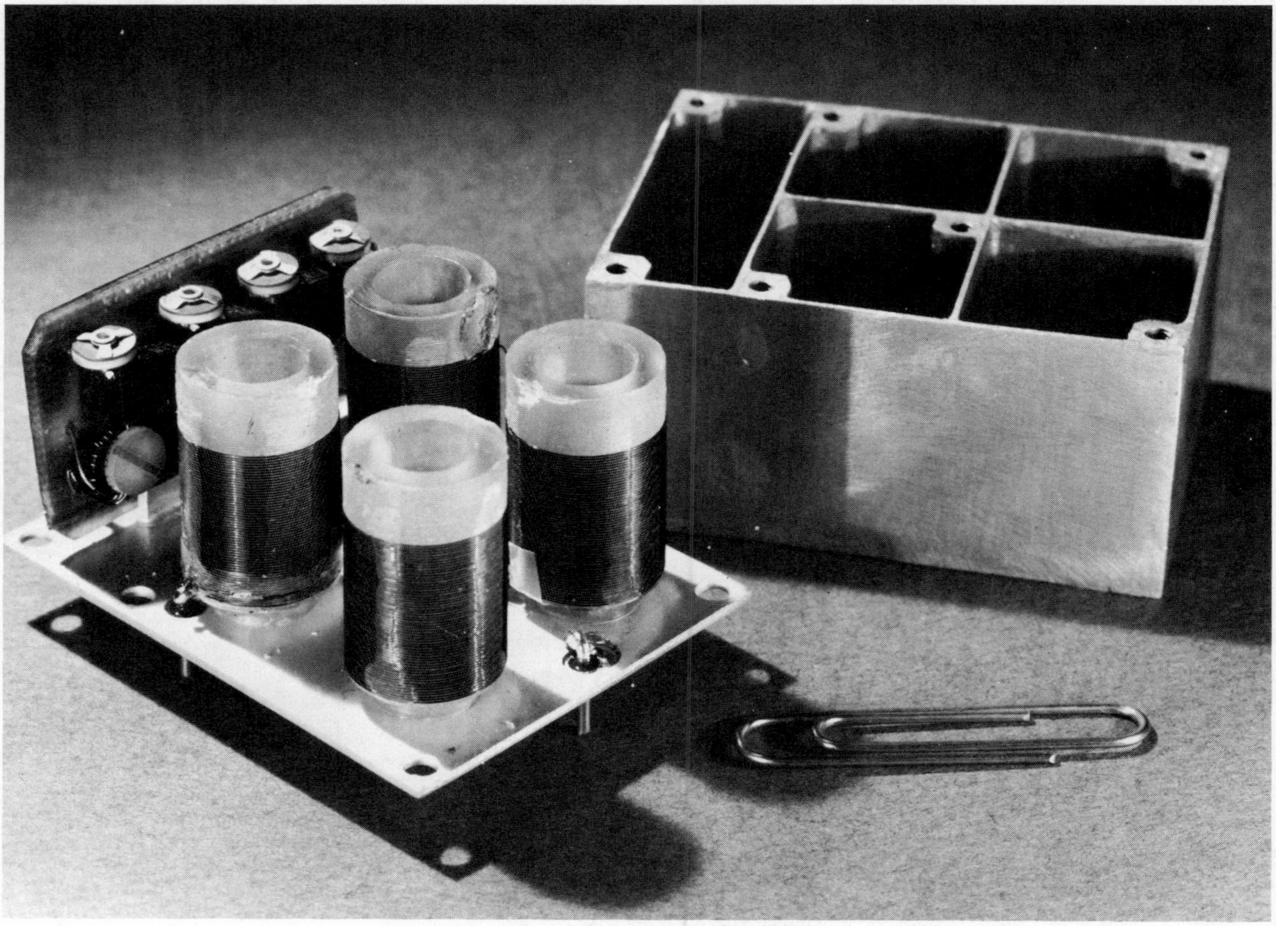

Fig. 9.29. 30-Mc helical and *LC* filter.

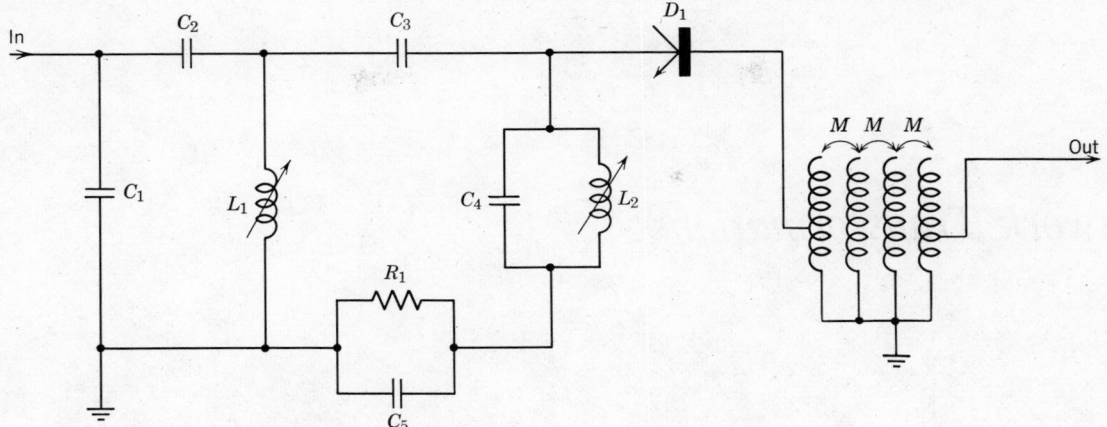

Fig. 9.30. Schematic diagram of 63 Mc × 6 = 378 Mc parametric multiplier.

The whole assembly can be designed in such a way that no spurious modes from the varactor will be passed through the high-frequency portion of the network. The fundamental frequency filter, in many cases can be a conventional *LC* network. The output from the varactor is connected to the input tap of the helical resonator filter. A schematic diagram of such a parametric multiplier is shown in Fig. 9.30. To eliminate unwanted outputs, the helical filter consists of several cavities. The high-quality helical filter adds only a negligible amount of loss to the multiplier network, in comparison to a conventional *LC* filter. A narrow passband is usually used since any noise in the initial stage of the harmonic multiplier or synthesizer is very damaging for the following stages at microwave frequencies. This arrangement with helical filters will cover the entire VHF and UHF bands, providing the cleanest harmonic output and the cleanest source of the above mentioned frequencies.

10

Network Transformations

A well-designed filter should satisfy not only the stated attenuation requirements but should also be designed for easy alignment in the manufacturing process. The filter must contain a reasonable number of components. This puts a large premium on basic ladder structures. The quality of such structures may be measured by several factors:

1. The number of components, and in particular, a minimum number of inductances.
2. The ratio of the largest to the smallest inductance in the circuit should be minimized.
3. The ratio of the largest to the smallest capacitance in the circuit should be minimized.
4. The connection of a capacitor in parallel with every inductance in the shunt arms of the ladder, thus having a provision for the absorption of the distributed capacity of the coil.
5. The connection of a capacitor from the "hot" node of any series-resonant circuit in the series arm of the filter to ground, thus avoiding difficulties in the realization of this type of circuit.
6. Proper distribution of tuning circuits which minimizes the sensitivity of the network with respect to terminal mismatches.

Let us assume that a filter has been designed by means of some arbitrary method of synthesis, and the schematic of the filter is given in the conventional form. There are numerous methods of modifying the schematic to gain some of the above mentioned conditions. It is possible to deal with the two-terminal (dipoles) within the network, providing a partial modification, or the entire four-terminal network may be analyzed. Looking for the possibility of modifying the entire network by such powerful means as Norton's transformation and the Star-Delta transformation can provide a variety of noncanonic network structures. In this field many ingenious

solutions have been developed with the intent to minimize the number of inductances in the ladder. Laurent introduced the Zig-Zag ladder formulated in the terms of image parameters. Cauer's treatment of this problem, involving the complicated concepts of group theory presents considerable conceptional difficulties. An equally promising procedure involves the use of approximate reactance transformations. A simple form of this makes use of capacitance inverters included in the ladder structures of bandpass filters without attenuation poles at finite frequencies. This structure was first suggested by Atiya. It was then used by S. Cohn for microwave filters. The apparent versatility and variety of realization procedures associated with the technique of filter synthesis has, however, one serious drawback. The complexity of the involved algebraic manipulation does not allow the direct control of the values of the branch elements— that is, the resulting element values. Personal experience is the most important time-saving factor at this step in design work. An opposite situation prevails in the older and simpler image-parameter technique. Thus, whereas the effective parameter theory provides a predictable and controlled response, it gives no direct insight into the values of the components. On the other hand, the image-parameter theory provides explicit expressions for element values in terms of the original design data, and yet gives no assurance of the response in the critical frequency range.

10.1 TWO-TERMINAL NETWORK TRANSFORMATIONS

In the design of filters, phase-splitting networks, as well as other networks, series and parallel combinations of inductors and capacitors are of the paramount importance. These combinations are commonly referred to as dipoles or two-terminal networks, and may contain two or more elements depending on the complexity of the network required.

It is often necessary for the circuit designer to change the configuration of the network, as we have previously illustrated. For example, change of configuration may be necessary to minimize the high voltages which appear across the elements of a series circuit at resonance.

Changing Configuration

The circuit designer is often required to change the element values of a network while preserving the configuration and the impedance characteristic. On the other hand, he may wish to also change the configuration while retaining the impedance characteristic.

The following tables are designed to minimize the time and effort for making such network transformations. Table 10.1 shows all possible transformations involving three- and four-element networks that use a minimum number of coils and capacitors to realize a particular impedance function. These are known as canonic networks. Below each figure are the necessary equations for performing the transformation.

Table 10.2 shows the various impedance functions which are characteristic of each of the networks. Finally, Table 10.3 lists the applicable equations for calculating the resonant frequencies associated with each network. A symbol key for Tables 10.1 through 10.3 is given in Table 10.4.

Use of the charts is straightforward, involving simple substitution into the various equations. This method, used as a substitute for the conventional methods of synthesis avoids the algebraic manipulations involved in expansion of partial fractions. The charts usually provide more choices of realizable dipole elements and hence can be regarded as a useful tool in everyday engineering practice.

Noncanonic Two-Terminal Networks

One basic and very useful three-element dipole is illustrated in Fig. 10.1a, and consists of a series-resonant circuit in shunt with a capacitor. Figure 10.1b is its equivalent circuit. Two conditions of equivalence imposed are: (1) The driving point impedance of the networks must be identical and (2) The basic configuration must be similar.

Figure 10.1b was derived by extracting a common capacitance from the network, and putting it in a different position. This transformation is often applied when the inductance L_2 in Fig. 10.1a is rather large; the inductance L_A in Fig. 10.1b will be considerably smaller.

The conditions stated were

1. $Z_1 = Z_2 = Z$.
2. L_A must be less than L_2.
3. The basic configuration must be preserved.

Setting $Z_1 = Z_2$ and solving for L_A, C_A, C_B, and C_M, the following are obtained:

$$L_A = \left(1 - \frac{C_3}{C_M}\right)^2 L_2 \tag{10.1.1}$$

$$C_A = \frac{C_2 C_M{}^2}{(C_M - C_3)(C_M - C_3 - C_2)} \tag{10.1.2}$$

$$C_B = \frac{C_M C_3}{C_M - C_3} \tag{10.1.3}$$

For an example, consider an application in crystal filter design, where the value of L_A can easily be produced by the crystal manufacturer. Then the ratio of L_A/L_2 is known. L_2 is the value of equivalent inductance and found to be of impractical magnitude. Using this ratio, Eqs. 10.1.1 through 10.1.3 can be represented in the following forms:

$$a = \frac{L_A}{L_2} \tag{10.1.4}$$

$$K = \frac{1 + \sqrt{a}}{1 - a} \tag{10.1.5}$$

$$C_M = K C_3 \tag{10.1.6}$$

$$C_A = \frac{K^2 C_2 C_3}{(K - 1)[(K - 1)C_3 - C_2]} \tag{10.1.7}$$

$$C_B = \frac{K C_3}{K - 1} \tag{10.1.8}$$

The resonant frequency of L_A and C_A is designated as f_A. With the aid of Eq. 10.1.7

$$f_A = \frac{1}{2\pi K} \sqrt{\frac{(K - 1)}{L_A}\left[\frac{K - 1}{C_2} - \frac{1}{C_3}\right]} = \frac{1}{2\pi\sqrt{L_A C_A}} \tag{10.1.9}$$

This frequency f_A represents the principal ordering information for the crystal manufacturer. The physical dimensions of both crystals in series and lattice arms can always, by this method, be brought to the same practical values.

Table 10.1 Dipole Transformation

EQUIVALENT NETWORKS	EQUIVALENT NETWORKS

1 — Z-CHARACTERISTIC I

(A)
$$L_1 = L_2\left(\frac{C_3}{C_3+C_4}\right)^2$$
$$C_1 = C_4(1+C_4/C_3)$$
$$C_2 = C_3+C_4$$

(B)
$$L_2 = L_1\left[(C_1+C_2)/C_2\right]^2$$
$$C_3 = C_2/(1+C_1/C_2)$$
$$C_4 = C_2/(1+C_2/C_1)$$

6 — Z-CHARACTERISTIC III

(A)
$$L_1 = W \quad L_2 = X$$
$$C_1 = Y \quad C_2 = Z$$
$$A = L_3 L_4 C_3 C_4$$
$$B = L_3 C_3 + L_4 C_4 + L_3 C_4$$
$$E = L_3 C_3 C_4 \quad D = C_4$$

(B)
$$L_3 = (L_1 C_1 - L_2 C_2)^2/(L_1+L_2)(C_1+C_2)^2$$
$$L_4 = L_1 L_2/(L_1+L_2)$$
$$C_3 = C_1 C_2(C_1+C_2)(L_1+L_2)/(L_1 C_1 - L_2 C_2)^2$$
$$C_4 = C_1+C_2$$

2 — Z-CHARACTERISTIC II

(A)
$$L_1 = L_4/(1+L_3/L_4)$$
$$L_2 = L_4/(1+L_4/L_3)$$
$$C_1 = C_2\left[(L_3+L_1)/L_4\right]^2$$

(B)
$$L_3 = L_2(1+L_2/L_1)$$
$$L_4 = L_1+L_2$$
$$C_2 = C_1\left[L_1/(L_1+L_2)\right]^2$$

7 — Z-CHARACTERISTIC III

(A)
$$L_1 = W \quad L_2 = X$$
$$C_1 = Y \quad C_2 = Z$$
$$A = L_3 L_4 C_3 C_4$$
$$B = L_3 C_3 + L_4 C_4 + L_4 C_3$$
$$E = C_3 C_4(L_3+L_4)$$
$$D = C_4$$

(B)
$$L_3 = \frac{L_1 L_2(L_1 C_1^2 + L_2 C_2^2)}{(L_1 C_1 - L_2 C_2)^2}$$
$$L_4 = (L_1 C_1^2 + L_2 C_2^2)/(C_1+C_2)^2$$
$$C_4 = C_1+C_2$$
$$C_3 = \frac{C_1 C_2(C_1+C_2)(L_1 C_1 - L_2 C_2)^2}{(L_1 C_1^2 + L_2 C_2^2)^2}$$

3 — Z-CHARACTERISTIC III

(A)
$$L_0 = L_2$$
$$C_0 = C_2 C_3/C_2+C_3$$
$$L_1 = L_3(1+C_3/C_2)^2$$
$$C_1 = C_2/(1+C_3/C_2)$$

(B)
$$L_2 = L_0$$
$$C_2 = C_1+C_0$$
$$L_3 = L_1\left[C_1/(C_1+C_0)\right]^2$$
$$C_3 = C_0(1+C_0/C_1)$$

8 — Z-CHARACTERISTIC III

(A)
$$L_0 = L_2 L_3/(L_2+L_3)$$
$$C_0 = C_3/1+(C_3/C_2)\left[L_3/(L_2+L_3)\right]^2$$
$$L_1 = \left[C_2(L_2+L_3)^2 + L_3^2 C_3\right]^2/(L_2+L_3)L_3^2 C_3^2$$
$$C_1 = C_3/1+(C_2/C_3)\left[(L_2+L_3)/L_3\right]^2$$

(B)
$$L_2 = L_0\left\{1+\left[L_0(C_1+C_0)^2/L_1 C_1^2\right]\right\}$$
$$L_3 = L_0+L_1\left[C_1/(C_1+C_0)\right]^2$$
$$C_3 = C_1 + C_0$$
$$C_2 = \frac{L_1^2 C_1^3 C_0(C_1+C_0)}{\left[L_0(C_1+C_0)^2 + L_1 C_1^2\right]^2}$$

4 — Z-CHARACTERISTIC III

(A)
$$L_0 = L_2+L_3$$
$$L_1 = L_2(1+L_2/L_3)$$
$$C_0 = C_2$$
$$C_1 = C_3\left[L_3/(L_2+L_3)\right]^2$$

(B)
$$L_2 = L_1 L_0/(L_1+L_0)$$
$$L_3 = L_0^2/(L_1+L_0)$$
$$C_2 = C_0$$
$$C_3 = C_1\left[(L_1+L_0)/L_0\right]^2$$

9 — Z-CHARACTERISTIC IV

(A)
$$L_3 = L_2(1+L_2/L_1)$$
$$L_4 = L_1+L_2$$
$$C_3 = C_1\left[L_1/(L_1+L_2)\right]^2$$
$$C_4 = C_2$$

(B)
$$L_1 = L_4\left[L_4/(L_3+L_4)\right]$$
$$L_2 = L_3 L_4/(L_3+L_4)$$
$$C_2 = C_4$$
$$C_1 = C_3\left[(L_3+L_4)/L_4\right]^2$$

5 — Z-CHARACTERISTIC III

(A)
$$L_1 = W \quad L_2 = X$$
$$C_1 = Y \quad C_2 = Z$$
$$A = L_4 L_3 C_4 C_3$$
$$B = L_4 C_4 + L_3 C_3 + L_4 C_3$$
$$E = L_3 C_4 C_3$$
$$D = C_4+C_3$$

(B)
$$L_4 = L_1 L_2/L_1+L_2$$
$$L_3 = (L_1^2 C_1 + L_2^2 C_2)^2/(L_1+L_2)(L_1 C_1 - L_2 C_2)^2$$
$$C_4 = C_1 C_2(L_1+L_2)^2/L_1^2 C_1 + L_2^2 C_2$$
$$C_3 = (L_1 C_1 - L_2 C_2)^2/L_1^2 C_1 + L_2^2 C_2$$

10 — Z-CHARACTERISTIC IV

(A)
$$L_4 = L_2$$
$$L_3 = L_1\left[(C_1+C_2)/C_2\right]^2$$
$$C_3 = C_2/(1+C_1/C_2)$$
$$C_4 = C_2/(1+C_2/C_1)$$

(B)
$$L_1 = L_3\left[C_3/(C_3+C_4)\right]^2$$
$$L_2 = L_4$$
$$C_1 = C_4(1+C_4/C_3)$$
$$C_2 = C_3+C_4$$

Table 10.1 (*continued*)

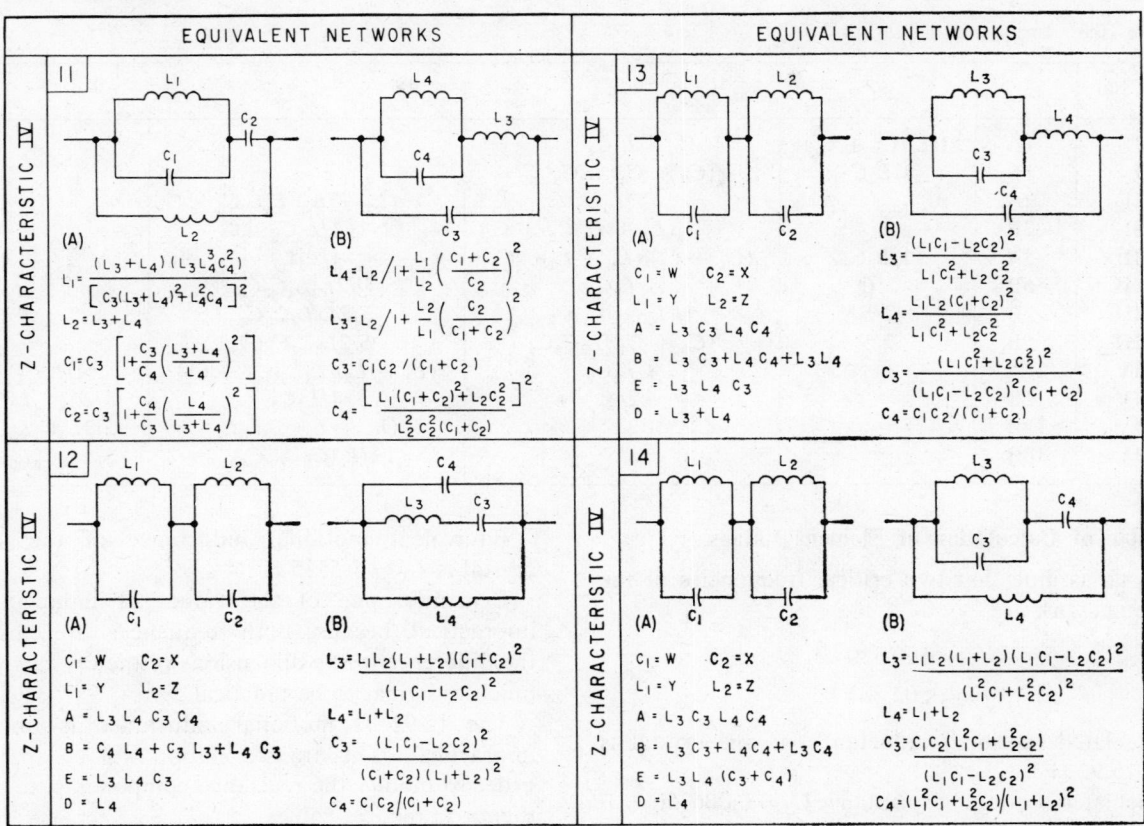

Table 10.2 Impedance Characteristics

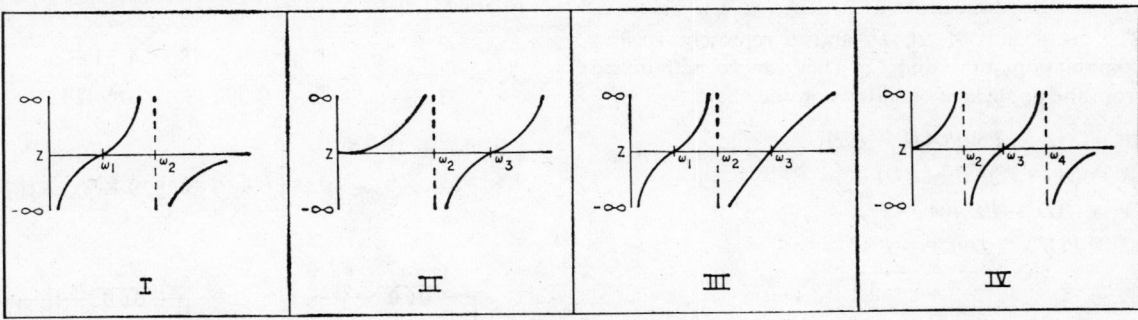

Table 10.3 Frequency Equations

Z-Char	Fig.	ω_1^2	ω_2^2	ω_3^2	ω_4^2
I	1A	$1/L_1(C_1 + C_2)$	$1/L_1C_1$		
I	1B	$1/L_2C_3$	$(C_3 + C_4)/L_2C_3C_4$		
II	2A		$1/L_1C_1$	$(L_1 + L_2)/L_1L_2C_1$	
II	2B		$1/(L_3 + L_4)C_2$	$1/L_3C_2$	
III	5A	$1/L_2C_2$	$(C_1 + C_2)/(L_1 + L_2)C_1C_2$	$1/L_1C_1$	
III	6B	$2/Q$	$1/L_3C_3$	$Q/2L_3L_4C_3C_4$	
III	7B	$2/S$	$1/(L_3 + L_4)C_3$	$S/2L_3L_4C_3C_4$	
III	5B	$2/S$	$(C_3 + C_4)/L_3C_3C_4$	$S/2L_3L_4C_3C_4$	
IV	12A		$1/L_2C_2$	$(L_1 + L_2)/L_1L_2(C_1 + C_2)$	$1/L_1C_1$
IV	12B		$2/S$	$1/L_4C_4$	$S/2L_4C_4L_3C_3$
IV	13B		$2/S$	$(L_3 + L_4)/C_3L_3L_4$	$S/2L_4C_4L_3C_3$
IV	14B		$2/S$	$1/L_3(C_3 + C_4)$	$S/2L_4C_4L_3C_3$

Example of Calculation of Element Values

Let us assume that two critical frequencies of the lattice network are

$$f_2 = 500.330 \text{ kc}$$
$$f_3 = 502.585 \text{ kc}$$

Calculated motional inductance of one crystal is $L_2 = 15.92$ H
  Calculated motional capacitance $C_2 = 0.006356 \ \mu\mu$F
  Interelectrode capacitance $C_3 = 0.593529 \ \mu\mu$F

Table 10.4 Symbol Key for Tables 10.1, 10.2, and 10.3

The equations describing the element values of some of the networks can be clarified by the use of more simplified notations. This has been done in the charts by substitution of symbols for some of the more complicated relationships. These symbols represent the following expressions.

(1) The coefficients A, B, D, and E represent various combinations of L and C. They can be determined from the applicable equations on the chart.

(2) $W = A(A - P^2)/P(AD - PE)$
  $X = (A - P^2)/(E - PD)$
  $Y = (AD - PE)/(A - P^2)$
  $Z = P(E - PD)/(A - P^2)$
  where $P = \dfrac{B + \sqrt{B^2 - 4A}}{2}$

(3) $Q = L_3C_3 + L_3C_4 + L_4C_4$
    $+ \sqrt{(L_3C_3 + L_3C_4 + L_4C_4)^2 - 4L_3L_4C_3C_4}$

(4) $S = L_3C_3 + L_4C_3 + L_4C_4$
    $+ \sqrt{(L_3C_3 + L_4C_4 + L_4C_3)^2 - 4L_3L_4C_3C_4}$

Equivalent motional inductance of the second crystal is $L_3 = 1.5535$ H

Evidently, one of the values of inductances is impractical, because both frequencies are close and the difference in the dimensions of the crystals will be much too great to be practical.

The 15.92 H motional inductance is obviously impractical. This transformation will be applied in order to modify the reactance components, to reduce them to practical values.

A crystal with an equivalent motional inductance of 1.5535 H can easily be manufactured. Therefore $L_A = 1.5535$ H.

From Eq. 10.1.4

$$a = \frac{L_A}{L_2} = \frac{1.5535}{15.92} = 0.09758$$

$$\sqrt{a} = 0.312378$$

From Eq. 10.1.5

$$K = \frac{1 + \sqrt{a}}{1 - a} = \frac{1 + 0.312378}{1 - 0.09758} = \frac{1.312378}{0.90242} = 1.4543$$

From Eq. 10.1.6

$$C_M = KC_3 = 1.4543 \times 0.593529 = 0.86317 \ \mu\mu\text{F}$$

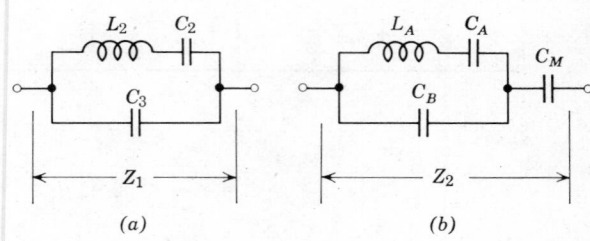

Fig. 10.1. Noncanonic transformation.

From Eq. 10.1.7

$$C_A = \frac{K^2 C_2 C_3}{(K-1)[(K-1)C_3 - C_2]} = 0.0667037 \, \mu\mu\text{F}$$

From Eq. 10.1.8

$$C_B = \frac{KC_3}{(K-1)} = \frac{1.4543 \times 0.593529}{0.4543} = 1.9 \, \mu\mu\text{F}$$

From Eq. 10.1.9

$$f_A = \frac{1}{2\pi\sqrt{L_A C_A}} = 494.413 \text{ kc}$$

This frequency f_A is very important because it determines the series resonance of the crystal to be ordered.

This frequency can be rapidly checked by the following formula:

$$f_A = f_2 \sqrt{1 - \frac{C_2}{C_3 K \sqrt{a}}} \qquad (10.1.10)$$

f_A is always less than f_2.

$$f_A = 500.330 \sqrt{1 - \frac{0.006356}{0.593529 \times 0.312378 \times 1.4543}}$$
$$= 494.398 \text{ kc}$$

The difference of 15 cps exists since the numbers were not carried out to a sufficient number of decimal places.

Equivalence of Two-Terminal Networks Having Mutual Coupling

The two-terminal network shown in Fig. 10.2*a* having mutual inductance between L_1 and L_2 can be transformed into Fig. 10.2*b*, if the coupled coils are regarded as a transformer substituted by its *T*-type equivalent circuit.

Part of the network between points 3 and 4 can be substituted by an equivalent two-terminal network with a parallel-resonant circuit instead of the series-resonant as in Fig. 10.2*b*. This network is shown in Fig. 10.2*c*.

Figure 10.2*d* shows the simplified schematic. The elements of the first circuit, Fig. 10.2*a*, and those in Fig. 10.2*d* are related in the following way:

$$C_2 = C_1 \frac{L_2}{L_6}$$

$$L_5 = L_1 - \frac{M^2}{L_2}$$

$$L_6 = L_2 \mp 2M + \frac{M^2}{L_2}$$

where

$$M = k\sqrt{L_1 L_2}$$

When $k = 1$, the value of $M^2 = L_1 L_2$ and therefore $L_5 = 0$. The equivalent circuit will be simpler and include only one coil and one capacitor as shown in Fig. 10.2*e*. Physically the exclusion of inductance L_5 from the circuit is explained by the fact that both constituent inductances in Fig. 10.2*c* are equal but opposite in sign.

If, for example, the value of capacitor C_1 in the circuit shown in Fig. 10.2*a* is known then

$$L_1 = \frac{L_6}{(1+n)^2}$$

$$L_2 = L_1 n^2$$

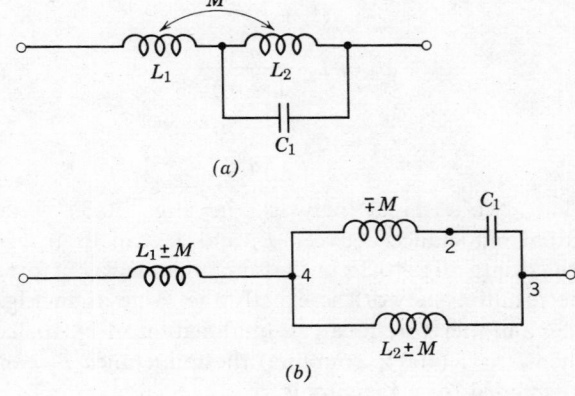

(a)

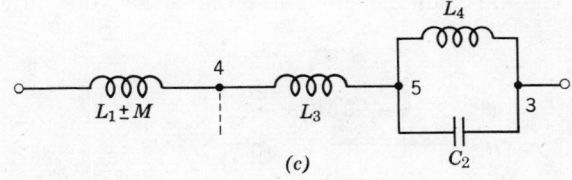

(b)

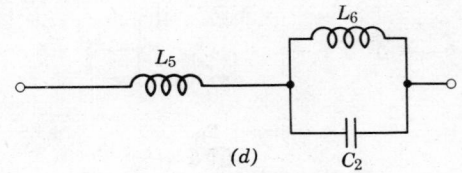

(c)

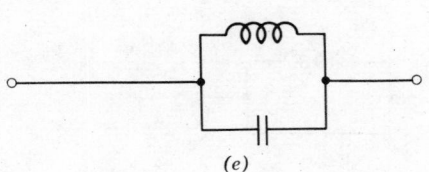

(d)

(e)

Fig. 10.2. Two-terminal network with mutual coupling.

where

$$n = \left(\sqrt{\frac{C_1}{C_2}} - 1 \right)^{-1}$$

If, on the other hand, the value L_1 is known then

$$L_2 = L_1 n^2$$

$$C_1 = C_2 \frac{L_6}{L_1}$$

where

$$n = +\sqrt{\frac{L_6}{L_1}} - 1$$

And finally, if $n = \pm\sqrt{(L_2/L_1)}$ then

$$L_1 = \frac{L_6}{(1 + n)^2}$$

$$L_2 = L_6 \frac{n^2}{(1 + n)^2}$$

$$C_1 = C_2 \frac{(1 + n)^2}{n^2}$$

The two-terminal network in Fig. 10.3a with mutual inductance between L_1 and L_2 can be transformed into Fig. 10.3b in the same manner as before. The resulting network is exactly the same as in Fig. 10.3b and therefore it can be modified into Fig. 10.3c. When $k = 1$ (100% coupling) the inductance L_3 can be excluded from the circuit.

The impedance of the two-terminal network, after

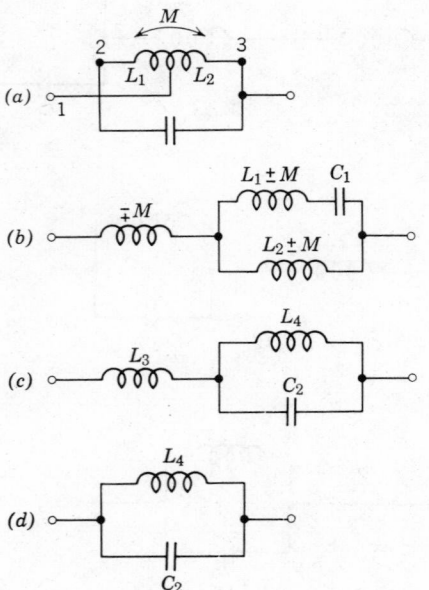

Fig. 10.3. Dipole transformation with mutual coupling.

its transformation into Fig. 10.3c, can be expressed by the following formula:

$$Z_a = \frac{j\omega L_1 n^2}{1 - \omega^2 L_1 C_1 (1 \pm n)^2}$$

where

$$n = \sqrt{\frac{L_2}{L_1}}$$

For the network shown in Fig. 10.3d

$$Z_d = \frac{j\omega L_4}{1 - \omega^2 L_4 C_2}$$

Both networks are equivalent when $Z_a = Z_d$. This implies the following:

$$L_4 = L_1 n^2$$

$$L_4 C_2 = L_1 C_1 (1 \pm n)^2$$

In complete analogy with the previous case in Fig. 10.2, the element values can be calculated by assuming one element as known:

1. Value of C_1 is known:

$$L_1 = \frac{L_4}{n^2}, \quad L_2 = L_4 \text{ with } n = \left(\sqrt{\frac{C_2}{C_1}} - 1 \right)^{-1}$$

2. The value of L_1 is known:

$$L_2 = L_4$$

$$C_1 = C_2 = \frac{n^2}{(1 + n)^2} \text{ with } n = \pm\sqrt{\frac{L_4}{L_1}}$$

3. The total inductance L_t in Fig. 10.3a is known:

$$L_1 = \frac{L_4}{n^2}; \quad L_2 = L_4; \quad C_1 = C_2 \frac{L_4}{L_t}$$

where

$$n = \left(\sqrt{\frac{L_t}{L_4}} - 1 \right)^{-1}$$

4. Ratio n is given:

$$n = \pm\sqrt{\frac{L_2}{L_1}}$$

then

$$L_1 = \frac{L_4}{n^2}$$

$$L_2 = L_4 C_1 = C_2 \frac{n^2}{(1 + n)^2}$$

10.2 DELTA-STAR TRANSFORMATION

The delta-star transformation is primarily used to modify reactive networks in order to achieve a

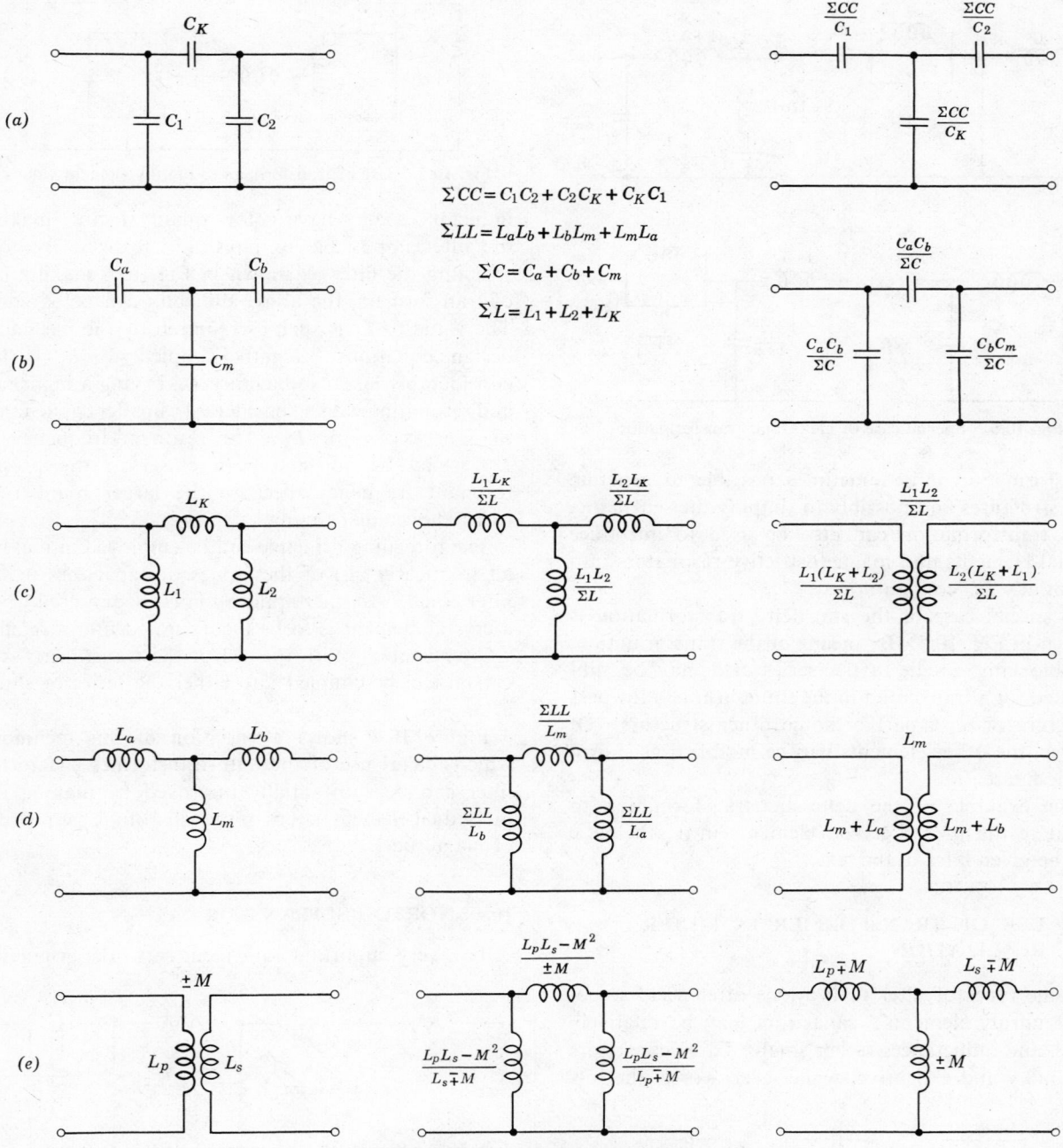

$$\Sigma CC = C_1 C_2 + C_2 C_K + C_K C_1$$

$$\Sigma LL = L_a L_b + L_b L_m + L_m L_a$$

$$\Sigma C = C_a + C_b + C_m$$

$$\Sigma L = L_1 + L_2 + L_K$$

Fig. 10.4. Star-Delta transformations.

different configuration of elements, or to develop a new relation of element values of some complicated networks. The delta (π) to star (T) transformation for capacitive elements is shown in line (a) of Fig. 10.4 where the element values of the star are given in terms of the element values of the delta. The reverse transformation, star-to-delta, for capacitors is given in line (b). In a similar manner lines (c), (d), and (e)

give the transformation for delta, star, and mutually coupled inductive circuits.

The transformations given can be used to change the coupling structure between resonators. In addition, a general star-delta transformation exists, where the elements of the schematic are not purely inductive or capacitive but may be of a complex impedance. With this general case of the star-delta

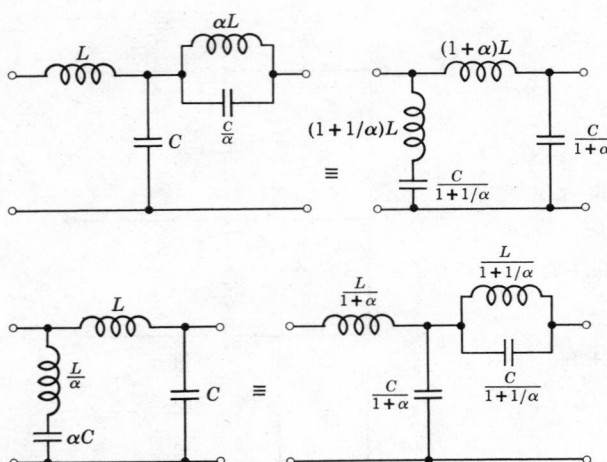

Fig. 10.5.   Special case of Delta-Star transformation.

transformation it is sometimes possible to generate new structures and possibly to simplify the schematic. This transformation can also be used to introduce crystal elements and magnetostrictive resonators into essentially *LC* configurations.

A special case of the star-delta transformation is shown in Fig. 10.5. By means of the transformation, the blocking circuit in the series arm may be substituted for a trap circuit in the shunt arm of a lowpass structure or as a part of some other structure. Of course, the other elements will be modified as shown in the figure.

The example of the delta-star transformation to facilitate the use of crystal elements in a schematic will be given later in the text.

10.3   USE OF TRANSFORMER IN FILTER REALIZATION

Some types of filter realizations often yield somewhat unruly elements; capacitors may be relatively large, and inductances rather small. Large capacitors are bulky and expensive, while, even worse, the low

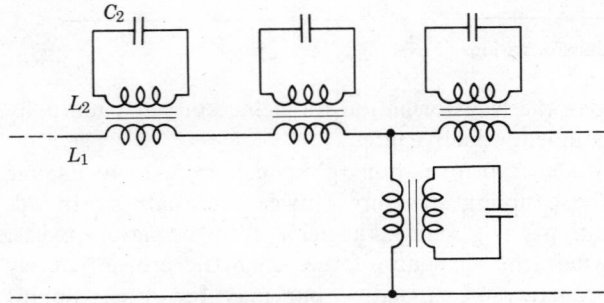

Fig. 10.6.   Use of transformers to increase Q

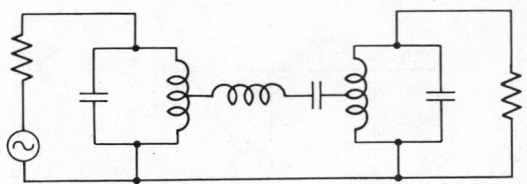

Fig. 10.7.   Use of transformers to modify element values.

inductances may have a low quality factor, making the filter impossible to physically realize. By constructing the filter as shown in Fig. 10.6 making use of transformers, the above difficulty can be avoided. The value of L_1 is such as required by the particular design equations. Tightly coupled to L_1 is the considerably larger inductance L_2, having a higher Q, and resonating with a considerably smaller capacitance than necessary for L_1. The resonant frequency of $L_2 C_2$ may be adjusted even closer to the proper resonant frequency, because the larger number of turns allows finer tuning.

The foregoing principle can be employed in making all (or nearly all) of the values of capacitors in the filter equal. Another application is encountered when a crystal element is to be incorporated into a regular *LC* schematic. With the aid of the transformer, the crystal can be coupled into either the series or shunt arm.

Figure 10.7 shows a variation of this technique which makes use of the auto-transformer. Here the filter can be substantially improved by making the individual element values of the filter of the same order of magnitude.

10.4   NORTON'S TRANSFORMATION

Two very important equivalent networks originated

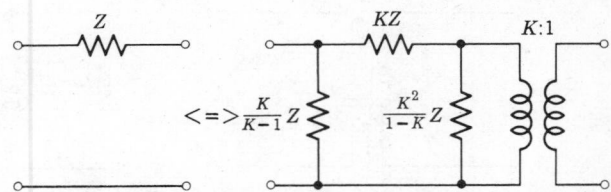

Fig. 10.8.   Norton's first transformation.

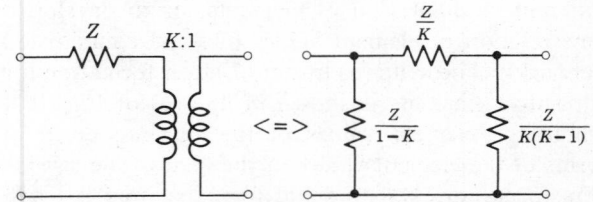

Fig. 10.9.   Another form of Norton's first transformation.

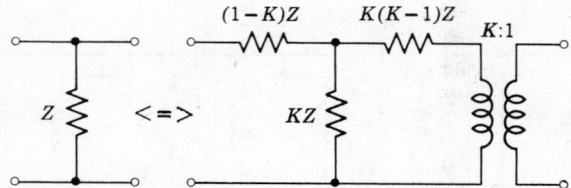

Fig. 10.10. Norton's second transformation.

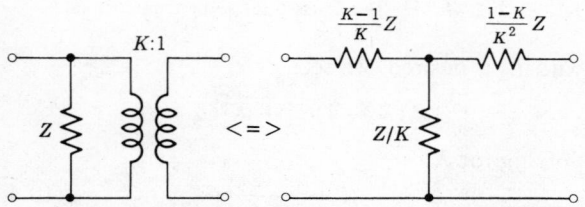

Fig. 10.11. Another form of Norton's second transformation.

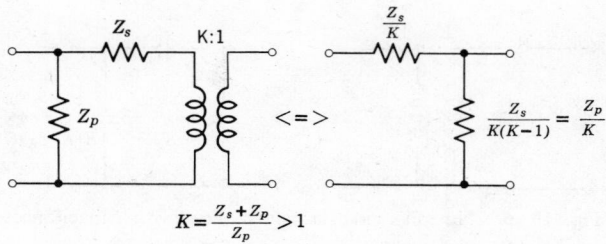

$$K = \frac{Z_s + Z_p}{Z_p} > 1$$

Fig. 10.13. Equivalent circuit derived from Fig. 10.9.

and capacitance free, and as having unity coupling coefficient and infinite winding inductances.

Norton's second transformation is shown in Fig. 10.10, and from it another equivalence, Fig. 10.11 which is more useful in practice, can be derived. As in the previous case, one of the lumped elements from each of the equivalent schematics is negative. It should be noted that in this second case a *T*-network results, whereas in the first case, the resulting network was of *π*-form.

From Norton's first transformation, Figs. 10.12 and 10.13 can be derived. Likewise, from Norton's second transformation, it is possible to obtain Figs. 10.14 and 10.15. These transformations are particularly important, since they can be used to change a *π*-input configuration to a *T*-input, and vice versa.

It should be noted that by the use of each of these transformations, the impedance of the given network is modified on one side of the transformation, all inductances and resistances are to be scaled by the factor $1/K^2$, and all capacitances are to be scaled by

by Norton can be used for network transformations which are extremely useful in the elimination of redundant elements or for the modification of element values. Figure 10.8 shows Norton's first transformation. By removing the transformer from the schematic on the right and placing it on the left, the remaining *π*-network must be modified by changing the element values as shown in Fig. 10.9 which gives Norton's first transformation in a more useful form.

Depending upon the value of *K*, the equivalent *π*-networks on the right side of Fig. 10.8 and Fig. 10.9 contain one negative element, which is indeed, unrealizable with passive components. However, it will be shown that in the application of this equivalence to certain forms of networks, it is possible to lump the negative element with an existing positive element, thus allowing the network to be practically realized for a certain range of *K* values.

The ideal transformer shown in these schematics has a finite transformation ratio of *K*, the turns ratio being numerically equal to $1/K$, by the usual definition. The transformer, being ideal, is considered loss free

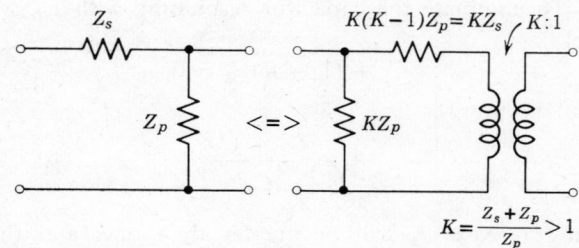

$$K = \frac{Z_s + Z_p}{Z_p} > 1$$

Fig. 10.14. Equivalent circuit derived from Norton's second transformation.

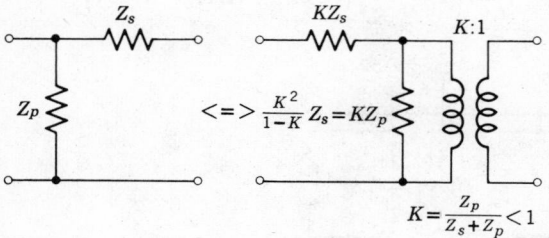

$$K = \frac{Z_p}{Z_s + Z_p} < 1$$

Fig. 10.12. Equivalent circuit derived from Norton's first transformation.

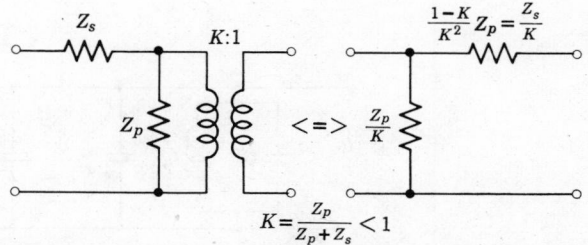

$$K = \frac{Z_p}{Z_p + Z_s} < 1$$

Fig. 10.15. Equivalent circuit derived from Fig. 10.11.

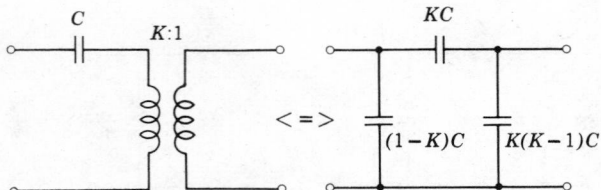

Fig. 10.16. Norton's first transformation with capacitances.

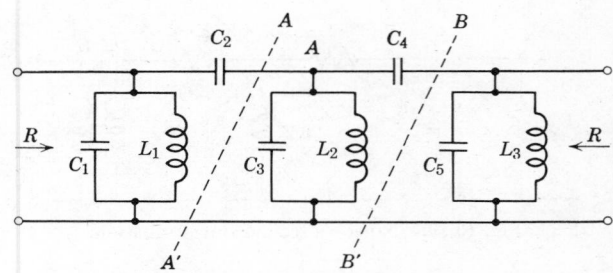

Fig. 10.17. Bandpass filter to be transformed.

K^2. With the aid of several transformations performed on various parts of a network, it will be possible to eliminate several redundant elements but finally end with an output impedance equal to the input impedance, if desired.

Figure 10.16 illustrates Norton's first transformation of Fig. 10.9 but with capacitances used for all impedances. This is obtained by using the relation $Z \propto (1/C)$. By use of this transformation, we can illustrate an example of the elimination of a redundant element.

Extreme Case of Transformation

Given the bandpass filter of Fig. 10.17, it is desired to economize the network by the elimination of a redundant capacitor. The transformation of Fig. 10.16 will be inserted in the position of the dotted lines AA' of Fig. 10.17 resulting in the schematic of Fig. 10.18. Note, that at this point, it is possible to eliminate either the capacitor resonating with L_1, or with L_2, or neither of these capacitors could be eliminated but the output impedance would be modified.

To eliminate the capacitor resonating with L_1, we set

$$C_1 + (1 - K)C_2 = 0$$

Solving for K,

$$K = \frac{C_1 + C_2}{C_2} \qquad (10.4.1)$$

In this case, K will be greater than one, and the transformation will be a step-down transformer; the impedance R/K^2 will be less than R.

When the elimination of the capacitor resonating

with L_2 is desired, we set

$$K(K - 1)C_2 + K^2C_3 = 0$$

Solving for K,

$$K = \frac{C_2}{C_2 + C_3} \qquad (10.4.2)$$

Now, K will be less than unity, the resulting transformation being a step-up transformer, and the impedance R/K^2 will be greater than R.

Note that it also is possible to eliminate either C_3 or C_5 from Fig. 10.17 by inserting the same transformation at the dashed lines BB'. Also, it would be possible to bring the output impedance of the filter back to the value of R, if we inserted a transformation at BB', and made

$$K_1 K_2 = 1 \qquad (10.4.3)$$

where K_1 is the transformation ratio performed at AA' and K_2 is the transformation ratio performed at BB'.

A most interesting consequence results in the use of Norton's transformation; the resonant circuits modified by its use will resonate at a frequency which is modified by the transformation. For example, if the filter of Fig. 10.17 is designed by the image-parameter approach, the resonant circuits L_1C_1, L_2C_3, and L_3C_5 all resonate at the upper cutoff design frequency f_2. By performing the transformation at AA' and eliminating C_1, then the resonant frequency of the center tank circuit will be

$$f = \left(2\pi \sqrt{\frac{L_2}{K^2}[K^2C_3 + K(K - 1)C_2]}\right)^{-1} \qquad (10.4.4)$$

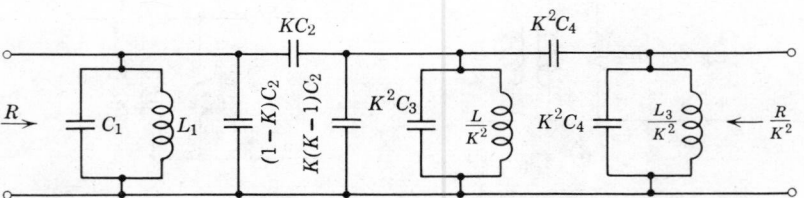

Fig. 10.18. Bandpass filter with transformation inserted.

Simplifying,

$$f = \left[2\pi \sqrt{L_2\left(C_3 + \frac{K-1}{K}C_2\right)} \right]^{-1}$$

Since

$$K = \frac{C_1 + C_2}{C_2}$$

then

$$f = \left[2\pi \sqrt{L_2\left(C_3 + \frac{C_1 C_2}{C_1 + C_2}\right)} \right]^{-1}$$

But, for the image-parameter design, $C_3 = C_2$. Then

$$f = \left[2\pi \sqrt{L_2 C_3 \left(\frac{C_3 + 2C_1}{C_3 + C_1}\right)} \right]^{-1}$$

or

$$f = f_2 \sqrt{\frac{C_1 + C_3}{2C_1 + C_3}} \tag{10.4.5}$$

Transformation Applied to Fundamental Type Section

The element values of the full fundamental bandpass section given in Fig. 10.19 may be expressed by:

$$L_1 = \frac{R}{\omega_2 - \omega_1} \qquad C_1 = \frac{\omega_2 - \omega_1}{R\omega_0^2}$$

$$L_2 = \frac{R(\omega_2 - \omega_1)}{2\omega_0^2} \qquad C_2 = \frac{2}{R(\omega_2 - \omega_1)} \tag{10.4.6}$$

By inspection of the foregoing design formulas it is apparent that, at narrow bandwidths, $C_2 \gg C_1$ and $L_2 \ll L_1$, making the filter difficult to realize. By use of the ideal transformer, however, this difficulty can be overcome.

Two ideal transformers will be inserted, such that the input and output impedances of the filter will be equal—that is, $K_1 K_2 = 1$. The first transformation will be inserted at the dashed line AA' in Fig. 10.19. The transformer inserted in Fig. 10.20a is replaced by its equivalent structure. The capacitor, $K_1(K_1 - 1)C_1$, must be negative, thus making $K_1 < 1$. By a unique choice of K, it is possible to eliminate the shunt capacitor entirely. Here, however, we will set $K_1 = 0.5$ and $K_2 = 2$.

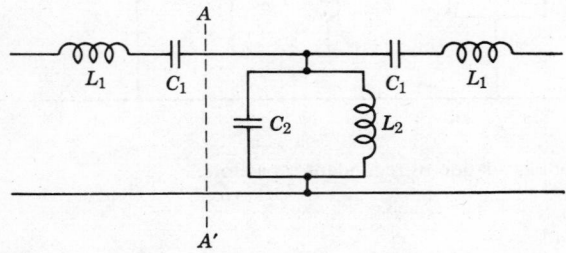

Fig. 10.19. Bandpass schematic.

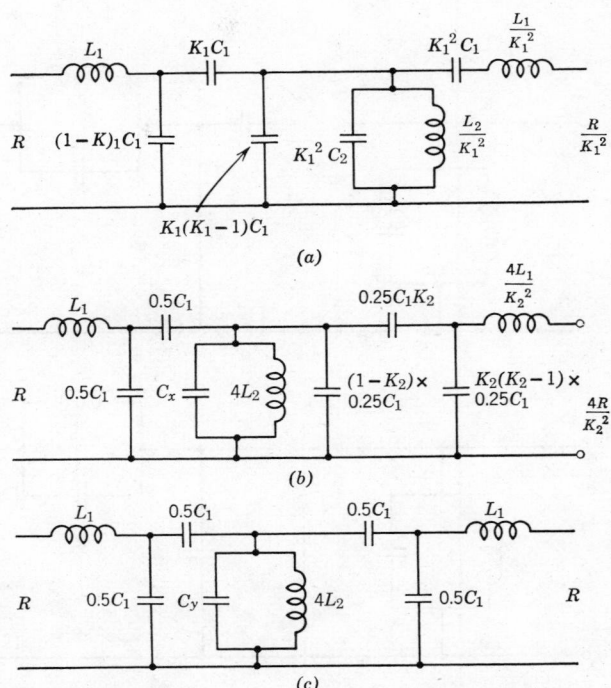

Fig. 10.20. Ideal transformer applied to fundamental type schematic.

The value of C_x, the tuned circuit capacitor of Fig. 10.20b, is changed by the transformation

$$C_x = K_1^2 C_2 + K_1(K_1 - 1)C_1$$
$$C_x = 0.25C_2 - 0.25C_1 = 0.25(C_2 - C_1)$$

The second transformer is to be inserted at a similar position on the right side between the coil and capacitor. This is shown in Fig. 10.20b. The value of K_2 to be used has been dictated; $K_2 = 2$. This will make the capacitor $(1 = K_2) 0.25C_1$ negative.

The new element values are shown in Fig. 10.20c where

$$C_y = 0.25C_2 - 0.25C_1 - 0.25C_1$$
$$C_y = 0.25(C_2 - 2C_1)$$

Elimination of Redundant Elements in the Coil-Saving Filter

The initial steps in filter design lead to a structure with a number of individual building blocks or partial networks. The next step is to connect the sections together and to eliminate redundant elements wherever possible. The network shown in Fig. 10.21 will be used to illustrate this final step. In Fig. 10.21a are four elementary filter sections which must be combined. It is assumed that the element values have been determined from Fig. 1.24. Two sections having peaks on opposite sides of the stopband are included. The

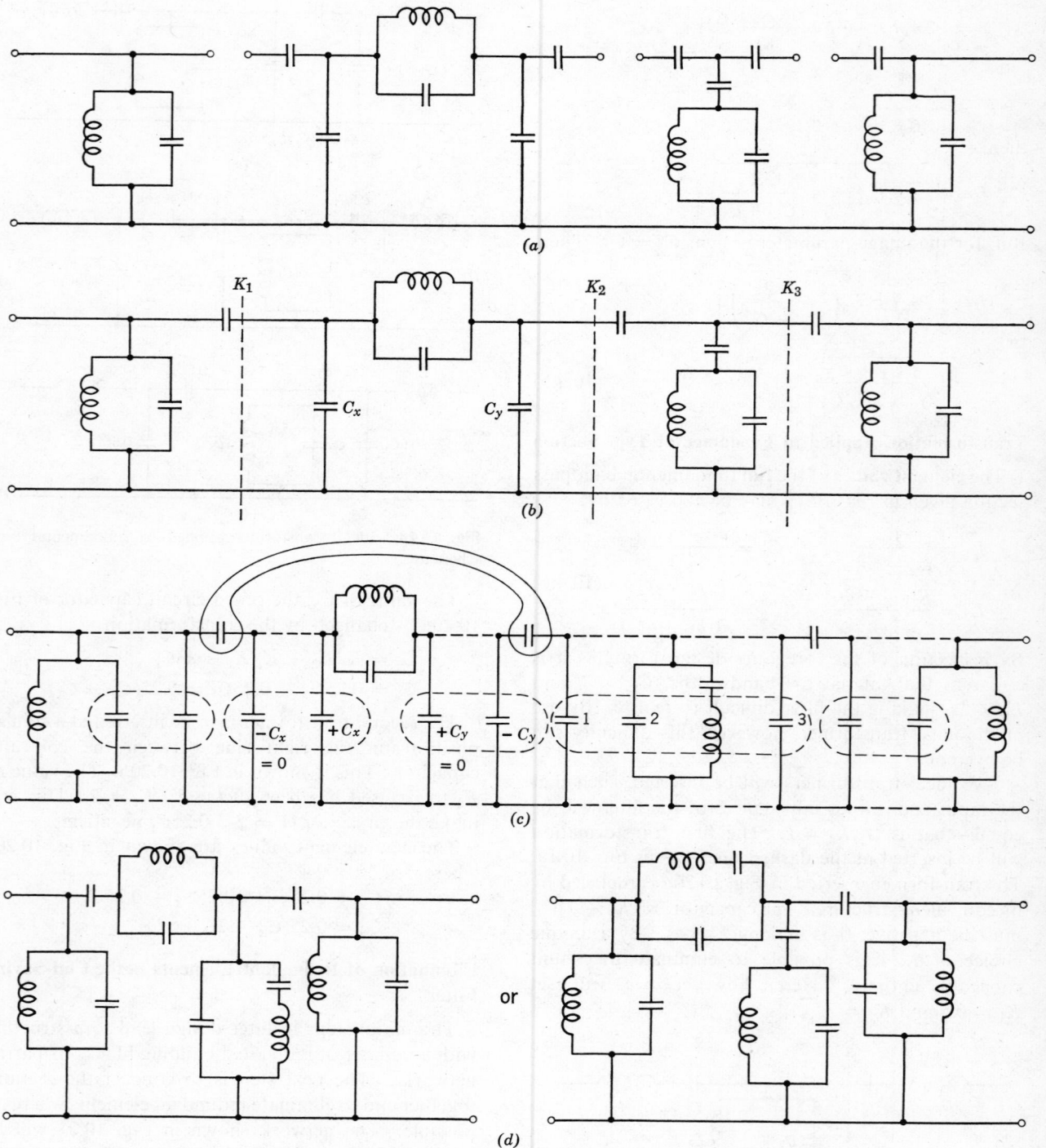

Fig. 10.21. Example of ideal transformer used for elimination of redundant capacitors.

input and output sections may be practically considered as impedance converters, because their contribution to the total attenuation response, especially in the proximity of the passband, is small. In Fig. 10.21*b* the series capacitors are merged together by the usual product-divided by sum rule. The filter now consists of four coils and ten capacitors. With the aid of Norton's first transformation, two redundant capacitors can be eliminated. For that purpose, an ideal transformer will be inserted in the position indicated by the letters K_1 and K_2. In the third row the elimination process is shown. Two capacitors, C_x and C_y, are eliminated by adding the negative capacitors from the ideal transformers. This permits us now to combine both series capacitors as shown in Fig. 10.21*c*. A third ideal transformer, inserted at K_3 serves only for establishing the desired impedance level in the output circuit of the completed filter. As the input and output impedances must be the same, the total transformation coefficient has to be equal to unity. That means

$$K_1 K_2 K_3 = 1$$

In row (*d*) the final schematics are shown, now consisting of seven capacitors and four coils. The schematic on the right side is electrically equivalent to the left schematic. The formulas for the dipole transformations necessary for this equivalence are given at the beginning of this chapter.

Ultimate Transformation of Bandpass Filters with Norton's Technique

In its final form, the coil-saving configuration of Fig. 10.21 is noncanonic in the sense that the number of reactors is not the absolute minimum. In its coil-saving form it still can be modified to generate a filter with the minimum number of capacitors. To illustrate the idea of ultimate transformation, let us consider the bandpass filter of Fig. 10.22 consisting of several sections having a minimum number of inductances.

By using the steps illustrated in this figure, the number of capacitors will be systematically reduced until a final stage is reached in which there are only two capacitors per section instead of the original three. Evidently, the input and output impedances are unequal.

The same procedure can be applied in the case of the filter, or part of a filter made up of different sections, such as the one shown in Fig. 10.23, which has a minimum number of inductors. The number of capacitors is again reduced step by step to two per section instead of the original three (ignoring, as before, the terminal elements). It should be noted that in the transformations illustrated in Figs. 10.22 and 10.23, the ideal transformers introduced are either step-up or step-down, and consequently each successive transformation tends to raise the nominal impedances of one end of the filter in relation to that of the other. If the number of sections is large enough, or the transformer ratios are high enough (which tends to be the case for a filter with a narrow passband) unrealizable element values can be generated and the final structures shown in Figs. 10.22 and 10.23 have no practical use. Thus one is forced to carry out only part of the transformation that is theoretically possible or to reverse the direction of each impedance at each successive transformation.

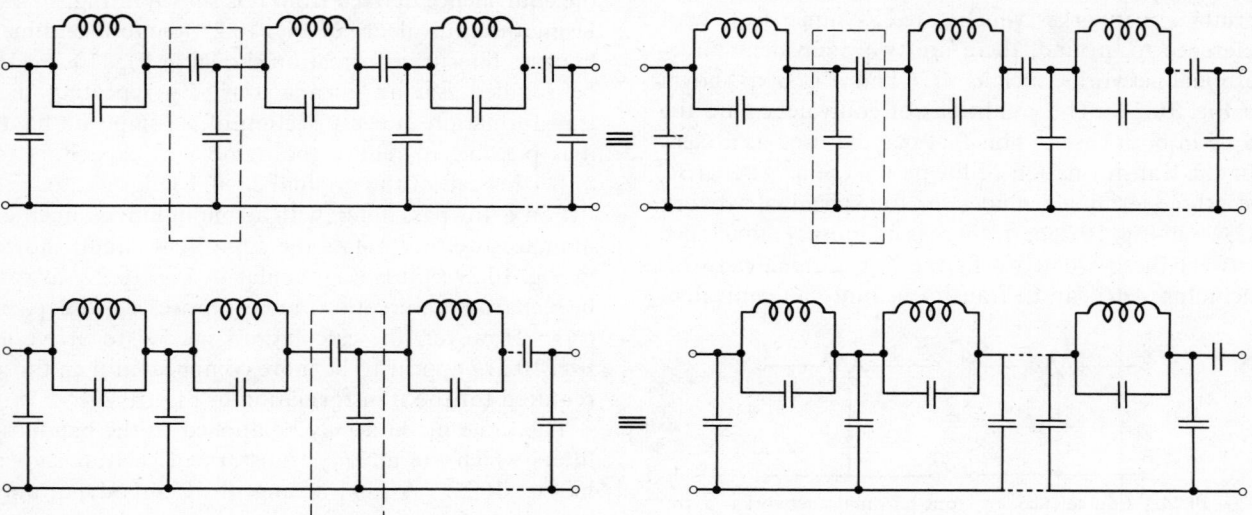

Fig. 10.22.   Ultimate transformation of bandpass filter with poles in the series arm.

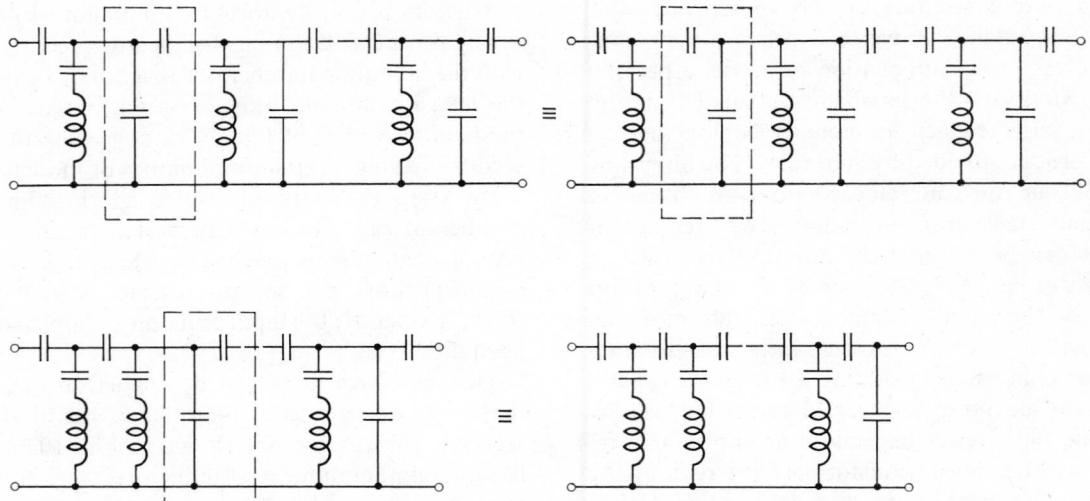

Fig. 10.23. Ultimate transformation of bandpass filter having poles in the shunt arm (series-resonant circuits).

The transformation illustrated in Figs. 10.22 and 10.23 therefore represents the theoretical limit of what is possible by the methods of Norton's transformation. The resultant structures obtained are in fact those of a lowpass or highpass filter to which Norton's transformations are not applicable.

10.5 APPLICATIONS OF MUTUAL INDUCTIVE COUPLING

Transformation of Lowpass Networks

The introduction of mutual inductances permits the reduction of the number of capacitors even in lowpass and highpass filters. To any two equivalent four-terminal networks which are asymmetrical with reference to ground, there are two equivalent three-terminal networks, such as ACE and $A'C'E'$ as shown in Fig. 10.24. The conditions of equivalence are the same in both cases. This fact may be used as a basis for the transformation of lowpass sections. Let ABC be a three-terminal included in a four-terminal network shown in Fig. 10.25a. If there is no mutual inductance between these two networks, the four-terminal network including ABC can be transformed into an equivalent

four-terminal network by using the equivalent three-terminal network shown in Fig. 10.25.

Considering the Norton's transformation, the following can be derived for the network in Fig. 10.26:

$$K = \frac{(C_a + C_b)}{C_a} = \frac{(C_1 + C_2)}{C_1}$$

$$C_1 = \frac{C_a}{K}$$

$$C_2 = \frac{C_b}{K} \tag{10.5.1}$$

Inductance L is used here as an auto-transformer, and the equivalence derived from it is shown in Fig. 10.27. Using the equivalence of Fig. 10.27, the four-terminal lowpass filter transformation shown in Fig. 10.28 can be fulfilled, saving a capacitor. By repeating this transformation in every section of a composite filter, it is possible to reduce the number of capacitors to $n + 1$ instead of the original $2n + 1$ (Fig. 10.29).

For a lowpass filter with a minimum number of inductors we thus obtain the equivalent circuit shown in Fig. 10.29d. It is very similar to Fig. 10.29c as can be seen from observation, and comprises $n + 1$ capacitors. However, the calculations needed to arrive at Fig. 10.29d appear to be more complicated than those required for the transformation in Fig. 10.29c.

The same method may be applied to the bandpass filters which are already transformed such as shown in Fig. 10.22. A final arrangement embodying only one capacitor per section, apart from the terminal elements, is shown in Fig. 10.30. This form represents

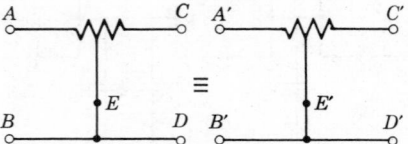

Fig. 10.24. Equivalence of four-terminal networks asymmetrical with respect to ground are also three-terminal equivalences.

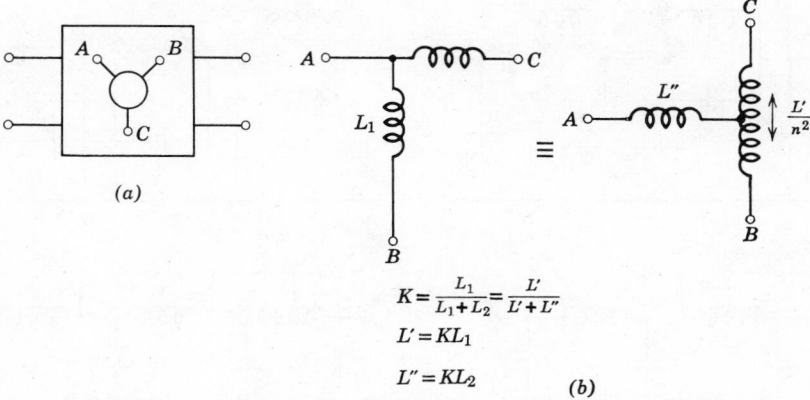

$$K = \frac{L_1}{L_1 + L_2} = \frac{L'}{L' + L''}$$

$$L' = KL_1$$

$$L'' = KL_2$$

Fig. 10.25. Three-terminal network enclosed in a four-terminal network.

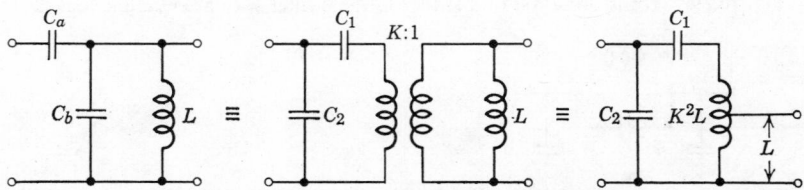

Fig. 10.26. Four-terminal network equivalence.

an absolute minimum number of elements in the bandpass filter.

Transformation of Highpass Networks

The method used to transform highpass sections differs from the method used for lowpass transformation. Let us examine the transformation of the network of Fig. 10.31*a* which is a highpass *T*-section having a single inductor. This network is equivalent to that in Fig. 10.31*b* where inductor *L* has been replaced by a tightly coupled transformer, with primary and secondary inductances equal to *L*. Points *A* and *B* are always at the same potential. The arrows indicate the direction of the induced voltages. Consequently, if *A* and *B* are connected, there is no change in the distribution of potentials and currents, so that the two circuits are equivalent. The schematic of Fig. 10.31*b* can be given in the form shown in Fig. 10.31*c* by interchanging the capacitor and transformer winding in each series branch; this obviously

affects neither the voltage at the terminals of the branch concerned nor the current through it.

The application of Norton's second transformation to Fig. 10.31*c* yields the network of Fig. 10.32. The transformer can be eliminated without changing the distribution of potentials and currents on the left side of it, provided all voltages on the right side are multiplied, and all currents on the right side divided by the transformation ratio *K*. This is shown in Figs. 10.33*a* and 10.33*b*.

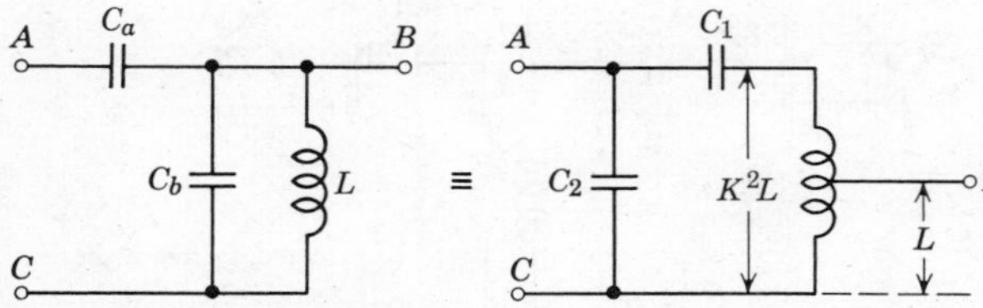

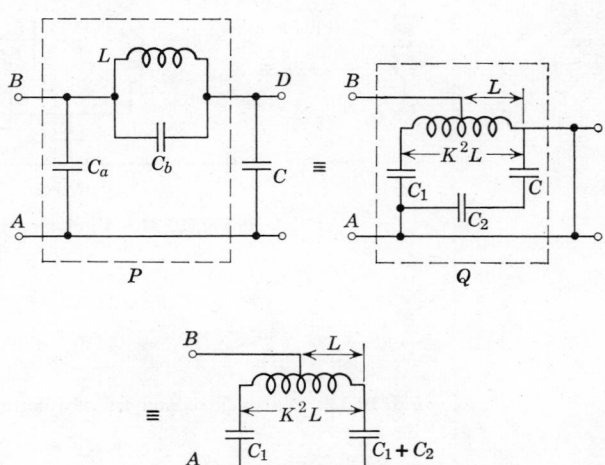

Fig. 10.28. Transformation of basic lowpass filter by converting three-terminal network *P* into network *Q*.

Fig. 10.27. Three-terminal equivalence.

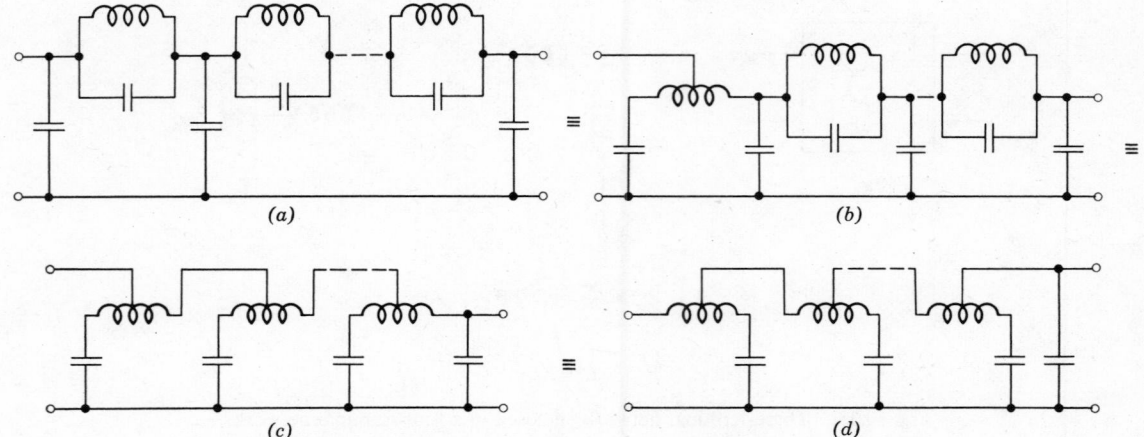

(a) (b)

(c) (d)

Fig. 10.29. Transformation of a ladder lowpass filter with attenuation peaks.

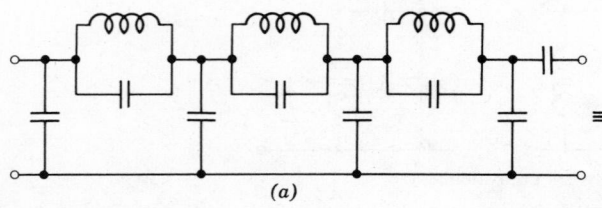

(a)

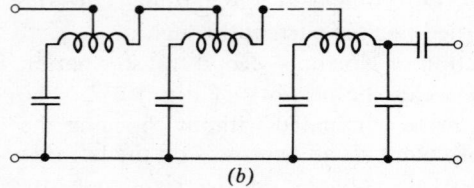

(b)

Fig. 10.30. Reduction of the number of capacitors in bandpass network without redundancy.

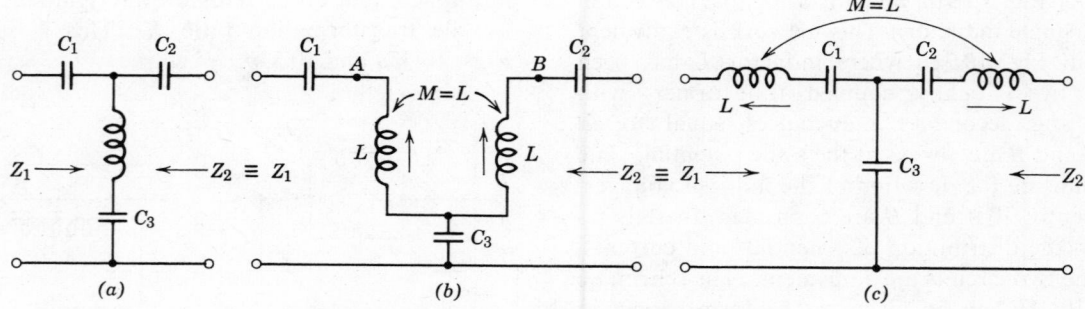

(a) (b) (c)

Fig. 10.31. Equivalence relating to highpass filters.

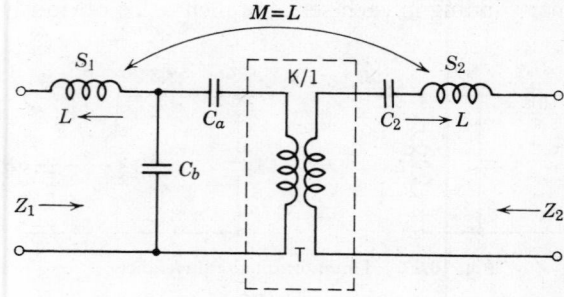

Fig. 10.32. Norton's second transformation.

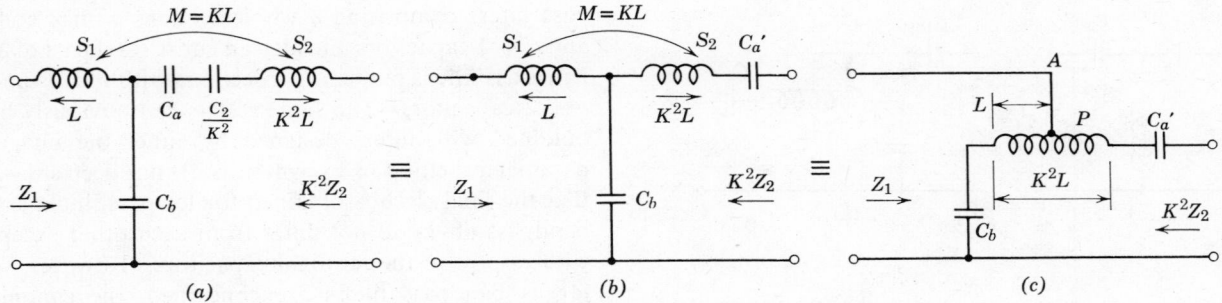

Fig. 10.33. Continuation of transformation of Fig. 10.31.

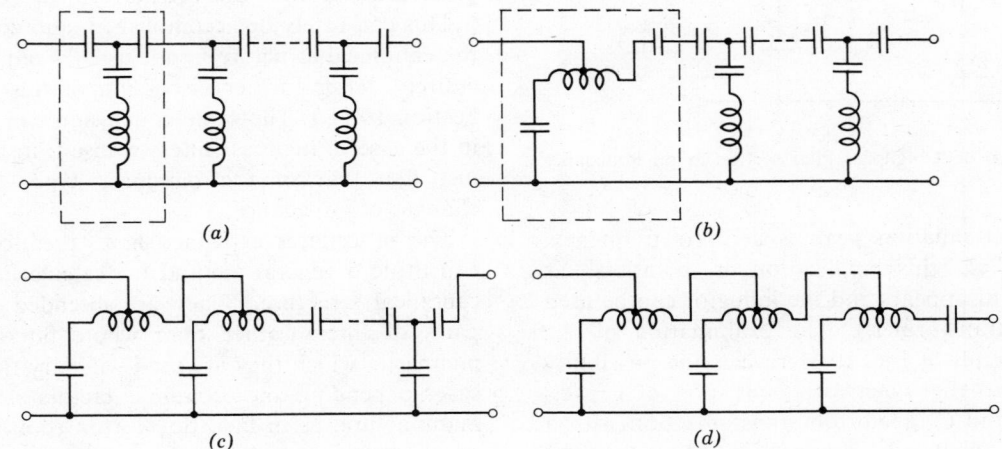

Fig. 10.34. Transformation of a composite highpass filter.

Since $K > 1$, there is a point P on inductor S of Fig. 10.33b having the property that the transformer ratio between S_1 and the part of S_2 between the coil input and P is unity. As the coupling is tight, P is always at the same potential as terminal A of winding S_1. It is therefore possible to join A and P without causing any other change. Thus the circuit in Fig. 10.33c can be obtained.

The equivalence of the circuits in Figs. 10.31a and 10.33c shows the possibility of obtaining a highpass section with one inductor and only two capacitors. Applied section by section to a filter comprising n sections each similar to the one before, this method makes it possible to design the filter having $(n + 1)$ capacitors instead of the original $2n + 1$. Figures 10.34a to 10.34d show the various stages in the transformation.

In the same manner as above, an already transformed bandpass filter can be most economically constructed having $(n + 2)$ capacitors instead of the original $2n + 2$. The original network is shown in Fig. 10.35a, and the transformed network in Fig. 10.35b. The transformation performed in Fig. 10.35

tends to raise the impedance from left to right, whereas the similar transformation of Fig. 10.30 tends to lower the impedance.

Highpass Networks as an Impedance Transformer

It is possible to use a highpass network, such as that in Fig. 10.31 as an impedance transformer.

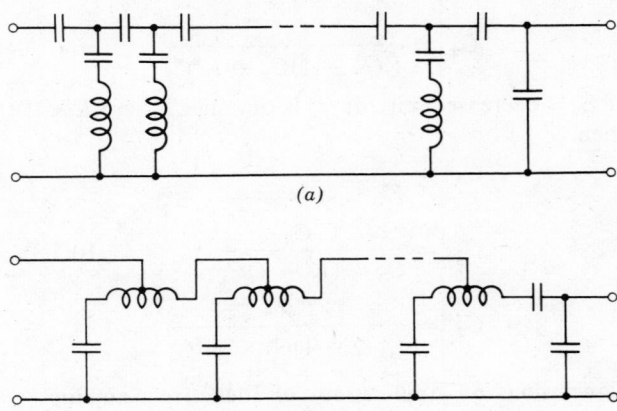

Fig. 10.35. Reducing the number of elements in a bandpass filter type shown in Fig. 10.23.

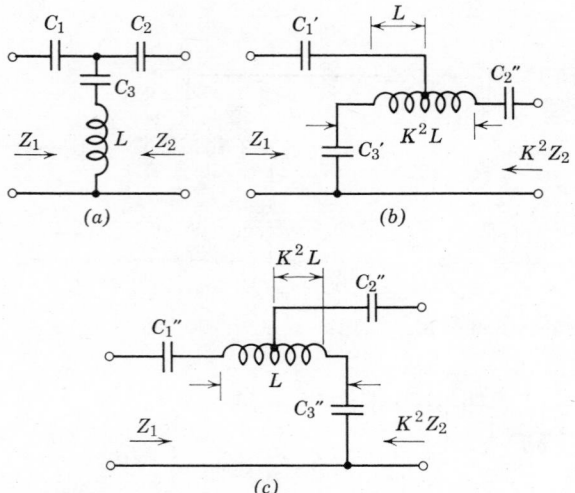

Fig. 10.36. Use of a highpass filter section as an impedance transformer.

(When the attenuation peak is at zero frequency, where $C_3 = \infty$, this application is pointless, as capacitor C_3 disappears and the inductor can be used as an auto-transformer.) The examination of the equivalent circuit of Fig. 10.31 reveals the possibility of applying to the T-network consisting of capacitors C_1, C_2, and C_3 a Norton transformation with a ratio chosen arbitrarily between $C_2/(C_2 + C_3)$ and $(C_1 + C_3)/C_1$. The impedance Z_2 can be modified within certain limits with the transformation shown in Fig. 10.36. If Z_2 is increased, the original circuit (a) modifies into circuit (b) where $K > 1$, then

$$C_3' = \frac{C_3}{K}$$

$$C_1' = \frac{C_1 C_3}{C_3 - (K - 1)C_1} \qquad (10.5.2)$$

$$C_2' = \frac{C_2 C_3}{K(K - 1)C_2 + K^2 C_3}$$

If Z_2 is decreased, circuit (c) is obtained where $K < 1$ then

$$C_3'' = \frac{C_3}{K}$$

$$C_1'' = \frac{C_1 C_3}{C_3 - (K - 1)C_1} \qquad (10.5.3)$$

$$C_2'' = \frac{C_2 C_3}{K(K - 1)C_2 + K^2 C_3}$$

Conclusions on Applications of Inductive Coupling

The various circuit transformations which have been described make it possible to design lowpass or high-

pass filters comprising n whole sections with n coils and $n + 1$ capacitors and also an entire (or a part of a) bandpass filter comprising n sections with n coils and $n + 2$ capacitors. The same results can obviously be obtained with filters designed by either the image-parameter method or by synthesis. It has been shown that the final circuits obtained for lowpass, highpass, bandpass filters do not differ from each other except with respect to the terminal capacitors. However, so far as bandpass filters are concerned, the canonic designs can only be obtained if all the preliminary Norton transformations are performed.

This can rarely be accomplished, and consequently the canonical structure is not usually obtained in its entirety. (For further discussion of this matter see Section 10.11.) The same conclusion can be applied in the case of highpass filters whose transformations, just like that of the bandpass type, necessitates changes of impedance.

The procedures explained here, therefore, do not constitute a general method for calculating filters of canonical structure. They are intended for use in parts of filters rather than whole filters, and the manner in which they are used will vary from case to case, depending on economic factors and case of manufacture, as in fact, do Norton transformations, of which they are a kind of generalization.

In the case of lowpass filters where transformation does not involve any change in impedances, the canonical form of Fig. 10.29 can almost always be obtained.

10.6 THE REALIZATION OF *LC* FILTERS WITH CRYSTAL RESONATORS

Consider the coil-saving *LC* bandpass filter of Fig. 10.37. The attenuation requirement was formulated such that *LC* components alone will not be sufficient to satisfy the design. The inband distortion will be excessive and the relative attenuation in the stopband will be insufficient unless elements with better quality factors are introduced.

Let us consider some possibilities of transformation in successive steps to obtain a desirable schematic in order to introduce the equivalent schematic of crystals. First, the impedance level of the crystal is usually much higher than the impedance level of the coil and capacitor and therefore a transformer should be incorporated to match the conventional element with a crystal resonator. In Fig. 10.38 the arrangement of the crystal that is to be used in the shunt arm of the filter is shown. Being inserted practically across the

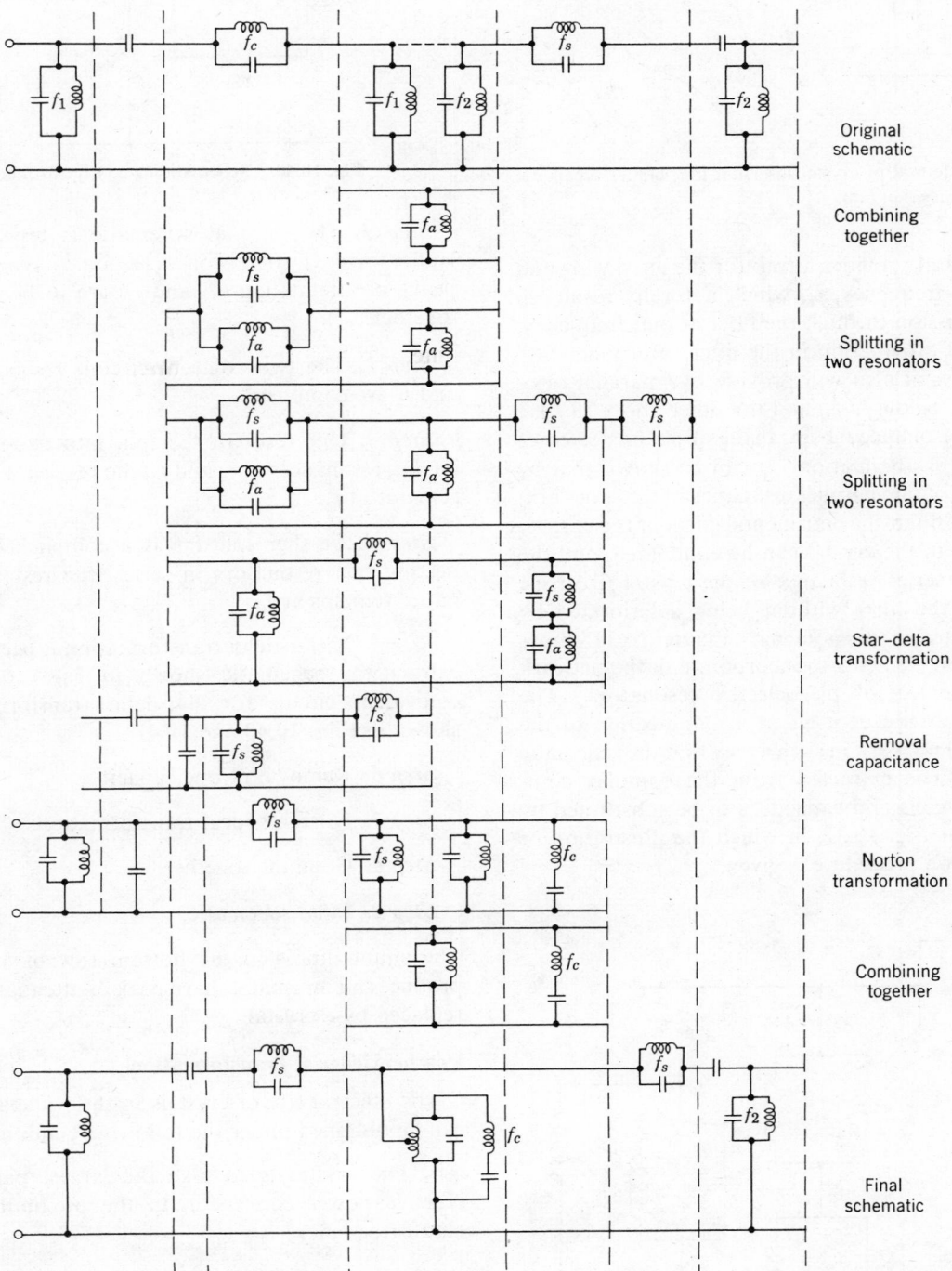

Original
schematic

Combining
together

Splitting in
two resonators

Splitting in
two resonators

Star–delta
transformation

Removal
capacitance

Norton
transformation

Combining
together

Final
schematic

Fig. 10.37. Design procedure for inserting crystal in *LC* filter.

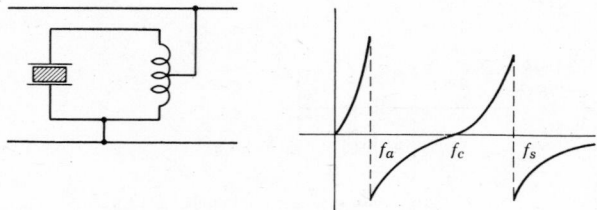

Fig. 10.38. Method of crystal insertion in a ladder schematic and the reactance diagram.

line, the crystal provides a trap for the energy around its resonant frequency, f_c, which naturally results in zero transmission through the filter at that frequency, or a peak of attenuation in the filter. But inherently the same crystal also will provide two parallel resonances: one below $f_c(f_a)$ and the other above $f_c(f_s)$.

This last resonance f_s being in the stopband decreases the amount of attenuation. It can be shown that by the addition of one parallel resonance in the series arm of the ladder filter, the detrimental effect of the parallel resonance f_s in the crystal can be eliminated, and the high-quality series resonance of the crystal f_c can be utilized for the filter without being deteriorated by other undesirable phenomena. Figure 10.39 shows the equivalent elements to incorporate in the network to allow the use of piezoelectric resonators. The problem now reduces itself to a modification of the conventional network in such a way that the schematic of Fig. 10.39 is extracted from the complex combinations of coils and capacitors to be substituted by the crystal of Fig. 10.37, in which the illustration of transformation procedure is given.

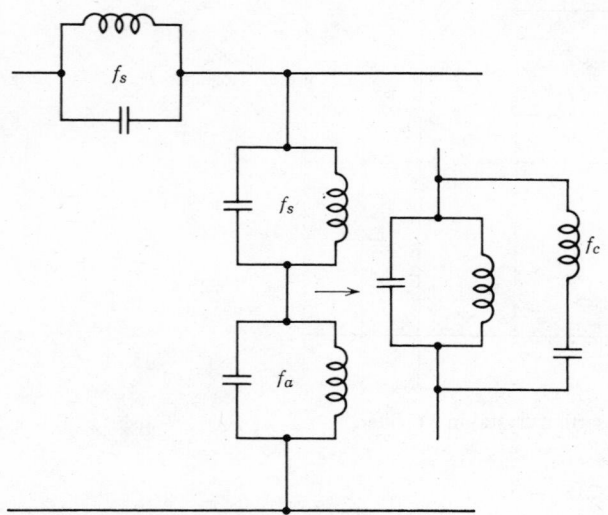

Fig. 10.39. Equivalent representation of crystal insert.

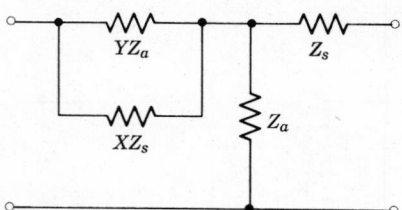

Fig. 10.40. Schematic to be transformed.

Step 1. The original schematic is represented in ladder form. Two networks (sections) having different peaks of attenuation (f_c and f_s) are to be connected together.

Step 2. The two adjacent circuits resonating at f_1 and f_2 are combined.

Step 3. One resonator is split into two resonators having resonances at f_s and f_a, the resulting resonance remains at f_c.

Step 4. Further splitting is accomplished by connecting two resonators in series, the resulting resonance remains at f_s.

Step 5. A star-delta transformation is performed to extract the schematic shown in Fig. 10.39. The equivalent circuit for star-delta transformation is shown in Figs. 10.40 and 10.41.

Step 6. Removal of one capacitor.

Step 7. Norton's ideal transformation.

Step 8. Combine together.

Step 9. Final schematic.

The shunt dipole in the bottom row of Fig. 10.37 produces an unusually sharp peak of attenuation when replaced by a crystal.

Poschenrieder's Transformation

The efficient use of crystals in the wideband filters can be obtained under the following conditions:

1. The crystal is to take the largest part of the reactive power conversion in the proximity of the cutoff frequency.

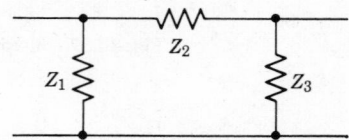

Fig. 10.41. Result of transformation of Fig. 10.40.

2. The critical and sharpest attenuation pole must be produced by a quartz resonator in the ladder schematic.

3. The bandwidth of the network must be as large as possible.

Much effort has been put on this problem in connection with requirements for the so-called super-group filters in carrier communication equipment. In Fig. 10.42a a part of an original schematic obtained by network transformation methods is shown. The section shown on the diagram includes a sharp attenuation pole at f_1 and another pole, less sharp, at f_2 below the passband. The transformed network is shown in Fig. 10.42b. It can be proved that both networks may be identical.

To achieve the necessary conditions for the given distribution of poles the following set of equations must be used in the network transformation.

$$\beta = \frac{L_1 C_1 - L_3 C_3}{L_2 C_2 - L_1 C_1}$$

$$\alpha = (L_1 C_1 - L_3 C_3)$$
$$\times \left[\frac{1}{L_2 C_2 - L_1 C_1} - \left(\frac{L_2 C_2 - L_3 C_3}{L_2 C_2 - L_1 C_1} \right)^2 \frac{1}{C_0 L_1'} \right]$$

$$L_2'' = \beta \cdot L_2$$

$$m = \frac{C_1 L_2 C_2 (L_1 + L_1 L_2'' / L_2 + L_2'')}{(L_2 C_2 - L_3 C_3)^2} - \frac{L_2 (C_2 - C_1)}{L_2 C_2 - L_1 C_1}$$

$$C_0' = \frac{C_0}{1 + \beta}$$

$$L_1' = (1 + \beta) \left(L_1 + \frac{L_1 L_2''}{L_2} + L_2'' \right) \qquad (10.6.1)$$

$$L_2' = \frac{L_2 (L_2 C_2 - L_3 C_3)}{L_2 C_2 - L_1 C_1}$$

$$L' = L_3 - L_2''$$

$$C_3' = \frac{C_0 (\beta - \alpha)(L_2 C_2 - L_1 C_1)^2}{L_2 C_1 (L_2 C_2 - L_3 C_3)}$$

$$L_1' C_1' = L_3 C_3$$

$$L_3' C_3' = L_1 C_1 \quad \text{(crystal frequency)}$$

$$L' C' = L_3 C_3$$

Both networks consist of the same number of reactive elements and the same number of meshes. The main goal of the transformation is to shift all the reactive power circulating within the schematic toward the cutoff, and also to obtain a realizable network which would include the equivalent circuit of a quartz resonator in the shunt arm. In the network of Fig. 10.42a a sharp pole is produced by $L_1 C_1$ in the series

arm. In Fig. 10.42b the sharpest pole is produced by $L_3' C_3'$ in the shunt arm. The capacitance ratio $C_p / C_s (C_2 / C_3)$ is (in the first approximation) inversely proportional to the distance between the pole frequencies f_1 and f_2. The quantity m determines the capacitance ratio which in the practical case would be greater than 140. The magnitude of m depends mainly on the ratio of the two tuning frequencies of the original ladder section. The shunt inductance L_2' in the transformed schematic is eminently suitable for increasing the impedance level of the crystal whose equivalent inductance value may be of the order of several henrys. The transformation may be readily applied to a suitable canonic ladder structure.

Introduction of Crystals into *LC* Lowpass Ladder Filters

If a lowpass filter, or part of a lowpass filter has a configuration as shown in Fig. 10.43 the following transformation can be used to facilitate the use of crystals, resulting in the schematic of Fig. 10.44. The condition to satisfy for L_3 is defined by

$$L_3 = \frac{L_1 C_1 - L_3 C_3}{C_2}$$

when $L_1 C_1$ and $L_3 C_3$ are given (see Fig. 10.43).

With the notation in Fig. 10.44 the element values will be expressed by

$$C_{\text{in}} = \frac{L_1 C_1 - L_3 C_3}{L_1 + L_3}$$

$$K_2 = \frac{C_1 (L_1 C_1 - L_3 C_3)}{L_3 C_3}$$

$$L_2 = L_1 + L_3$$

$$L_s C_s = L_1 C_1$$

$$C_p = C_{\text{in}} \frac{C_3}{C_1}$$

$$L_4 C_4 = L_3 C_3 = L_2 K_2$$

The ratio C_p / C_s for crystals is approximately 140 or higher, corresponding to a relatively small difference in the resonant frequency of $L_1 C_1$ and $L_3 C_3$. The use of this transformation permits this ratio to be increased by the use of an external capacitance to increase C_p. Two capacitors with values $+K_0$ and $-K_0$ may be put in parallel with C_p such that $(C_p + K_0)/C_s \geq 140$. Then, using a transformation having a negative capacitance in the structure with $-K_0$, L_4, C_4 and C_5, the negative capacitance can be absorbed.

If the absolute value of capacitor K_0 is larger than C_5, the modification is impossible and therefore an impedance transformation must be performed before

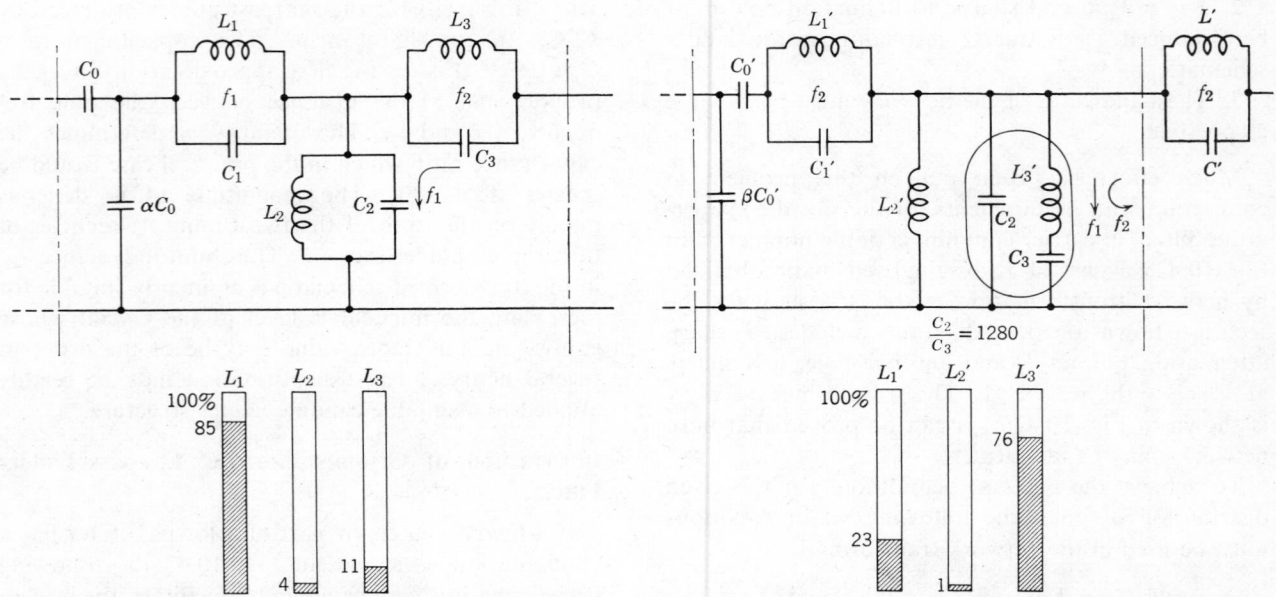

Fig. 10.42. Network transformation and reactive power distribution in the inductances of a filter (calculated by Poschenrieder).

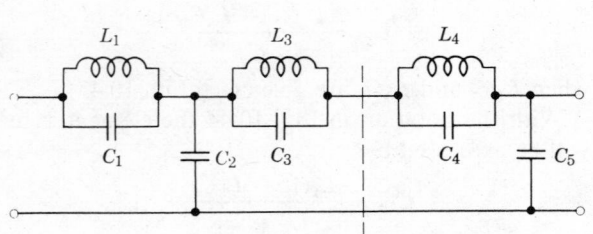

Fig. 10.43. Lowpass filter before crystal insertion.

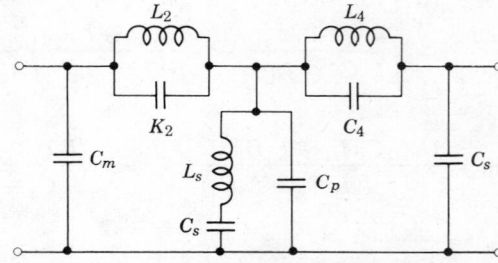

Fig. 10.44. Transformation of Fig. 10.43 adapted for crystal.

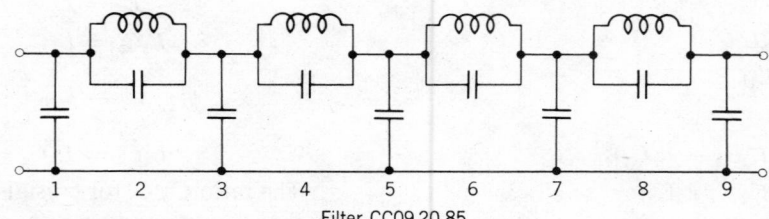

Filter CC09 20 85

Fig. 10.45. Filter CC09 20 85.

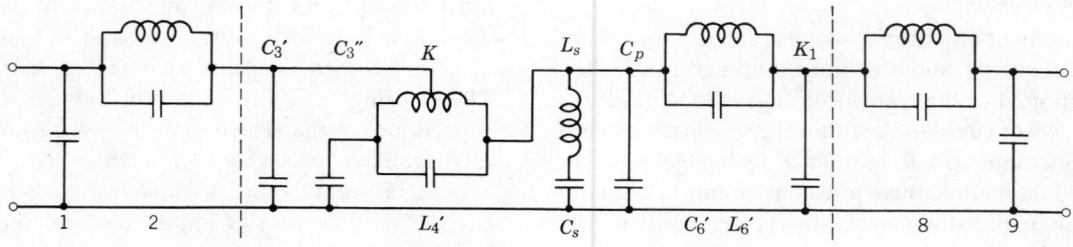

Fig. 10.46. Transformed schematic of Fig. 10.45.

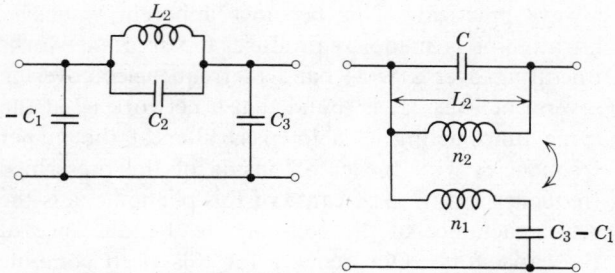

Fig. 10.47. Negative capacitor transformation.

the network is modified for crystal elements. The upper limit for the distance between peaks of attenuation which can be reached with this technique is approximately 200×10^{-4}.

The limitations of this technique are dictated by the value of motional inductance of the crystals; it is necessary to keep this value very large in comparison with the values of other inductances, otherwise the realization will lead to less interesting alternatives with transformers or with several identical crystals in parallel.

An example of this transformation is shown in Figs. 10.45 and 10.46. The normalized element values of Fig. 10.45 can be found in the catalog of Chapter 5. The modified element values are between the dashed lines of Fig. 10.46. In the original schematic, the closest peak of attenuation is produced by $L_6 C_6$, and is at 143×10^{-4} from the cutoff frequency.

10.7 NEGATIVE AND POSITIVE CAPACITOR TRANSFORMATION

As we know, besides the reduction of circuit elements in a given schematic, certain transformations can lead to new configurations in essentially a conventional network. The new network may then be especially suited for a variety of realizations, such as crystals or for ferramic resonators.

Let us call the equivalence of Fig. 10.47 the negative capacitor transformation. It permits an unrealizable four terminal network, with one negative capacitor to be substituted by another entirely realizable network employing an auto-transformer.

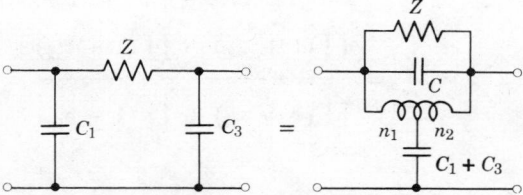

Fig. 10.48. Positive capacitor transformation.

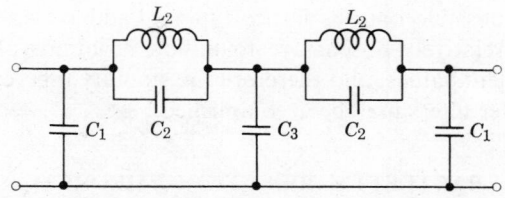

Fig. 10.49. Filter used in example of application.

A second transformation permits a network of ladder type to be substituted by a bridged-T network of the type shown in Fig. 10.48. This constitutes the positive capacitor transformation.

In the catalog of Cauer-Chebyshev filters of Chapter 5, it was noted that for filters having a very small value of passband ripple, the filter may exhibit a negative capacitor and is consequently unrealizable. The application of the negative capacitor transformation will permit the filter to be successfully realized.

Examples of Application

The filter of Fig. 10.49 was designed with the use of the catalog of normalized parameters to have a maximum passband ripple of 0.04 dB. The filter has been scaled to have its cutoff frequency at 308 kc and its input and output impedance equal to 1000 ohms. The original element values are:

$$C_1 = -155 \text{ pF}$$
$$C_2 = 14.935 \text{ pF}$$
$$C_3 = 514 \text{ pF}$$
$$L_2 = 0.0174 \text{ mH}$$

Figure 10.50 gives the transformed network having the following set of element values:

$$C_0 = 102 \text{ pF}$$
$$C = 2293 \text{ pF}$$
$$L = 0.11 \text{ mH}$$
$$L_2 = 0.0174 \text{ mH}$$

The negative capacitor transformation permits the realization of certain ladder filters which otherwise

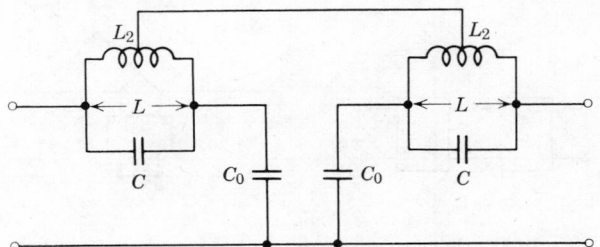

Fig. 10.50. Transformed schematic of Fig. 10.49.

are possible only as lattice types. Ladder responses are relatively unsensitive to minor variations of the element values, and therefore the stability inherent to ladder filters may be accomplished.

10.8 BARTLETT'S BISECTION THEOREM

It is known from the theory of four-terminal networks that for every symmetrical ladder section there corresponds an equivalent lattice schematic having equal transmission properties and the same characteristic impedances.

According to Bartlett's method, in order to obtain the equivalent lattice, the ladder network has to be bisected and the short-circuit impedance of the half-section is used as the series arms of the lattice sections. The open-circuit impedances are used as diagonal arms of the lattice. This procedure is illustrated in Fig. 10.51. Both *T*- and *π*-schematics can be transformed into their lattice equivalent. Conversely the lattice schematic can be, under certain conditions, transformed into *π*- or *T*-equivalent schematic. In Fig. 10.52, the illustration of Bartlett's theorem for reactive networks is shown. It is necessary to stress that under certain conditions the lattice cannot be transformed into a ladder because of the negative reactances involved in the transformation. On the other hand, every ladder has a lattice equivalent.

Transformation of the All-Pass Chain

Since the all-pass network can be invariably realized in lattice form (or its equivalent bridges) it will be shown that simple lattices in cascade may be transformed into a single equivalent lattice, which then may be transformed to its semilattice form, thus reducing the number of elements used.

The simple first-order networks, as shown in Fig. 1.8*a*, and even their simplest equivalent forms, are not

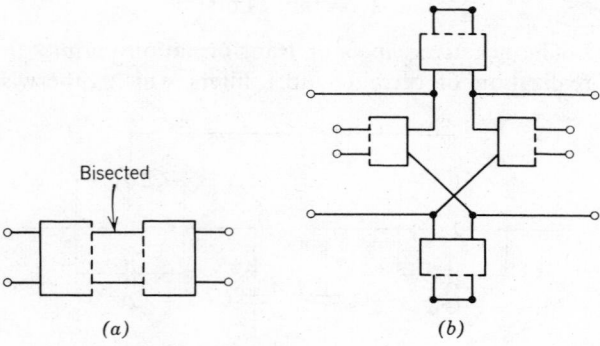

(a) *(b)*

Fig. 10.51. Symmetric ladder section and its equivalent lattice.

always practical. This becomes apparent whenever an attempt is made to produce a pair of networks operating over a wide band of frequencies covering several octaves. It is found that a network is, at the same time, acting as a lowpass filter at the higher frequencies with the cutoff inside of the prescribed frequency limits. The cause of this phenomena is the self-capacitance of the coils and lead inductance of the capacitors. One remedy for this is to combine pairs of the first-order networks together and realize them as a second-order network. It is recommended, also, to combine one high-frequency section with one low-frequency section to get a better practical lattice of the second-order. Next to be combined are the highest and lowest of what remains, and so on, until either one or none is left.

In Fig. 10.53 a chain of lattices is shown. In the case of lattices with equal characteristic impedances:

$$Z_a Z_b = Z_{an} Z_{bn} = Z_{ak} Z_{bk} = Z_0{}^2 \qquad (10.8.1)$$

where $k = 1, 2, \ldots, n$

n is the number of lattices

k is the number of lattices between 1 and n

$Z_a, Z_b =$ impedances of the lattice arm

Z_0, the characteristic impedance, is given by

$$Z_0{}^2 = Z_a Z_b \qquad (10.8.2)$$

The transmission constant g for any type of nonequal lattice with equal impedances will be expressed by:

$$g = g_1 + g_2 + \cdots + g_n = n(a + jb) \qquad (10.8.3)$$

The index associated with g shows the number of the lattice to which it refers. The transmission constant for any of the lattices will be expressed by:

$$g_k = 2 \tanh^{-1} \sqrt{\frac{Z_{ak}}{Z_{bk}}} = \ln \frac{1 + z_{ak}}{1 - z_{ak}} \qquad (10.8.4)$$

According to expression (10.8.3) for the transmission constant, a similar expression in a different form will be

$$g = \ln \frac{1 + z_a}{1 - z_a} = \ln \prod_{1}^{n} \frac{1 + z_{ak}}{1 - z_{ak}}$$

This expression can be solved to evaluate the normalized impedance z_a and z_b

$$z_a = \frac{Z_a}{Z_0} = \frac{\displaystyle\prod_{1}^{n}(1 + z_{ak}) - \prod_{1}^{n}(1 - z_{ak})}{\displaystyle\prod_{1}^{n}(1 + z_{ak}) + \prod_{1}^{n}(1 - z_{ka})}$$

$$z_b = \frac{Z_b}{Z_0} = \frac{1}{z_a}$$

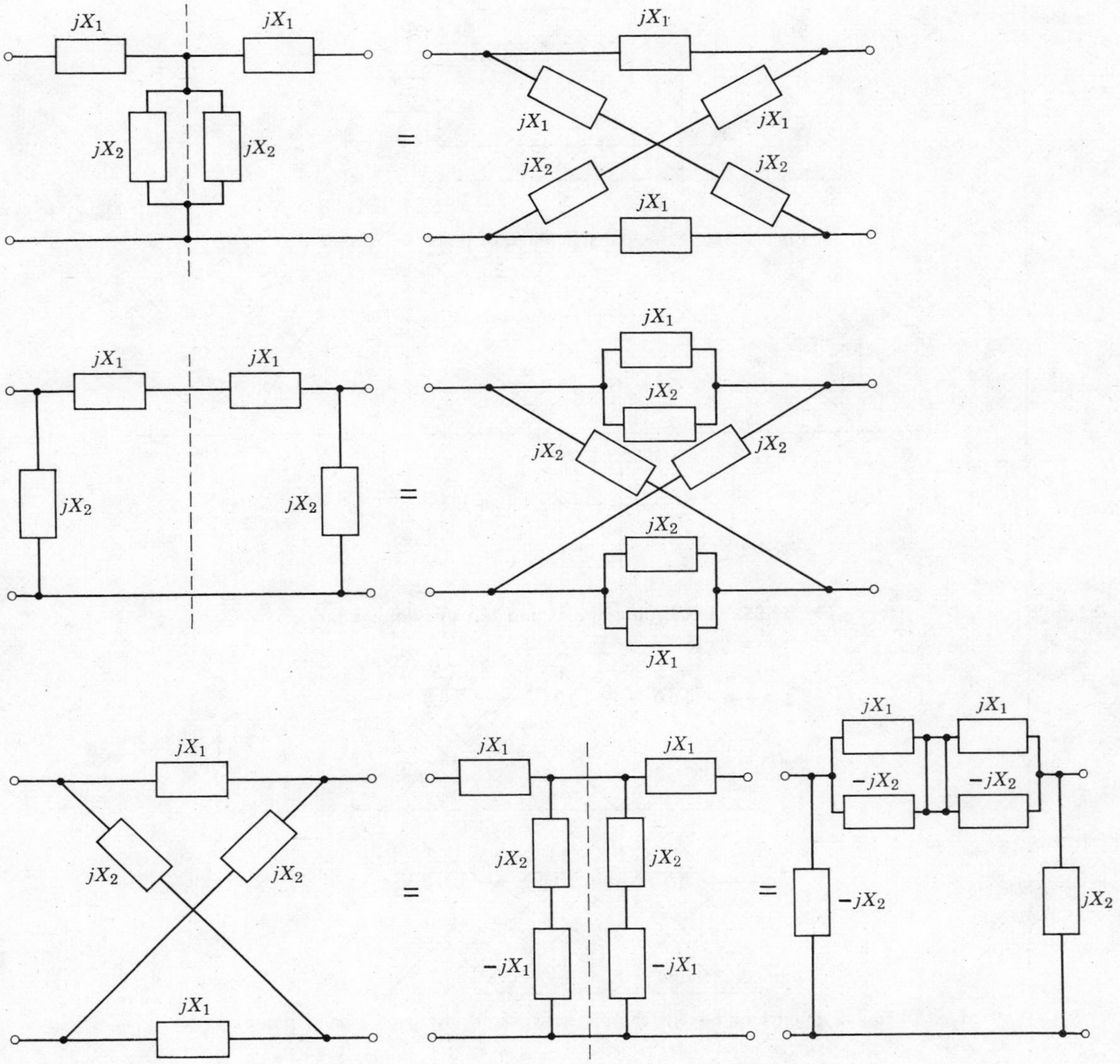

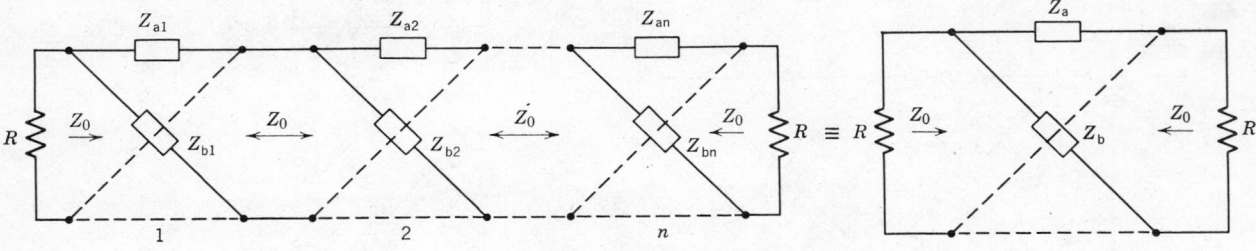

Fig. 10.52. Illustration of Barlett's theorem.

Fig. 10.53. Chain of lattices.

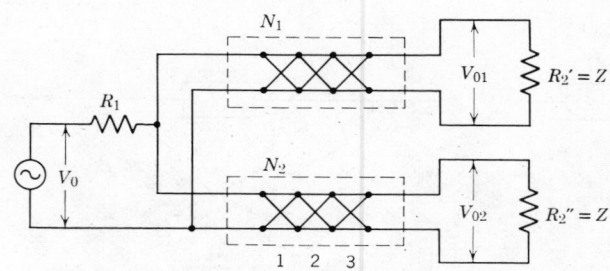

Fig. 10.54.   Principal schematic of phase networks.

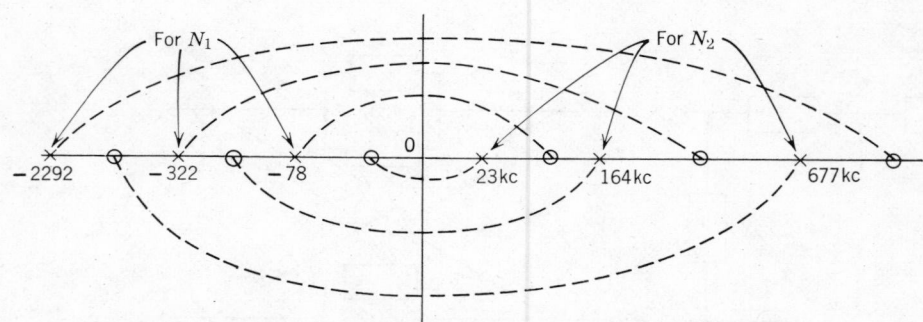

Fig. 10.55.   Location of poles and zeros of voltage ratio.

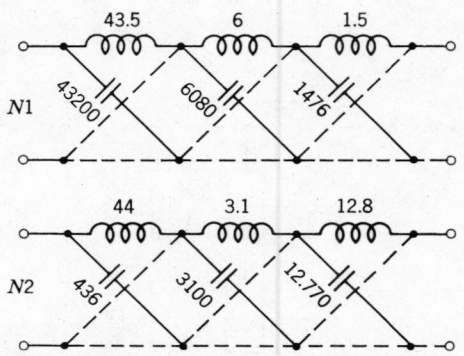

Fig. 10.56.   Elements of 6-transfer pole network for wide base-band phase splitter.

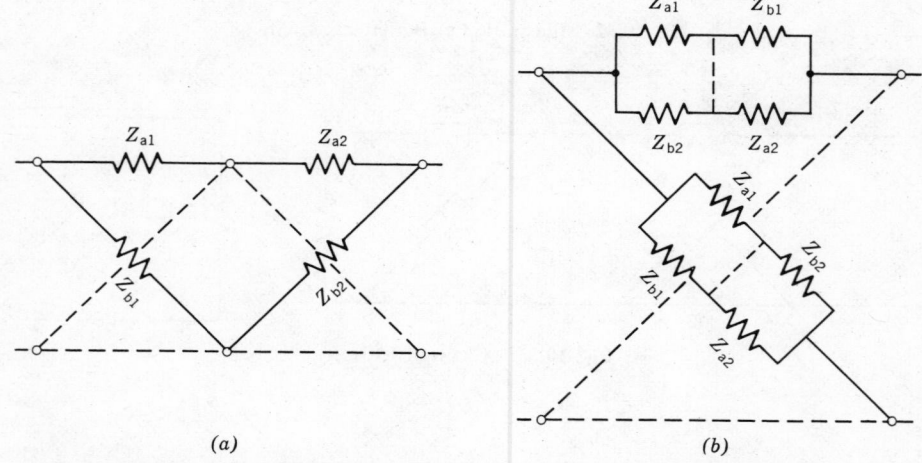

Fig. 10.57.   Cascade connections of two lattices and its one-lattice equivalent.

This means that z_a and z_b are reciprocal impedances. If we have equal lattices connected in cascade, as in the case of pulse-stretching and phase-correcting networks, the expressions for z_a and z_b will be slightly modified and will take the following forms:

$$z_a = \frac{(1 + z_{ak})^n - (1 - z_{ak})^n}{(1 + z_{ak})^n + (1 - z_{ak})^n}$$

$$z_b = \frac{1}{z_a}$$

Practical Steps in Lattice-Transformation Procedure

Let us assume that a phase splitter (such as that used in single sideband systems) shown in Fig. 10.54 has been synthesized by some nonelementary fashion. When the synthesis is accomplished, information has usually been obtained for symmetrical nonequal all-pass lattices of the first order such as shown in Figs. 10.55 and 10.56.

Figure 10.57a shows the cascaded connection of two lattices with equal characteristic impedances and the intermediate step in the transformation to combine two cascaded lattices into a single equivalent lattice is given in Fig. 10.57b. The dashed line between the impedance in each arm of Fig. 10.57b shows that at these points the potentials are equal and the points could be connected together.

The addition of three simple lattices can be done in a simple way. The first operation is to combine two of the lattices as has been done previously and then combine the resulting second-degree lattice with the remaining simple first-order lattice. The most practical procedure for combining the simple lattices is to combine, at first, two lattices having the lowest and highest transfer poles. This procedure is used because, in an extreme case, the poles may not be realized in the single-lattice form. Figure 10.58 shows the elements of a first-order and a second-order lattice which will be combined into a single third-order lattice. Both

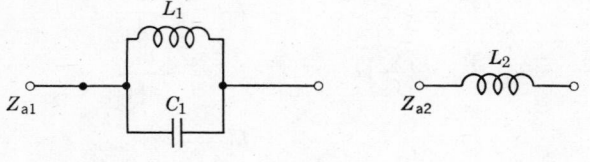

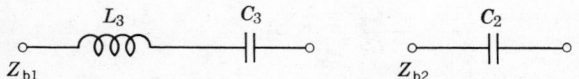

Fig. 10.58. Elements of second- and first-order all-pass lattices.

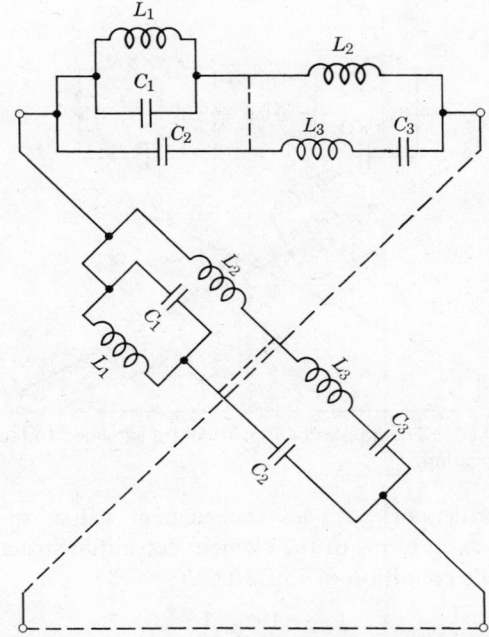

Fig. 10.59. Equipotential points connected for simplification.

lattices are ideal in the sense that no losses are involved.

In the reactances of Fig. 10.59, there are two resonant circuits which resonate at the same frequency. In the upper arm, the first anti-resonance will be produced when L_1 resonates with $(C_1 + C_2)$; the second one with L_2, L_3, and C_3. The second anti-resonance is hidden and must be "extracted." In the lattice arm of Fig. 10.59 the first anti-resonance is produced when C_1 resonates with $L_1 L_2 / L_1 + L_2$ and the second anti-resonance with C_2, C_3, and L_3. The second resonance in this circuit is also hidden and must also be "extracted."

The circuit, after the proper dipole transformation has been applied, is shown in Fig. 10.60. The two anti-resonant circuits in each arm of the lattice are resonated at the same frequency and may be combined by the simple addition of impedances.

Figure 10.61 is the equivalent schematic which is chosen for practical realization in semilattice form with a differential transformer and has two reciprocal branches at all frequencies.

10.9 CAUER'S EQUIVALENCE

In Figs. 10.62 and 10.63 two network equivalences introduced by Cauer are shown. The networks a and b are equivalent to network c, provided the conditions below the circuits a and b are fulfilled. The equations

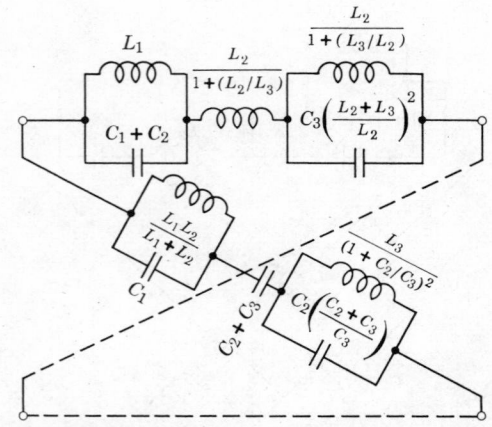

Fig. 10.60.   The dipoles of Fig. 10.59 transformed to facilitate simplification.

below network c gives the element values of that network in terms of the elements of initial structures.

If the condition of Fig. 10.62b

$$\frac{1}{C_1} + \frac{1}{C_2} + \frac{1}{C_3} = 0$$

is satisfied, then the elements of Fig. 10.62a in terms of 10.62b are

$$L_1 = MC_3$$

$$L_3 = MC_1$$

$$L_2 = \frac{L}{M}$$

$$C = MC_2$$

$$M = \frac{L}{C_1 + C_3}$$

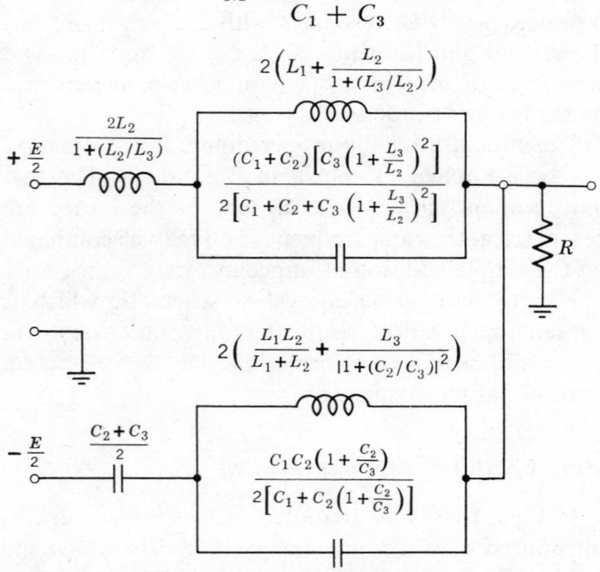

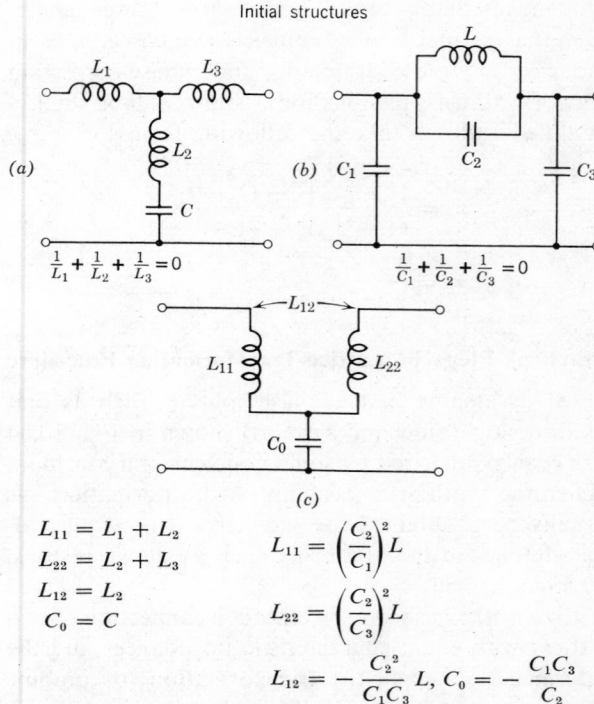

Fig. 10.62.   Initial structures and canonic equivalent.

$$L_{11} = L_1 + L_2 \qquad L_{11} = \left(\frac{C_2}{C_1}\right)^2 L$$

$$L_{22} = L_2 + L_3 \qquad L_{22} = \left(\frac{C_2}{C_3}\right)^2 L$$

$$L_{12} = L_2$$

$$C_0 = C \qquad L_{12} = -\frac{C_2{}^2}{C_1 C_3} L, \; C_0 = -\frac{C_1 C_3}{C_2}$$

Fig. 10.61.   Final schematic of the third-order all-pass semi-lattice.

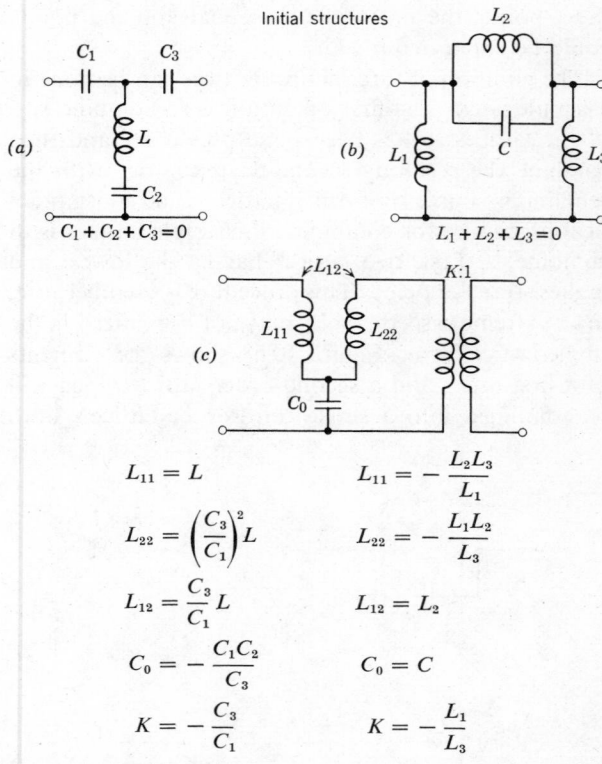

$$L_{11} = L \qquad L_{11} = -\frac{L_2 L_3}{L_1}$$

$$L_{22} = \left(\frac{C_3}{C_1}\right)^2 L \qquad L_{22} = -\frac{L_1 L_2}{L_3}$$

$$L_{12} = \frac{C_3}{C_1} L \qquad L_{12} = L_2$$

$$C_0 = -\frac{C_1 C_2}{C_3} \qquad C_0 = C$$

$$K = -\frac{C_3}{C_1} \qquad K = -\frac{L_1}{L_3}$$

Fig. 10.63.   Initial structures and equivalent form with mutual coupling.

For Fig. 10.63*b*, if the condition

$$L_1 + L_2 + L_3 = 0$$

is satisfied, then the elements of Fig. 10.63*a* in terms of 10.63*b* are

$$C_1 = ML_3$$
$$C_3 = ML_1$$
$$C_2 = ML_2$$

$$L = \frac{C}{M} \quad \text{with} \quad M = C \frac{L_1 + L_3}{L_1 \times L_3}$$

Transformation of Normalized *LP* for Canonic Realization

In the tables of Chapter 5 all filters are designed based on prototypes without mutual coupling between the coils. This is the most practical approach, since such structures are reasonably insensitive to minor variation of element values and can easily be adjusted. The disadvantage of the conventional structures is in the number of reactive components. For example, a lowpass of the third order used throughout Chapter 5 as an example, requires four elements, but theoretically three is sufficient as shown in Fig. 10.64. The schematics which include only that theoretical number of reactors (which is correspondent to the order of the numerator polynomial) are called canonic schematics. In the general case they always include elements with mutual coupling.

The canonic structure shown in Fig. 10.64 has evident practical advantages from the point of view of size, weight, price, and tuning time. The ideal transformer which is equivalent to stars of coils (one of which is negative) has lossless windings, strong coupling, and an infinite inductance. Practically, only 98% coupling can be realized and only finite inductance in a somewhat lossy transformer can be achieved. In Fig. 10.65 the equivalent schematics and the conditions for their equivalence are shown. In filter schematics usually the coils are wound in opposite

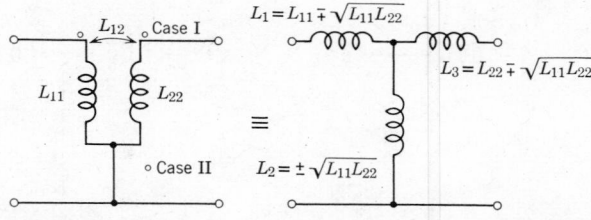

Case I—Opposite winding direction—L_1 or L_3 are negative (use upper sign).
Case II—Same winding direction—L_2 is negative (use lower sign).

Fig. 10.65. Ideal transformer and its equivalent.

directions (see case I in Fig. 10.65). In all-pass networks the direction of coil winding is the same (see case II). In both cases the necessary condition for the star of inductances is as in Fig. 10.62.

Numerical Example

It is assumed that the normalized lowpass network shown in Fig. 10.66 is given. The problem is to calculate the equivalent canonic lowpass network.

Since the voltage transfer function and short-circuit transfer function for bilateral networks are identical, the structure shown in Fig. 10.66*b* can be designed with the aid of the catalog.

For this example

$$\text{CC03 02 30} \quad \text{with} \quad K^2 = \infty$$

The corresponding normalized element values are

$$L_1 = -0.206$$
$$L_2 = 1.096$$
$$C_2 = 0.177$$
$$L_3 = 0.711$$

The problem is to substitute the negative star of inductances by mutually coupled coils and one separate coil. In this operation the unrealizable negative inductance L_1 will disappear.

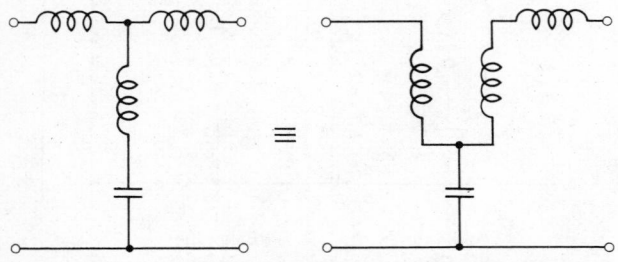

Fig. 10.64. Comparative realization in conventional and canonic forms for $n = 3$.

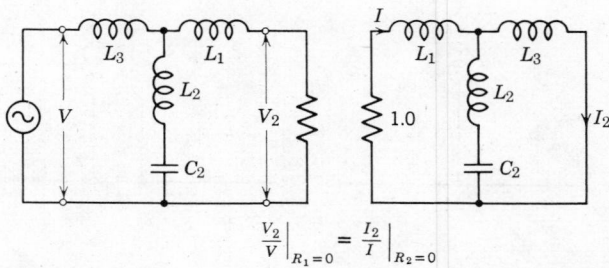

$$\left. \frac{V_2}{V} \right|_{R_1 = 0} = \left. \frac{I_2}{I} \right|_{R_2 = 0}$$

Fig. 10.66. Networks having equal transmission properties.

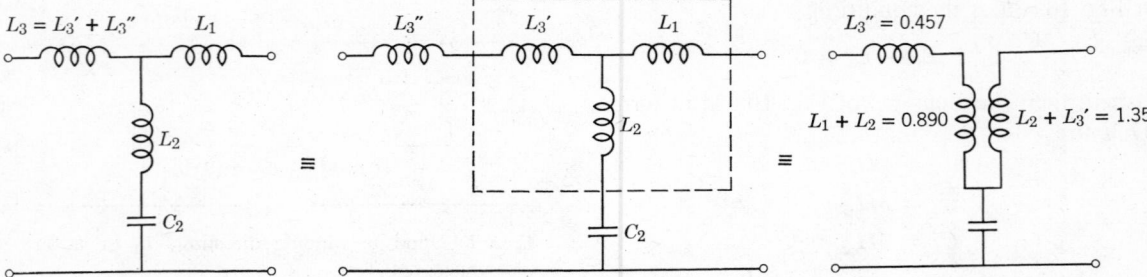

Fig. 10.67. Lowpass filter with $n = 3$.

From the necessary condition in Fig. 10.62 the value

$$L_3' = \frac{1}{L_1} - \frac{1}{L_2} = \frac{1}{0.254}$$

and

$$L_3' = L_3 - L_3' = 0.457$$

In Fig. 10.67c the final values of inductances are shown.

10.10   CANONIC BANDPASS STRUCTURES

The conventional lowpass-bandpass transformation essentially leads to the substitution of a series-resonant circuit for all coils and of a parallel-resonant circuit for every capacitor (see Chapter 5). But a complex lowpass filter can be transformed differently on the basis of whole sections. For example it can be subdivided into several partial four-terminal networks such as shown in Fig. 10.68, each transformed separately into bandpass structures and finally supplemented by partial networks which correspond to full removal at $\Omega = \infty$, realized by conventional substitution.

The final bandpass structure, therefore, consists of conventionally transformed coils and capacitors as given in Table 5.4, and the partial networks corresponding to finite poles which have to be connected together to correspond to a realizable lowpass structure. The disadvantage of the schematic in the table is that they include more network elements than is theoretically necessary.

In a canonic lowpass, the attenuation pole at finite frequency has a form shown in Fig. 10.68a. The network in that figure has no poles at zero or infinity and can be made from two reactive elements as has been shown in Fig. 10.65.

The coils of Fig. 10.68a must then satisfy the conditions written in Fig. 10.62a so that

$$L_3 = -\frac{L_1 L_2}{L_1 + L_2} \tag{10.10.1}$$

with

$$L_1 + L_2 > 0$$
$$L_3 + L_2 > 0$$
$$L_2 > 0$$

Figure 10.68b shows the basic form of the transformed filter from the lowpass prototype in Fig. 10.68a. Evidently this form is not realizable since some elements in the lowpass and bandpass are negative. With the help of Norton's equivalences (Fig. 10.11) which can be applied separately to the

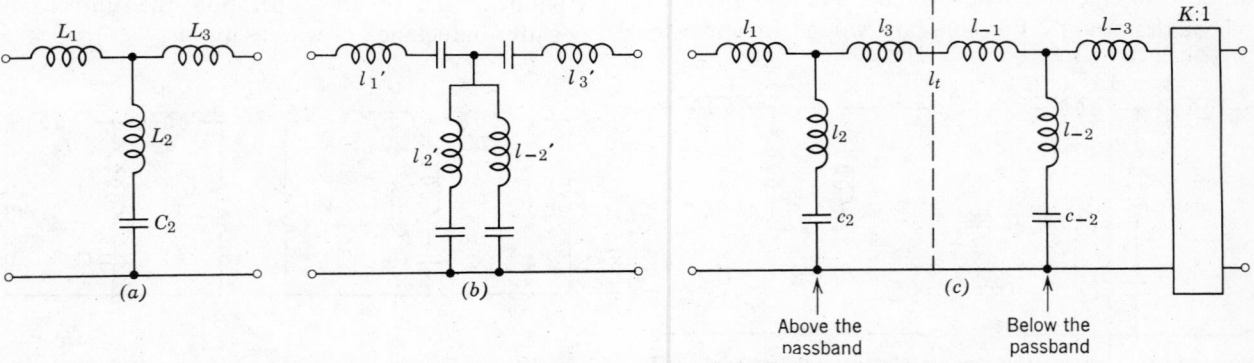

Fig. 10.68.   Canonic lowpass bandpass structures without poles at $\Omega = 0$ or $\Omega = \infty$.

series-resonant circuits in the shunt arm of the basic bandpass, the final schematic will take the equivalent form shown in Fig. 10.68c which preserves the original form of lowpass prototype.

The element values of this network are given in the following set of relations:

$$
\begin{align}
(1) \qquad & l_1 = L_1 b \\
(2) \qquad & l_t = l_3 + l_{-1} = -L_1 bd \\
(3) \qquad & l_2 = L_2 \Omega_{-1}{}^2 d \\
(4) \qquad & l_{-2} = L_2 dK \\
(5) \qquad & l_{-3} = L_1 \Omega_1{}^2 Kb \\
(6) \qquad & c_2 = \frac{1}{\Omega_1{}^2 l_2} \\
(7) \qquad & c_{-2} = \frac{1}{\Omega_{-1}{}^2 l_{-2}} \\
(8) \qquad & b = 1 - \Omega_{-1}{}^2 \\
(9) \qquad & d = 1 + m_1 + \Omega_1{}^2 \\
(10) \qquad & m_1 = \frac{L_1}{L_2} \\
(11) \qquad & K = 1 + m_1
\end{align}
\qquad (10.10.2)
$$

The values $\Omega_1{}^2$ and $\Omega_{-1}{}^2$ are identical with those in Table 5.4. The inductances l_3 and l_{-1} are evaluated with Eqs. 10.10.1 and 10.10.2 so that

$$
l_3 = \frac{l_1 \times l_2}{l_1 + l_2}; \qquad l_{-1} = l_t - l_3
$$

The T of coils of Fig. 10.68c satisfies Eq. 10.10.1 and can be realized with mutually coupled coils. The use of ideal transformers can be easily avoided when in the following partial network all inductances are multiplied, and all capacitors are divided by K^2. The completed network consists of n-partial networks with the transformation coefficient equal to K_t

$$
K_t = \prod_1^n K_1
$$

When in Eqs. 10.10.2, all m are positive, the coefficients K_t will be larger than unity. On the other hand, when all m are negative, K_t will be less than one.

A long chain (ladder) will often lead to an unfavorable ratio between the element values and to a high value of transformation coefficient. If the lowpass filter schematic has an attenuation pole at $\Omega = \infty$, full removal could be made in a different place and in this manner influence the value of the inductance ratio m. It is desirable to choose the position of the pole at $\Omega = \infty$ in the network in such a manner that the geometric mean of all transformation coefficients is

in the proximity of one. If the schematic has two or more attenuation poles at $\Omega = \infty$, there are many possibilities to control the impedance ratio (transformation coefficient).

The networks in Figs. 10.62 and 10.63 are dual and the duality relations are shown in the following set of equations.

$$
\begin{align}
& C_1 = \frac{1}{l_1 \Omega^2} \\
& C_3 = \frac{1}{l_3 \Omega^2} \\
& L = l_1 + l_3 \qquad (10.10.3) \\
& C_2 = \frac{1}{(l_1 + l_3)\Omega^2} \\
& \frac{1}{l_2 c} = \frac{1}{l c_2} = \Omega^2
\end{align}
$$

The canonic schematic which is shown in Fig. 10.68 has the biggest advantage in the fact that its realization requires the smallest number of elements.

The second advantage is that every lowpass ladder network can be transformed into a bandpass ladder of this form. The configuration also has disadvantages: even "strong" coupling is only an approximation to the necessary unity coupling. Nevertheless, the networks in canonical form are very useful in practice and find many applications.

10.11 BANDPASS LADDER FILTERS HAVING A CANONICAL NUMBER OF INDUCTORS WITHOUT MUTUAL COUPLING

Complex multielement ladder bandpass networks are designed, as a rule, with a minimum number of inductors. Consequently, the number of capacitors are slightly greater than in conventional filters. The lowpass-bandpass transformation procedure in this form is analogous to the previous transformation except that not only Norton's transformation, but the Cauer equivalences as well are used.

As a starting point in the discussion, we begin with a nonrealizable network, the dual to Fig. 10.68a, having no pole of attenuation at $\Omega = \infty$. This network is shown in Fig. 10.69a. It is possible to represent the entire filter as a combination of partial networks according to Fig. 10.69. One of the capacitors (C_1 or C_3) will consist of two capacitors, one positive and one negative, to satisfy the conditions in Fig. 10.62b. After the formal lowpass-bandpass transformation is performed, the network will be as shown in Fig. 10.69c. Using the relations shown

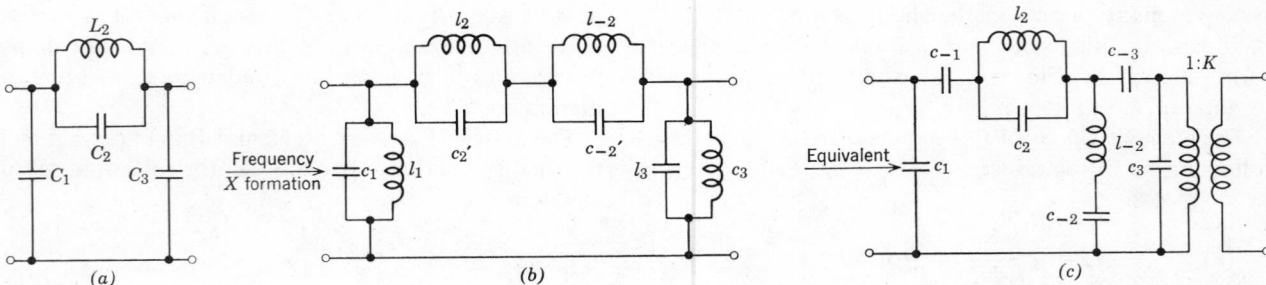

Fig. 10.69. Filter with canonic number of inductances.

below and the element values of the original lowpass filter, the element values of the coil-saving filter will be obtained.

$$c_1 = C_1(1 - \Omega_{-1}^2)$$
$$c_3 = C_3 K_1(1 - \Omega_{-1}^2)$$
$$l_2 = \frac{1}{\Omega_1^2 c_2}$$
$$c_2 = C_2 t_1 \Omega_{-1}^2$$
$$c_{-1} = -C_1 t_1(1 - \Omega_{-1}^2)$$
$$c_{-3} = -C_3 t_1(1 - \Omega_{-1}^2)K^2$$
$$c_{-2} = -m_1 C_3 t_1(1 - \Omega_{-1}^2)^2 \qquad (10.11.1)$$
$$l_{-2} = \frac{1}{\Omega_{-1}^2 c_{-2}}$$
$$C_3 = \frac{C_1 C_2}{C_1 + C_2}$$

where:

$$m_1 = \frac{C_1}{C_2}$$
$$t_1 = 1 + m_1 + \Omega_{+1}^2$$
$$K_1 = 1 + m_1 \Omega_{-1}^2$$

The values Ω_1^2 and Ω_{-1}^2 have a meaning similar to that in Chapter 5, Table 5.4.

$$\Omega_+^2 = \frac{1}{l_+ c_+}, \qquad \Omega_-^2 = \frac{1}{l_- c_-}$$

In the case when C_1 is positive in Fig. 10.69a, all element values in Fig. 10.69c are positive with the exception of c_{-1} and c_3.

Such a partial network alone cannot be realized with passive components. Two similar partial networks provide a situation which is extracted in Fig. 10.70. With the aid of Norton's equivalences, the network of Fig. 10.70a is reduced to 10.70c which is perfectly realizable.

The transformation procedure has been accomplished with the following set of relations.

$$c_t = C_3 + K^2 C_{(1+1)}$$
$$c_p = c_{-3}(1 - K_2)$$
$$c_{-t} = K_1^2 C_{-(1+1)}$$
$$c_{-p} = K_2^2 \frac{(c_3 + c_t)c_{-t}}{c_{-3} + c_t + c_{-t}}$$
$$K_2 = \frac{c_{-3}}{c_{-3} + c_t}$$
$$K_p = K_2 K_1$$

The final schematic has one ideal transformer with a transformation ratio $1:K$ on the end of the network. The total transformation coefficient of the entire composite structure which can involve many networks similar to the one in Fig. 10.69 will be equal to

$$K_t = \prod_1^n K_p$$

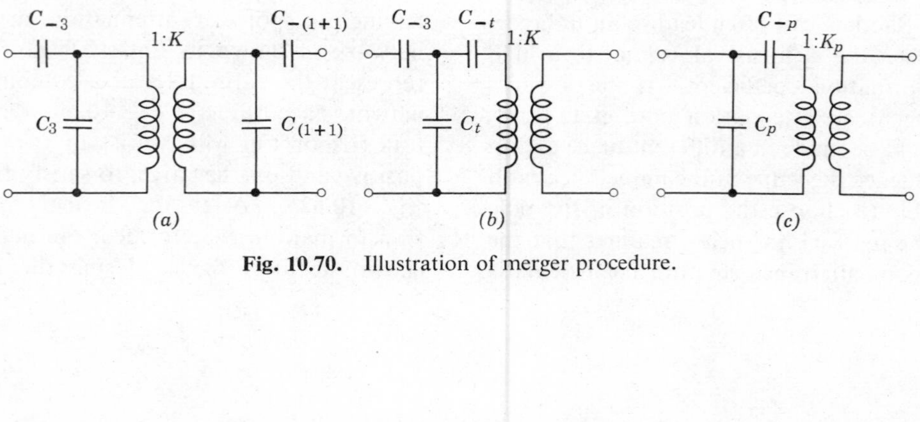

Fig. 10.70. Illustration of merger procedure.

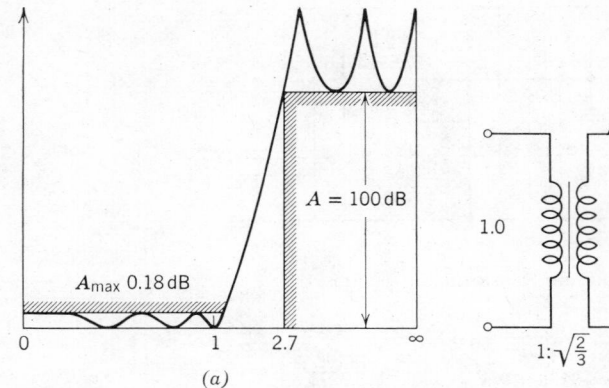

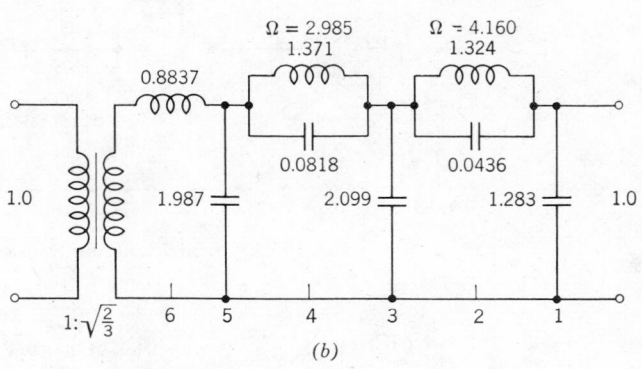

Fig. 10.71. Normalized lowpass *CC*06 20 21 *b*.

Every one of n partial networks according to Fig. 10.69c has in the beginning and at the end a negative capacitor, c_{-1} and c_3. The lowpass schematic must have a minimum of two poles of attenuation at $\Omega = \infty$. One pole must be removed on the input side and one on the output side of the filter in order to obtain the realizable network with the relations (10.11.2).

According to the above relations the values of c_p and c_{-p} can be negative. In this case a strictly passive network is unrealistic. The basic form of Table 5.4 with the same number of elements is perfectly valid for passive realization, because it permits a different order of attenuation pole removal. The change of the order of pole removal will lead to structures having mutual coupling.

The values c_p and c_{-p} are positive when

$$c_t \geq 0 \qquad (10.11.3)$$

and

$$c_{-t} \leq -(c_t + c_{-3})$$

The configuration of Fig. 10.69 although having a minimum number of coils without mutual coupling, still has disadvantages, considering the amount of calculation necessary with no real certainty that a realizable network will develop. From this point of view, the filters of Table 5.4 are more advantageous. In spite of the difficulties described, the form of filters in Fig. 10.69 is often used, since the advantages are comparatively larger than the difficulties mentioned.

Example of Antimetric Structure Transformation

The bandpass filter to be developed is to be inserted between a 500-ohm source and a 50-ohm output load. The center frequency is 468 kc with a bandwidth 35 kc. The lowpass model of Fig. 10.71 will be transformed into a coil-saving bandpass structure. The element values of the lowpass are shown in Fig. 10.71 normalized according to

$$L' = L\frac{\omega_r}{R_r}$$

$$C' = C\omega_r R_r$$

and are taken from the catalog of Chapter 5 for

$$CC06\ 20\ 21\ b$$

Step 1. Using complimentary normalization with respect to the bandwidth of the bandpass filter to be developed according to expressions

$$C'' = C_r\omega_r R_r a$$

$$L'' = L\frac{\omega_r}{R_r} a$$

the normalized lowpass will be as shown in Fig. 10.72 with a relative bandwidth $1/a = 35/468$.

Step 2. Representing the lowpass as a combination of partial networks shown in Fig. 10.73 will constitute the second step. The circuit now is prepared for partial network transformations.

Step 3. The equivalent bandpass form will be as shown in Figs. 10.74a and b, both including negative capacitors. The partial networks are now ready for

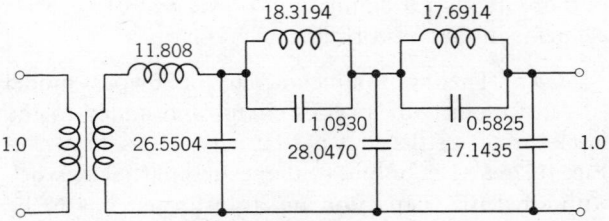

Fig. 10.72. Lowpass network normalized with respect to bandwidth.

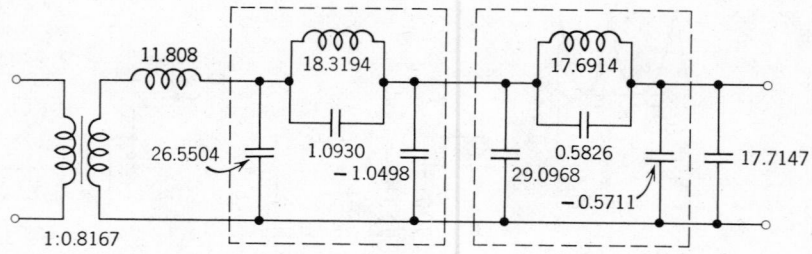

Fig. 10.73. Subdivision into transformable sections.

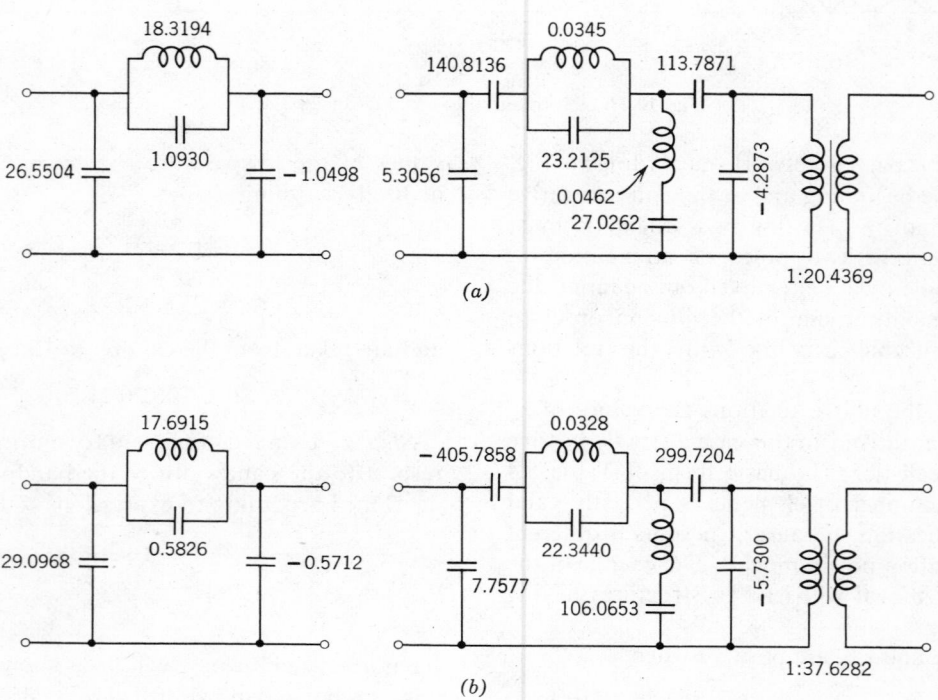

Fig. 10.74. Transformed partial networks.

combining through transformers at the end of the first partial network.

Step 4. The combining of both networks is performed and shown in Fig. 10.75. Two negative elements have vanished. The transformation coefficient is reduced. The transformation of input and output circuits for elimination of the rest of negative elements in the network has taken place.

Step 5. The input inductance in the lowpass model is transformed into a series circuit and added to the input circuit of the first partial network as shown in Fig. 10.76a. The output of the second partial network with a negative capacitor and transformer has to be supplemented by a parallel circuit which is a result of frequency transformation of a capacitor in a lowpass

model in Fig. 10.71. The situation is shown in Fig. 10.76b.

Step 6. Combine all ideal transformers into one transformer with transformation ratio 1:10. Transform the series-resonant circuits into parallel-resonant

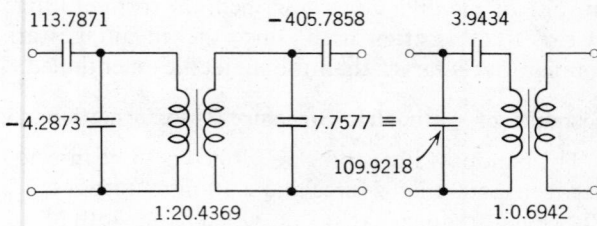

Fig. 10.75. Combining two partial networks and transformation into bandpass circuit.

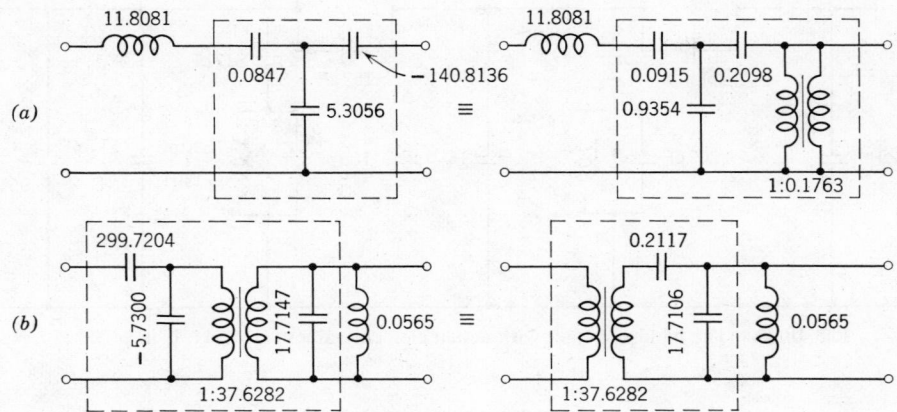

Fig. 10.76. Transformation of (*a*) input and (*b*) output circuits.

circuits to eliminate damaging distributed capacitances. The normalized network is shown in Fig. 10.77.

Step 7. The denormalizing procedure has taken place. The final schematic is shown in Fig. 10.78 with actual element values.

The transformation procedure of the original lowpass structure can be started with the network dual to 10.62*b* and shown in 10.62*a*. The resulting structure will be a canonic one with a canonic number of inductors and capacitors. Therefore a coil-saving structure does not mean the minimum number of inductive components, since the canonic form of a similar filter will require only six coils instead of seven. Moreover the number of capacitors in the canonic filter is seven. In the equivalent coil-saving filter the number is 10, after all possible transformations are performed.

To illustrate the transformation procedure of a lowpass filter into a canonic structure with mutual coupling, Fig. 10.79 summarizes the steps for a similar example.

Step 1. Normalized lowpass element values are obtained.

Step 2. The element values of the original lowpass filter are renormalized with respect to the bandwidth, which in this case is relatively wide, with $a = 2$.

Step 3. With aid of Cauer equivalents shown in Fig. 10.62 the partial networks are recalculated and the schematic is shown in Fig. 10.79. The pole at $\Omega = \infty$ is moved between the finite poles.

Step 4. The transformation of lowpass to bandpass structure is performed.

Step 5. Elements are combined and mutual coupling is introduced.

Summary on the Lowpass-Bandpass Transformation

If a lowpass filter has at $s = \infty$ at least one double pole (such as, for example, a type *CC* filter with even

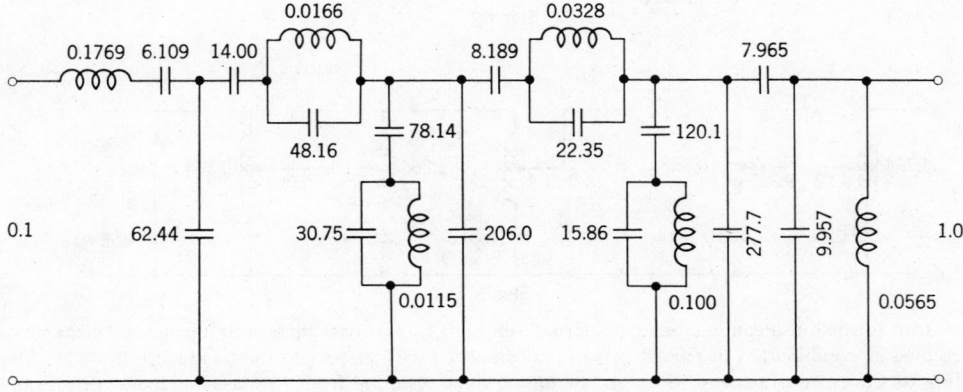

Fig. 10.77. Complete normalized network.

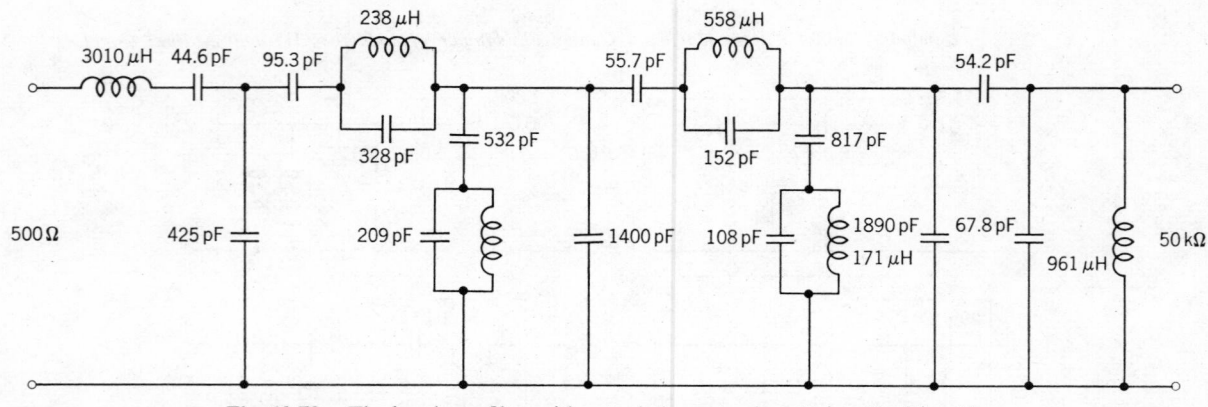

Fig. 10.78. The bandpass filter with actual element values (L in μH, C in ρF).

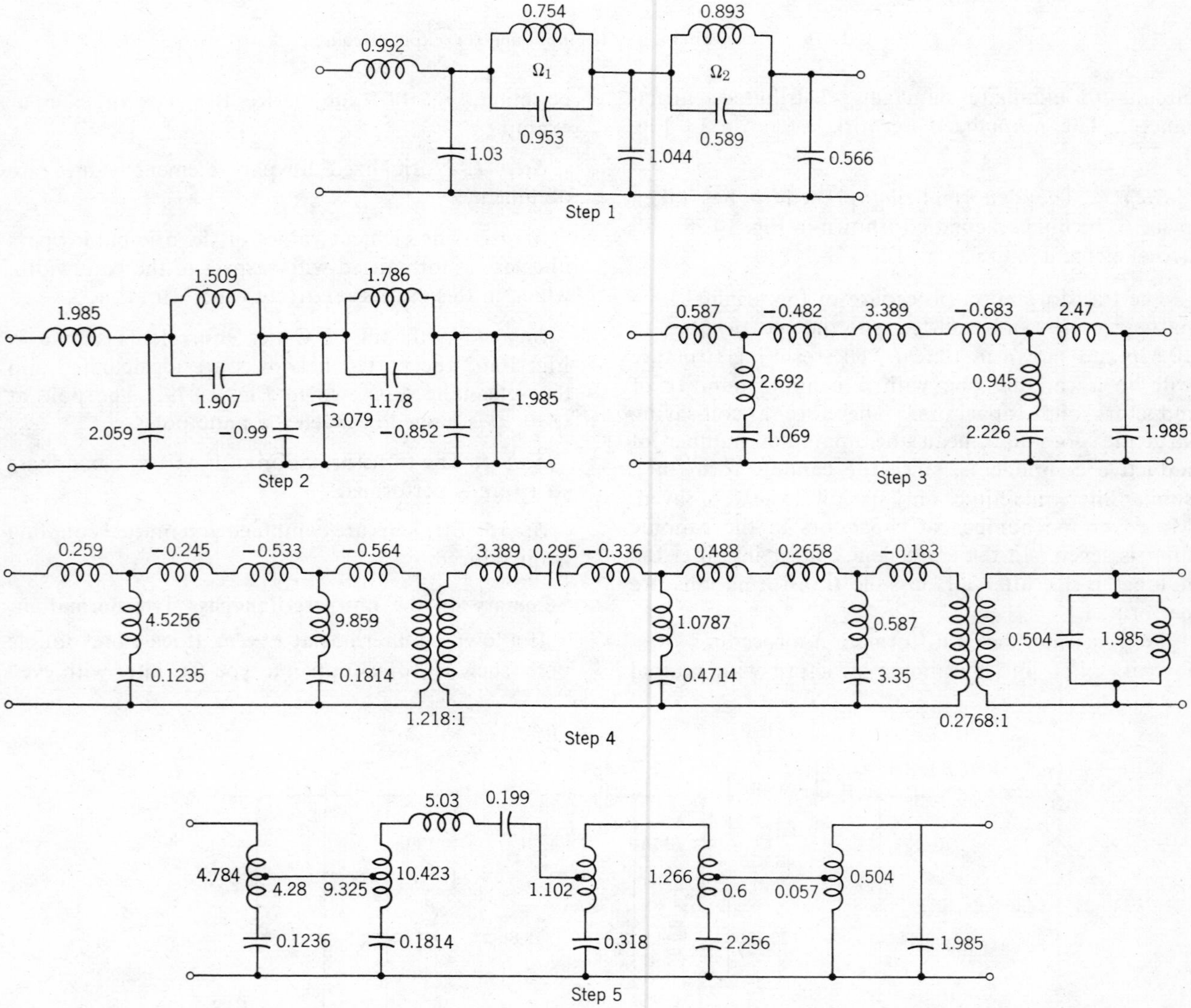

Fig. 10.79. Lowpass transformed into canonic bandpass structure. Step 1. Lowpass filter with normalized element values. Step 2. Lowpass filter represented as combination of partial networks normalized with respect to the bandwidth ($a = 2$). Step 3. Network transformed according to Cauer equivalences. Step 4. Bandpass filter resulting from lowpass-bandpass transformation. Step 5. Elements combined and mutual coupling introduced.

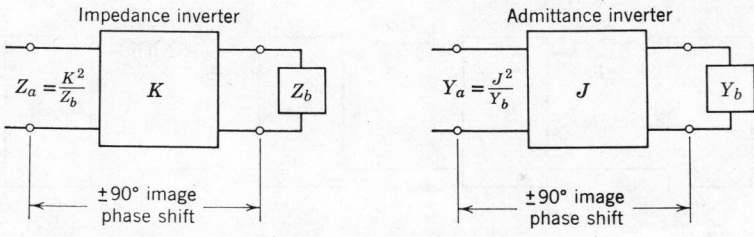

Fig. 10.80. Definition of inverters.

degree case b), the transformation may result in a filter having a minimum number of inductances. The calculation of elements may proceed in several steps. The original lowpass model is taken directly from the catalog. The partial transformation coefficient is to be determined between the sections, allowing the evaluation of the negative capacitor as part of Norton's transformer. The adjacent positive capacitor in parallel has to be equal to the original capacitor C_1 or C_3 across the line in the shunt arm of lowpass prototype. The series arm remains unchanged.

In the following step the most complicated operation takes place; several auxiliary quantities for each section have to be calculated, among them being the normalized resonant frequencies. With the aid of these quantities the normalized element values may be evaluated. Some additional circuit transformations may be performed to eliminate the redundant and negative elements from the schematic. In the final step the transformers are eliminated and the normalized element values are evaluated.

10.12 IMPEDANCE AND ADMITTANCE INVERTERS

In deriving design equations for certain types of bandpass and bandstop filters it is desirable to convert the prototypes that use both inductors and capacitors to equivalent forms that use either only inductors or only capacitors. This can be done with the aid of the idealized inverters which are symbolized in Fig. 10.80.

An idealized impedance inverter operates like a quarter-wavelength line of characteristic impedance K at all frequencies. Therefore if it is terminated in an impedance Z_b on one end, the impedance Z_a seen looking in at the other end is

$$Z_a = \frac{K^2}{Z_b} \qquad (10.12.1)$$

An idealized admittance inverter operates like a quarter-wavelength line of characteristic admittance J at all frequencies. Thus if an admittance Y_b is

attached at one end, the admittance Y_a seen looking in the other end is

$$Y_a = \frac{J^2}{Y_b} \qquad (10.12.2)$$

As indicated in Fig. 10.80, an inverter may have an image phase shift of either $\pm 90°$ or an odd multiple thereof.

Because of the inverting action indicated by Eqs. 10.12.1 and 10.12.2, a series inductance with an inverter on each side looks like a shunt capacitance from its exterior terminals. Likewise, a shunt capacitance with an inverter on both sides looks like a series inductance from its external terminals. Making use of this property the prototype circuits in Fig. 10.81 can be converted to either of the equivalent forms in Fig. 10.82 which have identical transmission characteristics to those prototypes in Fig. 10.81. As can be seen from Eqs. 10.12.1 and 10.12.2 inverters have the ability to shift impedance or admittance levels depending on the choice of the K or J parameters. For this reason in Fig. 10.82a the sizes of R_a, R_b, and the inductances L may be chosen arbitrarily and the response will be identical to that of the original prototype as in Fig. 10.81a, provided that the inverter parameters K are specified as indicated by the equations in Fig. 10.82a. The same holds for the circuit in

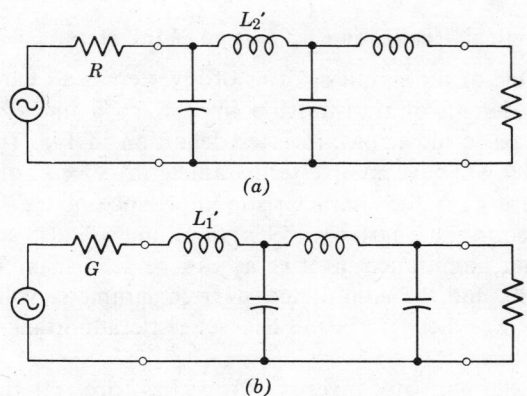

Fig. 10.81. Lowpass prototype circuits.

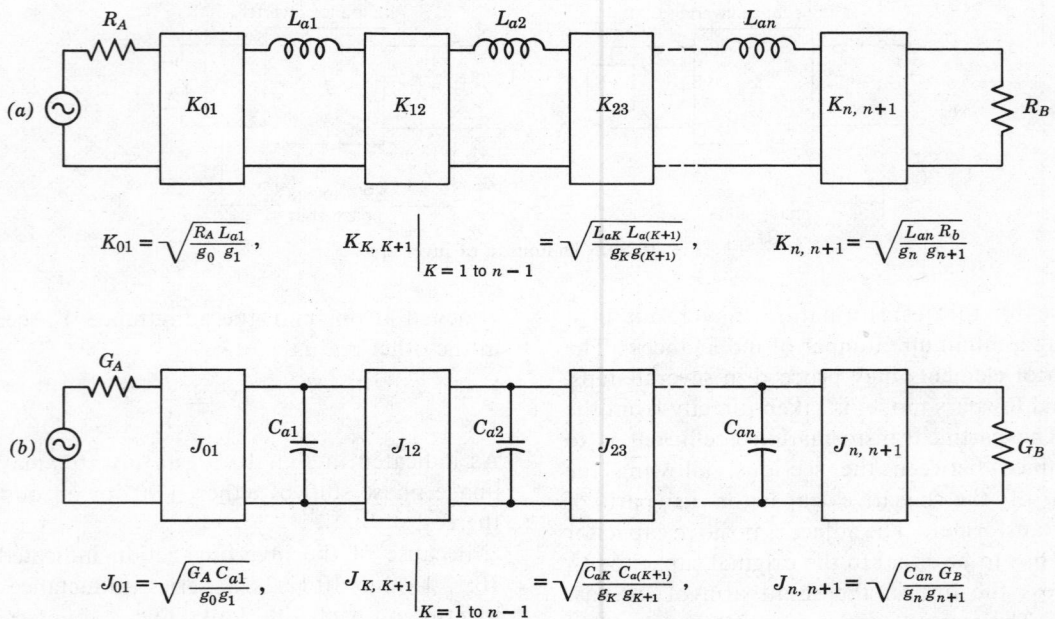

Fig. 10.82. Lowpass prototypes modified to include impedance or admittance inverters.

Fig. 10.82*b* only on the dual basis. Note that the g_k values referred to in the equations in Fig. 10.82 are the prototype element values.

A given circuit as seen through an impedance inverter looks like the dual of that given circuit. Thus the impedances seen from inductor L_{a1} in Fig. 10.82*a* are the same as those seen from inductance L_1' in Fig. 10.81*b* except for an impedance scale factor. The impedances seen from inductor L_{a2} in Fig. 10.82*a* are identical to those seen from inductance L_2' in Fig. 10.81*a*, except for a possible impedance scale change. In this manner the impedances in any point of the circuit in Fig. 10.82*a* may be quantitatively related to the corresponding impedances in the circuit in Fig. 10.81*b*.

Practical Realization of *K*- and *J*-Inverters

One of the simplest forms of inverters is a quarter-wavelength of transmission line. Such a line obeys the basic impedance-inverter definition in Fig. 10.80 and it will have an inverter parameter of $K = Z_0$ ohms where Z_0 is the characteristic impedance of the line. Of course, a quarter-wavelength of line will also serve as an admittance inverter as can be seen from Fig. 10.80, and the admittance inverter parameter will be $J = Y_0$ where Y_0 is the characteristic admittance of the line.

Although its inverter properties are relatively narrow-band in nature, a quarter-wavelength line can be used satisfactorily as an impedance or admittance inverter in narrowband filters. Thus if we have several identical cavity resonators, and if we connect them by lines which are a quarter-wavelength long at frequency ω_0, then by properly adjusting the coupling at each cavity it is possible to achieve a multiresonator Chebyshev response. Note that if all the resonators exhibit say, series-type resonances, and if they were connected together directly without impedance inverters, they would simply operate like a single series resonator. Some sort of inverters between the resonators are essential to obtain a multiple-resonator response if all of the resonators are of the same type—that is, if all exhibit a series-type resonance or all exhibit a parallel-type resonance.

Besides a quarter-wavelength line, there are numerous other circuits which operate as inverters. All necessarily produce an image phase of some odd multiple of $\pm 90°$, and many have good inverting properties over a much wider bandwidth than does a quarter-wavelength line. Figure 10.83 shows four inverting circuits which are of special interest for use as *K*-inverters (that is, inverters to be used with series-type resonators). Those shown in Fig. 10.83*a* and *b* are particularly useful in circuits where the negative *L* or *C* can be absorbed into adjacent positive series elements of the same type so as to give a resulting circuit having all positive elements. The inverters shown in Fig. 10.83*c* and *d* are particularly

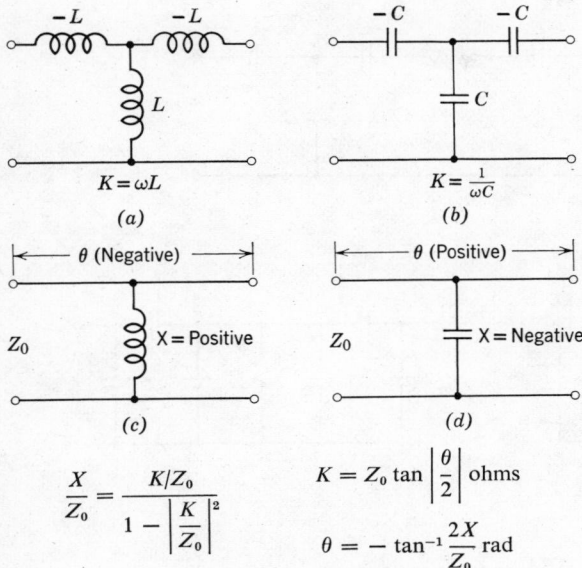

$$\frac{X}{Z_0} = \frac{K/Z_0}{1 - \left|\dfrac{K}{Z_0}\right|^2}$$

$$K = Z_0 \tan\left|\frac{\theta}{2}\right| \text{ ohms}$$

$$\theta = -\tan^{-1}\frac{2X}{Z_0} \text{ rad}$$

Fig. 10.83. *K* inverting circuits.

useful in circuits where the line of positive or negative electrical length θ shown in the figures can be added to or subtracted from adjacent lines of the same impedance. The circuits shown in Fig. 10.83a and c have an over-all image phase shift of $-90°$ while those in Fig. 10.83b and d have an over-all image phase shift of $+90°$. The impedance-inverter parameter K indicated in the figure is equal to the image impedance of the inverter network and is analogous to the characteristic impedance of a transmission line. The

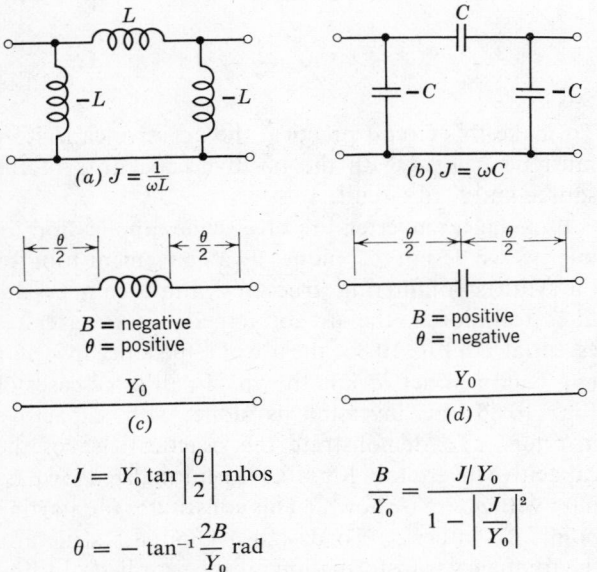

$$J = Y_0 \tan\left|\frac{\theta}{2}\right| \text{ mhos}$$

$$\theta = -\tan^{-1}\frac{2B}{Y_0} \text{ rad}$$

$$\frac{B}{Y_0} = \frac{J/Y_0}{1 - \left|\dfrac{J}{Y_0}\right|^2}$$

Fig. 10.84. *J* inverting circuits.

networks in Fig. 10.83 are much more broadband inverters than is a quarter-wavelength line.

Figure 10.84 shows four inverting circuits which are of special interest for use as *J*-inverters (that is, inverters to be used with parallel-type resonators). These circuits will be seen to be the duals of those in Fig. 10.83, and the inverter parameters *J* are the image admittances of the inverter networks.

Figure 10.85 shows yet another form of inverter composed of transmission line of positive and negative characteristic admittance. The negative admittances are in practice absorbed into adjacent lines of positive admittance.

Numerous other circuits will operate as impedance or admittance inverters, the requirements being that their image impedance be real in the frequency band of operation, and that their image phase be some odd multiple of $\pm\pi/2$.

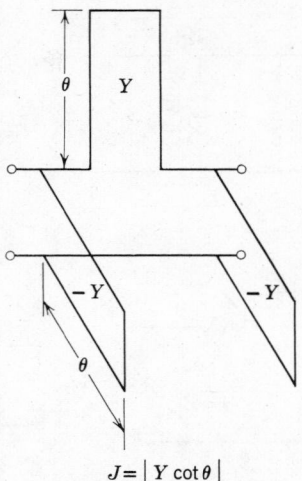

$$J = |Y \cot\theta|$$

Fig. 10.85. An admittance inverter formed from stubs of electrical length θ.

Example 1

Let us convert the shunt impedance Z_1 of Fig. 10.86a into the series impedance Z_2. A *K* inverter is used. To find the element values Eq. 10.12.1 and Fig. 10.83 can be employed:

$$Z_2 = \frac{K^2}{Z_1}$$

where

$$K^2 = \frac{1}{\omega_0^2 C^2}$$

ω_0 being the center frequency of the system. When Z_1 is given as a parallel-resonant circuit with the same center frequency ω_0, the equation for the impedance

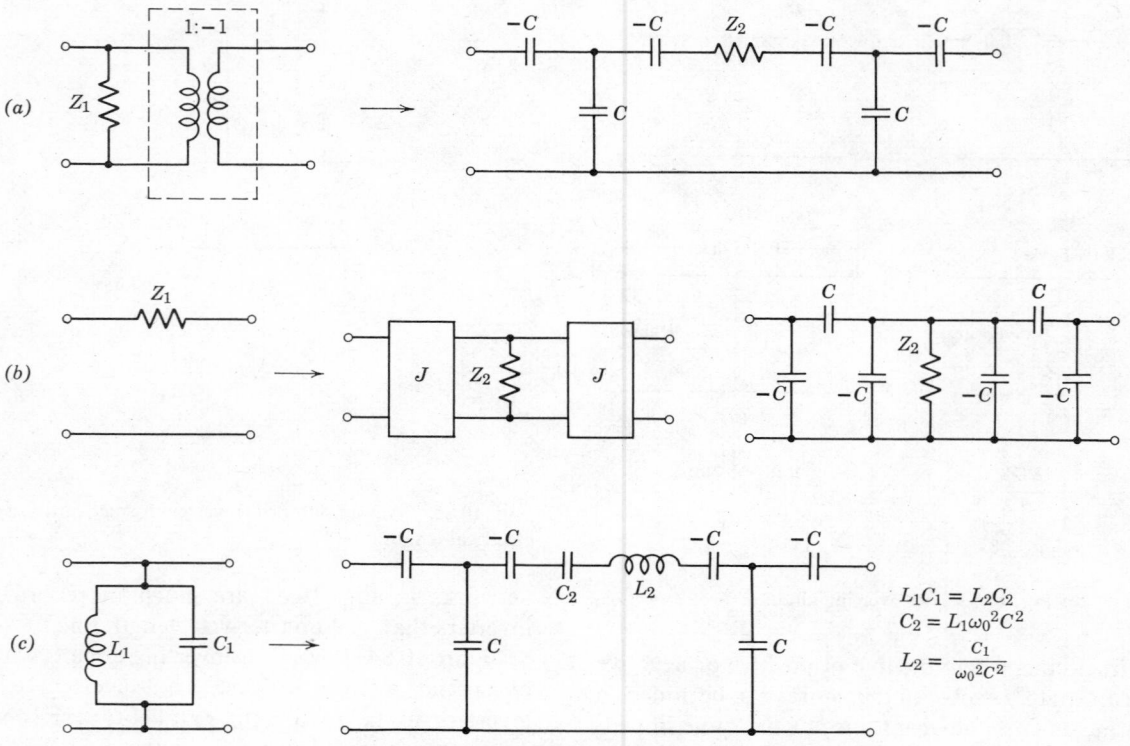

Fig. 10.86. Examples of impedance inversion.

Z_2 may be expressed in the form

$$\frac{1 + s^2 L_2 C_2}{s C_2} = \frac{1 + s^2 L_1 C_1}{s L_1}\frac{1}{\omega_0 C^2}$$

where the resonant frequency of both circuits is the same

$$L_1 C_1 = L_2 C_2$$

The elements of the series circuits are

$$C_2 = \frac{L_1}{K^2} = L_1 \omega_0 C^2$$

$$L_2 = C_1 K^2 = \frac{C_1}{\omega_0{}^2 C^2}$$

The negative capacitors between the shunt arms may be easily combined with positive C_2 to yield a realizable capacitor. The negative capacitors at the ends of the section are absorbed by the load being converted into a parallel combination of resistor and negative capacitor.

Example 2

The series impedance of Fig. 10.86b is to be converted into a parallel impedance as shown on the right side of the figure. Using the J-inverting circuit of Fig. 10.84 the actual network will be as shown in Fig. 10.86b. With the notation of impedances the equation for the circuit values may be expressed by

$$Z_2 = \frac{K^2}{Z_1}$$

$$K^2 = \frac{1}{\omega_0{}^2 C^2}$$

To make this circuit practical the negative capacitors must be merged with the positive capacitors of the source and load circuit.

Impedance inverters receive wide application in microwave design techniques as a convenient tool for the synthesis of filtering structures. But even in crystal filter techniques, the use of impedance inverters is essential. In Fig. 10.86c the use of impedance inverters in a bandpass network is shown. In all three cases of Fig. 10.86 the inversion is done with capacitive inverters. To demonstrate the practical use of the capacitive inverter, in Fig. 10.87a a normalized lowpass filter with $n = 5$ is shown. This constitutes the starting point of synthesis. To design a bandpass structure, the frequency transformation must be applied. In Fig. 10.87b the circuit of the bandpass structure is shown.

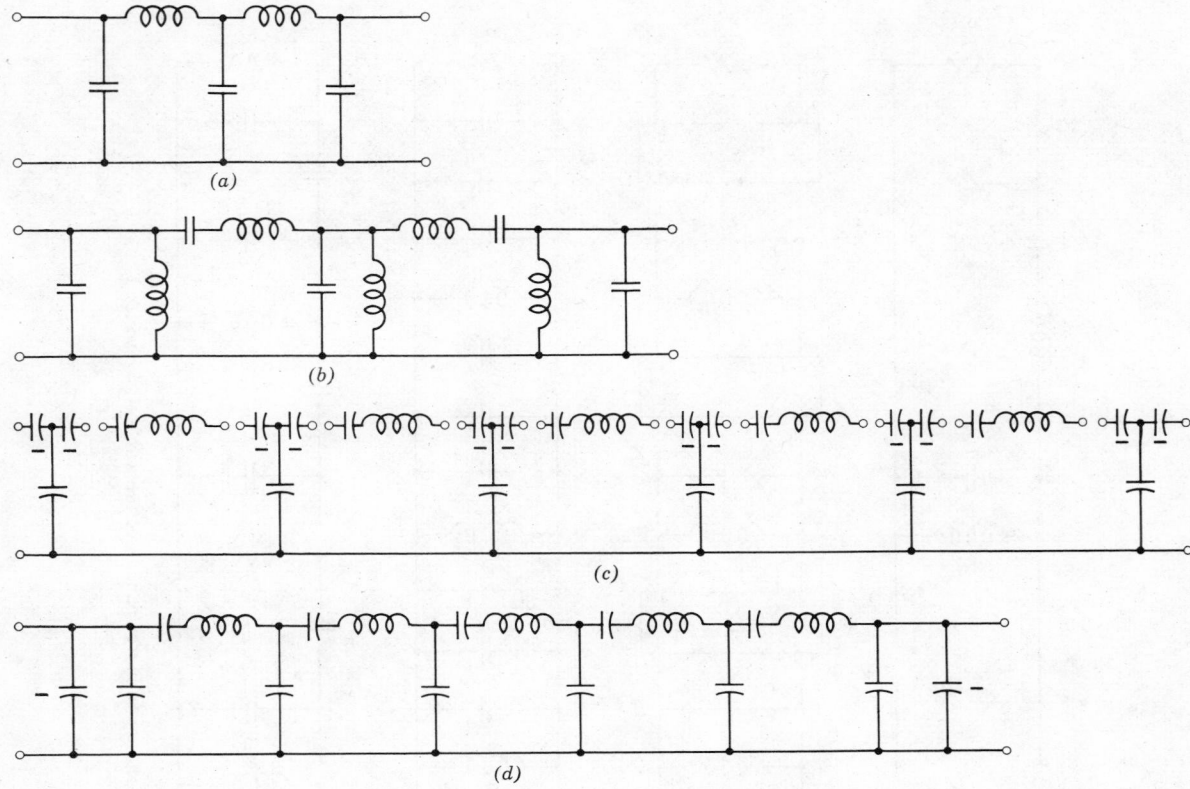

Fig. 10.87. (*a*) Normalized lowpass filter with $n = 5$. (*b*) After lowpass-bandpass transformation. (*c*) Using impedance converters the bandpass network is transformed. (*d*) After combination of the series elements.

In the following step, a frequency inverting structure will be used to remove parallel inductances from the above schematic. The resulting circuit is shown in Fig. 10.87*c*. Evidently some of the capacitors are negative. If we combine the series positive and negative capacitors in every section of the ladder, we obtain the resulting schematic of the filter which is equivalent to Fig. 10.87*b* and is shown finally in Fig. 10.87*d*. In this form the filter is ready for the next type of transformation based on Bartlett's bisection theorem in order to transform it into a convenient semilattice structure. The ladder of the type shown in Fig. 10.87*d*, can be synthesized to give a prescribed characteristic provided its lowpass prototype, such as in Fig. 10.87*a*, is a ladder with all transmission poles at infinity.

Lowpass-Bandpass Transformation of *LC* Filters with Inverters

With reference to Fig. 10.88*a* the series resonant circuit is equivalent to an ideal inverter followed by parallel-resonant circuit and terminated by another ideal inverter. Thus, Fig. 10.88*b* shows the second consecutive step of the procedure. In row *c* all

inverters are substituted by capacitors, and row *d* shows the network after combining positive and negative capacitors. The coefficients a_1, a_2, a_3, etc., are ladder coefficients of the lowpass prototype.

For the network in Fig. 10.88*b* the element values are

$$L_1 = \frac{J_1^2 B}{a_1} \qquad C_1 = \frac{A a_1}{J_1^2}$$

$$L_2 = \frac{J_2^2 H}{J_1^2 a_2} \qquad C_2 = \frac{J_1^2 D a_2}{J_2^2}$$

$$L_3 = \frac{J_1^2 J_3^2 B}{J_2^2 a_3} \qquad C_3 = \frac{J_2^2 A a_3}{J_1^2 J_3^2}$$

where J is defined as $J = 1/\omega C'$ (impedance inverter in this case). The output impedance can be found from

$$R' = \frac{J_2^2 J_4 \cdots}{J_1^2 J_3 \cdots} KR \quad \text{for } n\text{-odd}$$

$$R' = \frac{J_1^2 J_3^2 \cdots}{J_2^2 J_4} \frac{1}{KR} \quad \text{for } n\text{-even}$$

Considering the input side of the ladder, when the

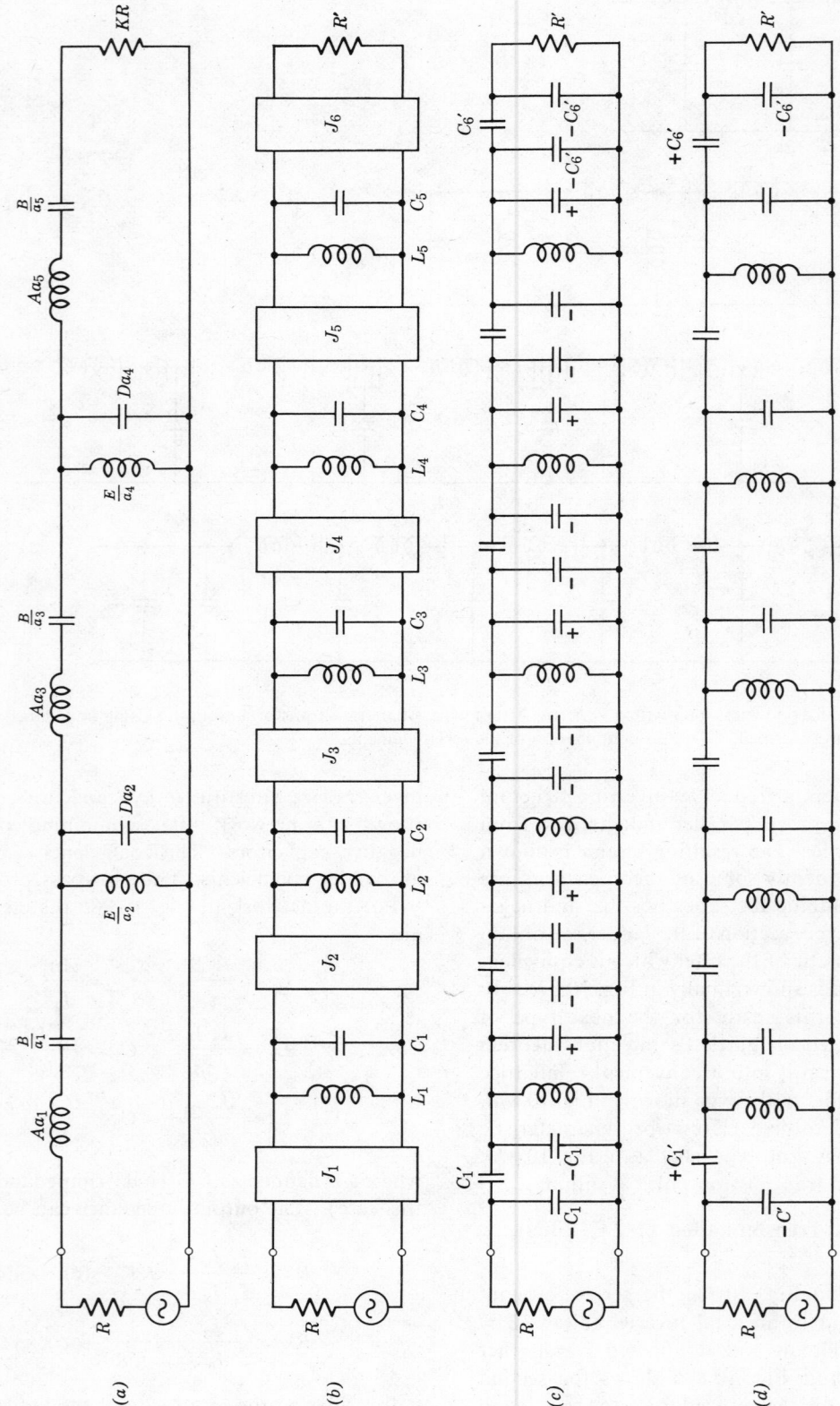

Fig. 10.88. Transformation of conventional bandpass structure into narrowband structure with aid of inverters.

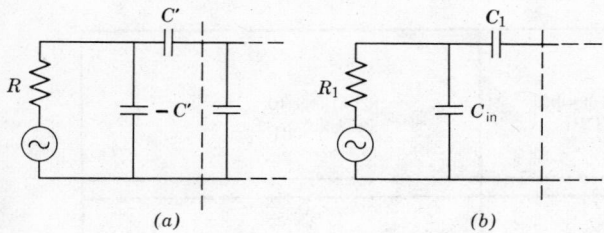

Fig. 10.89. Negative capacitance transformation.

first inverter J_1 is replaced by the corresponding set of capacitors, one of the negative capacitors cannot be absorbed. This situation is shown in Fig. 10.88d.

To overcome the inconveniences the circuit in Fig. 10.89a is replaced by the similar one in Fig. 10.89b in which both R_1 and C_1 are specified. Equating real and imaginary parts of the two impedances at the midband frequency ω_m

$$\frac{R}{1 + \omega_m^2 C R^2} = \frac{R_1}{1 + \omega_m^2 C_{in}^2 R_1^2}$$

and

$$\frac{1}{\omega_m C(1 + \omega_m^2)C^2 R^2} = \frac{\omega_m C_{in} R_1^2}{1 + \omega_m^2 C_{in}^2 R_1^2} + \frac{1}{\omega_m C_1}$$

where

$$\omega_m = 2\pi\sqrt{f_1 f_2}$$

and $f_1 f_2$ are limiting bandpass frequencies. Equating all the inductances to the specified value

$$L = L_1 = \frac{J_1^2 B}{a_1} = \frac{\Delta f}{f_m} \frac{1}{a_1 \omega_m^3 C_{in}^2 R}$$

The quantities which belong to the output circuit can be evaluated in like manner.

Direct-Coupled Bandpass Filters

Conventional frequency transformation yields the bandpass filter of Fig. 10.88a. The constants shown

in this figure are given by

$$A = \frac{R}{\Delta\omega} \qquad D = \frac{1}{R\,\Delta\omega}$$

$$B = \frac{\Delta\omega}{R\omega_m^2} \qquad E = \frac{R\,\Delta\omega}{\omega_m^2}$$

where the conventional transformation frequency is given by

$$\omega_m = 2\pi\sqrt{f_1 f_2}$$
$$\Delta\omega = 2\pi(f_2 - f_1)$$

An alternative transformation used by S. Cohn which partially corrects the frequency dependence of the inverting networks is

$$\omega_m = 2\pi[f_1 + f_2 - \sqrt{(f_2 - f_1)^2 + f_1 f_2}]$$

and

$$\Delta\omega = \frac{1}{a}\omega_0 = \frac{f_0(f_2 - f_1)}{f_1 f_2}\omega_m$$

For very narrow bandwidths these frequency definitions tend to the conventional definition.

When the ladder coefficients of lowpass prototype filters are given, the following quantities must be specified (see Fig. 10.90).

L—common inductance value (of the order of R_1/ω_m)

R_1, R_2—the required termination impedances

C_{in} and C_{out}—the required input and output capacitors

Then in terms of given parameters the auxiliary design resistance may be determined.

$$R = \frac{R_1}{1 + \omega_m^2 C_{in}^2 R_1^2 - (\Delta f/f_m)(R_1/a_1\omega_m L)}$$

The capacity of first inverter is

$$C_1' = \sqrt{\frac{\Delta f C_m}{f_m \omega_m R a_1}}$$

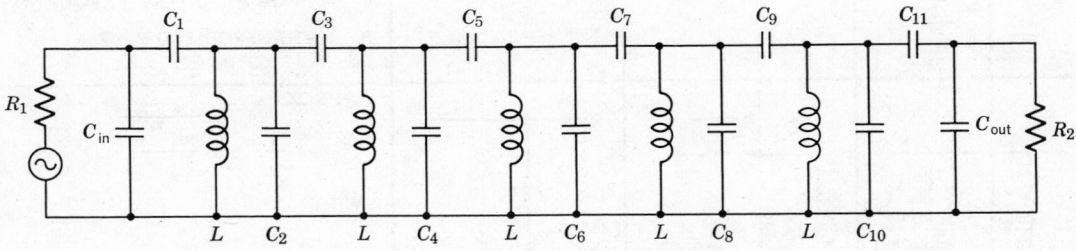

Fig. 10.90. Final schematic of transformed bandpass filter.

Table 10.5 Five methods of transformating a nonresonant generator and/or load into the first and/or last resonant circuit of the network.

IN ALL CIRCUITS BELOW $I = \sqrt{\dfrac{R_1}{Q_{OUT}X_2}}\dfrac{E_1}{R_1}$	IN ALL CIRCUITS BELOW $E = \sqrt{\dfrac{X_2}{Q_{OUT}R_1}}\,E_1$	TRANSFORMING CIRCUIT	RELATIONS TO OBTAIN Q_{OUT}
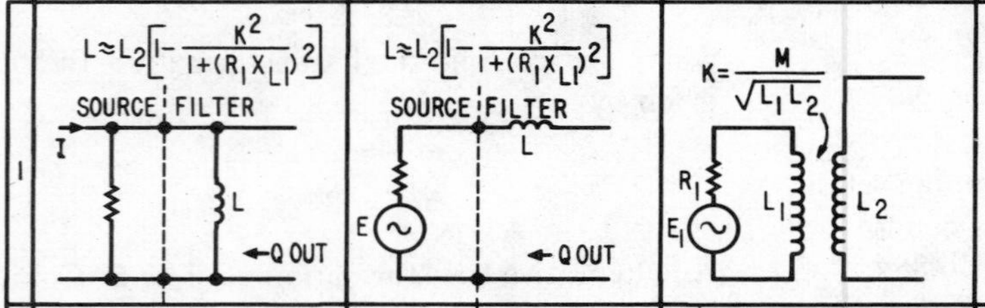 1 $L \approx L_2\left[1 - \dfrac{K^2}{1+(R_1 X_{L1})^2}\right]$ SOURCE FILTER, L, ← Q OUT	$L \approx L_2\left[1 - \dfrac{K^2}{1+(R_1 X_{L1})^2}\right]$ SOURCE FILTER, L, ← Q OUT	$K = \dfrac{M}{\sqrt{L_1 L_2}}$ R_1, E_1, L_1, L_2	**EXACT:** $\dfrac{\dfrac{R_1}{X_{L1}} + \dfrac{X_{L1}}{R_1}}{Q_{out}} \cdot \dfrac{1}{\dfrac{X_{L1}/R_1}{Q_{out}}}$
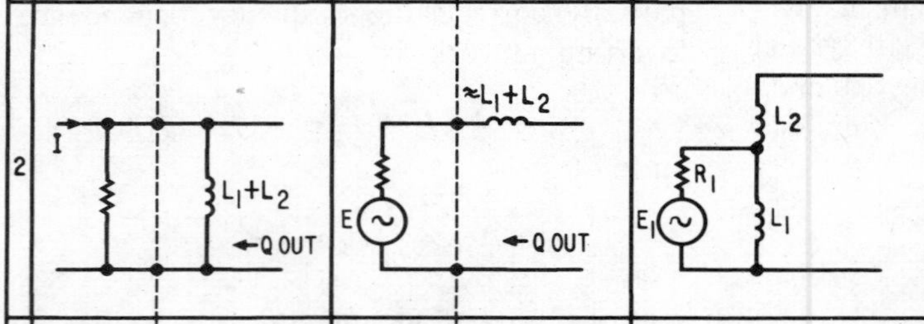 2 $L_1 + L_2$ ← Q OUT	$\approx L_1 + L_2$ ← Q OUT	R_1, E_1, L_2, L_1	**APPROX:** $\dfrac{L_1}{L_2} \approx \sqrt{\dfrac{R_o}{Q_{out}X_2}}$ When $\dfrac{L_1}{L_2} \ll 1$ $\dfrac{R_1}{X_{L1}} \gg 1$
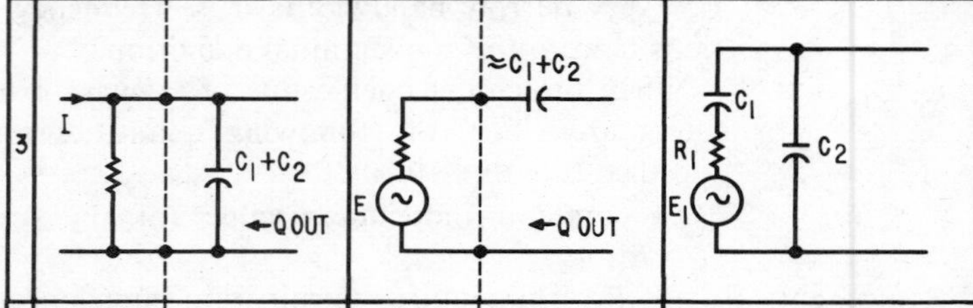 3 $C_1 + C_2$ ← Q OUT	$\approx C_1 + C_2$ ← Q OUT	R_1, E_1, C_1, C_2	**APPROX:** $\dfrac{C_1}{C_2} \approx \sqrt{\dfrac{X_2}{Q_{OUT}R_1}}$ When $\dfrac{C_1}{C_2} \ll 1, \quad \dfrac{R_1}{X_{C1}} \ll 1$
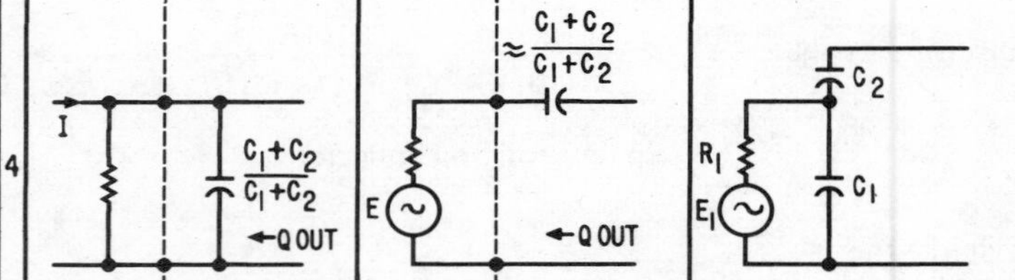 4 $\dfrac{C_1 + C_2}{C_1 + C_2}$ ← Q OUT	$\approx \dfrac{C_1 + C_2}{C_1 + C_2}$ ← Q OUT	R_1, E_1, C_2, C_1	**APPROX:** $\dfrac{C_1}{C_2} \approx \sqrt{\dfrac{Q_{OUT}X_2}{R_1}}$ When $\dfrac{C_1}{C_2} \gg 1, \quad \dfrac{R_1}{X_{C1}} \gg 1$ — — — — **EXACT** $C_1 = \sqrt{\dfrac{C_a Q_{out}}{\omega_o R_1}}$ $\dfrac{C_1}{C_2} = \sqrt{\dfrac{R_x}{R_1}} - 1 \quad C_a = \dfrac{C_1 C_2}{C_1 + C_2}$ ω_o = NODE RESONANT FREQUENCY R_x = TRANSFORMED R AT ω_o
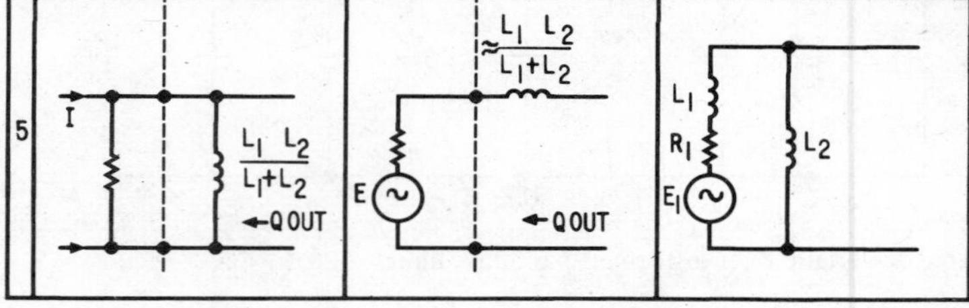 5 $\dfrac{L_1 L_2}{L_1 + L_2}$ ← Q OUT	$\approx \dfrac{L_1 L_2}{L_1 + L_2}$ ← Q OUT	L_1, R_1, E_1, L_2	**APPROX:** $\dfrac{L_1}{L_2} \approx \sqrt{\dfrac{Q_{out}R_1}{X_2}}$ WHEN $\dfrac{L_1}{L_2} \gg 1, \quad \dfrac{R_1}{X_{L1}} \ll 1$

where

$$C_m = \frac{1}{\omega_m{}^2 L}$$

The series arm capacities are then given by the formulas

$$C_1 = \frac{C_1{}'(1 + \omega_m{}^2 C_{\text{in}}^2 R_1{}^2)(1 + \omega_m{}^2 C_1{}'^2 R^2)}{(1 + \omega_m{}^2 C_{\text{in}}^2 R_1{}^2) - \omega_m{}^2 C_{\text{in}} C_1{}' R_1{}^2 (1 + \omega_m{}^2 C_1{}'^2 R^2)}$$

$$C_3 = \frac{\Delta f}{f_m} \frac{C_m}{\sqrt{a_1 a_2}}$$

$$C_5 = \frac{\Delta f}{f_m} \frac{C_m}{\sqrt{a_2 a_3}}$$

$$C_7 = \frac{\Delta f}{f_m} \frac{C_m}{\sqrt{a_3 a_4}}$$

$$C_9 = \frac{\Delta f}{f_m} \frac{C_m}{\sqrt{a_4 a_5}}$$

$$C_{11} = \frac{C_6{}'(1 + \omega_m{}^2 C_{\text{out}}^2 R_2{}^2)(1 + \omega_m{}^2 C_6{}'^2 R'^2)}{(1 + \omega_m{}^2 C_{\text{out}}^2 R_2{}^2) - \omega_m{}^2 C_{\text{out}} C_6{}' R_2{}^2 (1 + \omega_m{}^2 C_6{}' R'^2)}$$

The design output resistance in terms of given parameters when order of transmission function n is odd is

$$R' = \frac{R_2}{(1 + \omega_m{}^2 C_{\text{out}}^2 R_2{}^2) - \dfrac{\Delta f}{f_m} \dfrac{k R_2}{a_n \omega_m L}}$$

where k is an arbitrary impedance correction factor. The capacity of last inverter is

$$C_{n+1}' = \sqrt{\frac{k C_m \Delta f}{a_n \omega_m R' f_m}}$$

When n is even,

$$R' = \frac{R_2}{(1 + \omega_m{}^2 C_{\text{out}}^2 R_2{}^2) - \dfrac{\Delta f}{f_m} \dfrac{R_2}{k a_n \omega_m L}}$$

and

$$C_{n+1}' = \sqrt{\frac{\Delta f}{f_m} \frac{C_0}{a_n \omega_m R'}}$$

The values of shunt capacitors across the inductors may be given in terms of the above parameters:

$$C_2 = C_m - C_1' - C_3$$

$$C_4 = C_m - C_3 - C_5$$

$$\cdot$$
$$\cdot$$
$$\cdot$$

$$C_{10} = C_m - C_6' - C_9$$

There is a restriction of choice of the common value of the inductor L; too high a value may yield some of the shunt capacitors negative or some of the series capacitors too small to be practical (for very narrow bandwidth filters).

There is also an upper limit to the value of the input and output capacitors but this is unlikely to be approached in practice. The restriction on the termination ratio, is likewise not met in practice. Thus nonequally terminated filters can readily be designed from an equally terminated model and conversely.

10.13 SOURCE AND LOAD TRANSFORMATION

The problem of internal transformation in the filter is closely associated with the problem of source and load transformation. The filter has to *see* a certain amount of resistance in the load circuit and a certain amount of internal resistance in the generator or the source. If the filter realization is for optimum performance their constituent elements are not adaptable to the value of outside resistances; the impedance of the filter (in case of image-parameter design) has to be modified with Norton's transformation. If on the other hand the filter design has a certain amount of losses in the input and output circuits these circuits must be modified accordingly to provide the proper operating condition from the point of view of Q-factor. Therefore the problem now is that of obtaining the proper transformation of the generator and load to produce the required Q_1 and Q_n. Table 10.5 shows five types of transforming circuits and their design equations. Let us consider case four. This is the most

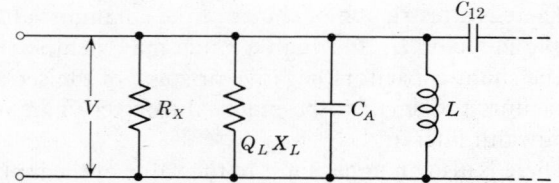

Fig. 10.91. Illustration of source transformation of type 4 of Table 10.5.

practical for many designs because the shunt capacitor may be used to absorb the distributed capacitance usually associated with the input and output circuit. Let us assume that the filter capacitor is C_a. Therefore,

$$\frac{C_1 C_2}{C_1 + C_2} = C_a \tag{10.13.1}$$

and the ratio between two capacitors will be given by

$$\frac{C_1}{C_2} = \sqrt{\frac{R_x}{R_1}} - 1 \tag{10.13.2}$$

where R_x is the transformed value of R_1 is required to produce the specified Q_{out}. Incidental coil dissipation (neglecting capacitor losses) is accounted for in the following way:

$$\frac{1}{R} = \frac{1}{R_x} + \frac{1}{Q_L X_L}$$

In this equation, R is the value of the nodal shunt resistance required to attain the specific Q_{out}. The output transformation is similar to the input transformation and the final circuit of the completed filter

will include on both sides one supplementary capacitor. Let us consider the following example:

$$C_a = 31.32 \text{ pF}$$

The measured Q-factor of the coil associated with the circuit at f_0 is 260. The equivalent schematic of input circuit will be as shown in Fig. 10.91 where L and C are assumed to be perfect elements. The required quality factor, according to reference tables is $Q_{out} = 7.66$. The reactance of the coil at $f_0 = 100$ Mc is

$$X_L = 2\pi \times 100 \times 0.075 = 47.1 \text{ ohms}$$

The value of transformed resistor from the source will be:

or

$$R_x = \frac{Q_{out} Q_L X_L}{Q_L - Q_{out}}$$

$$R_x = \frac{7.66 \times 47.1 \times 260}{260 - 7.66}$$

The ratio between the two capacitors (substituting for the input capacitor) will be

$$\frac{C_1}{C_2} = \sqrt{\frac{R_x}{R_1}} - 1 = \sqrt{7.2} - 1 = 1.64$$

Using Eq. 10.13.1 for the capacitance ratio and substituting it into the original expression for the sum of two capacitors, the following will be the solution for C_2:

$$C_2 = \frac{31.32 \times 2.64}{1.64} = 50.2 \text{ pF}$$

and therefore C_1 is equal to 82.4 pF.

Bibliography

CHAPTER 1. FILTERS IN ELECTRONICS

Belevitch, V., Recent Developments in Filter Theory, *IRE Trans. Circuit Theory*, CT-5, pp. 236–252, December 1958.

Cohn, S., Direct-Coupled-Resonator Filters, *Proc. IRE*, **45**, pp. 187–196, February 1957.

Fritzsche, G., Ist die Wellengrössentheorie noch aktuell? *Wissenschaftliche Z. der Tech. Universität Dresden*, 14H6, 1965.

Matthaei, G. L., Interdigital Band-Pass Filter, *IRE Trans. PGMTT-10*, pp. 479–491, November 1962.

Matthaei, G. L., Comb-Line Band-Pass Filters of Narrow or Moderate Bandwidth, *Microwave J.*, **6**, pp. 82–91, August 1963.

Matthaei, G. L., Young, L., and Jones, E., *Microwave Filters, Impedance-Matching Network and Coupling Structures*, McGraw-Hill, New York, 1964.

Oswald, J., Les filtres elementares, *C. & T.* N4, 1953.

Zobel, O., Theory and Design of Electric Wave Filters, *Bell System Tech. J.*, January 1923.

Zverev, A., and Blinchikoff, H., Application of Network Synthesis to Broadbanding of Low-Frequency Antennas, *IEEE Trans. Communication Systems*, CS-12, pp. 237–250, June 1964.

Zverev, A., Introduction to Filters, *Electro-Technology*, **73**, pp. 61–90, June 1964.

Zverev, A., Microwave-Filter Design, *Electro-Technology*, **74**, pp. 46–56, December 1964.

Zverev, A., Composite Filters, *Electro-Technology*, **75**, pp. 48–53, February 1965.

Zverev, A., Unconventional Filter Design, *Electro-Technology*, **75**, pp. 40–46, March 1965.

Zverev, A., The Golden Anniversary of Electric Wave Filters, *IEEE Spectrum*, **3**, pp. 129–131, March 1966.

CHAPTER 2. THEORY OF EFFECTIVE PARAMETERS

Bader, W. Polynomvierpole vorgeschriebener Frequenzabhängigkeit, *Arch. Elektrotech.*, **34**, pp. 181–209, April 1940.

Bosse, G., *Synthese Elektrischer Siebschaltungen*, S. Hirzel Verlag, Stuttgart, 1963.

Cauer, W., *Theory der Linearen Wechselstromschaltungen*, Academie Verlag, Berlin, 1954.

Darlington, S., Synthesis of Reactance-Four-Poles which Produce Prescribed Insertion Loss Characteristics, *J. Math. Phys.*, **30**, pp. 257–353, September 1939.

Darlington, S., Network Synthesis Using Tchebyshev Polynomial Series, *Bell System Tech. J.*, **31**, pp. 613–665, July 1952.

Fetzer, V., Die numerische Berechnung von Filterschaltungen mit allgemeinen Parametern nach der modernen Theorie unter besonderer Berücksichtigung der Cauerschen Arbeiten, *Arch. elekt. übertr.*, **5**, pp. 499–508, November 1951; **6**, pp. 419–431, October 1952.

Fritzsche, G., Die Pole-Nullstellen-Method, *Wissenschaftliche Z. der Tech. Universität Dresden*, 11H1, 1962.

Haase, K. H., Zur Aufstellung der charakteristischen Funktion und Berechnung der Betriebsdämpfung symmetrischer und antimetrischer Betriebsparameter-Filter, *Frequenz*, **6**, pp. 168–179, May 1952.

Herrero, J., and Willoner, G., *Synthesis of Filters*, Prentice Hall, 1966.

Laurent, T., Le filtre zig-zag, *Ericsson Tech.*, **4**, pp. 3–28, January 1936.

Ming, N., Verwirklichung von linearen Vierpolschaltungen vorgeschriebener Frequenzabhängigkeit unter Berücksichtigung übereinstimmender Verluste aller Spulen und Kondensatoren, *Arch. Elektrotech.*, **39**, pp. 452–471, July 1949.

Piloty, H., Beiträge zur Berechnung von Wellenfiltern., *Elekt. Nachrtech.*, p. 137, 1938.

Piloty, H., Kanonische Kettenschaltungen für Reaktanzvierpole mit vorgeschriebenen Betriebseigenschaften, *Telegr.-Fernspr.-Tech.*, **29**, pp. 249–258, 279–290, 320–325, September–November 1940.

Ufelman, A., A General Rule for the Distribution of the roots of a Characteristic Polynomial, *Electrosvyaz* N7, 1959.

Van Valkenberg, M. E., *Introduction to Modern Network Synthesis*, Wiley, New York, 1960.

CHAPTER 3. FILTER CHARACTERISTICS IN THE FREQUENCY DOMAIN

Belevitch, V., Tschebyshev filters and amplifier network, *Wireless Engr*, **29**, pp. 106–110, April 1952.

Humpherys, D., Minimum Insertion Loss Filters of Even Order, *IEEE Trans. Circuit Theory*, CT-11, pp. 414–416, September 1964.

Humpherys, D., Rational Function Approximation of Polynomials to Give an Equiripple Error, *IEEE Trans. Circuit Theory*, CT-11, pp. 479–486, December 1964.

Lubkin, Y. J., Papoulis Filters, *Electron. Des.*, **11**, pp. 55–56, 1963.

Orchard, H. J., Formulae for ladder filters, *Wireless Engr*, **30**, pp. 3–5, January 1953.

Rumpelt, E., *Über den Entwurf elektrischer Wellenfilter mit vorgeschriebenem Betriebverhalten*, doctoral dissertation, Tech. Hoch., Munich, 1947.

Storch, L., Synthesis of Constant-Time-Delay Ladder Networks using Bessel Polynomials, *Proc. IRE*, **42**, pp. 1666–1675, November 1954.

Szentirmai, G., The Design of Arithmetically Symmetrical Band-Pass Filters, *IEEE Trans. Circuit Theory*, CT-10, pp. 367–375, September 1963.

Zverev, A., Filters in Electronics, *Westinghouse Engr*, **23**, pp. 59–64, March 1963.

CHAPTER 4. ELLIPTIC FUNCTIONS AND ELEMENTS OF REALIZATION

Bedrosian, S. D., Elliptic Functions in Network Synthesis, *J. Franklin Inst.*, **271**, pp. 12–30, January 1961.

Bosse, G., and Nonnenmacher, W., Einheitliche Formel für Filter, *Frequenz*, **13**, pp. 33–44, February 1959.

Grossman, A., Synthesis of Tchebyscheff Parameter Symmetrical Filter, *Proc. IRE*, **45**, pp. 454–473, April 1957.

Kurth, C., Übersicht über die Berechnung von Filtern, *Frequenz*, **10**, pp. 391–396, 1956; **11**, pp. 12–19, pp. 43–53, 1957.

CHAPTER 5. THE CATALOG OF NORMALIZED LOWPASS FILTERS

Christian, E., and Eisenmann, E., *Filter Design Tables and Graphs*, Wiley, New York, 1966.

Fritzsche, G., and Buchholz, G., Filter Katalog, *Nachrichten-technik*, 14H4, 1964.

Geffe, P. *Simplified Modern Filter Design*, John F. Rider, Inc., New York, 1963.

Glowatzki, E., Katalogisierte filter, *Nachrtech. Z.*, **9**, pp. 508–513, November 1956.

Humpherys, D., Synthesis of Even Order Filters with Chebyshev Magnitude Response, *Electron. Equip. Engng.*, **12**, pp. 55–58, March 1964.

Kawakami, M., Nomographs for Butterworth and Chebyshev Filters, *IEEE Trans. Circuit Theory*, CT-10, pp. 288–289, June, 1963.

Saal, R., and Ulbrich, E., On the Design of Filters by Synthesis, *IRE Trans. Circuit Theory*, CT-5, December, 1958.

Saal, R., *Der Entwurf von Filtern mit Hilfe des Kataloges Normierter Tiefpasse*, Telefunken GMBH, Backnang, W. Germany, 1963.

Ulbrich, E., and Piloty, H., Über den Entwarf von Allpassen Tiefpassen und Bandpassen mit eines in Tschebyscheffschen Sinne approximeten Konstanten Gruppenlaufzeit, *Arch. elekt. Übertr.*

CHAPTER 6. DESIGN TECHNIQUE FOR POLYNOMIAL FILTERS

Atiya, F. S., Theorie der maximal-geebneten und quasi-Tschebyscheffschen-Filter, *Arch. elekt. übertr.*, pp. 441–450, September 1963.

Dishal, M., Design of Dissipative Bandpass Filters Producing Desired Exact Amplitude-Frequency Characteristics, *Proc. IRE*, **37**, pp. 1050–1069, September 1949.

Feldtkeller, R., *Hochfrequenz-Bandfilter*, S. Hirzel Verlag, Stuttgart, 1953.

Green, E., *Amplitude-Frequency Characteristics of Ladder Networks*, Marconi's Wireless Telegraph Co., Chelmsford, Essex, England, 1954.

Reference Data for Radio Engineers, 4th ed., International Telephone & Telegraph Corp., New York, 1956.

CHAPTER 7. FILTER CHARACTERISTICS IN THE TIME DOMAIN

Craig, J. W., *Theory and Design Data for Uniformly Dissipative, Doubly Terminated Bandpass and Lowpass Filters*, Lincoln Laboratory Technical Report 411, February 1966.

Dishal, M., Gaussian-Response Filter Design, *Elect. Commun.*, **36**, pp. 3–26, 1959.

Eggen, C. P., McAllister, A. S., Modern Low Pass Filter Characteristics, *Electro-Technol.*, **78**, pp. 50–55, August 1966.

Fritzsche, G., Anwendungen des Filterkataloges, *Nachrichten-technik*, **14**, H4-12, 1964; **15**, H1-2, 1965.

Henderson, K. W., and Kautz, W. H., Transient Responses of Conventional Filters, *IRE Trans. Circuit Theory*, CT-5, pp. 333–347, December 1958.

Humpherys, D., A Relationship Between Impulse Response Overshoot and Phase Ripple, *IEEE Trans. Circuit Theory*, CT-10, pp. 450–452, September 1963.

Humpherys, D., *Rational Function Approximation of Polynomials with Equiripple Error*, doctoral dissertation, University of Illinois, 1963.

CHAPTER 8. CRYSTAL FILTERS

Dishal, M., Modern Network Theory Design of Single-Sideband Crystal Ladder Filters, *Proc. IEEE*, **53**, pp. 1205–1216, September 1965.

Haas, W., Die Verwendung von Quarzen in Netzwerken die nach der Betriebsparameter Theory berechnet werden, *Frequenz*, No. 5, March 1962.

Herzog, W., *Siebschaltungen mit Schwingkristallen*, DVB, Wiesbaden, 1949.

Humpherys, D., Synthesis of Narrow Band Crystal Filters, *Electro-Technology*, **76**, pp. 36–46, November 1965.

Humpherys, D., Insertion Loss Synthesis of RLCX Networks, submitted for publication in *IEEE Trans. Circuit Theory*.

Indjoudjian, D., and Andrieux P., *Les filtres a cristaux Piezo-electriques*, Gauthier-Villars, Paris, 1953.

Mason, W. P., *Electromechanical Transducers and Wave Filters*, Van Nostrand, Princeton, New Jersey, 1948.

Poschenrieder, J., Die Wellen Parameter Theory als Einfaches, Hilfsmittel zur Realizierung von Quartz Band Filters, *Nachrtech. Z.* **12**, pp. 132–138, 1959.

Poschenrieder, Steile Quartz Filter Grossere Bandbreite in Abzweigschaltung, *Nachrtech. Z.*, **9**, pp. 561–565, 1956.

CHAPTER 9. HELICAL FILTERS

Dishal. M., Alignment and Adjustment of Synchronously Tuned Multiple-Resonant-Circuit Filters, *Elec. Commun.*, pp. 154–164, June 1952.

Fubini and Guillemin, Minimum Insertion Loss Filters, *Proc. IRE*, **47**, pp. 37–41, January 1959.

Macalpine, W. and Schildknecht, R., Coaxial Resonator with Helical Inner Conductor, *Proc. IRE*, **47**, pp. 2099–2105, December 1959.

Zverev, A. and Blinchikoff, H., Realization of Filters with Helical Components, *IRE Trans. Component Parts*, CP-8, pp. 99–110, September 1961.

CHAPTER 10. NETWORK TRANSFORMATIONS

Colin, J. E., De l'introduction de cristaux piézoélectriques en des filtres passe-bas et passe-haut en échelle (avec extension aux filtres passe-bande), *Câbles & Transmission*, **16**, No. 2, pp. 85–93, April 1962.

Gillard, J., Reducing the Number of Elements in Ladder Networks by the Use of Auto-Transformers, *Phillips Telecommun. Rev.*, Vol. 24, May 1963.

Norton, E. L., U.S. Patent NR 1681554.

Norton, E. L., U.S. Patent NR 1708950.

Watanabe, H., On the Circuit with a Minimum Number of Coils, *IRE Trans. Circuit Theory*, CT-7, pp. 77–78, March 1960.

Zverev, A. and Blinchikoff, H., Network Transformations for Wave Filter Design, *Electronics*, **32**, pp. 52–54, June 26, 1959.

Zverev, A., Practical Design of All-Pass Networks, *Electron. Ind.*, **24**, pp. 77–80, February 1965.

Index